MRI AND CT OF THE SPINE

MRI AND CT OF THE SPINE

KRISHNA C. V. G. RAO, M.D.
Clinical Professor of Radiology
Uniformed Services University of Health Sciences
and
Neuroradiologist
Department of Radiology
Walter Reed Army Medical Center
Washington, D.C.

J. POWELL WILLIAMS, M.D.
Professor of Radiology
Department of Radiology
University of South Alabama
Mobile, Alabama

BENJAMIN C. P. LEE, M.D.
Director of Neuroradiology
Division of Neuroradiology
Washington University Medical Center
St. Louis, Missouri

JOHN L. SHERMAN, M.D.
Director
MRI of Colorado
Colorado Springs, Colorado

SANS TACHE

Williams & Wilkins

BALTIMORE • PHILADELPHIA • HONG KONG
LONDON • MUNICH • SYDNEY • TOKYO

A WAVERLY COMPANY

Editor: Timothy H. Grayson
Project Managers: Raymond E. Reter and Lauren Byrne
Copy Editor: Jim Madru
Designers: Norman W. Och and Wilma Rosenberger
Illustration Planner: Marty Tenney
Cover Designer: Wilma Rosenberger

Accurate indications, adverse reactions, and dosage schedules for drugs are provided in this book, but it is possible that they may change. The reader is urged to review the package information data of the manufacturers of the medications mentioned.

Printed in the United States of America.

Chapter reprints are available from the publisher.

Library of Congress Cataloging-in-Publication Data

MRI and CT of the spine / Krishna C. V. G. Rao . . . [et al.] ; with contributions by Patricia C. Davis, Solomon Batnitsky, R. Stephen Williams.
p. cm.
Includes bibliographical references and index.
ISBN 0-683-07133-5
1. Spine—Magnetic resonance imaging. 2. Spine—Tomography. I. Rao, Krishna C. V. G.
[DNLM: 1. Spine—pathology. 2. Spinal Diseases—diagnosis. 3. Magnetic Resonance Imaging—methods. 4. Tomography, X-Ray Computed—methods. WE 725 M9385 1994]
RD768.M74 1994
617.3'750754—dc20
DNLM/DLC
for Library of Congress 93-22969
CIP

93 94 95 96 97
1 2 3 4 5 6 7 8 9 10

PREFACE

The availability of cross-sectional imaging of the spine in recent years has been a quantum leap over earlier imaging modalities. Methods of examining the spine over the past 12 years have changed drastically with water-soluble contrast material, computed tomography, and magnetic resonance imaging. Each of these methods has undergone many changes and improvements in this time period.

Although this textbook has been in development over the last 10 years, the information presented is current and relevant. Our particular concern is the timeliness of the information relating to the selection of the ideal imaging modality given the imaging options that are available.

Several textbooks are available that deal with magnetic resonance imaging of the spine or with computed tomography of the spine. No other text, however, relates to the decisions that have to be made in practice—selecting the ideal imaging technique that can provide the desired clinical information, with the least invasive features, while considering the cost-effectiveness. Cost-effectiveness is a focus in the health care industry, and it is particularly important in the context of spine imaging. This textbook is useful in providing an in-depth review of spine imaging.

We believe that current imaging techniques have plateaued; as such, our book is on the "cutting edge." Throughout this text, we have stressed the clinical aspects of imaging with profuse illustrations while the technical aspects of imaging are briefly reviewed with appropriate references.

We thank the authors and numerous other contributors for their forbearance and their enthusiasm. Several radiology colleagues have helped us by providing their clinical material, without which we would not have accomplished this text.

We would like to thank Williams & Wilkins headed by Mr. Tim Grayson and Will Passano for encouraging us throughout this period to continue our efforts, and to their editorial staff of Ray Reter, Lauren Byrne, and numerous others who helped us in finally delivering the book.

Each of us would like to thank our wives and families for bearing with our idiosyncrasies during this period.

Krishna C. V. G. Rao
J. Powell Williams
B. C. P. Lee
John L. Sherman

ACKNOWLEDGMENTS

This book is the result of the advice, cooperation and the sustained enthusiasm of our numerous associates in radiology, orthopedics and the neurosciences, as well as our own learning experience from our diagnostic miscues.

Several of our present and former Residents, Fellows and staff members in neuroradiology have contributed their time in manuscript preparation, proof reading, especially Fouad E. Gellad, John Joslyn, Sudhir Arora. Jeffery Baka and Bharat Mehta from Henry Ford hospital have provided additional material for certain chapters.

Any book on neuro-imaging depends on the variety and depth of clinical material. This has been accomplished with many of the illustrations contributed by our present and former associates, including Douglas C. Brown, Mary L. Turkiewicz, Carl A. Geyer, Hector Robles. Additional material through the courtesy of AJNR was provided by Nick Patronas, Randy Jenkins, Joseph M. Levy and Alex C. Mamourian. They are acknowledged in the appropriate sections.

Ed Chandler and others of the photography department at Walter Reed Army Medical Center, as well as others from University of Southern Alabama, Mallinckrodt Institute and Emory University hospitals were very helpful in working with us in the production of the illustrative material.

We also acknowledge the great support provided by the publishing house of Williams & Wilkins and their staff. Finally none of this could have been achieved without the continued forbearance and support of our family members.

Krishna C. V. G. Rao
J. Powell Williams
Benjamin C. P. Lee
John L. Sherman

CONTRIBUTORS

Solomon Batnitzky, M.D.
Professor of Diagnostic Radiology
Chief, Section of Neuroradiology
University of Kansas Medical Center
Kansas City, Kansas

Patricia C. Davis, M.D.
Associate Professor of Radiology
Director of Pediatric Neuroradiology
Emory University School of Medicine
Atlanta, Georgia

Donald A. Eckard, M.D.
Assistant Professor of Diagnostic Radiology
Chief, Section of Interventional Neuroradiology
Department of Diagnostic Radiology
University of Kansas Medical Center
Kansas City, Kansas

Hilton I. Price, M.D.
Neuroradiologist
Christian Hospital Northeast
St. Louis, Missouri

R. Stephen Williams, M.D.
Neuroradiologist
Department of Radiology
University of South Alabama
Mobile, Alabama

CONTENTS

CHAPTER / 1

Technical Considerations

B. C. P. Lee

A basic knowledge of the technical principles involved in imaging of the spine with computed tomography (CT) and magnetic resonance (MR) allows selection of appropriate parameters, avoidance of pitfalls in interpretation, and an understanding of the artifacts that may mimic pathologic process or degrade the studies. With CT, imaging is based on the electron density of the tissue. Bone and soft tissues can be well differentiated. However, since the density characteristics of the various tissues within the spinal canal are almost similar, identification of the spinal cord and nerve roots within the thecal space depends on the volume of surrounding cerebrospinal fluid. In most instances, delineation of the spinal cord and nerve roots requires intrathecal contrast material, whereas evaluation of the epidural contents to a limited extent requires intravenous contrast material.

MR imaging (MRI) of the spine is based on the proton density of the tissues. As a result, MRI is excellent in evaluating the intraspinal contents. Utilizing a combination of pulse sequences, the soft-tissue anatomy and pathologic changes can be identified, providing an added dimension not available with other imaging modalities. Artifacts are more common with MRI than with CT. As a result, proper selection of the image-acquisition parameters is more important with MRI compared with CT. This chapter provides a basic understanding of the imaging principles of MRI and CT, as well as technical considerations and imaging strategies that apply to spine imaging.

Magnetic Resonance Imaging

PHYSICAL PRINCIPLES

Magnetic resonance images are created from electrical signals generated from protons, which are then localized in three-dimensional space. Three components are required for excitation of protons: the presence of protons, a strong static magnetic field (magnet), and application of radio frequencies (RF). Localization of protons in three-dimensional space is achieved by introducing gradients within the magnetic field in the three orthogonal planes.

Proton Excitation/Relaxation. Protons in the body are randomly aligned and are in constant motion. When a patient is placed within the field of a giant powerful magnet, the protons in that patient's body are all aligned in the direction of the strong magnetic field and continue to rotate, or *precess*, along the axis in this (longitudinal) direction. A radio frequency (RF) applied at a right angle to the magnetic field causes the protons to deflect and continue to precess in the (transverse) direction of the RF. When this RF pulse is switched off, the protons relax along the transverse direction (T2 relaxation), as well as back to the longitudinal direction of the magnetic field (T1 relaxation) (Fig. 1.1). The rates of relaxation depend on many factors, such as the physical state, size, and chemical structure of the compound to which the proton is attached. T2 relaxation is always shorter than T1 relaxation. The strength of the RF signal detected depends on these rates of relaxation [1–3].

Localization. The magnetic field within a magnetic resonance imager is uniform and homogeneous; thus all protons precess at the same rate irrespective of location. Special gradient coils create a magnetic field that is stronger at one end and weaker at the opposite end. Protons in the region of stronger field precess faster and generate stronger signals than those in the lower end of the field (Fig. 1.2). Such gradients are used (*a*) to define the orientation, location, and thickness of the section and (*b*) to localize individual protons within the section (Fig. 1.3). By electronically changing the orientation of these gradients, cross-sectional images can be created in orthogonal or oblique planes [2].

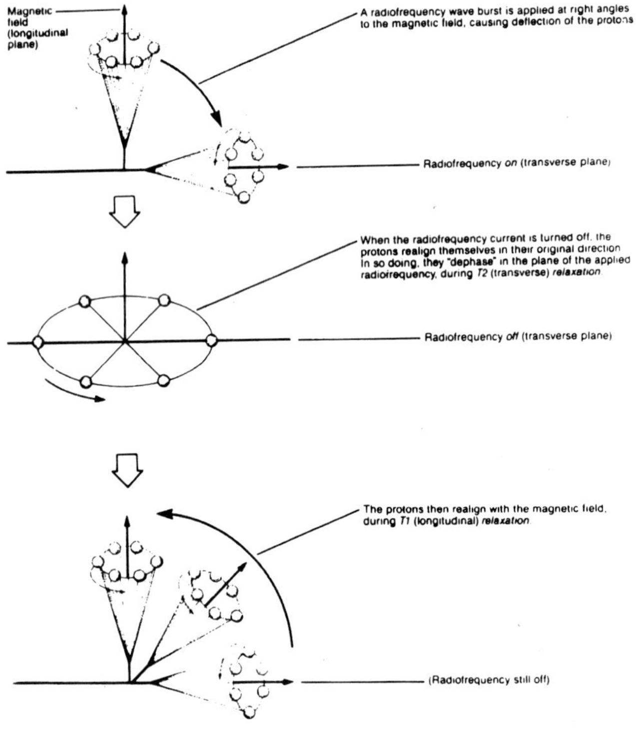

Figure 1.1. Effect of application and cessation of RF at right angles to direction of precession.

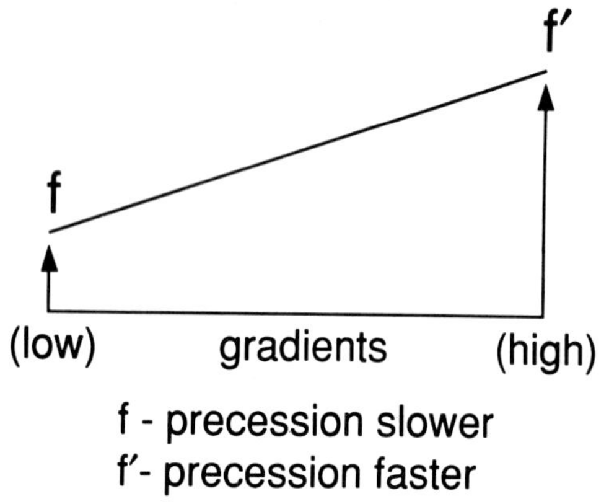

Figure 1.2. Application of a gradient creates differences in the magnetic field; protons in the stronger field precess faster than those in the weaker field.

Slice Selection. After applying a gradient along the axis of a slice (*z* axis), the location of the section is selected by shifting the frequency range.

Localization Within the Section. This is achieved by phase encoding and frequency readout. The *phase* is the degree of rotation, or precession, at a particular time after beginning the RF excitation. It is to be expected that the phase angle is greater with protons that precess faster than with those that precess more slowly (Fig. 1.4). By encoding or assigning a line in the *y* axis for each degree of phase, the signal generated by proton relaxation along this line may be determined or read in the *y* axis (Fig. 1.5).

TECHNICAL AND HARDWARE CONSIDERATIONS

MRI diagnosis of spinal abnormalities relies on anatomic changes and alterations in signal intensity in the abnormal regions. MRI is the method of choice for demonstrating abnormalities of the spinal cord

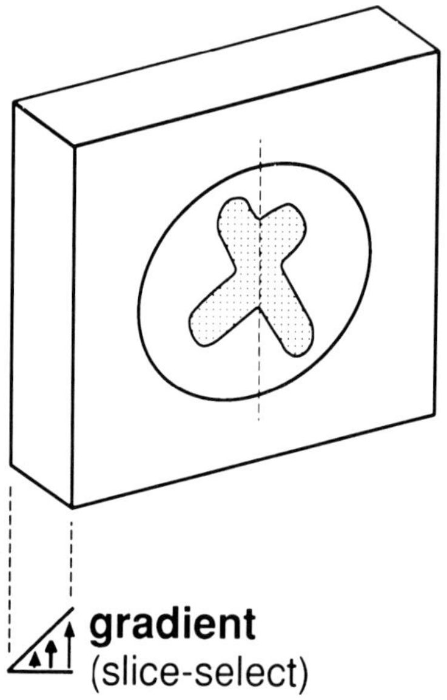

Figure 1.3. A gradient is used to select a slice for scanning. A strong gradient produces a thin slice.

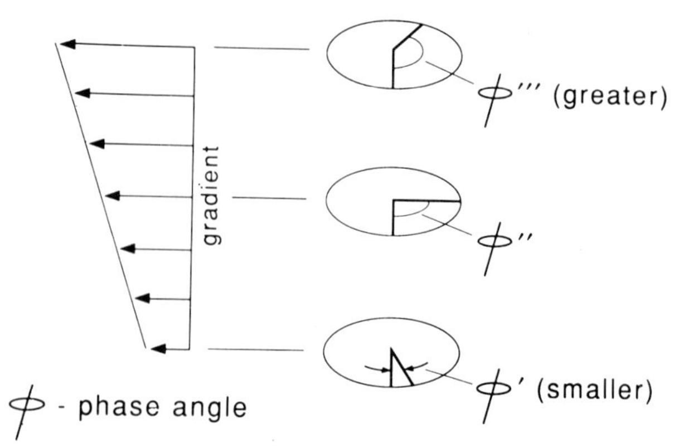

Figure 1.4. The phase angle at a particular time is related to the speed of precession. Each of these phase angles is encoded.

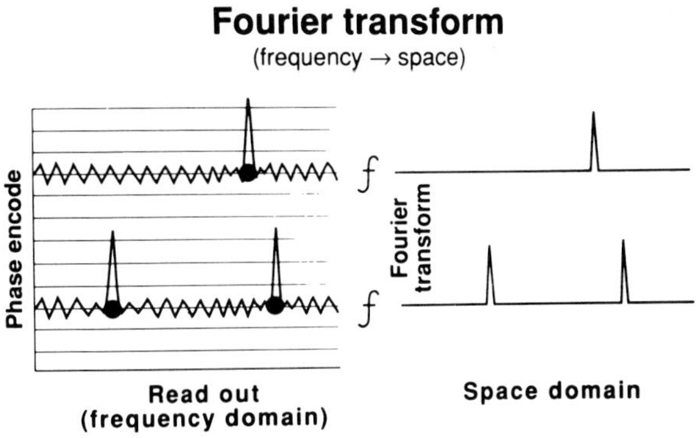

Figure 1.5. The signal along each of the phase-encoding lines is read out (frequency direction) and Fourier transformed into peaks of signal.

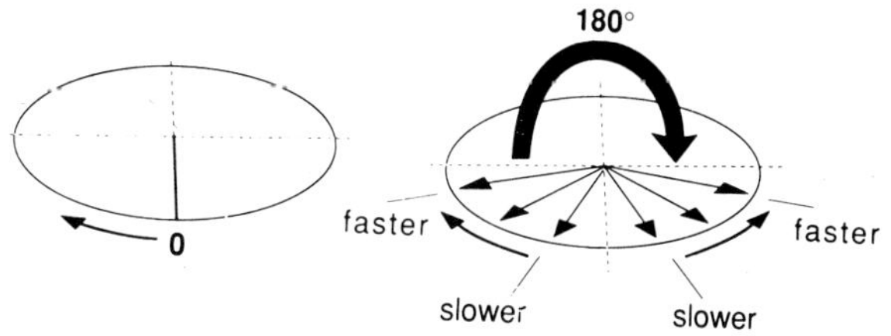

Figure 1.6. Spin echo. After an initial 90-degree RF pulse, a second 180-degree pulse is applied at a predetermined time. The direction of precession is reversed; the fast and slow protons "catch up" and dephase. At this point, signal is received. The time between each 90-degree RF pulse is the repetition time (TR). The echo time (TE) is twice the interval between the 90- and 180-degree RF pulses.

and canal but is limited in its ability to evaluate bony abnormalities. Spin-echo and gradient-refocused echo techniques are used routinely, but newer techniques utilizing rapid spin echoes shorten scanning time without compromising image quality [4]. The technical factors used are not universal, and they need to be optimized for specific field strengths as well as for scanners of individual vendors. Before discussing details of the imaging protocols, it is important to understand some of the basic principles relevant to resolution and artifacts in MR images of the spine.

Scanning Time

The scan time is dependent on the repetition time (TR), number of gradient-encoding steps, and signal averages:

$$\text{Scanning time} = \text{TR} \times \text{gradient step} \times \text{signal average}$$

This equation is applicable to all scanning techniques, and the scan time should be optimized for individual clinical problems.

Spatial and Contrast Resolution

Spatial resolution is dependent on the pixel size and the signal-to-noise (S/N) ratio of the image [3]. The ideal scan of the spine should have the maximal spatial and contrast resolution and be conducted in the briefest time possible. A short scanning time avoids patient movement and improves efficiency in use of the MRI scanner.

Spatial Resolution. The matrix size is determined by field of view (FOV) divided by the number of gradient steps utilized for phase encoding and readout; that is,

$$\text{Matrix size} = \text{FOV/gradient steps}$$

Readout steps may be altered by electronically increasing the amount of sampling in the frequency readout direction without increasing scanning time. Phase-encoding gradients are acquired sequentially, and the scanning time increases with the number of these steps utilized.

Signal-to-Noise (S/N) Ratio. The S/N ratio is related to the matrix size, magnetic field strength, bandwidth of the transmitted RF, efficiency of the receiver coil, and its proximity to the spine [5]. The signal changes logarithmically with the matrix size: Decreasing the matrix by half reduces the signal strength to a quarter and requires the scan time to be quadrupled. The signal is linearly related to the field strength so that a scanner of 1.5 T has three times the signal of a 0.5-T unit. The signals emanating from the spine and adjacent structures are within a relatively narrow range of frequencies. Signals outside this range are mainly due to noise. By transmitting an RF signal of narrow bandwidth, only clinically relevant signal is detected, and the S/N ratio is improved. Surface and Helmholz coils placed directly on the skin surface improve RF reception because they are close to the spine compared with body coils, which are within the gantry of the scanner.

The S/N ratio is also affected by the scanning technique and parameters such as the repetition, echo time, and flip angle. The signal is stronger with spin-echo than gradient-echo techniques and in general improves with longer repetition and shorter echo times. The way in which signal is affected by the flip angle is complex.

Radio-Frequency (RF) Coils

Radio-frequency (RF) coils are used to transmit and receive radio-frequency waves used for imaging. Body coils are located within the gantry of the scanner, close to the magnet, while surface coils may be placed directly on the body surface close to the region of interest. Some coils have a dual transmit and receive function, while others receive RF signals only.

Body Coils. Body coils are installed within the gantry of the scanner and are used for both transmission and reception of RF signals. These coils are used for examinations of the entire spinal column, but they have limited resolution. Reducing the field of view and increasing gradient phase-encoding steps to improve spatial resolution leads to marked deterioration of the S/N ratio, which necessitates an exponential increase in scan time. This prolonged scan time is usually impractical, and alternative methods of improving signal strength without a time penalty are preferable when a limited region of the spine is evaluated.

Surface Coils. Surface coils are used for receiving RF signals transmitted by the whole-body coil. Most of these coils are circular in shape, but rectangular coils are available for imaging longer segments of the lumbar and thoracic spine. The FOV is determined by the diameter and width of the coil and with the routine number of gradient steps (i.e., 256) results in images with excellent spatial resolution. The coils are placed on the skin surface close to the spine. Because of its proximity to the region of interest, the signal is strong, and it is even possible to use small matrixes to further improve spatial resolution. The RF signal received is strongest near the skin surface and diminishes progressively until it becomes undetectable beyond the radius of the coil. Because of this drawback, deep-seated abnormalities beyond the spinal column cannot be evaluated.

Helmholz Coils. These coils are used to examine the neck and consist of pairs of surface coils oriented parallel to each other. The region of interest is placed between the coils, which receive signals from both the anterior and posterior surfaces. These coils are ideal for examining the cervical region because structures both within and anterior to the spine are visible.

Phased-Array Coils. This system consists of four to six surface coils, each with its own receiver and memory board. The coils are arranged in a linear manner but are coupled together to provide a larger field of view equal to the length of the coils and within the homogenity of the magnet—thus increasing the FOV to 48 cm. By using the four separate coils the S/N ratio is increased from the spatial information received by the receiver coils. Usually, only four of the six coils are used in any given situation. Phased-array coils allow examination of the entire spine without repositioning the patient, thus reducing the total time of the examination. In the

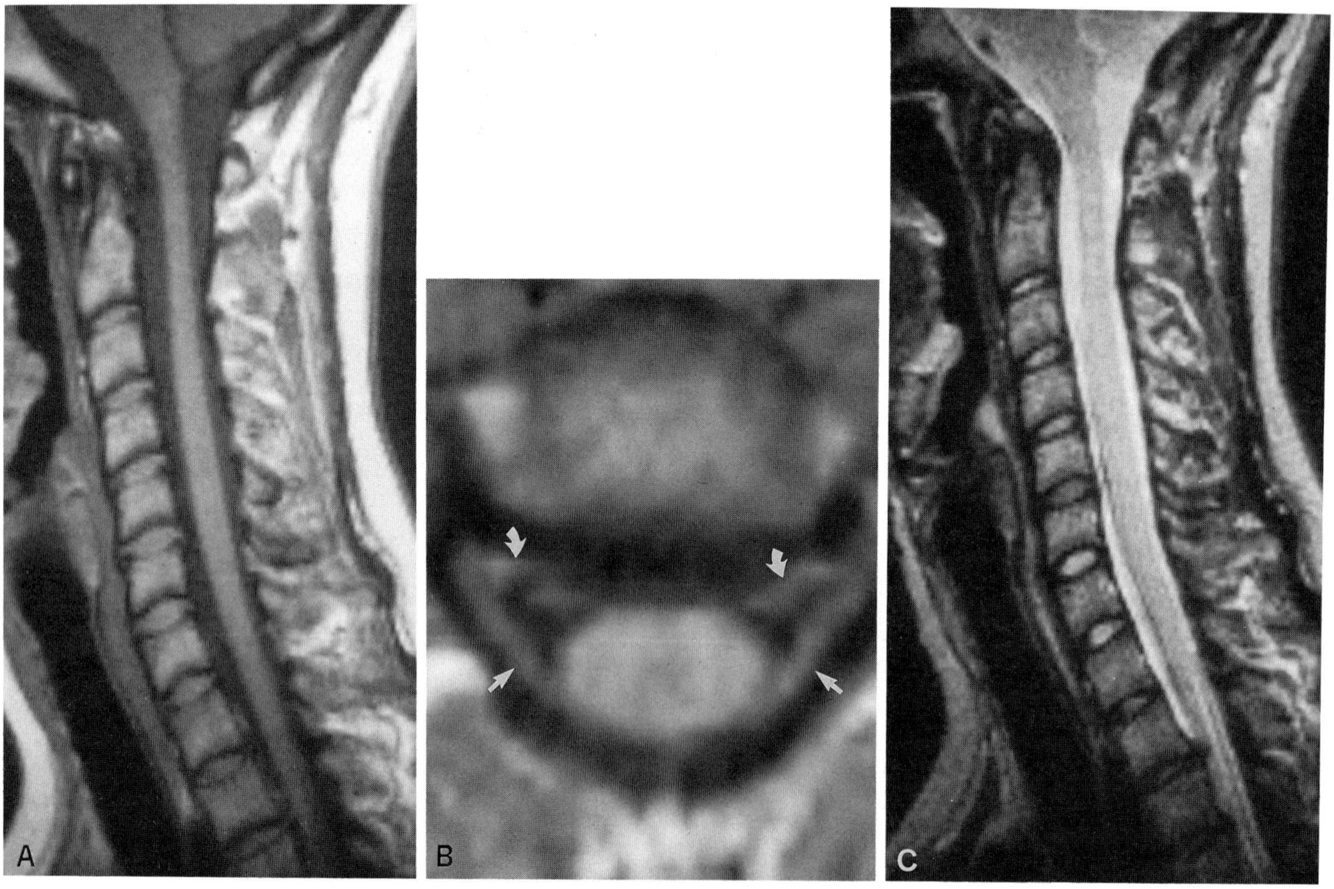

Figure 1.7. Spin-echo image of the cervical spine (**A**) and T1-weighted image (**B**). The dorsal (*arrows*) and ventral roots (*curved arrows*) are visible in the axial image. (**C**) T2-weighted image.

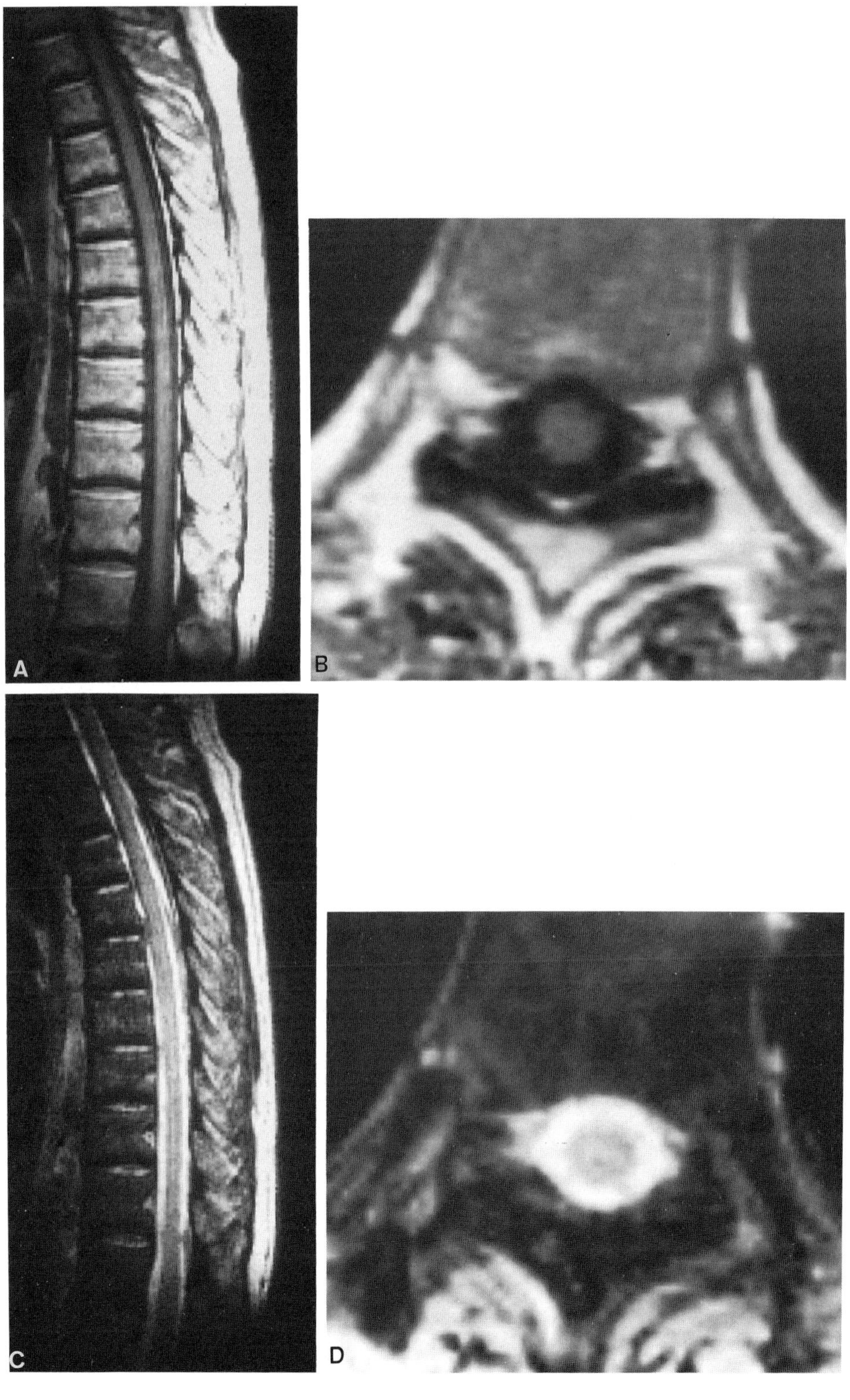

Figure 1.8. Spin-echo images of the thoracic spine (**A, B**), T1-weighted image (**C**), and T2-weighted image (**D**).

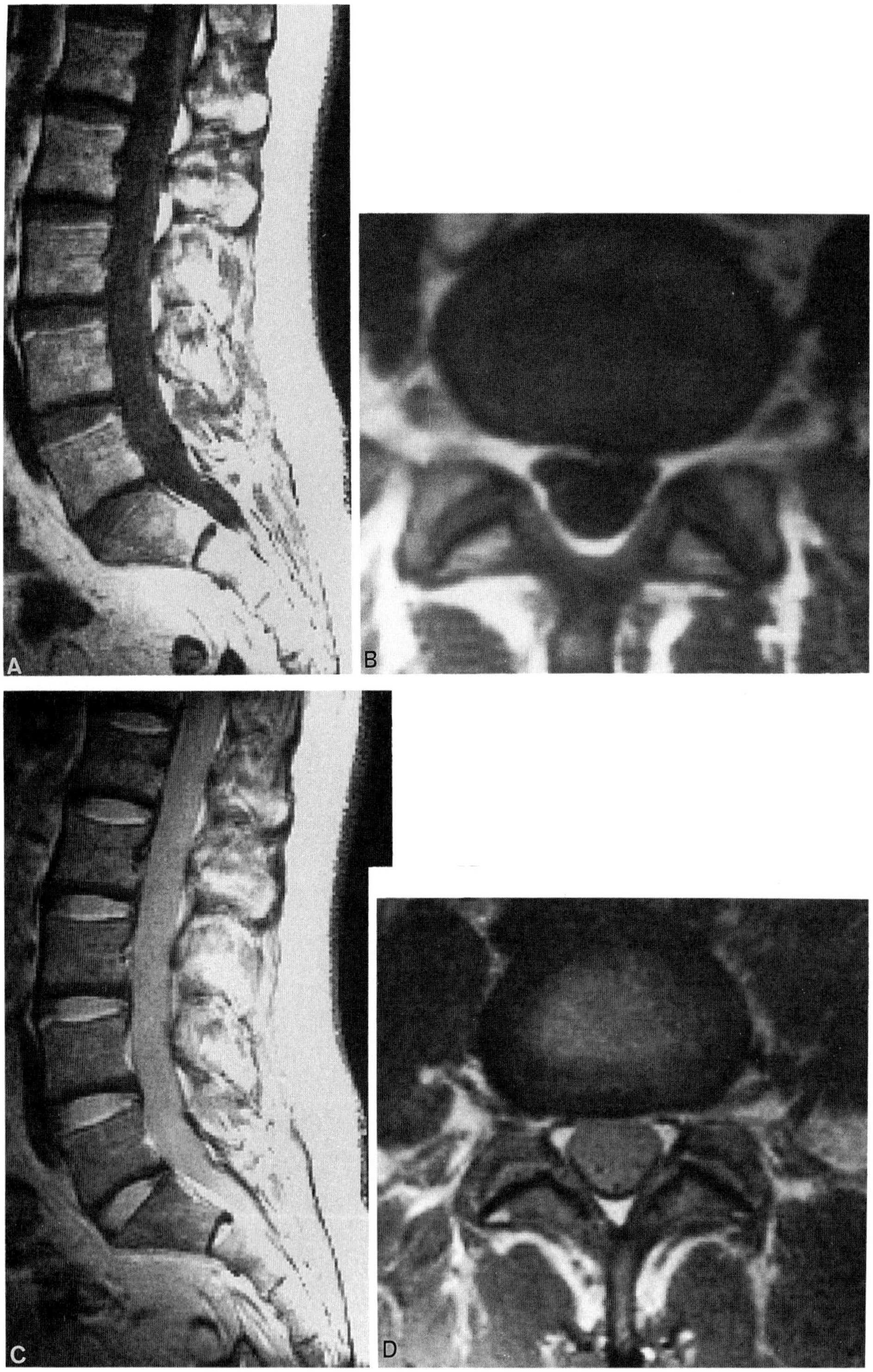

Figure 1.9. Spin-echo images of the lumbar spine. (**A, B**) T1-weighted images, (**C, D**), proton density, and T2-weighted image (**E, F**).

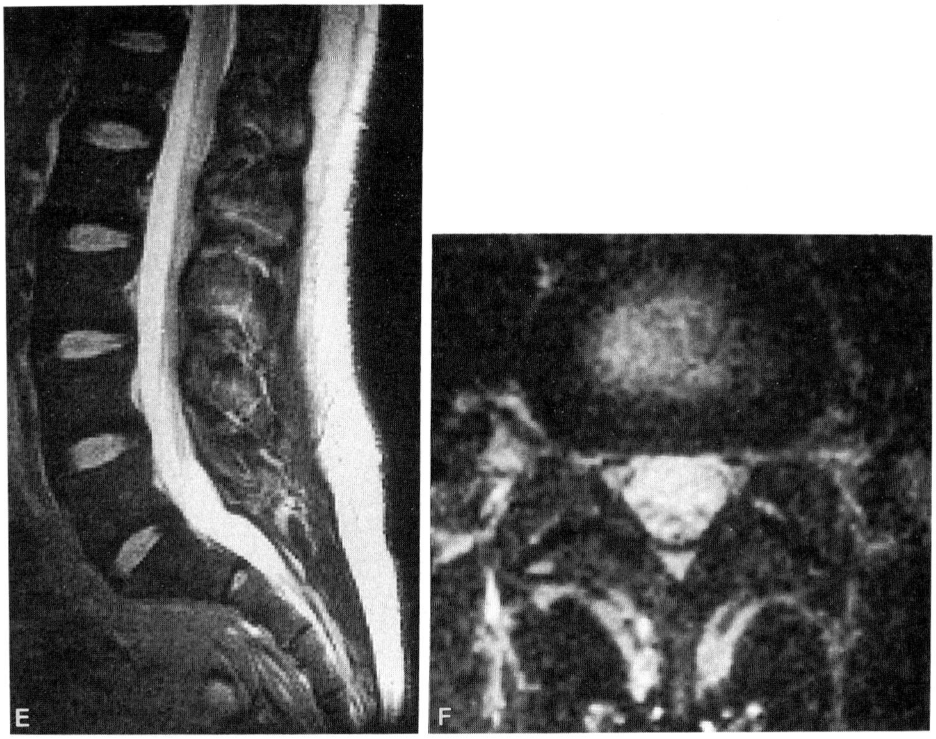

Figure 1.9–*continued*

phased-array arrangement, the cross-talk between adjacent coils is decreased, which also improves the S/N ratio. Each coil-receiver-memory system produces its own region of image. These are then combined, leading to the large FOV. Phased-array coil systems require acquisition in 512 × 512 matrix. The resultant increased processing time is corrected by simultaneously processing the data while the next level of imaging data is being acquired. This requires more powerful computers.

Hardware Improvements

Newer hardware has resulted in improved image quality and faster scanning times. Although the scanning time depends on the parameters used and remains unchanged, more powerful computers have made loading of the scanning programs and reconstruction of the images almost instantaneous. Better and stronger gradient coils allow thinner sections and shorter echo times, resulting in better spatial and contrast resolutions. Phase-array coils, as they become available, will allow for imaging of the entire spinal axis with the same quality as segmental surface coils and a reduction of the total scanning time.

Scanning Techniques

The aim of MRI of the spine is twofold: (*a*) to determine anatomic distortion and displacement of the spinal cord and nerves and (*b*) to detect abnormal signals within lesions and the deformed normal spinal cord and nerves. The first is akin to the changes observed in myelography and intrathecal CT examinations; the latter observations are unique to MRI.

Spin Echo. Protons are excited individually by repetitive RF pulses. The time between these excitations is known as the *repetition time* (TR). Spin-echo technique is used to enhance the signal generated by proton relaxation and is used routinely in MRI. After the initial 90-degree pulse, an additional 180-degree RF pulse is applied, and the resulting signal is received at a time identical to that between the two pulses; the interval between the initial 90-degree pulse and reception is known as the *echo time* (TE) (Fig. 1.6). The TR or TE may be adjusted to receive predominantly T1, T2, or proton-density signals. *T1 relaxation* is defined as the signal present after two-thirds of the longitudinal relaxation has occurred; spin-echo studies using TR and TE are predominantly T1-weighted. With a long TR, the protons relax almost completely; a long TE allows dephasing in the transverse plane, and the signal is predominantly T2-weighted. Multiple short and long TEs are possible when long TRs are used; the shorter TEs of these sequences are primarily proton-density-weighted [7]. T1-weighted images are obtained with short TRs and TEs; T2-weighted images are obtained with long TRs and dual-echo TEs. Images of the first echo of the T2-weighted images provide proton-density information.

In addition to multiple echo (TE), most spin-echo scans are acquired in multiple sections. Since the TR is up to several seconds, an RF may be initiated in multiple sections consecutively; the number of

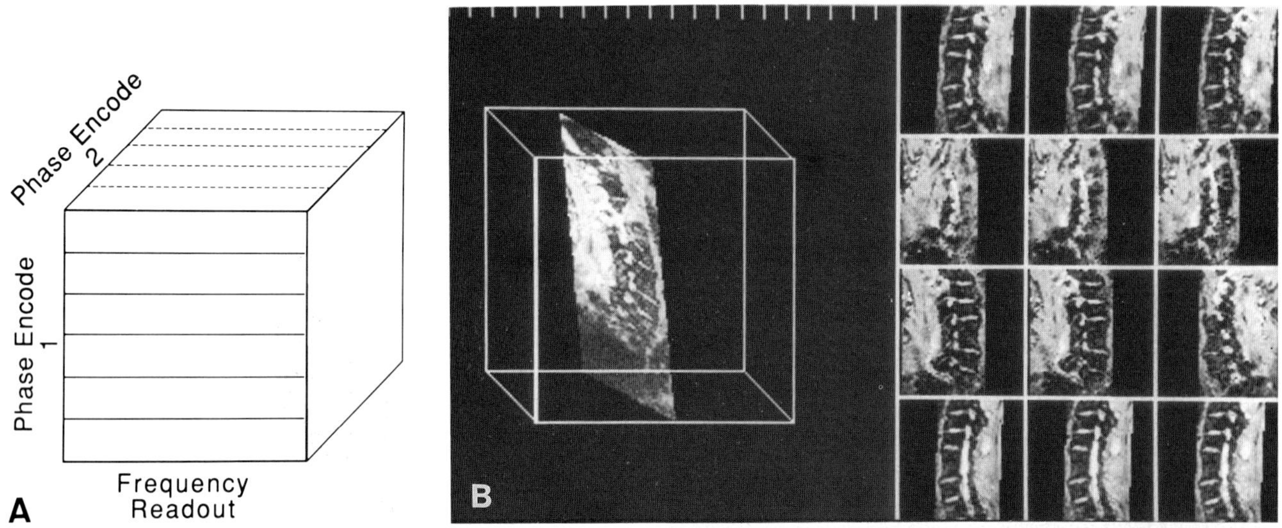

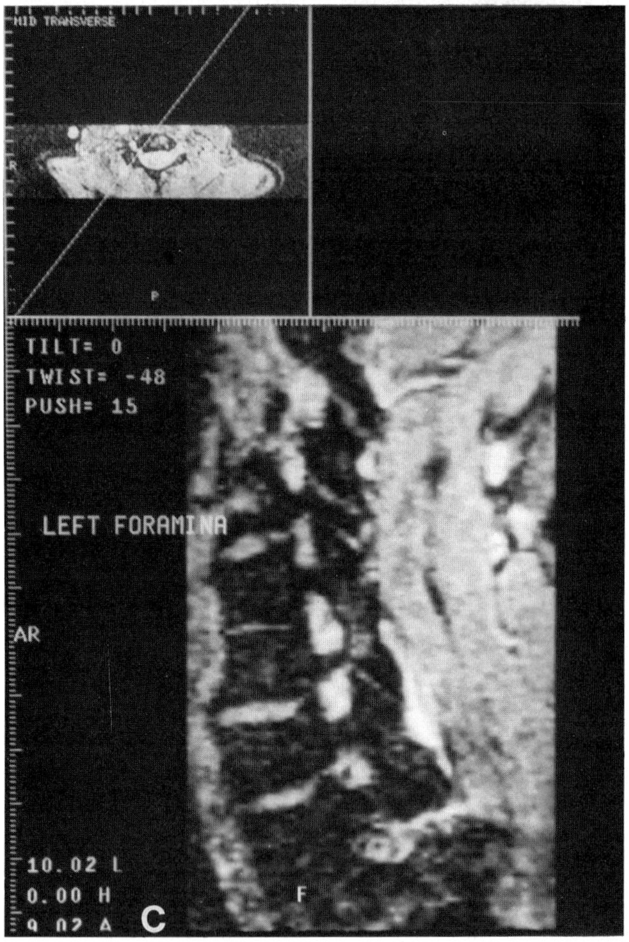

Figure 1.10. (**A**) Three-dimensional or volume scan is performed by applying a phase-encoding gradient at right angles to the initial encoding direction. (**B**) Application of this principle in the cervical spine with a series of sections of 2-mm thickness with no interslice gap. (**C**) Oblique sagittal image of the cervical neoroforamen generated from the preceding technique.

sections that may be acquired is determined by the longest TE used. Total scanning time is thus considerably shortened by this technique (Figs. 1.7 through 1.9).

Gradient Echo. Instead of using full 90- or 180-degree RF pulses, small flip angles are used, coupled with very short TRs. The precise T1 and T2 information received by these techniques is complex and depends on the flip angle as well as the TR and TE used [4, 7–9]. In two-dimensional sections, the slice thickness and location are determined by gradients in this direction. When phase encoding is applied along this slice-selection direction, true three-dimensional data are acquired (Fig. 1.10). Multiplanar sections may then be reconstructed from the three-dimensional data. This is particularly useful in the cervical region, since this allows evaluation of the neural foramen en face in the sagittal oblique plane. Gradient-echo techniques are used mainly for myelographic-type anatomic delineation of the spinal cord and nerves, and the signal observed has no real relevance to T1 and T2 relaxations (Fig. 1.11).

Fast Spin Echo. In conventional spin-echo technique, an initial 90-degree pulse is applied. It is followed at a predetermined time (1/2 TE) by an 180-degree RF pulse, and the signal is received at the same time interval following this latter refocusing RF. The signal data are received sequentially, line by line, along the phase-encoding direction. When a long TR is used, multiple echoes may be produced by repeating the 180-degree refocusing RF pulse. Fast spin echo (FSE) is a modification of the rapid acquisition with relaxation enhancement (RARE) technique, which takes advantage of the signal available in these repeated echoes. Instead of acquiring one line of phase-encoded data from a single echo, data are collected from several echoes during the same TR [10–12]. For example, when four echoes are produced, signal from each of the echoes is collected along separate phase-encoding lines, thus shortening the scan time by a quarter. Up to 16 lines may be used to acquire as many echoes within a single TR, and the entire image may be synthesized in one-sixteenth the time of the conventional technique. Unlike conventional spin echo, where the echo time is exactly at the time of reception of the RF, the echo time of FSE is a composite of several echoes. The nominal echo time is determined by the time the initial echo is collected, and the rest of the data are "filled in" by the signal from the subsequent echoes. The exact role of this technique is not yet determined, but preliminary experience suggests that the S/N ratio of these images is superior to that produced by conventional T2 spin-echo methods (Fig. 1.12).

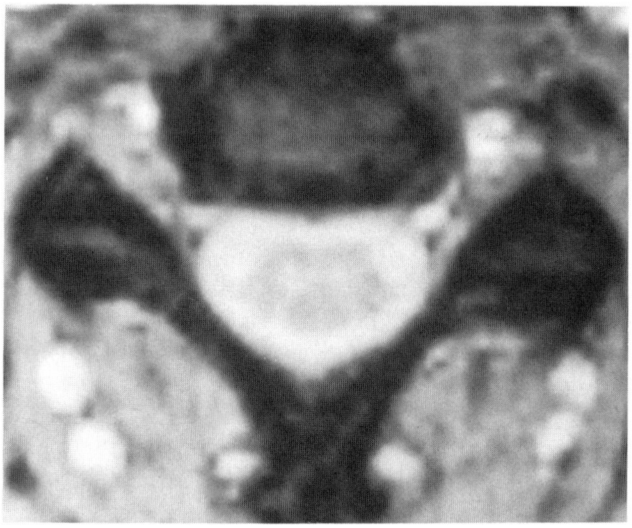

Figure 1.11. Gradient-echo image of the cervical spine.

Intravenous Contrast. Contrast agents rely on the T1 shortening of rare metals. Gadolinium is highly toxic, but it is rendered innocuous by chelation with DTPA. Intravenous gadolinium-DTPA is the contrast agent used with T1-weighted techniques. In general, enhancement within the spinal cord indicates increased vascularity or breakdown of the blood–spinal cord barrier [13, 14]. Epidural, meningeal, and subarachnoid space enhancement is due to increased vascularity. The specific use of this agent will be described in subsequent chapters.

Artifacts

Artifacts not only degrade the quality of the MR images but may be misinterpreted as true abnormalities. Some of these artifacts may be corrected by software programs installed in the scanner, but others require judicious implementation of scanning

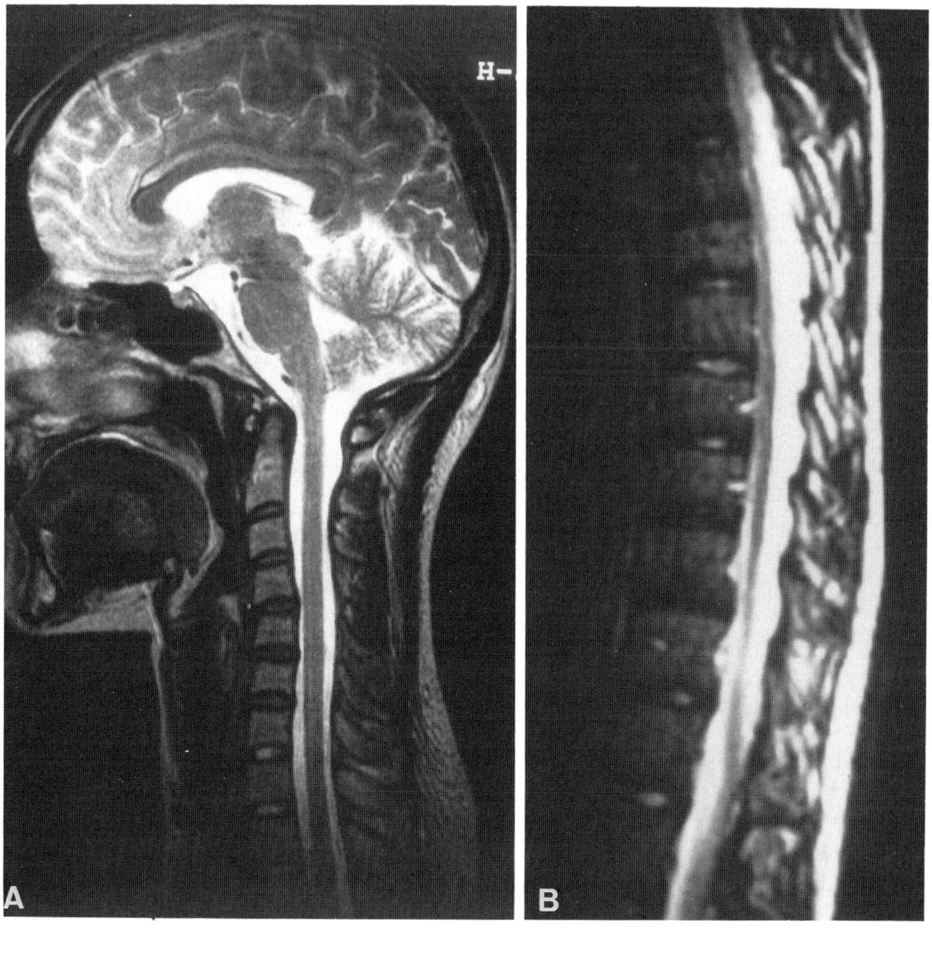

Figure 1.12. Fast spin-echo image of the cervical spine.

parameters. It is important to try to understand the physical basis of the common artifacts so that they may be corrected by changing imaging parameters. MRI artifacts may be divided into those related to movement, sampling, chemical shift, and susceptibility effects [15–17]. Artifacts due to electronic or engineering errors require assistance from service engineers and will not be discussed here.

Physiologic Movement Artifacts. Involuntary physiologic movement artifacts include cardiac motion, pulsation of vessels outside the spine and within the subarachnoid space, as well as respiration and bowel peristalsis. These physiologic movements may cause misregistration during the short time interval between phase encoding and frequency readout. The resulting artifact is always in the phase-encoding direction [18] (Fig. 1.13). Since this interval is proportional to the echo time, such artifacts are more pronounced in T2-weighted images, where this time is longer than in T1-weighted sequences. Artifacts secondary to pulsations within the subarachnoid space are confined to the spinal canal, but pulsations outside the canal, respiratory and bowel movement artifacts, are "transmitted" throughout the region imaged and cause considerable degradation in the images of the spinal canal.

Four methods are available for correction of phase-encoding movement-related artifacts. First, artifacts arising from pulsations within the spinal canal may be reduced by application of appropriate correction gradients, known variously as *gradient-motion nulling, gradient-motion reduction,* (GMR), *flow void,* and *motion artifact suppression technique* (MAST) [19]. These techniques utilize gradients that are opposite in polarity and have an equivalent phase angle to those of protons in motion. The positive phase is canceled by the negatively applied gradient and will thus eliminate or reduce the artifacts caused by signal from moving protons (Fig. 1.14). Second, phase encoding is normally arranged in an anteroposterior direction for routine scanning of the spine, and artifacts arising from vascular and cardiac pulsations and respiratory movement are present anterior to the spinal canal. These artifacts may be suppressed by application of a presaturation pulse in this region without obscuring the region of interest in the spine [20]. This 90-degree RF pulse applied immediately prior to the imaging pulses has the effect of saturating the protons in this region so that they cannot emit signals to cause artifacts. The presaturation pulse may be varied in thickness and may be placed in any location and direction (Fig. 1.15). When properly placed, such a pulse is extremely effective in suppressing movement artifacts, but it fails where the region of movement cannot be completely covered. This occurs most commonly in patients with kyphosis, in whom the region between the presaturation slab and the curved thoracic spine is not included. Phase-encod-

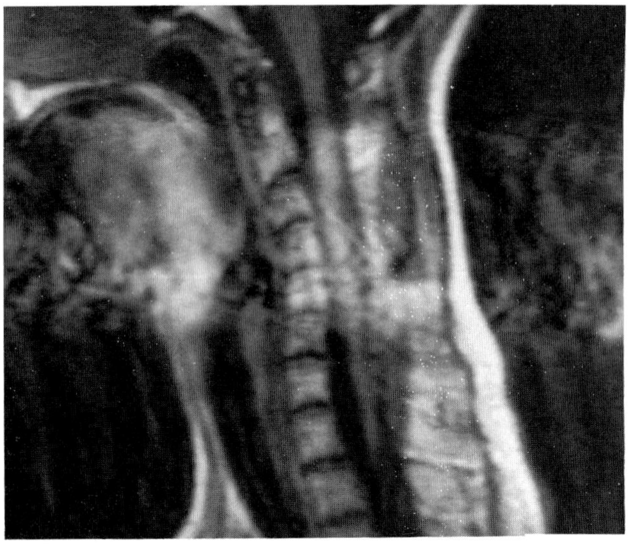

Figure 1.13. Movement artifact due to respiration and swallowing is in the phase-encode direction.

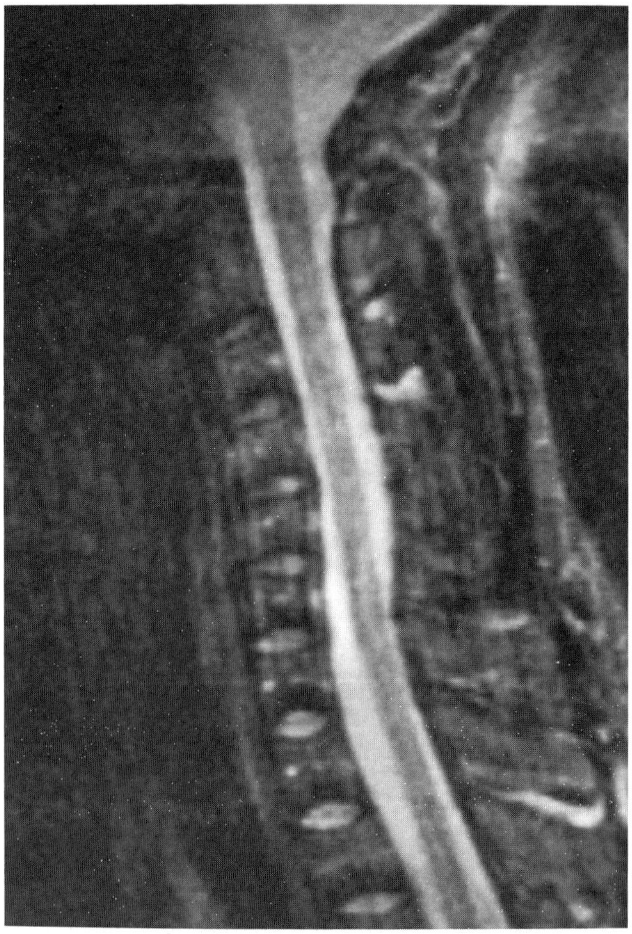

Figure 1.14. Pulsation artifacts within the subarachnoid space are suppressed by the gradient-motion correction technique. Respiratory motion outside the spinal canal is uncorrected.

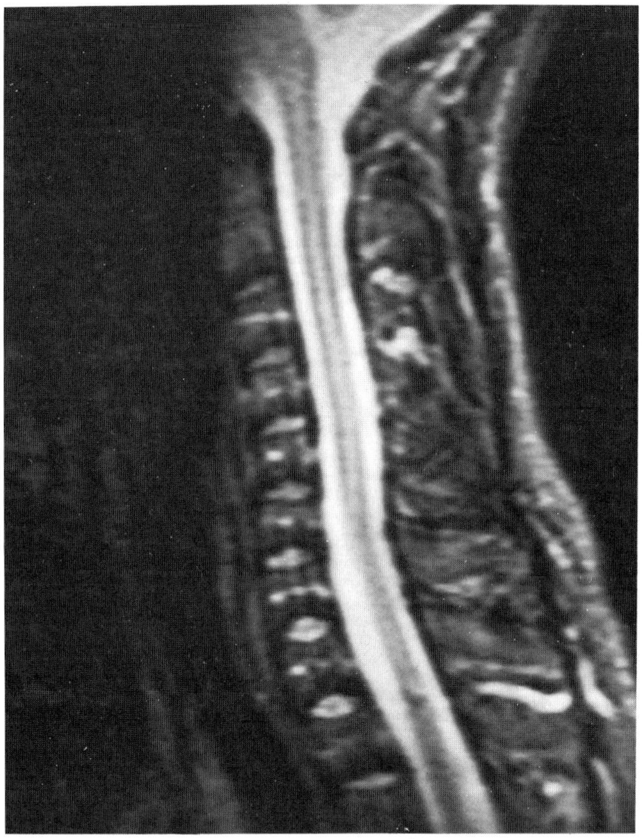

Figure 1.15. Respiratory and swallowing movement is suppressed by application of a presaturation pulse anterior to the spine. Note the pseudosyrinx due to truncation artifact (see text).

ing movement artifacts that are present in more than one direction may be suppressed by additional presaturation in these locations. Third, artifacts caused by movement in the anteroposterior direction may be eliminated by switching the phase-encoding to the superoinferior direction. This technique, unfortunately, results in "wrap around" phase-related artifacts in the superoinferior direction and will be considered below. Finally, cardiac gating and triggering the RF pulse to the ECG or finger pulse plethysmograph improve pulsation artifacts within as well as outside the spinal canal but fail to correct artifacts from other types of physiologic movement.

Gradient motion correction techniques are installed in most scanners and usually are applied routinely for all spine scanning. Presaturation pulses must be implemented by scanning personnel using a "scout" scan as a guide to placement. Changing the direction of phase encoding may be effective in some cases, but pulsatile gating is time-consuming and not totally effective and is seldom used except in cases recalcitrant to other methods of motion suppression.

Sampling Artifacts. *Truncation artifacts* are due to undersampling of data points, which results in "aliasing" and failure to resolve anatomic structures that are in close proximity. Spatial resolution on MRI scanning is dependent on the frequency of sampling; i.e., resolution increases with the frequency. When the sampling frequency is less than that of the structure imaged, the margins between structures are not resolved [21] (see Fig. 1.15). These artifacts are seen commonly when the body coil is used. They are sometimes seen in surface-coil images of small children, in whom the spine is small relative to the coil.

A truncation artifact may be corrected by decreasing the field of view (i.e., using a surface coil instead of a body coil) and increasing both the frequency and the phase-encoding steps. Increasing frequency encoding, for example, from 128 to 256 to 512 steps is achieved without a time penalty. A similar increment in phase encoding doubles or quadruples the scanning time. In practice, use of 256 phase-encoding steps constitutes the best compromise for time and spatial resolution. A truncation artifact from the spinal cord–cerebrospinal fluid interface is common because this boundary is in the same orientation as the phase-encoding gradient, which is more likely to be undersampled. This truncation or Gibb artifact is seen as parallel dark and bright lines within the spinal canal which may be mistaken for a syrinx on T1- and T2-weighted images.

Spatial Localization Artifact. This "wrap around" artifact is in phase-encoding directions and is due to failure to localize an RF signal situated outside the region of interest. This artifact is seen when the FOV used is smaller than the torso. This problem is particularly troublesome with the body coil, necessitating a large FOV. When surface coils are used, signal anterior to the region of interest of the surface coils may be "wrapped" posteriorly by orientating the phase encoding in the anteroposterior direction. Such unwanted signal will then be located behind the back and thus will not be superimposed on the spine (Fig. 1.16).

Chemical Shift Artifact. This artifact is the result of slight differences in precession frequencies between fat and water in tissues. This difference, which becomes progressively more marked with increasing field strengths, may be noticed in images obtained at 1.5 T and is accentuated by narrowing of the frequency bandwidth [22]. Fat and water give rise to two distinct signals which are spatially encoded and displayed at slightly different locations. The artifact is present only in the frequency-encoding direction and should not be mistaken for an abnormality (Fig. 1.17).

Electronics and Engineering Artifacts. T2 and GRE images may be suboptimal due to poor signal of low-field magnets and inhomogeneity of the magnetic field. The signal may be improved by increasing signal averaging, but this may be impractical be-

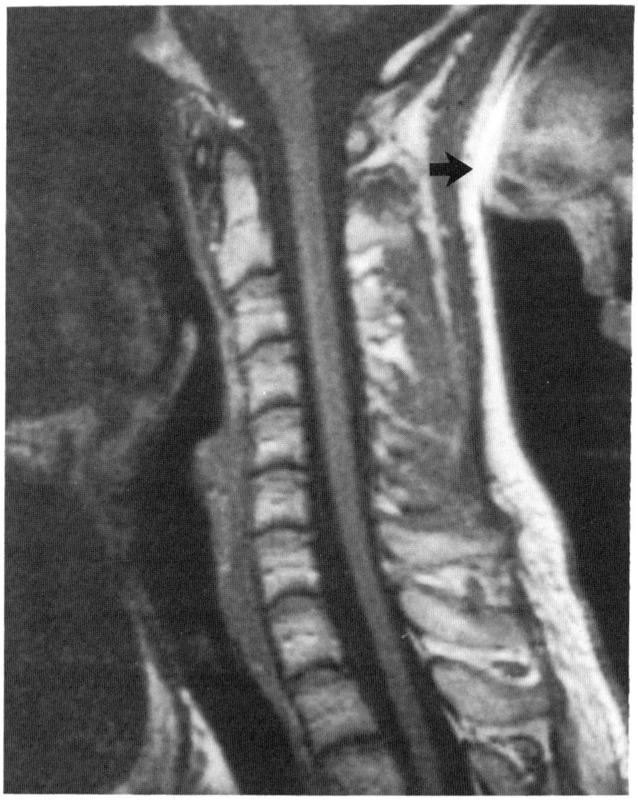

Figure 1.16. Wrap-around artifact. The nose (*arrow*) is wrapped to the posterior part of the image because of the small field of view.

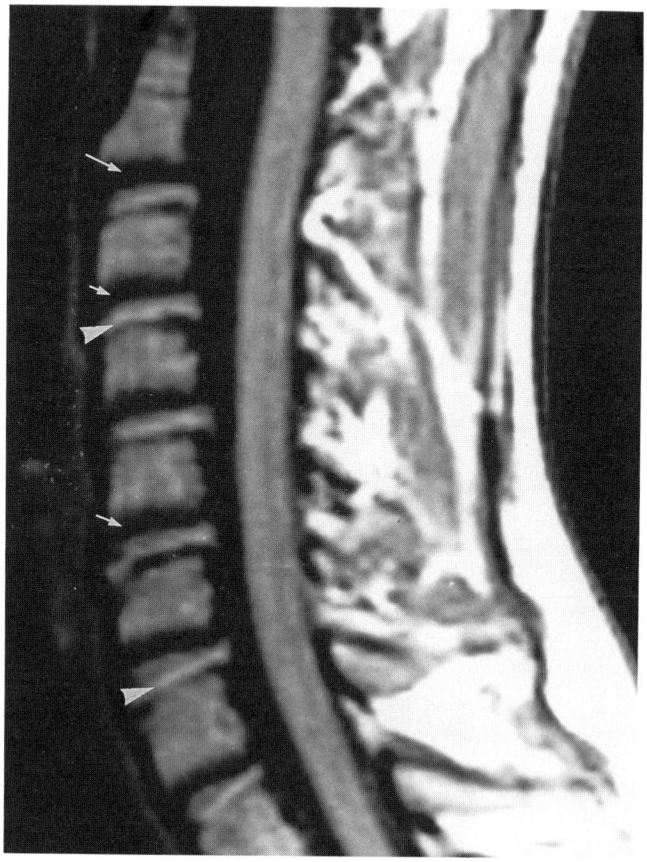

Figure 1.17. Chemical shift artifact. Lateral T1-weighted MRI of the cervical spine demonstrating the hypointense disk endplate margin along the superior aspect of the disk space (*arrow*) and bright signal along the opposite side of the disk endplate (*arrowhead*) resulting from signal-encode direction.

cause of prolongation of scanning time. Artifacts which are not recognized as one those described above are usually due to defective electronics and engineering problems. They are usually responsible for excessively noisy images and have to be corrected by the field engineer.

Magnetic Susceptibility Artifacts. Retained metallic fragments following spinal surgery and trauma cause severe signal loss far in excess of the size of the metallic object. Such an artifact is caused by extreme shortening of relaxation due to the metal [23] and is currently uncorrectable.

CLINICAL SCANNING PROTOCOLS

The techniques for imaging of the spine using both body and surface coils will be discussed. Protocols for evaluation of specific diseases will be considered in subsequent chapters. The technical factors employed in this chapter are based on experience with the Siemens and GE scanners. Since the specific technical settings vary somewhat among scanners from different vendors, those presented here should be used as a guide only [24] (Tables 1.1 and 1.2).

T1-weighted, proton-density (also referred to as *balanced* or *spin-density images*), T2-weighted, and gradient-echo (GRE) imaging sequences are used in imaging of the spine. Sagittal T1 sequences are used routinely in all scans, but the choice of the other sequences depends on the disease process under evaluation. In general, lesions of the spinal cord and canal require T2-weighted images, but degenerative and disk diseases may be imaged by either GRE or T2-weighted techniques. Both these techniques produce a myelographic effect useful for assessing anatomic distortions of the subarachnoid space. GRE scans take less time than T2-weighted techniques, but signal changes around the intervertebral disks are poorly characterized. The choice among these techniques for evaluation of degenerative and disk diseases depends on the time available and cooperation of the patient. Images of the first echo of a T2-weighted (spin-density) double-echo sequence are superior to GRE for evaluation of lumbar disk and degenerative diseases. The selection of sequences and projections is summarized in Table 1.1.

Multi-echo spin and gradient-echo sequences are acquired with surface, Helmholtz, and body coils with multi-slice techniques. A 256 phase-encode and either a 256 or a 512 frequency-encode matrix are used with the surface and Helmholtz coils; a 512 $\times$ 512 matrix should be obtained with the body coil.

GRE images are proton or T2*-weighted images which produce a myelographic effect with bright cerebrospinal fluid signal. (In T2*-weighted MRI, the asterisk represents effective relaxation time following free induction decay within heterogeneous tissue. It is not true T_2 and is shorter in time than T_2). Sections are 4 mm or less in thickness and are separated by no more than 10% interslice gap or are contiguous using an interleave technique. Orthogonal sagittal and axial projections are used, except for evaluation of cervical and lumbar disk disease, where angled sections parallel to the disk spaces are more sensitive in detecting subtle herniations. Gradient-motion compensation techniques are routinely applied. Phase encoding is in the anteroposterior direction, and physiologic movements are

Table 1.1.
Degenerative and Disk Disease

	Cervical	Thoracic	Lumbar	Whole Spine
T1				
Sagittal	+	+	+	+
Axial	(+) (optional)	+	+	+
T2				
Sagittal	+	+	+	+
Axial		−	+	−
GRE				
Sagittal	−	−	(+)	−
Axial	+	+	−	−
Contrast Material (postoperative)				
Axial	+	+	+	−
Sagittal	(+)	(+)	(+)	(+)
Coil	Helmholz	Surface	Surface	Body
FOV*	230/280	230/400	250/400	500
Matrix†	256 × 256o	256 × 256o	256 × 256o	512 × 512o
Thickness‡				
Sagittal	3–4 mm	3–4 mm	4–5 mm	3–5 mm
Axial	4–5 mm	4–5 mm	4–5 mm	4–5 mm
Coronal	4–5 mm	4–5 mm	4–5 mm	4–5 mm

*Axial/sagittal and coronal.
†o = oversampling in the readout direction, e.g., 256o = 512.
‡No gap between section is preferred, up to 20% permissible.
Note: Techniques for T1W1: TR500–800, TE15–22
T2W1: TR2500, TE20, 90
GRE: TR300, TE11, flip angle 30°
Gradient motion, nulling routinely.
Presaturation anterior to spine for all T2-weighted images.

Table 1.2.
Other Diseases, Including Neoplasms

	Cervical	Thoracic	Lumbar	Whole Spine
T1				
Sagittal	+	+	+	+
Axial	+	+	+	+
T2				
Sagittal	+	+	+	+
Axial	(+) (optional)	(+)	(+)	(+)
Contrast Material				
Axial	+	+	+	+
Sagittal	(+)	(+)	(+)	(+)
Coronal	(+)	(+)	(+)	(+)
Coil	Helmholz	Surface	Surface	Body
FOV*	230/280	230/400	250/400	500
Matrix†	256 × 256o	256 × 256o	256 × 256o	512 × 512o
Thickness‡				
Sagittal	3–4 mm	3–4 mm	4–5 mm	3–5 mm
Axial	4–5 mm	4–5 mm	4–5 mm	4–5 mm
Coronal	4–5 mm	4–5 mm	4–5 mm	4–5 mm

*Axial/sagittal and coronal.
†o = oversampling in the readout direction, e.g., 256o = 512.
‡No gap between section is preferred, up to 20% permissible.
Note: Techniques for T1W1: TR500–800, TE15–22
T2W1: TR2500, TE20, 90
Gradient motion, nulling routinely.
Presaturation anterior to spine for all T2-weighted images.

suppressed by placing a selective presaturation pulse anterior to the spine. Where such a technique is unavailable, the phase encoding is in the superoinferior direction, and a phase-wrap correction technique is utilized. Cardiac and respiratory gating is rarely necessary.

ANALYSIS OF MRI SCANS

Anatomic Delineation

Optimal anatomic delineation is achieved when there are sharp differences in contrast between the cerebrospinal fluid, spinal cord, nerve roots, and epidural spaces. Cerebrospinal fluid is dark and epidural fat appears bright on T1-weighted images, and this technique is optimal for evaluating compression and displacement of the spinal cord and nerve root. However, because of similarities in the signal of the subarachnoid space, posterior longitudinal ligament, and intervertebral disk margins, it is often difficult to differentiate between these structures. T2-weighted and gradient-echo techniques produce a myelographic effect which is especially effective in demonstrating subtle compression of the subarachnoid space and nerve roots.

Contrast Changes

T1 techniques are useful in demonstrating contrast changes due to hematomas and fat and may detect decreased signal in some lesions. T2 techniques are in general more sensitive in demonstrating signal changes caused by abnormalities of the spinal cord and vertebrae. They are particularly sensitive to subtle changes in water content within intervertebral disks caused by degenerative disease and disk herniations. Gradient-echo images are dependent on proton density and T2* effect and are less reliable than T2-weighted techniques in detection of pathologic changes of signal.

Computed Tomography

Computed tomography (CT) involves sectional imaging of the spine, primarily in the axial plane. The basic principle in CT imaging involves passage of a thin source of x-ray beam perpendicular to the longitudinal axis of the body. After its passage through the body with its different tissue characteristics, the attenuated x-ray beam is picked up by detectors. The thin beam may be fan-shaped or a pinpoint. The attenuated data are collected in all 360 degrees by an array of detectors. The analog data are digitized, computed, and subsequently displayed as a cross-sectional or tomographic sectional image of the region of interest. The final image is the result of application of extensive mathematical computation, which has been dealt with extensively by others [26]. In imaging of the spine, CT scanning is excellent in demonstrating the osseous anatomy and, to a limited extent, the soft-tissue anatomy.

Evaluation of the soft-tissue anatomy within the spinal canal is augmented by utilizing intravenous or intrathecal contrast material. Intravenous contrast material opacifies the vascular channels and tissues with increased neovascularity. The vascular enhancement

in the extradural space provides additional contrast between the osseous and ligamentous structures,

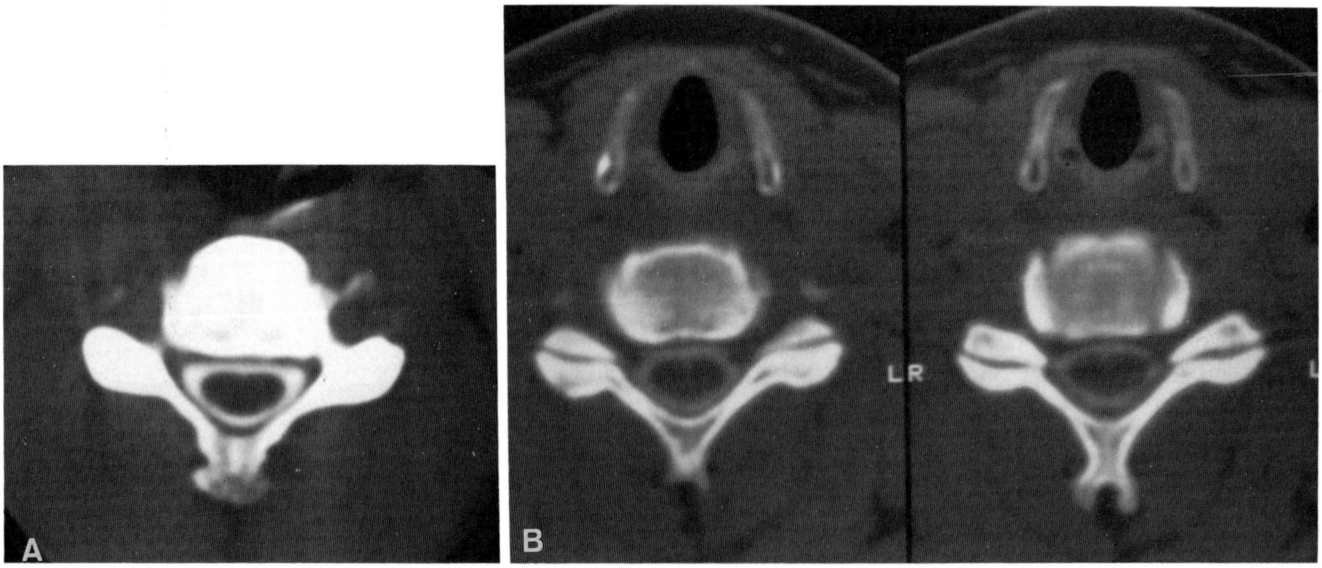

Figure 1.18. Axial CT sections in (**A**) soft tissue and (**B**) bone window algorithm demonstrate the difference in soft-tissue detail and contrast. To a limited extent, these also can be achieved by changing the window level and contrast setting of individual slices.

the epidural fat, and the spinal cord and nerve roots. Intrathecal contrast material allows opacification of the cerebrospinal fluid spaces, providing differentiation of intradural anatomy from the extradural compartment.

Image quality depends on numerous factors, including the anode tube–detector system configuration, the attenuation coefficiency of the x-ray beam, the heat capacity of the anode tube, the sensitivity of the detector system, the thickness of the slice that can be imaged, and the spatial resolution.

SCANNING TECHNIQUE

CT scans are all acquired in the axial plane. Conventional scanners acquire individual scans sequentially, but newer scanners have the capability of acquiring scan data in a continuous fashion. This is possible by virtue of the "slip ring" design, whereby the anode rotates continuously as the table moves in increments. The parameters which may be varied are angulation, thickness of the section, amperage, and the reconstruction algorithm. In addition to display of images in the original axial plane, two- and three-dimensional images in multiple planes (MPR) can be reconstructed from the original data set.

Angulation and Slice Thickness

In general, the gantry is angled parallel to the intravertebral disk spaces when herniation and degenerative diseases are suspected. Scan sections should cover the disks and at least half the adjacent vertebral bodies. Orthogonal sections are useful in evaluation of other spinal disorders. For routine scanning, slice thickness varies between 3 and 5 mm, preferably with 1 to 2 mm of overlap between contiguous slices. When finer detail is desired or when three-dimensional or MPR reconstructions are contemplated, 2-mm-thick sections are used.

When scans are acquired using the "spiral" mode, the thickness of the sections during one revolution of scanner rotation is determined by the speed of the table movement. Sections as thin as 1 mm are possible with table speeds of 1 mm/s. This method is idcal for three-dimensional and MPR reconstructions.

Radiographic Factors

A power setting of 120 kV is used in most scanners and is usually fixed. Between 400 and 600 mA is used for evaluation of soft tissue and bone. A lower amperage is permissible for bone scanning

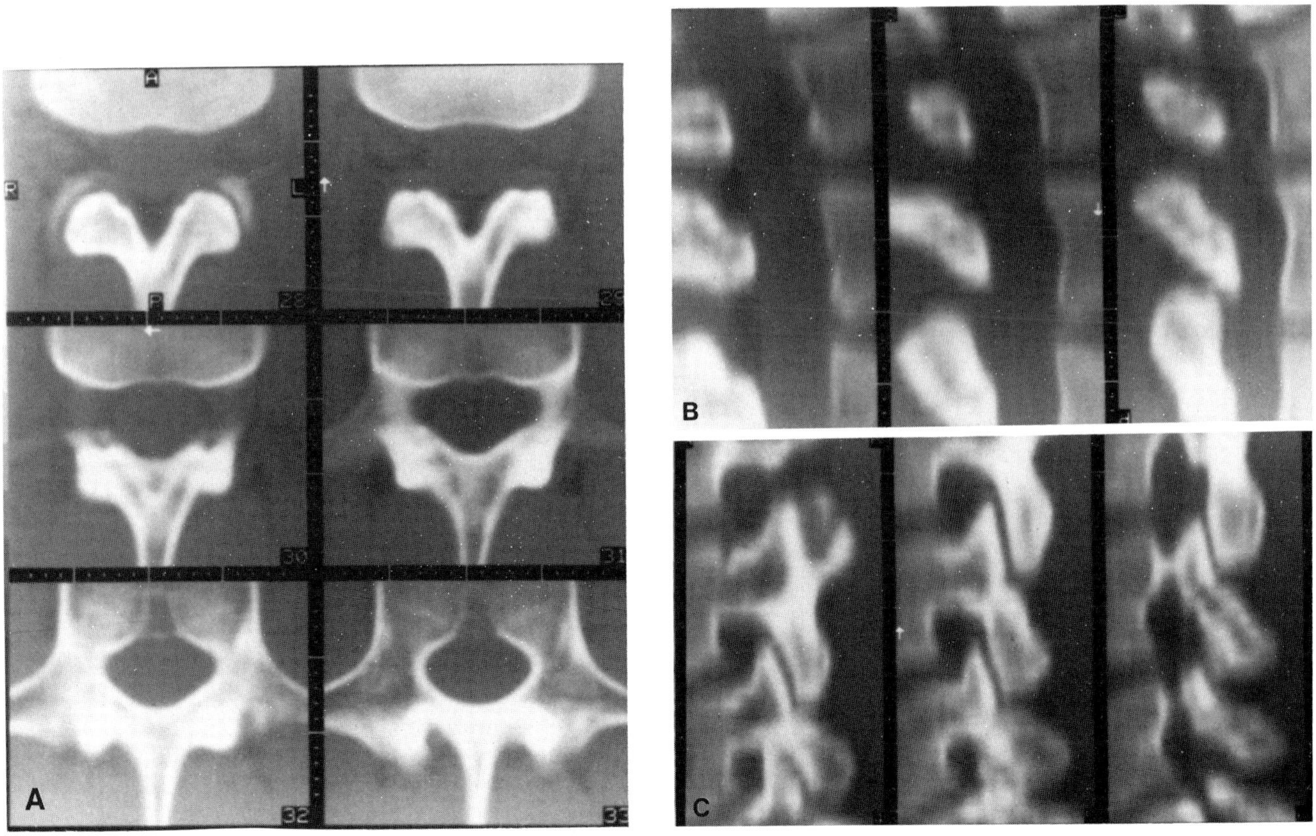

Figure 1.19. MPR images obtained by stacking series of (**A**) contiguous axial sections producing 1-mm sagittal tomographic sections in the (**B**) midline and (**C**) to show the facet joints and neural foramen.

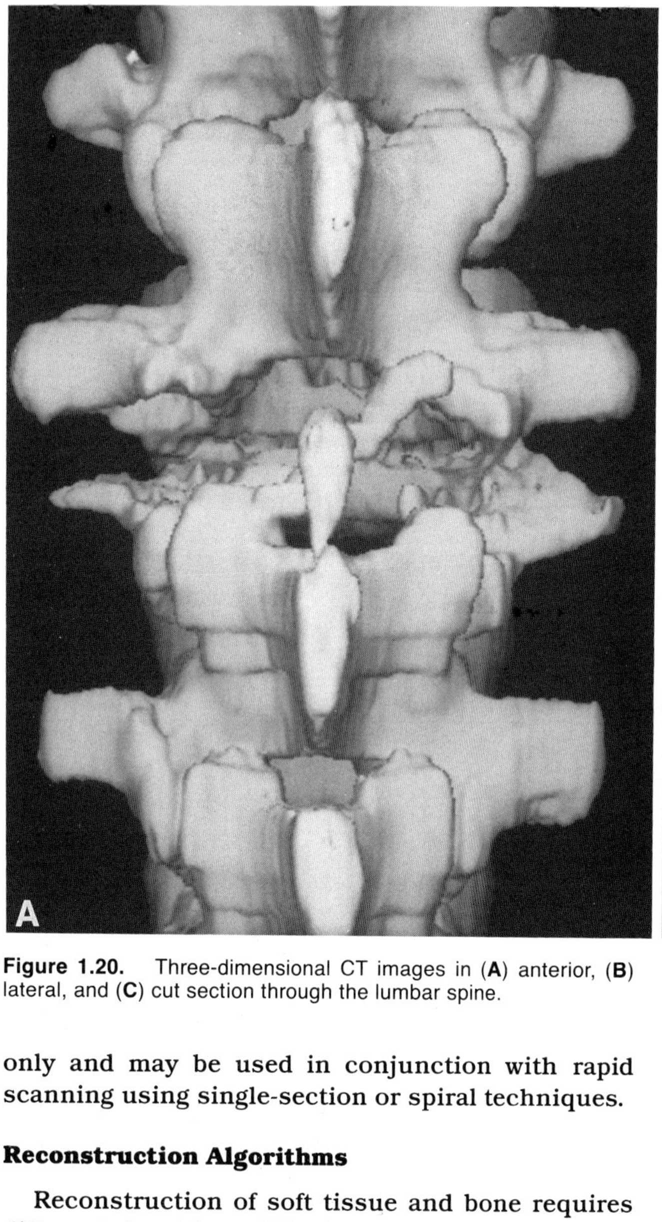

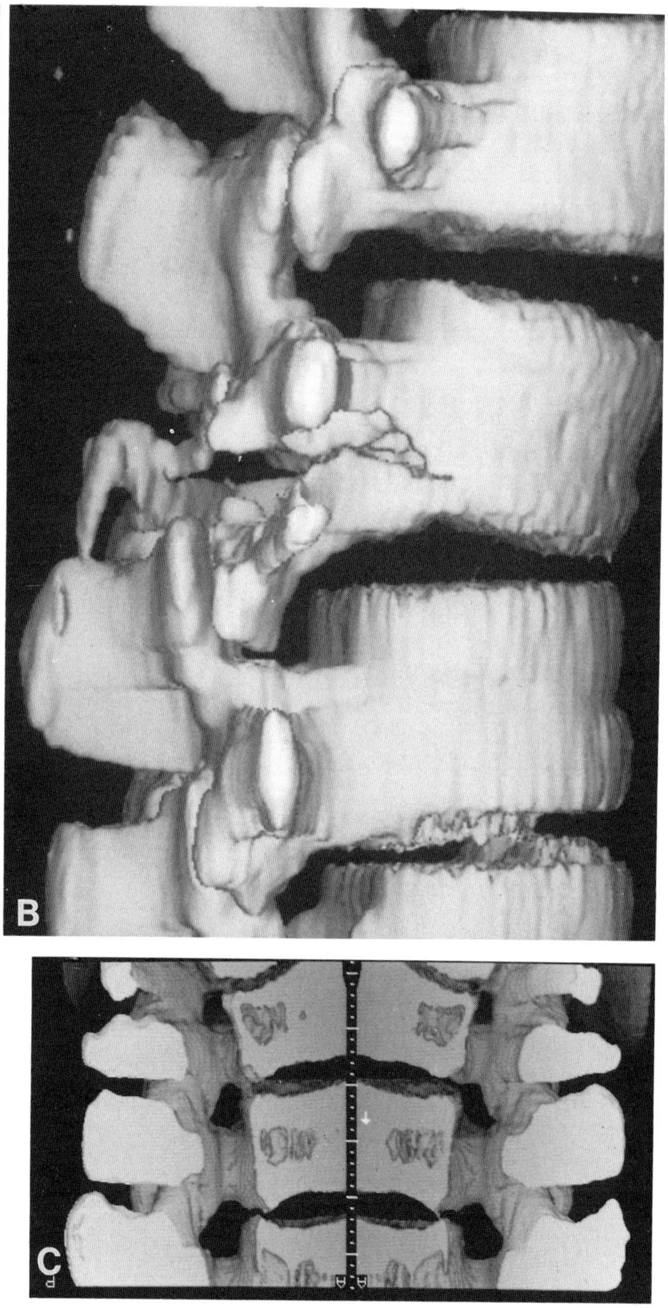

Figure 1.20. Three-dimensional CT images in (**A**) anterior, (**B**) lateral, and (**C**) cut section through the lumbar spine.

only and may be used in conjunction with rapid scanning using single-section or spiral techniques.

Reconstruction Algorithms

Reconstruction of soft tissue and bone requires different algorithms. The differences in densities between soft tissues are small, and reconstruction methods that enhance such subtle differences are required. Bone has very high density compared with other tissues, and algorithms that enhance and sharpen the edges are desirable, at the expense of poor visualization of surrounding soft tissues. Most scanners have a choice of these two types of algorithms, with reconstruction characteristics in between (Fig. 1.18).

MPR and Three-Dimensional Reconstruction

Multiplanar re-format (MPR) is used for visualizing the spine in the sagittal, coronal, and oblique planes. It preserves the soft-tissue and bone algorithms of the original studies. They are useful in determining the relationship of lesions relative to the disk space and the vertebral bodies (Fig. 1.19). Three-dimensional reconstruction is another method for evaluating the relationship of lesions to different portions of the vertebrae and is a valuable adjunct to the original axial and MPR scans (Fig. 1.20). More experience is required before its true utility can be realized.

CT ARTIFACTS

Image degradation in CT is caused by artifacts resulting from patient motion, the presence of metallic objects within the region being scanned, detector calibration errors in the equipment, or errors in data processing (25). Patient motion-related artifacts result from respiratory, cardiac, or bowel motion or an inability to hold still due to the patient's physical condition. Physiologic motion is usually not a major concern in spine imaging. Artifacts also may result from metallic objects, dense calcified masses, or large pockets of air. These may create streak artifacts or artifacts due to edge enhancement between two objects of significant difference in attenuation values. Such artifacts can be corrected by changing the orientation of the scan plane or by viewing the image in a wider window setting.

Artifacts may result from partial volume averaging, usually in a thick section which consists of two or more tissues with significantly different attenuation values, such as a herniated disk and an adjacent bony spur. This can be corrected by obtaining a thinner section.

Calibration errors within the detectors also may result in circular artifacts. They often occur at the same level in different sections but may not be appreciated except when overlying soft-tissue structures. For a more detailed review of this subject, the reader is referred to the discussion by Villifana [26].

References

1. Balter S. An introduction to the physics of magnetic resonance imaging. Radiographics 1987;7:371–384.
2. Pavlicek W. Magnetic resonance instrumentation and imaging formation. Radiographics 1987;7:809–820.
3. Fullerton GD. Magnetic resonance imaging signal concepts. Radiographics 1987;7:579–594.
4. Wehrli FW. Introduction to fast-scan magnetic resonance. Milwaukee: General Electric, 1986.
5. Kanal E, Wehrli FW. Signal-to-noise ratio, resolution and contrast. In: Werhli FW, Shaw D, Kneeland JB, eds. Biomedical magnetic resonance imaging: principles, methodology and applications. New York, VCH Publishers, 1988:47–114.
6. Harms SE. Multiple spin echo magnetic resonance imaging of the brain. Radiographics 1986;6:117–123.
7. Haase A, Frahm J, Matthaei D, et al. FLASH imaging: rapid NMR imaging using low flip angle pulses. J Magn Res 1986;67:258–266.
8. Mills TC, Ortendahl DA, Hylton NM, et al. Partial flip angle MR imaging. Radiology 1987;162:531–539.
9. Winkler ML, Ortendahl DA, Mills TC, et al. Characteristics of partial flip angle and gradient reversal MR imaging. Radiology 1988;166:17–26.
10. Hennig J, Nauteth A, Friedburg H. RARE imaging: a fast imaging method for clinical MR. Magn Reson Med 1986;3:823–833.
11. Melki PS, Mulkern RV, Panych M, et al. Comparing the FAISE method with conventional dual-echo sequences. JMRI 1991;1:319–326.
12. Jolesz FA. Fast spin-echo technique extends versatility of MR. Diagn Imaging 1992;3:78–87.
13. Sze G. Gadolinium-DTPA in spinal disease. Radiol Clin North Am 1988;26:1009–1024.
14. Valk J. Gd-DTPA in MR of spinal lesions. AJNR 1988; 9:345–350.
15. Bellon EM, Haacke EM, Coleman PE, et al. MR artifacts: a review. AJNR 1986;147:1271–1281.
16. Harris R, Wesby G. Artifacts in magnetic resonance imaging. In: Kressel KY, ed. Magnetic resonance annual 1988. New York: Raven Press, 1988.
17. Johnson BA, Kelly WM. Common MRI artifacts: an overview. MRI Decisions 1989;(March/April):17–23, (May/June):26–32.
18. Ehman RL, McNamara MT, Brasch RC, et al. Influence of physiologic motion on the appearance of tissue in MR imaging. Radiology 1986;160:795–801.
19. Pattany PM, Phillips JJ, Chiu LC, et al. Motion artifact suppression technique (MAST) for MR imaging. J Comput Assist Tomogr 1987;11:369–377.
20. Felmlee JP, Ehman RL. Spatial presaturation: a method for suppressing flow artifacts in magnetic resonance imaging without gating. Radiology 1987;164:559–564.
21. Czervionke LF, Czervionke JM, Daniels DL, et al. Characteristic features of MR truncation artifacts. AJR 1988;151:1219–1228.
22. Soila KP, Viamonte M, Starewicz PM. Chemical shift misregistration effect in magnetic resonance imaging. Radiology 1985;153:819–820.
23. Wendt RE III, Wilcott MR III, Nitz W, et al. MR imaging of susceptibility induced magnetic field inhomogeneities. Radiology 1988;168:837–841.
24. Heiken JP, Brown JJ, eds. Manual of clinical magnetic resonance imaging. New York: Raven Press, 1992.
25. Joseph PM. Artifacts in computed tomography. In: Newton TH, Potts DG, eds. Radiology of the skull and brain: technical aspects of computed tomography, vol. 5. St. Louis: Mosby, 1981.
26. Villafana T. Physics and instrumentation. In: Lee SH, Rao KCVG, eds. Cranial MRI and CT. 3d ed. New York: McGraw-Hill, 1992.

CHAPTER / **2**

Anatomy

K. C. V. G. Rao

Conventional radiography is often adequate in evaluating the osseous anatomy of the spine. Such studies usually consist of anteroposterior, lateral, and oblique views. Computed tomography (CT) provides an additional dimension not available even with pleuridirectional tomography, demonstrating the axial anatomy of the osseous component. To a limited extent, CT also demonstrates the soft tissues both within and surrounding the spine. In CT, evaluation of the soft-tissue contents of the spinal canal is facilitated by the use of intravenous contrast (IVCT), and differentiation of the various compartments is facilitated by the use of intrathecal contrast (CTM). This limitation of CT has been eliminated to a great extent with magnetic resonance imaging (MRI). In the majority of circumstances, MRI also has eliminated the need for conventional myelography and CT following myelography (CTM).

In general, the osseous anatomy can be evaluated best by CT. In evaluating the ligaments and joints, both CT and MRI are equally applicable, whereas MRI is the ideal imaging modality when evaluating the soft tissue and cerebrospinal fluid spaces within the spinal canal. Knowledge of the normal anatomic landmarks and usual variants of the spine and its contents on CT and MRI studies is indispensable when evaluating for pathologic processes involving the spine and its contents. This chapter deals with the normal anatomic structures seen on CT and MRI.

The spinal column consists of 33 vertebrae joined together by multiple ligaments, synovial joints, and cartilaginous intervertebral disks. The vertebrae throughout the spine have a similar general configuration, with distinctive features, however, that characterize the cervical, thoracic, lumbar, and sacral segments of the spine. The craniocervical junction, which includes the axis and atlas vertebrae has a unique characteristic. Anatomic landmarks in the base of the skull need to be recognized because they form part of the craniocervical junction and are often involved in various pathologic processes.

Craniocervical Junction

The anatomy of the craniocervical junction is more complex than that of the rest of the spine. This junction consists of the occipital condyles of the basiocciput on either side of the foramen magnum and atlas (C1) and axis (C2) of the cervical vertebrae. The osseous structures and ligaments in this region have to conform to motion in flexion, extension, and rotation [1] (Fig. 2.1).

The *basiocciput* forms the posteroinferior aspect of the skull. It consists of the foramen magnum, the occipital condyles, the hypoglossal canal, the jugular tubercles, and the clivus (Fig. 2.2). The foramen magnum is the largest foramen of the skull. It is bounded by four ossified centers of the occipital bone: anteriorly by the basioccipital segment, laterally by two exooccipital segments containing the occipital condyles, and posteriorly by the supraoccipital or squamosal segment. These various segments fuse between the age of 2 and 4 years. The foramen magnum is oval in shape and averages 35 mm in the sagittal and 30 mm in the transverse directions. Narrowing of the foramen as in achondroplasia or enlargement as in the Chiari malformation can be best evaluated by axial CT or MRI in sections perpendicular to the anterior and posterior lips of the foramen magnum.

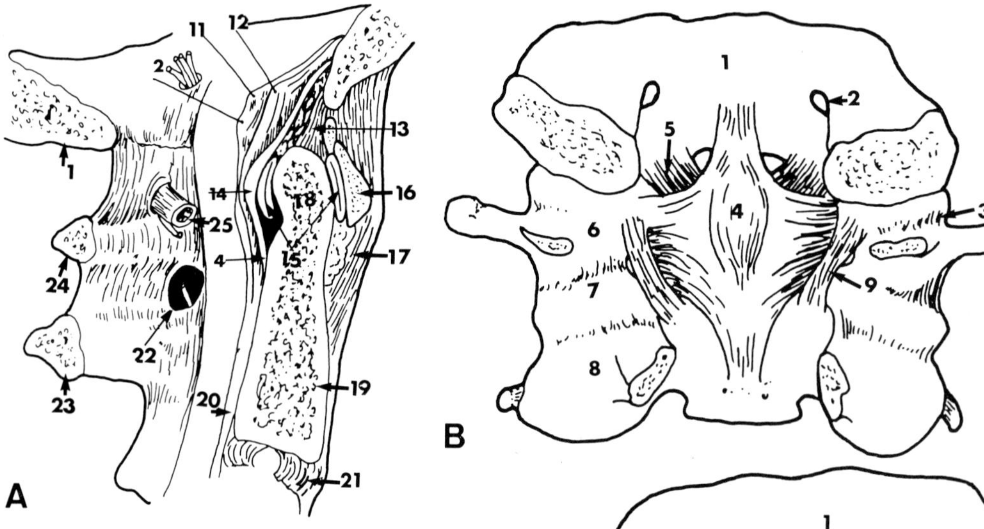

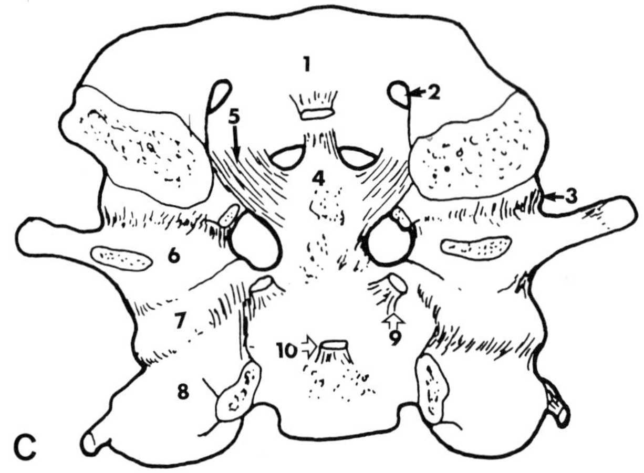

Figure 2.1. Schematic drawing of craniovertebral junction. A, Sagittal section through midline. **B,** Coronal section viewed from back after removing the posterior elements. **C,** Viewed from back after removal of the cruciate, transverse, and atlantoaxial ligaments. (1) Occipital bone, (2) hypoglossal canal, (3) atlanto-occipital joint, (4) attachment of cruciate ligament to dens, (5) alar ligament, (6) atlas, (7) atlantoaxial joint, (8) axis, (9) accessory atlantoaxial ligament, (10) cruciate ligament remnants, (11) superficial layer of tectorial membrane, (12) deep layer of tectorial membrane, (13) apical ligament of dens, (14) vertical portion of cruciate ligament, (15) synovial space around dens, (16) anterior arch of atlas, (17) anterior longitudinal ligament, (18) dens, (19) body of axis, (20) posterior longitudinal ligament, (21) intervertebral disk, (22) intervertebral foramen, (23) arch of axis, and (24) arch of atlas.

The *occipital condyles* appear reniform in cross section and have a convex surface that articulates with the articular masses of the atlas. The atlanto-occipital articulation is best evaluated in the coronal plane on CT [2], and in coronal and sagittal sections on MRI (Fig. 2.3). Congenital variations of the atlanto-occipital articulation include condylar hypoplasia (resulting in flattened condyles) and variable ossification between the occipital bone and the lateral masses or transverse process of the atlas. Although coronal CT or MRI may show the osseous abnormality, this is one region where conventional tomography is probably superior to these imaging studies [3].

The *hypoglossal canal* courses through the basiocciput in an anterolateral direction. The canal is best demonstrated on both CT and MRI in the axial or coronal plane utilizing thin sections with minimal interslice gap (see Fig. 2.2). The hypoglossal nerve (CN XII) and a branch of the ascending pharyngeal artery and its accompanying veins pass through the canal. These structures can on occasion be identified on MRI studies. The jugular tubercle forms the roof of the hypoglossal canal.

The lower part of the clivus forms the anterior boundary of the foramen magnum and is separated from the upper half of the clivus (which arises from the sphenoid bone) by the spheno-occipital synchondrosis. This usually fuses by the age of 15 years.

Flexion-extension of the head takes place at the C1–occipital condyle level, whereas rotary motion occurs between the C1 and C2 level. To accommodate these movements, strong ligaments attach the basiocciput to the bodies of the atlas and axis and continue down as the *anterior* and *posterior longitudinal ligaments* along the anterior and posterior surfaces of the vertebral bodies [4] (see Figs. 2.1 and 2.3). Similarly, the ligamentum nuchae attaches between the external occipital protuberance and adjacent crest to the spine of the seventh cervical vertebra. Additional ligaments connect both the atlas and the axis with the basiocciput. These include the apical ligament of the dens and the alar ligaments. Rotary motion at the atlantoaxial joint is achieved by the cruciate ligament attached to the posterior surface of the odontoid and laterally to the junction of the anterior arch with lateral mass of C1 (Fig. 2.4). In axial sections, CT and MRI may not show a completely closed bony ring of the atlas unless the sections are obtained perpendicular to the longitudinal axis of the posterior neural arch.

The first two cervical vertebrae are unique and dif-

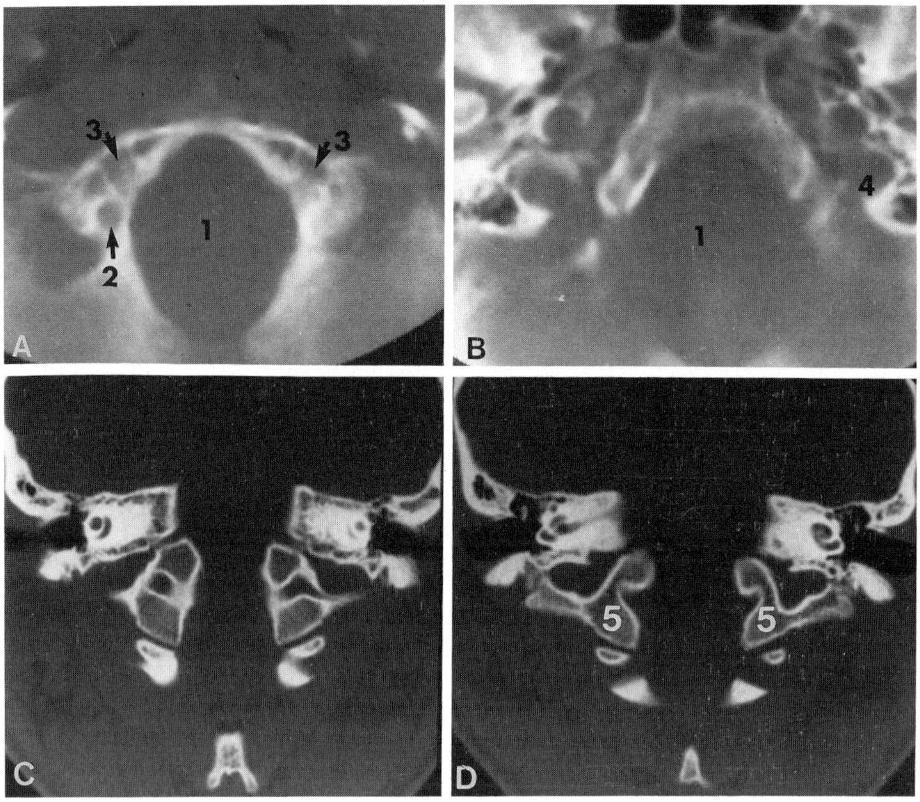

Figure 2.2. **Base of skull. A, B,** Contiguous axial section. **C, D,** Adjacent coronal section. (1) Foramen magnum, (2) hypoglossal canal, (3) groove for vertebral artery, and (4) jugular foramen. Note the relationship of the (5) occipital condyle and lateral mass of the atlas.

ferent in their appearance from the rest of the cervical spine. They have additional ligaments which can occasionally be recognized on the CT or MRI studies, in addition to ligaments which are common to the rest of the spinal column.

ATLAS

The atlas (C1) vertebra originates from the C1 sclerotome. It is in the form of a ring, made up of two lateral masses connected by an anterior and posterior arch (Fig. 2.4 and 2.5). A small protuberance is seen in the outer surface of the anterior arch, the anterior tubercle, which is the site of attachment of the longismus colli muscle. The inner surface of the ring is smooth and articulates with the odontoid process of the second cervical vertebra (the axis). The posterior arch has a rudimentary spinous process, the posterior tubercle. Occasionally, the posterior arch of the atlas fails to unite, and the ring is incomplete. This should not be confused with a fracture. Fusion of the posterior arch may be delayed until up to 4 years of age [5]. The lateral masses on either side consist of the articular surfaces and the transverse process. The superior facet articulates with the occipital condyles. The inferior articular facet has a concave to flat surface and slants upward and medially when viewed in coronal plane on CT or MRI (see Fig. 2.3). The transverse processes are broad and are perforated by the foramen transversarium for passage of the vertebral artery. The artery also makes a groove in the superior surface of the posterior arch of the atlas at its junction with the lateral mass (see Fig. 2.5). A slight irregularity and a small tubercle on the medial surface of the lateral mass of the atlas (C1) may be seen, which is the site of attachment of the transverse ligament (see Figs. 2.4 and 2.5).

AXIS

The second cervical vertebra is distinguished by the odontoid process. The main body of the vertebra originates from the C2 sclerotome, while the dens originates from the C1 sclerotome. This embryologic peculiarity of the C1–C2 region explains the various congenital anomalies occurring at this level, the most common being the os odontoideum.

The odontoid process extends upward from the superior surface of the body of the axis (see Figs. 2.3, 2.4, and 2.5). A dark signal band exists at the junction of the dens with the body of C2 until the age of 4 to 6 years, best seen on sagittal MRI (Fig. 2.6). In cross section, the dens appears circular and slightly

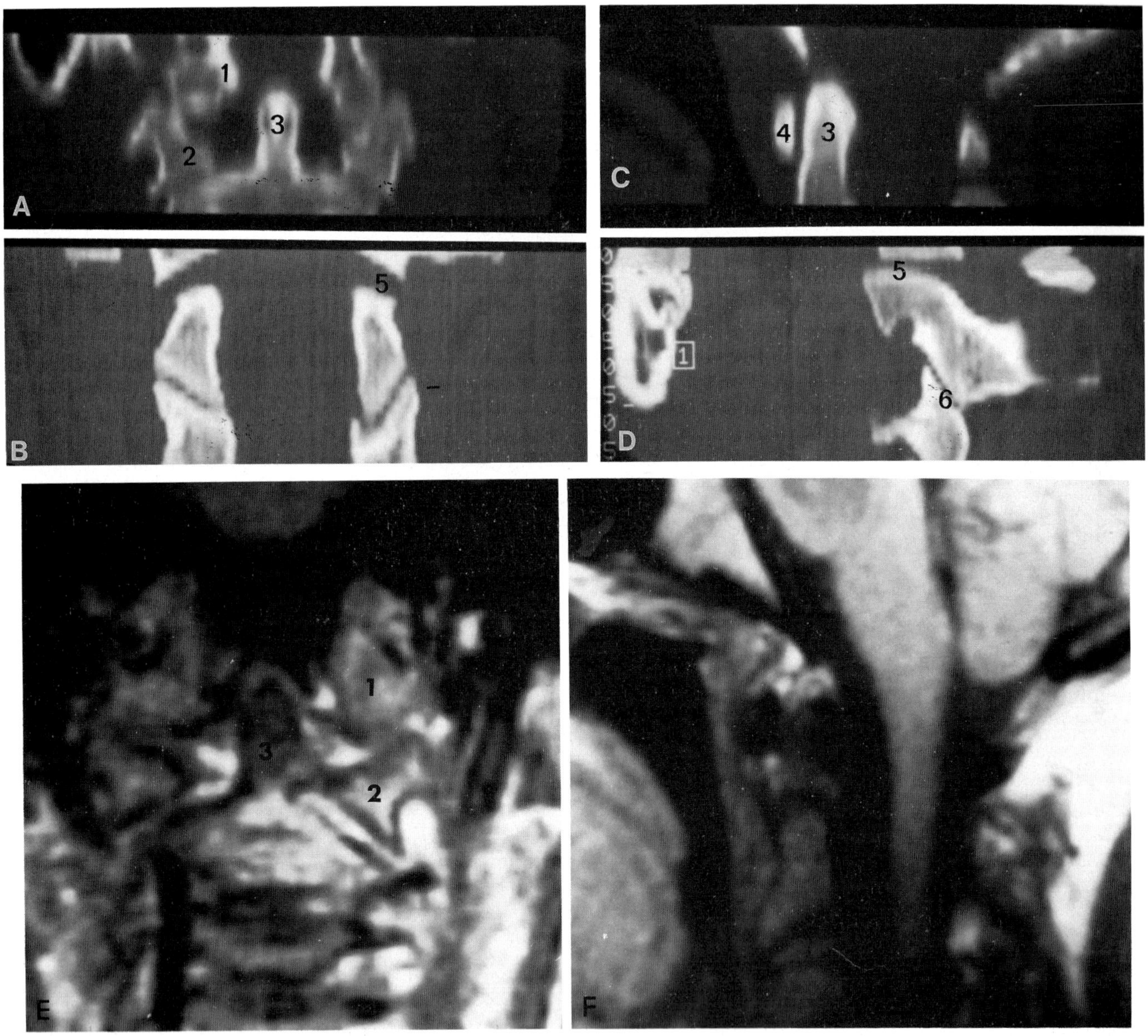

Figure 2.3. Craniovertebral junction. Re-formatted CT in coronal view: (**A**) anterior level, (**B**) posterior level. Re-formatted CT sections in (**C**) midsagittal and (**D**) parasagittal planes. T1-weighted MRI in (**E**) coronal and (**F**) midsagittal planes. (1) Occipital condyles, (2) lateral mass of atlas, (3) odontoid process, (4) anterior arch of atlas, (5) atlanto-occipital joint, (6) atlantoaxial facet joint, (7) foramen magnum, (8) apical ligament, (9) fat in between the apical and alar ligament, and (10) posterior longitudinal ligament.

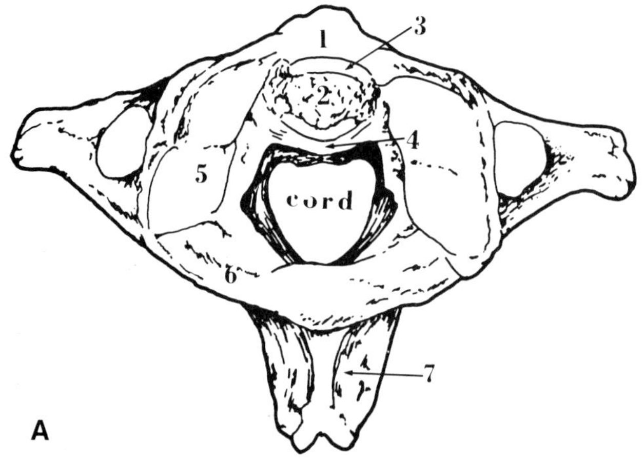

Figure 2.4. Atlantoaxial joint. A, Schematic diagram showing the ligaments and relationship of dens to the anterior arch of C1. (1) Anterior arch of C1, (2) odontoid process, (3) anterior synovial membrane, (4) transverse ligament, (5) lateral condyles, (6) posterior arch of C1, and (7) spinous process. **B,** Contiguous axial CT sections. **C,** Re-formatted midline section of C1–2 region. **D, E,** T1-weighted adjacent axial MRI images. (1) Anterior arch of atlas, (2) odontoid process, (3) cruciate ligament, (4) foramen transversorium, (5) body of axis, (6) synovial space, (8) apical fat, and (9) spinal cord.

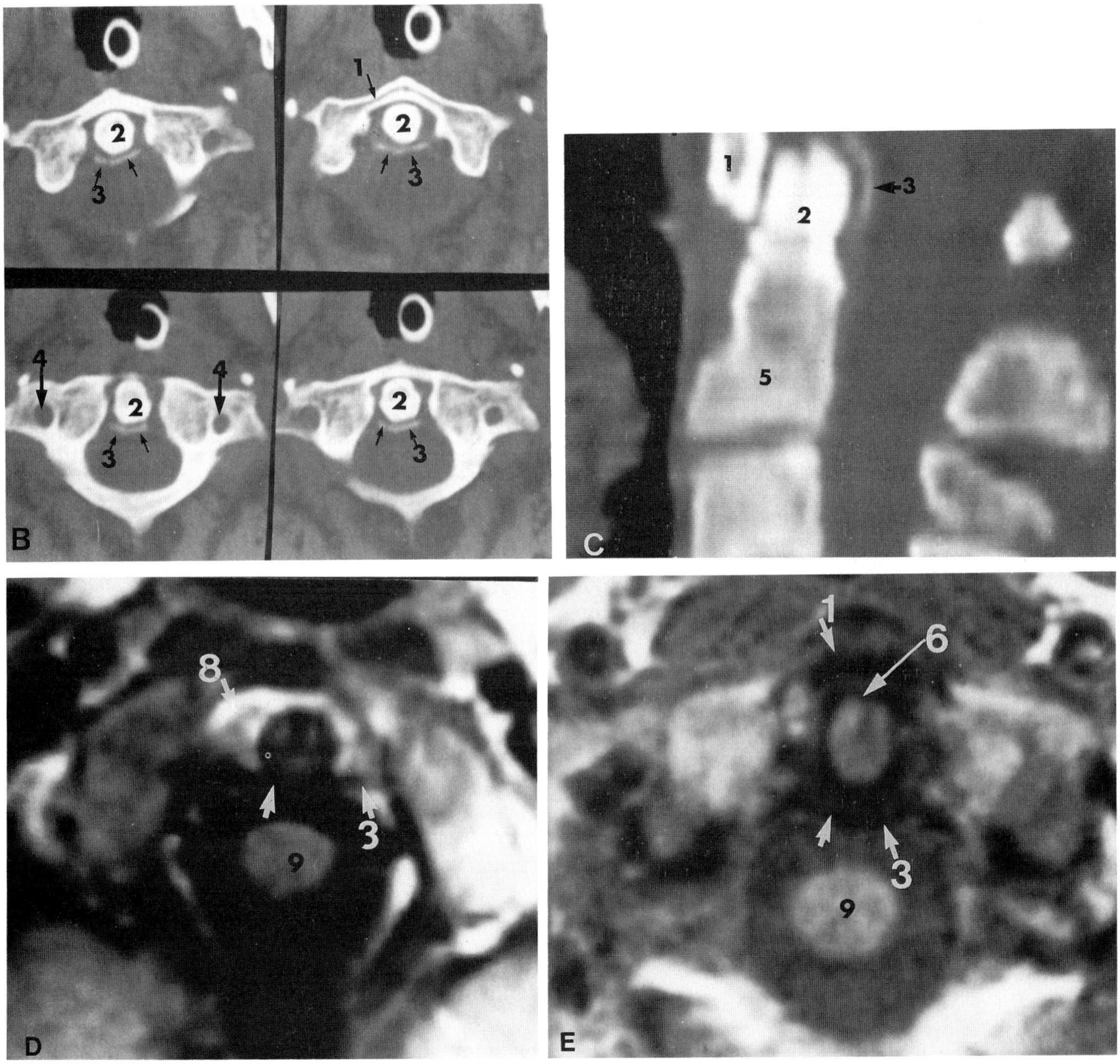

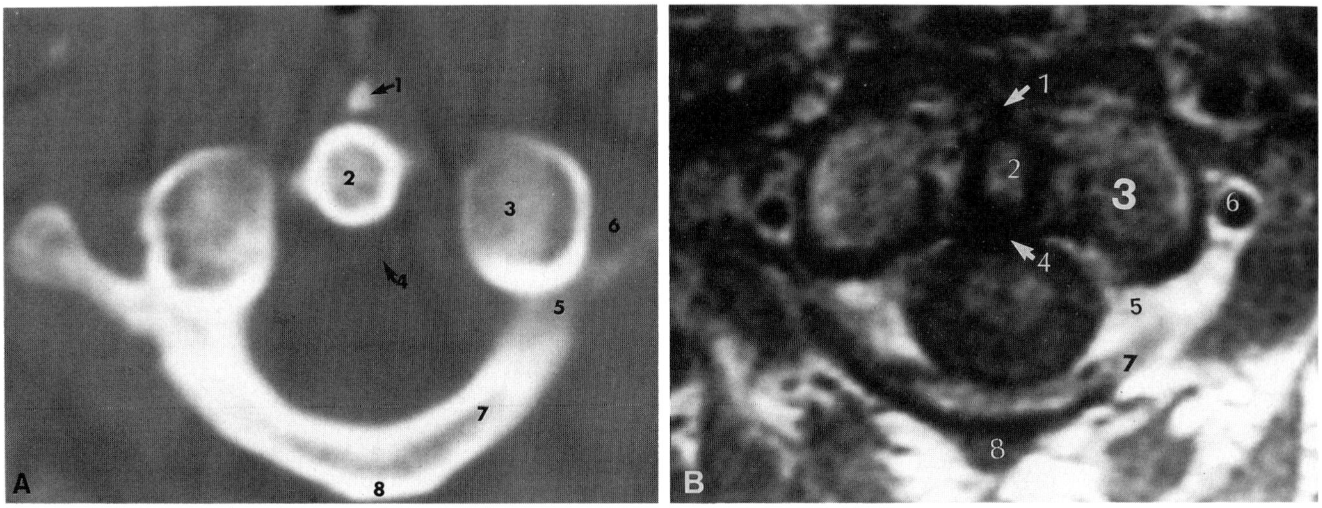

Figure 2.5. **A,** Axial CT scan through the atlas vertebrae. **B,** Axial T1-weighted MRI. (1) Anterior arch of atlas, (2) odontoid process, (3) lateral condyles, (4) transverse ligament, (5) groove for vertebral artery, (6) vertebral artery, (7) posterior ring, and (8) rudimentary spinous process.

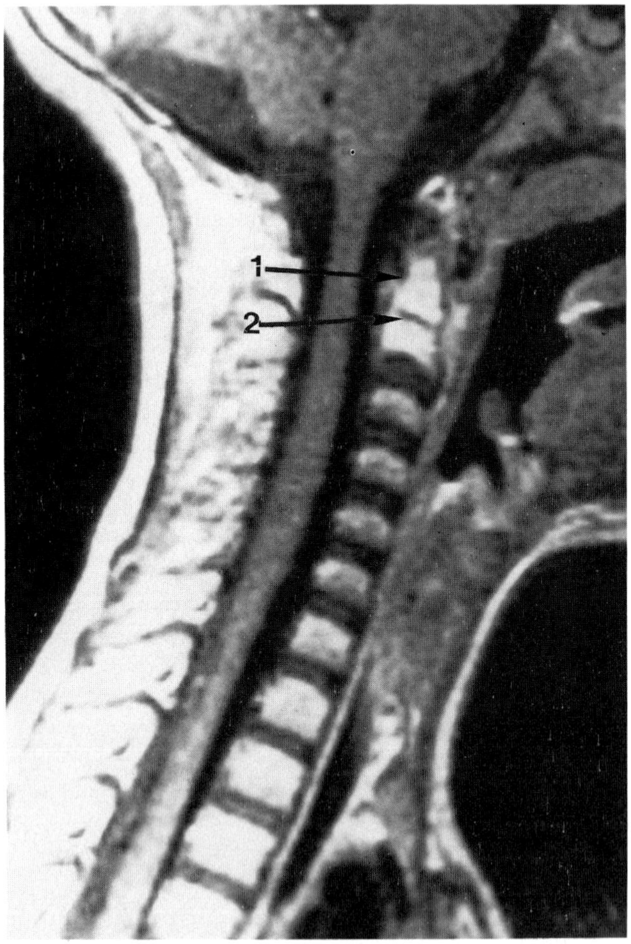

Figure 2.6. T1-weighted sagittal MRI in a 2-year-old child. (1) Tip of odontoid process, and (2) hypointense signal band between the base of the odontoid process and the body of the C2 vertebra.

flattened anteriorly as it articulates with the anterior arch of C1. On axial sections, the atlantodens space should not exceed 2 mm (see Figs. 2.4 and 2.5). A space larger than 3 mm in adults is abnormal, suggesting either torn ligaments in cases of trauma or joint pathology such as that seen in rheumatoid arthritis [6]. In infants and children, this space may be as wide as 5 mm.

On either side of the body of C2 are the lateral masses which contain the articular surfaces and the short, blunt transverse processes (Fig. 2.7). The superior facets are horizontal, facing upward and outward, and articulate with the inferior facets of C1. The atlantoaxial joints permit rotary movement of the head and are best evaluated by CT or MRI in the coronal plane (see Fig. 2.3). Axial sections through the atlantoaxial joint usually show the facets of C1 lateral to the body of C2. This appearance should not be mistaken for a fracture of C2. The inferior articular facets face downward and forward and lie posterior to the facet of C3. They are aligned in the typical fashion of the rest of the lower cervical spine. The transverse processes are small and are perforated by the *foramen transversarium*, which is directed slightly superolaterally, unlike the vascular foramina in the rest of the cervical vertebrae. The spinous process is large and often bifid. The pedicles, lamina, and spinous process form the lateral and posterior walls of the spinal canal and a closed ring with the vertebral body (Fig. 2.7).

Lower Cervical Spine

The five remaining cervical vertebrae are fairly similar in appearance. They consist of a body, pedicles, lamina, transverse processes, articular pillars,

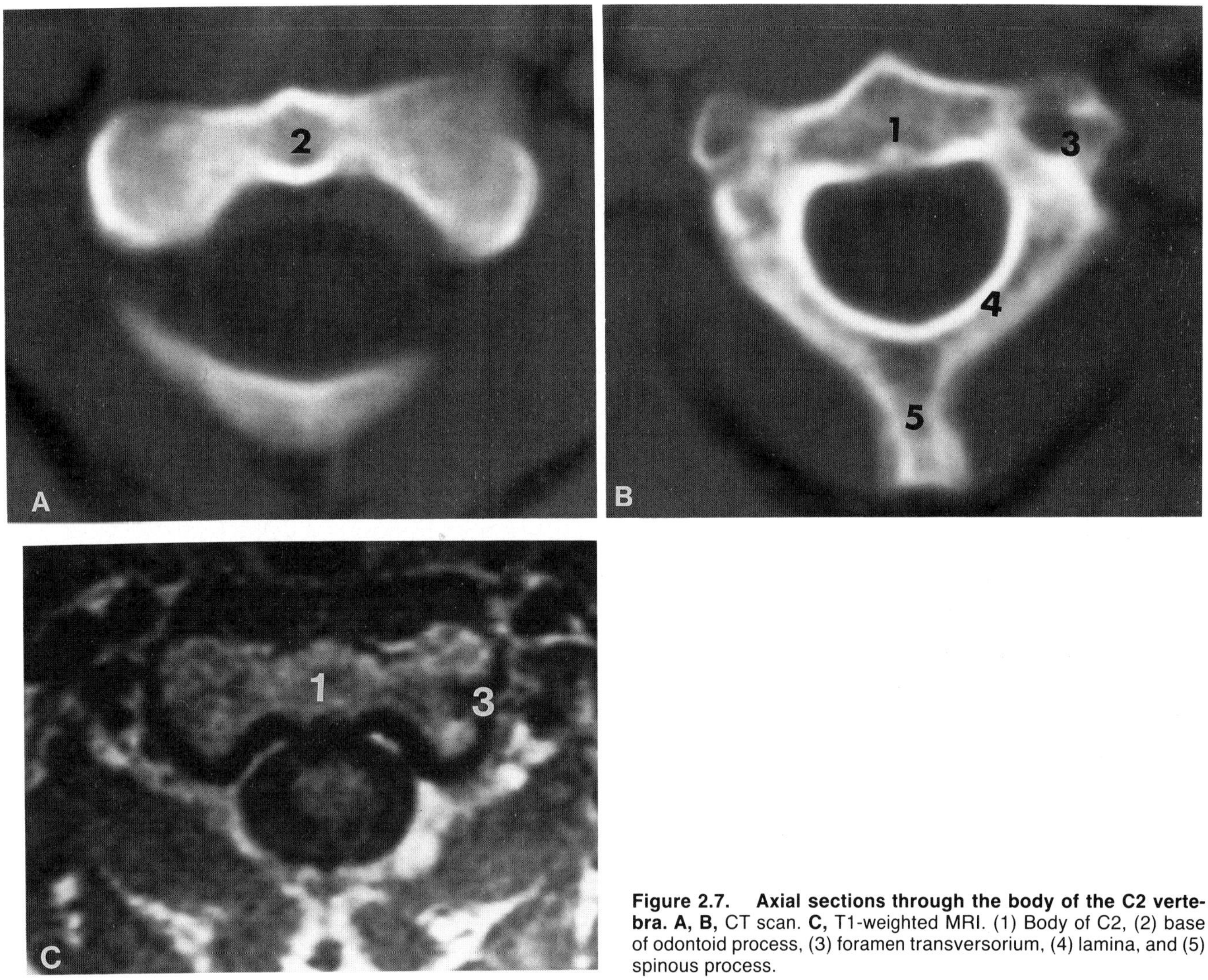

Figure 2.7. Axial sections through the body of the C2 vertebra. A, B, CT scan. **C,** T1-weighted MRI. (1) Body of C2, (2) base of odontoid process, (3) foramen transversorium, (4) lamina, and (5) spinous process.

and spinous processes (Figs. 2.8 through 2.13). The cervical spinal canal is triangular in cross section. It has a fairly uniform dimension from C3 to C7. The average anteroposterior diameter at C1 is 25 mm, at C2 it is 22 mm, and from C3 to C7 it is 18 mm. Sagittal diameter of 12 mm or lower is indicative of spinal stenosis.

The vertebral bodies consist of a cortical outer layer and a cancellous inner portion. The height of the vertebral body averages 15 mm. The transverse diameter of the body is greater than the anteroposterior diameter. The superior surface of the body ends on each side in an upturned lip which forms the uncinate processes. The uncinate process articulates with the inferolateral aspect of the vertebra above to form the uncovertebral joints. The uncinate processes indent the posterolateral surface of the intervening disk and form the anteromedial boundary of the intervertebral foramen (see Figs. 2.8 and 2.9). On CT, the uncinate process is seen as dense, compact bone bordering the posterolateral margin of the disk. On axial MRI, the signal depends on the extension of bone marrow within the uncinate process (see Fig. 2.10). In children and young adults, the uncinate process has an intermediate to bright signal relative to the adjacent disk material (see Fig. 2.11). In adults, depending on the extent of degenerative changes in the adjacent disk, the uncinate process forms the uncovertebral joint (joint of Luschka) and has a decreased signal with all pulse sequences. The anterior and posterior longitudinal ligaments attached to the disk margins and along the surfaces of the vertebral bodies have similar attenuation values as the annular fibers of the disk on CT, whereas on both T1- and T2-weighted MRI they have decreased signal (see Fig. 2.11). On MRI, similar decreased signal is also seen at the site of the intervertebral venous entry zone. The sagittal T1- and T2-weighted images help in differentiating the posterior longitudinal ligament/annular fibers from the cerebrospinal fluid space within the spinal canal.

The neural arches include the pedicles which con-

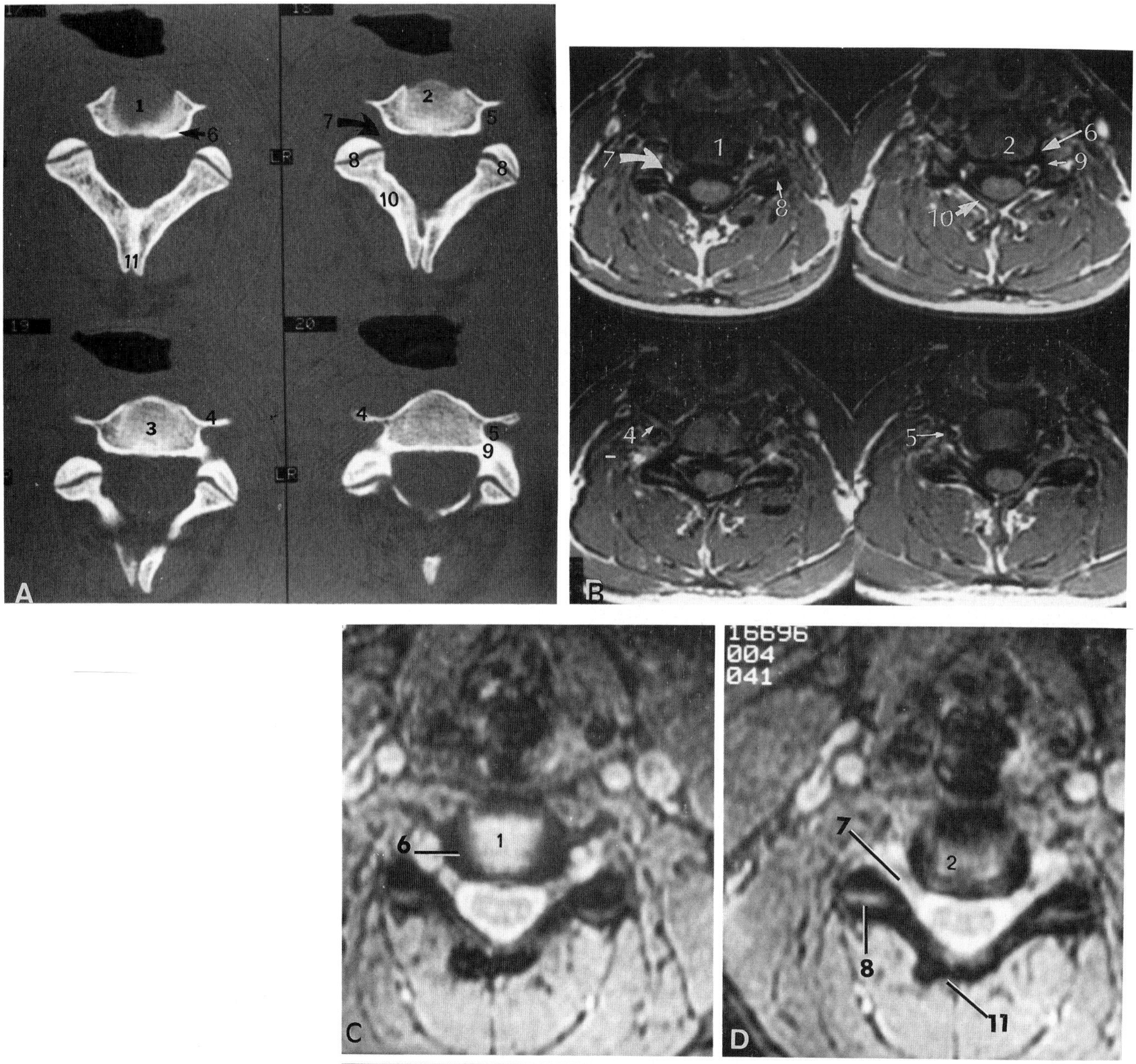

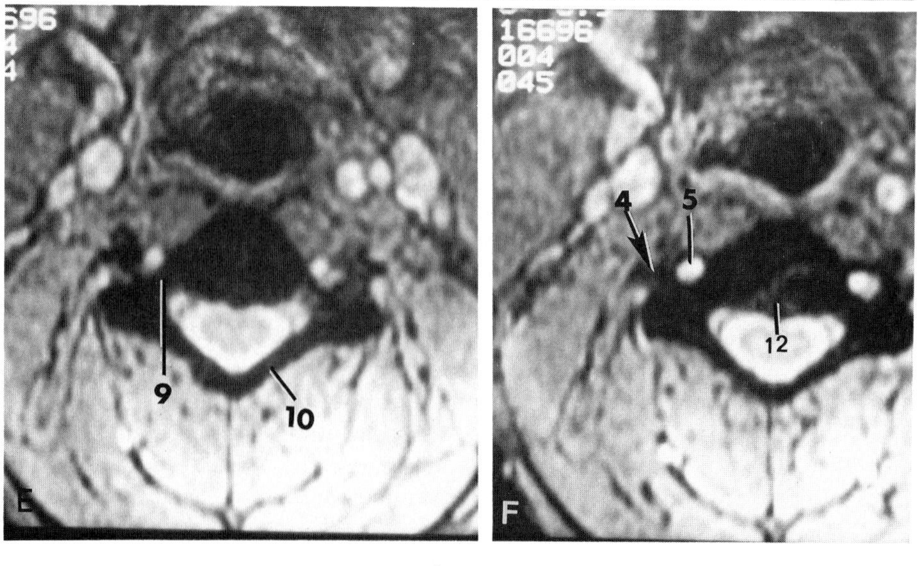

Figure 2.8. **A,** Contiguous axial CT scan. **B,** Axial T1-weighted MRI. **C–F,** Axial T2-weighted MRI. (1) Disk, (2) disk and adjacent endplate of the vertebral body, (3) vertebral body, (4) transverse process, (5) foramen transversorium, (6) uncinate process, (7) neural canal, (8) facet joint, (9) pedicle, (10) lamina, (11) bifid spinous process, and (12) venous entry zone.

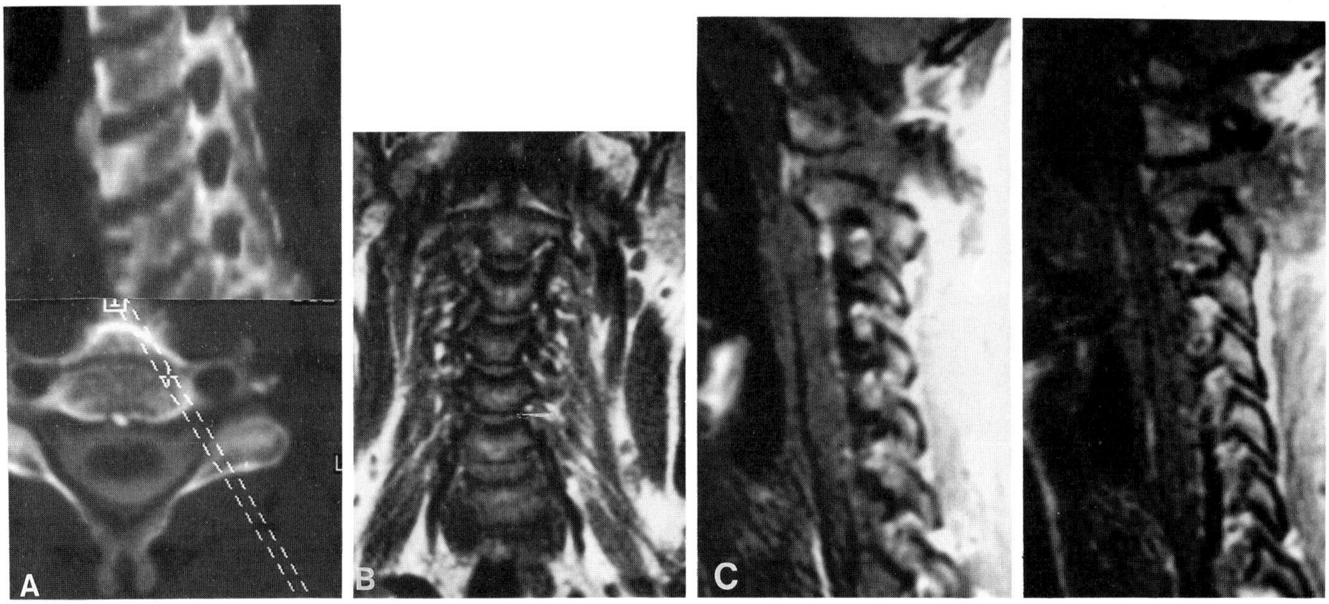

Figure 2.9. **A,** Parasagittal re-formatted CT scan. Coronal (**B**) and parasagittal (**C**) T1-weighted MRI. (1) Vertebral body, (2) disk space, (3) uncinate process, (4) inferior facet, (5) superior facet, and (6) neural foramen.

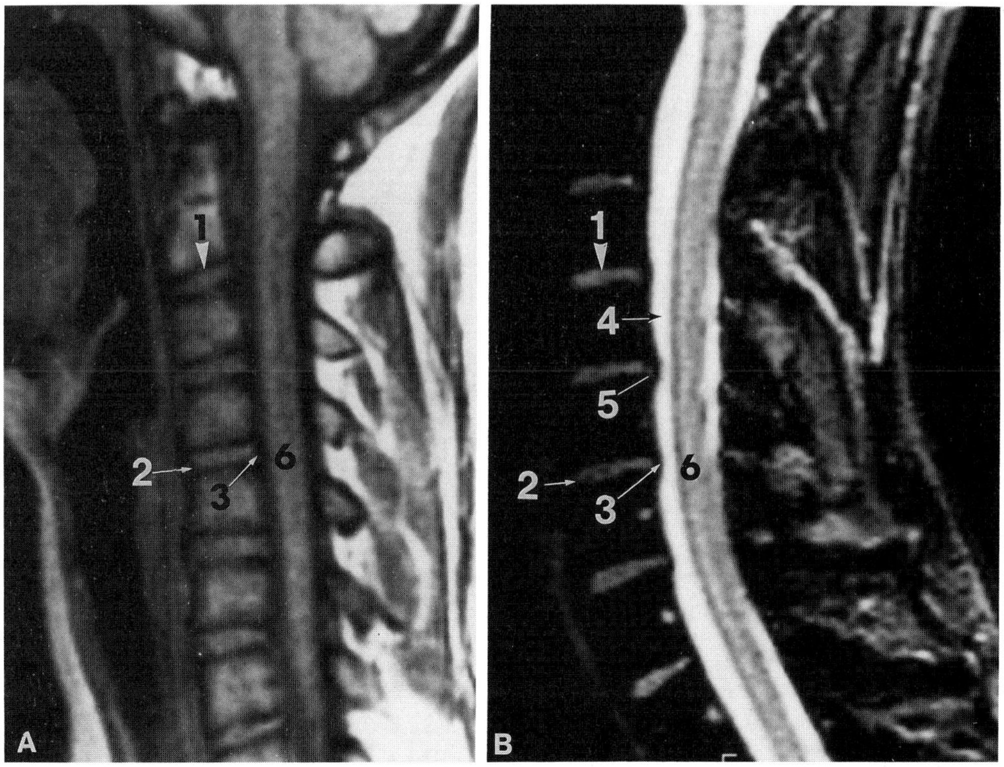

Figure 2.10. Midline sagittal MRI through the cervical spine. **A,** T1-weighted image. **B,** T2-weighted image. (1) Nucleus pulposus has intermediate signal on the T1-weighted image and bright signal on the T2-weighted section; (2) the annulus has hypointense signal in both pulse sequences, merging with the ligaments; (3) posterior longitudinal ligament; (4) CSF spaces identified from the ligaments and dura on T2-weighted sections; (5) bulging disk; and (6) spinal cord.

Figure 2.11. Axial T1-weighted MRI images in a 2-year-old child demonstrating the bright signal in the uncinate process (*arrow*).

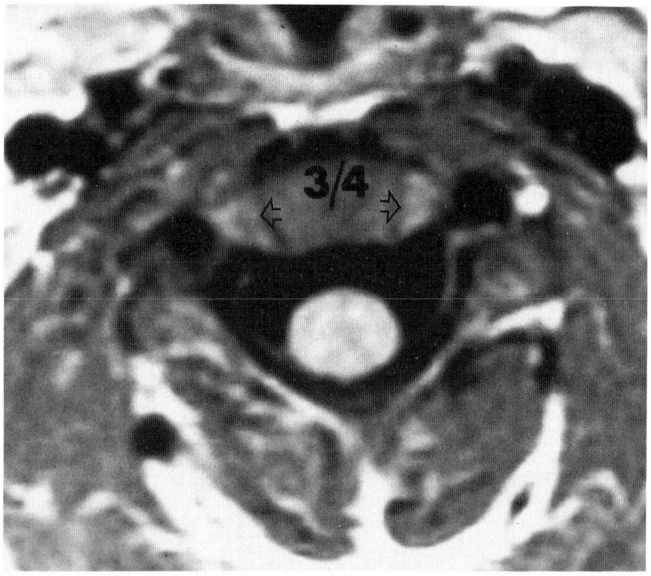

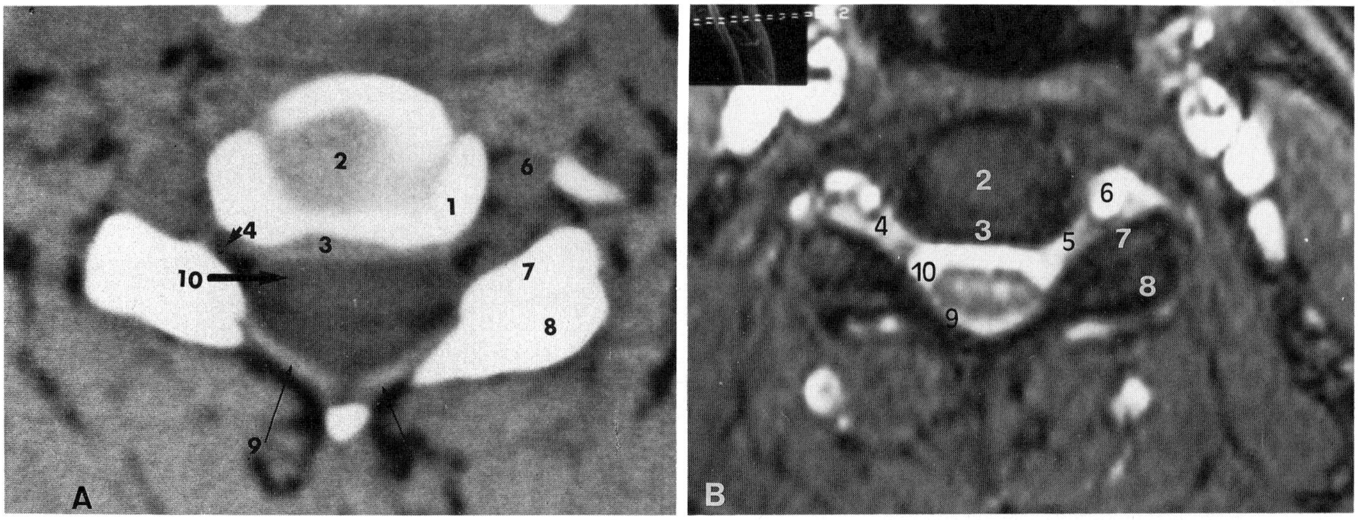

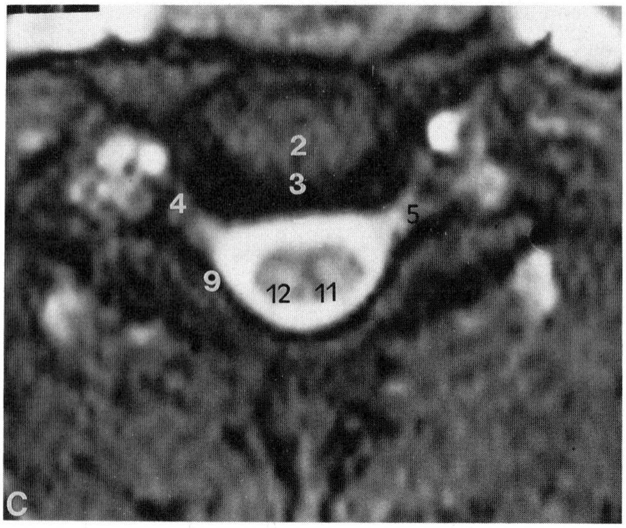

Figure 2.12. **A,** Axial CT scan through cervical disk space level. **B, C,** Contiguous axial T2-weighted MRI images. (1) Uncinate process, (2) nucleus pulposus of disk, (3) confluence of annular fibers, posterior longitudinal ligament, and dura, (4) neural canal, (5) nerve roots, (6) vertebral artery in foramen transversorium, (7) superior facet, (8) inferior facet of subjacent superior vertebrae, (9) ligamentum flavum, (10) CSF, (11) spinal cord white matter, (12) gray matter.

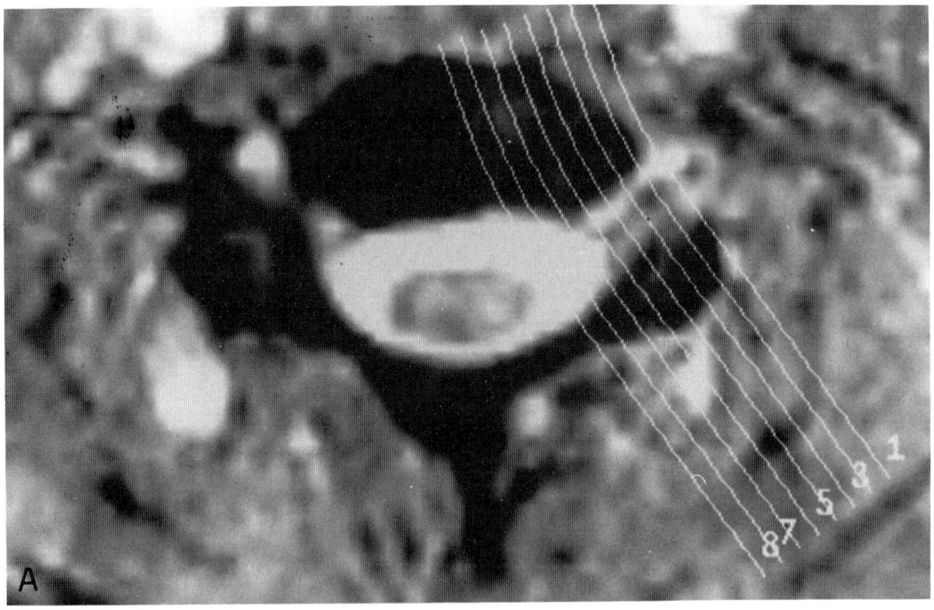

Figure 2.13. **A**, Axial section from three-dimensional FT volume acquisition with lines demarcating the plane of parasagittal oblique images. **B**, Similar to CT re-formatted images obtained to evaluate the neuralforamen en face along its entire length (1–4).

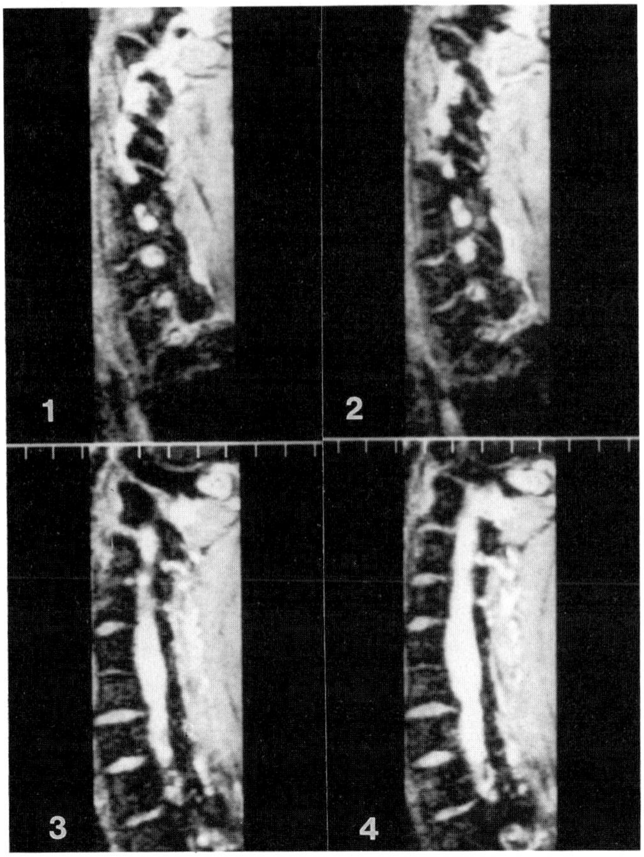

nect the vertebral body to the lamina. The thin and narrow laminae join posteriorly to form the spinous process. The spinous processes are usually short and bifid except at the C7 level, which is usually prominent. The pedicles, laminae, and spinous process form the posterolateral wall of the spinal canal, creating a closed axial ring.

Projecting laterally from the junction of the pedicles and laminae are the articulating facets, which have a superior and inferior surface. The pars interarticularis or isthmus is the ridge of bone connecting the articular pillars with the junction of the laminae and pedicles. Each articular process consists of a superior and inferior facet. The superior facet is directed upward and backward, while the inferior facet is directed downward and forward (see Figs. 2.9 and 2.12). The superior facet lies anterior to the inferior facet of the subjacent vertebrae.

Just anterior to the articular pillars are the transverse processes, which are perforated by the foramen transversarium, except at the C7, where they are either small or nonexistent. The vertebral artery arises from the subclavian artery and enters the foramen transversarium at the C6 level. The artery on each side extends upward to the atlas, where it turns behind the lateral mass of C1 to pierce the dura and enter the cranial base through the foramen magnum. On the lateral aspect of the transverse processes are the anterior and posterior tubercles. Between these two tubercles there is a groove for the emerging spinal nerves.

The intervertebral foramen (*neural foramen*) is a canal, with the internal and external orifices seen en face in a 45-degree oblique projection in the sagittal plane. On CT and MRI, the length of the canal is best evaluated in axial images performed with a 10-degree angulation posteroanteriorly from the sagittal plane (see Figs. 2.9 and 2.12). The boundaries of the neural foramen are formed anteriorly by the uncinate process and the adjacent vertebral body and the posterolateral portion of the intervertebral disk. The posterior wall is formed by the superior facet joint and the articular pillar. The superior and inferior walls are formed by the pedicles of the adjacent vertebral body. The neural canal is 12 to 15 mm in length, whereas the width and height of the neural foramen are 4 mm and 7 to 8 mm, respectively. The dorsal and ventral roots occupy the lower half of the foramen, the upper half being filled with fat and small veins. On CT, the osseous boundaries of the cervical neural foramen are evaluated with thin axial sections and subsequent re-formatted images. MRI in the axial and sagittal planes is often adequate in demonstrating the soft-tissue contents within the neural canal and foramen, although sagittal oblique views allow evaluation of the foramen en face (see Fig. 2.13).

Thoracic Spine

The 12 thoracic vertebrae increase in their axial diameter from above downward and are recognized by minor differences in the cervical or lumbar regions (Figs. 2.14 through 2.18). They are larger than the cervical vertebrae but smaller than the lumbar vertebrae. They are recognized by articular facets for the rib heads. The articular facets are located at the superoposterolateral angle of the body encroaching on the lower posterior angle of the body above, except for the facet of T11 and T12, which are located on the pedicle.

The superior and inferior surfaces of the thoracic vertebrae are flat. The anteroposterior and transverse diameters of the vertebral body are about the same. The height of the vertebral body averages 20 mm. The pedicles, laminae, and spinous processes form the boundaries of the spinal canal, which is almost circular (see Figs. 2.14 and 2.15). The transverse processes arise from the articular pillars and are directed laterally, backward, and upward. On their anterior aspect is the costotransverse facet, which articulates with the neck and tubercle of the corresponding rib.

The articular processes are located at the junction of the pedicle and lamina. The superior articular facets lie posterior to the pedicle, facing mainly backward (see Fig. 2.16). The inferior articular facet is in front of the lamina, facing forward. The facet joints project in a coronal plane in the thoracic region, whereas they project in an oblique plane in the cervical region and in the sagittal plane in the lumbar region. The facet joints extend up slightly above the level of the intervertebral disk. The spinous processes are long and extend posteriorly and downward.

The intervertebral foramina are directed laterally and are bounded by the vertebral body anteriorly, the facet joints posteriorly, and the adjacent pedicles superiorly and inferiorly. The intervertebral foramina are best seen in parasagittal projections in MRI or reformatted CT images. The exiting nerve roots and dorsal root ganglion are located in the upper middle portion of the neural foramen (see Figs. 2.16 and 2.17).

The thoracic spinal canal is elliptical in its upper

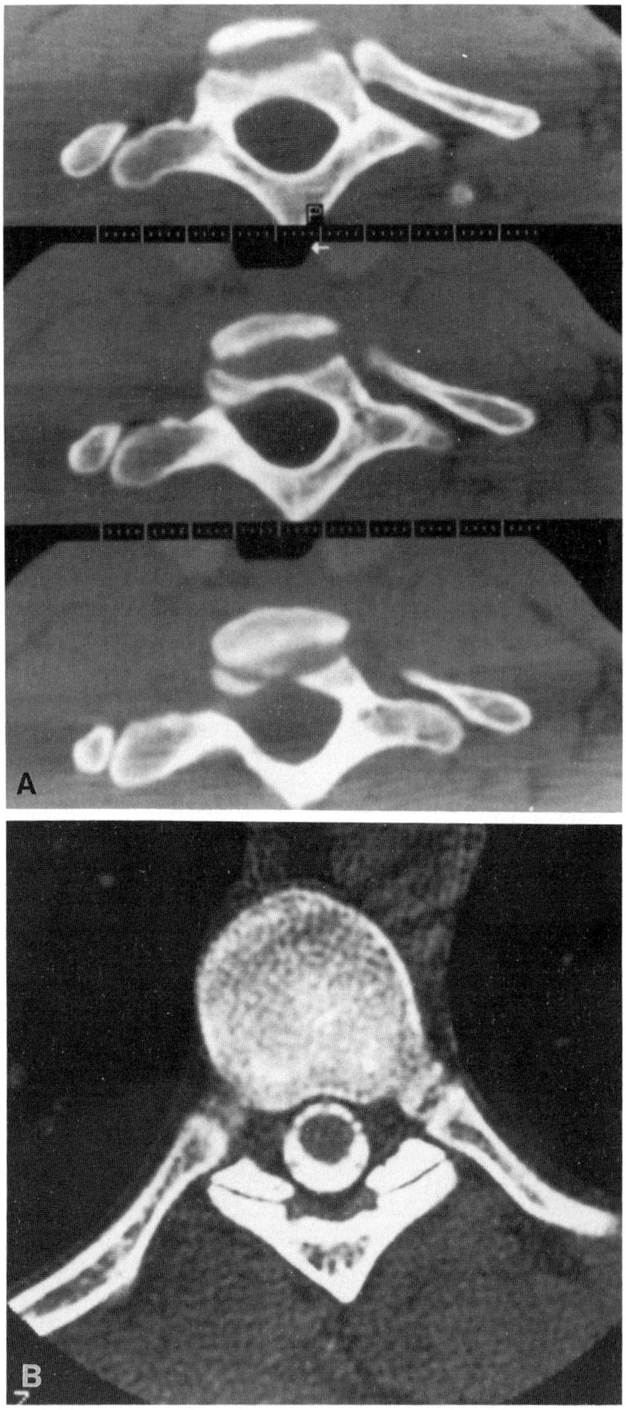

Figure 2.14. A, Contiguous axial CT sections through the C7–T1 level demonstrates the attachment of the articular facet for the rib to the posterolateral margins of the vertebral body. **B,** Axial CT scan through T11 vertebra. The articular facet for the rib is with the pedicle.

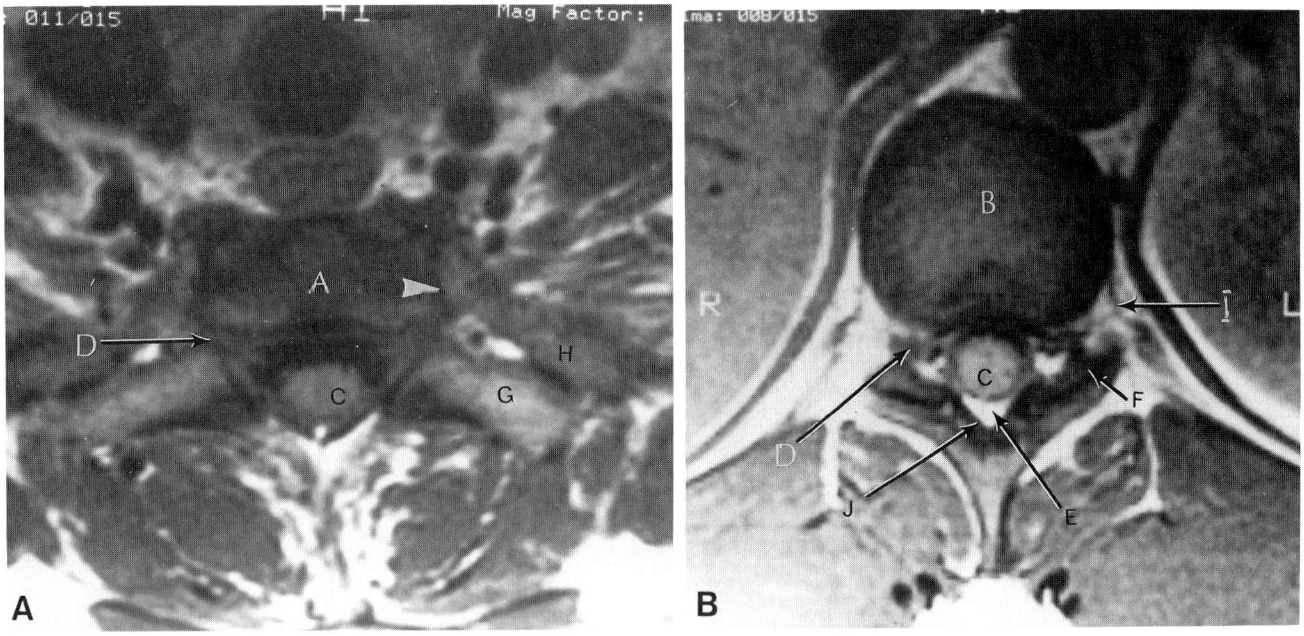

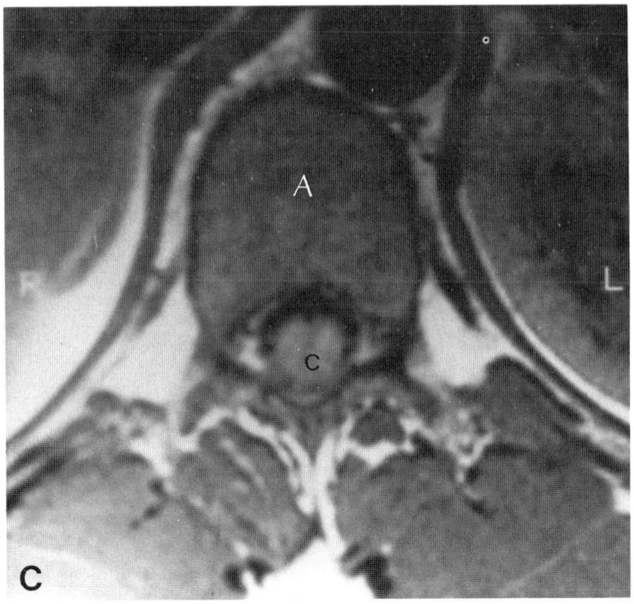

Figure 2.15. **A,** T1-weighted MRI through disk and adjacent T1 vertebra with articular facet along posterolateral surface (*arrow head*). Note the almost rectangular appearance of the vertebral body. **B,** Axial section through midthoracic disk level. **C,** Axial section through the vertebral body level. Note transition of the spinal cord and the vertebral body between the proximal and midthoracic spine (*A* = vertebral body; *B* = disk; *C* = spinal cord; *D* = neurovascular bundle in articulating surface of rib; *I* = radiculomedullary artery).

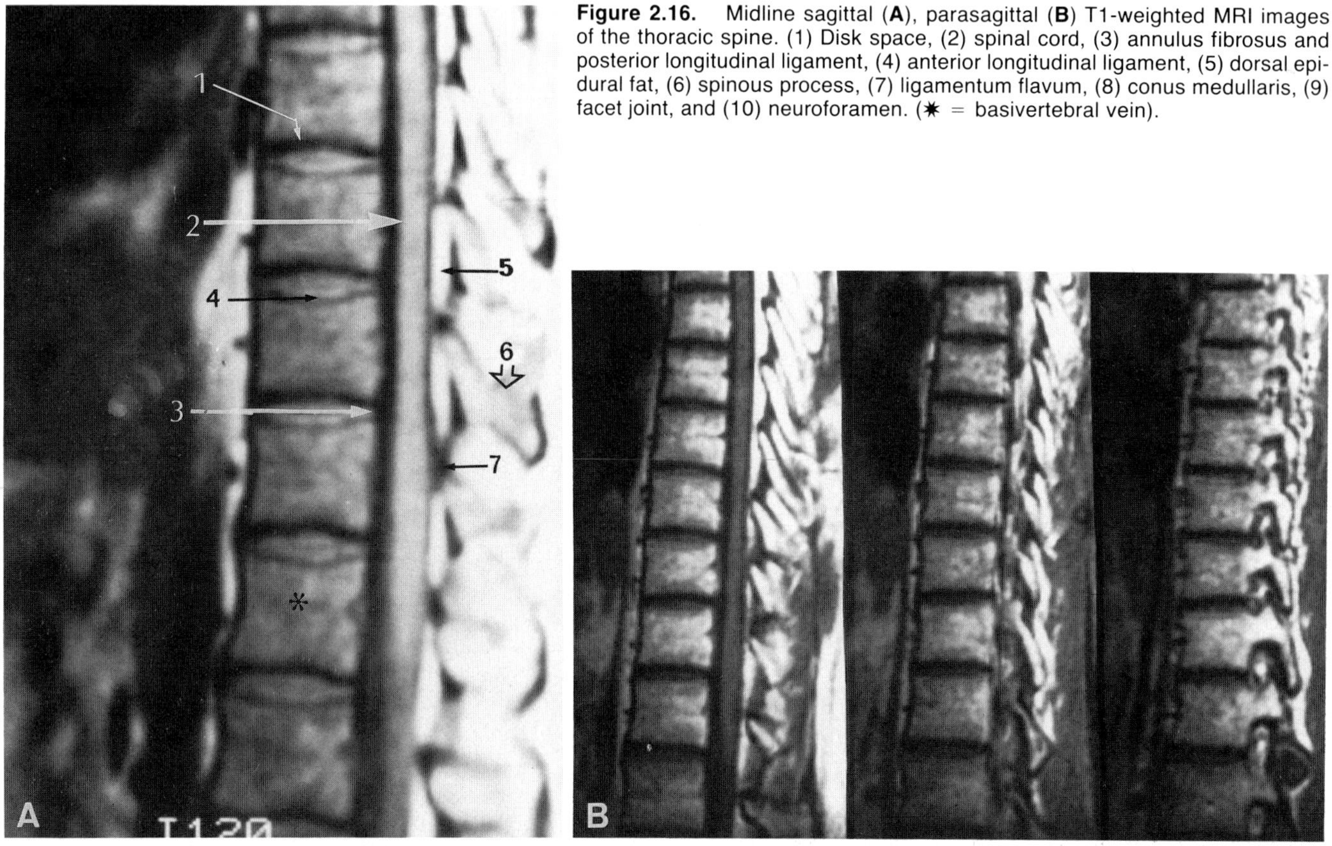

Figure 2.16. Midline sagittal (**A**), parasagittal (**B**) T1-weighted MRI images of the thoracic spine. (1) Disk space, (2) spinal cord, (3) annulus fibrosus and posterior longitudinal ligament, (4) anterior longitudinal ligament, (5) dorsal epidural fat, (6) spinous process, (7) ligamentum flavum, (8) conus medullaris, (9) facet joint, and (10) neuroforamen. (✱ = basivertebral vein).

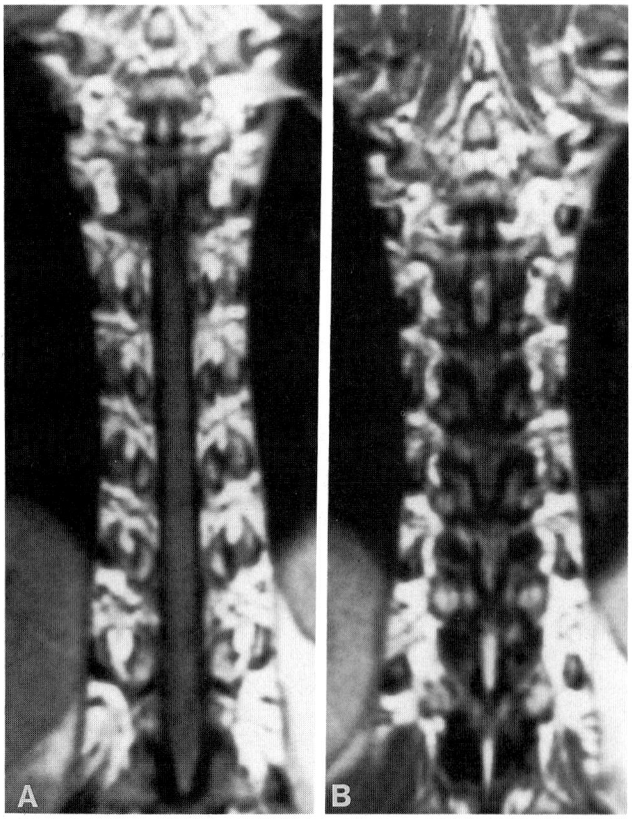

Figure 2.17. Coronal T1-weighted MRI images of the thoracic spine. A, Midline section shows the spinal cord and surrounding CSF space.The neural canal is directed laterally. **B,** Posterior aspect of the spinal canal with the laminae closer than in the cervical or lumbar region.

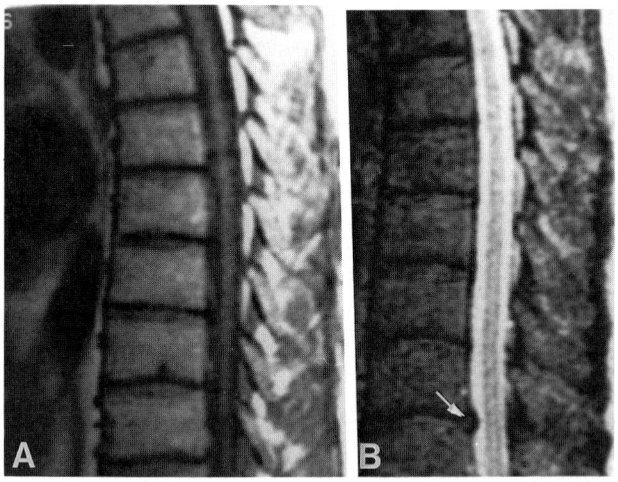

Figure 2.18. Thoracic spinal canal in sagittal plane. T1-weighted (**A**) and T2-weighted (**B**) MRI sections through the midline show the fair amount of CSF surrounding the spinal cord. A small asymptomatic disk protrusion is seen (*arrow*).

portion and round in its middle and lower portions (see Figs. 2.14 and 2.15). The subarachnoid space almost entirely fills the canal, with a small amount of surrounding epidural fat. Unless a pathologic process is identified or suspected, sagittal MRI utilizing a short TR spin-echo sequence followed by a low flip angle gradient-echo or T2*-weighted image is adequate in demonstrating the length of the thoracic spine and contents (see Fig. 2.18).

Lumbar Spine

The five lumbar vertebrae are the heaviest and largest of the vertebrae and are recognized by their size and absence of articular facets on their vertebral body (Figs. 2.19 through 2.22). The vertebral body consists of a cortical outer layer and a cancellous inner portion (see Figs. 2.19 and 2.20). The body is traversed by the basivertebral veins, which emerge from its dorsal surface near the midline. The height of the lumbar vertebrae averages 25 mm. The transverse diameter of the body is greater than its anteroposterior diameter.

The neural arches formed by the pedicles, laminae, and spinous processes form the boundaries of the spinal canal, which is oval in the upper lumbar and triangular in the lower lumbar region (see Fig. 2.21). The normal sagittal diameter of the canal measures between 15 and 25 mm, the normal interpediculate distance measures between 20 and 30 mm, gradually increasing in distance between L1 and L5, and the surface area of the vertebrae is between 2 and 3.5 cm^2.

The transverse processes extend laterally and slightly posteriorly from the junction of the pedicle and laminae, but in the fourth and fifth vertebrae, they arise more anteriorly from the pedicle.

The superior articular facets are directed medially and backward whereas the inferior articular facets are directed laterally and slightly forward (see Fig. 2.21). Unlike the thoracic facet joints, which are in the coronal plane, the lumbar zygoapophyseal joints are in the sagittal plane. However, variations in angulation and the plane of the lumbar facet joint surfaces are common. The variation in angulation of the facet joint can be better evaluated in the transaxial plane with CT than with MRI (see Fig. 2.21). Asymmetry in the joint angulation between the right and left facet joints is not uncommon [7, 8]. The more coronal the orientation of the facet joint surfaces, the greater is the incidence of facet joint disease [9].

The intervertebral foramen (neural foramen) is directed laterally (see Figs. 2.21 and 2.22). It is bounded superiorly and inferiorly by the posterior margins of the adjacent vertebral bodies and their pedicles. The anterior boundary of the foramen is the posterolateral margin of the adjacent vertebral bodies and the intervertebral disk. The posterior boundary is formed by the superior articular process of the subjacent vertebral body forming the facet joint. The nerve root and the posterior root ganglion lie in the upper portion of the foramen. The neural foramen is oval shaped, widest along its superoinferior plane. It can be visualized both in the sagittal and axial planes in both CT and MRI studies. The lateral recess, which represents the entrance to the neural foramen, is bounded anteriorly by the intervertebral disk and laterally by the superior articular process and the facet joint (see Fig. 2.21) and is best evaluated in the axial plane.

Sacrum

The sacrum is composed of five fused vertebral bodies. It has a triangular shape with its apex pointing inferiorly and dorsally. The sacrum articulates

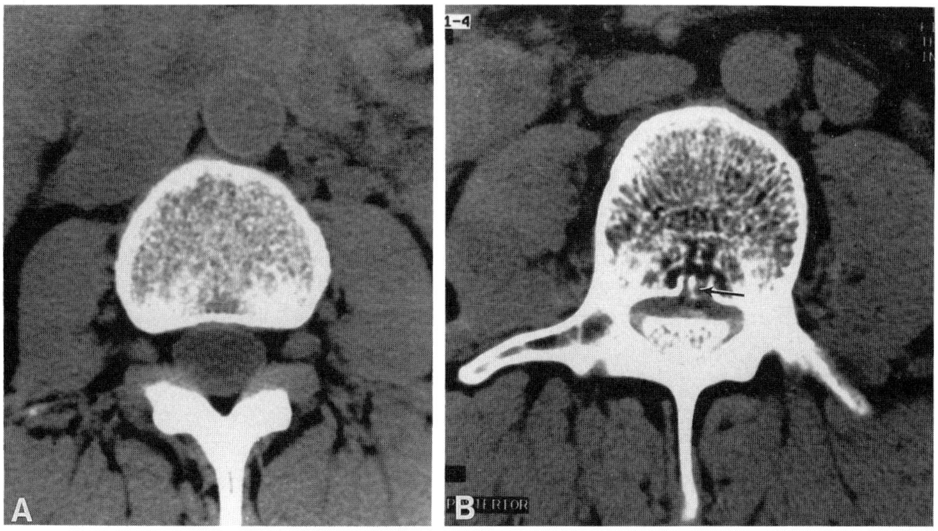

Figure 2.19. Axial CT scan through lumbar vertebrae demonstrating (**A**) the concave posterior margin of the L3 vertebra and (**B**) the straight posterior margin of the L4 vertebra. The entry zone of the basivertebral vein is associated with a small bony ridge (*arrow*). The transverse process is prominent.

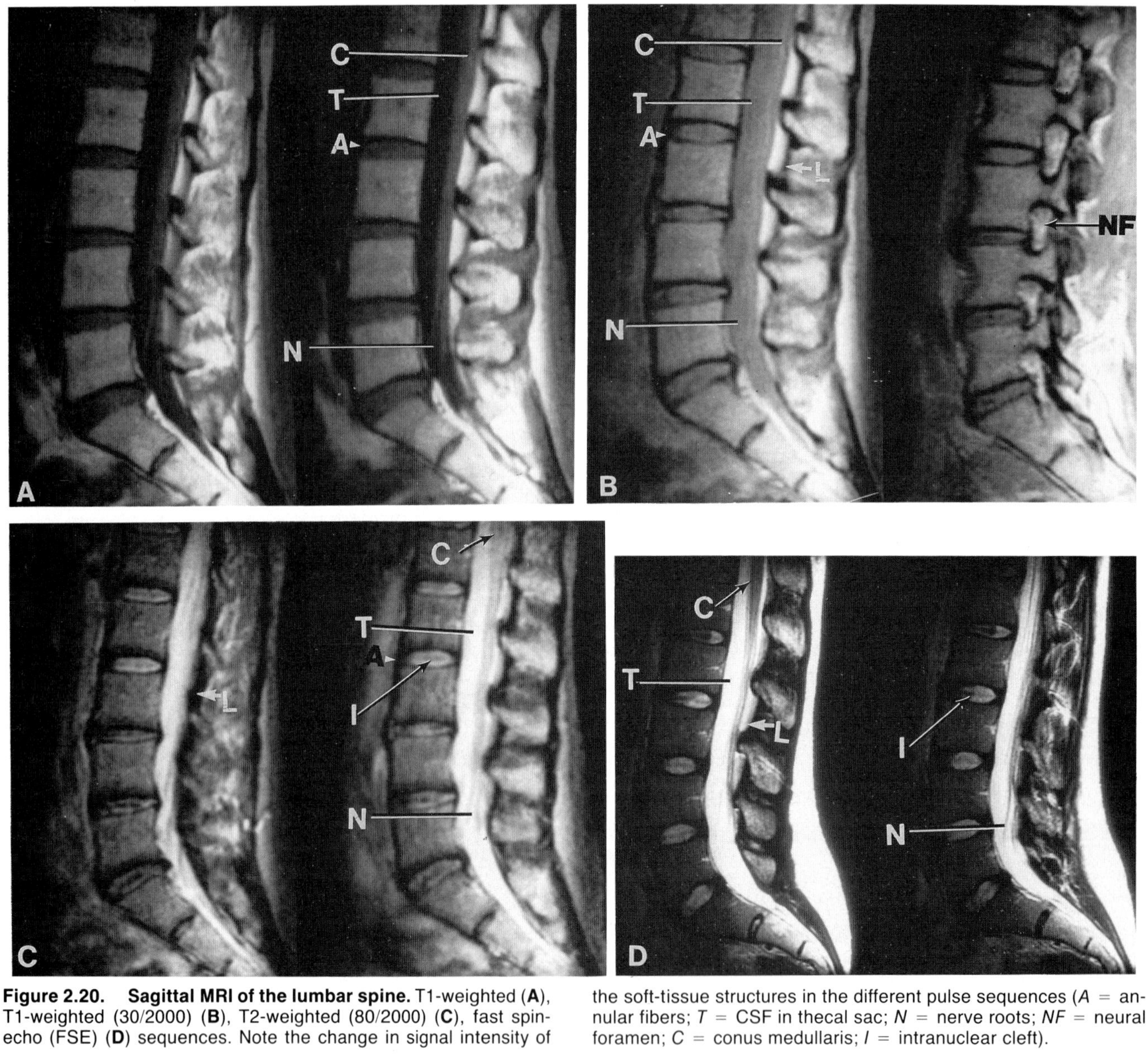

Figure 2.20. Sagittal MRI of the lumbar spine. T1-weighted (**A**), T1-weighted (30/2000) (**B**), T2-weighted (80/2000) (**C**), fast spin-echo (FSE) (**D**) sequences. Note the change in signal intensity of the soft-tissue structures in the different pulse sequences (*A* = annular fibers; *T* = CSF in thecal sac; *N* = nerve roots; *NF* = neural foramen; *C* = conus medullaris; *I* = intranuclear cleft).

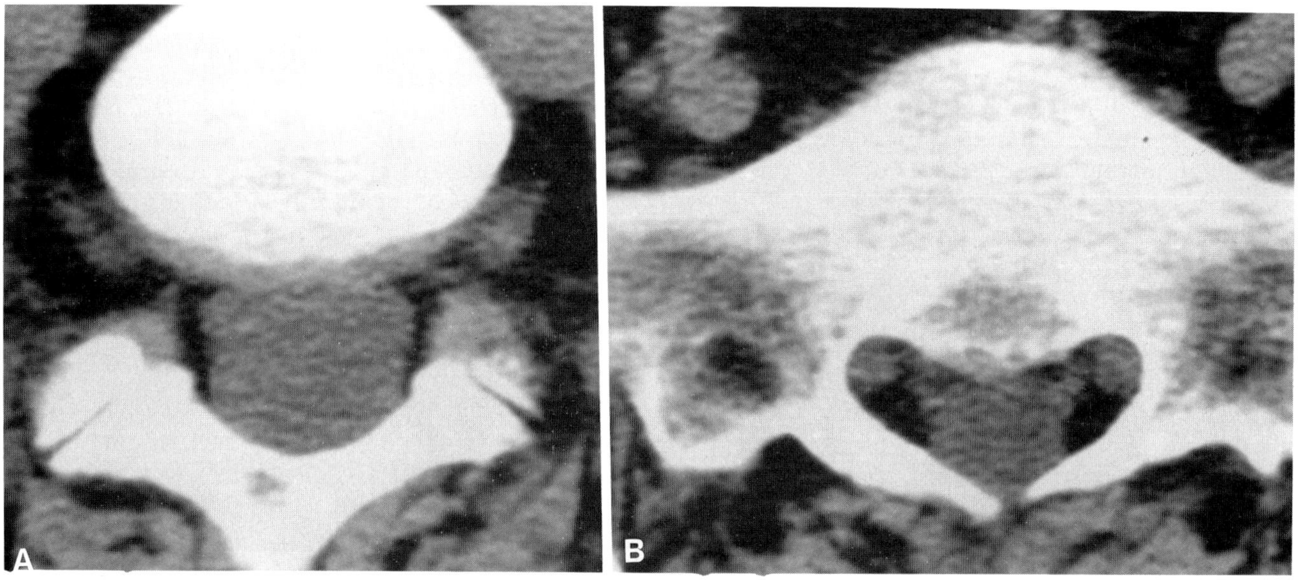

Figure 2.21. Axial CT sections show (**A**) the oval shape of the spinal canal in the upper lumbar region and (**B**) the triangular shape in the lower lumbar region.

Figure 2.22. Lumbar neural foramen. A, T1-weighted sagittal MRI. **B–D,** Re-formatted CT scans. The neural foramen is pear-shaped, with the lumbar nerve roots in the upper portion of the foramen. (1) Superior articular process, (2) isthmus, and (3) inferior articular process. (✱ = neural foramen; ↗* = more medially the neural canal or subarticular canal).

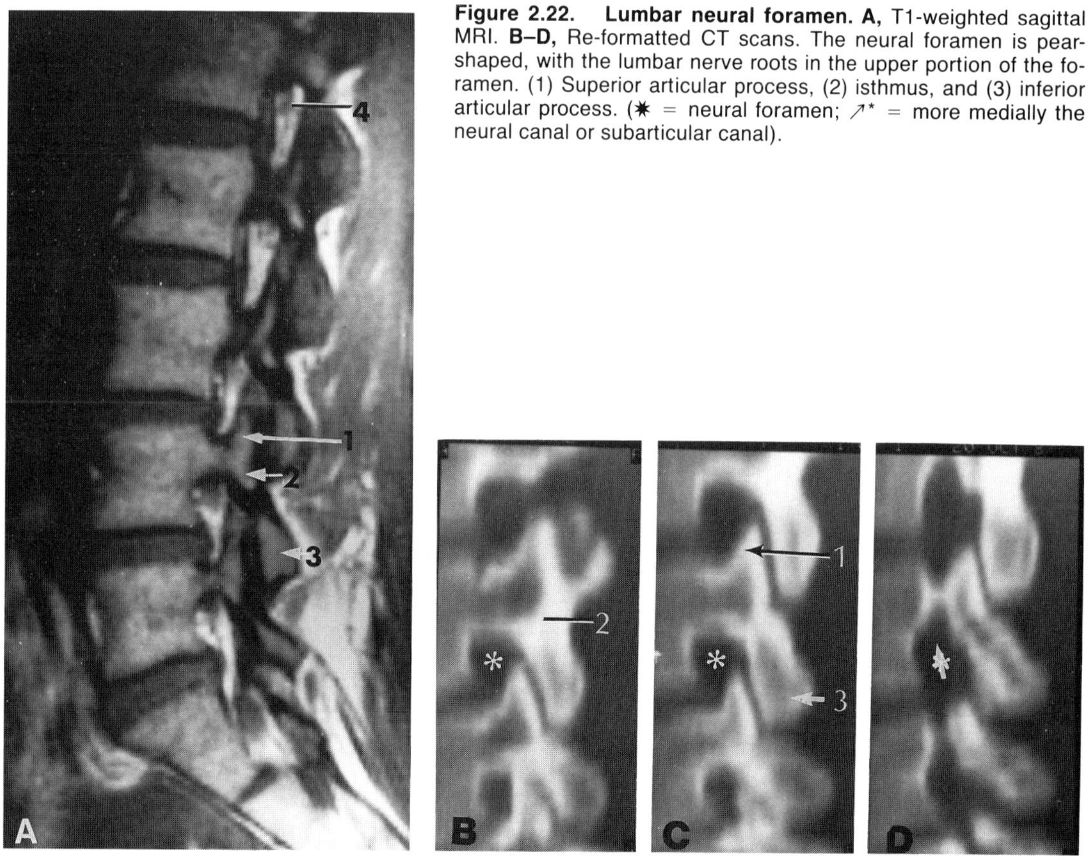

with the fifth lumbar vertebra superiorly and with the ileum laterally. The anterior surface of the sacrum reveals four transverse ridges which are the remnants of the intervertebral disks. The ventral sacral foramina, from which the ventral divisions of the sacral nerve emerge, are located on each side of the ridge. The dorsal sacral foramina are located on the posterior surface of the sacrum lateral to the sacral crests, which represent the fused facets. The sacral canal is usually incomplete in its lower portion, forming the sacral hiatus. Varying degrees of concavity of the sacrum are encountered. Because of its curvature, the sacral neural foramen is best imaged in the axial or coronal plane [10] (Fig. 2.23). The entire sacral canal is best evaluated on sagittal MRI sections.

The advantage of MRI and, to a lesser extent, CT is the ability to evaluate contiguous vertebral segments and the disk spaces in all orthogonal planes, unlike plain radiography. Signal changes within the vertebrae also can be evaluated. In MRI, the signal characteristics of the osseous spine depend on the age of the patient and the constituents of the bony spine. In the neonatal period, the vertebral body primarily consists of red marrow, which is gradually replaced by the yellow marrow such that by age 30

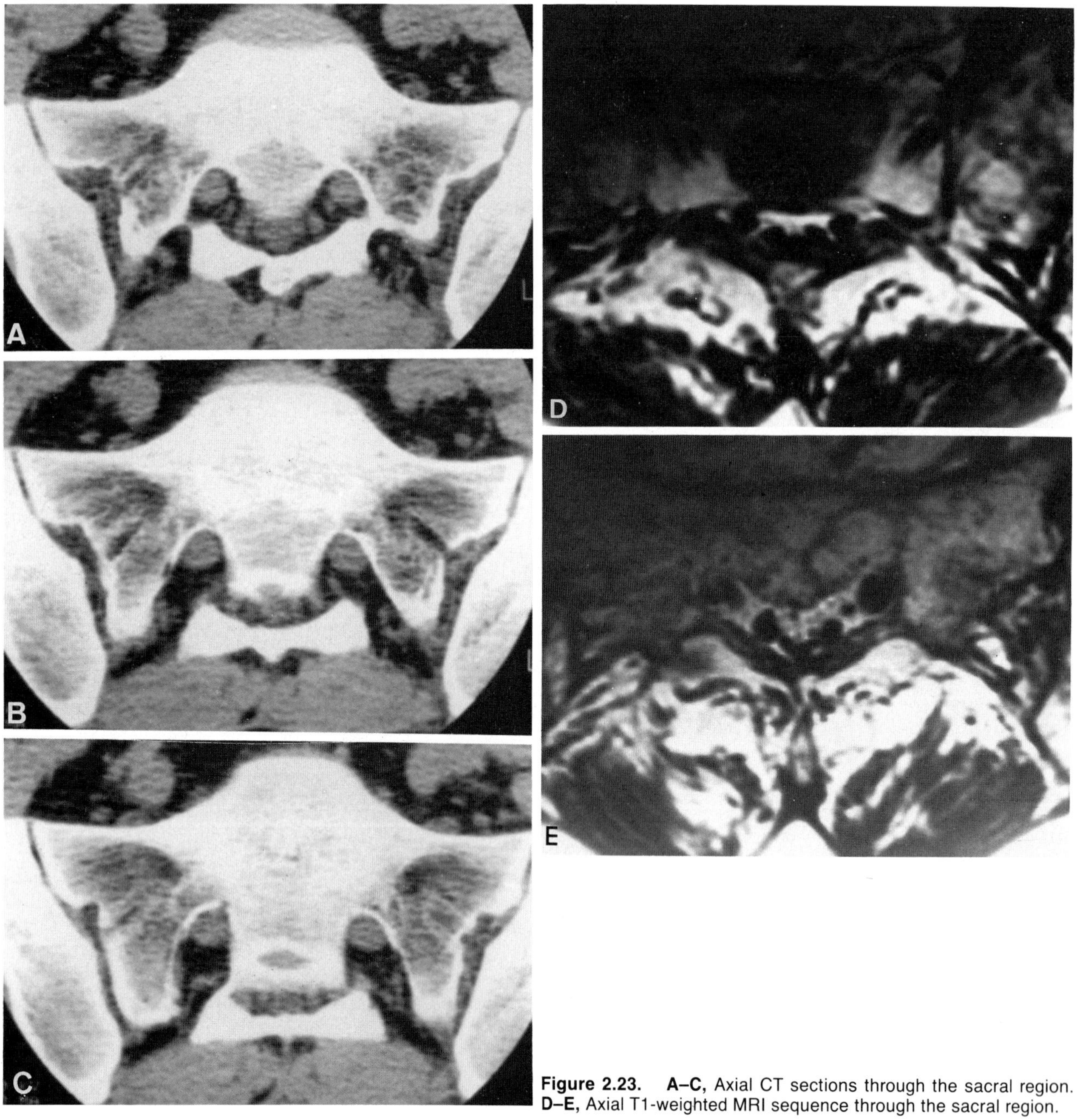

Figure 2.23. **A–C,** Axial CT sections through the sacral region. **D–E,** Axial T1-weighted MRI sequence through the sacral region.

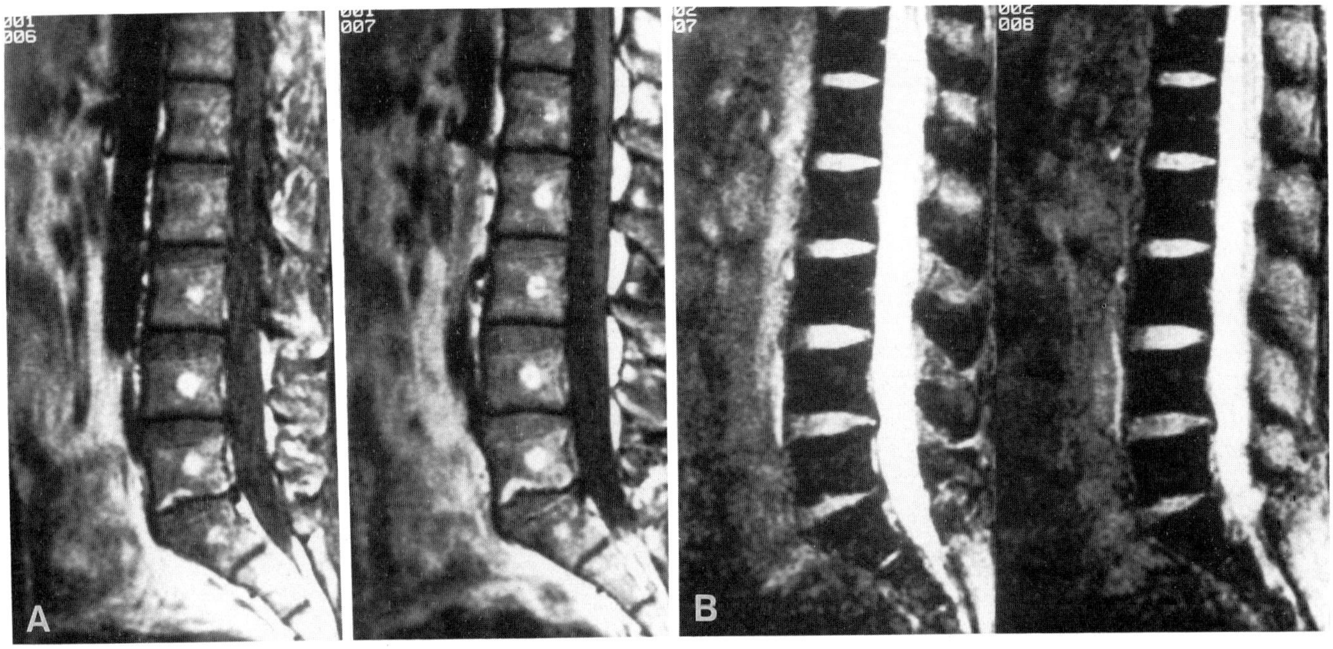

Figure 2.24. **A,** T1-weighted MRI. Fatty replacement in vertebral marrow seen as round bright signal in midline adjacent to the venous channels. **B,** T2-weighted MRI. These are isointense on T2-weighted MRI. The isointensity results from fat replacing the red marrow, often seen in the older population.

nearly 60% is yellow marrow. In addition, with aging, focal areas of fatty infiltration also may occur within the stroma of the yellow marrow. On MRI studies, the water and marrow in the vertebrae result in bright signal intensities in T1-weighted images in infants (see Fig. 2.6), isointense signals during adolescence, and heterogeneous signals beyond the third decade. On both T2 and gradient-echo sequences, the vertebral water and marrow have a decreased signal intensity. The normal vertebra appears relatively homogeneous, without a trabecular pattern, and has a low signal intensity. This is due to a combination of volume averaging and the dampening effect of the trabecular matrix over the relative volume of the marrow within the vertebral body. The signal intensity of marrow is better evaluated with spin-echo sequences than with gradient-echo sequence. This is due to loss of signal in gradient echo, resulting from greater magnetic susceptability. Focal areas of hyperintense signal lesions on T1-weighted images which are less intense on T2-weighted sequences are seen as a normal process of aging. These focal bright lesions represent fatty replacement following ischemic changes in the marrow space [11] (Fig. 2.24). The dense cortical bone at the margin of the vertebral body has a low signal intensity on both T1 and T2-weighted images. A band of low signal along the inferior cortical margin of the vertebra on sagittal MRI sections is the result of chemical shift artifact resulting from marrow (fat)–disk (water) interface (Fig. 2.25).

In axial sections approximately midway between

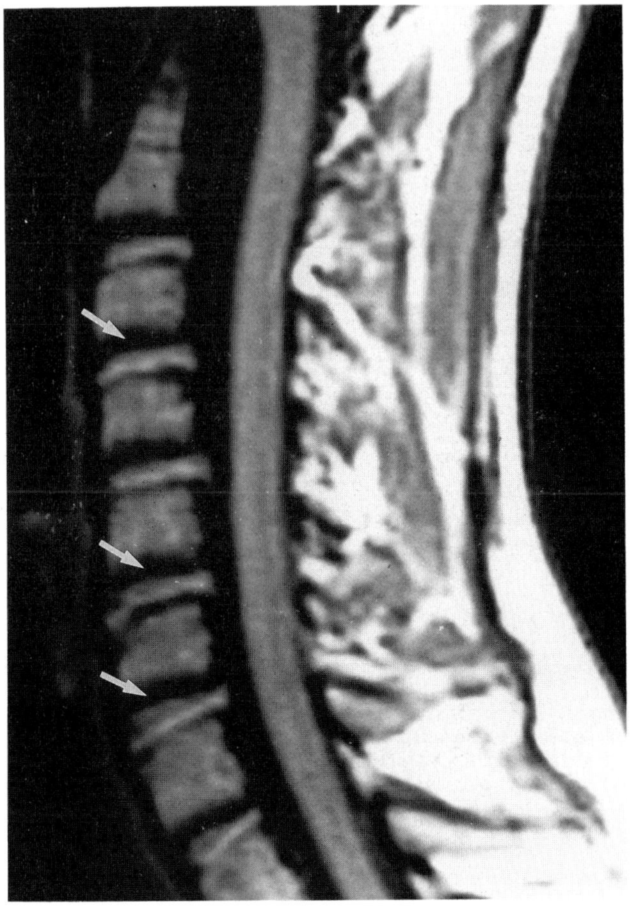

Figure 2.25. Chemical shift artifacts seen on T1-weighted MRI as thick, hypointense bands (*arrowheads*) along the interface of the disk-vertebra margins should not be confused with the vertebral end plates. Note linear bright signal on the opposing surface of the disk margin.

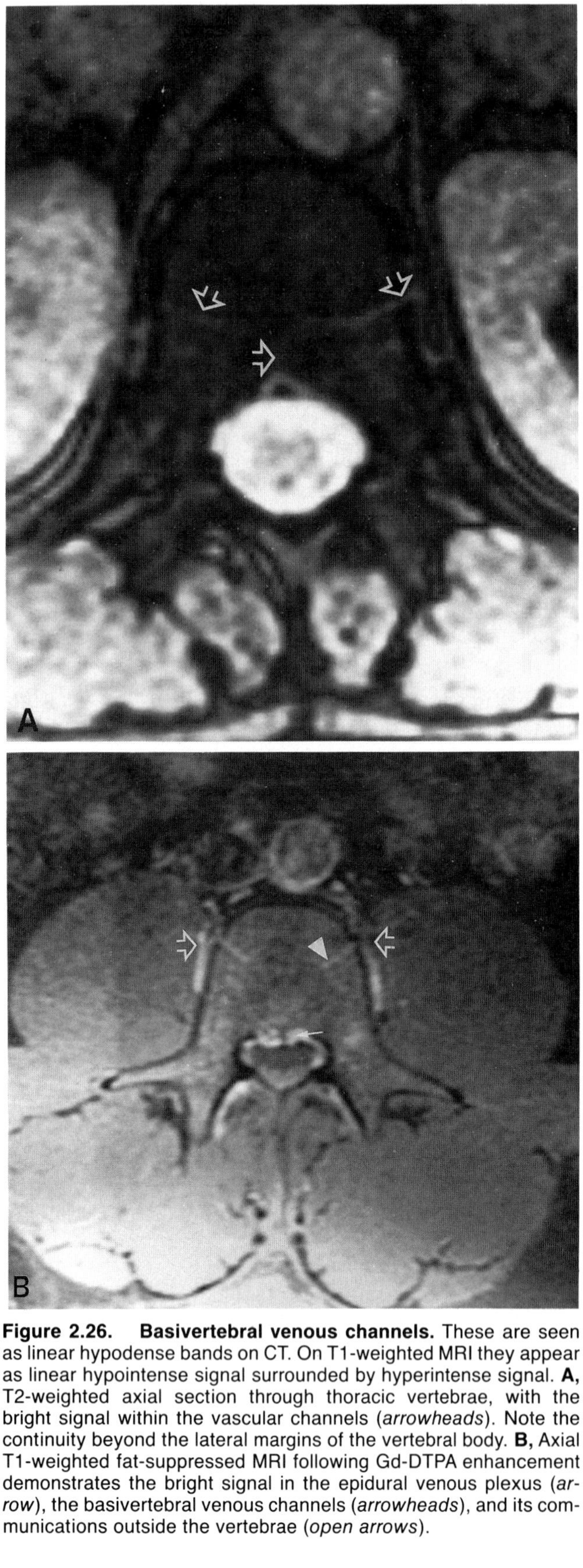

Figure 2.26. **Basivertebral venous channels.** These are seen as linear hypodense bands on CT. On T1-weighted MRI they appear as linear hypointense signal surrounded by hyperintense signal. **A,** T2-weighted axial section through thoracic vertebrae, with the bright signal within the vascular channels (*arrowheads*). Note the continuity beyond the lateral margins of the vertebral body. **B,** Axial T1-weighted fat-suppressed MRI following Gd-DTPA enhancement demonstrates the bright signal in the epidural venous plexus (*arrow*), the basivertebral venous channels (*arrowheads*), and its communications outside the vertebrae (*open arrows*).

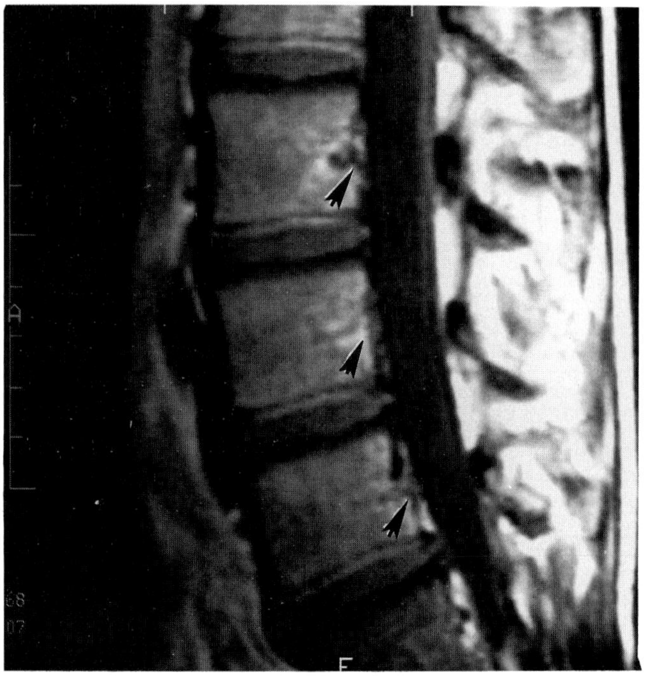

Figure 2.27. Midline sagittal MRI sections demonstrate the entry zone (*arrowhead*) of the basivertebral veins and its relationship to that of the epidural veins.

the endplates, one or several lucencies often may be seen crossing the vertebral body. These correspond to the basivertebral veins which interrupt the bony trabeculae [12, 16, 2.19]. The basivertebral veins exit from the dorsal surface of the body. At their exit points, a region of increased bone density on CT (see Figs. 2.19 and 2.21) and curvilinear low signal on MRI is seen (Fig. 2.26) which corresponds to the impacted bone. The basivertebral entry zones as well as the basivertebral venous channels within the vertebral bodies are best seen in the lumbar vertebrae, although they are also present in the cervical and thoracic regions (see Fig. 2.26). The entry zones of the venous channels are often seen as linear foci of hypointense signal on T1-weighted sections and as bright signal in sagittal T2-weighted as well as on gadolinium-enhanced T1-weighted MRI sections (Fig. 2.27). The bright signal is due to the slow venous flow and some amount of surrounding epidural fat. On CT and sometimes on MRI, the venous channels within the vertebral bodies may be mistaken for fractures [12, 14]. Differentiation from fracture is based on their location in the midaxial plane, distinct osseous walls, and in MRI the linear bright signal due to the combination of slow venous flow and surrounding lipid material.

Normal Variants

Variations in the anatomy of the spine are not unusual. The two most common of these are bifid spinous processes and transitional vertebrae. *Spina bifida occulta* is nonfusion of the lamina due to an anomaly in closure of the neural tube. It is often an incidental finding, usually in the lumbar region, but it also may involve the thoracic or cervical spine. An intradural anomaly should be excluded when spina bifida is associated with a skin dimple or a hairy patch.

When there are more than five non-rib-bearing vertebrae in the lumbar region, the additional "lumbar" vertebra or *transitional vertebra* is most often due to lumbarization of the first sacral vertebrae. Less often, six non-rib-bearing vertebrae may be the result of hypoplasia or absence of the twelfth rib. Occasionally, only four non-rib-bearing lumbar vertebrae may be present, resulting from "sacralization" of the fifth lumbar vertebrae. Recognition of these anatomic variants is important when evaluating the lumbosacral spine for a herniated disk. The level of the disk can be erroneously interpreted in cross-sectional imaging if the transitional anomaly is not recognized. Anteroposterior views of the thoracolumbar spine are necessary in this evaluation.

Spondylolysis is due to a cleft involving the pars interarticularis or isthmus of the vertebrae. It is most often seen in the lumbar spine, involving the L4 or L5 vertebra. The clefts may involve the pedicle (retrosomatic cleft) [15], or the lamina (retroisthmic cleft). In the lumbar region, spondylolysis is seen in 5% of the general population. In majority of instance, spondylolysis is the result of fracture, most often from repeated minor stress trauma.

On axial CT sections, spondylolysis appears as an irregular linear defect in the pars interarticularis (Fig. 2.28). The defect in the pars may mimic a zygapophyseal joint; however, the normal facet joints can be seen above and below the pars defect. The facet joints are best seen in axial sections through the disk space, whereas the pars defect is seen in axial section through the midportion of the vertebral body. Parasagittal and oblique re-formatted sections provide additional information not readily seen on the axial sections. Spondylolysis is often associated with spondylolisthesis. Identification of the pars defect on MRI may occasionally be difficult even when it is associated with spondylolisthesis [16], since it may have an appearance similar to a degenerative spur of the superior facet, osteoblastic metastasis, or partial facetectomy. On MRI, spondylolysis should be suspected when there is inability to resolve the marrow signal in the pars and decreased signal adjacent to the facet joint in axial views and between the superior and inferior facets of the pars on parasagittal sections. Because of sclerosis at the site of the defect, there is decreased signal intensity in all pulse sequences. Occasionally, a focal or linear bright signal within the dark region can be identified. This represents soft tissue or marrow interposition, often seen when associated with spondylolisthesis. In rare instances, spondylolysis also may involve the cervical vertebrae [17]. This is usually congenital rather than a result of occult trauma. Spondylolysis in the cervical region most often involves the C6 vertebral segment. Plain radiographs complemented with CT are ideal in identifying the bony abnormality. Other anomalies of the osseous spine include fusion anomalies of the vertebral bodies and posterior elements. Fusion anomalies are discussed in Chapter 4.

Facet Joints

The facet joints are formed by the apposing surfaces of the superior and inferior articular facets of the adjacent vertebrae (Figs. 2.9, 2.17, and 2.22). They thus constitute the posterolateral joint complex of the motion segment of the spine [9, 18–20]. The orientation of the superior and inferior articular facets differs in the different regions of the spine [21, 22]. Facet joints are diarthrodial joints with a synovial lining. The joint surfaces are covered by articular cartilage of varying thickness in the different regions. The cartilage is thickest in the central portion of the joint. The joint capsule is made of a thin fibrous layer loosely attached to the joint margins and reinforced by the attachment of the adjacent ligamentum flavum [20]. On CT they are best evaluated on axial sections with bone algorithm (Fig. 2.29), although slightly oblique parasagittal re-formatted studies demonstrate not only their alignment but also encroachment of the neural foramina in degenerative changes involving the spine. However, this necessitates contiguous thin or overlapping CT sections for proper evaluation. The facet joints also can be well visualized on MRI studies. The joint space, consisting of the synovial membrane, fluid, and cartilage, has an intermediate signal on T1-weighted sequence and a hyperintense signal on T2-weighted images, whereas the cortical-cartilage interface and the menisci appear hypointense (Fig. 2.30). Variations in the extent of the joint capsule and the menisci can be demonstrated on MRI [21, 22], although the clinical significance in evaluating facet disease is not clear. Occasionally, a focal area of bright signal, representing the apical synovial fat pad, is noted within the lumbar facet joint on sagittal section [23]. With degenerative changes in the facet joints, this bright signal is not identified.

Intervertebral Disks

Intervertebral disks form the articulations between the vertebral bodies. They vary in thickness, size, and shape in the different regions of the spine.

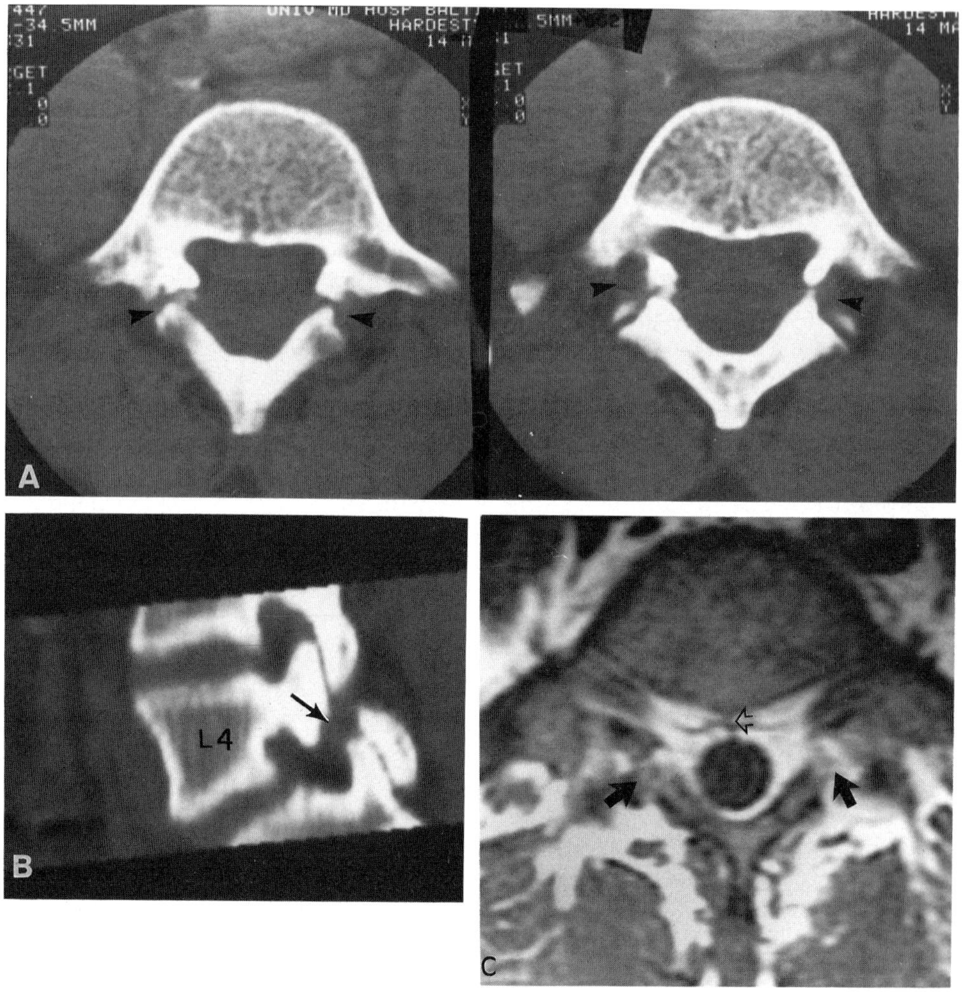

Figure 2.28. Spondylolysis and spondylolesthesis. A, Axial CT section. **B,** Re-formatted sagittal view on CT. Bilateral spondylolysis is present (*arrowheads*). The pars defect (*arrow*) at the junction of isthmus and neck of inferior articular facet. Note increased sagittal diameter of the spinal canal. **C,** Axial T1-weighted MRI.

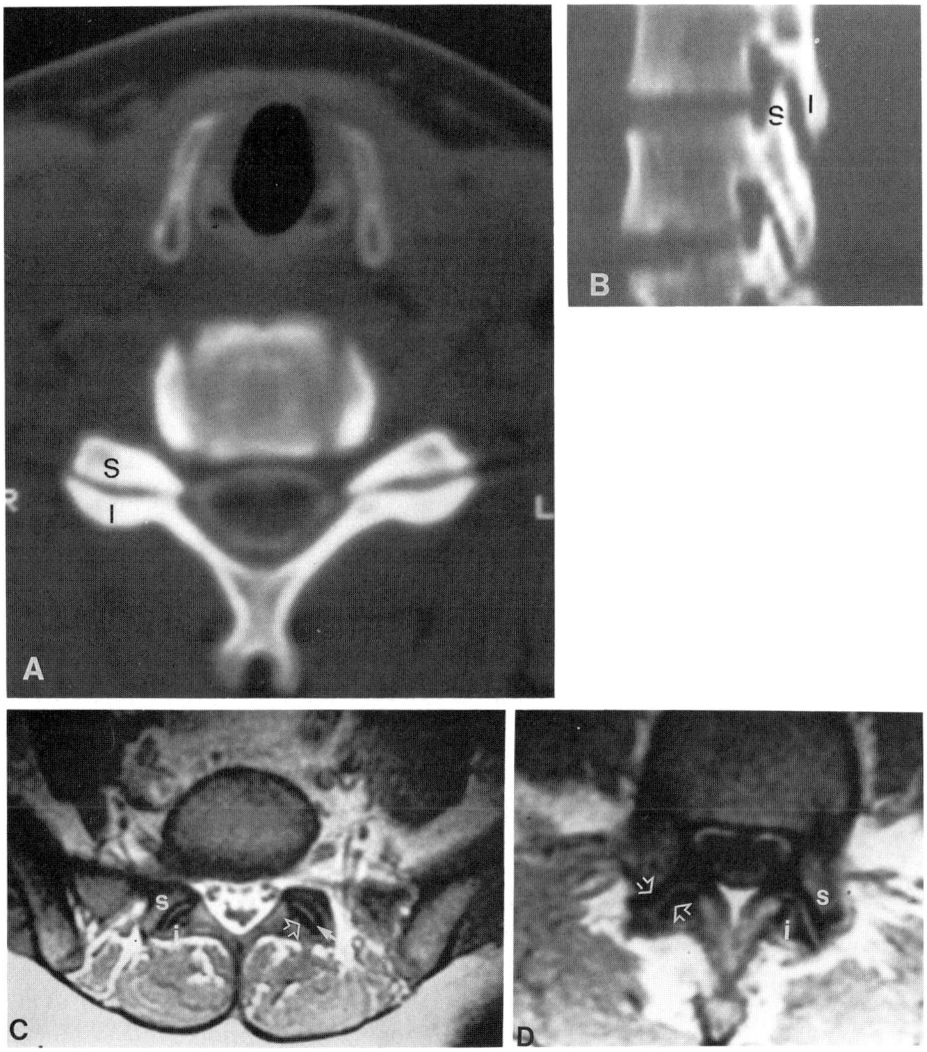

Figure 2.29. Facet joints. These are diarthroidal joints, the superior facet forming the posterior border of the neural foramen. The anterior articulating surface is the superior facet of the vertebra below, whereas the posterior articulating surface is from the inferior facet of the vertebra above. Their articulation planes differ in the cervical and lumbar regions. **A,** Axial CT section through cervical region. **B, C,** Sagittal re-formatted CT scan showing the superior (*S*) and inferior (*I*) articular facets of the joint. Axial T1-weighted MRI through lumbar region. The synovial fluid and menisci has an intermediate to bright signal. The articular cartilage has a dark or hypointense signal. **D,** With aging and hypertrophy of the cartilage (*open arrow*), the synovial space and menisci are difficult to identify.

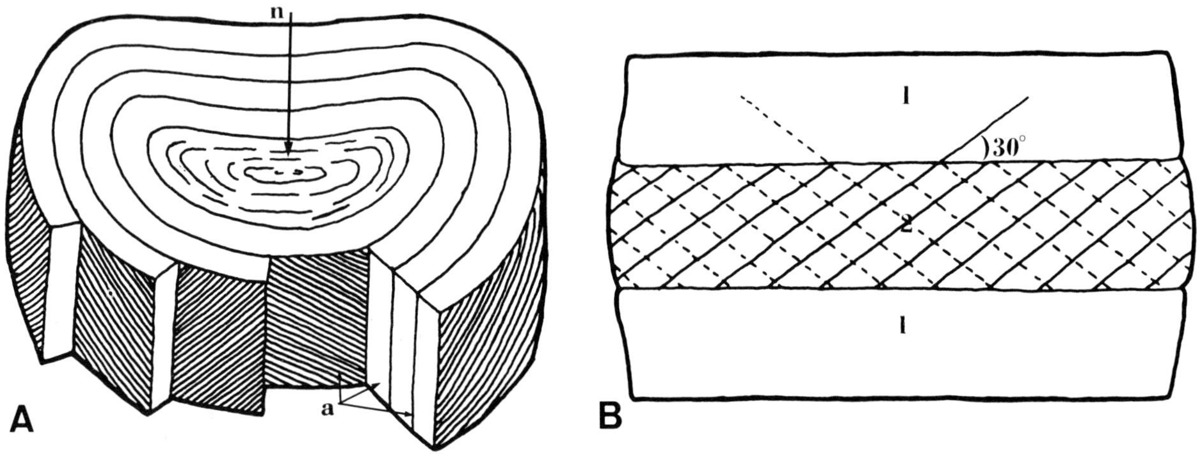

Figure 2.30. Schematic drawing of the disk. A, Concentric layers of the annulur fibers surround the nucleus pulposus in adjacent layers. The fibers are oriented in opposite directions at an approximately 30 degrees to the vertebral body endplate 1 (*N* = nucleus pulposus; *AF* = annular fibers). **B,** Frontal view of the annular fibers of the intervertebral disk. Note the 30-degree orientation of the fibers in relation to the endplates (1 = vertebral body; 2 = annular fibers).

The axial cross-sectional area of the thoracic and lumbar disks corresponds to the surfaces of the vertebral bodies between which they are located. Cervical disks are slightly smaller than their adjacent vertebral bodies because of the uncinate processes. The anterior margins of all intervertebral disks are convex. The posterior margins of all cervical and thoracic disks are concave. The first three lumbar disks have concave posterior margins, but the degree of concavity is much less than in the thoracic spine. The fourth lumbar disk usually has a slight concavity, but it may be normally flat. The fifth lumbar disk normally is flat or mildly convex posteriorly. Demonstration of the disk without including the adjacent vertebral body is easiest in the lumbar region, when appropriate slice thickness (3 to 5 mm) is utilized with both CT and MRI. In the thoracic and cervical regions, demonstration of the disk without interfering bone density requires a section thickness of between 1.5 and 3 mm.

Intervertebral disks account for approximately 25% of vertebral column length (excluding the sacrum and coccyx) [24, 25]. However, the disks are not distributed evenly. Lumbar disks, as one may suspect, are thickest both in absolute measure and relative to the height of the adjacent vertebral body. Thoracic disks are thinnest relatively, since their weight-bearing load is lighter and stability of the thoracic column is aided by the ribs and their musculature. The intervertebral disks may vary in thickness between regions of the same disk space. Lumbar and cervical disks are thicker anteriorly, which contributes to the lordosis of these regions. The most wedge-shaped disk is at the L5–S1 level. The normal height of the disk at the L5–S1 level on plain radiographs is 15 mm anteriorly and 6 mm posteriorly in females and 20 and 6 mm, respectively, in males. Thoracic disks are of nearly uniform thickness; the natural thoracic kyphosis is due almost entirely to the shape of the vertebral bodies, with a slight anterior decrease in the height of the disk compared with the posterior margin. The intervertebral disk is a bicovex structure. It serves as a shock absorber, and it distributes the mechanical and rotary stresses applied to the vertebral bodies. The intervertebral disk consists of the central nucleus pulposus, surrounded by the annulus fibrosus and peripheral cartilaginous endplates. These merge with the cortical surface of the adjacent vertebral bodies.

NUCLEUS PULPOSUS

The nucleus pulposus is a remnant of the embryonic notochord composed of collagen fibrils embedded in a mucoprotein and mucopolysaccharide gel. The polysaccharide gives the nucleus a water-imbibing ability which decreases with age [25]. This loss of hydration plays a significant role in disk degeneration, as more and more stresses are shifted from the failing nucleus pulposus to the annulus fibrosus.

The nucleus pulposus in cervical and lumbar disks occupies a larger percentage of the disk space than in their thoracic counterparts. The nucleus pulposus is located centrally in the cervical disks but is situated between the middle and posterior third in the lumbar disks.

ANNULUS FIBROSUS

The annulus fibrosus encases the nucleus pulposus and forms the outer boundary of the intervertebral disk. The outer peripheral portion is mostly made up of fibrous tissue, whereas the more central component has cartilaginous elements. The fibers of the annulus are laminated, with each layer slanting

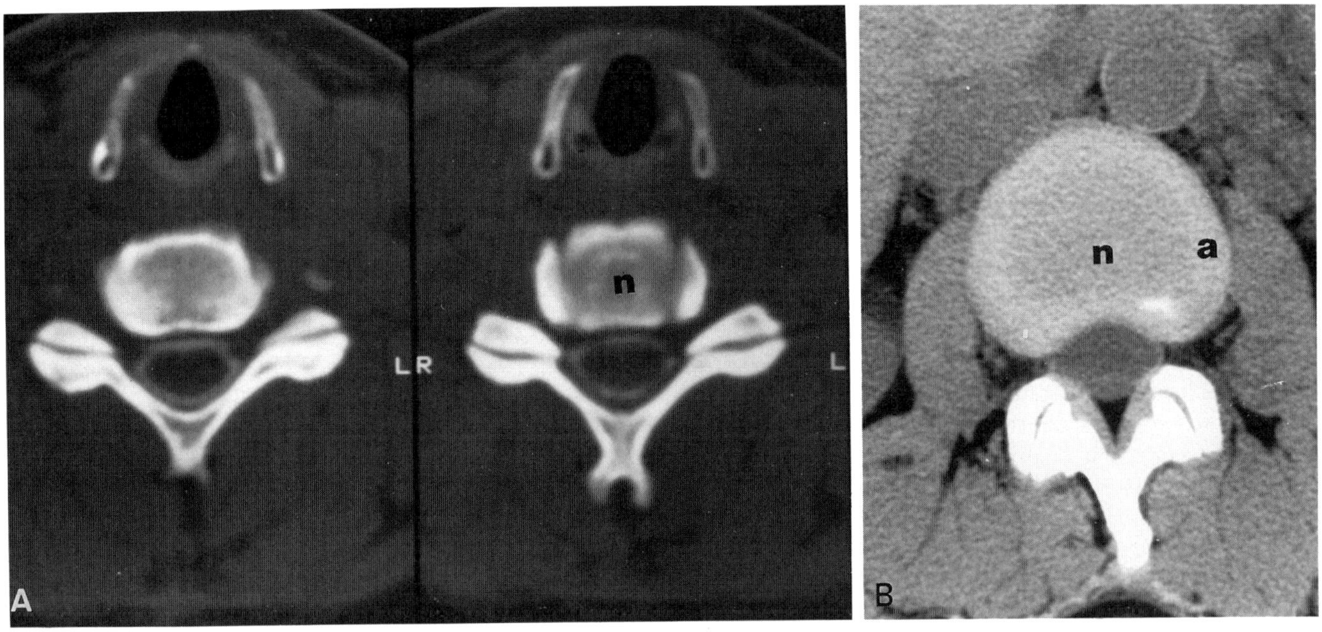

Figure 2.31. Axial CT through (**A**) cervical and (**B**) lumbar disk space. The annular fibers (*a*) at the periphery are hyperdense compared with the more central fibers and the nucleus pulposus (*n*).

in opposite directions (see Fig. 2.25). The fibers are angled approximately 60 degrees to the axis of the vertebral body. The centrally located fibers of the annulus fibrosis insert into the hyaline cartilage endplates (Sharpey's fibers) of the vertebral bodies. This provides the pathway for diffusion of metabolites and fluids through the vertebral endplates. Peripheral fibers pass over the edge of the vertebral body to unite with the cartilaginous endplate. These peripheral insertions are much stronger than the central ones. Sharpey's fibers, when abnormally stressed by disk degeneration, can ossify, thus forming osteophytes.

On CT sections, the normal intervertebral disk has a hyperdense peripheral region that represents the annulus and a central mildly hypodense nucleus pulposus (Fig. 2.31). If the section involves the nucleus pulposus along with the cartilaginous portion of the annulus and the adjacent cortical bone, it usually has a hyperdense appearance. This is common in the cervical and thoracic regions due to the small height of the disk space and partial volume averaging of all the elements. In the lumbar region with thin axial sections, the posteromedially situated nucleus pulposus is relatively more hypodense. The annulus is thicker and wider anterolaterally. The nucleus pulposus has a ground-glass appearance relative to the cortical bone. This limited evaluation of the disk by CT is due to the inherent limitations of x-ray attenuation and beam-hardening effect. The signal characteristics of the disk on MRI depend on the physiologic and biochemical changes resulting from aging, as well as other factors such as stress (Fig. 2.32).

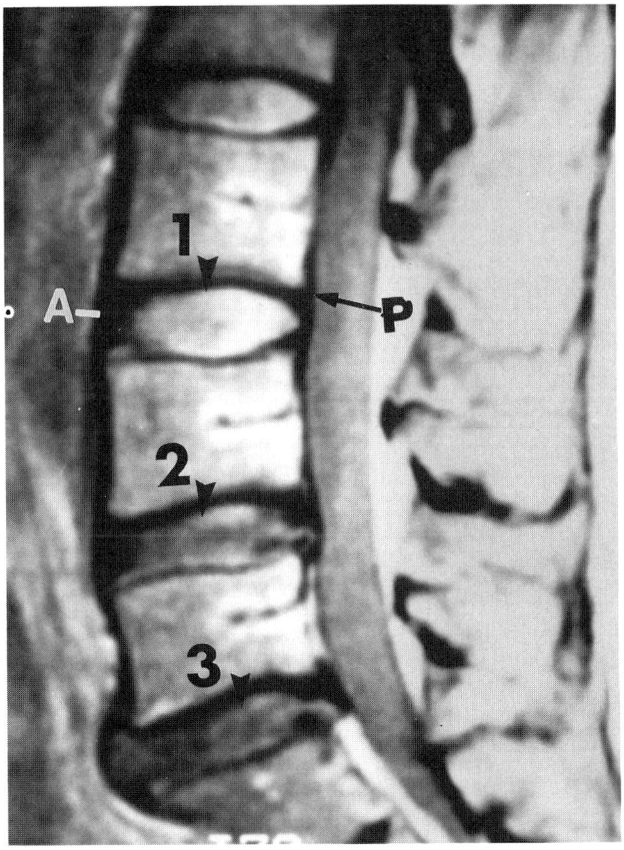

Figure 2.32. Sagittal T2-weighted (30/2000) MRI images of lumbar spine demonstrating regressive changes within the disk space. (1) normal hydrated disk; (2) decreased signal within the disk with thinning of the posterior annular fibers; the linear bright signal probably indicates protrusion of nucleus pulposus within the transverse tear of the annular fibers; and (3) circumferential bulging of the annulus fibrosus with some loss in height of the disk space and marked decreased signal within the disk material (*A* = anterior longitudinal ligament, *P* = posterior longitudinal ligament).

Collagen and elastic fibers form the matrix of the nucleus pulposus [26, 27]. The collagen within the matrix consists of highly proteinaceous solution of complex carbohydrates such as chondromucoprotein, which itself is made up of chondroitin sulfate, keratosulfate with a mucoprotein complex [28]. With aging, the mucoprotein complex within the disk decreases in quantity, along with an increase in the fibrous component. Simultaneously, there is a decrease in the total water content. This results in the nucleus changing from a gelatinous substance to a desiccated fibrous mass, whereas peripherally, the cartilaginous elements increase within the annulus fibrosus. Eventually, there is a complete loss of differentiation of the annulus from the nucleus pulposus, resulting in complete disk degeneration. The degenerated disk consists of necrotic tissue with calcification, fibrocartilage, and gelatinous residue.

Although these are the biochemical end results, the physiologic changes detected by MRI presumably are related to the loss in hydration as the natural process of aging. This is manifested on MRI studies as a progressive decrease of the bright signal within the nucleus pulposus, as well as the loss in height of the disk space [29, 30]. In the normal hydrated disk, the nucleus pulposus has an intermediate signal intensity on T1-weighted images and hyperintense signal on T2- and T2*-weighted images. The annulus fibrosus and the cartilaginous endplate have a hypointense signal in both the T1- and T2-weighted images (see Fig. 2.20).

In the newborn infant's spine, the nucleus pulposus is more prominent. In addition, the collagen within the fibrocartilaginous component of the nucleus is minimal [31], contributing to the larger area of bright signal on MRI. With aging, the annulus fibrosus gradually increases its fibrous component from the periphery toward the nucleus pulposus in a radial fashion, depending on the stress applied to the disk space and the body contour. There is an increase in fibrous content in the region of the equator within the nucleus pulposus, the site of the notochordal element also referred as the *intranuclear cleft* [32] (see Fig. 2.20). The intranuclear cleft is often seen beyond the second decade, extending from the midline posteriorly in T2-weighted sagittal MRI images. With the regressive changes of aging, the hypointense signal of the intranuclear cleft merges with the hypointense signal of the rest of the disk material (Fig. 2.33). These regressive changes involve the vertebral bodies, the facet joints, and the disk spaces. The characteristic anatomic and biochemical changes in the discovertebral complex associated with aging have been correlated extensively with MRI studies [31–38] (Table 2.1) (see also Chap. 5).

Table 2.1.
MRI in Physiological Changes in Disk with Aging

Disk Type and Age	Anatomic Characteristics	MRI Signal Characteristics
Type 1 *(0–9 years)*	Disk contains notochordal elements. The fibrocartilage of the disk is sharply demarcated from hyline cartilage of the vertebral body. There is sharp demarcation between the fibrocartilage of the annulus and the nucleus pulposus.	Low signal on T1-weighted and bright signal on T2-weighted MRI. Sharp demarcation from the hyline cartilage of the vertebral body.
Type 2 *(10–20 years)*	Fifty percent of disk space is occupied by anulus fibrosus. The nucleus is located posteriorly in the upper lumbar region and anteriorly in the lower lumbar region. The notochordal element is replaced by fibrous band. Margin between anulus and nucleus pulposus is ill defined.	Disk has bright signal on T2-weighted MRI. The peripheral portion of the anulus forming the Sharpey's fibers has low signal intensity.
Type 3 *(20 years and beyond)*	Transition between anulus and nucleus pulposus is ill defined. Radial and vertical tears in the inner fibers of the anulus occur with brown pigmentation. The nucleus becomes more fibrous with decreased hydration. The endplates show cracks. Loss in height of the disk space.	Anular tears shows linear bands of dark signal with areas of bright signal from fluid within the cracks.
Type 4 *(40 years and beyond)*	With aging there is disintegration of the anulus and the nucleus pulposus. There is marked narrowing of the disk space, with eburnation and sclerosis of the endplates.	Loss in height. Decreased signal on both T1- and T2-weighted images. Focal linear areas of bright signal depending on the presence of fluid.

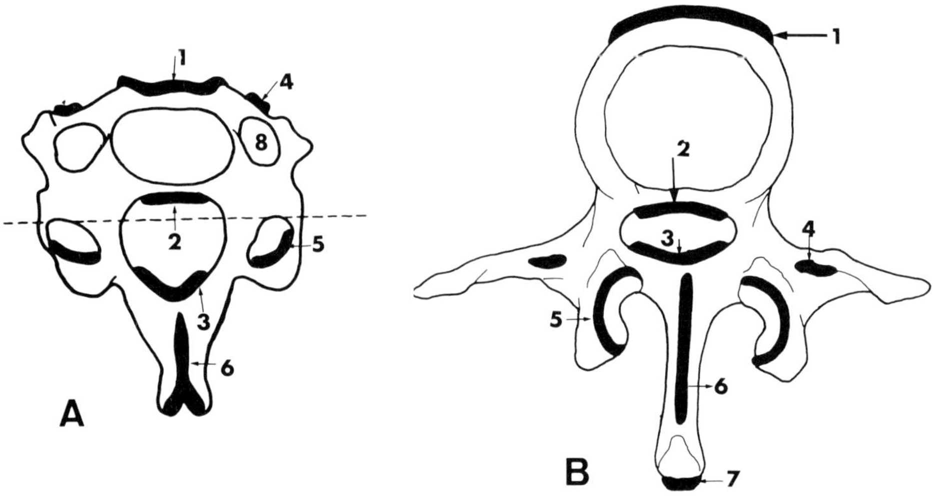

Figure 2.33. Schematic drawing of the spinal ligaments in the axial plane. A, Cervical spinc. **B,** Lumbar spine. (1) Anterior longitudinal ligament, (2) posterior longitudinal ligament, (3) ligamentum flavum, (4) intertransverse ligament, (5) capsular ligament, (6) interspinous ligament, (7) supraspinous ligament, and (8) foramen transversorium.

Ligaments

The vertebral bodies are linked together by the in tervertebral disks and the anterior and posterior longitudinal ligaments. The facet joints, lamina, and spinous and transverse processes all possess ligamentous attachments. The basic attachments of the ligaments below the C2 level are relatively constant throughout the vertebral column (Fig. 2.33). The thoracic vertebrae have additional ligaments connecting the ribs.

ANTERIOR LONGITUDINAL LIGAMENT

The anterior longitudinal ligament courses along the anterior aspect of the vertebral bodies from C2 to the sacrum (see Figs. 2.33 and 2.34). It is narrowest in the cervical region and becomes broader as it descends. The deepest fibers are firmly bound to the annulus of the disk and to the edges of the adjacent vertebral bodies. The deep fibers span only one disk space. Intermediate fibers span two to three vertebral bodies, and the superficial fibers extend over four or five levels. The ligament is firmly attached to the edges of the vertebral bodies; however, the ligament is thickest opposite the midvertebral body. It tends to fill the concavity of the vertebral body and forms the periosteam of the underlying bone. The anterior longitudinal ligament is thicker and stronger than its posterior counterpart. The normal ligament is not seen on CT. On MRI studies, the ligament has decreased signal intensity in all pulse sequences (Fig. 2.34). On sagittal images, the ligament merges with the anterior cortical surface of the vertebral body, and at the disk level, it merges with the annulus fibrosus.

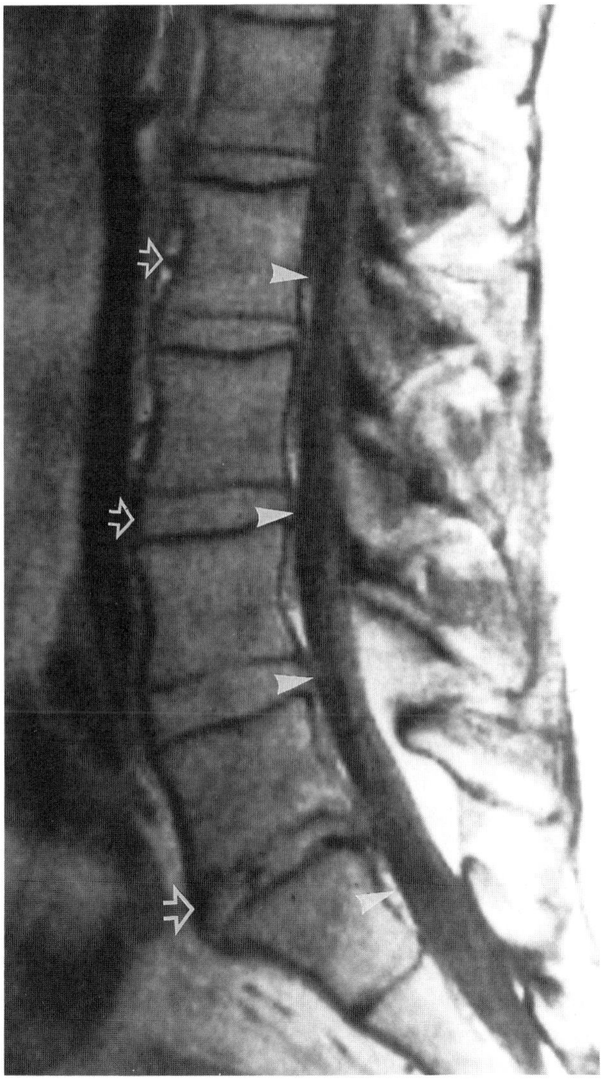

Figure 2.34. Sagittal T1-weighted MRI. Anterior longitudinal ligament (*open arrows*) at the disk and vertebral body levels. Posterior longitudinal ligament (*arrowheads*) is attached to the annulus and adjacent Sharpey's fibers but is strung like a bow across the vertebral body. The potential space is filled with fat and epidural veins. The ligament is deficient laterally.

POSTERIOR LONGITUDINAL LIGAMENT

The posterior longitudinal ligament extends from the posterior aspect of C2 to the coccyx (see Figs.

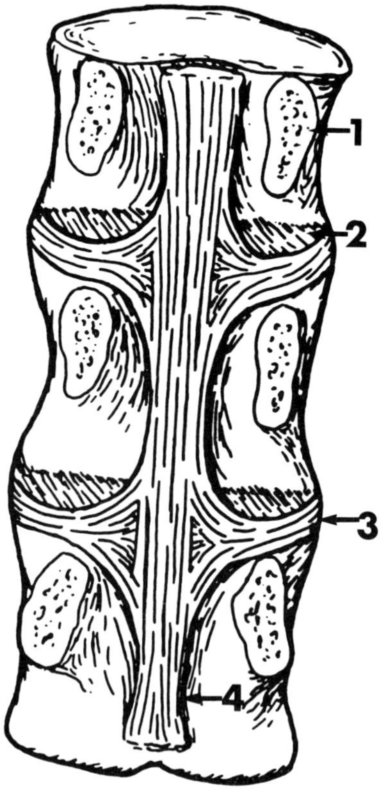

Figure 2.35. Schematic view of the posterior longitudinal ligament from behind with the pedicles and the posterior elements cut away. The ligament flares laterally and fuses with the annular fibers of each intervertebral disk. (1) Pedicle, (2) annular fibers of the intervertebral disk. (3) lateral extension of the posterior longitudinal ligament, and (4) posterior longitudinal ligament.

2.33–2.37). The ligament is wider in the cervical region and diminishes in breadth caudally. It is thickest in the thoracic region. In the thoracic and lumbar spine, the deepest fibers of the ligament flare out over the annulus fibrosus laterally and terminate in the proximal neural foramen. Laterally, the ligament is attached firmly to the annulur fibers (Fig. 2.36). At each disk level the deep fibers are attached to the annulus and the adjacent vertebral body. The superficial fibers extend over two or three vertebral bodies and are attached loosely to the vertebrae. Viewed laterally, the ligament appears bowstrung over the concave posterior aspect of the vertebral body. The intervening space is filled with the basivertebral venous plexus, loose areolar tissue, and fat. On CT, the posterior longitudinal ligament cannot be differentiated from the annulus in sections through the disk, and at the vertebral body level, they both blend with the dural tube. The intervening space consists of basivertebral venous plexus, loose areolar tissue, and perivascular fat. On MRI, the posterior longitudinal ligament has a decreased signal intensity and cannot be separated from the annulus at the disk level and from the dural sac at the vertebral body level. In both sagittal and axial images, the intervening space between the cortical margin of the vertebral body and the posterior longitudinal ligament has a bright or heterogeneous signal intensity resulting from the basivertebral venous plexus and the perivascular fat (Figs. 2.34 and 2.37).

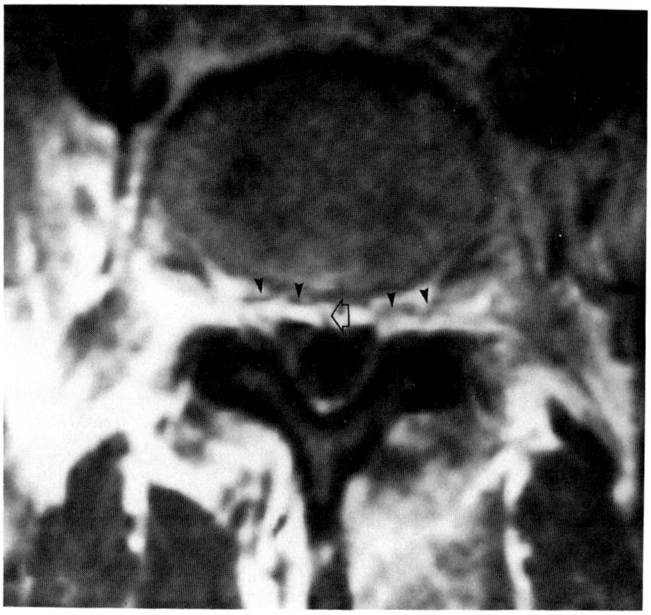

Figure 2.36. Axial T1-weighted MRI image demonstrating the epidural venous plexus (*arrowheads*) between the vertebral body and the posterior longitudinal ligament. The ligament cannot be identified separate from the dura. Epidural plicae (*open arrow*) connecting the dura to the posterior aspect of the vertebral body.

LIGAMENTUM FLAVUM

The ligamentum flavum spans the individual space between adjacent laminae from C2 to S1 to form a smooth posterior wall of the bony spinal canal (Fig. 2.37). The ligament courses from one neural foramen, across the midline, to the contralateral neural foramen [39, 40]. In the cervical region, the ligament fuses with the medial portion of the facet joint capsule (Fig. 2.38). In the thoracic and lumbar regions (Figs. 2.39 and 2.40), the ligament forms the posterior margin of the neural foramen and turns dorsally to fuse with the thin capsule of the articular facet on the lateral surface of the superior articulating process of the vertebra below [41] (Fig. 2.40). The ligament stabilizes the facet joint and prevents displacement of the articular process into the vertebral canal. The ligament is yellow in color because of its high elastin content. The elasticity aids in restoring upright posture after bending and in maintaining a smooth posterior wall. The thickness of the ligament varies, measuring approximately 1.5 mm in the cervical region, around 2 mm in the thoracic region, and between 4 and 6 mm in the lumbar region [40]. The ligament can be as thick as 9 mm at the L4–5 level but is usually thinner at the L5–S1 level. Medially, the ligament from each side fuses with the interspinous ligament posteriorly (see Figs. 2.37 and 2.40).

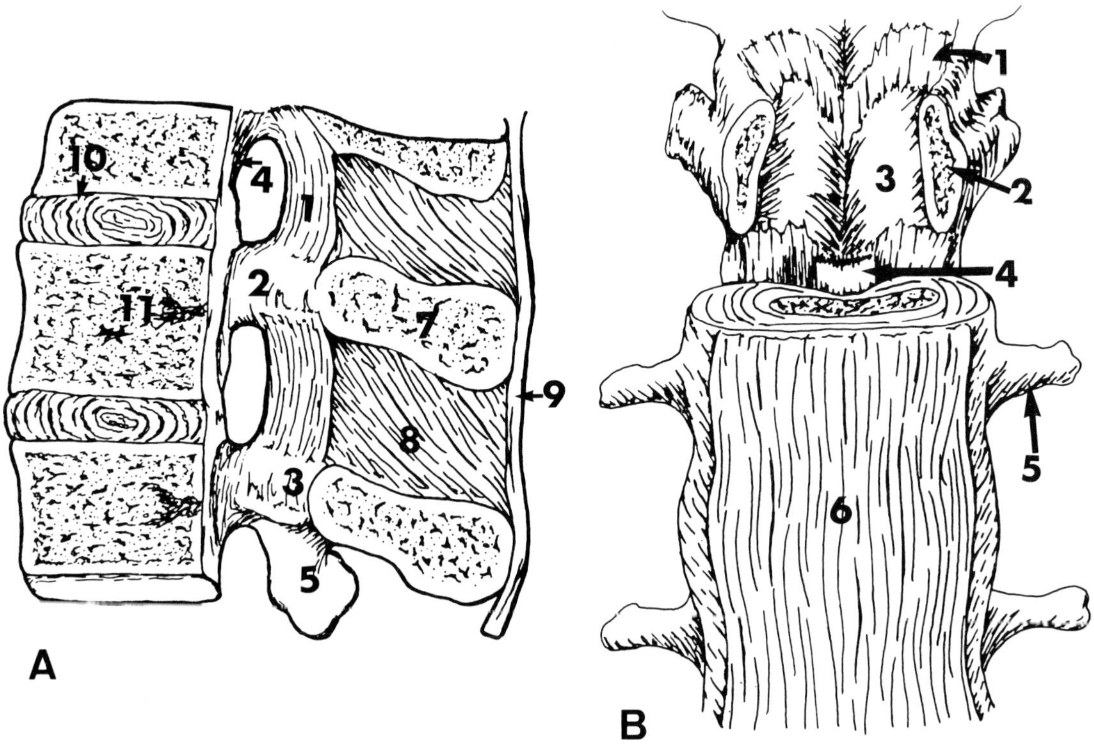

Figure 2.37. Schematic drawing of the vertebral column viewed from (**A**) the front with one vertebral body cut away and (**B**) the lateral aspect of the spinal canal. (1) Ligamentum flavum, (2) pedicle, (3) lamina, (4) posterior longitudinal ligament, (5) transverse process, (6) anterior longitudinal ligament, (7) spinous process, (8) interspinous ligament, (9) supraspinous ligament, (10) annular fibers, and (11) vertebral body with basivertebral veins.

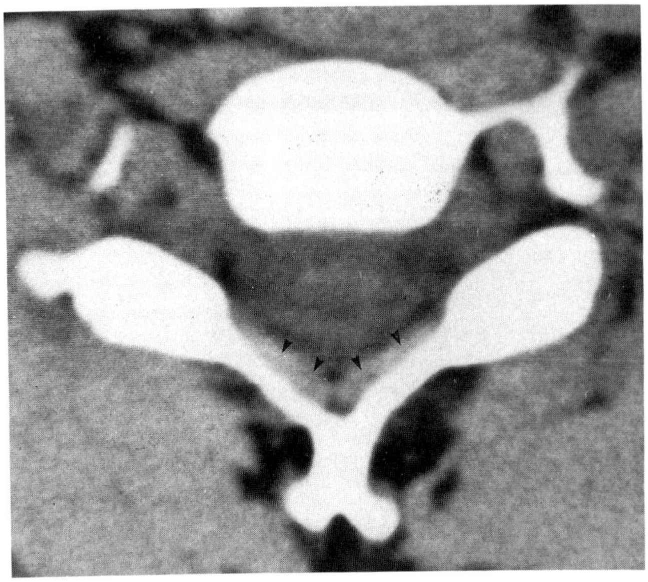

Figure 2.38. Ligamentum flavum (*arrowheads*) on axial CT scan in the cervical region. It has a similar density to the paraspinal muscles.

The annulus fibrosus is further strengthened by ligamentous bands which are attached laterally to the pedicle and extend posteriorly to be attached to the superior articular process. These fibrous bands continue posteriorly to merge with the ligamentum flavum [42]. They are difficult to separate from the cortical margin of the inferior wall of the intervertebral foramen in sagittal MRI sections.

On axial CT, the ligament has a soft-tissue density similar to muscle. It appears thinner in front of the lamina, due to partial volume averaging [38, 43]. When utilizing a bone window setting to determine spinal stenosis, one should be cautious to set the cursor anterior to the ligamentum flavum, since it forms the posterior wall of the canal. Although the ligament can be visualized in the coronal and sagittal planes [44, 45], it is best evaluated in axial sections (see Figs. 2.38–2.40). It has an intermediate signal intensity, except at its site of attachment, where is has a decreased signal intensity. Occasionally, ossification of the ligament may result in formation of a bony spur within the neural foramen [87] (Fig. 2.42). This may be mistaken for a fracture or disk.

OTHER LIGAMENTS

The interspinous, supraspinous, and intertransverse ligaments form attachments between the spinous and transverse processes [46]. The interspinous and supraspinous ligaments are located between adjacent spinous processes (see Fig. 2.37). The interspinous ligament spans the space between adjacent spinous processes from their roots to their apexes. It is best developed in the lumbar region. The ligament restricts excessive flexion between adja-

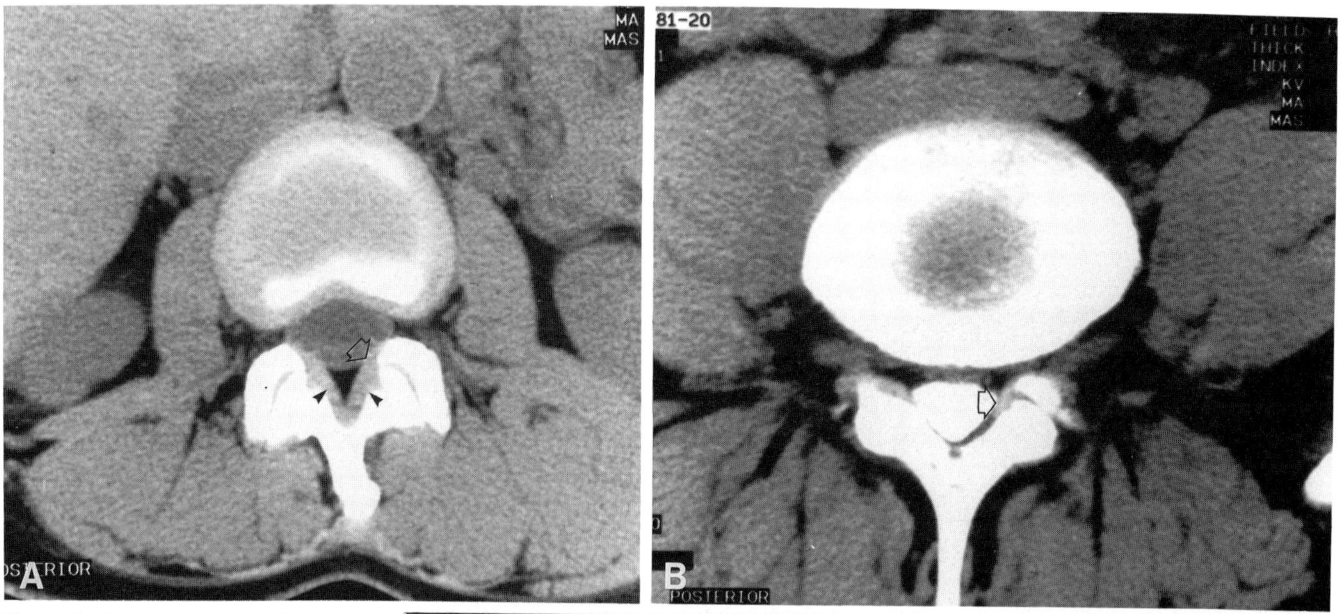

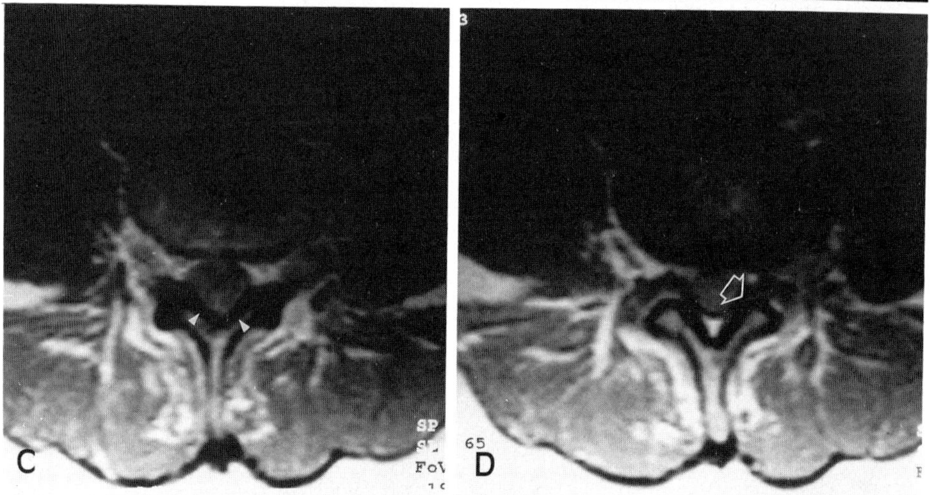

Figure 2.39. Ligamentum flavum in the lumbar region. A,B, Axial CT scans. **C,D,** Axial T1-weighted MRI images. The ligament is mildly hyperdense on CT and hypointense on MRI (*arrowheads*). It merges anteriorly with the facet joint capsule (*open arrow*). The annular fibers of the disk are connected to the ligament by small connecting fibers.

cent vertebrae and tends to distribute movement over several segments. The ligament is thin and weak in the thoracic region and thick and strong between the lumbar spinous processes. Paired interspinous muscles on either side of the ligament insert on the transverse processes. The supraspinous ligament connects the apices of the spinous processes, which usually terminate at the L4 level and occasionally at the L3 or L5 level. Superiorly, it extends beyond C7 to be attached to the external occipital protuberance and evolves into the ligamentum nuchae, a much stronger and wider ligament. The intertransverse ligament is more prominent in the thoracic spine. In the lumbar region it is thin and membranous, appearing more as lumbar fascia than ligament. The back musculature is attached to these ligaments, especially in the thoracic region. Blood vessels and nerves pass through these ligaments to supply the facet joints and the posterior structures. The ileolumbar ligament, frequently seen in axial CT and MRI sections, courses posterolaterally from the transverse process of the L5 vertebra to the iliac crest.

Spinal Cord and Nerves

The spinal cord is the continuation of the medulla oblangata. Its proximal end is demarcated at the level of the foramen magnum. It terminates in adults at the level of T12–L1 disk interspace as the *conus medullaris*. In infants, the tip of the conus medullaris is at the L3 level and gradually attains the "adult" position before the age of 5 years [47]. The *cauda equina* is the extension of the nerve roots caudal to the conus. The *filum terminale* is a fibrous band extending from the tip of the conus to the coc-

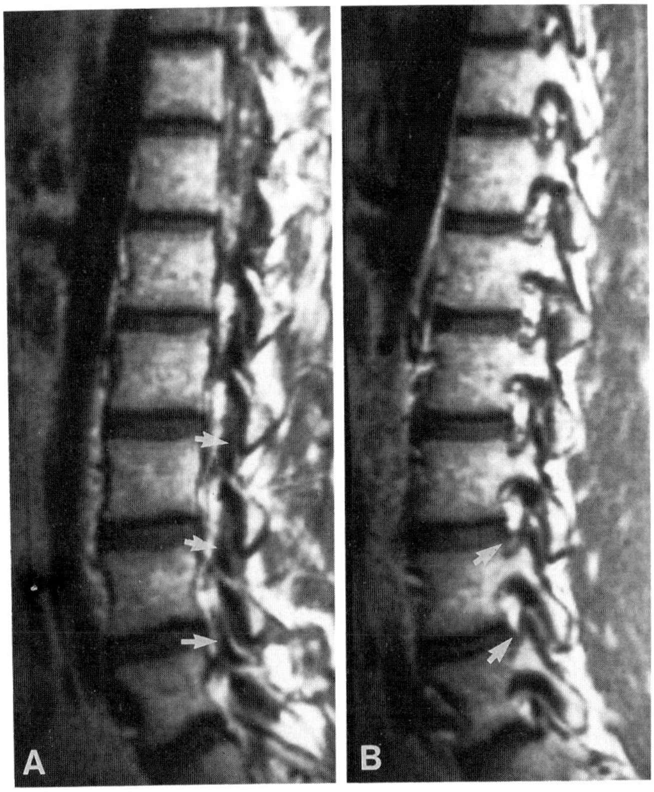

Figure 2.40. Sagittal T1-weighted MRI sequence through (**A**) the subarticular canal and (**B**) the adjacent neural foramen. The ligament (*arrows*) forms the posteromedial attachment to the facet joint capsule and extends along the inferior margins of the neural foramen to merge with the annular fibers of the disk.

cyx. The spinal cord has an anterior median fissure and a shallower posterior median sulcus (Fig. 2.43). Posterolateral sulci run on either side of the posterior median sulcus. The spinal cord is also invested by the pia mater with a loosely attached arachnoid membrane similar to the brain. The dura surrounds the cord, forming the thecal sac or dural tube. Inferiorly, the distal end of the thecal sac usually ends at a variable distance between the first and second sacral vertebral segments. The septum posticum anchors the cord posteriorly. Anteriorly, the distal end of the thecal sac is anchored to the posterior longitudinal ligament by a midline fascial septum. The septum often can be seen on axial MRI sections at the L5–S1 level due to tapering of the thecal sac and variable amounts of fat between the dural tube and the posterior longitudinal ligament [48] (see Fig. 2.36).

The subarachnoid space, which is filled with cerebrospinal fluid (CSF), surrounds the cord. It is prominent between the foramen magnum and the C2 level. Beyond this level and up to the level of the conus, the subarachnoid space is variable, but it again becomes prominent below the level of L1 or distal to the conus medullaris. The arachnoidal sleeves with its contained CSF surround the nerve roots for a variable distance within the neural foramen. The CSF has pulsatile motion, which is maximal in the cervical region, gradually diminishing in the lumbosacral region.

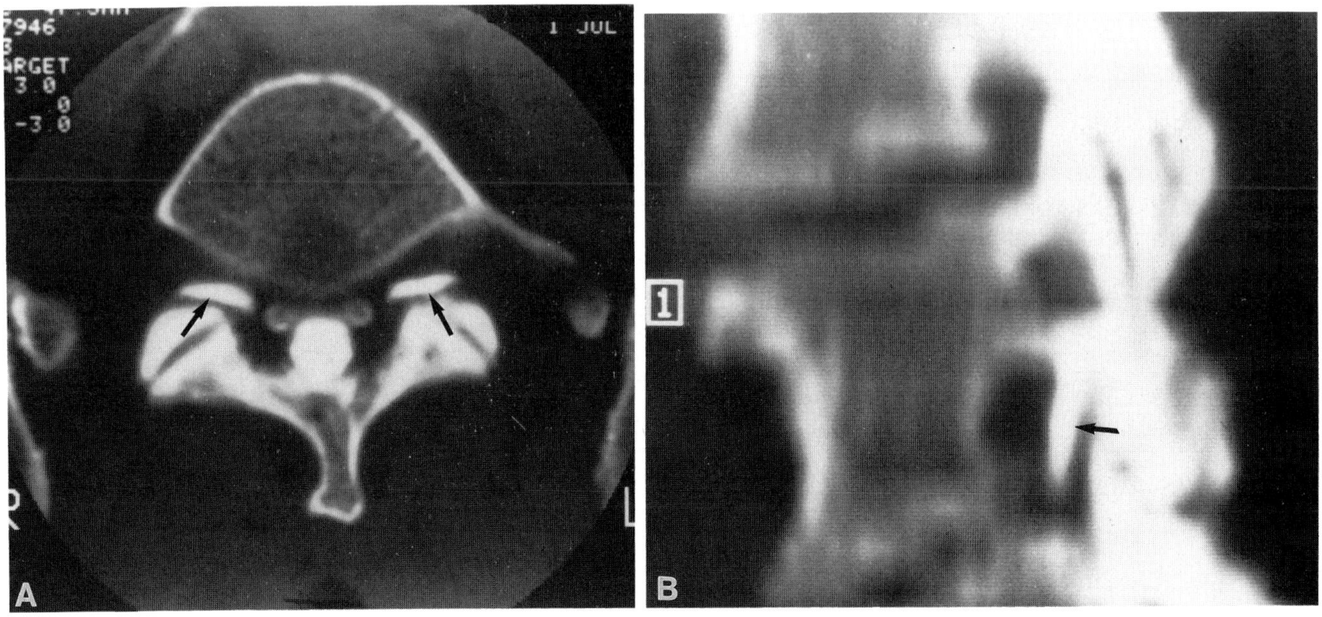

Figure 2.41. Axial (**A**) and re-formatted sagittal (**B**) CT scans demonstrating bilateral foraminal spur (*arrow*). This is a normal variant resulting from ossification of the ligamentous band along the pedicle which connects the ligamentum flavum with the annular fibers of the disk.

The spinal cord has two focal areas of increased girth (Figs. 2.16 and 2.43). The proximal area of increased girth in the cervical region is between C4 and T1, with its widest diameter opposite the C5–C6 disk space. This corresponds to the cervical enlargement to accommodate the brachial plexus nerves. Overall in the cervical region the cord is twice as wide in the coronal plane as compared with the sagittal plane (Fig. 2.43). The dentate ligament is attached to the cervical cord along its lateral surface midway between the anterior (ventral) and posterior (dorsal) nerve roots. In the thoracic region, the cord is oval in shape. Proceeding caudally, opposite the T10 vertebral body and extending down to the conus is another fusiform enlargement corresponding to the lumbosacral plexus.

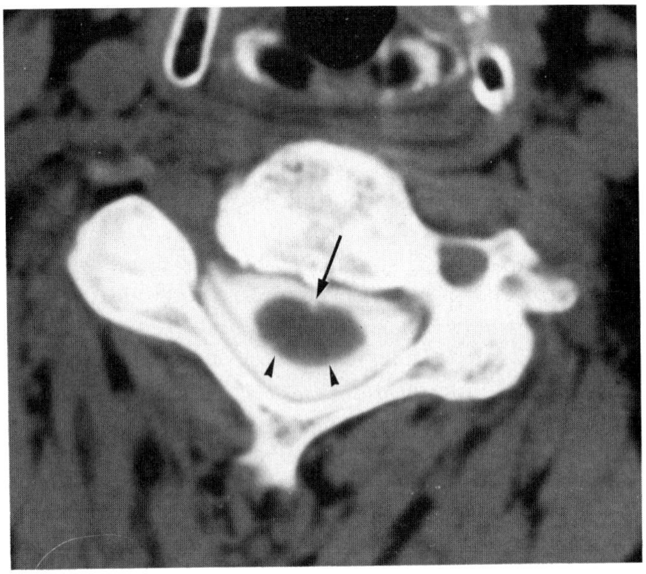

Figure 2.42. Axial CT scan of the cervical cord following intrathecal contrast showing its gross morphology. (1) Anterior median fissure (*arrow*), and (2) posterior lateral sulci (*small arrows*).

The spinal cord consists of central gray and peripheral white matter. An ependymal lined central canal runs throughout the length of the spinal cord which communicates cephalad with the craniospinal subarachnoid space at the level of the obex. There are 31 pairs of spinal nerves. Each pair consists of an anterior motor and a posterior sensory root (Fig. 2.44). The roots of each pair are invested within the dural sheath which extends laterally to the neural foramina. The anterior and posterior roots join beyond the dural sheath to form the spinal nerves. The first to the seventh cervical spinal nerves leave the spinal canal through the intervertebral foramen above that vertebral level. The eighth cervical nerve passes through the foramen between C7 and T1. In the thoracic region, the nerve roots exit and descend in a downward direction. As a result, in the upper thoracic region the cord segment is two levels lower than the corresponding vertebrae. In the lower thoracic region, this is further accentuated, there being a difference of three levels between the cord segment level and the vertebrae. In the lumbar region, the terminal portion of the cord, the tip of the conus med-

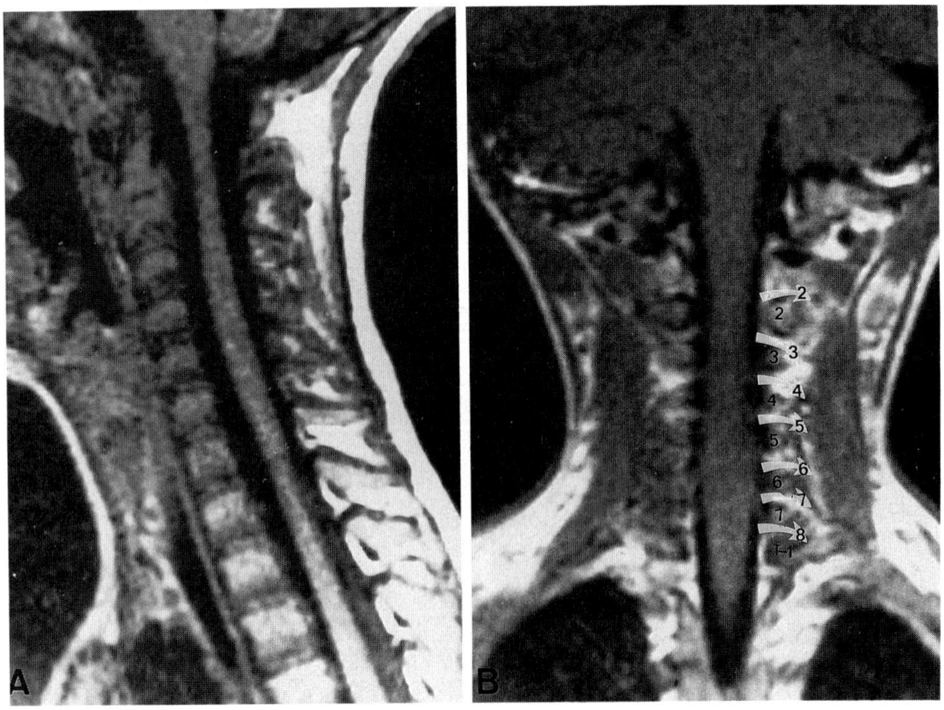

Figure 2.43. Sagittal (**A**) and coronal (**B**) T1-weighted MRI images of the entire cervical region demonstrating the circumferential widening of the spinal cord between C4 and T1 and the relationship of the cervical nerves to their neural foramen.

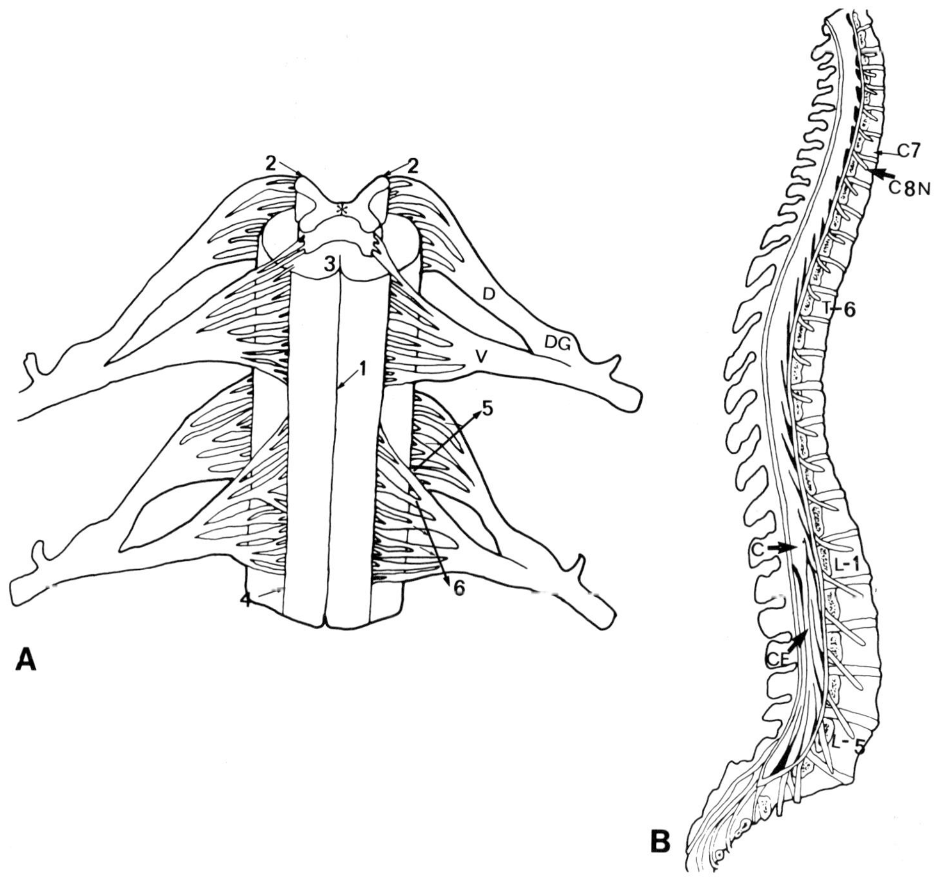

Figure 2.44. **A,** Spinal cord section viewed from the front showing spinal nerve and spinal nerve root origins (*D* = dorsal nerve root; *DG* = dorsal nerve root ganglion; *V* = venteral nerve root; *1* = anterior median fissure; *2* = gray matter; *3* = white matter; *4* = central canal; *5* = fila of posterior root; *6* = fila of anterior root). **B,** Lateral cut section of the spinal column to show the relationship of vertebrae to the nerve root (*C8N* = the C8 nerve root that exits through the foramen between C7 and T1; *C* = conus medullaris; *CE* = cauda equina).

ullaris, is at the L1–2 disk level (Fig. 2.45). The conus is oval shaped with a rounded eminence posteriorly. The lumbar and sacral nerves descends downward below the conus as the cauda equina, packed into anterolateral and posterolateral bundles, located along the dependent portion of the thecal sac in a crescentric manner (Fig. 2.46). As in the cervical region, the anterior and posterior nerve roots are encased in the dural sheath which extends slightly beyond the neural foramina.

IMAGING OF THE SPINAL CORD AND NERVES

The spinal cord, the nerves, and the surrounding CSF are preferably imaged by MRI, but they can be evaluated by CT following intrathecal (CTM) or intravenous contrast. Because MRI is noninvasive, it is ideal in evaluating the spinal canal and its contents. MRI utilizing an appropriate pulse sequence is sensitive in demonstrating the cord and its surroundings in the three orthogonal planes.

Several technical factors influences the imaging characteristics, however. In most instances, a combination of spin-echo or gradient-echo sequences allows for evaluation of the anatomic structures. MRI related artifacts, however, may result in an abnormal configuration or signal changes. This is particularly common in the cervical region and to a lesser extent in the thoracic or lumbar region. The artifacts may result from data truncation, vascular or CSF pulsation, chemical shift, even echo rephasing, magnetic field distortion, and patient motion [49–52]. The most important of these are related to the pulsatile characteristics of CSF [52, 53]. Signal loss may be mistaken for vascular structures. In a similar fashion, an increased signal within the CSF compartment may mimic a disk or other lesion, usually in the last slice of multislice axial T1-weighted sections. Ghost shadows may superimpose on the normal structures, usually in the phase-encoding axis, and may be mistaken for cord lesions. They are often seen with long-TR spin-echo sequences. Truncation (Gibb's) artifact is a curvilinear high signal band superimposed on the spinal cord which occurs with two-dimensional FT imaging, usually seen on both axial and sagittal images [52].

Some of these artifacts can be corrected by utilizing a smaller field of view (FOV), changing the matrix, filtering, and using gating and motion-suppression techniques [54]. New acquistion methods, such as volume acquisition (3DFT) [55] or utilizing phase-array coils, allow for higher contrast, thinner sections,

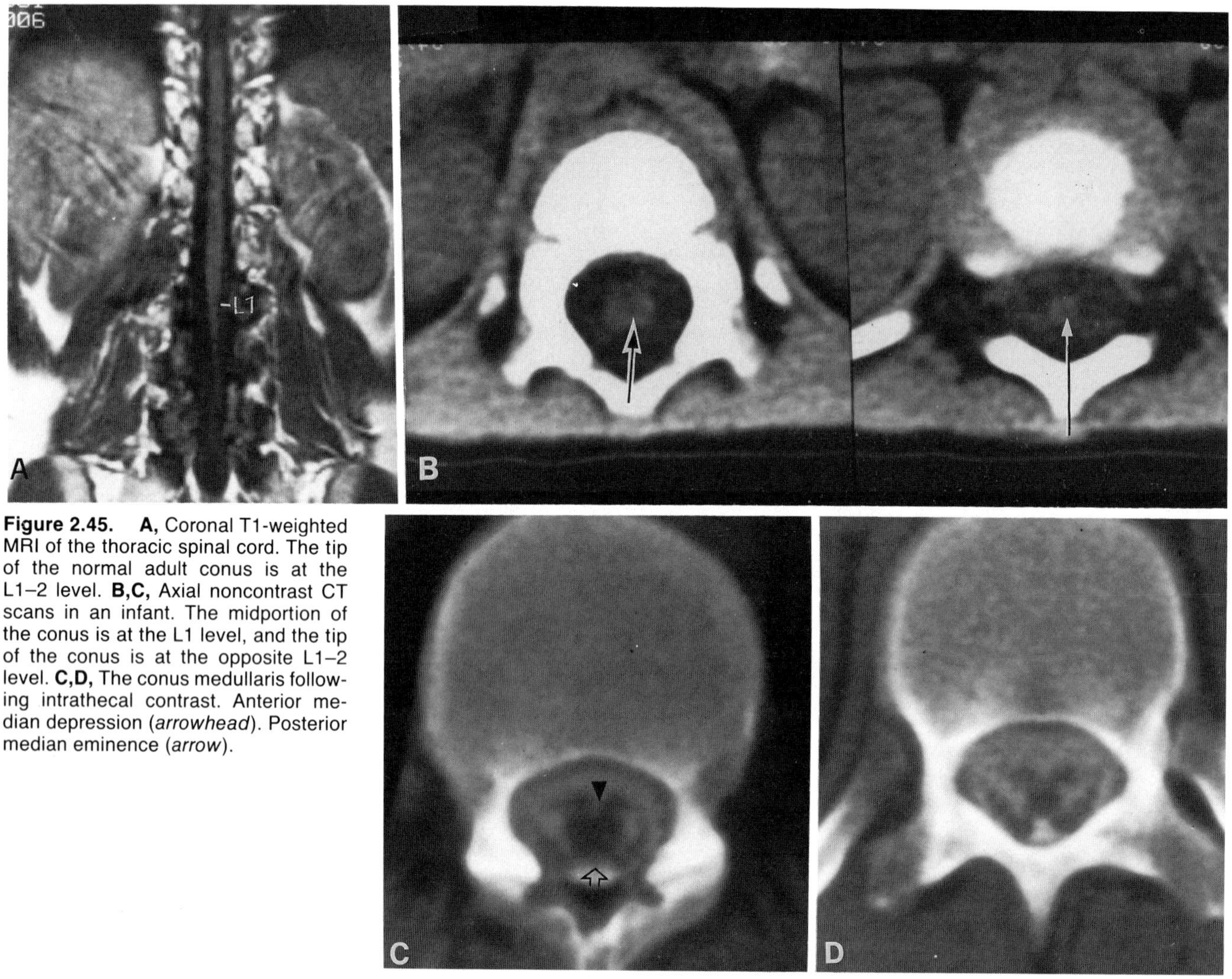

Figure 2.45. **A,** Coronal T1-weighted MRI of the thoracic spinal cord. The tip of the normal adult conus is at the L1–2 level. **B,C,** Axial noncontrast CT scans in an infant. The midportion of the conus is at the L1 level, and the tip of the conus is at the opposite L1–2 level. **C,D,** The conus medullaris following intrathecal contrast. Anterior median depression (*arrowhead*). Posterior median eminence (*arrow*).

and minimal artifacts (aliasing artifact at both ends of the excited slab of tissue). These were discussed in greater detail in Chapter 1.

Measurement of cord diameter with conventional myelography and CTM has been attempted to define the normal indices at different levels of the cord [56–59]. These measurements are subject to wide variation. On CTM, the measurements depend on numerous factors, including partial volume averaging, absorption of contrast material in the pial layer of the cord, observer reliability [57, 60], and utilizing appropriate window levels and settings [56, 61]. MRI is a more sensitive and relatively accurate modality in evaluating cord diameter, provided an appropriate pulse sequence is utilized. The total length of the cord in the sagittal and coronal planes can be evaluated, allowing for observation of focal or diffuse cord swelling or atrophy. T1-weighted sagittal images allow evaluation the cord along its entire length. As such, actual measurement is often not necessary in the majority of cases. On T1-weighted sequences, the cord appears bright compared with the surrounding CSF, and similar to the conventional myelogram, it appears dark against the brighter CSF on T2-weighted sequences [62].

In the cervical region, the elliptical shape of the cord, as well as the major fissures, is seen consistently on axial T1-weighted MRI. The dorsal and ventral nerve rootlets, as well as the osseous elements, are also clearly visualized (Fig. 2.47). On T1-weighted MRI, the CSF, ligaments, and dural tube, as well as the vascular structures, all appear hypointense relative to the intermediate signal of the spinal cord. On T2-weighted spin-echo or gradient-echo images, the CSF is bright, whereas the spinal cord and dural tube have a hypointense signal. The epidural fat has a bright signal. The epidural veins, depending on the flow, may appear bright. Visualization of the internal architecture of the cord is difficult with the preceding pulse sequences [63]. The gray and white matter within the cord is best evaluated on gradient-echo sequences or with long acquisition times and a

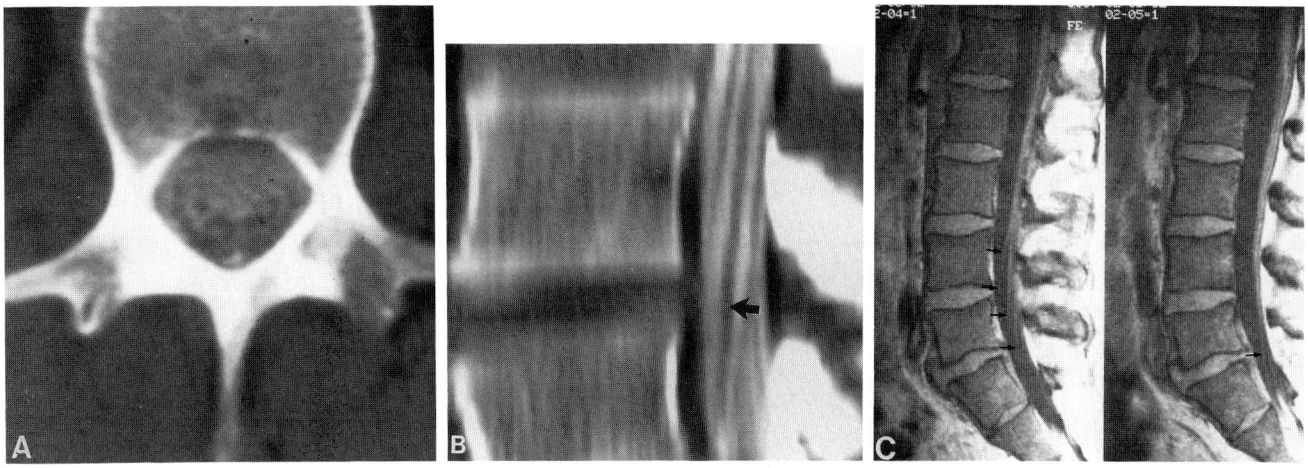

Figure 2.46. Cauda equina. After intrathecal contrast, (**A**) axial and (**B**) re-formatted sagittal CT images. **A,** The nerve roots lie along the dorsal aspect of the thecal sac. **B,** The filum terminale is seen as a thick band (*arrow*) surrounded by lumbosacral nerve roots. **C,** Sagittal T1-weighted MRI demonstrating the conus and the cauda equina (*arrow*).

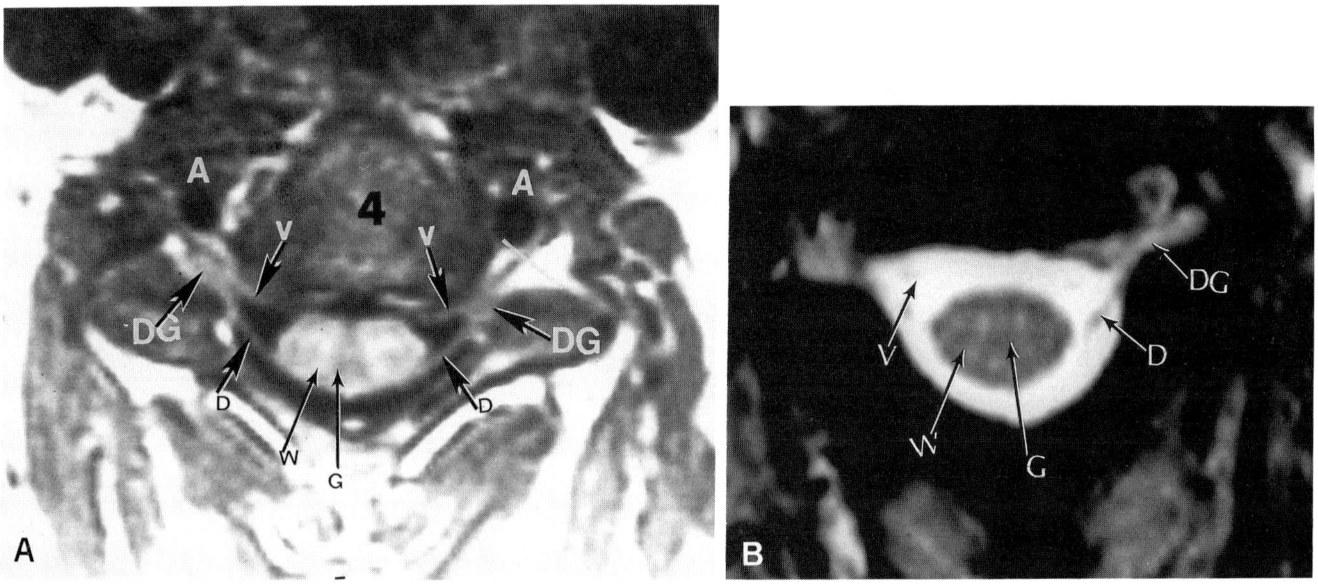

Figure 2.47. Axial (**A**) T1-weighted and (**B**) T2-weighted MRI images demonstrating the cervical spinal cord and nerve roots 1 (*V* = ventral nerve root; *D* = dorsal nerve root; *DG* = dorsal root ganglion; *W* = white matter; *G* = gray matter; *A* = vertebral artery).

smaller matrix. In caudal axial sections through the cervical region the gray matter is located in the anterior aspect of the spinal cord, whereas in the cephalad segment the gray matter is more posterolateral. The anterior and posterior horns are best seen in axial sections through the caudal segment of the cervical cord. The nerve rootlets that eventually form nerve roots can be routinely evaluated on MRI (see Fig. 2.47). The first seven cervical nerves are numbered according to the vertebral pedicle below the exiting nerve, with the exception of the eighth cervical nerve, which is above the pedicle of the first thoracic vertebrae (see Fig. 2.43). In the remainder of the spine, the nerves are numbered according to the vertebral pedicle above the exiting nerve.

On CT, the contents of the spinal canal and the neural foramina require intrathecal contrast and contiguous 1.5- to 2-mm thin sections extending between the pedicles of the adjacent vertebral body (Fig. 2.48). Depending on the lateral extension of the dural sheath, the nerve can be followed within the midportion of the neural foramen [64, 65]. In the lateral portion of the neural foramen, the nerve root may be identified depending on the perineural fat and epidural veins. On IVCT, the hypodense nerve root is seen as a homogeneous soft-tissue density surrounded by the hyperdense epidural veins. Occasionally, the dorsal root ganglion can be identified situated along the posterolateral aspect of the neural foramen.

The cervical neural foramen is routinely evaluated on axial (Figs. 2.47 and 2.48) and sagittal images.

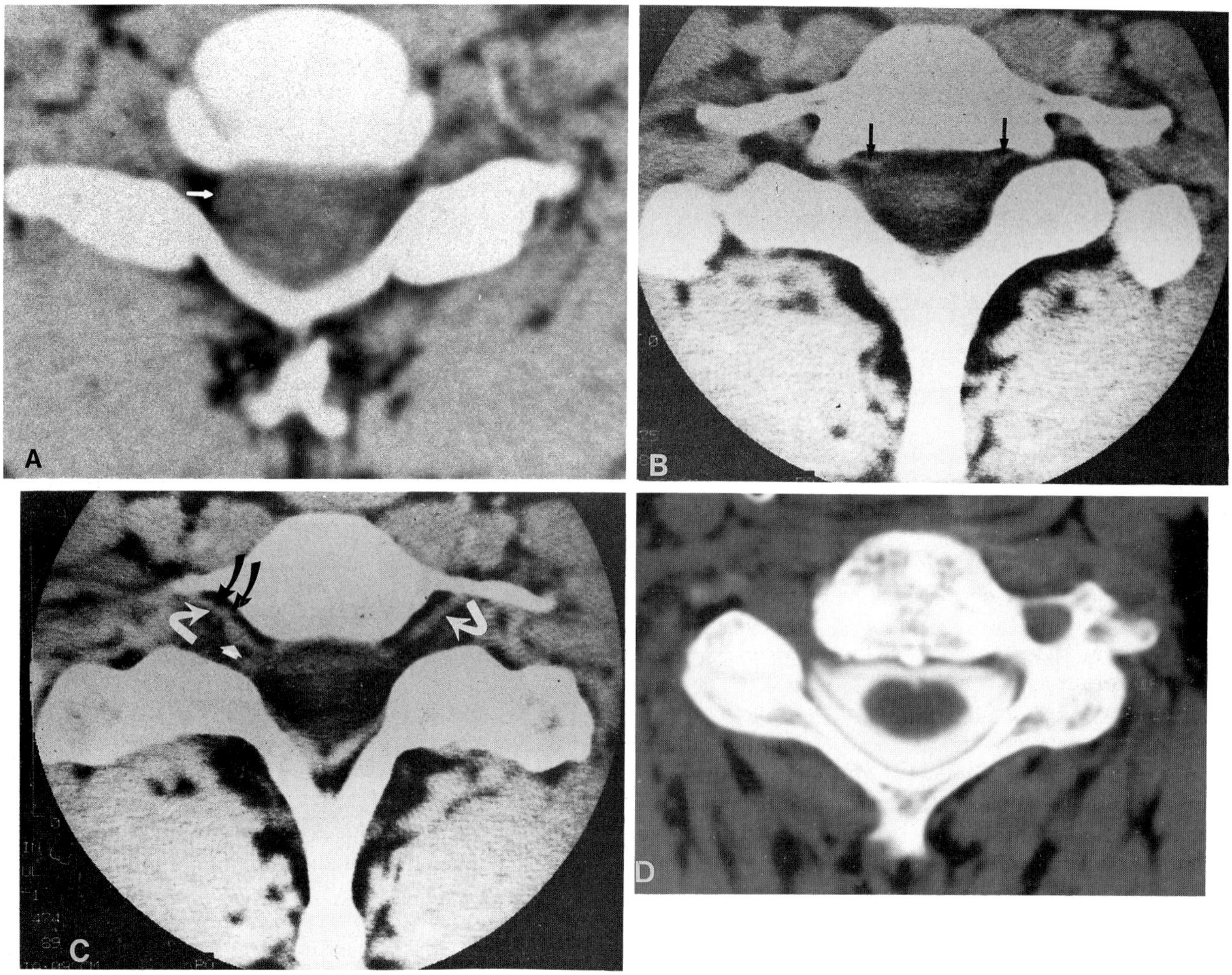

Figure 2.48. Cervical spinal cord. A, Noncontrast CT scan demonstrating the thecal sac and the mild lateral outpouching which demarcates the proximal origin of the nerve root sleeve (*arrow*). **B,** Bilateral nerve roots (*arrows*) surrounded by the epidural fat. **C,** The right dorsal ganglion (*arrow*) is noted within the neural foramen. Bilateral cervical nerves (*curved white arrow*) and the epidural vein (*black arrow*). **D,** CT following intrathecal contrast demonstrates the spinal cord (*c*), the dorsal nerve root (*d*), and the thecal sac and nerve root sleeve filled with contrast material.

Parasagittal oblique views allow evaluation of the neural foramen en face [64, 66–68]. Similar information without sacrificing time also can be obtained utilizing three-dimensional FT gradient-echo imaging with subsequent 1-mm re-formatted images in the axial plane and 45-degree oblique sagittal plane on each side to visualize the neural foramen en face (see Fig. 2.13). The nerve roots are located in the lower portion of the foramen, between the uncinate process and the superior articular process [64]. On T2-weighted images, the nerve roots appear hypointense, surrounded by the bright signal of the perineural fat and the surrounding veins. Similar results also can be obtained using Gd-DTPA–enhanced T1-weighted studies [64, 69, 70].

In the thoracic region, visualization of the contents of the spinal canal by CT is facilitated by CTM; on the other hand, the spinal cord and CSF space are well seen on MRI. The thoracic spinal cord is round to oval in shape. Occasionally, the anteromedian fissure and the posterior sulci can be identified. There is an overall increase in diameter of the thoracic cord between T10 and L1, with the widest segment between T11 and T12. The cord tapers to terminate as the conus at the L1–2 level (see Fig. 2.45). In the thoracic region, the rootlets form a single trunk before exiting the canal. Within the neural foramen, the trunk is still invested with dura which fuses with the epineurium just distal to the posterior root ganglion. The thoracic nerve roots occupy the central portion of the neural foramen, surrounded by neurovascular bundle and perivascular fat.

In the lumbar region, the intrathecal contents consist of the conus, the cauda equina, and the filum terminale. Sagittal T1-weighted MRI is ideal in demonstrating the position and diameter of the conus.

On MRI studies, the conus reaches the adult position between the L1–2 disk space in a majority of patients before age 1 [71]. In axial sections, the conus has an oval or, less often, round appearance with a small rounded eminence along its posterior surface. On both CT and MRI, two characteristic features are useful in identifying the conus [72, 73]. These are a small groove in the anterior aspect of the cord, which is a continuation of the anterior median fissure, and a conical projection at the same level along the dorsal aspect [72] (Fig. 2.49). Obliteration of these two landmarks should raise suspicion of an intramedullary lesion. A fibrous filament extends from the conus to the distal end of the dural sac. On CT and MRI, the filum cannot be separated from the descending lumbar and sacral nerve roots. Rarely, in MRI, due to the presence of a small amount of fat, the filum may have a bright signal, which should not be mistaken for a neurofibroma (Fig. 2.50). The lumbar and sacral nerve roots which form the cauda equina appear crowded in the proximal portion, but individual nerve roots can be identified in the distal thecal sac. On axial sections they are layered in a crescentric manner along the dependent surface of the thecal sac (see Fig. 2.46).

The course of the ventral and dorsal roots becomes more oblique the more caudal the segment. The lumbar and sacral roots can be identified for a variable distance, each within their nerve root sheath, after they emerge from the thecal sac. The L5, S1, and sometimes L4 roots emerge at or slightly higher than the disk space just superior to their neural foramen [74] (Fig. 2.51). This means that a lateralized hertniated disk at the L4–5 disk space can impinge on the L5 nerve root, but a lateralized L2–3 herniated disk rarely impinges on the L3 nerve root unless significant vertical migration of the herniated disk has occurred. Because of the descending oblique course of the lumbar nerve roots, they exit within the superior portion of the neural foramen, above the disk space (Fig. 2.52). On axial section the nerve roots and the thecal sac are surrounded by a variable amount of fat in the anterolateral gutter of the spinal canal. Unlike the cervical region, identification of the dorsal and ventral nerve roots within the thecal sac or the nerve root sheath is not possible, although they have been described based on cadaver specimens [75]. The posterior root ganglion of the nerve is usually located directly beneath the pedicle under which it exits. The ganglion can be identified on both the CT and MRI due to the ample surrounding fat [76]. Occasionally, the anterior ramus of the spinal nerve may be identified coursing around the vertebral body.

NORMAL VARIANTS WITHIN THE SPINAL CANAL

Minor variations in the appearance of thecal sac can be seen on imaging studies and can be confused as abnormal. The distal end of the thecal sac may extend beyond the S1–2 segment. The dural sheath along with its contained nerve root extends up to the midportion of the neural foramen. Occasionally, it may extend beyond the lateral margin of the neural foramen. The size of the dural sheath can be variable in isolated cases. Large dural sheath extensions may be an isolated anomaly, forming a cystic outpouching called *Tarlov's cyst*. Tarlov's cysts may occur at any level but are common in the lumbosacral region. When very large and located in the sacral canal, they may cause focal enlargement of the canal and neural foramen.

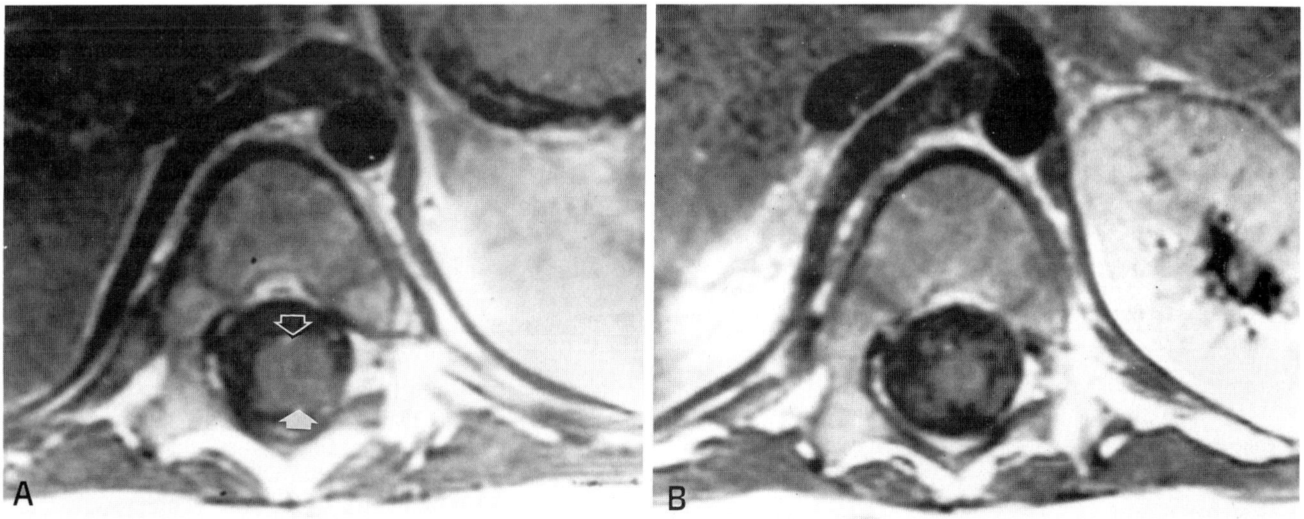

Figure 2.49. T1-weighted MRI images of (**A**) conus medullaris demonstrating the anterior median sulcus (*small arrow*) and the posterior median eminence (*arrowhead*) and (**B**) lower portion of the conus (see also Fig. 2.45).

Occasionally, two or, less often, three nerve roots of different spinal segments may share a common nerve root sheath for a short distance within the spinal canal, separating into their individual nerve roots just proximal to the neural foramen. This most often involves the L4–5 or L5–S1 nerve root, resulting in a conjoined nerve root sheath [77]. It has a characteristic appearance on CT and MRI (Figs. 2.53 and 2.54). A conjoined nerve root sheath is almost always unilateral and has the appearance of an elongation of the dural sheath on one side.

Epidural Space

The epidural space separates the thecal sac from the bony canal. It is composed of loose connective tissue, fat, and a plexus of veins. The fat content varies among individuals, being most abundant in the lumbar and sacral regions as a result of the large canal diameter and narrow thecal sac. This helps in visualizing the contents of the epidural space by CT and MRI. Large epidural veins should be recognized so that they are not confused with extruded disk fragments. Occasionally by CT and often in MRI, epidural plicae in the lower lumbosacral region can be identified (see Fig. 2.36). The plicae are thin, inconsistent, midline linear bands that connect the thecal sac to the bony canal. This explains the paracentral location of extruded disks in the lumbosacral region [48].

EPIDURAL VEINS

An extensive venous plexus lies in the epidural space anteriorly and courses along the entire vertebral column. The venous plexus consists of medial and lateral veins on either side of the posterior longitudinal ligament (Fig. 2.56) coursing roughly parallel to the ligament. At the disk levels, where the ligament flares laterally, the veins lie lateral to the ligament and the thecal sac. At the level of the midvertebral body, the veins are confluent, anastomosing with the basivertebral vein as it exits the vertebral body. These form the retrocorporeal venous

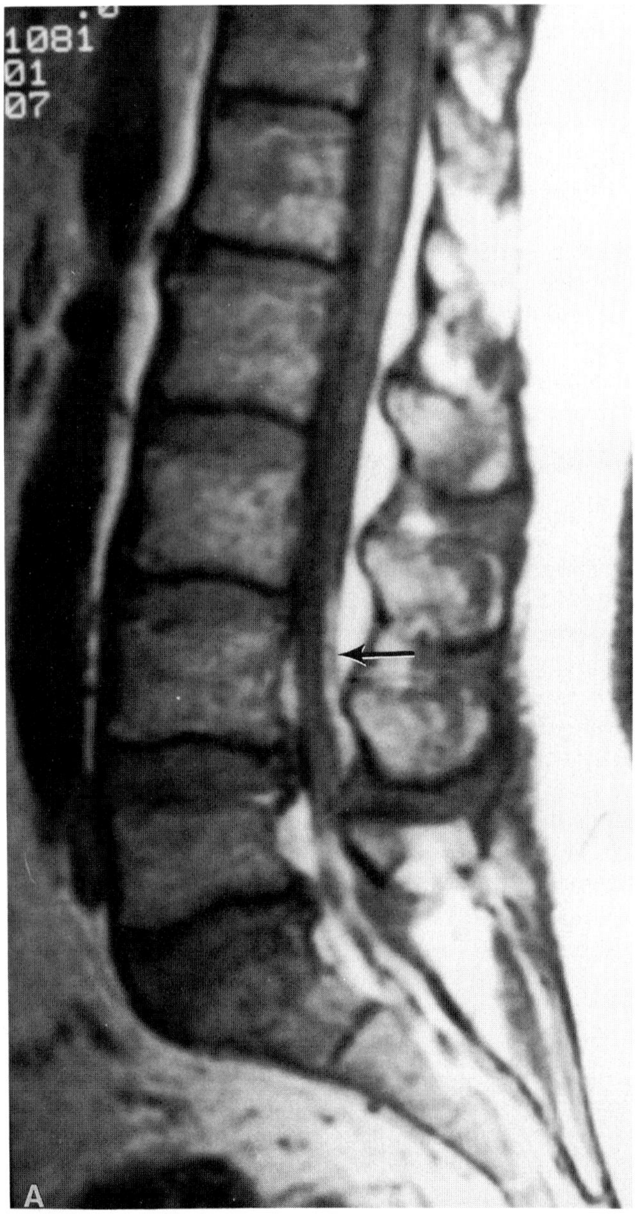

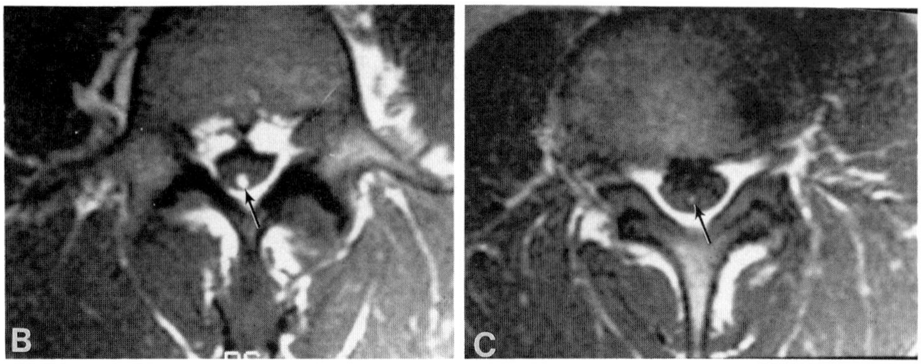

Figure 2.50. Sagittal (**A**) and contiguous axial (**B,C**) T1-weighted MRI images. Bright signal (*arrow*) of the filum terminale is due to fatty infiltration.

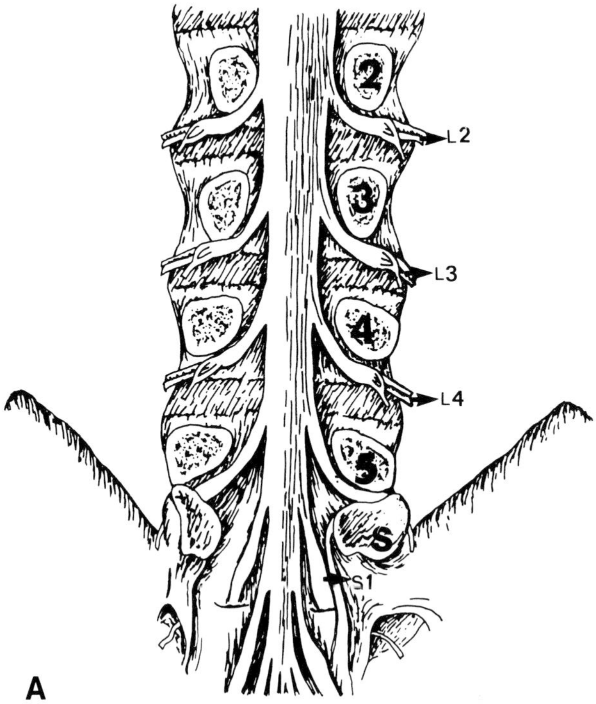

Figure 2.51. A, Lumbar nerve roots viewed from behind with posterior elements removed. Note that the upper lumbar nerve roots exit the thecal sac below the intervertebral disk (i.e., the L3 nerve root exits the thecal sac below the L2–3 disk space. B–D, Contiguous axial CT sections demonstrating the relationship of the thecal sac and nerve root sheath within the spinal canal.

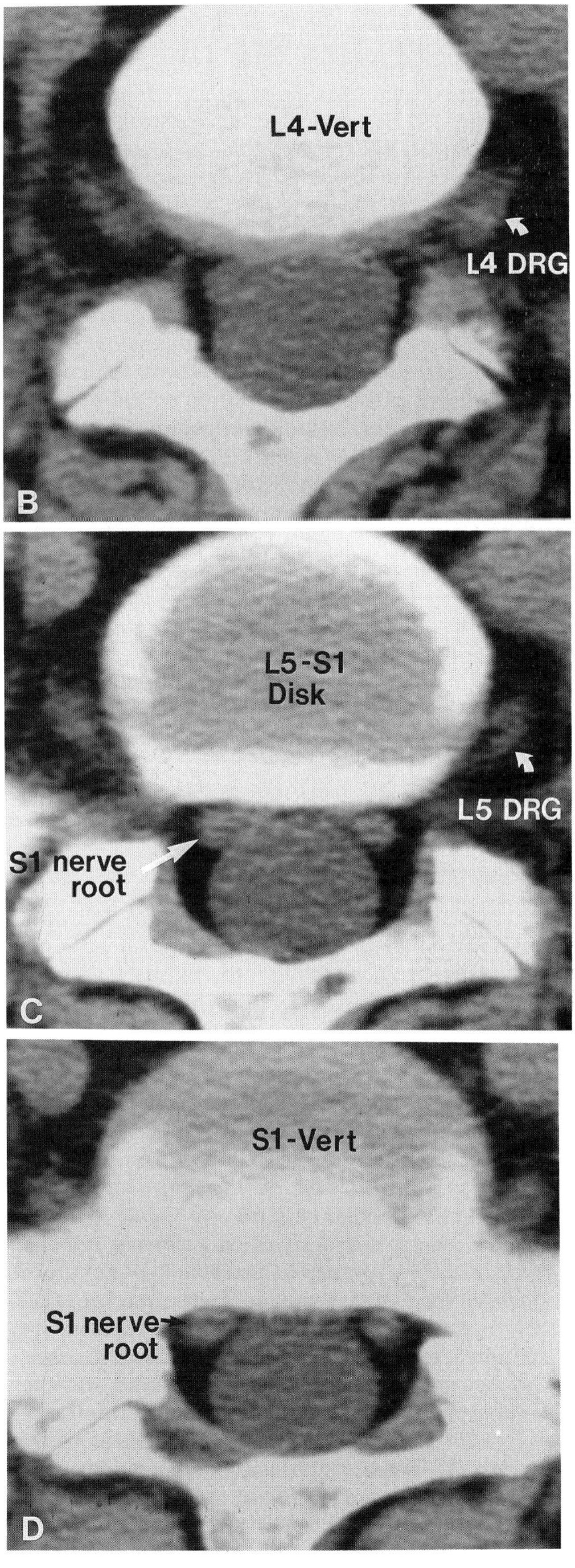

channels (Batson's plexus) located between the vertebral body and the posterior longitudinal ligament. The epidural veins can be identified on IVCT or MRI. Epidural venography is necessary in demonstrating the venous tributaries within the spinal canal as well as their anastomotic channels [78]. The veins exit the canal through the veins of the lateral foramen along with the nerve root. The veins anastomose with the sacral, ascending lumbar, intercostal, or vertebral plexus depending on the region of the spinal column. With large herniated or extruded disks, some of the epidural mass most likely represents the dilated obstructed veins, which can be recognized as enhancing components or surrounding the disk margin. These probably contribute to the enhancing disk seen on IVCT and Gd-DTPA–enhanced MRI studies [69, 70, 79] (Fig. 2.56). The veins in the epidural space within the thoracic or cervical spinal canal are difficult to visualize, except in the neural foramen. This is due to the smaller size of the veins, relative paucity of epidural fat, and larger space occupied by the cord.

Six longitudinal venous channels drain the spinal cord. The most constant venous channels run along the anterior median fissure and posterior median sulcus [80]. The others are often incomplete and variable. The midline posterior vein is the largest

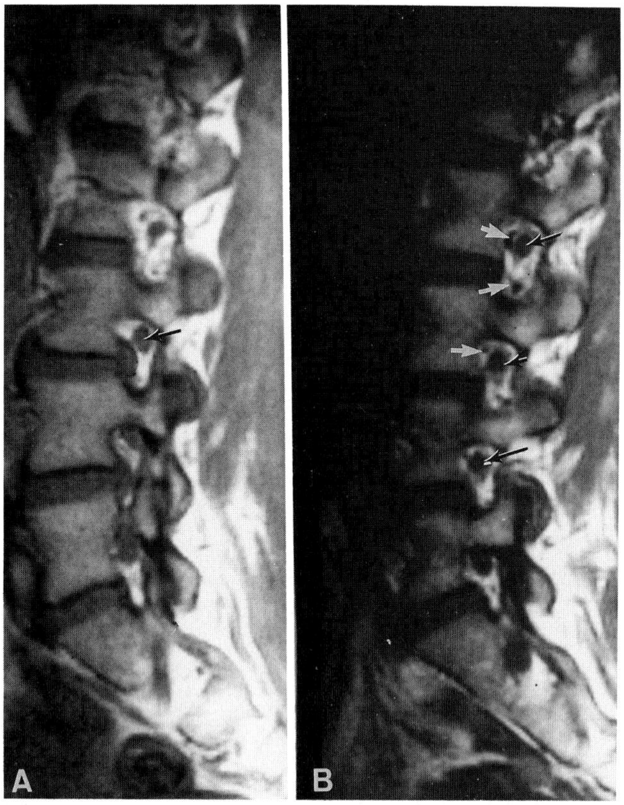

Figure 2.52. **A,B,** Sagittal T1-weighted MRI images of the lumbar spine. Relationship of nerve root sheath to the disk within the neural foramen. The nerve root sheath with the ventral and dorsal nerve root lies in the superior portion of the neural foramen (*arrowhead*). The radiculomedullary artery and vein lie anterior to the sheath (*white arrow*).

and drains the posterior two-thirds of the spinal cord. The venous tributaries anastomose extensively. They drain cephalad into the inferior petrosal sinuses and by a series of randomly distributed radiculomedullary vein into the intervertebral veins. There are no direct transdural connections to the epidural veins. The normal veins of the spinal cord are difficult to visualize on CTM or MRI.

ARTERIAL SUPPLY

The vertebral, costocervical, intercostal, and lumbar arteries give off paired small spinal branches which enter the spinal canal through the neural foramen to supply the vertebral canal and its contents. Except for the vertebral arteries, which can be identified in the cervical region, the normal arterial branches are difficult to identify on CT or MRI. A ventral branch supplies small twigs to the outer lamellae of the disk in children. The ventral branch through its caudal and cranial extension supplies the two adjacent vertebral bodies, the posterior longitudinal ligament, and the ventral dura before entering the basivertebral foramen of the vertebral body. Anastomoses exist between the paired spinal arteries and the adjacent spinal arteries above and below [81].

Arteriolar tufts in the cartilaginous endplates seen in the fetal spine disappear with aging and are not seen in the adult spine [82]. Thus, except for the outermost layer of the annulus fibrosus, the metabolically active intervertebral disk is avascular and depends on diffusion of nutrients through the cartilaginous endplates. As aging reduces the permeability of the endplates, there is compromise of the nutrition, leading to degeneration.

The spinal arteries give off branches to the ventral and dorsal roots named the *anterior* and *posterior radicular arteries*, respectively. Most of the posterior radicular arteries continue as the posterior radiculomedullary arteries and supply the posterior spinal artery. Only a few (usually four to nine) unilateral anterior radiculomedullary arteries feed the anterior spinal artery [81, 83]. The spinal cord is supplied by one anterior and two posterior spinal arteries which originate from the anterior and posterior spinal rami of the intradural vertebral arteries. The anterior spinal artery supplies the anterior two-thirds of the spinal cord, including the corticospinal tracts, and courses in or on the median fissure. The two posterior spinal arteries form interconnecting longitudinal arteries, each one lying in front of the other, behind the attachment of the dorsal roots (Figs. 2.55 and 2.56).

The upper cervical cord is supplied by the anterior spinal artery. This is derived from either the cervical segment of the vertebral artery or from branches of the thyrocervical or costocervical trunk (Fig. 2.57). The upper and middle thoracic cord is supplied by the anterior spinal arteries formed by branches from the intercostal arteries derived from the aorta (see Figs. 2.58 and 2.59).

The lower thoracic and lumbar segments of the cord are fed by the largest of the radiculomedullary arteries (artery of Adamkiewicz), which originates from intercostal or lumbar arteries between T6 and L2, often from the left side, most commonly at the T9 level [83]. This artery may supply up to two-thirds of the entire spine. There are ascending and descending branches which anastomose with other radiculomedullary branches. The anterior spinal artery anastomoses with the posterior spinal artery at the conus medullaris. Sacral and lumbar radicular arteries supply the cauda equina and the filum terminale. They have potential collateral channels with the anterior spinal artery. The posterior spinal arteries anastomose freely and in short segments are replaced by a vascular plexus. Because of the extensive anastomotic network, occlusion of one posterior radiculomedullary artery has little effect, unlike the anterior spinal artery, which is the vessel most often involved in ischemic episodes.

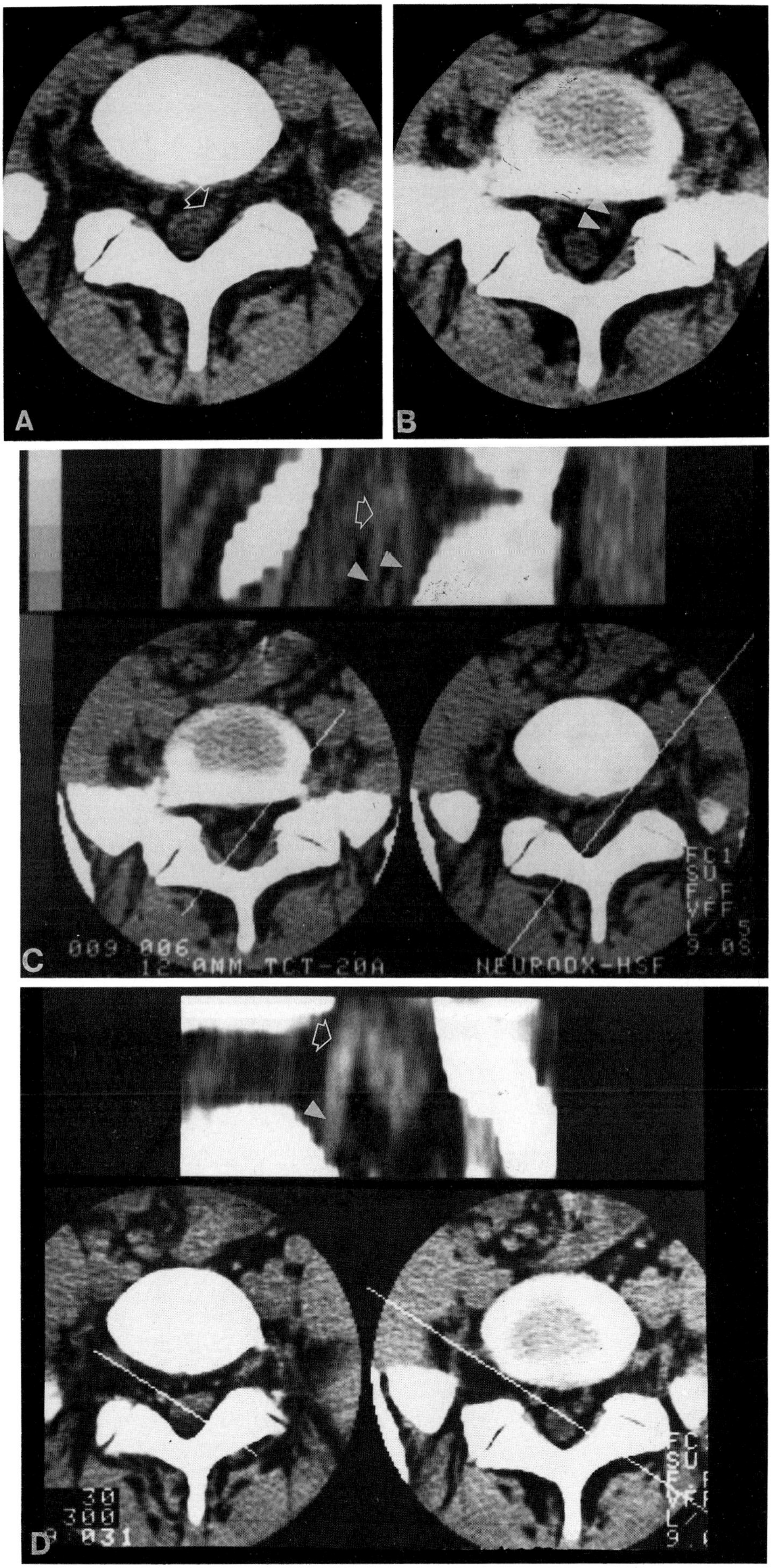

Figure 2.53. Conjoint nerve roots. **A,B,** Axial CT scans demonstrate an outpouching of the thecal sac on the left (*arrow*). In the next lower section two nerve roots (*arrowheads*) are noted. **C,** Sagittal re-formatted image demonstrating the two nerve roots (*arrowheads*) from a single nerve root sheath (*arrows*). **D,** The normal right nerve root (*arrowhead*) and sheath (*open arrow*).

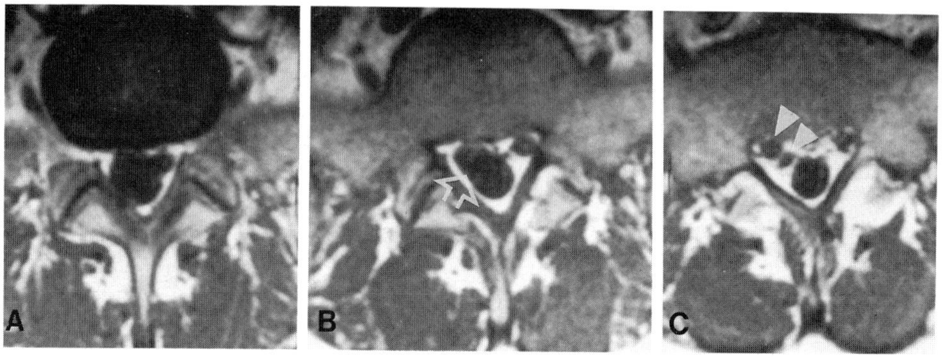

Figure 2.54. Conjoint nerve roots. A,B,C, Axial T1-weighted MRI images demonstrate a conjoint nerve root sheath (*open arrow*) with two separate nerves at the level below (*arrowheads*).

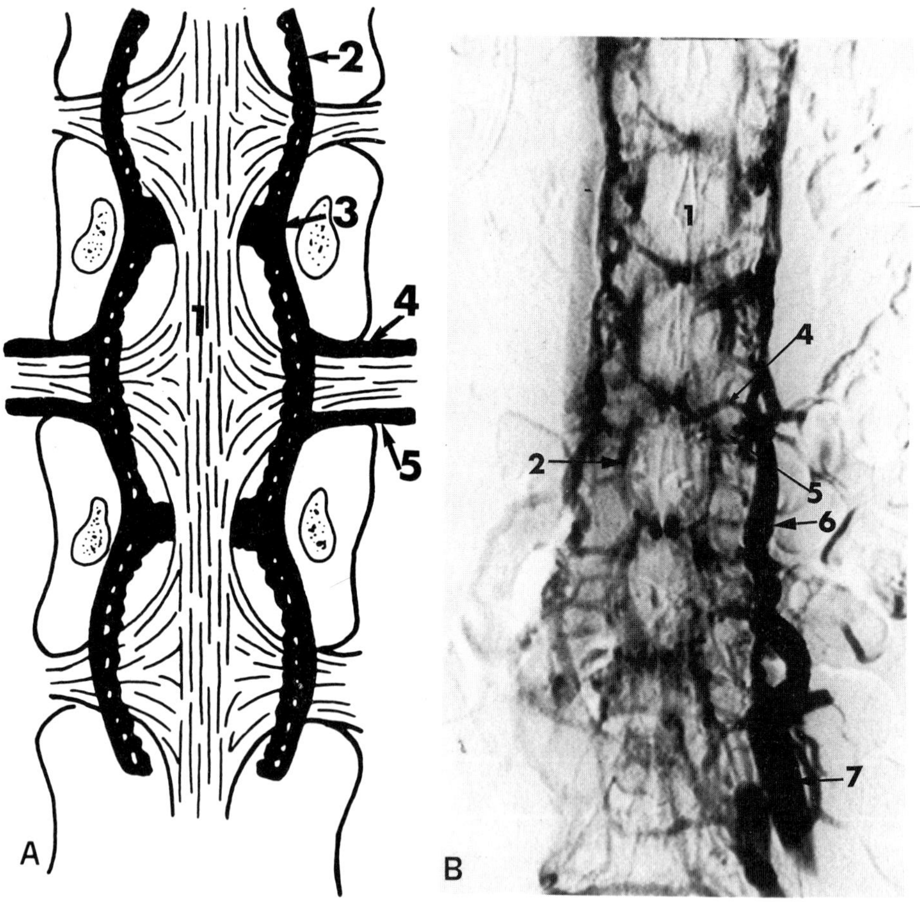

Figure 2.55. The epidural venous anatomy and drainage pathways. A, Lumbar spine viewed from behind after removal of posterior elements removed. **B,** Lumbar venogram demonstrating the venous anatomy. (1) Posterior longitudinal ligament, (2) medial and lateral epidural veins, (3) retrocorporeal venous anastomosis, (4) superior vein of the lateral foramen, (5) inferior vein of the lateral foramen, (6) ascending lumbar vein, and (7) lateral sacral vein.

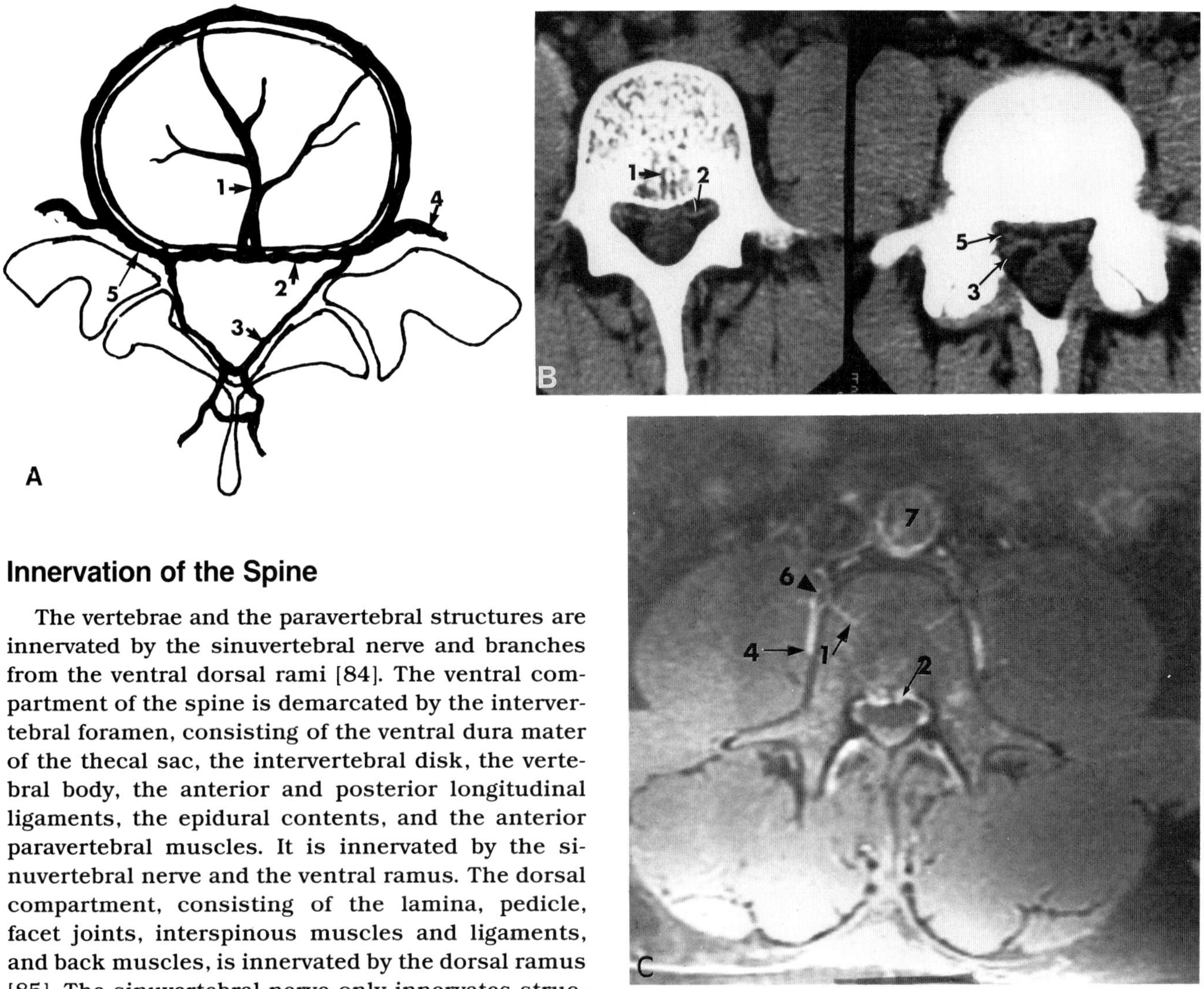

Figure 2.56. **A,** Schematic diagram of the epidural vein in axial view. **B,** Noncontrast CT scan demonstrating the epidural veins (*arrows*). **C,** Gd-DTPA–enhanced fat-suppressed T1-weighted MRI image. (1) Basivertebral vein, (2) anterior epidural vein, (3) posterior epidural veins, (4) ascending lumbar vein, (5) foraminal vein, (6) anterior external plexus, and (7) inferior vena cava.

Innervation of the Spine

The vertebrae and the paravertebral structures are innervated by the sinuvertebral nerve and branches from the ventral dorsal rami [84]. The ventral compartment of the spine is demarcated by the intervertebral foramen, consisting of the ventral dura mater of the thecal sac, the intervertebral disk, the vertebral body, the anterior and posterior longitudinal ligaments, the epidural contents, and the anterior paravertebral muscles. It is innervated by the sinuvertebral nerve and the ventral ramus. The dorsal compartment, consisting of the lamina, pedicle, facet joints, interspinous muscles and ligaments, and back muscles, is innervated by the dorsal ramus [85]. The sinuvertebral nerve only innervates structures within the spinal canal [84]. Innervation of the lateral and anterior aspects of the ventral compartment is derived from branches of the ventral ramus and sympathetic nervous system (Fig. 2.60).

The annulus fibrosus in its different portion receives innervation from the ventral ramus and gray ramus communicans. The anterior longitudinal ligament and the vertebral periosteum are also innervated by the gray ramus of the sympathetic trunk (Fig. 2.60). The exact nature of the many nerve fibers to the vertebral soft tissues is uncertain. Proprioceptive and nociceptive (pain) functions are part of their functions. The exact contribution of these pain fibers in patients with low back pain is controversial, as is the contribution of the sympathetic innervation.

These small nerves cannot be demonstrated on CT or MRI. However, understanding the innervation and the complex distribution should lead to a better appreciation of the complexities in evaluating patients with back pain syndromes [86].

The Paraspinal Muscles

The paraspinal muscles are paired structures throughout the spinal column which control flexion-extension of the spine (erector spinae muscles) and stabilize the spine (posture muscles). These muscles are routinely demonstrated on cross-sectional imaging studies of the spine, although little attention is paid to them, in part because of their infrequent pathology. By virtue of its multiplanar imaging capabilities and superior soft-tissue contrast, MRI is ideally suited to the evaluation of these muscles. T1-weighted images best demonstrate the fat and fascial planes necessary for anatomic definition.

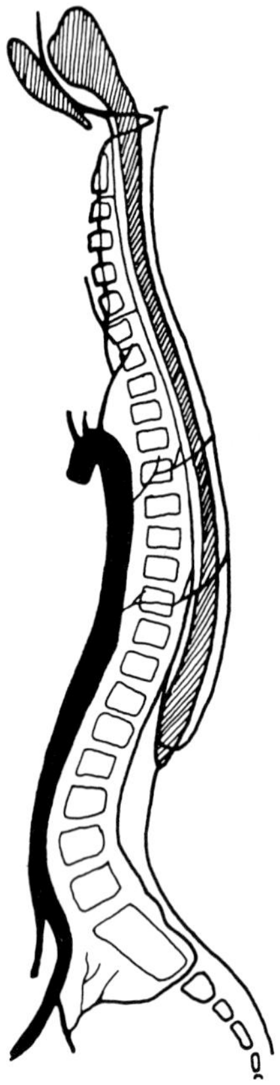

Figure 2.57. Lateral view of the anterior radiculomedullary arteries of the spinal cord. (1) Cervical anterior radiculomedullary arteries, (2) posterior spinal artery, (3) anterior spinal artery, (4) upper thoracic radiculomedullary artery, (5) artery of Adamkiewicz, and (6) sacral radicular artery.

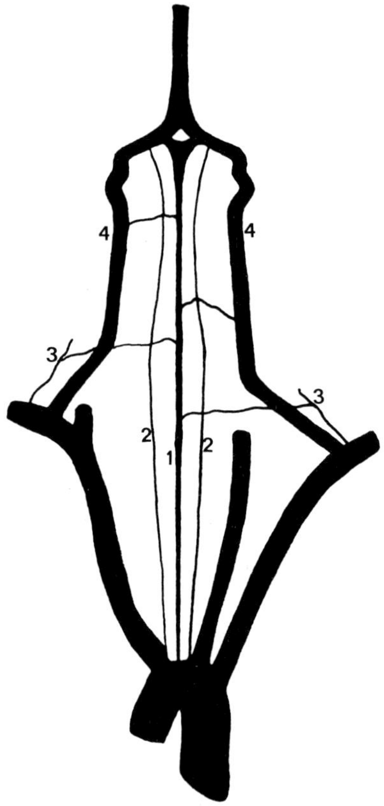

Figure 2.58. Major arterial feeders in the cervical spinal cord. (1) Anterior spinal artery, (2) posterior spinal arteries, and (3) anastomotic branches from (*i*) thyrocervical trunk, and (*ii*) vertebral arteries.

The deep muscles of the back are divided into a superficial, or transversocostal, group and a deep, or transversospinal, group [88–91]. A third group of muscles, the suboccipital muscles, helps control movements at the craniocervical junction. The superficial back muscles including the trapezius, latissimus dorsi, levator scapuli, and rhomboids, are primarily involved in movements of the shoulder girdle.

THE SUPERFICIAL GROUP

The superficial, or transversocostal, group includes the splenius capitis, the splenius cervicis, and the erector spinae. The splenius muscles originate from the spinous processes of the cervical and upper thoracic vertebrae and the nuchal ligament and insert laterally on the occipital bone and mastoid process (splenius capitis) and the upper cervical transverse processes (splenius cervicis). Working together, the splenius muscles extend the neck and retract the head. Separately, they draw the head to one side and rotate it ipsilaterally in concert with the contralateral sternocleidomastoid muscle. The splenius capitis lies immediately deep to the trapezius and is consistently demonstrated on axial images of the neck (see Figs. 2.61 and 2.62). The splenius cervicis is variably present, lying inferior, deep, and lateral to the splenius capitis.

The erector spinae is a much larger muscle, present in all segments of the spine, consisting of three separate muscle bundles, the medially located spinalis muscles (spinalis cervicis and spinalis thoracis), the intermediately located longissimus muscles (longissimus capitis, longissimus cervicis, and longissimus thoracis), and the laterally located iliocostalis muscles (iliocostalis cervicis, iliocostalis thoracis, and iliocostalis lumborum), flanking the spinal column on each side. In the lumbosacral region, the erector spinae originates from a broad, thick tendon (the sacrospinalis tendon) originating from the medial sacral crest, the lumbar and eleventh and twelfth thoracic spinous processes, and the dorsal aspect of the iliac crest. The sacrospinalis tendon divides superiorly into the three muscle bundles. As their name implies, the spinalis muscles originate from vertebral spinous processes and insert on spinous processes several segments cephalad. The spinalis cervicis is small, may be absent, and is not readily distinguished on imaging studies. The spinalis thoracis is apparent on axial scans, blending with the more lateral longissimus thoracis

Figure 2.59. Schematic drawing of arterial supply in the lumbar spine. (1) Radicular artery, (2) segmental lumbar artery, (3) aorta, and (4) segmental branches within marrow of the vertebrae.

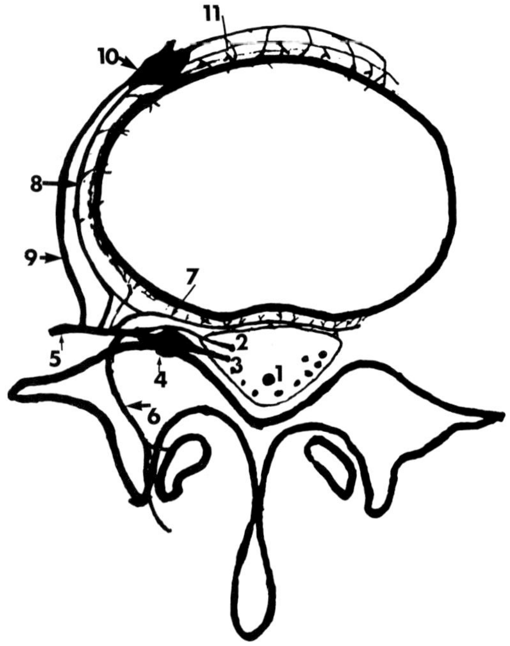

Figure 2.60. Schematic drawing of innervation of the lumbar spine. (1) Filum terminale, (2) ventral n. root, (3) dorsal n. root; (4) dorsal root ganglion, (5) ventral ramus of spinal nerve, (6) dorsal ramus of spinal nerve, (7) sinuvertebral nerve, (8) gray rami communicans, (9) white rami communicans, (10) paraspinal sympathetic ganglion, and (11) anterior sympathetic branches.

and the more anterior semispinalis (see also Figs. 2.63 and 2.64). The longissimus thoracis is the largest bundle of the erector spinae, originating from the sacrospinalis tendon and vertebral transverse processes and inserting on transverse processes of more cephalad vertbrae. It is readily demonstrated on axial (see Fig. 2.64 and 2.66) and coronal (see Figs. 2.63 and 2.65*B*). The longissimus cervicis and longissimus capitis are demonstrated on both axial and coronal images of the neck (Figs. 2.61 and 2.62). The longissimus capitis lies between the longissimus cervicis and the semispinalis capitis (see also Figs. 2.61*A* and 2.61*B*). The most lateral group, the iliocostalis muscles (Figs. 2.65 and 2.66), originate from the sacrospinalis tendon and the angles of ribs, inserting on more cephalad ribs and on cervical transverse processes. Together the erector spinae are extensors of the spinal column.

THE DEEP GROUP

The deep, or transversospinal, group includes the semispinalis muscles (semispinalis thoracis, semispinalis cervicis, and semispinalis capitis), the multifidus, and the rotatores (rotatores cervicis, rotatores thoracis, and rotatores lumborum). These muscles, as their name implies, originate from vertebral transverse processes, ascending obliquely and medially to insert on adjacent and distant spinous processes. The semispinalis thoracis and cervicis extend the spine and rotate it contralaterally. The semispinalis capitis and more medial spinalis capitis extend the neck and rotate the face contralaterally. The semispinalis capitis lies deep to the splenius capitis and is a constant finding on axial and coronal scans (see Figs. 2.61 and 2.62), lying between the medial and deeper semispinalis cervicis and the more lateral longissimus capitis and cervicis. The semispinalis thoracis blends with the more superficial spinalis thoracis and is difficult to distinguish from the deeper multifidus on routine imaging studies (Figs. 2.64).

The multifidus lies in a groove on each side of the spinous processes from the sacrum to the axis, its most caudal fascicles originating from the erector spinae aponeurosis, the dorsal sacroiliac ligaments, and the posterior superior iliac spine. The individual fascicles vary in length, ascending medially to the adjacent vertebrae (deepest fascicles), the second or third above (intermediate fascicles), and as far as the fourth above (most superficial fascicles). The multifidus is largest and most well-formed in the lumbosacral region, where it is easily seen on axial images immediately adjacent to the spinous processes, medial and anterior to the erector spinae (Fig. 2.66). On coronal scans (Fig. 2.63 and 2.65*A*), the bundles can be seen ascending medially from transverse processes toward more cephalad spinous processes. The multifidus rotates the spine contralaterally. The

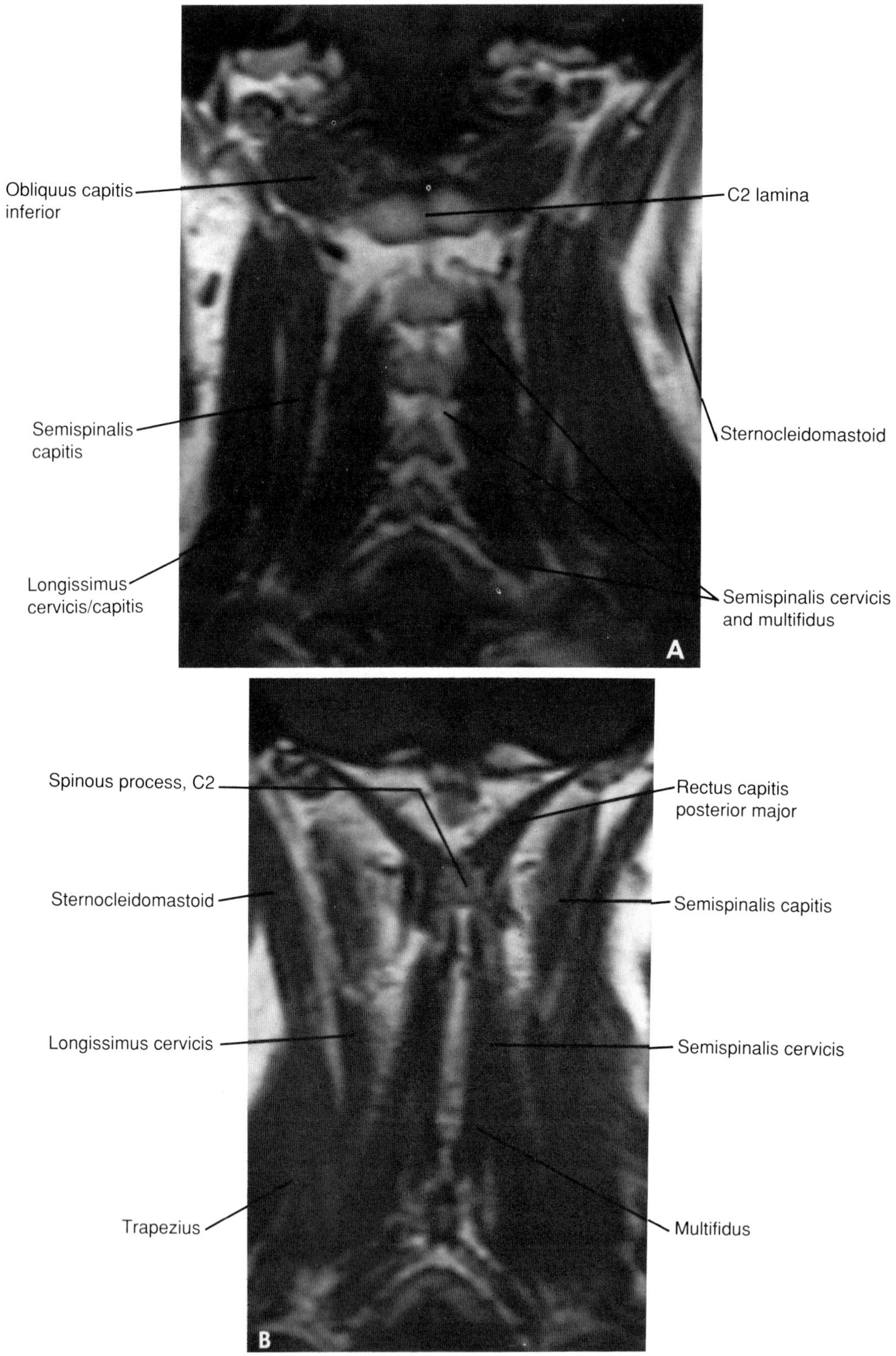

Figure 2.61. Coronal T1-weighted images of the neck at the level of the vertebral laminae (**A**) and just posterior to the laminae (**B**).

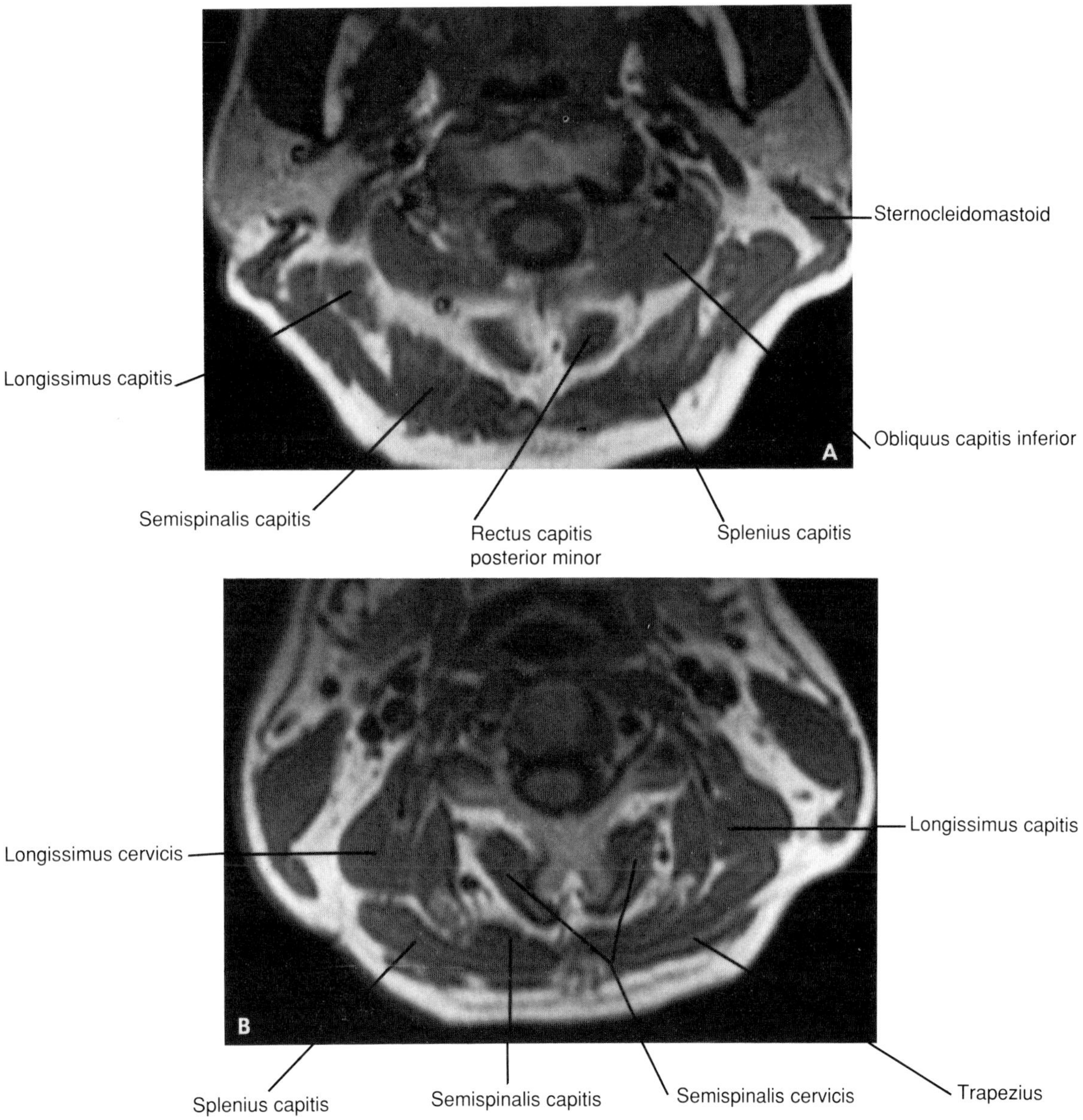

Figure 2.62. Axial T1-weighted images of the neck at the level of the body of C2 (**A**) and through a mid-cervical vertebra (**B**).

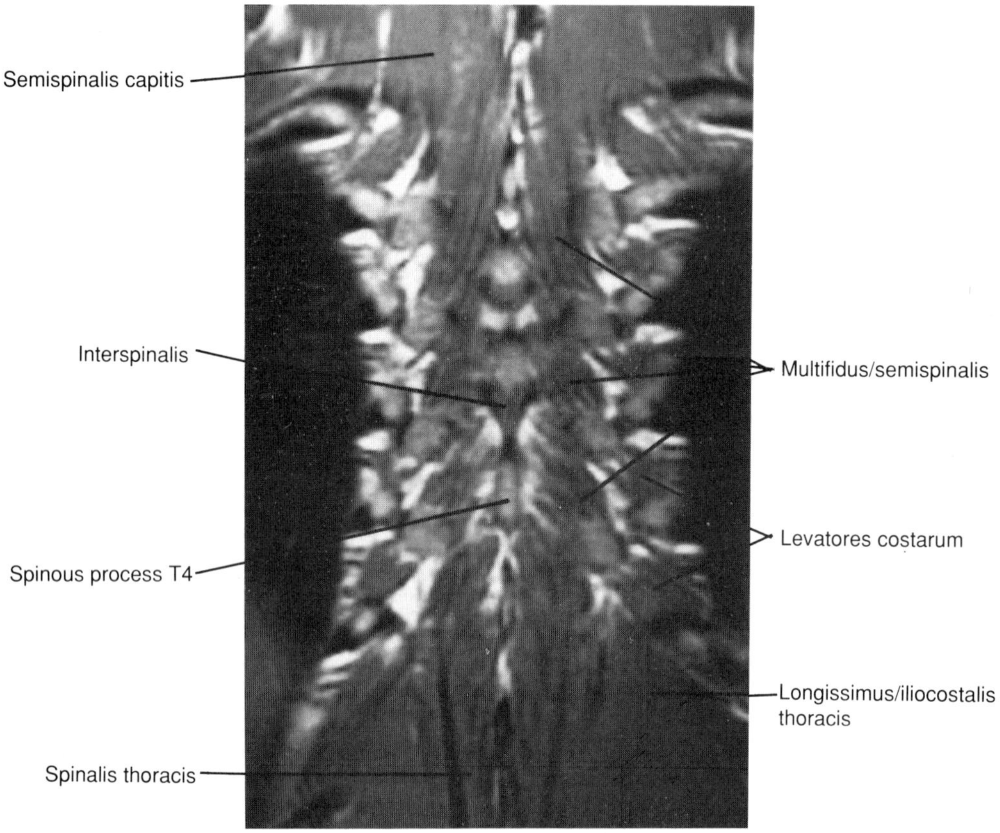

Figure 2.63. Coronal T1-weighted image of the thoracic region at the level of the vertebral spinous processes.

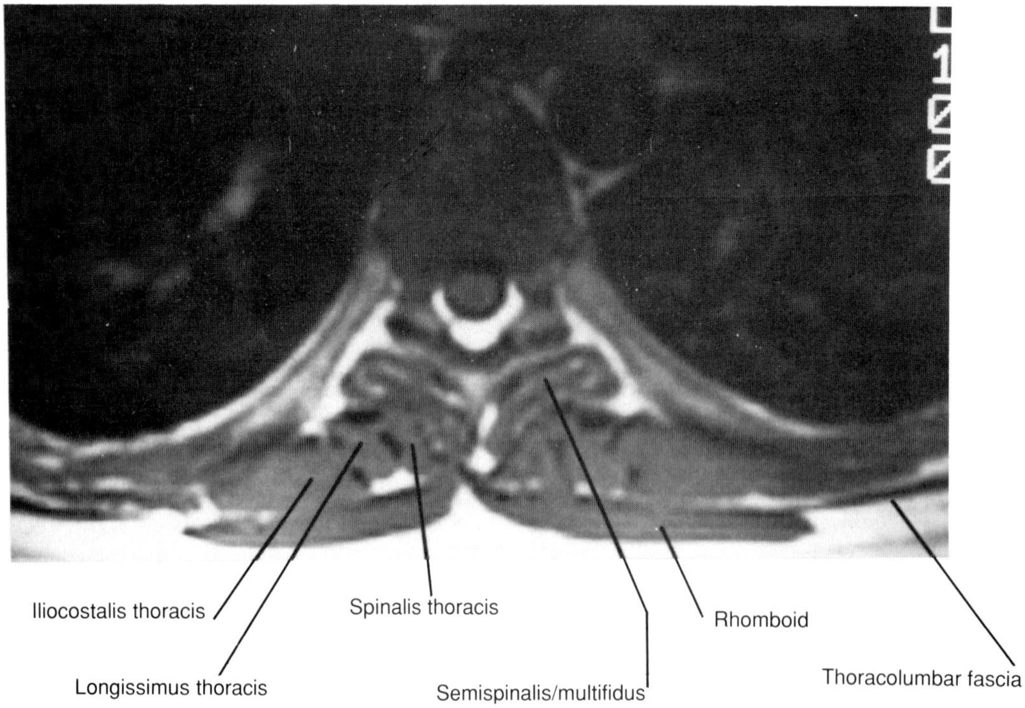

Figure 2.64. Axial T1-weighted image through the midthoracic spine.

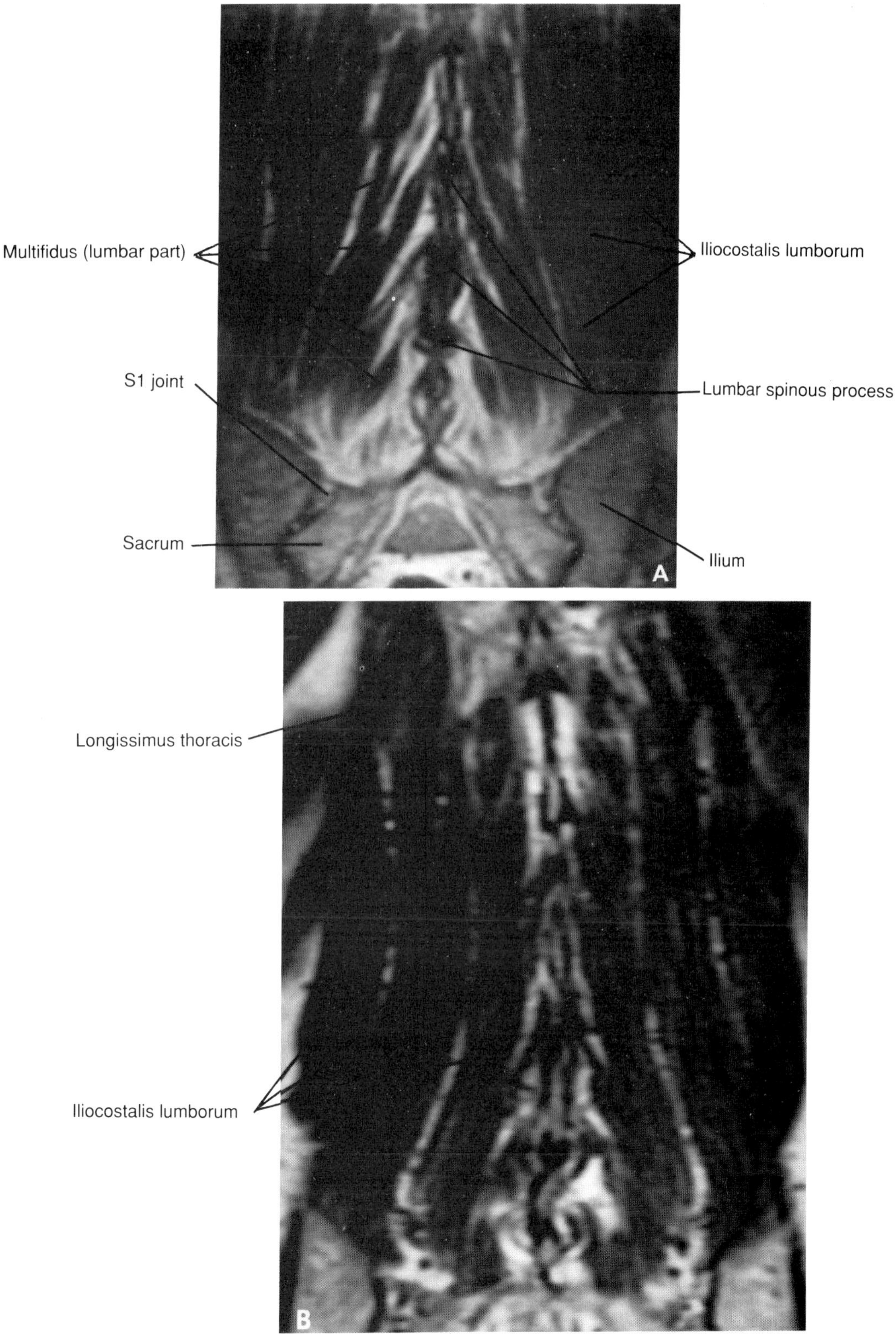

Figure 2.65. Coronal T1-weighted images of the lumbosacral region. Part **A** is slightly more anterior than part **B,** demonstrating the multifidus muscles.

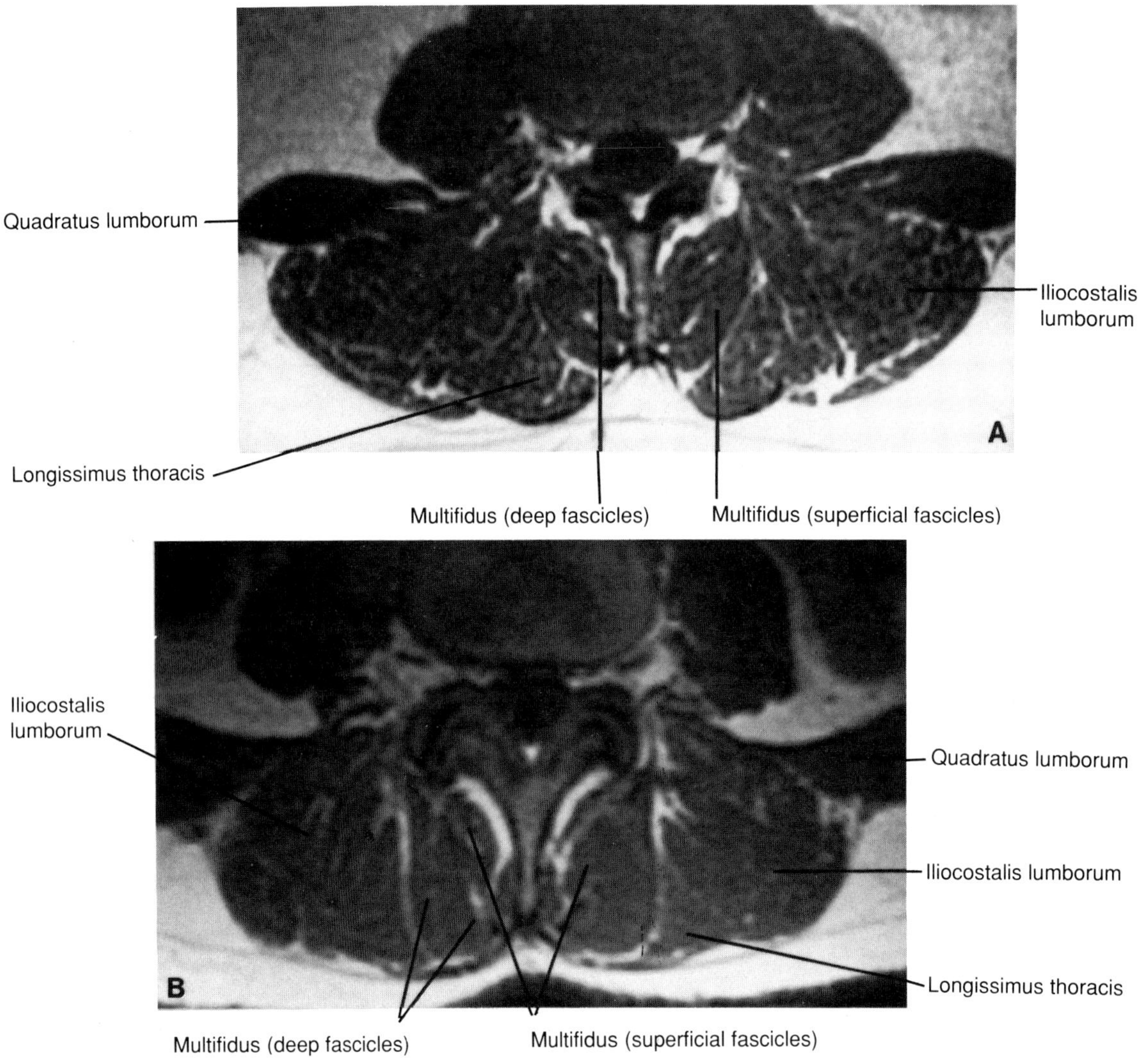

Figure 2.66. Axial T1-weighted images through the upper (**A**) and lower (**B**) lumbar spine.

rotatores, so named because of their action, lie deep to the multifidus, extending from transverse processes to the spinous processes of the next vertebra above (rotatores brevis) or two vertebrae above (rotatores longis), also rotating the spine to the opposite side. They are not readily distinguished on imaging studies.

Innervation of the superficial and deep paraspinal muscles is derived from the dorsal rami of spinal nerves.

THE SUBOCCIPITAL MUSCLES

The suboccipital muscles extend the head at the atlanto-occipital articulation (rectus capitis posterior major and minor) and rotate the head at the atlantoaxial joint (obliquus capitis superior and inferior), lying deep to the splenius capitis and semispinalis capitis. The rectus capitis posterior major ascends laterally from the axial spine to insert on the inferior nuchal line and the occipital bone below, and is best seen on coronal images (see Fig. 2.61*A*). In the triangle between the diverging rectus capitis posterior major muscles lie the rectus capitis posterior minor muscles, ascending from the atlantal tubercle to the inferior nuchal line, medial to the rectus capitis posterior major (see Fig. 2.62*A*). The obliquus capitis inferior extends from axial spine to the transverse process of the atlas and can be seen on both axial (see Fig. 2.61) and coronal (see Fig. 2.62*A*) images. The obliquus capitis superior extends from the transverse process of the atlas to the occipital bone between the superior and inferior nuchal lines, lateral to the semispinalis capitis, and overlapping the rectus capitis posterior major. Innervation to the suboccipital muscles is derived from the dorsal rami of the first cervical spinal nerves.

References

1. McRea DL. Craniovertebral junction. In: Newton TH, Potts DG, eds. Radiology of the skull and brain, part 1: The skull. St Louis: Mosby, 1971:260–271.

2. Daniels DL, Williams AL, Haughton VM. Computed tomography of the articulations and ligaments of the occipito-atlantoaxial region. Radiology 1983;146:709–716.
3. Smoker WRK, Keyes WD, Dunn VD, et al. MRI vs conventional radiologic evaluation of the craniovertebral and cervicomedullary junction. Radiographics 1986;6:953–994.
4. Burget JL, Sick H, Dirheimer Y, et al. CT of the main ligaments of the cervico-occipital hinge. Neuroradiology 1985;27:112–118.
5. Calvy TM, Segall HD, Gilles FH, et al. CT anatomy of the craniovertebral junction in infants and children. AJNR 1987;8:489–494.
6. Bandschuh CV, Modic MT, Kearney F, et al. Rheumatoid arthritis of the cervical spine: Surface coil MR imaging. AJNR 1988;9:565–571.
7. Lindt M. Biomechanics of the lumbar spine. In: Frankel VH, Nordin M, eds. Basic biomechanics of the skeletal system. Philadelphia, Lea & Febiger, 1980:255–290.
8. Morris JM. Biomechanics of the lumbar spine. In: Finnison BE, ed. Low back pain. Philadelphia: Lippincott, 1980:27–43.
9. Van Schalk JPJ, Verbeist H, Van Schalk FDJ: The orientation and shape of the lower lumbar facet joints: A computed tomographic study of the variation in 100 patients with low back pain, and a discussion of their possible clinical importance. In: Donavan Post MJD, ed. Computed tomography of the spine. Baltimore: Williams & Wilkins, 1984:495–505.
10. Whalen MA, Gold RP. Computed tomography of the sacrum: normal anatomy. AJNR 1982;3:547–554.
11. Hajek PC, Baker LL, Goober JE, et al. Focal fat deposition in axial bone marrow: MR characteristics. Radiology 1987;
162:245–249.
12. Saratoris DJ, Resnick D, Guerra J Jr. Vertebral venous channels: CT appearance and differential considerations. Radiology 1985;155:745–749.
13. Lee BCP, Kazam E. Computed tomography of the spine and spinal cord. Radiology 1978;128:95–102.
14. Dorwart RH, DeGroot J, Sauerland EH, et al. Computed tomography of the lumbosacral spine: normal anatomy, anatomic variants and pathologic anatomy. Radiographics 1982;2:459–498.
15. Johansen JG, McCarty JD, Haughton VM. Retrosomatic cleft: computed tomographic appearance. Radiology 1983;148:447–448.
16. Johansen DW, Farnum GN, Latchaw RE, et al. MR imaging of the pars interarticularis. AJNR 1988;9:1215–1220.
17. Schwartz AM, Wechsler RJ, Lamdy DM, et al. Posterior arch defect of the cervical spine. Skeletal Radiol 1982; 8:135–139.
18. Carrera GF, Haughton VM, Syvertson A, et al. Computed tomography of the lumbar facet joints. Radiology 1980; 134:145–148.
19. Carrera GF, Williams AL. Current concepts in evaluation of lumbar facet joints. CRC Crit Rev Clin Imaging 1984;21:83–104.
20. Hadley LA. Anatomico-roentgenographic studies of the posterior spinal articulations. AJR 1961;86:270–276.
21. Xu GL, Haughton VM, Carrera GF. Lumbar facet joint capsule: appearance at MR imaging and CT. Radiology 1990;177:415–420.
22. Yu S, Sether LA, Haughton VM. Facet joint menisci of the cervical spine: Correlative MR imaging and cryomycotomy study. Radiology 1987;164:79–82.
23. Schellinger D, Weiner L, Ragsdale BD, et al. Facet joint disorders and their role in the production of low back pain and sciatica. Radiographics 1987;7:923–944.
24. Coventry MB. Anatomy of the intervertebral disk. Clin Orthop 1969;67:9–15.
25. Coventry MD, Ghormley RK, Kernohan JW. The intervertebral disk: its microscopic anatomy and pathology: 1. Anatomy development and physiology. J Bone Joint Surg 1945;27A:105–112.
26. Eyring EJ. The biochemistry and physiology of the intervertebral disc. Clin Orthop 1969;67:16–23.
27. Ghosh P, Bushell GR, Taylor TFK, et al. Collagen, elastin, and noncollagenous protein of the intervertebral disk. Clin Orthop 1977;12:124–132.
28. Lipson SJ, Muir H. Proteoglycans in experimental intervertebral disk degeneration. Spine 1984;6:194–210.
29. Pech P, Haughton VM. Lumbar intervertebral disk: correlative MR and anatomic study. Radiology 1985;156: 699–701.
30. Modic MT, Pavlicek M, Weinstein MA, et al. Magnetic resonance imaging of the intervertebral disk disease. Radiology 1984;152:103–111.
31. Ho PSP, Yu S, Sether LA, et al. Progressive and regressive changes in the nucleus pulposus: 1. Neonate. Radiology 1988;169:87–91.
32. Aguila LA, Piraino DW, Modic MT, et al. The intranuclear cleft of the intervertebral disk: Magnetic resonance imaging. Radiology 1985;155:155–158.
33. Yu S, Haughton VM, Lynch KL, et al. Fibrous structure in the intervertebral disc: Correlation of MR appearance with anatomic sections. AJNR 1989;10:1105–1110.
34. Yu S, Haughton VM, Ho PSP, et al. Progressive and regressive changes in the nucleus pulposus: 2. Adult. Radiology 1988;169:93–97.
35. Sether LA, Yu S, Haughton VM, et al. Intervertebral disk: normal age related changes in MR signal intensity. Radiology 1990;177:385–388.
36. Aoki J, Yamamoto I, Kitamura N, et al. Endplate of the discovertebral joints: Degenerative changes in the elderly adult. Radiology 1987;164:411–414.
37. Modic MT, Steinberg PM, Ross JS, et al. Degenerative disk disease: assessment of changes in vertebral body marrow with MR imaging. Radiology 1988;166:193–199.
38. DeRoos A, Kressel H, Spitzer C, et al. MR imaging of marrow changes adjacent to end plates in degenerative lumbar disk disease. AJR 1987;149:531–534.
39. Yong-hing K, Reilly J, Kirkaldy-Willis WH. The ligamentum flavum. Spine 1976;1:226–234.
40. Zarzur E. Anotomic studies of the human lumbar ligamentum flavum. Anesth Analg 1984;63:499–502.
41. Ramsey RH. The anatomy of the ligamentum flavum. Clin Orthop 1966;44:129–140.
42. Nowicki BH, Haughton VM. Neural foraminal ligaments of the lumbar spine: appearance at CT and MR imaging. Radiology 1992;183:257–264.
43. Haughton VM, Syvertstein A, Williams AL. Soft tissue anatomy within the spinal canal as shown on computed tomography. Radiology 1980;134:649–655.
44. Ho PSP, Yu S, Sether LA. Ligamentum flavum: appearance on sagittal and coronal MR images. Radiology 1988;168:469–472.
45. Grenier JP, Kressel HY, Schiebler ML, et al. Normal and degenerative posterior spinal structures: MR imaging. Radiology 1987;165:517–525.
46. Heylings DJA. Supraspinous and interspinous ligaments of the human spine. J Anat 1978;125:127–131.
47. Harwood-Nash DCF, Fitz CR. Neuroradiology in infants and children, vol 3. St. Louis: Mosby, 1976.
48. Schellinger D, Manz HJ, Vidic B, et al. Disk fragment migration. Radiology 1990;175:831–836.
49. Bronskill MG, McVigh ER, Kucharcyzk W, et al. Syrinx-

like artifacts on MR images of the spinal cord. Radiology 1988;166:485–488.
50. Levy LM, DiChiro G, Brooks RA, et al. Spinal cord artifacts from truncation errors during MR imaging. Radiology 1988;166:479–483.
51. Lufkin RB, Pusey E, Stark DD, et al. Boundary artifacts due to truncation error in MR imaging. AJR 1986;147: 1283–1287.
52. Rubin JB, Enzmann DR. Imaging of spinal CSF pulsation by 2DFT MR: significance during clinical imaging. AJNR 1987;8:297–306.
53. Rubin JB, Enzmann DR. Harmonic modulation of proton MR precessional phase by pulstile motion: origin of spinal CSF flow phenomenon. AJNR 1987;8:307–318.
54. Pattany P, Phillips J, Chiu L, et al. Motion artifact suppression technique (MAST) for MR imaging. J Comput Assist Tomogr 1987;11:369–377.
55. Tsurudu JS, Norman D, Dillon W, et al. Three-dimensional gradient recalled MR imaging as a screening tool for the diagnosis of cervical radiculopathy. AJNR 1990; 10:1263–1271.
56. Gellad F, Rao KCVG, Joseph PM, et al. Morphology and dimensions of the thoracic cord by computer assisted metrizamide myelography. AJNR 1983;4:614–617.
57. Nordqvist L. The sagittal diameter of the spinal cord and subarachnoid space in different age group: a roentgenographic postmortem study. Acta Radiol (Diagn) (Stockh) 1964;227:1–96.
58. Thijssen HOM, Keyser A, Horstink MWM, et al. Morphology of the cervical spinal cord on computed tomography. Neuroradiology 1979;18:57–62.
59. Seibert CE, Barnes JE, Dreisbach JN, et al. Accurate CT measurement of the spinal cord using metrizamide: physical factors. AJNR 1981;2:75–78.
60. Koehler PR, Anderson RE, Baxter B. The effect of computed tomography viewer controls on anatomical measurements. Radiology 1979;130:189–194.
61. Rosenbloom S, Cohen WA, Marshall C, et al. Imaging factors influencing spine and cord measurements by CT: a phantom study. AJNR 1983;4:646–649.
62. Daniels DL, Hyde JS, Kneeland JB, et al. The cervical nerves and foramina: local coil MR imaging. AJNR 1986; 7:129–133.
63. Czervionke LF, Daniels DL, Ho PSP, et al. Cervical neural foramina: correlative anatomic and MR imaging study. Radiology 1988;169:753–759.
64. Czervionke LF, Daniels DL, Ho PSP, et al. The MR appearance of gray and white matter in the cervical spinal cord. AJNR 1988;9:557–562.
65. Pech P, Daniels DL, Williams AL, et al. The cervical neural foramina: correlation of microtomy and CT anatomy. Radiology 1985;155:143–146.
66. Modic MT, Masaryk TJ, Ross JS, et al. Cervical radiculopathy: value of oblique MR imaging. Radiology 1987; 163:227–231.
67. Yenerich DO, Haughton VM. Oblique plane MR imaging of the cervical spine. J Comput Assist Tomogr 1986; 10:823–826.
68. Flannagan BD, Lufkin RB, McGlade C, et al. MR imaging of the cervical spine: neural vascular anatomy. AJNR 1987;8:27–32.
69. Berger RK, Williams AL, Daniels DL, et al. Contrast enhancement in spinal MR imaging. AJNR 1989;10:633–637.
70. Ross JS, Modic MT, Masaryk TJ, et al. Assessment of extradural degenerative disease with Gd-DTPA enhanced MR imaging: correlation with surgical and pathological findings. AJNR 1989;10:1243–1249.
71. Wilson DA, Prince JR. MR imaging determination of the location of the normal conus medullaris throughout childhood. AJNR 1989;10:259–262.
72. Monjati A, Wayne WS, Raushing W, et al. MR of the cauda equina. AJNR 1987;8:893–900.
73. Grogan JP, Daniels DL, Williams AL, et al. The normal conus medullaris: CT criteria for recognition. Radiology 1984;151:661–664.
74. Calliet R. Low back syndrome. 3d ed. Philadelphia: FA Davis, 1981.
75. Kostelic JK, Haughton VM, Sether LA. Lumbar spinal nerves in the neural foramen: MR appearance. Radiology 1991;178:837–839.
76. Kostelic J, Haughton VM, Sether L. Proximal lumbar spinal nerves in axial MR imaging, CT, and anatomic sections. Radiology 1992;183:239–242.
77. Helms CA, Dorwart RH, Gray MB. CT appearance of conjoined nerve roots and differentiation from herniated nucleus pulposus. Radiology 1982;144:803–807.
78. Theron J. Spinal plebography. New York: Springer-Verlag, 1978.
79. Russel EJ, D'Angelo CM, Zimmerman RD, et al. Cervical disk herniation: CT demonstration after contrast enhancement. Radiology 1984;152:703–712.
80. Crock HV, Yoshhizawa H, Kame SK. Observations on the venous drainage of the human vertebral body. J Bone Joint Surg 1973;55B:528–533.
81. Djindjian R, Merland JJ, Djindjian M, et al. Angiography of spinal column and spinal cord tumors. Stuttgart: Theime-Stratton, 1981.
82. Bernick S, Calliet R. Vertebral end-plate changes with aging of human vertebrae. Spine 1982;7:97–102.
83. Doppman JL, DiChiro G, Ommaya AK. Selective arteriography of the spinal cord. St. Louis: WH Green, 1969.
84. Pederson HE, Blunck CFJ, Gardner E. The anatomy of lumbosacral posterior rami and meningeal branches of spinal nerves (sinu-vertebral nerves) J Bone Joint Surg 1956;38A:377–391.
85. Bogduk N. The innervation of the lumbar spine. Spine 1983;8:286–292.
86. Jinkins JR, Whittemore AR, Bradley WG. The anatomic basis of vertebrogenic pain and autonomic syndrome associated with lumbar disk extrusion. AJNR 1989;10: 219–231.
87. Helms CA, Sims R. Foraminal spurs: a normal variant in the lumbar spine. Radiology 1986;160:153–154.
88. Pansky B. Review of gross anatomy. 4th ed. New York: Macmillan, 1979:152–159.
89. Ferner H, Staubesand J, eds. Sobotta's atlas of human anatomy. vol. 2, 10th ed. Baltimore: Urban & Schwarzenberg, 1983:225–237.
90. Williams PL, Warwick R, Dyson M, Bannister LH, eds. Gray's anatomy. 37th ed. New York: Churchill Livingstone, 1989: 587–591.
91. Cahill DR, Orland MJ. Atlas of cross-sectional anatomy. Philadelphia: Lea and Febiger, 1984:116–130.

CHAPTER / **3**

The Craniovertebral Junction

J. Sherman

The evaluation of the craniovertebral junction is directed primarily at determining the adequacy of the canal space for the spinal cord and medulla (the neuraxis), making allowance for the normal ranges of movements in flexion, extension, and rotation. Once it is assured that the neuraxis is not compromised and the craniocervical junction is stable, one can proceed to the task of recognition and diagnosis of specific disease processes and anatomic variations.

Normal Anatomy

The craniovertebral junction is a complex transition area which includes the posterior central region of the base of the skull and the upper two cervical segments (Figs. 3.1 and 3.2). The basilar part of the occipital bone is the basiocciput. The clivus is the sloping part of the basiocciput. The foramen magnum lies posterior to the basiocciput and is part of the occipital bone. It is typically ovoid in shape with the long axis parallel to the midsagittal plane. CT and MRI are both suitable for evaluation of this region.

The anterior third of the foramen magnum is usually angular in shape. The anterior edge of the foramen magnum is slightly thickened and lies between the anterior ends of the occipital condyles and posteroinferior tip of the clivus. The midpoint of the anterior lip of the foramen magnum is termed the *basion*. The posterior edge of the foramen magnum is typically thin and rounded, and the midpoint is termed the *opisthion*. Since the anterior midpoint of the foramen magnum is defined by the clivus, the plane and the length of the foramen magnum depend on the length of the clivus.

In normal adults, the length of the foramen magnum (anteroposterior diameter) measures 38 mm (ranging from 31 to 43 mm), and the width is 30 mm (ranging from 26 to 34 mm) [1] (Table 3.1). These measurements of the bony margins of the foramen magnum do not include ligaments, dural bands, fibrous membranes, and synovial cavities which also encroach on the space allotted for passage of the neuraxis and vertebral arteries.

The cranium is joined to the spine by the atlanto-occipital joints, the anterior and posterior atlanto-occipital membranes, and the occipitoaxial ligaments. The anteroposterior relationships of the spine to the skull base can be assessed using Wackenheim's line [2], which is drawn along the posterior margin of the clivus. The inferior extension of Wackenheim's line should fall tangent to the posterior tip of the odontoid process (Fig. 3.3).

The occipital condyles and the superior articular facets of the atlas form the atlanto-occipital joints and allow head "nodding" forward and backward and limited lateral flexion of the head on the neck. Complex rotation is achieved by the atlantoaxial joints, which include the paired lateral joints formed by the superior facets of C2 and the inferior facets of C1 and the median joints which form a "pivot point" along the anterior and posterior surfaces of the dens.

The anterior median joint is a synovial cavity between the anterior arch of C1 and the anterior margins of the dens. The posterior median joint is between the posterior margin of the dens and the transverse ligament of the atlas. The transverse lig-

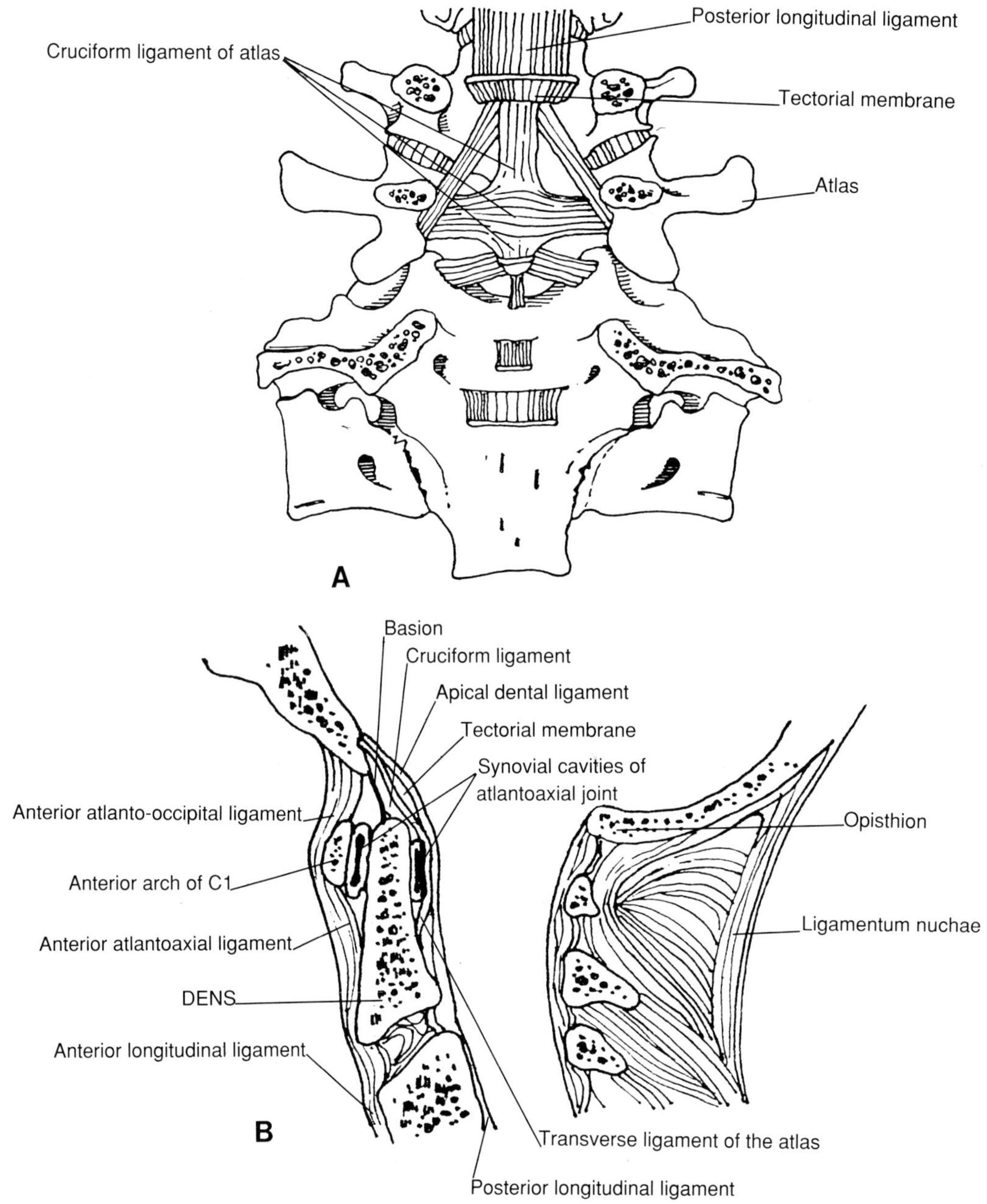

Figure 3.1. Line drawings of craniovertebral junction anatomy. **A,** Midline-sagittal diagram. **B,** Coronal diagram.

ament is a strong band that runs between the lateral masses of the atlas, passing behind the dens. The superior cruciate ligament extends upward from the middle of the transverse ligament to the basion. The inferior cruciate ligament extends downward to the back of the body of the axis. The membrana tectoria is a broad sheet of fibrous tissue continuous with the posterior longitudinal ligament which attaches to the endocranial surface of the basion and merges laterally with accessory atlantoaxial ligaments, which, in turn, are attached to the atlantoaxial joint capsules and run along the lateral masses. The apical ligament of the dens stretches from the apex of the dens to the basion. The alar ligaments of the dens are short bands that run from the side of the dens upward to the medial side of the occipital condyles.

The foramen magnum and upper cervical spinal canal form a "funnel" in which the cross-sectional area is largest at the foramen magnum and progressively narrows to the level of approximately C3, below which the dimensions are fairly constant.

MRI has replaced myelopolytomography, myelography, and CT-assisted myelography as the modality of choice in the evaluation of the craniovertebral junction. It is noninvasive and is able to image the bones, neural structures, and associated soft-tissues in multiple planes without artifact [3–5]. Conventional radiographic methods are, however, recommended for evaluation of bone details [6], par-

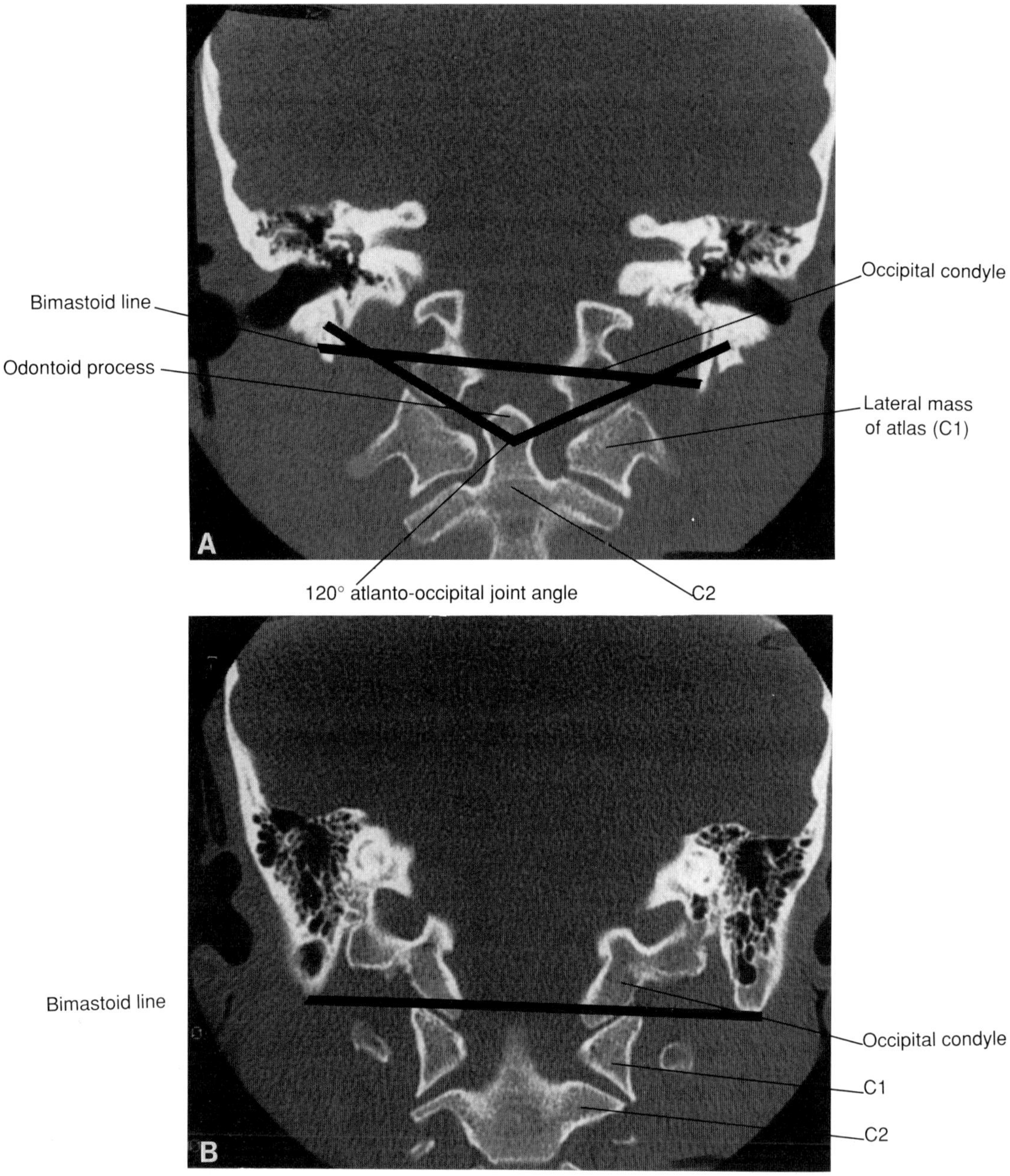

Figure 3.2. Normal coronal CT of craniovertebral junction. Note normal angulation of atlanto-occipital and atlantoaxial joints. There is slight rotation of the axis, which causes asymmetrical distance between the lateral mass of the atlas and the odontoid process.

ticularly for lesions involving the lateral masses, facet joints, and spinous processes of the upper cervical spine.

Measurements obtained from MRI usually correlate well with their CT and radiographic counterparts [7] (Table 3.2). It should be noted that various measurements (Chamberlain's line, McGregor's line, Wackenheim's line, McRae's line, Boogaard's line, atlanto-occipital joint axes angle, bimastoid line, digastric line) (Table 3.3) were used in the past to determine indirectly the effect of bony pathologic processes because the spinal cord and cervicomedullary junction could not be visualized using conventional radiographic techniques (Figs. 3.3 through 3.5). These criteria were most useful in the evaluation of basilar invagination and atlanto-occipital dislocations.

Specific measurements are not as important on MRI images because the bone and spinal cord are both seen, allowing precise evaluation of pathoanatomic relations.

Clinical Presentation

The clinical signs and symptoms of diseases or malformations of the craniovertebral junction are due to compression or disturbance of the lower brainstem, high cervical cord, and rarely, vertebro-

Table 3.1.
Foramen Magnum Measurements

Age (years)	Sex	Width	Length (AP diameter)
1–3	F	31 ± 3	39 ± 4.5
1–3	M	31 ± 2.5	40 ± 5.4
3–7	F	34 ± 2.4	39 ± 4.8
3–7	M	34 ± 2.4	43 ± 3.7
>7	F	35 ± 2.8	44 ± 3.5
>7	M	35 ± 2.8	45 ± 3.5

Source: Kruyff and Jeffs [114].

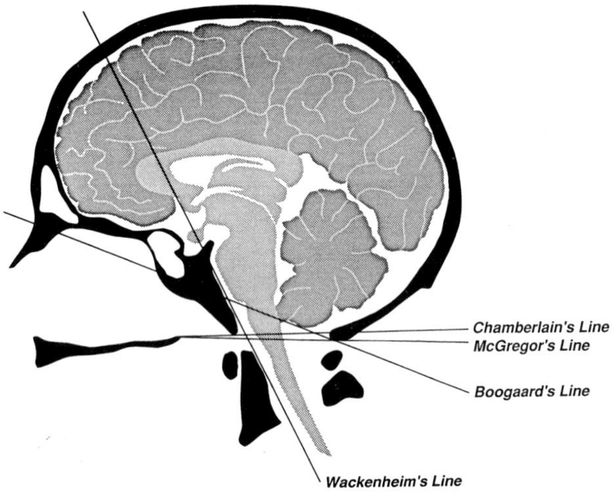

Figure 3.3. Diagram of sagittal craniovertebral junction relationships showing Chamberlain's, McGregor's, Boogaard's, and Wackenheim's lines.

Table 3.2.
Normal Measurements of Craniocervical Junction

Atlantodental interval	< 2.1 mm
Basion-dental interval	4 mm
Cervicomedullary angle	135–175 degrees (mean 155.2)
Basal angle	123–152 degrees (mean 137)
C1 subarachnoid space (AP)	16 mm (mean 12–21) [112]
C2 subarachnoid space (AP)	14 mm (mean 11–18)
C3 subarachnoid space (AP)	12 mm (mean 11–14)
C1 spinal canal (AP)	25 mm (mean 19–32) [113]
C2 spinal canal (AP)	22 mm (mean 16–27)
C3 spinal canal (AP)	20 mm (mean 15–25)

Table 3.3.
Craniovertebral Junction Measurements

Chamberlain's line extends from the posterior edge of the hard palate to the opisthion (i.e., the posterior border of the foramen magnum) [103]. Normally, Chamberlain's line lies above the dens. Basilar invagination should be suspected if the tip of the odontoid projects more than 2.5 mm above this line and confirmed if this distance exceeds 6.6 mm.

McGregor's line is a modification of Chamberlain's line in which the line extends from the hard palate to the undersurface of the occiput [104]. The dens should lie no more than 4.5 mm above McGregor's line.

McRae's line measures the anteroposterior dimension of the foramen magnum between the basion and the opisthion [105]. The dens should not project above McRae's line. This criteria has the advantage of being independent of facial structures.

Fischgold's bimastoid line connects the mastoid tips. The dens measures from 3 mm below to 10 mm above this line [106].

Fischgold's digastric line extends between the two digastric grooves, which lie medial to the base of the mastoid processes. The distance between this line and the middle of the atlanto-occipital joint is less than 10 mm in basilar impression.

The *atlanto-occipital joint angle* is determined by drawing intersecting lines along the surfaces of the atlanto-occipital joints on each side. Basilar invagination is suspected if the angle is greater than 150 degrees [107].

Wackenheim's line is drawn along the posterior surface of the clivus. This line should lie tangential to the posterior tip of the dens [108].

Power's ratio is used to assess craniovertebral dislocation but determining the ratio between basion–posterior C1 arch line and the opisthion–anterior C1 arch line. This ratio is greater than one in patients with atlanto-occipital dislocation [109].

Klaus's height index is the distance from Twining's line to the tip of the odontoid process and should not exceed 30 mm. Measurements less than 30 mm confirm basilar invagination [110,111].

basilar insufficiency. Patient complaints are variable. There is no clinical grouping of symptoms which is characteristic of any specific disorder. The most common symptom is dull neck and occipital headache. *Press headache* is a headache that occurs when abdominal pressure is raised by sneezing, defecating, or coughing (*cough headache*). This type of headache is often seen in patients with cerebellar tonsillar ectopia (Chiari malformations) as is downbeating nystagmus. Other symptoms include vertigo, paralysis, disturbance of equilibrium, and sensory disturbances. Long tract signs with hyperflexia, Babinski response, spasticity, and weakness are frequent. Sensory disturbances are common but not very distinct. The Lhermitte's sign is common and indicates sensory disturbances of an electric shock character mostly in the arms or along the vertebral column caused by certain movements of the head and neck. Dissociated sensory disturbance should raise the possibility of associated syringomyelia. Cerebellar ataxia and lower cranial nerve dysfunction can occur, causing dysarthria and dysphagia in some instances. Compression of the vertebral arteries may cause altered consciousness and periods of confusion.

Considering the wide range of symptoms, it is not surprising that a diagnosis of multiple sclerosis is frequently made before the true nature of the problem is determined.

Occipital Bone and Foramen Magnum Anomalies

MRI is excellent for evaluation of the foramen magnum. Axial images can be placed through the true plane of the foramen magnum using a sagittal

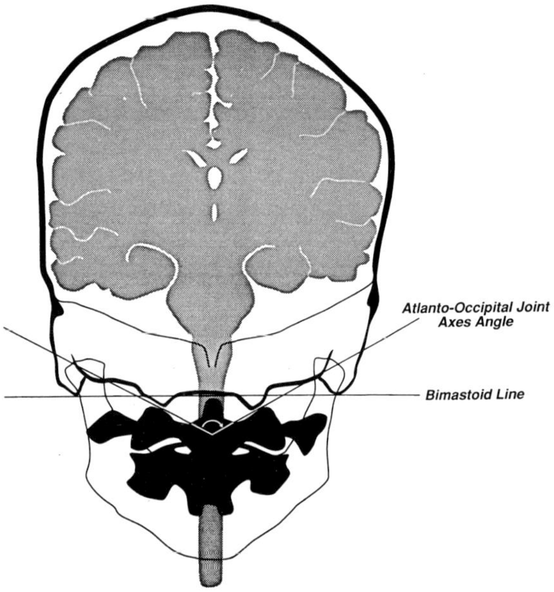

Figure 3.4. Diagram of coronal craniovertebral junction relationships showing atlanto-occipital joint axes angle and bimastoid line.

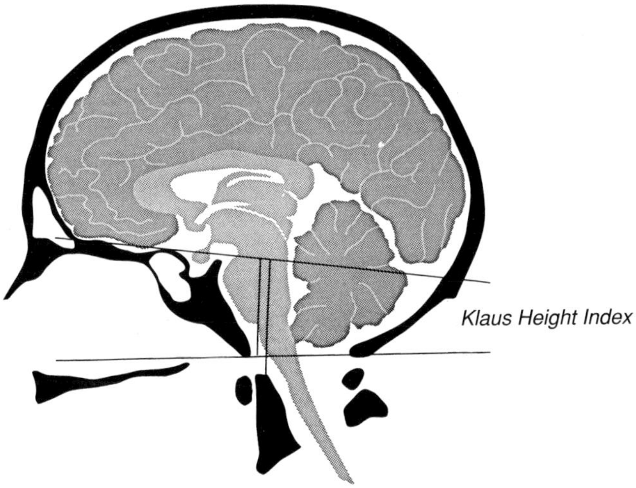

Figure 3.5. Diagram of sagittal craniovertebral junction relationships showing Klaus height index.

image as a "pilot." If the cortical margins are not adequately visualized using T1-weighted techniques, then T2-weighted gradient-echo or spin-echo sequences are recommended. Although the bony margins of the foramen magnum are clearly seen on axial CT scans, the size may be overestimated if the plane of the section is just slightly above the true foramen magnum. This error may be avoided by multiplanar CT re-formations. In general, CT scans with sagittal re-formations are superior to MRI in evaluating bony abnormalities.

OCCIPITAL VERTEBRA

Basiocciput anomalies anterior to the foramen magnum develop when there is persistence of primitive elements of the preatlantal vertebrae. These anomalies include bony excrescences involving the anterior margin of the foramen magnum, an anterior bony arch (hypocondylar arch), and a median or paramedian third occipital condyle which articulates with the dens or anterior arch of the atlas. Various anomalous pseudoarticulations have been reported between the occipital bone and the transverse processes of the atlas [8, 9]. These anomalies usually require CT examination for full evaluation.

FORAMEN MAGNUM STENOSIS AND ACHONDROPLASIA

Foramen magnum stenosis may be developmental or acquired. Paget's disease and disorders of bone metabolism can cause progressive thickening and sclerosis of the skull and may lead to obliteration of basal foramina and foramen magnum stenosis [10]. Stenosis of the foramen magnum may occur as part of a localized anomaly of the craniovertebral junction including axis and atlas development or in association with achondroplasia (Fig. 3.6). Achondroplasia is the most common form of congenital dwarfism. This autosomal dominant disorder is characterized by a decreased rate of endochondral ossification with normal membranous ossification, which results in foramen magnum stenosis. Spinal stenosis is the result of abnormally low vertebral bodies and short pedicles. Since the spinal cord remains of normal size, the subarachnoid space is diminished, and the surrounding epidural fat and connective tissue are almost entirely absent.

The foramen magnum is almost always below age-corrected mean values, but severe stenosis with high cervical cord compression occurs in less than 1% of achondroplastic patients [11]. Symptomatic patients who undergo surgical decompression typically have foramen magnum measurements that are 3 standard deviations or more below the mean [12, 13]. Neurologic deficit is invariably seen if the sagittal diameter of the foramen magnum is 19 mm or less [14, 15]. Stenosis is often most marked in the transverse (side-to-side) orientation and is exacerbated by dysplastic bony expansion of the bony margin of the foramen magnum, especially the opisthion [16]. The cisterna magna is typically not visualized, and the cerebellar tonsils are pressed against the medulla, exacerbating the foramen magnum stenosis in some patients.

Severe stenosis causes spinal cord compression and chronic myelopathy. MRI reveals focal constriction of the spinal cord and loss of cross-sectional cord area. The vascular supply to the cord may be compromised. Occult trauma superimposed on the compressed cord may cause focal myelomalacia, softening, demyelination, and gliosis. These changes can be recognized on T2-weighted MRI images as areas of increased intensity which are most

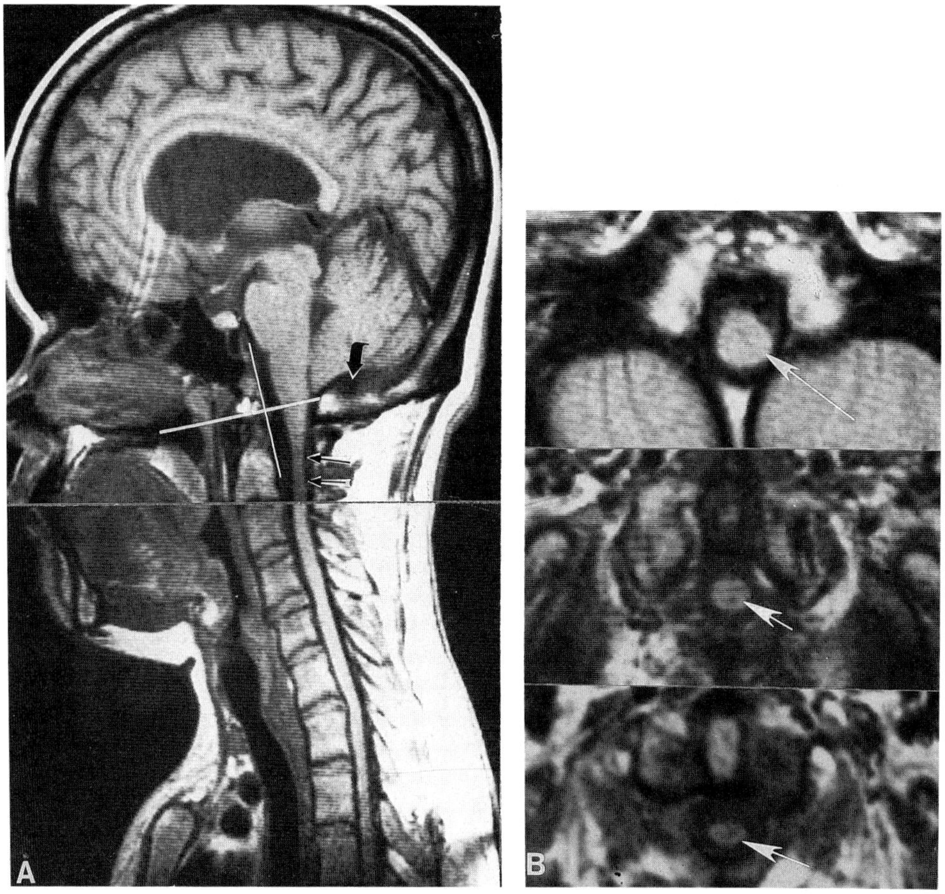

Figure 3.6. Achondroplasia. A 26-year-old woman. Some typical features of achondroplasia are demonstrated, including foramen magnum stenosis, bone hypertrophy, chronic upper cord atrophy/myelopathy, and cervical spine stenosis. **A,** Sagittal image. Note marked hypertrophy of posterior margin of the foramen magnum (opisthion) (*large arrow*). McGregor's and Wackenheim's lines are normal. There is atrophy of upper cervical cord (*arrows*) due chronic foramen magnum stenosis. **B,** Axial images (collage) showing spinal cord (*arrows*). Atrophy of cord is apparent on lowest axial image.

severe in the posterior gray matter and dorsal columns similar to *traumatic central cord syndrome* (Fig. 3.7).

Respiratory problems may be the result of occult spinal cord compression at the cervicomedullary junction [17, 18]. If sedation is administered, the patient must be observed closely and monitored physiologically during and after the procedure.

Odontoid dysplasia in achondroplasia is rare, but C1–2 subluxation due to odontoid hypoplasia may occur. The craniocervical junction is uniformly narrowed [19, 20], and the normal "funnel" shape of this area is lost.

BASILAR INVAGINATION AND PLATYBASIA

Platybasia is a term used to describe an abnormally flattened basal angle. There are no signs or symptoms that can be attributed to platybasia alone. Basilar invagination and platybasia are related findings that often coexist. *Primary* basilar invagination is congenital, and *secondary* basilar invagination (basilar impression) is acquired. Primary basilar invagination can be separated into two types based on the development of the basiocciput. Anterior basilar invagination is associated with basioccipital hypoplasia of variable severity [21]. The clivus is short, and there may be persistence of the spheno-occipital synchondrosis (Fig. 3.8). In the second type of congenital basilar invagination (Dejerine type, convexobasia), the foramen magnum and paramedian occiput are upwardly convex and together with the upper cervical spine invaginate into the posterior fossa [22–24] (Fig. 3.9). Basilar invagination traditionally has been determined radiographically by observing that the odontoid process projects more than 2.5 mm above Chamberlain's line or 4.5 mm above McGregors's line [25–28]. The tip of the dens should not extend above McRae's line (plane of the foramen magnum) [29]. The Klaus height index is reduced to less than 30 mm in basilar invagination [30]. The normal basal angle is less than 140 degrees. This angle is measured at the intersection of lines defined by the nasion, tuberculum sella, and basion (Welcker's basal angle).

Primary basilar invagination may be associated with platybasia, occipitalization of the atlas, condylar hypoplasia, Klippel-Feil anomaly, atlantoaxial dislocation, Chiari malformation [31, 32], syringomyelia, and syringobulbia [33]. MRI is less helpful than CT and other radiographic methods for determining the presence of associated bone anomalies such as partial or complete occipitalization of the atlas [34].

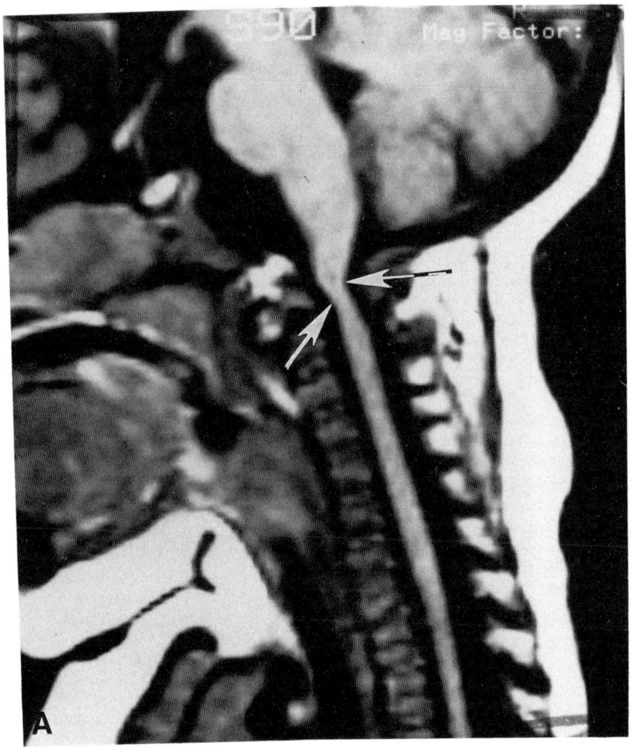

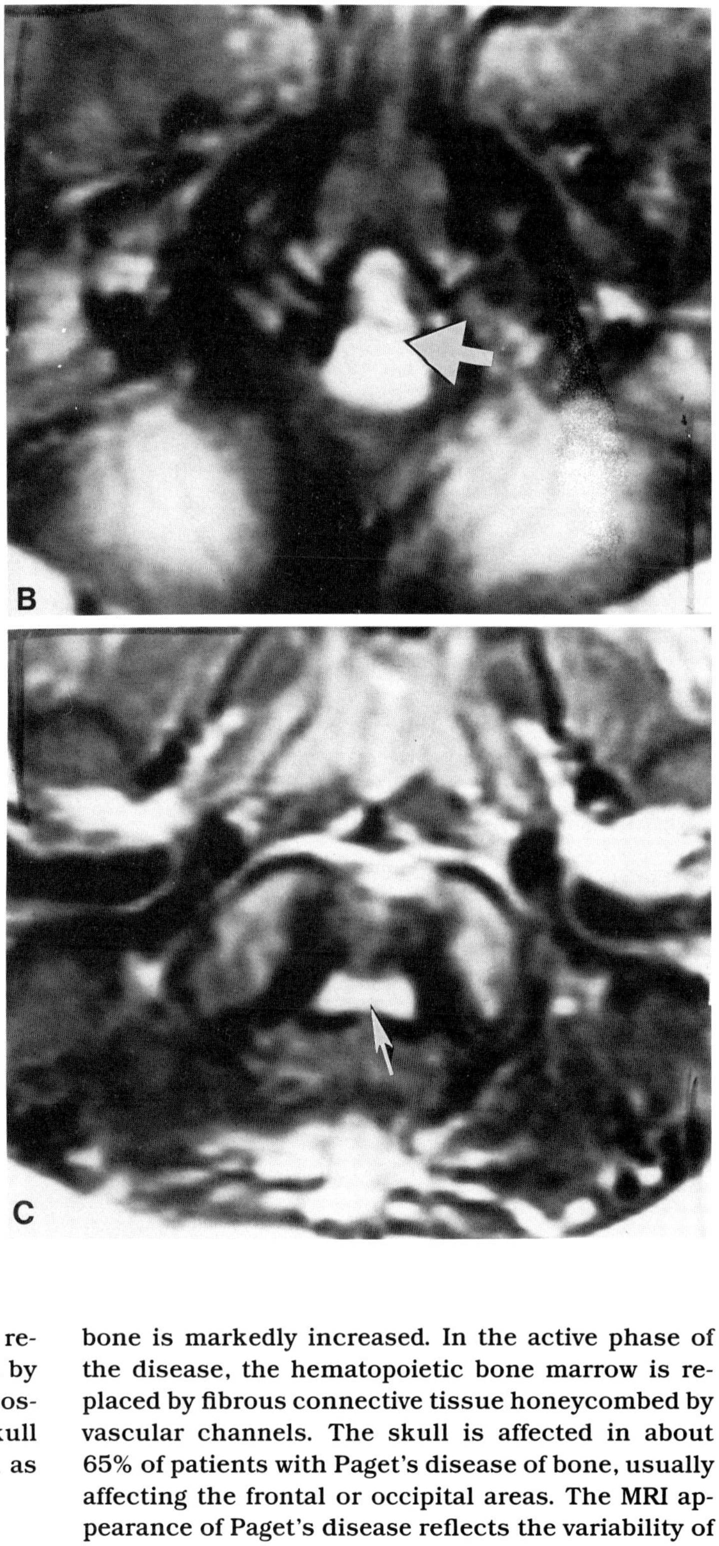

Figure 3.7. **Achondroplasia.** A 6-month-old girl with delayed motor development. Severe foramen magnum stenosis with cervicomedullary compression and myelopathy. **A,** Sagittal image. Note compressive deformity (*arrows*) of spinal cord and cervicomedullary junction. **B,** Axial image through foramen magnum showing keyhole-shaped foramen magnum stenosis and cervicomedullary compression (*arrow*). **C,** Axial image through C1 level. Note stenotic canal in anteroposterior orientation and focal compression of spinal cord (*arrow*).

Secondary basilar invagination is sometimes referred to as *basilar impression.* It can be caused by Paget's disease (Fig. 3.10), rheumatoid arthritis, ossification disorders, fractures of the posterior skull base, and acquired metabolic bone disease such as hyperparathyroidism and osteomalacia.

PAGET'S DISEASE

Paget's disease of the bone is a chronic disorder of the adult skeleton in which there is increased osteolytic and osteoblastic activity and excessive bone remodeling leading to enlargement and softening of the affected bones [35]. The cause of Paget's disease of bone is unknown, but total blood flow through bone is markedly increased. In the active phase of the disease, the hematopoietic bone marrow is replaced by fibrous connective tissue honeycombed by vascular channels. The skull is affected in about 65% of patients with Paget's disease of bone, usually affecting the frontal or occipital areas. The MRI appearance of Paget's disease reflects the variability of the different phases of the disease. Hypointense marrow on short-TR/TE and long-TR/TE images may simulate metastatic disease and lymphoma. High signal areas may reflect hypervascularity in the affect bone. Distinguishing features of Paget's disease include the presence of cortical thickening with a pattern of prominent trabeculation [36].

Secondary basilar invagination due to pagetic re-

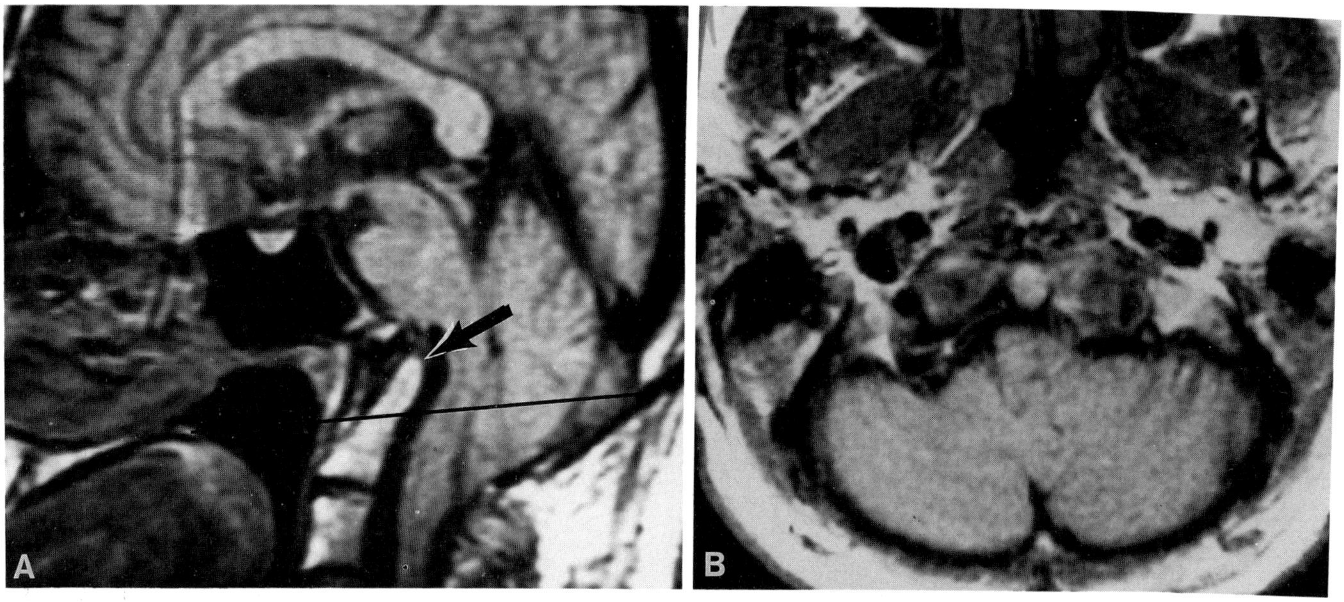

Figure 3.8. Congenital basilar invagination due to hypoplasia of the basiocciput. **A,** Sagittal image. Note short clivus and high anterior arch of C1. The odontoid process is situated well above Chamberlains line (*arrow*) and causes impingement of brainstem at pontomedullary junction. **B,** Axial image.

Figure 3.9. Severe basilar invagination due to osteogenesis imperfecta tarda in a 26-year-old man. **A,** Sagittal image showing calvarial thickening, convex deformation of the skull base, impingement of pons by odontoid process (*wavy arrows*), and cervical syringomyelia (*arrow*). **B,** Coronal image showing marked convexity of skull base.

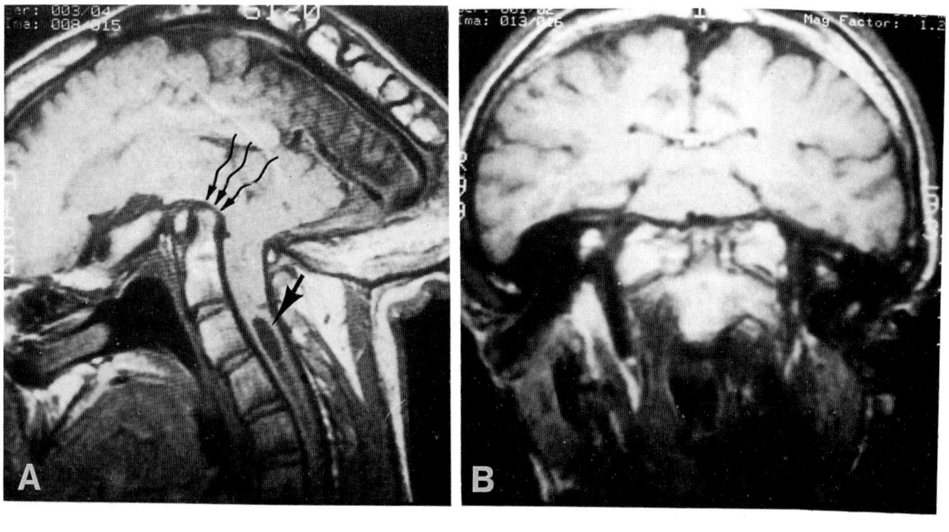

Figure 3.10. Secondary basilar invagination due to Paget's disease. A 76-year-old woman with severe involvement of cranium. Note markedly thickened calvarium (*white arrows*). **A,** Sagittal image demonstrates convex skull base due to bone softening from Paget's disease. Note brainstem impingement by dens (*arrow*). **B,** Coronal image.

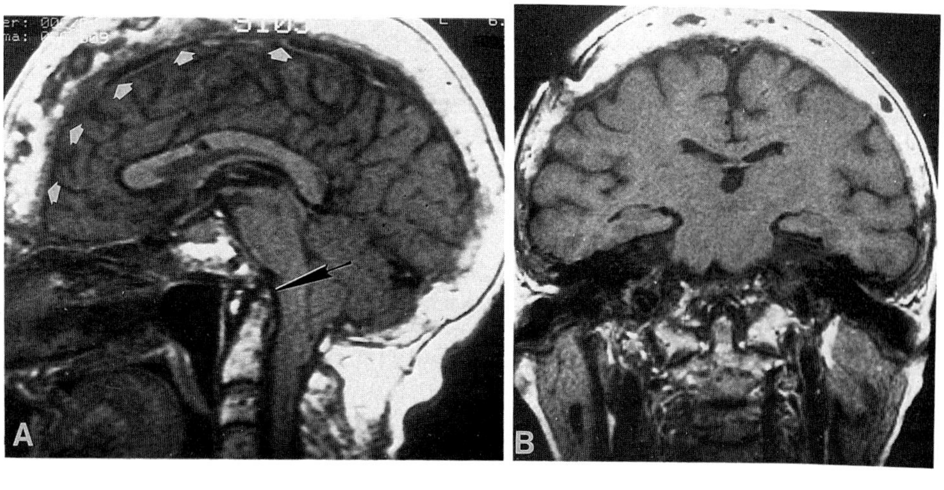

modeling and bone softening is a well-known development occurring in up to one-third of patients with Paget's diseases of the skull [37].

Atlas Anomalies

OCCIPITALIZATION OF THE ATLAS

Occipitalization of the atlas (atlas assimilation) is a failure of segmentation between the skull and the atlas (Fig. 3.11). This is one of the most common congenital anomalies of the craniovertebral junction and is quite variable, ranging from complete incorporation of the atlas into the occiput to partial fusion of one of the arches or lateral masses. CT with sagittal and coronal reformations is optimal for evaluation of this anomaly, and MRI is optimal for the associated anomalies involving the cervicomedullary junction.

This anomaly is usually asymptomatic but is frequently associated with other abnormalities such as basilar impression, Klippel-Feil syndrome, occipital vertebra, Chiari I malformation, and syringomyelia [38]. However, there may be ligamentous incompetency, allowing atlantoaxial subluxation. This is seen particularly in patients in whom there is associated fusion of C2 and C3. Approximately 50% of these patients gradually develop atlantoaxial subluxation [39].

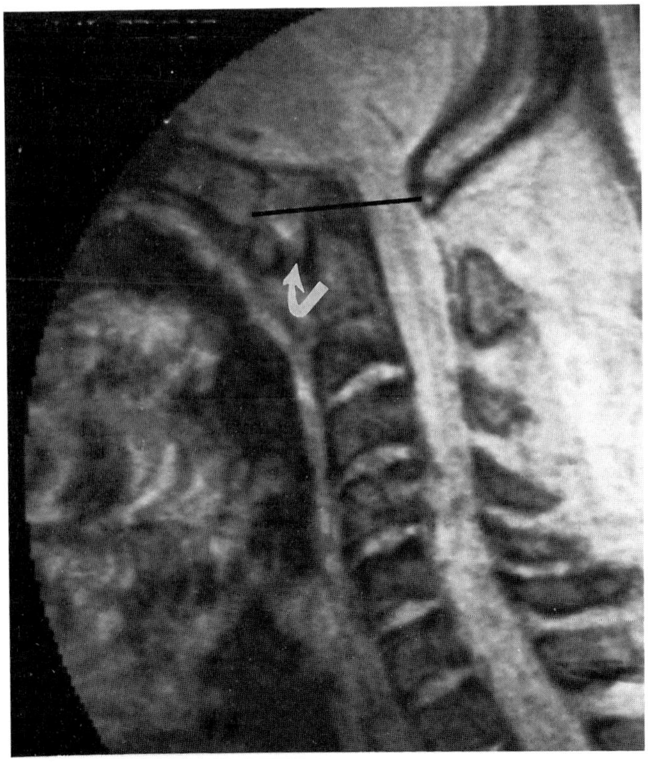

Figure 3.11. Occipitalization of C1 with short clivus, basilar invagination, and foramen magnum stenosis (note McRae's line). C1–2 subluxation (*curved arrow*) due to incompetency of the transverse ligament. Note absence of posterior arch of C1. Sagittal image.

ATLAS RACHISCHISES

There may be aplasia or hypoplasia of the arch. Segmentation defects (rachischises) may affect the posterior arch, anterior arch, or both (split atlas).

ATLANTOAXIAL FUSION

This involves congenital fusion of the anterior arch of the atlas with the dens. This is the equivalent of "block vertebrae" that are typically seen at mid-cervical levels and may be very variable in extent [40] (Figs. 3.12 and 3.13).

PONTICULUS POSTICUM

The ponticulus posticum is a semilunar ossification projecting posteriorly from the articular edge of the atlas superior articular facet. The anomaly is also known as *Kimmerle's deformity*. The projection is variable and may unite with the adjacent neural arch of the atlas to produce the "arcuate foramen" through which the vertebral artery passes.

Odontoid Anomalies and Atlantoaxial Subluxation

Maldevelopment of the odontoid process is most frequently identified as an isolated phenomenon such as os odontoideum. However, odontoid dysplasia has been found in at least eight different syndromes, including Morquio syndrome and other mucopolysaccharidoses, spondyloepiphyseal dysplasia congenita, spondylometaphyseal dysplasia, metatropic dysplasia, pseudoachondroplasia, diastrophic dwarfism, and Conradi's disease [41]. These syndromes become significant when they cause atlantoaxial instability. The criteria for diagnosis of atlantoaxial subluxation are given in Table 3.4.

Atlantoaxial subluxation may be due to developmental or acquired incompetence of the ligaments (transverse, alar, and accessory ligaments) (Fig. 3.14) or loss of integrity of the odontoid process or anterior arch of the atlas. Several conditions are associated with atlantoaxial subluxation, including trauma, inflammatory arthritis (especially rheumatoid arthritis), bacterial inflammations, retropharyngeal abscess [42]. Down syndrome, os odontoideum, tumors, and bone dysplasias [43–46].

Atlantoaxial instability may not be apparent at the time of examination [47], and the neurologic deficit which results from atlantoaxial dislocation is variable, depending on the rate of development of the dislocation [48]. Chronic dislocation may lead to intermittent or low-grade constant cord compression. In some patients, cord compression is caused by a fibrous pseudotumor that forms at the C1–2 junction secondary to chronic atlantoaxial subluxation

Figure 3.12. Congenital partial fusion of C1 and C2 and partial occipitalization of C1. **A,** Plain radiographs in flexion in the lateral view, and in frontal projection. **B,** Coronal T1-weighted MRI. The lateral mass of C1 on one side is fused with the occipital condyle (*arrowheads*). On the opposite side the deformed odontoid process (*wavy arrows*) is fused with the arch and lateral mass of C1 (*arrows*). **C,** Axial T1-weighted MRI demonstrating the fusion anomaly between C1 and C2 (*arrow*). **D,** CT in axial plane better demonstrates the fusion of the odontoid process with the arch and lateral mass of C1 (*arrows*).

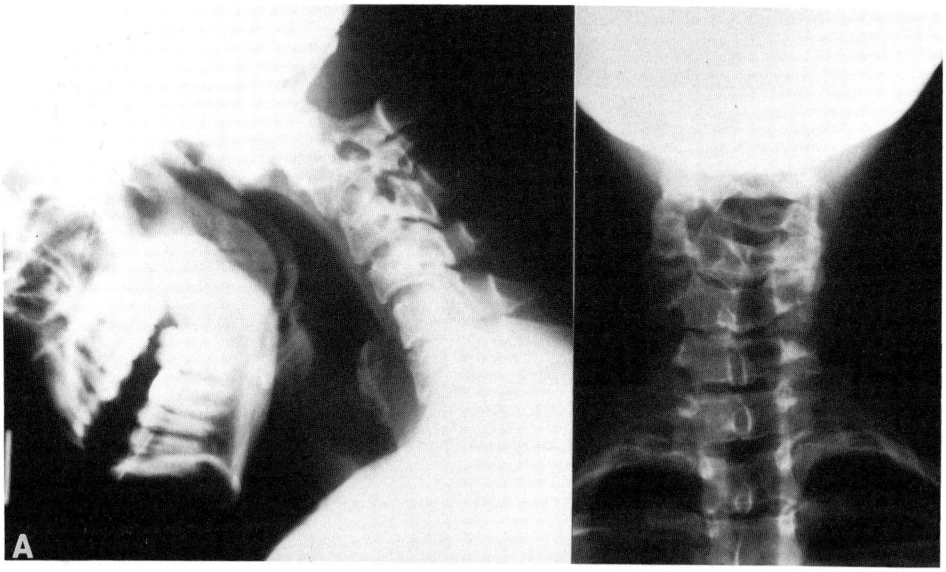

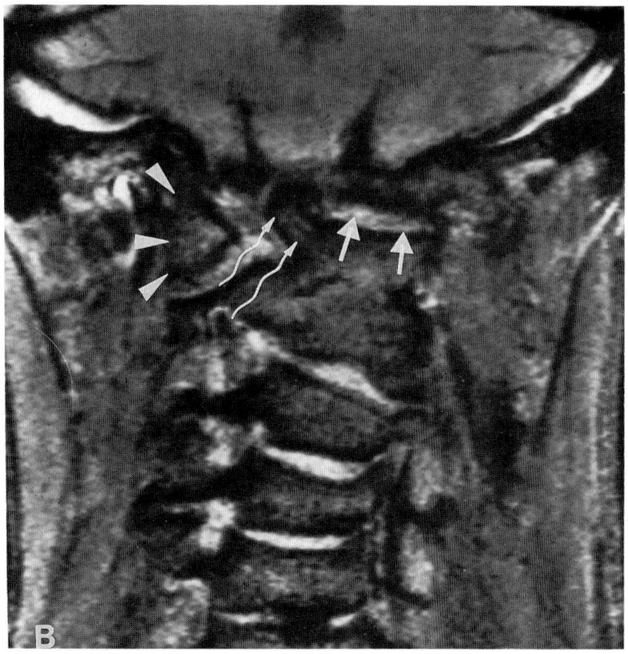

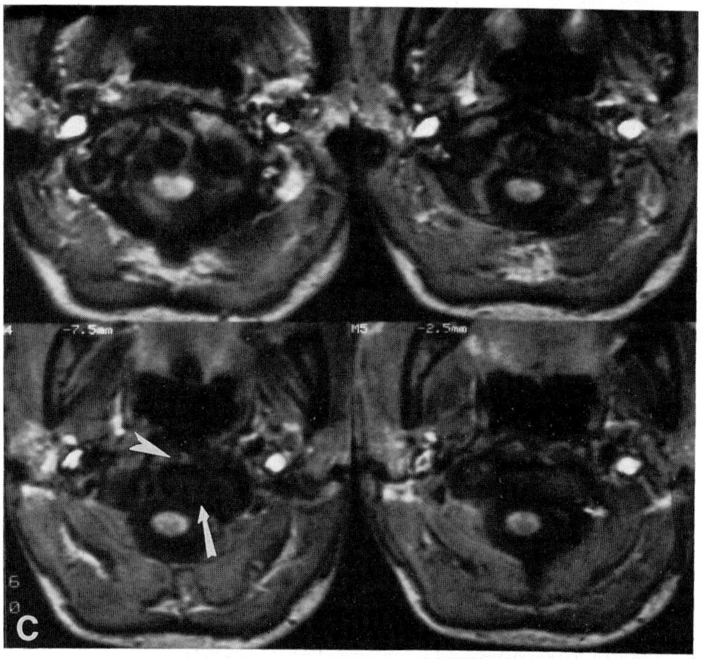

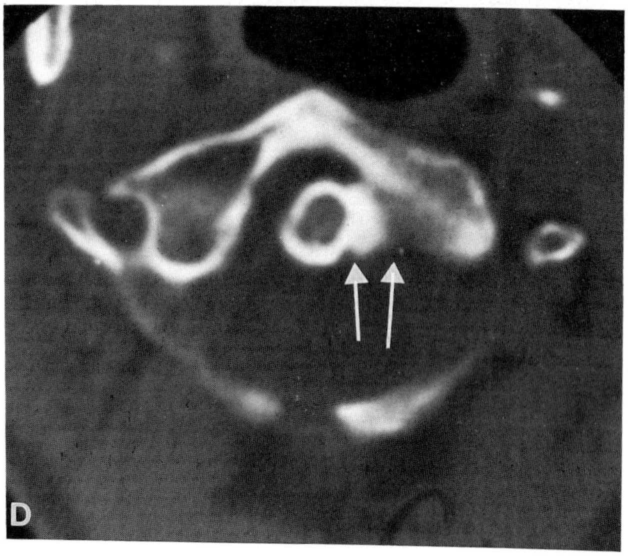

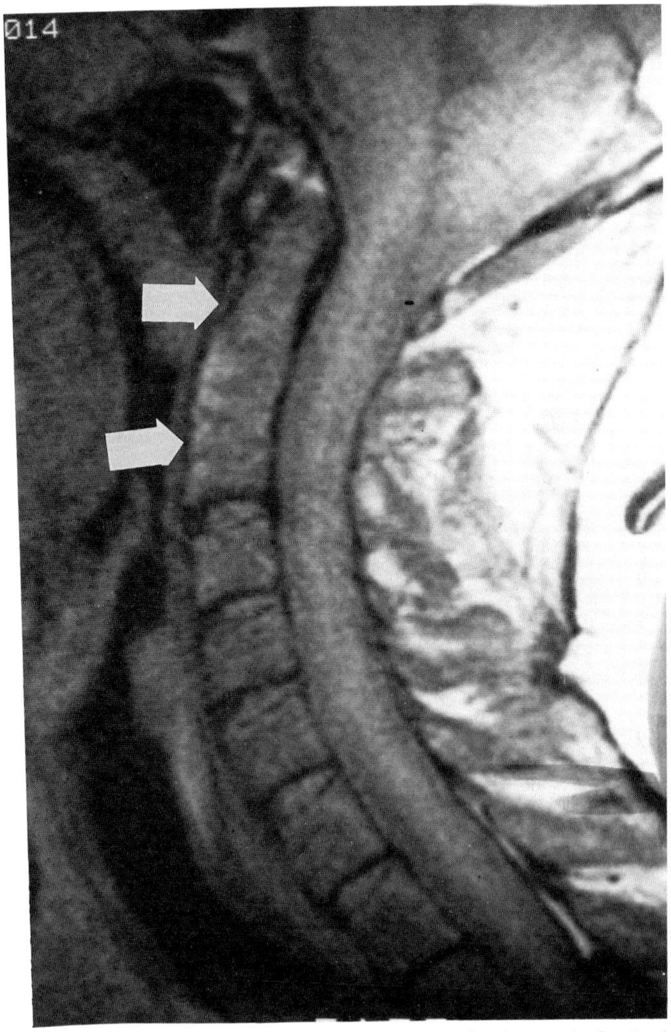

Figure 3.13. Congenital fusion of C2 and C3 (*arrows*). Sagittal image.

[49]. The chronic fibrous mass is of intermediate intensity on T1-weighted images and generally low in signal intensity of T2-weighted images. The appearance of the fibrous pseudotumor may simulate rheumatoid arthritis. Not all patients with atlantoaxial instability have myelopathy [50]. Images in flexion and extension are required to allow demonstration of the full range of instability and to identify positionally dependent cord encroachment [51]. Dislocation is more apt to occur in flexion, resulting in anterior displacement of the atlas in relationship to the axis. The spinal cord may be indented ventrally by the body of C2 and compressed posteriorly by the arch of the atlas.

OSSICULUM TERMINALE

A persistent ossiculum terminale (Bergman's ossicle) is formed by failure of fusion of the terminal ossicle of the odontoid process. This minor anomaly is not associated with instability and is not considered clinically significant.

Table 3.4.
Subluxation Criteria

Anterior atlantoaxial subluxation is present when the atlantodental interval is greater than 2.1 mm.
Subaxial subluxation is present when there is an offset between adjacent vertebrae of more than 1 mm or more than 15% of the anteroposterior diameter of the vertebral body [115].
Rotatory atlantoaxial subluxation is present when there is unilateral offset of the lateral mass of C1 in relation to the articular pillar of C1.
Vertical atlantoaxial subluxation (cranial settling, atlantoaxial impaction, acquired basilar impression) is present if the superior aspect of the intact dens is even with or above the McCrae line (foramen magnum). The anterior arch of C1 is more closely related to the base of C2 [116].

OS ODONTOIDEUM

Os odontoideum is the most common odontoid anomaly (Figs. 3.15 through 3.17) and represents either a failure in development and fusion of the primary or secondary ossification center of the odontoid or the sequelae of trauma to the odontoid in early childhood [52–54]. Most patients with os odontoideum have instability and cervical pain, but only one-third of the patients have signs or symptoms indicative of neural deficits [55]. Anterior atlantoaxial instability is much more common than posterior instability [56]. Progressive cervical myelopathy is the most serious of clinical symptoms and signs in this condition. It is thought to be caused by compression of the spinal cord due to atlantoaxial instability [57]. Chronic instability leads to hypertrophic changes in the ligaments and other soft tissues around the odontoid, stenosis of the adjacent spinal canal, and in some patients, chronic myelopathic changes in the spinal cord.

The typical configuration of this abnormality allows differentiation from acute fracture of the odontoid, but it is much more difficult to exclude a chronic distracted nonunited odontoid fracture. The lesion consists of an ovoid or round ossicle with a smooth, well-corticated margin separate from the axis. It is of variable size and is usually located in the position of the normal odontoid tip or near the basion. The gap between the os odontoideum and the lower part of the C2 vertebral body is wide and lies above the level of the superior articular facet of the axis. The fragment is usually about one-half the normal size of the odontoid process. The anterior arch of C1 is typically hypertrophic, a useful observation when it is necessary to exclude acute fracture.

ODONTOID HYPOPLASIA

Aplasia of the odontoid is an extremely rare anomaly, whereas hypoplasia of the odontoid is more common and represents failure of development of the

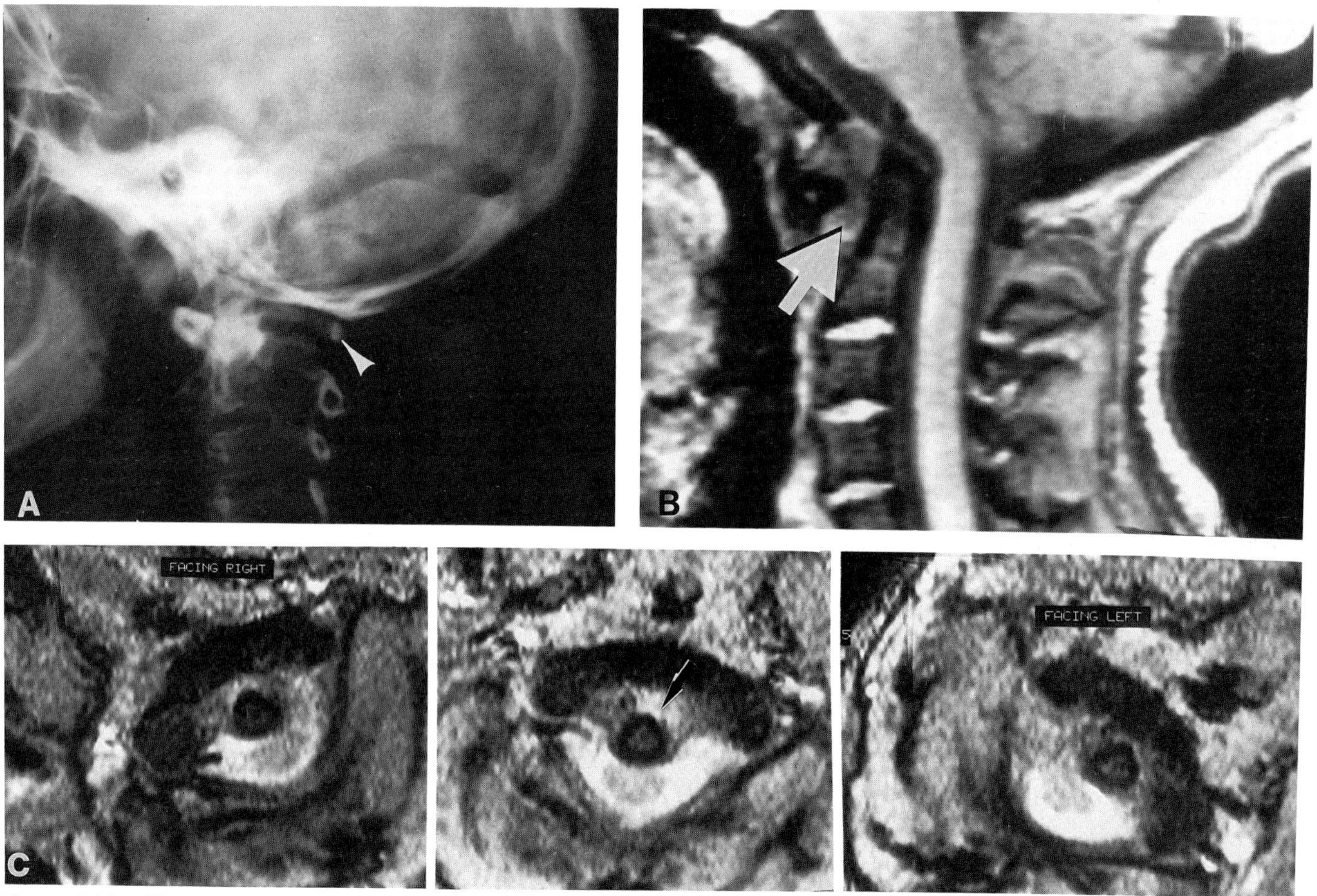

Figure 3.14. Hypoplasia of C1 with incompetency of transverse ligament causing C1–2 subluxation. **A,** Lateral radiograph shows hypoplastic posterior arch (*arrowhead*). **B,** Sagittal image shows C1–2 subluxation (*large arrow*) and synovial pseudomass anterior to odontoid process. **C,** Axial kinematic study demonstrates C1–2 subluxation (*arrow*).

Figure 3.15. Os odontoideum. A 33-year-old woman without a history of trauma. Note detached tip of dens (*arrow*), "squared of" hypertrophic anterior arch of C1 (*arrowheads*), and hypertrophy of periarticular tissues. **A,** Extension image. **B,** Flexion sagittal image. Note subluxation of C1–2.

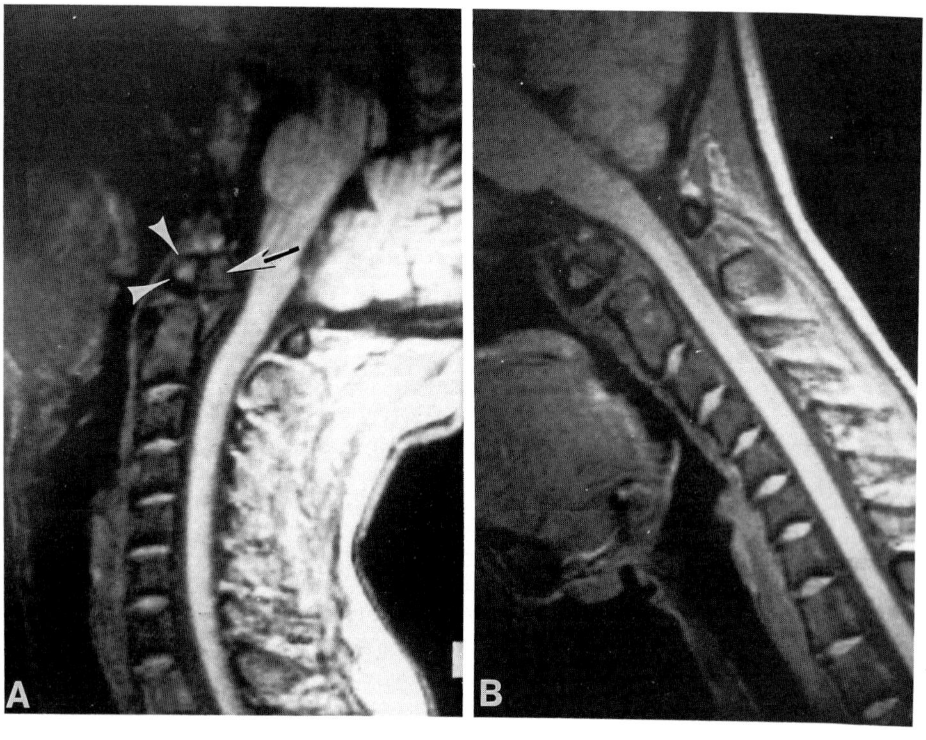

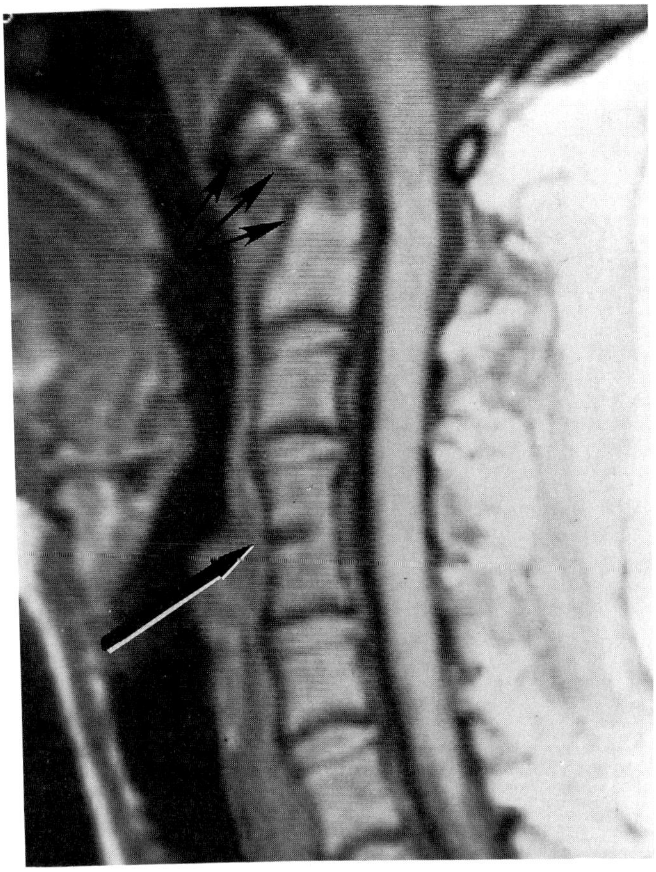

Figure 3.16. Os odontoideum with subluxation (*small arrows*) and fusion of C4 and C5 (*large arrow*). A 25-year-old woman without a history of trauma. Note large anterior arch of C1 and poor definition of tip of odontoid.

dens. Odontoid hypoplasia/dysplasia is frequently seen in patients with mucopolysaccharidosis (especially Morquio syndrome), spondyloepiphyseal dysplasia congenita, and occasionally, other skeletal dysplasia syndromes.

MORQUIO SYNDROME (MUCOPOLYSACCHARIDOSIS IV)

Osseous deformity is found in all types of the mucopolysaccharidoses (Table 3.5) and is frequently manifested by atlantoaxial instability. Patients with Morquio syndrome are the most likely patients with any of the mucopolysaccharidoses to be evaluated for neuraxis compression (Fig. 3.18). In this autosomal recessive disorder there is a deficiency of galactoside-6-sulfatase, which leads to accumulation of keratin sulfate in susceptible tissues. There are two variations of the Morquio syndrome: type A and type B. Morquio B patients are less likely to have an anomalous odontoid. In Morquio A, the odontoid is hypoplastic or absent, and atlantoaxial instability is worsened by ligamentous laxity [58, 59].

Atlantoaxial instability with myelopathy of the upper cervical cord is the most severe complication of Morquio syndrome, resulting in frequent paraplegia [60]. Instability is caused by the combination of the aplastic odontoid and severe ligamentous laxity. Upper cervical canal stenosis is often present.

Progressive myelopathy can be arrested and some improvement achieved by reduction of the atlantoaxial subluxation and posterior occipitocervical fusion. Surgery is indicated when myelopathy is confirmed [61].

SPONDYLOEPIPHYSEAL DYSPLASIA CONGENITA

This is a variably expressive autosomal dominant disorder characterized by retarded ossification of the spine and proximal femora. Hypoplasia of the odontoid and ligamentous laxity with atlantoaxial instability (Fig. 3.19) are fairly constant findings in spondyloepiphyseal dysplasia congenita (SEDc). Cervical myelopathy due to instability appears much earlier in SEDc than in the Morquio syndrome. Respiratory symptoms of myelopathy may occur in the newborn or infant [62]. Instability is frequently treated by performing an occipital–C1–2 posterior fusion (Fig. 3.20).

Miscellaneous Skeletal Dysplasias

There are a variety of uncommon skeletal dysplasias associated with craniovertebral maldevelopment other than those detailed above. They include spondylometaphyseal dysplasia, pseudoachondroplasia, Conradi disease, Kniest dysplasia, metatrophic dwarfism, diastrophic dwarfism, Scott's syndrome, cartilage-hair syndrome, and neurofibromatosis type II [63–65] (Figs. 3.20 through 3.24). Camurati-Englemann disease may cause brainstem compression due to foramen magnum stenosis [66].

Inflammatory Disease

Atlantoaxial dislocation may be caused by inflammatory processes around the craniovertebral region. Pyogenic infections, tuberculosis, and the inflammatory arthridites have all been implicated. Infectious subluxation of the atlantoaxial joint, known as *Grisel's syndrome*, may present as torticollis. It usually presents in children but has been reported in adults [67]. The infection usually starts in the tonsils, pharynx, or middle ear and then spreads to the cervical vertebrae. Neurologic complications occur in approximately 15% and include readiculopathy, myelopathy, and sudden death.

RHEUMATOID ARTHRITIS

Rheumatoid arthritis is a systemic inflammatory disease of unknown etiology characterized by a chronic inflammatory and proliferative reaction of the synovial lining of the bursae and articular capsules. The inflamed and hypertrophic tissue

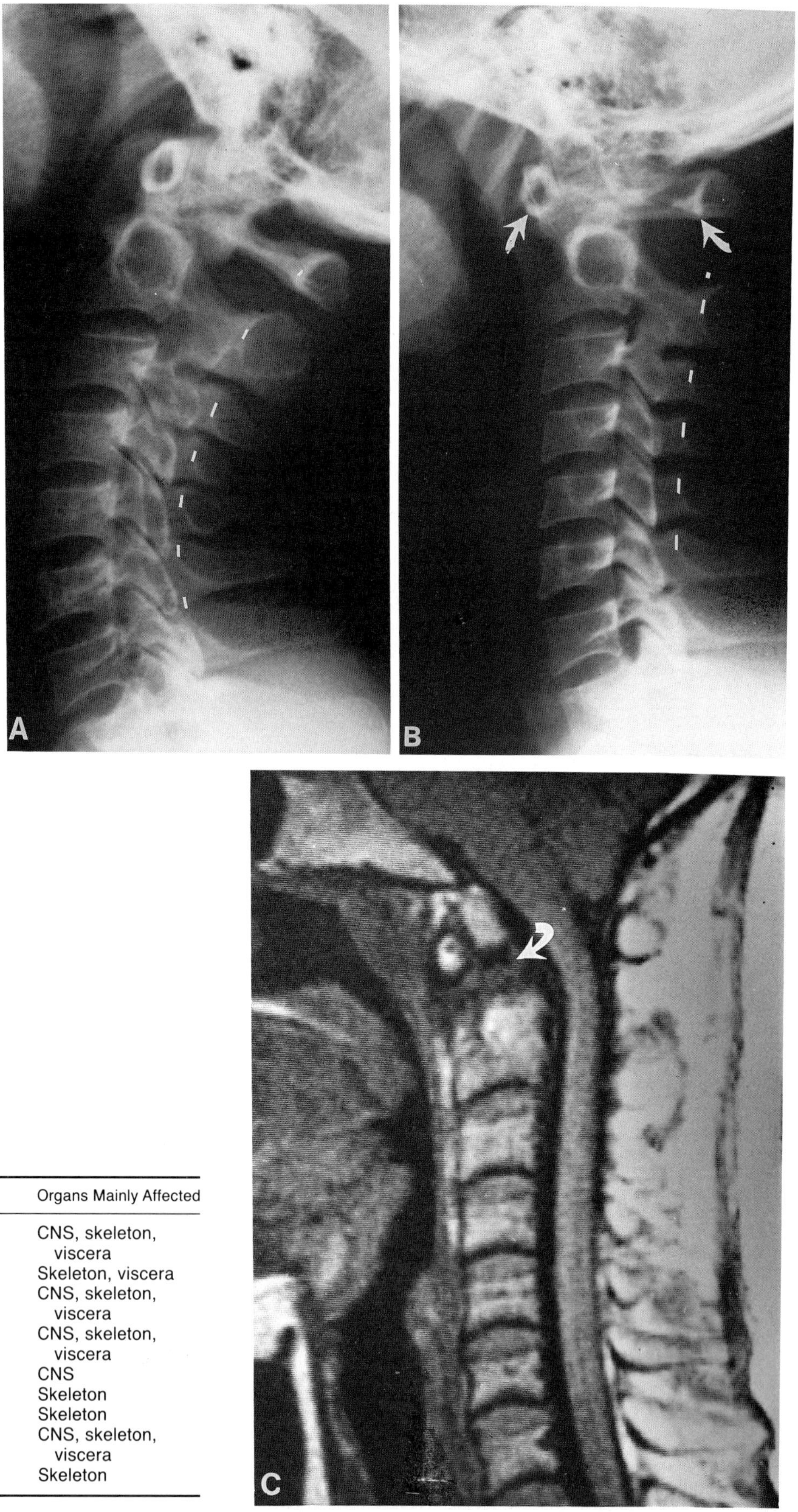

Figure 3.17. Os odontoideum with subluxation in a 39-year-old woman. **A,** Extension lateral radiograph showing enlarged anterior arch of C1 and poor definition of odontoid. Dashed-line is spinolaminar line. **B,** Flexion lateral radiograph demonstrating anterior subluxation of C1. **C,** Sagittal MRI image showing separation of tip of odontoid from base of ondontoid (*curved arrow*) and enlarged anterior arch of C1.

Table 3.5.
The Mucopolysaccharidoses

Mucopolysaccharidosis	Designation	Organs Mainly Affected
Hurler	MPS 1-H	CNS, skeleton, viscera
Scheie	MPS 1-S	Skeleton, viscera
Hurler/Scheie	MPS I H/S	CNS, skeleton, viscera
Hunter	MPS II	CNS, skeleton, viscera
Sanfilippo (A–D)	MPS III (A–D)	CNS
Morquio (A,B)	MPS IV (A,B)	Skeleton
Maroteaux-Lamy	MPS VI	Skeleton
Sly	MPS VII	CNS, skeleton, viscera
DiFerrante	MPS VIII	Skeleton

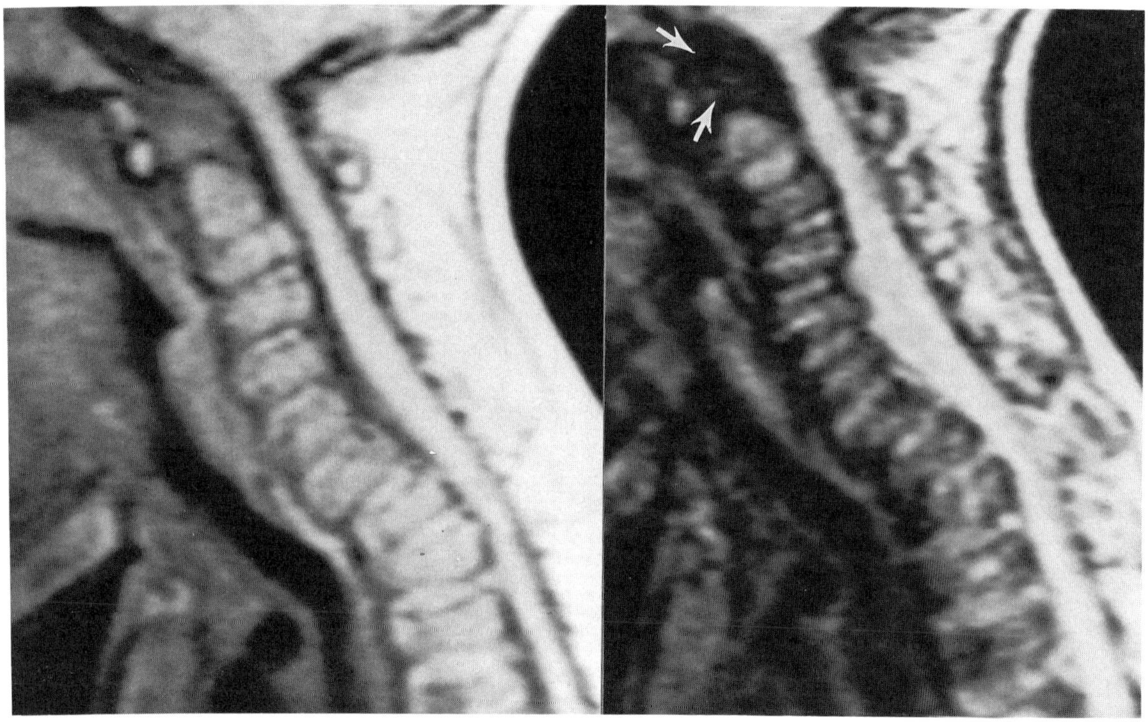

Figure 3.18. Odontoid hypoplasia/dysplasia in Morquio syndrome in a 7-year-old boy. Note hypointense appearance of dysplastic ondontoid cartilage on T2-weighted image (*arrows*). There is stenosis of cervicocranial junction down to C3 level.

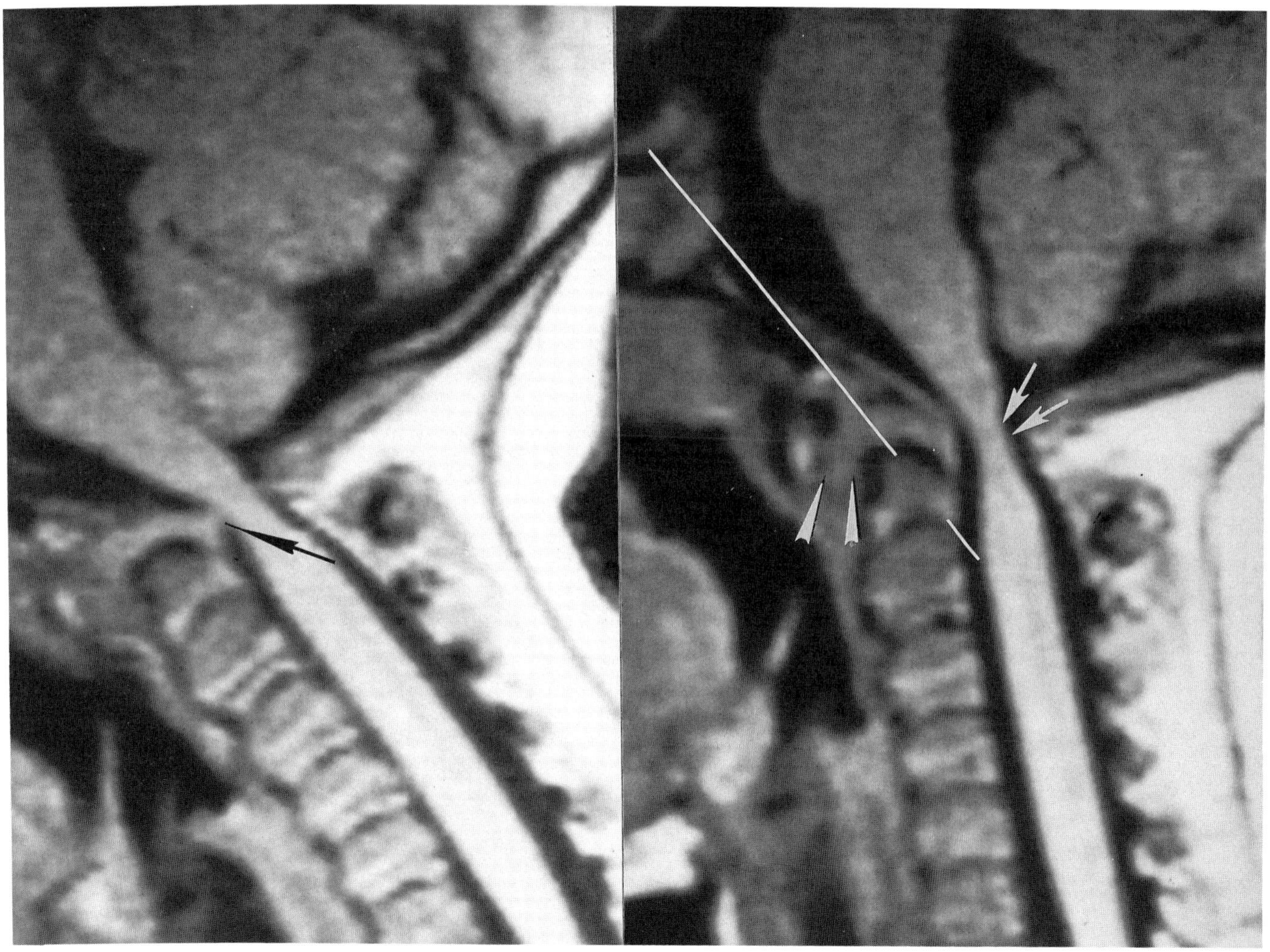

Figure 3.19. Odontoid hypoplasia in spondyloepiphyseal dysplasia congenita in a 4-year-old girl with chronic myelopathy. Note "waistlike" narrowing of spinal cord (*arrows*), abnormal relationships of Wackenheim's line (*white line*), and C1–2 subluxation (*arrowheads*). The subluxation has caused stenosis of the cervicocranial junction. **A,** Extension sagittal image. **B,** Flexion sagittal image demonstrates increased subluxation and cord impingement (*black arrow*).

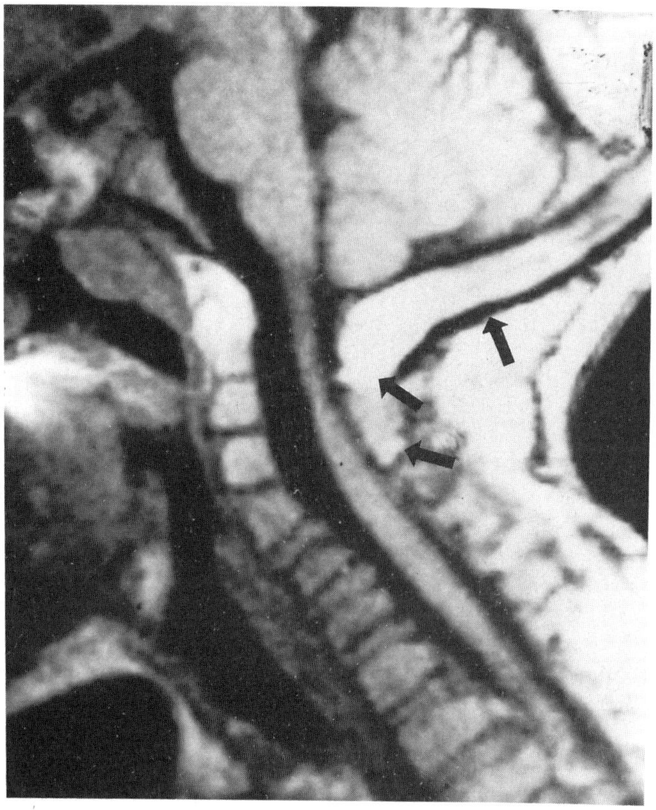

Figure 3.20. Treated odontoid hypoplasia in a 15-year-old girl with spondyloepiphyseal dysplasia congenita and severe chronic myelopathy, stable since age 3. An occipitocervical fusion has been performed (*arrows*). Note chronic cord atrophy.

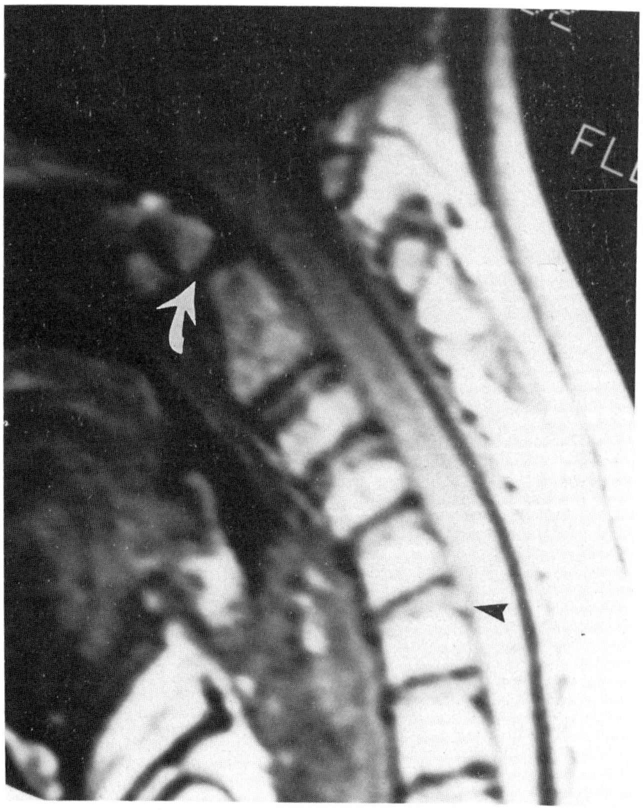

Figure 3.21. Pseudoachondroplasia with os odontoideum in a 33-year-old woman. Note os odontoideum with minimal instability (*curved white arrow*). Symptoms of neck and arm pain are thought to be due to disc protrusion at C5–6 (*arrowhead*).

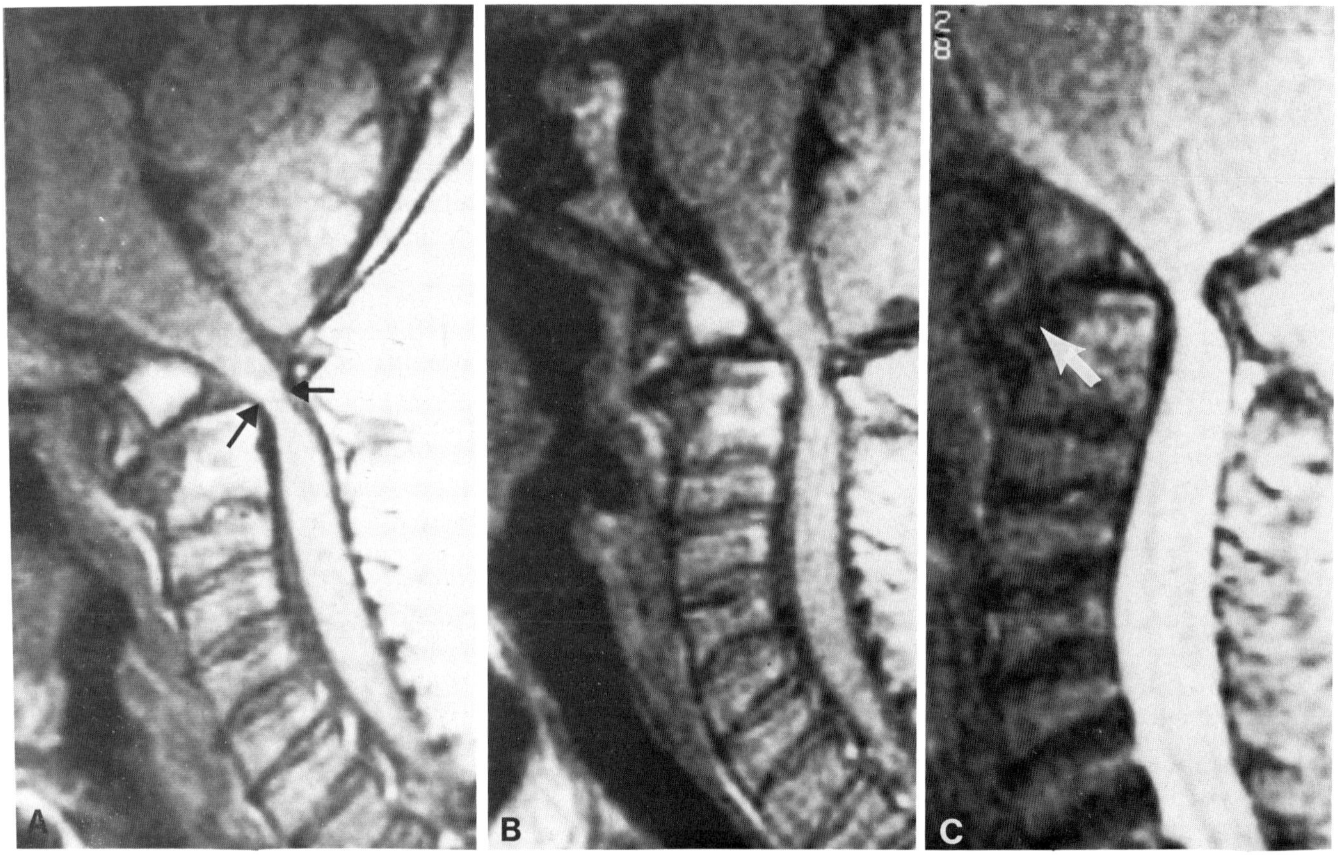

Figure 3.22. Spondylometaphyseal dysplasia in a 27-year-old woman with unstable os odontoideum. Note difference between flexion and extension images. MRI finding of focal cord narrowing suggests chronic myelopathy, but surprisingly, this patient has only mild hyperreflexia. **A,** Flexion T1-weighted image. Cord appears compressed (*arrows*). **B,** Extension T1-weighted image. Note partial reduction of subluxation. **C,** Neutral T2-weighted image. Chronic fibrous pseudomass anterior to odontoid appears hypointense (*white arrow*).

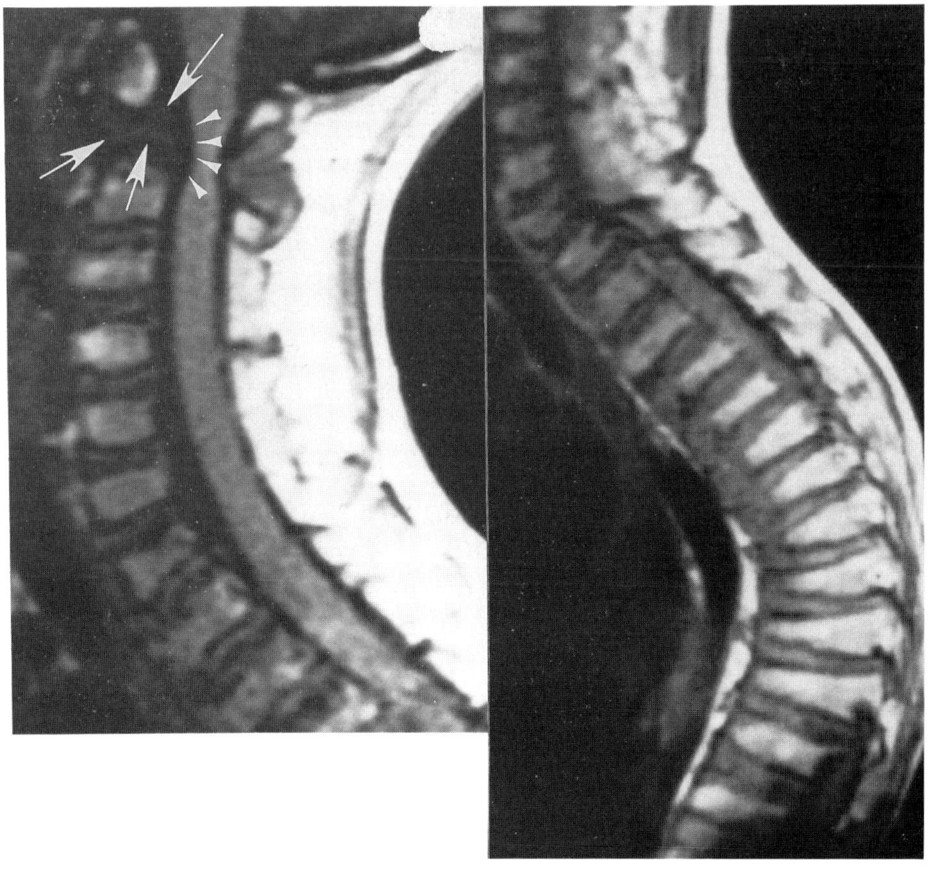

Figure 3.23. Metatrophic dwarfism in a 7-year-old boy with odontoid hypoplasia (*arrows*), C1–2 instability, and significant compression of spinal cord (*arrowheads*). Symptoms primarily related to C2 level, although thoracic stenosis is also present. Note broad, anteriorly flattened vertebrae.

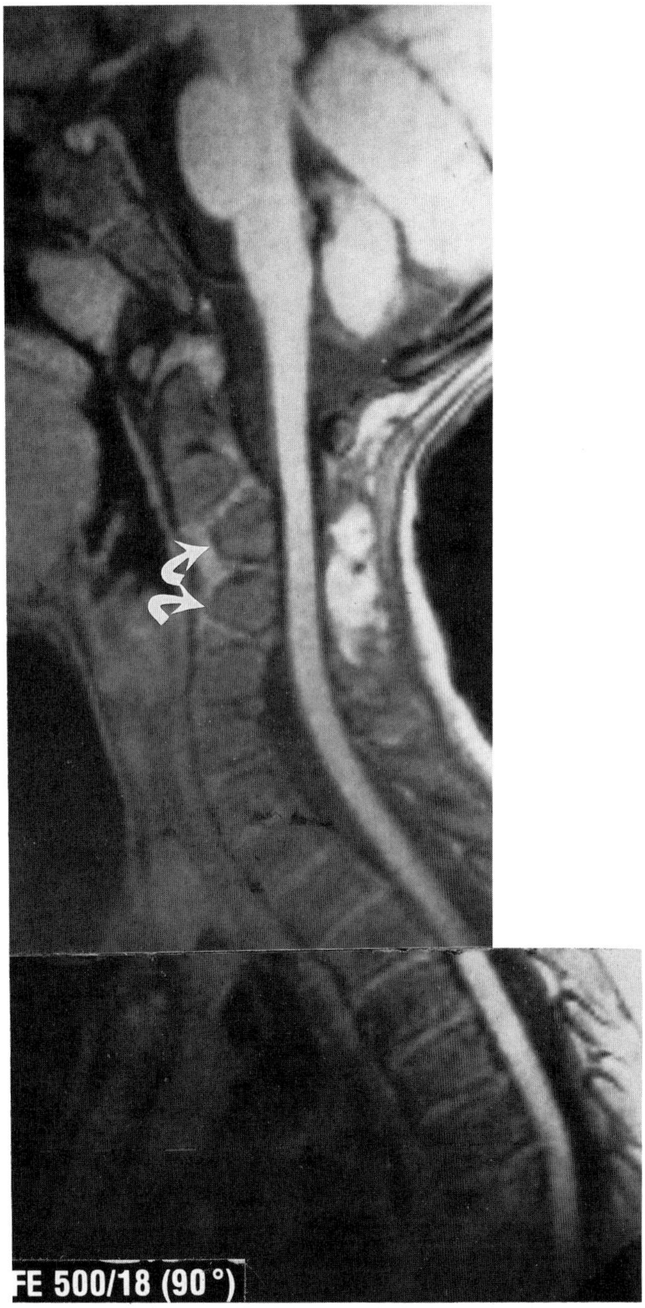

Figure 3.24. Diastrophic dwarfism in a 4-year-old boy. Odontoid is incompletely formed. The deformity at C3 and C4 (*curved arrows*) is a typical finding in diastrophic dwarfism and causes focal stenosis, but there were no signs of instability and no clinical evidence of a myelopathy.

(pannus) is often associated with destruction of articular cartilage and subchondral bone. Eroding pannus and hyperemic effects cause ligamentous laxity commonly leading to instability and atlantoaxial subluxation [68, 69] (Figs. 3.25 through 3.28).

Pannus may form in the retrodental, predental, supradental, and laterodental areas [70]. Some patients with chronic atlantoaxial subluxation develop fibrous granulation tissue and hypertrophy of connective tissue elements as an abnormal response to chronic stresses and friction [71]. This periodontoid fibrous pannus has been shown to decrease in size when the subluxation is stabilized by cervical fusion [72]. Granulomatous or fibrous pannus can prevent full reduction of subluxation by occupying the predental space. Erosion and collapse of the occipitocervical and atlantoaxial facet joints may cause cranial subluxation (cranial settling), allowing the dens to invaginate into the foramen magnum [73].

The goal of imaging in rheumatoid arthritis is (*a*) to assess the effect of the disease on the spinal cord and cervicomedullary junction, (*b*) to determine if instability is present, and (*c*) to identify the areas involved by the disease and document the effects of the disease on articular surfaces and soft tissue for treatment planning and prognosis. MRI of the odontoid is frequently superior to radiographic evaluation because superimposition of bony structures and osteopenia obscure bony detail on the radiographs. High-resolution CT is optimal for cortical bone detail but is usually restricted to axial sections (supplemented by computer reformations). Patients with significant pathology involving the craniovertebral junction are usually not good candidates for direct coronal CT because of the risk and difficulty of the hyperextended position.

Dislocations which may occur between C1 and C2 (atlantoaxial) or below C2 (subaxial) cause compression of the spinal cord and/or medulla oblongata. There are seven synovial spaces at the craniocervical junction that can be affected by rheumatoid arthritis: the paired atlanto-occipital and lateral atlantoaxial joints, the predental and postdental synovial joints, and, variably present, a small bursa between the membrana tectoria and the median part of the transverse ligament of the atlas [74]. Approximately 30% of all rheumatoid patients develop a subluxation of the cervical spine [75, 76]. Five types of subluxation are possible at the C1–2 level (Table 3.6).

Flexion and extension views demonstrate the full range of instability and may reveal subluxation in flexion which reduces completely in extension [77]. Flexion/extension sagittal MRI, may be recommended to help determine whether focal cord narrowing represents reversible cord compression or irreversible chronic myelopathic atrophy.

Mechanical compression, arteritis, rheumatoid pachymeningitis, or constricting epidural pannus have direct effects on the spinal cord and also may lead to microvascular compression or major vessel occlusion [78]. In mild cord compression, areas of demyelination have been seen in the lateral corticospinal tracts [79]. With increasing severity of compression, demyelination is more marked, there is neuronal loss, gliosis, microcyst formation, necrosis, and cavitation. Secondary vascular disturbances occur in the central gray matter and ventral poste-

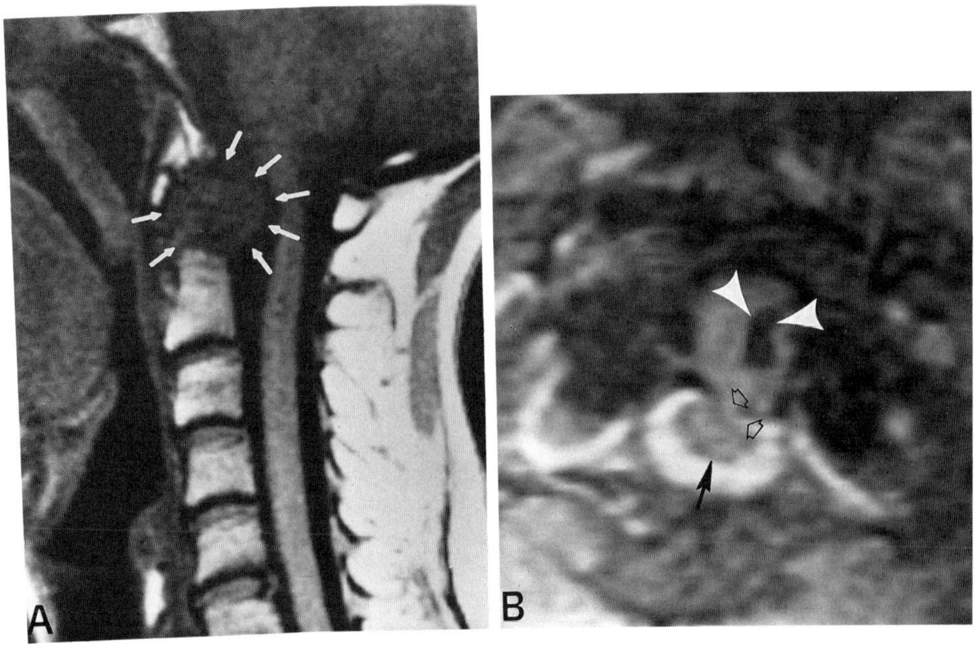

Figure 3.25. Rheumatoid arthritis in a 54-year-old woman. **A,** Sagittal image shows marked erosion of odontoid process and pannus causing a chronic synovial pseudomass which compresses the spinal cord. **B,** T2*-weighted axial MRI image demonstrates erosion of odontoid process (*arrowheads*) and compression of spinal cord (*arrow*) by panus (*small arrows*).

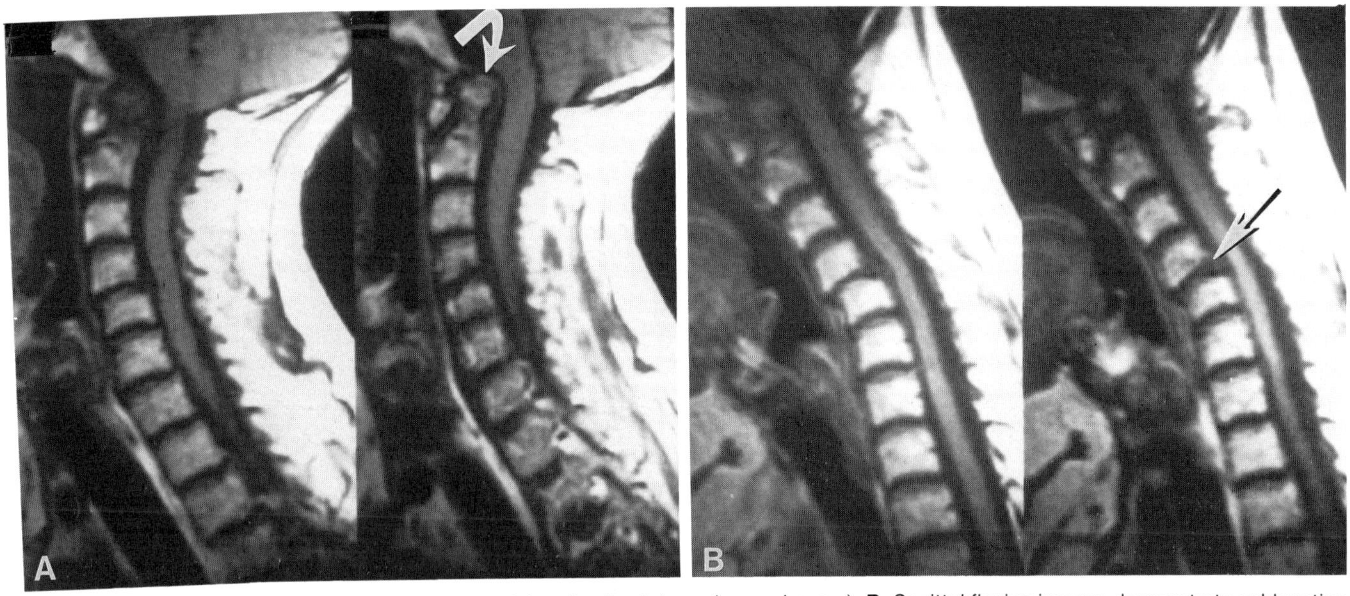

Figure 3.26. Rheumatoid arthritis with atlantoaxial and subaxial subluxation in a 63-year-old woman. **A,** Sagittal extension images show vertical subluxation odontoid process impinging on brainstem (*curved arrow*). **B,** Sagittal flexion images demonstrate subluxation at C1–2 and C4–5 (*arrow*) levels.

rior columns, i.e., in the terminal supply of the anterior spinal artery [80]. High signal intensity in the spinal cord on T2-weighted or proton density spin-echo images is seen in some patients with severe degrees of spinal cord compression. A hyperintense lesion is seen more frequently in severely clinically impaired patients and is considered an indication of poor prognosis [81]. Cervicomedullary compression may be caused by cranial subluxation of the odontoid into the foramen magnum or basilar impression.

Bundschuh et al. [82] reported that all patients with a cervicomedullary angle less than 135 degrees were neurologically abnormal. All rheumatoid patients with an abnormal cervicomedullary angle and/or cervicomedullary junction compression should undergo careful neurologic evaluation. The MRI appearance of spinal cord compression is usually but not always associated with clinical signs of a myelopathy.

Degenerative Arthritis

Radiographic features typical of osteoarthritis are often observed in or around the atlantoaxial joint

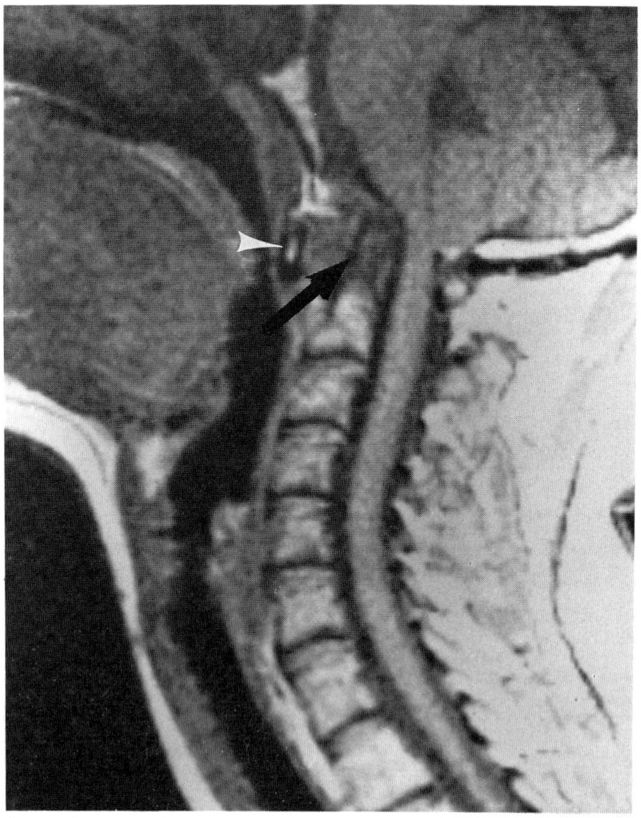

Figure 3.27. Rheumatoid arthritis with severe posterior subluxation of odontoid process. Note low-signal intensity in tip of odontoid (*arrow*). Anterior arch of C1 (*arrowhead*). Sagittal extension image demonstrates that the C1–2 subluxation is "fixed" by the predental pannus, preventing reduction in extension. There is severe cervicomedullary compression.

and odontoid process [83, 84]. The most common abnormality is a small bony excresence or osteophyte projecting from the apex of the dens [85]. Abnormalities seen best on CT include osteophytes of the median facet of the atlas, cysts, small ossicles, and transverse ligament calcification.

Chondrocalcinosis

Hypertrophic changes and calcification in the transverse ligament and adjacent soft tissue may be seen in calcium pyrophosphate dihydrate deposition disease (CPPD) [86, 87] (Figs. 3.29 and 3.30). Hyperostosis of the odontoid tip may occur. The cortical margins of the odontoid and atlas may become irregular, and small cysts or lacunae develop. With severe involvement, this process can cause an epidural inflammatory "mass" and can affect the atlantoaxial joint, leading to subluxation and spinal cord compression. Tumoral calcinosis may be seen. Calcifications in the hypertrophic ligamentous pseudomass are detected on CT but not on MRI.

Other Arthritides

Diffuse idiopathic skeletal hyperostosis (DISH) may cause irregularity of the cortical margins of the axis or atlas as part of the diffuse process of calcification and osteophyte formation that typically occurs at ligamentous attachments. Other inflammatory arthritides that affect the craniocervical

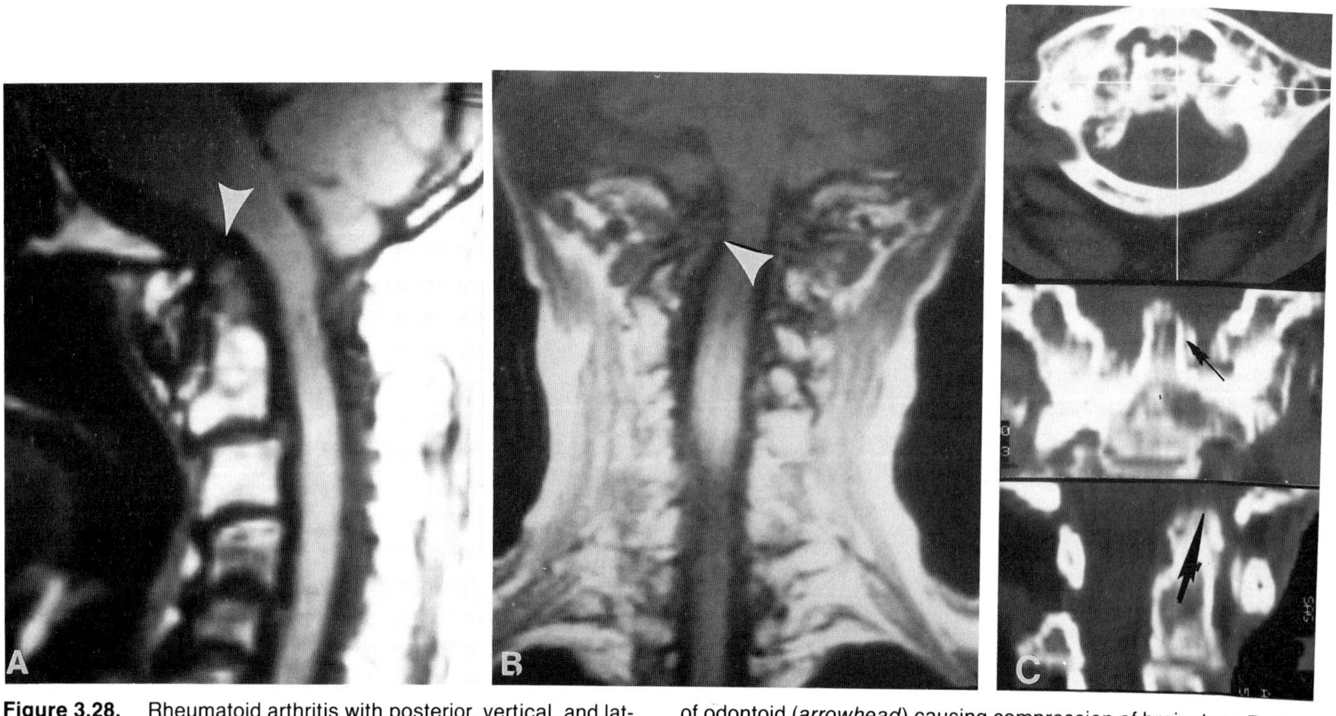

Figure 3.28. Rheumatoid arthritis with posterior, vertical, and lateral subluxation of odontoid in a 66-year-old woman with facial paresthesias. **A,** Sagittal MRI image demonstrates increased atlantoaxial joint space, irregularity of odontoid, and vertical subluxation of odontoid (*arrowhead*) causing compression of brainstem. **B,** Coronal MRI image shows lateral subluxation (*arrowhead*) and cord deviation. **C,** CT scan also demonstrates odontoid subluxations (*arrows*) and irregularity.

junction include juvenile rheumatoid arthritis, psoriatic arthritis [88], and rarely, ankylosing spondylitis.

Odontoid Pseudomass on MRI

Occasionally, sagittal MRI images appear to reveal a soft-tissue pseudomass posterior to the odontoid process. This appearance is caused by volume averaging of the posterior odontoid cortex and the adjacent synovium and ligaments, particularly when the image is slightly parasagittal in location. This appearance is most likely to be seen in patients with slight scoliosis and usually can be recognized by comparison of corresponding axial images [89].

Table 3.6.
Subluxations in Rheumatoid Arthritis

Anterior atlantoaxial
Vertical altantoaxial (cranial)
Lateral atlantoaxial
Rotatory atlantoaxial
Subaxial

Chiari Malformations

MRI is the procedure of choice for demonstrating the Chiari malformations. Sagittal T1-weighted images are the most helpful [90]. Chiari (1891, 1895) and Arnold (1894) described a complex deformity of the brainstem and cerebellum. These malformations have been divided into four types [91]:

Type I—displacement of the cerebellar tonsils and occasionally the part of the inferior vermis into the cervical spinal canal (Fig. 3.31).

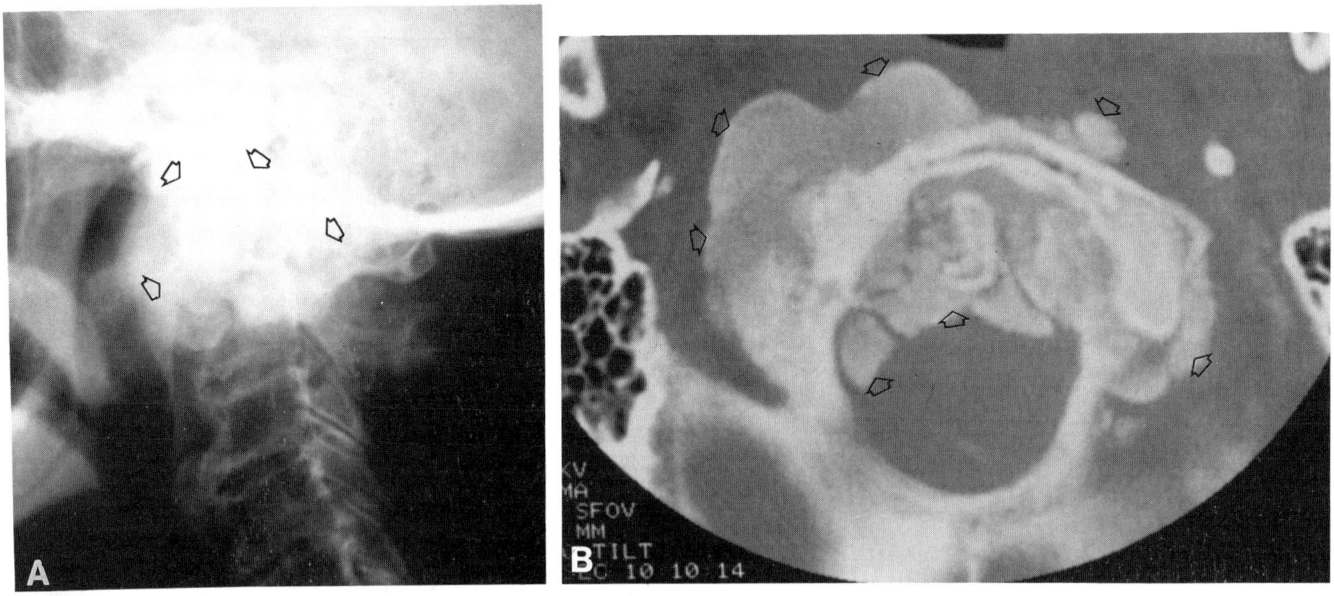

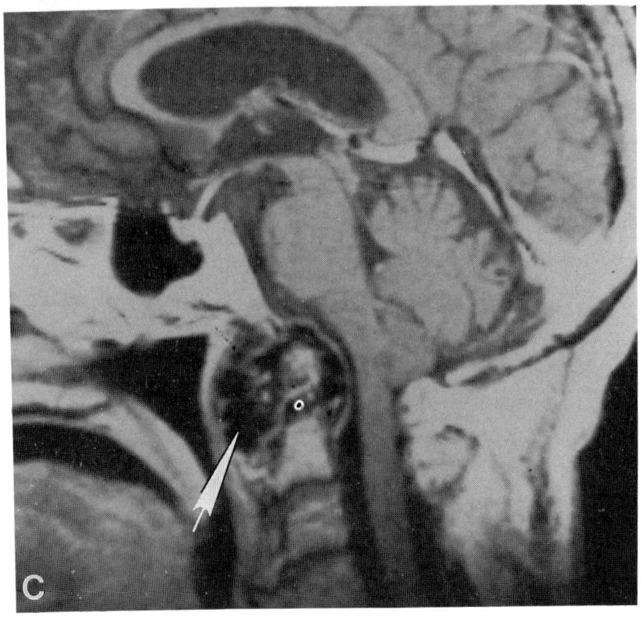

Figure 3.29. Calcium pyrophosphate deposition disease (chondrocalcinosis) with tumoral calcinosis. **A,** Lateral radiograph shows diffuse increased density around odontoid process (*arrows*). **B,** CT scan shows diffuse calcifications in soft tissues (*arrows*) around odontoid (tumoral calcinosis) and irregularity of dens. **C,** Sagittal MRI image demonstrates low-signal intensity tumoral calcinosis (*arrow*) around dens. There is C1–2 subluxation and brainstem impingement. (Case courtesy of Wendy Smoker, M.D.)

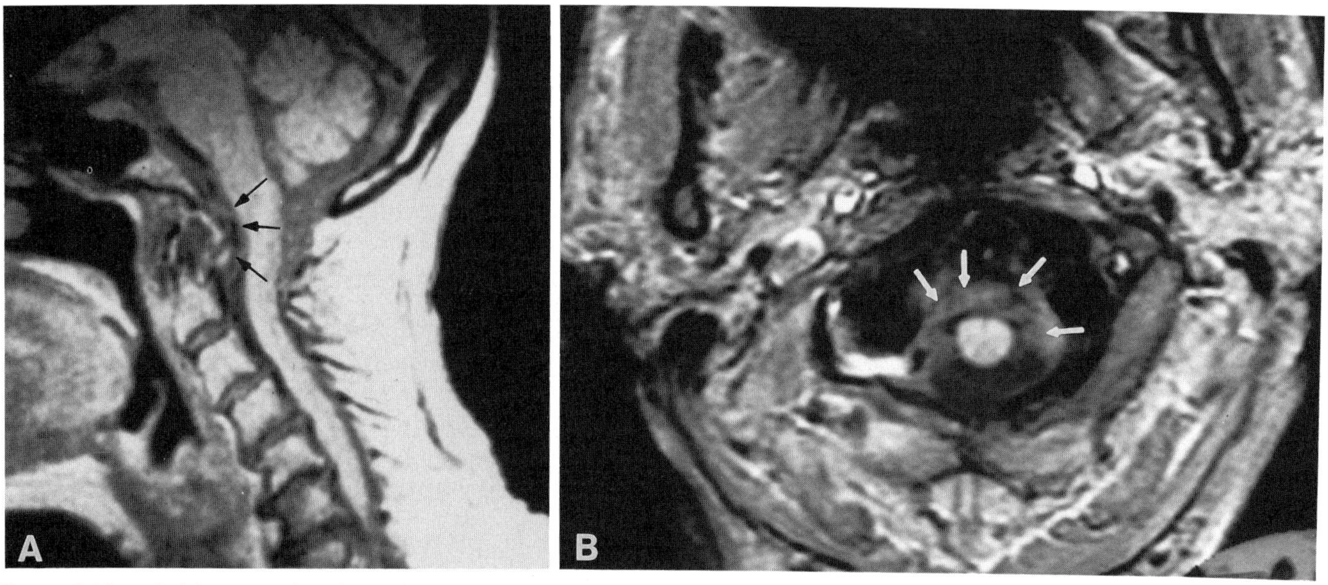

Figure 3.30. Calcium pyrophosphate deposition disease with chronic hypertrophic soft tissue pseudomass causing stenosis at craniocervical junction. There was no subluxation. **A,** Sagittal MRI image. Note pseudomass causing deviation of cervicomedullary junction (*arrows*). **B,** Axial MRI image demonstrates mass effect of thickened soft tissues (*arrows*) on the spinal cord.

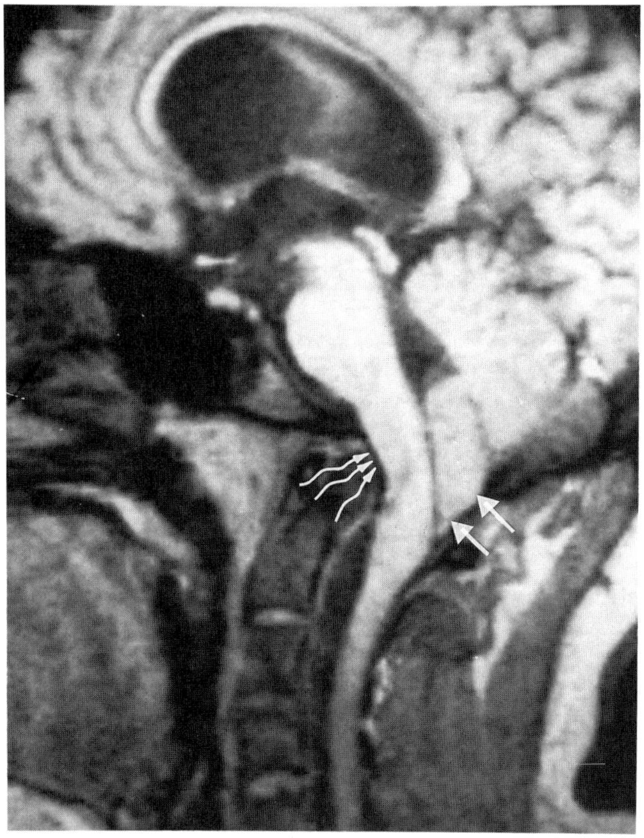

Figure 3.31. Chiari I malformation and mild basilar invagination. The brainstem is impinged by the dens (*wavy arrows*). The cerebellar tonsils (*arrows*) are pointed, compressed, and extend 15 mm below McRae's line (plane of the foramen magnum).

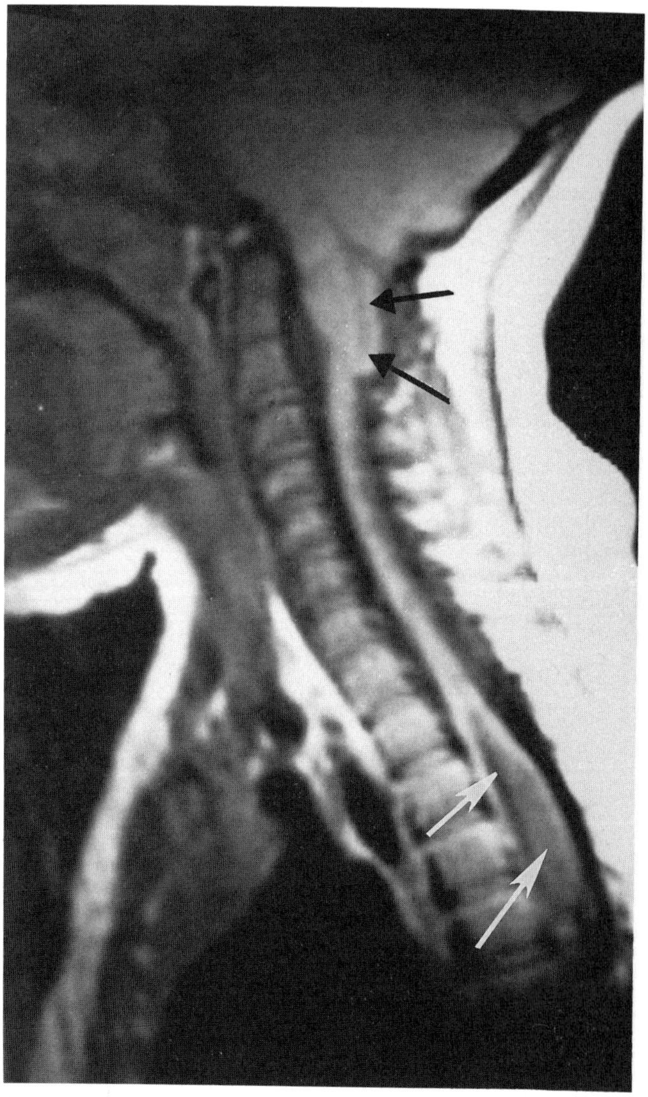

Figure 3.32. Chiari II malformation and syringohydromyelia (*white arrows*) in a 3-year-old boy. Note the extension of the cerebellar tonsils (*black arrows*) and part of inferior vermis into spinal canal.

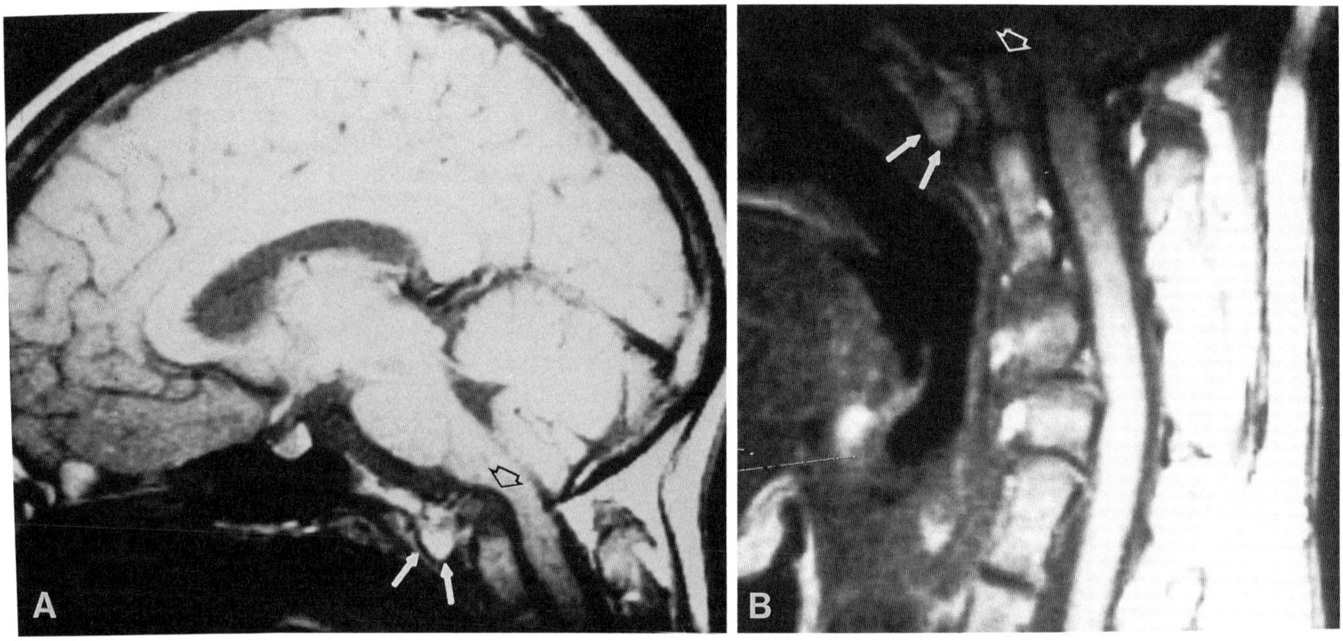

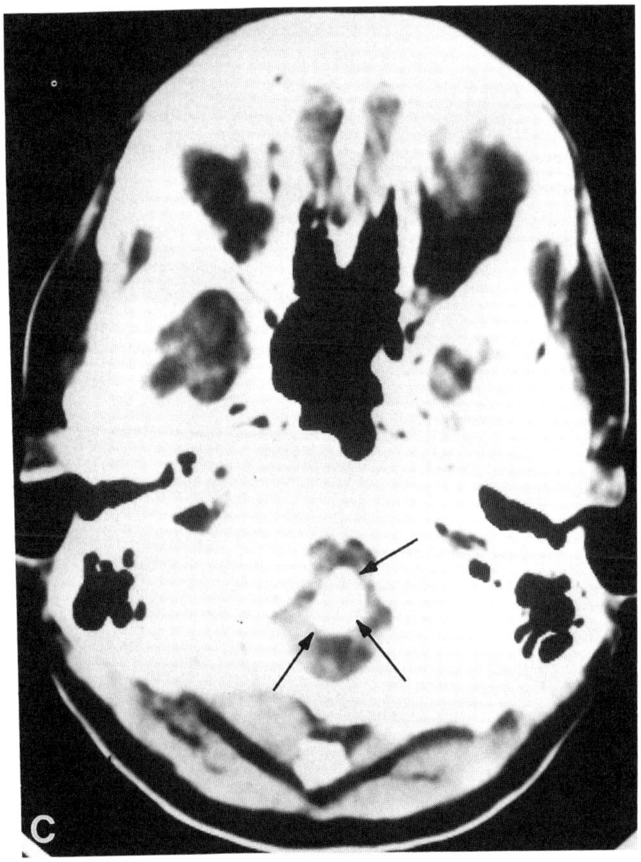

Figure 3.33. Klippel-Feil syndrome with anomalies including occipitalization of C1, flairing of the basiocciput (accessory condyle formation) (*small arrows*), basilar invagination (*open arrow*), partial fusion at C2–3, fusion at C5–6, and T1–2, and vertebral artery hypoplasia. **A,** Sagittal T1-weighted brain image showing basilar invagination and basiocciput flaring. **B,** Sagittal T1-weighted cervical spine image. **C,** CT scan through skull base reveals abnormal central location of odontoid process in foramen magnum (*arrows*).

Type II (Arnold-Chiari malformation)—displacement of the brainstem and lower cerebellum into the cervical canal. There is usually associated cranial or spinal dysraphism (Fig. 3.32).

Type III—displacement of the brainstem and lower cerebellum into the cervical canal and dorsal herniation through a cervical spina bifida (myelocerebellomeningocele).

Type IV—cerebellar hypoplasia.

The Chiari I malformation is most common and may exist as an isolated clinically insignificant MRI finding. However, it is also associated with craniovertebral anomalies such as basilar invagination, Klippel-Feil syndrome, occipitalization of the atlas, and syringomyelia. There is no association with spina bifida or myelomeningocele [92]. Sagittal MRI images are most helpful for depiction of this anomaly. Coronal MRI images are at times misleading. In contrast to the Chiari II malformation, the position of the fourth ventricle is normal. The cisterna magna is not visualized, the cerebellar tonsils frequently appear "pointed" due to compression within the foramen magnum, and cervicomedullary "kinking" is sometimes present. Tonsillar protrusion below the plane of the foramen magnum of more than 3 mm is most often associated with clinical symptomatology [93]. In the majority of patients with communicating syringomyelia, the tonsils average 8 to 9 mm below the foramen magnum and appear compressed or pointed in all patients [94]. MRI demonstration of less than 2 mm of tonsillar ectopia is probably of no clinical significance in the absence of syringomyelia.

The Chiari I malformation is clinically extremely variable in expression. The clinical findings tend to be manifested in early to middle adulthood, sometimes simulating multiple sclerosis [95]. The most frequent complaints include suboccipital, neck, and/or upper extremity pain (72%), altered sensation (47%), unsteady gait (42%), headache (42%), and weakness (32%). Downbeating nystagmus may be the only clinical manifestation [96]. The symptoms most likely result from arachnoidal scarring and adhesions increasing pressure effects on the brainstem and tonsils. Altered or obstructed cerebrospinal fluid flow leads to hydrocephalus and/or syringomyelia in some patients.

Chiari II malformation (Arnold-Chiari malformation) includes dysplastic changes involving the cerebellum, midbrain, ventricular system, falx, and cranium. It is often associated with other spinal anomalies recognized in infancy or childhood. The medulla, lower pons, fourth ventricle, and inferior vermis may extend into the cervical spinal canal dorsal to the spinal cord. The tentorium cerebelli and with it the transverse sinus are attached very low on the occipital bone, near the foramen magnum. "Kinking" or "buckling" of the caudally displaced cervicomedullary junction is usually seen. A cerebellar vermian "tail" extends well into the cervical canal, and tonsillar ectopia is usually present but is not required for diagnosis [97]. A defect in the posterior arch of the atlas is often present, and the foramen magnum is usually enlarged [98]. The ventricle may be slitlike throughout its length, or it may present with caudal cystic or diffuse dilatation extending rostrally into the aqueduct. The variable nature of the fourth ventricular abnormality associated with the Chiari II malformation is important for planning a surgical approach [99].

Chiari III and IV malformations are rare and usually obvious at birth.

Klippel-Feil Syndrome

In this condition there is fusion of one or more cervical vertebrae producing a "block" vertebra. These patients frequently have a short or asymmetrical neck, fixed head position, unilateral high scapula, kyphosis or scoliosis, and miscellaneous other anomalies. The Klippel-Feil syndrome is associated with other vertebral segmentation abnormalities such as occipitalization of the atlas and hemivertebrae (Fig. 3.33).

Rarely, there is compromise of the vertebral circulation with head turning, which may be manifested as vertigo, nausea, visual scotomas, and unconsciousness [100]. Secondary spondyloarthropathy develops due to the altered biomechanical relationships in which more stress is distributed to disks and vertebrae above and/or below the block vertebra.

Chiari I malformation and syringomyelia occur with increased frequency in the Klippel-Feil syndrome. Other developmental anomalies that may accompany the Klippel-Feil syndrome include urogenital anomalies, cleft palate/lip, cardiovascular anomalies, and syndactly [101, 102].

References

1. Coin CG, Malkasian DR. Foramen magnum. In: Newton TH, Potts DG, eds. Radiology of the skull and brain, Vol 1: The skull. St Louis: Mosby, 1971.
2. Wackenheim A. Roentgen diagnosis of the craniovertebral region. New York: Springer-Verlag, 1974.
3. Walker HS, Dietrich RB, Flannigan BD, Lufkin RB, Peacock WJ, Kangarloo H. Magnetic resonance imaging of the pediatric spine. Radiographics 1987;7:1129–1152.
4. Pettersson H, Larsson EM, Holtas S, Cronquist S, Egund N, Zymunt S, Brattstrom H. MR imaging of the cervical spine in rheumatoid arthritis. AJNR 1988; 9:573–577.
5. Reynolds H, Carter SW, Murtagh FR, Rechtine GR. Cervical rheumatoid arthritis: value of flexion and extension views in imaging. Radiology 1987;164:215–218.
6. Smoker WRK, Keyes WD, Dunn VD, Menezes AH. MRI versus conventional radiologic examinations in the evaluation of the craniovertebral and cervicomedullary junction. Radiographics 1986;6:953–994.
7. Bundschuh C, Modic MT, Kearney F, Morris R, Deal C. Rheumatoid arthritis of the cervical spine: surface-coil MR imaging. AJNR 1988;9:565–571.
8. von Torklus D, Gehle W. The upper cervical spine: regional anatomy, pathology and traumatology. A system radiological atlas and textbook. New York: Grune & Stratton, 1972.
9. Naidich TP, McLone DG, Harwood-Nash DC. Malformations of the craniocervical junction. In: Newton TH, Potts EG, eds. Computed tomography of the spine and spinal cord. San Anselmo: Clavadel Press, 1983:355–366.
10. Applegate LJ, Applegate GR, Kemp SS. MR of multiple cranial neuropathies in a patient with Camurati-Englemann Disease: case report. AJNR 1991;12:557–559.
11. Langer LO Jr, Romoin DL. Achondroplasia. In: Bergsma D, ed. Birth defects compendium, 2d ed. New York: Liss, 1979:10–11.
12. Wang H, Rosenbaum AE, Reid CS, Zinreich SJ, Pyeritz RE. Pediatric patients with achondroplasia: CT evaluation of the craniocervical junction. Radiology 1987; 164:515–519.

13. Hecht JT, Nelson FW, Butler IJ, Horton WA, et al. Computerized tomography of the foramen magnum: achondroplastic values compared to normal standards. Am J Med Genet 1985;20:355–360.
14. List CF. Neurologic syndromes accompanying developmental anomalies of occipital bone, atlas and axis. Arch Neurol Psychiatry 1941;45:577–616.
15. Menezes AH, VanGilder JC, Graf CJ, et al. Craniocervical abnormalities: a comprehensive surgical approach. J Neurosurg 1980;53:444–455.
16. Wang H, Rosenbaum AE, Reid CS, Zinreich SJ, Pyeritz RE. Pediatric patients with achondroplasia: CT evaluation of the craniocervical junction. Radiology 1987; 164:515–519.
17. Reid CS, Pyeritz RE, Kopits SE, Maria BL, et al. Cervicomedullary compression in young patients with achondroplasia: value of comprehensive neurologic and respiratory evaluation. J Pediatr 1987;110(4):522–530.
18. Stokes DC, Phillips JA, Leonard CO, Dorst JP, et al. Respiratory complications of achondroplasia. J Pediatr 1983;102(4):534–541.
19. Marin-Padilla M, Marin-Padilla TM. Developmental abnormalities of the occipital bone in human chondrodystrophies (achondroplasia and thantophoric dwarfism). Birth Defects 1977;13(3D):7.
20. Hecht JT, Nelson W, Butler IJ, et al. Computerized tomography of the foramen magnum: achondroplastic values compared to normal standards. Am J Med Genet 1985;20:355–360.
21. Wackenheim A. Hypoplasia of the basi-occipital bone and persistence of the spheno-occipital synchondrosis in a patient with transitory supplementary fissure of the basi-occipital. Neuroradiology 1985;27:226–231.
22. Dejerine J. Dystrophie osseuse par aplasie de la substance spongieuse du corps basilaire de l'occipital. Rev Neurol 1926;2:281–300.
23. McRae DL. The significance of abnormalities of the cervical spine. AJR 1960;84:3–25.
24. Wackenheim A. The pathogenesis of two distinct cervico-occipital malformations. In: Voth D, Glees P, eds. Diseases in the craniocervical junction: anatomic and pathologic aspects and detailed clinical accounts. New York: de Gruyter, 1987.
25. Chamberlain WE. Basilar impression (platybasia). Yale J Biol Med 1938;11:487.
26. McRae DL. Bony abnormalities in the region of the foramen magnum: Correlation of the anatomic and neurologic findings. Acta Radiol 1953;40:335–355.
27. VonTorklus D, Gehle W. The upper cervical spine: regional anatomy, pathology and traumatology. In: Georg Thieme Verlag, ed. A systemic radiological atlas and textbook. New York: Grune & Stratton, 1972.
28. Dolan KD. Cervicobasilar relationships. Radiol Clin North Am 1977;(2):155–166.
29. McRae DL. The significance of abnormalities of the cervical spine. AJR 1960;84:3–25.
30. Dolan KD. Cervicobasilar relationships. Radiol Clin North Am 1977;(2):155–166.
31. Bewermeyer H, Dreesbach HA, Hunermann B, Heiss WD. MR imaging of familial basilar impression. J Comput Assist Tomogr 1984;8:953–956.
32. Wadia NH. Myelopathy complicating congenital atlanto-axial dislocation. Brain 1967;90:449–472.
33. Sherman JL, Citrin CM, Barkovich AJ. MR imaging of syringobulbia. J Comput Assist Tomogr 1987;11:407–411.
34. Smoker WRK, Keyes WD, Dunn VD, Menezes AH. MRI versus conventional radiologic examinations in the evaluation of the craniovertebral and cervicomedullary junction. Radiographics 1986;6:953–994.
35. Barry HC. Paget's disease of bone. Edinburgh: E&S Livingstone, 1969.
36. Roberts MC, Kressel HY, Fallon MD, Zlatkin MB, Dalinka MK. Paget's disease: MR imaging findings. Radiology 1989;173:341–345.
37. Resnick D, Niwayama G. Paget's disease. In: Resnick D, Niwayama C, eds. Diagnosis of bone and joint disorders. 2d ed. Philadelphia: Saunders, 1987:2127–2170.
38. Foster JB, Hudgson P. The radiology of communicating syringomyelia. In: Barnett HJM, Foster JB, Hudgson P, eds. Syringomyelia. Philadelphia: WB Saunders, 1973: 50–63.
39. McRae DL, Barnum AS. Occipitalization of the atlas. AJR 1953;70:23–46.
40. Wackenheim A. C½ block vertebra, fusion of the anterior arch of the atlas with the axis, follow-up of the fusion in a child. Neuroradiology 1983;16:416–417.
41. Kopits SE, Perovic MN, McKusick VA, Robinson RA, Bailey JA. Congenital atlantoaxial dislocations in various forms of dwarfism. J Bone Joint Surg 1972;54A: 1349.
42. Chen YM, Elster AD. MR imaging of fatal atlantoaxial dislocation caused by retropharyngeal abscess. AJNR 1990;11:992.
43. VanGilder JC, Menezes AH, Dolan KD. Developmental/acquired abnormalities. In: The craniovertebral junction and its abnormalities, 1987. Mount Kisco, N.Y.: Futura, 1987.
44. Gilles RH, Biona M, Sotrel A. Infantile atlanto-occipital instability: the potential danger of extreme extension. Am J Dis Child 1979;133:30–37.
45. Hreidarsson S, Magram G, Singer H. Symptomatic atlantoaxial dislocation in Down syndrome. Pediatrics 1982;69:568–571.
46. Pueschel SM, Scola FH. Atlantoaxial instability in individuals with Down syndrome: epidemiologic, radiographic, and clinical studies. Pediatrics 1987;80:555–560.
47. Lipson SJ. Dysplasia of the odontoid process in Morquio's syndrome causing quadriparesis. J Bone Joint Surg 1977;59A:340–344.
48. Blaw ME, Langer LO. Spinal cord compression in Morquio-Brailsford's disease. J Pediatr 1969;74:593–600.
49. Sze G, Brant-Zawadzki MN, Wilson CR, et al. Pseudotumor of the craniovertebral junction associated with chronic subluxation: MR imaging studies. Radiology 1986;7:985–989.
50. Kopits SE, Perovic MN, McKusick V, Robinson RA, Bailey JA. Congenital atlantoaxial dislocations in various forms of dwarfism. J Bone Joint Surg 1972;54A:1349–1350.
51. Sherman JL, Kopits SE, Citrin CM: MR imaging of extrinsic compression of the neuraxis in skeletal disease. AJNR 1988;9:1007.
52. Fielding JW, Mensinger RN, Hawkins RJ. Os odontoideum. J Bone Joint Surg 1980;62A:376–383.
53. Hensinger RN, Fielding JW, Hawkins RJ. Congenital anomalies of the odontoid process. Orthop Clin North Am 1978;9:901–912.
54. Michaels L, Prevost MJ, Crang DF. Pathological changes in a case of os odontoideum (separate odontoid process). J Bone Joint Surg 1969;50A:965–972.
55. Fielding JW, Mensinger RN, Hawkins RJ. Os odontoideum. J Bone Joint Surg 1980;62A:376–383.
56. Shirasaki N, Okada K, Oka S, et al. Os odontoideum with posterior atlantoaxial instability. Spine 1991;7: 706–715.

57. Minderhoud JM, Braakman R, Penning L. Os odontoideum: clinical, radiological and therapeutic aspects. J Neurol Sci 1969;8:521–544.
58. Blaw ME, Langer LO. Spinal cord compression in Morquio-Brailsford's disease. J Pediatr 1969;74:593–600.
59. Langer LO, Carey LS. The roentgenographic features of the KS mucopolysaccharidosis of Morquio (Morquio-Brailsford's disease). AJR 1966;97:1–20.
60. Stevenson RE. Mucopolysaccharidosis IV. In: Bergsma D, ed. Birth defects compendium, 2d ed. New York: Liss, 1979:732.
61. Kopits SE. Orthopedic complications of dwarfism. Clin Orthop 1976;114:153–179.
62. Kopits SE. Orthopedic complications of dwarfism. Clin Orthop 1976;114:153–179.
63. Kopits SE, Perovic MN, McKusick V, et al. Congenital atlantoaxial dislocations in various forms of dwarfism. J Bone Joint Surg 1972;54A:1349–1350.
64. Perovic NM, Kopits SE, Thompson RC. Radiologic evaluation of the spinal cord in congenital atlantoaxial dislocation. Radiology 1973;109:713–716.
65. Sherman JL, Kopits SE, Citrin MC. MR imaging of extrinsic compression of the neuraxis in skeletal disease (a). AJNR 1988;9:1007.
66. Applegate LJ, Applegate GR, Kemp SS. MR of multiple cranial neuropathies in a patient with Camurati-Engelmann disease: case report. AJNR 1991;12:557–559.
67. Mathern GW, Batzdorf U. Grisel's syndrome. Clin Orthop 1989;244:131–146.
68. Heywood AWB, Learmonth ID, Thomas M. Cervical spine instability in rheumatoid arthritis. J Bone Joint Surg 1988;70B:702–797.
69. Lipson SJ. Rheumatoid arthritis in the cervical spine. Clin Orthop 1989;239:121–127.
70. Bundschuh C, Modic MT, Kearney F, Morris R, Deal C. Rheumatoid arthritis of the cervical spine: surface-coil MR imaging. AJNR 1988;9:565–571.
71. Sze G, Brant-Zawadzki MN, Wilson CR, Norman D, Newton TH. Pseudotumor of the craniovertebral junction associated with chronic subluxation: MR imaging studies. Radiology 1986;161:391–394.
72. Larsson E, Hotas S, Zygmunt S. Pre- and postoperative MR imaging of the craniocervical junction in rheumatoid arthritis. AJNR 1989;10:89–94.
73. Slatis P, Santavirta S, Sandelin J, et al. Cranial Subluxation of the Odontoid Process in Rheumatoid Arthritis. J Bone Joint Surg 1989;71A:189–195.
74. Walmsley R. Joints. In: Romanes GJ, ed. Cunningham's textbook of anatomy. 11 ed. London: Oxford University Press, 1972:207–257.
75. Mathews JA. Atlantoaxial subluxation in rheumatoid arthritis. Ann Rheum Dis 1969;28:260–265.
76. Bundschuh C, Modic MT, Kearney F, Morris R, Deal C. Rheumatoid arthritis of the cervical spine: surface-coil MR imaging. AJNR 1988;9:565–571.
77. Reynolds H, Carter SW, Murtagh FR, Rechtine GR. Cervical rheumatoid arthritis: value of flexion and extension views in imaging. Radiology 1987;164:215–218.
78. Kudo H, Iwano K, Woshizawa H. Cervical cord compression due to extradural granulation tissue in rheumatoid arthritis. J Bone Joint Surg 1984;66B:426–430.
79. Ogino H, Tada K, Okada K, et al. Canal diameter, anteroposterior compression ratio, and spondylotic myelopathy of the cervical spine. Spine 1983;8:1–15.
80. Hashizume Y, Iijima S, Kishiimoto H, Yanagi T. Pathology of spinal cord lesions caused by ossification of the posterior longitudinal ligament. Acta Neuropathol 1984;63:123–130.
81. Takahashi M, Yamashita Y, Sakamoto Y, Kojima R. Chronic cervical cord compression: clinical significance of increased signal on MR images. Radiology 1989;173:219–224.
82. Bundschuh C, Modic MT, Kearney F, Morris R, Deal C. Rheumatoid arthritis of the cervical spine: surface-coil MR imaging. AJNR 1988;9:565–571.
83. Shore L. On osteo-arthritis in the dorsal intervertebral joint: a study in morbid anatomy. Br J Surg 1935; 22:833–849.
84. Harata S, Tohno S, Kawagishi T. Osteoarthritis of the atlantoaxial joint. Int Orthop 1981;5:277–282.
85. Genez BM, Willis JJ, Lowrey CE, et al. CT findings of degenerative arthritis of the atlanto-odontoid joint. AJR 1990;154:315–318.
86. Dirheimer Y, Wackenheim C, Dietemann JL. Calcifications of the transverse ligament in calcium dihydrate deposition disease (CPPD). Neuroradiology 1985; 27:87.
87. Weingarten K, Kelly AB, Zimmerman RD. Compression of the cervicomedullary junction due to calcium pyophosphate deposition disease (a) AJNR 1989;10:886–892.
88. Kellebrew G, Gold RH, Sholkoff SD. Psoriatic spondylitis. Radiology 1973;108:9–16.
89. Smoker WRK, Keyes WD, Dunn VD, Menezes AH. MRI versus conventional radiologic examinations in the evaluation of the craniovertebral and cervicomedullary junction. Radiographics 1986;6:953–994.
90. Wolpert SM, Anderson M, Scott RM, Kwan ESK, Runge VM. Chiari II malformation: MR imaging evaluation. AJNR 1987;8:783–792.
91. Friede FL. Dysplasia of cerebellar hemispheric organization. In: Friede RF, ed. Developmental pathology. Vienna: Springer-Verlag, 1975:314–326.
92. Naidich TP, McLone DG, Harwood-Nash DC. Malformations of the craniocervical junction. In: Newton TH, Potts DG, eds. Computed tomography of the spine and spinal cord. San Anselmo: Clavadel Press, 1983:355–366.
93. Barkovich AJ, Wippold FJ, Sherman JL, Citrin CM. Significance of cerebellar tonsillar position of MR. AJNR 1986;7:795–799.
94. Sherman JL, Barkovich AJ, Citrin CM. The MR appearance of syringomyelia: new observations. AJNR 1986;7: 985–995.
95. Banerji NK, Millar JHD. Chiari malformation presenting in adult life. Brain 1974;97:157–168.
96. Saez RJ, Onofrio BM, Yanagihar T. Experience with Arnold-Chiari malformation, 1960 to 1970. J Neurosurg 1976;45:416–422.
97. Naidich TP, McLone DG, Harwood-Nash DC. Malformations of the craniocervical junction. In: Newton TH, Potts DG, eds. Computed tomography of the spine and spinal cord. San Anselmo: Clavadel Press, 1983:355–366.
98. Kruyff E, Jeffs R. Skull abnormalities associated with the Arnold Chiari malformation. Acta Radiol (Diagn) 1966;5:9–24.
99. Venes JL, Black KL, Latack JT. Preoperative evaluation and surgical management of the Arnold-Chiari II malformation. J Neurosurg 1986;64:363–370.
100. Faintness and syncope. In: Adams RD, Victor M, eds. Principles of neurology. New York: McGraw-Hill, 1981: 248–257.
101. MacEwen D. The Klippel-Feil syndrome. J Bone Joint Surg 1975;57B:261–267.
102. Nagib MG, Maxwell RE, Chou SN. Klippel-Feil syndrome

in children: clinical features and management. Childs Nerv Syst 1985;1:255–263.
103. Chamberlain WE. Basilar impression (platybasia): bizarre developmental anomaly of occipital bone and upper cervical spine with striking and misleading neurologic manifestations. Yale J Biol Med 1939;11:487–496.
104. McGregor J. The significance of certain measurements of the skull in the diagnosis of basilar impression. Br J Radiol 1948;21:171–181.
105. McRae DL. The significance of abnormalities of the cervical spine. AJR 1960;84:3–25.
106. Hinck VC, Hopkins CE, Savara BS. Diagnostic criteria of basilar impression. Radiology 1961;76:579.
107. Dolan KD. Cervicobasilar relationships. Radiol Clin North Am 1977;15:155–166.
108. Wackenheim A. Roentgen diagnosis of the craniovertebral region. New York: Springer-Verlag, 1974.
109. Powers B, Miller BD, Kramer RS, et al. Traumatic anterior atlanto-occipital dislocation. Neurosurgery 1979; 4(1):12–17.
110. Klause E. Roentgen diagnosis of platybasia and basilar impression: further experiences with a new examination method. Fortsch Geb Rontgenstrahlen 1957; 86:460.
111. Dolan KD. Cervicobasilar relationships. Radiol Clin North Am 1977;(2):155–166.
112. Nordqvist L. The sagittal diameter of the spinal cord and subarachnoid space in different age groups: A roentgenographic post-mortem study. Acta Radiol Suppl. (Stockh) 1964;227:1–96.
113. Bojsen E. The cervical spinal canal in intraspinal expansive processes. Acta Radiol (Stockh) 1954;42:101–115.
114. Kruyff E, Jeffs R. Skull abnormalities in the Arnold-Chiari malformation. Acta Radiol [Diagn] 1966;5:9–24.
115. Anderson RJ. The diagnosis and management of rheumatoid synovitis. Orthop Clin North Am 1975;6(3):629–639.
116. Dirheimer Y, Babin E. Upward atlantoaxial dislocation in rheumatoid arthritis. Neuroradiology 1974;7:229–236.

CHAPTER / **4**

Congenital Anomalies

B. C. P. Lee / K. C. V. G. Rao

Congenital anomalies of the spine may involve the osseous elements or the spinal cord and meninges or elements of both components. The more obvious anomalies where all elements may be involved usually present at birth and less often during infancy. Imaging studies in this group of patients usually are called for to demonstrate the extent of the intrinsic soft-tissue anomalies as well as the associated cranial involvement following surgical correction of the open midline defect.

In most congenital anomalies of the spine evaluated beyond the neonatal period there is either an occult or obvious skeletal anomaly. Such patients are usually in their infancy or adolescence and are investigated for evaluation of progressive scoliosis; cutaneous stigmata such as skin dimples, skin tags, nevi, and the midline patch of hypertrichosis, or manifestations of neurologic disorders. Rarely, the intrinsic spinal cord anomaly may not be detected until later in life, when such patients are investigated for progressive or acute cord compression or radiculopathy. A combination of imaging studies may be required to demonstrate the suspected pathologic process. Sonography, myelography, computed tomography (CT) following myelography (CTM), and magnetic resonance imaging (MRI) are all useful in imaging the spine and its contents. At the present time, MRI is the imaging study of choice because it is noninvasive and provides the maximum diagnostic information. The other imaging studies, except sonography in neonates, are usually employed as ancillary studies.

In infants and young children, appropriate sedation is required to obtain the desired imaging studies without motion-related artifacts. If CT is contemplated, the study should be complemented with prior intrathecal contrast material to evaluate the total extent of the spinal canal, including the craniovertebral junction. Depending on the age of the patient, anywhere from 3 to 15 ml of contrast may be required. Contiguous thin (1.5 to 3 mm) or overlapping CT sections allow for subsequent re-formatted images in the sagittal or coronal plane.

MRI is excellent in evaluating spinal cord anomalies because of its inherent sensitivity in differentiating the various soft-tissue components within the spinal canal. Rapid evaluation of the entire spine is possible with appropriate pulse sequences, surface coil selection, and field of view without the need for contrast studies, except in rare circumstances.

Spinal Cord and Vertebral Relationship

The conus medullaris of the spinal cord is situated at the lumbosacral level at birth. The vertebral column grows more rapidly than the spinal cord; myelographic and cadaver data suggest that the conus medullaris ascends to the L2–3 level by 5 years of age, reaching the adult location of T12/L1 shortly after 12 years [1, 2]. A more recent MRI report on normal children, however, suggests that the conus reaches the L2–3 level as early as 2 years of age and in many cases ascends to the adult location shortly after this age [3].

Each segment of the spinal cord gives rise to a pair of nerves that exit through the intervertebral foramina. However, the relation of the spinal nerves to the vertebrae in the cervical region is different from that in the rest of the spinal cord. The first cervical (C1) nerve exits between the occipital bone and the atlas, and the C2 to C7 nerves exit between the corresponding vertebrae and the one immediately above. However, since there are eight cervical nerves and only seven cervical vertebrae, the C8 nerve exits in the foramen between the C7 and T1 vertebrae. As a result of this discrepancy, the nerves of the thoracic and lumbar cord exit immediately below rather than above the corresponding vertebrae. Because of the unequal growth of the vertebrae and spinal cord, the spinal nerves assume progressively more oblique downward courses as they exit from the neural foramina. The cauda equina nerves ultimately become almost vertical in their course and exit through neural foramina several vertebral levels below the conus medullaris.

Embryology

Embryologically, the spinal cord has a dual origin: The cranial and main portions of the spinal cord arise from cells of the central nervous system, and the conus medullaris and filum terminale arise from a separate cell mass [4–6]. The central nervous system and the spinal cord proper are formed by *neurulation.* In this process, a specialized portion of the ectoderma known as the *neural plate* gives rise to the neural tube. The midline neural plate grows into a groove or furrow in the dorsal aspect of the ectoderm. The margins of the plate fold and fuse in the midline to form the neural tube (Fig. 4.1). This closure begins at 2 to 3 weeks in the midthoracic region and extends cranial and caudally until the last openings (the cranial anterior and caudal posterior neuropores) are closed by 24 to 26 days of gestation, thus sealing off any communication with the external surface of the ectoderm. At the same time, as the neural tube is formed, the spinal cord is divided into segments corresponding to the individual vertebral bodies which arise from the primitive notochord. During this period, neural crest cells form between the neural tube and the skin surface. These give rise eventually to the dorsal roots, autonomic ganglia, and peripheral nerves at each segmental level.

The most caudal portion of the spinal cord arises at 28 to 32 days, after closure of the posterior neuropore, from a separate group of ectodermal cells known as the *caudal cell mass.* These cells enlarge and undergo vacuolation to form many vesicles which then coalesce by the process of canalization into a single cavity known as the *ventriculus terminales.* Retrogressive differentiation takes place, by which the central cavity communicates with the central canal of the spinal cord, the proximal portion of the caudal cell mass becomes the conus medullaris, and the distal portion involutes to be the film terminalis (Fig. 4.2).

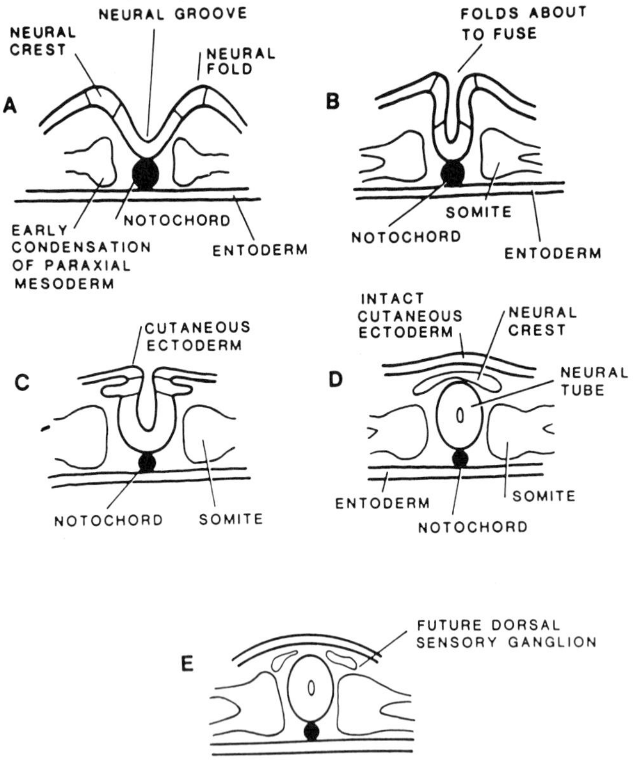

Figure 4.1. **A,** The spinal cord is derived from the midline portion of the ectoderm known as the neuroectoderm which becomes the neural groove adjacent to the notochord. **B–E,** The groove folds and fuses dorsally starting from the thoracic region, proceeding cranial and caudally until the anterior and posterior neuropores are closed before the first month of gestation. A portion of the neuroectoderm between the neural tube and the cutaneous ectoderm is separated and forms the dorsal ganglia (neurocrest).

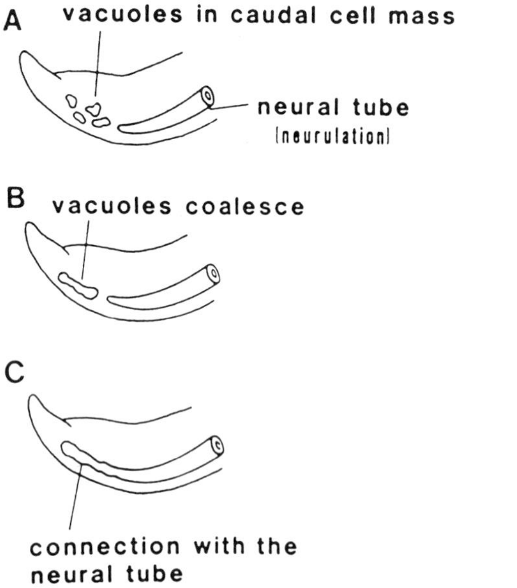

Figure 4.2. **A,** The distal portion of the spinal cord is derived from a caudal cell mass that is separate from the neural tube. This mass initially enlarges and then undergoes vacuolation. **B,C,** The vacuoles coalesce by a process of canalization into a single cavity which eventually communicates with the central canal.

Congenital Abnormalities

Abnormalities of the spinal cord or spinal dysraphism may be divided into open and closed defects—spina bifida aperta and occult spinal dysraphism, respectively. Open defects are generally secondary to failure of neural tube closure and communicate with the exterior. Occult dysraphisms include defects of the caudal cell mass (lipomeningomyelocele, tethered spinal cord), spinal lipoma, syringomyelia, diastematomyelia, and neurenteric cyst [4–6]. Open defects are seldom imaged because they are repaired surgically shortly after birth to prevent infection of the meninges and central nervous system. The term *spinal dysraphism* is used in the radiologic literature to describe both open and occult defects, since such a distinction can no longer be made after surgical closure of the former type [7–12].

Lack of closure of the neural tube results in meningoencephalocele and meningomyelocele at the cranial or caudal end and less commonly at other locations of the spinal cord. Defective involution of the caudal mass results in a tethered cord, which is usually associated with lipomas and other congenital masses (lipomyelomeningocele). Syringomyelia is associated with both defects, as well as with diastematomyelia and tethered cord. Defects secondary to abnormal development of the dural sac give rise to simple meningoceles and are not usually associated with abnormalities of the spinal cord.

Neural Tube Closure Defects

SPINA BIFIDA APERTA

The most severe form of closure defect involving the entire spinal cord, rachischisis, is generally incompatible with life. Less severe defects are due to persistence of the anterior or posterior neuropore and occasionally to failure of closure of intermediate portions of the neural tube. Although it has been suggested that the defects result from reopening of the neuropore and neurotube after they have closed, there is new evidence to indicate that the defects occur because of primary failure of closure of these structures [13–15]. The involved portion of the spinal cord is always exposed to the exterior in these anomalies—spina bifida aperta.

Cervical Meningocele and Meningomyelocele

Persistence of the anterior neuropore at the craniocervical junction beyond 24 to 26 days of gestation results in a meningocele or meningoencephalocele depending on whether neurologic tissues is present. These defects are always associated with bony abnormalities. The portion of the occipital bone involved with the meningocele is defective, the posterior arches of the involved vertebrae are bifid or absent, and the spinal canal is frequently enlarged. MRI is used to determine whether neurologic tissue is contained within the meningocele, and this is best seen on T1-weighted images. T2-weighted images confirm the presence of cerebrospinal fluid (CSF) within the meningocele in older children but are unreliable in those under 1 year of age because the signal of the white matter is identical to that of CSF in this age group.

Cervical meningomyelocele may present as an isolated lesion but is commonly associated with other congenital anomalies such as the Arnold-Chiari anomaly. The anomaly is sometimes referred to as *Chiari III*, in which the cerebellar tonsils and portions of the hemispheres are contained within the meningomyelocele (Fig. 4.3). Infants with these anomalies invariably have severe neurologic defects of the brain. In the Dandy Walker anomaly, the enlarged fourth ventricle is sometimes continuous with the meningocele, and there may be herniation of portions of the cerebellum or occipital lobe into this cavity.

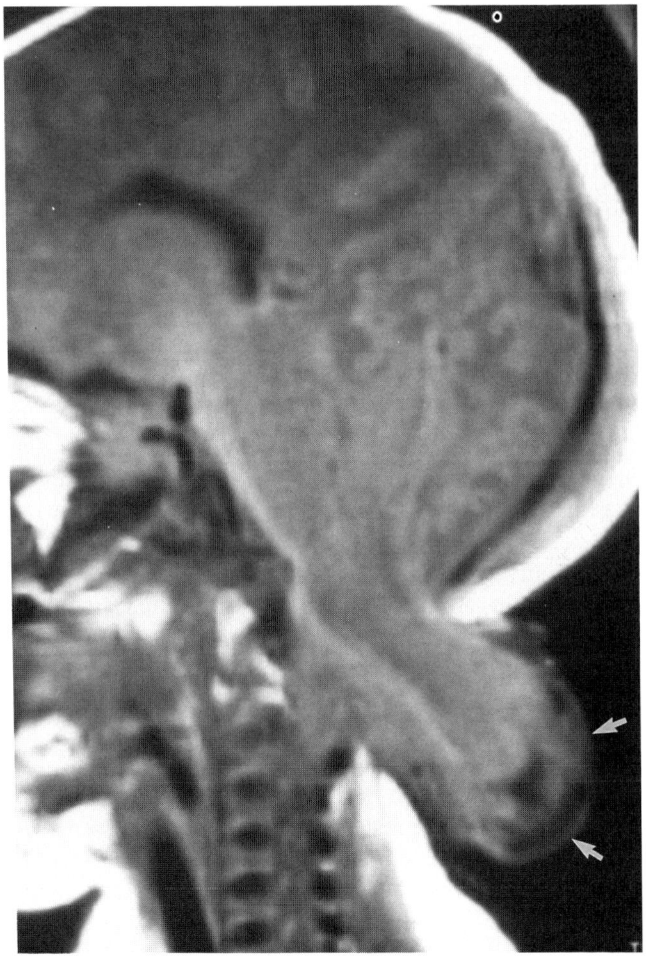

Figure 4.3. Chiari III malformation consists of the Arnold-Chiari malformation in which the ectopic cerebellar contents are herniated as a cervico-occipital meningomyelocele (*arrows*).

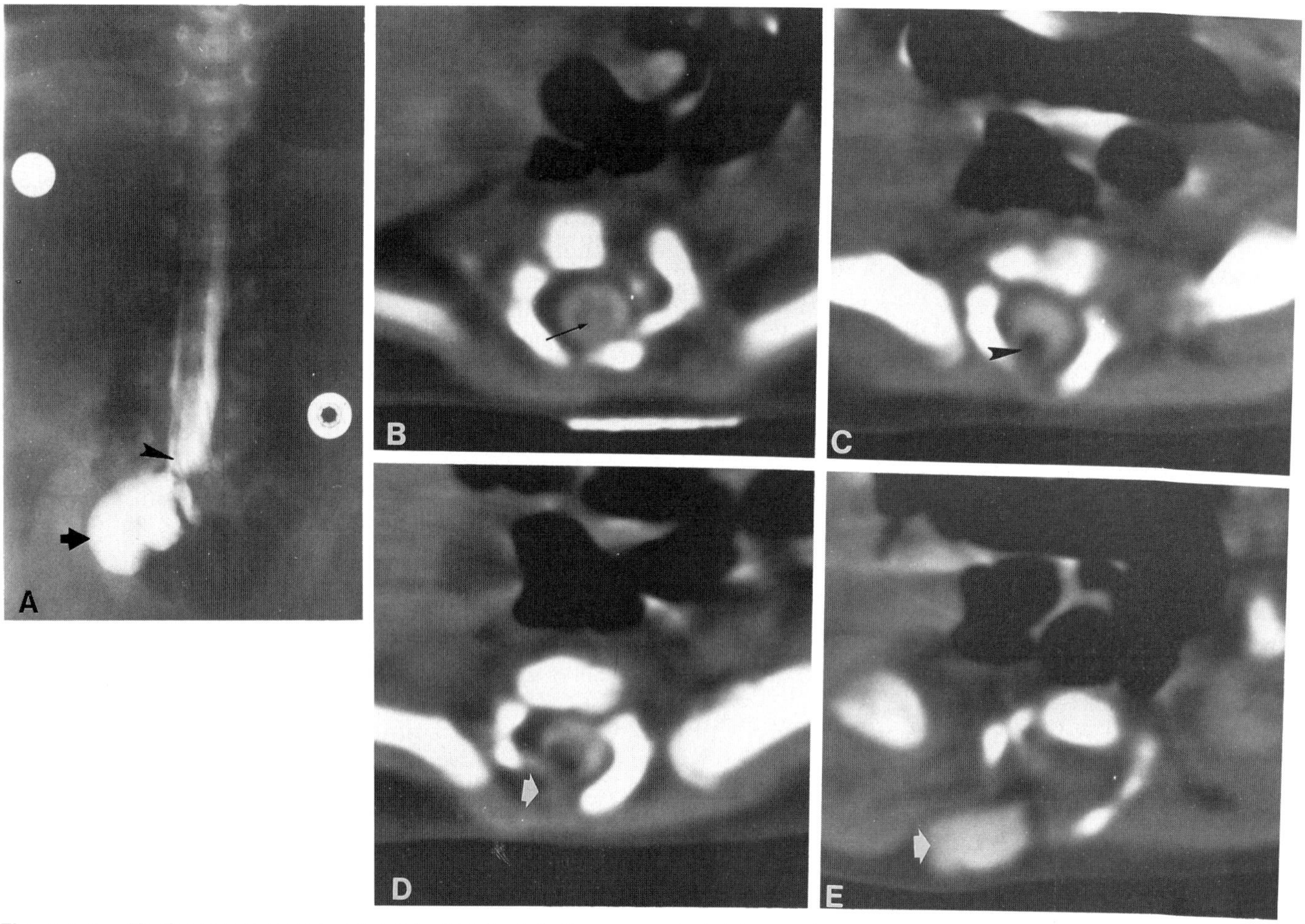

Figure 4.4. Meningomyelocele in a 3-week-old infant. A, Frontal view of myelogram demonstrates the tethered cord (*arrowhead*) and the large meningocele (*arrow*). **B–E,** CTM sections demonstrating the wide spinal canal, the syrinx within the cord (*arrow*), the lipoma (*arrowhead*), and the large bony defect with the meningocele (*white arrow*) under the skin surface. **F,** Repaired meningocele. Note the enlarged subarachnoid space anterior to the placode where the spinal cord is tethered. Syringomyelia are noted at several locations in the spinal cord (*arrows*).

Lumbosacral Meningomyelocele

These defects result from persistence of the posterior neuropore and must be differentiated from defects of caudal neural tube formation, which will be discussed below [6, 10]. Lumbosacral meningomyelocele may occur alone but is frequently associated with the Arnold-Chiari malformation, hydrocephalus, and other severe neurologic abnormalities. Bony spina bifida, which is always present, may be quite extensive, and the spinal canal is usually markedly enlarged. The distal spinal cord is abnormal and exposed posteriorly to the exterior. The site where the abnormal spinal cord is attached posteriorly is surrounded by a network of abnormal vessels and is known as the *zona medullovasculosa* or *placode*. The region around this is the *zona epitheliosa*, which has the capacity to epithelialize and enclose the meningocele from the exterior. The subarachnoid space anterior to the placode may be enlarged and give rise to a true meningomyelocele. Unlike the meningocele of occult dysraphism, where the meningocele is posterior and surrounds the abnormal spinal cord, the enlarged subarachnoid space in this condition is anterior to the placode (Fig. 4.4). Since these lesions are open to the exterior, they are potential sources of CSF leakage and infection and have to be surgically corrected urgently.

Lumbosacral meningomyelocele is frequently associated with the Chiari II malformation. Other spinal cord abnormalities, such as diastematomyelia and syringomyelia, also may be present. The syringomyelia often extends directly into the placode. The spinal cord may be tethered at the site of the placode or by a thickened filum terminale caudally. Lipomas may be present adjacent to the filum terminale rather than at the placode. The spinal cord may become tethered after surgical correction of the meningomyelocele [16–19]. Other postoperative complications include inclusion dermoid and epidermoid masses at the site of surgery.

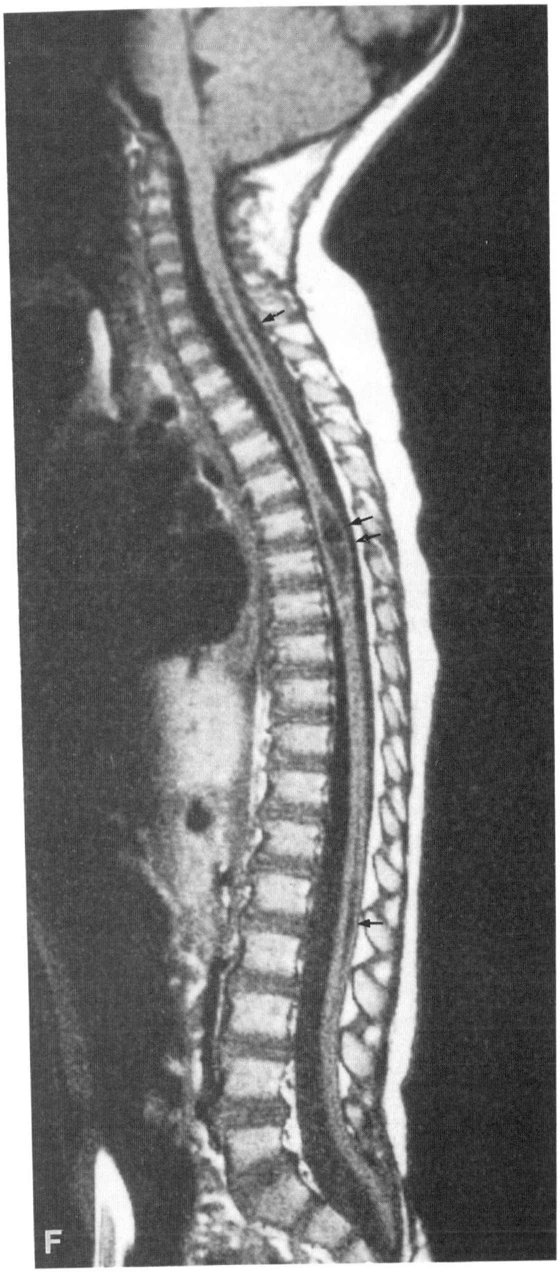

Figure 4.4–*continued*

Occult Spinal Dysraphism

CAUDAL TUBE DEFECTS

Lipomeningomyelocele

Defects are caused by failure of the caudal cell mass to differentiate into the conus medullaris and filum terminale [20–24]. As the caudal cell mass develops after closure of the posterior neuropore, such defects are closed on the skin surface. Cutaneous evidence of the abnormality is inconspicuous, but occasionally, a hair tuft, a hemangioma, or a soft-tissue swelling may be visible [25].

On CT, lipomeningomyelocele demonstrates a well-defined spina bifida defect, a large spinal canal, hypodense subcutaneous fat, and a contiguous lipoma within the spinal canal along the dorsal surface of the spinal cord. This can be better appreciated on CTM (Fig. 4.5). The distal portion of the conus attached to the cord may have a sharp, distinctive cleavage plane; at other times this is not so clear. The cleavage plane is better evaluated on MRI studies compared with CT. On MRI, the tethering of the conus medullaris to the posterior aspect of the dural sac and the presence of an associated lipoma can be well defined on sagittal and axial sections (see Fig. 4.5). The lipoma often extends extradurally and is frequently accompanied by a meningocele, justifying the term *lipomeningomyelocele* (Fig. 4.6). The meningocele, unlike that of open spina bifida, which is anterior to the placode, surrounds the tethered cord and the associated lipoma (Fig. 4.7). The tissue at the site of tethering is totipotential and may develop into a teratoma [26] (Fig. 4.8), or other cell types of primitive neuroectodermal tumor (PNET).

Tethered Spinal Cord

Although the spinal cord is most commonly tethered at the site of a lipoma or teratoma as part of the lipomeningomyelocele defect, tethering may be caused by a short, thick filum terminale [27, 28] (Fig. 4.9). This occurs because of defective regression following canalization of the caudal cell mass. A lipoma also may be present concurrently, and hence this entity may be considered a mild manifestation of lipomeningomyelocele. Although severe motor and sensory weakness is sometimes present, the neurologic deficits and bladder dysfunctions are usually mild and are often not detected until the child undergoes a growth spurt. The neurologic defects are produced because the spinal cord is anchored and cannot ascend as the vertebral column grows. Occasionally, symptoms of a tethered cord are not apparent until adulthood.

Plain radiographs may be normal, although spina bifida occulta is a common finding. Both CTM and MRI demonstrate widening of the lumbosacral spinal canal, with the thick filum stretched taut along the dorsal aspect of the dural tube (see Fig. 4.7). The associated fibrous band or lipoma at the fixed site of the filum also can be identified on these studies. On MRI, the diagnosis of tethered cord secondary to thickened fixed filum terminale should not be made purely on the basis of sagittal studies, since in most sagittal MRI studies the filum as well as the descending nerve roots may be seen as a composite band along the posterior surface of the thecal sac (Fig. 4.10). Axial sections usually confirm the presence of the thickened filum when it measures more than 2 mm in diameter. Other features to look for include the bowstring appearance of the spinal cord and

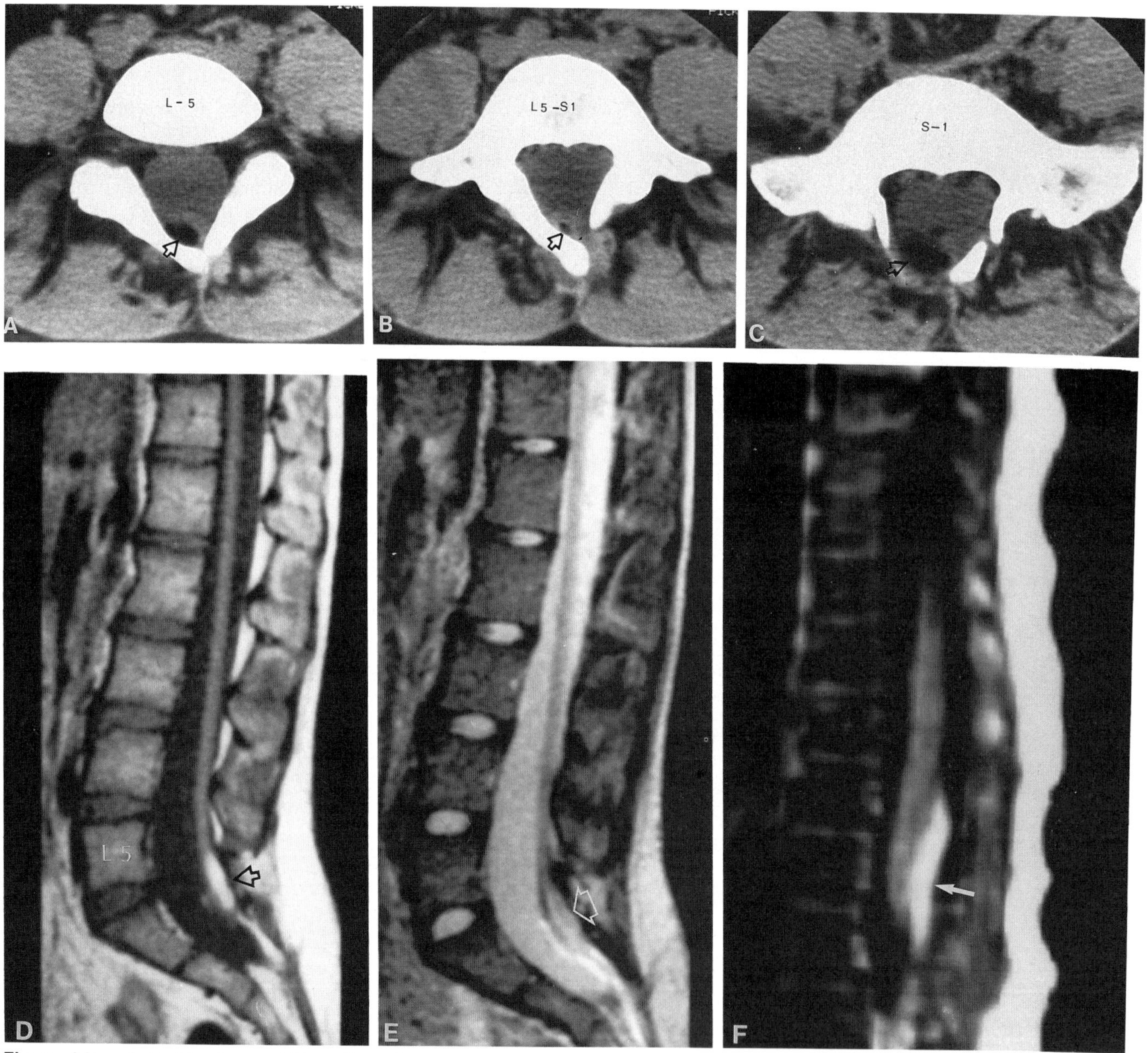

Figure 4.5. **A–C,** Contiguous CT sections demonstrating the spina bifid and associated bony defect, the hypodense lipoma (*arrowhead*). **D,** T1-weighted sagittal MRI image. **E,** T2-weighted sagittal MRI image. The spinal cord is fixed and has a "bow string" appearance. The placode is attached to a lipoma (*arrow*). The subarachnoid space is enlarged. **F,** Sagittal image in another patient demonstrating the conus medullaris tethered posteriorly by a lipoma (*arrow*).

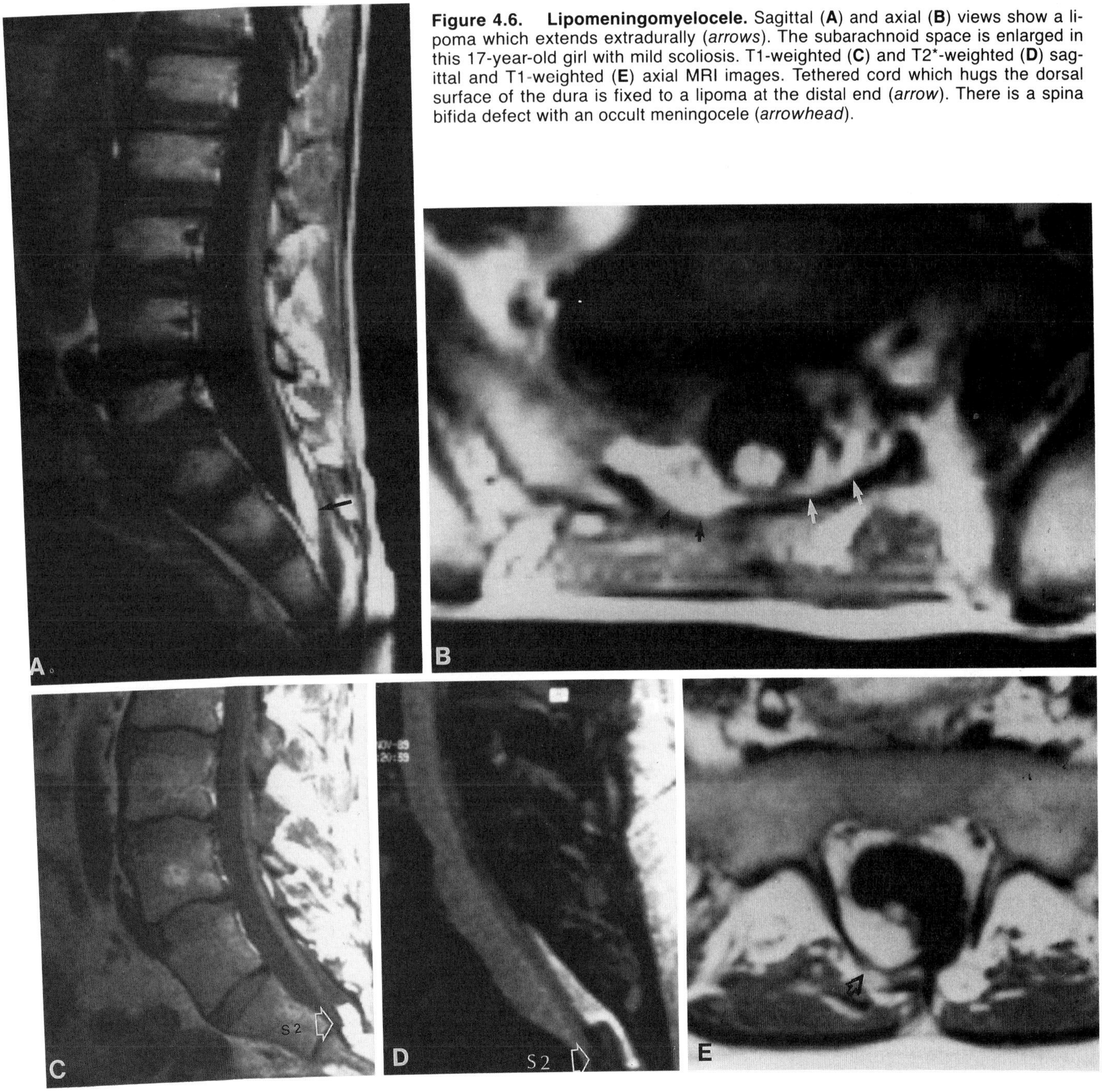

Figure 4.6. Lipomeningomyelocele. Sagittal (**A**) and axial (**B**) views show a lipoma which extends extradurally (*arrows*). The subarachnoid space is enlarged in this 17-year-old girl with mild scoliosis. T1-weighted (**C**) and T2*-weighted (**D**) sagittal and T1-weighted (**E**) axial MRI images. Tethered cord which hugs the dorsal surface of the dura is fixed to a lipoma at the distal end (*arrow*). There is a spina bifida defect with an occult meningocele (*arrowhead*).

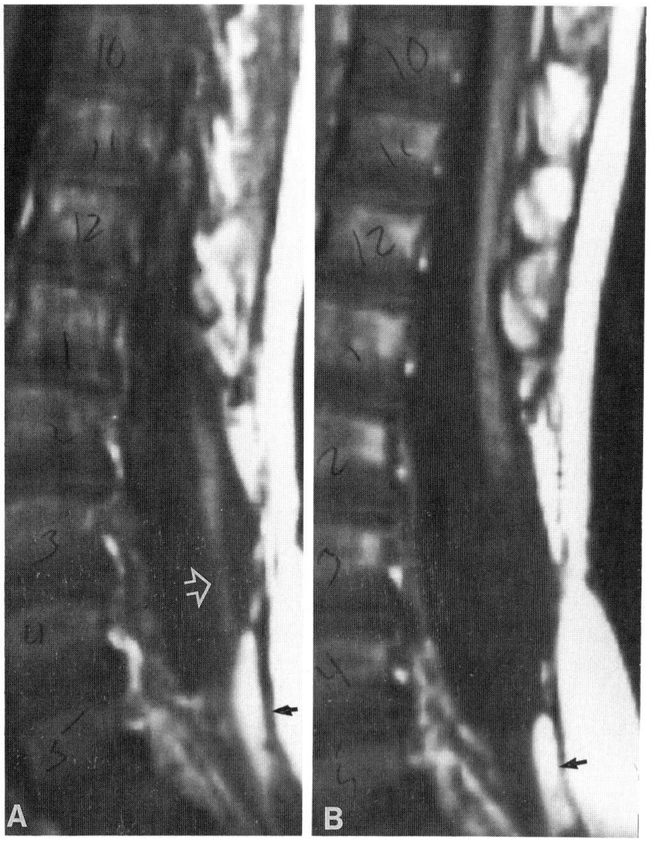

Figure 4.7. Lipomeningomyelocele. Lateral views show lipoma (*arrow*) at the caudal cord and an enlarged subarachnoid space anterior as well as posterior to the tethered spinal cord (*open arrow*).

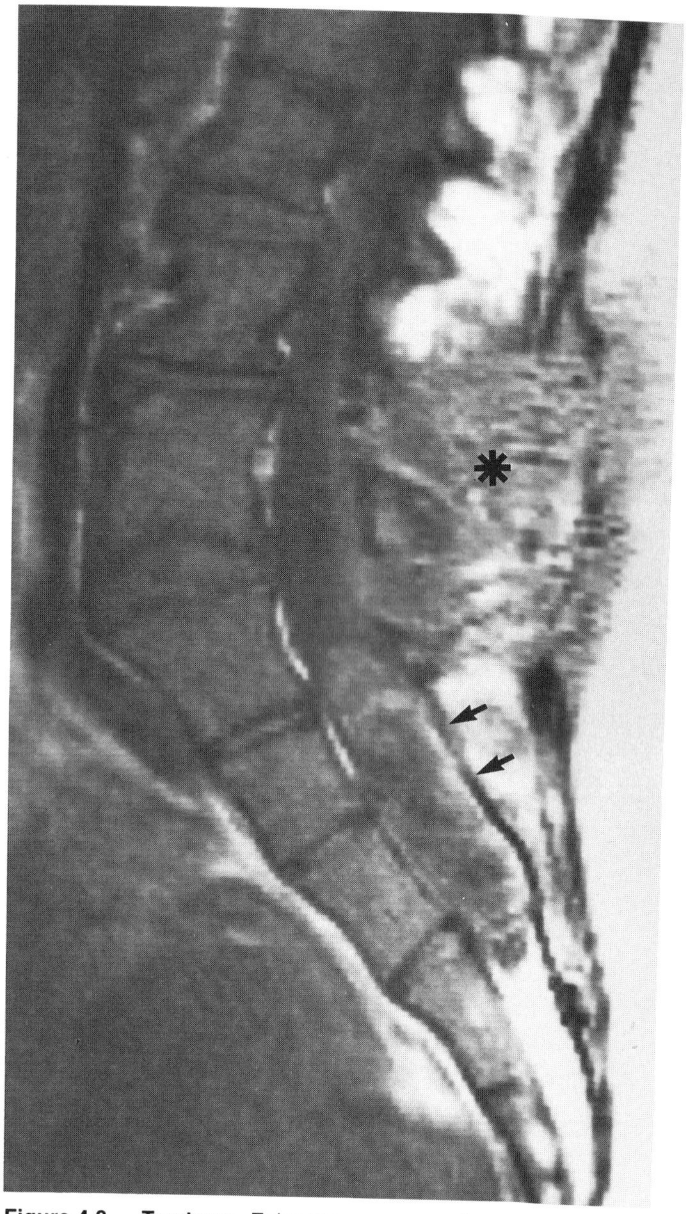

Figure 4.8. Teratoma. Enhancing mass causing tethering of the spinal cord (*arrow*) has been partly excised. Operative changes (*).

the presence of a fibrous or lipomatous component along its distal end.

Dermal Sinus

The spinal cord may be tethered at the site of a dermal sinus, which is visible on the skin surface. When the neural fold fuses to form the neural tube, the cutaneous ectoderm also fuses and is separated from the neuroectoderm. When this separation is incomplete, a sinus tract remains (Fig. 4.11). The misplaced ectodermal cells may develop into inclusion epidermoid or dermoids [6, 15]. They may become infected and act as a source for recurrent meningitis [29]. A dermal sinus is usually encountered in young patients. The skin opening of the sinus may be covered by a hairy patch. It is usually in the midline, often in the lumbosacral region, although dermal sinuses may occur at other levels, including the craniocervical region. Myelography followed by CT is useful on occasion for demonstrating the osseous anomaly or tethering of the dura where the sinus tract is fixed to the skin surface. The presence of an associated epidermoid or dermoid is seen on CT as a filling defect either dorsal or ventral to the deformed spinal cord. The exact nature of the cyst is better appreciated on MRI studies. The MRI appearances of these lesions are nonspecific; they may be solid or cystic, the signal of the latter depending on the contents (Fig. 4.12). Enhancement should raise the suspicion of secondary infection (Fig. 4.13).

Myelocystocele

This rare abnormality consists of a cystic dilatation of the conus medullaris and is thought to be due to defective regression of the caudal cell mass [27, 28]. This abnormality is covered by intact skin and is associated with bladder and cloacal extrophy and severe vertebral defects such as sacral agenesis. The mild form of this abnormality is known as *ventriculus terminalis*. The intermedullary location distin-

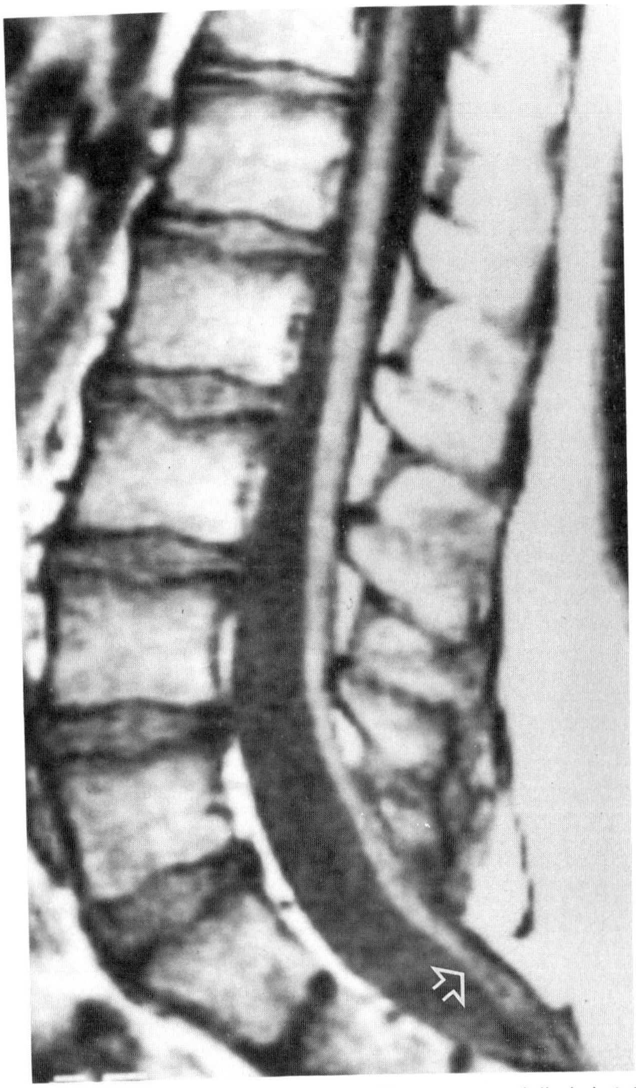

Figure 4.9. Tethered spinal cord. The conus medullaris is tethered and continuous with a thick, short filum terminale (*arrows*). Note enlarged subarachnoid space without lipoma.

guishes these cystic cavities from meningomyelocele.

SPINAL LIPOMA

Spinal lipomas are collections of fat tissue derived from the mesenchyme. They are closely attached to the spinal canal and may be intradural or partially extradural, and typically they are located along the dorsal aspect of the spinal canal. Extradural lipomas are often attached to the subcutaneous tissue, forming a soft mass under the skin.

Apart from those associated with spinal cord tethering, isolated congenital lipomas may occur at other locations within the spinal canal. Small incidental lipomas are usually asymptomatic [29]. On CT they appear hypodense (Fig. 4.14). On MRI they are recognized by the characteristic signal—hyperintense (bright) on T1- and hypointense on T2-weighted images—and by their lack of mass effect on the spinal cord (see Fig. 4.5). Lipomas may be present alongside the filum terminale without causing tethering [30] (Fig. 4.15). Widespread lipomatosis is commonly extradural but also may be intradural and involve the nerve root sheaths as well as the spinal canal (Figs. 4.16 and 4.17). Congenital extradural lipomatosis is occasionally seen to involve most of the spinal canal (Fig. 4.18) and has to be distinguished from lipomas secondary to Cushing's disease or prolonged steroid therapy. These secondary lipomas are more likely to be symptomatic [31–34] (Fig. 4.19).

DIASTEMATOMYELIA AND DIPLOMYELIA

In diastematomyelia, the spinal cord is split in the midline by a fibrous band or bony spur between the posterior margin of the vertebral body and the posterior arch. The divided cord may be enclosed within a single or separate dural sacs [35]. Diplomyelia is a true duplication of the spinal cord in which each cord is contained within individual dural sacs and bony canals and has a full complement of paired spinal nerves exiting through individual intervertebral foramina. Unfortunately, the radiologic descriptions are not critical in their distinction between diastematomyelia and diplomyelia. Using these strict criteria, most of the reported cases are probably diastematomyelia. Diastematomyelia is present in the lumbar and lower thoracic spine and occasionally involves the upper thoracic spine. Occasionally, diastematomyelia may be present at more than one level. This condition is more common in females than in males. Cutaneous stigmata are often present [66]. Scoliosis is a common feature.

Several mechanisms have been suggested for the embryogenesis of these conditions [6]: (*a*) *Split notochord theory*—The notochord is split before or after it has been formed and allows the endoderm to herniate dorsally to become adherent to the cutaneous ectoderm. A fistula between the amniotic and yolk sac forms and divides the spinal cord as well as the vertebrae into two. According to this theory, persistence of the dorsal and ventral portions results in diastematomyelia, and persistence of the ventral portion alone gives rise to an enterogenous cyst [36] (Fig. 4.20). (*b*) *Abnormal neurulation theory*—As a result of overgrowth, instead of meeting in the midline the neural folds fuse with the neural plate on their own sides and form two separate tubes. The mesenchyme is thought to fill in the midline gap between the neural tubes secondarily and subsequently form the dura and septum of a diastematomyelia [14, 37, 38]. (*c*) Diastematomyelia also has been attributed to duplication of the caudal cell mass during the canalization stage into separate

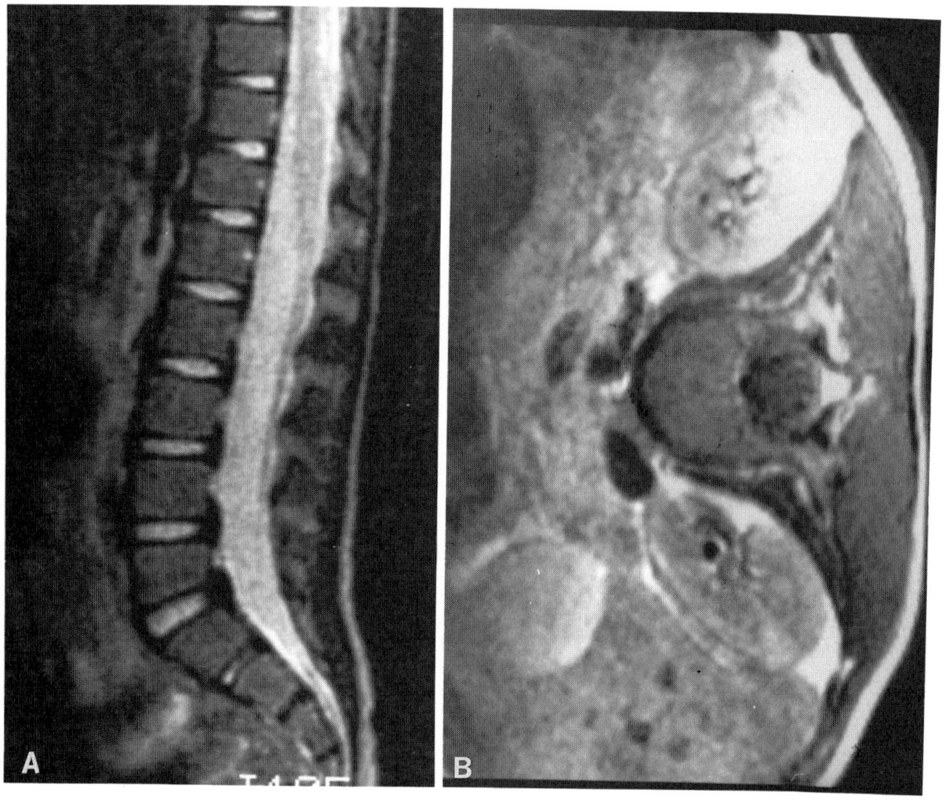

Figure 4.10. A 2-year-old boy. **A,** Sagittal MRI shows apparent thick filum suggestive of a fixed tethered cord with prominent anterior subarachnoid space. **B,** Axial image demonstrates normal appearance of the cauda equina.

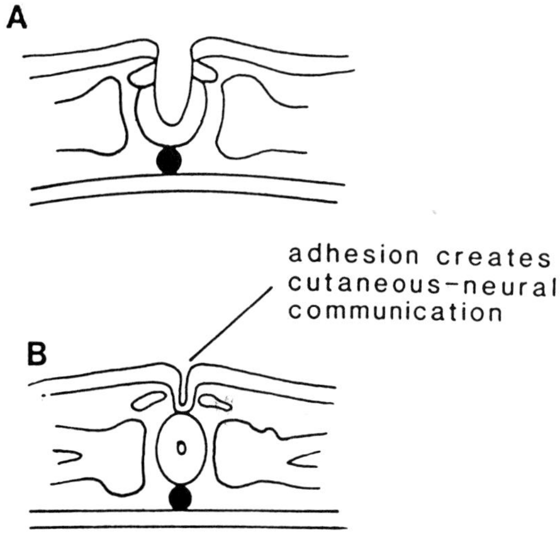

Figure 4.11. When a remnant of the cutaneous ectoderm becomes adherent to the spinal cord during neural tube closure, a dermal sinus and cyst may form.

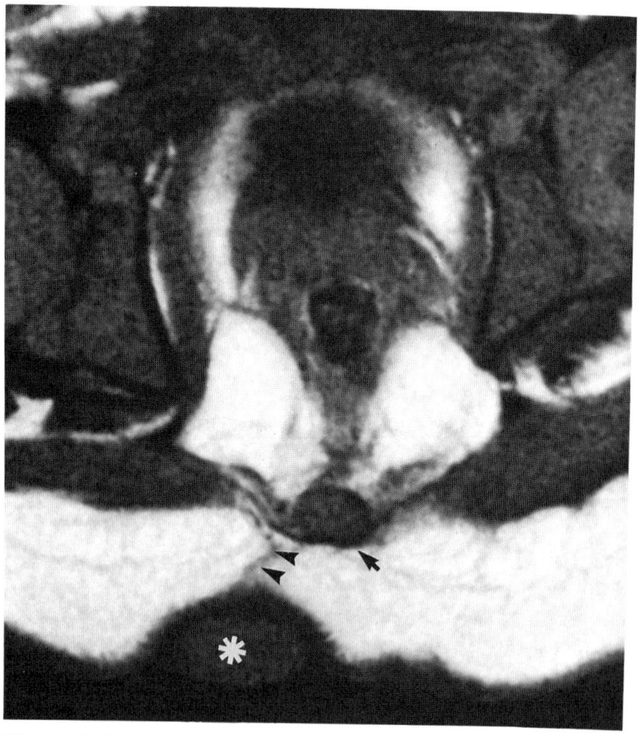

Figure 4.12. Dermal sinus. Axial view shows sinus tract (*arrowheads*) and cyst (*arrow*). A vitamin E tablet is placed at the external site of the sinus tract (*).

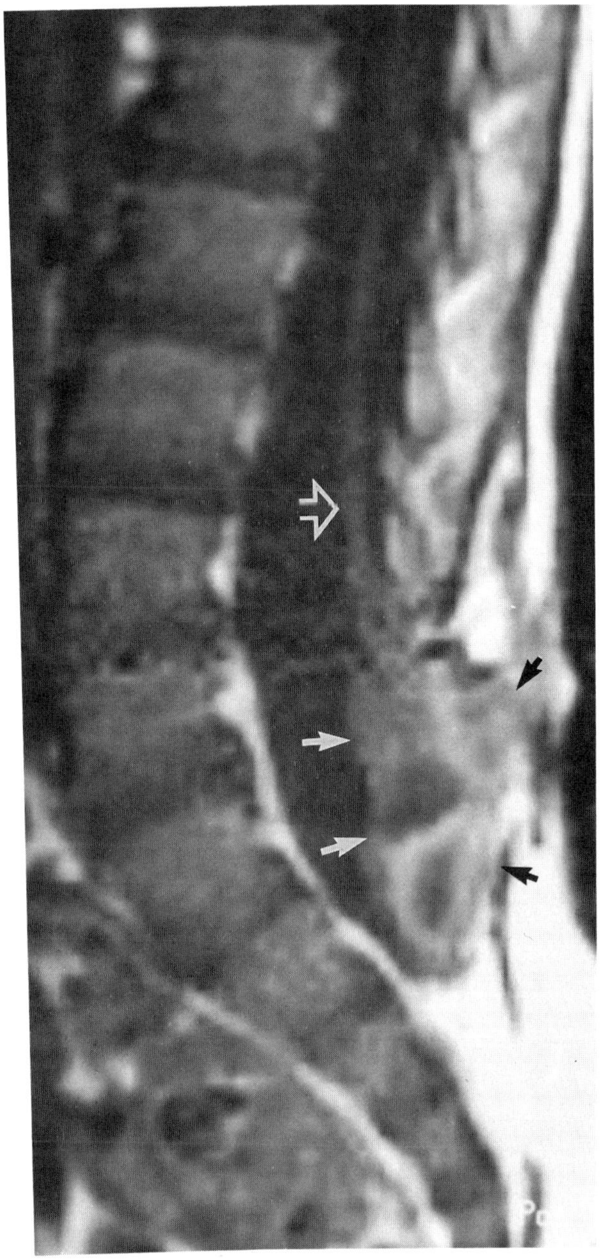

Figure 4.13. **Infected dermal cyst.** Multiloculated enhancement is shown (*arrows*). Note tethered spinal cord (*open arrow*).

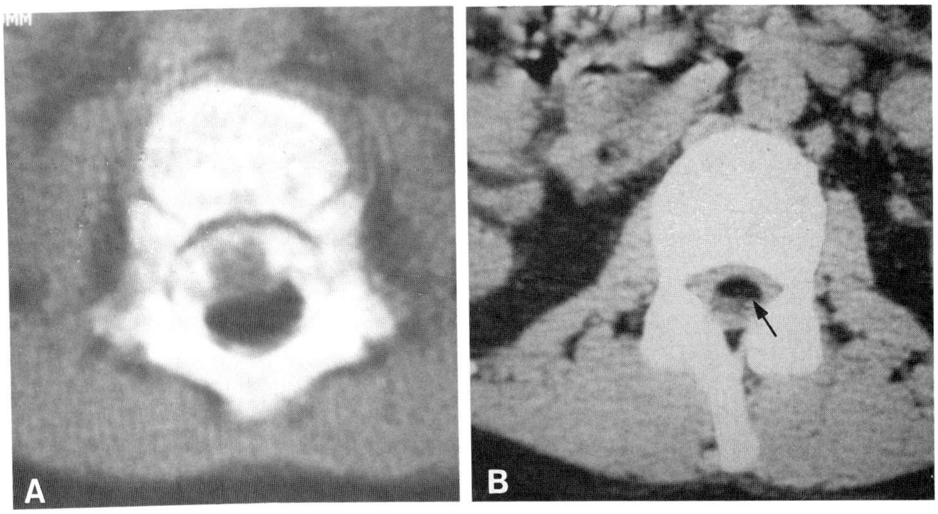

Figure 4.14. **A,** Axial CT in a child. **B,** CT in an adult. Isolated asymptomatic intradural lipoma which has decreased attenuation.

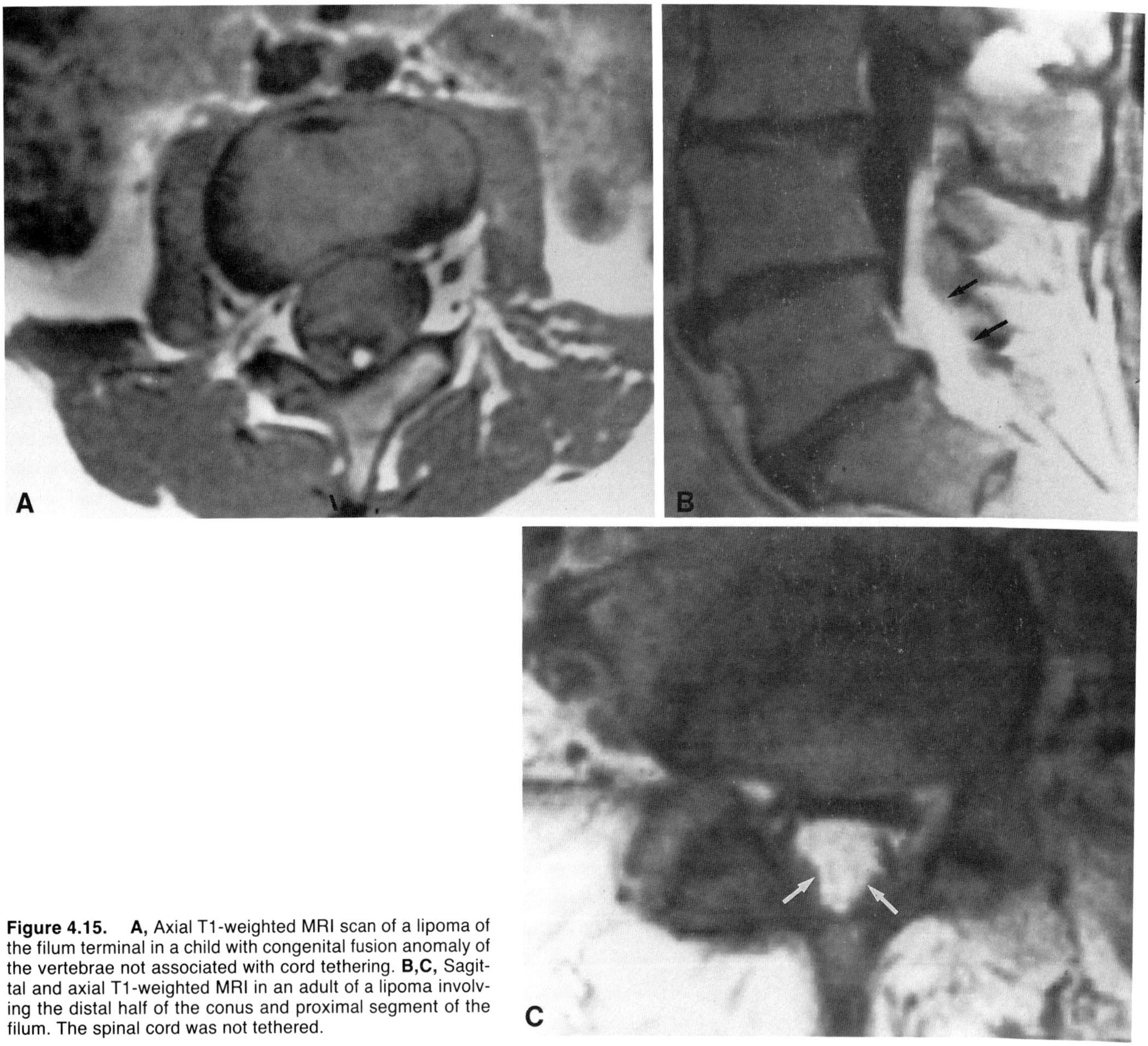

Figure 4.15. A, Axial T1-weighted MRI scan of a lipoma of the filum terminal in a child with congenital fusion anomaly of the vertebrae not associated with cord tethering. **B,C,** Sagittal and axial T1-weighted MRI in an adult of a lipoma involving the distal half of the conus and proximal segment of the filum. The spinal cord was not tethered.

Figure 4.16. Lipomatosis. A, A contiguous axial CT. Sagittal (**B**) and coronal (**C**) reformatted images in another patient with intradermal lipoma.

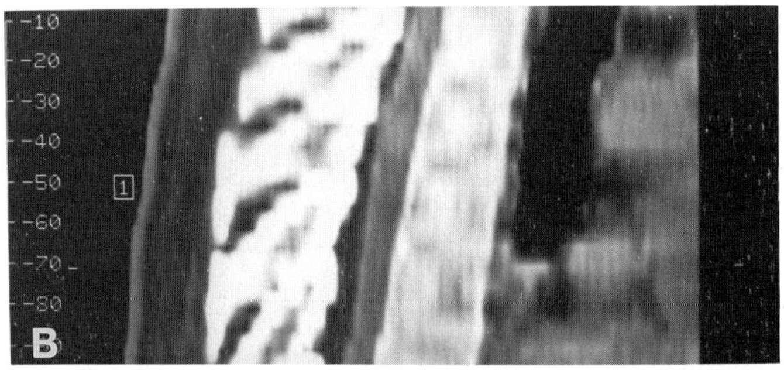

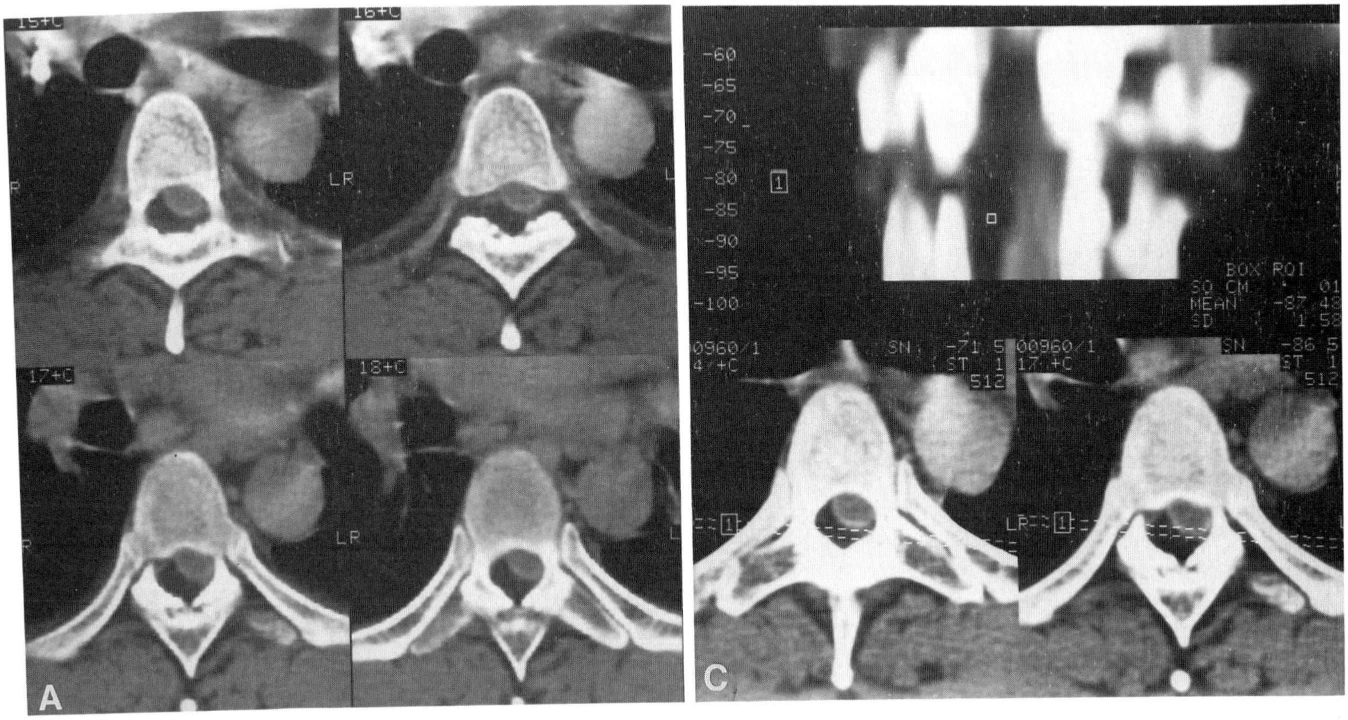

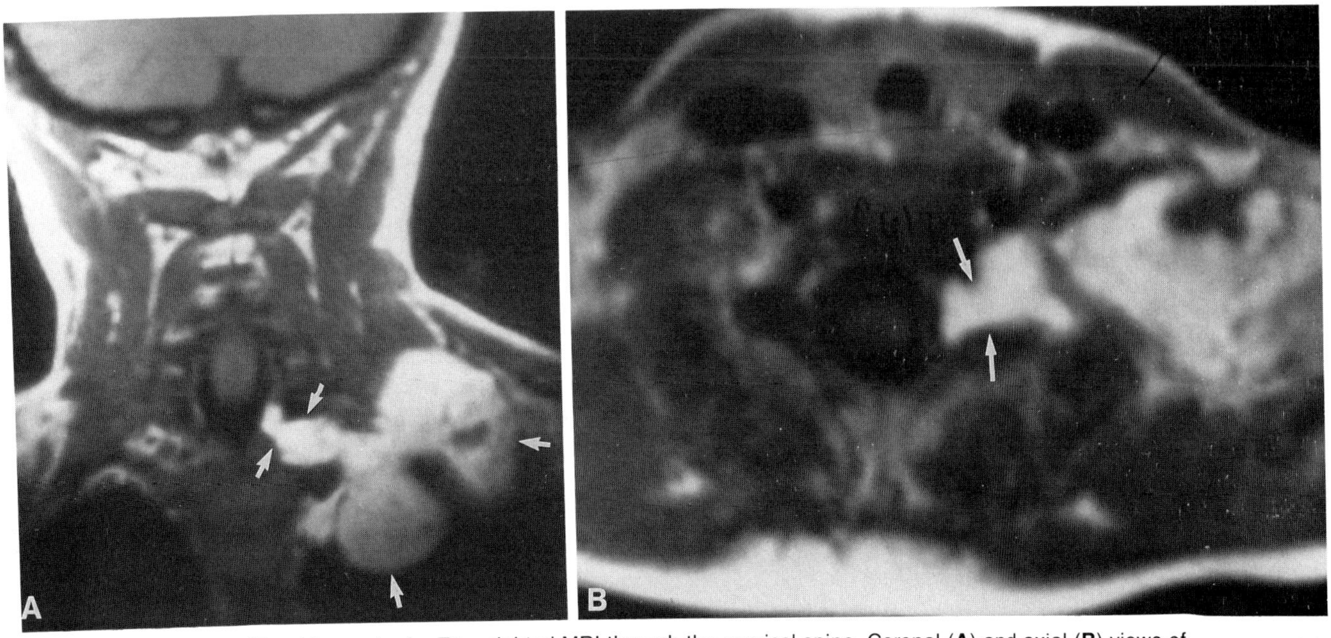

Figure 4.17. Lipomatosis. T1-weighted MRI through the cervical spine. Coronal (**A**) and axial (**B**) views of a lipoma extending through the neural foramina into the paraspinal region (*arrows*).

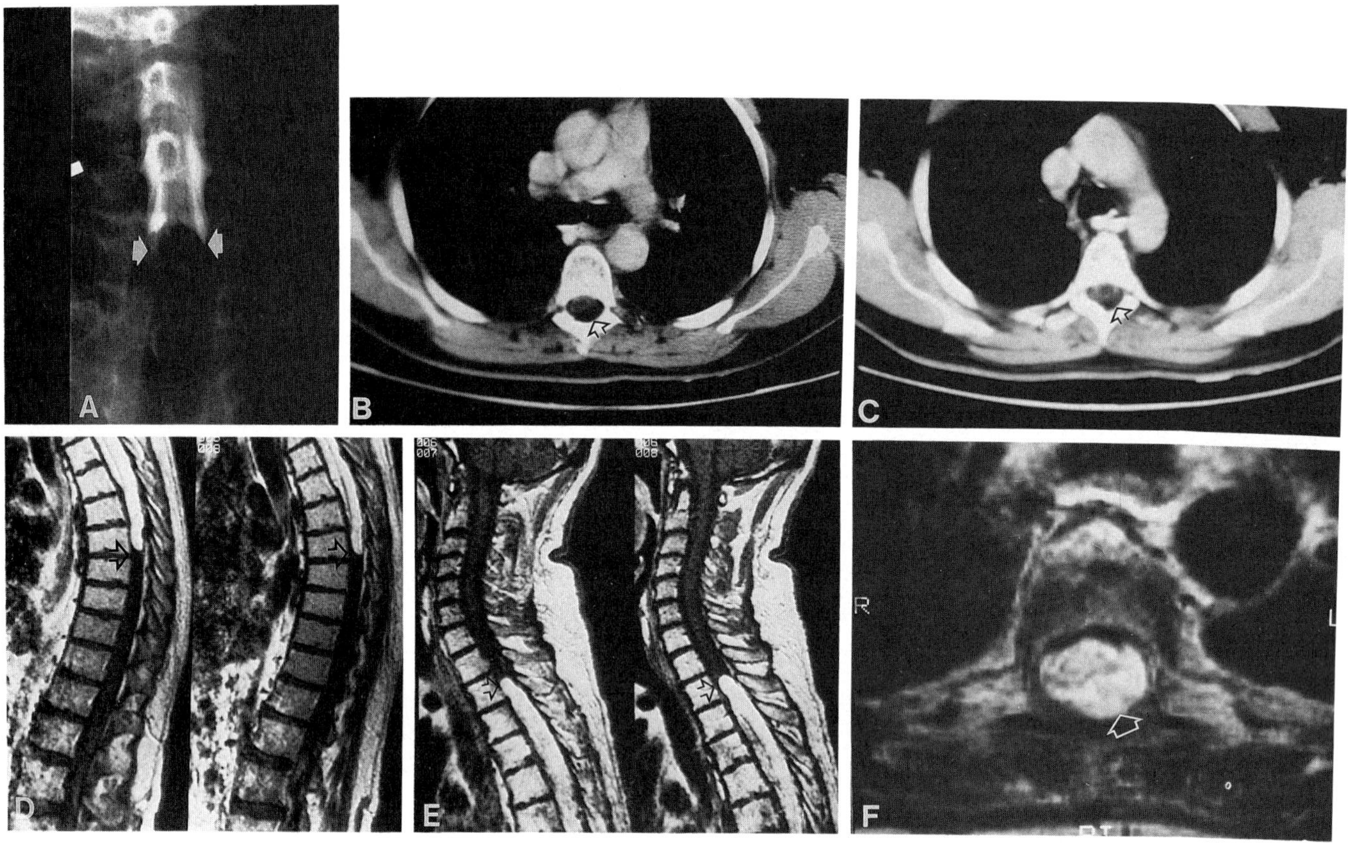

Figure 4.18. **Extradural lipoma. A,** Frontal view of the myelogram demonstrating an expansile intradural mass. **B,C,** Contiguous axial CT with the lipoma (*arrow*) seen as a hypodense mass. **D,E,** Sagittal T1-weighted MRI demonstrating the hyperintense lipoma (*arrow*) along its superior and inferior extent within the spinal canal. **F,** Axial T1-weighted MRI. The spinal cord is displaced anteriorly by the large lipoma (*arrow*).

cords [39, 40]. This theory may account for diastematomyelia of the conus medullaris but cannot explain involvement of the thoracic cord, which does not arise from this precursor [4].

On plain radiographs, a bony spur is seen in nearly 50% of cases (see Fig. 4.21). Other osseous anomalies associated with diastematomyelia includes hemivertebrae, spina bifida, partial or fused vertebral bodies, decreased height of the vertebrae, and a sagittally narrow but wide spinal canal. The widening of the canal seen as increased interpediculate distance is partly due to thinning of the pedicles. These changes are most prominent at the site of the bony spur or fibrous band but may extend one or two vertebral segments above and below the level of the split cord. Rarely, diastematomyelia may occur at more than one level (Fig. 4.22).

CT is useful in demonstrating the bony spur, but CTM is ideal in demonstrating the bony spur or fibrous band and the contents of the thecal sac. On MRI, the fibrous band of diastematomyelia is visible as a decreased or isointense signal that does not change between T1- and T2-weighted images (Fig. 4.23). The bony spur often has decreased signal but occasionally has the typical increased signal of marrow on T1-weighted images (Fig. 4.24). Butterfly and hemivertebrae may be present at several levels and may account for scoliosis and spinal deformities. The affected spinal cord may either appear single, with a midline split at one or several levels, or may be frankly divided into two separate cords, with discernible dural sacs. The appearances of true diplomyelia, using the strict criteria mentioned above, have not, in our view, been documented adequately. The chords may be anchored by the midline fibrous and bony lesion or by actual tethering of the conus medullaris (Fig. 4.25).

CAUDAL REGRESSION SYNDROME

This syndrome is associated with varying anomalies such as agenesis of the caudal spinal elements, absence of one or both sides of the pelvis, and fusion of both lower extremities. The agenesis of the caudal spinal elements may be as high as the L1 or T12 vertebra [67]. The syndrome includes nonneurologic elements such as congenital anomalies of the lung parenchyma and the cardiac and renal systems. In those who survive, nearly 30% are mentally retarded. The anomaly is presumed to result from developmental arrest before the fourth week of gestation.

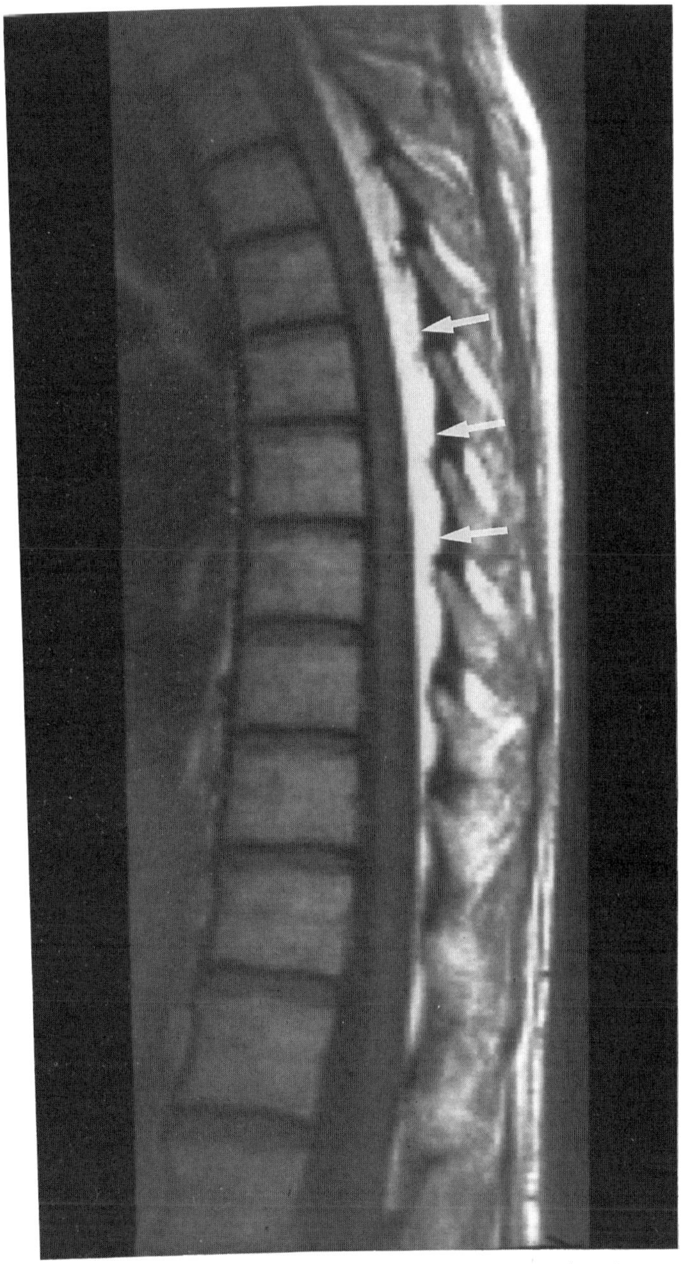

Figure 4.19. **Lipomatosis.** Epidural lipoma secondary to prolonged steroid therapy (*arrows*).

On imaging studies, the lumbar vertebrae up to the L1 level may be absent. When the caudal element is down to the L5 vertebrae, there is often sacral agenesis. The spinal canal below the lowest intact vertebra is often hypoplastic, narrow, or in the form of a fibrous band.

Imaging studies are necessary to identify the extent of the osseous abnormality and the severity of the lumbosacral agenesis. MRI provides a noninvasive evaluation of the extent of the spinal cord within the identifiable spinal canal [68] (Fig. 4.26). MRI is also useful in identifying the level of the conus and the presence of other anomalies at a higher level. Determination of the tethered cord and presence of

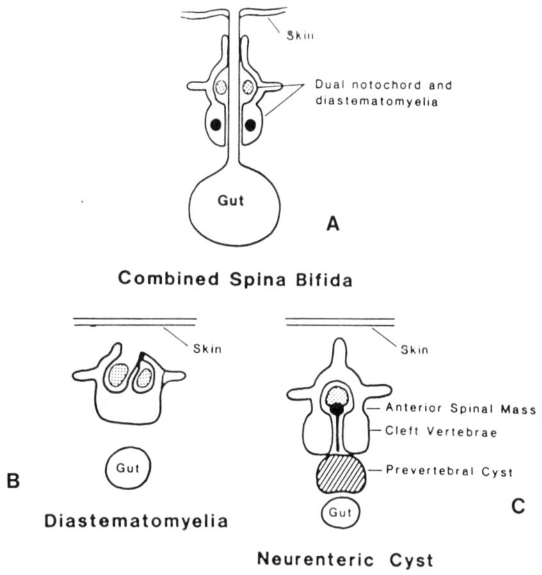

Figure 4.20. Split notochord. A, Where there is a persistent communication between the ectoderm (skin) and the endoderm (gut), the neural tube and notochord (vertebrae and disks) are split. **B–C,** Diastematomyelia or neuroenteric cyst (and butterfly vertebral body) is present depending on where the posterior or anterior portion of the communication persists.

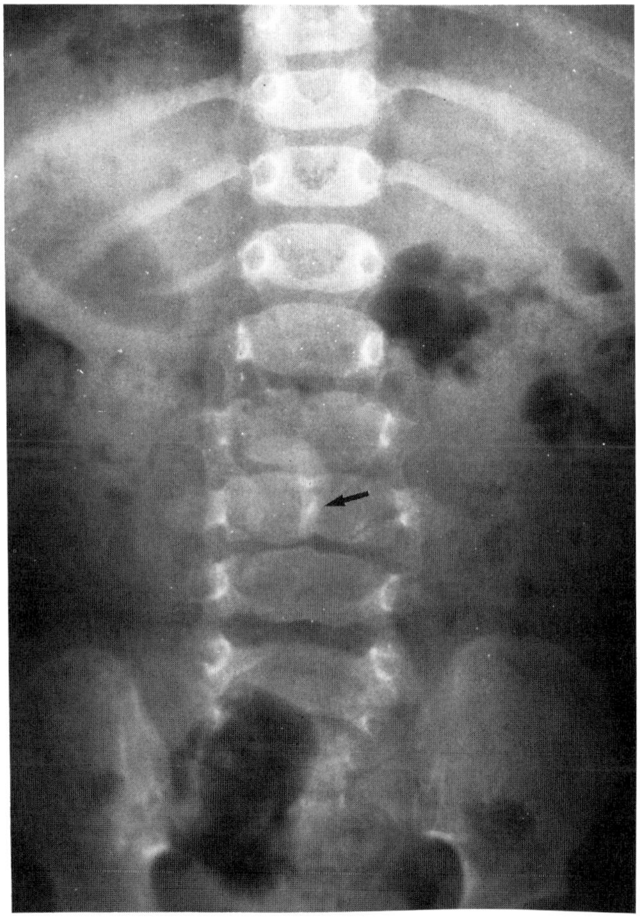

Figure 4.21. Diastometomyelia. Frontal view of spine shows a midline bony spur (*arrow*) at the L3 level. There is thinning of the pedicle from the L1 to the L5 level with increased interpediculate distance.

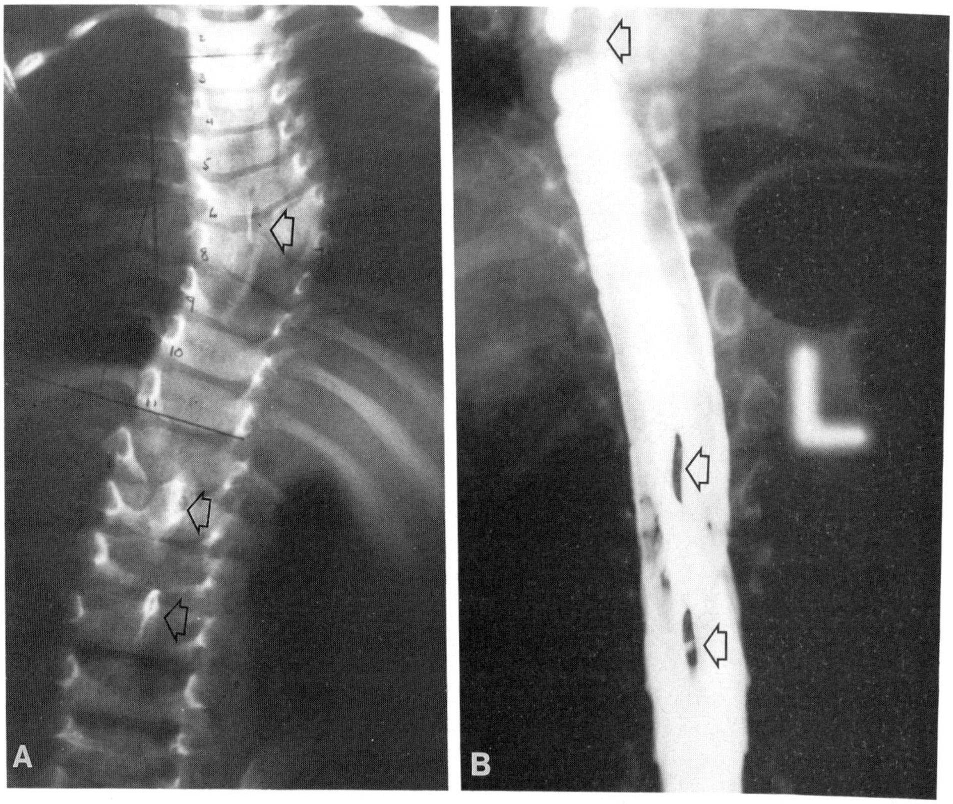

Figure 4.22. Diastometomyelia at multiple levels. A, Frontal view of the thoracolumbar spine demonstrates midline bony spurs at three levels (*arrows*). There is scoliosis associated with a hemivertebrae (T7). **B,** Oil-based myelogram demonstrates the tethered cord from the bony spurs.

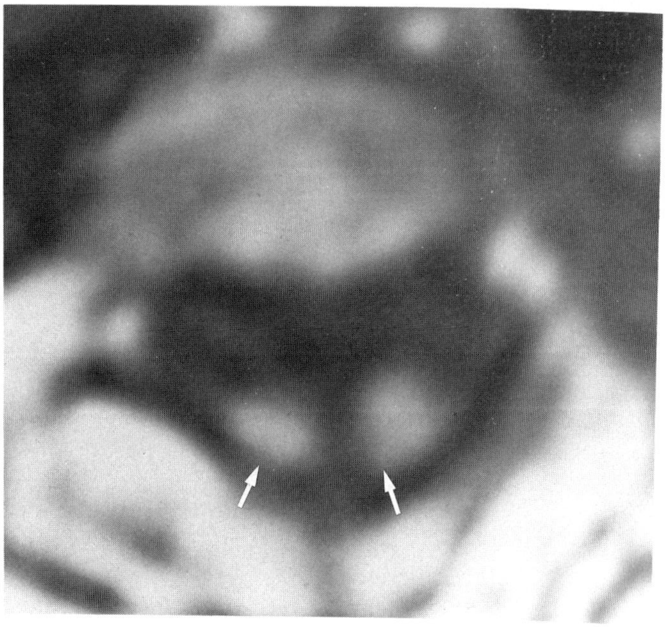

Figure 4.23. Diastematomyelia. Axial T1-weighted MRI shows the divided spinal cord (*arrows*).

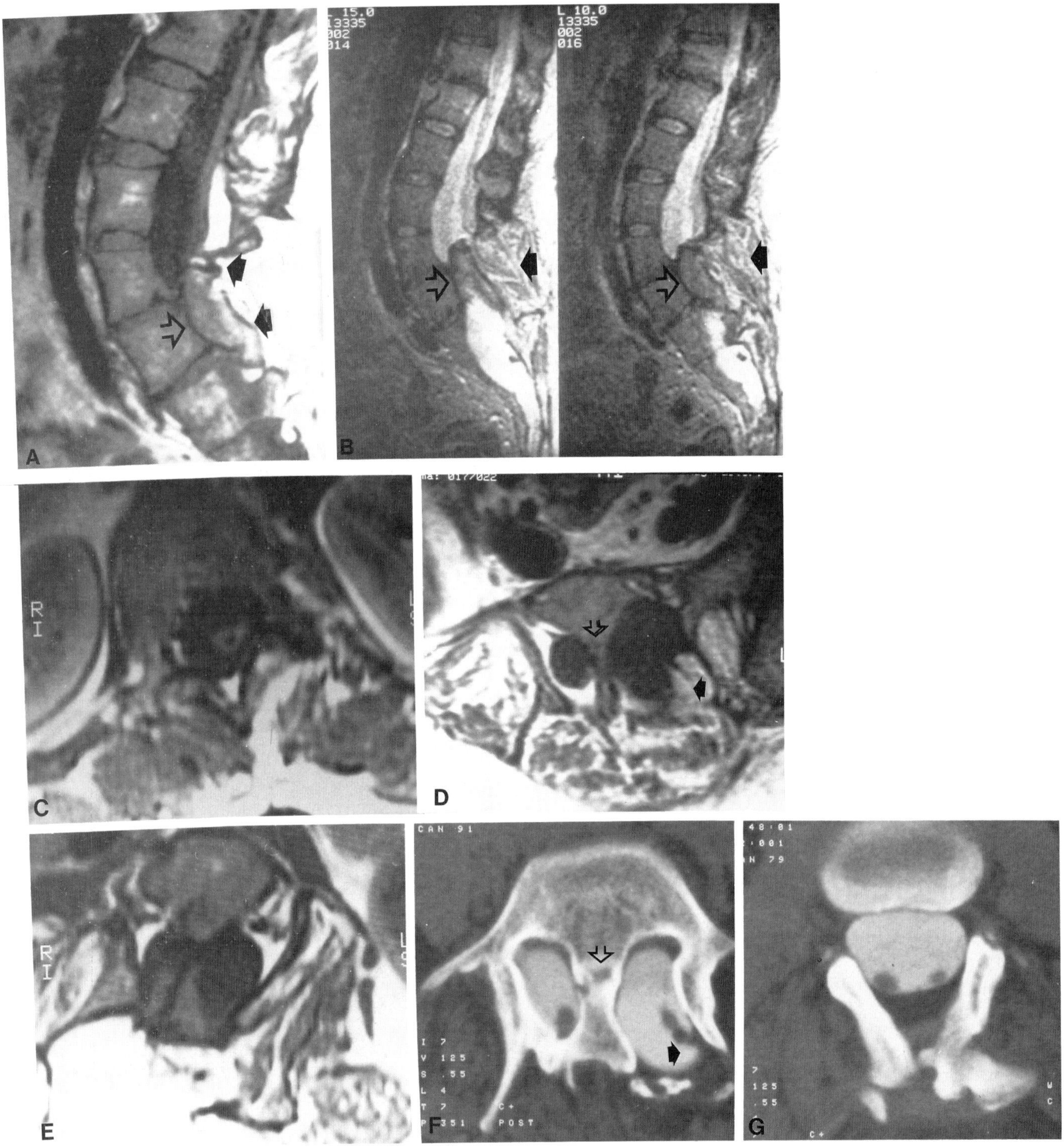

Figure 4.24. Diastometomyelia. T1-weighted (**A**) and T2-weighted (**B**) sagittal MRI images. Tethered cord fixed to an elongated, broad-based bony spur (*open arrow*) at the L5–S1 levels. There is absence of the normal spinous process as a result of the bony fusion anomaly (*arrow*). **C–E,** Axial MRI sections show the divided thecal sac, the bone spur with marrow signal (*open arrow*), and the syrinx within the attached conus. **F,G,** Axial CTM demonstrating the thick bony spur with the conus attached within the midline bony mass (*open arrow*), the duplication of the spinal canal and thecal sac, and at a higher level the enlarged thecal sac.

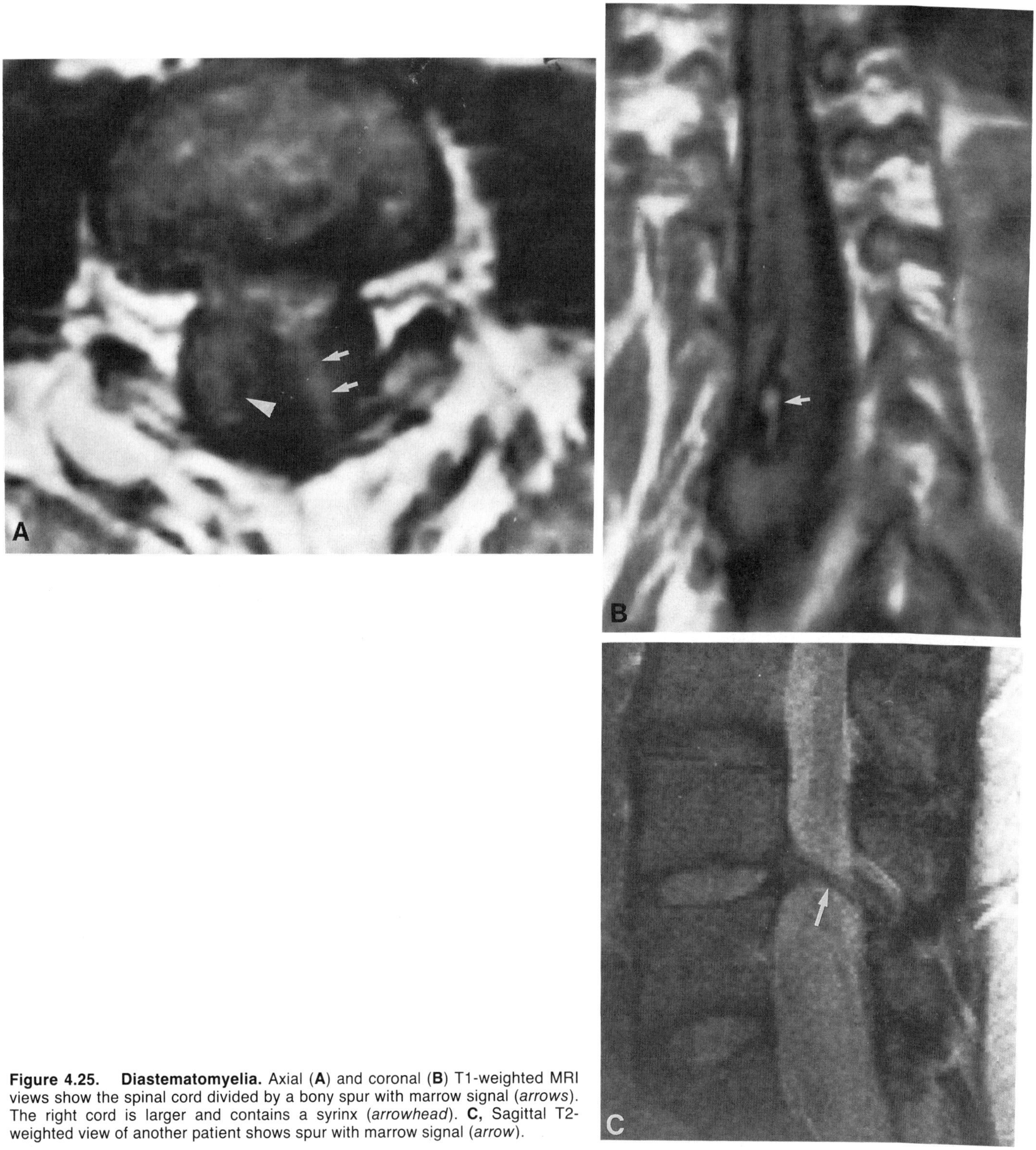

Figure 4.25. Diastematomyelia. Axial (**A**) and coronal (**B**) T1-weighted MRI views show the spinal cord divided by a bony spur with marrow signal (*arrows*). The right cord is larger and contains a syrinx (*arrowhead*). **C,** Sagittal T2-weighted view of another patient shows spur with marrow signal (*arrow*).

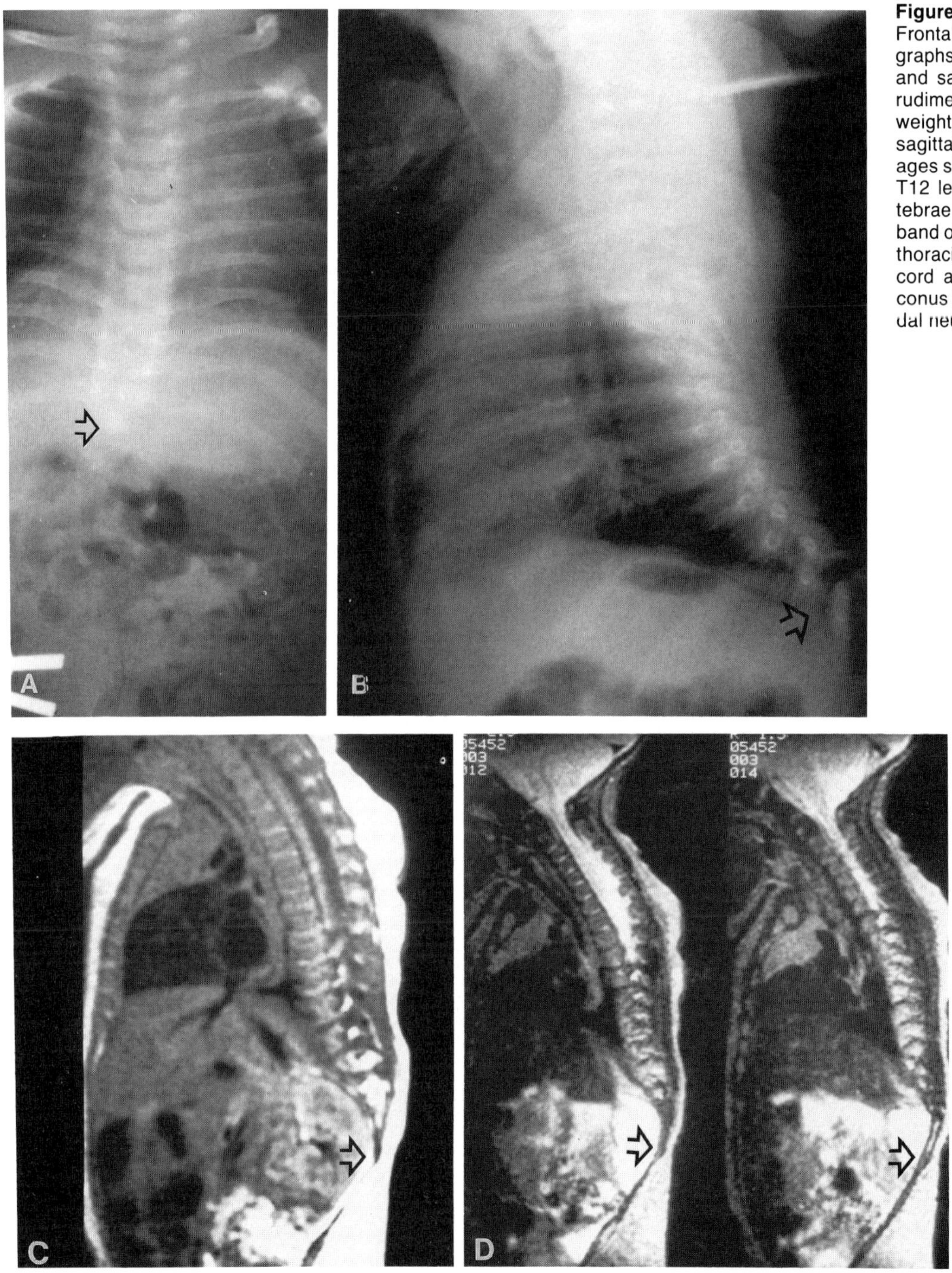

Figure 4.26. Caudal regression. Frontal (**A**) and lateral (**B**) plain radiographs show absence of the lumbar and sacral vertebrae in an infant with rudimentary lower extremities. T1-weighted sagittal (**C**) and T2-weighted sagittal and parasagittal (**D**) MRI images show the spinal cord ending at the T12 level. The lumbar and sacral vertebrae are replaced with a thick fibrous band of tissue (*arrow*). The cervical and thoracic vertebrae as well as the spinal cord are normal, with absence of the conus indicating regression of the caudal neuropore.

fibrous bands may be difficult due to the varying signal characteristics of the abnormal soft-tissue elements.

SYRINGOMYELIA

Syringomyelia is frequently present in all types of spinal dysraphism. The pathogenesis and MRI appearances have been discussed previously. The syringomyelia may involve a part or the entire spinal cord and become continuous with a dilated ventriculus terminalis of the conus medullaris (Figs. 4.4 and 4.27). Syrinxes may be present in one or both tubes of the divided spinal cords [41] (Figs. 4.24*B* and 4.28).

SCOLIOSIS

Scoliosis is common in all types of occult spinal dysraphism. Tethering of the conus medullaris, at the site of diastematomyelia (see Fig. 4.22), neurologic dysfunction secondary to syringomyelia, and actual mechanical deformity from bony lesions such as hemivertebrae may all contribute to the deformity. It is therefore imperative to visualize the entire spinal cord and conus medullaris in all cases.

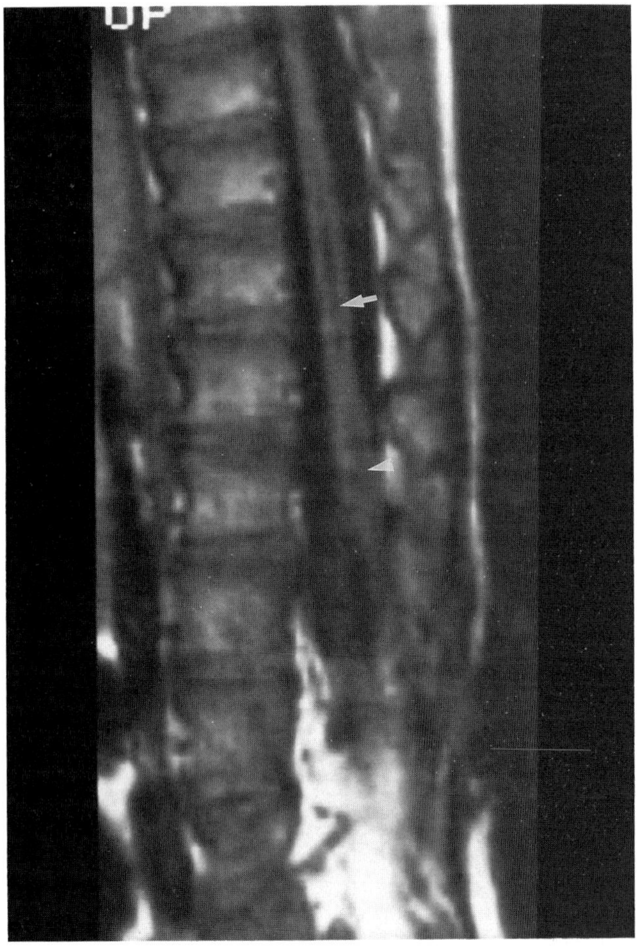

Figure 4.27. Syringomyelia. Syringomyelia in the lower spinal cord (*arrows*) is present adjacent to a similar cavity (ventriculus terminalis) in the conus medullaris (*arrowheads*). Note large lipoma caudal to the conus.

NEUROENTERIC CYSTS

Neuroenteric cyst is a rare congenital anomaly with a cystic mass usually located in the intradural extramedullary compartment of the spinal canal. Neuroenteric cysts are the result of interposition of an endodermal cell rest between the primitive notochord and the neuroectoderm (see Fig. 4.20). Occasionally, the cyst may communicate through a butterfly vertebra or hemivertebra, with the extraspinal component of the cyst lining the gastrointestinal (GI) tract within the mediastinum or upper abdomen. More often the anteriorly located intradural cyst is attached to the vertebra or the GI tract by a fibrous band. The intradural cyst is usually located in the cervical region, less often in the lower thoracic region adjacent to the conus. The cyst usually becomes symptomatic as a result of an increase in size from mucinous secretions by the columnar mucin-secreting epithelium of the cyst wall. Patients usually present with symptoms of cord compression between the second through fourth decades.

Plain radiographs often demonstrate fusion anomalies of the vertebral bodies, such as hemivertebrae, butterfly vertebrae, or partially fused vertebrae. On myelography and CTM, a smooth, lobulated extramedullary mass, usually opposite to the bony changes, is present. Differentiation from an anterior intra- or extradural arachnoid cyst is difficult. On plain CT, the mucinous secretions of the neuroenteric cyst have a higher density than the surround-

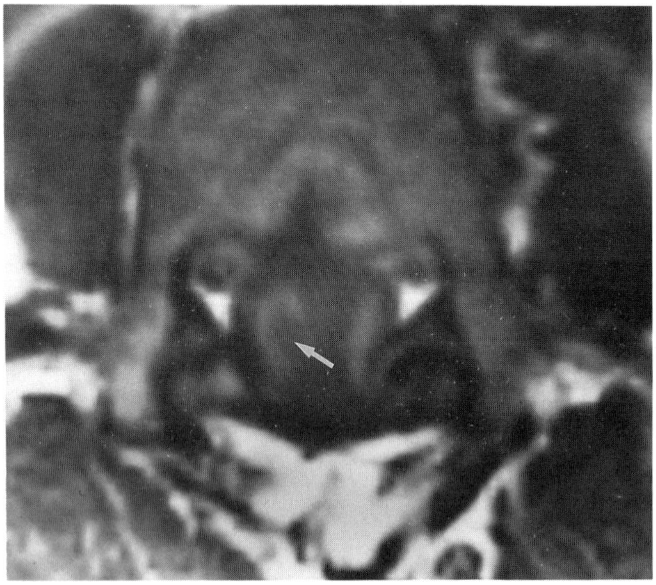

Figure 4.28. Syringomyelia. Syrinx is present in the right cord of diastematomyelia (*arrow*).

ing CSF. MRI is ideal in demonstrating the location of the cyst, its contents, and the associated cord compression without the need for invasive studies (Fig. 4.29). On MRI, the cyst may have the same signal characteristics as CSF, although more often it is mildly hyperintense on T1-weighted images. The cyst does not enhance with contrast material and often compresses the spinal cord [42–47] (Fig. 4.30). Enterogenous cysts have to be differentiated from arachnoid cysts, which are common in the thoracic region and are usually located posteriorly.

DURAL SAC ABNORMALITIES

Sacral and Thoracic Meningoceles

Unlike the dorsal midline cervical and lumbosacral meningomyeloceles, these anomalies are not defects of neural closure or caudal tube development. Anterior sacral meningocele is an absence of a portion of the sacral vertebra which results in herniation of the spinal contents anteriorly [48, 49]. It is not associated with an abnormal spinal cord, nor with tethering of the conus medullaris. Sagittal MRI scans are ideal for evaluating the extent of this abnormality (Fig. 4.31). Lateral thoracic meningocele is associated with enlargement of the neural foramen which allows formation of a meningocele. This abnormality usually occurs by itself but may be part of a more generalized dural ectasia of neurofibromatosis [50]. Solitary cysts are thought to result from repeated episodes of subarachnoid pressure or trauma [51, 52]. The MRI signal of CSF within lateral meningoceles distinguishes these meningoceles from neurofibromas which cause similar enlargements.

Dural Ectasia and Lumbar Meningoceles

Dural ectasia consists of dilatation of the spinal canal and nerve root sheaths and may be part of the generalized spinal dysplasia of neurofibromatosis or may occur alone without other associated development abnormalities [53–55]. Enlargement of the bony spinal canal and neural foramina should not be confused with that caused by spinal cord and nerve tumors, since this abnormality is associated with abnormalities of the spinal cord or nerves (Fig. 4.32). Dilatation of the lumbar and sacral nerve root sheaths may be present alone (Fig. 4.33) or may be part of a generalized dural ectasia. Commonly seen dilatations of the nerve root sleeves known as *Tarlov cysts* are of no significance (Fig. 4.34) but may be quite striking in patients with severe ankylosing spondylitis [56–60]. The mechanism of formation in this condition is unclear, but it may be associated with increased pulsatile pressure of the CSF or adhesions. Dural ectasia and dilated root cysts generally are asymptomatic and should not be confused with pseudomeningoceles secondary to trauma, which often require surgical correction [61].

VERTEBRAL ANOMALIES

The notochord is intimately related to the development of the spinal cord and vertebral column. It is not surprising that a dysraphic spinal cord is often associated with bony vertebral abnormalities. The connections between spina bifida, ectasia, hemivertebrae and butterfly vertebrae with meningomyelocele, lipomeningomyelocele, and diastematomyelia were discussed earlier. Failure of the segmentation involving several vertebrae seen as block vertebrae also may be associated with abnormalities of the spinal cord, whereas fusion of one or two vertebrae is often asymptomatic.

Each vertebral segment develops in a cephalocaudad fashion from the mesodermal element, which separates into the ventromedial sclerotome. Each sclerotome subsequently undergoes cartilage

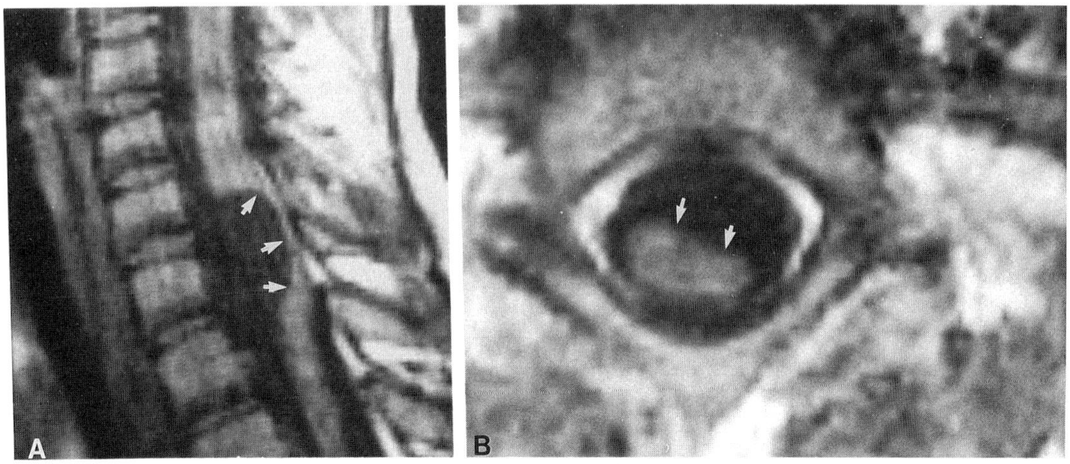

Figure 4.29. Neuroenteric cyst. Sagittal (**A**) and axial (**B**) views show cyst anterior to and compressing the spinal cord (*arrows*).

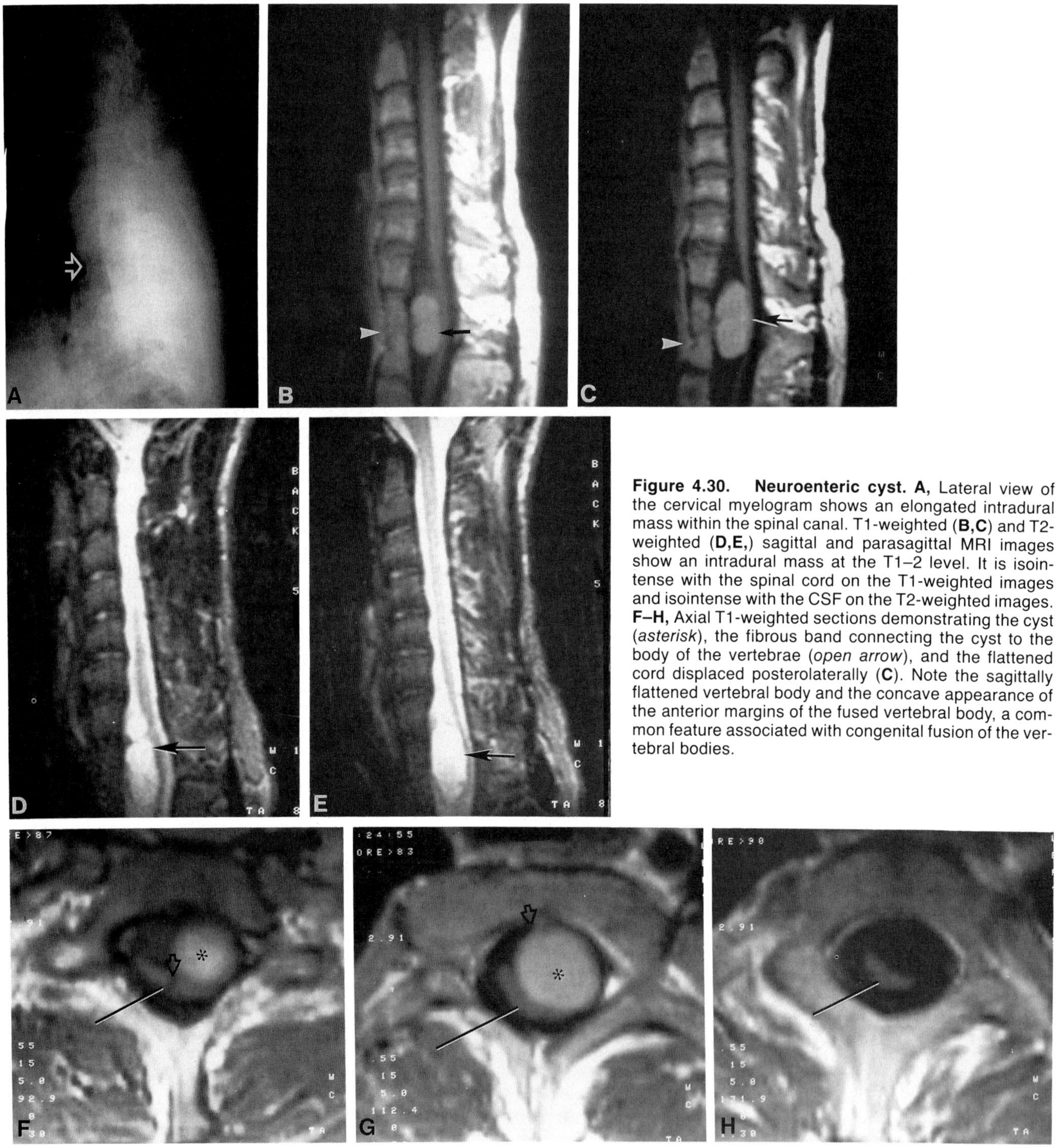

Figure 4.30. Neuroenteric cyst. A, Lateral view of the cervical myelogram shows an elongated intradural mass within the spinal canal. T1-weighted (**B,C**) and T2-weighted (**D,E,**) sagittal and parasagittal MRI images show an intradural mass at the T1–2 level. It is isointense with the spinal cord on the T1-weighted images and isointense with the CSF on the T2-weighted images. **F–H,** Axial T1-weighted sections demonstrating the cyst (*asterisk*), the fibrous band connecting the cyst to the body of the vertebrae (*open arrow*), and the flattened cord displaced posterolaterally (**C**). Note the sagittally flattened vertebral body and the concave appearance of the anterior margins of the fused vertebral body, a common feature associated with congenital fusion of the vertebral bodies.

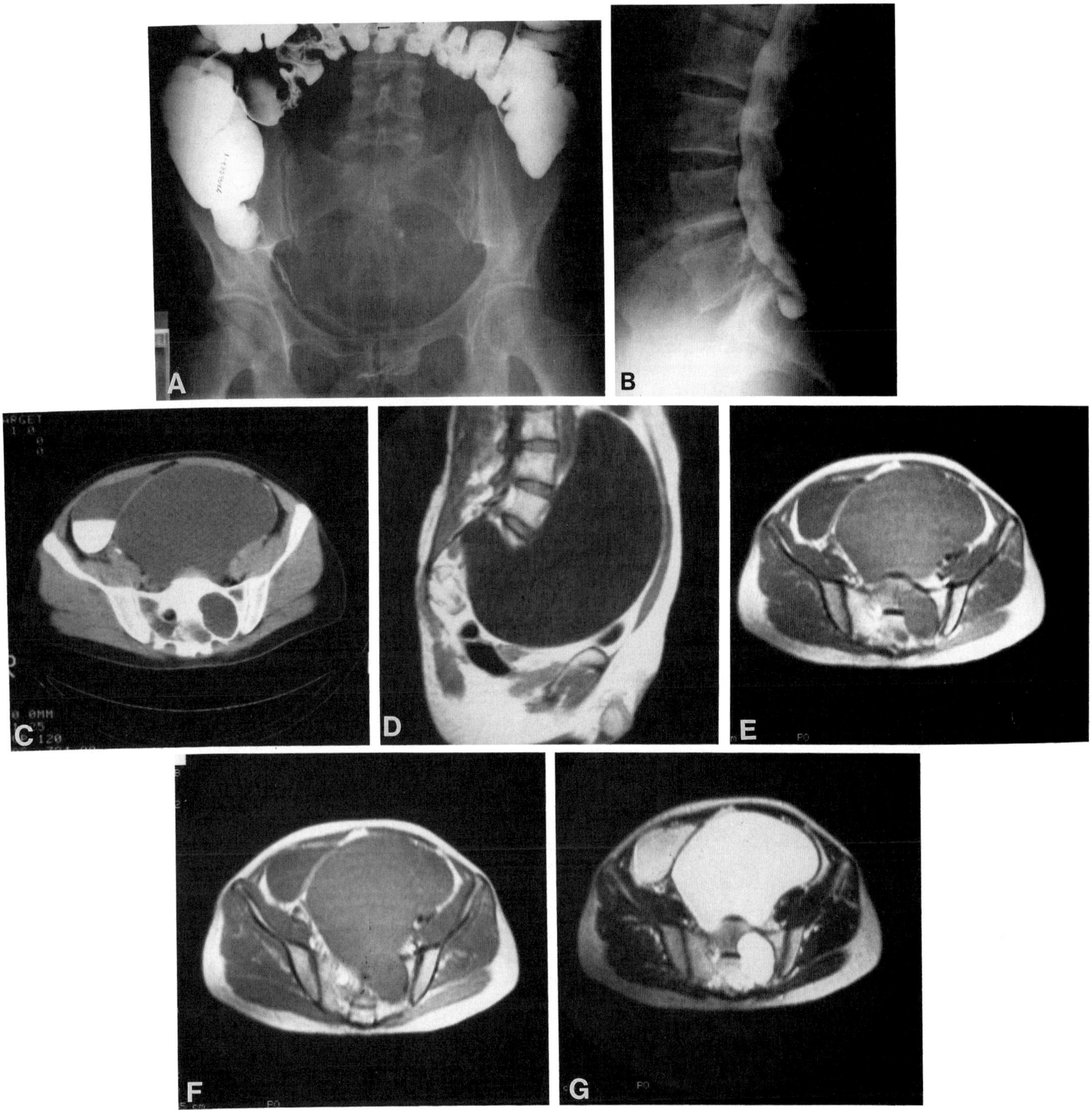

Figure 4.31. Anterior meningocele. A, Frontal view of the abdomen shows elevation of the barium-filled colon from a large pelvic mass. **B,** Lateral view of the lumbar myelogram demonstrates contrast material in an enlarged intradural cyst anterior to the thecal sac. **C,** CT section through the lower pelvis shows marked enlargement of the sacral neural foramen (*arrow*) associated with presacral cystic mass (*arrowheads*). **D,** T1-weighted parasagittal MRI shows a large cystic mass extending between the sacral foramen and the pelvic cavity. **E,F,** Axial MRI balanced images demonstrating the extension of the pelvic mass into the enlarged sacral foramen. **G,** Axial T2-weighted MRI identifying the anterior sacral meningocele.

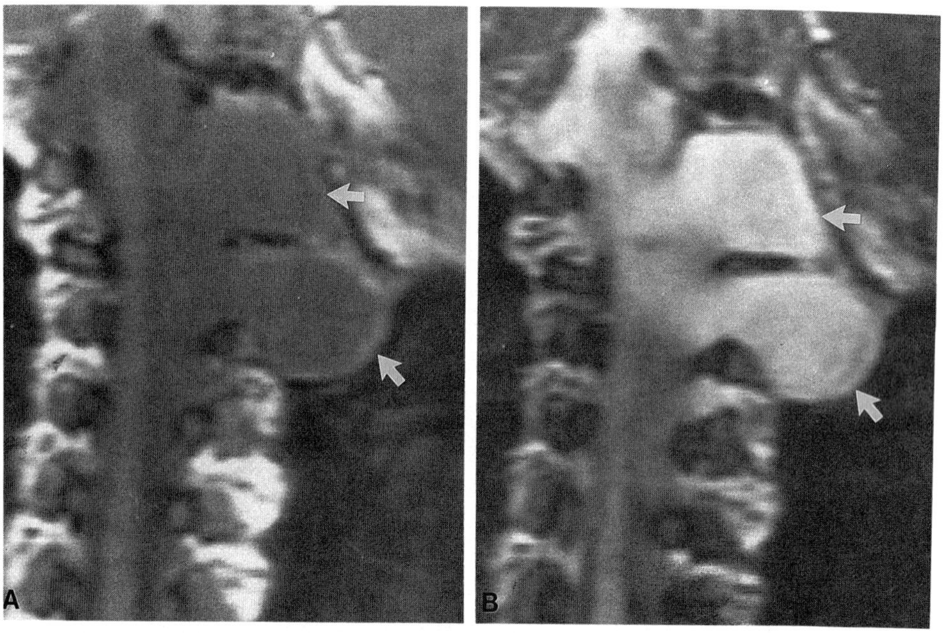

Figure 4.32. Dural ectasia. Proton-density (**A**) and T2-weighted (**B**) images show enlarged upper cervical nerve root sheaths (*arrows*). The lack of neural tissue or mass differentiates them from neurofibromas.

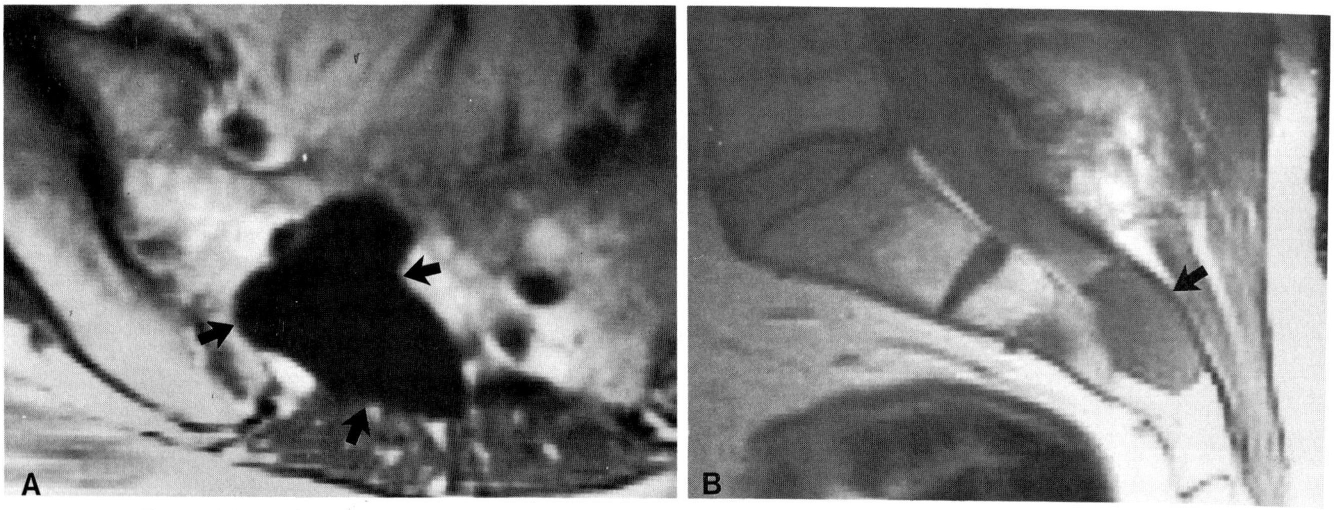

Figure 4.33. Nerve root cysts. A, Axial view shows a very enlarged sacral root sheath (*arrows*) with CSF signal characteristics. **B,** Sagittal view shows a smaller cyst in another patient (*arrow*).

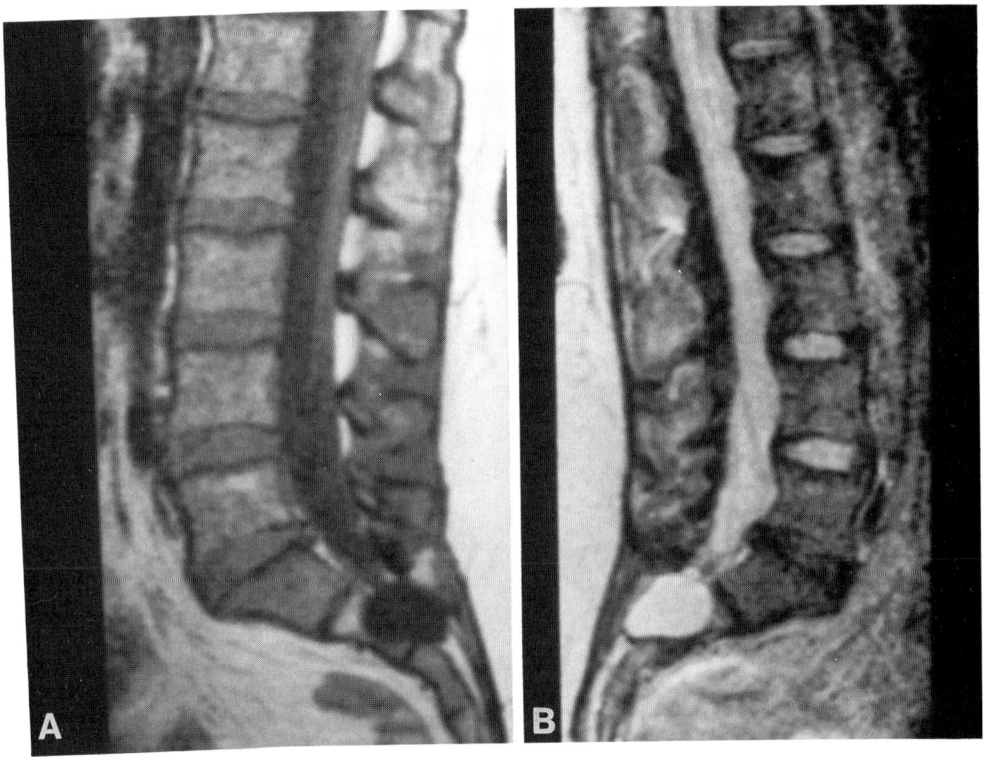

Figure 4.34. Tarlov cyst. T1-weighted (**A**) and T2-weighted (**B**) sagittal MRI sections show an incidental perineural cyst causing scalloping of the sacral vertebral body. Note the brighter signal intensity of the cyst content compared with the CSF in the thecal sac.

deposition and subsequently ossification. Osseous anomalies are the result of arrest during the process of differentiation and fusion of the ossification centers of each vertebra. Although the majority of these anomalies can be identified on plain radiographs, they are best evaluated by CT.

The most common of these anomalies is spina bifida occulta. The anomaly is the result of nonfusion of the adjacent neural arches posteriorly. It is more common in females than in males. The vertebrae most often involved include L5, S1, C1, and C7. In the thoracic region, this anomaly is variable but common.

Hemivertebrae result from persistence of the ventral and dorsal ossification center for each centrum which fail to unite in the midline. This is common in the upper thoracic region but may involve the cervical vertebrae (Fig. 4.35). Unilateral absence or hypoplasia of a pedicle of the vertebra is due to a complete or incomplete cleft in the vertebral arch. The affected vertebra demonstrates a posterior location of the lateral mass and absence of the transverse process. The contralateral pedicle is usually thicker and sclerotic (Fig. 4.36).

Spondylolysis of the vertebrae refers to a defect involving the pars interarticularis or isthmus. The cleft usually divides the neural arch into an anterosuperior segment consisting of the pedicle, superior facet, and transverse process and a posteroinferior segment consisting of the inferior facet and the lamina. Most of these occur in the lower lumbar region, involving the L5 and to a lesser extent the L4 vertebrae. Spondylolysis is considered to be acquired anomaly due to fractures during childhood which may not have been recognized initially. Two variations in the position of the clefts have been identified: (*a*) a retroisthmic cleft, wherein the deficit is at the junction of the isthmus and the spinous process (this is probably an acquired anomaly), and (*b*) retrosomatic cleft, which is anterior and close to the junction of the pedicle with the vertebral body. This more likely is the result of fusion anomaly and often is associated with spondylolisthesis. Spondylolyses are often asymptomatic, seen in nearly 5% of the population. In later life, as a result of aging and continued stress, they are prone to result in degeneration. With the associated degeneration involving the facet joints and laxity of the facet ligaments, they often result in spondylolisthesis. Although rare, spondylolysis also can involve the cervical region. In the cervical region the C6 vertebra is often involved (Fig. 4.37).

Block vertebrae are the result of an absent or rudimentary disk space. The combined vertebrae may have a height equivalent to the two vertebrae, but occasionally they are taller. They are often associated with a narrow sagittal diameter of the vertebral body. The anterior margin of the block vertebrae is usually concave. The adjacent vertebrae above and below the block vertebrae may show other anomalies such as

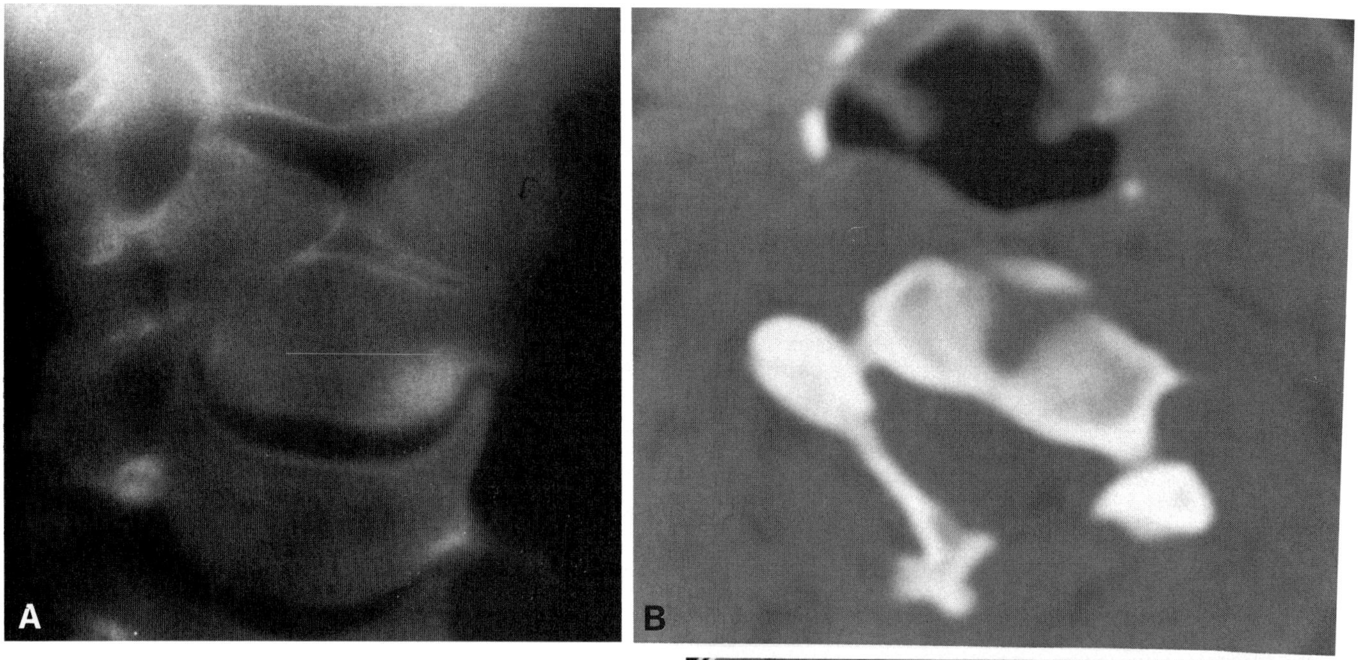

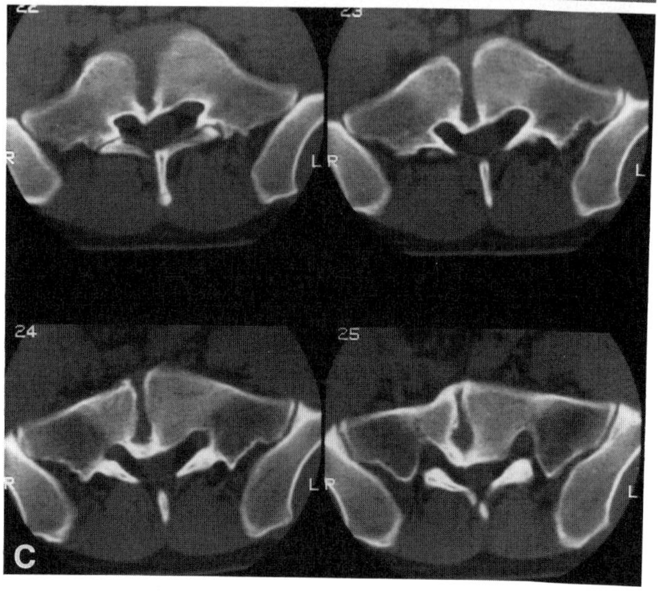

Figure 4.35. Hemivertebrae. A, Frontal tomogram in the cervical region shows the butterfly vertebra which is fused with the adjacent distal vertebra. **B,** Axial CT scan shows the anomaly better. **C,** Contiguous axial section through the sacral region demonstrates a similar anomaly.

a hemivertebra, and they may be associated with posterior fusion anomalies (Fig. 4.38).

Anomalies of the craniovertebral junction resulting from faulty segmentation between the occipital condyles and the superior facet of the atlas may result in unilateral or bilateral fusion of the occipital condyles with the lateral masses of the atlas or between the neural arch of the atlas and the occiput. These anomalies are best resolved with thin-section CT with re-formatted images or by tomography (Fig. 4.39). MRI is useful in demonstrating cervicomedullary compression resulting from the osseous elements or the associated soft-tissue components.

Klippel-Feil Anomaly

In this abnormality, two or more segments of the cervical vertebral column fail to segment into the individual vertebral bodies and posterior elements and result in a block vertebra. The patient usually presents with short neck, low-lying hairline, and limited cervical motion. The spinal canal is deformed as well as enlarged, and the neural foramina are grossly disorganized. The spinal cord and spinal nerves are likewise abnormal and cause severe neurologic deficits (Fig. 4.40). This abnormality is usually associated with other abnormalities of the craniocervical junction such as platybasia, basilar invagination, and occipitalization of the atlas.

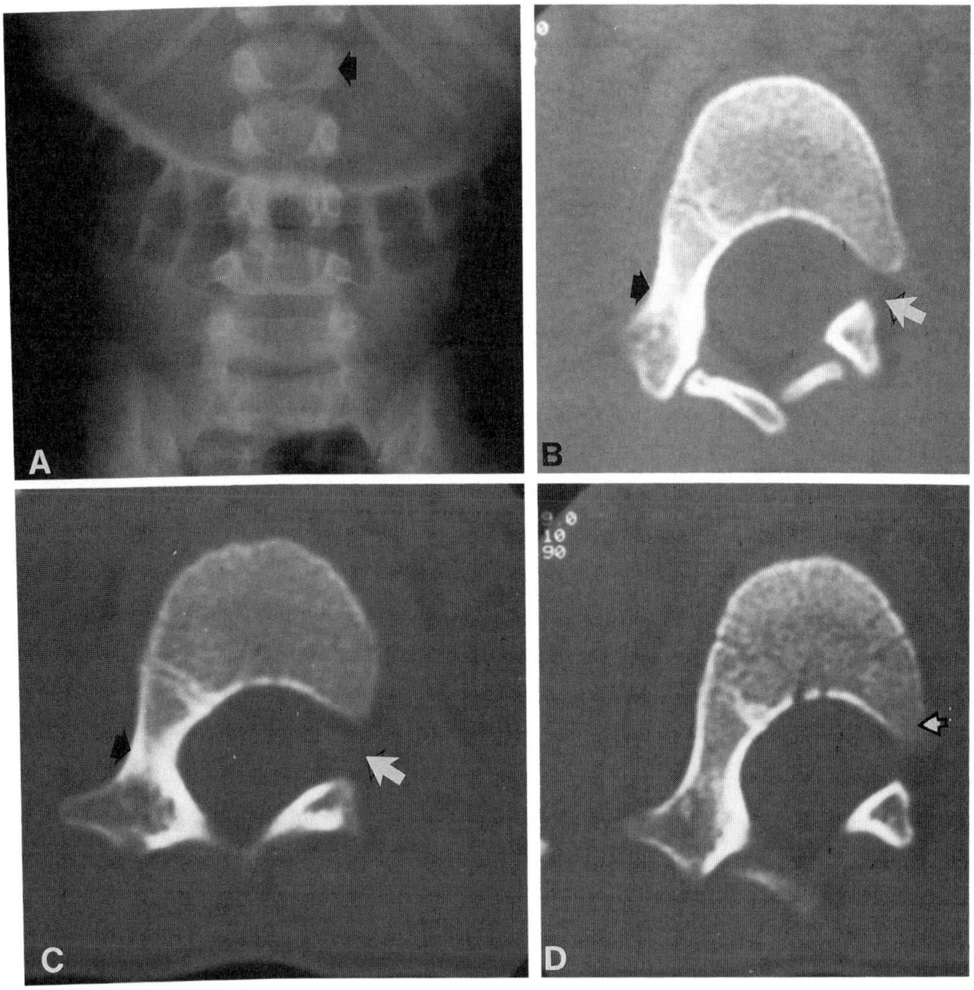

Figure 4.36. Absence or hypoplasia of the pedicle. A, Frontal view shows absence of the pedicle (*arrow*) of the L1 vertebra. **B–D,** Contiguous axial CT sections. The left pedicle is hypoplastic (*arrow*). The opposite pedicle (*black arrow*) is elongated, thickened, and sclerotic. The hypoplasia of the pedicle is associated with a deformed and rudimentary isthmus and facet.

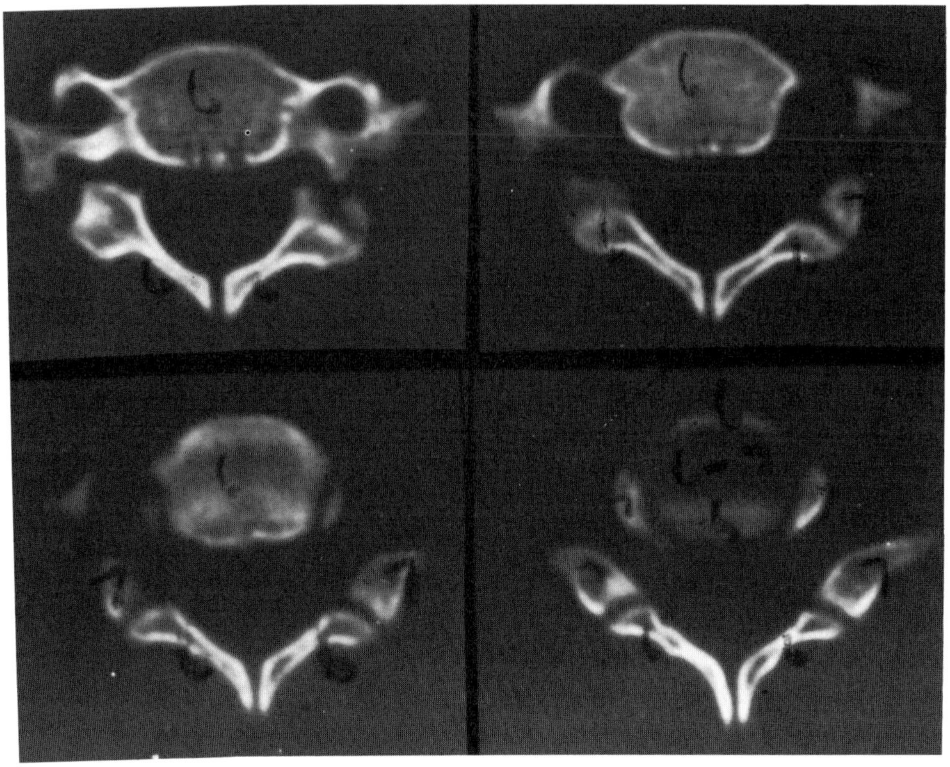

Figure 4.37. Spondylolysis of the C6 vertebrae. Contiguous axial sections demonstrating retrosomatic spondylolysis of the C6 vertebrae.

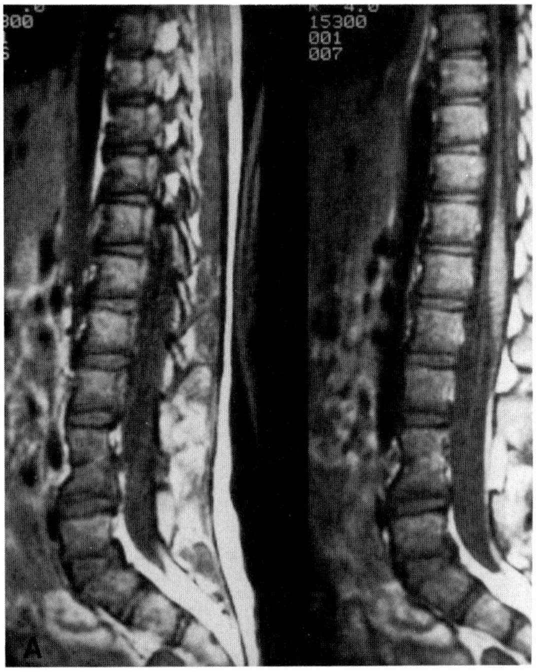

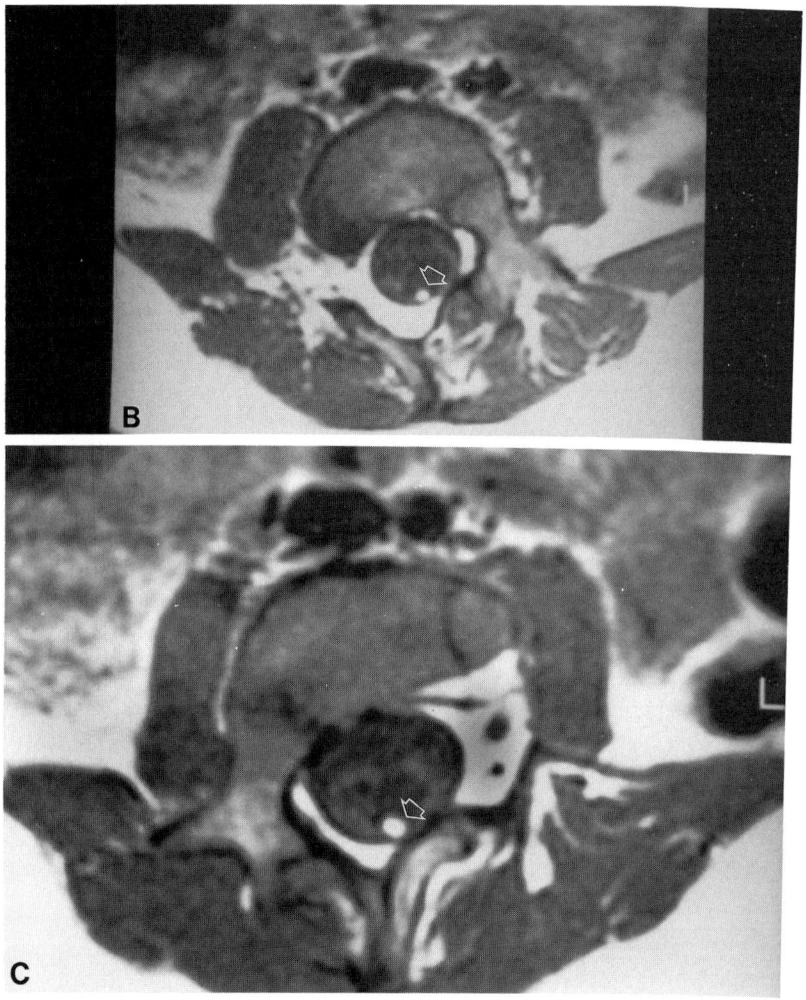

Figure 4.38. Block vertebrae with hypoplasia of the pedicle. A, Sagittal MRI shows fusion of the L3–4 vertebrae with concavity of the anterior vertebral margin. The subarachnoid space is prominent. The conus is at the L1–2 level. **B,C,** Hypoplasia of the pedicles (*arrows*) on opposite sides of the L3 and L4 vertebrae. Generous amount of epidural fat is present surrounding the thecal sac. Filar lipoma is seen as a bright signal within the thecal sac (*white arrows*).

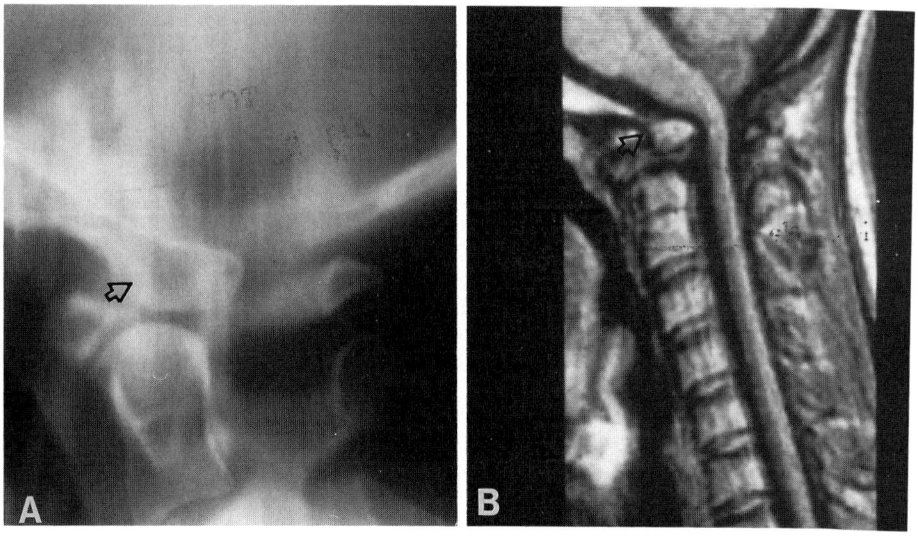

Figure 4.39. Assimilation of the dens with the occiput. A, Midline sagittal tomogram shows the dens fused to anterior lip of foramen magnum (*open arrow*). **B,** Midline sagittal T1-weighted MRI shows resulting kinking of the cervicomedullary junction.

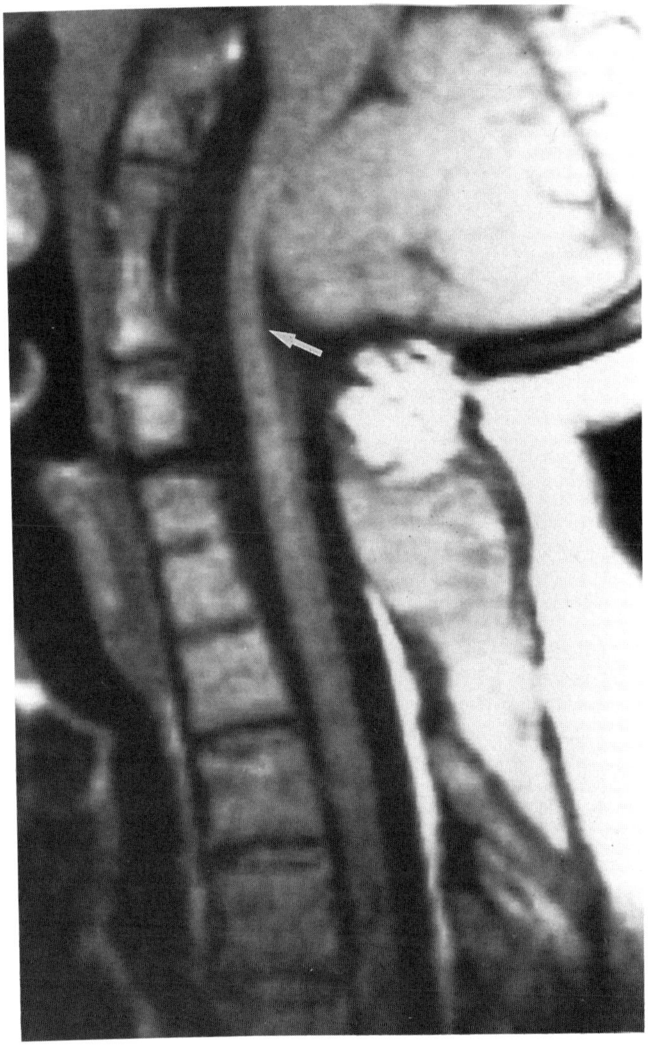

Figure 4.40. **Klippel-Feil abnormality.** Sagittal view shows fused deformed vertebrae. Note abnormal foramen magnum with compression of the upper cervical cord by the cerebellar tonsils (*arrow*).

References

1. Harwood-Nash DC, Fitz CR. Myelography. In: Neuroradiology in infants and children. St. Louis: Mosby, 1976: 1133.
2. Kirks DR. Spine and contents. In: Practical pediatric imaging. Boston: Little, Brown, 1984:134.
3. Wilson DA, Prince JR. MR imaging determination of the location of the normal conus medullaris throughout childhood. AJNR 1989;10:259.
4. Lemire RJ, Loesser JD, Leech RW, et al. Normal and abnormal development of the human nervous system. New York: Harper & Row, 1975.
5. O'Reilly R, Muller F. The normal and abnormal development of the nervous system in the early human embryo. J Pediatr Neuroscience 1986;2:89–100.
6. French BN. The embryology of spinal dysraphism. Clin Neurosurg 1982;30:295–340.
7. Barnes PD, Lester PD, Yamanashi WS, et al. Magnetic resonance imaging in infants and children with spinal dysraphism. AJNR 1986;7:465–472.
8. Altman NR, Altman DH. MR imaging of spinal dysraphism. AJNR 1987;8:533–538.
9. Brunberg JA, Latchaw RE, Kanal E, et al. Magnetic resonance imaging of spinal dysraphism. Radiol Clin North Am 1988;26:181.
10. Jaspan T, Worthington BS, Holland IM. Comparative study of magnetic resonance imaging and computed tomography–assisted myelography in spinal dysraphism. Br J Radiol 1988;61:445.
11. Duthoy MJ, Lund G. MR imaging of the spine in children. Eur J Radiol 1988;8:188.
12. Rindahl MA, Colletti PM, Zee CS, et al. Magnetic resonance imaging of pediatric spinal dysraphism. Magn Reson Imaging 1989;7:217–224.
13. Warkany J. Morphogenesis of spina bifida. In: McLaurin RL, ed. Myelomeningocele. New York: Grune & Stratton, 1977:31–39.
14. Padget DH. Spina bifida and embryonic neuroschisis—a causal relationship. Definition of the postnatal conformations involving a bifid spine. Johns Hopkins Med J 1968;128:233–252.
15. Padget DH. Neuroschisis and human embryonic maldevelopment: new evidence on anencephaly, spina bifida and diverse mammalian defects. J Neuropathol Exp Neurol 1970;29:192–216.
16. Heinz ER, Rosenbaum AE, Scarff TB, et al. Tethered spinal cord following meningomyelocele repair. Radiology 1979;131:153–160.
17. Nelson MD Jr, Bracchi M, Naidich TP, McLone DG. Natural history of repaired myelomeningocele. Radiographics 1988;8:695.
18. Williamson MR, Glasier CM, Chadduck WM, et al. MRI in the evaluation of spinal bifida patients in the remote period after meningomyelocele repair. Pediatr Radiol 1989;19:442–441.
19. Just M, Schwarz M, Ludwig B, et al. Cerebral and spinal MR-findings in patients with postrepair myelomeningocele. Pediatr Radiol 1990;20:262–266.
20. Naidich TP, McLone DG, Mutluer S. A new understanding of dorsal dysraphism with lipoma (lipomyeloschisis): radiologic evaluation and surgical correction. AJNR 1983;4:103–106.
21. Roos RAC, Vielvoye GJ, Voormolen JHC, et al. Magnetic resonance imaging in occult spinal dysraphism. Pediatr Radiol 1986;16:412–416.
22. Cecchini A, Locatelli D. Lipomyelomeningoceles: a neuroradiological approach. J Neuroradiol 1988;15:49–61.
23. Davis PC, Hoffman JC Jr, Ball TI, et al. Spinal abnormalities in pediatric patients: MR imaging findings compared with clinical, myelographic and surgical findings. Radiology 1988;166:679.
24. Tortori-Donati P, Cama A, Rosa ML. Occult spinal dysraphism: neuroradiological study. Neuroradiology 1990; 31:512–521.
25. Albright AL, Gartner JC, Wiener ES. Lumbar cutaneous hemangiomas as indicators of tethered spinal cords. Pediatrics 1989;83:977–980.
26. Monajati A, Spitzer RM, Wiley JL, et al. MR imaging of a spinal teratoma. J Comput Assist Tomogr 1986;10:307–310.
27. Levy LM, Di Chiro G, McCullough D, et al. Fixed spinal cord: diagnosis with MR imaging. Radiology 1988; 169:773.
28. Merx JL, Bakker-Niezen SH, Thijssen HOM, et al. Tethered spinal cord syndrome: a correlation of radiological features and perioperative findings in 30 patients. Neuroradiol 1989;31:63–70.
29. Walker AE, Bucy PC: Congenital dermal sinuses: a

source of spinal meningeal infection and subdural abscesses. Brain 1934;57:401–402.
30. Barkovich AJ, Raghavan N, Chuan SH, et al. Wedge-shaped cord terminus: a radiographic signa of caudal regression. AJNR 1989;10:1223–1228.
31. Peacock WJ, Murovic JA. Magnetic resonance imaging in myelocystoceles: report of two cases [Abstract]. Radiology 1990;174:589.
32. Taviere V, Brunelle F, Baraton J, et al. MRI study of lumbosacral lipoma in children. Pediatr Radiol 1989; 19:316–320.
33. Okumura R, Minami S, Asato R, et al. Fatty filum terminale: assessment with MR. J Comput Assist Tomogr 1990; 14:571.
34. Quint DJ, Boulos RS, Sanders WP, et al. Epidural lipomatosis. Radiology 1988;169:485–490.
35. Arroyo IL, Barron KS, Brewer EJ Jr. Spinal cord compression by epidural lipomatosis in juvenile rheumatoid arthritis. Arthritis Rheum 1988;31:447.
36. Gero BT, Chynn KY. Symptomatic spinal epidural lipomatosis without exogenous steroid intake: report of a case with magnetic resonance imaging. Neuroradiology 1989;31:190.
37. Doppman JL. Epidural lipomatosis. Radiology 1989;171: 581–582.
38. Naidich TP, Harwood-Nash DC. Diastematomyelia: hemicord and meningeal sheaths; single and double arachnoid dural tubes. AJNR 1983;4:633–636.
39. Bentley JFR, Smith JR. Developmental posterior enteric remnants and spinal malformations: the split notochord syndrome. Arch Dis Child 1960;35:76–86.
40. Herren RY, Edwards JE. Diplomyelia (duplication of the spinal cord). Arch Pathol 1940;30:1203–1214.
41. Patten BM. Overgrowth of the neural tube in young human embryos. Anat Rec 1952;113:381–393.
42. Lemire RJ, Loesser JD, Leech RW, et al. Normal and abnormal development of the human nervous system. New York: Harper & Row, 1975:78–80.
43. Gardner WJ. Diastematomyelia and Klippel-Feil syndrome. Cleve Clinic Q 1964;31:19–44.
44. Schlesinger AE, Naidich TP, Quencer RM. Concurrent hydromyelia and diastematomyelia. AJNR 1986;7:473–477.
45. Wilkins RH, Odom GL. Spinal intradural cysts. In: Handbook of clinical neurology. Vol. 20. 1976:55–102.
46. Aoki S, Machida T, Sasaki Y, et al. Enterogenous cyst of cervical spine: clinical and radiological aspects (including CT and MRI). Neuroradiology 1987;29:291–293.
47. Pierot L, Dormont D, Oueslati S, et al. Gadolinium-DTPA enhanced MR imaging of intradural neurenteric cysts. J Comput Assist Tomogr 1988;12:762–764.
48. Gandon Y, Hamon D, Carsin M, et al. Radiological features of intradural epidermoid cysts: contribution of MRI to the diagnosis. Neuroradiology 1988;15:335.
49. Geremia GK, Russell EJ, Clasen RA. MR imaging characteristics of a neurenteric cyst. AJNR 1988;9:978.
50. Pierot L, Dormont D, Oueslati S, et al. Gadolinium-DTPA enhanced MR imaging of intradural neurenteric cysts. J Comput Assist Tomogr 1988;12:762–764.
51. VK, Gupta RK, Sharma BS, et al. Craniospinal enterogenous cyst. J Comput Assist Tomogr 1990;14:470–472.
52. Martin B, Boyer de Latour F. MR imaging of anterior sacral meningocele. J Comput Assist Tomogr 1988; 12:166–167.
53. Willinsky RA, Grosman H, Cooper PW, et al. Radiology of sacral cysts. J Can Assoc Radiol 1988;39:21–25.
54. Miles J, Pennybacker J, Sheldon P. Intrathoracic meningocele: its development and association with neurofibromatosis. J Neurol Neurosurg Psychiatry 1969;32:99–110.
55. Cross GO, Reavis JR, Saunders WW. Lateral intrathoracic meningocele. J Neurosurg 1949;6:423–432.
56. Maiuri F, Corriero G, Giampaglia F, Simonetti L. Lateral thoracic meningocele. Surg Neurol 1986;26:409–412.
57. Laws JW, Pallis C. Spinal deformities in neurofibromatosis. J Bone Joint Surg [Br] 1963;45:674–682.
58. Heard G, Payne EE. Scalloping of the vertebra bodies in von Recklinghausen's disease of the nervous system (neurofibromatosis). J Neuro Neurosurg Psychiatry 1956;25:345–350.
59. So CB, Li DKB. Anterolateral cervical meningocele in association with neurofibromatosis: MR and CT studies. J Comput Assist Tomogr 1989;13:692–695.
60. Tarlov IM. Perineural cysts of the spinal nerve roots. Arch Neurol Psychiatry 1938;40:1067–1074.
61. Larsen JL, Smith D, Fossan G. Arachnoidal diverticula and cystlike dilatations of the nerve-root sheaths in lumbar myelography. Acta Radiol 1980;21:141–145.
62. Abello R, Rovira M, Sanz MP, et al. MRI and CT of ankylosing spondylitis with vertebral scalloping. Neuroradiology 1988;30:272–275.
63. Rubenstein DJ, Alvarez O, Ghelman B, et al. Cauda equina syndrome complicating ankylosing spondylitis: MR features. J Comput Assist Tomogr 1989;13:511–513.
64. Mitchell MJ, Sartoris DJ, Moody D, et al. Cauda equina syndrome complicating ankylosing spondylitis. Radiology 1990;175:521.
65. Verstraete KLA, Martens F, Smeets P, et al. Traumatic lumbosacral nerve root meningoceles: the value of myelography, CT and MRI in the assessment of nerve root continuity. Neuroradiology 1989;31:425–429.
66. Hilal SK, Marton D, Pollack E. Diastematomyelia in children: radiographic study of 34 cases. Radiology 1974; 112:609–614.
67. Winberg 1974.
68. Tunnel WP, Austin JC, Barnes PD, et al. Neuroradiologic evaluation of sacral abnormalities in imperforate anus complex. J Pediatr Surg 1987;22:58–67.

CHAPTER / **5**

Degenerative Disk and Vertebral Disease

K. C. V. G. Rao / J. P. Williams

Degenerative disease of the spine is a common medical problem encountered by physicians in all specialties. The degenerative changes may involve any region of the spine, although they most often involve the lumbosacral region, followed in order of severity and clinical presentation by the cervical region and less often the thoracic region. Diagnosis and treatment, as well as the resulting loss of time during the prime years, result in significant economic loss to the individual and to society. Next to respiratory infections, this is the most common source of chronic disability.

The degenerative changes, in a sense, are a result of aging, which primarily involves the disk and subsequently or simultaneously affects the various components of the discovertebral complex, such as the adjacent vertebral bodies, the ligaments, the facet joints and capsule, and the surrounding paraspinal muscles [1, 2]. As a consequence of the degenerative changes, disk herniation is more common in the younger patient, whereas in older patients involvement of the other elements of the discovertebral complex leads to spinal stenosis. It is not uncommon that beyond age 40, nearly 85 to 90% of adults demonstrate degenerative changes in one or more components of the discovertebral complex on imaging studies. However, there is minimal correlation between the severity of the degenerative changes evidenced on imaging studies and the severity of the disability in symptomatic patients [3, 4].

The exact pathophysiology and biochemical changes in degenerative disk disease are not clearly understood. Various contributory factors such as body habitus, rotational stress on the annulus fibrosus in the lower lumbar vertebrae, anomalies of the spine, diet, working conditions, genetic and metabolic factors, and trauma are known to accelerate the process. Similarly, although autoimmune mechanisms have been implicated, their role in degenerative changes of the disk substance has not been established.

Even though the degenerative process starts in the disk, since the various components of the spine work as a functional unit, primary damage to one segment of this functional unit produces secondary changes in the other components, facet arthropathy, hypertrophic changes in the ligaments and capsules, spinal stenosis, muscular atrophy, and motion instability. The symptom complex of low back pain is due to a combination of one or more of the preceding elements. The clinical presentation of degenerative changes in the spine results from the primary or secondary effects on the neural tissue, either within the spinal canal or as the nerve roots exit from the neural foramina, as well as from innervation of the functional components of the spine, including the disk, the facet joints, the joint capsule, ligaments, and the paraspinal musculature [5–7].

Pathophysiology of Discovertebral Degeneration

The intervertebral disks allow for the movement of adjacent vertebral bodies, at the same time retaining their axial relationship. The disks also serve to transmit and distribute the weight between the larger surface area of the vertebral bodies. This accounts for the increased height and volume of the vertebral bodies and the disk material in the lumbar region relative to the cervical region.

Degenerative changes involve all three elements of the disk: the outer annular fibers, the central nucleus pulposus, and the adjacent fibrocartilaginous endplates. With aging, the tensile strength of the annulus fibrosus with its obliquely oriented fibers normally arranged in concentric lamellae and slanting in opposite directions is lost. Axial rotational and shearing forces result in radial and vertical tears within the annulus [8–10]. The center of the disk similarly shows biochemical changes. The nucleus pulposus, which is in the form of a hydrophilic sponge consisting of mucoprotein gel within a collagen fibril network composed of proteoglycans, keratin sulfate, hyaluronic acid, and chondroitin sulfate, loses its hydrostatic ability with a decrease in the mucopolysaccharide content and an increase in collagen content [11]. The combined effect of weight bearing and the axial rotational stresses forces the nucleus pulposus into the fissures created by the tears within the annulus. In the younger individual, the rotational and shearing forces lead to herniation of the nucleus pulposus, whereas with aging the degenerative changes within the disk lead to a decrease in the height of the disk, along with circumferential bulging of the disk. Degenerative changes also affect the endplates of the adjacent vertebral bodies as well as the subchondral bone marrow and trabeculae. There is disruption of the fibrocartilage in the endplates with attempts at regeneration and formation of granulation tissue. The reparative process results in osteophytes and a reactive bone with thick trabeculae. Adjacent to the endplates, the red marrow is replaced by fibrous tissue with some amount of neovascularity. With continued insult there is further progression of new or reactive bone formation [12, 13]. Within the vertebral body the red marrow is replaced by lipid component or the yellow marrow. The amount of yellow marrow as well as the fibrosis in the bone can be variable and depends on the continued stress or trauma to the discovertebral endplates [12, 13].

Simultaneous with the degenerative process in the disk, capsular and facet arthritic changes result in facet hypertrophy and laxity of the facet joint capsule, leading to instability. Other changes include microfractures of the facet, synovial hypertrophy, and new bone formation. Similarly, there is hypertrophy, calcification, or ossification of the ligaments, with loss of its elasticity. All the preceding leads to varying degrees of spinal stenosis.

Clinical Manifestations

With minor variations, the degenerative process leading to disk herniation is similar in the various segments of the spine. The clinical manifestations, however, are dissimilar. In the lumbosacral region, degenerative disk disease and disk herniation may present as a low back pain localized to a vertebral segment or as radiculopathy. The low back pain without motor deficits is related to the sensory fibers innervating the spine mediated through the ventral ramus of the somatic spinal root. This may present as localized pain or as radiating or referred pain resulting from the somatic distribution of the afferent fibers of the spinal nerve consisting of sympathetic fibers which join with the sensory fibers of the spinal nerve root, the recurrent meningeal nerve [5–7]. The sympathetic fibers innervate the periphery of the disk, as well as the discovertebral complex over the entire length of the spine (paravertebral autonomic neural plexus). This anatomic basis for vertebrogenic pain [5, 6] often accounts for the discrepancy between the clinical manifestations and the extent of degenerative changes identified on imaging studies. Radicular pain is due to compression of a nerve root segment within either the spinal canal or the neural foramen. This is usually due to focal compression of the nerve root, as may occur with a herniated disk. *Radiculopathy* refers to radiated pain in the distribution of the irritated nerve root segment with both sensory and motor symptoms. Occasionally, the symptoms may mimic vascular claudication, consisting of pain dysesthesia and muscle spasm. This is often seen with spinal stenosis. In the thoracic and cervical regions, degenerative disk disease may present with symptoms of pain of dermatomal distribution, radiculopathy, or myelopathy. Myelopathy should be suspected when in addition to pain there is weakness and spasticity. This usually indicates spinal cord involvement, either from extrinsic compression or intrinsic cord pathology. The role of imaging studies is to demonstrate the morphologic changes that may account for patients' clinical symptoms. In the cervical region, since there are eight cervical nerves and seven cervical vertebrae, the herniated disk usually compresses the nerve root corresponding to the adjacent superior vertebral body (Table 5.1). In the lumbar region, since the conus medullaris is at the L1–2 level and the nerve roots have to travel a longer distance within the thecal sac and nerve root sheaths, depending on the size and location of the herniated disk, more than one segmental nerve root may be involved. Often, the nerve root involved is one segment lower than the disk space involved (Table 5.2). In most instances where patients present with both

Table 5.1.
Clinical Localization in Cervical Radiculopathy

Nerve Root Segment & Disk Level	Sensory	Motor	Reflex	Remarks
C3 C2–3 Disk level	Pain back of neck mastoid region and pinna of ear.	None	Nil	
C4 C3–4 Disk level	Pain or hypalgesia back of neck, scapular region, and anterior chest	Weakness levator scapulae	Nil	
C5 C4–5 Disk level	Hypalgesia and pain lower neck, shoulder and anterior arm	Paresis of biceps brachii and deltoid	Impaired or absent biceps reflex	Impaired shoulder motion
C6 C5–6 Disk level	Pain in shoulder, medial border of scapula, lateral aspect of arm and dorsal forearm	Paresis of biceps	Impaired or absent biceps reflex	Impaired sensory changes in the thumb and index finger
C7 C6–7 Disk level	Pain shoulder and scapular region, lateral aspect of arm and dorsal aspect of forearm	Atrophy and weakness of triceps, pronator teres	Impaired or absent triceps reflex	Hypalgesia and sensory changes involving the index and middle fingers
C8 C7–T1 Disk level	Pain medial border of scapula, arm and forearm	Atrophy small muscles of hand and hypothenar eminence	Nil	Hypalgesia and sensory changes involves the ring and little finger

Table 5.2.
Clinical localization in Lumbar Radiculopathy

Nerve Root Segment	Sensory	Motor	Reflex	Remarks
L3 L2–3 Disk level	Greater trochanter to medial side of thigh	Quadriceps and anterior tibial paresis	Loss of knee jerk	Intact saphenous nerve distribution
L4 L3–4 Disk level	Outer part of thigh across patella to anterior medial aspect of lower leg and medial aspect of foot	Paresis of quadriceps and anterior tibial muscle	Diminished knee jerk	
L5 L4–5 Disk level	From back of thigh above knee over lateral condyle to anterior lateral quadrant of lower leg and big toe	Paresis of extensor muscles	Absent posterior tibial reflexes	Impaired foot dorsiflexion
S1 L5–S1 Disk level	From back of thigh, back of leg and lateral malleolus to little toe	Paresis of peroneal muscles, and gluteal muscles	Absent ankle jerk	
Cauda equina syndrome	Loss of sensation in saddle area (perianal, gluteal region and back of upper thigh)	Paresis of gluteal muscle, knee flexors, flexors and extensors of the foot. Knee extensor is intact	Absent ankle jerk, but knee jerk is preserved	Pain on coughing. Bladder and bowel sphincter disturbance

motor and sensory findings, the associated morphologic changes within the spine and its contents can be identified.

Imaging Studies

Degenerative disk disease and its sequelae are ideally evaluated with MRI and CT scans. Other imaging studies often utilized for diagnostic purposes include plain radiography, myelography, and less often discography and radionuclide scanning.

PLAIN RADIOGRAPHY

X-rays of the spine are not useful in demonstrating the early changes of disk degeneration or the presence of disk herniation. The progressive changes of disk degeneration and their direct or indirect effects on the adjacent vertebral bodies can be suspected when there is narrowing of the disk space and occasionally calcification within the disk. In the younger patient, x-rays are useful in identifying osseous or discovertebral anomalies leading to degenerative changes at an earlier age, which may account for the low back pain. These include fusion anomalies, spondylolysis, spondylolisthesis, and congenital narrowing of the spinal canal. With progression to more severe degenerative changes in the disk and adjacent vertebral segments, plain x-rays may demonstrate the osteophytes arising from the endplates, calcification or ossification of the longitudinal ligaments, and accumulation of pockets of gas within the disk space as well as within the facet joints. Plain radiography is often obtained prior to surgery, since this is an ideal means of identifying transitional vertebrae.

MYELOGRAPHY

Prior to the availability of CT or MRI, the role of myelography in degenerative disk disease was to demonstrate the extradural disease that may compress the thecal sac and its contents, including the spinal cord and nerve roots. Although it has a high degree of accuracy in the evaluation of disk herniation and spinal stenosis, myelography is insensitive in demonstrating foraminal and far lateral disk herniation and in certain percentages of disk herniation at the L5–S1 level when associated with a large epidural space. Myelography is a relatively invasive procedure. At the present time, myelography is combined with CT when evaluating spinal stenosis secondary to degenerative disk disease and in patients with severe kyphoscoliosis.

COMPUTED TOMOGRAPHY

Computed tomography (CT) allows direct visualization of the bony contours of the spinal canal, the intervertebral disk, and disk herniation as well as the soft tissues within the canal. Disk herniation is seen on CT as focal regions of high attenuation, and this is especially useful in the lumbar region because of the presence of epidural fat. The epidural fat provides a low-attenuation contrast, separating the herniated disk from the thecal sac (Fig. 5.1). CT is excellent in demonstrating the osseous and cartilaginous abnormalities associated with degenerative disk disease [14–18]. CT is equal to MRI in evaluating for herniated disk in the lumbar region, since the large amount of epidural fat provides adequate low-density contrast from the higher density of the herniated disk. CT is also more cost-effective compared with MRI studies. MRI should be reserved for those cases where the herniated disk cannot be separated from the thecal sac due to a lack of epidural fat, as may occur in the upper lumbar, thoracic, and cervical regions. When evaluating for disk herniation or spinal stenosis in the cervical or thoracic region, CTM is usually obtained following noninvasive evaluation with MRI. This may follow conventional myelography or a small amount of intrathecal contrast material (CTM) [19, 20]. The intrathecal contrast material allows appreciation of the herniated disk or the extent of spinal stenosis. CT following intrave-

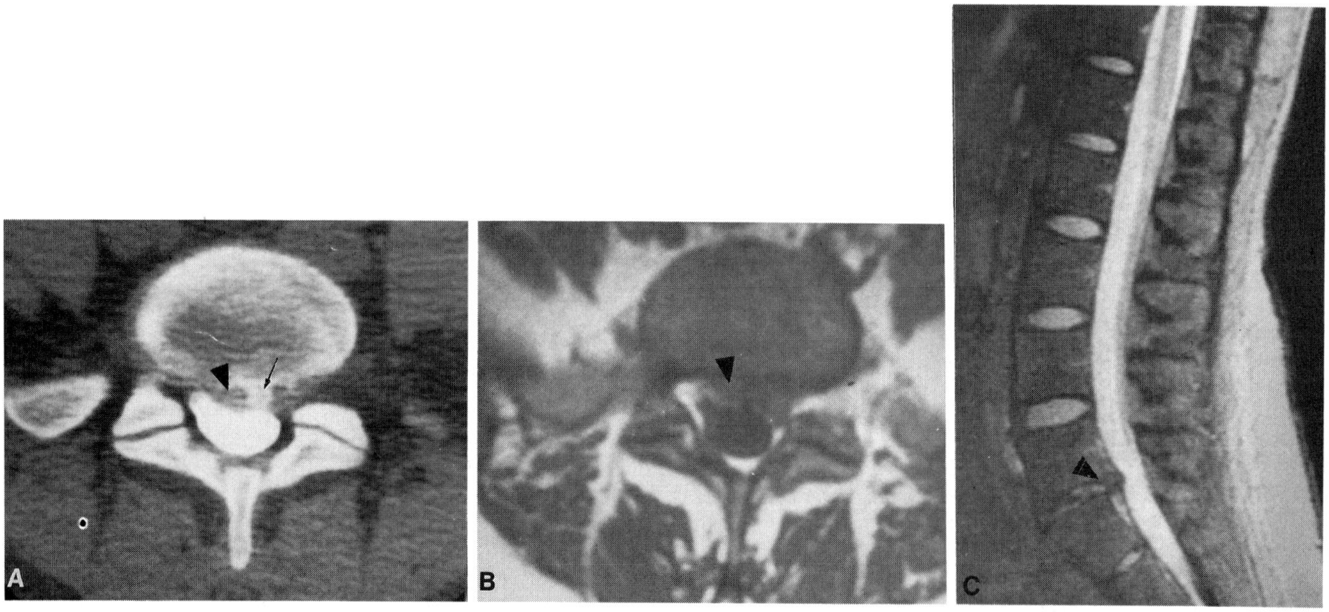

Figure 5.1. Disk herniation on CT. A, Axial CT section through L5–S1 disk level. The left paracentral herniated disk (*arrowhead*) has a density similar to the parent disk. Focal hyperdensity within the disk (*arrow*) represents calcification or osteophyte. Contrast material within the thecal sac shows deformity of the thecal sac and nonfilling of the nerve root sleeve on the left side. **B,** Axial T1-weighted MRI section. The herniated disk has a similar signal as the parent disk. The CSF within the thecal sac has a lower signal intensity than the herniated disk. The epidural fat has a bright signal which allows separation of the disk material from the thecal sac. **C,** Sagittal T2-weighted MRI section demonstrates the entire length of the lumbar spinal canal with loss in height and signal change within the L5–S1 disk space. The herniated disk (*arrowhead*) is still bound by the thin annular fibers. The posterior longitudinal ligament and the dura are displaced posteriorly. Sagittal images allows assessment of the normal position of the conus and nerve roots within the thecal sac.

nous contrast material (IVCT) is considered equal if not superior to MRI in evaluating cervical disk disease [21–24]. However it is not popular, because a fair volume of intravenous contrast material needs to be utilized to opacify the epidural veins. Overall, CT provides a rapid, noninvasive examination. The main drawback of CT is its limitation in differentiating the soft tissues within the canal and the nonavailability of sagittal and coronal planes of orientation, often necessary in proper diagnosis. Present-day high-resolution, fast-scanning and instant multiplanar re-formation has further improved the usefulness of CT.

MAGNETIC RESONANCE IMAGING

Magnetic resonance imaging (MRI) combines the multiplanar imaging capability of CT with the higher contrast sensitivity of the soft tissues. With refinement in technology, such as improved surface coils [25–28] and pulse sequences [29–35], CSF or cardiac gating [36–37], and faster volume acquisition [35], MRI is the primary imaging modality in the diagnosis of degenerative disk disease and disk herniation, having proven equal to, if not better than, CT or CTM [25–27, 38, 39]. Routine MRI depends on proton density and the relaxation time within tissues. Imaging sequences that provide maximum information when evaluating disk and vertebral disease may vary depending on individual preferences. A combination of T1- and T2-weighted images in the axial and sagittal planes is usually employed (Fig. 5.2). In the lumbar region, CSF pulsation is not a factor; however, in the cervical and thoracic regions, CSF pulsations may degrade image quality. Shorter imaging times without loss of resolution are being achieved with fast spin-echo (FSE) sequences or with a variable-flip-angle gradient-refocused echo (GRE) sequence. Three-dimensional volume acquisition GRE imaging allows thin axial sections with minimal image degradation. This allows for contiguous slices without a large interslice gap [35]. Appropriate phase and frequency encoding axes are important so as to avoid the potential pitfall of missing a small herniated disk, especially in the thoracic region [40]. The imaging studies provide some insight regarding not only the morphologic changes but also to a limited extent the biochemical and pathologic changes in the disk as well as within the adjacent vertebral bodies [41–43].

Other imaging studies utilized in evaluating back pain include *discography*, which involves injection of contrast material into the intervertebral disk space [45]. About 1 to 2 ml of contrast material is injected into the nucleus pulposus of the disk to evaluate the location and contour of the nucleus pulposus. When there is degeneration of the disk with fissures within the annulus or herniation of disk material, the contrast injection elicits pain similar to that experienced by the patient, in addition to distortion of the disk space either due to degenerative changes within the disk or from the herniated disk material. Tears in the annulus not appreciated by

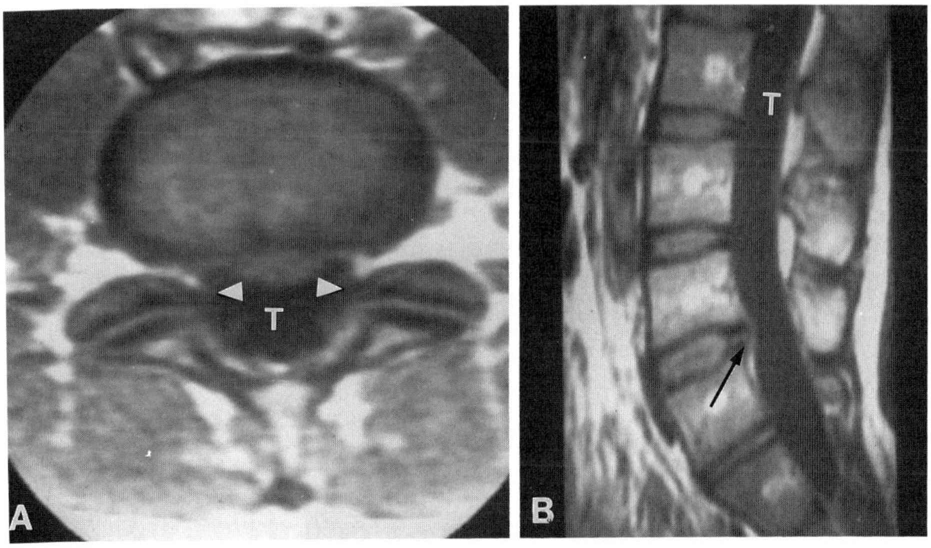

Figure 5.2. Disk herniation on MRI. Axial (**A**) and midsagittal (**B**) T1-weighted MRI sequences. The central herniated disk has a similar signal intensity as the parent disk. The axial and sagittal images show deformity of the thecal sac. The thecal sac with its contained CSF is hypointense, as are the Sharpey's fibers and thin annular fibers demarcating the herniated disk material. The small amount of epidural fat has a hyperintense signal. The conus is in the normal position. Focal round bright signal within the vertebral body represents fatty replacement of the red marrow. The moderate-sized right paracentral herniated disk has a hypointense signal (*arrowhead*), still partly within the confines of the posterior longitudinal ligament (*arrow*). The CSF within the thecal sac (T) has a lower signal compared with the herniated disk. The disk is demarcated from the thecal sac by the bright signal of fat in the epidural space.

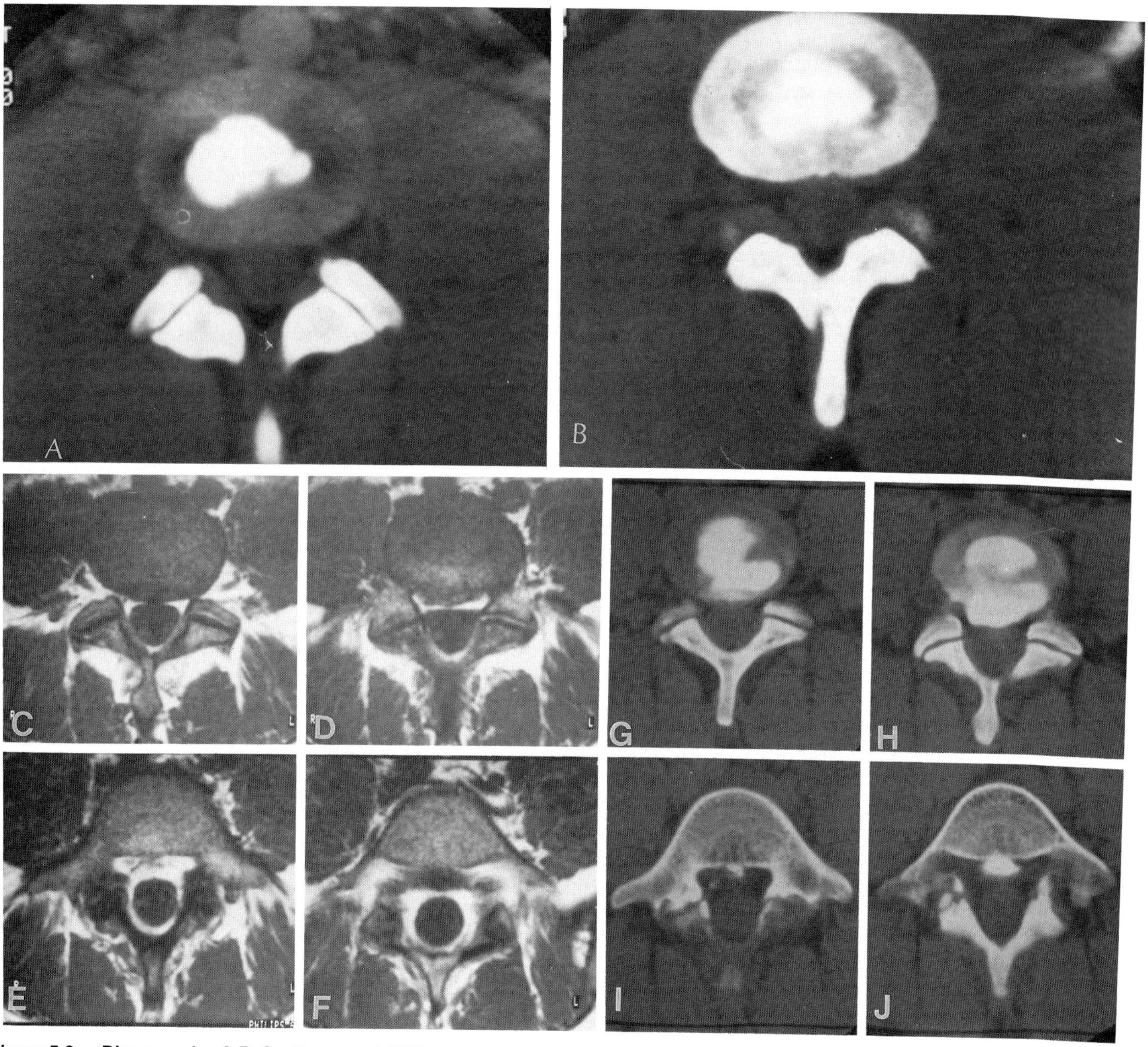

Figure 5.3. Discography. A,B, Contiguous axial CT sections with the contrast material within the confines of the disk space. Irregular outer margin of the contrast column indicates fissures within the inner annular fibers. **C–F,** Axial T1-weighted MRI sequences demonstrate the spondylolisthesis with adequate central canal and lateral recess. **G–J,** Discograms in same patient demonstrate the extravasation of contrast material through the annular tear.

MRI or conventional CT as well as disk herniations can be demonstrated when discography is combined with CT [9] (Fig. 5.3). Although there are advocates of this procedure [46A, 46B], since these findings also can be demonstrated in asymptomatic patients, we believe that it has limited clinical application.

Degenerative Disk Disease

LUMBOSACRAL REGION

Degenerative disk disease is one of the most common causes of low back pain. Among the numerous causes of low back pain, 10 to 15% of symptomatic patients have a herniated disk, other than bulging of the disk. Of those with herniated disks in this region, 85% of disk herniations involve the L4–5 and L5–S1 disk levels. Disk herniation may occur at more than one level in 10% of these patients [47] (Fig. 5.4). The peak incidence of disk herniation is in the late thirties in males and slightly later in females [48].

In most instances of symptomatic disk herniation there is an antecedent history of some form of trauma or unusual motion of the spinal axis. Because of the ongoing degenerative process affecting the annulus fibrosus, the viable nucleus pulposus protrudes within or through the fissures or tears. The shearing injury and the axial loading may progress to circumferential bulging of the annulus fibrosus beyond the confines of the adjacent vertebral bodies (*bulging disk*) or to focal protrusion of the nucleus pulposus (*herniated disk*). In most symptomatic patients this herniation occurs in a posterior or posterolateral direction. Disk herniation also can occur in an anterior or anterolateral direction or centrally into the adjacent vertebral body (*intraver-*

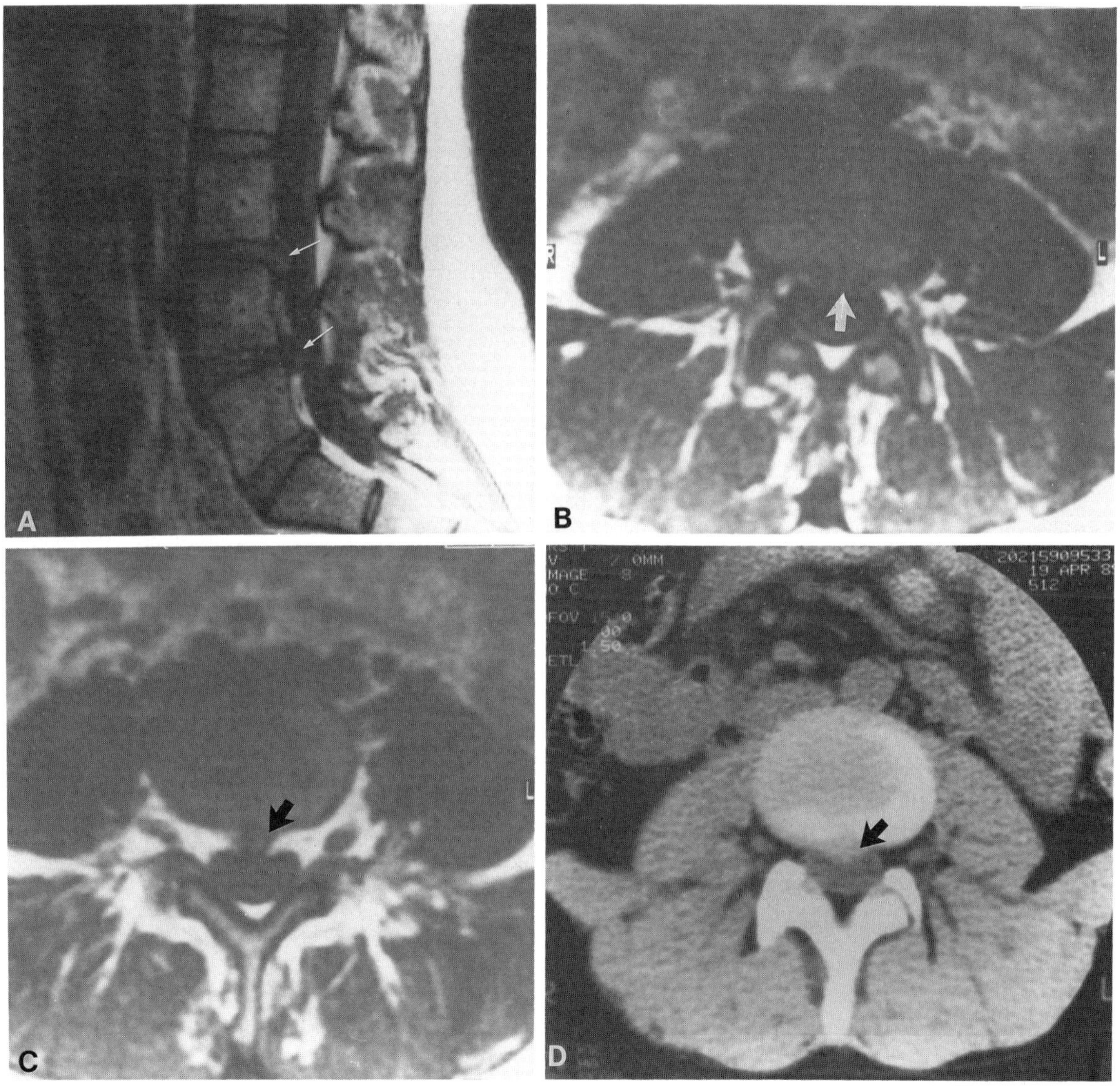

Figure 5.4. Multiple-level disk herniation. A, T1-weighted sagittal MRI demonstrates disk herniations (*arrows*) at L3–4 and L4–5 levels with disk bulge at the L5–S1 level. **B,** Axial T1-weighted MRI demonstrating a moderate-sized herniated disk at the L3–4 level (*arrow*). Axial T1-weighted MRI at the L4–5 level (**C**) and CT at corresponding level (**D**) demonstrate a small central herniated disk.

Figure 5.5. Vacuum disk. Axial CT (**A**) and sagittal re-formatted section (**B**) demonstrate the vacuum disk with nitrogen gas (*arrow*) within the disk space. **C,** Lateral spine x-ray following myelogram. **D,E,** MCT at the L4–5 level. Vacuum phenomenon (*arrowhead*) within both the disk space and the epidural space. There is an associated left paracentral herniated disk. The disk space at L4–5 is markedly narrow. Sagittal (**F**) and axial (**G**) T1-weighted MRI images showing Schmorl's node (*arrowhead*) and adjacent nitrogen gas (*open arrow*). Both have hypointense signal.

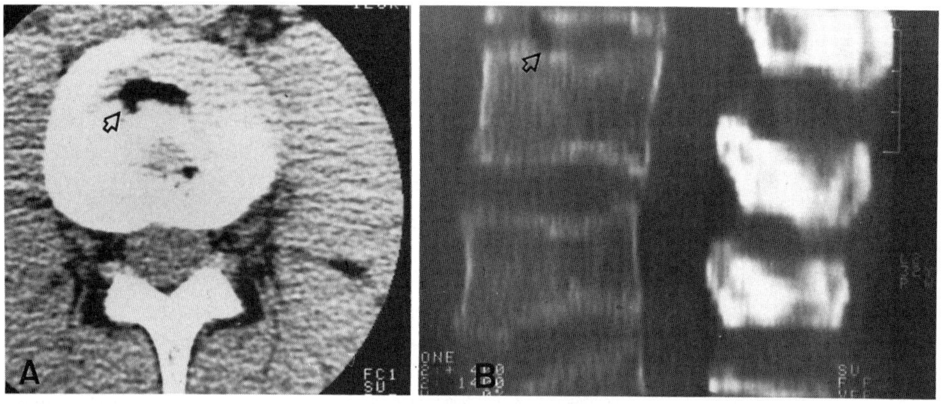

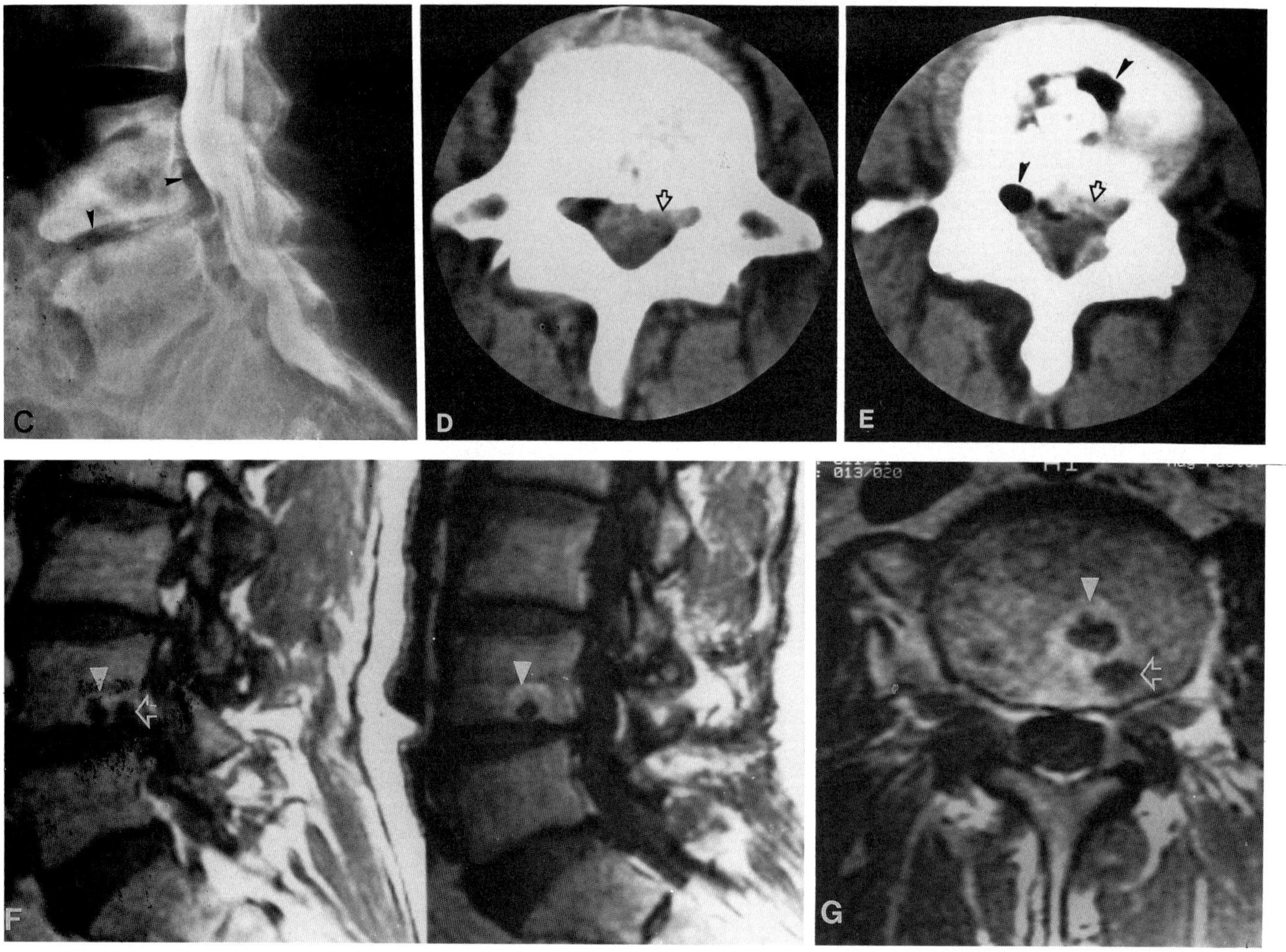

tebral disk herniation, Schmorl's node). As degenerative changes progress beyond the fourth or fifth decade, disk herniation is less prevalent. This is due to loss of hydration and biochemical changes within the nucleus leading to varying degrees of calcification, fibrosis, and accumulation of gas within the desiccated nucleus pulposus. The nitrogen gas from the vacuum disk may extend within the herniated disk and behind the posterior longitudinal ligament or within the epidural space with extruded and sequestered disk herniation (Fig. 5.5).

Schmorl's Nodes

Schmorl's nodes are intradiscal herniations into the adjacent vertebral bodies through the cartilaginous endplates [49, 50]. They are common in the lumbar region, although they are also seen in the thoracic and cervical regions. In the younger age group with symptoms of low back pain, these have been reported in the upper lumbar vertebrae [50]. On conventional radiographs of the spine, Schmorl's nodes appear as focal extensions of the disk space within the surface of the adjacent vertebral body. They have a sclerotic bony margin and are often associated with the vacuum disk phenomenon [50, 51]. The combination of narrow disk space, sclerosis of the adjacent vertebral endplates, vacuum disk phenomenon, and focal defects in the bone is also referred as *intervertebral osteochondrosis* [49, 50]. On CT, a Schmorl's node appears as a sharply marginated sclerotic density with a central hypodense region (Fig. 5.6). On MRI, a Schmorl's node appears as a focus of isointense to slightly hyperintense signal surrounded by a margin of low signal intensity on T1-weighted images and marked hypointensity on T2-weighted images. The hypointensity is further accentuated in the GRE sequence. The disk space is usually narrowed with some loss of the normal disk signal in the T2-weighted images [52]. The presence of Schmorl's nodes alone may be an incidental finding; however, intradiscal herniation may be associated with low back pain.

Anterior Disk Herniation

Although the annulus is much thicker anterolaterally and is firmly attached to the anterior longitudinal ligament and at the edges of the vertebral body merging with the Sharpey's fibers, anterior and anterolateral disk herniation is not uncommon. It usually involves the anterosuperior aspect of the vertebral body and may mimic an avulsion fracture. This is more common in the adolescent population. In one series, anterior disk herniations accounted for 29% of the herniated disks, presenting with symptoms of vertebrogenic low back pain [5]. *Spondylosis deformans* refers to anterior disk herniation associated with osteophyte formation. It is seen in more than 50% of patients with discovertebral degeneration. This is more often the case if the discal herniation occurs anteriorly at the junction of the cartilaginous endplate and the bony rim. The osteophytes may bridge the anteriorly herniated disk. On CT, the disk material extending in an oblique plane between the endplates and the bony margin can be demonstrated on axial and re-formatted sagittal images. The apophyses appear sclerotic (Fig. 5.7). On MRI,

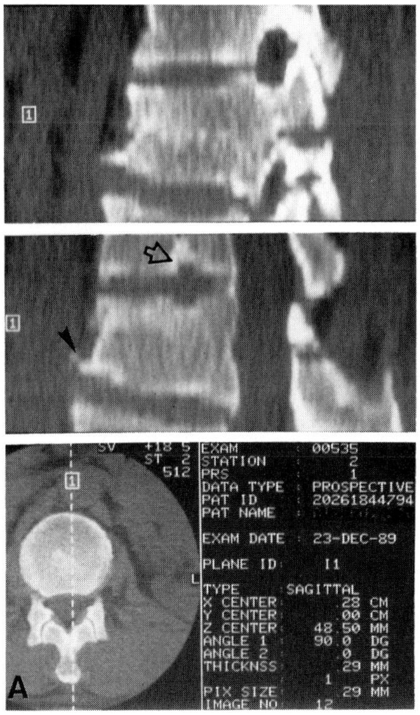

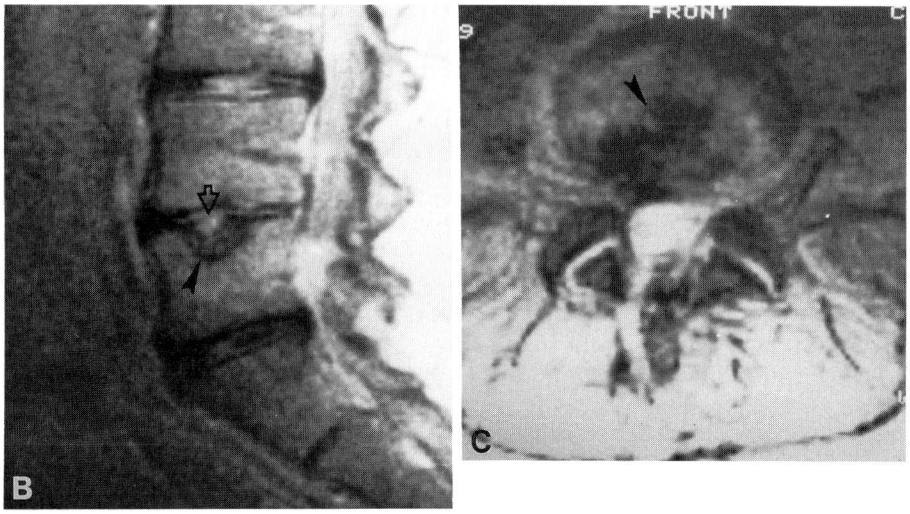

Figure 5.6. Schmorl's node. A, Re-formatted sagittal CT demonstrates intravertebral disk herniation (Schmorl's node) (*open arrow*) associated with spondylosis deformans. Anterior osteophyte (*arrowhead*). Sagittal (**B**) and axial (**C**) T1-weighted MRI images in another case demonstrates marked narrowing of the disk space (*open arrow*) with adjacent Schmorl's node (*arrowhead*).

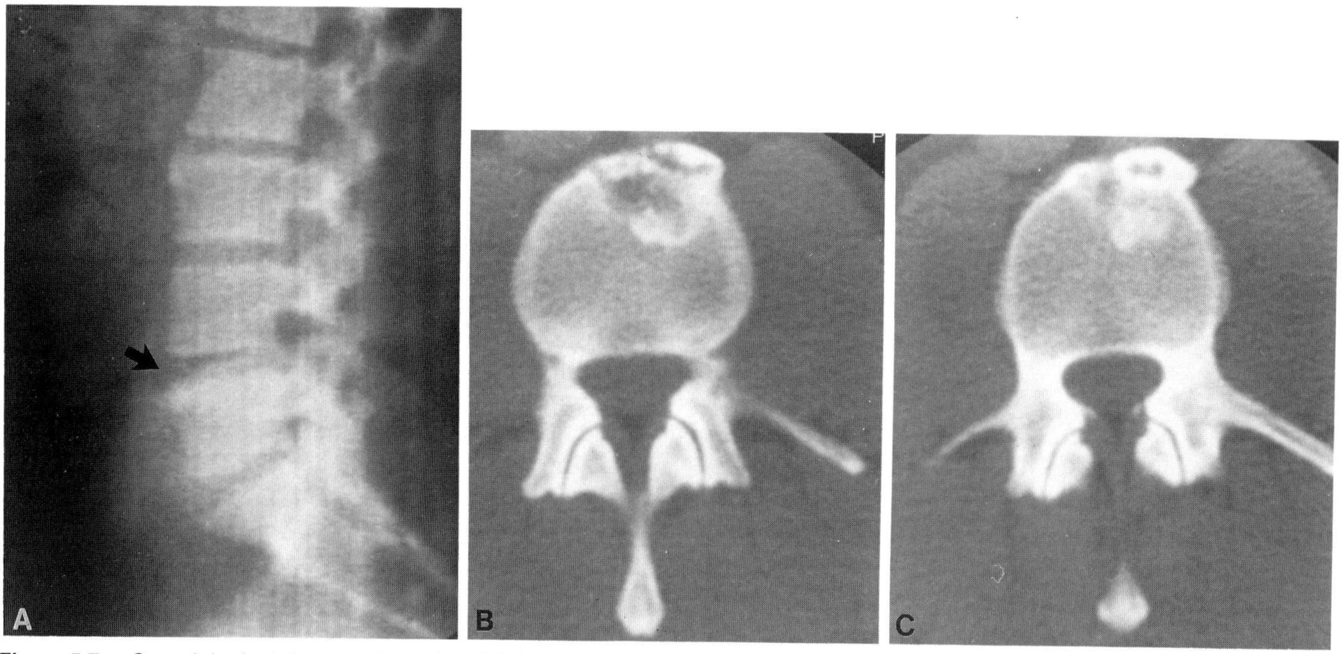

Figure 5.7. Spondylosis deformans (anterior disk herniation). **A,** Lateral plain x-ray of lumbar spine with a sclerotic bone density and mild loss in height of the anterosuperior margin of the L5 vertebral body (*arrowhead*). **B,C,** Contiguous axial CT scans show an irregular sclerotic bone density surrounding a hypodense focus along the anterior lip of the vertebral body and adjacent endplate. The anterior cortical surface of the bone is avulsed and sclerotic. This is characteristic of herniation of disk material following disruption of annular fibers as they merge with the Sharpey's fibers.

the anterior intradiscal herniation follows the signal characteristic of the disk, with the anterior bone fragment having a hypointense signal (Fig. 5.8). These herniations are more common in the lower lumbar region, although spondylosis deformans also can involve the thoracic and cervical regions.

Disk Herniation

Disk herniation is the most common cause of back pain and is often associated with radicular symptoms due to compression of the nerve roots within the spinal or neural canal. In the cervical and thoracic region, disk herniation may result in cord compression. Other pathologic processes may respond with similar symptoms, so accurate diagnosis is important for appropriate treatment. In this context, the purpose of imaging studies is to identify if the low back pain is due to discovertebral changes, disk herniation, or some other pathologic entity.

The terminology utilized in describing the sequelae of disk disease varies not only between the different medical specialties but also among radiologists. A common nomenclature which signifies the pathologic process accurately and is accepted by all specialties is important [53].

1. *Bulging annulus or bulging disk*—when there is symmetrical extension of the annulus beyond the confines of the adjacent vertebral endplates secondary to degenerative changes involving the annulus (Fig. 5.9).
2. *Herniated nucleus pulposus*—refers to thinning of the annulus with focal prolapse or herniation of the nucleus pulposus beyond the margins of the endplates of the vertebral body (see Fig. 5.9). It is still subligamentous and often confined within the thin outer annulur fibers and posterior longitudinal ligament. In the sagittal plane, it is still at the disk level. In the axial plane, the herniated disk material may be central or paracentral in location. Other terms include *disk protrusion* and *disk prolapse*.
3. *Extruded disk*—when the herniated nucleus pulposus is beyond the confines of the annulus (see Fig. 5.9). The tear in the annulus may be radial or transverse. Transverse annular tears usually occur where the annular fibers merge with Sharpey's fibers near the endplates. The extruded disk material is still attached to the rest of the nucleus pulposus within the disk space by a narrow pedicle. The extruded disk may migrate in a cephalad or caudad direction.
4. *Herniated disk with free fragment (sequestered)*—when a portion of the disk fragment is entirely separate from the parent disk, (see Fig. 5.9). A fragmented disk may penetrate the posterior longitudinal ligament and present as a discrete mass in the epidural space or rarely penetrate the dura. The separated or "free" fragment may migrate in a cephalad, caudad,

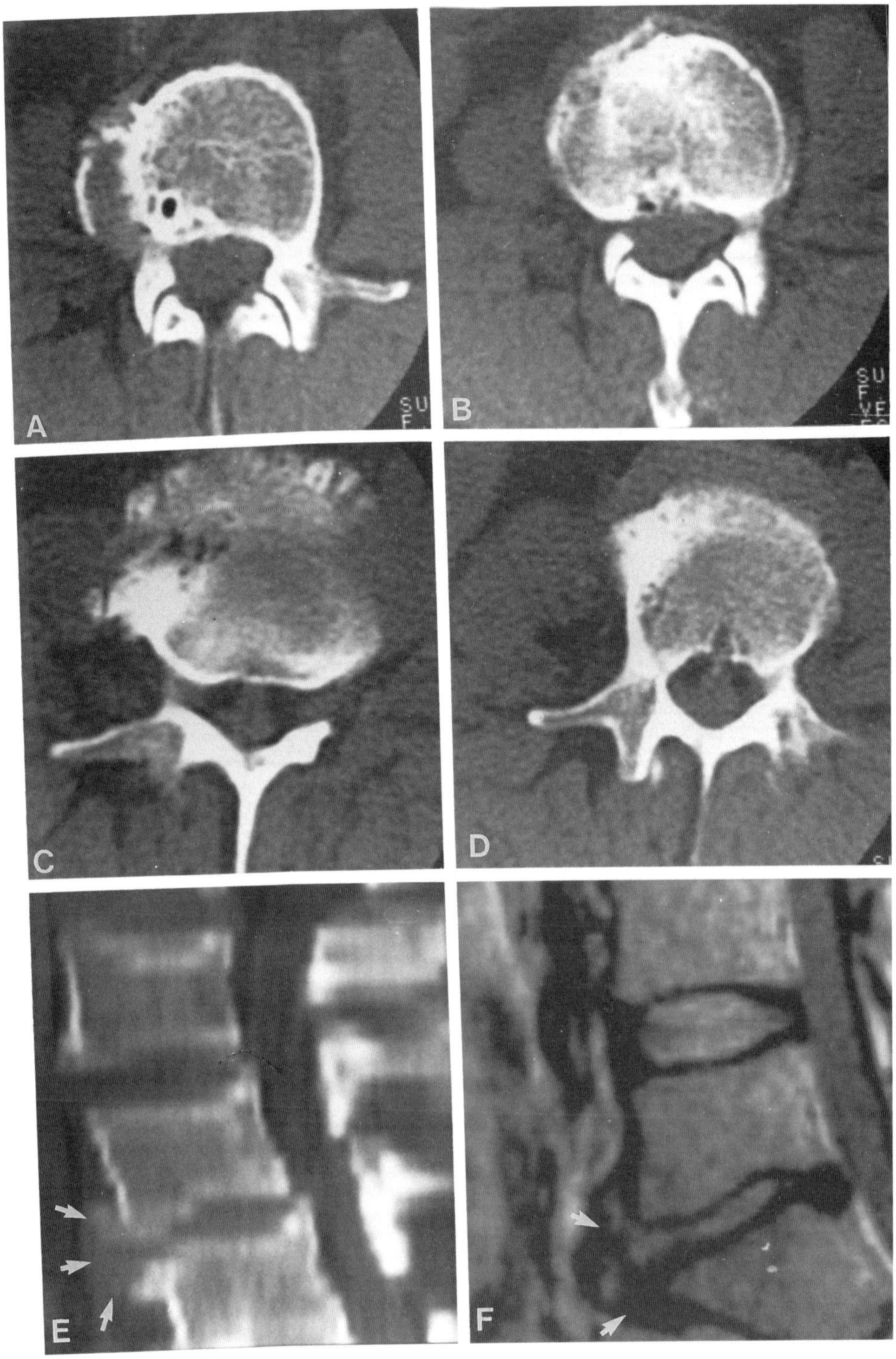

Figure 5.8. Spondylosis deformans. Contiguous axial CT sections (**A–D**) and sagittal re-formatted midline section (**E**). A large anterior and anterolateral herniated disk (*white arrows*) with avulsion of the endplates of adjacent vertebral bodies is present. Dense cortical bone bridges the herniated disk. (**F–H**) MRI studies of another patient with anterior disk herniation. Sagittal T1-weighted image (**F**) and first echo (**G**) and second echo (**H**) of T2-weighted images demonstrate the moderate-sized anterior disk herniation (*arrowheads*) at the L5–S1 disk level. It follows the signal characteristic of the disk. There is narrowing of the disk space. The marrow of the adjacent vertebral bodies shows bright signal on T2-weighted sequences and intermediate signal on T1-weighted sequences, indicative of type 1 marrow changes. A small posterolateral herniated disk (*open arrow*) is also present, better evaluated on the axial T1-weighted image (**I**).

or lateral direction and is often sequestered within the canal at a different location.

Degenerative changes within the disk space are a continuous process of aging whether this results in a bulging disk or herniation of disk material. This is best evaluated on MRI in the sagittal plane (Fig. 5.10). Loss in height of the disk space with linear irregular bands of decreased signal reflects a combination of decreased hydration and fissures or tears in the inner fibers of the annulus fibrosus. On sagittal images, varying degrees of indentation of the thecal sac are common. Distinguishing a bulging disk from a herniated disk requires evaluation in additional planes other than just the sagittal plane.

Bulging Disk

A bulging disk or bulging annulus indicates generalized or symmetrical extension of the annulus beyond the margins of the vertebral bodies due to its

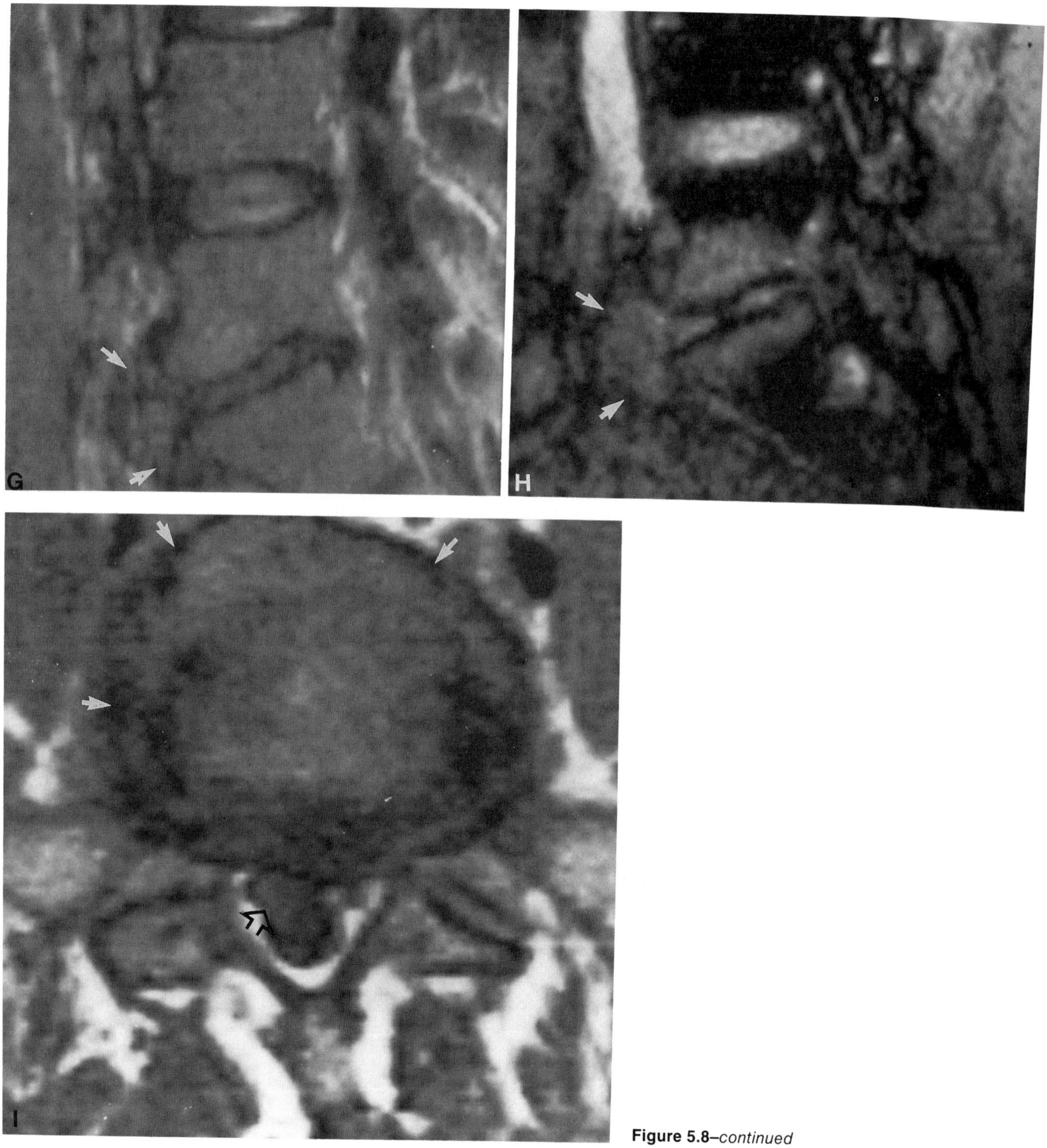

Figure 5.8–*continued*

laxity or redundancy. A bulging disk occurs when the nucleus pulposus desiccates and fissures or tears develop in the annulus fibrosus. It is often associated with loss of disk space height. The bulging of the disk may be asymmetrical in patients with scoliosis or subluxation. When it occurs in isolation, a bulging disk can be a cause of back pain, but not necessarily associated with radiculopathy. In the later stages of the continued degenerative process, bulging disks are often associated with bridging osteophytes. When associated with degenerative changes involving the facet joints, hypertrophy, and thickening of the capsule and ligaments, it often leads to spinal stenosis. Distinguishing between

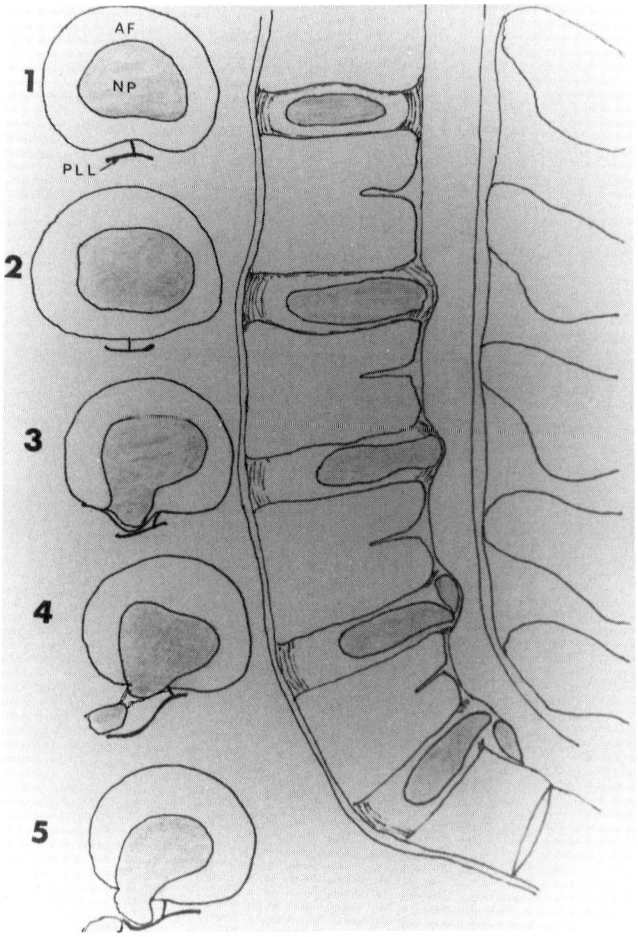

Figure 5.9. Diagrammatic representation of disk degeneration. 1 = Normal appearance of nucleus pulposus (NP) and annular fibers (AF) in relation to the vertebral bodies. 2 = Bulging disk associated with thin but intact annulus extending beyond confines of the vertebral body. Usually bulge is circumferential. 3 = Herniated disk when there is disruption or a tear of annular fibers and the nucleus pulposus herniates into this defect. It is usually focal. Usually, the annular fibers are not completely torn, and the tear is within the confines of the posterior longitudinal ligament. 4 = Extruded disk, when the nucleus pulposus has extruded through the torn annulus but is still connected with the disk space. 5 = Extruded disk with free fragment. The free fragment may become sequestered.

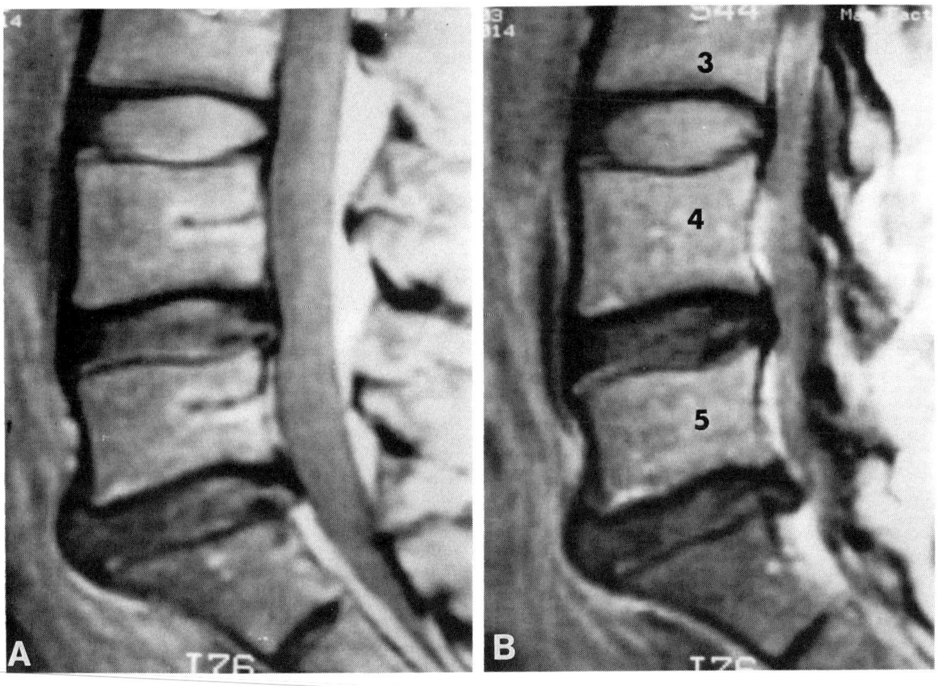

Figure 5.10. (A,B). Degenerative changes in disk space. Sagittal proton-weighted MRI demonstrating (*a*) normal bright signal in L3–4 disk space, (*b*) hypointense signal with bulging of the disk at L4–5 disk level, and (*c*) loss in height of disk space in addition to hypointense signal within the disk space, with both anterior and posterior disk herniations at the L5–S1 disk level. The progressively decreased signal within the disk space reflects decreased hydration and fissures in the inner layers of the annulus fibrosus. The wide band of decreased signal along the inferior margin of the vertebral bodies represents chemical shift artifact.

bulging disk and herniated disk is important because discectomy or chemonucleolysis is not the appropriate treatment.

On CT, a bulging disk is seen as a smoothly outlined extension of soft-tissue density, with an attenuation value less than that of cortical bone but slightly higher than that of the thecal sac or nerve root sleeves [54] (Fig. 5.11). Unlike the annulus of the normal disk, which follows the contour of the vertebral body, bulging disks usually extend beyond the outer margin of the vertebral body with a convex outer contour. An asymmetrical bulge on CT or MRI occasionally may be mistaken for a herniated disk. This is most often related to patient positioning within the gantry, section orientation, facet degeneration more on one side than on the other, or scoliosis (Fig. 5.12). Evaluation of contiguous sections is helpful in the differentiation. On MRI, a bulging disk usually has a hypointense signal on T1-weighted images. Fissures within the inner portion of the annulus are not seen on spin-echo sequences, but occasional linear bright signals may be seen in the sagittal long-TR or GRE sequences. On sagittal images, bulging disks may extend beyond the margins of the endplates, with some loss of disk height. On axial sections, the nucleus pulposus and inner fibers of the annulus appear spread out circumferentially. Depending on the extent of the degenerative

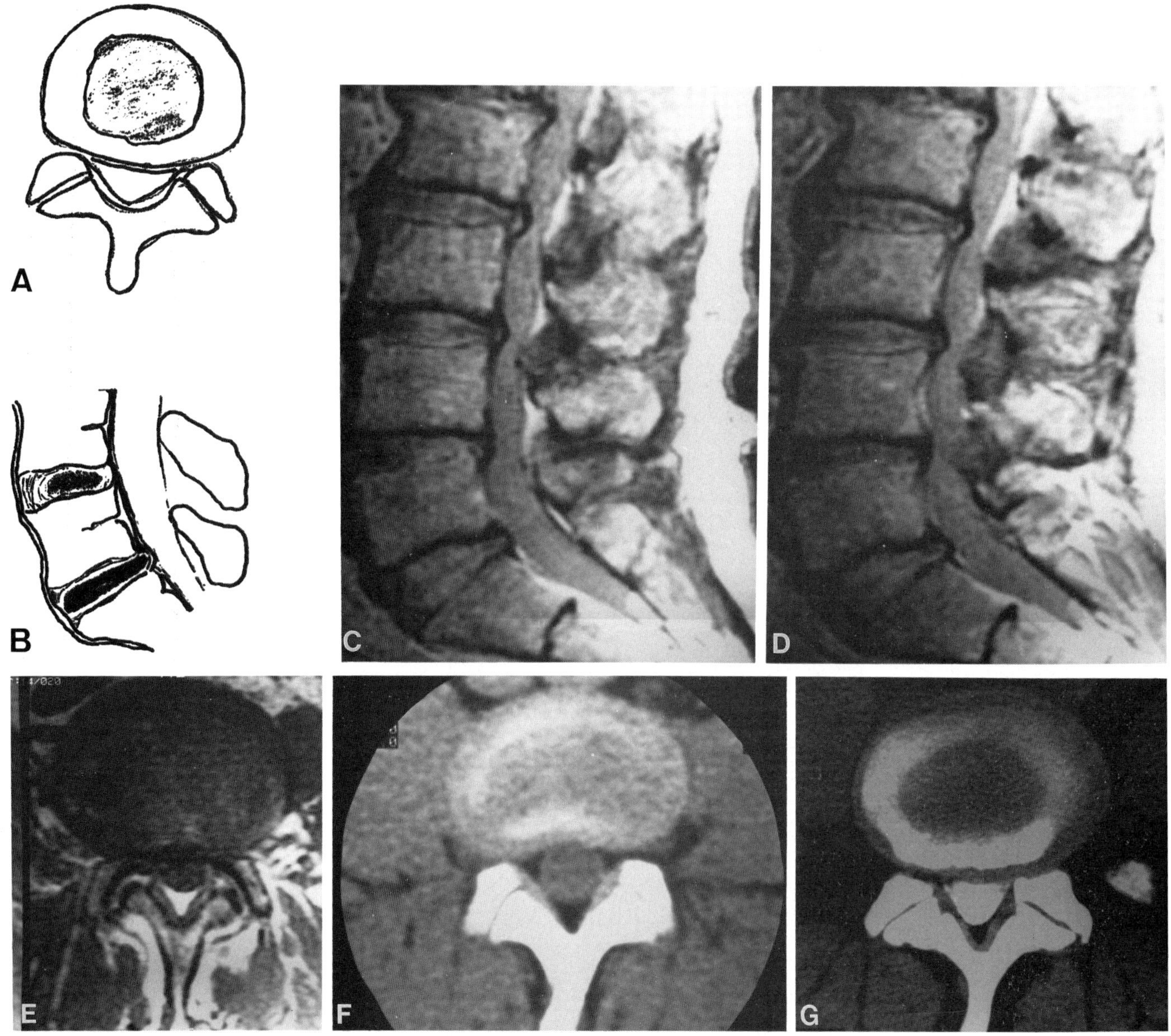

Figure 5.11. Bulging disk. Line drawing of a bulging disk in axial (**A**) and sagittal (**B**) planes. **C,D,** Sagittal MRI sequences demonstrate loss in height of the disk space at multiple levels with bulging disks. Axial T1-weighted MRI (**E**), axial noncontrast CT (**F**), and MCT (**G**) show symmetrical bulging of the disk extending beyond the cortical margin of the vertebra. The thecal sac and nerve root sheath appear normal.

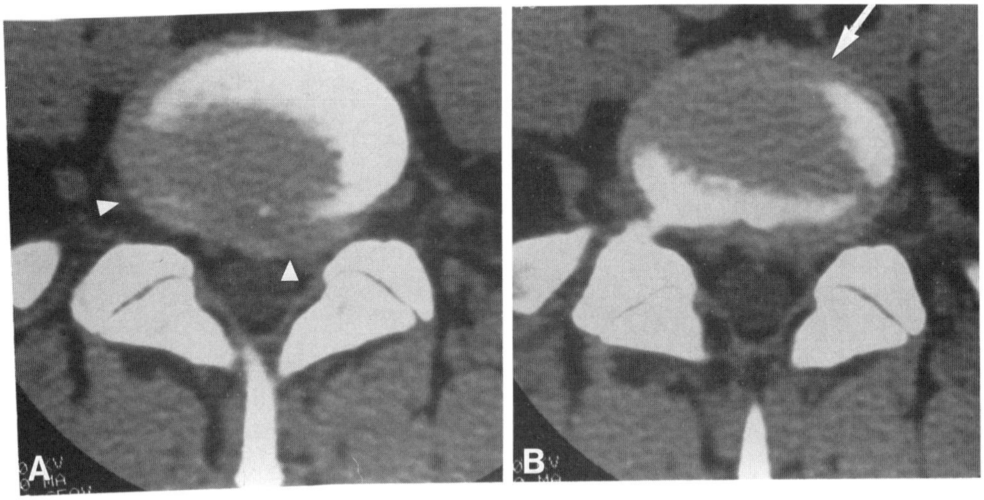

Figure 5.12. Herniated disk with asymmetrical bulge on axial CT. A,B, Contiguous axial CT sections demonstrating a broad-based extension of disk material (*arrowheads*) beyond the cortical margin associated with a left paracentral herniated disk (*open arrow*). This is caused by mild scoliosis, resulting in the section demonstrating portions of the pedicle on the right side and the foramen on the opposite side.

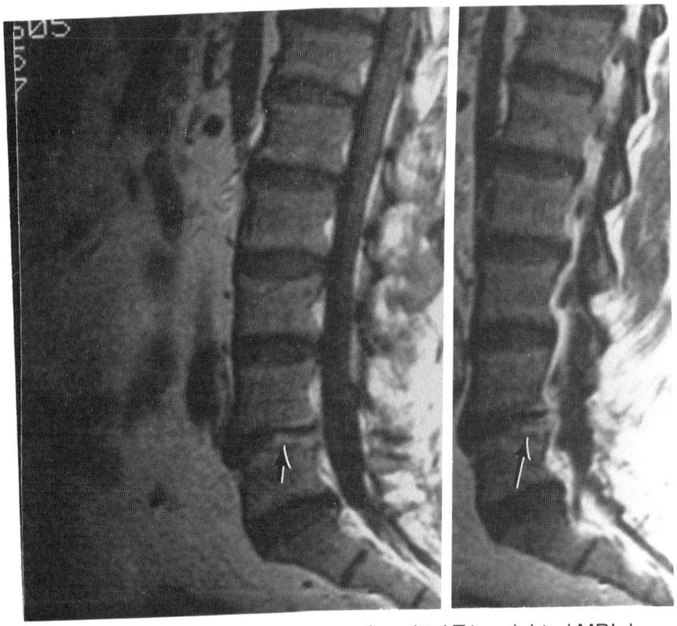

Figure 5.13. Gd-DTPA–enhanced sagittal T1-weighted MRI demonstrating linear band of enhancement within the disk space, probably representing a healing process within the inner annular fibers of the annulus fibrosus.

process, the nucleus may have varying signal intensity on T2-weighted and GRE images. The signal intensity can be bright due to focal fluid accumulation [33, 55] or have varying degrees of hypointensity almost similar to the outer annulus. The loss of signal is presumably due to the decrease in water content [36]. Gd-DTPA–enhanced MRI has been utilized to demonstrate the fissures within the deeper fibers of the annulus [56]. Enhancement is better appreciated on the sagittal images (Fig. 5.13). Enhancement is presumed to coincide with the healing process. Its recognition in the management of low back pain is at the present time of doubtful value. Bulging disks probably cause low back pain on the basis of stretching and irritation of the somatic neural plexus [5].

Herniated Nucleus Pulposus

In the adult population, the nucleus pulposus in the lumbar region is located slightly posteriorly within the disk space, the annulus being wider and thicker anteriorly. Due to a combination of thinner annulus fibrosus posteriorly, loose attachment to the posterior longitudinal ligament, and lack of ligament support in its posterolateral aspect, the nucleus pulposus herniates more often posteriorly than anteriorly. The posteriorly herniated nucleus pulposus (HNP) may be central, lateral, or more often paracentral in location owing to the poor support offered by the posterior longitudinal ligament. Focal small central HNPs may not be symptomatic, but large central and smaller paracentral and laterally located HNPs are always associated with low back pain, usually unilateral, radiating to the buttock and/or down the leg.

Sensory symptoms consist of loss of power with absent reflexes. The dermatomal level often assists in localizing the nerve root involved (see Table 5.2). In the lumbosacral region, the compromised nerve root is usually posterolateral to the herniated nucleus pulposus. The nerve root involved is usually one segment lower than the involved disk level (Fig. 5.14). For example, HNP at the L4–5 disk level would compress the L5 nerve root, and this would occur in a similar manner at other levels in the lumbosacral region. Since the lumbosacral nerves travel for a fair distance within the thecal sac, disk herniations at higher levels may not match the clinical findings. Patients may find relief with conservative therapy, with disappearance or decrease in the size of the HNP. However, when conservative therapy fails, these patients often benefit from surgical treatment. This is also the group of patients who most often respond to chemonucleolysis [57] or percutaneous discectomy [58].

CT diagnosis of herniated disk is dependent on identifying a focal hyperdense central or lateral soft-

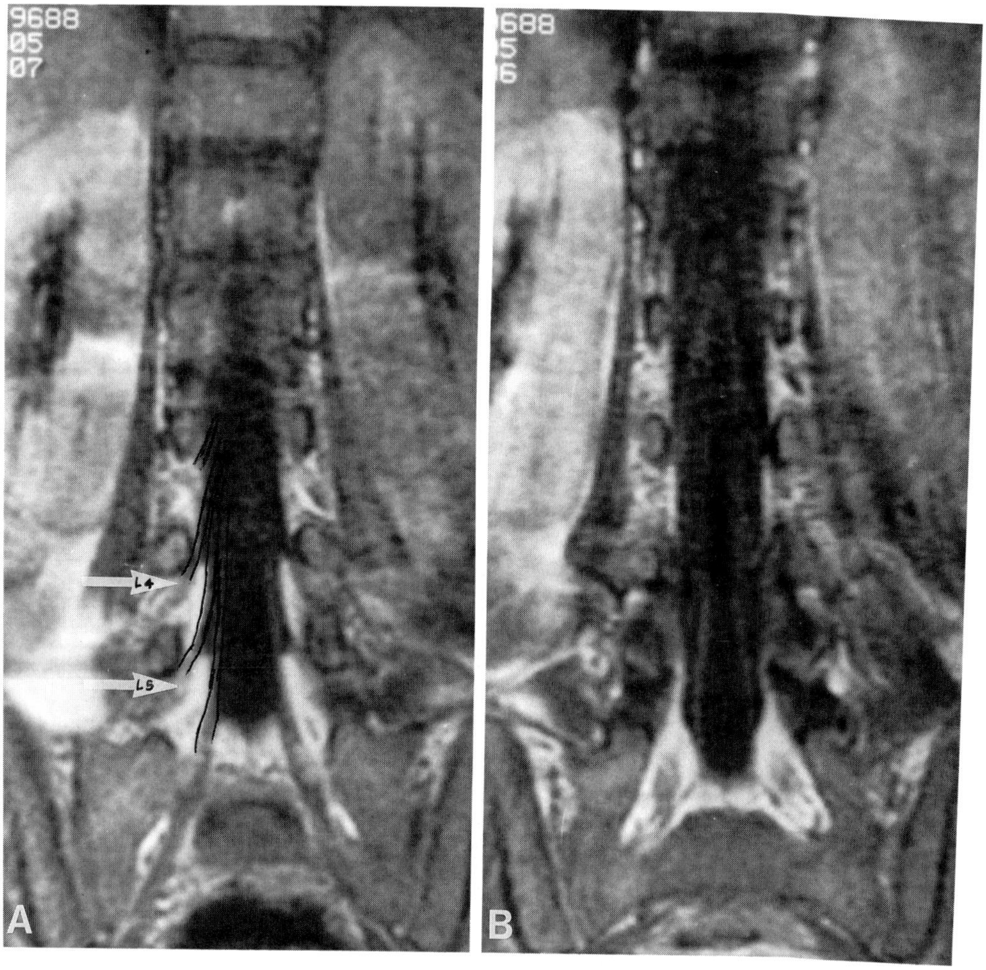

Figure 5.14. Coronal MRI sections through the lumbar spine modified to demonstrate the longer distance that the nerve root travels within the thecal sac and nerve root sheath prior to exiting the spinal canal. The L4 nerve root exits under the L4 pedicle and the L5 nerve root under the L5 pedicle. Consequently, a herniated disk within the spinal canal at the L4–5 level will affect the L5 nerve root, whereas an extraforaminal disk at the L4–5 level will affect the L4 nerve root.

tissue mass beyond the confines of the posterior margin of the vertebral body [59, 60] (Fig. 5.15). The attenuation value of the herniated disk is higher than that of the thecal sac or the nerve root and its sheath. Loss of the adjacent epidural fat is often seen. The low attenuation of the epidural fat provides the contrast that separates the disk from the thecal sac and the nerve root sheath within the spinal canal. Depending on the size of the herniated disk, the nerve root sheath may be displaced posteriorly or may not be visualized. Deformity of the thecal sac is common, unless a fair amount of epidural fat is present. The hyperdensity of the herniated disk on CT is due to volume averaging of the fibrocartilage and the mucoprotein components of the disk material. Speckled or punctate calcification may occasionally be seen within the herniated disk (Figs. 5.1 and 5.16). Calcification also may be present in the posterior longitudinal ligament, which may sharply demarcate the disk. Rarely, the herniated disk may be markedly hyperdense, a so-called hard or calcified disk. Sometimes these are difficult to differentiate from bony spurs. In most instances, CT without intrathecal or intravenous contrast material is adequate to demonstrate herniated disk in the lumbar region. CTM may be indicated in those rare cases in which the attenuation value of the herniated disk is similar to that of the thecal sac or when there is complete absence of epidural fat (see Fig. 5.16).

On MRI, the morphologic changes are similar to those seen with CT. MRI allows evaluation in both axial and sagittal planes. Evaluation is based on a combination of T1- and T2-weighted images in the axial and sagittal planes. The axial images are obtained in a plane parallel to the disk spaces. The T2-weighted images are either spin-echo or GRE sequences (Fig. 5.17). On T1-weighted sections, the herniated disk has an intermediate to hypointense signal, similar to the signal intensity within the disk space, with the soft-tissue mass extending posteriorly beyond the margins of the vertebral body. The posterior margin of the herniated disk is demarcated by the thinned out annulus and the posterior longitudinal ligament. The signal intensity of a herniated disk on T2-weighted images can be variable. Most often it is hypointense, but the signal intensity may range from intermediate to hyperintense. This variability of the signal depends on the water content of the herniated nucleus pulposus and the extent of degeneration within the annular–nucleus pulposus

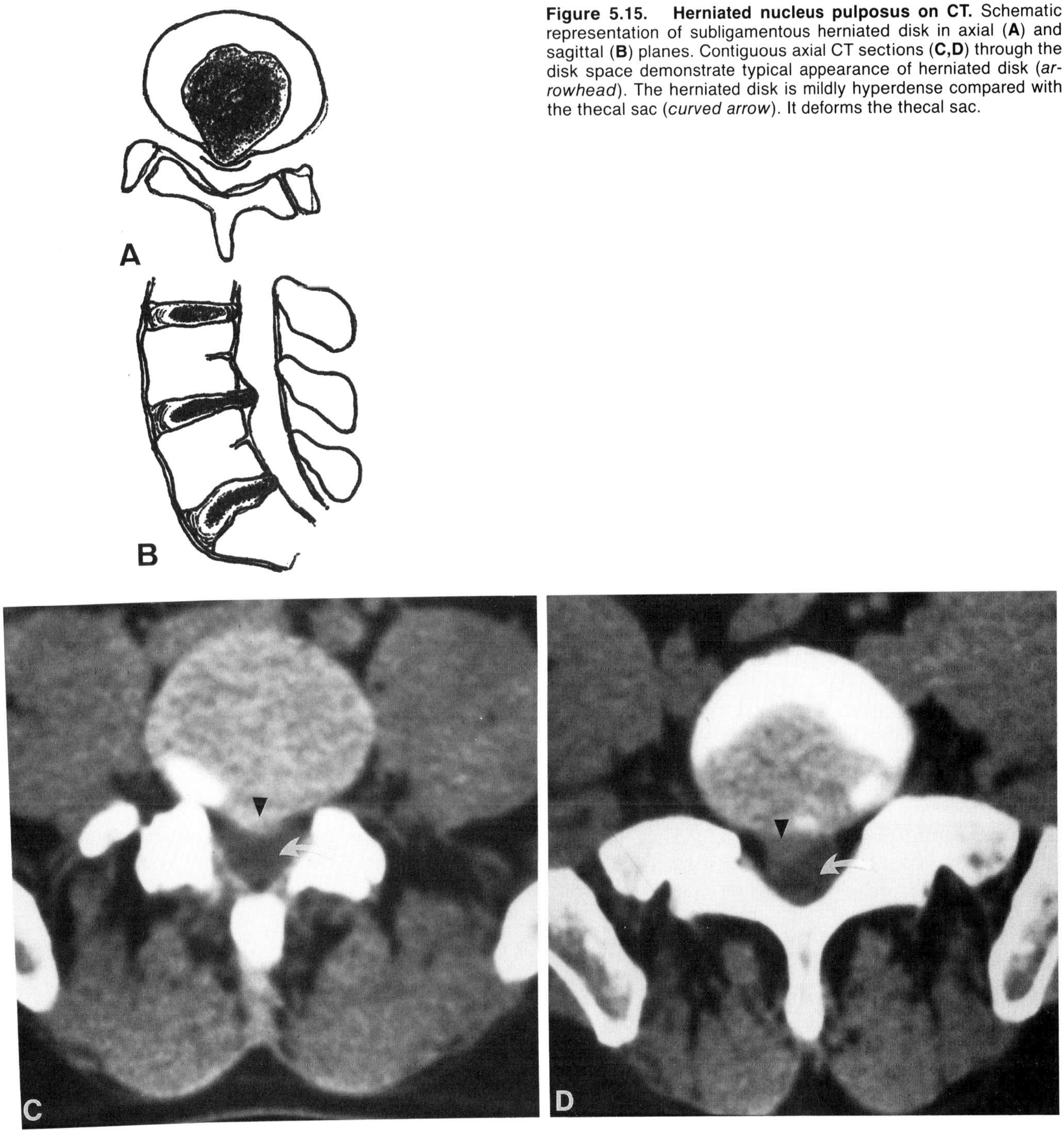

Figure 5.15. Herniated nucleus pulposus on CT. Schematic representation of subligamentous herniated disk in axial (**A**) and sagittal (**B**) planes. Contiguous axial CT sections (**C,D**) through the disk space demonstrate typical appearance of herniated disk (*arrowhead*). The herniated disk is mildly hyperdense compared with the thecal sac (*curved arrow*). It deforms the thecal sac.

component. The hyperintense signal within the herniated disk is probably related to biochemical changes, but it also might be related to edema. These signal characteristics are best evaluated on the T2-weighted images. With large herniated disks, the epidural space above and below the herniated disk is increased, consisting of some fat but more often of dilated epidural veins. This may be confused with an extruded disk, especially in the acute stage of disk herniation (Fig. 5.18). Usually the thinned out annulus and the posterior longitudinal ligament form a sharp hypointense posterior border demarcating the herniated disk from the adjacent epidural space. The diagnosis of subligamentous herniated disk is based on (*a*) focal extradural disk material within the confines of the posterior longitudinal ligament, (*b*) loss of epidural fat, (*c*) nonvisualization of the adjacent nerve root, and (*d*) associated loss in disk height and signal changes within the disk space. Sagittal oblique projections [61] or Gd-DTPA–enhanced MRI [62] is not usually necessary except in unusual circumstances, such as differentiation of a herniated disk from postoperative scar or foraminal and extraforaminal disk herniation from other conditions.

Lateral and Extraforaminal HNP

A herniated nucleus pulposus (HNP) may be located beyond the confines of the spinal canal, either within the nerve root canal or at the outer margin of the canal (Fig. 5.19). Both of these locations are difficult to demonstrate on myelography due to the variable lateral extension and filling of the nerve root sheath with contrast material. Both CT [63, 64] and MRI [65], because of their axial image orientation, are excellent in demonstrating extraforaminal or lateral herniated disks. There is a higher incidence of lateral and extraforaminal disk herniation in males, often with a history of long-term inadequate treatment for low back pain. Since the far lateral disk is located above the pedicle and may migrate cephalad, root involvement is typically one segment higher than with intraspinal herniated disks [65]. Foraminal and extraforaminal herniated disks are more common at the L2–3 and L3–4 disk spaces, unlike the usual intraspinal HNP at lower levels.

On imaging studies (Figs. 5.20 and 5.21), foraminal and extraforaminal HNPs have density or intensity characteristics similar to that of a central or paracentral herniated nucleus pulposus. Far lateral herniated disks usually have a smooth, oval contour, with deformity or obliteration of the paraspinal fat planes. The laterally placed disk may be partly foraminal in location. Parasagittal or sagittal oblique views are excellent in demonstrating the disk within the neural foramen. Because of their strategic location, being partly within the neural canal or foramen,

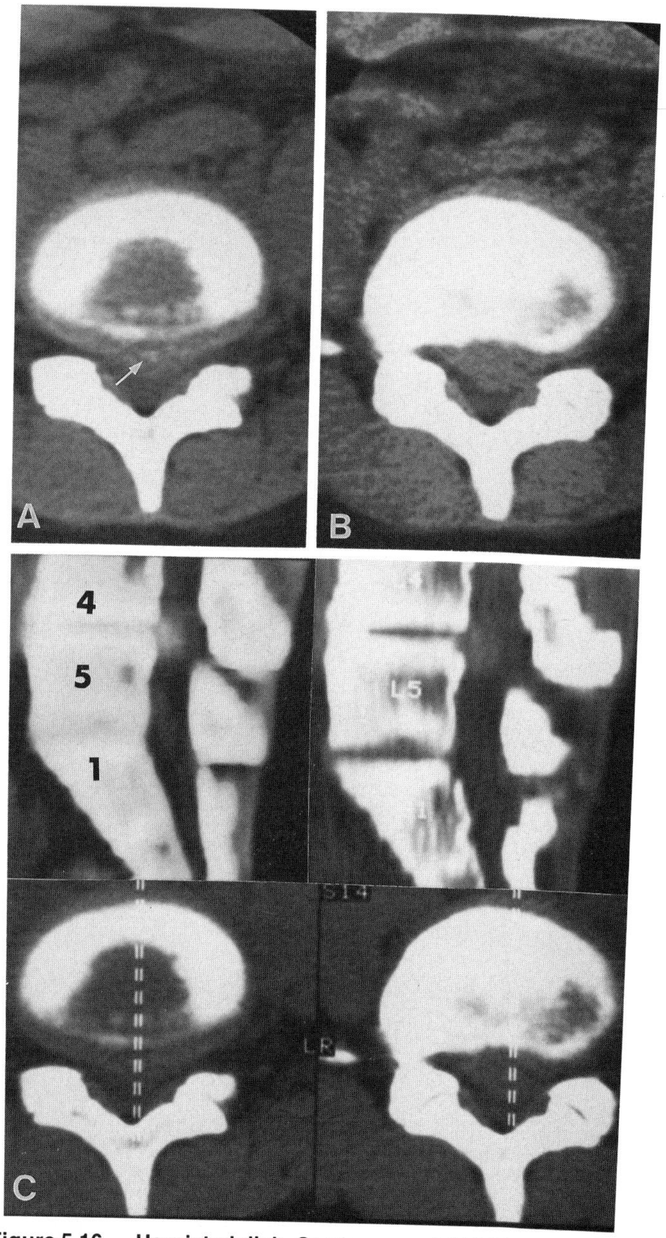

Figure 5.16. Herniated disk. Contiguous axial (**A,B**) and sagittal (**C**) re-formatted, noncontrast CT scans demonstrate a moderate size herniated disk. The herniated disk is opposite the narrowed L4–5 disk space with deformity of the thecal sac. **D–F,** Contiguous sagittal MRI sequences in another patient with L4–5 herniated disk. T1-weighted (**D**), proton-density (**E**), and T2-weighted (**F**) images demonstrate the signal characteristics of the herniated disk (*arrowhead*) on the different pulse sequences and encroachment of the neural canal.

both the herniated disk material and the nerve root may be inflamed. Both foraminal and extraforaminal herniated disks may have bright signals on T2-weighted images. Compression of the venous plexus is not uncommon. Enhancement of the disk and the adjacent venous plexus may be confused with a nerve sheath tumor (Fig. 5.22). Gd-DTPA–enhanced MRI may be necessary for demonstrating an extraforaminal herniated disk. If a Gd-DTPA–enhanced

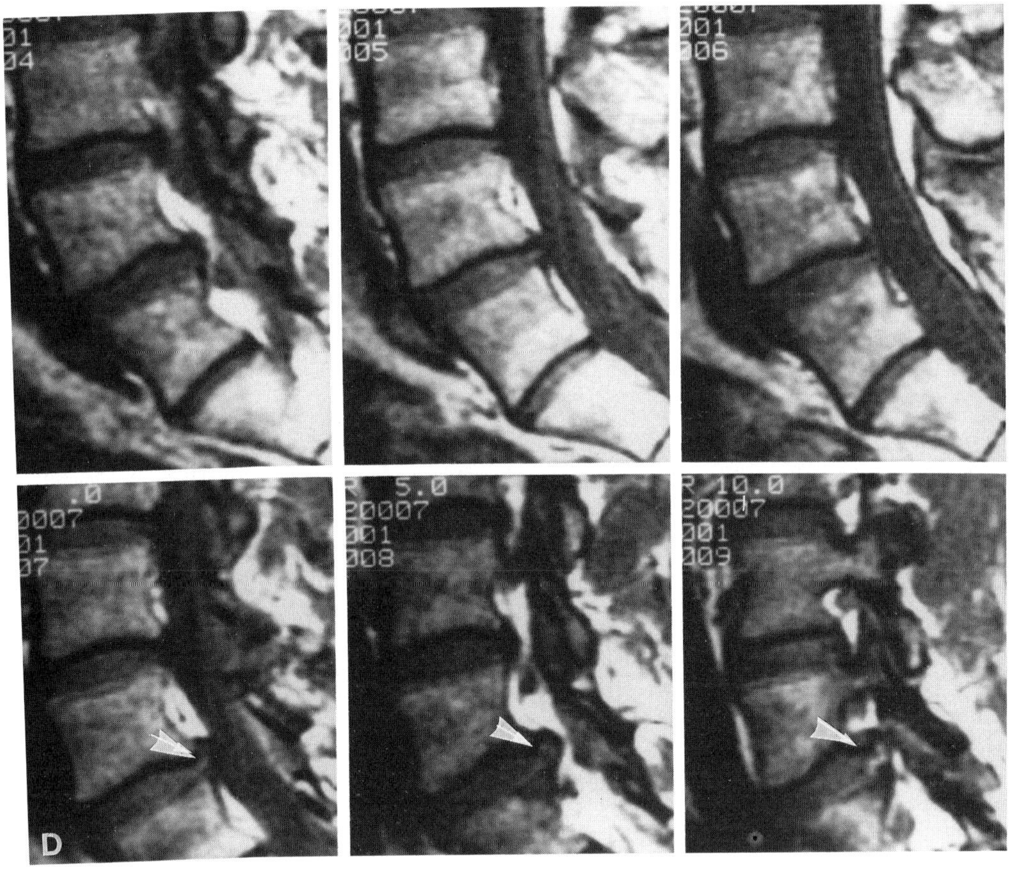

Figure 5.16–*continued*

study is contemplated, use of a fat-suppression sequence may allow appreciation of subtle enhancement without interference from the epidural fat.

Extruded Disk

The diagnosis of an extruded disk is based on demonstrating an extradural soft-tissue mass that is connected to the parent nucleus pulposus by a narrow pedicle or on the ability to demonstrate an absence of annulur fibers in a broad-based large herniated nucleus pulposus (Fig. 5.23). The extruded disk may migrate in a cephalad or caudad direction. It is usually paracentral in location at the L5–S1 level and to a lesser extent at the L4–5 level. Cephalad or caudad migration occurs when there is a transverse tear of the annulus, usually at its point of attachment between the superior or inferior endplate of the vertebral body, respectively. The paracentral location in relation to the vertebral body is due to lateral thinning of the posterior longitudinal ligament and the presence of a sagitally aligned thin septum between the posterior aspect of the vertebral body and the posterior longitudinal ligament [66]. Because of the weakness of the ligament laterally and the potential space between the body of the disk and the posterior longitudinal ligament, an extruded disk and "free fragments" migrate into this space or laterally beyond the confines of the ligament.

On CT, an extruded disk should be suspected if the herniated disk material is identified both at the disk level and with a contiguous cephalad or caudad component opposite the vertebral body. The tear or rupture of the annulus is not seen, since this often occurs at the attachment of the annulus to the endplates. Similarly, the midline septum between the posterior margin of the vertebral body and ligament [66] may not be seen due to the partial volume averaging and higher attenuation of the herniated disk. Occasionally, the extruded disk may have a slightly lower attenuation value similar to the thecal sac. This probably represents edema or fluid within the extruded disk (Figs. 5.23 and 5.24). IVCT may be useful in demonstrating the tear in the annulus and the attached posterior longitudinal ligament.

On MRI, an extruded disk should be suspected when, apart from the findings similar to CT, absence or disruption of the annulus/posterior longitudinal ligament can be identified. On axial and sagittal MRI sections, unlike the subligamentous herniated disk (HNP), the extruded disk material is not completely bound by the hypointense signal, which normally represents the thinned out annulus and the posterior longitudinal ligament. This is best appreciated

Figure 5.16–*continued*

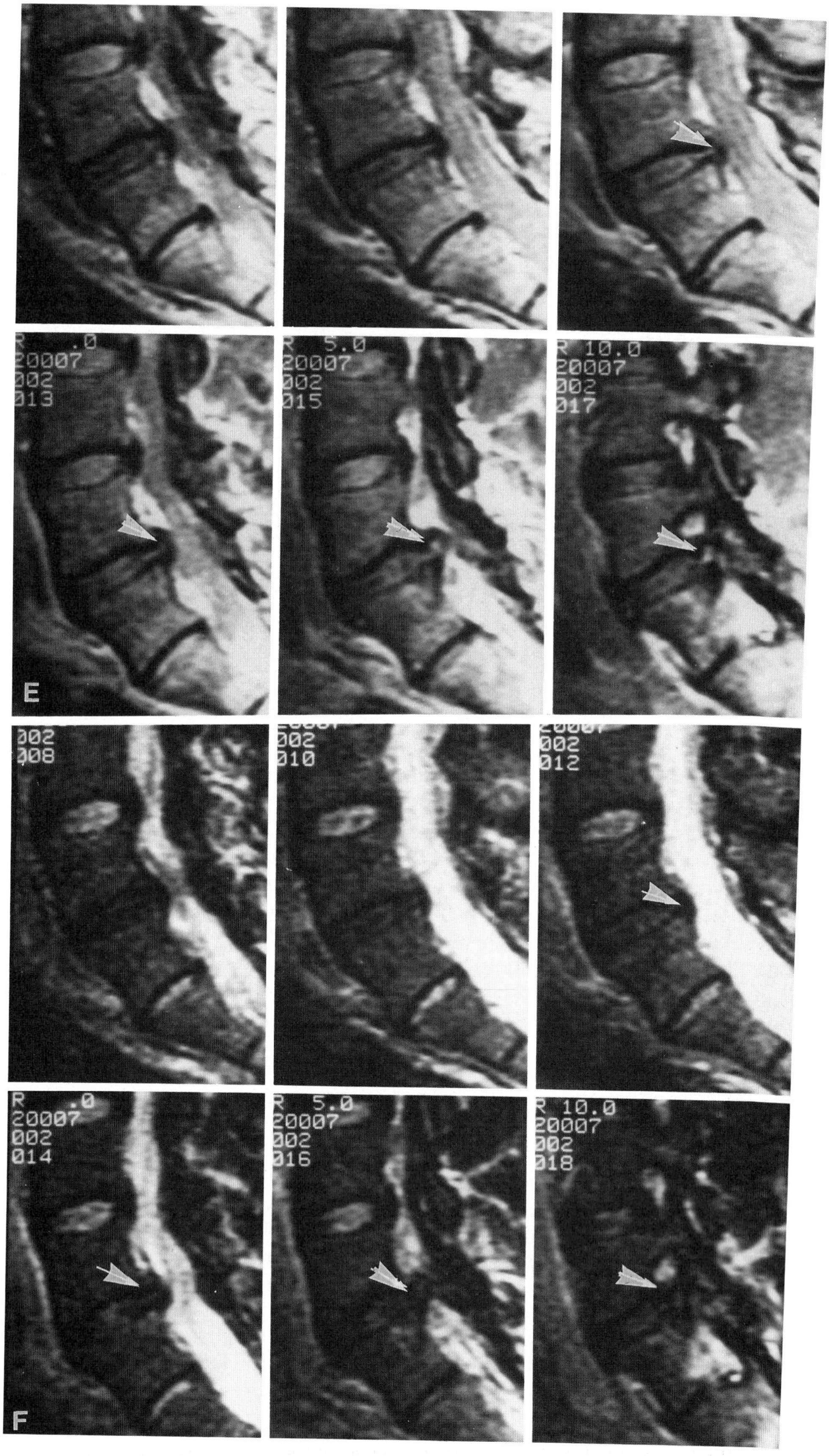

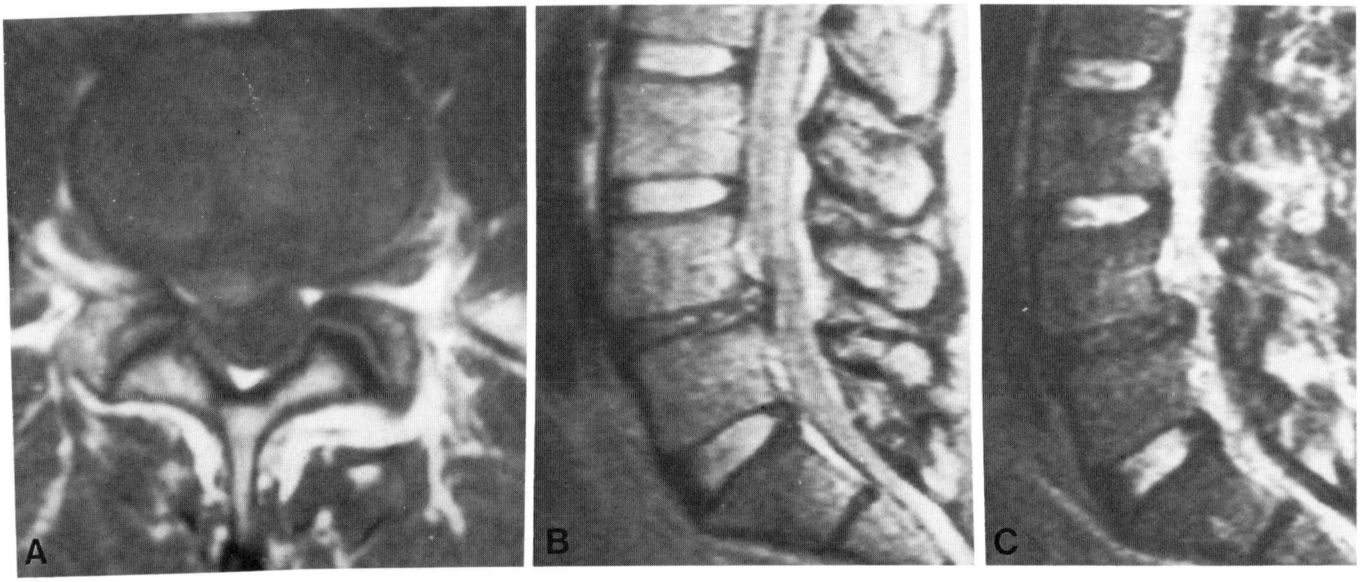

Figure 5.17. Herniated disk MRI. A, Axial T1-weighted MRI of a right paracentral herniated disk (*arrowhead*) has a slightly brighter signal than the parent disk and the thecal sac. **B,C,** Sagittal MRI sections. Proton-density (1800/30) (**B**) and T2-weighted (1800/80) (**C**) sections demonstrate the narrow disk space with the herniated disk material within the confines of the annular fibers and posterior longitudinal ligament. The proton-density image demonstrates the slightly hyperintense signal of the herniated disk better than the T2-weighted section.

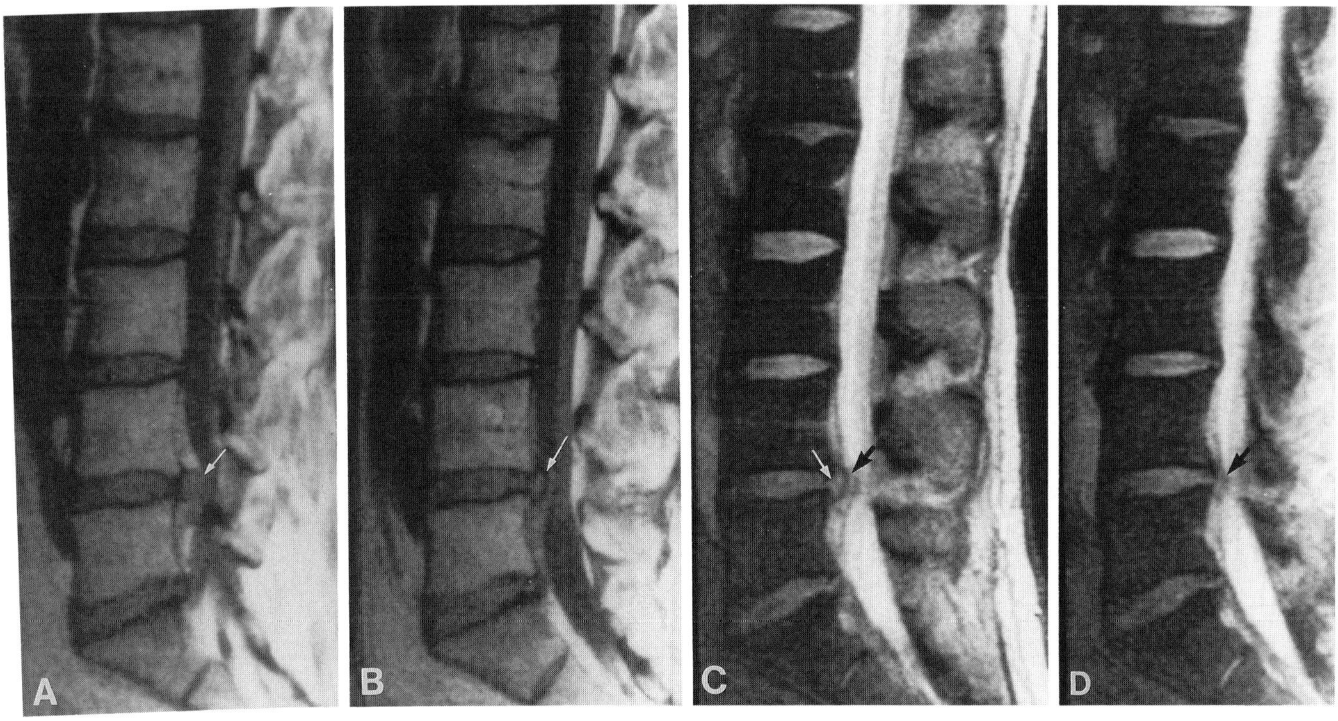

Figure 5.18. T1-weighted (**A,B**) and T2-weighted (**C,D**) sagittal MRI sequences with the subligamentous herniated disk within the thin annular fibers (*white arrow*) and the posteriorly displaced dura (*dark arrow*). The heterogeneous signal within the prominent epidural space (*arrowhead*) consists of dilated venous channels and epidural fat.

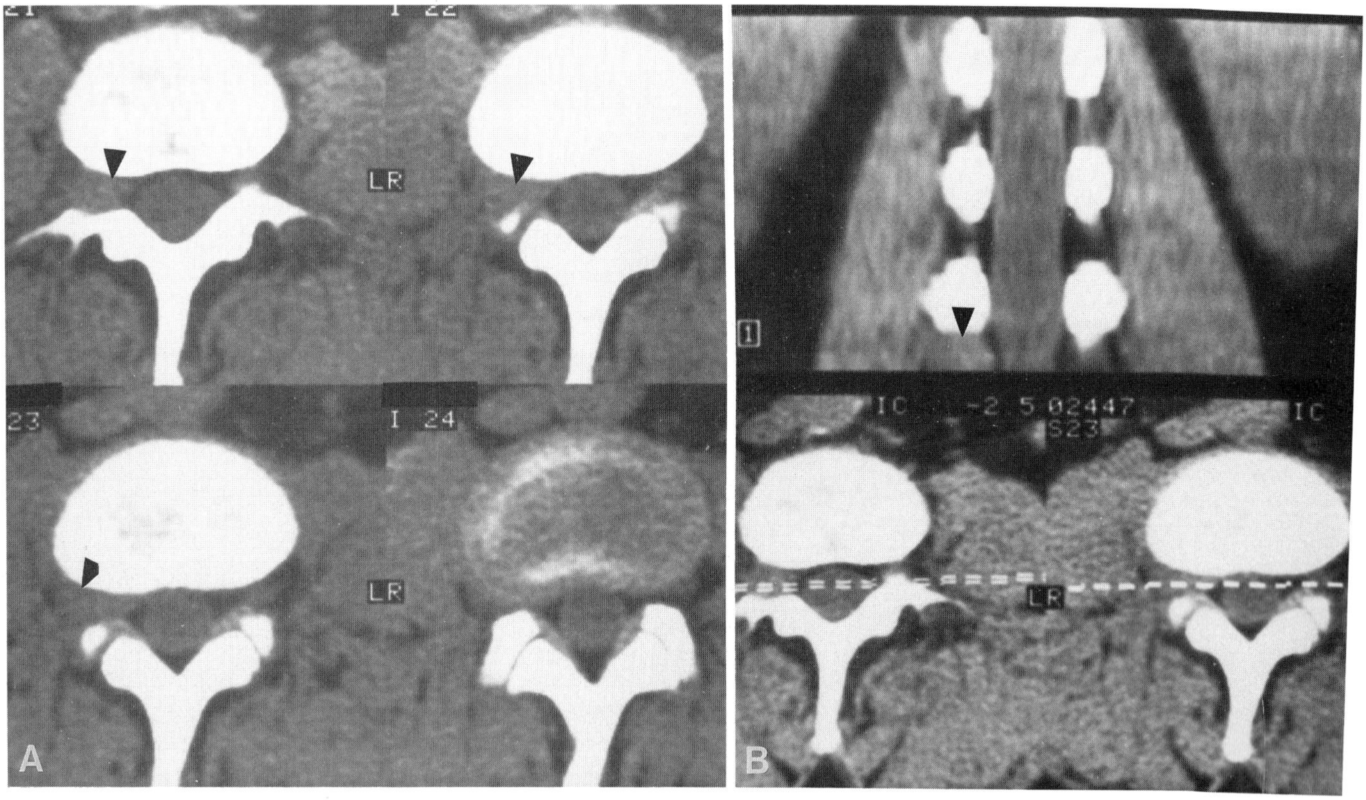

Figure 5.19. Foraminal herniated disk at the L3–4 level. Contiguous noncontrast axial CT (**A**) and coronal re-formatted CT (**B**) demonstrate the mildly hyperdense herniated disk (*arrowhead*) within the neural foramen on the right side.

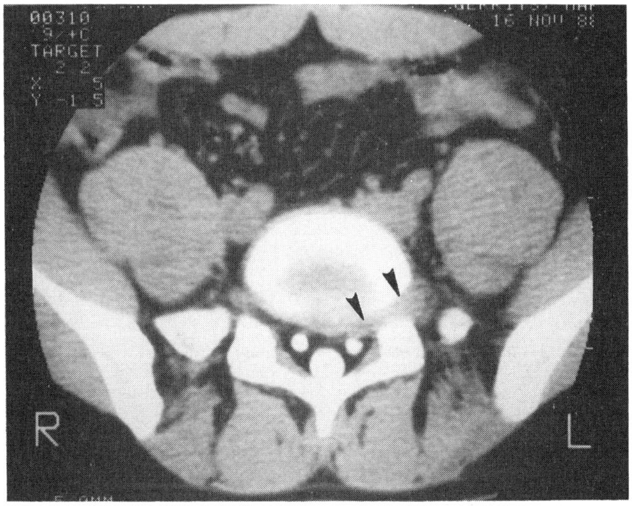

Figure 5.20. Foraminal disk herniation at L4–5. Axial CTM demonstrates the lateral herniated disk extending beyond the neural foramen (*arrowhead*). The L5 nerve root sheath is not compressed because it is still within the spinal canal.

in the T2-weighted or GRE images (Figs. 5.24 and 5.25). The pedical between the parent disk and the herniated disk is best demonstrated on sagittal T2-weighted images (see Fig. 5.2). The extruded disk often has a hypointense or intermediate signal on T1-weighted images and variable degrees of bright signal intensity on T2-weighted and GRE images. Occasionally, Gd-DTPA–enhanced MRI may be indicated as a means of demonstrating a torn or disrupted posterior longitudinal ligament [62]. Distinction between subligamentous herniated disk and extruded disk may have a clinical value if non-surgical treatment is being considered, since these patients are not candidates for treatment with chemonucleolysis or percutaneous aspiration.

Free Fragments and the Sequestered Disk

A detached free fragment from an extruded disk may migrate in the potential space between the vertebral body and the posterior longitudinal ligament, or the fragmented disk material may be located laterally within the neural canal. It thus becomes "sequestered" by migrating cephalad, caudad, or laterally (Fig. 5.26). It is not unusual for patients with extruded disks to find some transient relief of symptoms following fragmentation of the herniated disk. Subsequently, their symptoms may recur, but at a different level. If the disk fragment has broken through the posterior longitudinal ligament or, as is more common, the disruption of the annulus is paracentral, the free fragment usually migrates caudad.

On CT, a herniated disk with a free fragment (sequestered disk) appears as a large soft-tissue mass, usually oval in configuration. It can have a higher or lower attenuation than the dural sac or the nerve root sheath. If the sequestered disk is adjacent to a

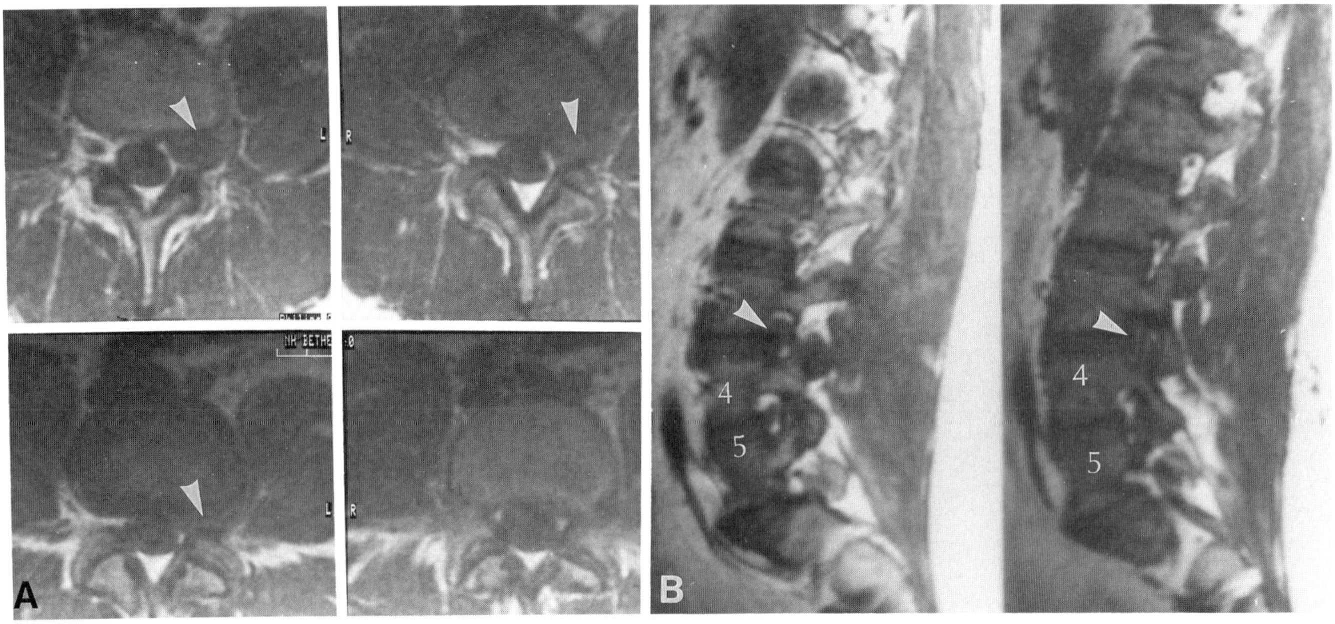

Figure 5.21. MRI in foraminal disk herniation at L3–4. Contiguous axial (**A**) and adjacent left parasagittal (**B**) T1-weighted MRI sequences. Large foraminal herniated disk (*arrowhead*) within the left neural foramen.

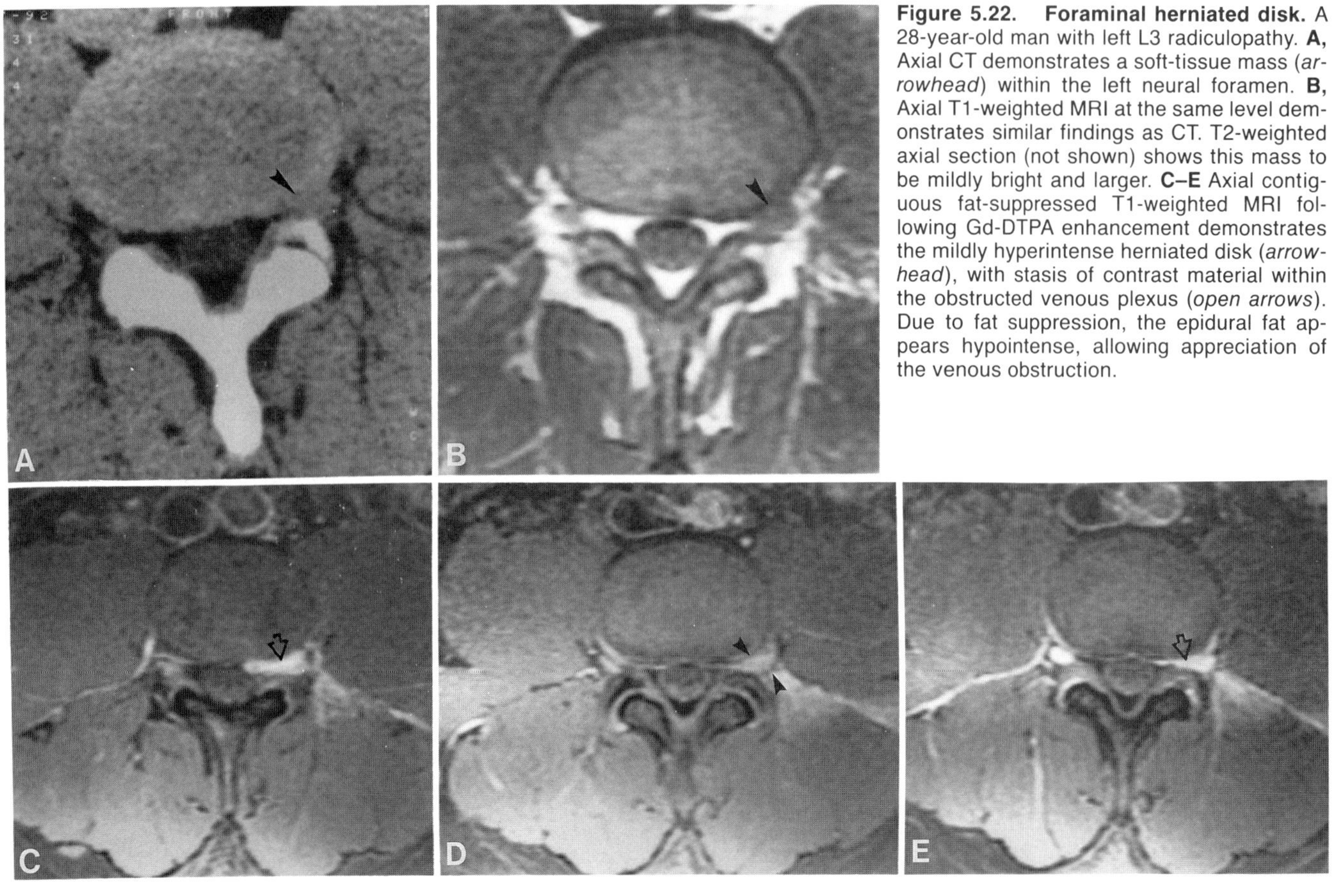

Figure 5.22. Foraminal herniated disk. A 28-year-old man with left L3 radiculopathy. **A,** Axial CT demonstrates a soft-tissue mass (*arrowhead*) within the left neural foramen. **B,** Axial T1-weighted MRI at the same level demonstrates similar findings as CT. T2-weighted axial section (not shown) shows this mass to be mildly bright and larger. **C–E** Axial contiguous fat-suppressed T1-weighted MRI following Gd-DTPA enhancement demonstrates the mildly hyperintense herniated disk (*arrowhead*), with stasis of contrast material within the obstructed venous plexus (*open arrows*). Due to fat suppression, the epidural fat appears hypointense, allowing appreciation of the venous obstruction.

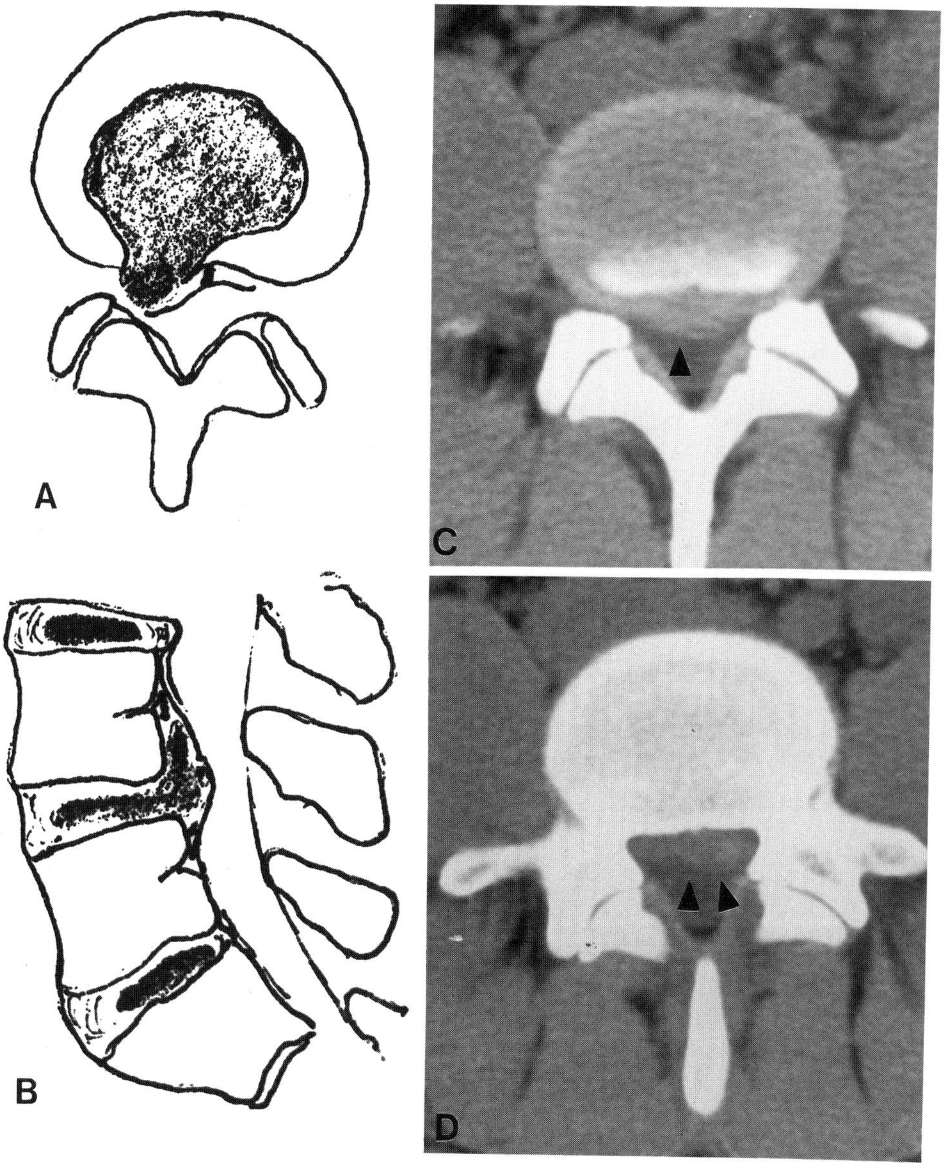

Figure 5.23. Schematic drawing of an extruded disk. Axial (**A**) and sagittal (**B**) views. Contiguous axial CT sections demonstrate a focal central herniated disk within the spinal canal (**C**) at the disk level with caudal extension of the herniated disk material opposite the L5 vertebral body (arrowheads) (**D**). CT diagnosis is based on this characteristic finding.

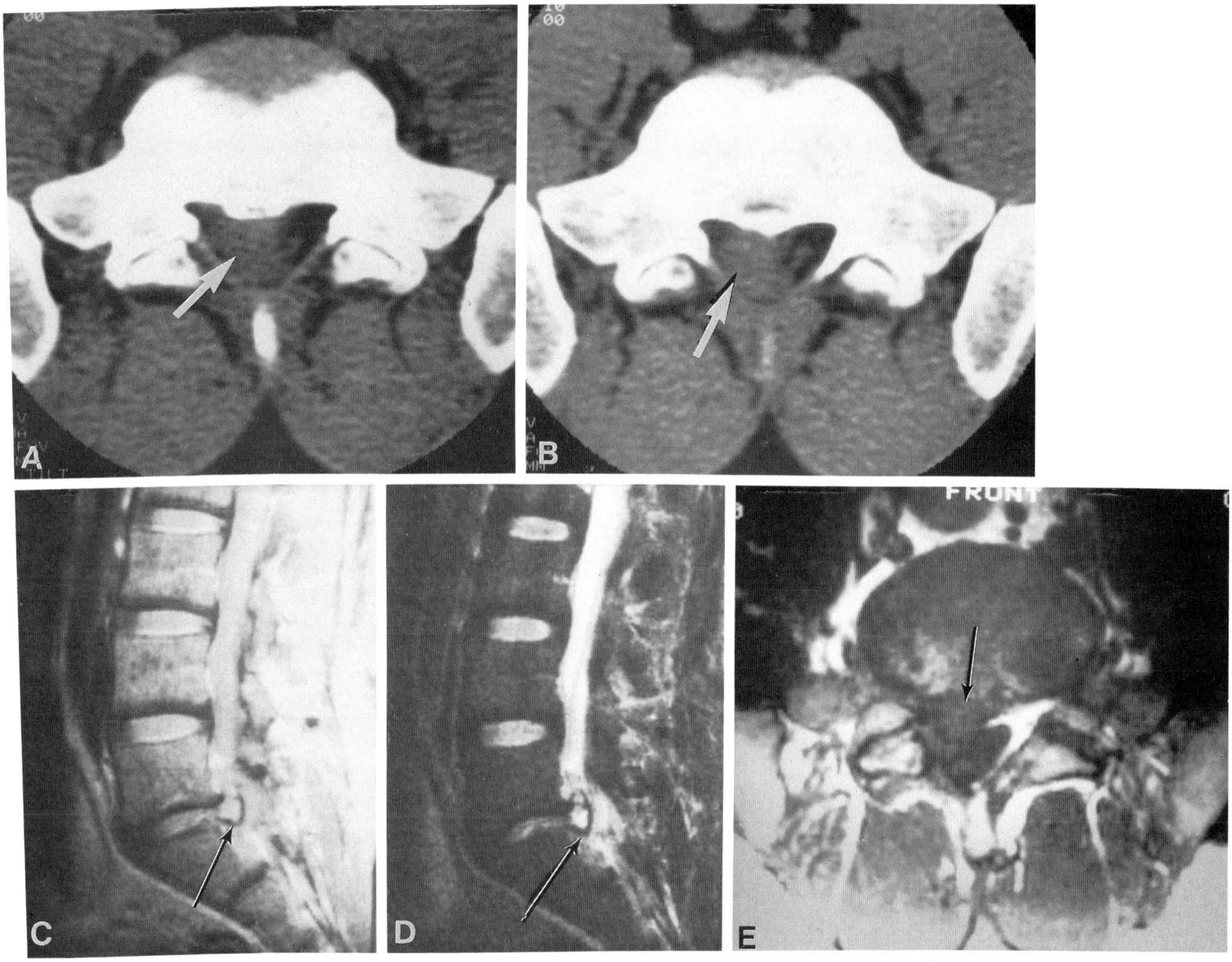

Figure 5.24. **Extruded disk. A,B,** Contiguous axial CT sections demonstrate an extruded right paracentral herniated disk (*arrowhead*) with similar density as the thecal sac. T2-weighted sagittal MRI sections—(**C**) (2000/30) and (**D**) (2000/80)—demonstrate the extruded disk (*arrow*) with disruption of the annular fibers along its superior margin. The extruded disk has a bright signal compared to the parent disk. **E,** The mildly hyperintense extruded disk (*arrow*) on axial T1-weighted MRI.

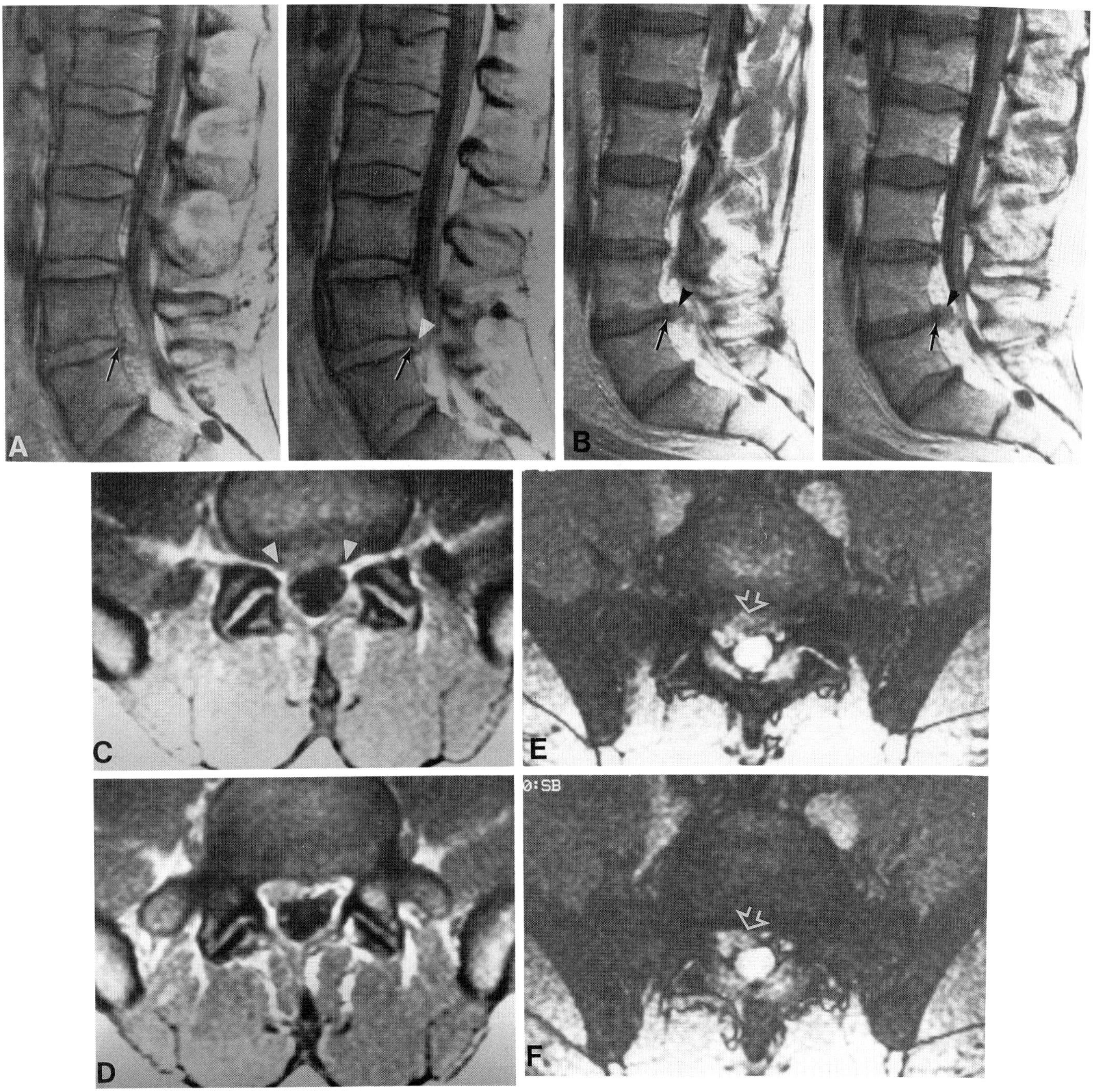

Figure 5.25. MRI of extruded disk. A, T1-weighted MRI demonstrates the narrow pedicle (*arrow*) connecting the parent disk with the extruded disk (*arrowhead*). **B,** T1-weighted Gd-DTPA–enhanced MRI demonstrates sharper delineation of the extruded disk due to contrast pooling in the epidural venous plexus. **C,D,** Contiguous axial T1-weighted MRI sections demonstrate the slightly hyperintense signal of the extruded disk and disruption of the annular fibers (*arrowhead*) **E,F,** Gradient-echo contiguous axial sections demonstrate the extruded disk (*arrowheads*).

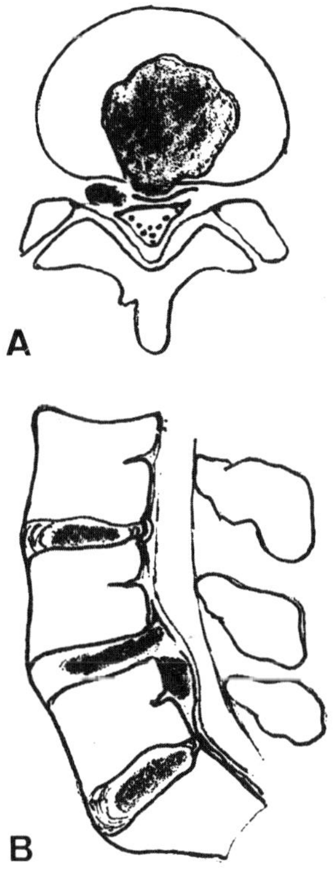

Figure 5.26. Schematic drawing of a sequestered disk. Axial (**A**) and sagittal (**B**) views. A fragmented disk not attached either to the parent disk or to the extruded disk is defined as a "free" fragment. The free fragment may migrate in any direction. Most often such fragments are sequestered beyond the disk level, often within the neural canal (lateral recess).

nerve root sheath, it may mimic a prominent thecal sac, tumor, conjoined nerve, or cystic mass. These are best evaluated with CTM [67, 68].

MRI diagnosis of a free disk fragment is based on T2-weighted studies. On T1-weighted sequences, the herniated disk as well as the free fragment may have an intermediate or hypointense signal. The involved disk space is decreased in height, with loss of signal in the sagittal views. On axial views, the sequestered disk material appears as an irregular or oval extradural soft-tissue mass adjacent to the suspected disk level or as a mass contiguous with the nerve root sheath in the adjacent neural canal or, rarely, at a slightly different level. The disk fragment has a bright signal on the T2-weighted studies (Fig. 5.27). The bright signal is presumed to be due to the higher water content [69]. Other factors which may account for the bright signal include tissue edema from the surrounding venous plexus or, less likely, punctate hemorrhage.

Intradural Disk Herniation

Intradural herniation of disk material is rare, with an incidence of 0.13% [70]. There is usually a history of low back pain with periods of exacerbation and progressive neurologic deficit. Occasionally, patients may have had previous back surgery not necessarily related to disk disease. Intradural disk herniation is presumed to result from chronic irritation with subsequent adhesions between the posterior longitudinal ligament and the dura. The disk material herniates directly through the weakened dura instead of displacing the dura backwards. It is often seen in the post-surgical patient.

Differentiation of the disk material from the thecal contents is difficult on NCCT. On myelogram and MCT, the disk material presents as an intradural mass with complete or incomplete block of the thecal sac. On IVCT, the periphery of the disk material enhances as a result of the inflammatory response or meningeal reaction (Fig. 5.28). Differentiation from other intradural lesions such as neurofibromas, meningiomas, metastases, and large extradural granulomas may be difficult [71, 72]. Intradural disk herniation has an isointense or hypointense signal on both T1- and T2-weighted images, with mildly hyperintense margins [72, 73] (Fig. 5.29). On Gd-DTPA–enhanced MRI, there is always peripheral enhancement similar to that seen with IVCT (Fig. 5.30). The clinical history, the narrowing of the adjacent disk space, and the CT or MRI appearance usually help in differentiation from other intradural lesions.

A very large herniated disk (giant disk) may result in a complete block. Large herniated disks at the L1–2 level may present with the cauda equina syndrome. The herniated disk may be mistaken for an intra- or extradural tumor. Diagnosis depends on the enhancement characteristics on both CT and MRI and the extradural origin of the mass on myelography and CTM.

Disk Regression

Regression of a herniated disk spontaneously or following conservative therapy is not uncommon [74–78]. Relief of symptoms is associated with reduction in the size or volume of the herniated disk (Fig. 5.31). Regression of the herniated disk is not unusual with subligamentous disk herniation. Regression of the disk material has been shown with CT [74, 75] and on MRI studies with extruded disk herniations [76]. The reduction in disk volume is probably due to inflammatory response and autologous absorption of the extruded disk. The decrease in disk volume also may be attributed to dehydration of the herniated disk material and may account for the change from an initially bright signal to a hypointense signal on follow-up MRI studies. MRI is useful as a noninvasive study in evaluating conservative treatment of disk herniation.

THORACIC SPINE

Degenerative changes involve the thoracic region in a manner similar to the rest of the spine, but the

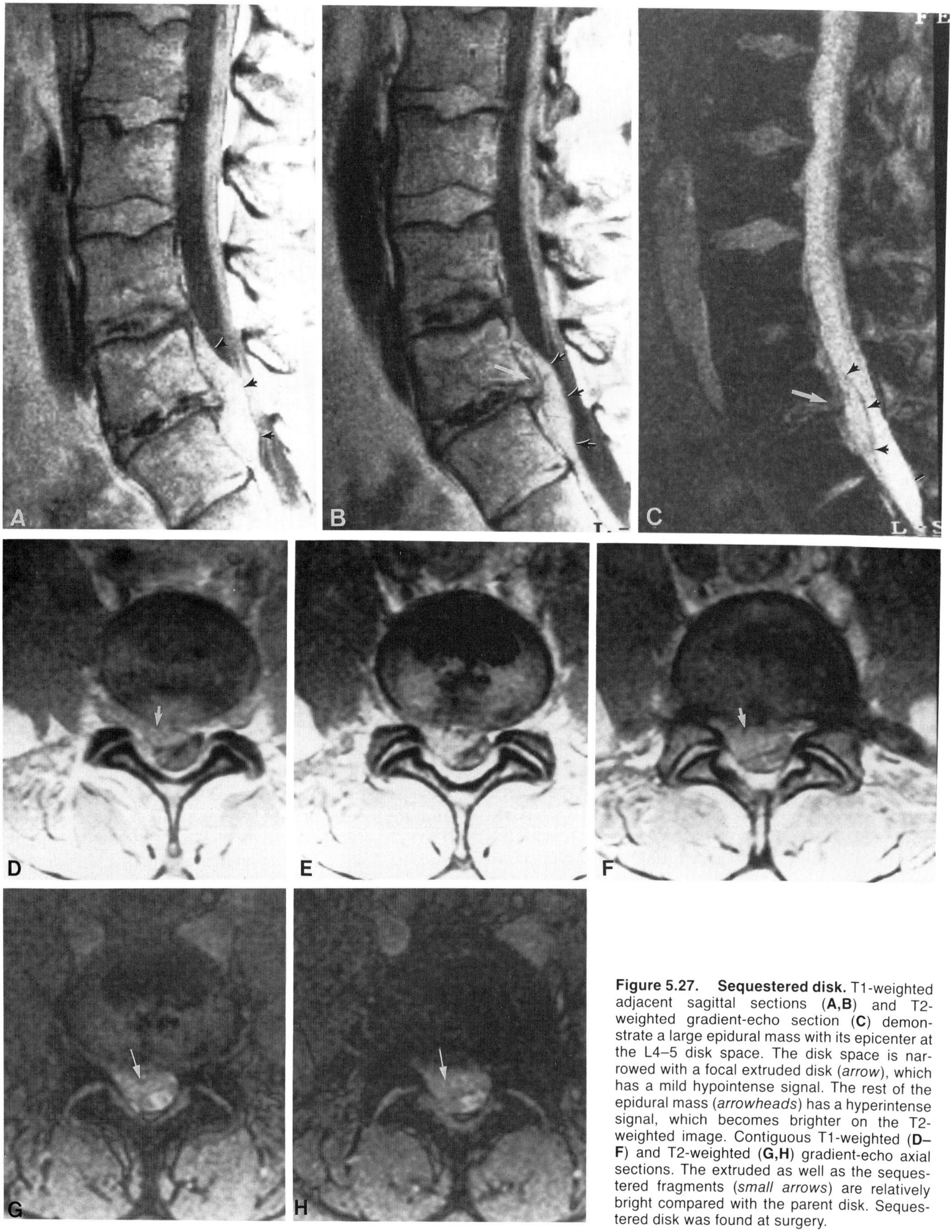

Figure 5.27. Sequestered disk. T1-weighted adjacent sagittal sections (**A,B**) and T2-weighted gradient-echo section (**C**) demonstrate a large epidural mass with its epicenter at the L4–5 disk space. The disk space is narrowed with a focal extruded disk (*arrow*), which has a mild hypointense signal. The rest of the epidural mass (*arrowheads*) has a hyperintense signal, which becomes brighter on the T2-weighted image. Contiguous T1-weighted (**D–F**) and T2-weighted (**G,H**) gradient-echo axial sections. The extruded as well as the sequestered fragments (*small arrows*) are relatively bright compared with the parent disk. Sequestered disk was found at surgery.

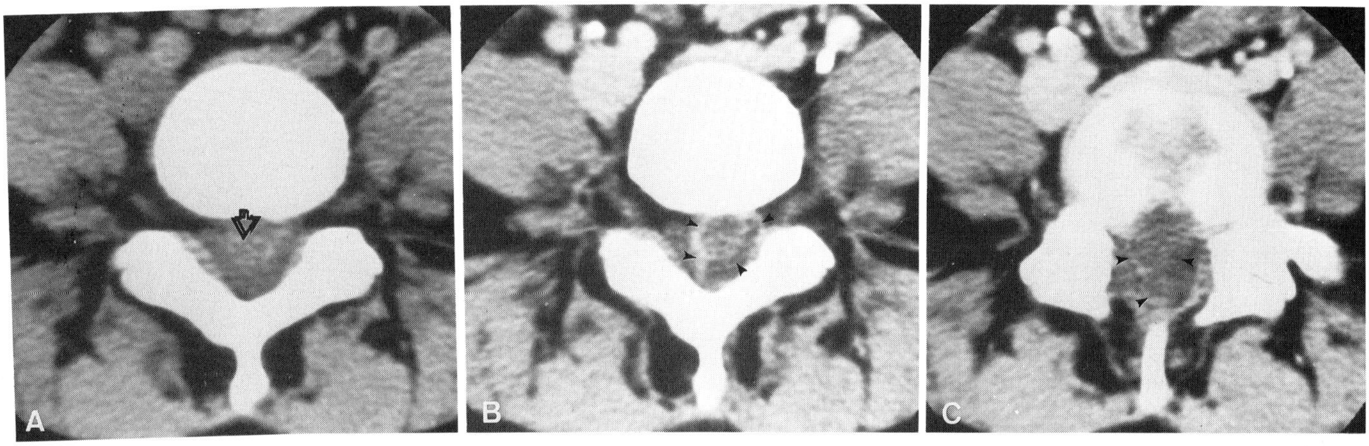

Figure 5.28. CT of intradural disk herniation. A, Noncontrast axial CT demonstrates a large hyperdense (*open arrow*) mass within the thecal. **B,C,** Contiguous axial IVCT scans demonstrate enhancement of the margins of the intraspinal mass and adjacent dura of the thecal sac. At surgery this was an extruded disk which had penetrated the dura.

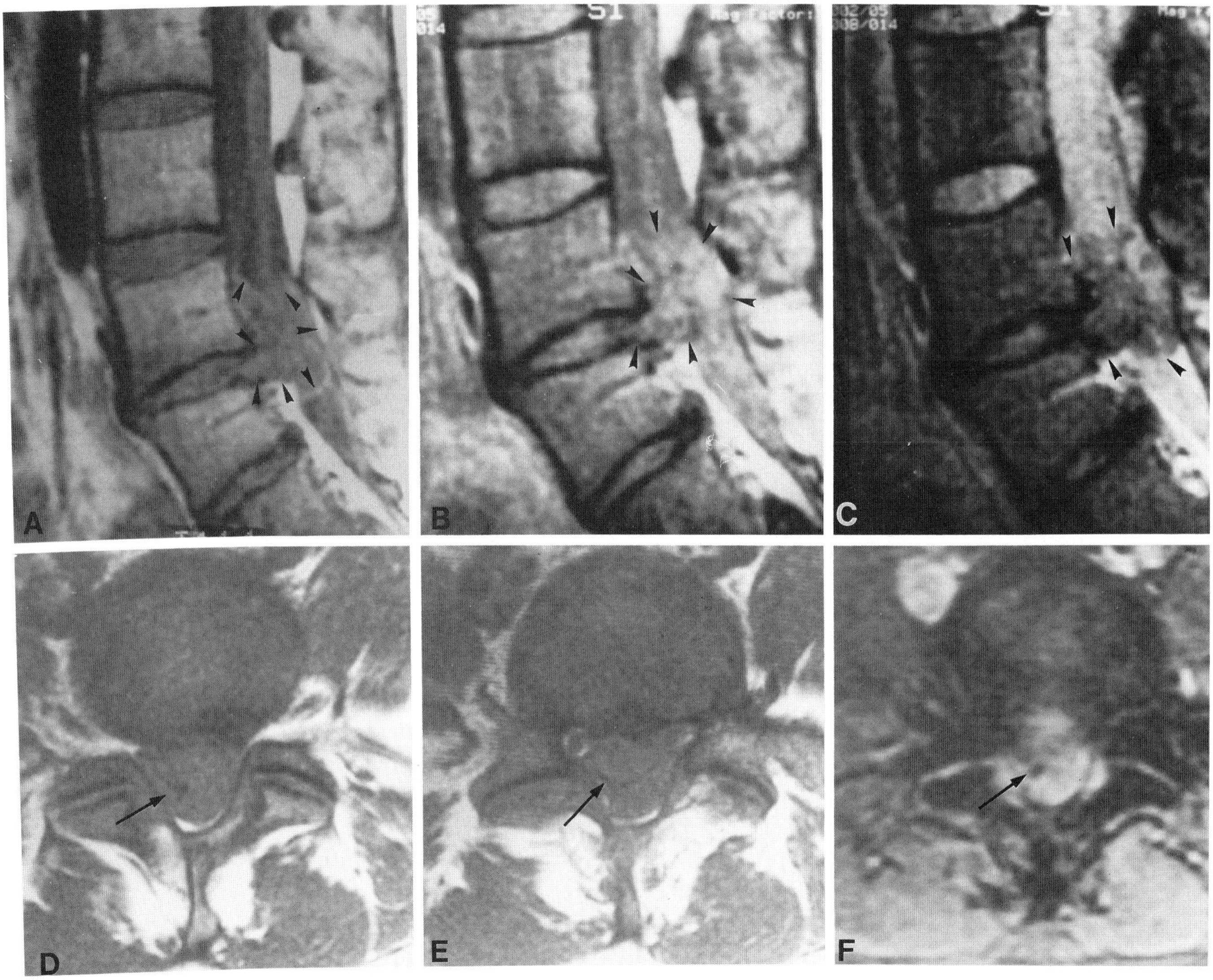

Figure 5.29. MRI of intradural disk herniation. A, T1-weighted sagittal MRI demonstrates a large intradural mass (*arrowheads*) contiguous with the epidural space and the adjacent disk space of L4–5. **B,** Proton-density sagittal section shows an intra- and extradural mass with bright signal compared with the CSF and areas of heterogeneous signal. **C,** T2-weighted section demonstrates more marked heterogeneous signal in the intra- and extradural mass. In all three sequences, the hypointense signal of the annular fiber–posterior longitudinal ligament is not seen. **D,E,** Contiguous axial MRI sections demonstrate the similar signal intensity of the intradural mass and CSF. **F,** T2-weighted axial section shows the intradural herniated disk with ill-defined margin (*arrowheads*) within the thecal sac.

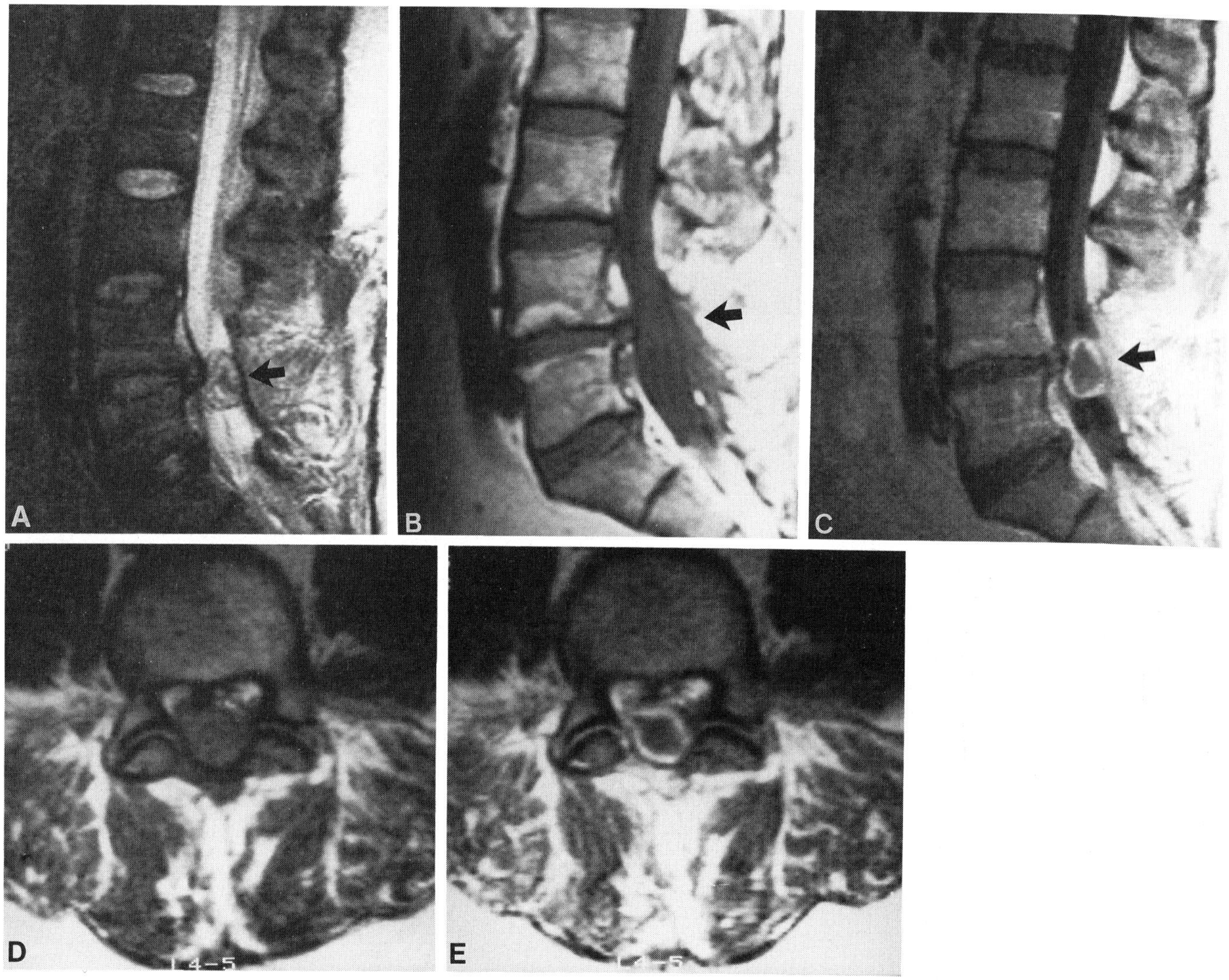

Figure 5.30. MRI in intradural disk herniation. A, T2-weighted sagittal image demonstrates a heterogeneous intraspinal mass (*arrow*) adjacent to a herniated disk. **B,** T1-weighted sagittal section shows the intraspinal mass to be isointense compared with CSF. The endplates of the adjacent vertebral bodies show type 1 changes. **C,** T1-weighted Gd-DTPA–enhanced image demonstrates peripheral enhancement around the intradural disk fragment. T1-weighted axial section (**D**) and Gd-DTPA–enhanced T1-weighted image (**E**) demonstrate the peripheral enhancement around the intradural disk fragment. (From Wasserstrom R, Mamourian AC, Black JF, et al.: Intradural lumbar disk fragment with ring enhancement on MR. AJNR 1993; 14:401–404.)

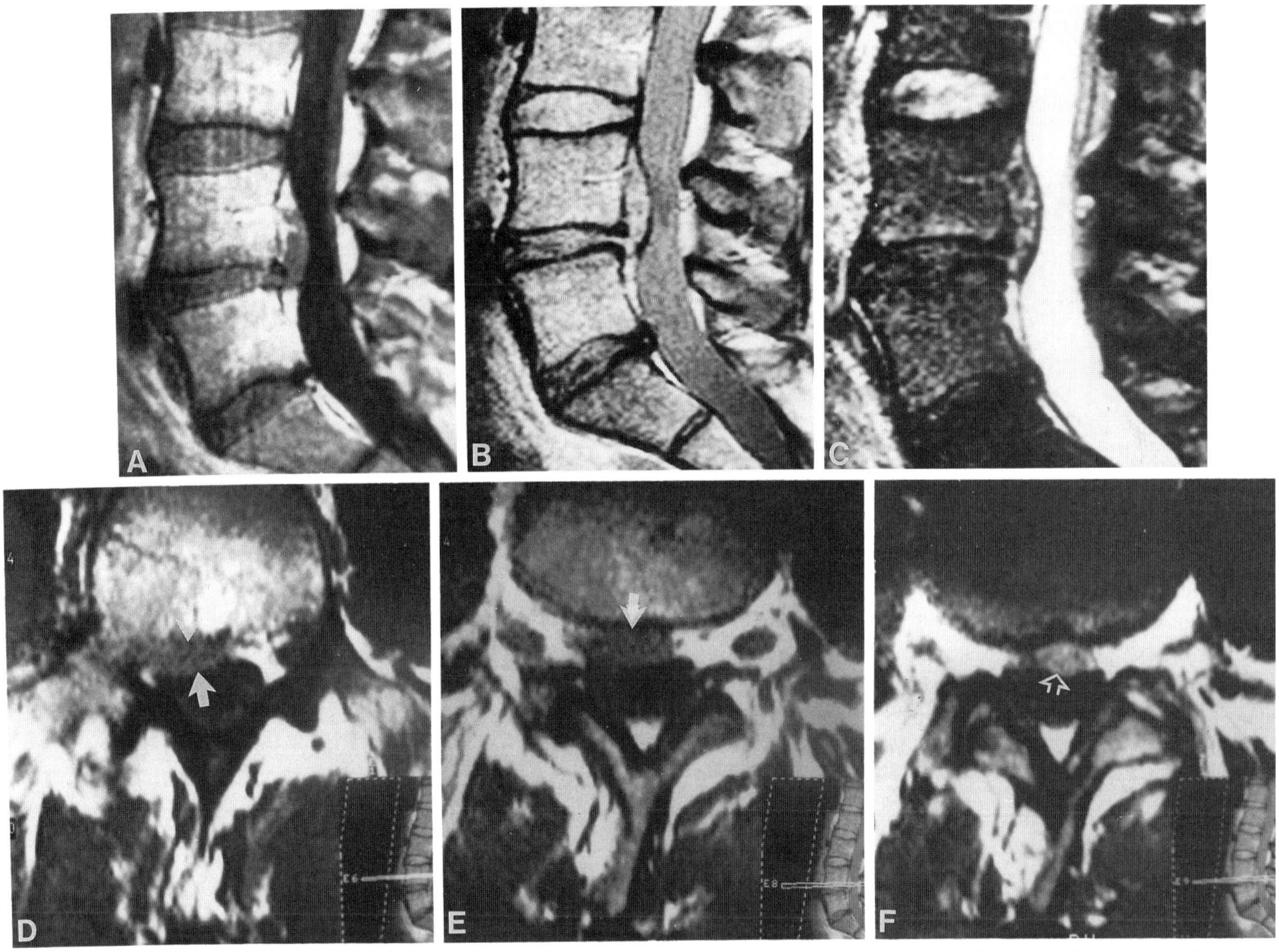

Figure 5.31. Disk regression. A–C, Sagittal T1-weighted, proton-density, and T2-weighted sagittal MRI sequences demonstrate an extruded disk with a sequestered fragment in the extradural space. It has a mild hypointense signal on the T1-weighted image, bright signal on the proton-density image, and a heterogeneous signal on the T2-weighted image. **D–F,** Contiguous axial section demonstrates the mild hypointense signal of the sequestered disk (*arrowhead*), with a hyperintense signal (*open arrow*) probably representing venous stasis. **G,H,** Twelve days following the initial study, T1-weighted contiguous axial sections before contrast (**G**) and following Gd-DTPA enhancement (**H**). The sequestered disk fragment shows peripheral enhancement. The size of the extradural disk also appears smaller. **I,** Fifty-six days following initial study, T1-weighted Gd-DTPA–enhanced axial section demonstrates further regression of the disk material, although there is persistent intense enhancement surrounding the disk. **J–L,** Follow-up study 410 days following the initial study. T1-weighted contiguous sagittal section (**J**), T2-weighted sagittal section (**K**), and T1-weighted axial section (**L**). There is complete regression of the disk material. The disk space is narrowed. Type 2 endplate changes are seen in adjacent vertebral bodies. Intravertebral disk herniation (Schmorl's node) not present initially is demonstrated.

Figure 5.31–*continued*

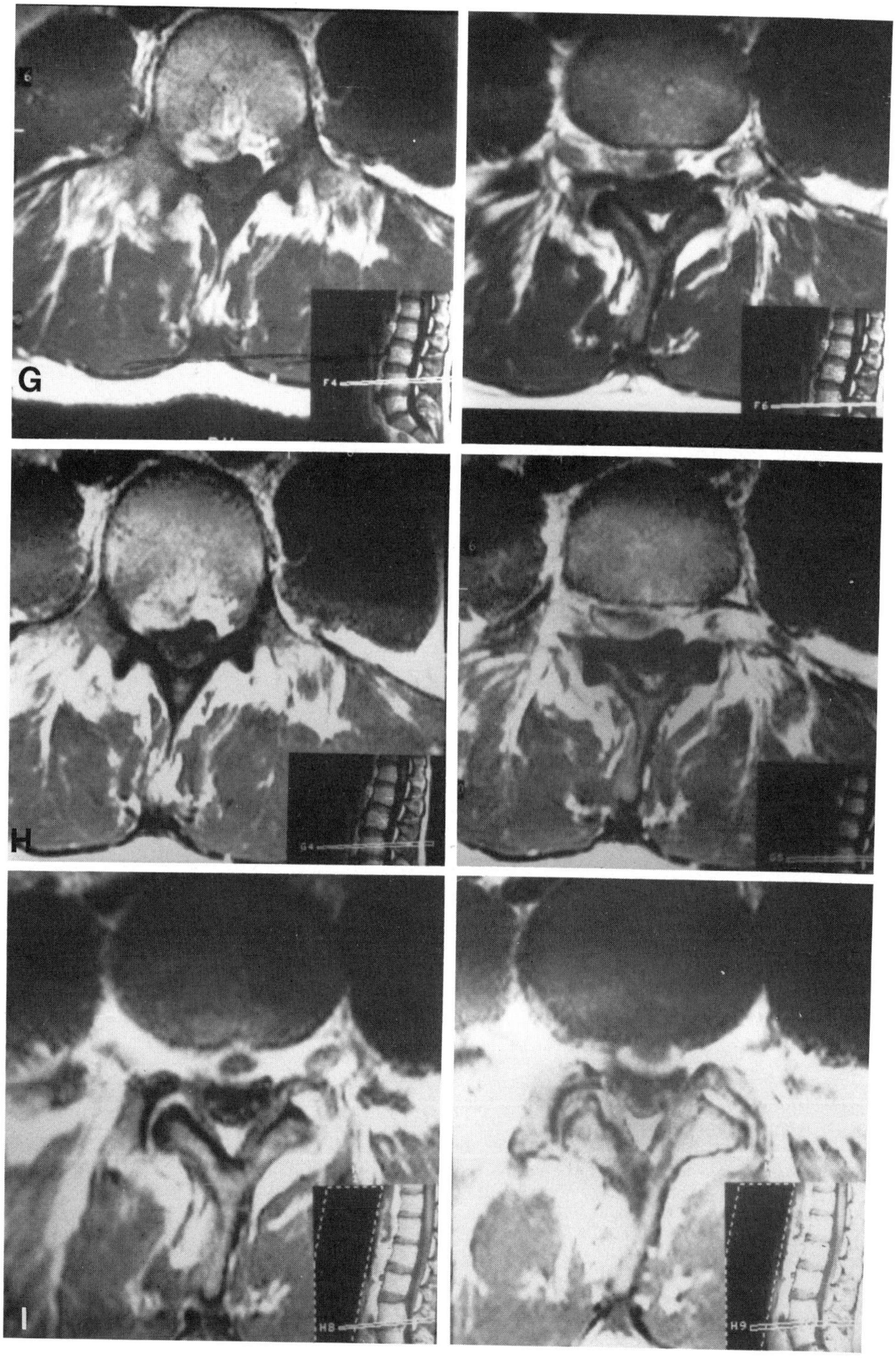

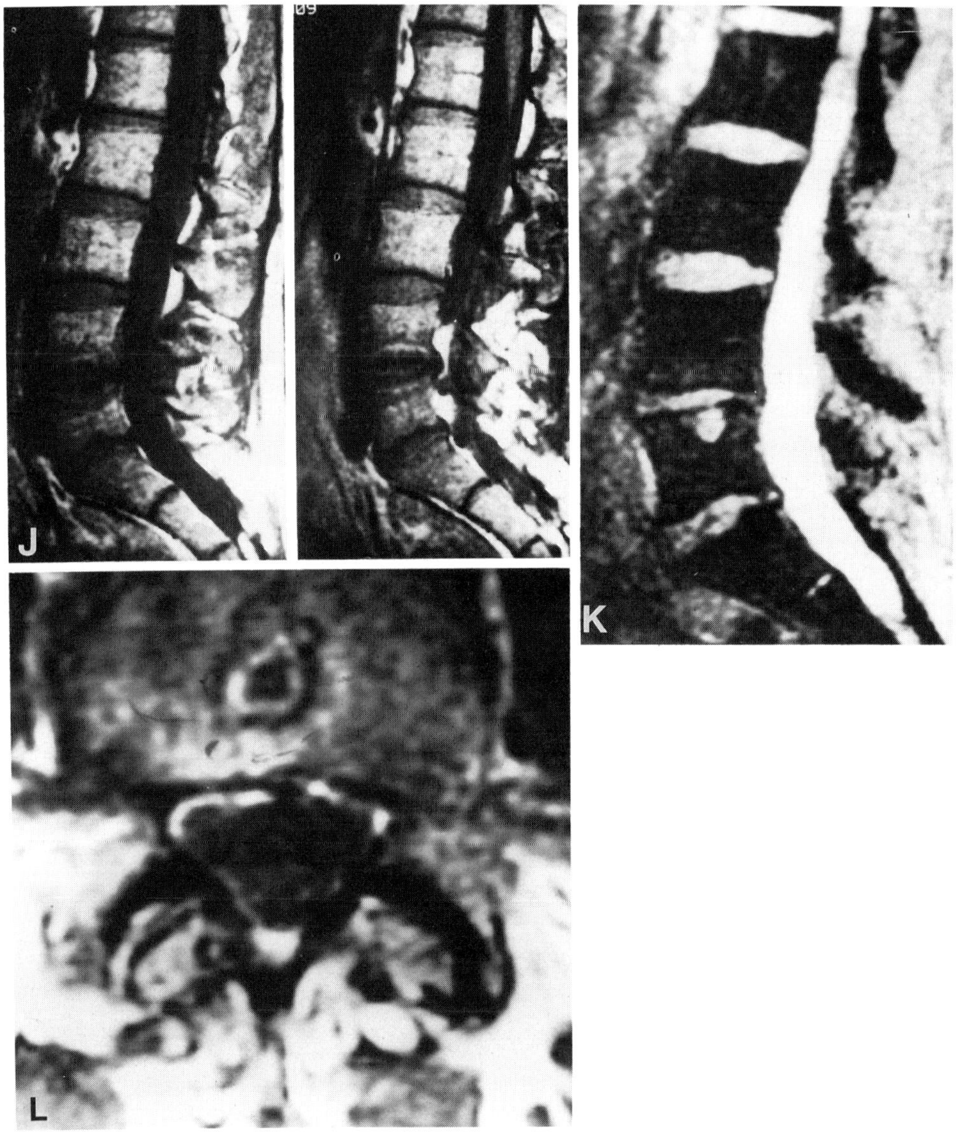

Figure 5.31–*continued*

process is slower than in the lumbar and cervical regions. This is due to the greater support provided by the thoracic rib cage and the less significant rotary component, with the greatest motion being flexion and a limited amount of extension. Age-related narrowing of the disk space with degenerative bulging of the disk is not uncommon [79]. Thoracic disk herniation is more common in males, unlike meningiomas, which are common in females. Disk degeneration with calcification and endplate eburnation is routinely observed in plain radiographs. It is conceivable that in many instances this degeneration may be associated with asymptomatic disk herniation. Such degeneration has been identified on CT studies [79, 80], and a higher incidence could be uncovered as a result of the greater utilization of MRI in evaluating the thoracic spine [81]. Thoracic disk herniation may involve multiple levels [82] (Fig. 5.32). The incidence of symptomatic disk herniation, however, is estimated at less than 0.1%. Thoracic disk herniation usually involves the lower thoracic region. The symptomatic disk usually compresses the spinal cord and the adjacent nerve root, resulting in a symptom complex of paresthesia below the level of the disk herniation associated with paraplegia or paraparesis. There is loss of both superficial and deep sensation [83]. These clinical findings may be associated with any extradural or intradural lesion, however. An antecedent history of torsion and flexion injury is common in the patient with a symptomatic herniated thoracic disk. Most symptomatic thoracic disk herniations occur between the third and fifth decades.

Plain radiographic findings in evaluating thoracic disk herniations are nonspecific. They consist of narrowing of the disk space and the presence of Shmorl's node and linear calcification. Both MRI and CTM are ideal imaging studies. Conventional myelography often demonstrates the extradural lesion if the disk herniation is significant, but a small central or paracentral disk could easily be missed by the density of the contrast material in the lateral gutters. Similarly, small but symptomatic disk herniations can be missed on plain CT due to lack of epidural fat and partial volume averaging.

CTM is necessary in demonstrating the extradural lesion (Fig. 5.33). Ideally, once the level is identified on myelography, contiguous 3-mm sections or overlapping sections with subsequent re-formation help in precise localization of the lesion and also in differentiating the disk herniation from a meningioma. CT is also useful in demonstrating the calcification commonly associated with thoracic disks. In most cases, the calcification within or peripheral to the thoracic disk is the result of avulsion and fragmentation of the endplate. This fragmentation may extend into the disk space and represent degeneration with calcification of the ruptured annulus and adjacent endplate. It has been suggested that in many instances small asymptomatic thoracic disks readily identified on CT are posterior epidural extensions of the intradiscal herniation (Schmorl's nodes) [84].

MRI is ideal as a screening test in evaluating for thoracic disk herniation [84–86], since the extradural herniated disk and any evidence of cord compression can be identified easily on sagittal images (Fig.

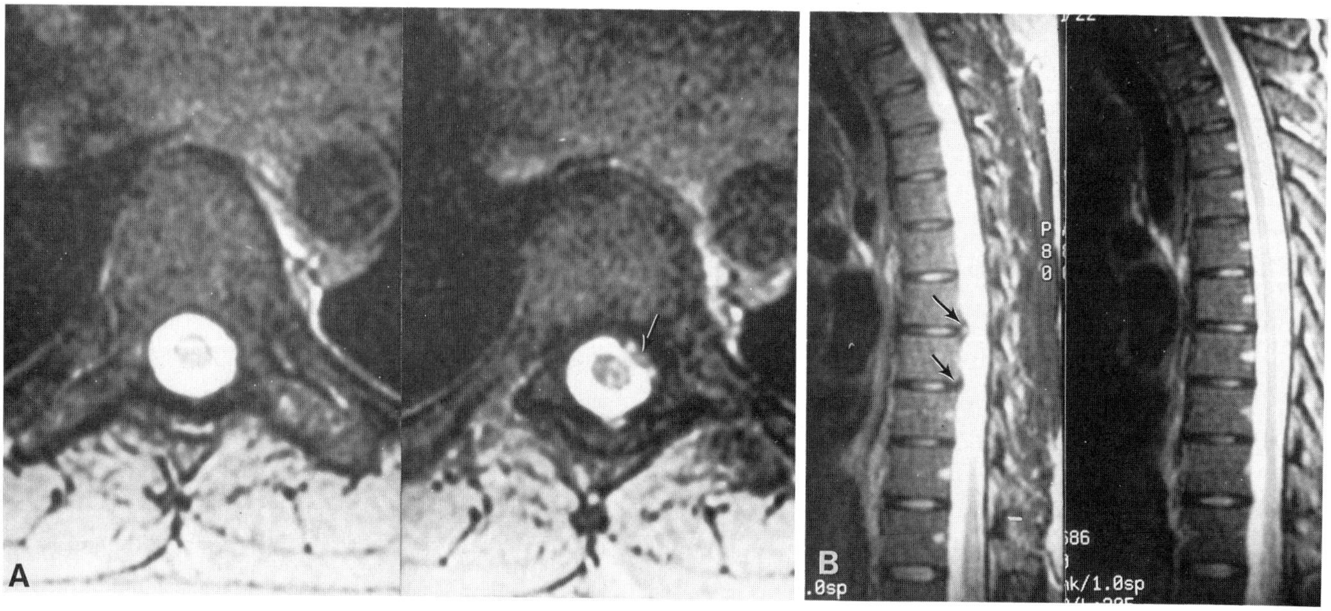

Figure 5.32. Asymptomatic thoracic disk herniation. Axial (**A**) and sagittal (**B**) T2-weighted FSE images of the thoracic spine demonstrate an incidental thoracic disk herniation (*arrow*).

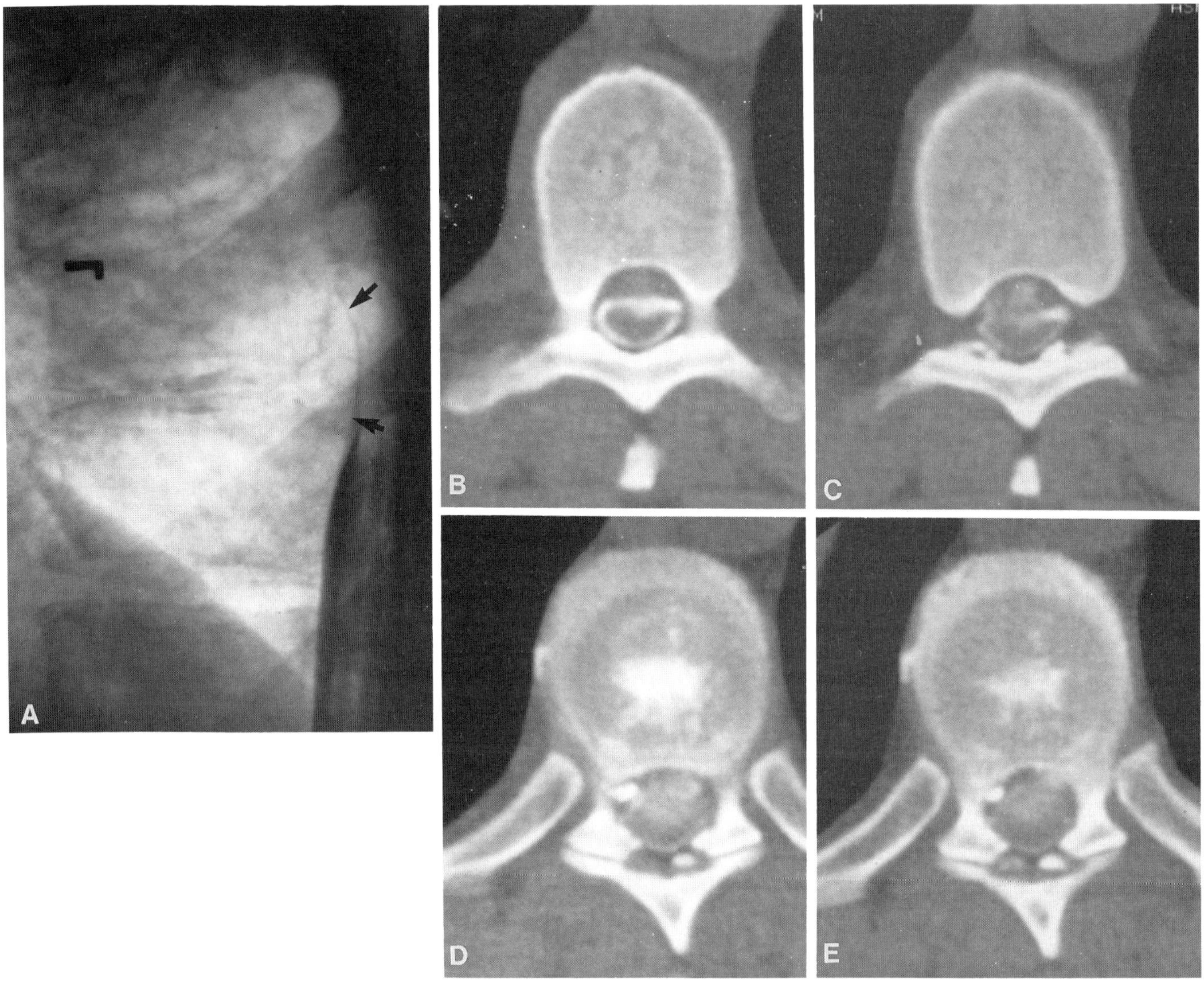

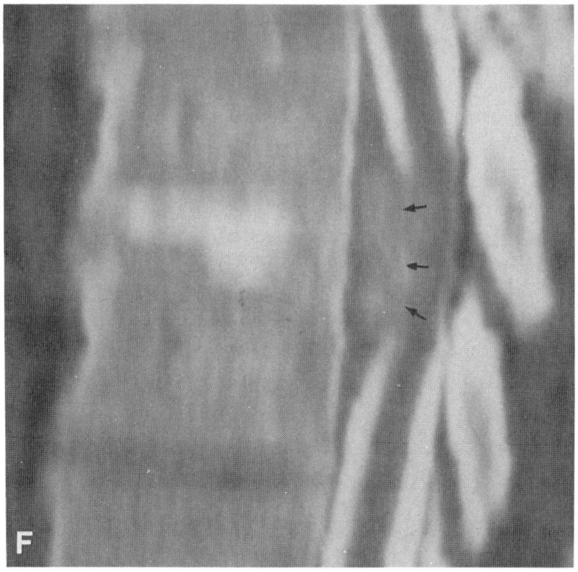

Figure 5.33. Thoracic disk herniation on CT. A, Lateral view of myelogram demonstrates a focal extradural defect (*arrows*) with partial obstruction of the spinal canal. Contiguous axial CTM scans (**B–E**) and sagittal re-formatted image (**F**) from the CTM demonstrate the partially calcified large herniated disk (*arrows*) with compression of the spinal cord.

5.34). T1- and balanced T2-weighted images help in differentiating a herniated disk from osteophytes or a meningioma. Unless there is calcification, a herniated disk has similar signal as the rest of the disk material. Axial images, particularly a GRE or T2*-weighted image, are necessary to demonstrate the laterality of the herniated disk.

Certain technical factors should be borne in mind when utilizing MRI for thoracic disk herniation:

1. Slice thickness should be kept less than 5 mm, with a minimal interslice gap.
2. Since cartilage and osteophytes have similar signal intensity, a calcified disk may be missed unless one of the pulse sequences is a T2*-weighted image or low-flip-angle GRE image.
3. Small herniated disks can be missed if the frequency-encoding axis is perpendicular to the spine [40].
4. Herniated disks can occasionally mimic meningiomas. Gd-DTPA–enhanced MRI helps in differentiating these two entities. Herniated disks will not show enhancement, whereas even small meningiomas demonstrate intense enhancement.

Unlike in the lumbosacral region, extruded and sequestered disk herniation is extremely rare. This is due to the smaller volume of the disk material and more central location of the nucleus pulposus.

CERVICAL DISK DISEASE

Disk herniation is less common in the cervical region compared with the lumbar region. This is in part due to the lesser compressive forces, the protection offered by the posterolaterally situated uncinate process, and the more lateral extension of the posterior longitudinal ligament. Since the maximum range of motion in the cervical region involves the C4 to C7 segments, disk herniation in the cervical region usually involves the disk spaces between C4

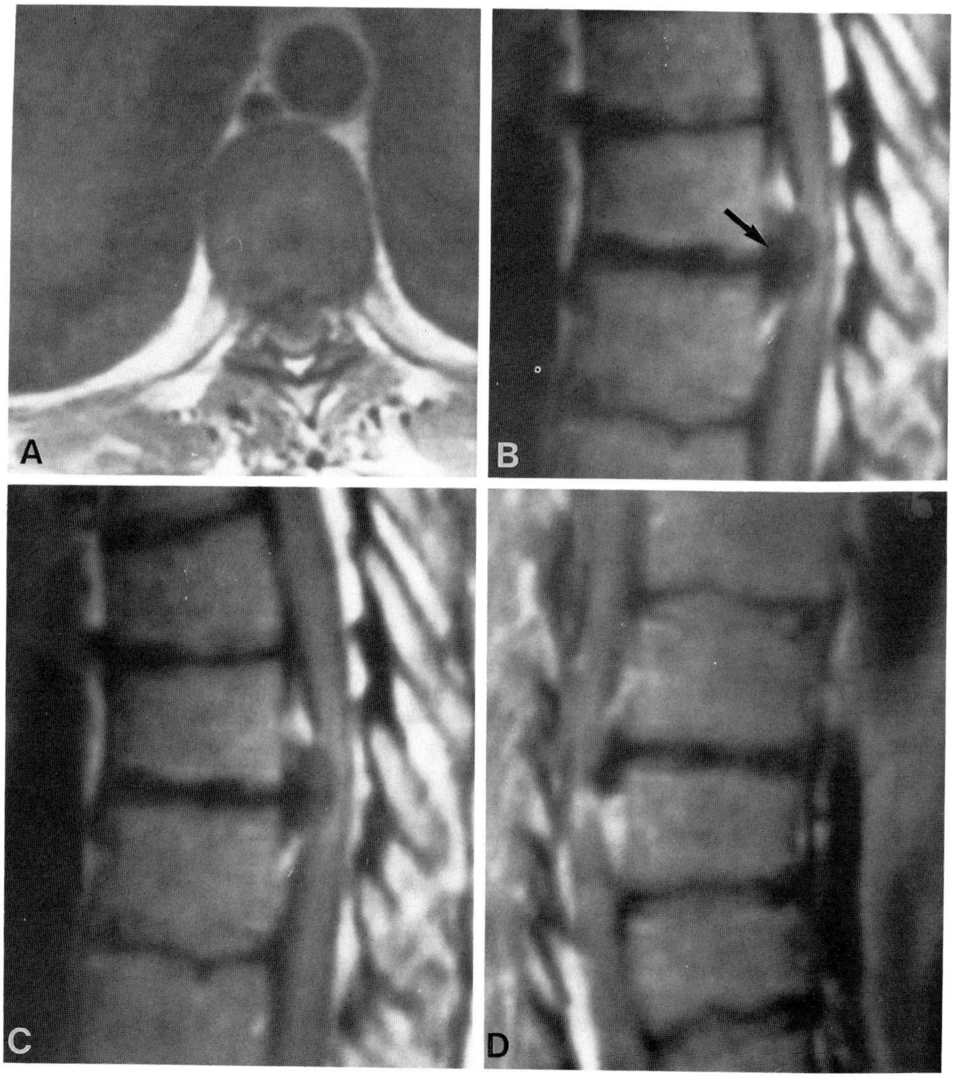

Figure 5.34. Thoracic disk herniation on MRI. T1-weighted axial (**A**) and sagittal (**B**) MRI sequences. The herniated disk has a low signal intensity (*arrow*). The disk space is narrowed. A prominent epidural space is seen above and below the herniated disk. Note the cord compression and bright signal within the cord. **C,D,** Gd-DTPA–enhanced contiguous sagittal T1-weighted MRI sequences. Enhancement of the epidural venous plexus is seen above and below the herniated disk. The disk itself does not show enhancement, thus differentiating from a meningioma. The main advantage of MRI over CT is its ability in demonstrating the cord ischemia.

and C7 [87]. There is often a history of minor trauma, usually with a rotary component. Disk herniation is more common in males than in females, often between the third and fifth decades.

The mechanism of disk degeneration and herniation is similar to that in the rest of the spine. Degeneration and disk herniation may result in a bulging disk, a posterolateral or central HNP, or disk extrusion. Unlike in the lumbar region, despite the narrower disk height, because of the relative circumferential diameter of the spinal cord and anatomic configuration of the vertebral bodies, even small disk herniations can be symptomatic. Bulging disks are not symptomatic unless associated with congenital narrowing of the spinal canal or age-related degenerative spinal stenosis. Focal central disk herniation may present with paraparesis or paraplegia due to cord compression and myelopathy, whereas posterolateral disk herniation may present with radiculopathy, myelopathy, or a combination of the two. The clinical presentation may be acute, subacute, or chronic as a result of compression or irritation of one or more nerve roots and/or the spinal cord. The symptoms may be nonlocalized, presenting as headaches; neck, scapula, or shoulder pain; or pain localized to a dermatomal distribution (see Table 5.1).

Given a choice among the existing imaging modalities, because it is noninvasive, MRI should be the preliminary screening test in evaluating cervical disk disease. Even though both CTM and IVCT are considered superior to MRI because of their spatial resolution, they should be utilized only if the MRI study is equivocal or when disk herniation is part of the spectrum of discovertebral disease.

In evaluating cervical disk disease by IVCT or CTM, certain technical factors should be considered. Since the height of the normal disk space is less than in the lumbar region and the herniated disk material is smaller in volume, the section thickness of each slice should be between 1.5 and 3 mm. Ideally, the section should be parallel to the disk space. An adequate cross-sectional area to cover the pedicles of the adjacent vertebral bodies should be included at each level. Intrathecal contrast material is always necessary in patients with short necks or broad, muscular shoulders when evaluating the C6–7 and C7–T1 disk levels.

Both plain CT and CTM studies are deficient in demonstrating the nerve root compression from small lateral cervical disks or hypertrophied ligaments. IVCT helps in demarcating the herniated disk from soft tissues within the cervical spinal canal [21–24]. It is useful in demonstrating lower cervical disk herniation (C5–6, C6–7), especially when lateral disk herniation needs to be differentiated from osteophytes. The herniated disk is identified by the peripheral marginal enhancement of the epidural soft tissues, including the epidural and intervertebral foraminal venous channels (Fig. 5.35).

MRI requires an approach that is similar to that in the lumbar region. More so than in the lumbar region, it is important to obtain axial, sagittal, and occasionally parasagittal oblique views [88–90]. Osteophytes and hypertrophy of the uncovertebral joints as well as the posterior longitudinal ligament may result in compromise of the spinal and neural canals. This may mimic a calcified annulus or a herniated disk. Distinguishing calcified and cartilaginous structures from herniated disk material can be accomplished by obtaining images in different pulse sequences in at least one plane. Different protocols have been suggested. The basic intent is to separate the signal intensity of the CSF from that of the ligamentous structures and to differentiate calcific and cartilaginous structures from herniated disk. T1-weighted axial and sagittal sections should be obtained routinely. To avoid the long acquisition time and associated CSF pulsation artifacts, the T2-weighted image is usually replaced with a gradient-echo (GRE) or fast spin-echo (FSE) sequence [30–32, 34, 91, 92]. This not only provides the myelographic appearance of the CSF but also allows differentiation of the soft herniated disk from posterior osteophytes, ligamentous hypertrophy, or a calcified disk (Fig. 5.36). Improvement in image quality and reliability of MRI in evaluating foraminal encroachment by a lateral herniated disk, osteophytes, or a hypertrophied uncinate process can be achieved by combining the GRE or FSE sequence with 3DFT imaging [35, 91, 92]. This technique allows for thin 2- to 3-mm sections in any plane, including parasagittal oblique views, which are often necessary in evaluating for foraminal encroachment (Fig. 5.37). The images may not have the spatial resolution of T1- or T2-weighted images, and they are not useful in detecting intrinsic cord abnormalities. If myelopathy is suspected, this is best evaluated with CSF-gated T2-weighted studies. In most routine screening of the cervical spine, a 2DFT gradient-echo sequence is adequate in demonstrating the extradural disease (Fig. 5.38).

On MRI, small central herniated disks may be present at multiple levels and often represent age-related degenerative processes (Fig. 5.39). Lateral herniated disks of moderate size often can be seen on both axial and sagittal images. However, lateral herniated disks may be difficult to recognize in the presence of hypertrophy of the uncinate process, since both have a similar signal intensity on T1-weighted images. On gradient-echo sequence, the herniated disk may be masked by the bright signal of the engorged epidural veins. Gd-DTPA–enhanced MRI with either 2DFT or 3DFT has been advocated [93]. The principle is similar to IVCT. Enhancement

Figure 5.35. IVCT in cervical disk herniation. Contiguous axial CT sections demonstrate the epidural veins (*white arrow*) demarcating the herniated disk (*arrowhead*) from the thecal sac.

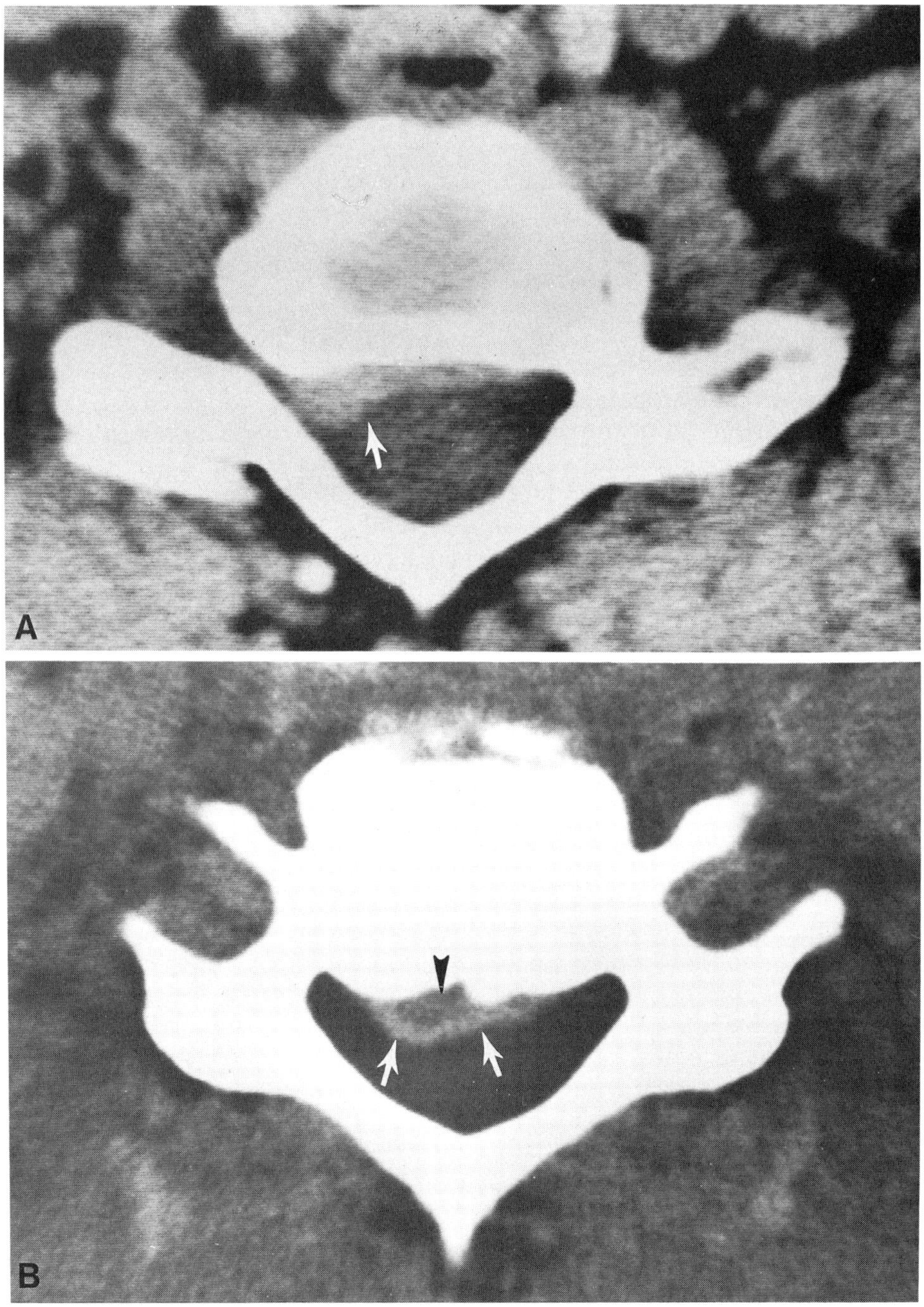

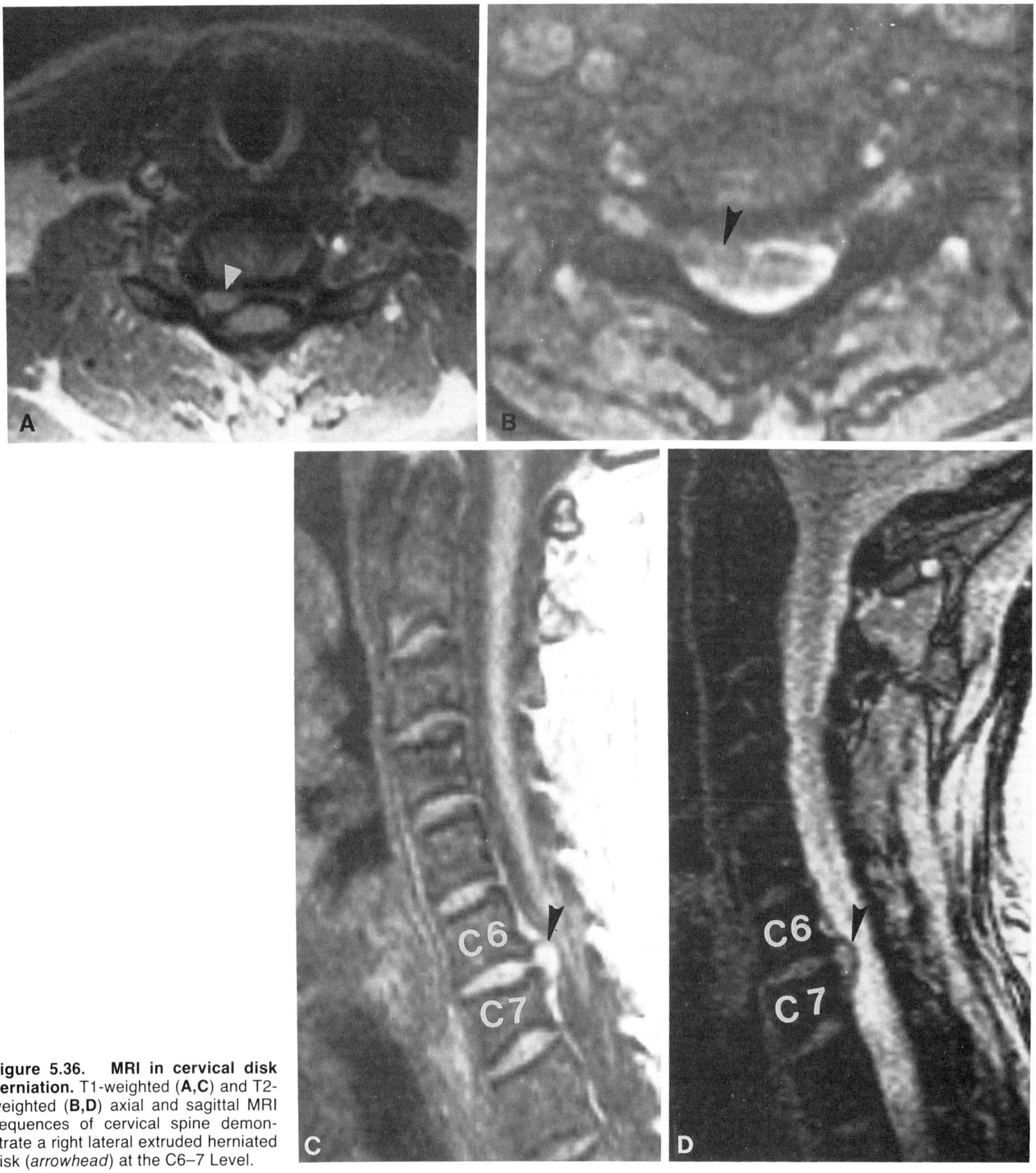

Figure 5.36. MRI in cervical disk herniation. T1-weighted (**A,C**) and T2-weighted (**B,D**) axial and sagittal MRI sequences of cervical spine demonstrate a right lateral extruded herniated disk (*arrowhead*) at the C6–7 Level.

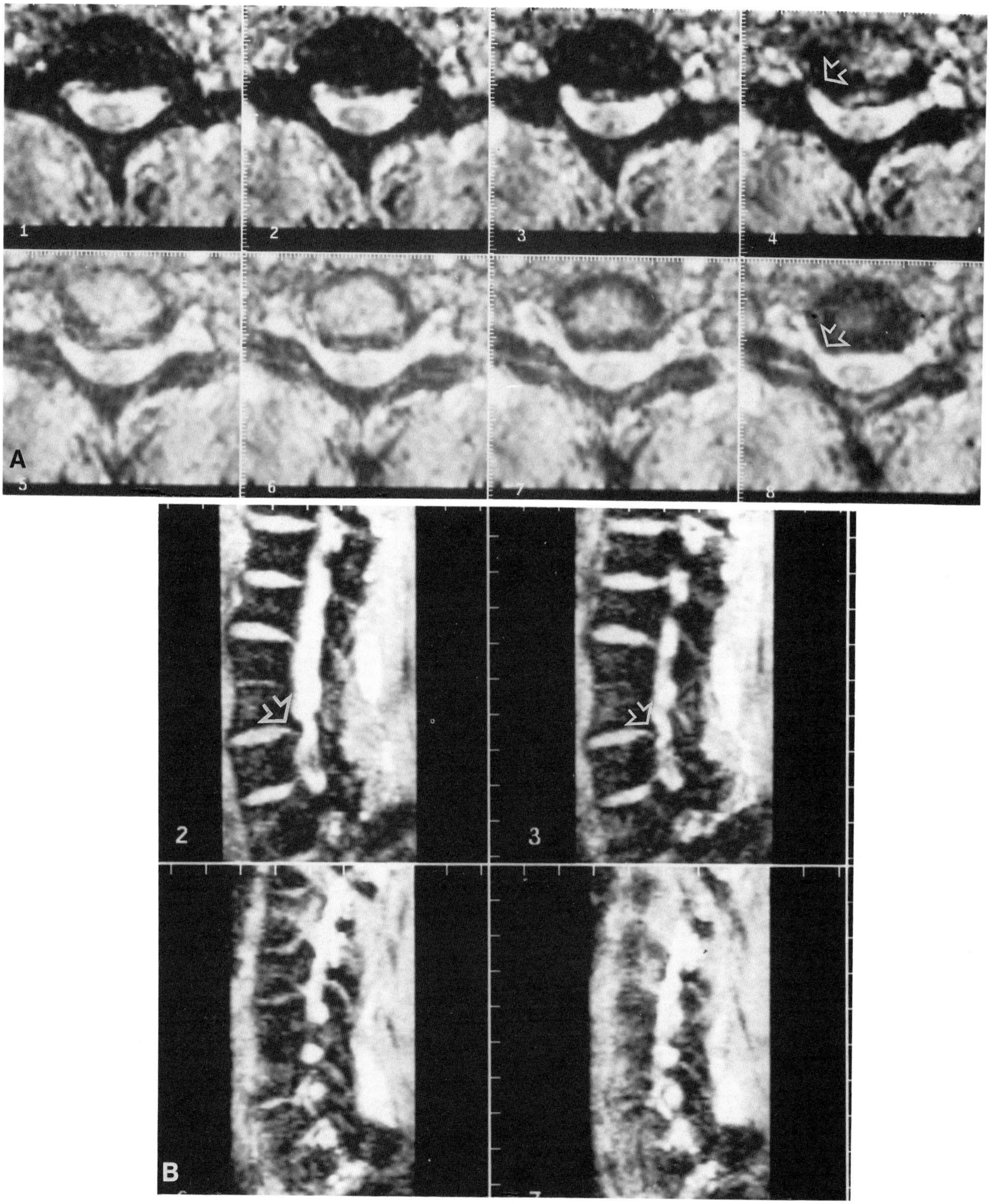

Figure 5.37. **A,** A 3DFT fast spin-echo T2-weighted MRI of the cervical spine with subsequent 2-mm partition demonstrates a bony osteophyte encroaching on the right neural canal (*open arrow*). **B,** Sagittal oblique re-formatted images from the volume acquisition can be useful in evaluating the neural canal and its contents.

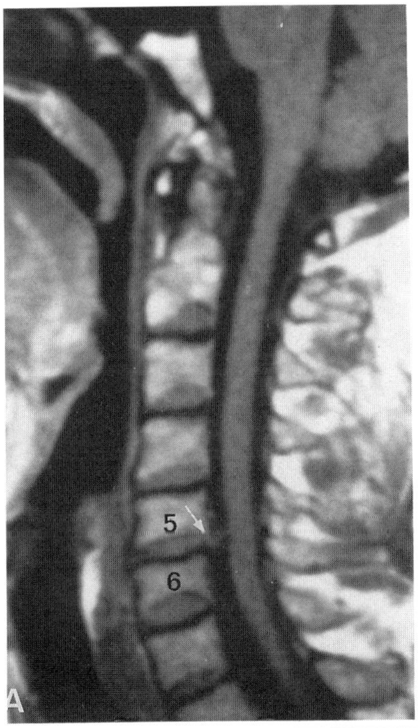

Figure 5.38. Cervical herniated disk. Sagittal T1-weighted (**A**) and sagittal (**B**) and axial (**C**) T2-weighted gradient-echo MRI images. Focal right lateral herniated disk at C5–6 disk space is well visualized. In most instances, gradient-echo images are adequate when screening for cervical degenerative disk disease. Another example of where fast spin-echo and gradient-echo are combined for evaluating cervical disk disease. Fast spin-echo sagittal T1-weighted (**D**) and T2-weighted (**E**) images show focal indentation of the spinal cord at C5–6 level (*arrow*). The signal characteristics are suggestive of osteophyte or ligamentous hypertrophy.

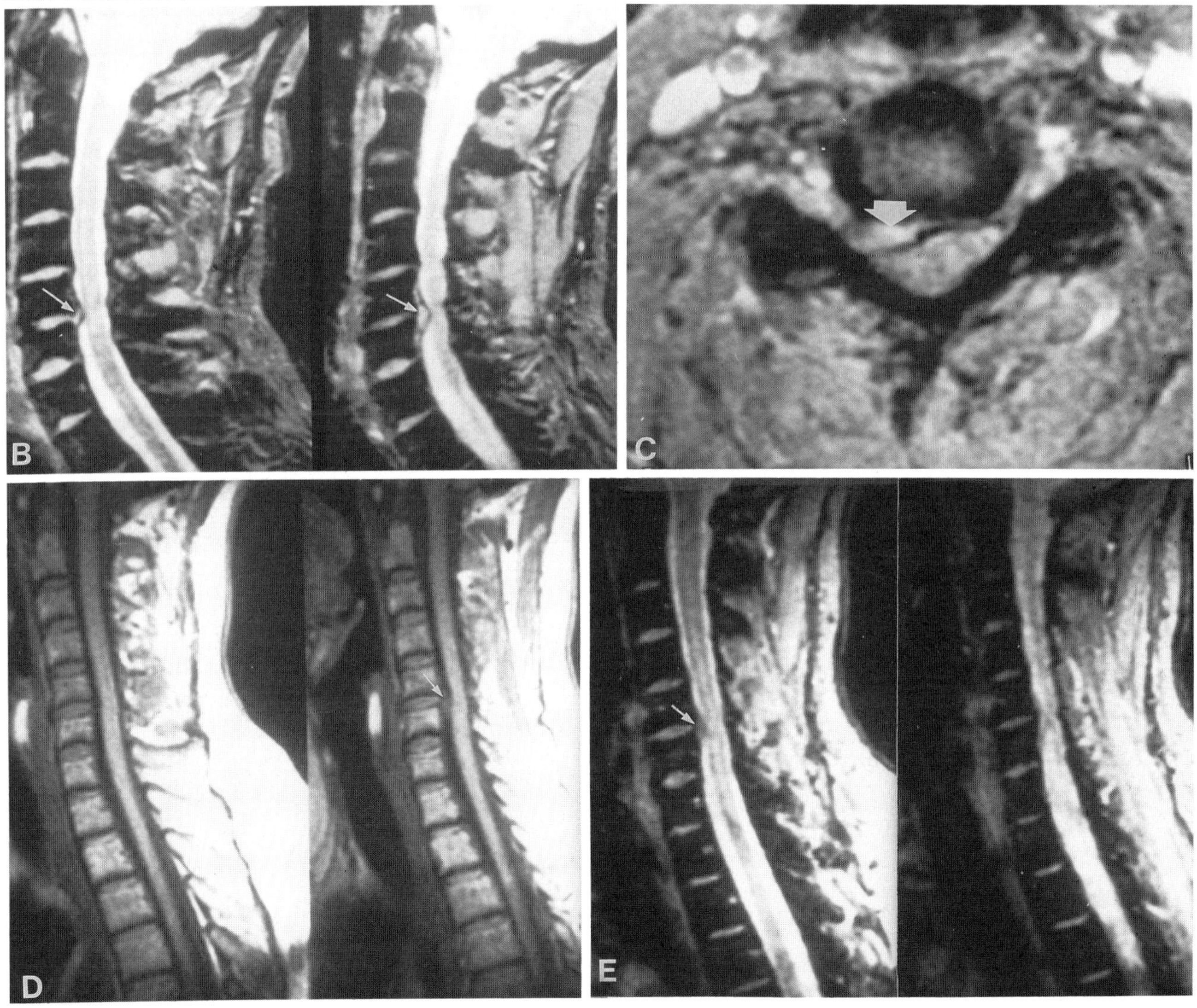

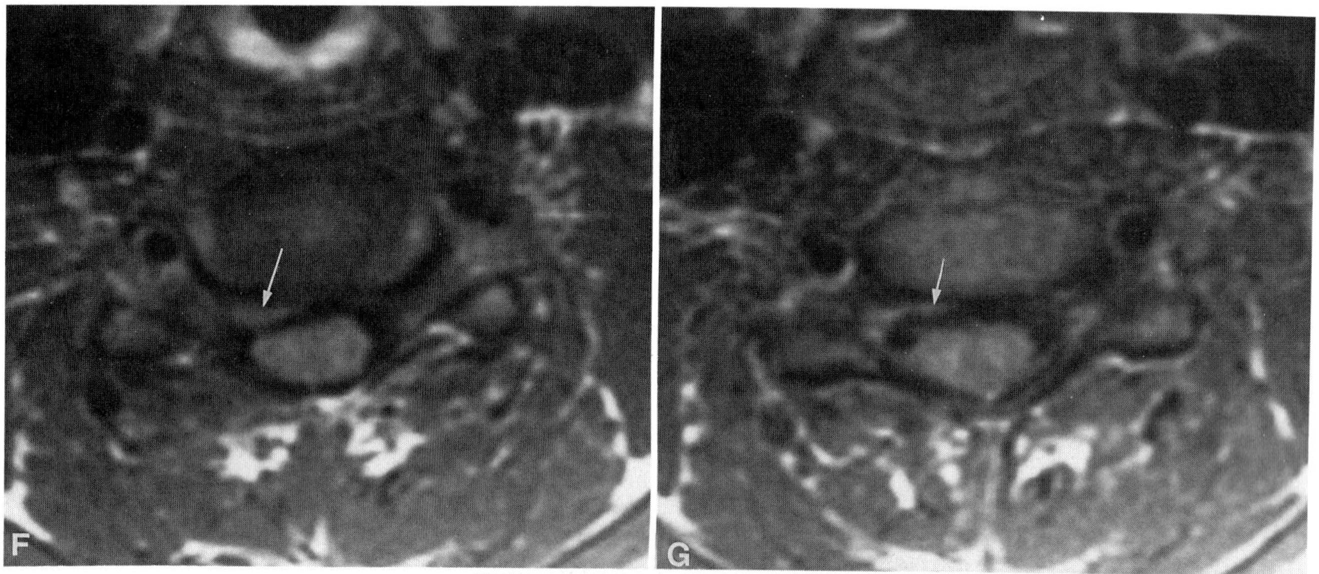

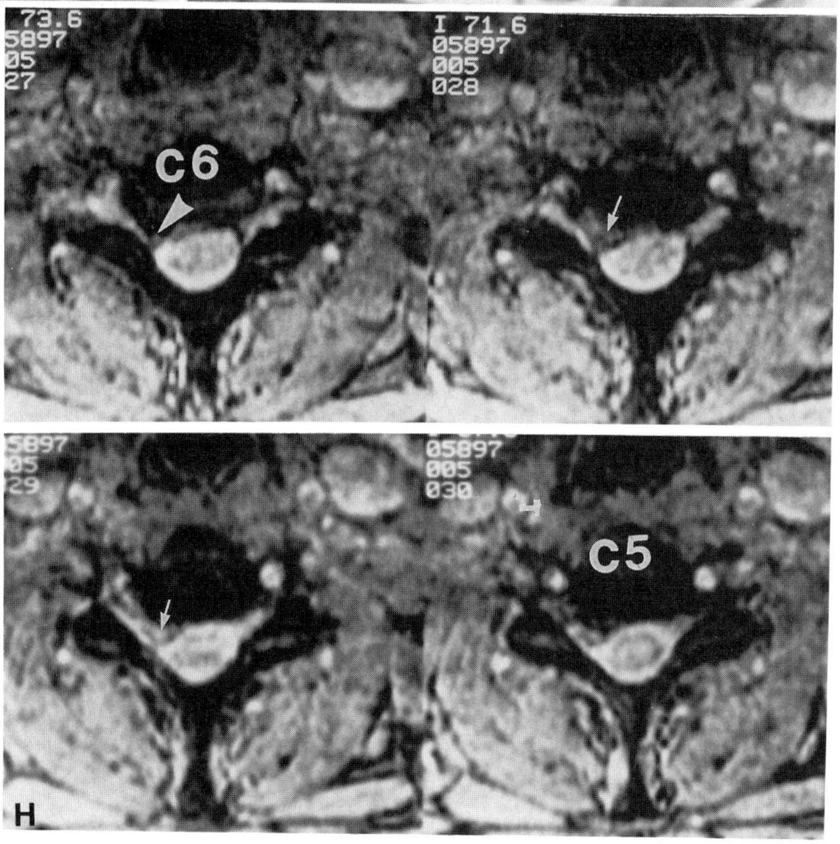

Figure 5.38–*continued* **F,G,** Axial T1-weighted images demonstrate a lateral herniated disk (*arrow*). **H,** Contiguous 2DFT T2-weighted gradient-echo image demonstrates a lateral herniated disk (*arrow*) and underlying uncinate hypertrophy (*open arrow*).

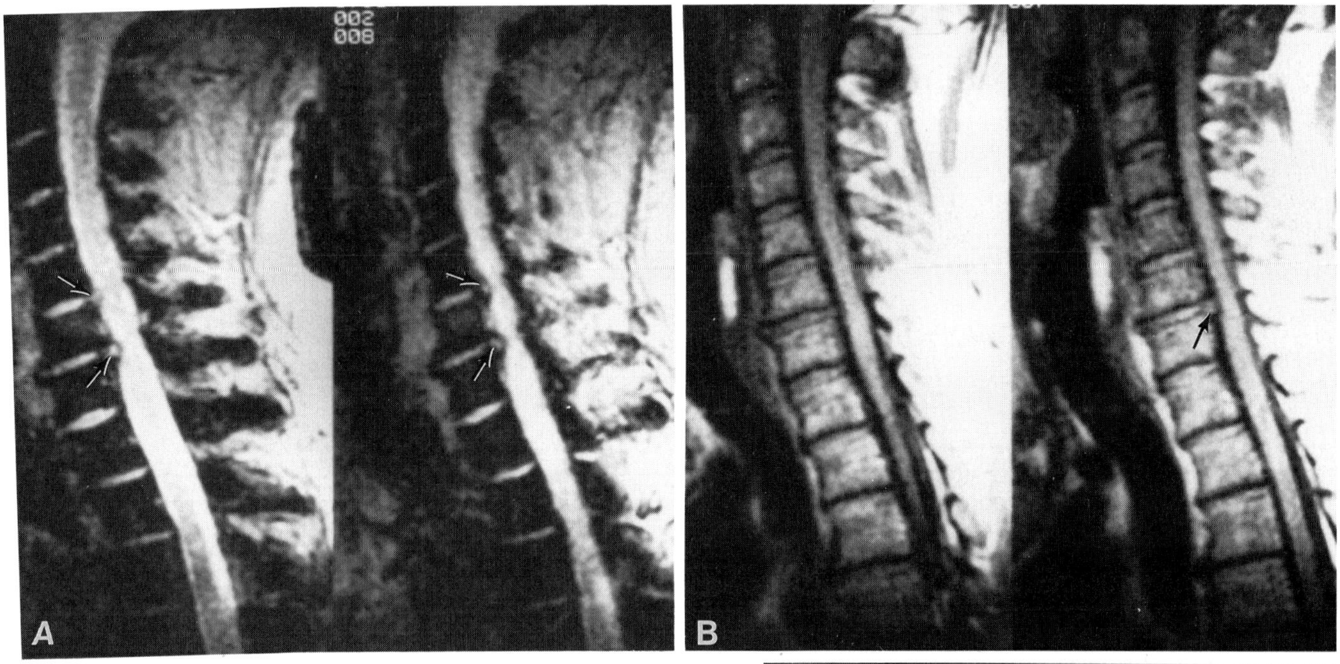

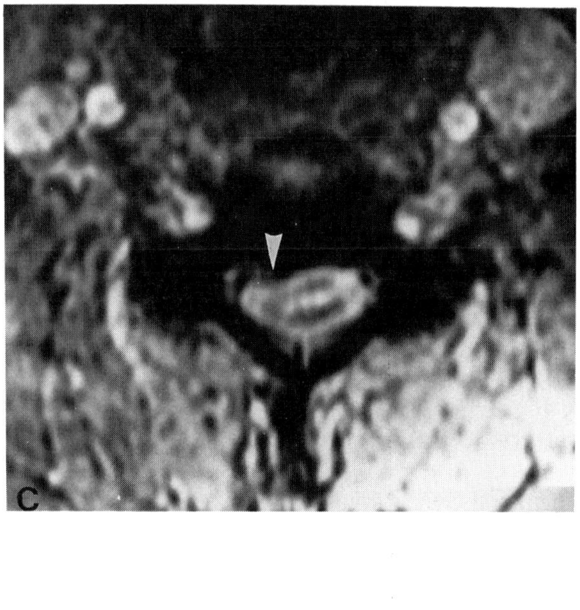

Figure 5.39. Multiple disk herniation. A, Adjacent sagittal T1-weighted MRI. **B,** Sagittal T2-weighted MRI. Disk herniation is seen at C5–6 and C6–7 levels (*arrows*). **C,** Axial T2-weighted gradient-echo image shows a lateral herniated disk.

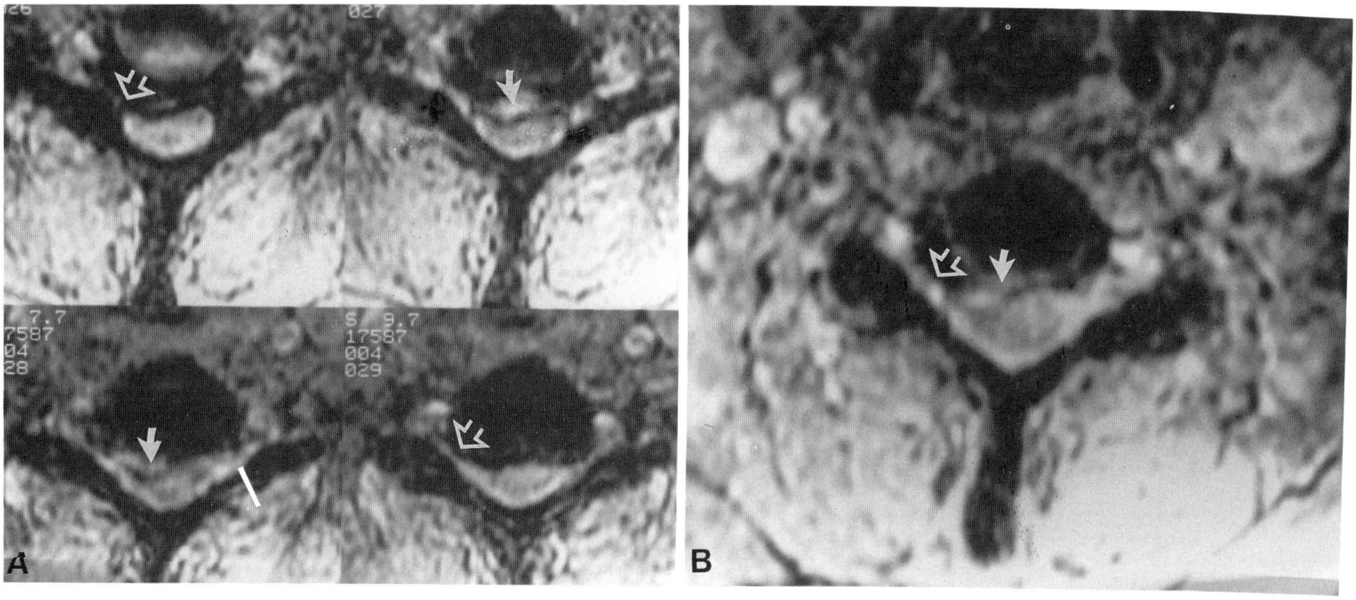

Figure 5.40. Patient with C7 radiculopathy. Contiguous axial MRI section (**A**) through C6–7 disk level and (**B**) 2DFT gradient-echo. T2-weighted MRI demonstrate right C6–7 foraminal encroachment by a combination of herniated disk (*arrow*) and uncinate hypertrophy (*arrowhead*). Cervical myelogram, frontal (**C**) and oblique (**D**) views, demonstrates nonfilling of the right C7 nerve root sleeve (*arrow*). **E,** Contiguous CTM section through this region demonstrates spinal stenosis from herniated disk as well as neural canal narrowing from osteophyte formation of right uncinate process.

of the epidural venous plexus and periphery of the disk margin on T1-weighted images helps in demonstrating the herniated disk, even if it is partly calcified. Myelography followed by CT can be utilized if the MRI findings are equivocal (Fig. 5.40).

Pitfalls and Differential Diagnosis

There are several pitfalls one must be aware of when evaluating CT or MRI studies for herniated disk as a cause of back pain. Pitfalls related to technique may result from (*a*) imaging just through the disk level without including the adjacent vertebral bodies, (*b*) obtaining only a single slice through the disk space, (*c*) asymmetry in the height of the individual vertebrae or scoliosis resulting in portions of the bulging disks being imaged in each individual section, (*d*) patient motion during the study, resulting in a suboptimal study or the sections not matching the levels posted on the reference or scout image, (*e*) not recognizing transitional vertebrae in the lumbosacral region, and (*f*) presence of spondylolisthesis. In addition, on MRI studies, other factors also may be contributory, such as the field strength of the magnet, availability of proper surface coils, and utilization of appropriate pulse sequences. These have been dealt in Chapter 1.

In most instances, the diagnosis of disk herniation is not difficult. In occasional cases other extradural intraspinal lesions may mimic a herniated disk. In most instances, this occurs with an extruded disk or fragment which is in an unusual location. The lesions which may be mistaken for a herniated disk include the following (Table 5.3).

Table 5.3.
Lesions Which May Be Confused with Herniated Disk

1. Asymmetrical bulging disk due to mild scoliosis.
2. Pseudo disk from spondylolisthesis.
3. Synovial cyst.
4. Conjoined nerve root sheath.
5. Perineural cyst.
6. Neurofibroma and schwanomma.
7. Epidural metastases.
8. Prominent epidural venous plexus.
9. Postoperative scar.

SYNOVIAL CYST

Synovial cysts usually occur in association with degenerative changes involving the facet joints. The cyst wall is a continuation of the synovial membrane of the facet joint. Rarely, the cyst wall may not appear in continuity with the joint space. The cyst is usually covered by the thinned-out ligamentum flavum, whose fibers are attached to the capsule of the facet joint and the adjacent pedicle. Most synovial cysts consist of mucinous fluid. The wall of the cyst may contain flecks of calcium.

Patients usually present with back pain [94]. However, radicular symptoms are not uncommon if the cyst extends anteriorly within the spinal canal and compresses the nerve root within the neural foramen [94, 95]. Clinical differentiation of low back pain from facet arthropathy and that from a herniated disk may be difficult. Most synovial cysts are encountered in the lower lumbar region, most often at the L4–5 level (Fig. 5.41). Synovial cysts also may occur in the cervical region [96, 97].

On CT, synovial cysts are often seen as soft-tissue

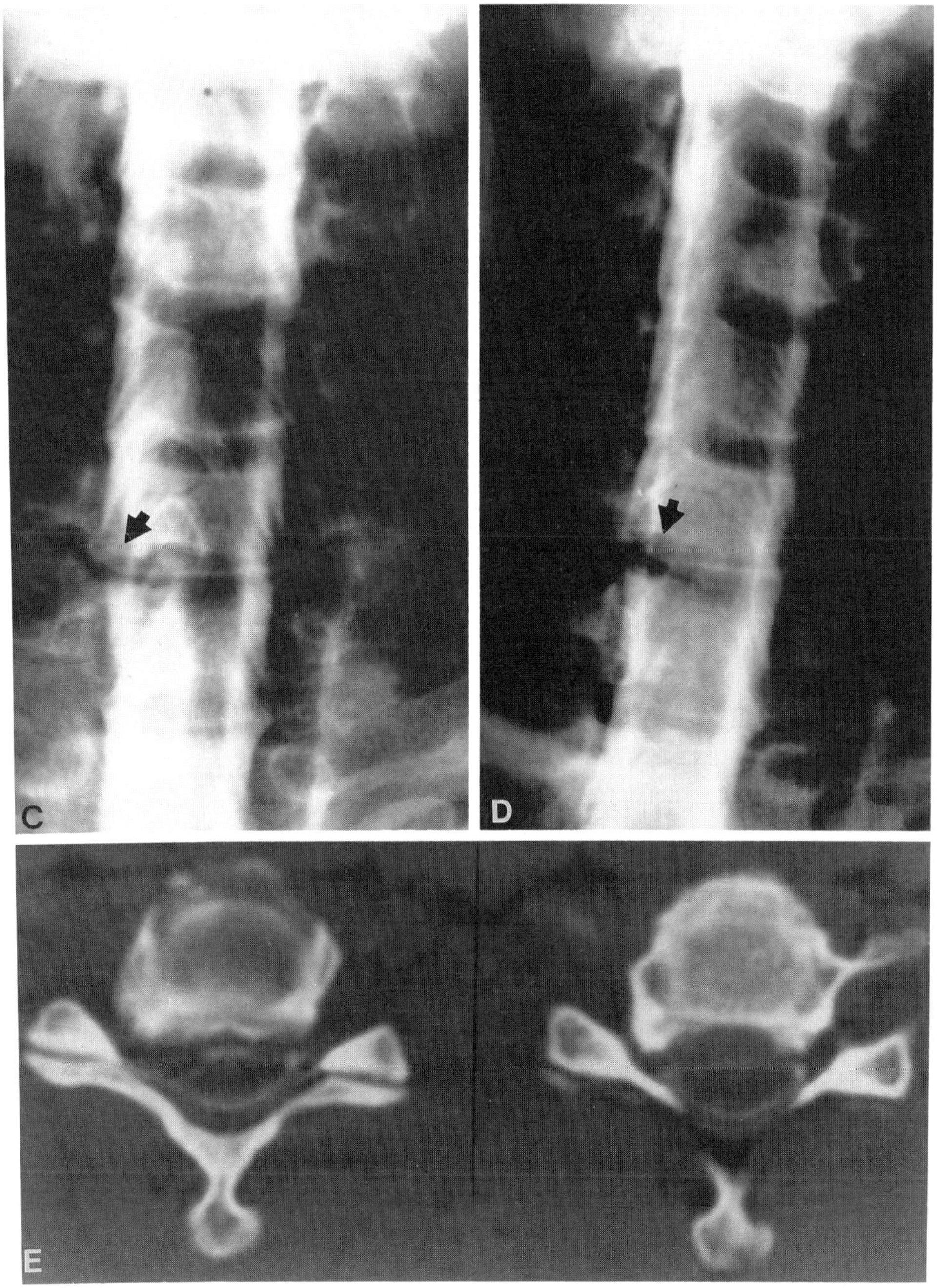

Figure 5.40–*continued*

masses that are hypodense or isodense compared with the adjacent thecal sac. They are usually located laterally, adjacent to the facet joints, but rarely they may extend anterolaterally, adjacent to the pedicle. The facet joints normally show irregular contour with osteophytosis. Presence of calcification within the cyst wall helps in the diagnosis [94–96, 98, 99] (see Fig. 5.41). Occasionally, the cyst wall may erode the adjacent bone [99]. Differentiation from a herniated disk can be made on the basis of CT and MRI studies (Fig. 5.42). On MRI, a synovial cyst is isointense or hypointense on T1-weighted sequences and hyperintense on GRE or T2-weighted images [100–102]. The cyst wall is hypointense in all pulse sequences. The diagnosis is thus made on the basis of the location of the cyst and the signal characteristics of the contents and cyst wall. Synovial cysts in the cervical region are usually associated with severe degenerative disk and facet disease, whose appearance on MRI is similar to a synovial cyst in the lumbar region (Fig. 5.43).

PIGMENTED VILLONODULAR SYNOVITIS

Pigmented villonodular synovitis (PVNS) is an inflammatory proliferative reaction of the synovium. The synovial lined villous fronds consist of giant cells, fibroblasts, lipid, and hemosiderin. PVNS pre-

Table 5.4.
Differential Diagnosis of Nonneoplastic Vertebral Body Lesions

Diagnosis	Vertebral Body Signal Intensity on T1	T2	Morphology	Signal Within the Disk on T2-Weighted MRI	Comments
Degenerative disk:					
Type 1	↓	↓	Adjacent vertebrae	↓	Normal plain x-rays
Type 2	↑	↑ min.	Adjacent vertebrae	↓	Minimal endplate sclerosis
Type 3	↓	±	Adjacent vertebrae	↓	
Diskovertebral sclerosis	↓	↑	Focal usually along anterior aspect	↓	Visible on plain x-rays
Schmorl's node	↓	±	Within vertebral body, adjacent to the endplate	Rim ↓ Center ±	Narrow disk space Center of node has similar signal as disk
Compression fracture	↓	↑	Signal changes may be minimal adjacent to fracture line	Normal	May be associated with soft-tissue swelling Plain x-rays Osteoporosis
Infection	↓	↑	Affects the disk space	±	Adjacent soft-tissue mass
Paget's disease	↓	±	Involves the vertebral body in its entirety	Normal	
Hemangioma	↑	↑	Focal or involves the entire vertebral body	Normal	Shows increased trabecular pattern
Fatty replacement	↑	±	Focal, usually central, adjacent to basivertebral vein	Normal	Normal plain x-rays
Metastases	↓	↑	Patchy or diffuse	Normal	May enhance with Gd-DTPA

dominantly involves the larger joints, although it may involve the synovium with extraarticular extension within the smaller joints, such as the facet joints [103–105]. PVNS presents as tumor-like nodules adjacent to the facet joint, often associated with erosive changes in the joint. On CT, PVNS appears as a focal hyperdense extradural mass adjacent to the facet joint [105]. Calcification within the mass is less common compared with synovial cysts. On MRI, PVNS has a heterogeneous signal due to the presence of hemosiderin within the soft-tissue mass [104] (Fig. 5.44).

CONJOINED NERVE ROOT

Conjoined nerve roots are common anomalies wherein two segmental nerve roots are located within a common nerve root sheath in its proximal course within the neural canal. Conjoined nerve roots within the sheath may be confused with a lateral herniated disk. Conjoined nerve roots often involve the L5 and S1 nerve roots. By themselves, they are not a cause of low back pain or radiculopathy.

On CT, the conjoined nerve roots appear as soft-tissue densities deforming the thecal sac on one side or located within the neural canal (lateral recess) deforming the epidural fat [106, 108]. The conjoined nerve root sheath is usually seen at a level slightly above the disk interspace and has the same density as the thecal sac (Fig. 5.45). Conjoined nerve roots may result in focal scalloping of the adjacent vertebral body and pedicle [107]. On MRI, the conjoined nerve root and its sheath are easy to identify because they have a similar signal intensity as rest of the thecal sac. There is a characteristic outpouching of the thecal sac on the side of the enlarged nerve root sheath. On adjacent lower sections, the separate nerve roots can be identified before they enter their respective neural foramina. Occasionally, conjoined nerve roots may present bilaterally. A herniated disk may be present at the same level as the conjoined nerve root. In most instances, these can be easily differentiated on MRI based on the signal characteristics and their appearance. In the occasional cases where this is difficult, myelography or CTM will confirm the diagnosis [108].

PERINEURAL CYST

Perineural cysts (Tarlov's cysts) are fusiform dilatations of the nerve root sheaths usually extending distal to the junction of the dorsal root with the posterior root ganglion. Occasionally, they may be continuous with the nerve root sheath and the thecal sac. The cyst wall is lined with pia-arachnoidal tissue similar to the thecal sac. Such cysts may occur at any level but are common in the lumbar and sacral regions. They may be bilateral. They are often associated with scalloping of the adjacent bony compartment and widening of the neural foramen. A large Tarlov's cyst in the sacral region may mimic a sacral meningocele. On CT, the larger cysts may be con-

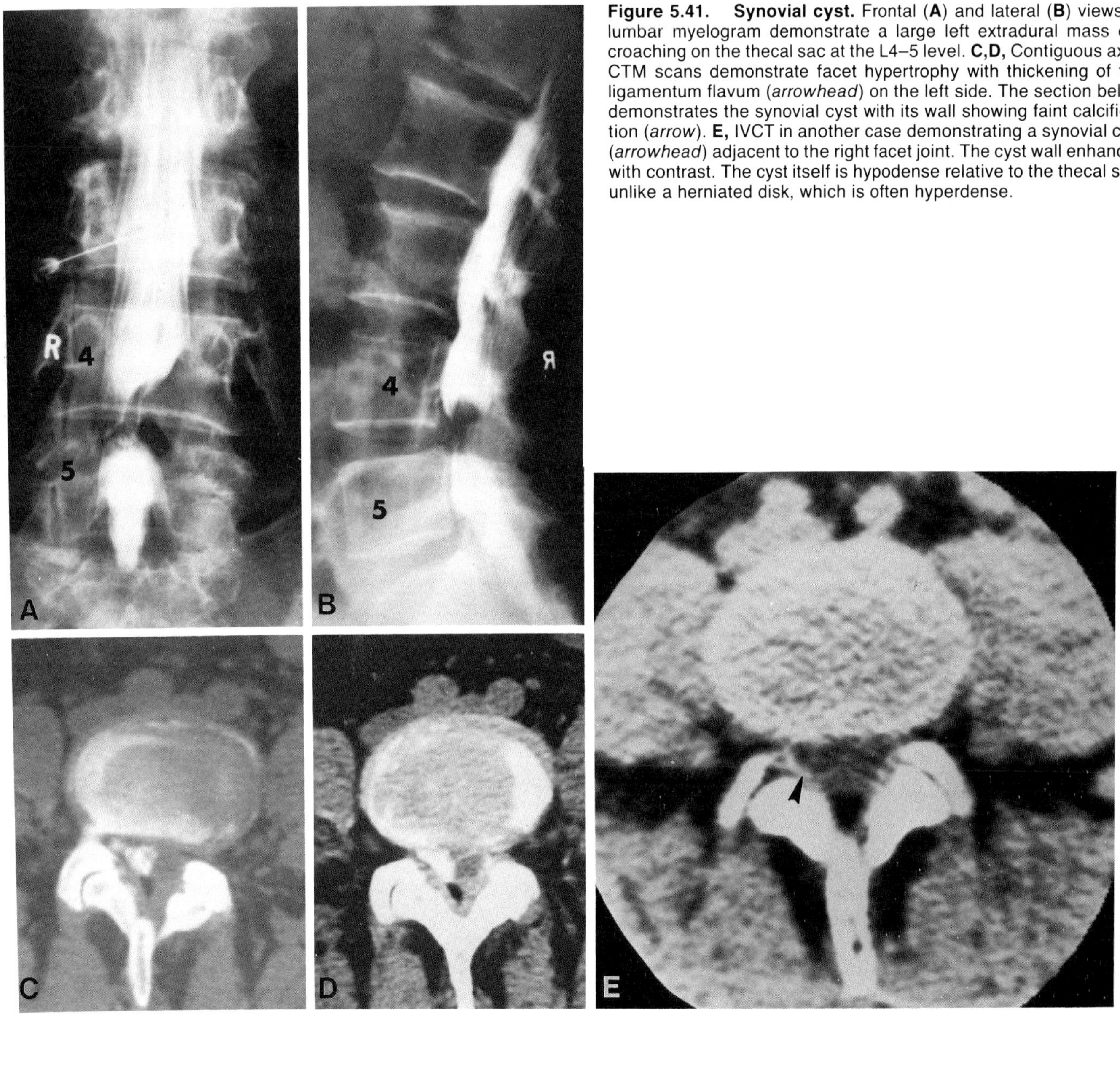

Figure 5.41. Synovial cyst. Frontal (**A**) and lateral (**B**) views of lumbar myelogram demonstrate a large left extradural mass encroaching on the thecal sac at the L4–5 level. **C,D,** Contiguous axial CTM scans demonstrate facet hypertrophy with thickening of the ligamentum flavum (*arrowhead*) on the left side. The section below demonstrates the synovial cyst with its wall showing faint calcification (*arrow*). **E,** IVCT in another case demonstrating a synovial cyst (*arrowhead*) adjacent to the right facet joint. The cyst wall enhances with contrast. The cyst itself is hypodense relative to the thecal sac, unlike a herniated disk, which is often hyperdense.

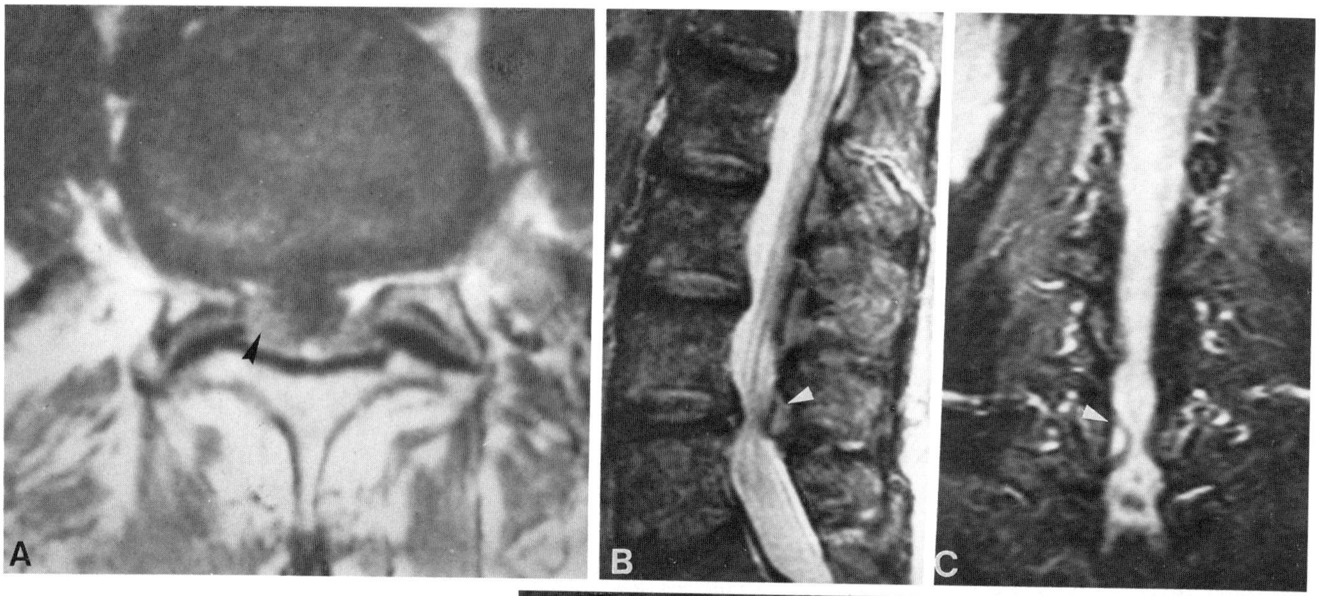

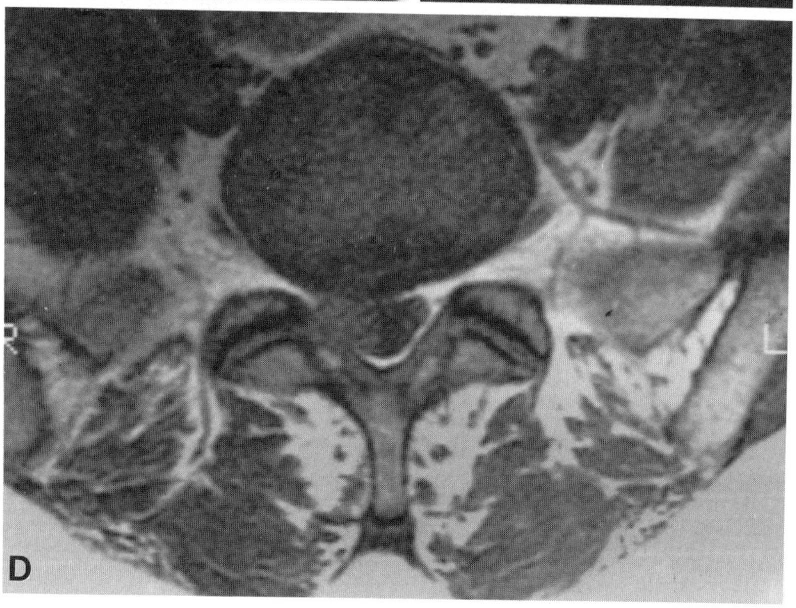

Figure 5.42. Synovial cyst on MRI. Axial T1-weighted (**A**), and sagittal (**B**) and coronal (**C**) T2-weighted MRI sequences. The synovial cyst adjacent to the right L4–5 facet joint has an intermediate signal (*arrowhead*) on the T1-weighted axial image and an intensely bright signal (*arrow*) similar to CSF on the T2-weighted images. The cyst wall has a dark signal, allowing separation from the thecal sac. **D,** T1-weighted axial section of a right lateral extruded disk has a similar signal as the synovial cyst in **A.** Differentiation is based on identifying epidural fat between the herniated disk and the ligamentum flavum. This is not seen in **A.**

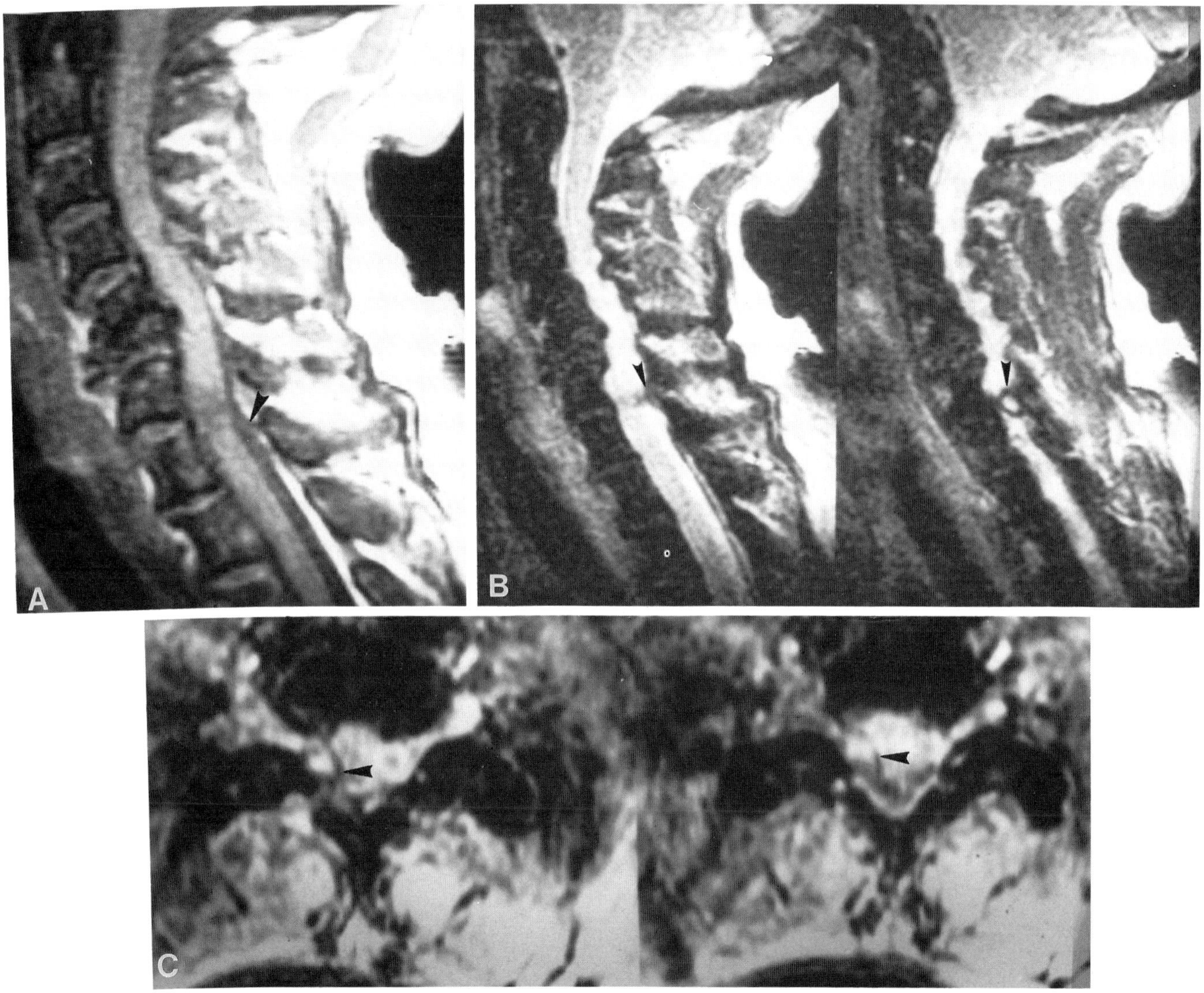

Figure 5.43. Synovial cyst in cervical region. T1-weighted sagittal MRI (**A**), adjacent sections of T2-weighted sagittal MRI (**B**), and contiguous axial T2-weighted section (**C**). Synovial cyst (*arrowhead*) has a bright signal within a hypointense cyst capsule adjacent to the facet joint. Although the level of the cyst is demarcated, it is not seen on the T1-weighted midsagittal section.

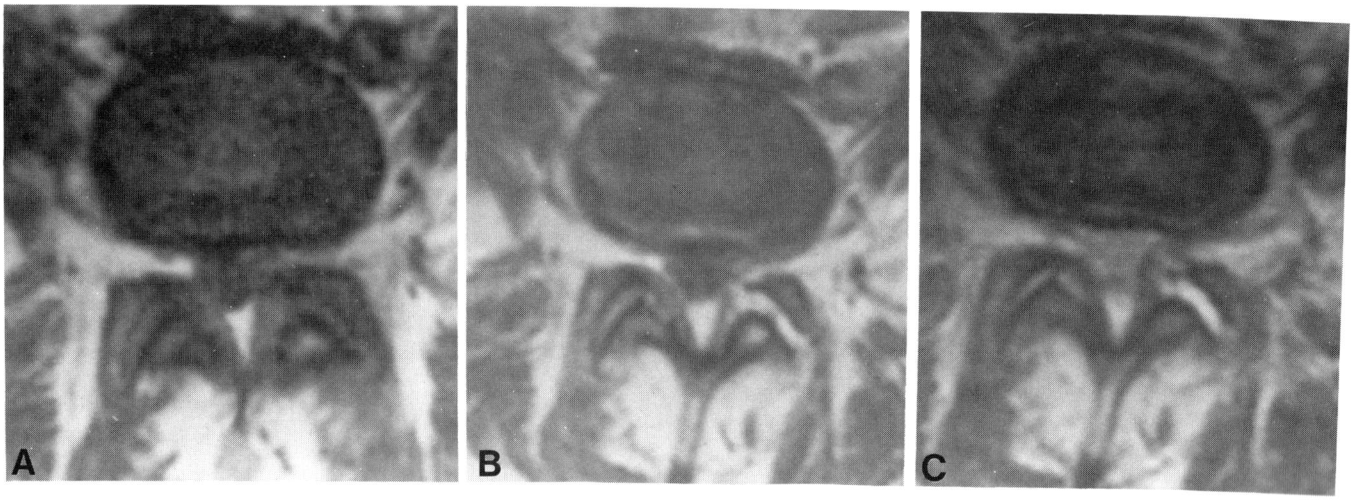

Figure 5.44. Pigmented villonodular synovitis. A,B, Contiguous axial T1-weighted MRI sequences. The left facet joint demonstrates erosive changes with bright signal within the facet joint capsule and an adjacent mass with intermediate signal (*arrow*). **C,** T2-weighted (1800/30) MRI demonstrates a heterogeneous signal within the mass. Facetectomy was performed. The mass consisted of synovial hypertrophy with hemorrhagic areas.

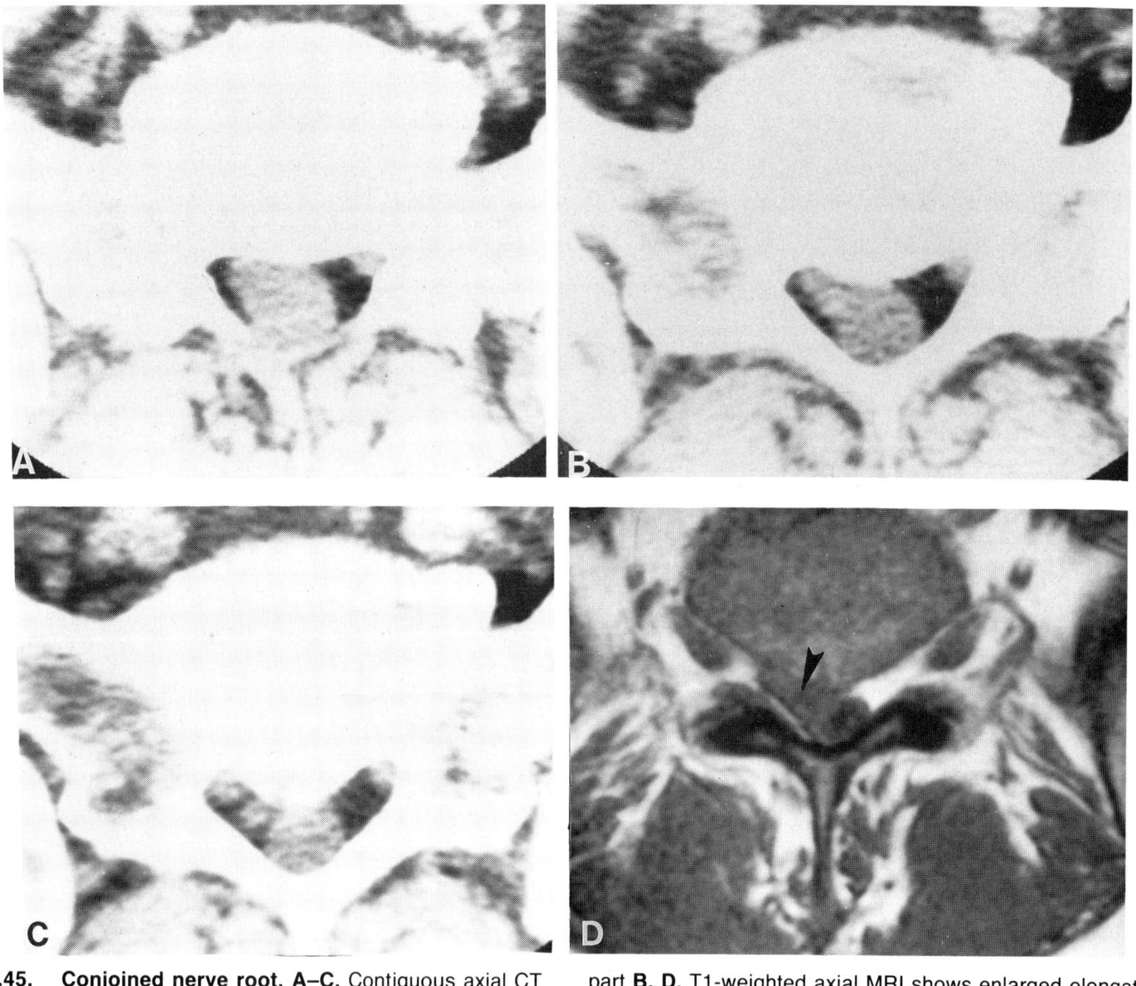

Figure 5.45. Conjoined nerve root. A–C, Contiguous axial CT sections demonstrate a conjoined nerve root on the right side. This may be mistaken for an extruded disk when seen in isolation, as in part **B. D,** T1-weighted axial MRI shows enlarged elongated nerve root sheath (*arrowhead*) from conjoined L5 and S1 nerve roots on the right side.

fused with tumors, whereas the smaller perineural cysts may be confused with lateral foraminal herniated disks. Their diagnosis and differentiation from herniated disks can be made with CTM or MRI. On MRI, perineural cysts follow the signal of the CSF. They may be brighter than CSF on T2-weighted images, which is explained on the basis of increased protein content or absence of CSF pulsation.

NEOPLASM

Nerve sheath tumors isolated within the lateral recess of the spinal canal or neural foramen may be mistaken for a herniated disk [109]. Such tumors can be differentiated from their characteristic appearance on both CT and MRI. On CT, nerve sheath tumors have decreased attenuation compared with herniated disks. Scalloping of the posterior margin of the vertebral body and adjacent pedicle is common. Most of these tumors enhance on IVCT. On MRI, the tumor may mimic a disk on T1-weighted images, but it has a hyperintense signal on T2-weighted images and enhances homogeneously on T1-weighted Gd-DTPA–enhanced images. The difficulty in diagnosis usually involves foraminal and extraforaminal herniated disks (Fig. 5.46).

Metastases involving the nerve roots or within the epidural space may mimic a herniated disk (Fig. 5.47). Clinical evidence of a primary site of neoplasm and CT and MRI evidence of bony metastasis help in the differentiation. The diagnosis can be confirmed by the characteristic enhancement of the lesion on contrast-enhanced studies.

POSTOPERATIVE SCAR

Differentiation of postoperative scar from recurrent disk herniation is based on the enhancement

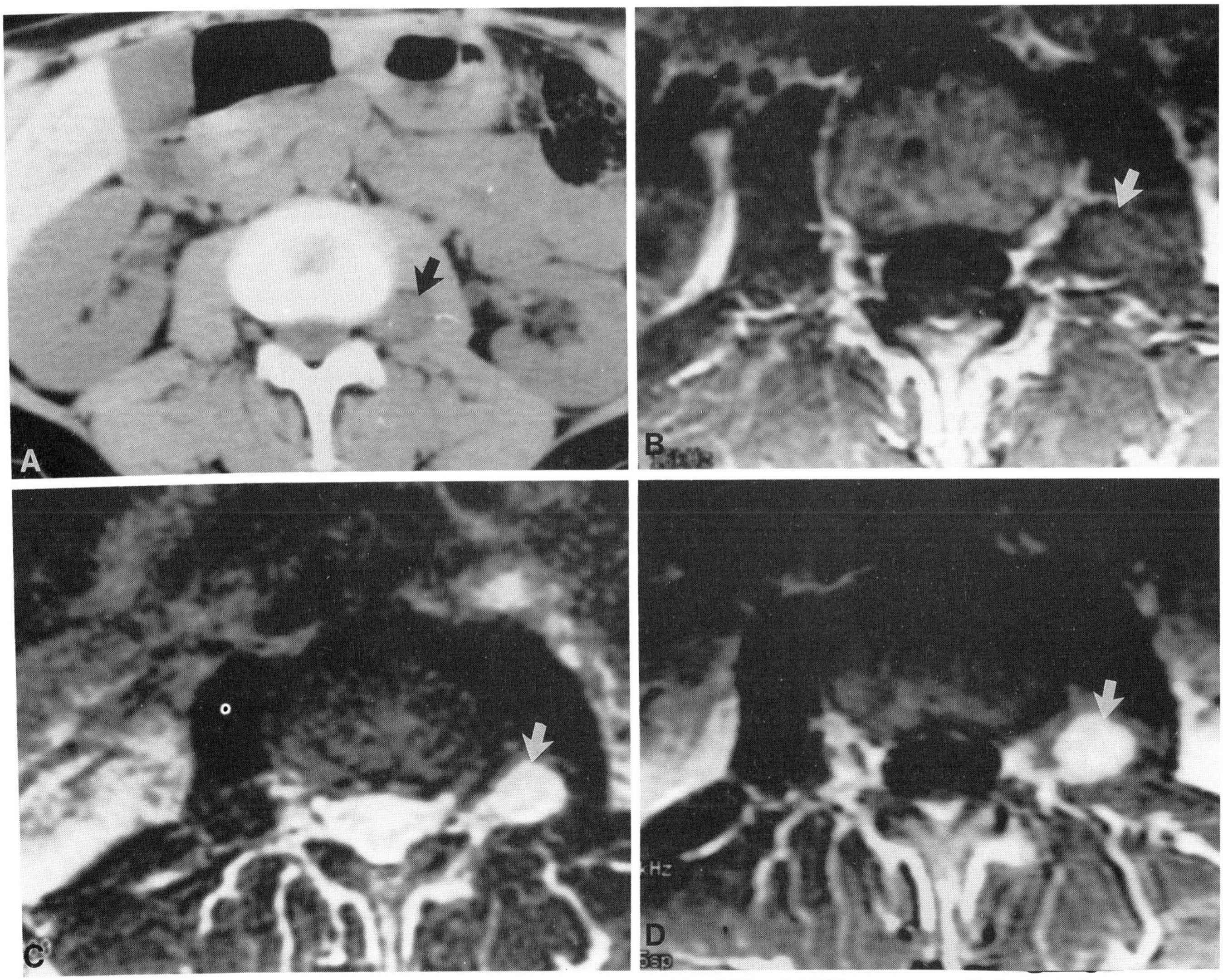

Figure 5.46. **A**, Axial CT demonstrates an extraforaminal neurofibroma, which could be mistaken for an extraforaminal disk. T1-weighted (**B**), T2-weighted (**C**), and Gd-DTPA–enhanced T1-weighted (**D**) axial MRI sequences with the typical appearance of the neurofibroma.

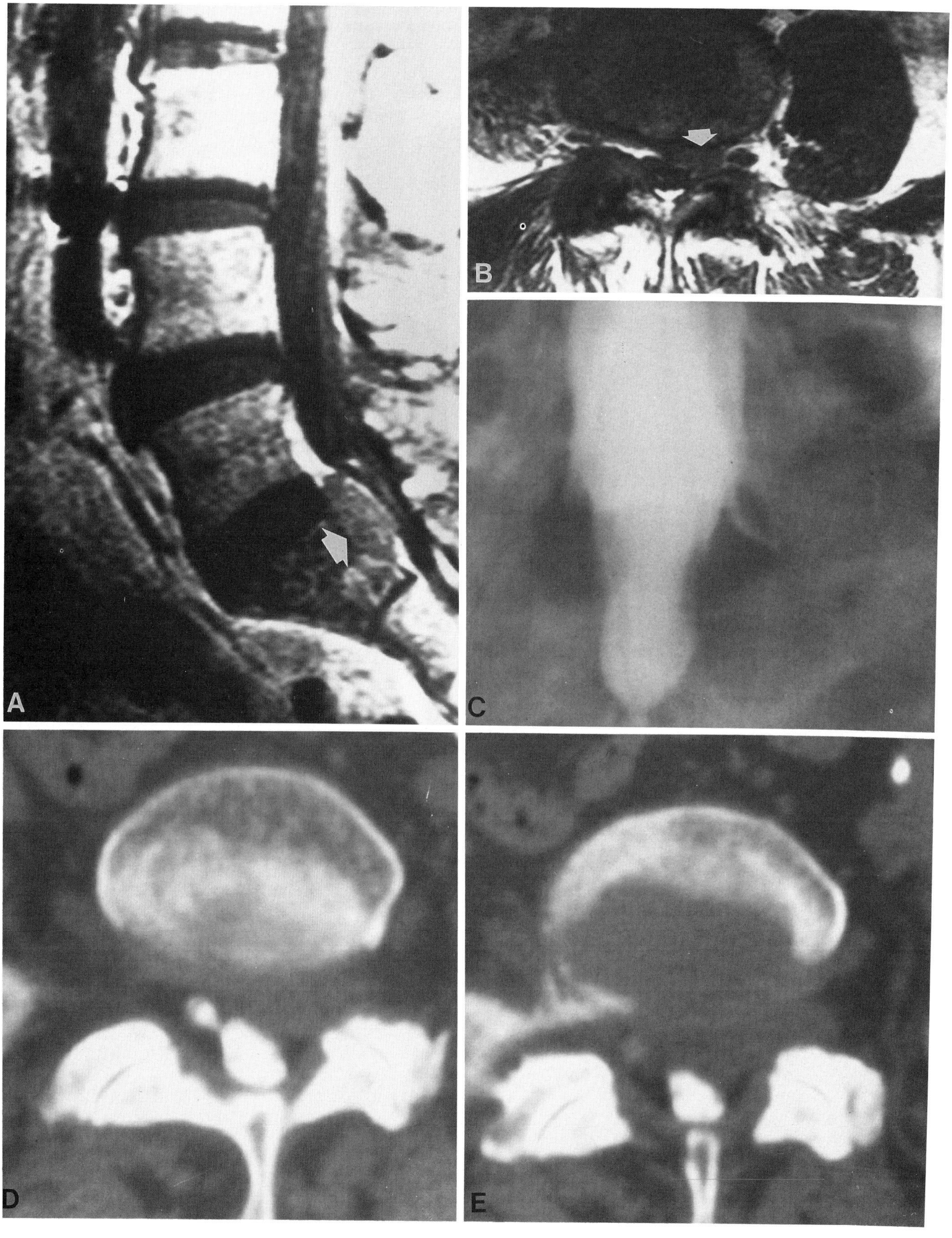
A
B
C
D
E

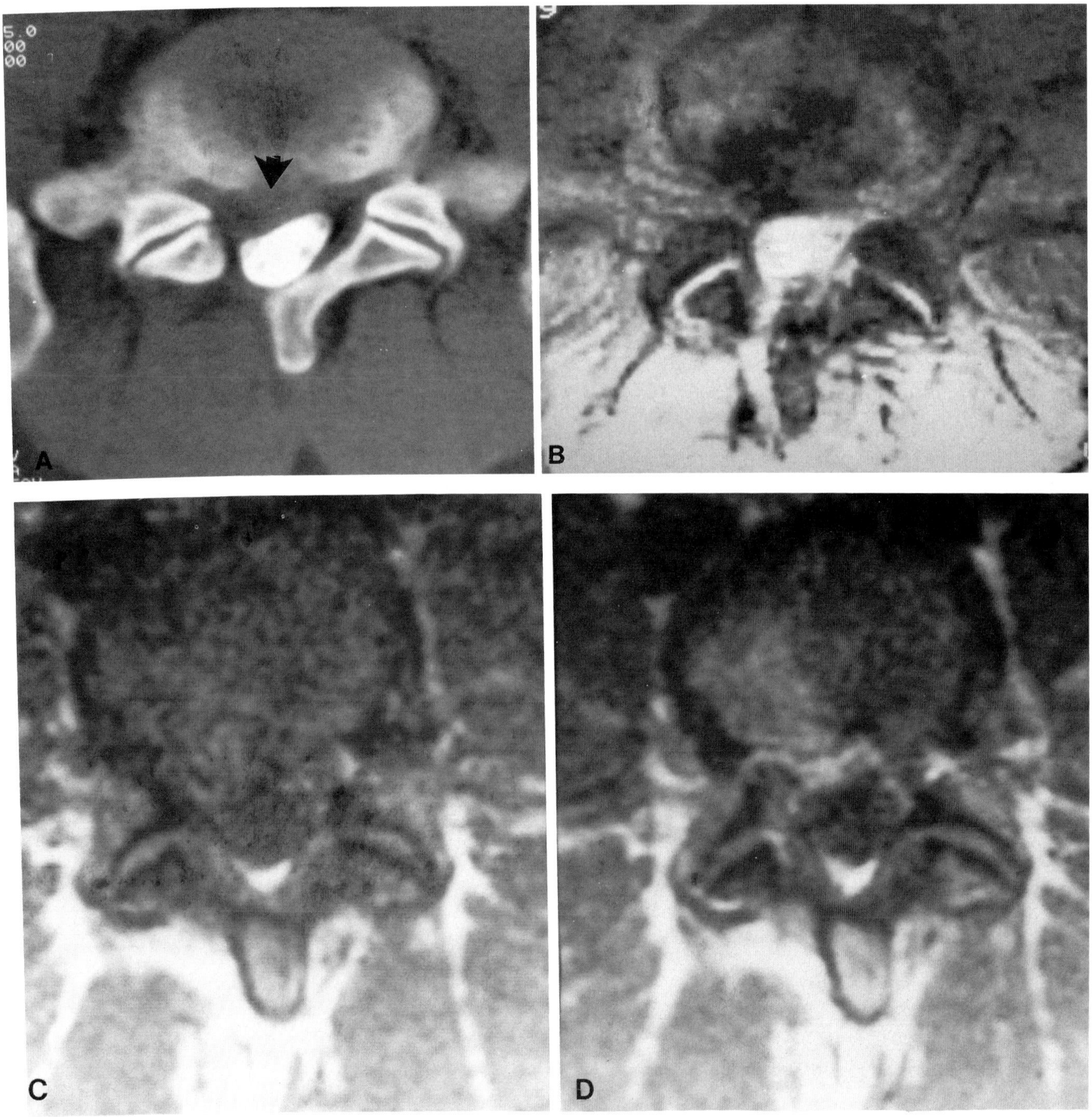

Figure 5.48. Postoperative scar. A, Axial CTM section demonstrates a large right paracentral soft-tissue mass deforming the thecal sac. The right nerve root sheath is not visualized. A recurrent or residual disk cannot be excluded. **B,** T2-weighted axial MRI section demonstrates a similar appearance (*arrowhead*). Precontrast (**C**) and post-Gd-DTPA–enhanced (**D**) T1-weighted MRI images demonstrate the enhancement of the epidural postoperative fibrous tissue (*arrow*). **E–H,** Postoperative extruded disk in another patient: T1-weighted (**E**) proton-density (**F**) sagittal section. T1-weighted axial section before (**G**) and after (**H**) Gd-DTPA enhancement. Unlike epidural scar tissue, the extruded herniated disk (*arrowhead*) does not enhance. Note absence of the lamina (*arrow*) on the left side.

Figure 5.47. Neoplasm. Patient with known cancer presented with low back pain and neurologic deficit. T1-weighted sagittal (**A**) and axial (**B**) MRI sequences demonstrate a focal extradural soft-tissue mass opposite the L5–S1 disk space and S1 vertebral body (*arrow*). It has the appearance of a herniated disk. The disk space at L5–S1 has normal height. **C,** Frontal view of the myelogram demonstrates nonfilling of the left S1 nerve root. **D,E,** Axial CTM sections demonstrate the large extradural mass to the left of midline. At surgery, this turned out to be epidural mets.

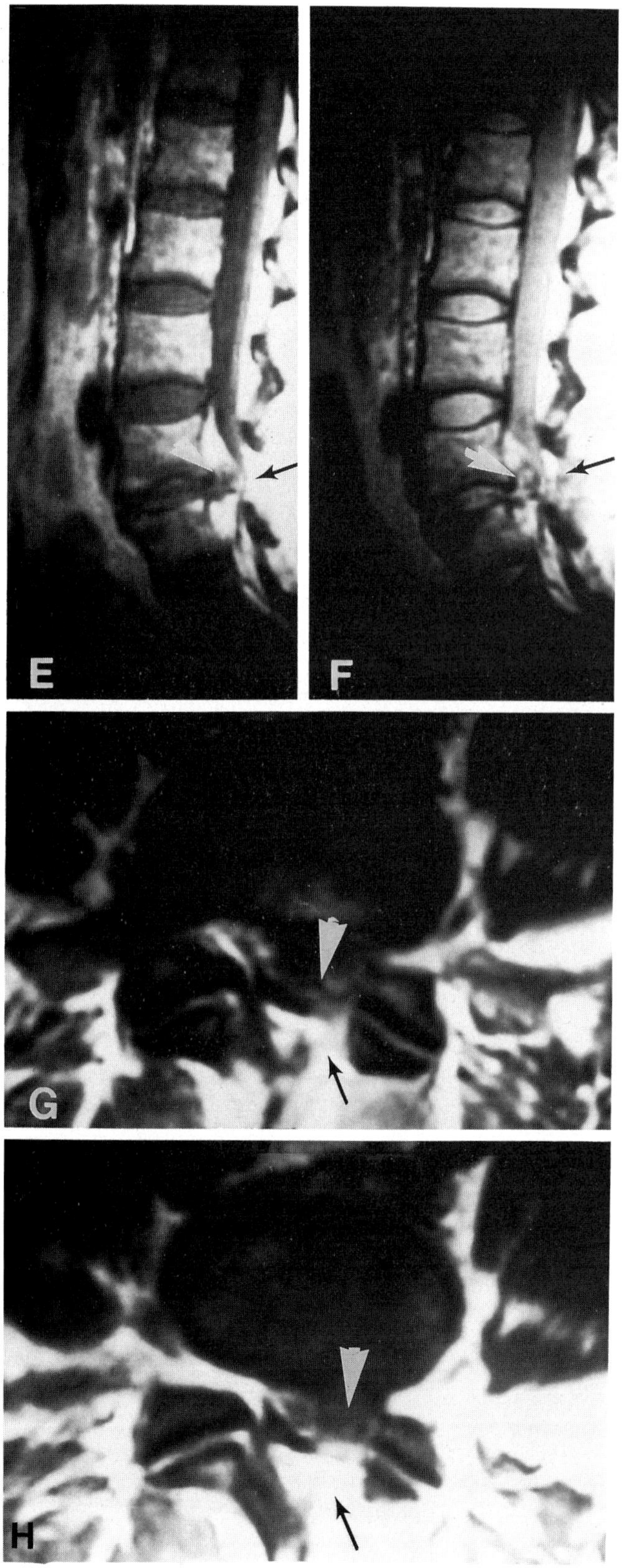

Figure 5.48–*continued*

characteristics of scar tissue. Most anterior epidural scars demonstrate enhancement with Gd-DTPA (Fig. 5.48), whereas enhancement of recurrent disk herniations with contrast material is uncommon (see Chap. 6).

DISK SPACE INFECTION

Disk space infection can present with low back pain and radicular symptoms. In the early stages of disk space involvement, plain radiographs and CT may not demonstrate the changes in the endplates or within the marrow space. Narrowing or destruction of the disk material is difficult to appreciate on CT, except in the later stages of the disease process. On MRI, the disk space infection may show disk herniation or more often bulging of the disk. Loss in height of the disk space, endplate changes, and alterations in the marrow signal are common. The appearance is similar to the pattern seen with type 1 discovertebral degenerative changes. Diagnosis of discitis is based on demonstrating narrowing of the disk space with signal changes within both the disk space and the marrow of adjacent vertebral bodies (Fig. 5.49). There is often edema of the prevertebral soft tissue. Gd-DTPA–enhanced T1-weighted MRI shows varying degrees of enhancement of the remnant of the disk space, the marrow space of the adjacent vertebral bodies, and the prevertebral soft tissues.

PROMINENT VASCULAR PLEXUS

A large epidural venous plexus, especially at the L5–S1 level, may mimic a central or paracentral disk on CT and T1-weighted MRI sequences. Differentiation from a herniated disk can be achieved with IVCT and on T2-weighted MRI sequences. The vascular plexus is hyperdense on IVCT and hyperintense on T2-weighted MRI. An unusual cause of prominent venous plexus may be obstruction of the inferior vena cava (Fig. 5.50).

Degenerative Changes in Vertebral Bodies

Degenerative changes secondary to aging and trauma not only affect the contents of the disk space but also involve the endplates of the vertebral bodies adjacent to the degenerative disk. These need to be differentiated from other nonneoplastic vertebral body lesions (see Table 5.4). These are best evaluated on MRI studies [41–43]. Three types of signal-intensity changes in the bone marrow adjacent to the endplates of degenerated disks have been observed on MRI [43]. Type 1 changes consist of a band of hypointense signal on T1-weighted images and hyperintense signal on T2-weighted spin-echo sequence (Fig. 5.51). This most likely represents

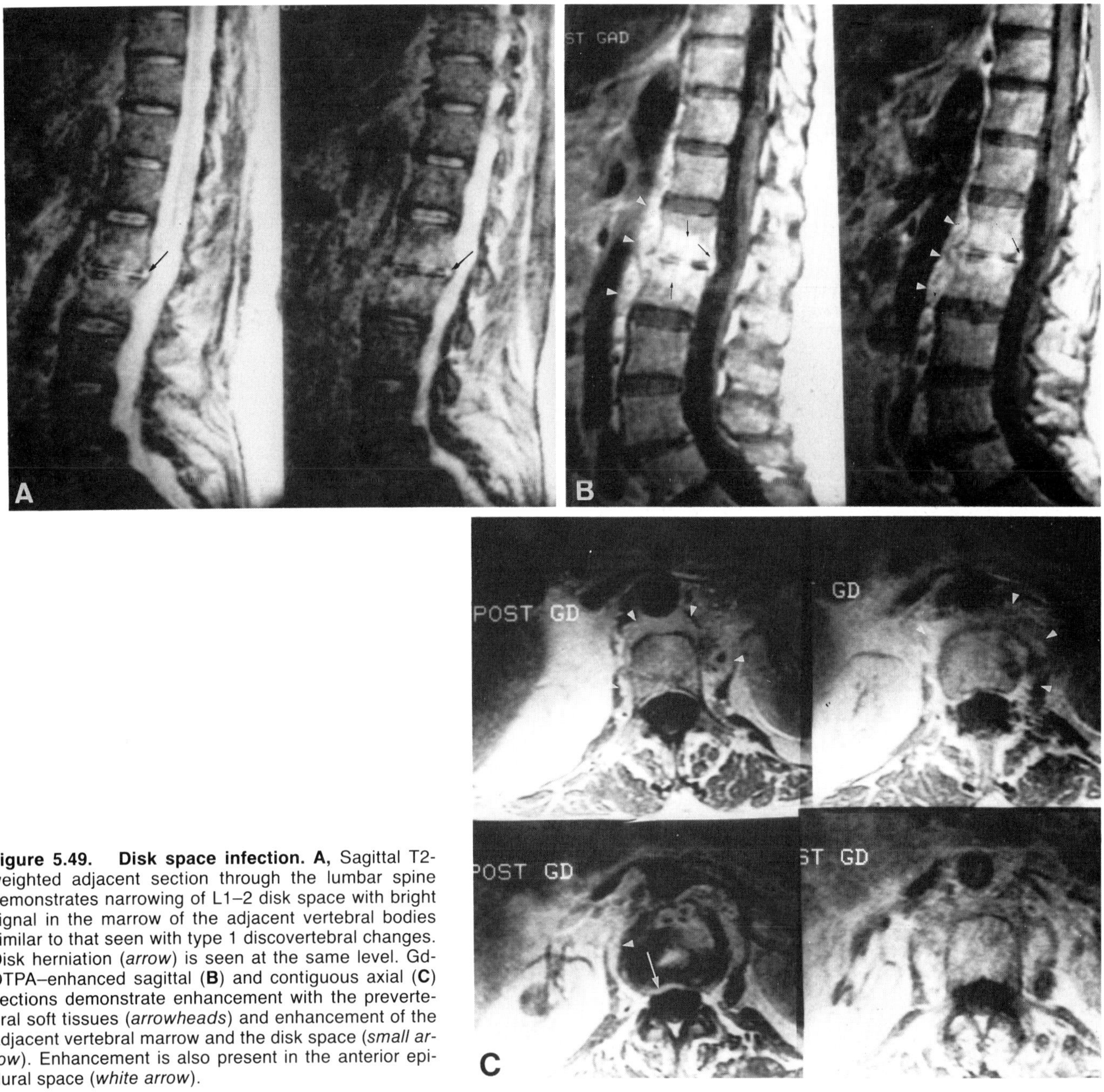

Figure 5.49. Disk space infection. A, Sagittal T2-weighted adjacent section through the lumbar spine demonstrates narrowing of L1–2 disk space with bright signal in the marrow of the adjacent vertebral bodies similar to that seen with type 1 discovertebral changes. Disk herniation (*arrow*) is seen at the same level. Gd-DTPA–enhanced sagittal (**B**) and contiguous axial (**C**) sections demonstrate enhancement with the prevertebral soft tissues (*arrowheads*) and enhancement of the adjacent vertebral marrow and the disk space (*small arrow*). Enhancement is also present in the anterior epidural space (*white arrow*).

edema of the adjacent marrow due to fissuring and disruption within the endplates and the annulus or local replacement of normal cellular marrow with vascularized fibrous tissue. The linear band may be focal adjacent to the anterior aspect of the endplate and probably is a sign of acute disk degeneration. Similar signal-intensity changes have been described with chymopapaine-treated disks [110]. Type 1 changes are often reversible.

Type 2 marrow changes appear as hyperintense signal on T1-weighted images and as isointense or mildly increased signal on T2-weighted images. Type 1 lesions over a period of time may convert to type 2 signal abnormalities. However, type 2 signal changes remain permanent and are not reversible. Type 2 signal changes most likely represent replacement of red marrow with yellow, fatty marrow. The mild increase in signal seen on T2-weighted images is probably due to a combination of yellow marrow and fibrous tissue. Type 2 changes are always associated with loss in disk height, as well as decreased signal intensity within the disk space (Fig. 5.52).

Type 3 marrow signal changes are characterized by hypointense signal in both T1- and T2-weighted images. They are less common than the other two types and are often associated with almost complete

Figure 5.50. Prominent epidural venous plexus. A, Frontal view of myelogram demonstrates nonfilling of the nerve root sheaths at the L4–5 level. **B,C,** Axial contiguous CTM sections demonstrate an anterior extradural mass mimicking a large sequestered disk. The vertebrae demonstrate numerous enlarged basivertebral venous channels. Axial (**D–F**) and sagittal (**G**) T1-weighted MRI images demonstrate enlarged epidural venous channels as well as numerous enlarged basivertebral channels (*arrows*). The venous channels on the left side in the paraspinal region demonstrate stasis or thrombosis. **H,** Frontal view of the venogram demonstrates aneurysmal dilatation of Batson's plexus and the basivertebral and epidural venous plexus around the L4 level.

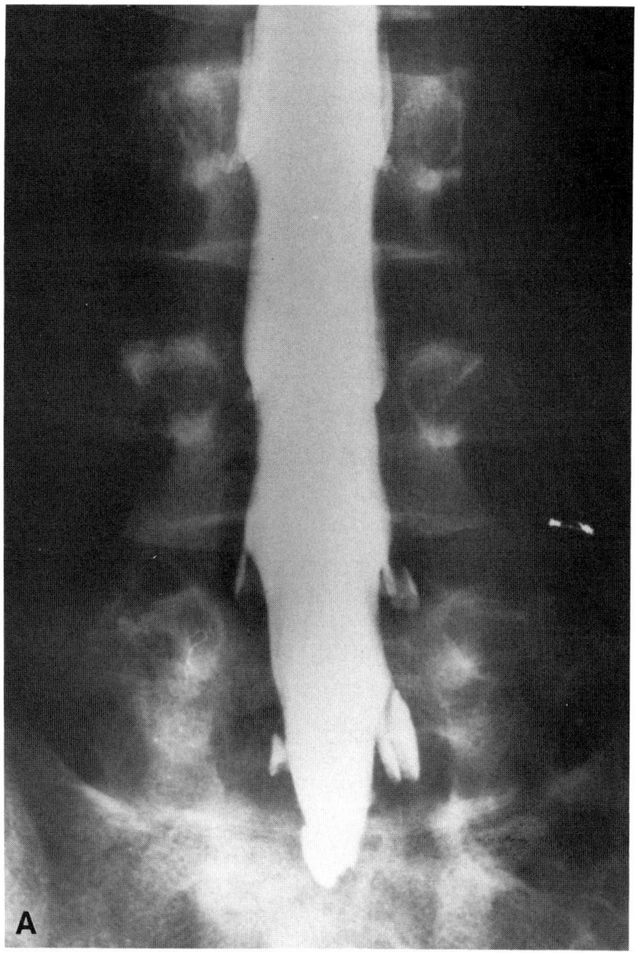

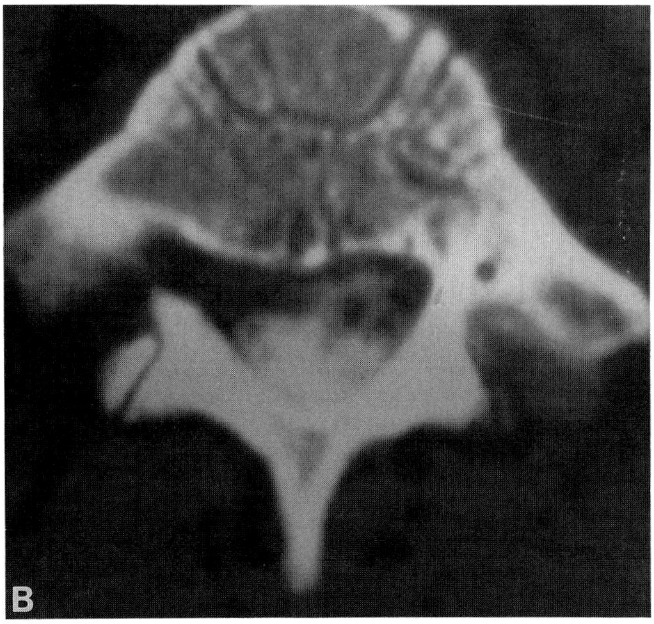

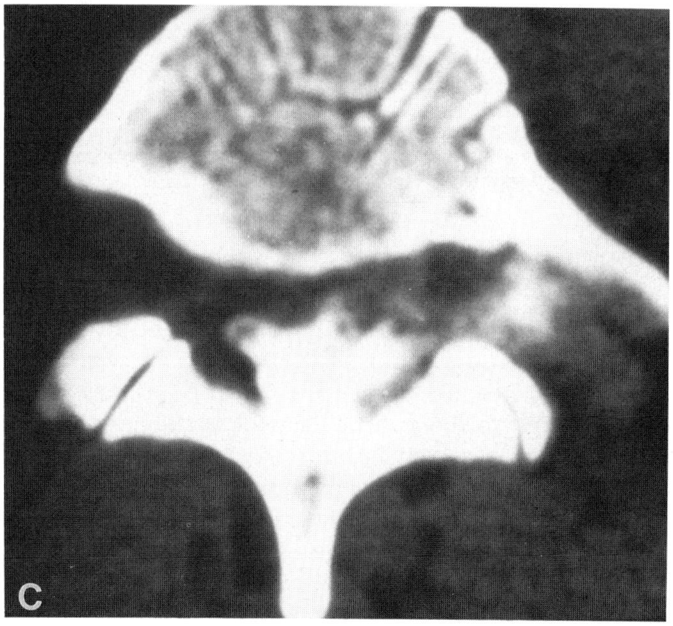

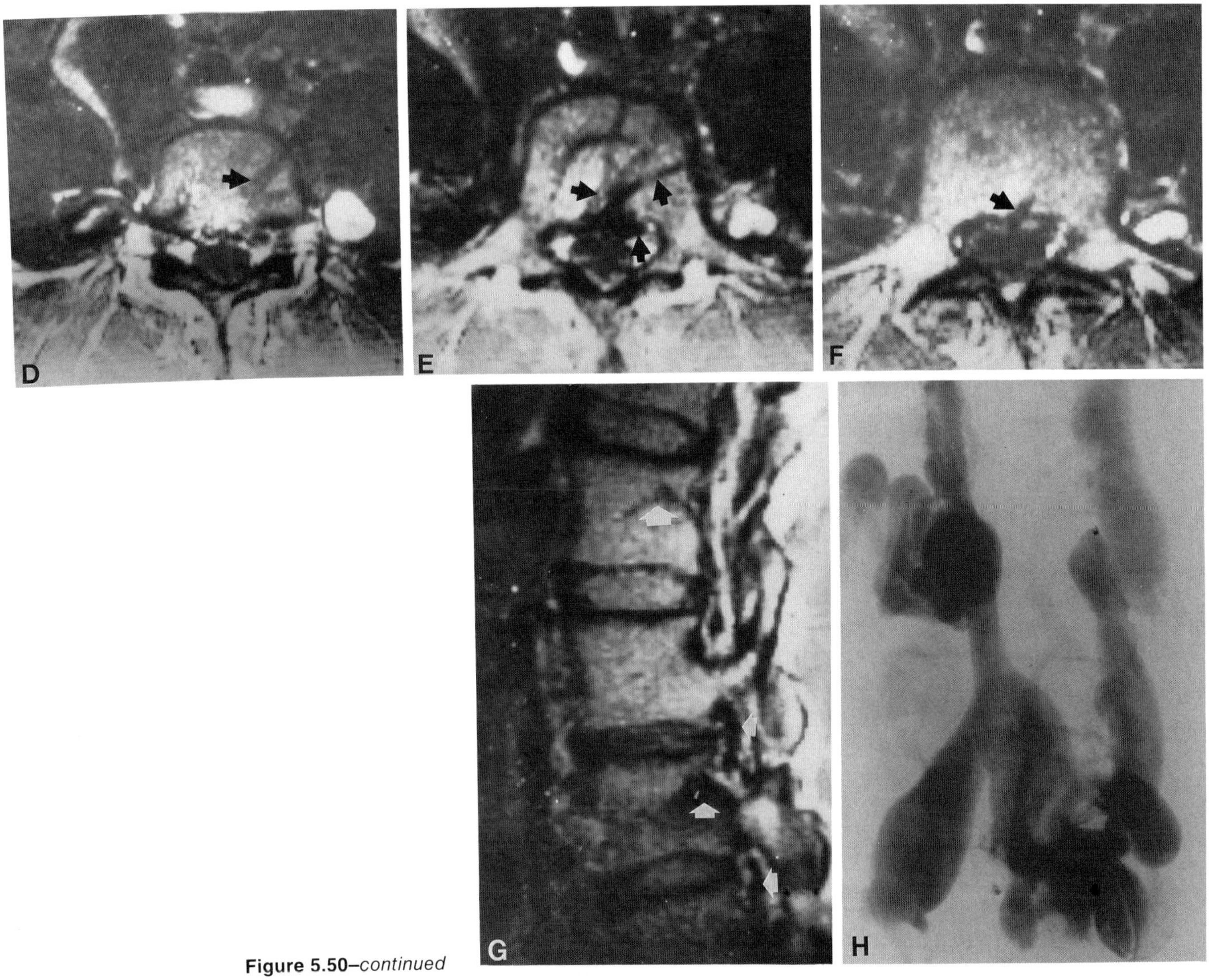

Figure 5.50–*continued*

disappearance of the disk space. On plain radiographs, this may be seen as dense sclerosis of the endplates with marginal osteophytosis (Fig. 5.53).

The endplate and adjacent marrow changes described above are most common in the lumbar region, although similar changes can be identified in the cervical and thoracic regions. Type 1 and 2 changes may not be appreciated on plain radiographs or CT scans, since, unlike MRI, where signal characteristics are dependent on tissue hydration, imaging depends on electron density. The significance of the bony changes seen on MRI and their therapeutic implications are not completely understood. Type 1 changes are similar to those seen with acute discitis or metastatic disease. Clinical history and other imaging characteristics are usually helpful in the diagnosis.

Discovertebral sclerosis or *hemispherical spondylosclerosis* is focal sclerosis of the anteroinferior aspect of the vertebral body associated with narrowing of the disk space and irregular sclerotic margin of the adjacent endplate. This often involves the L4 vertebral body, although it may involve the L5 vertebra and to a lesser extent other vertebrae. Plain radiographs have a characteristic appearance. Discovertebral sclerosis has a hypointense signal on T1-weighted images and mild hyperintensity on T2-weighted MRI images [111] (Fig. 5.54). This may be confused with metastatic disease. It usually affects middle-aged females, often presenting with low back pain.

Signal changes as a result of aging and discovertebral degeneration should be differentiated from other nonneoplastic process which may affect the vertebrae [112] (see Table 5.3).

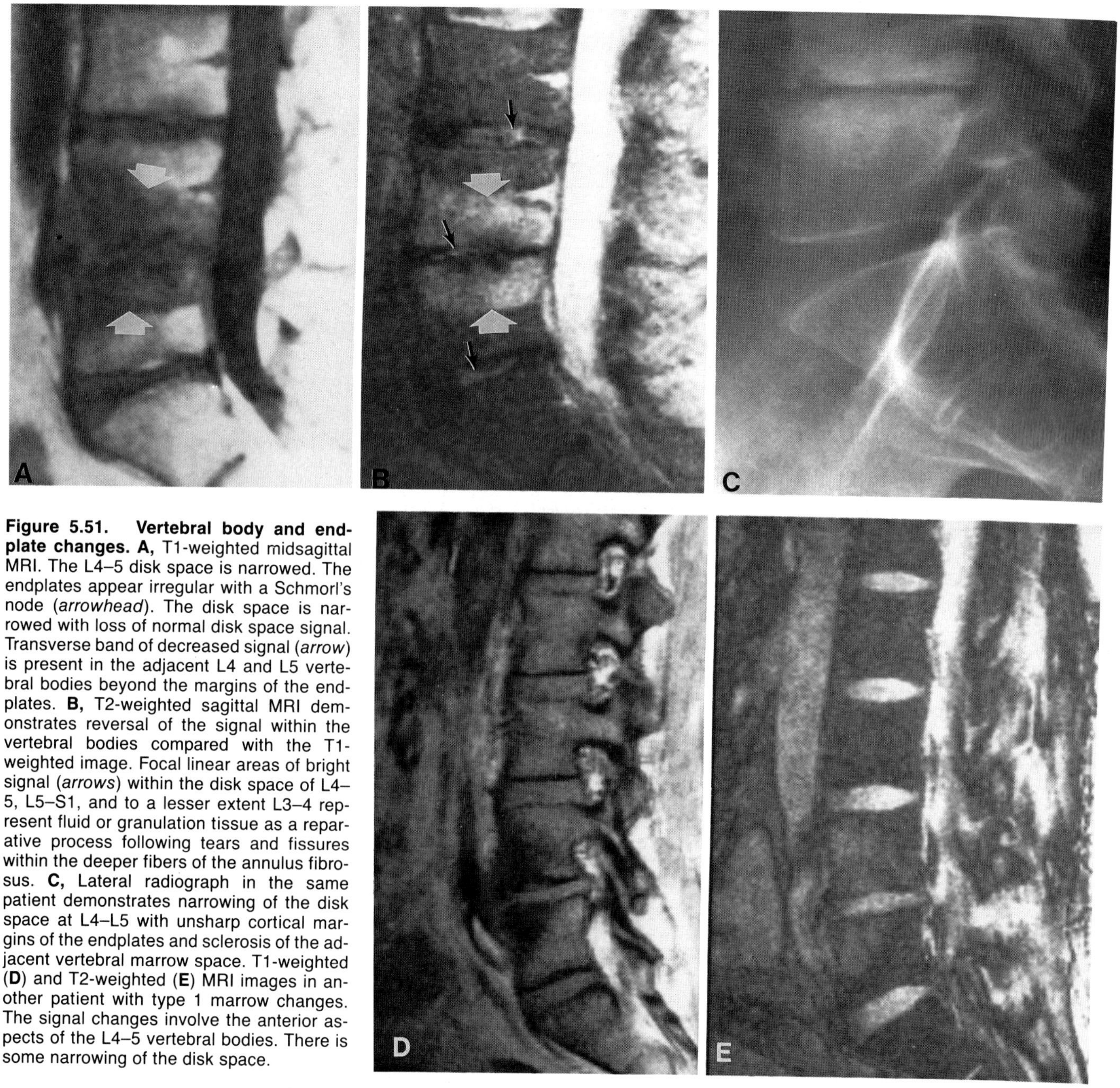

Figure 5.51. Vertebral body and endplate changes. A, T1-weighted midsagittal MRI. The L4–5 disk space is narrowed. The endplates appear irregular with a Schmorl's node (*arrowhead*). The disk space is narrowed with loss of normal disk space signal. Transverse band of decreased signal (*arrow*) is present in the adjacent L4 and L5 vertebral bodies beyond the margins of the endplates. **B,** T2-weighted sagittal MRI demonstrates reversal of the signal within the vertebral bodies compared with the T1-weighted image. Focal linear areas of bright signal (*arrows*) within the disk space of L4–5, L5–S1, and to a lesser extent L3–4 represent fluid or granulation tissue as a reparative process following tears and fissures within the deeper fibers of the annulus fibrosus. **C,** Lateral radiograph in the same patient demonstrates narrowing of the disk space at L4–L5 with unsharp cortical margins of the endplates and sclerosis of the adjacent vertebral marrow space. T1-weighted (**D**) and T2-weighted (**E**) MRI images in another patient with type 1 marrow changes. The signal changes involve the anterior aspects of the L4–5 vertebral bodies. There is some narrowing of the disk space.

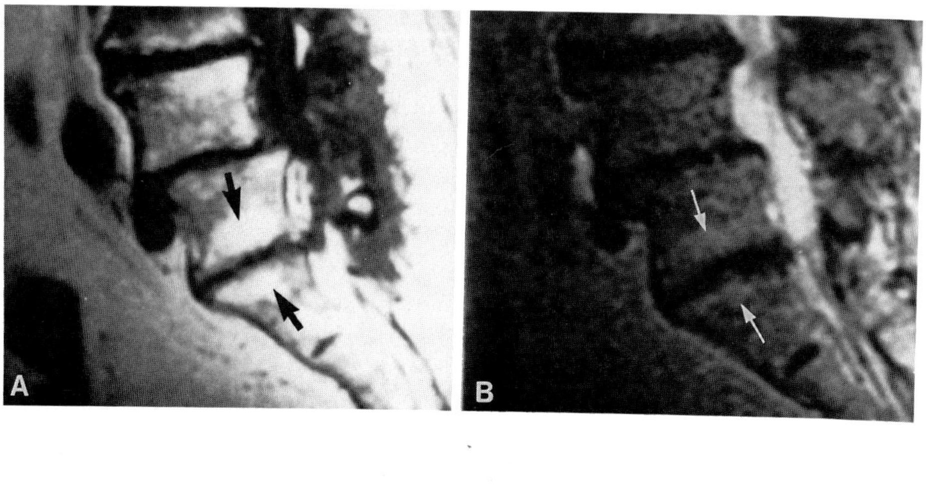

Figure 5.52. Type 2 marrow changes. A, T1-weighted MRI. Focal linear bright signal band in adjacent vertebral bodies of L5–S1 (*arrow*). The disk space is narrowed. **B,** T2-weighted MRI demonstrates isointense signal in the same region.

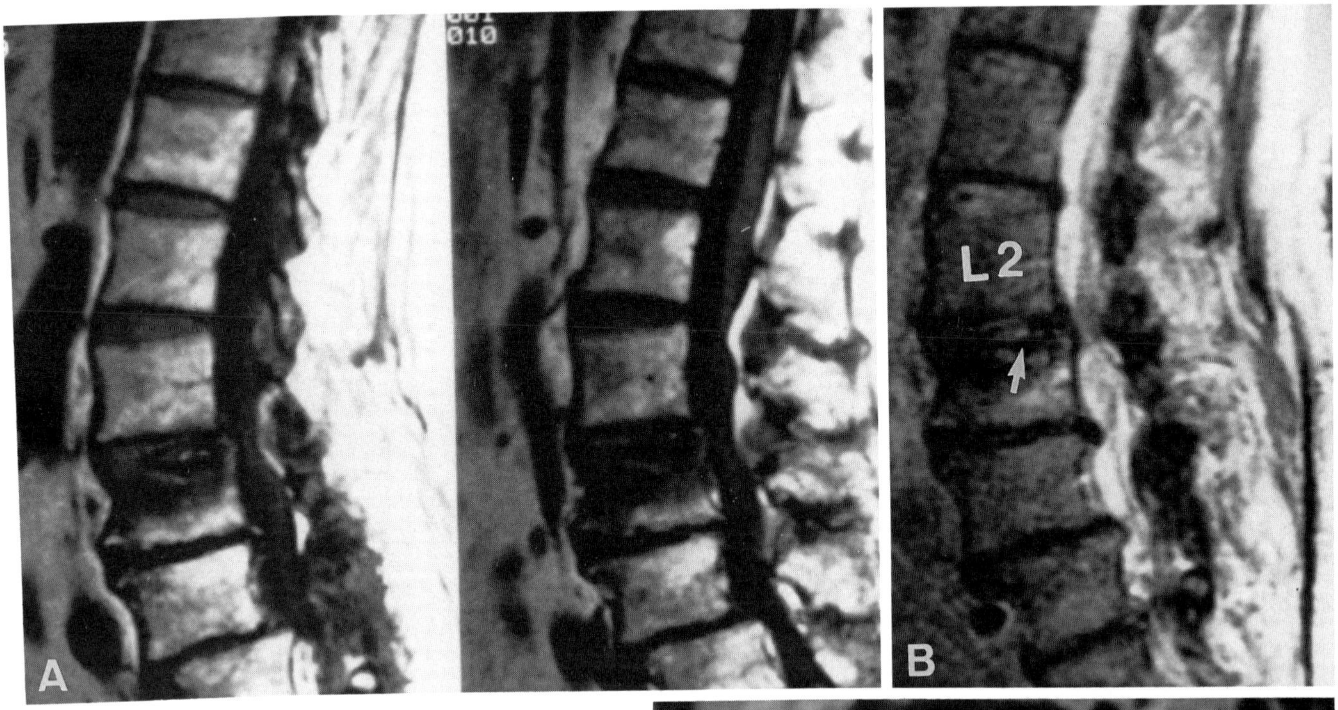

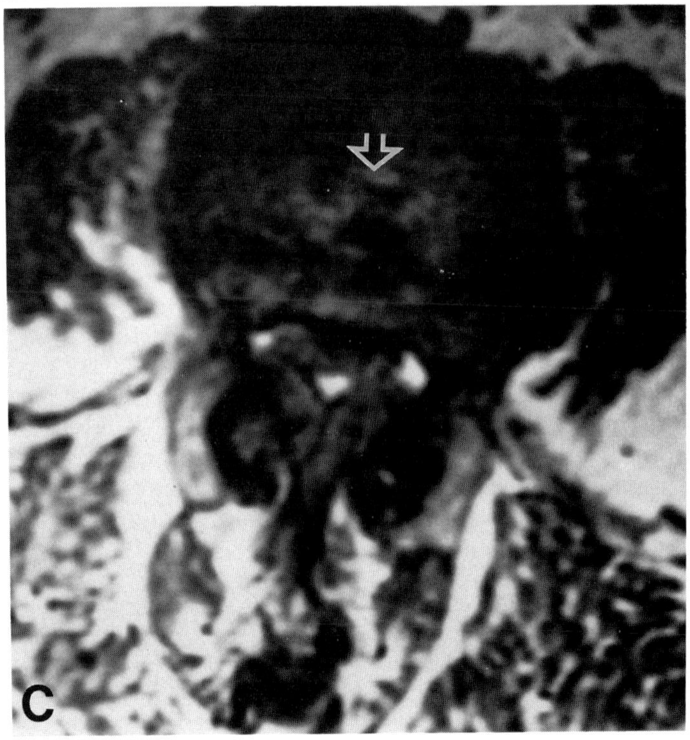

Figure 5.53. Type 3 marrow changes. A, T1-weighted adjacent sagittal MRI sections. Note the decreased signal along the upper half of the L1 vertebral body, with narrowing of the L1–2 disk space. **B,** On T2-weighted MRI, the signal remains hypointense. A Schmorl's node (*arrow*) is present within the endplate of the L2 vertebral body. **C,** Axial T1-weighted MRI image through the disk and endplate demonstrates a Schmorl's node (*arrow*).

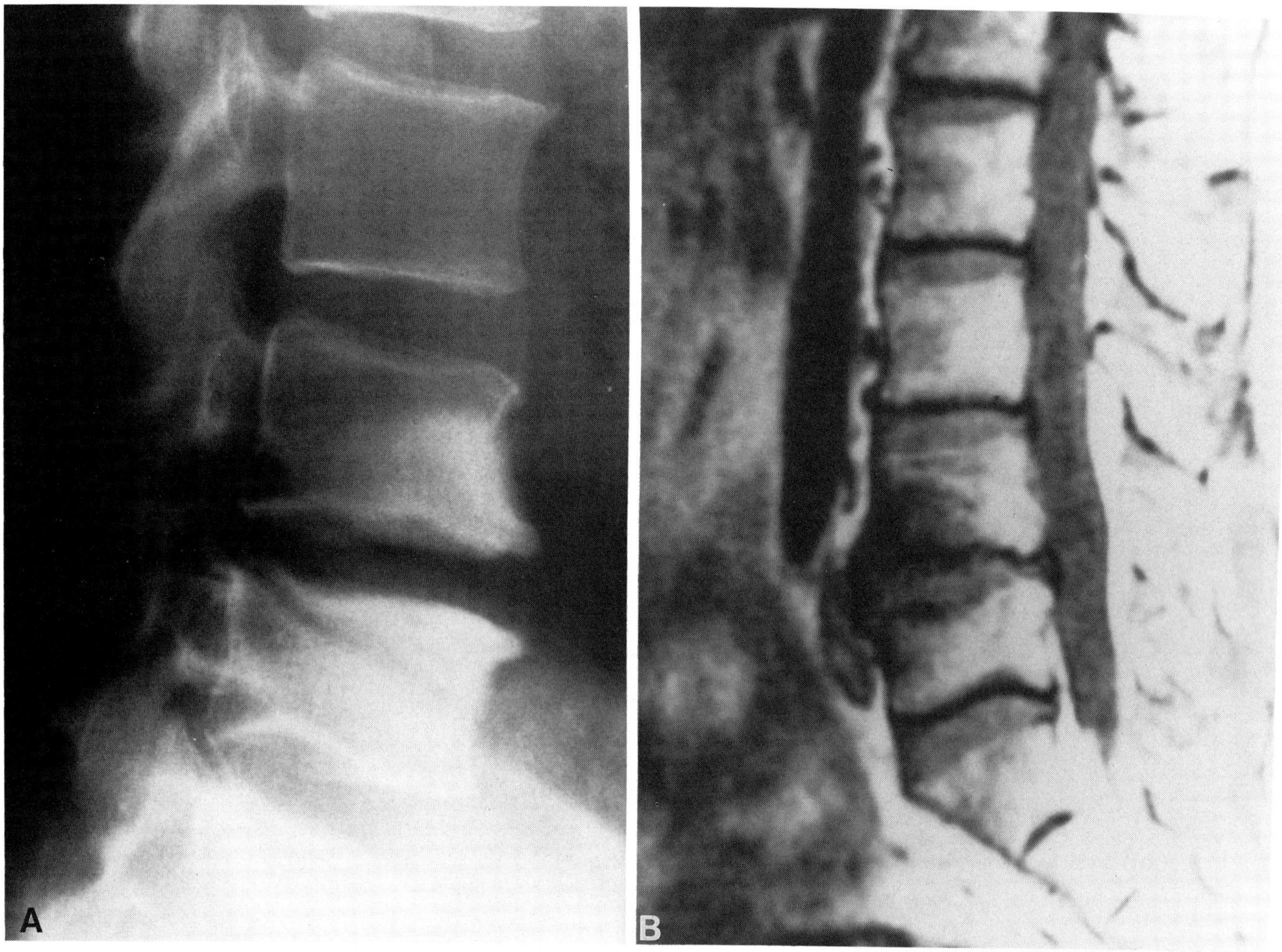

Figure 5.54. Discovertebral sclerosis. A, Lateral x-ray of the lumbar spine demonstrates marked sclerosis of the anteroinferior margin of the L4 vertebral body. **B,** Sagittal MRI section demonstrates hypointense signal in the same area. Although the disk space is not narrowed, there are degenerative osteophytes along the anterior aspects of the adjacent vertebral bodies.

Degenerative Spinal Disease

Other than disk herniation, back pain also can result from degenerative changes involving the other elements that form the "motor unit." The degenerative process in the disk adversely affects the related spinal structures. This results in narrowing of the spinal canal—spinal stenosis, facet arthropathy, ligamentous hypertrophy, spinal instability, and muscle atrophy.

SPINAL STENOSIS

Spinal stenosis may involve the central spinal canal, its lateral recess (neural canal), the intervertebral foramina, or a combination of all three components. Spinal stenosis may be developmental or, as is more common, an acquired disease from numerous causes (Table 5.5).

The *congenital* or *developmental* form of spinal stenosis is rare. It may be part of a systemic process such as in *achondroplasia* or be limited to the lumbar or cervical segments of the spine. Congenital spinal stenosis should be suspected when the angle between the two laminae is less than 90 degrees, the lamina are smaller, and the superior and inferior articulating facets are facing anteroposteriorly. These osseous anomalies result in greater stress on the disk spaces and the facet joints, resulting in early degenerative changes and patients presenting with symptoms at an earlier age.

Congenital lumbar stenosis is more often seen in males than in females, usually in their second or third decade. Patients present with neurogenic claudication [113], consisting of radicular pain and motor and sensory deficits that develop when walking or standing but which are relieved with rest. Such patients may or may not have low back pain. Imaging findings are similar to those seen with acquired spinal stenosis, except that the degenerative osseous and ligamentous components are less prominent.

Tablc 5.5.
Spinal Stenosis

1. CONGENITAL:
   - Idiopathic
   - Achondroplasia
   - Morquio's disease
   - Bone dysplasias
2. ACQUIRED:
   - Degenerative changes in disk
   - Spondylolysis
   - Spondylolesthesis
   - Facet arthropathy
   - Posttraumatic
   - Postsurgical
   - Ankylosing spondylitis
   - Ossification posterior longitudinal ligament (OPLL)
   - Ossification or hypertrophy of ligamentum flavum
   - Diffuse idiopathic skeletal hyperostosis (DISH)
   - Metabolic causes:
     - Paget's Disease
     - Acromegaly
     - Epidural lipomatosis
     - Flourosis

(Combination of one or more of the above entities may acount for the spinal stenosis)

The stenosis usually involves the L1 to L5 segments of the spinal canal and spares the sacral region (Fig. 5.55).

Congenital cervical spinal stenosis is rare but should be suspected when symptoms of radiculopathy present in a person, especially after minor trauma. A true lateral radiograph of the cervical spine demonstrates a narrow spinolaminar distance. On cross-sectional imaging studies, congenital stenosis is suspected when the canal width in the sagittal plane is decreased, the spinal cord is flattened anteroposteriorly, and there is a decrease in CSF space surrounding the spinal cord. Even minimal degenerative changes may accentuate the neurologic deficits.

Lumbar spinal stenosis is the result of extension of the degenerative process of aging to other elements of the discovertebral complex beyond the disk space. The primary change starts in the disk space with fissuring of the annulus, leading to bulging of the annulus and loss of disk height. Disruption of the endplates leads to calcium deposition and the development of bridging periosteum, which subsequently forms osteophytes. There is hypertrophy and/or calcification of the spinal ligaments. Loss of synovial fluid and degenerative changes in the synovium leads to laxity of the capsule of the facet joints. Continued stress results in new but disorganized bone formation at the margins of the facet joints. Similar stress on the lamina and pedicles results in increased sclerosis. The decreased motion from the pain leads to disuse atrophy of the paraspinal muscles. These are progressive events.

The resulting stenosis may demonstrate elements of more than one of the preceding components at one or more levels. It most often involves the L4–5 and L5–S1 levels. The spinal stenosis may be diffuse, indicating an overall decrease in diameter as well as a narrowing of the neural canal and foramen, often identified as *central spinal stenosis*. This extends between the L1 and L5 level. Patients present with a variety of symptoms depending on the neural element involved and the vascular compression.

The clinical history is of persistent progressive low back pain associated with radicular pain involving both lower limbs and buttocks and a variety of sensory symptoms which are exacerbated by walking and standing. These symptoms mimic those of vascular claudication and, if not recognized, usually lead to extensive vascular imaging studies. The symptom complex defined as *neurogenic claudication* [113] should be suspected when symptoms become worse as the day progresses and are often associated with standing for a long period.

Stenosis may involve the lateral recess (subarticular canal) [18] which is formed anteriorly by the intervertebral disk and the posterior margins of the adjacent vertebral bodies and posteriorly by the medial aspect of the superior articular facet and pars interarticularis. Parts of the lateral wall are formed by the adjacent pedicle and extension of the ligamentum flavum, and the medial surface is made up of the soft dural sac. Hypertrophic changes involving the disk, facet joints, and ligaments, as well as osteophytic projections, narrow the subarticular canal and result in entrapment of the nerve root. This presents clinically with pain and radiculopathy.

Foraminal encroachment results in symptoms similar to those of subarticular stenosis. The neural or intervertebral foramen is a short segment bounded by the pedicles of the adjacent vertebral bodies, anteriorly by the posteroinferior margin of the adjacent vertebrae and posterolateral margin of the disk, and posteriorly by the lateral aspect of the facet joint. Identification of stenotic processes involving these various segments of the discovertebral complex on imaging studies helps in the surgical management of symptomatic patients.

Imaging Studies

The osseous and cartilaginous changes associated with spinal stenosis are best evaluated with CT; however, since a herniated disk is always a consideration, MRI should be the primary examination. In most instances, this may be adequate. If the symptoms cannot be explained on the basis of MRI studies, and if surgery is a consideration, CT combined with myelography (CTM) is often indicated. Axial CT sections may need to be complemented with re-for-

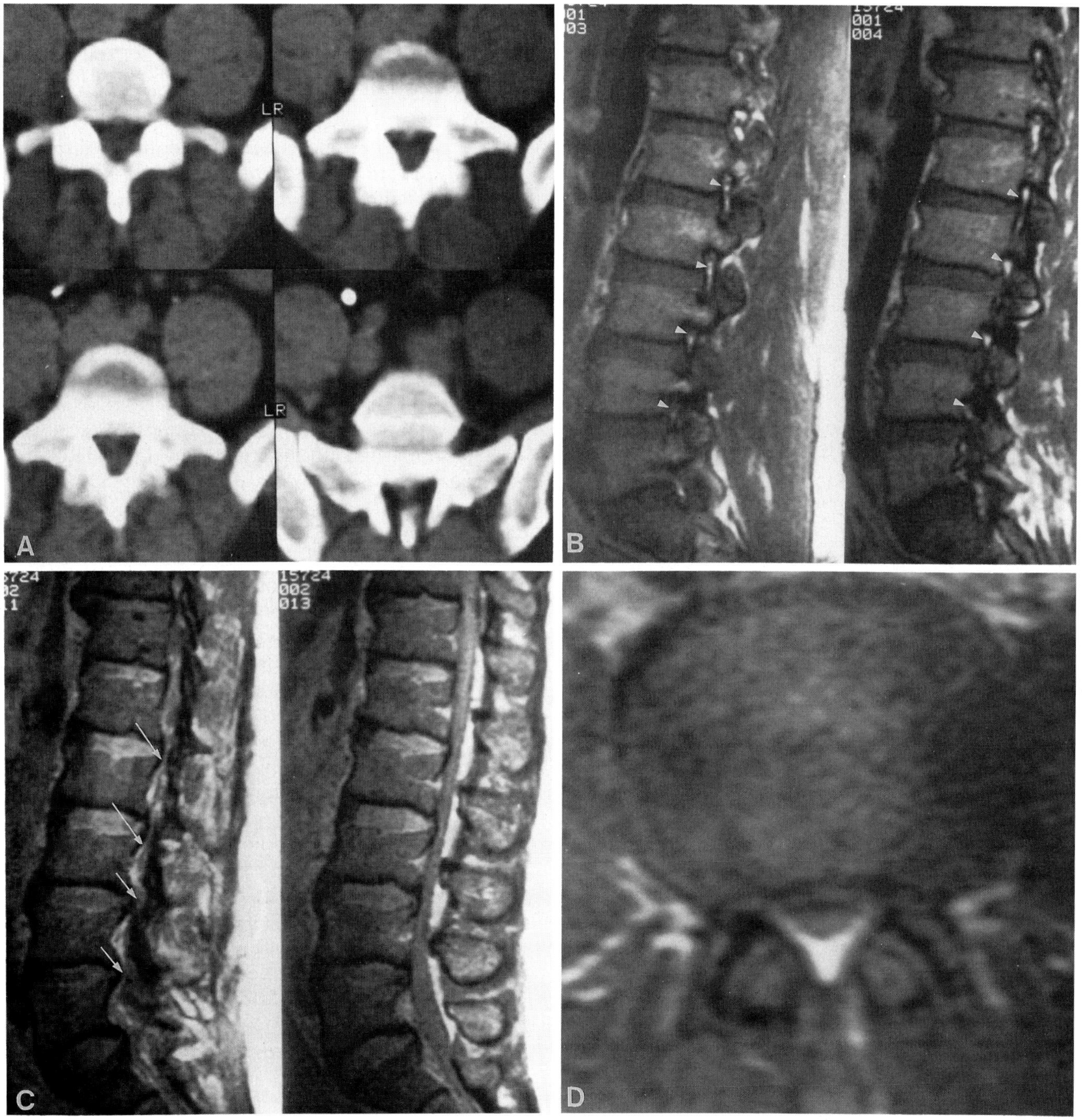

Figure 5.55. Congenital lumbar stenosis. A 28-year-old woman with symptoms of neurogenic claudication. **A,** Contiguous axial CT scan. The spinal canal is triangular in shape, with marked thickening of the laminae. The laminae are short, contributing to the spinal stenosis. **B,C,** Adjacent sagittal MRI images from the left neural foramen to the midline. The neural foramen (*small arrowhead*) and subarticular canal (*small arrow*) show stenosis in the sagittal plane. The thecal sac is narrowed between the L1 and L5 level. **D,** Axial T1-weighted MRI at the L3–4 disk level demonstrates the short, thick laminae. There is symmetrical bulging of the disk, which with the short pedicle leads to a narrow subarticular canal.

matted sagittal and coronal images to demonstrate the overall extent of the stenosis.

Central canal stenosis results in decreased sagittal and transverse diameter of the spinal canal. This is most often seen in association with degenerative disk and facet joint disease. Syndesmophytes at the margins of attachment of the annulus fibrosus along with similar changes involving the capsular attachments around the facet joints encroach on the central canal. Depending on the extent of the facet hypertrophy, bulging disk, and hypertrophy of the spinal ligaments, the canal has an exaggerated trefoil appearance on axial sections and a corrugated appearance on the sagittal images (Fig. 5.56). This is always associated with lateral recess stenosis and foraminal stenosis. The stenosis may be accentuated by preexisting congenital narrowing, achondroplasia, Paget's disease, or diffuse idiopathic skeletal hyperostosis (DISH). Postoperative changes and trauma also may result in spinal stenosis, although these tend to be focal. MRI evaluation is based on both sagittal and axial sections in both T1- and T2-weighted sequences (Fig. 5.57). In most instances, the conventional T2-weighted spin-echo sequence is replaced with a GRE or FSE sequence. Calcification of the facet joints and prominence of the ligaments are seen as hypointense signal on both the T1- and T2-weighted images. Bony osteophytes may appear hypointense or isointense depending on the amount of marrow and fibrous tissue. The thecal sac is flattened anteroposteriorly. A fair amount of epidural fat may be interposed dorsally between the hypertrophied thickened ligamentum flavum and the thecal sac. In severe central stenosis, the nerve roots may be pinched within the thecal sac. These can be assessed on MRI by the increased brightness of the CSF signal distal to the stenosis compared with the CSF signal in the rest of the thecal sac [114]. The brightness of the CSF signal distal to the block is presumed to be due to stasis and decreased movement of CSF (Fig. 5.58). The hypertrophy of the ligaments as well as the osteocartilaginous component in central spinal stenosis is best evaluated with CT, often following myelography (CTM). Since the stenosis often involves more than one level, contiguous thin sections (5 mm) should be obtained. This allows for subsequent re-formatted images in other orthogonal planes (Fig. 5.59). Although canal measurements have been described, they have limited value because the actual encroachment and narrowing of the thecal sac and the nerve root sheath can be identified directly on both MRI and CTM. Canal measurements also may be misleading if the stenosis results from soft-tissue hypertrophy, such as stenosis resulting from a bulging disk and thickened ligamentum flavum. These can be defined clearly on axial CT and MRI images.

The severity of the clinical symptoms depends to some extent on the severity of the stenosis. In most instances, central stenosis in the lumbar region is focal at the disk levels and may extend laterally to involve the lateral recess. Stenosis is common between L4–5 and L5–S1 unless associated with a preexisting disease process, in which case other discovertebral segments, including the thoracolumbar segments, may be affected.

Lateral recess stenosis may occur with an adequate central spinal canal. This is more often seen at the L4–5 and L5–S1 disk levels. Lateral recess stenosis results from bony osteophytes, a calcified bulging disk, and hypertrophy of the facet joints and ligaments; however, the most common cause is a combination of hypertrophic changes involving the superior articular facet and the facet joint capsule (Fig. 5.60). Narrowing of the lateral recess should be suspected when the width of the canal is less than 3 mm on axial CT sections [115]. Compression of the nerve root within the canal is best evaluated on CT and MRI in axial sections through the pedicle.

Foraminal stenosis results from hypertrophic osteophytes from facet joints and endplates of the vertebral body as well as a bulging disk compromising the foramina. Foraminal stenosis may occur as a result of degenerative spondylolisthesis. Laxity of the facet joint capsule leads to forward displacement of the superior facet, which compresses the nerve roots within the foramina. Although axial CT and MRI may demonstrate the foraminal narrowing, this is best evaluated on sagittal MRI and re-formatted sagittal CT images. Unlike in cervical region, it is rarely necessary to obtain parasagittal oblique MRI sections, although rarely a combination of parasagittal oblique imaging following Gd-DTPA enhancement may be necessary to identify associated disk herniation [62].

Back pain also may be secondary to mechanical instability of the spine from *spondylolysis* and subsequent *spondylolisthesis*, as well as the resulting spinal stenosis. *Spondylolysis* refers to a defect in the pars interarticularis. This usually involves the L5 vertebra and less often the L4 vertebra. With aging, there is both bony and soft-tissue proliferation, resulting in narrowing of the lateral recess. This may extend further laterally to involve the neural foramen, leading to foraminal stenosis. The pars defect and the stenosis are easily recognized on CT (Fig. 5.61). The pars defect may go unnoticed on axial MRI sections, since both the pars defect and the hypertrophic changes of the adjacent facet joints have hypointense signal on T1- or T2-weighted MRI studies (Fig. 5.62). Identification of spondylolysis is facilitated by evaluating both axial and sagittal images and if there is associated spondylolisthesis.

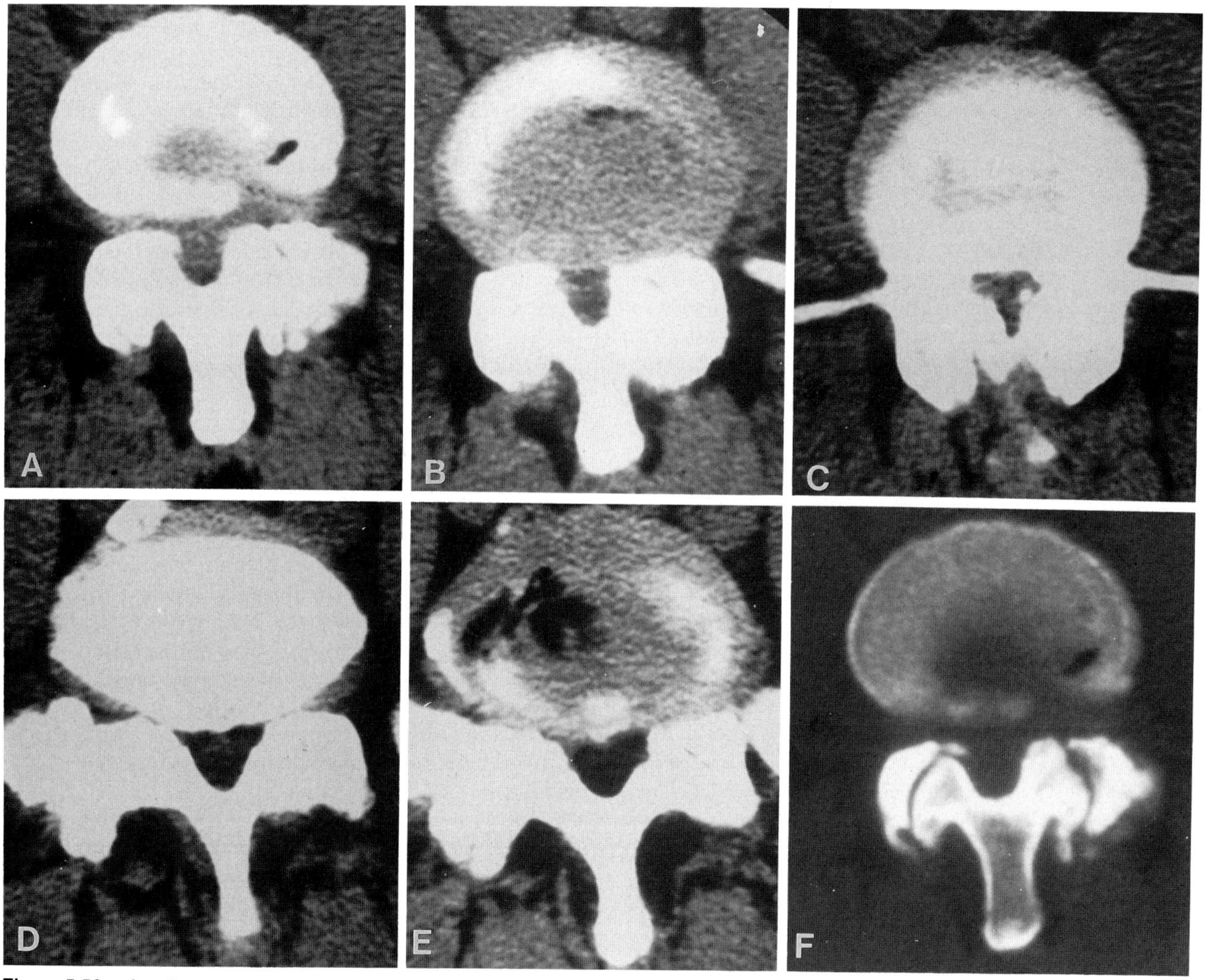

Figure 5.56. Lumbar central canal stenosis. A–E, Axial CT sections. In soft-tissue setting, the osseous and cartilaginous structures appear markedly dense. The central spinal canal appears markedly narrow, resulting from marked hypertrophy of the facet joints, short laminae, bulging disk, and calcification and thickening of the ligaments. **F–J,** Axial sections in bone algorithm demonstrate the osteocartilaginous elements contributing to the stenosis to better advantage.

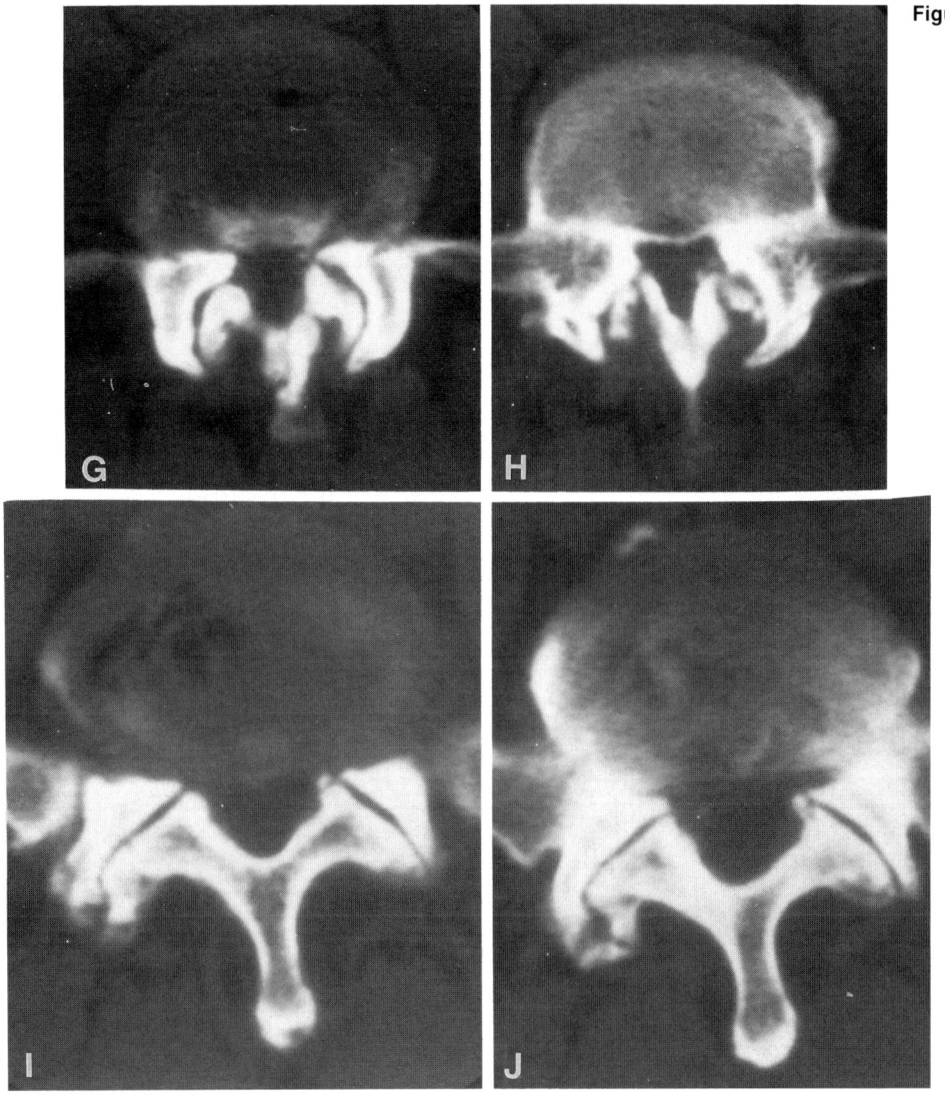

Figure 5.56–*continued*

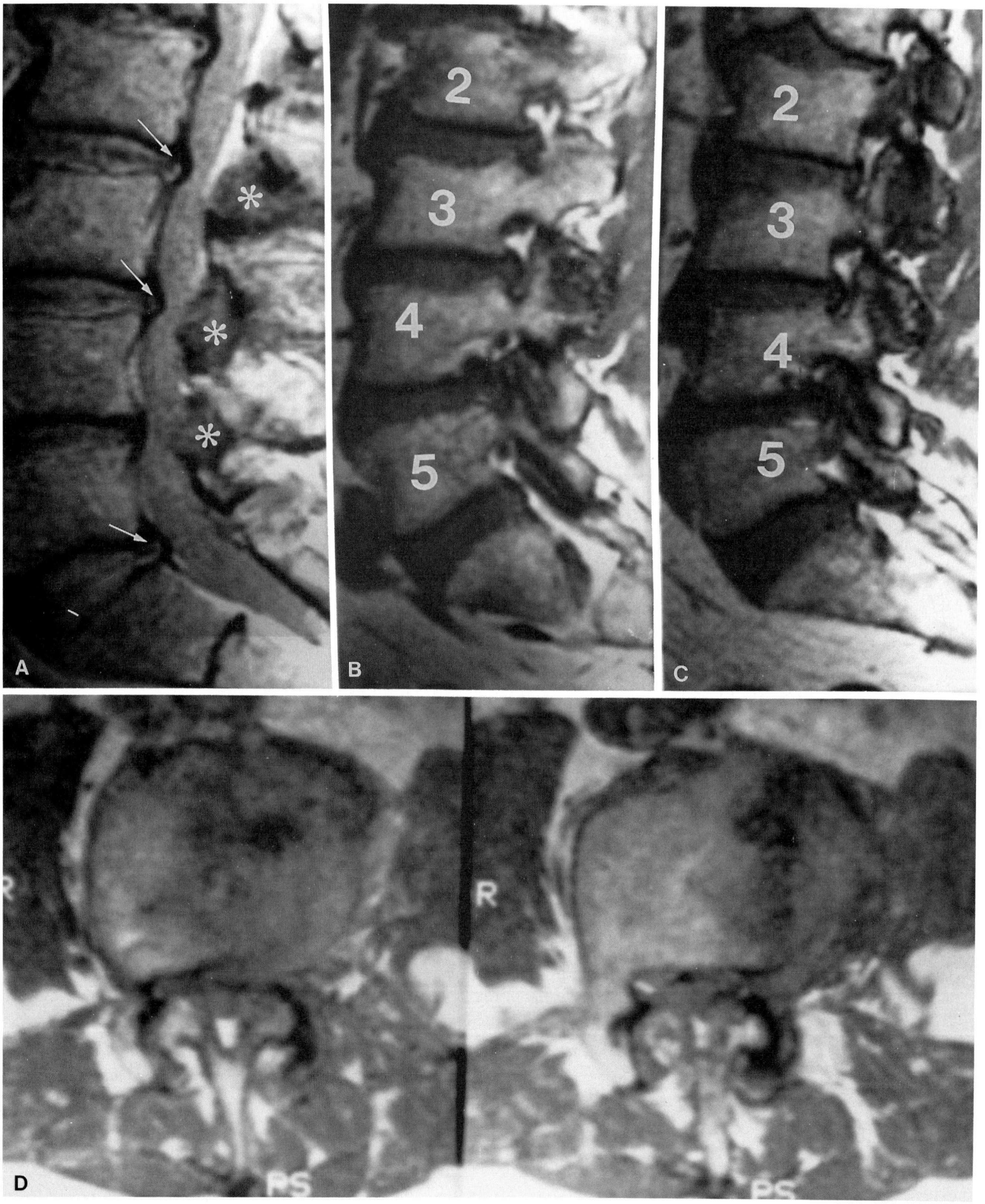

Figure 5.57. Lumbar spinal stenosis. A, Sagittal midline MRI shows lumbar spinal canal stenosis at each disk level anteriorly from herniated disk (*arrow*) and posteriorly from thickening of the ligamentum flavum (✱). Parasagittal MRI sections through the right (**B**) and left (**C**) foraminal levels. Foraminal encroachment is demonstrated at multiple levels between the L2 and L5 level. **D,** Contiguous axial T1-weighted MRI demonstrates both central canal and neural foramen stenosis from the bulging disk, marked degenerative facet, and ligamentous changes.

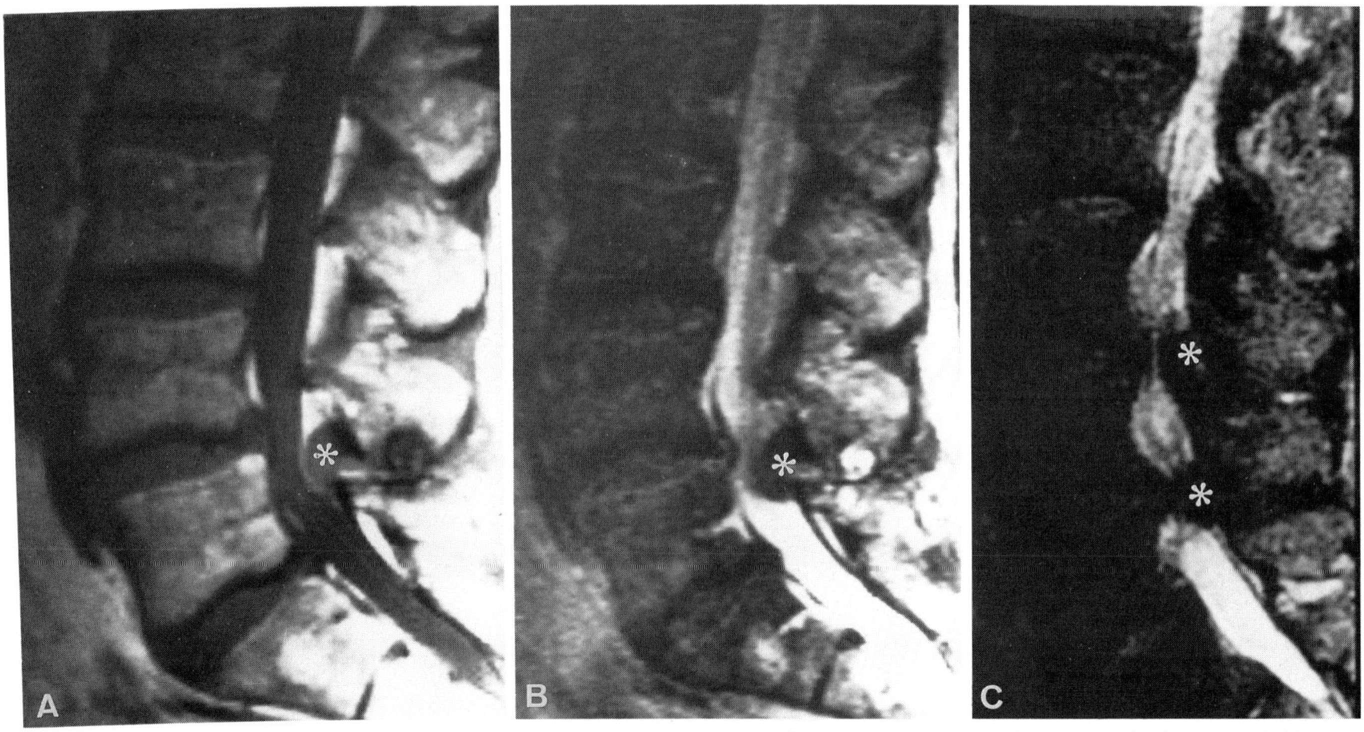

Figure 5.58. Spinal stenosis secondary to soft-tissue thickening. A, Sagittal T1-weighted MRI shows focal stenosis at the L4–5 disk level from combination of degenerative disk disease and focal thickening of the ligamentum flavum (✱). **B,** Sagittal T2-weighted MRI demonstrates severe focal stenosis with hyperintense CSF signal distal to the obstruction. The increased CSF signal is presumed to be due to absent CSF pulsation. **C,** Sagittal T2-weighted MRI in another patient demonstrates severe spinal stenosis at both the L3–4 and L4–5 levels, accentuating the hyperintense signal of CSF below the obstruction.

Spondylolisthesis is subluxation of an adjacent vertebral body, usually due to bilateral spondylolysis or as a result of degenerative disk and facet disease. The latter is also referred as *pseudospondylolisthesis*. The symptoms are due to mechanical instability resulting from stresses applied to these joint surfaces, as well as from compression of the neurovascular bundle within the narrowed foramen. The spinal canal may appear normal or capacious in sagittal images. With pseudospondylolisthesis, central canal stenosis results from slippage of the vertebrae and compression of the thecal sac between the pseudo-disk herniation and the neural arch of the anteriorly displaced vertebrae (Fig. 5.63). This type of stenosis is often seen in females and involves the L4–5 level [14, 116]. The symptoms become worse after laminectomy. Flexion and extension lateral views may show instability and motion, which are difficult to assess by CT or MRI. These plain x-rays should be obtained if surgery is contemplated.

THORACIC SPINE

Stenosis of the thoracic spine is less prevalent or symptomatic when compared with that of the lumbar or cervical region. This is probably due to the available canal diameter and the diameter of the spinal cord, as well as the less mechanical stress of the joint surface. Symptomatic stenosis is usually the result of thickening, calcification, or ossification of the posterior longitudinal ligament (OPLL), bony osteophytes compressing the spinal cord, or compression of the nerve roots within the foramen. OPLL is more common in the Japanese population [117], although it is not uncommon in Caucasians. Among both races it more often involves the cervical spine. OPLL is often associated with diffuse idiopathic skeletal hyperostosis (DISH). Plain radiographs may not show the calcification of the ligament. CT is useful in demonstrating the calcification or ossification of the ligament as well as the extent and severity of the cord compression [118] (Fig. 5.64). Ideally, the study should consist of overlapping contiguous sections such that the extent of the hypertrophic changes and the ossification or calcification can be evaluated in the sagittal plane.

On MRI, spinal stenosis due to thickening and calcification of the ligament is seen as a thick, irregular band of hypointense signal along the posterior margin of the vertebral bodies on both T1- and T2-weighted images [119]. Cord ischemia or myelopathy should be suspected in the presence of hyperintense signal within the spinal cord on sagittal T2-weighted images (see Fig. 5.64). This may not be a consistent finding due to motion-related artifacts from both CSF and structures outside the spinal canal. Focal

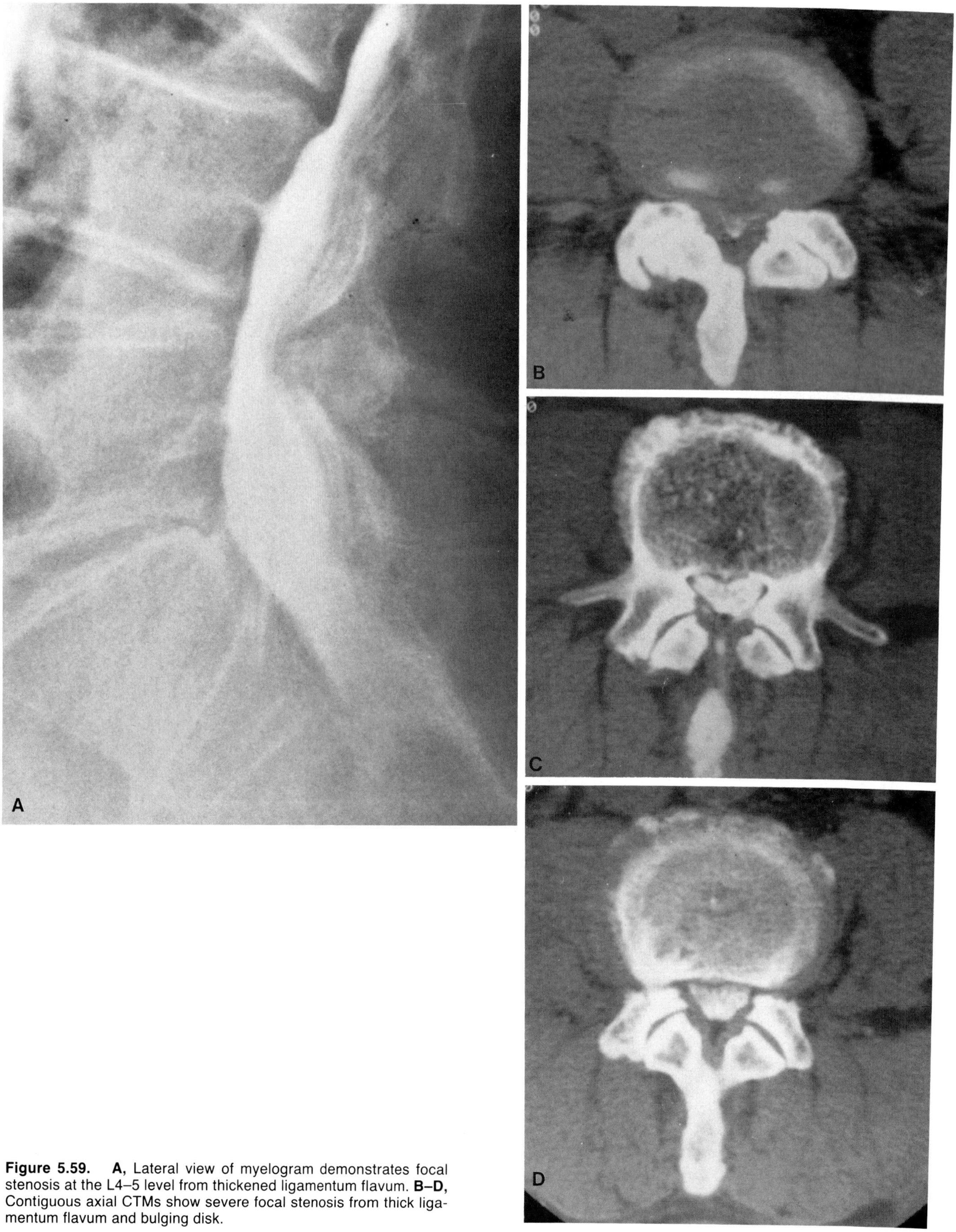

Figure 5.59. A, Lateral view of myelogram demonstrates focal stenosis at the L4–5 level from thickened ligamentum flavum. **B–D,** Contiguous axial CTMs show severe focal stenosis from thick ligamentum flavum and bulging disk.

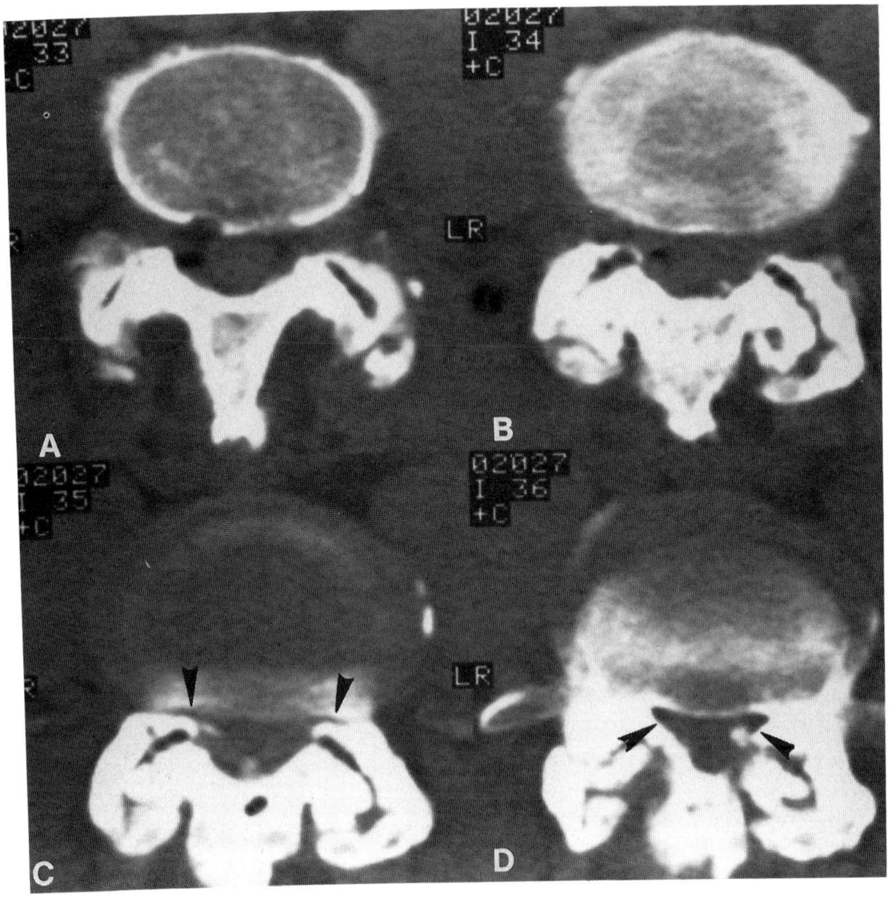

Figure 5.60. Lateral canal stenosis. Contiguous axial CT scans show stenosis of the subarticular canal (*arrow*) resulting from hypertrophy of the articular facets and hypertrophy and calcification of the facet joint capsule. Vacuum phenomenon is present in the facet joint on both sides.

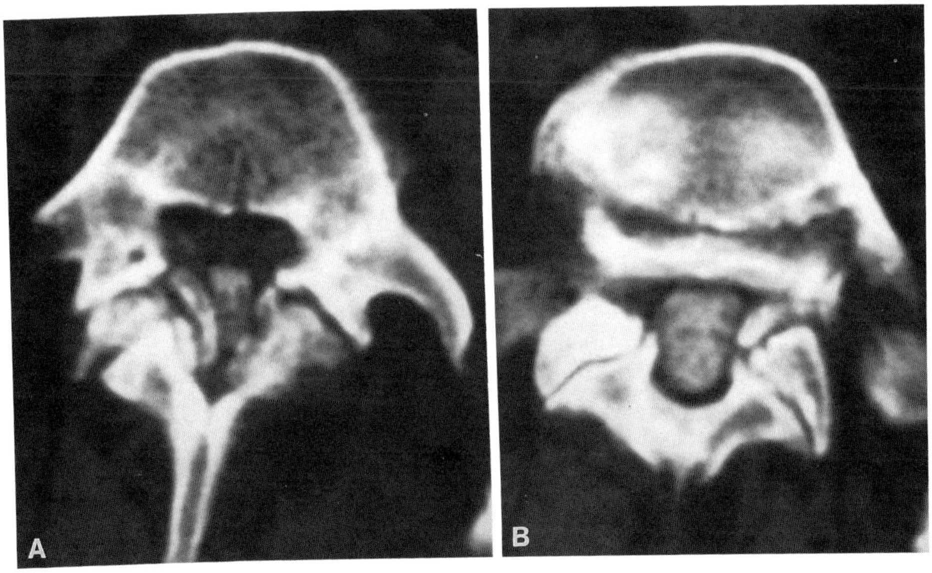

Figure 5.61. Axial CT sections through the L5–S1 level. Bilateral spondylolisis of the L5 posterior neural arch (*arrowhead*). Anterior subluxation of L5 over S1 leads to degenerative changes in the facet joint capsule and hypertrophic osteophytes leading to foraminal stenosis even though the central canal is capacious.

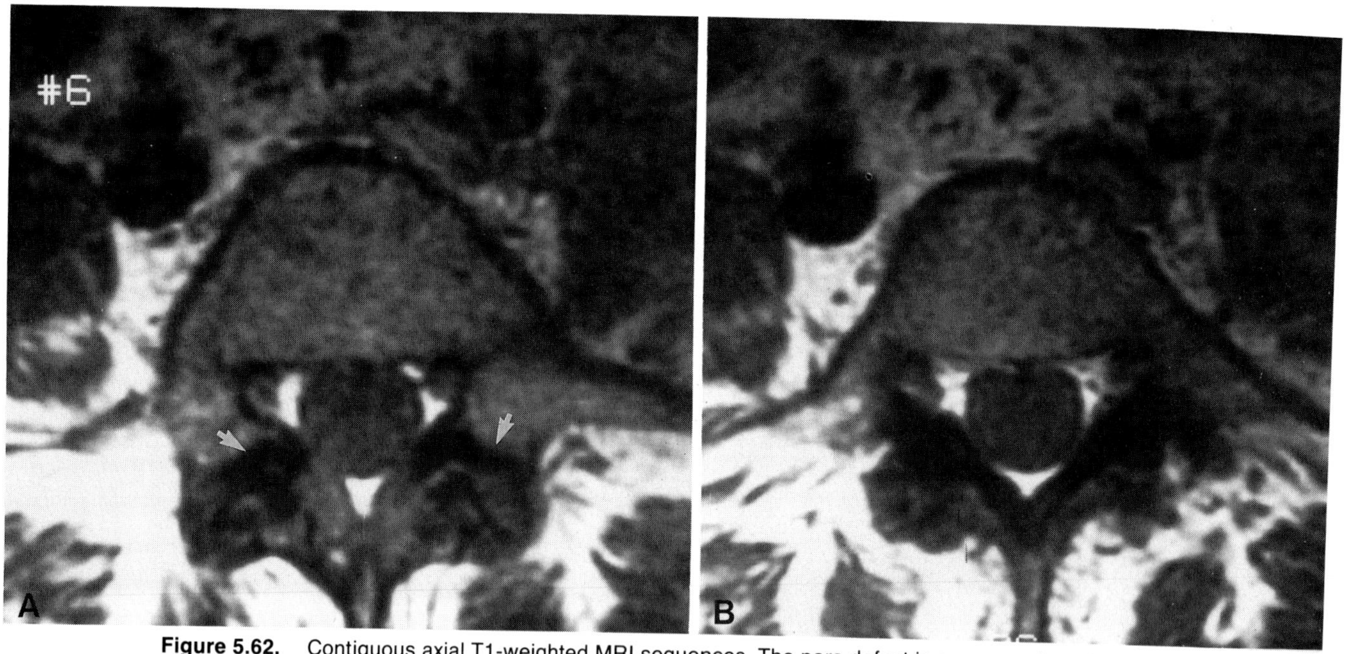

Figure 5.62. Contiguous axial T1-weighted MRI sequences. The pars defect is seen as a hypointense signal anterior to the superior articular facet (*open arrow*).

spinal stenosis from cartilaginous hypertrophy can be confused with meningioma and will often require a Gd-DTPA–enhanced MRI study to exclude this entity.

Similar to OPLL as a cause of spinal stenosis, thickening and ossification of the ligamentum flavum also can result in compressive cord symptoms. Due to similarity in signal characteristics of CSF and calcified ligaments, a focal thickened or ossified ligamentum flavum can be missed on T1-weighted MRI studies. This is best evaluated with CTM [120].

CERVICAL REGION

In the cervical region, stenosis of the spinal canal as a result of degenerative disk disease is more common than disk herniation. This is due to the greater range of motion and less weight bearing compared with the thoracic or lumbar region. Similar to the lumbar region, the degenerative process involves both the discovertebral complex and the apophyseal joints. In most instances it involves the C3 to C6 discovertebral segments, with C4–5 and C5–6 being involved more often than others. Degenerative changes leading to varying severity of spinal stenosis is part of the aging process, usually seen beyond the fifth decade. The process may be accentuated by trauma, certain occupations, preexisting borderline spinal stenosis, or association with OPLL and DISH. Cervical stenosis at a level beyond C3 and including the craniovertebral junction is often seen with congenital anomalies, including Morquio's syndrome, Down syndrome, and a variety of bone dysplasias (Fig. 5.65) (see chap. 3).

Symptoms of myelopathy result from cord compression by the degenerated disk material along with a thickened calcified or ossified posterior longitudinal ligament and from similar changes in the ligamentum flavum. Symptoms are accentuated when the neck is hyperextended. Radicular pain with both motor and sensory changes results from neural canal stenosis, from a combination of hypertrophy of the uncinate process and its capsule, from degenerative changes of the disk material, and from similar changes in the facet joints.

Both CTM and MRI are equally effective in demonstrating the morphologic anatomy in evaluations of the spinal cord and nerve root impingement. Each has its own advantages.

MRI evaluation of cervical spinal stenosis, although more complicated and time-consuming, allows a more detailed evaluation of the soft-tissue anatomy. Imaging in more than one plane is necessary, primarily when evaluating neural canal stenosis. Flow and pulsation artifacts are more common, thus necessitating faster imaging techniques, dedicated surface coils, and techniques to suppress or decrease the flow-related artifacts. In the T1-weighted image, the cord morphology can be evaluated. A focal or diffuse decrease in CSF around the cord is suggestive of spinal stenosis. Since the thickened ligaments, osteophytes, and calcium deposits within the ligaments all appear hypointense, similar to CSF, this needs to be separated from the CSF utilizing a T2-weighted sequence. Whether a conventional spin-echo (SE), gradient-echo (GRE), or fast-spin-echo (FSE) sequence is utilized depends on the desired clinical information, since each has its

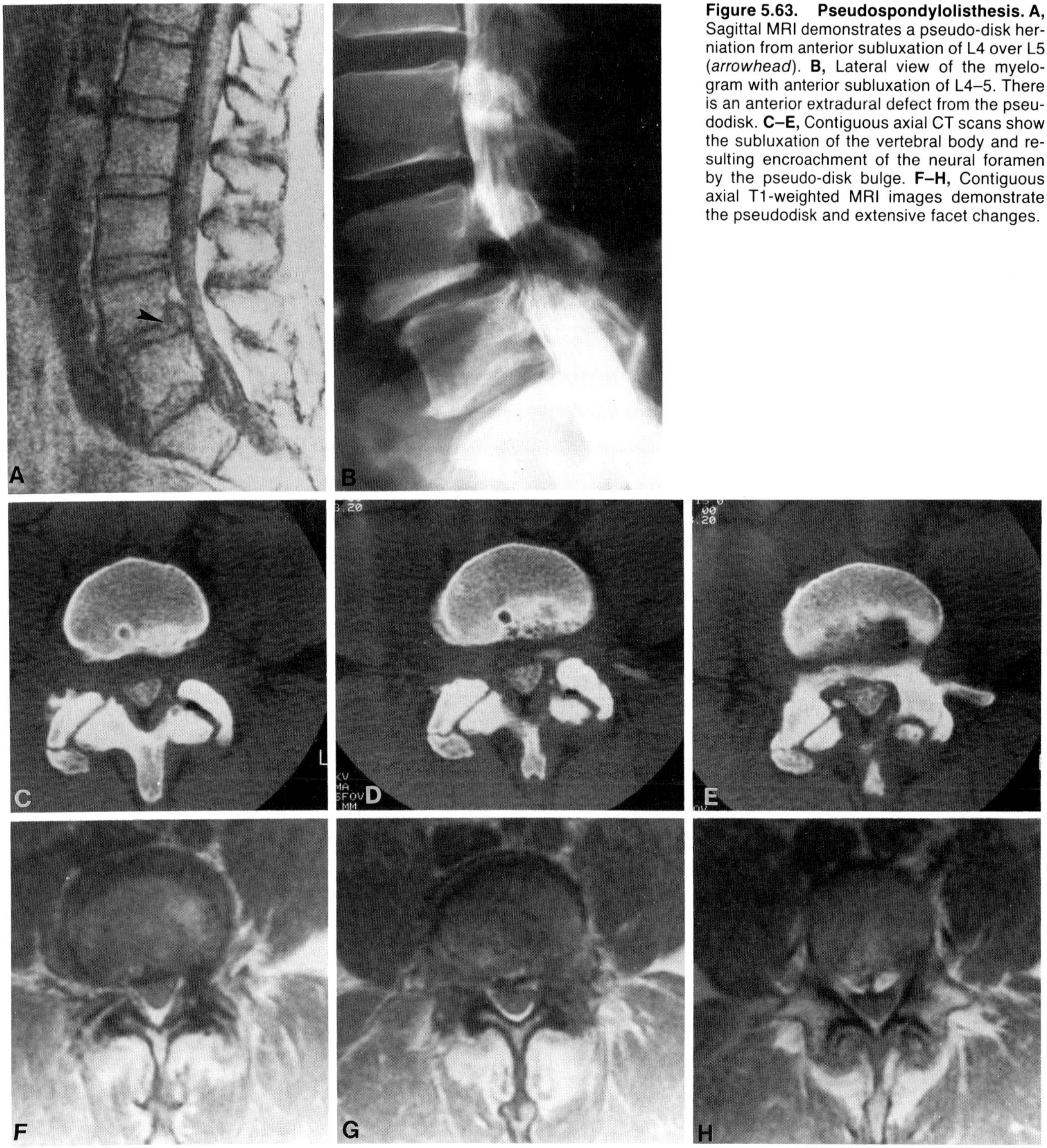

Figure 5.63. Pseudospondylolisthesis. A, Sagittal MRI demonstrates a pseudo-disk herniation from anterior subluxation of L4 over L5 (*arrowhead*). **B,** Lateral view of the myelogram with anterior subluxation of L4–5. There is an anterior extradural defect from the pseudodisk. **C–E,** Contiguous axial CT scans show the subluxation of the vertebral body and resulting encroachment of the neural foramen by the pseudo-disk bulge. **F–H,** Contiguous axial T1-weighted MRI images demonstrate the pseudodisk and extensive facet changes.

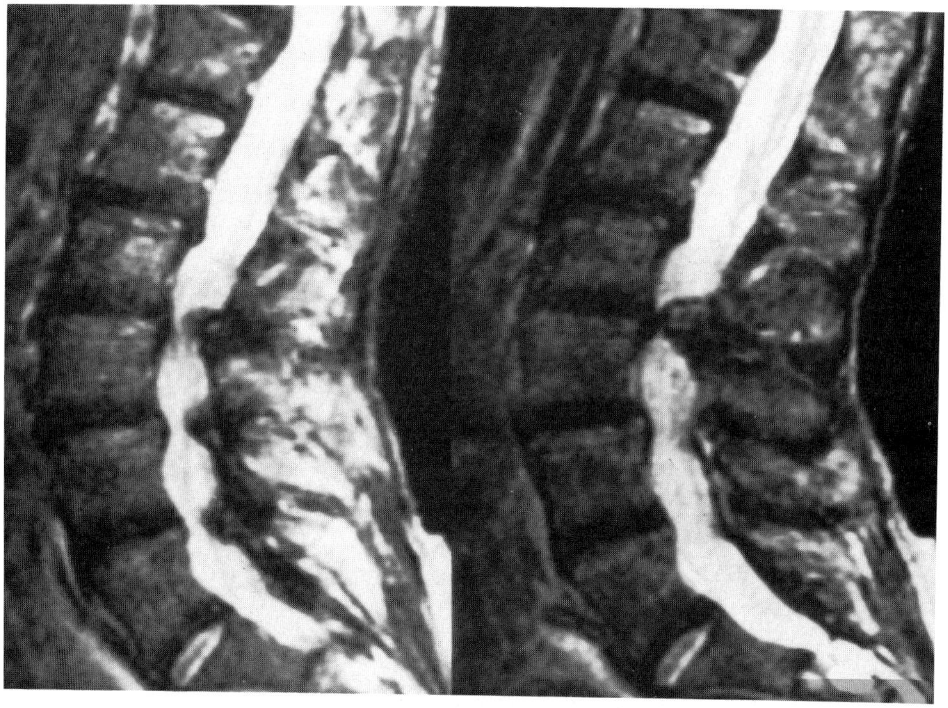

Figure 5.64. Ossification of ligament resulting in spinal stenosis. A, Contiguous T2-weighted sagittal MRI demonstrates ossification of the ligamentum flavum. **B,** Sagittal MRI demonstrates narrow spinal canal with degenerative osteophytes.

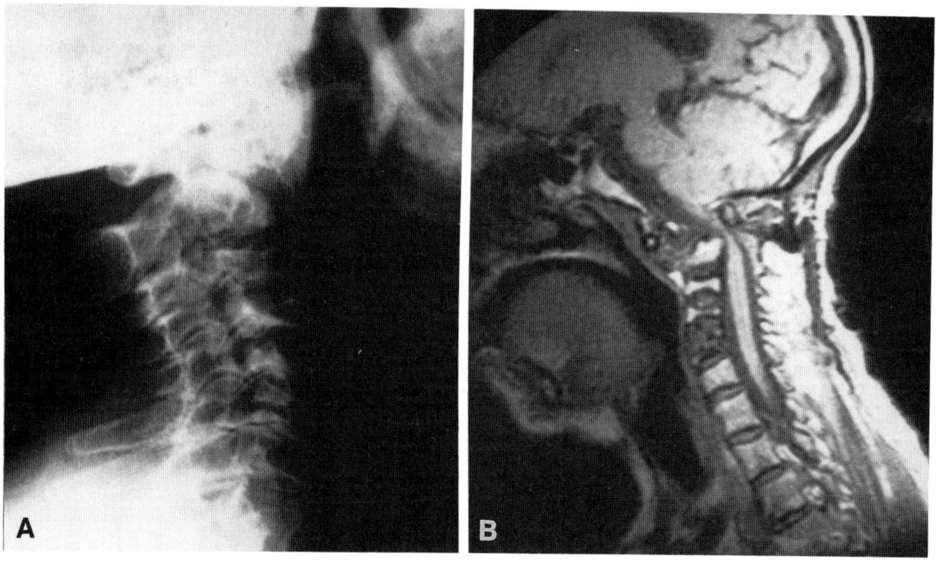

Figure 5.65. A, Lateral x-ray of the cervical spine in Morquio's syndrome with deformity of the cervical spine at the C1–2 level. **B,** Sagittal T1-weighted MRI shows stenosis of the craniovertebral junction. Vertebral body changes involve the upper cervical vertebrae.

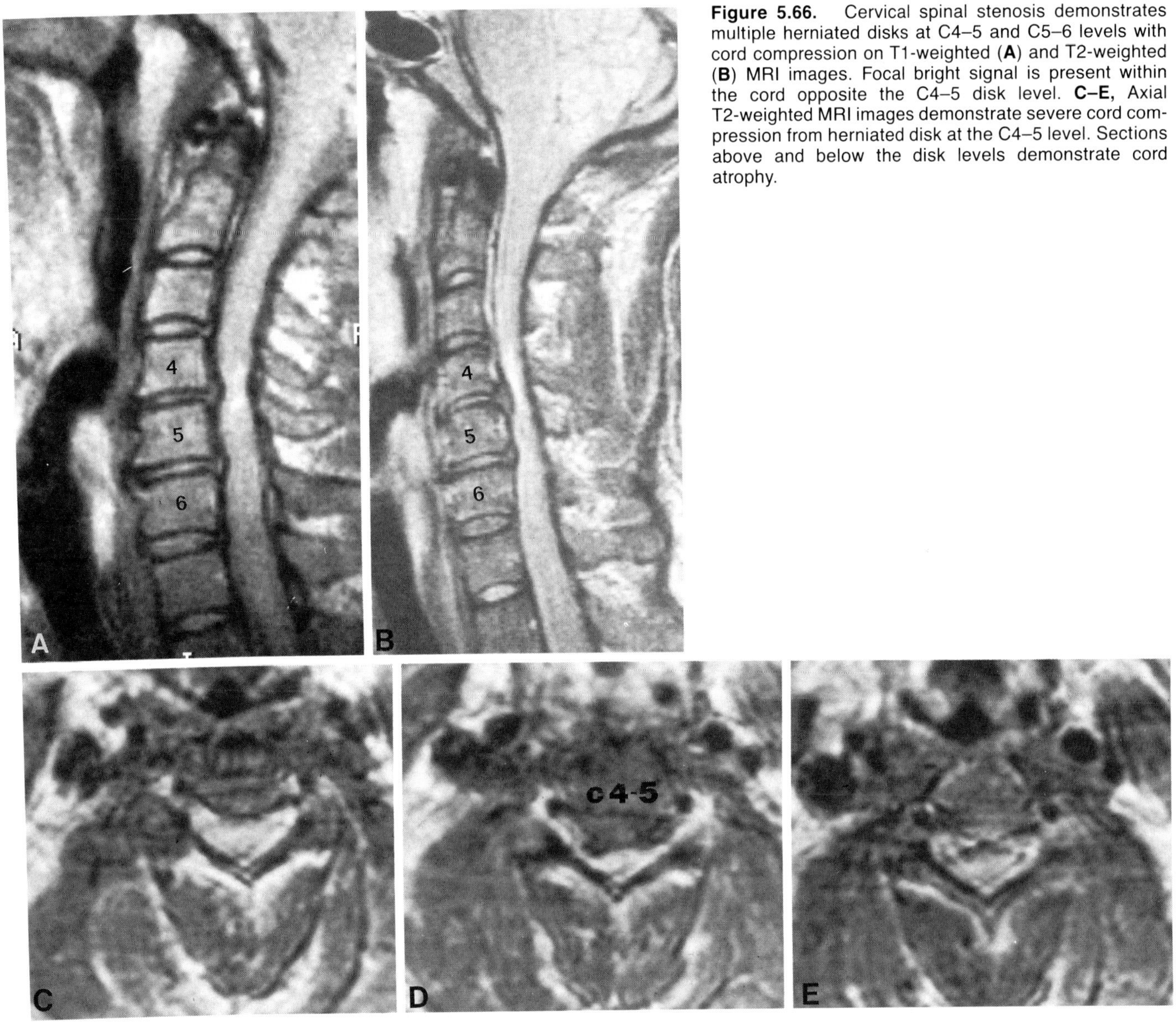

Figure 5.66. Cervical spinal stenosis demonstrates multiple herniated disks at C4–5 and C5–6 levels with cord compression on T1-weighted (**A**) and T2-weighted (**B**) MRI images. Focal bright signal is present within the cord opposite the C4–5 disk level. **C–E,** Axial T2-weighted MRI images demonstrate severe cord compression from herniated disk at the C4–5 level. Sections above and below the disk levels demonstrate cord atrophy.

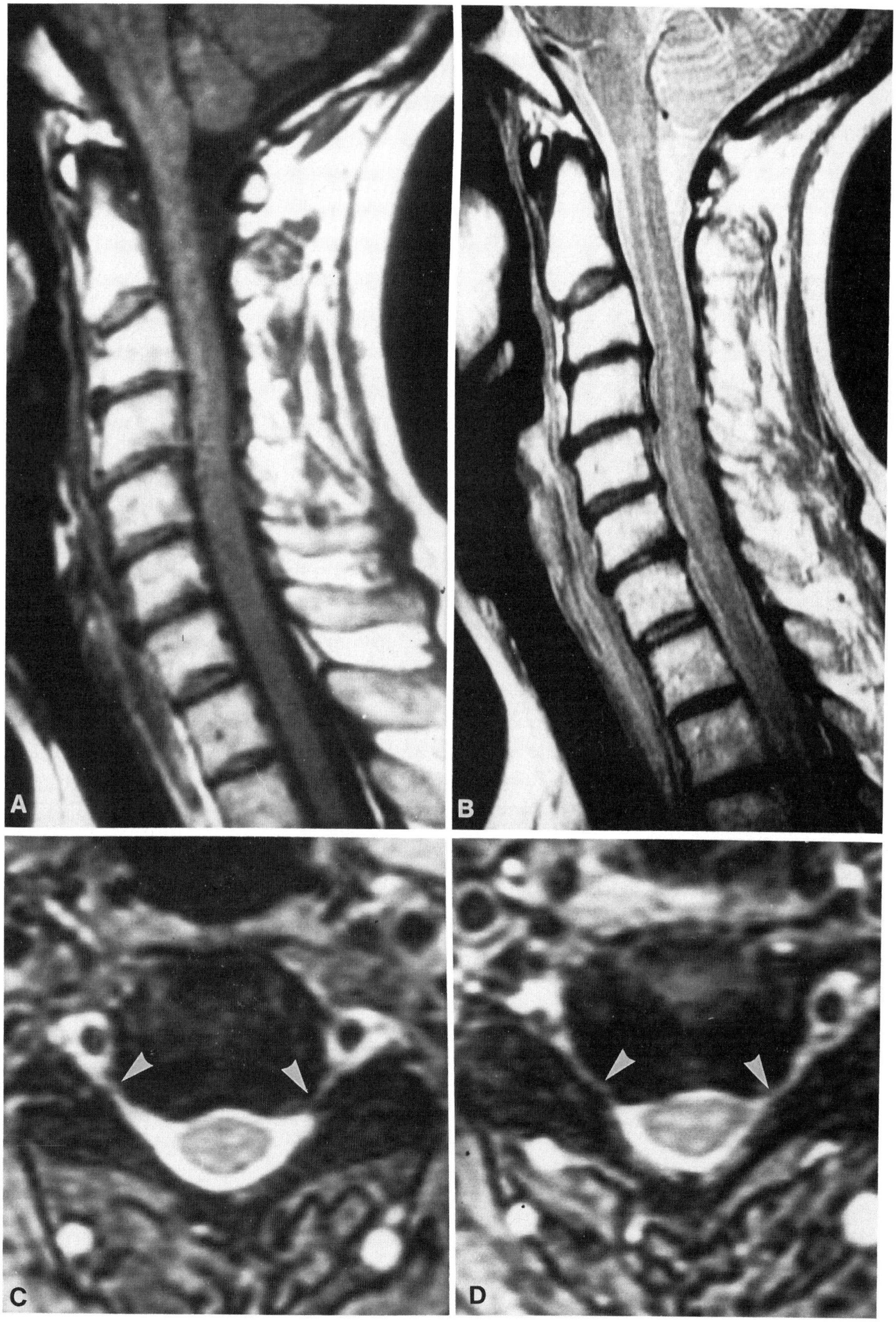

Figure 5.67. Sagittal T1-weighted (**A**) and T2-weighted (**B**) MRI images demonstrate cervical spondylitic disease with mild indentation of the thecal sac at multiple levels between C3 and C7. **C,D,** Contiguous T2-weighted MRI images through the C3–4 disk level. Bilateral neural canal stenosis from degenerative changes of the facet joints and hypertrophy of the uncinate process (*arrowhead*).

own limitations. Routine screening for cervical spinal stenosis usually consists of T1-weighted sagittal and axial sections and a gradient-echo sequence in either the sagittal or axial plane. The axial sections are usually acquired with a volume slab such that thin 2-mm sections can be obtained subsequently. This allows evaluation of both the degenerative osteophytes and the associated disk herniation. Cord compression can be assessed by the decrease in the amount of surrounding CSF and the flattened appearance of the spinal cord. Myelomalacia should be suspected in the presence of focal hyperintense signal within the spinal cord on T2-weighted spin-echo images (Fig. 5.66). The focal hyperintense signal foci represent a combination of ischemia and edema, although small syrinx or cavitation may have a similar appearance [121]. Gd-DTPA–enhanced MRI may be useful in differentiating ischemic necrosis of the cord from cavitary lesions. Since these focal bright foci within the cord are adjacent to the site of severe stenosis, they should not be mistaken for other intrinsic cord lesions.

Foraminal stenosis is evaluated in a similar manner as for a lateral herniated disk with 3DFT gradient-echo imaging with subsequent partition in the axial and parasagittal oblique planes [35, 91]. Foraminal encroachment by a hypertrophied uncinate process or the facet joints appears as hypointense signal (Fig. 5.67). Small lateral disks compressing the nerve roots in the entry zone may be missed. In the appropriate clinical setting, this can be excluded with Gd-DTPA–enhanced MRI [93].

The osteocartilaginous elements contributing to cervical spinal stenosis are best evaluated with CTM. This usually follows myelography or instillation of small amount of contrast material into the thecal space. Contiguous thin (2 to 3 mm) or overlapping slices without tilting the gantry allow for evaluation in both the axial and re-formatted sagittal planes. This is necessary because the height of the disk spaces is smaller, and the neural foramen can be evaluated in its entirety without volume averaging of the adjacent pedicles. Stenosis of the canal from osteophytes or thickened ligaments can be demonstrated. Because of excellent contrast resolution between the soft tissue and the osteocartilaginous elements, CT is excellent in demonstrating the severity and extent of the cord compression, foraminal encroachment, and nerve root impingement (Fig. 5.68). However, a small, laterally placed disk can be missed unless subsequent MRI or IVCT studies are obtained.

Cord compression is suspected when there is marked thinning of the contrast within the subarachnoid space, with flattening or deformity of the spinal cord in both the axial and sagittal planes, similar to MRI. The neural canal may or may not be compromised, depending on the extent of the soft-tissue and bony involvement (Fig. 5.69). Intravasation of contrast material within the cord has been observed on delayed CT studies. It is indicative of cavitation [122] or cystic necrosis [123] secondary to compressive myelopathy resulting from severe cervical spondylitic disease. Faint focal enhancement within the cord in patients with severe cord compression which reverses itself following decompressive laminectomy and improvement in neurologic function also has been observed [124].

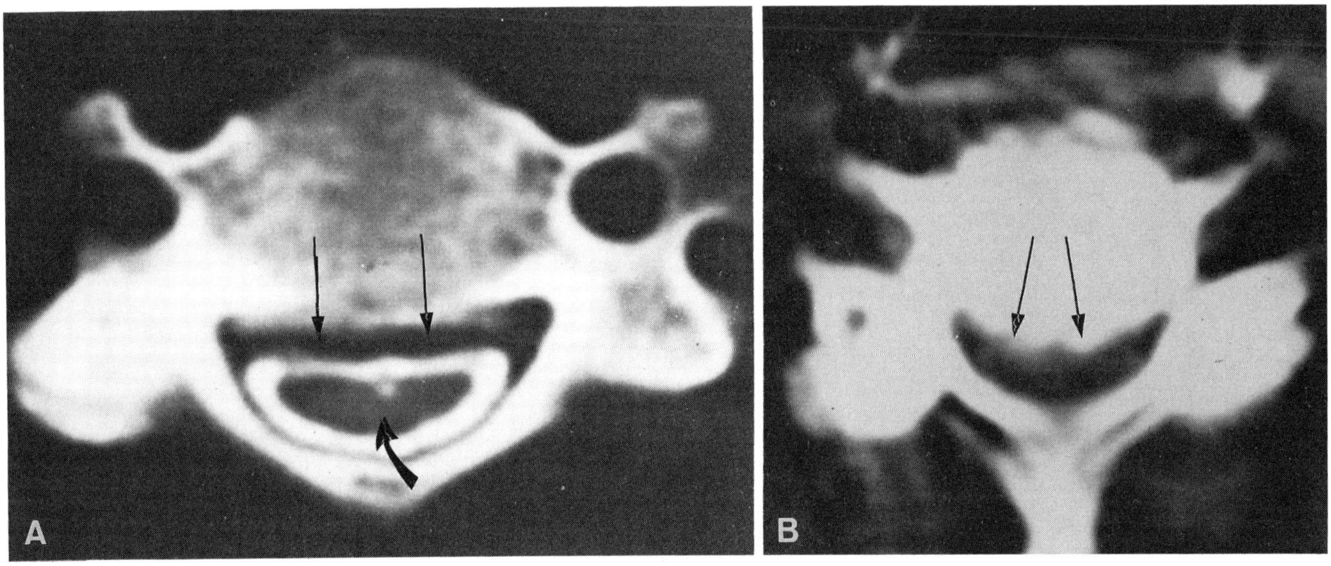

Figure 5.68. **A,** Axial CTM demonstrates cord compression secondary to hypertrophy of the posterior longitudinal ligament (*arrows*). The spinal cord is compressed (*curved arrow*). **B,** Cervical spinal stenosis from osteocartilaginous hypertrophy.

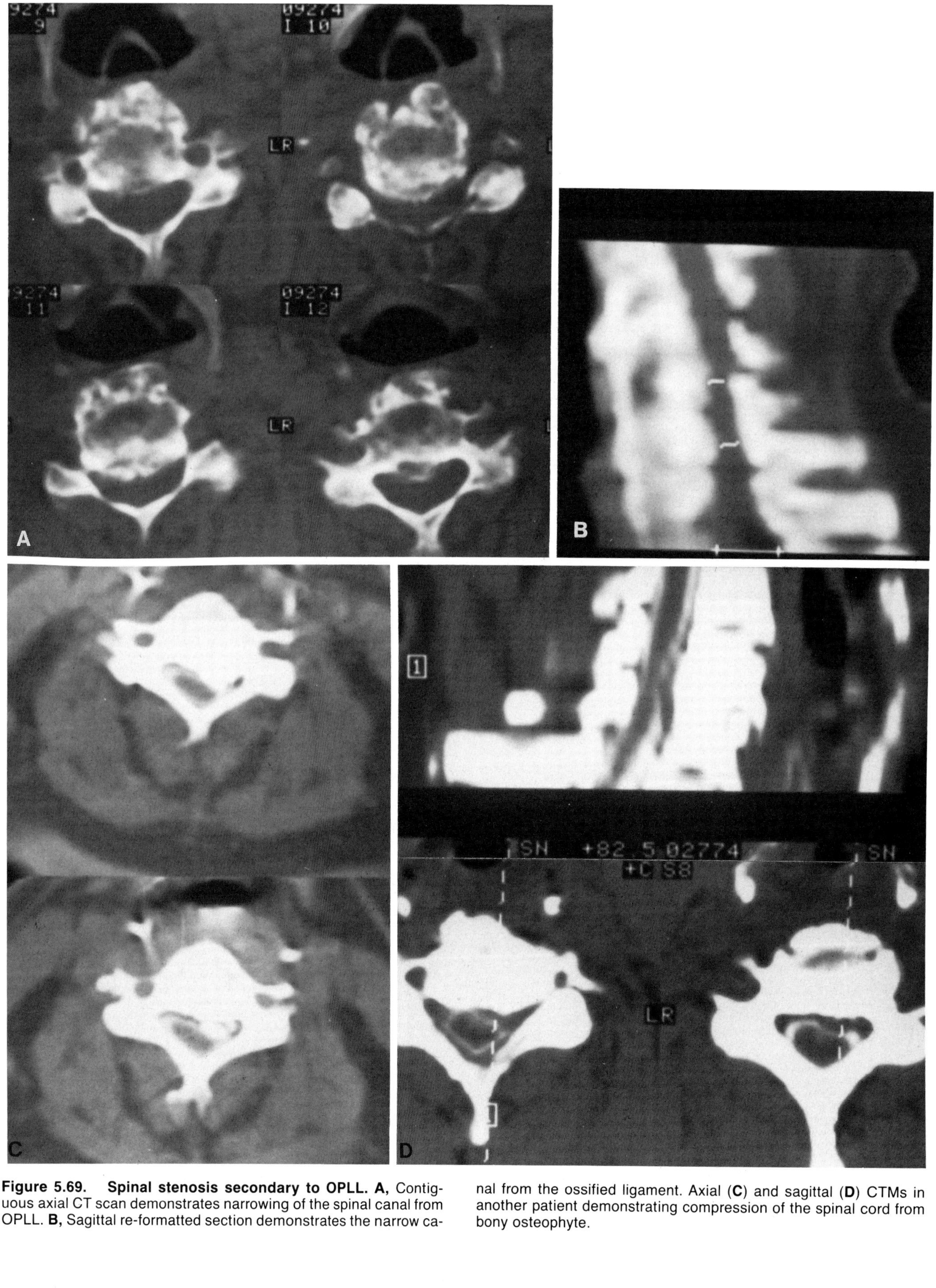

Figure 5.69. Spinal stenosis secondary to OPLL. A, Contiguous axial CT scan demonstrates narrowing of the spinal canal from OPLL. **B,** Sagittal re-formatted section demonstrates the narrow canal from the ossified ligament. Axial (**C**) and sagittal (**D**) CTMs in another patient demonstrating compression of the spinal cord from bony osteophyte.

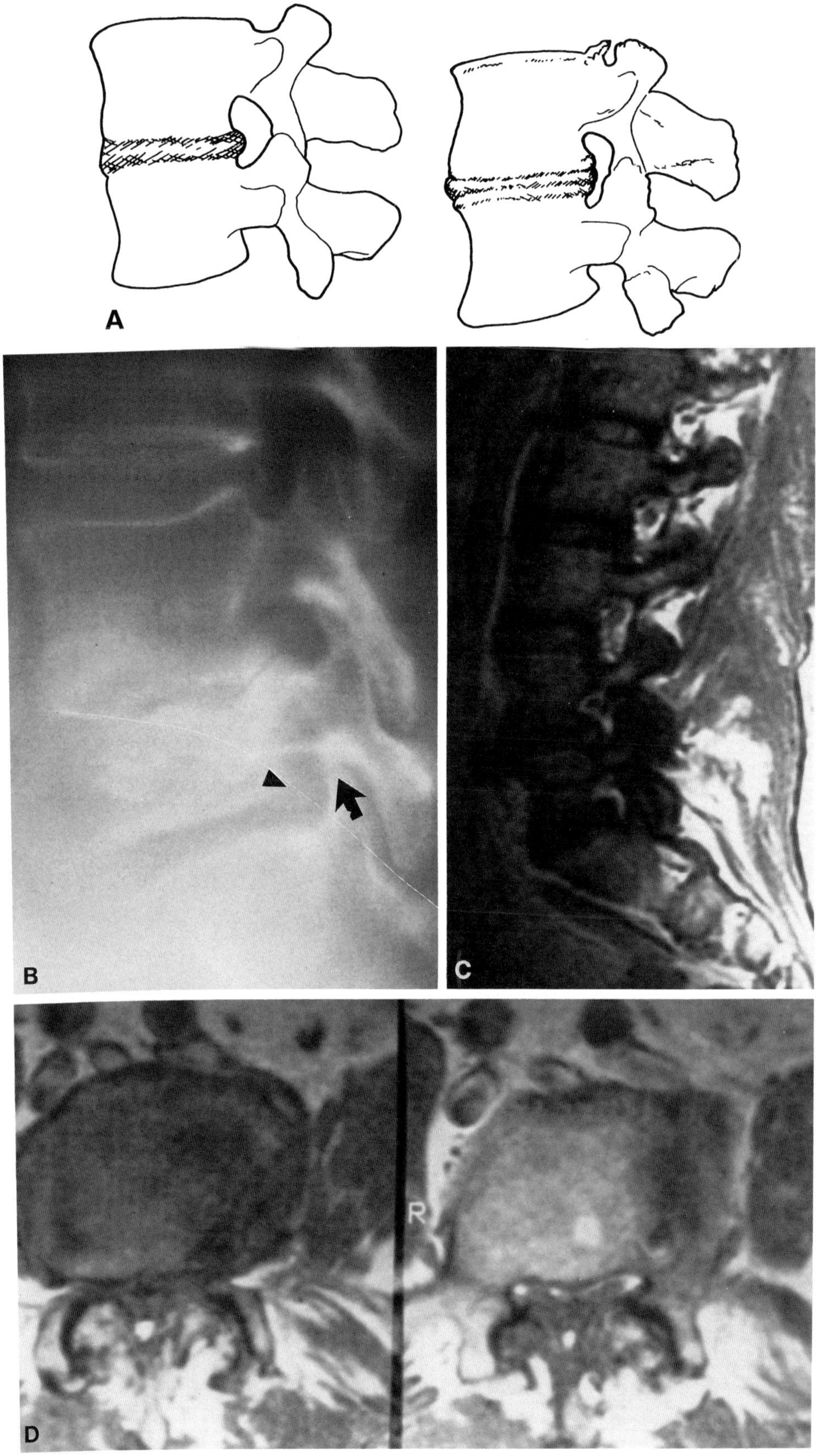

Figure 5.70. **A,** Schematic drawing demonstrating the progressive narrowing of the foramen from degenerative bulging disk and facet joint hypertrophy. **B,** Lateral tomogram of the lumbar spine demonstrates the foraminal stenosis from anterior slippage of the superior articular facet (*arrow*), loss in disk height, and posterior osteophyte (*arrowhead*). Parasagittal (**C**) and contiguous axial (**D**) T1-weighted MRI images demonstrate foraminal and lateral recess stenosis from facet hypertrophy and bulging disk.

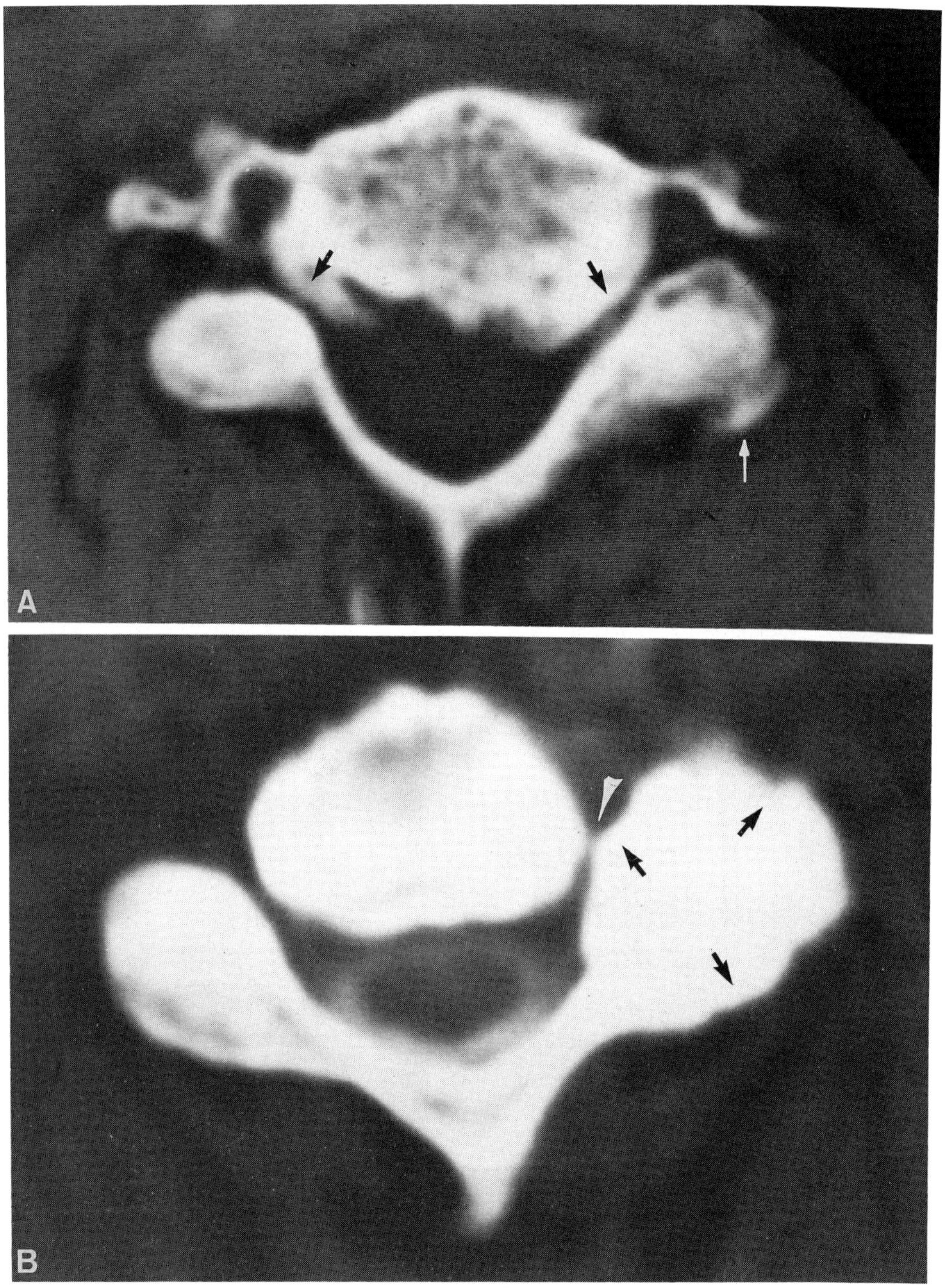

Figure 5.71. **A,** Axial CT scan demonstrates marked hypertrophy of the facet joint (*arrow*) with narrowing of the neural canal (*arrowhead*). **B,** Bilateral foraminal stenosis from facet joint hypertrophy (*white arrow*) and hypertrophy of the uncinate process (*arrow*).

FACET JOINT DISEASE

Facet joint disease is usually a part of the degenerative process leading to spinal stenosis. However severe degenerative facet joint changes with minimal disk disease are a major cause of low back pain [125]. Since the facet joints are synovial lined joints, degenerative changes in the cartilage result in calcification and osteochondral deposits, cyst formation, and entrapment of the synovium and its proliferation. These changes compress the nerve root in the lateral recess (Fig. 5.70). Continued inflammation and fluid buildup within the capsule result in direct irritation of the sensory plexus around the joint, causing focal or diffuse back pain and muscle spasm. Hypertrophic changes in the facet joint primarily involve the superior articular facet and contribute to the lateral recess syndrome. Unusual occupational stresses initially may involve one facet joint and subsequently may implicate the opposite joint. Facet arthropathy and narrowing of the lateral recess on the opposite side are often seen following unilateral facetectomy, foraminotomy, or laminectomy.

Facet arthropathy is the most common cause of foraminal stenosis. This may result from bony osteophytes, anterior slippage of the inferior articulating facet, or buckling of the joint capsule into the foramen, and it leads to narrowing of the upper half of the foramen, compressing the exiting nerve root.

The severity and extent of facet joint arthropathy can be evaluated with CT. The soft-tissue mass and osteocartilaginous component leading to the spinal canal and lateral recess stenosis are easily identified (Fig. 5.71). The presence of vacuum phenomena within the facet joint is evidence of cartilage and synovial degeneration. On MRI, facet hypertrophy is seen as an isointense mass with hypointense signal within and surrounding it representing cartilaginous changes and the capsular ligament, respectively. Fluid within the joint space as well as subchondral cysts appear as bright signals on T2-weighted images. The vacuum phenomenon is seen as a hypointense signal on both T1- and T2-weighted images.

MUSCLE ATROPHY

Atrophy of the paraspinal muscle is often seen in association with degenerative disk disease. This is seen as atrophy of the paraspinal muscles with increased fatty infiltration on both CT and MRI. The atrophic muscles, by not providing adequate support, may contribute to motion instability and to further progressive changes within the ligaments and the discovertebral complex. The role of muscle tone or muscle bulk in degenerative disk disease is not clearly defined.

References

1. Pritzker KPH. Ageing and degeneration in the lumbar intervertebral disk. Orthop Clin North Am 1977;8:65–67.
2. Brown MD. The pathophysiology of disc degeneration. In: Orthop Clin North Am 1970;2:359–370.
3. Teresi LM, Lufkin RB, Reicher MA, et al. Asymptomatic degenerative disk disease and spondylosis of the cervical spine: MR imaging. Radiology 1987;164:83–88.
4. Penning L, Wilmink JT, van Woerden HH, et al. CT myelographic findings in degenerative disorders of the cervical spine: clinical significance. AJNR 1986;7:119–124.
5. Jinkins JR, Whittemore AR, Bradley WG. Anatomic basis of vertebrogenic pain and autonomic syndrome associated with lumbar disk extrusion. AJNR 1989; 10:219–233.
6. Jackson HC, Winkelman RK, Bickel WH. Nerve endings in the human lumbar spinal column and related structures. J Bone Joint Surg 1966;48A:1272–1281.
7. Pederson HE, Blumck CP, Gardner E. The anatomy of the lumbosacral posterior rami and meningeal branches of the spinal nerves (sinuvertebral nerves). J Bone Joint Surg 1956;38A:377–391.
8. Yu S, Haughton VM, Sether LA, et al. Annulus fibrosus in bulging intervertebral disks. Radiology 1988;169: 761–763.
9. Yu S, Sether LA, Ho PSP, et al. Tears of the annulus fibrosus: correlation between MR and pathologic findings in cadavers. AJNR 1988;9:367–370.
10. Yu S, Haughton VM, Ho PSP, et al. Progressive and regressive changes in the nucleus pulposus: II. The adult. Radiology 1988;169:93–97.
11. Lipson SJ, Muir H. Proteoglycans in experimental intervertebral disk degeneration. Spine 1984;6:194–210.
12. Martel W, Seegar JF, Wicks JD, et al. Traumatic lesions in the discovertebral junction in the lumbar spine. AJR 1976;127:457–460.
13. Kirkaldy-Willis WH, Yong-Hing K, Reilly J. Pathology and pathogenesis of lumbar spondylosis and stenosis. Spine 1978;3:319–328.
14. Dorwart RH, Vogler JB, Helms CA. Computed tomography of spinal stenosis. Radiol Clin North Am 1983;21: 301–321.
15. Haughton VM, Syvertsen A, Williams AL. Soft tissue anatomy within the spinal canal as seen on computed tomography. Radiology 1980;134:649–655.
16. Heithoff KB. High resolution computed tomography of the lumbar spine. Postgrad Med 1981;70:193–213.
17. Daniels DL, Grogan JP, Johansen JG, et al. Cervical radiculopathy: computed tomography and myelography compared. Radiology 1984;151:109–113.
18. Ciric I, Mikhael MA, Tarkington JA, et al. The lateral recess syndrome. J Neurosurg 1980;53:433–437.
19. Larsson EM, Holtas S, Cronqvist S, et al. Comparison of myelography, CT myelography and magnetic resonance imaging in cervical spondylosis and disk herniation: pre- and post-operative findings. Acta Radiol 1989;30:233–238.
20. Badami JP, Norman D, Barbaro NM, et al. Metrizamide CT myelography in cervical myelopathy, radiculopathy: correlation with conventional myelography and surgical findings. AJNR 1986;6:59–64.
21. Russell EJ, D'Angelo CM, Zimmerman RD, et al. Cervical disk herniation: CT demonstration after contrast enhancement. Radiology 1984;152:703–712.
22. Russell EJ. Cervical disk disease. Radiology 1990;177: 313–325.
23. Baleriux D, Noterman J, Ticket L. Recognition of cervical soft disk herniation by contrast enhanced CT. AJNR 1983;4:607–611.
24. Krause D, Drape JL, Maitrot D, Woerly B, Tongio J, eds. CT and MRI of disk herniations. Heidelberg: Springer-Verlag, 1991.
25. Modic MT, Masaryk T, Boumphrey F, et al. Lumbar herniated disk and canal stenosis: prospective evaluation by surface coil MR, CT and myelography. AJNR 1986; 7:709–717.
26. Modic MT, Masaryk TJ, Mulopulos GP, et al. Cervical radiculopathy: prospective evaluation with surface coil MR imaging, CT with metrizamide and metrizamide myelography. Radiology 1986;161:753–759.
27. Brown BM, Schwartz RH, Frank E, et al. Preoperative evaluation of cervical radiculopathy and myelopathy by surface coil MR imaging. AJNR 1988;9:859–866.
28. Axel LL. Surface coil magnetic resonance imaging. J Comput Assist Tomgr 1984;8:381–384.
29. Edelman RR, Atkinson DJ, Silver MS, et al. FRODO pulse sequences: a new means of eliminating motion, flow and wrap around artifacts. Radiology 1986;166: 231–236.
30. Enzmann DR, Rubin JB. Cervical spine MR imaging with a partial flip angle gradient refocussed pulse sequence: I. General considerations and disk disease. Radiology 1988;166:467–471.

31. Haacke EM, Lenz G. Improving MR image quality in the presence of motion using rephasing gradient. AJR 1987; 148:1251–1258.
32. Hedberg MC, Drayer BP, Flom RA, et al. Gradient echo (GRASS) MR imaging in cervical radiculopathy. AJNR 1988;9:145–151.
33. Modic MT, Pavlicek W, Weinstein MA, et al. Magnetic resonance imaging of intervertebral disk disease: clinical and pulse sequence considerations. Radiology 1984;152:103–111.
34. Vandyke C, Ross JS, Tkach J, et al. Gradient-echo MR imaging of the cervical spine: evaluation of extradural disease. AJNR 1989;10:627–632.
35. Tasurda JS, Norman D, Dillon W, et al. Three dimensional gradient-recalled MR imaging as a screening tool for the diagnosis of cervical radiculopathy. AJNR 1989;10:1263–1271.
36. Enzmann DR, Rubin JB, Wright A. Use of cerebrospinal fluid gating to improve T2-weighted images: I. Spinal cord. Radiology 1987;162:763–767.
37. Rubin JB, Wright A, Enzmann DR. Lumbar spine: motion compensation for cerebrospinal fluid on MR imaging. Radiology 1988;166:225–231.
38. Maravilla KR, Lash HP, Weinreb JC, et al. Magnetic resonance imaging of the lumbar spine with CT correlation. AJNR 1985;6:237–245.
39. Pech P, Haughton VM. Lumbar intervertebral disk: correlative MR and anatomic study. Radiology 1985;156: 699–701.
40. Enzmann DR, Griffin C, Rubin JB. Potential false negative MR images of the thoracic spine in disk disease with switching of phase and frequency-encoding gradients. Radiology 1987;165:635–639.
41. Aoki J, Yamamoto I, Kitamura N, et al. Endplate of the discovertebral joint: degenerative changes in the elderly adult. Radiology 1987;164:411–414.
42. deRoos A, Kressel H, Spritzer C, et al. MR imaging of marrow changes adjacent to end plates in degenerative lumbar disk disease. AJR 1987;149:531–534.
43. Modic MT, Steinberg PM, Ross JS, et al. Degenerative disk disease: assessment of changes in vertebral body marrow with MR imaging. Radiology 1988;166:193–199.
44. Yu S, Haughton VM, Sether LA, et al. Comparison of MR and diskography in detecting radial tears of the annulus: a postmortem study. AJNR 1989;10:1077–1081.
45. Abdelwahab IF, Gould ES: The role of diskography after negative post myelographic CT scans: retrospective view. AJNR 1988;9:187–190.
46a. Gibson MD, Buckley J, Maukinney R, et al. Magnetic resonance imaging and discography in the diagnosis of disc degeneration. Comparative study of 50 discs. J Bone Joint Surg 1986;68:364–373.
46b. Schneidermann G, Flannigan B, Kingston S, et al. Magnetic resonance imaging in the diagnosis of disk degeneration: correlation with discography. Spine 1987; 12:276–291.
47. Shapiro R. Myelography. 3d ed. Chicago: Year Book Publishers, 1975.
48. Frymoyer JW: Back pain and sciatica. N Engl J Med 1988;318:291–300.
49. Resnick D, Niwayama G. Intervertebral disk herniations: cartilagenous (Schmorl's) nodes. Radiology 1978; 107:89–97.
50. Resnick D. Degenerative disease of the vertebral column. Radiology 1985;156:3–14.
51. Knutsson F. The vacuum phenomenon in the intervertebral disks. Acta Radiol 1942;23:173–179.
52. Grenier N, Grossman RL, Schiebler ML, et al. MR imaging spectrum of vacuum phenomena in degenerative lumbar disk disease: pitfalls and usefulness. Radiology 1987;164:861–865.
53. Taveras JM. Herniated intervertebral disks: A plea for a more uniform terminology [Editorial]. AJNR 1989;10: 1283–1284.
54. Williams AL, Haughton VM, Meyer GA, et al. Computed tomographic appearance of the bulging annulus. Neuroradiology 1982;142:403–408.
55. Yu S, Haughton VM, Sether LA, et al. Criteria for classifying normal and degenerated intervertebral disks. Radiology 1989;170:523–526.
56. Ross JS, Modic MT, Masaryk TJ, et al. Tears of the annulus fibrosus: assessment with Gd-DTPA enhanced MR imaging. AJNR 1989;10:1251–1254.
57. McCulloch JA, McNab I. Selection of patients for chemoneucleolysis: In McCulloch JA, McNab I, eds. Sciatica and chymopapain. Baltimore: Williams & Wilkins, 1983:103–127.
58. Onik G, Helms CA, Ginsberg L, et al. Percutaneous lumbar diskectomy using new aspiration probe. AJR 1985;144:1137–1140.
59. Williams AL, Haughton VM, Syvertsen A. Computed tomography in the diagnosis of herniated nucleus pulposus. Radiology 1980;135:95–99.
60. Williams AL, Haughton VM, Daniels DL, et al. CT diagnosis of lateral herniated disk. AJNR 1982;3:95–99.
61. Grenier N, Greselle JF, Vtal JM, et al. MR features of foraminal and extraforaminal lumbar herniated disks: surgical correlation in 26 cases. Presented at ASNR 1989 [Abstract]. AJNR 1989;10:890.
62. Ross JS, Modic MT, Masaryk TJ, et al. Assessment of extradural disk disease with Gd-DTPA enhanced MR imaging: correlation with surgical and pathological findings. AJNR 1989;10:1243–1250.
63. Bonneville JF, Runge M, Catlin F, et al. Extraforaminal disk herniation: CT demonstration of Sharpey's fibers avulsion. Neuroradiology 1989;31:71–74.
64. Novetsky GJ, Berlin L, Epstein AJ, et al. Extraforaminal herniated disk: value of computed tomography. AJNR 1982;3:653–656.
65. Osborne AG, Hood RS, Sherry RG, et al. CT/MR spectrum of far lateral and anterior lumbo-sacral disk herniations. AJNR 1988;9:775–778.
66. Schellinger D, Manz HI, Braislav V, et al. Disk fragment migration. Radiology 1990;175:831–836.
67. Firooznia H, Benjamin V, Kricheff II, et al. CT of lumbar spine disk herniation: correlation with surgical findings. AJNR 1984;5:91–96.
68. Dublin AB, McGahan JP, Reid MH. The value of computed tomographic metrizamide myelography in the neuroradiological evaluation of the spine. Radiology 1983;146:79–86.
69. Masaryk TJ, Ross JS, Modic MT, et al. High resolution MR imaging of sequestered lumbar intervertebral disks. AJNR 1988;9:351–358.
70. Graves VB, Finney HL, Mailander J. Intradural lumbar disk herniation. AJNR 1985;7:495–499.
71. Hodge CJ, Binet EF, Keiffer SA. Intradural herniation of lumbar intervertebral disk. Spine 1978;3:346–350.
72. Holtas S, Nordstrom CH, Larsson EM, et al. Case reports: MR imaging of intradural disk herniation. J Comput Assist Tomogr 1987;11:353–355.
73. Wasserstrom R, Mamourian AC, Black JF, et al. Intradural lumbar disk fragment with ring enhancement on MR. AJNR 1993;14:401–404.
74. Guinto FC Jr, Hashim H, Stumer M: CT demonstration

of disk regression after conservative therapy. AJNR 1984;5:632–633.
75. Teplick JG, Haskin ME. Spontaneous regression of herniated nucleus pulposus. AJNR 1985;6:331–335.
76. Bazzao A, Gallucci M, Masciocchi C, et al. Lumbar disk herniation: MR imaging assessment of natural history in patients treated without surgery. Radiology 1992; 185:135–141.
77. Sal JA, Saal JS, Herzog RJ. The natural history of lumbar intervertebral disc extrusion treated nonoperatively. Spine 1990;15:683–686.
78. Pradatschar K, Fiore DL, Barbiero A. The natural history of lumbar disc herniations assessed by a CT follow-up study. Neuroradiology 1991;33(suppl 1):84–85.
79. Ryan RW, Lally JF, Kozic Z. Asymptomatic calcified thoracic disk: CT recognition. AJNR 1988;9:363–369.
80. Arce CA, Dohrmann GJ. Thoracic disc herniation: improved diagnosis with computed tomographic scanning and a review of the literature. Surg Neurol 1985; 23:356–361.
81. Ross JS, Perez-Rayes N, Masaryk TJ, et al. Thoracic disk herniation: MR imaging. Radiology 1987;165:511–515.
82. Alvarez O, Roque CT, Pampati M, et al. Multilevel thoracic disk herniations: CT and MR studies. J Comput Assist Tomogr 1988;12:650–655.
83. Benson MKD, Byrnes DF. The clinical syndromes and surgical treatment of thoracic intervertebral disk prolapse. J Bone Joint Surg 1975;57(A)(4):471–477.
84. Awwad EE, Martin DS, Smith KR Jr. The nuclear trail sign in thoracic herniated disk. AJNR 1992;13:137–143.
85. Francavilla TL, Powers A, Dina T, et al. MR imaging of thoracic disk herniations. J Comput Assist Tomogr 1987;11:1062–1065.
86. Parizel PM, Rodesch G, Baleriaux D, et al. Gd-DTPA enhanced MR in thoracic disk herniation. Neuroradiology 1989;31:75–78.
87. Epstein BS. The spine: a radiological text and atlas. 4th ed. Philadelphia: Lea & Febiger, 1967.
88. Edelman RR, Stark DD, Saini S, et al. Oblique plane of section in MR imaging. Radiology 1986;159:807–810.
89. Haughton VM, Fenerich DO, Oblique plane MR imaging of the cervical spine. J Comput Assist Tomogr 1987; 10:823–826.
90. Modic MT, Masaryk TJ, Ross JS, et al. Cervical radiculopathy: value of oblique MR imaging. Radiology 1987; 163:227–231.
91. Youssem DM, Atlas SW, Goldberg HI, et al. Degenerative narrowing of the cervical spine neural foramina: evaluation with high resolution 3DFT gradient echo MR imaging. AJNR 1991;12:229–236.
92. Sze G, Merriam M, Oshio K, et al. Fast spin-echo imaging in evaluation of intradural disease of the spine. AJNR 1992;13:1383–1392.
93. Ross JS, Ruggieri JP, Tkach JA, et al. Gd-DTPA enhanced 3D MR imaging of cervical degenerative disk disease: initial experience. AJNR 1992;13:127–136.
94. Hemminghytt S, Daniels D, Williams AL, et al. Intraspinal synovial cysts: natural history and diagnosis by CT. Radiology 1982;145:375–376.
95. Casselman ES. Radiologic recognition of symptomatic spinal synovial cysts. AJNR 1985;6:971–974.
96. Patel SC, Sanders WP. Synovial Cyst of the cervical spine: case report and review of the literature. AJNR 1988;9:602–603.
97. Quaghebaur G, Jeffree M. Synovial cyst of the high cervical spine causing myelopathy. AJNR 1992;13:981–983.
98. Mercader J, Gomez JM, Cardenal C. Intraspinal synovial cyst: diagnosis by CT. Neuroradiology 1985;27: 346–348.
99. Gorey NT, Hyman RA, Black KS, et al. Lumbar synovial cysts eroding bone. AJNR 1992;13:161–163.
100. Jackson DE, Atlas SW, Mani JR, et al. Intraspinal synovial cysts: MR imaging. Radiology 1989;170:527–530.
101. Liu SS, Williams KD, Drayer BP, et al. Synovial cysts of the lumbosacral spine: diagnosis by MR imaging. AJNR 1989;10:1239–1242.
102. Selbergleit R, Gebarski SS, Brunberg JA, et al. Lumbar synovial cysts: correlation of myelographic, CT, MR, and pathologic findings. AJNR 1990;11:777–779.
103. Retrum CE, Schmidlin TM, Taylor WK, et al. CT myelography of extradural pigmented villonodular synovitis. AJNR 1987;8:727–729.
104. Titlebaum DS, Rhodes CH, Brooks JS, et al. Pigmented villonodular synovitis of a lumbar facet joint. AJNR 1992;13:164–166.
105. Savitz MH, Katz SS, Goldstein HB, et al. Hypertrophic synovitis of the facet joint forming a paraarticular mass in cases of herniated intervertebral disk. Spine 1987;12:509–510.
106. Helms CA, Dorwart RH, Gray M. The CT appearance of conjoint nerve roots and differentiation from a herniated nucleus pulposus. Radiology 1982;144:803–807.
107. Hoddick WK, Helms CA. Bony spinal changes that differentiate conjoined nerve roots from herniated nucleus pulposus. Radiology 1985;154:119–120.
108. Bouchard JM, Copty M, Langelier R. Pre-operative diagnosis of conjoined nerve anomaly with herniated lumbar disk. Surg Neurol 1978;10:229–231.
109. Yang WC, Zappula R, Malis L. Neurilemmoma in intervertebral foramen. J Comput Assist Tomogr 1981;5: 904–906.
110. Masaryk TJ, Boumphrey F, Modic MT, et al. Effects of the chemonucleolysis demonstrated by MR imaging. J Comput Assist Tomogr 1986;10:917–923.
111. Sobel D, Zyroff J, Thorne RP. Diskogenic vertebral sclerosis: MR imaging. J Comput Assist Tomogr 1987; 11:855–857.
112. Hayes CW, Jensen ME, Conway WF. Non-neoplastic lesions of the vertebral bodies: findings in magnetic resonance imaging. Radiographics 1989;9:883–903.
113. Verbeist H. Neurogenic intermittent claudication in cases with absolute and relative stenosis of the lumbar vertebral canal (ASLC & RLSC), in cases with narrow lumbar intervertebral foramina, and in cases with both entities. Clin Neurosurg 1962;20:204–214.
114. Quint DJ, Patel SC, Sanders WP, et al. Importance of absence of CSF pulsation artifacta in the MR detection of significant myelographic block at 1.5 T. AJNR 1989; 10:1089–1095.
115. Lee BCP, Kazam E, Newman AD. Computed tomography of the spine and spinal cord. Radiology 1978;128:95–102.
116. Epstein BS, Epstein JA, Jones MD. Degenerative spondylolisthesis with an intact neural arch. Radiol Clin North Am 1977;15:275–292.
117. Hiramatsu Y, Nobechi T. Calcification of the posterior longitudinal ligament of the spine in the Japanese. Radiology 1971;100:307–312.
118. Murkhani J, Russell WJ, Hayabuchi, et al. Computed tomography of posterior longitudinal ossification: its appearance and diagnostic value with special reference to thoracic lesions. J Comput Assist Tomogr 1982; 11:41–50.

119. Leutehans JJ, Coughlin BF, Weinstein MR. Ossification of the posterior longitudinal ligament by MR. AJNR 1987;8:924–925.
120. Stollman A, Pinto R, Benjamin V, et al. Radiologic imaging of symptomatic ligamentum flavum thickening with and without ossification. AJNR 1987;8:991–994.
121. Al-Mefty O, Harkey LB, Middleton TH, et al. Myelopathic cervical spondylytic lesions demonstrated by magnetic resonance imaging. J Neurosurg 1988;68:217–222.
122. Iwasaki Y, Tashiro K, Kikuchi S, et al. CT myelography with intramedullary enhancement in cervical spondylosis. J Neurosurg 1985;63:363–368.
123. Jenkins JR, Bashir R, Al-Mefty O, et al. Cystic necrosis of the spinal cord in compressive myelopathy: demonstration by iopamidol CT-myelography. AJNR 1986; 7:693–701.
124. Chakera T. Personal communication, 1985.
125. Schellinger D, Weber L, Ragsdale BD, et al. Facet joint disorders and their role in production of back pain and sciatica. Radiographics 1987;7:923–944.

CHAPTER / 6

Postoperative Spine

K. C. V. G. Rao / J. P. Williams

Imaging of the postoperative spine is usually indicated when there is persistence or recurrence of preexisting symptoms. The symptoms may persist in the immediate postoperative period, within the first few weeks or months, or may be delayed, presenting months or years following the surgery. Selection of the appropriate imaging modalities depends on the region of the spine involved, the presenting symptoms, and suspected clinical diagnosis, as well as the presence of metallic prosthetic devices or materials that may be present. The most common reason for imaging studies is the failed back surgery syndrome (FBSS) following surgery for disk herniation or spinal stenosis (Table 6.1) and less often for all the other early and delayed complications of spine surgery. These include infection, arachnoiditis, dural tears, pseudomeningocele, spinal stenosis, or spinal instability. If spine surgery has been performed for intradural lesions, imaging studies are required to evaluate for tethering of the cord, residual or recurrent intradural lesions, and arachnoiditis.

Failed Back Surgery Syndrome

The *failed back surgery syndrome* (FBSS) refers to patients who have recurrence of their debilitating symptoms following previous back surgery either for spinal stenosis or herniated disk. Symptoms develops months to years later following their initial surgery. More than one-third of these patients have persistence or recurrence of their symptoms following the surgery [1], and many of them are functionally incapacitated, requiring further investigations. Although the surgical procedure may differ, the symptom complex in most such patients is similar. It has been presumed that in most cases, FBSS is the result of an inadequate initial diagnosis due to the limited diagnostic capabilities in the past or of inadequate or improper selection of the surgical procedure [2]. Diagnosis has improved with direct visualization of the spine and its contents on high-resolution CT and MRI, as well as the use of nonionic intrathecal contrast material with CTM, and intravenous Gd-DTPA–enhanced MRI. Images can be evaluated not only in the axial cross-sectional plane but also in other orthogonal planes. On CT, this is achieved with multiplanar re-formatted images [3], whereas MRI allows imaging in any orthogonal plane. CT or CT following myelography (CTM) allows evaluation of most postoperative alterations of the

Table 6.1.
The Postoperative Spine

Early Complications
1. Epidural hematoma
2. Residual disk herniation
3. Infection
4. Dural tear
Late Complications
1. Epidural fibrosis
2. Recurrent disk herniation
3. Recurrent or residual disk herniation
4. Spinal stenosis
5. Arachnoiditis
6. Facet joint arthropathy
7. Postfusion syndrome
Miscellaneous
1. Pseudomeningocele
2. Pseudoarthrosis
3. Cord cavitation
4. Cord atrophy

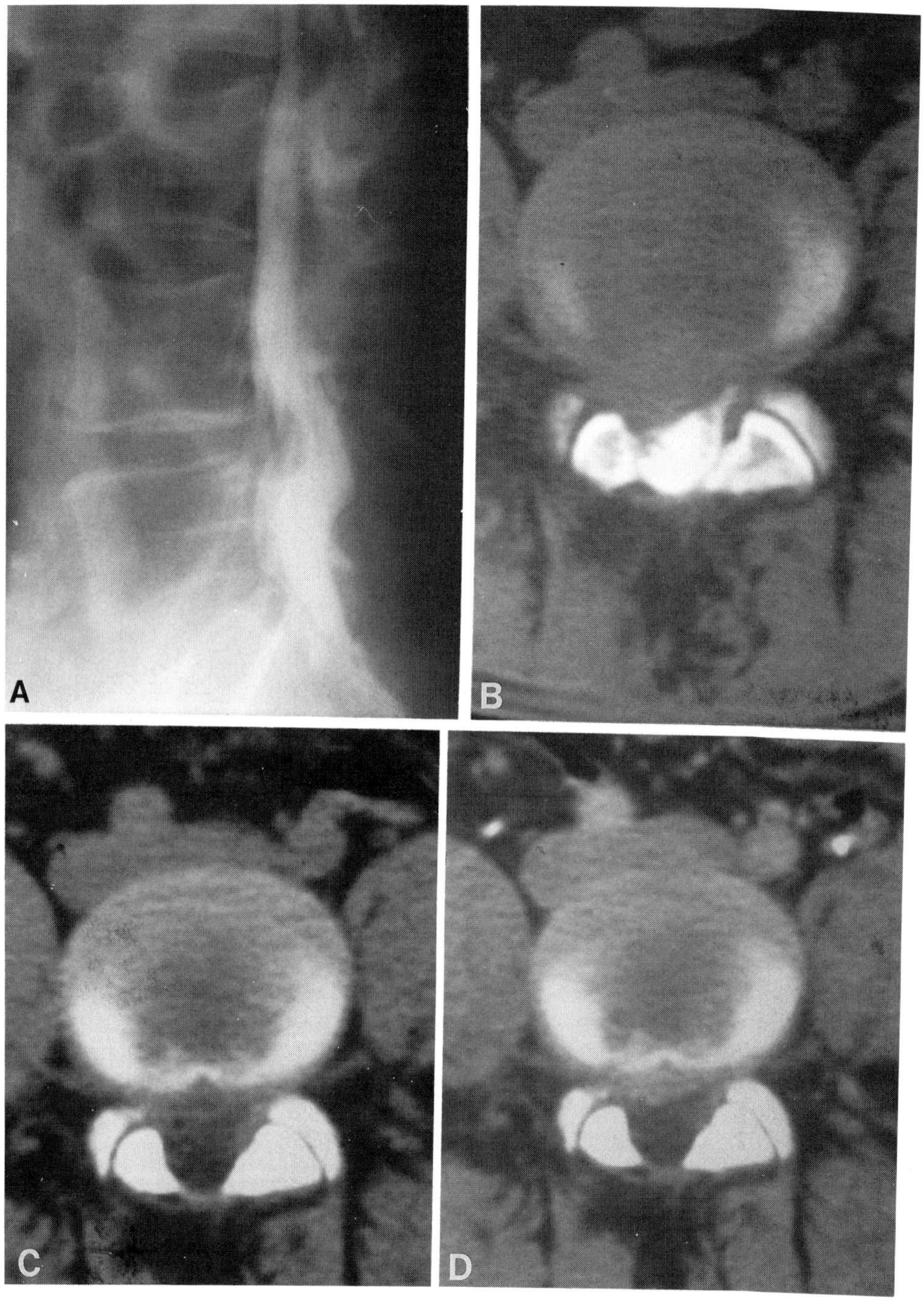

Figure 6.1. Appearance of epidural scar on different imaging modalities in a 41-year-old patient with recurrence of symptoms following back surgery for disk herniation. **A,** Lateral view of lumbar myelogram. A large extradural soft-tissue component displaces the contrast-containing thecal sac at the level of previous laminectomy (L4–5). This could result from disk herniation or epidural scar. The disk space is narrowed. **B,** Axial MCT through this region demonstrates the mildly hyperdense extradural mass displacing the thecal sac and the nerve root sleeve. The densities of the extradural mass and the parent disk are similar. Thus differentiation of herniated disk from epidural scar is difficult. Non-contrast-enhanced and (**C**) intravenous contrast-enhanced (**D**) CT scans at a similar level demonstrate homogeneous enhancement of the previously hyperdense extradural mass. The posterior margin of the enhancing scar is irregular and sharply demarcated from the thecal sac. The scar extends into the neural foramen. This is the characteristic appearance of the majority of the postoperative scars on IVCT. **E,F,** Precontrast axial T1-weighted MRI images. The epidural scar and the CSF-containing thecal sac have a similar hypointense signal. Lack of normal epidural fat does not allow separation of the thecal sac from the epidural scar. **G,** Axial Gd-DTPA–enhanced T1-weighted MRI image with enhancement of the epidural scar.

spine [4, 5]. However, differentiation of epidural fibrosis from recurrent or residual disk herniations is difficult (Fig. 6.1). Prior to the availability of MRI, this was resolved with CT following intravenous contrast enhancement (IVCT) [6–11]. MRI [12–16] and augmentation of MRI with paramagnetic contrast material (Gd-DTPA) [17–19] have replaced IVCT in this evaluation.

In evaluating the postoperative spine, preliminary frontal and lateral views in the upright position are usually indicated. Plain radiographs allow evaluation of the level and type of surgical procedure initially performed. In addition, flexion and extension views in the lateral plane may be indicated to exclude mechanical instability resulting from either the surgical procedure or facet degeneration. Unlike CT and MRI, which are performed in the supine position, plain radiographs can be obtained in the upright weight-bearing position, allowing for evaluation of alignment of the vertebral bodies and the presence of anterior or posterior subluxation as well as scoliosis. Radiographic studies are also useful in evaluating large posterolateral bone grafts and are essential for the accurate interpretation of interbody

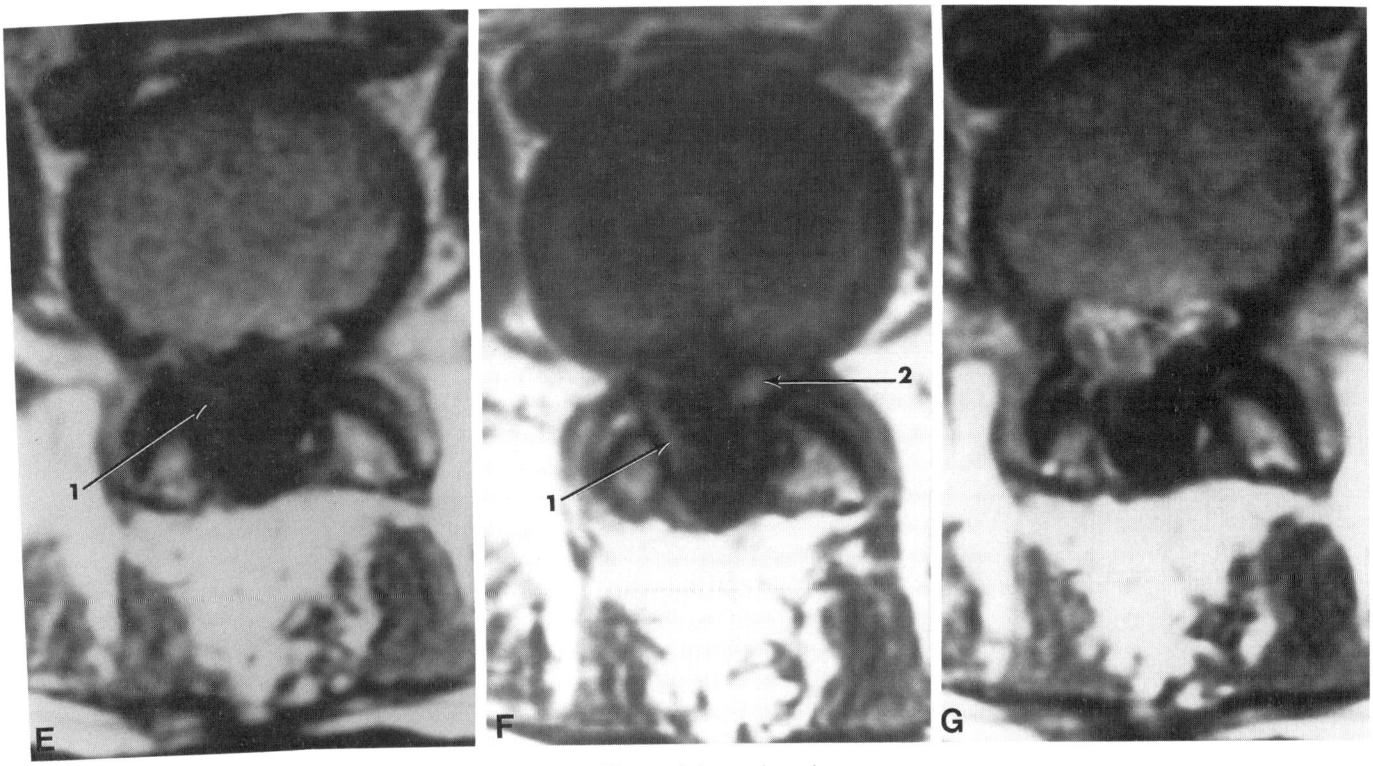

Figure 6.1–*continued*

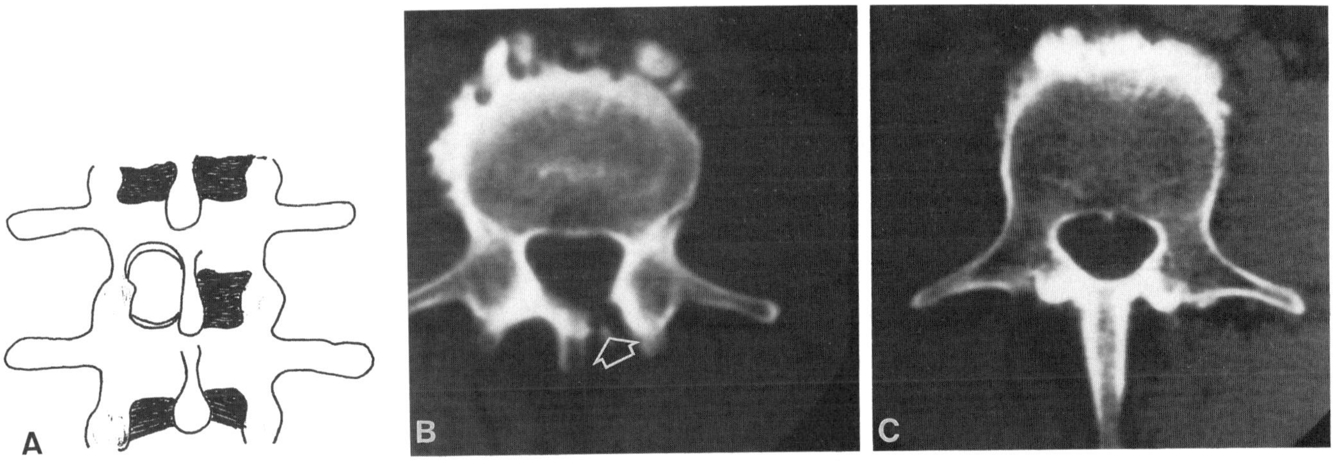

Figure 6.2. **A,** Schematic drawing of spinal laminotomy. **B,C,** Contiguous axial CT sections demonstrate the laminotomy (*arrow*) with preservation of the upper portion of the laminae.

fusion, particularly in the lumbar region. Destruction of the bony endplates of the vertebral bodies, whether due to surgery or infection, although detectable on CT and MRI studies, is best determined by preliminary evaluation of plain x-rays.

Postoperative Changes

In the lumbar region, the most common surgical treatment for intervertebral disk herniation is discectomy. The exposure to the disk space is obtained by a posterior approach between the two laminae. The interlaminal space may be increased by partially removing the undersurface of the lamina-laminotomy (Fig. 6.2) or by complete removal of the lamina on the side of the disk-hemilaminectomy (Fig. 6.3). Depending on the preoperative evaluation, hemilaminectomy may be combined with partial removal of the medial aspect of the superior facet to create more space for the passage of the nerve root (Fig. 6.4). Bilateral removal of laminae at one or more levels is usually seen in patients operated on for central spinal stenosis (Fig. 6.5), whereas widening of the foramina by facetectomy is usually indicated when

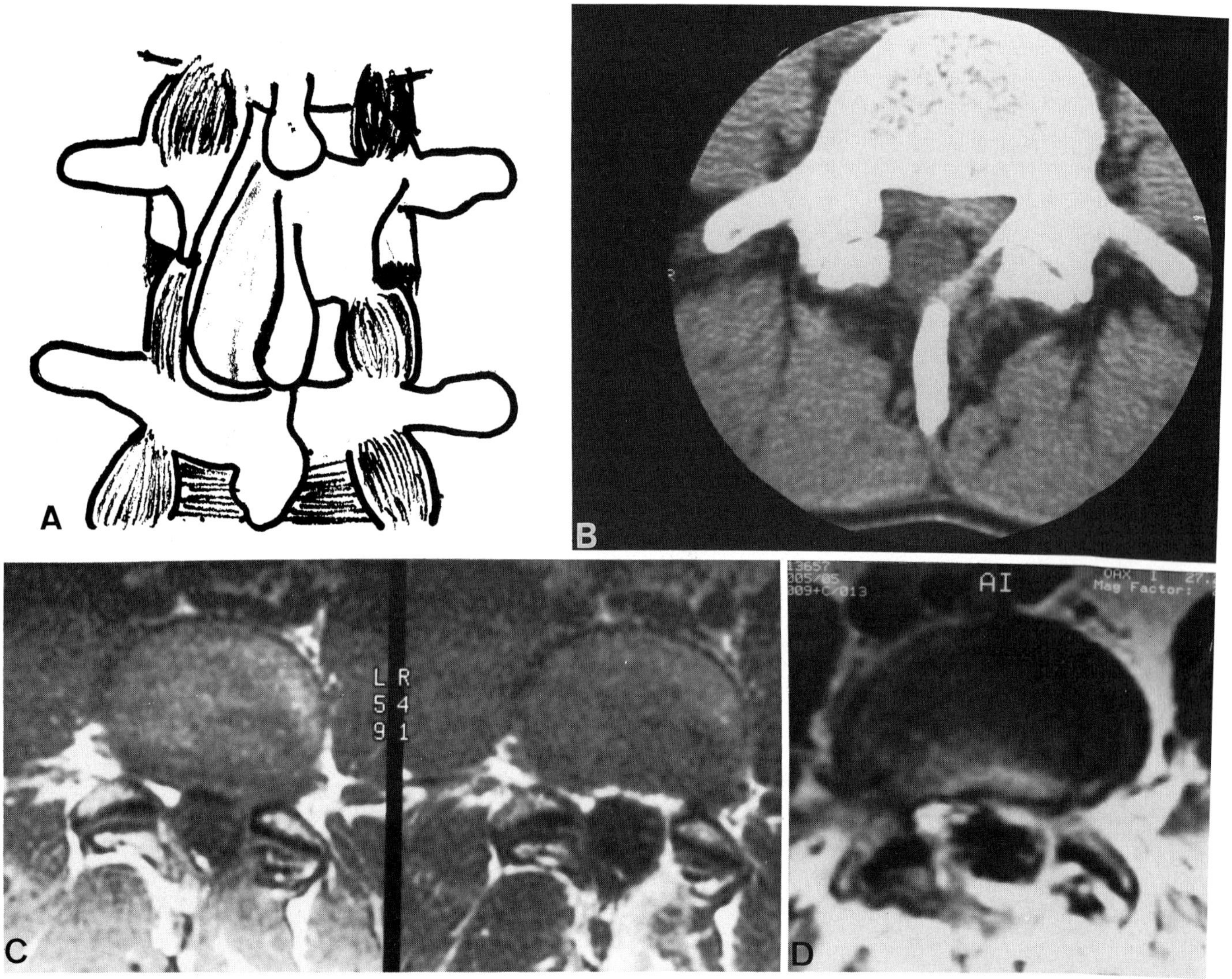

Figure 6.3. **A,** Schematic line drawing showing complete removal of the lamina on one side. This allows better exposure to the anterior epidural space. **B,** Axial CT section with absence of the lamina and ligamentum flavum on the right side. Precontrast (**C**) and Gd-DTPA–enhanced (**D**) T1-weighted MRI images in another patient demonstrate the absent lamina on the left side with the epidural scar extending posteriorly beyond the laminectomy defect. There is homogeneous enhancement of the epidural posterolateral fibrous tissue and within the facet joint.

there is foraminal or neural canal narrowing. The ligamentum flavum is sacrificed. The thecal sac and the nerve root are separated from the herniated disk, and the disk material both within the spinal canal and within the intervertebral space is curettaged and removed. Following hemostasis, occasionally a fat graft may be placed around the released nerve root [20–22]. This is usually to prevent compression of the nerve root from subsequent scar tissue.

During the process of discectomy there is a variable amount of epidural venous bleeding, which probably accounts for the variation in the subsequent epidural fibrosis and scar tissue formation. The amount of the epidural fibrous tissue and scarring also depends on the type of the surgical exploration and laminectomy [23]. In the normal course of events, healing and repair occur with the blood products being absorbed, followed by fibroblasts and macrophages replacing the clot, usually within 48 to 72 hours. In the next 5 to 7 days, immature capillaries develop, and at the same time collagen is laid in the matrix, forming the granulation tissue. Simultaneous contraction of the myofibroblasts with gradual decrease in the proliferating vasculature results in a firm scar within a few months. The scar is relatively avascular. However, if for any reason an irritant focus persists in the region, whether due to mechanical causes, such as further disk herniation, biochemical alteration, or infection, the neovascularity within the scar tissue persists or may recur at a later stage. Discrete calcification within the scar tissue or in close proximity to the junction of the scar tissue and the enchondral bone margin is not unusual. Normal scar tissue has minimal water content, since it

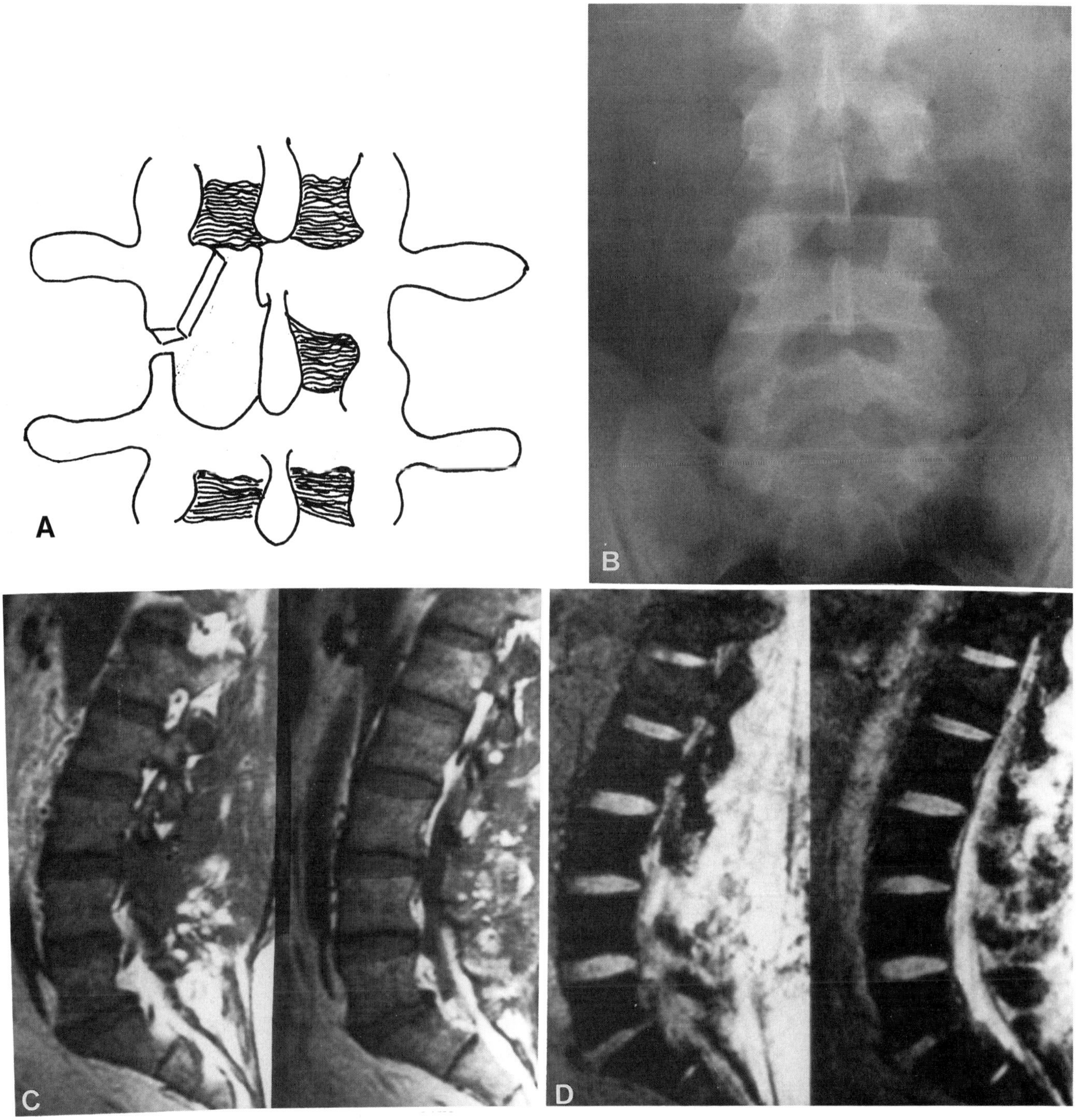

Figure 6.4. **A,** Schematic drawing of hemilaminectomy with facetectomy. **B,** Frontal view of plain x-rays with absence of the lamina and facet (*arrow*) T1-weighted MRI section demonstrates hemilaminectomy with facetectomy for exposure to the herniated disk within the neural foramen. T1-weighted (**C**), T2-weighted (**D**) sagittal, and T1-weighted axial (**E**) precontrast MRI sequences. Sagittal (**F**) and axial (**G**) sections following Gd-DTPA. Enhancement is noted within the epidural scar and along the surgical site. The remnant of the annular fibers has a hypointense signal. Minimal enhancement is also noted within the residual foraminal herniated disk (*open arrow*).

consists of closely packed collagen tissue, whereas granulation tissue has a higher water content. Enhancement of the epidural scar or fibrous tissue with intravenous contrast material depends on various factors, including the maturity of the scar tissue and the neovascularity within the scar tissue. In most instances, a herniated disk, since it is avascular, does not show this characteristic homogeneous enhancement. This difference in the attenuation value or signal characteristics of the disk on CT and MRI, respectively, is used in differentiating herniated disk from epidural fibrosis.

Facetectomy with or without laminectomy is often performed when there is narrowing of the lateral re-

Figure 6.4–*continued*

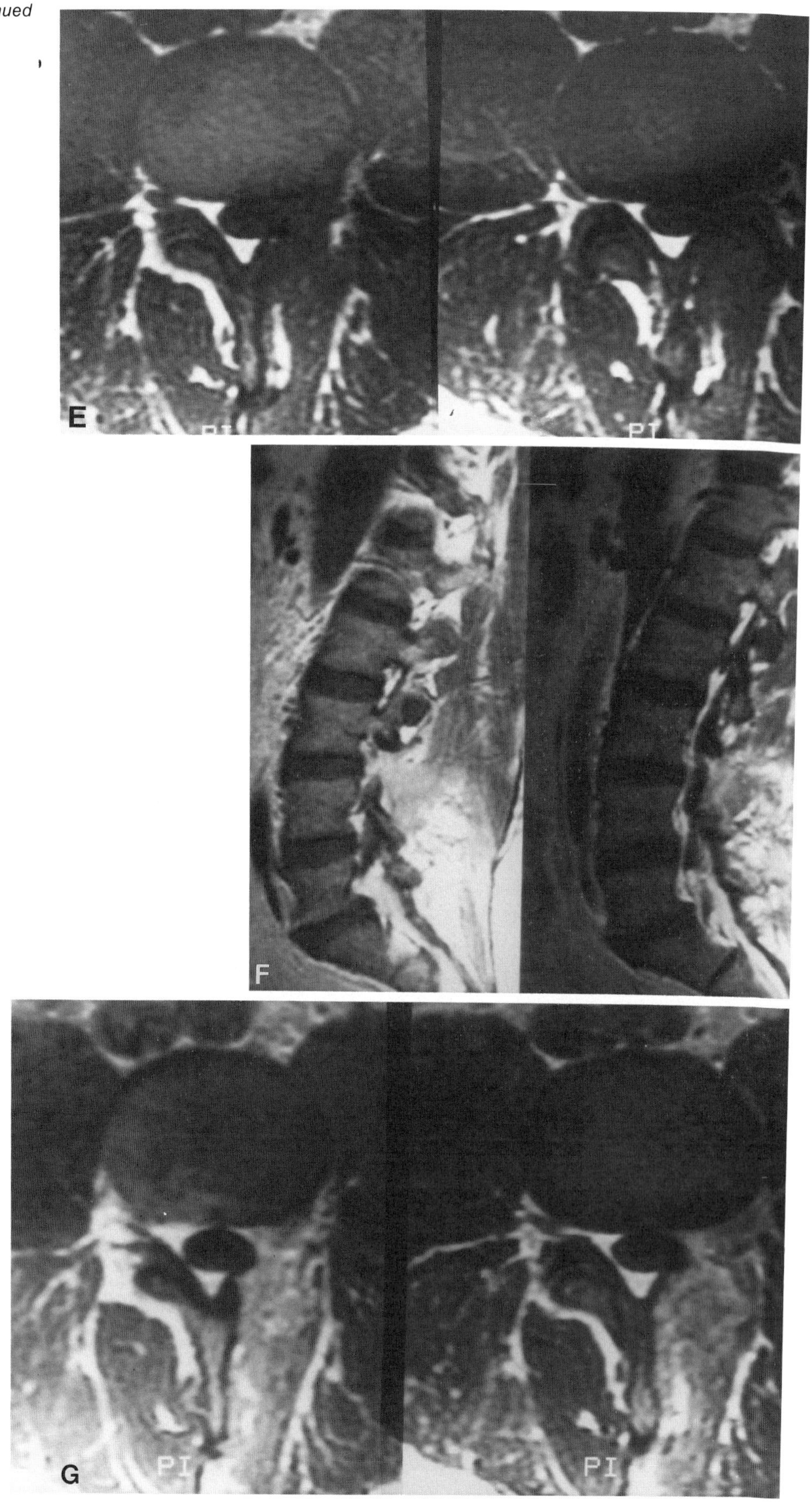

cess, most often involving the facet joint, or when the herniated disk is within the neural foramina. The laminectomy is extended with removal of the medial surface of the superior facet of the involved facet joint (see Fig. 6.4).

Fusion of the adjacent vertebral bodies to provide stability includes anterior interbody fusion utilizing either bone grafts (Fig. 6.6) or, less often, metal plates or posterolateral paraspinal fusion with bone struts between the adjacent pars and transverse processes. When more than one level needs to be stabilized, posterolateral fusion is performed utilizing bone grafts or other prosthetic devices.

In the thoracic region, postoperative changes following discectomy are different from those in the lumbar region. Thoracic discectomy is performed utilizing the lateral approach following costotransversectomy. This involves removal of the pedicle or posterolateral portion of the adjacent vertebral bodies, including the disk space (Fig. 6.7). This surgical approach allows for a stable facet joint and intact connecting ligaments between the lamina and spinous process, preventing subsequent kyphosis. Epidural scarring is less common. Multilevel laminectomies may be performed when dealing with intradural lesions such as neoplasms or vascular anomalies. Metallic prosthetic devices are often utilized in the treatment of scoliosis or following spinal trauma. These may hinder evaluation with CT or MRI.

In the cervical region, discectomy may involve either an anterior or a posterior approach, although it is usually performed by the anterior approach with or without bone grafts [24] (Fig. 6.8). On plain x-rays

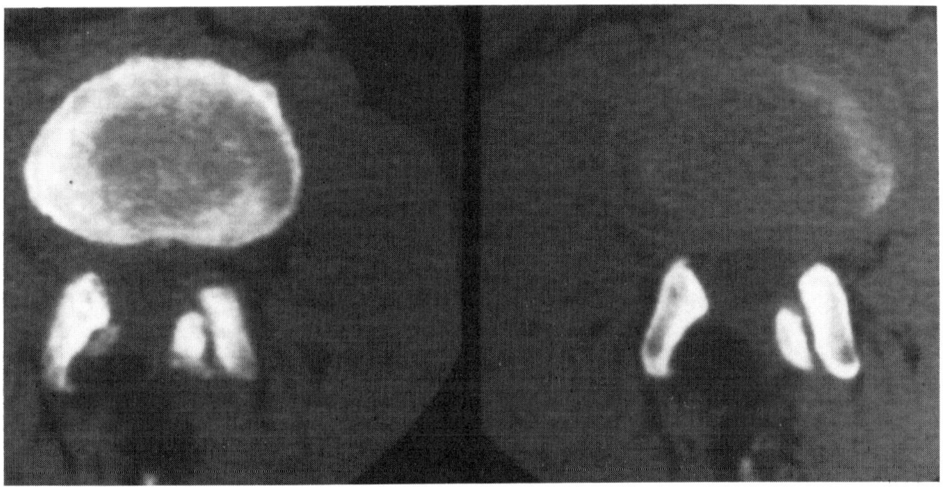

Figure 6.5. Contiguous axial CT scan with bilateral removal of the laminae and the spinous process with partial facetectomy (*arrow*) in a patient with spinal stenosis.

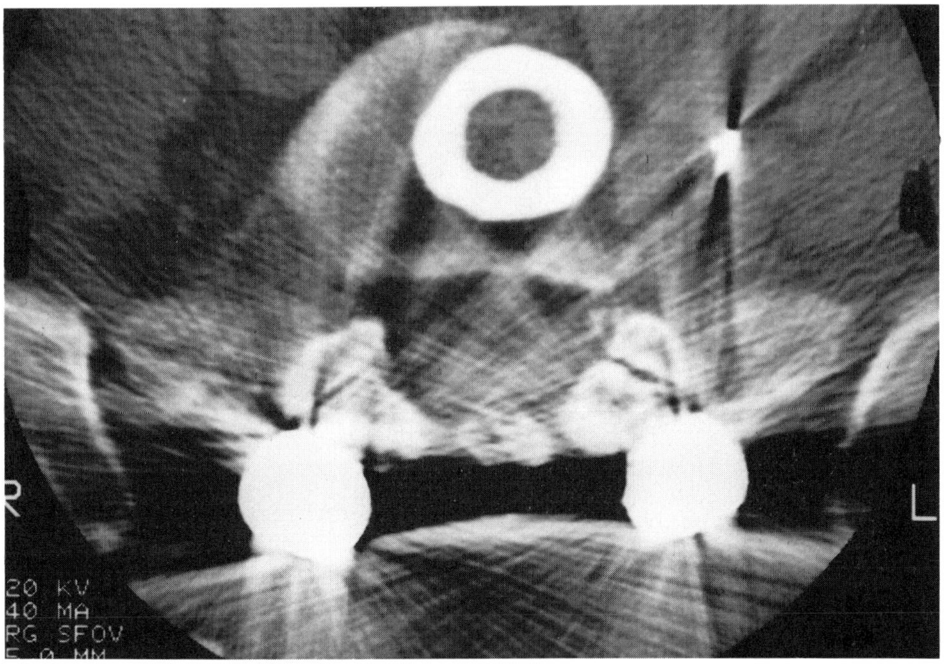

Figure 6.6. Anterior interbody fusion. Axial CT section demonstrates a bone graft placed between the vertebral bodies and posterior stabilization with rods. CT performed 10 days following surgery to identify the location of the graft within the vertebral space.

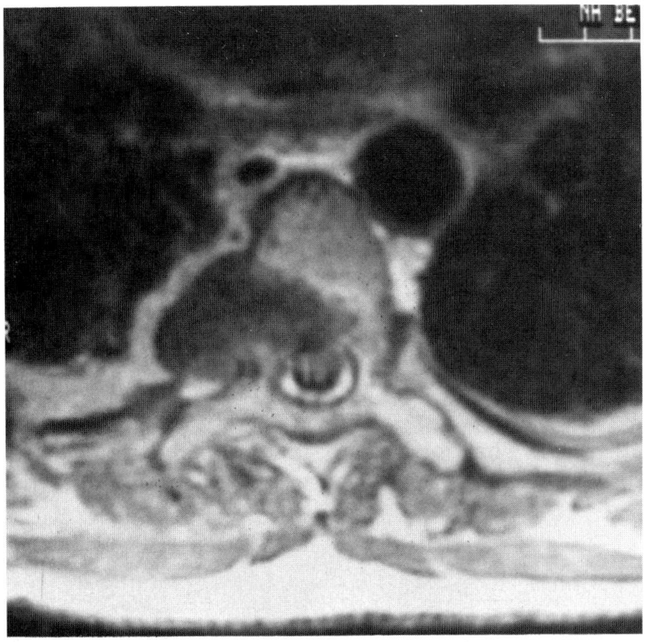

Figure 6.7. MRI of the thoracic region following costotransversectomy. The pedicle portion of the vertebral body and the medial portion of the adjacent rib are removed to provide exposure to the anterior epidural space.

the disk space appears narrowed, although the cortical bone appears intact. MRI is useful in identifying the postoperative changes. Anterior interbody fusion is usually performed in the cervical spine for the treatment of cervical spondylitic disease associated with narrowing of the spinal canal as well as adjacent neural foramina. It is indicated when the stenosis is focal and involves one level. The primary purpose is to decrease motion and prevent further hypertrophic changes at the adjacent disk margins as well as the neural foramina. Anterior interbody fusion may consist of placing a rounded bone plug after drilling a hole in the disk space and adjacent vertebral body (Cloward's procedure) [25]or of removing the disk and anterior endplates of adjacent vertebrae and placing a square plug of bone in the rectangular space (Smith-Robinson procedure) [26]. When cervical spondylitic disease involves more than one level, either multilevel laminectomy or anterior fusion with strut grafts may be utilized. The most common postoperative complications of these procedures include dislodgment of the graft, infection, nonacceptance of the graft, and increased incidence of spondylitic changes above and below the graft. Less

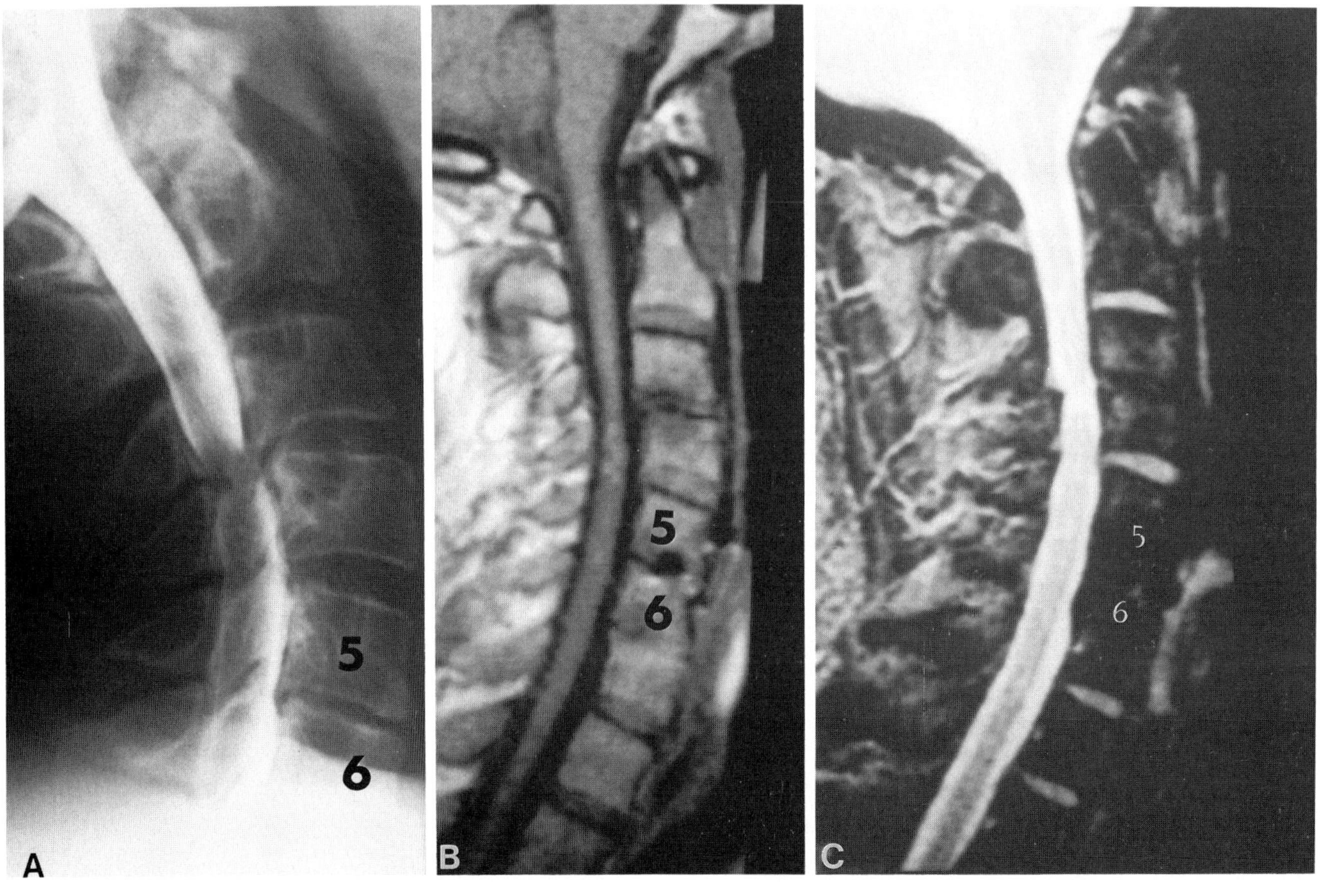

Figure 6.8. Anterior cervical discectomy. A, Lateral view of cervical myelogram demonstrates narrowing of the C5–6 disk space following anterior cervical discectomy. Focal spinal stenosis is present at a higher level involving the C3–4 disk space. Sagittal T1-weighted MRI (**B**) and T2-weighted MRI (**C**) demonstrate complete obliteration of the C5–6 disk space. Focal stenosis is identified at the C3–4 level similar to the myelogram.

common complications include cord compression and cord atrophy.

Imaging Considerations

Most of the osseous changes can be evaluated on plain radiographies. CT and MRI are useful in demonstrating the associated soft-tissue changes within the spinal canal. Artifacts may degrade the CT or MRI studies. Metallic hardware such as wire sutures and metal plates or rods often result in artifacts both on CT and MRI. In such patients, plain x-rays or myelography followed by conventional tomography may be necessary (Fig. 6.9). In most cases, CTM with imaging in bone algorithm to a limited extent reduces the metallic streak artifacts. Similarly, magnetic susceptibility artifacts on MRI can be reduced with fast spin-echo (FSE) imaging and by decreasing the spacing between the echoes.

LUMBAR SPINE

Evaluation of the postoperative lumbar spine can be done either with CT or MRI. On both CT and MRI, absence of the ligamentum flavum is a good indication of previous surgical exploration [23]. Laminotomy and partial facetectomy defects, which may be difficult to recognize on plain x-rays, can be appreciated on CT when thin sections are obtained. On MRI, the laminectomy site is identified by the absence of the cortical bone and the hyperintense marrow signal. On sagittal images, heterogeneous signal from the scar tissue and particulate metallic or bone fragments within the soft tissues often identifies the surgical site.

Soft-tissue changes following surgery are best evaluated with CT or MRI augmented with contrast material. Wide variability in the asymptomatic "normal" postoperative spine can be identified. This depends on the amount of the epidural hematoma, the surgical procedure, the rate of absorption of the blood products, and the rate of formation of granulation tissue. These factors in general also account for the extent of subsequent epidural fibrosis and scar tissue. There is, however, no correlation between the extent of epidural fibrosis and the severity of the patient's symptoms [27]. The epidural fibrosis may be similar in both symptomatic and asymptomatic patients observed on clinical evaluation of FBSS [28].

Imaging of the postoperative spine is most often performed for differentiation of epidural fibrosis or scar from residual or recurrent herniated disk. Both CT and MRI following intravenous contrast material have been used in this evaluation. Studies comparing the relative accuracy of IVCT and Gd-DTPA–enhanced MRI in differentiating epidural fibrosis (scar) from disk material have shown MRI to have higher sensitivity compared with CT [12–19, 29]. IVCT has an accuracy of nearly 75% [6, 18, 29], whereas MRI is consistently better in differentiating the two entities [17, 18, 29]. On gadolinium-enhanced MRI, the reliability in detecting enhancement of the scar tissue

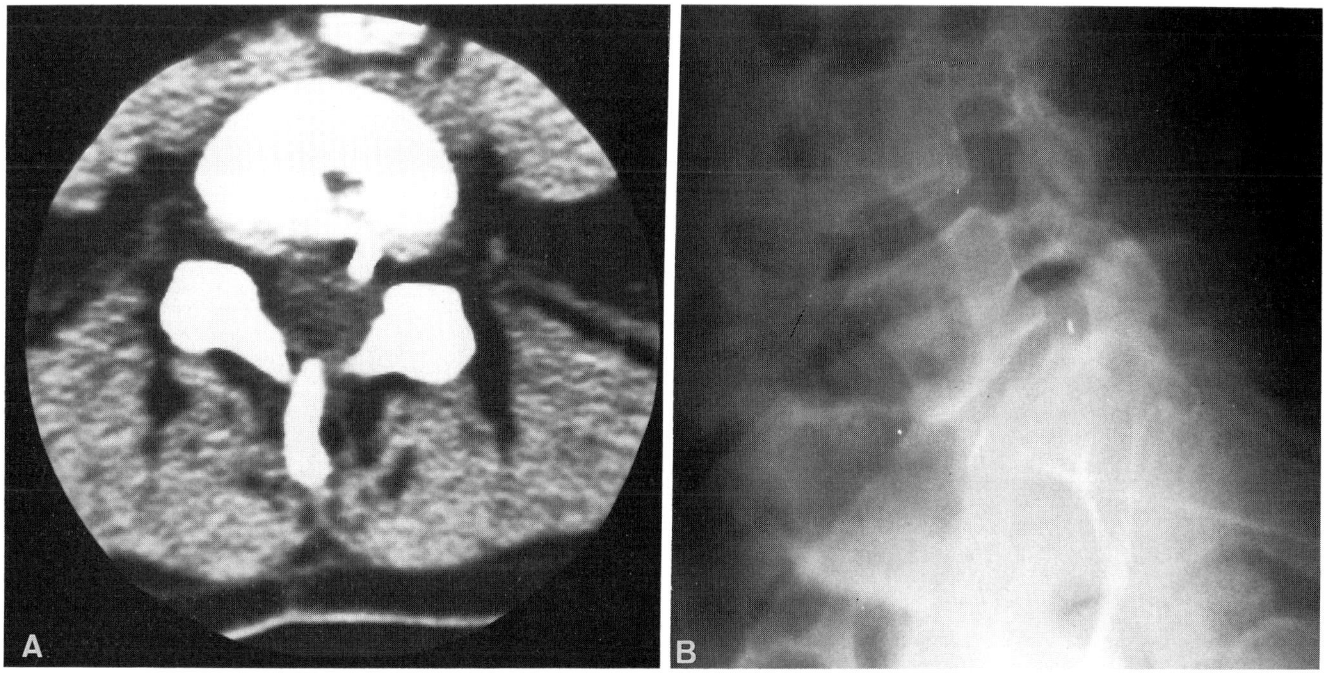

Figure 6.9. **A,** Non-contrast-enhanced axial CT demonstrates a focal linear hyperdensity compressing the thecal sac, simulating a bony osteophyte. **B,** Lateral view of the lumbar spine demonstrates a metallic clip within the spinal canal causing the artifact seen on CT.

has been reported to be as high as 90% [17]. The increased detection capability is related to the higher spatial and contrast resolution of MRI compared with CT.

At the present time, Gd-DTPA–enhanced MRI has replaced IVCT in the differentiation of epidural fibrosis from residual or recurrent disk herniation. Enhancement of epidural fibrosis with IVCT [7, 30] and with Gd-DTPA–enhanced MRI [15, 17–19] is due to extravascular pooling of contrast material. IVCT enhancement decreases with maturity of the scar unless the scar is inflamed as a result of irritation from bony spurs or residual disk herniation [7] and usually requires a larger bolus of contrast material [11]. The enhancement, however, depends on the presence of the vascularity within the scar tissue. Extravasation of contrast material from the vascular compartment into the extravascular space usually occurs within a few minutes following intravenous injection and lasts for a maximum period of 30 minutes [31]. Time and density curve studies in an animal model have shown enhancement of the granulation tissues following Gd-DTPA injection reaching a peak in about 5 minutes [32]. Since nearly 50% of GD-DTPA is cleared in 15 to 20 minutes [33], the enhanced study should be obtained within 5 to 10 minutes following injection of contrast material. This is of importance because residual herniated disk will not show enhancement during this period. Delay in scanning may result in enhancement of the periphery of the disk or the entire disk, leading to errors in diagnosis. Even though there may be variability in the degree of enhancement of the scar tissue following Gd-DTPA injection, enhancement is noted irrespective of the age in the inflamed scar tissue [15–19].

A recent animal experimental study has shown that peak enhancement of scar tissue depends on the maturity of the scar tissue, the dose of the contrast (0.3 mmol/kg) and imaging within 15 minutes of injection [53]. In our experience, this decreased enhancement of the scar in postoperative patients is often seen in patients with more mature scar tissue beyond 12 to 14 months old. Studies of enhancement in experimental scar in animal models have shown that enhancement with Gd-DTPA is maximum 4 weeks after surgery, with a rapid decrease in enhancement by 4 months [34]. It is conceivable that the lack of enhancement beyond 4 months in the experimental posterior epidural tissue, as compared with enhancement in anterior epidural scar, is related to the absence of persistent mechanical irritation at the surgical site. Absence of detectable enhancement in the aged asymptomatic epidural scar tissue is probably due to decreased vascularity of the mature scar, chondrification, and osteophyte formation. Beyond 6 months, the epidural fibrous tissue gradually converts into mature scar, with a decrease in its vascular component.

In the immediate postoperative period (1 to 3 weeks), invariably a soft-tissue epidural mass is present at the level of the discectomy. It consists of localized hematoma from variable accumulations of blood products and edema in the adjacent tissue. On CT, this has a variable density ranging from mild hypodensity to hyperdensity when compared with the adjacent thecal sac. The difference in density of the epidural mass depends on the resorption of the blood products and the amount of fibrous tissue within the mass. The fat planes surrounding the thecal sac are not clearly visualized (Fig. 6.10). The endplate and the site of curettage following disk removal are difficult to identify due to marrow edema. In some case this appearance may last beyond the 3-week period [7]. Heterogeneous enhancement of the epidural mass is often seen on IVCT. During this pe-

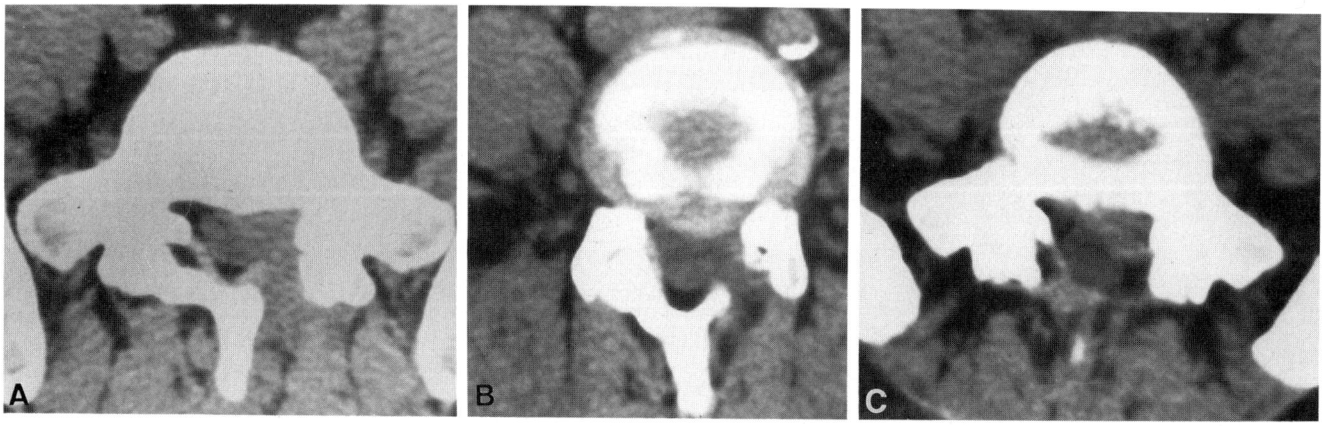

Figure 6.10. **A,** Axial non-contrast-enhanced CT performed 7 days following disk surgery demonstrates mildly hyperdense epidural soft-tissue mass with loss of fat planes. The soft-tissue mass extends beyond the laminectomy defect. The mild hyperdensity may represent residual disk and hematoma. **B,C,** Contiguous axial IVCTs demonstrate ill-defined peripheral enhancement of an anterior epidural mass in another patient studied 7 days following surgery. In both instances, the patients were asymptomatic.

Figure 6.11. Sagittal T2-weighted MRI, first echo (**A**) and second echo (**B**), images following discectomy at the L3–4 and L4–5 levels. Focal epidural mass is present with disruption of the annular fibers at both these levels, almost similar to preexisting herniated disk. Focal intermediate to bright signal is seen within the epidural space and the site of laminectomy. Precontrast (**C**) and postcontrast Gd-DTPA–enhanced (**D**) T1-weighted MRI images demonstrate the hematoma with enhancement of the granulation tissue, the nerves, and the margins of the thecal sac.

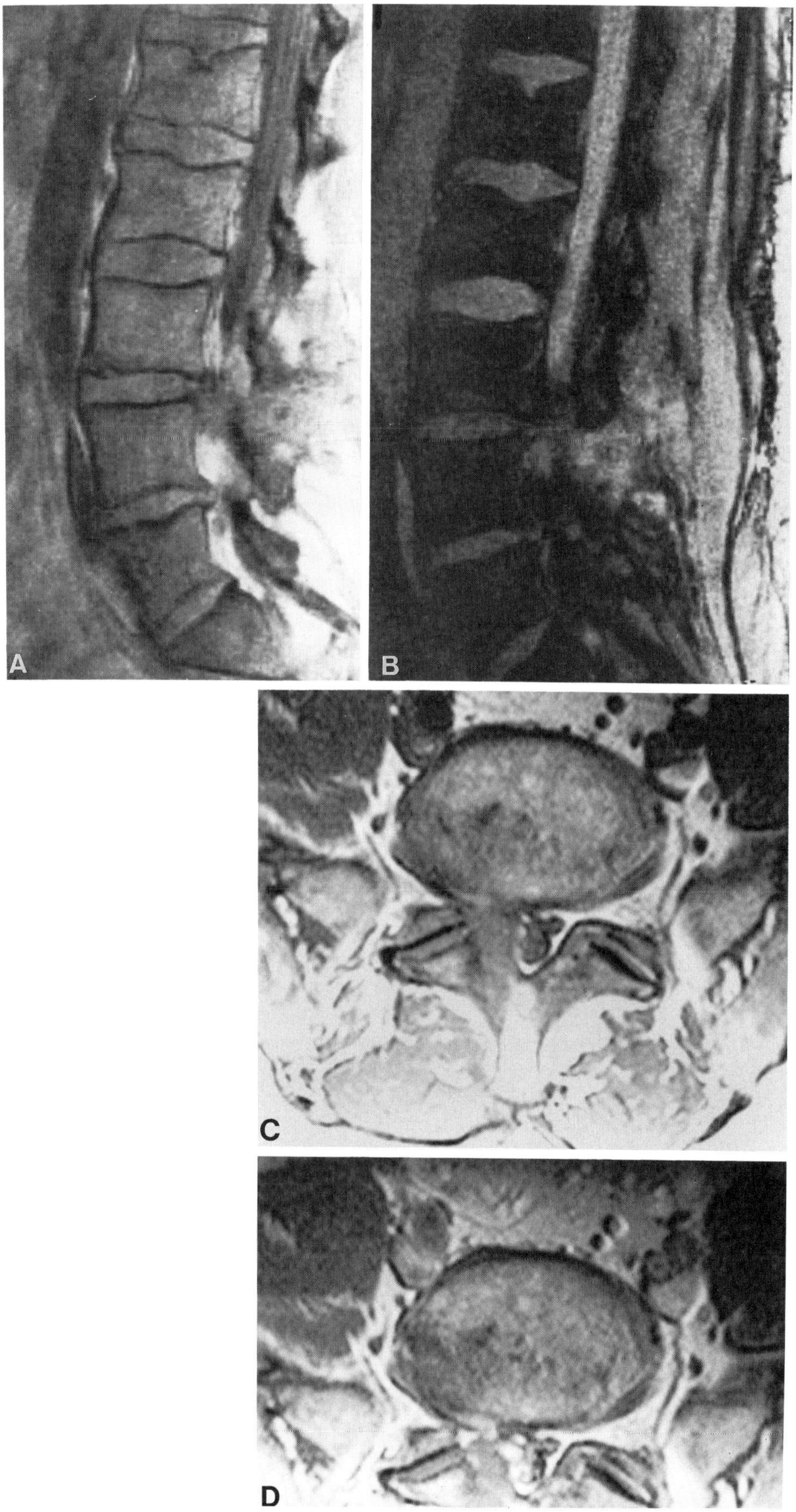

riod, differentiation of residual disk herniation from epidural fibrosis is difficult. On T1-weighted MRI sequences, the epidural soft-tissue mass has an intermediate to hypointense signal depending on the organization of the hematoma and the formation of granulation tissue. Deformity of the thecal sac is often present and has the appearance of the herniated disk prior to surgery. On T2-weighted images, the epidural mass has an intermediate to bright signal (Figs. 6.11 and 6.12). The disk space may show decreased height. The endplates and marrow of the adjacent vertebral bodies appear mildly hyperintense on T2- or T2*-weighted images and hypointense on T1-weighted images similar to type 1 degenerative changes. Even though there are mild marrow signal changes, enhancement of the marrow space adjacent to the discectomy is not a feature of the normal postoperative spine. Enhancement, if present beyond 4 to 6 weeks after surgery, is usually indicative of discitis. On axial images, hyperintense signal within the curettage site of the disk space with enhancement is often seen at this time (Fig. 6.13).

In the initial months following discectomy, IVCT and Gd-DTPA–enhanced T1-weighted MRI images show heterogeneous enhancement of the epidural space, the laminectomy site, the adjacent facet joints, and the nerve root sheath, as well as within the muscle-subcutaneous space. The epidural fibrous tissue is usually less prominent. The thecal sac appears normal or may show retraction toward the scar [35] (Figs. 6.14 and 6.15). The disk space is narrowed with decreased signal on both T1- and T2-weighted MRI images. There is consistent enhancement of the epidural scar with encasement of the nerve root.

In most instances, 4 to 6 months following discectomy, there is minimal epidural scar tissue causing a mass effect. Both CT and MRI demonstrate a decrease in size with retraction of the extradural fibrous tissue and ipsilateral traction on the thecal sac. The epidural fibrous tissue may appear as a hyperdense linear band or as an extradural mass with ill-defined margins compared with the smooth appearance of the residual or recurrent disk [15]. On

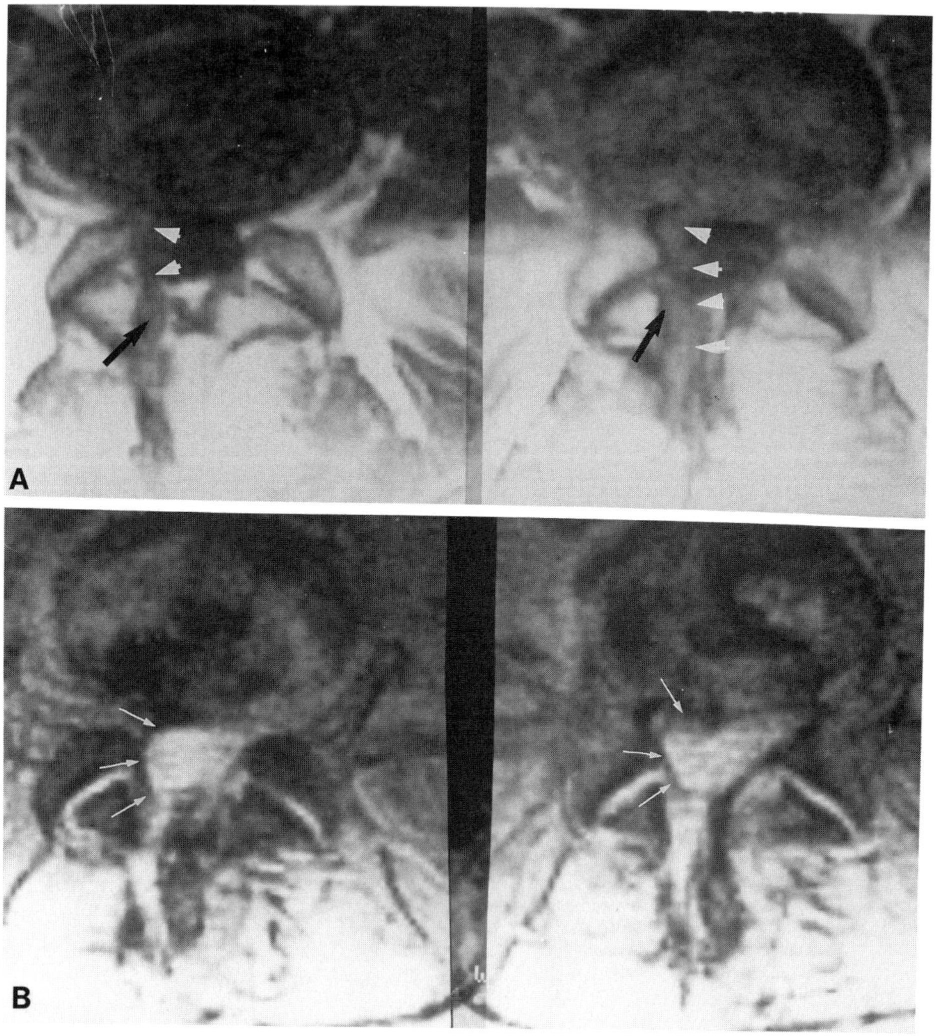

Figure 6.12. Postoperative MRI in an asymptomatic patient 10 days following back surgery. **A,** Contiguous axial T1-weighted MRI demonstrates the intermediate signal of the epidural soft-tissue mass most likely representing hematoma. **B,** T2-weighted MRI shows that the hematoma (*arrows*) has slightly brighter signal than CSF.

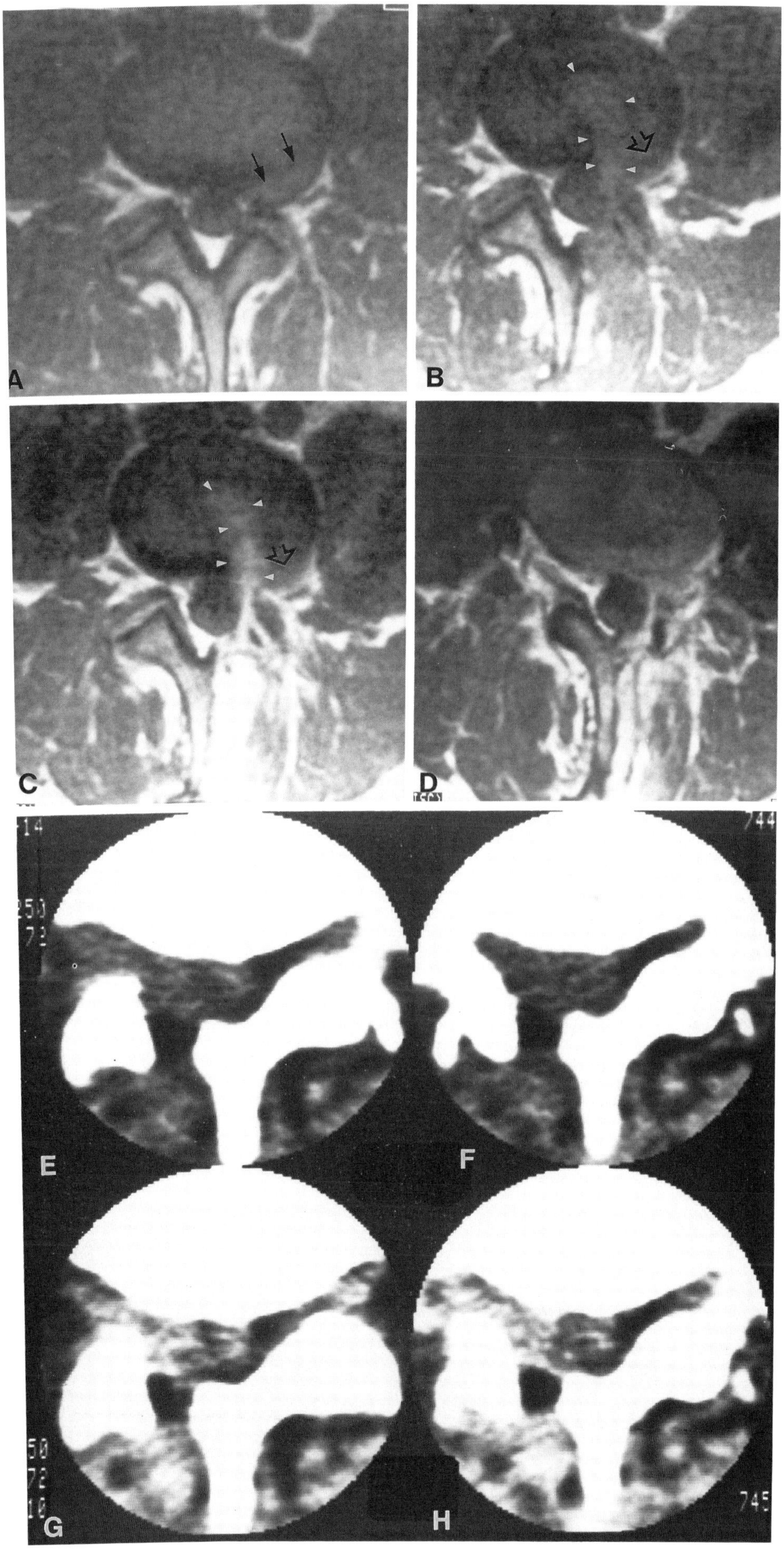

Figure 6.13. Postoperative MRI 4 weeks after back surgery with recurrence of symptoms. **A,** T1-weighted axial MRI prior to surgery demonstrates an extraforaminal disk herniation (*arrow*). **B,** Non-contrast-enhanced T1-weighted axial MRI. There is persistence of portions of the extraforaminal herniated disk (*open arrow*). An ill-defined, irregular, mildly hyperintense signal is seen within the disk space at the curettage site (*arrowheads*). **C,** Gd-DTPA–enhanced MRI demonstrates enhancement of this site (*arrowheads*), as well as the epidural scar tissue. **D,** In the section below the disk level, there is enhancement of the granulation tissue (*arrow*) surrounding the remnant of the disk. Contiguous non-contrast-enhanced axial CT scans (**E,F**) and IIVCT scans (**G,H**) 3 weeks following back surgery in a patient with persistent symptoms. Enhancement of the epidural soft-tissue mass extending into the noural foramen is seen. There is some enhancement of the thecal sac. The enhancing epidural mass can represent organizing hematoma or granulation tissue. Note the position of the fat graft (*arrow*) within the laminectomy site.

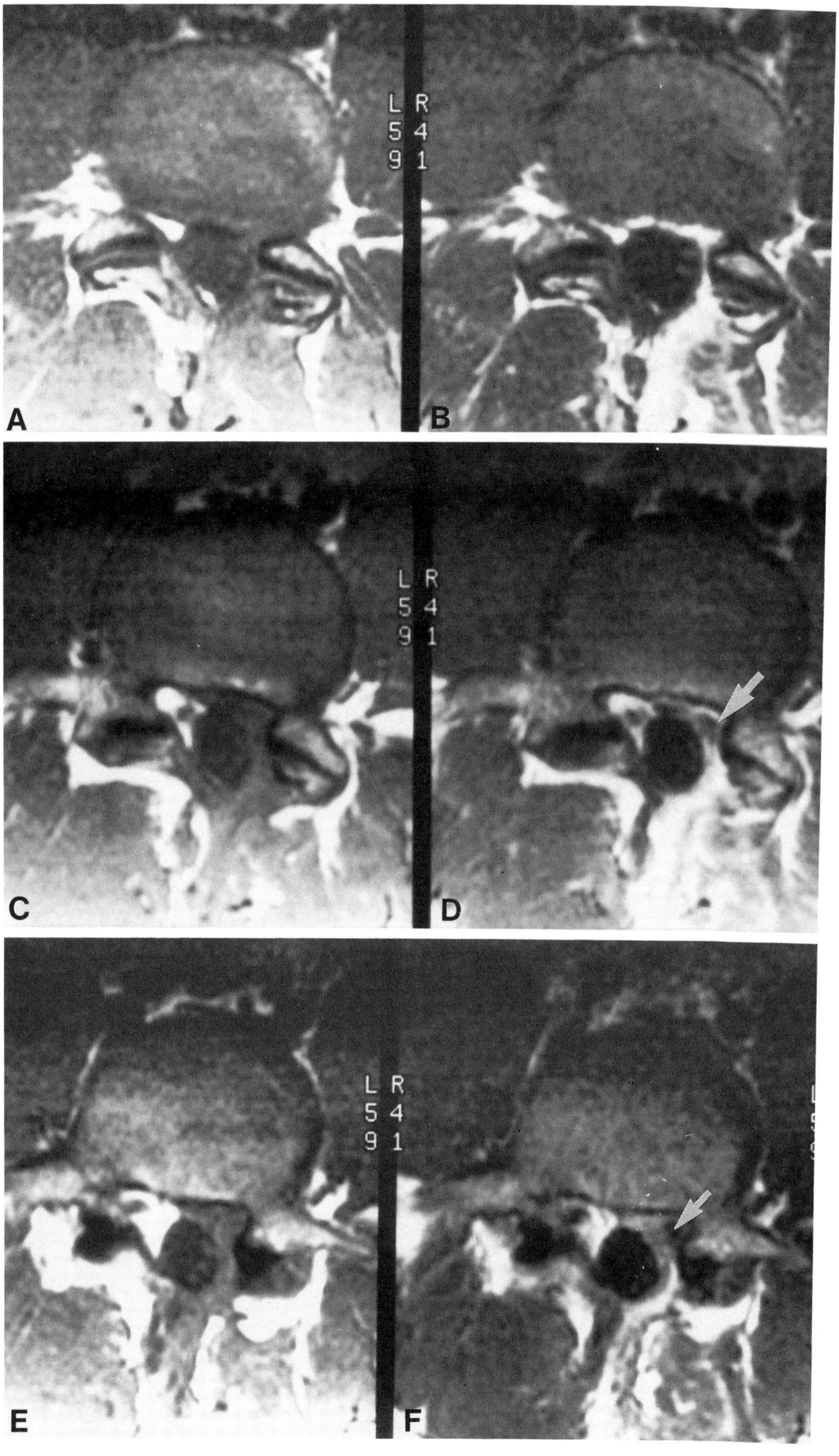

Figure 6.14. Postoperative lumbar spine 3 months following surgery demonstrates the decrease in epidural fibrosis following contraction and conversion to scar tissue in an asymptomatic patient. Contiguous axial non-contrast-enhanced T1-weighted MRI images (**A,C,E**) and Gd-DTPA–enhanced T1-weighted MRI images (**B,D,F**) demonstrate homogeneous enhancement of the epidural scar tissue. The nerve root within the scar does not enhance (*arrow*). Retraction of the thecal sac toward the scar is noted.

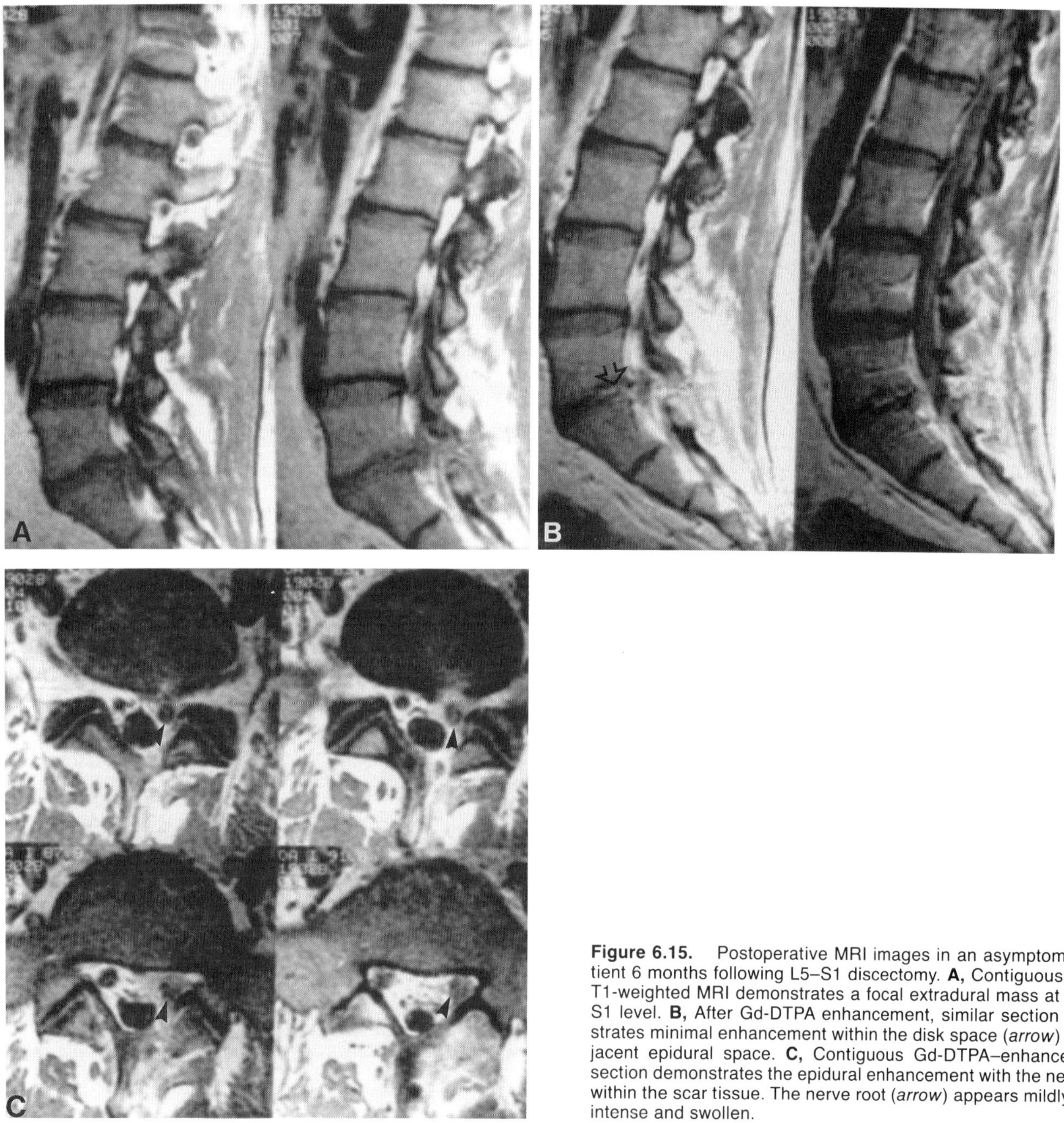

Figure 6.15. Postoperative MRI images in an asymptomatic patient 6 months following L5–S1 discectomy. **A,** Contiguous sagittal T1-weighted MRI demonstrates a focal extradural mass at the L5–S1 level. **B,** After Gd-DTPA enhancement, similar section demonstrates minimal enhancement within the disk space (*arrow*) and adjacent epidural space. **C,** Contiguous Gd-DTPA–enhanced axial section demonstrates the epidural enhancement with the nerve root within the scar tissue. The nerve root (*arrow*) appears mildly hyperintense and swollen.

CT, the epidural fibrous tissue appears hyperdense with a variable increase in density on IVCT (Fig. 6.16). On T1-weighted MRI images, epidural fibrosis has a hypointense signal. Enhancement is variable, but epidural fibrosis consistently enhances on Gd-DTPA–enhanced studies (see Fig. 6.17).

Recurrent or residual disk is differentiated from epidural scar by its lack of enhancement, smooth contour, and occasional enhancement along the periphery of the herniated disk (Fig. 6.18). This is similar in appearance to the enhancement seen with regressive disk changes [36]. In majority of the instances, absence of homogeneous enhancement and continuity of the herniated disk with the parent disk help in differentiating from scar tissue. These two diagnostic features may not be present when the residual herniated disk is sequestered. A sequestered disk may show homogeneous enhancement. Similar problems in differentiation may be encountered when epidural fibrosis is associated with calcifica-

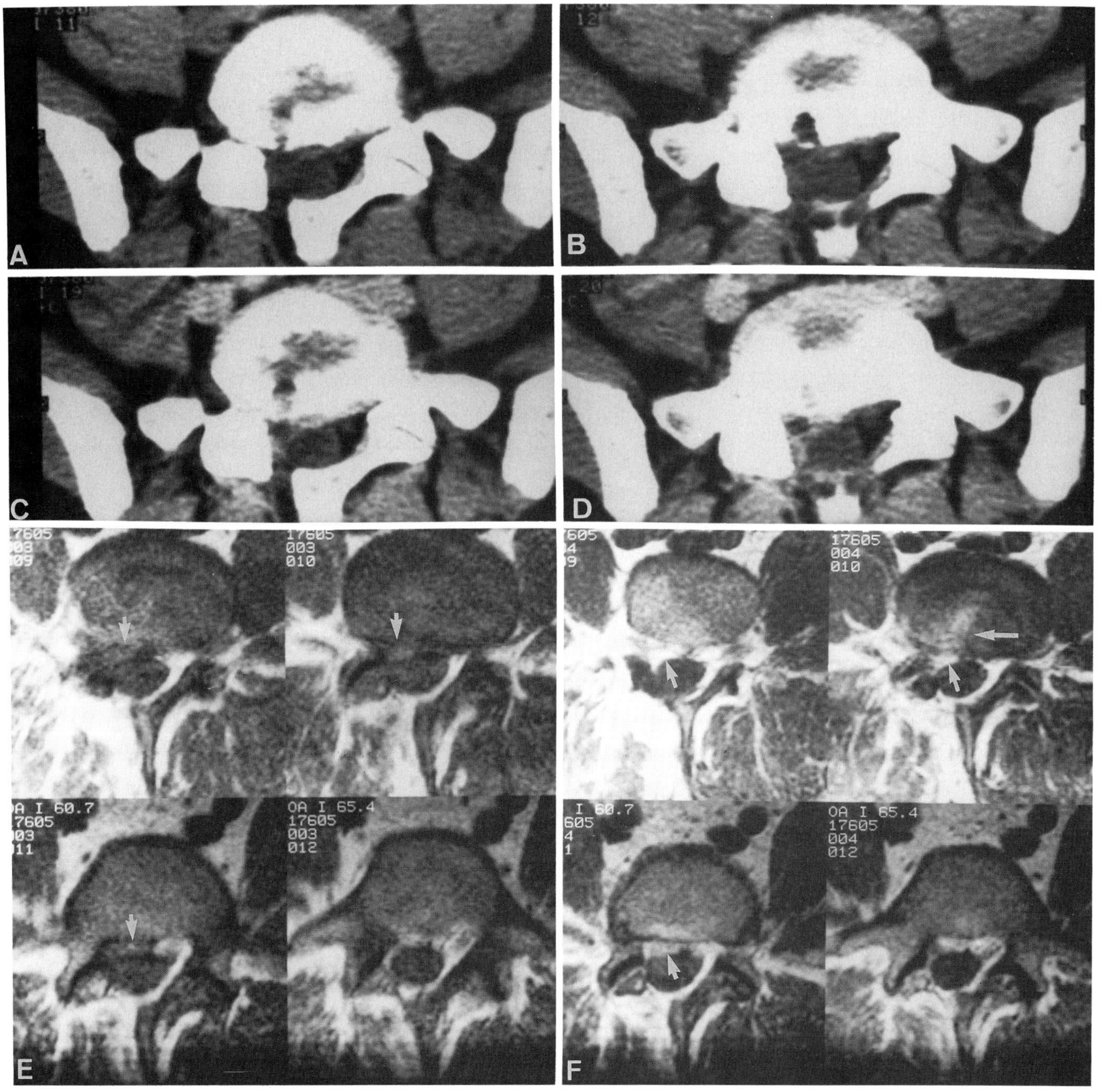

Figure 6.16. Postoperative MRI images 12 months following previous back surgery demonstrating consistent but variable enhancement of the epidural scar on IVCT. Precontrast (**A–C**) and post-IVCT (**B,D**) scans demonstrate minimal enhancement of the epidural scar. This is the pattern in most cases beyond 12 to 14 months on CT. An example of MRI during the same time period following surgery demonstrates on axial T1-weighted precontrast (**E**) and (**F**) Gd-DTPA–enhanced (**B**) MRI sequences consistent enhancement of the epidural scar (*arrows*).

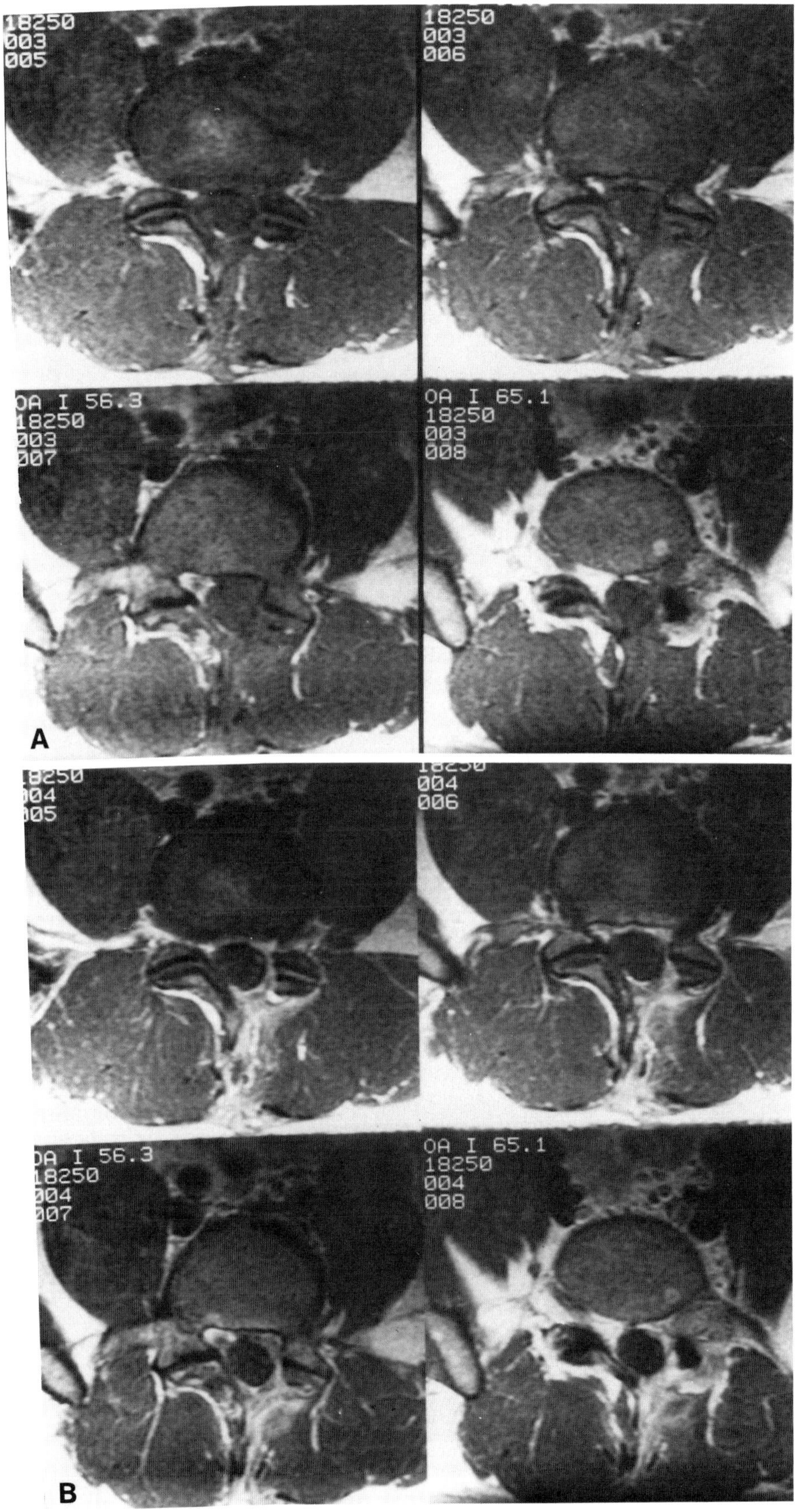

Figure 6.17. Postoperative MRI beyond the 18-month period in an asymptomatic patient. **A,** Contiguous axial T1-weighted MRI demonstrates epidural scar involving the anterolateral and posterior aspects of the thecal sac on the left side. The laminectomy defect is not clearly seen. **B,** Gd-DTPA–enhanced MRI shows consistent but variable intensity of enhancement within the scar tissue.

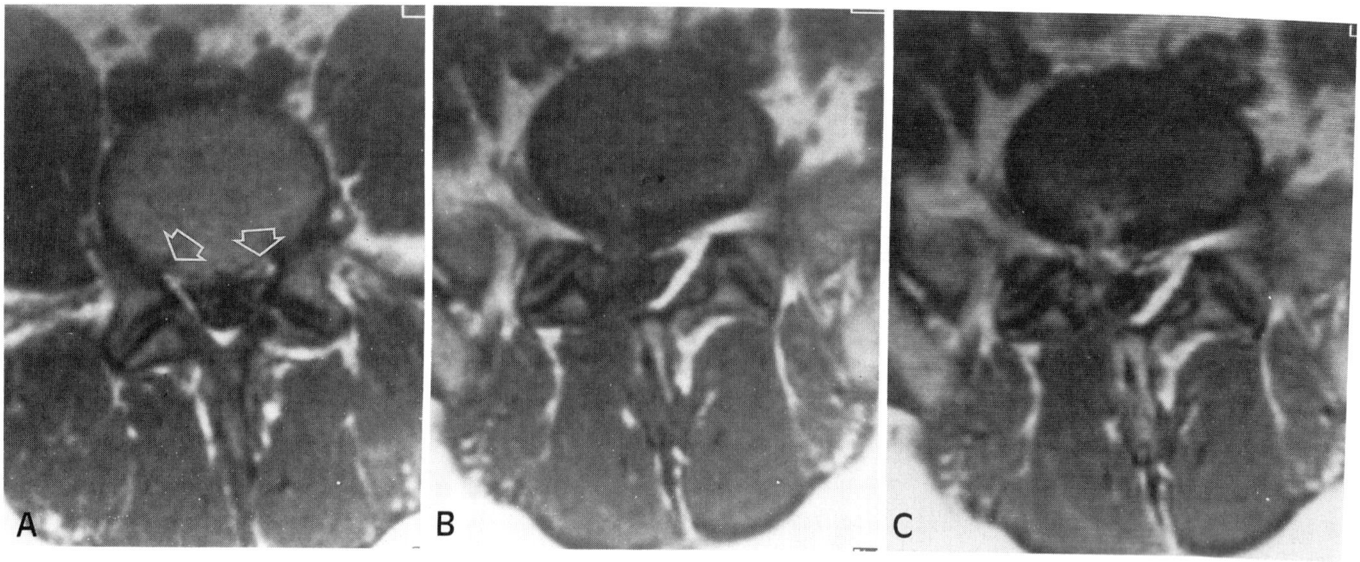

Figure 6.18. Recurrent herniated disk. A,B, Contiguous axial T1-weighted MRI sequences demonstrate a laminectomy on the right side with an anterior epidural mass. On the lower section (**B**) there is focal extension of the disk material beyond the confines of the annular ligament (*arrows*). **C,** Gd-DTPA–enhanced MRI demonstrates enhancement adjacent and surrounding the residual disk. This was proven to be a herniated disk at surgery.

tion of the scar tissue or only the peripheral margins of the scar tissue enhance. This pattern is seen in nearly 10% of postoperative spines. Combination of T1- and T2-weighted sagittal images prior to Gd-DTPA–enhanced MRI are helpful in demonstrating the hypointense signal of osteophytes. In most instances, if conservative treatment fails, such patients need to be explored for possible residual herniated disk.

Residual disk herniation as a complication of back surgery is not uncommon. This often presents as recurrence of symptoms and neurologic deficits during the recuperative stage following early ambulation. It is usually the result of incomplete discectomy. This is more common with herniated disk within the neural foramen or with extraforaminal disk herniation (see Fig. 6.13). The imaging characteristics of the remnant foraminal and far lateral disks are similar to those of the preoperative studies. Recurrent disk herniation should be suspected when the postoperative imaging studies demonstrate a central or lateral herniated disk within the spinal canal, usually months or years following the initial discectomy (Fig. 6.19). More often the disk herniation involves one level above or below the site of previous discectomy (see Fig. 6.19).

THORACIC SPINE

Postoperative evaluation of the thoracic spine can be accomplished with either CT or MRI, except when metallic hardware is present. When metallic hardware is present, MRI may not be feasible or may be difficult to evaluate from the "blooming" artifacts from the metal. CT studies also can degraded by metallic streak artifacts. Plain radiography or polytomography is often necessary to complement these studies.

Apparent enlargement of the neural foramen following discectomy is a common finding. Depending on the surgical approach, whether the costotransversectomy was posterolateral or anterolateral and extrapleural, soft-tissue changes in the paravertebral region can be identified (Fig. 6.20). This often represents postoperative edema, which usually resolves within a few weeks. A small amount of extrapleural fluid collection may be identified on sagittal and axial MRI images. The fluid collection appears to be contiguous with the spinal CSF space and gradually resolves with time. This should not be confused with postoperative pseudomeningocele. Enhancement of the scar tissue is similar or slightly less prominent than in the lumbar region. Focal epidural scar is less common. Differentiation of focal postoperative scar from meningioma may be difficult, since both these may enhance on IVCT or gadolinium-enhanced MRI.

CERVICAL SPINE

In the cervical spine, postoperative complications are related to the surgical approach. Epidural fibrosis is seen more often following laminectomy or foraminotomy than following anterior discectomy. Postoperative complications in the cervical region include cord compression from bone grafts, injury to the cervical nerve roots during foraminotomy, and less often, epidural hematoma. Between the first and third postoperative week, the complications to be aware of include rejection of the bone graft used in

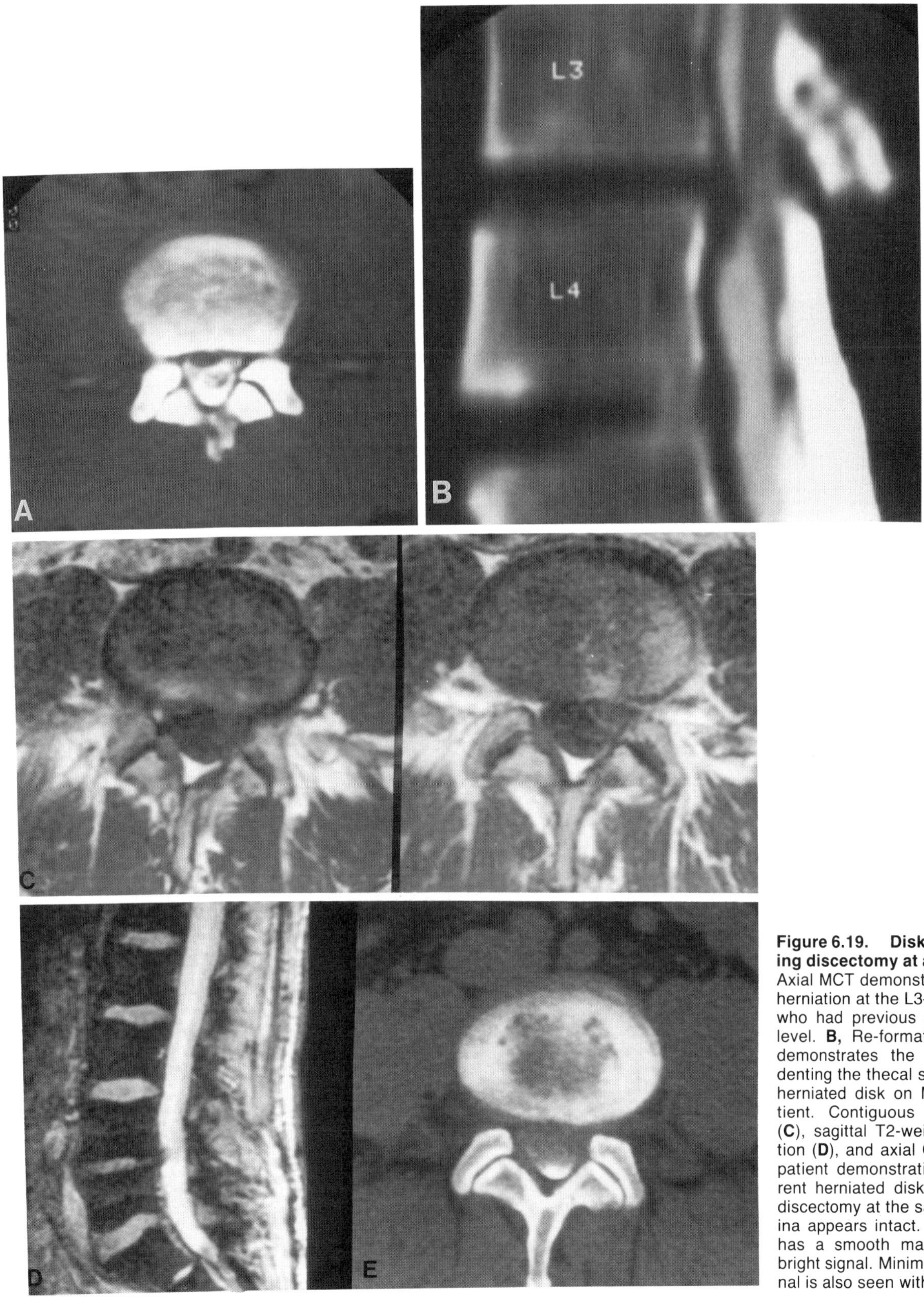

Figure 6.19. Disk herniation following discectomy at a different level. A, Axial MCT demonstrates a central disk herniation at the L3–4 level in a patient who had previous surgery at a lower level. **B,** Re-formatted sagittal image demonstrates the herniated disk indenting the thecal sac. **C–E,** Recurrent herniated disk on MRI in another patient. Contiguous axial T1-weighted (**C**), sagittal T2-weighted midline section (**D**), and axial CTM (**E**) in another patient demonstrating a lateral recurrent herniated disk following previous discectomy at the same level. The lamina appears intact. The herniated disk has a smooth margin, with a mildly bright signal. Minimal hyperintense signal is also seen within the disk space.

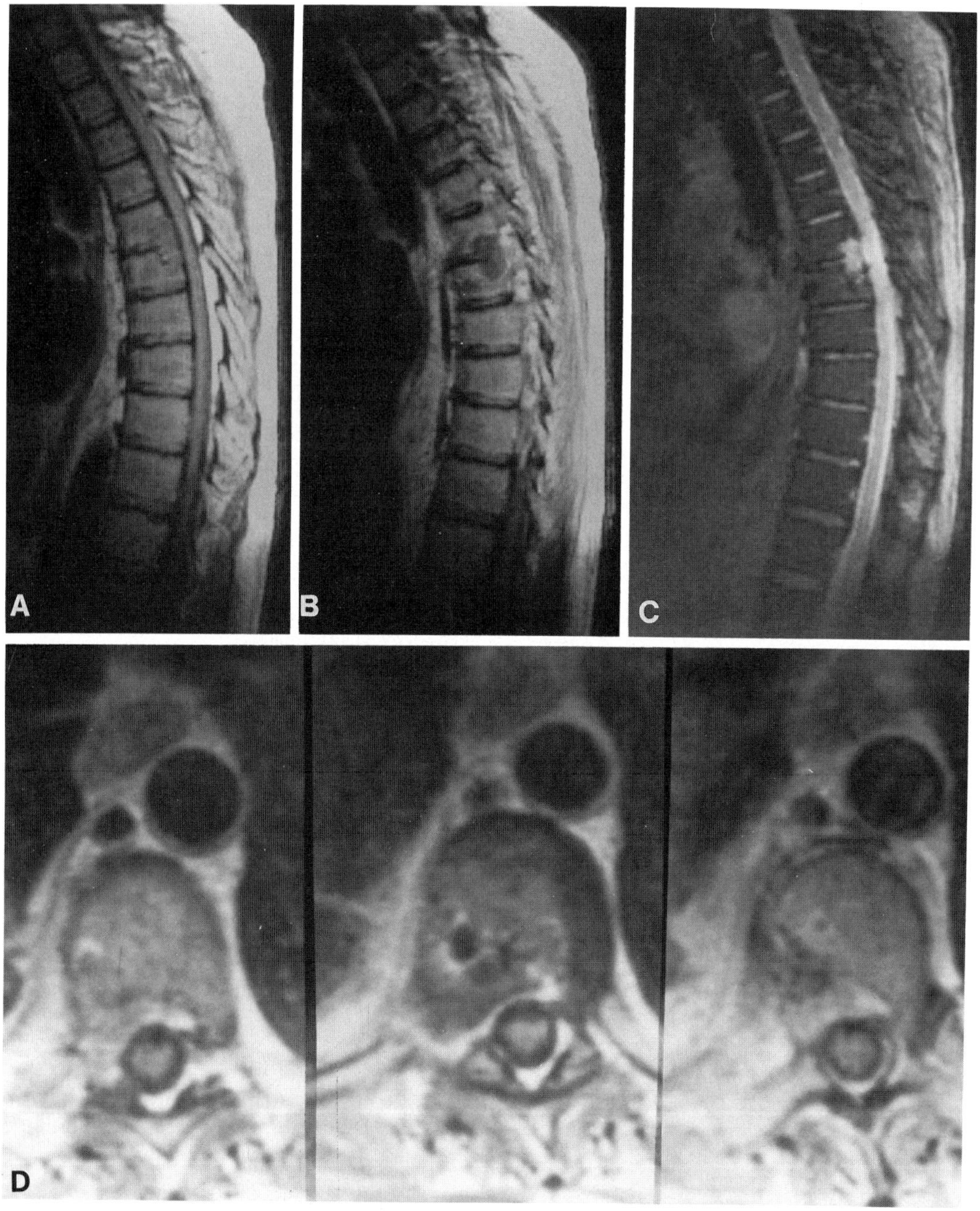

Figure 6.20. Postoperative changes in the thoracic spine following discectomy. Sagittal and parasagittal T1-weighted (**A,B**) and parasagittal T2-weighted (**C**) MRI sequences demonstrate enlarged foramen with CSF signal density (*arrow*). Contiguous axial T1-weighted precontrast (**D**) and Gd-DTPA–enhanced (**E**) MRI images demonstrate the hematoma, which is mildly hypodense, surrounded by the enhancing granulation tissue within the epidural space and extending into the neural foramen. There is enhancement of the prevertebral soft tissue extending along the extrapleural space.

interbody fusion following anterior discectomy, infection, and if laminectomy has been performed, lack of stability of the facet joints. Beyond this period, the delayed complications are usually associated with spinal stenosis from further degenerative changes or disk herniation at a level above or below the prior fusion.

Plain radiographs are adequate in evaluating the bone graft placed within the disk space in anterior interbody fusion. CT and MRI allow identification of the extent of bony fusion as well as the adjacent soft-tissue changes.

On CT, the bone fragments and graft have a density similar to that of the adjacent vertebral bodies. In the normal healing process, the graft is incorporated into the adjacent vertebral bodies. Nonfusion should be suspected when there is increased sclerosis of the graft with focal regions of heterogeneous hypodensity in the adjacent vertebral bodies. CTM is useful in identifying the position of the graft in relation to the intervertebral space. Thin axial sections with subsequent sagittal re-formatted images allow evaluation of the bone graft and impingement, if any, of the spinal cord. The viability of the graft is best assessed by the absence of edema in the prevertebral space. The presence of edema in the prevertebral space beyond 7 to 10 days should raise the possibility of avascular necrosis or superimposed infection. Within the spinal canal, an increase in the epidural space should raise the suspicion of impending epidural infection and abscess formation. This is best evaluated on CTM or IVCT (Fig. 6.21). On IVCT, the posteriorly displaced dura demonstrates intense enhancement. The graft and adjacent vertebral endplates changes are best evaluated on bone window settings (Fig. 6.22). Nonunion and pseudoarthrosis should be suspected when there is dense sclerosis of the bone graft or linear lucent zones are identified. Plain radiographs in flexion and extension are usually adequate in evaluating subluxation.

On MRI, depending on the nature of the graft material, the signal characteristics may change. It is important to know the nature of the graft material used, since allografts appear hypointense on both T1- and T2-weighted images, whereas autografts, when healing, tend to be hyperintense on T1-weighted images and isointense or hypointense on T2-weighted images; however, when fusion is associated with incorporation of the graft into adjacent vertebral bodies, the presence of marrow results in signal characteristics similar to the vertebral bodies.

Early Complications

Although epidural fibrosis and recurrent disk are the major considerations in the postoperative spine, other postoperative complications may account for the persistence or recurrence of symptoms. Those encountered in the immediate postoperative period, usually within the initial 1 to 3 weeks following surgery, include epidural hematoma, postoperative infection, and dural tear. They are relatively uncommon.

EPIDURAL HEMATOMA

A small amount of epidural hematoma from the epidural venous plexus is not uncommon after back surgery despite hemostasis. It is most often identified in the lower lumbosacral region due to the increased epidural space, often at the L5–S1 disk level. Such hematomas are usually asymptomatic and do not need evacuation. They may be identified on CT as mildly hyperdense soft tissue displacing the thecal sac [37]. Well-organized, focal hematomas may be mistaken for a fragmented disk. In the initial 24 to 48 hours, the serosanguineous exudate on MRI has a hypointense signal on T1-weighted images and mildly isointense to hyperintense signal on T2-weighted images (Fig. 6.23). The presence of isointense signal on T1-weighted images and focal hypointense signal on T2-weighted images is suggestive of hematoma in the deoxyhemoglobin phase. More often the siderosis along the dural tube has a hypointense signal on T2-weighted or gradient-echo images, similar to the signal intensity of the dura and posterior longitudinal ligament.

Focal hematoma in the thoracic or cervical region as an incidental finding is not unusual. Surgical intervention is usually not necessary unless the hematoma is large or there is clinical evidence of progressive neurologic deterioration.

POSTOPERATIVE INFECTION

Postoperative infection is uncommon after spine surgery. This may manifest as disk space infection or osteomyelitis with epidural abscess. Diagnosis is usually delayed because the findings on imaging studies may be attributed to changes in the immediate postoperative period.

Discitis

Discitis is an unrecognized complication of back surgery, since the persistent back pain is often considered to be postoperative nerve root irritation or, rarely, residual disk herniation. Discitis is usually the result of a nonspecific inflammatory response, since in many cases a pathogen may not be isolated. Discitis as a complication of back surgery has an incidence of 1 to 3% [38, 39]. Elevation of the erythrocyte sedimentation rate (ESR) beyond the second to fourth week after surgery should raise the clinical suspicion of discitis. Plain radiography may not show the early marrow and endplate changes

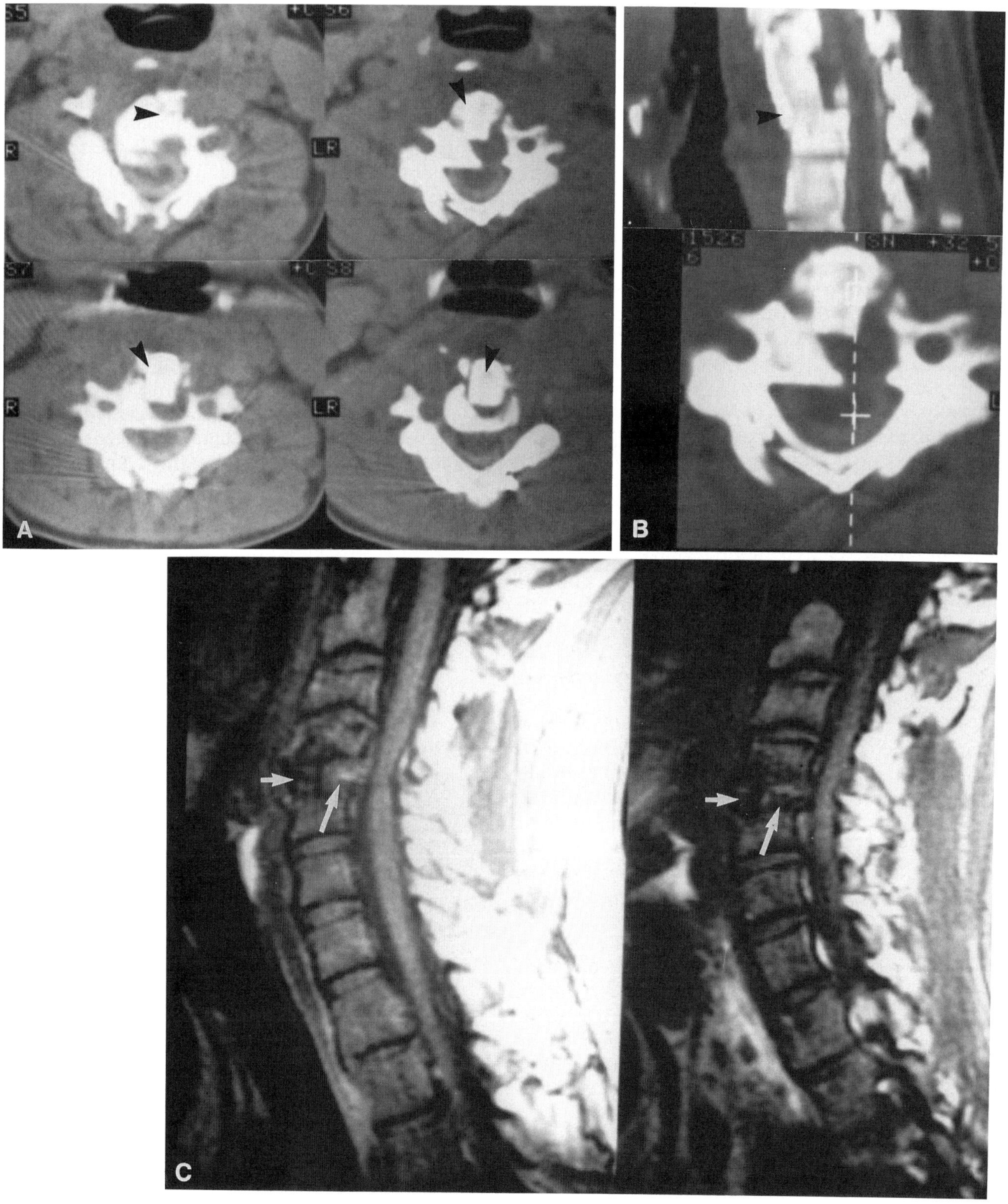
A
B
C

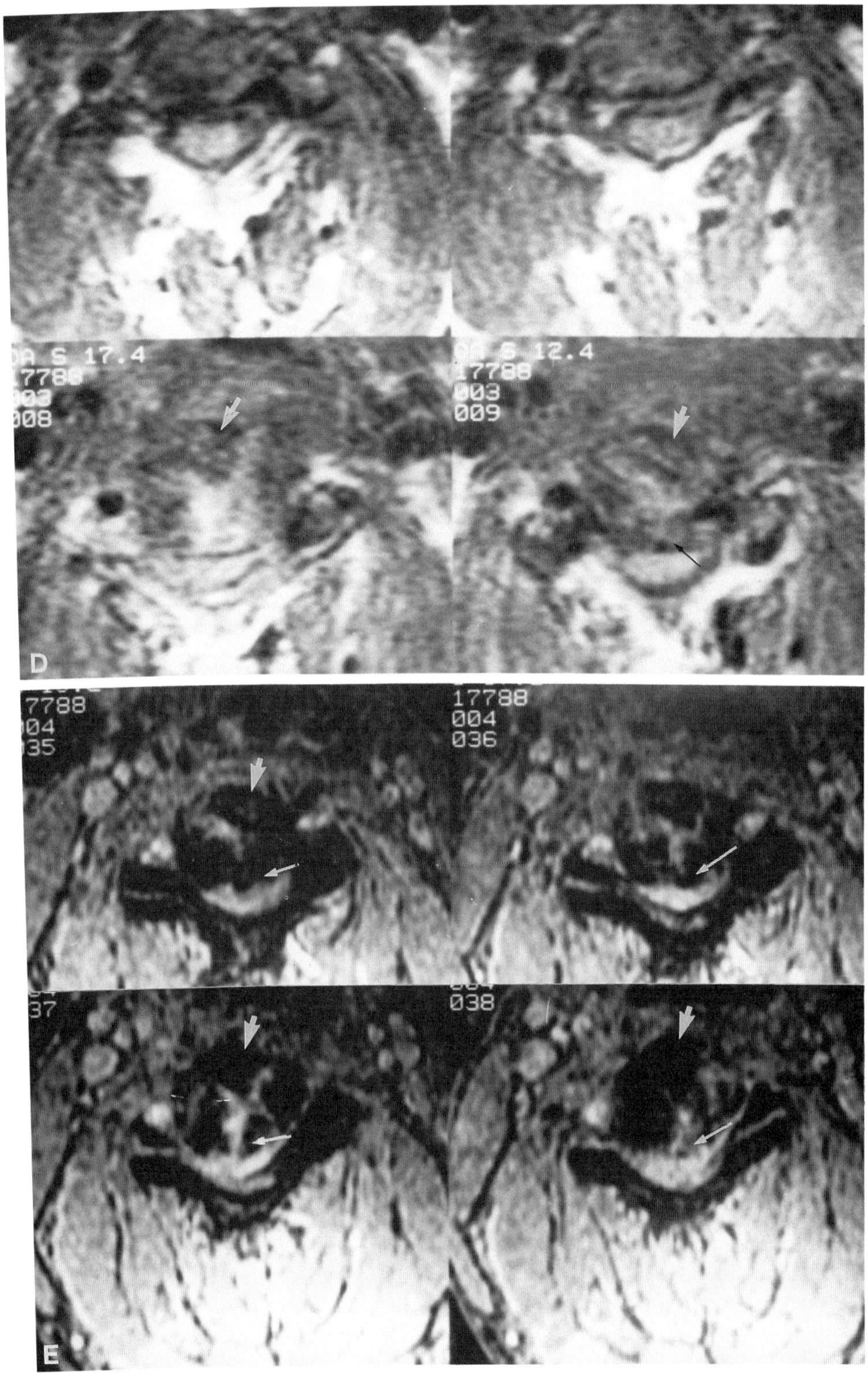

Figure 6.21. Postoperative CT/MRI in cervical disk following dislodgment of the graft. **A,** Contiguous axial CTM demonstrates anterior displacement of the graft (*arrowhead*). There is mild cord compression with effacement of the contrast material in the ventral aspect of the thecal sac. **B,** Re-formatted sagittal midline section demonstrates the partial dislodgment of the graft (*arrowhead*). The cord is compressed from the alignment of the vertebral bodies. **C–E,** MRI sequences in another patient with persistent symptoms following anterior discectomy with a bone graft between the C4–5 disk space. **C,** Contiguous sagittal T1-weighted MRI demonstrates the bone graft (*arrow*) with enhancing residual disk (*large arrow*). There is focal spinal stenosis with ill-defined focal bright signal from ischemia. Contiguous axial T1-weighted (**D**) and T2-weighted (**E**) MRI sections demonstrate persistent spinal stenosis. There is focal bright signal surrounding the graft (*arrow*), with the residual herniated disk (*small arrow*) compressing the thecal sac and spinal cord. The stenosis is exaggerated on the T2-weighted gradient-echo image.

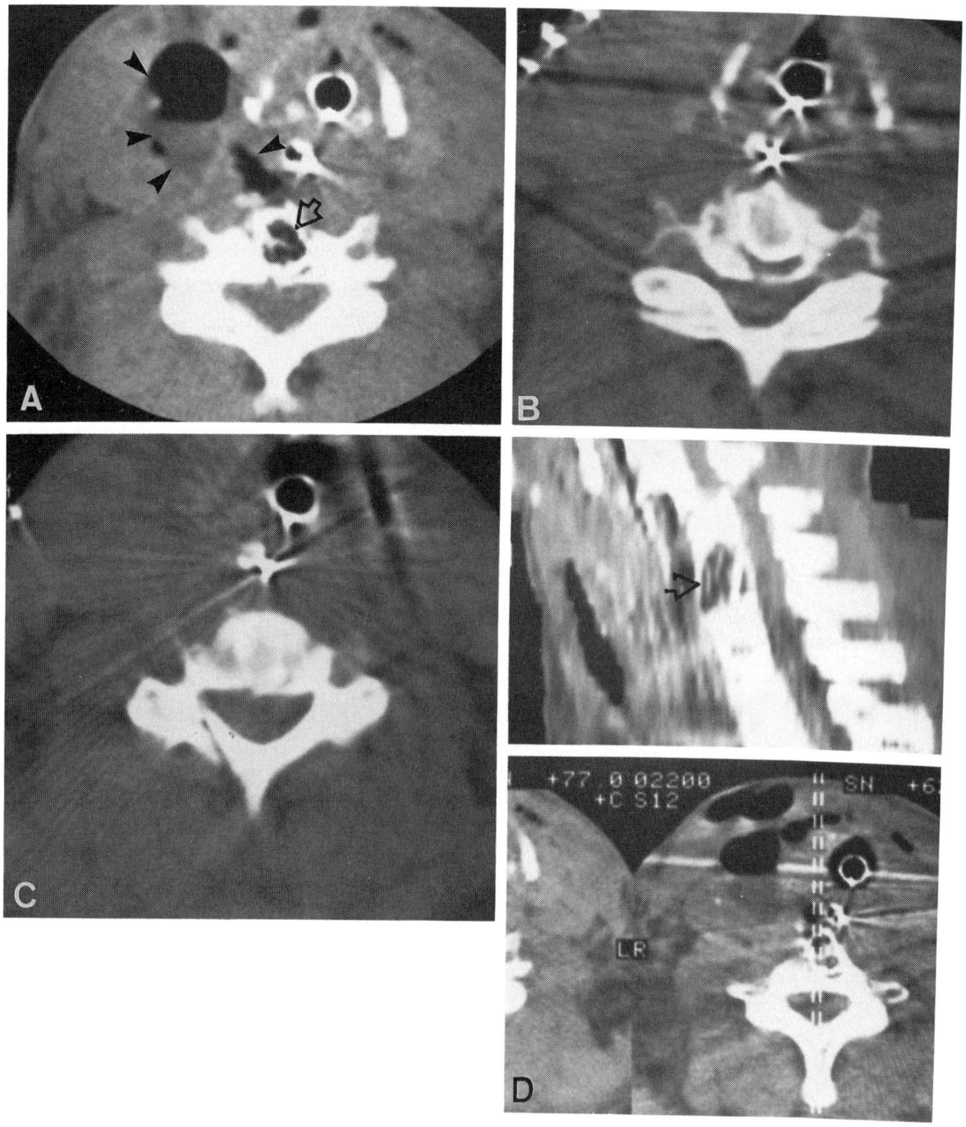

Figure 6.22. **A–C,** Contiguous axial CT scans demonstrate hypodensity within the allograft (*open arrow*) following anterior discectomy and placement of a bone graft. There is extensive prevertebral swelling with a large abscess cavity (*arrowhead*). **D,** sagittal re-formatted CT scan demonstrates necrosis within the graft (*open arrow*) and extensive prevertebral swelling as well as the abscess cavity.

resulting from disk space infection. Both CT and radionuclide studies are unreliable in differentiating normal postoperative changes from early disk space infection [39]. The early morphologic and biochemical changes within the disk space and adjacent vertebral bodies are best evaluated with MRI [40, 41] (Fig. 6.24). Fat-suppressed images allow better evaluation of the bone marrow spaces. Sagittal images demonstrate loss of disk height. The intranuclear cleft is not visualized [42, 43]. The disk space, as well as the marrow of the adjacent vertebral body, has a hypointense signal on the T1-weighted images and a relatively bright signal on T2-weighted images. The signal change within the marrow is seen as a horizontal band, although this may be localized to the posterior half of the adjacent vertebral bodies. There may be associated prevertebral edema in addition to the epidural postoperative changes. On Gd-DTPA–enhanced T1-weighted MRI sequences, the previously hypointense marrow shows significant enhancement. Enhancement is also noted within the disk space and the extradural soft-tissue mass when compared with the precontrast studies. The disk space and adjacent vertebral body changes have an appearance similar to those seen with type 1 degenerative vertebral body changes [44]; however, enhancement of the marrow spaces is uncommon in the normal postoperative patient or with type 1 degeneration. Signal abnormalities and enhancement with Gd-DTPA, as with discitis, are also seen following chymopapaine injection [45, 46].

Discitis, if not recognized, may lead to destruction of the endplate and necrosis, resulting in epidural abscess. This may spread along the posterior or anterior longitudinal ligament beyond one discovertebral segment. The most common organism involved is *Staphylococcus aureus*, although other organisms may be responsible. The infection usually spreads through the hematogenous route. Most of these patients have preexisting diabetes or vascular

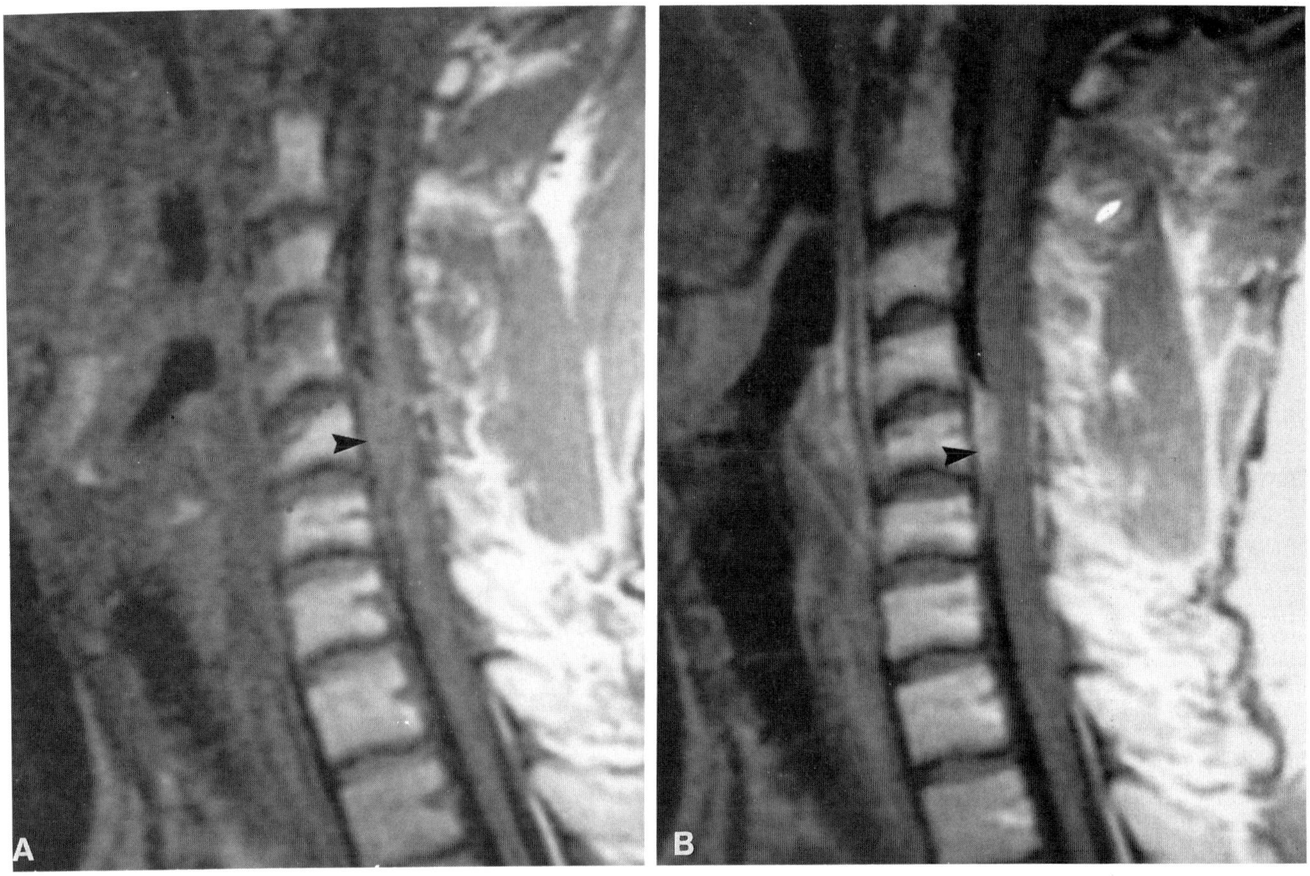

Figure 6.23. Cervical epidural hematoma. A, Sagittal midline section demonstrates an irregular, elongated soft-tissue epidural mass (*arrowhead*) mildly compressing the cord in a patient with persistent neurologic deficit following extensive cervical laminectomy. **B,** Gd-DTPA–enhanced MRI demonstrates enhancement of this focal mass (*arrowhead*). This disappeared on follow-up MRI study. This was considered to represent focal hematoma.

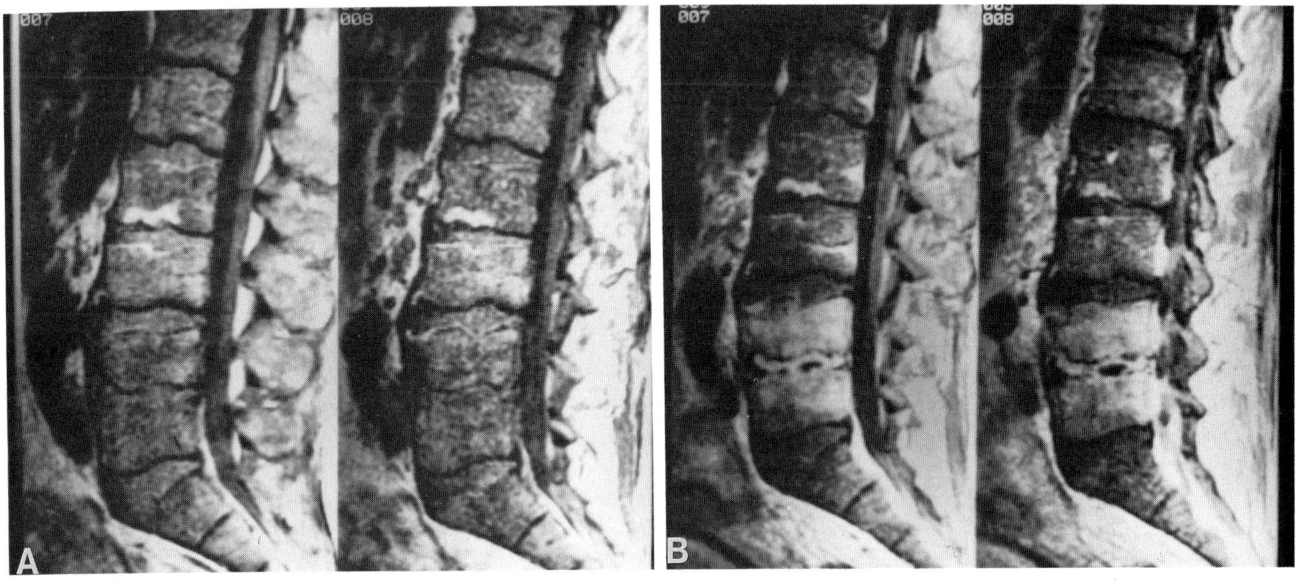

Figure 6.24. Discitis. A, Contiguous sagittal T1-weighted MRI demonstrates narrowing of the L4–5 disk space with ill-defined decreased signal of the marrow in the adjacent vertebral bodies. **B,** Gd-DTPA–enhanced sagittal T1-weighted MRI demonstrates intense enhancement of the disk space as well as the adjacent vertebral body marrow. The patient is 12 days after discography, with persistent back pain, elevated ESR, and low-grade temperature. Similar changes may be seen following discitis after surgery.

insufficiency. Infection should be suspected when there are clinical signs of inflammation with local tenderness beyond the initial recuperative period, usually beyond 7 to 10 days. In the cervical region, this is usually associated with soft-tissue swelling of the neck and torticollis.

Epidural Abscess

The disk space infection, if not recognized, may progress to prevertebral and epidural abscess. This is common when extensive surgery was done, such as anterior interbody fusion, extensive laminectomy, or anterior cervical discectomy with placement of allograft. Epidural abscess also may follow discography or chymopapaine injection. Although the most common organism involved is *S. aureus*, other organisms also may be implicated. This is more often the case with cervical epidural abscess resulting from hematogenous and contiguous spread of infection from nasopharyngeal and oropharyngeal spaces. The epidural abscess is usually ventral in the lumbar and cervical regions and dorsal in the thoracic region. However, the epidural abscess may occur at the level of the disk surgery or at a slightly different level. Discitis often precedes epidural abscess and most likely represents the stage of cellulitis.

On CT, soft-tissue swelling is identified by the absence of the fascial and fat planes. Bone destruction can be suspected by the irregular cortical bone margins, the presence of spicules of bone fragments, and variable but heterogeneous sclerosis of the vertebrae. CTM is necessary to demonstrate the epidural extension of the inflammation or abscess and the presence of cord compression (Fig. 6.25). Although IVCT may demonstrate the soft-tissue enhancement demarcating the necrotic abscess components, MRI is better in this evaluation.

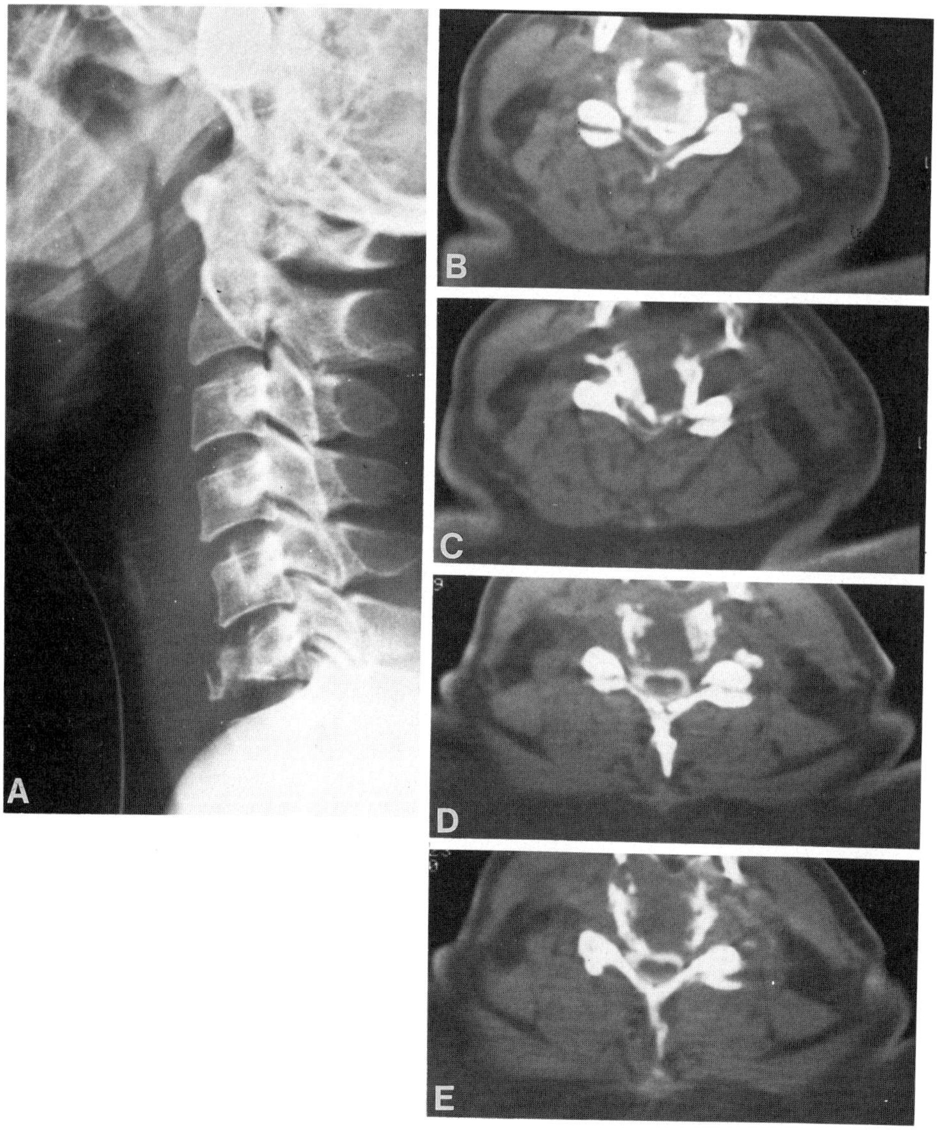

Figure 6.25. Epidural abscess following anterior discectomy. A, Lateral plain x-ray of the cervical spine demonstrates extensive prevertebral soft-tissue swelling with destruction of the anterior margin of the C6 vertebral body. **B–E,** Contiguous axial CTMs demonstrate the epidural extension of the abscess with cord compression. **F,G,** Axial IVCTs in another patient presenting with neurologic deficit, low-grade temperature, and paraplegia developing 2 weeks following surgery. Plain radiograph showed mild soft-tissue swelling at the surgical site. IVCTs demonstrate the epidural abscess with an enhancing ring contralateral to the laminectomy site in the cervical region.

On MRI, epidural abscess and soft-tissue inflammation are seen as hypointense signal on T1-weighted images and as bright signal on T2-weighted and gradient-echo sequences. Similar signal changes also can be identified within the vertebral bodies due to marrow edema (Fig. 6.26). Bone fragments and necrotic osseous and ligamentous components appear as low signal in both T1- and T2-weighted sequences. In most instances, there is intense enhancement of the dural tube and the abscess capsule following Gd-DTPA–enhanced MRI. Both CT and MRI are useful in evaluating the healing process and follow-up.

DURAL FISTULA

Dural fistula with persistent drainage of cerebrospinal fluid (CSF) is a rare complication following back surgery. It usually results from inadvertent tear of the dura and arachnoid. In most instances, the CSF drainage into the soft tissue is self-limiting, usually healing with fibrosis and occasionally resulting in pseudomeningocele.

Rarely, the fistula may persist, with the CSF collection forming a soft-tissue mass in the back and presenting with spinal headache. Pockets of air within the subcutaneous tissue may be identified on plain radiographs, although these can be identified easily on CT (Fig. 6.27). CSF leak from dural tear can be misdiagnosed, unless clinically suspected. On MRI, CSF leaks from dural tears can be difficult to appreciate because the heterogeneous signal in the soft tissue may be mistaken for soft-tissue edema and microshrapnel of bone and metallic fragments from the surgical procedure (Fig. 6.28). Direct visualization of the CSF leak and the site of the dural tear is best achieved with CTM. Thin axial CT sections through the region of interest with sagittal re-formatted images allow precise localization.

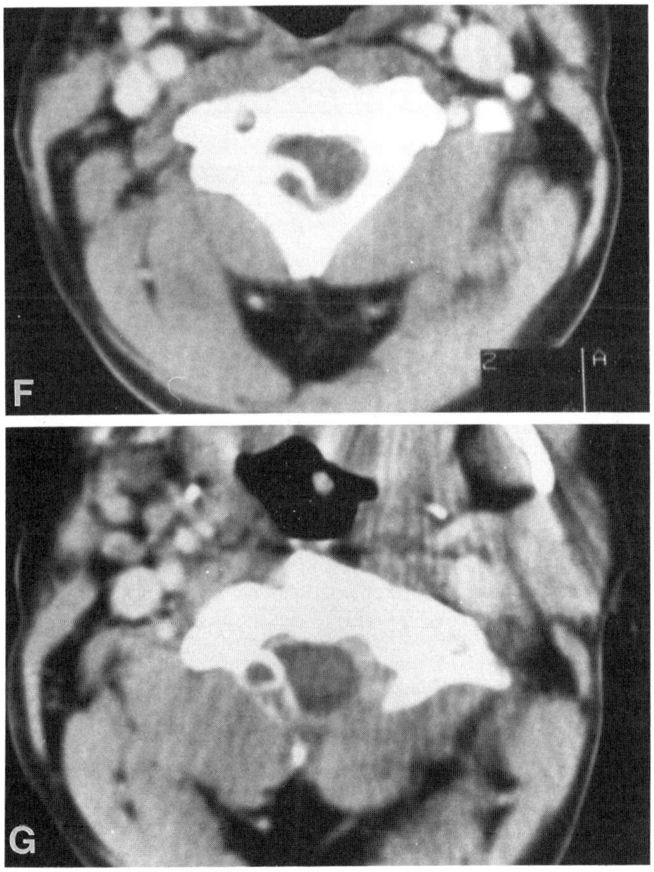

Figure 6.25–*continued*

Delayed Complications

The three most common causes of FBSS are epidural fibrosis (18%), recurrent disk herniation (12%), and spinal stenosis (60 to 70%). Diagnosis of epidural fibrosis and residual disk herniation is accomplished with Gd-DTPA–enhanced MRI. Spinal stenosis usually manifests months or years following back surgery. Other less common delayed postoperative complications include arachnoiditis, pseudomeningocele, and pseudoarthrosis.

SPINAL STENOSIS

Spinal stenosis following back surgery is the result of a combination of progressive degenerative changes of aging and the added mechanical stress resulting from surgery on the discovertebral complex. The stenosis often involves the neural foramen and the neural canal (lateral recess). This is the result of a combination of facet hypertrophy, forward displacement of the superior facet, and bulging of the remnant of the annulus due to loss in disk height [47]. The symptoms may be further aggravated by the associated epidural fibrosis encasing the nerve root.

Less often, postoperative stenosis may result in overall narrowing of the central canal due to hypertrophic changes in the facet joints, as well as the ligamentum flavum. Since during disk surgery the ligamentum flavum is sacrificed at the level of the herniated disk, central stenosis more often involves one level above and below the operated disk space (Fig. 6.29).

The extent of the stenosis is best evaluated on sagittal MRI in long-TR or gradient-echo images (Fig. 6.30). The degenerated disk, hypertrophied ligament, and osteophytes all appear hypointense. T1-weighted images are useful in differentiating the ligamentous and osseous components, except when there is marked bony sclerosis. CTM is more useful in this evaluation because the bone and calcified ligaments can be separated easily from the thickened ligaments. In a similar manner, a degenerated, bulg-

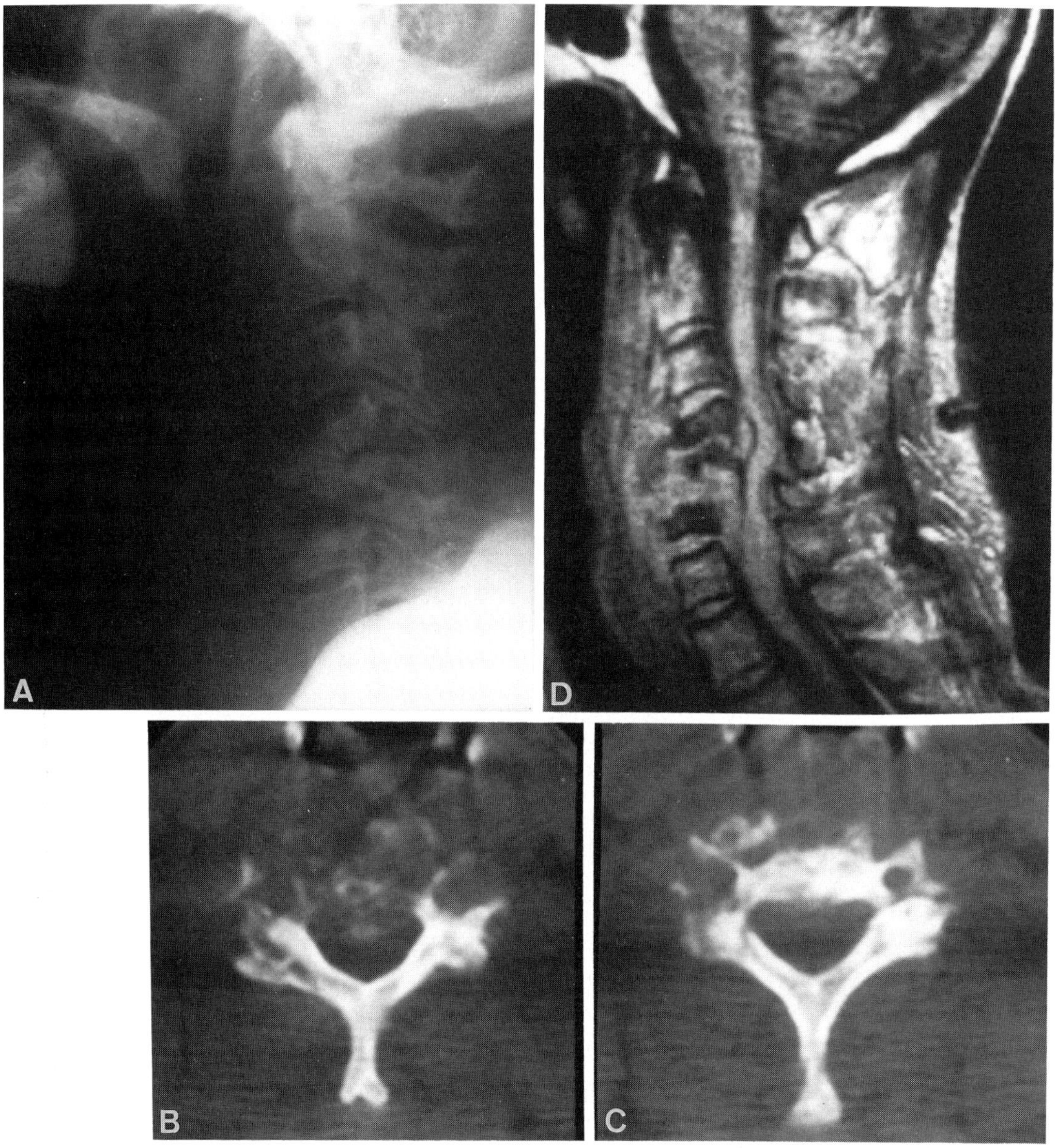

Figure 6.26. Anterior epidural abscess on MRI in a patient 2 months following anterior cervical discectomy and Cloward's procedure. **A,** Plain x-ray demonstrates extensive prevertebral swelling with destruction and partial fusion of the anterior aspects of the C4 and C5 vertebrae. **B,C,** Contiguous CT sections demonstrate the bone destruction with extension into the spinal canal. **D** Sagittal T1-weighted MRI. The remnant of the vertebral body and endplates has ill-defined margins with hypointense signal. The necrotic graft material has a heterogeneous bright signal and projects beyond the anterior and posterior aspects of the vertebral bodies. The epidural abscess is still within the confines of the dura and posterior longitudinal ligament, compressing the spinal cord.

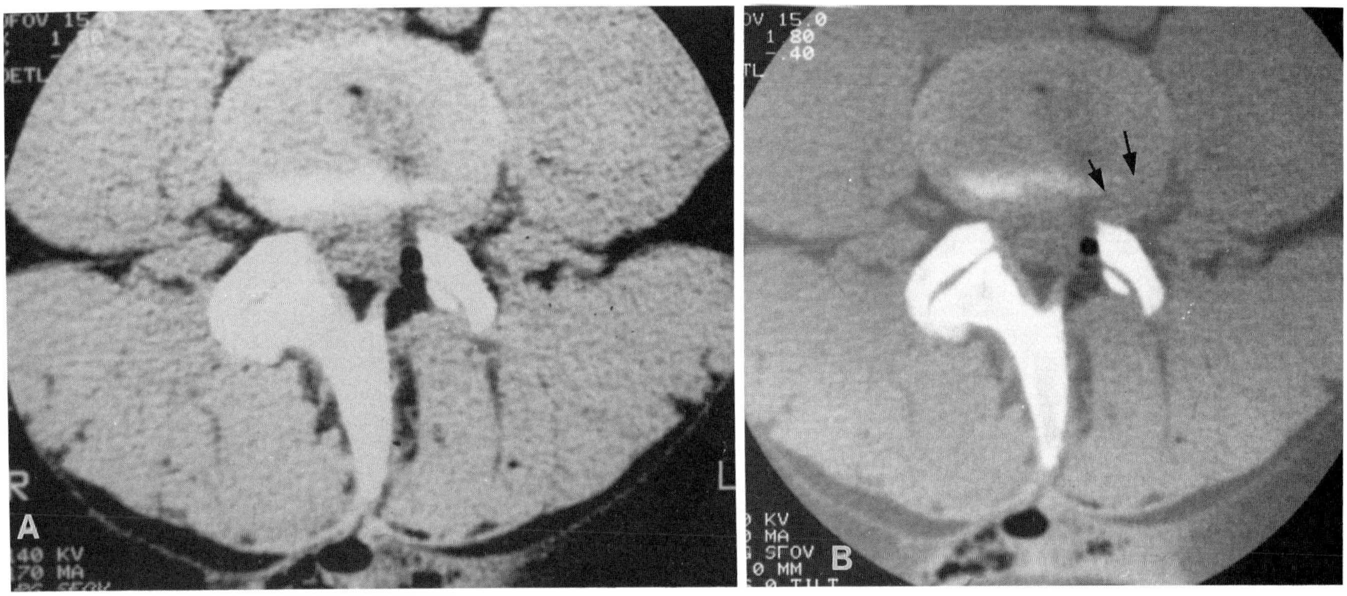

Figure 6.27. Dural fistula. CT scans in a patient with persistent low back pain, soft-tissue swelling, and serous discharge 7 days following back surgery. Axial CT scans with soft-tissue (**A**) and bone (**B**) window settings. Collection of air is present within the subcutaneous tissue, the spinal canal, and the disk space. Residual foraminal disk herniation (*arrows*).

ing annulus can be separated from vertebral end-plate sclerosis (see Chap. 5).

ARACHNOIDITIS

Spinal arachnoiditis is a common postoperative complication. In one large series, spinal arachnoiditis accounted for 6 to 16% of the patients presenting with FBSS [26]. Factors prior to or during surgery may be responsible for the serous meningitis usually resulting in arachnoiditis. These include intrathecal injection of anesthetics or steroids, intrathecal injection of contrast agents, and postoperative infection. In the past, oil-based intrathecal contrast agents were associated with arachnoiditis [48, 49], especially when this material was introduced into the thecal sac admixtured with blood. The incidence of arachnoiditis with presently available nonionic contrast materials is extremely rare [50].

Diagnosis of postoperative arachnoiditis with CT or MRI is difficult and often requires myelography for confirmation. This is particularly true of postoperative arachnoiditis in the cervical and thoracic regions. In the lumbar region, several myelographic features have been described. Myelographic studies may demonstrate adhesions of the cauda equina and nerve roots to the inner surfaces of the meninges, usually along the posterior aspect of the thecal sac. Arachnoiditis also may present as filling defects, narrowing, shortening, or occluding the thecal sac. The latter is often associated with a more severe form of arachnoiditis. Associated epidural scarring is often present.

On CTM, arachnoiditis should be suspected when the sac appears "empty," resulting from the posterior adhesions of the nerve roots. The nerve root sleeves may not be visualized. In the severe form of arachnoiditis, the clumped and deformed nerve roots may be seen within the thecal sac and mistaken for a nerve root tumor. The margins of the thecal sac appear irregular on both CTM and the prior myelogram (Fig. 6.31). MRI diagnosis is based on combination of T1-weighted, T2-weighted, and gradient-echo sequences. Sequential axial and sagittal images show findings similar to those seen with CTM, such as the empty thecal sac sign, the nodular tumor–like clumping, and the foreshortened and narrow thecal sac on the sagittal images. Gd-DTPA–enhanced MRI may show in the occasional patient enhancement of the nerve roots. Enhancement of the pial surface of the nerve roots, especially when it has a nodular appearance, is more suggestive of intradural metastases than chronic adhesive arachnoiditis.

Postoperative arachnoidal adhesions may result in the formation of subarachnoid cysts, which may not communicate with the rest of the subarachnoid space. On delayed CTM studies, contrast material may diffuse into the subarachnoid cyst, while the contrast material in the rest of the subarachnoid space has diluted due to absorption (Fig. 6.32).

Diagnosis of subarachnoid cyst on MRI is difficult. On T2-weighted images, the loculated nonturbulent CSF within the cyst may have a brighter signal compared with the normal CSF space. CT is better in evaluating spinal subarachnoid cyst, whatever its

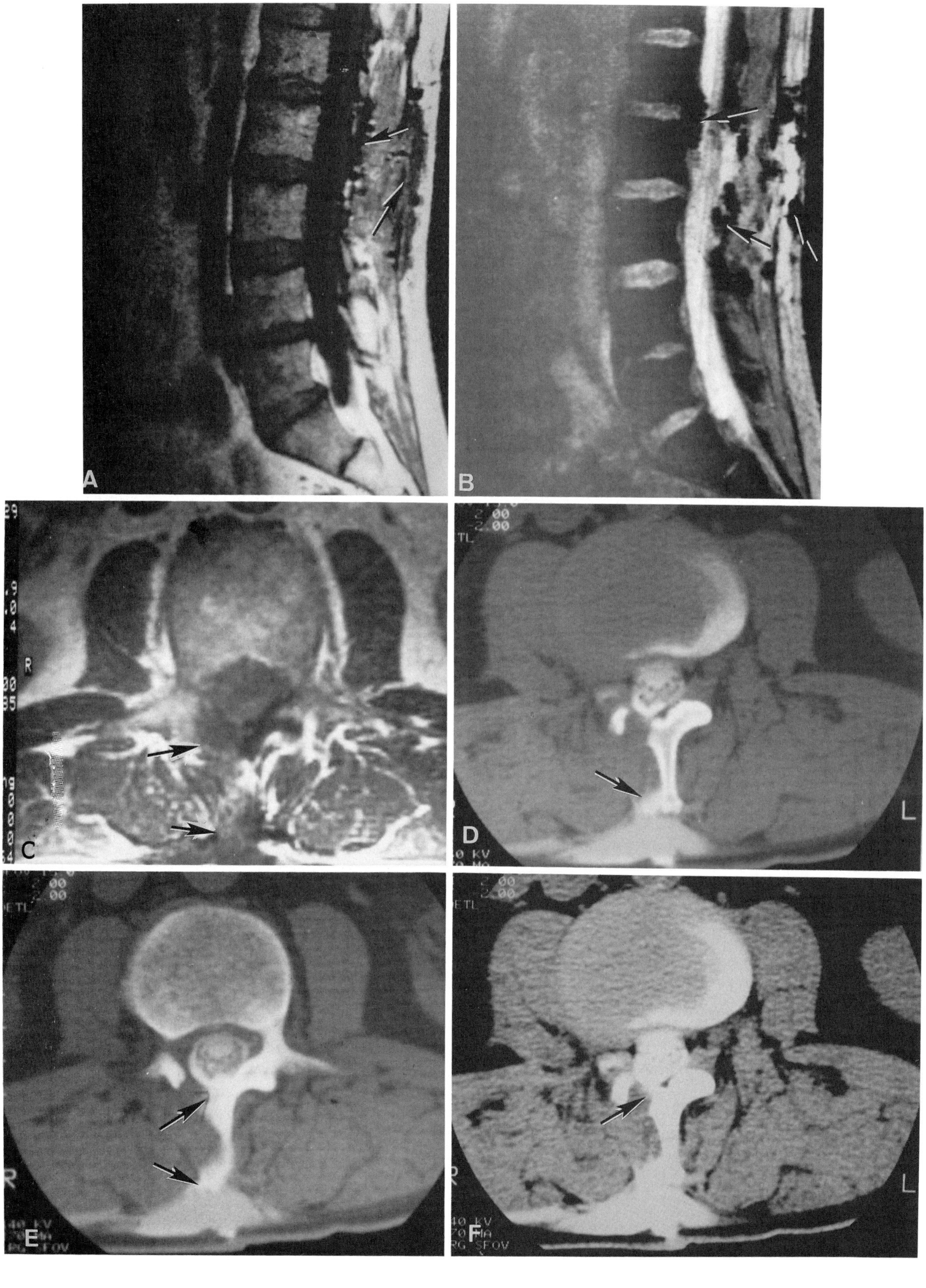
A
B
C
R
D
L
E
R
F
R
L

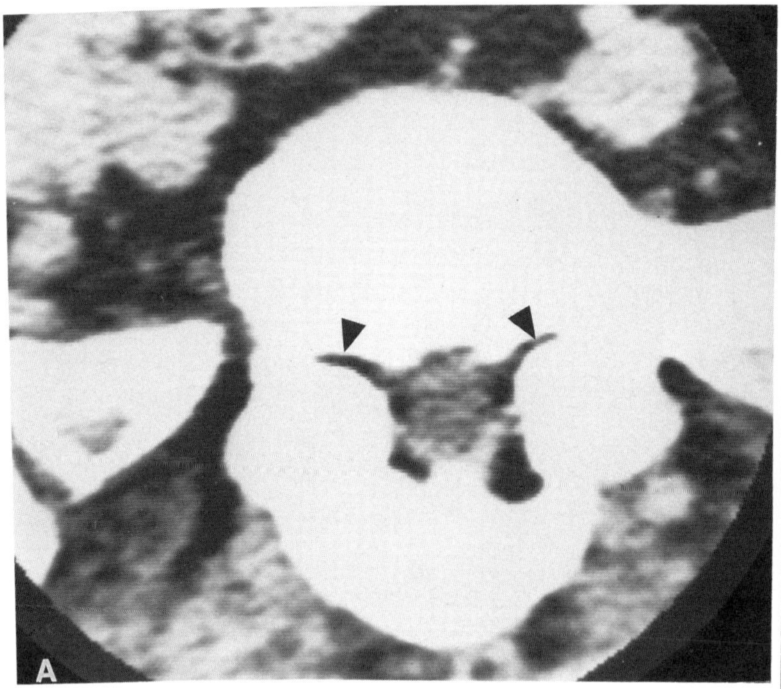

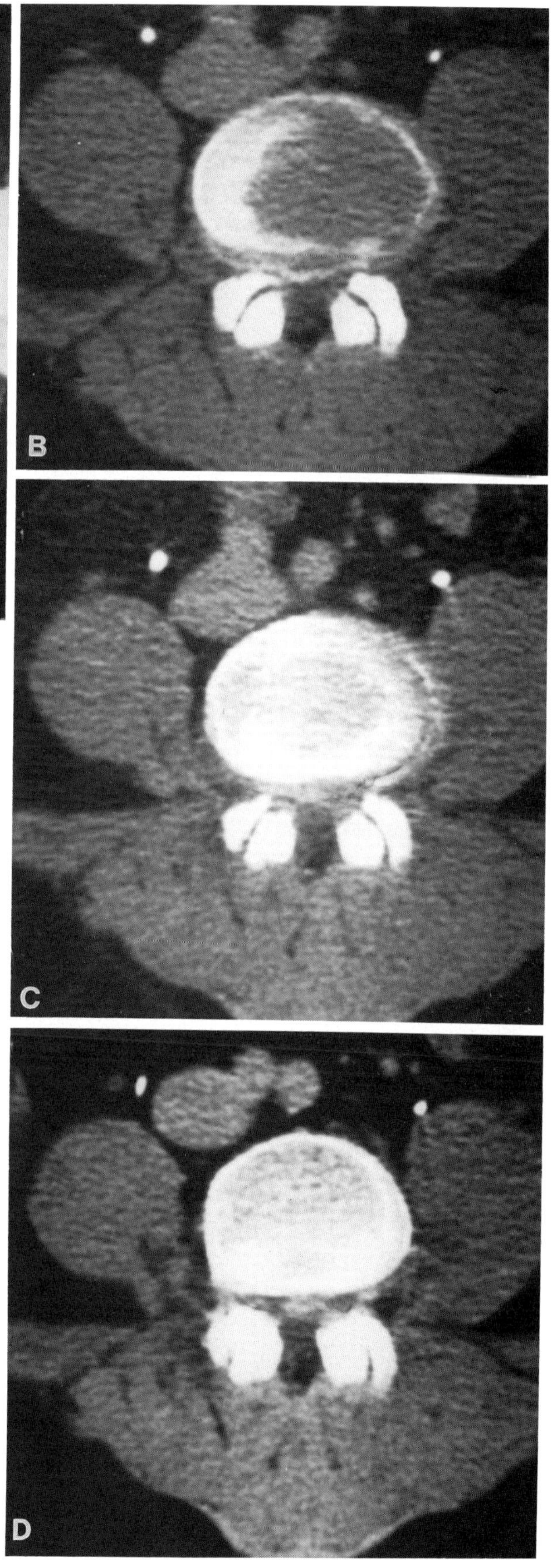

Figure 6.29. Postoperative stenosis. A, Axial CT scan demonstrates bilateral subarticular canal stenosis (*arrows*) from bilateral facet hypertrophy. Patient also had posterior spinal fusion for spinal instability. **B–D,** IVCT scans in another patient following laminectomy for spinal stenosis demonstrating further narrowing of the spinal canal from degenerative disk, enhancing epidural scar and facet hypertrophy.

Figure 6.28. Dural fistula. A 23-year-old man operated on for extraforaminal disk herniation with persistent low CSF pressure headache following disk surgery. MRI sequences obtained 2 weeks following surgery. T1-weighted (**A**) and T2-weighted (**B**) sagittal MRI images demonstrate multiple low-signal-intensity foci within the anterior epidural space as well as in the site of the laminectomy and adjacent subcutaneous tissue. Focal areas of bright signal are also seen in the posterior aspect of the surgical site. **C,** Axial T1-weighted MRI demonstrates mildly hypointense signal along the laminectomy site extending into the subcutaneous space (*arrows*). Contiguous axial CTMs in bone (**D,E**) and soft-tissue (**F**) window settings. The extravasation of the intrathecal contrast material tracking between the dural fistula (*arrows*) and the subcutaneous space is clearly demonstrated.

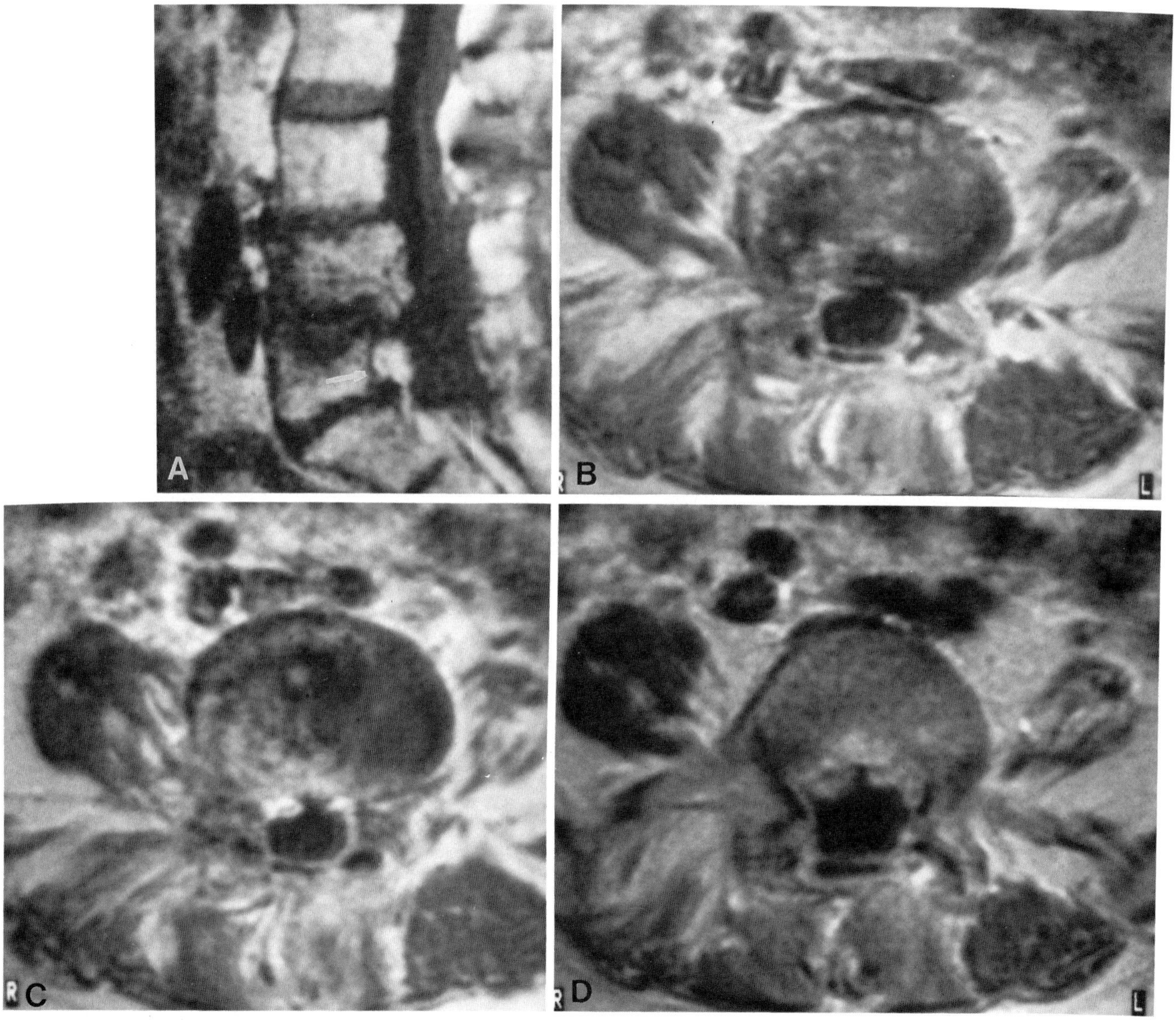

Figure 6.30. Postoperative stenosis. Laminectomy performed 7 years back with persistent debilitating low back pain. **A,** Gd-DTPA–enhanced T1-weighted sagittal midline section demonstrates subluxation of L4 vertebrae. The L3–4 disk space is narrowed. The L4–5 disk space and adjacent endplates demonstrate decreased marrow signal. Enhancement of the epidural venous plexus is seen opposite the L5 vertebral body (*arrow*). Axial sections through L4–5 (**B**) and L5–S1 (**C,D**) levels demonstrate minimal epidural scar enhancement. **E,** Frontal view of lumbar myelogram demonstrates nonfilling of the nerve root sleeve on the right side. **F,G,** MCT scans through the L5–S1 level demonstrate the anterior epidural scar, with deformity of the thecal sac. The nerve root sheath on the right side is not well defined. The postoperative changes are much more clearly demonstrated. Atrophy of the paraspinal muscles can be appreciated on both the MRI and CT studies.

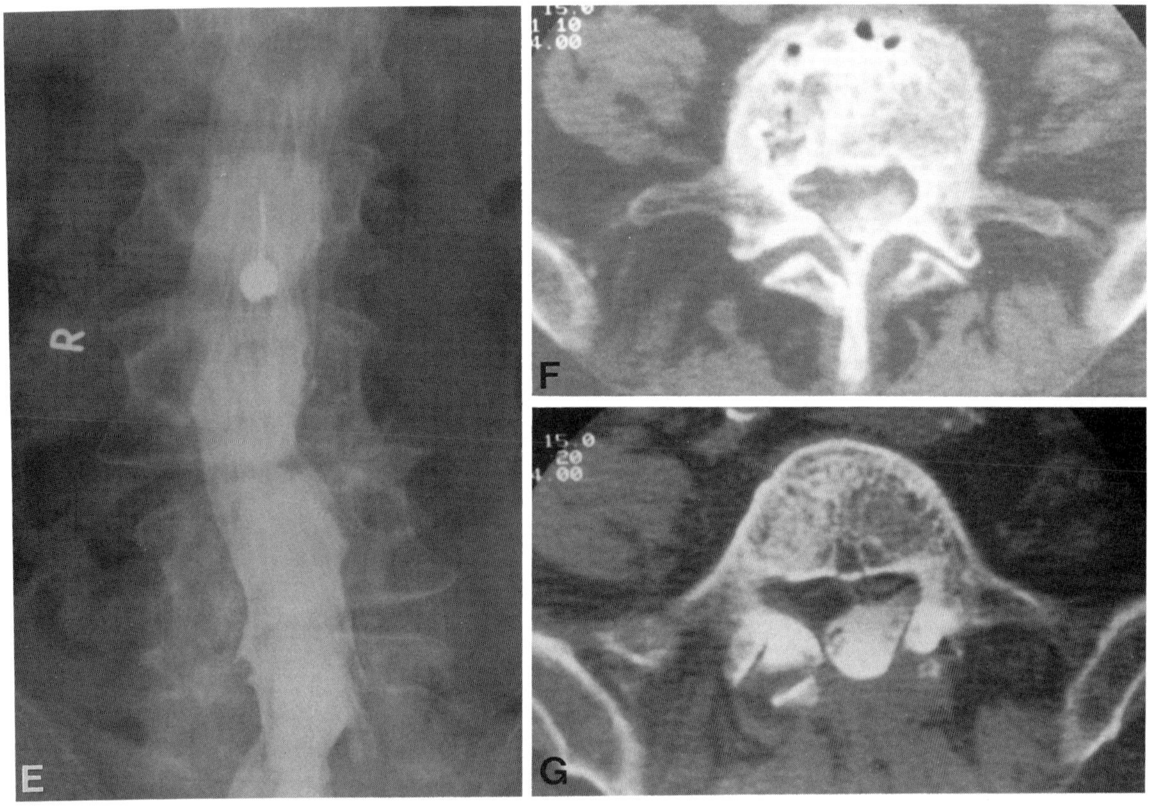

Figure 6.30–*continued*

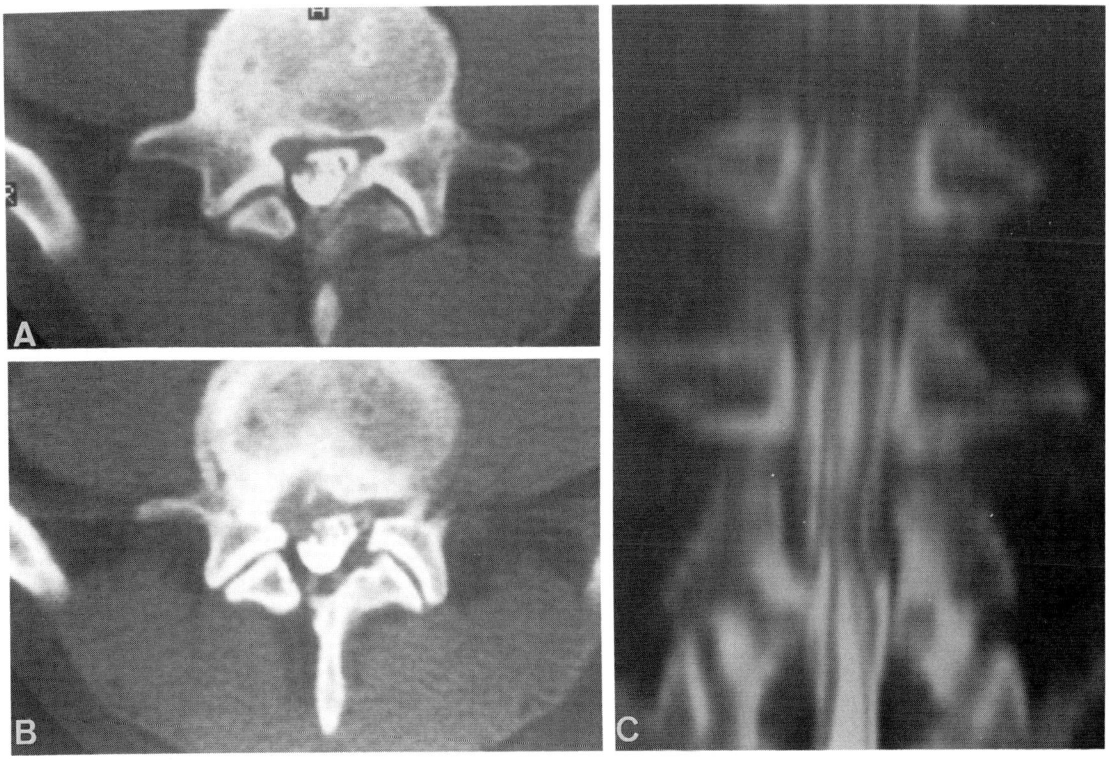

Figure 6.31. Spinal arachnoiditis. A,B, Axial CTM sections demonstrate the thick nerve roots with deformity of the nerve root sheath from epidural scarring and arachnoiditis. **C,** Coronal re-formatted image of the same patient demonstrating the thickening of the nerve roots within the thecal sac. These findings are indicative of arachnoiditis.

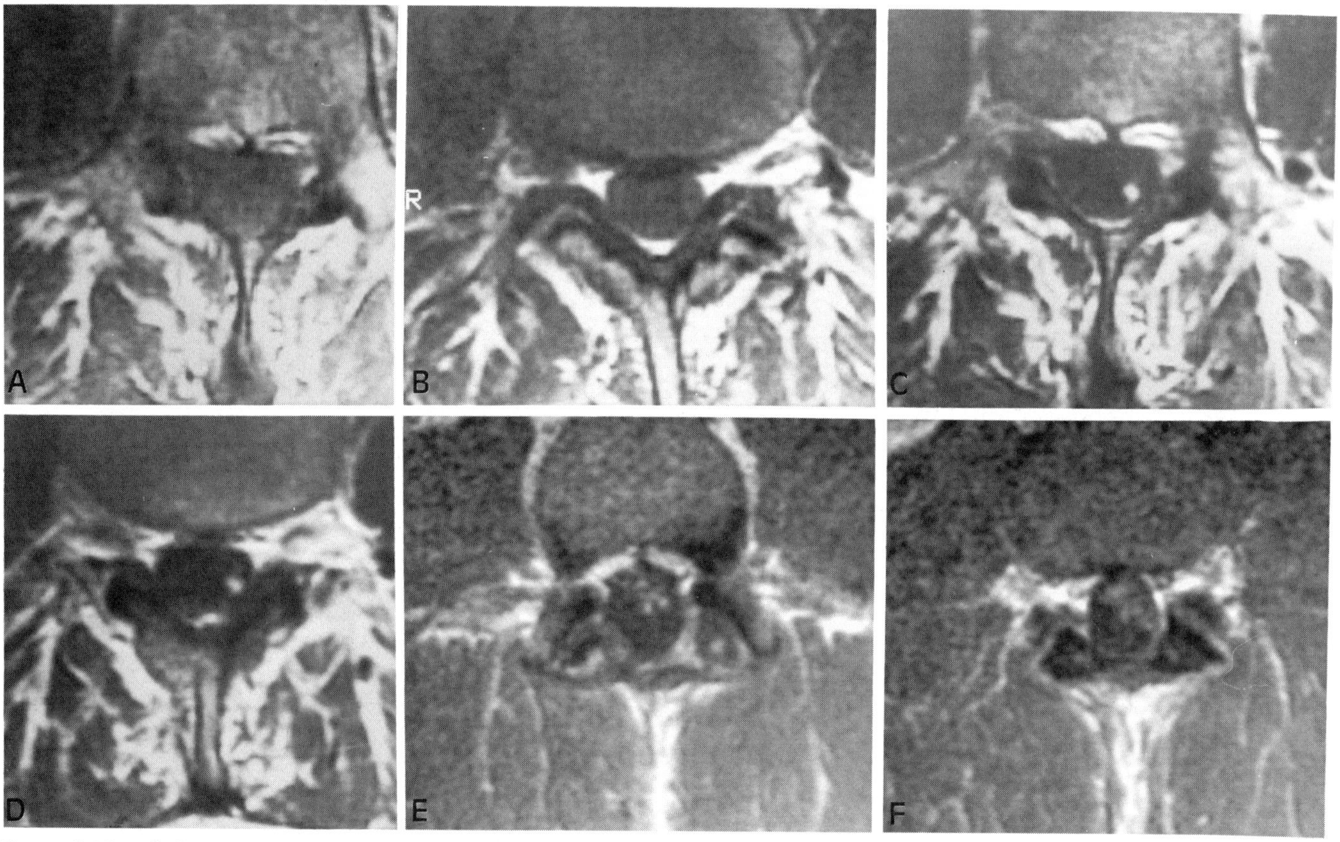

Figure 6.32. Spinal arachnoiditis following surgery. Precontrast (**A,B**) Gd-DTPA–enhanced (**C,D**) MRI sequences demonstrating enhancement of the nerve root in a patient suspected of arachnoiditis. Similar pattern may be seen with intradural metastasis and requires clinical correlation. **E,F,** Gd-DTPA–enhanced axial MRI sections in another patient with arachnoiditis demonstrating previous laminectomy with enhancing clumped nerve roots within the thecal sac.

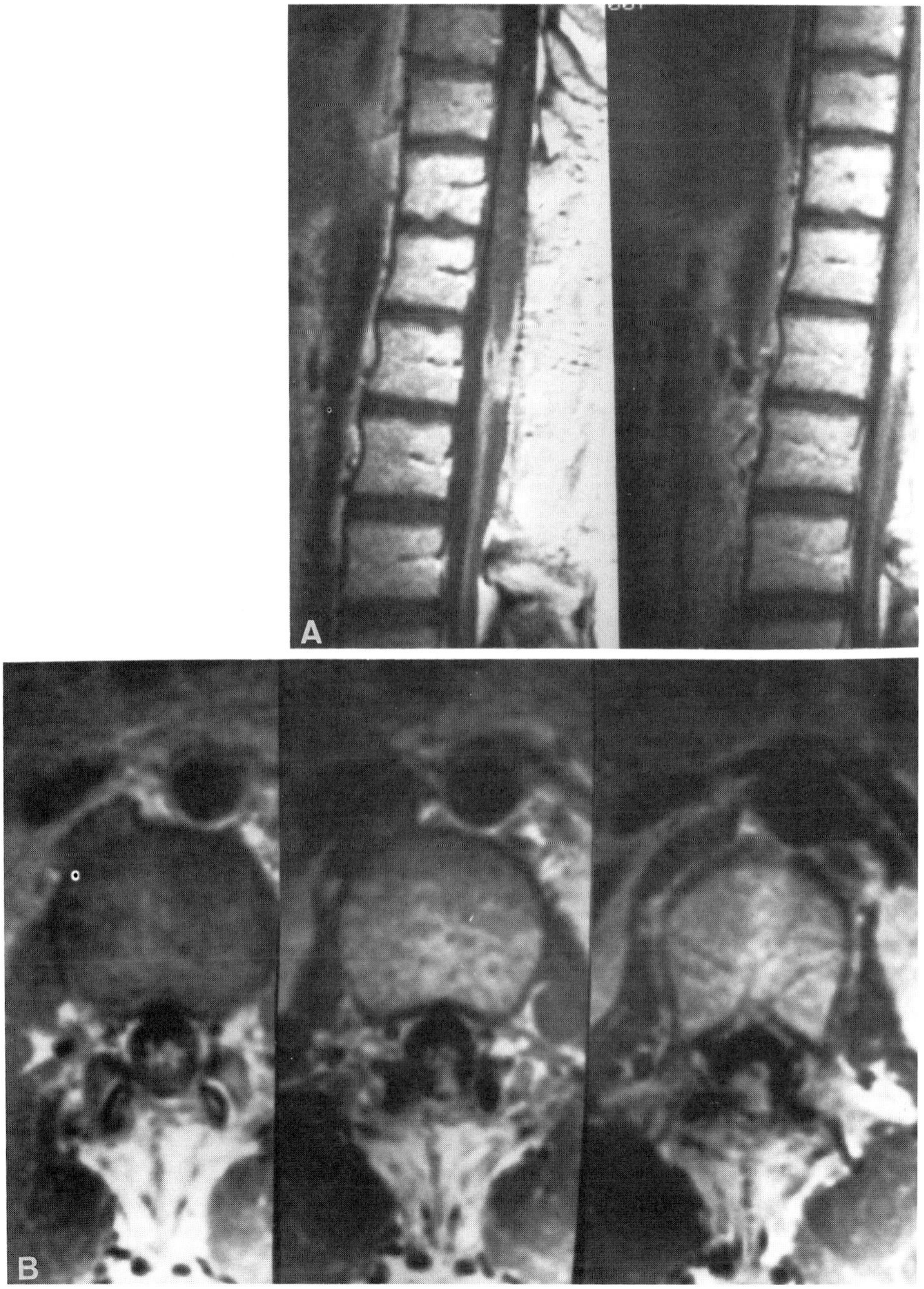

Figure 6.33. Postoperative dural scarring. Precontrast (**A**) and Gd-DTPA–enhanced (**B**) sagittal MRI sequences demonstrate nerve roots adherent to the dorsal aspect of the dura. Patient is being followed for recurrence of tumor. Determination of recurrent or residual neoplasm from postoperative scar enhancement is difficult. This can only be assessed with change in the character and extent of enhancement on sequential MRI studies.

etiology (see Fig. 6.32). Chronic adhesive arachnoiditis in the cervical and thoracic regions may result in cavitation within the cord. The appearance is similar to posttraumatic syrinx.

DURAL SCARRING

Postoperative tethered cord is often seen following invasion of the intradural space for surgery of intradural neoplasms or vascular anomalies. Evaluation of the postoperative tethered cord is by Gd-DTPA–enhanced MRI studies (Fig. 6.33). These are more often indicated for evaluating the presence of a residual intradural lesion. Often this assessment is difficult unless the enhancing intradural lesion is similar or larger than the original lesion. Most often this assessment requires follow-up studies.

PSEUDOMENINGOCELE

Pseudomeningocele is the result of herniation of the arachnoidal membrane following a tear in the dura during surgery [51]. Pseudomeningoceles should be differentiated from small lateral outpouchings of the thecal sac, which are a common finding following laminectomy (Fig. 6.34). Pseudomeningocele may follow wide laminectomy [52]. The pseudomeningocele is connected to the CSF space by a narrow neck (Fig. 6.35). Expansion of the pseudomeningocele over a period of time results from a weak arachnoidal membrane or fibrous tissue expanding from the transmitted CSF pulsation. The outpouching of the dural tube is due to a lack of support posteriorly, since the dura is attached to a weakened scar tissue. The nerve roots may become adherent to the arachnoidal pouch. Although this may be an incidental finding on imaging studies, occasionally patient symptoms have been attributed to the stretching of the adherent nerve roots. Pseudomeningocele is often seen in the cervical region following suboccipital craniotomy (Fig. 6.36), but it may occur in any region of the spine.

On CT, pseudomeningocele is observed as a hypodense CSF-containing sac extending posteriorly beyond the confines of the laminectomy site. On CTM, the contents within the sac communicate with the rest of the CSF space. The nerve roots adherent to the wall of the sac are seen as nodular filling defects. On MRI, the contents of the sac follow the signal intensity of the CSF in the different pulse sequences, hypointense on T1-weighted images and hyperintense on T2-weighted images.

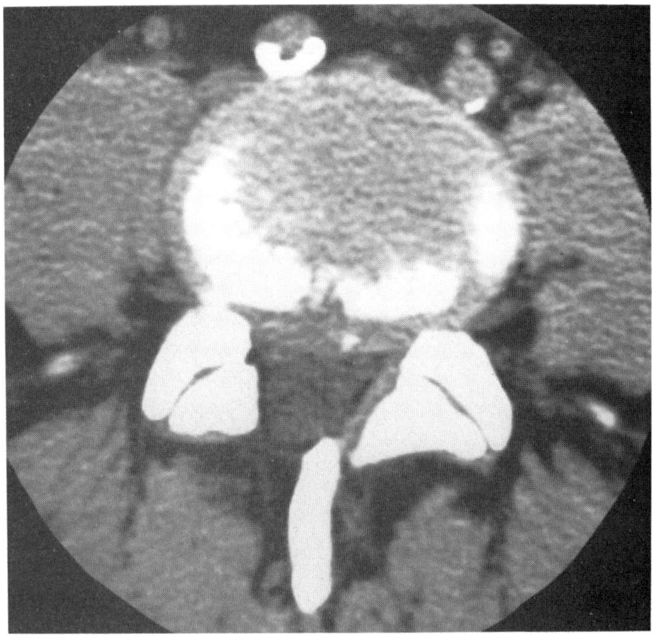

Figure 6.34. Pseudomeningocele. Axial IVCT demonstrates minimal posterior extension of the thecal sac through the laminectomy defect due to absence of the ligamentum flavum and traction of the thecal sac by the scar tissue. There is further bulging of the disk. This is a normal postoperative course.

Pseudomeningoceles should be differentiated from postoperative fluid collections in the soft tissues posteriorly within the muscle planes. They do not communicate with the CSF spaces. Differentiation of an abscess from these serous collections may be difficult. Marginal enhancement on contrast studies and loss of fat planes are probably suggestive of cellulitis prior to abscess formation.

PSEUDOARTHROSIS

Pseudoarthrosis refers to failure of solid fusion of the adjacent bony surfaces after posterior spinal fusion or interbody fusion resulting in mechanical instability. Pseudoarthrosis is best evaluated by plain radiography. On CT, the absence of solid fusion can be suspected when hypodense linear areas are seen on axial CT sections within the regions of the bone grafts with sclerotic hyperdense bone grafts. On MRI, these appear as hypointense signal on both T1- and

Figure 6.35. Pseudomeningocele. Sagittal T1-weighted (**A**), T2-weighted (first echo) (**B**), and T2-weighted (long TR) (**C**) images demonstrate a large pseudomeningocele extending into the subcutaneous tissue through an L4–S1 laminectomy defect. The CSF within the meningocele has a slightly brighter signal compared with the CSF within the spinal canal. This is due to increased protein content and/or lack of pulsatile motion within the pseudomeningocele. **D–F,** Gd-DTPA–enhanced contiguous axial sections demonstrate the neck of the pseudomeningocele (*arrowhead*). There is enhancement of the epidural soft tissues as well as of the nerve roots (*small arrow*) and nerve root sheath (*open arrow*) from arachnoiditis.

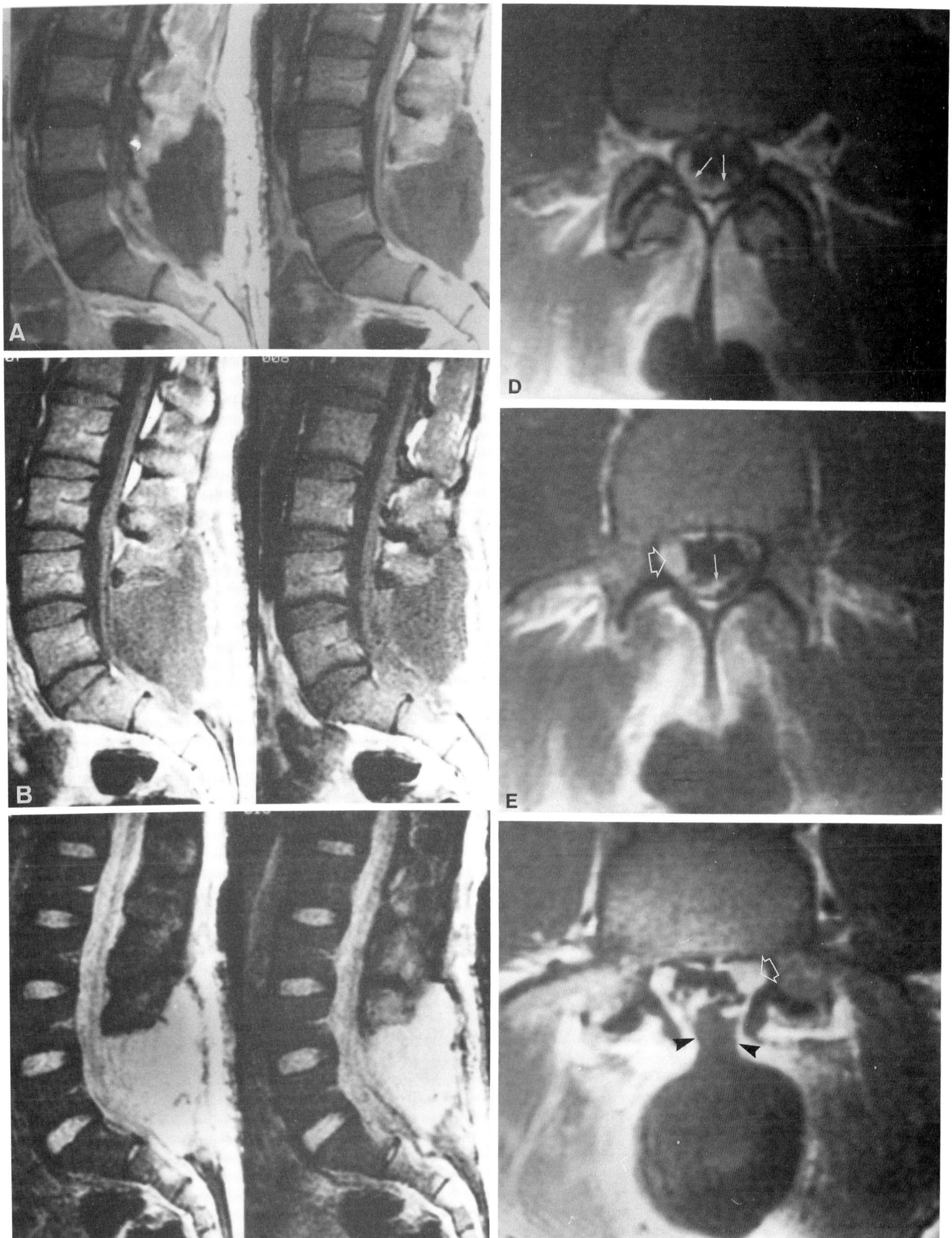
A
B
C
D
E
F

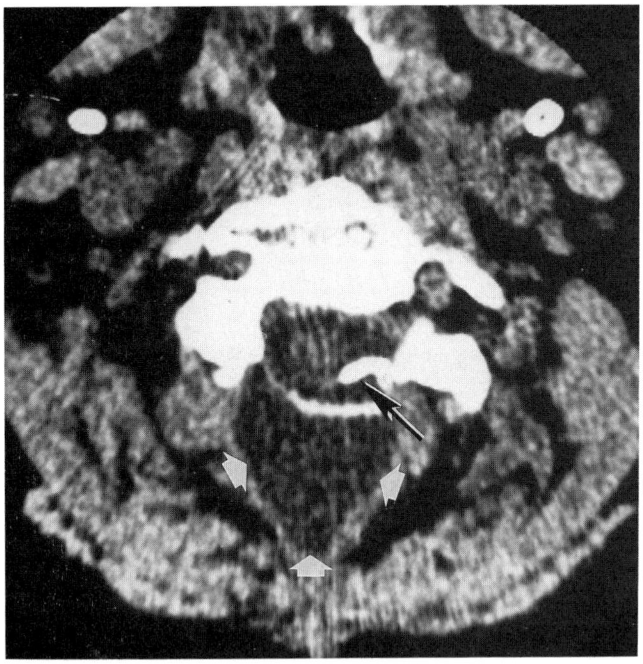

Figure 6.36. Pseudomeningocele following cervical laminectomy. Pseudomeningocele (*arrowhead*) in the cervical region following cervical laminectomy. A shunt catheter (*arrow*) is seen within the spinal canal.

T2-weighted images. MRI has limited value in evaluating for the presence of pseudoarthrosis.

POSTOPERATIVE SPINAL CORD ATROPHY OR CAVITATION

Postoperative atrophy or cavitation within the cord may occur following cervical or thoracic spine surgery for extradural disease such as disk herniation or spinal stenosis. Whether these are manifestations of preexisting cord ischemia or the result of vascular compromise during surgery is difficult to ascertain. Arachnoiditis may result in a similar appearance. MRI is ideal in demonstrating these cord abnormalities; however, even with Gd-DTPA–enhanced MRI, it may be difficult to define the etiologic cause for this type of cord cavitation. CT requires prior administration of intrathecal contrast material for demonstration of the atrophic cord and intravasation of contrast material within the cavitating lesion (see Chap. 12).

References

1. Burton CV, Kirkaldy-Willis WH, Young-Hing K, et al. Causes of failure of surgery on the lumbar spine. Clin Orthop 1981;157:191–199.
2. Fager CA, Freidberg SR. Analysis of failures and poor results of lumbar spine surgery. Spine 1980;5:87–94.
3. Glenn WV, Rhodes ML, Altschuler EM, et al. Multiplanar display of computerized tomography applications in the lumbar spine. Spine 1979;4:282–352.
4. Mali JC, Kaiser JA. The usual appearance of the postoperative lumbar spine. Radiographics 1987;7:245–269.
5. Meyer JD, Latchaw RE, Roppolo HM, et al. Computed tomography and myelography of the postoperative lumbar spine. AJNR 1982;3:223–228.
6. Braun IF, Hoffman JC, Davis PC, et al. Contrast enhancement in CT differentiation between recurrent disk herniation and post-operative scar: prospective study. AJR 1985;145:785–790.
7. Rao KCVG, Williams JP. Contrast enhanced CT in differentiating postoperative scar from recurrent disk herniation. Symp Neuroradiol 1983.
8. Schubiger O, Valabanis A. Postoperative lumbar CT: technique, results, and indications. AJNR 1983;4:595–597.
9. Schubiger O, Valabanis A. CT differentiation between recurrent disk herniation and post-operative scar formation: the value of contrast enhancement. Neuroradiology 1982;22:251–254.
10. Teplick JG, Haskin ME. Intravenous contrast enhanced CT of the postoperative lumbar spine, improved identification of recurrent disk herniation, scar, arachnoiditis and diskitis. AJNR 1984;5:373–383.
11. Yang PJ, Seeger JF, Dzioba RB, et al. High dose IV contrast in CT scanning of the postoperative lumbar spine. AJNR 1986;7:703–707.
12. Ross JS. Magnetic resonance imaging of the postoperative spine. Top Magn Reson Imag 1988;1:39–52.
13. Ross JS, Masryk TJ, Modic MT, et al. Lumbar spine: postoperative assessment with surface-coil MR imaging. Radiology 1987;164:851–860.
14. Ross JS, Masaryk TJ, Modic MT. Postoperative cervical spine: MR assessment. J Comput Assist Tomgr 1987; 11:955–962.
15. Bundschuh CV, Modic MT, Ross JS, et al. Epidural fibrosis and recurrent disk herniation in the lumbar spine: MR imaging assessment. AJNR 1988;9:169–178.
16. Hochhauser L, Kieffer SA, Cacoryian ED, et al. Recurrent post discectomy back pain: MR-surgical correlation. AJNR 1988;9:769–774.
17. Hueftle MG, Modic MT, Ross JS, et al. Lumbar spine: postoperative MR imaging with Gd-DTPA. Radiology 1988;167:817–824.
18. Bundschuh CV, Stein L, Slusser JH, et al. Distinguishing between scar and recurrent herniated disk in postoperative patients: value of contrast enhanced CT and MR imaging. AJNR 1990;11:949–958.
19. Ross JS, Masaryk TJ, Schrader M, et al. MR imaging of the postoperative lumbar spine: assessment with gadopentate dimeglumine. AJNR 1990;11:771–776.
20. Langenskiold A, Kiviluoto O. Prevention of epidural scar formation after operations on the lumbar spine by means of free fat transplant. Clin Orthop 1976;115:92–95.
21. Gill GG, Sakovich L, Thompson E. Pedicle fat grafts for the prevention of scar formation after laminectomy: an experimental study in dogs. Spine 1979;4:176–186.
22. Jacobs RR, McClain O, Neff J. Control of postlaminectomy scar formation. Spine 1980;5:223–227.
23. Teplick JG, Haskins ME: CT of the postoperative lumbar spine. Radiol Clin North Am 1983;21:395–421.
24. Jacobs S. Anterior cervical fusion. Surg Annu 1976; 8:413–446.
25. Cloward RB. The anterior approach for removal of ruptured cervical disks. J Neurosurg 1958;15:601–617.
26. Robinson RA, Smith GW. Anterolateral cervical disc removal and interbody fusion for cervical disc syndrome. J Johns Hopkins Hosp 1955;96:223–231.
27. Long DM, Filtzer DL, Ben Debra M, et al. Clinical features of the failed-back syndrome. J Neurosurg 1988;69:61–71.

28. Cervilline P, Curri D, Volpin N, et al. Computed tomography of epidural fibrosis after discectomy: a comparison between symptomatic and asymptomatic patients. Neurosurgery 1988;23:710–713.
29. Sotiropoulos S, Chafetz NI, Lang P, et al. Differentiation between postoperative scar and recurrent disk herniation: prospective comparison of MR, CT and contrast enhanced CT. AJNR 1989;10:639–643.
30. Wilmink JT, Roukema JG. Effects of IV contrast administration on intraspinal and paraspinal tissues: a CT study: 1. Measurement of CT attenuation numbers. AJNR 1987;8:703–709.
31. Brasch RC, Weinmann HJ, Wesket GE. Contrast enhanced MR imaging: animal studies using Gd-DTPA complex. AJR 1984;142:625–630.
32. Ross JS, Delamarter R, Aikawa M, et al. Gadolinium-DTPA–enhanced MR imaging of the postoperative lumbar spine: time course and mechanism of enhancement. AJR 1989;152:825–834.
33. Weinman HJ, Brasch RC, Press WR, et al. Characteristics of gadolinium-DTPA complex: a potential NMR contrast agent. AJR 1984;142:619–624.
34. Ross JS, Blaser S, Masaryk TJ, et al. Gd-DTPA enhancement of posterior epidural scar: an experimental model. AJNR 1989;10:1083–1088.
35. Boden SD, Davis DO, Dina TS, et al. Contrast-enhanced MR imaging performed after successful lumbar disk surgery: prospective study. Radiology 1992;182:59–64.
36. Bozzao A, Gallucci M, Masciocchi C, et al. Lumbar disk herniation: MR imaging assessment of natural history in patients treated without surgery. Radiology 1992;185: 135–141.
37. Zilkha A, Irwin GA, Fagleman D. Computed tomography of spinal epidural hematoma: AJNR 1983;4:1073–1075.
38. Lindholm TS, Pylkkanen P. Discitis following removal of intervertebral disc. Spine 1982;7:618–622.
39. Puranen J, Makela J, Lahde S. Postoperative intervertebral discitis. Acta Orthop Scand 1984;55:461–465.
40. Boden SD, Davis DO, Dina TS, et al. Postoperative discitis: distinguishing early MR imaging findings from normal postoperative disk space changes. Radiology 1992;184:765–771.
41. Forster A, Pothmann R, Winter K, et al. Magnetic resonance imaging in nonspecific discitis. Pediatr Radiol 1987;17:162–163.
42. Ross JS. Magnetic resonance assessment of the postoperative spine: degenerative disk disease. Radiol Clin North Am 1991;29:793–808.
43. Hyman RA, Gorey MT. Imaging strategies for MR of the spine. Radiol Clin North Am 1988;26:505–533.
44. Modic MT, Steinberg PM, Ross JS, et al. Degenerative disk disease: assessment of changes in vertebral body marrow with MR imaging. Radiology 1988;166:193–199.
45. Masaryk TJ, Boumphrey F, Modic MT, et al. Effects of chemoneucleolysis demonstrated by MR imaging. J Comput Assist Tomgr 1986;10:917–923.
46. Deeb ZL, Schmiel S, Daffner RH, et al. Intervertebral disc-space infection after chymopapain injection. AJR 1986;144:671–674.
47. Brodsky AE. Post-laminectomy and post-fusion stenosis of the lumbar spine. Clin Orthop 1976;115:130–139.
48. Hackney DB, Grossman RI, Zimmerman RA, et al. MR characteristics of iophendylate (pantopaque). J Comput Assist Tomogr 1986;10:401–403.
49. Quencer RM, Tenner M, Rothman L: The postoperative myelogram. Radiology 1975;123:667–679.
50. Haughton VM, Eldevik OP, Ho KC, et al. Arachnoiditis from experimental myelography with aqueous contrast media. Spine 1978;3:65–69.
51. Miller PR, Elder FW. Meningeal pseudocyst (meningocele spurious) following laminectomy. J Bone Joint Surg 1968;50A:268–276.
52. Teplick JG, Peyster RG, Teplick SK, et al. CT identification of postlaminectomy pseudomeningocele. AJNR 1983;4:179–182.
53. Nguyen CM, Haughton VM, Ho KC, et al. MR contrast enhancement: An experimental study in postlaminectomy epidural fibrosis. AJNR 1993;14:997–1002.

CHAPTER / 7

Infection

D. A. Eckard / S. Batnitzky / H. I. Price

Spinal infection is an uncommon but important clinical problem. If recognized early and treated appropriately, a full recovery often can be expected. Unfortunately, unrecognized vertebral osteomyelitis can still be the cause of substantial neurologic disability or death.

The spectrum of spinal infection includes osteomyelitis and discitis (infectious spondylitis), epidural abscess, transverse myelitis, and meningeal infections. A number of imaging modalities are available for the evaluation of spinal infection. These include plain films of the spine, computed tomography (CT), radionuclide scintigraphy, myelography, CT following myelography (MCT), and magnetic resonance imaging (MRI). MRI is generally considered the best imaging modality to evaluate spinal infection [1]. These various types of spinal infection and imaging findings will be discussed in this chapter.

Infectious Spondylitis

Infections of the bone (osteomyelitis) and infections of the disk (discitis) usually occur in conjunction with one another, and thus the term *infectious spondylitis* is used in this chapter when referring to these types of infections [2]. Infectious spondylitis accounts for approximately 5% of all cases of pyogenic osteomyelitis [3].

Pathogenic organisms reach the spine by hematogenous spread through the arteries, hematogenous spread through the veins from Batson's plexus, or direct inoculation via a diagnostic or surgical procedure [4]. The most common sources of hematogenous spread are the genitourinary tract, skin, and respiratory tract, with most pelvic sources of infection such as urinary tract infections, enteric fistulas, surgery, and the postpartum state inoculated via Batson's plexus [5, 6]. In approximately 37% of cases, no source of infection can be identified [7].

The blood-borne bacteria usually localize in the subchondral portions of the vertebral body via the nutrient equatorial artery and metaphyseal arteries. Anastomoses are present between the equitorial and metaphyseal arteries in children younger than 15 years old but not in adults. Because bone infarct is believed necessary for pyogenic osteomyelitis to occur, the more vascular nature of the vertebral endplates makes infectious spondylitis in children much less common [8].

After the infection settles into the subchondral space, it usually spreads contiguously into the disk space, causing both an osteomyelitis and a discitis [9]. The infection can then spread across the disk into the adjacent contiguous endplate. Lower-grade infections such as tuberculosis and fungal infections often pass in a subligamentous manner to the adjacent vertebral body and thus do not involve the disk space [2]. This lack of disk space narrowing is characteristic of tuberculosis and fungal infections. It is rare for one vertebra to be involved alone in either pyogenic or nonpyogenic infectious spondylitis, occurring in less than 1% of patients [2]. Infection rarely spreads into the pedicles or posterior elements. When present, it should suggest tuberculosis. Tuberculosis also should be suspected if there is extensive paravertebral abscess formation.

With treatment or host response, various degrees of regenerative processes occur. Earlier treatment leads to less damage. The major response is that of bone sclerosis in both pyogenic and nonpyogenic infections. If treatment is delayed, lysis of the bone can cause complete bony collapse of the affected vertebral bodies. In tuberculosis, anterior wedging causing a gibbus deformity is not uncommon.

Pyogenic spondylitis is most commonly caused by *Staphylococcus aureus*, but almost any organism can cause it, including *Streptococcus* and gram-negative organisms [6]. Nonpyogenic spondylitis is most commonly caused by tuberculosis but also can be caused by brucellosis and a variety of fungi, including *Coccidioides*, *Aspergillus*, and *Candida*. Direct aspiration or tissue biopsy is usually necessary to identify the causative organism.

Affected patients are usually in the fifth to seventh decades of life, with an average age of 50 years. Males predominate 2:1 over females [7]. Symptoms vary depending on the site, the individual host response, and the type of organism. Most patients present with back pain of several weeks' duration, which may be constant or intermittent, mild or severe. Limitation of motion is commonly seen, and the patient may or may not be febrile. The development of neurologic symptoms related to the infection is a late finding and is felt to be due to either an associated epidural abscess or compression deformity of the involved vertebral bodies with retropulsion of fragments and infectious debris. The white blood cell (WBC) count and erythrocyte sedimentation rate (ESR) are often elevated, but this is not always seen. Predisposing factors are identified in only a minority of cases and include diabetes mellitus, blunt trauma, alcoholism, immunosuppression, human immunodeficiency virus (HIV) infection, intravenous drug abuse, infectious endocarditis, and history of a recent surgical or invasive diagnostic spinal procedure [10].

Nonpyogenic osteomyelitis presents in a similar fashion, although usually with a more prolonged clinical course lasting from months to years. Tuberculous spondylitis in North America is most prevalent in adults, with a mean age of approximately 40 years, which is younger than pyogenic spondylitis [11]. Eighty percent of patients with vertebral tuberculosis do not have evidence of pulmonary infection [12].

Although infectious spondylitis usually involves both the vertebral body and the disk, this may not always be the case in children because a direct hematogenous route to the disk space exists [13, 14]. Infants and children with discitis may present with back pain, hip pain, abdominal pain, or reluctance to ambulate. The ESR is usually elevated, and the WBC count may or may not be elevated. As in adults, the most common cultured organism is *S. aureus*. In children, discitis is usually a benign self-limited condition that does well with or without antibiotic therapy.

RADIOGRAPHIC EVALUATION

Of the many radiographic procedures available to evaluate infectious spondylitis, MRI is best. Modic et al. [15] studied 23 patients with the microbiologic and histologic diagnosis of spinal osteomyelitis and found that MRI had a sensitivity of 96%, a specificity of 92%, and an accuracy of 94%. Typically, T1-weighted images show narrowing of the involved disk space, low signal intensity within the adjacent vertebral bodies, subligamentous or epidural soft-tissue masses which are generally isointense to the disk, and cortical erosions of the endplates [16] (Figs. 7.1 through 7.6). Rarely, the T1-weighted images appear normal. T2-weighted images usually show narrowing of the disk space, high signal intensity of the adjacent vertebral bodies, high-signal-intensity epidural or subligamentous masses, and erosions of the cortical endplates [16]. The disk often shows increased signal intensity on T2-weighted scans but may show normal or decreased signal intensity (see Figs. 7.1 through 7.5).

Gadolinium-enhanced scans may be helpful (see Figs. 7.1 through 7.6). Postcontrast scans can show (*a*) thick, dense enhancement at the periphery of the disk, (*b*) thin peripheral disk enhancement, (*c*) broad, dense enhancement of most of the disk space, (*d*) patchy nonconfluent areas of uptake, and (*e*) thin and sometimes subtle areas of uptake in a small portion of the disk periphery [17]. In the paravertebral soft-tissue masses, inhomogeneous enhancement is often seen. Contrast-enhanced images also allow for better visualization and delineation of associated epidural abscesses. Contrast material often obliterates the abnormal low signal intensity seen within the vertebral bodies on T1-weighted images. Therefore, it is important that unenhanced images be obtained prior to giving contrast material.

Fat and water MRI scans, T2* and STIR scans, also have been evaluated in patients with infectious spondylitis [16]. Fat (water signal–suppressed) sequences correlated closely with T1-weighted scans. Water (fat signal–suppressed) sequences correlated closely with T2-weighted images, showing increased signal intensity within involved marrow. T2*-weighted images are similar to T2-weighted images but have less contrast. STIR images are also similar to T2-weighted images but have better contrast between normal and abnormal tissue. In general, T1- and T2-weighted images should be obtained. Contrast-enhanced scans are occasionally useful. STIR scans, because of their high contrast, are very helpful, and a fat scan could be of benefit in questionable cases, since it is very sensitive to marrow replacement [16].

Infants and children with infectious spondylitis show changes similar to adults. However, the abnormality on T1-weighted images within the marrow is

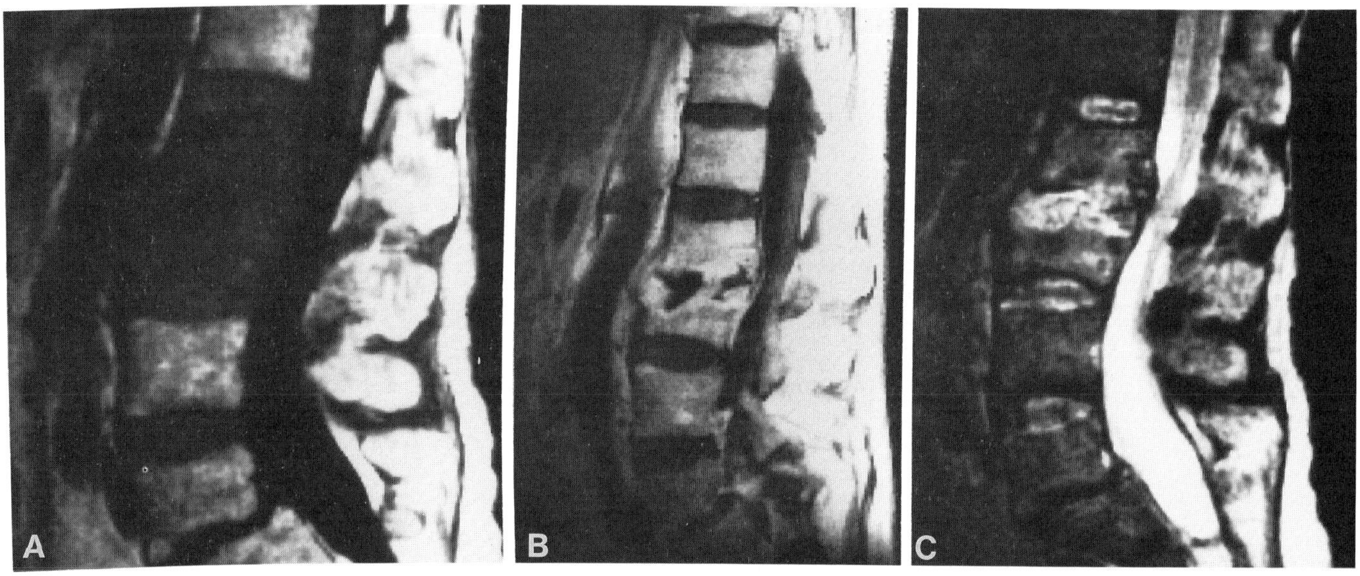

Figure 7.1. ***Escherichia coli*** **infectious spondylitis. A,** T1-weighted scan (SE 500/17) shows typical findings of infectious spondylitis with decreased signal intensity involving the L2 and L3 vertebral bodies. The disk space is difficult to visualize. **B,** T1-weighted scan (SE 500/17) following Gd-DTPA. Following contrast administration, there is enhancement of the L2 and L3 vertebral bodies, which obscures the abnormality in signal intensity seen on the unenhanced T1-weighted images. The disk space is seen to be irregular and narrowed and shows patchy rim enhancement. There is irregularity of the endplates at L2 and L3. Mild anterior extradural deformity is present from mild retropulsion. **C,** T2-weighted scan (SE 2000/90) shows increased signal intensity involving the L2 and L3 vertebral bodies. The disk shows markedly increased signal intensity.

less obvious because of the low-signal-intensity marrow normally found in children. In infants, the abnormal area in the vertebral bodies may actually appear as increased signal intensity relative to the adjacent normal marrow [18]. Another difference between adults and children is that the disk usually shows decreased signal intensity on the T2-weighted images in children [18].

MRI may not be as useful in evaluating for resolution of the infection. The abnormalities seen on MRI can persist for some time. However, with successful treatment, it has been noted that contrast enhancement in the disk, the adjacent vertebral bodies, and the paravertebral soft tissues progressively decreases (see Fig. 7.5) with resolution despite persistent changes on unenhanced images [17].

Between 1 and 2 years following the infection, end-stage changes are seen which usually consist of narrowing or obliteration of the disk space and occasionally fusion of the involved vertebral bodies across the disk space. Often, the marrow fat which has been destroyed by the infection is not replaced, but the acute inflammatory changes in the vertebral bodies and the surrounding soft tissues resolve [2].

Disk space changes are often not present or occur later in tuberculous spondylitis, which may confuse the diagnosis of an infectious process and suggest metastatic disease [1]. Thrush and Enzmann [16] evaluated three patients with tuberculous spondylitis and found areas of increased and decreased signal intensity within adjacent vertebral bodies on T1-weighted images (Figs. 7.7 and 7.8). These areas of high and low signal intensity on T1-weighted images were all of high signal intensity on T2-weighted images. Focal cortical bone erosion is common. The disk spaces and signal intensities are generally well preserved even with advanced bony disease. Large anterior and lateral paraspinous masses are usually present. They are isointense on T1-weighted images and increased in signal intensity on T2-weighted images. Epidural masses may be present. Involvement of the posterior elements favors tuberculosis spondylitis over pyogenic spondylitis, being quite common in the former and very rare in the latter. More than two vertebral bodies are involved in from 50 to 65% of cases [12].

Plain films in infectious spondylitis can show loss of definition of endplate margins, narrowing of disk height, associated soft-tissue swelling, endplate erosions, or vertebral body destruction (see Fig. 7.2). However, plain film findings are often subtle and do not become abnormal for 2 to 3 weeks after the onset of symptoms [6, 19]. The differentiation of osteomyelitis from degenerative disk disease is also very difficult. The presence of intradiscal gas, the *vacuum phenomenon*, formerly was thought to exclude the presence of infection, but this has been disproved [20, 21].

CT findings in infectious spondylitis are mainly related to the adjacent soft-tissue changes with swelling and obliteration of adjacent fat planes (see Fig. 7.4). Fragmentation and/or erosive changes in

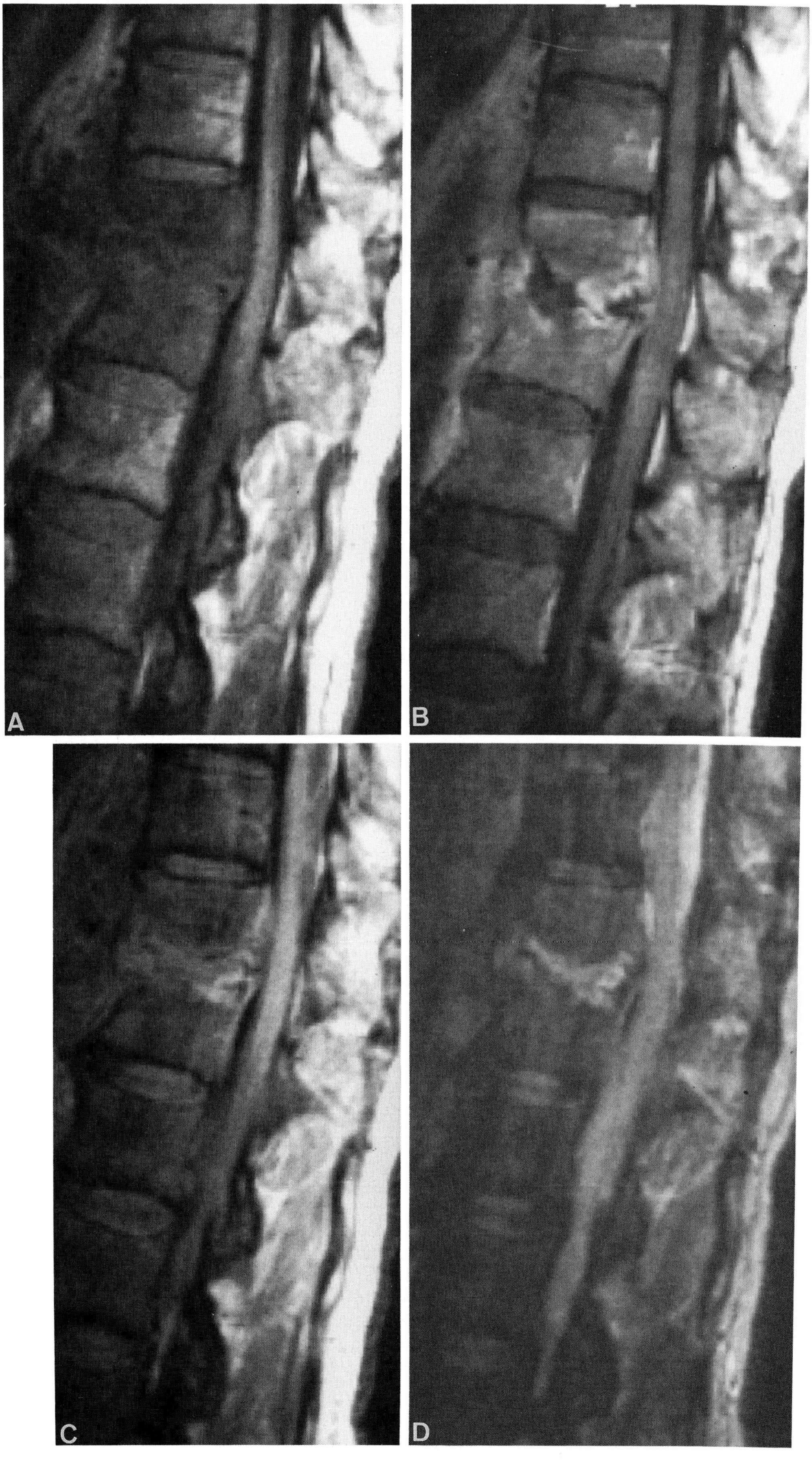
A
B
C
D

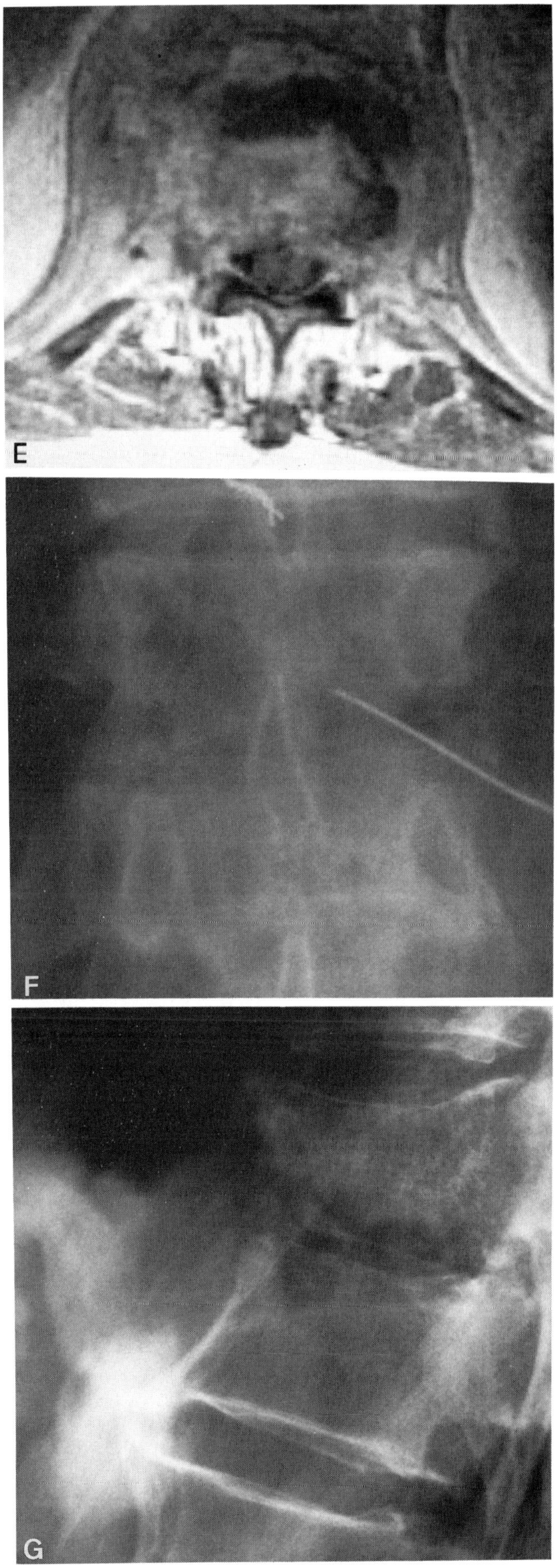

Figure 7.2. *Staphylococcal aureus* infectious spondylitis and epidural abscess in a 69-year-old man with a 4-month history of increasing back pain. **A,** Sagittal T1-weighted scan (SE 500/17) shows decreased signal intensity in the T11 and T12 vertebral bodies. There is mild retropulsion and possible epidural mass. The disk is not well seen. **B,** Sagittal T1-weighted scan (SE 500/17) following Gd-DTPA shows narrowing and destruction as well as irregular enhancement of the disk. Enhancement of the vertebral bodies obscures the decreased signal intensity seen on the unenhanced T1-weighted images. There is enhancement of a small epidural abscess. Irregularity and destruction of the endplates of the vertebral bodies are seen. **C,** Sagittal proton-density scan (SE 2000/25) shows subtle increased signal of the disk that is better seen on T2-weighted images. **D,** Sagittal T2-weighted scan (SE 2000/90) shows increased signal of the disk and subtle increased signal in the T11 and T12 vertebral bodies. **E,** Axial T1-weighted image (SE 800/17) following Gd-DTPA shows narrowing of the spinal canal from mild retropulsion and an epidural abscess. There is swelling of the perivertebral soft tissues. **F,G,** Plain films of the spine obtained at the time of needle biopsy show narrowing of the T11 and T12 disk space as well as irregularity and decreased density of the endplates, especially the anteroinferior portion of the T11 vertebral body.

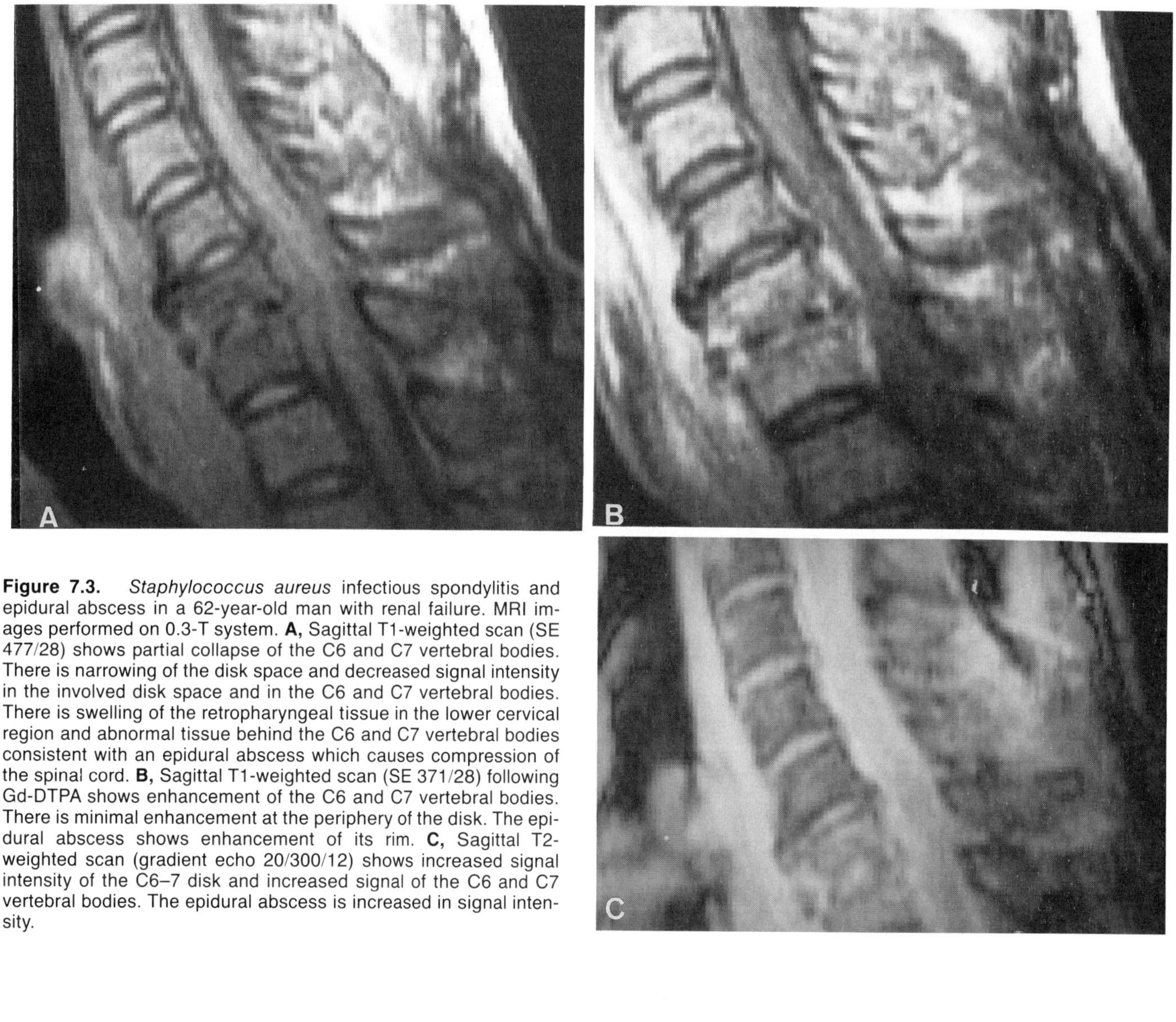

Figure 7.3. *Staphylococcus aureus* infectious spondylitis and epidural abscess in a 62-year-old man with renal failure. MRI images performed on 0.3-T system. **A,** Sagittal T1-weighted scan (SE 477/28) shows partial collapse of the C6 and C7 vertebral bodies. There is narrowing of the disk space and decreased signal intensity in the involved disk space and in the C6 and C7 vertebral bodies. There is swelling of the retropharyngeal tissue in the lower cervical region and abnormal tissue behind the C6 and C7 vertebral bodies consistent with an epidural abscess which causes compression of the spinal cord. **B,** Sagittal T1-weighted scan (SE 371/28) following Gd-DTPA shows enhancement of the C6 and C7 vertebral bodies. There is minimal enhancement at the periphery of the disk. The epidural abscess shows enhancement of its rim. **C,** Sagittal T2-weighted scan (gradient echo 20/300/12) shows increased signal intensity of the C6–7 disk and increased signal of the C6 and C7 vertebral bodies. The epidural abscess is increased in signal intensity.

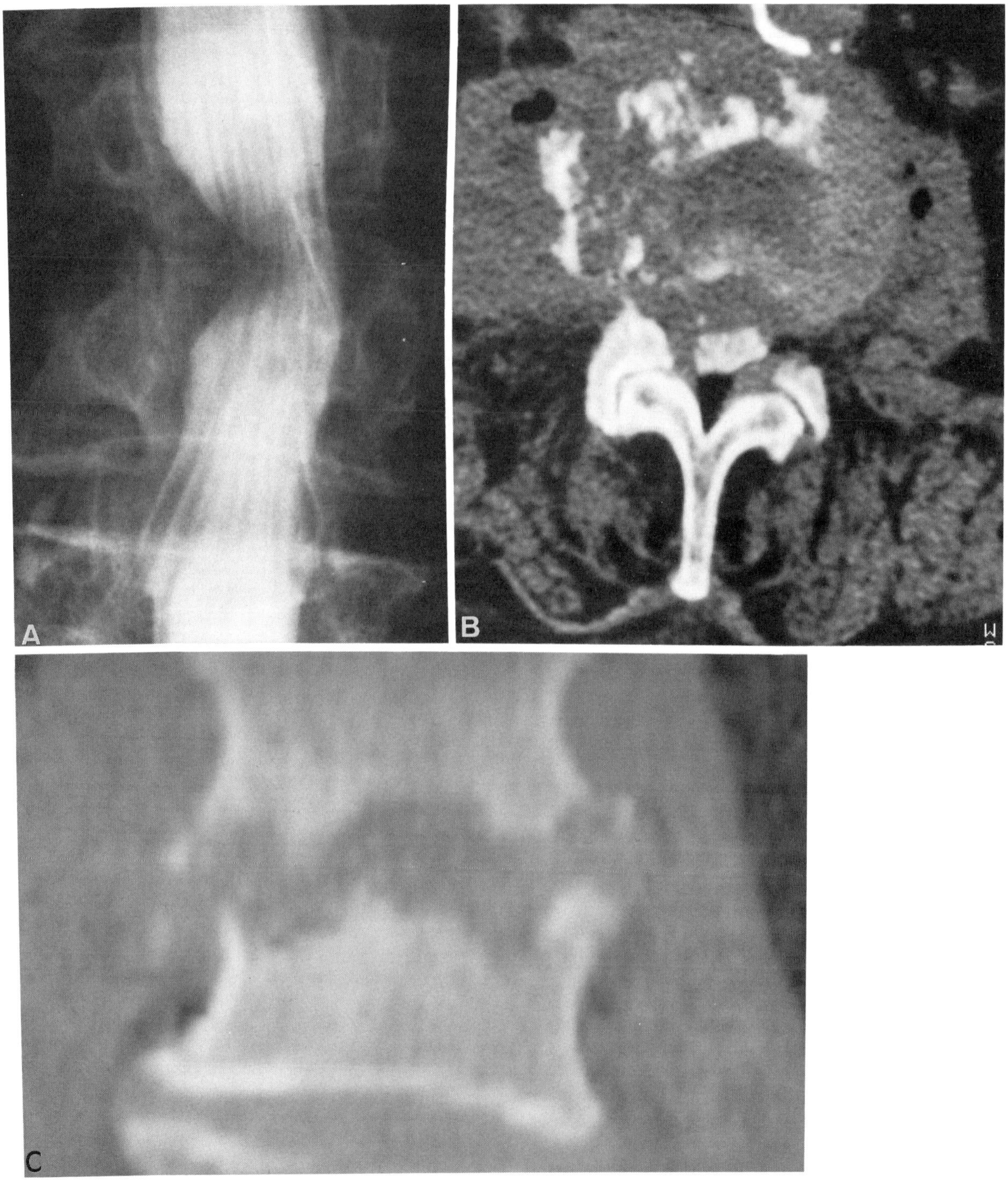

Figure 7.4. Infectious spondylitis and epidural abscess in an 82-year-old woman who developed back pain following cholecystectomy. **A,** Lumbar myelogram shows extradural deformity at the L3–4 level on the right consistent with an epidural mass lesion. The L4 root sleeve on the right is not seen. **B,** Postmyelography CT (MCT) at the L3–4 level shows an epidural mass located anteriorly and laterally. Bony fragmentation and partial destruction of the lower L3 vertebral body are seen. There is swelling of the paravertebral soft tissues and gas within the tissues. The findings are due to infectious spondylitis and an associated epidural abscess. **C,** Coronal re-formatted CT images show to better advantage the irregularity and destruction of the L3 and L4 endplates. MRI following surgical decompression and drainage of the epidural abscess shows typical features of infectious spondylitis. **D,** T1-weighted scan (SE 500/20) shows decreased signal of the L3 and L4 vertebral bodies. The disk is not well seen. There is an extradural mass which is isointense to slightly hypointense to cord. **E,** T1-weighted scan (SE 500/20) following Gd-DTPA shows thick, irregular enhancement of the disk and diffuse enhancement of the epidural collection. The epidural enhancement could represent residual/recurrent abscess or postoperative granulation tissue. There is partial compression and irregularity of the L3 and L4 vertebral bodies. **F,** T2-weighted scan (SE 2500/90) shows slightly increased signal of the L3 and L4 vertebral bodies. The disk is increased in signal intensity. This finding is quite striking given the decreased signal intensity of the other disk spaces consistent with degenerative change in this older patient. The epidural collection is not markedly increased in signal intensity in this case, suggesting that it represents postoperative granulation tissue.

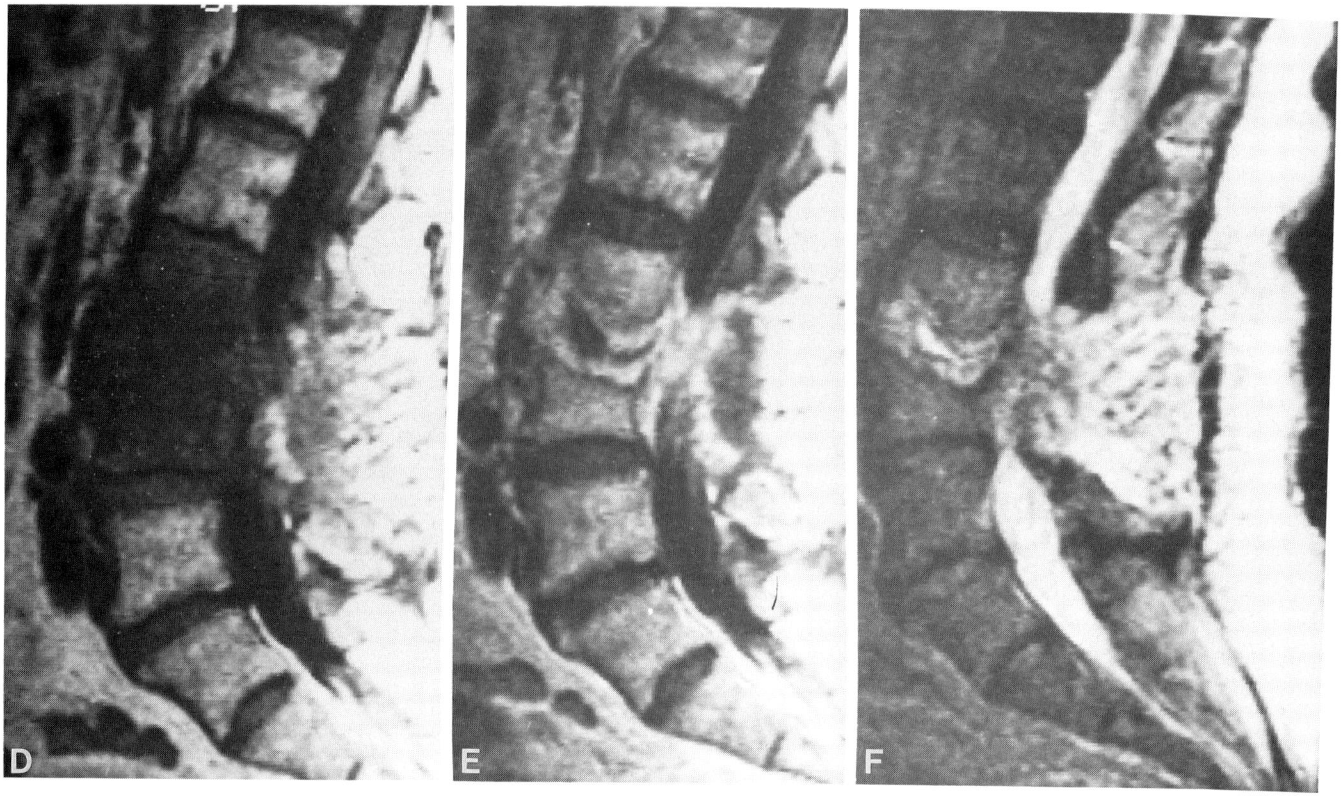

Figure 7.4–*continued*

Figure 7.5. Postoperative *S. aureus* infectious spondylitis in a 66-year-old man who developed recurrent low back pain and right lower extremity pain 6 weeks after L4–5 laminectomy. Initial scans show changes of early infectious spondylitis. **A,B,** Sagittal T1-weighted images (SE 750/25) show narrowing of the L4–5 disk space and subtle decreased signal intensity in the right half of the lower L4 and upper L5 endplates. A soft-tissue mass is seen behind the L4–5 disk space. Postoperative changes are noted in the soft tissues posterior to the spine. **C,D,** Sagittal T1-weighted scans (SE 750/25) following Gd-DTPA show enhancement of the areas in the vertebral bodies which were seen to be decreased in signal intensity on the unenhanced images. There is enhancement of the soft-tissue mass behind L4–5 which could represent scar tissue or an epidural abscess. The cystic area behind the spinal canal represents a pseudomeningocele from a dural tear at the time of surgery. **E,F,** Sagittal T2-weighted images (SE 2000/90) again show narrowing of the disk space. There is increased signal intensity within the disk. There is increased signal intensity in the superoposterior aspect of the L5 vertebral body, where there was decreased signal intensity on the T1-weighted images. The patient's pain persisted and got worse, so repeat MRI was performed 4 weeks later, which now shows more obvious infectious spondylitis. **G,H,** Sagittal T1-weighted scans (SE 750/25) now show more extensive decreased signal intensity in the L4 and L5 disks. There also has been enlargement of the epidural "mass." Disk space narrowing and postoperative changes are again seen. **I,J,** Sagittal T1-weighted scans (SE 750/25) after Gd-DTPA show enhancement of the involved portions of the vertebral bodies which again obscures the abnormality of the vertebral bodies. There is enhancement of most of the epidural "mass." The small area that does not enhance probably represents liquid pus. There is mild enhancement of the posterior portion of the disk that was not seen on the previous MRI study. **K,L,** Sagittal T2-weighted scans (SE 2000/90) show increased signal intensity in the L4 and L5 vertebral bodies which is much more apparent than on the previous study. There is increased signal intensity in the L4–5 disk. The epidural "mass" is markedly increased in signal intensity (*arrows*). **M,N,** Axial T1-weighted images (SE 700/20) before (**M**) and after (**N**) Gd-DTPA show the liquid portion of the epidural abscess to better advantage (*curved arrows*). The sagittal enhanced images were obtained 30 minutes after the axials. Some of the contrast material has seeped into the abscess over time on the sagittal images. The cystic collection behind the spinal canal is the pseudomeningocele (*arrowhead*). Symptoms resolved after surgical drainage of the abscess, appropriate antibiotics, and repair of the dural tear. Follow-up scans were obtained 5 months following treatment. **O,P,** Sagittal T1-weighted scans (SE 700/15) show continued narrowing of the L4–5 disk space. The decreased signal intensity in the L4 and L5 vertebral bodies is no longer seen. **Q,R,** Sagittal T1-weighted scans (SE 700/15) following Gd-DTPA show a small amount of residual enhancement within the disk and in the epidural space. The changes are most likely secondary to scar. **S,** Sagittal T2-weighted scan (SE 3000/90) shows resolution of the abnormal increased signal in the vertebral bodies and disk. The bright epidural mass seen previously has resolved. The little remaining epidural tissue shows decreased signal intensity. **T,U,** Axial T1-weighted scans before (**R**) and after (**S**) Gd-DTPA show minimal epidural tissue which enhances following contrast administration and probably represents scar. The large epidural abscess is no longer seen.

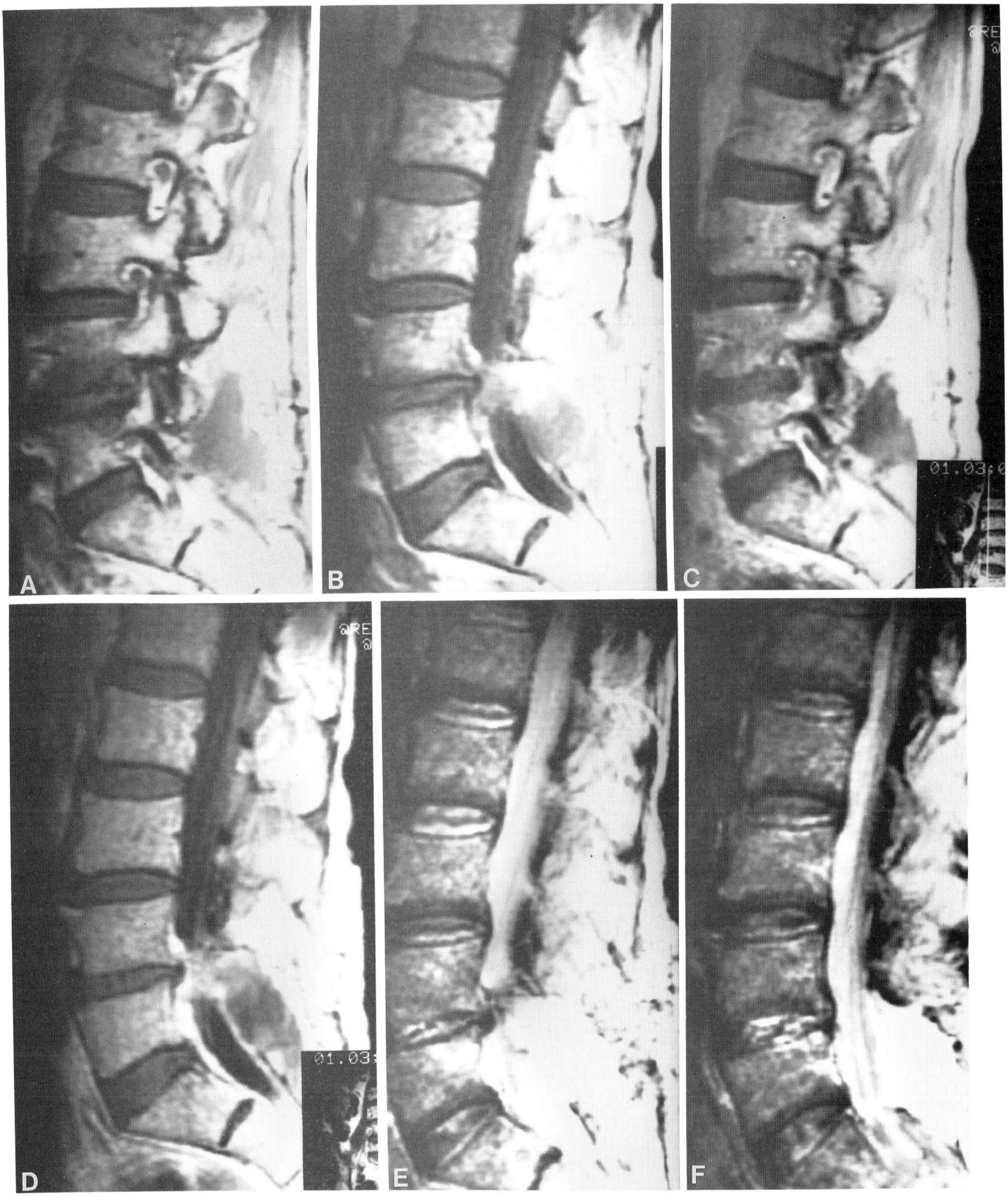
A
B
C
D
E
F

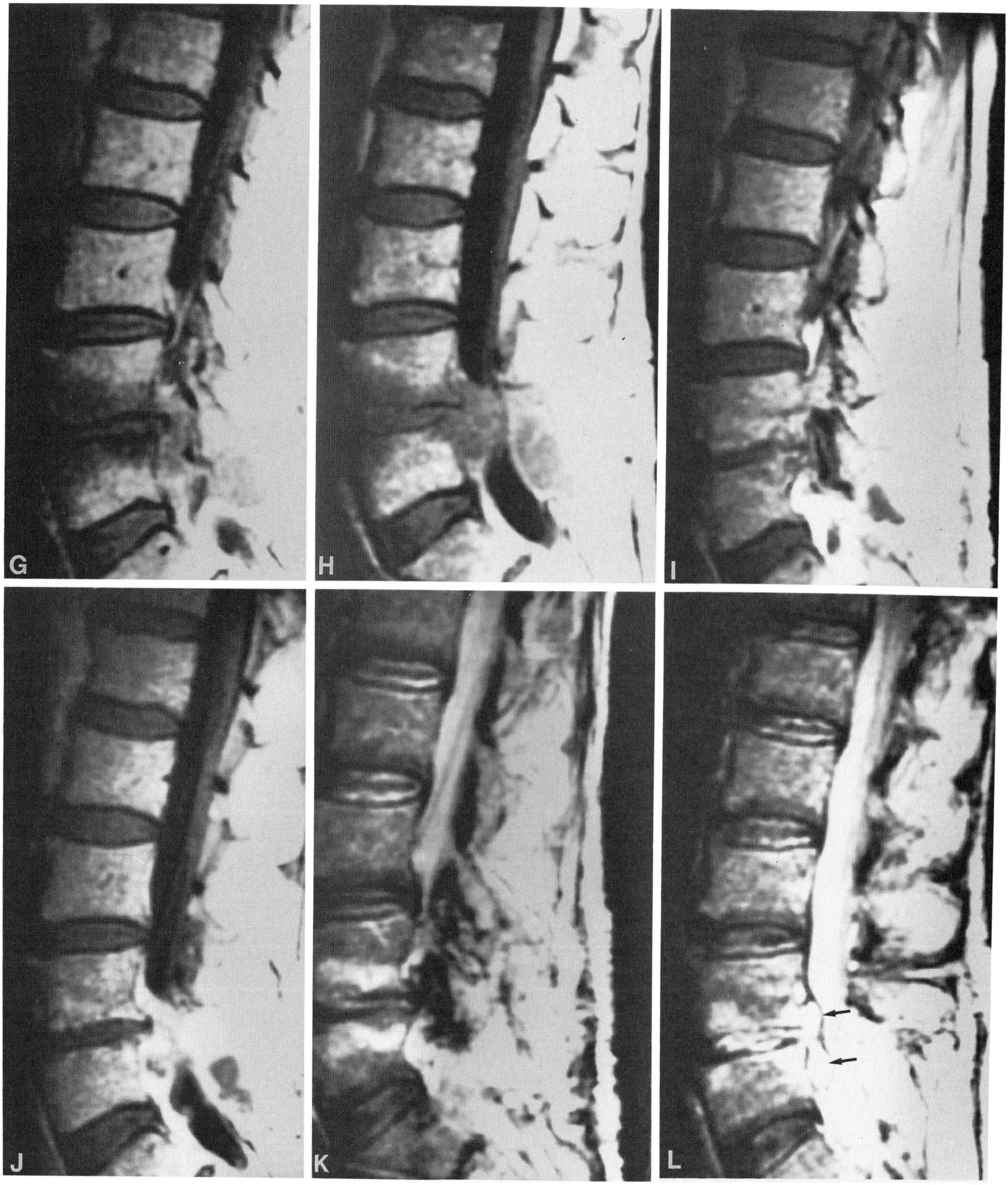

Figure 7.5–*continued*

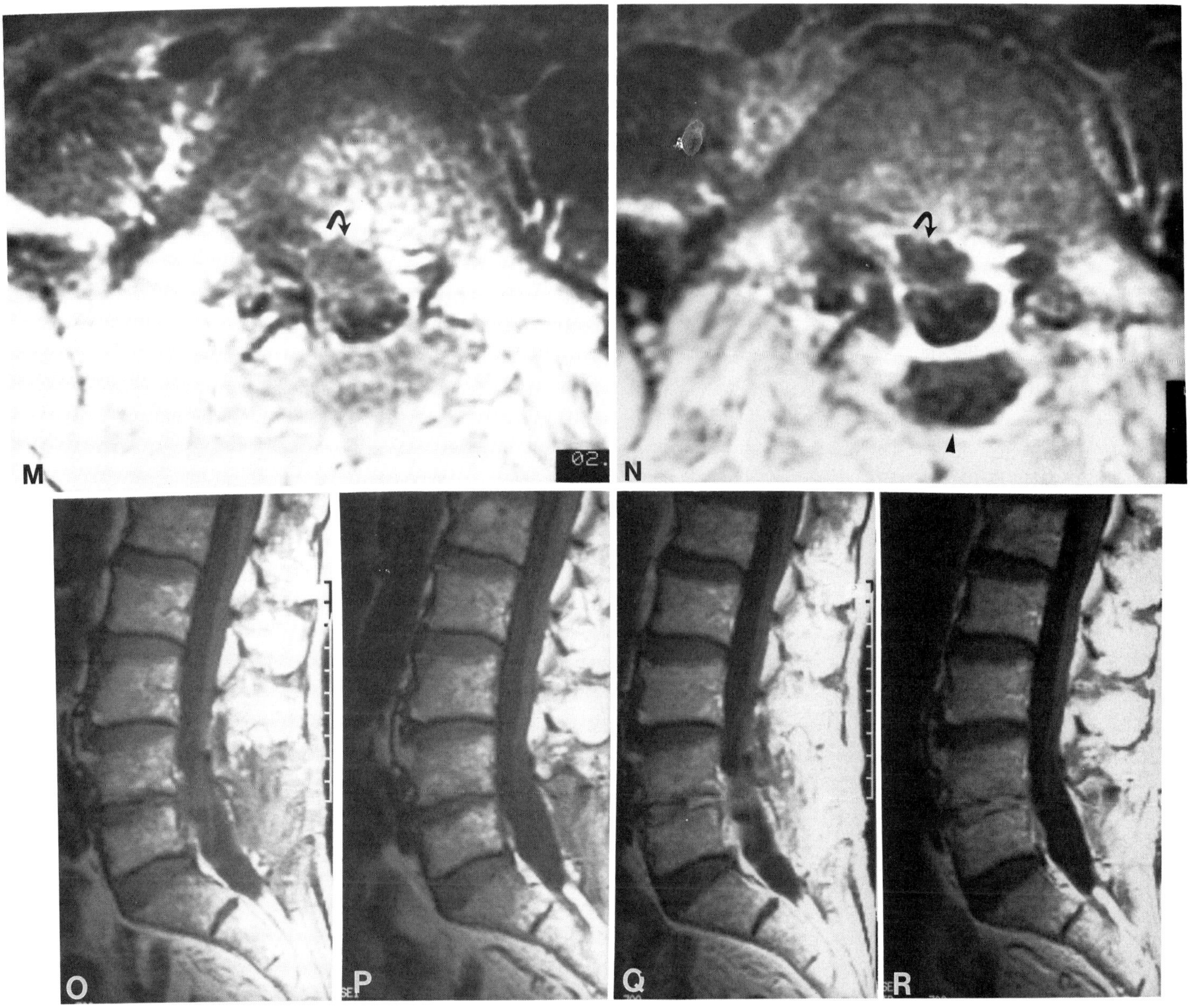

Figure 7.5–*continued*

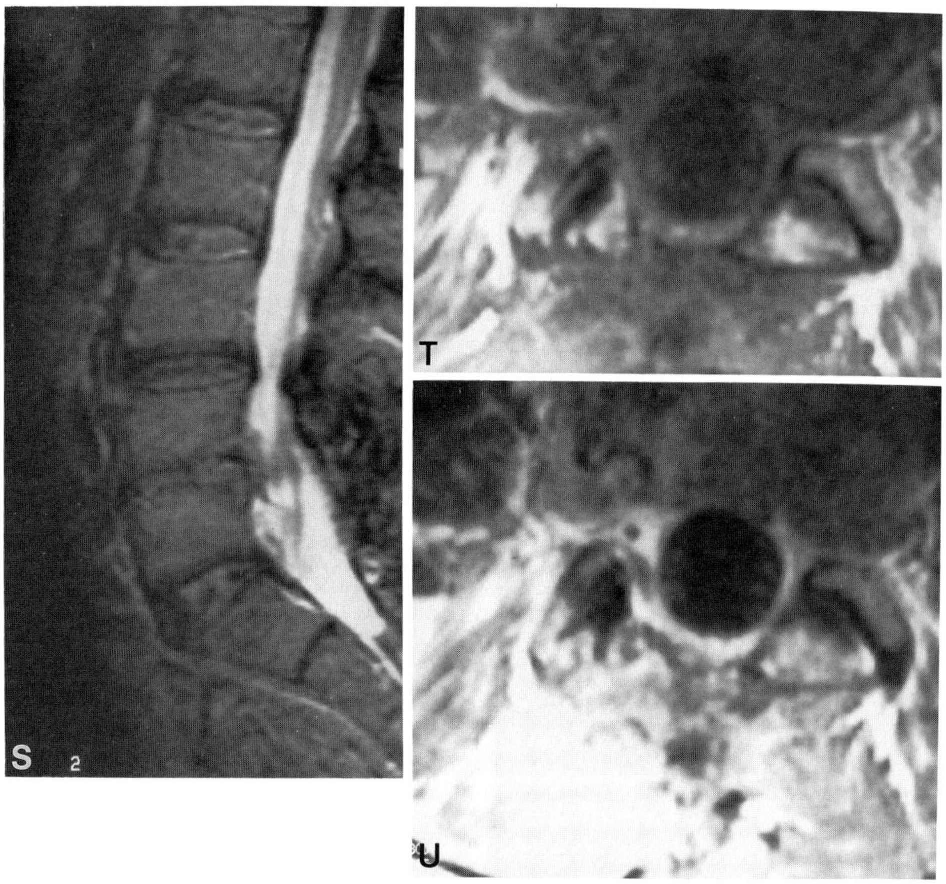

Figure 7.5–*continued*

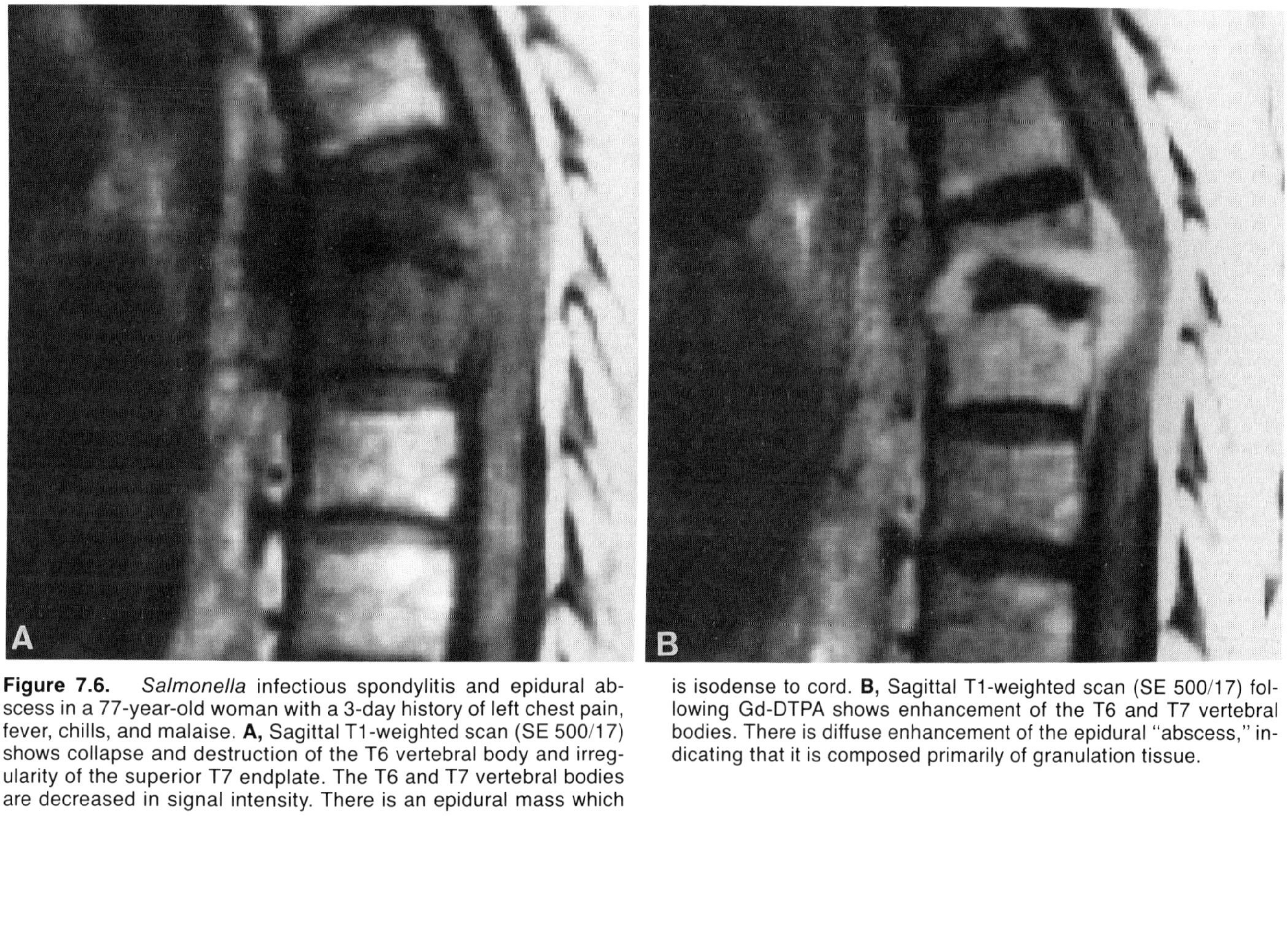

Figure 7.6. *Salmonella* infectious spondylitis and epidural abscess in a 77-year-old woman with a 3-day history of left chest pain, fever, chills, and malaise. **A,** Sagittal T1-weighted scan (SE 500/17) shows collapse and destruction of the T6 vertebral body and irregularity of the superior T7 endplate. The T6 and T7 vertebral bodies are decreased in signal intensity. There is an epidural mass which is isodense to cord. **B,** Sagittal T1-weighted scan (SE 500/17) following Gd-DTPA shows enhancement of the T6 and T7 vertebral bodies. There is diffuse enhancement of the epidural "abscess," indicating that it is composed primarily of granulation tissue.

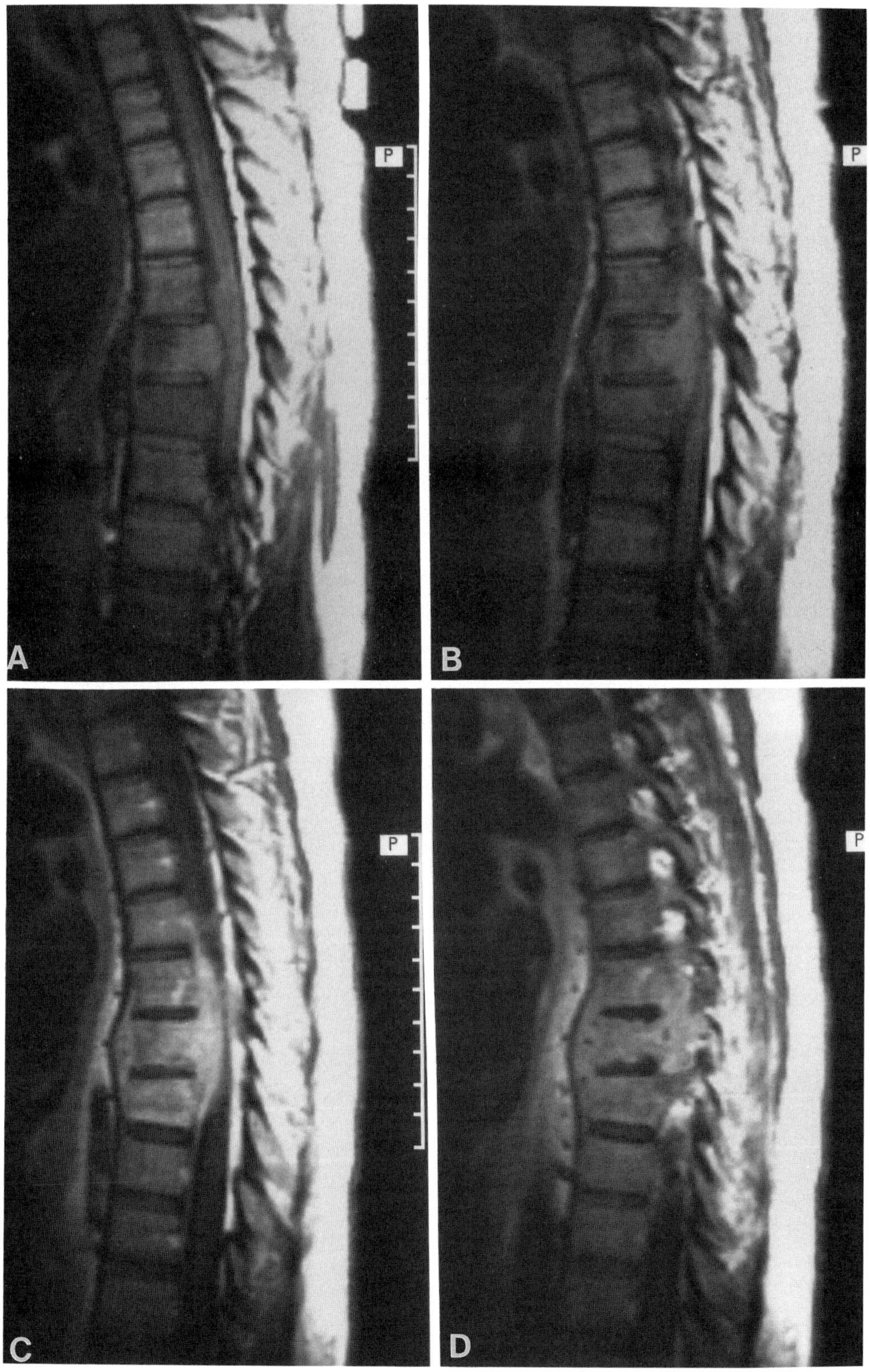

Figure 7.7. Tuberculosis osteomyelitis in a 28-year-old woman with right-sided chest pain and intermittent fevers. **A,B,** T1-weighted scans (SE 500/15) show areas of mixed increased and decreased signal intensity in the T6, T7, and T8 vertebral bodies and an epidural mass. There is relative sparing of the disk spaces. Multilevel involvement, mixed signal intensity within the vertebral bodies, and relative preservation of the disk spaces favor a process such as TB. **C,D,** T1-weighted scans (SE 500/15) show mild enhancement of the involved vertebral bodies and enhancement of the prevertebral and epidural masses. There is not abnormal enhancement of the disks. Note is made of irregularity of the inferior endplate at T7. **E,** Axial T1-weighted scan (SE 900/15) shows the large left paraspinal mass and the epidural abscess. Involvement of the pedicle on the left favors TB over a pyogenic process. There is also noted to be pneumonia/atelectasis in the left lung. **F,** Axial T1-weighted scan (SE 900/15) after Gd-DTPA shows enhancement within the epidural abscess and the paraspinal mass. **G,** CT following nonionic contrast material shows partial obstruction of the T7 vertebral body and confirms destruction of the left pedicle with a large associated paraspinal mass. There is atelectasis/pneumonitis in the adjacent lung and a left pleural effusion.

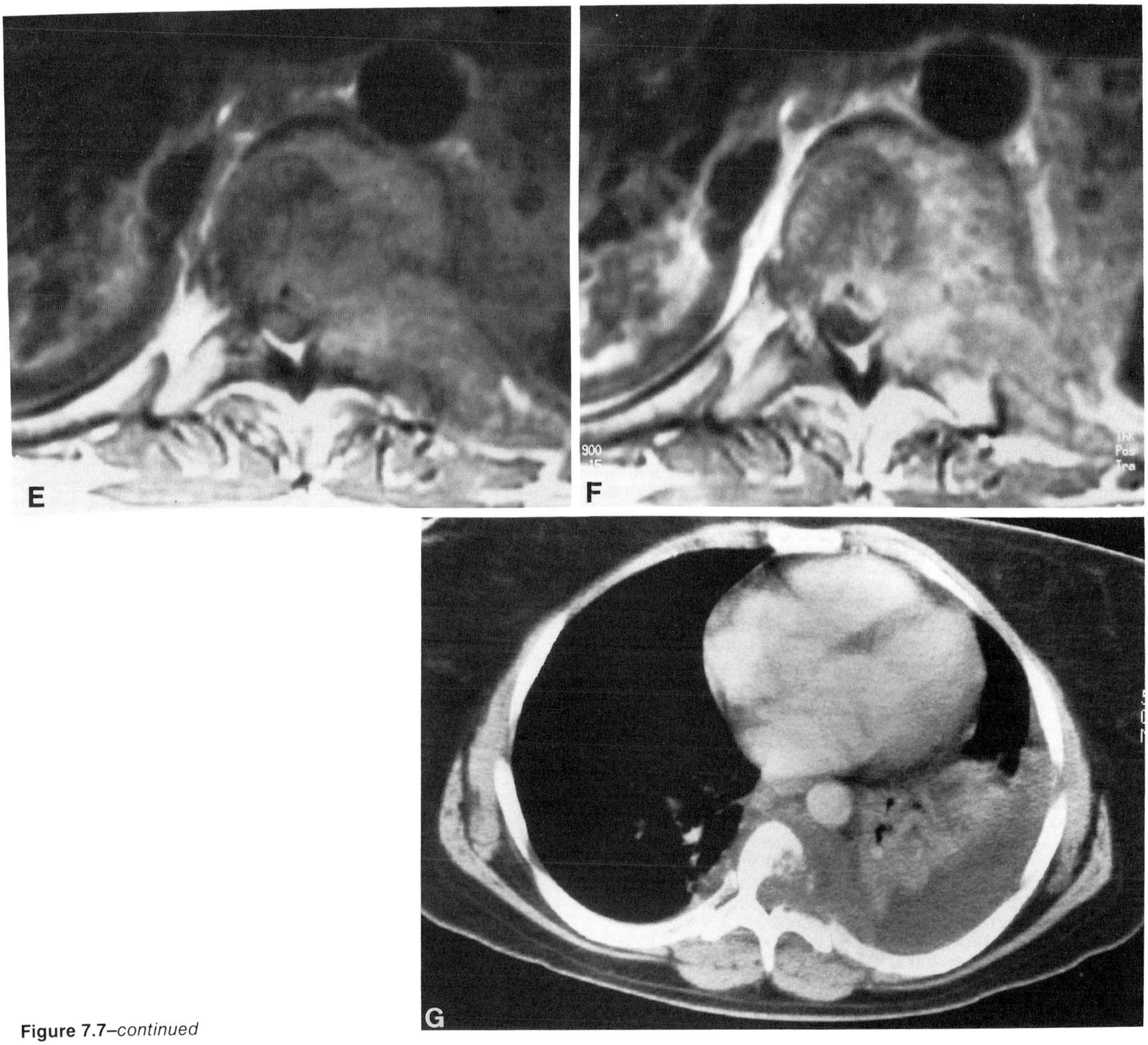

Figure 7.7–*continued*

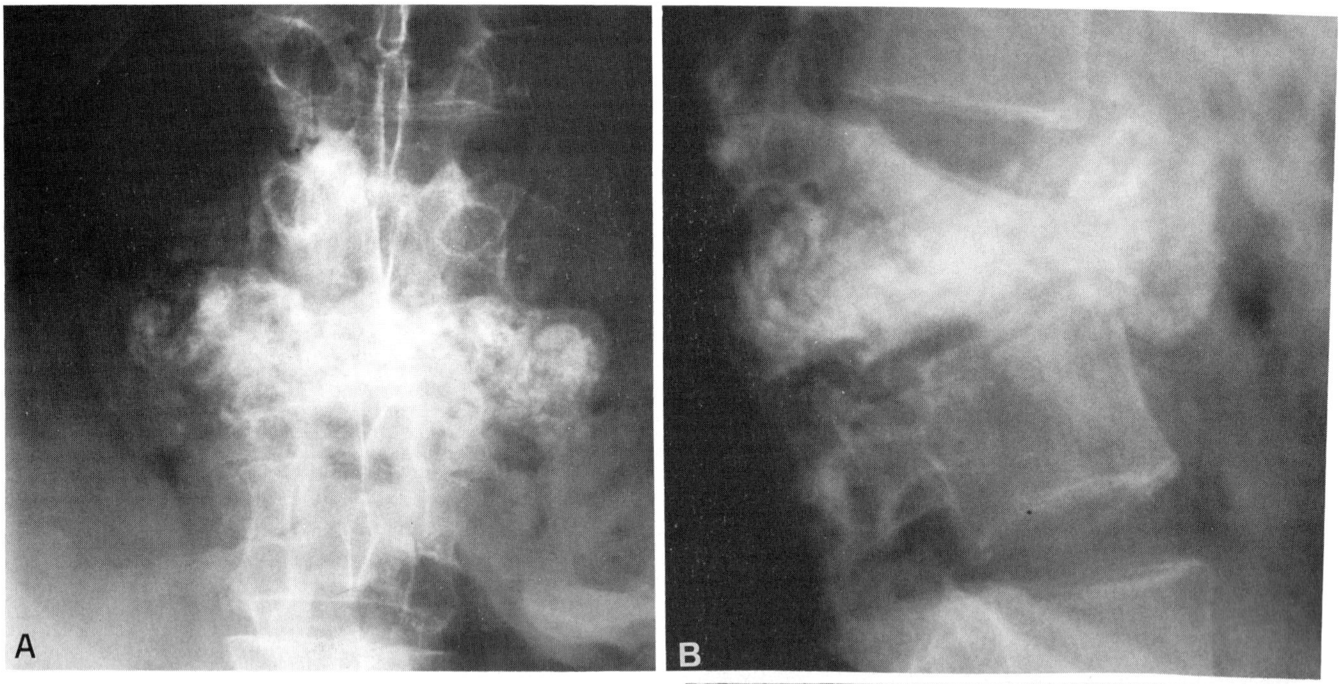

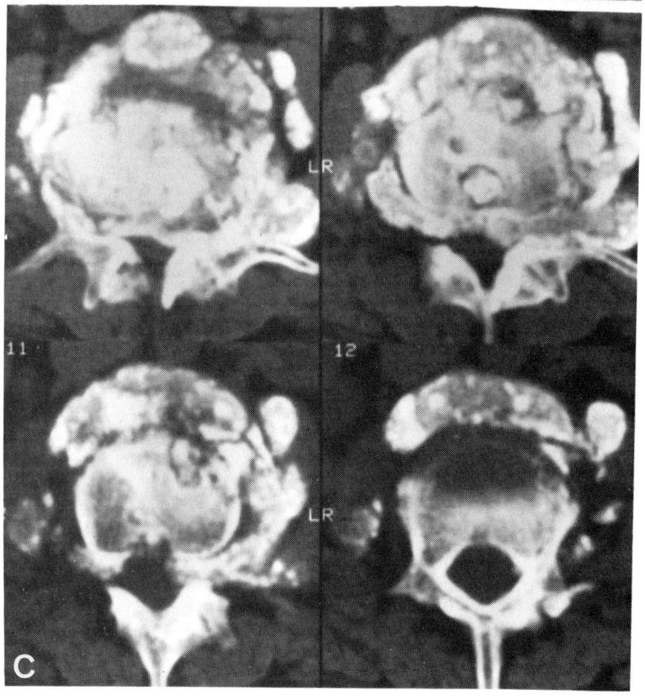

Figure 7.8. Tuberculous spondylitis in a 27-year-old man who presented with weight loss, low back pain, and weakness and numbness of the lower extremities. **A,B,** Plain films of the spine reveal a gibbus deformity and exuberant callus formation typical of end-stage tuberculosis. **C,** CT shows exuberant callus and marked narrowing of the spinal canal. **D,** Sagittal T1-weighted MRI (SE 500/17) shows areas of mixed increased and decreased signal intensity. There is marked narrowing of the spinal canal with compression of the cauda equina. **E,** Sagittal T2-weighted scan (SE 2100/80) shows increased signal intensity in the L3 and L4 vertebral bodies. The L3–4 disk space is largely destroyed.

the vertebral endplates may be seen but are often difficult to appreciate and differentiate from other pathologies on axial images.

Radionuclide scintigraphy using bone scanning agents is sensitive in detecting infectious spondylitis but is not specific. Specificity can be increased by using indium-labeled leukocytes, but false-negative studies can occur with chronic infection and false-positive studies can occur in patients with chronic inflammatory conditions such as rheumatoid arthritis and fracture [8]. Scintigraphic studies are most useful if the MRI findings are not diagnostic.

Epidural Abscess

Infection in the spinal epidural space is unusual, but it is an important clinical entity because treatment significantly improves outcome. In most cases (78%), a source of infection can be identified [8]. The source of infection may be a distant site with hematogenous spread or may be a local site with direct

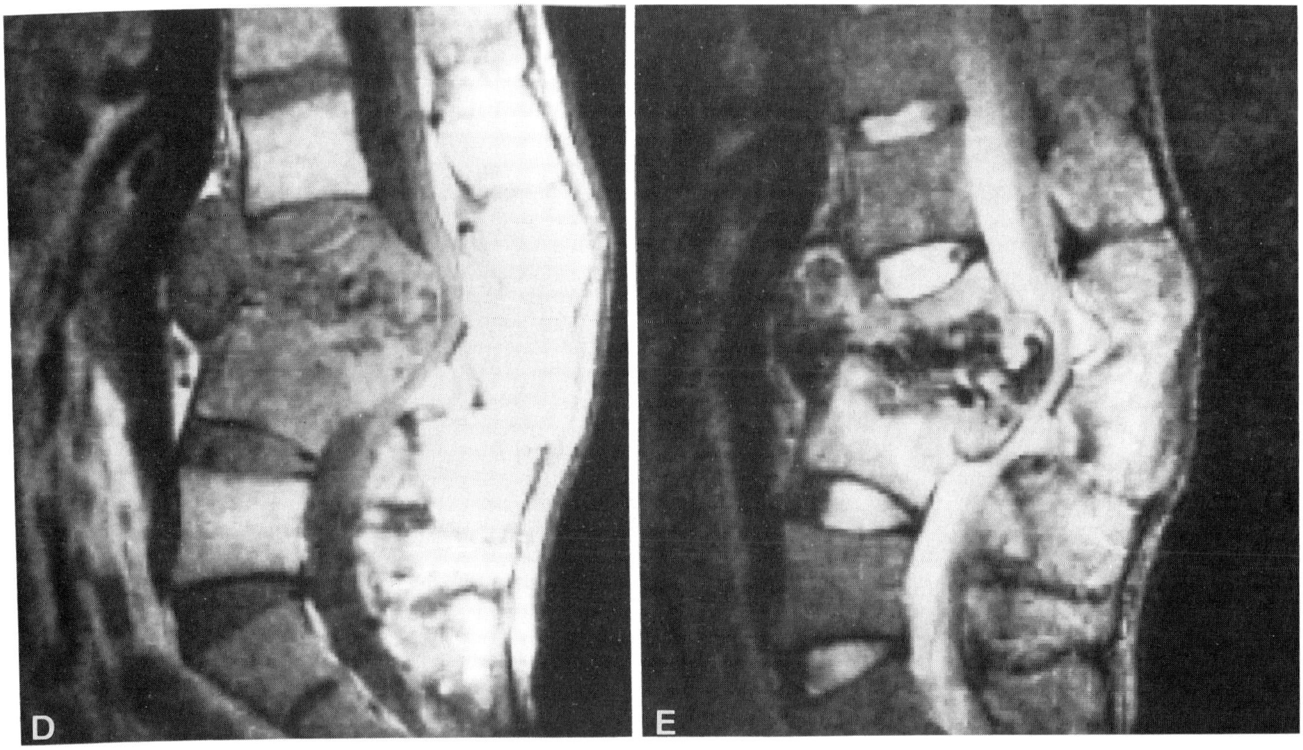

Figure 7.8–*continued*

spread, such as in patients with infectious spondylitis, or from adjacent tonsillar and nasopharyngeal infections. Direct inoculation also can occur secondary to spinal surgery or penetrating trauma. *S. aureus* is the most common infecting organism, but other gram-positive and gram-negative organisms are fairly common [22–24]. Risk factors include diabetes mellitus, IV drug abuse, alcohol abuse, and chronic renal failure.

Symptoms of spinal epidural abscess may occur in an acute form with onset over an interval of days or in a chronic form with onset over weeks to months [22, 23]. Admitting symptoms most often include back pain, radicular pain, and fever with weakness and paralysis appearing later. Stiffness and cramping are often the earliest symptoms [23]. Weakness and paralysis are not usually entirely reversible [23]. Fever may not be prominent, or it may be absent entirely, especially in those with more chronic symptoms. WBC count and ESR are usually elevated. Spinal epidural infections in children are rare [25]. When present, they are more difficult to diagnose because of atypical presentations.

Neurologic deterioration is usually related to direct compression of neural tissue by the infected tissue. Deterioration can be seen without mechanical compromise and is thought to be related to ischemic damage [8].

Although the term *epidural abscess* is used to describe infection in the spinal epidural space, it is often a misnomer. In a recent surgical review, 35% of patients were found to have liquid pus, 13% had granulation tissue alone, and 52% had both granulation tissue and pus [26]. No significant difference in clinical presentation, organism involved, or outcome could be identified when comparing the groups [26].

The management of spinal epidural abscess generally includes decompressive laminectomy and operative debridement followed by antibiotics. Treatment with percutaneous drainage or antibiotics alone also has been reported [4, 27–30]. It is expected that this will prove to be acceptable treatment in selected patients with lumbar abscesses and no evidence of neural compression [26].

RADIOGRAPHIC EVALUATION

MRI is the procedure of choice in evaluating patients for suspected epidural abscess (Figs. 7.2 through 7.6 and 7.9). In general, epidural infections are hypointense or isointense with the spinal cord on T1-weighted images and of high signal intensity on proton-density and T2-weighted images [1]. In most cases, the extradural location of the mass is readily apparent with a convex margin toward the spinal canal. If there is compression or infarction of the cord, it also may show increased signal intensity on T2-weighted images. Gadolinium-enhanced MRI is often helpful in better delineating the extent and

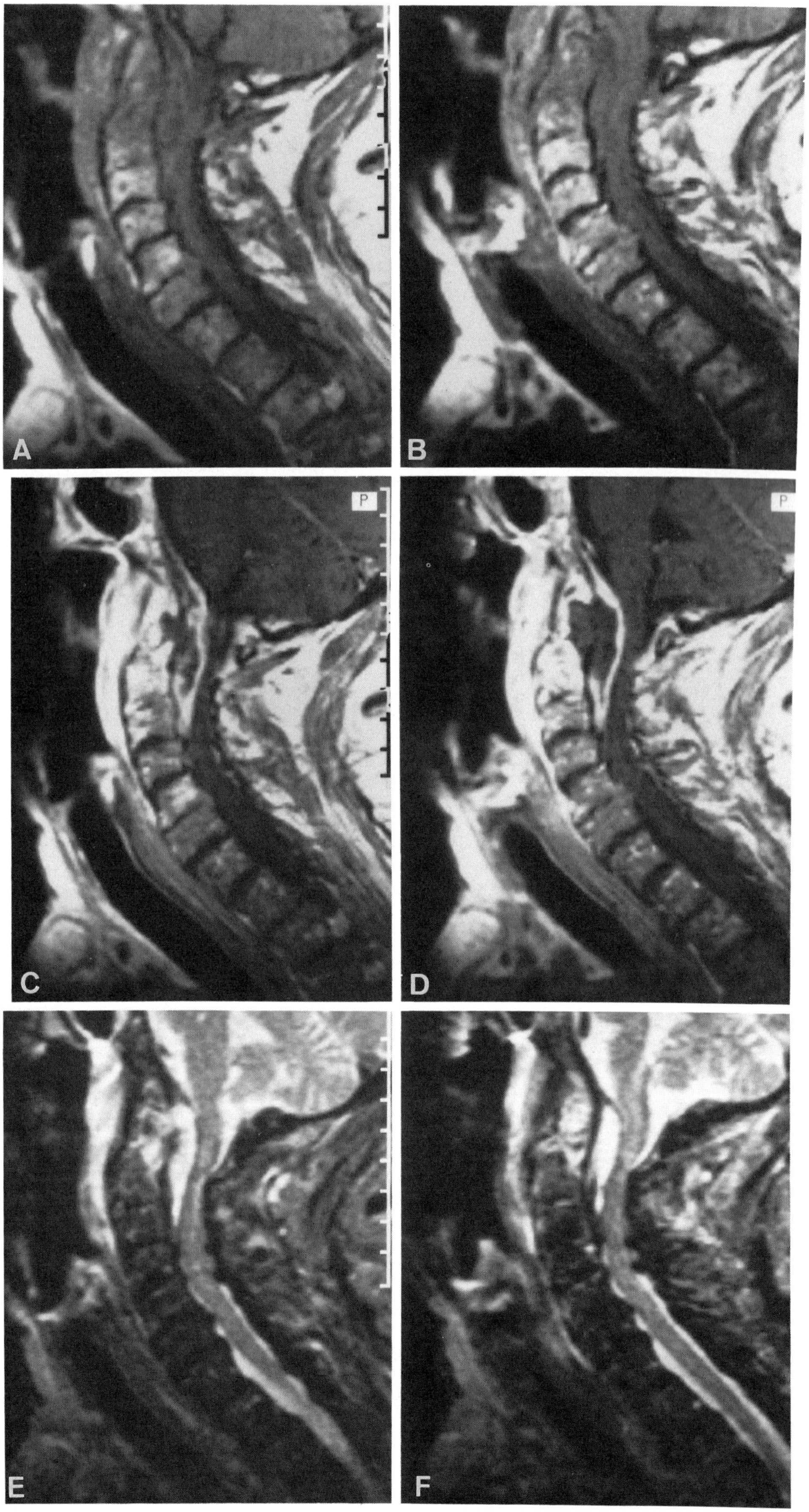

Figure 7.9. Osteomyelitis and epidural abscess in a 78-year-old woman who presented with progressive upper and lower extremity weakness. Five weeks prior to admission, the patient had fallen with a resultant nondisplaced fracture of the dens. **A,B,** Sagittal T1-weighted scans (SE 520/15) show decreased signal within the dens. There is an epidural mass which is in part isointense and in part decreased in signal intensity compared with the spinal cord. The cord is displaced posteriorly and compressed. The prevertebral tissues are swollen. **C,D,** Sagittal T1-weighted scans (SE 520/15) after Gd-DTPA show enhancement of the swollen prevertebral tissues and of the dens. There is enhancement around the periphery of the epidural abscess which suggests that is is largely liquid. The posterior of the dens is destroyed. **E,F,** Sagittal T2-weighted scans (SE 2587/80) show increased signal intensity of the epidural collection. There is increased signal intensity of the cord behind the dens, indicating tissue damage. **G,** Axial T1-weighted scan (SE 693/15) shows low-signal-intensity isointense epidural abscess and paravertebral mass with loss of normal fat planes. **H,** Axial T1-weighted scan (SE 693/15) after contrast much more clearly delineates the abscess (*arrow*) and shows diffuse enhancement of the prevertebral soft tissues.

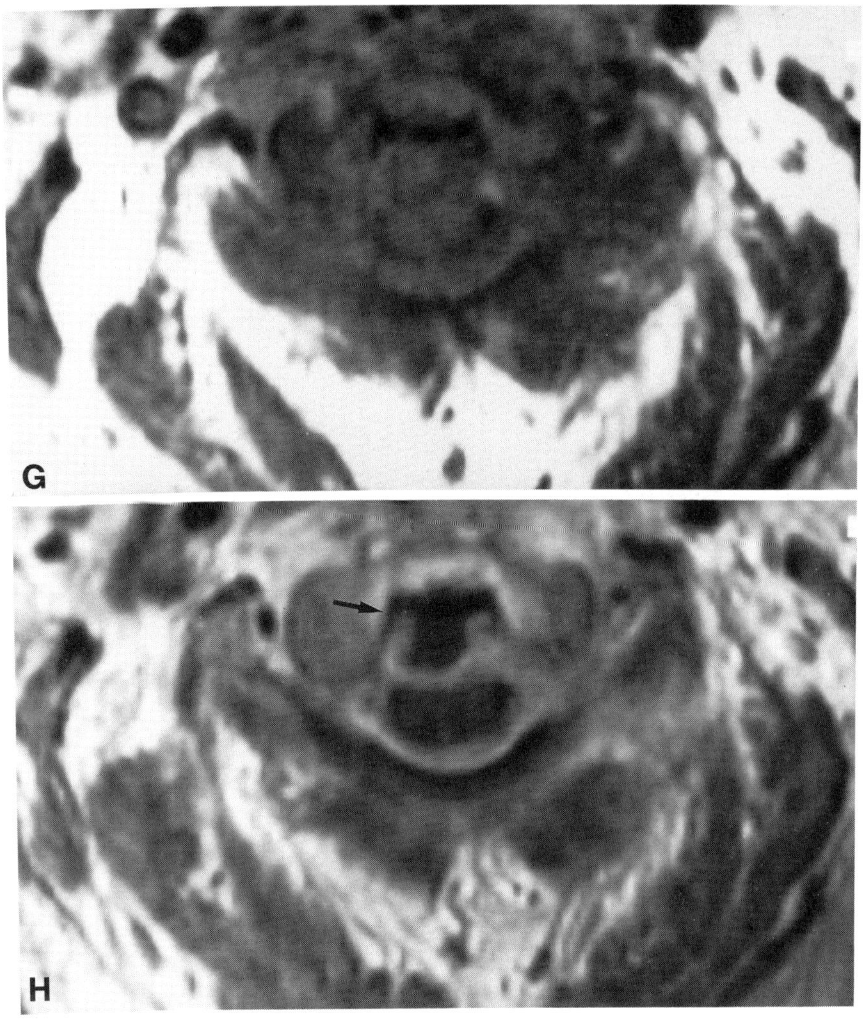

Figure 7.9 *continued*

anatomic location of the abscess. An epidural abscess may show diffuse homogeneous enhancement, inhomogeneous enhancement, or thin peripheral enhancement [1, 31].

In some patients, the epidural abscess may show inhomogeneous mixed signal intensity on T1- and T2-weighted images which can be difficult to distinguish from the signal changes of an adjacent meningitis [1]. In these cases, contrast-enhanced images may be helpful. If the infection has a hemorrhagic component, possibly from rupture of epidural veins, it could be mistaken for a subdural or epidural hematoma [2].

The abscess collection may be localized or can be quite extensive, especially in the thoracic region. The infection can be located ventrally, which is more common in the cervical and lumbar regions, or it can be located dorsally or circumferentially, both of which are more common in the thoracic region.

Epidural infections in the high cervical region can insinuate around the vertebral bodies and into the intervertebral foramina causing loss of normal fat planes and epidural fat. On T1-weighted images, the inflammatory tissue is usually isointense to muscle. On T2-weighted images, there is generally increased signal intensity (see Fig. 7.9).

If after an MRI examination (without and with IV contrast material) the diagnosis is still uncertain, myelography with high-resolution CT is recommended. Myelography typically shows an area of total or subtotal block with signs of an epidural mass lesion (Figs. 7.4 and 7.10). CT scanning following the myelogram better localizes the extent of the abscess and confirms its extradural location (Figs. 7.4, 7.10, and 7.11). Now that MRI is readily available in most locations, the need for myelography in evaluating patients with suspected epidural abscess should be reserved for difficult cases or patients with contraindications to MRI because of the potential of contaminating additional spinal compartments.

CT without myelography may not be diagnostic. If there is concomitant infectious spondylitis, paraspi-

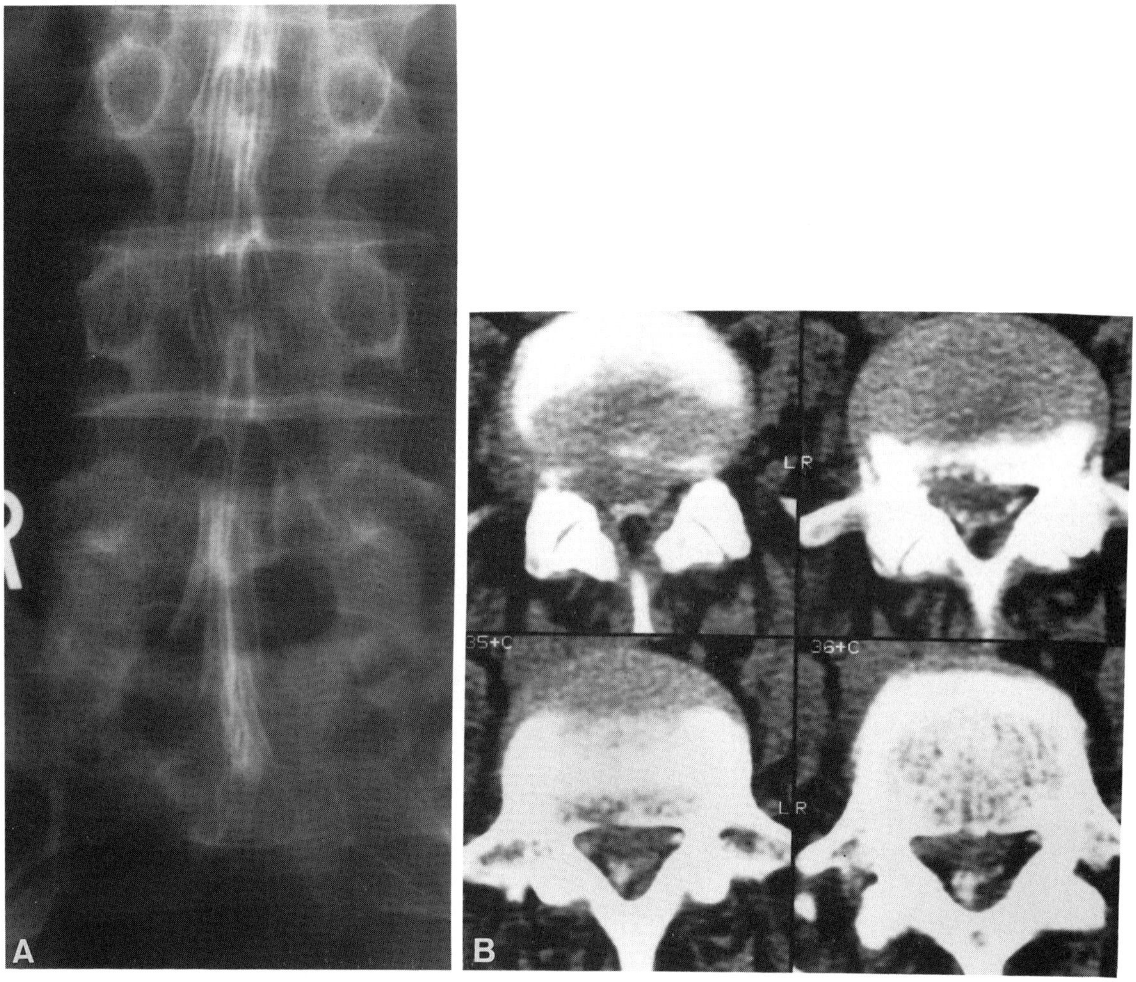

Figure 7.10. Streptococcus epidural abscess in a 63-year-old patient with a 22-day history of back pain. There was no history of fever, and the WBC count was normal. Clinically, it was thought the patient had a herniated disk. **A,** Lumbar myelogram demonstrates a very small spinal canal secondary to extradural compression of the thecal sac. (Note the distance of the thecal sac from the pedicles.) There is poor filling of the root sleeves at multiple levels, most pronounced at L5 on the right and S1 on the left. No bony destructive lesions are identified. **B,** Postmyelography CT (MCT) shows an epidural mass which was seen to extend over multiple lumbar levels. No definite bony destruction was seen.

nal soft-tissue swelling and bony fragmentation may be seen. If IV contrast material is given, an enhancing inflammatory mass or rim may be seen within the spinal canal. Lucent extradural collections may be seen on unenhanced scans. However, like myelography, CT of the spine should take a back seat to MRI in most patients with suspected epidural abscess.

The prognosis for patients with spinal epidural abscess has improved dramatically since the early descriptions of the disease [26]. Dandy's initial report [32] in 1926 indicated an 83% mortality. Following the introduction of antibiotics and improved surgical techniques, mortality decreased. Baker et al. [22] reported 18% mortality in 1975. Mortality today is generally much less than 18%, with Curling et al. [26] reporting 6.9% in their 1990 series. Continued improvement in the prognosis of this disease is expected because of better diagnostic modalities such as MRI as well as newer antibiotics and neurosurgical techniques.

Infections of the Cord

INTRAMEDULLARY SPINAL CORD ABSCESS

Primary infections of the spinal cord are extremely rare. They are seen most commonly in immunocompromised patients such as those with HIV infection or in IV drug abusers. Patients generally present with rapid deterioration of neurologic function. Pain may or may not be present. Patients with cord abscesses due to *Pseudomonas cepacia, Toxoplasma,* and *Mycobacterium tuberculosis* have been described recently [33–35].

On MRI, most infections will show focal mass effect with decreased signal intensity on T1-weighted images and increased signal intensity on T2-weighted images (Fig. 7.12). Contrast-enhanced scans show

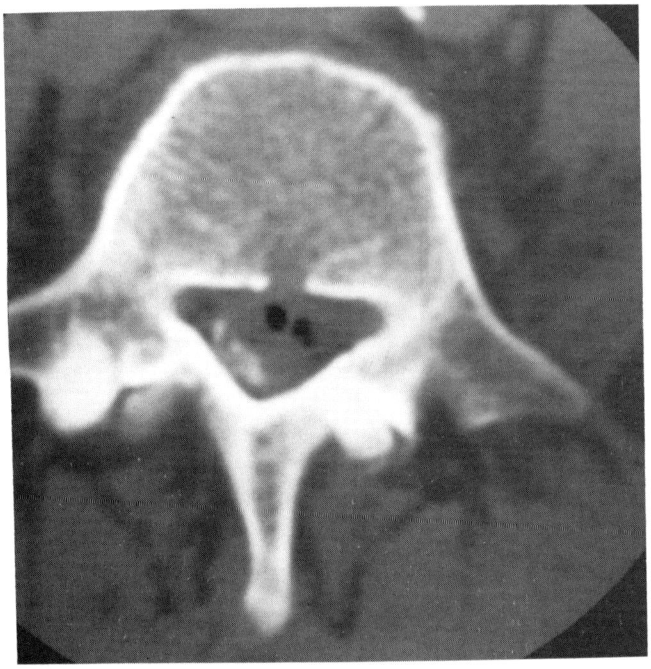

Figure 7.11. Epidural abscess. Postmyelography CT (MCT) shows compression of the thecal sac by a soft-tissue mass which contains gas. Differential diagnosis would include epidural abscess and herniated disk. MRI showed findings of infectious spondylitis and epidural abscess.

nodular areas of enhancement or ring enhancement on T1-weighted images. Contrast-enhanced scans allow for differentiation of the abscess from surrounding edema. The enhancement pattern on MRI is similar to that from tumor.

Myelography often shows cord enlargement which is often associated with a spinal block. Postmyelography CT confirms the cord enlargement but is not specific.

MYELITIS

Myelitis represents an inflammatory process of the spinal cord which may be caused by a variety of conditions. Viral infection, allergic reaction related to a previous infection or vaccination, a paraneoplastic-type process, sarcoidosis, and vascular insufficiency have all been described as causes [2, 36]. Specific infectious agents that have been described as causing the disease include herpes simplex virus, cytomegalovirus, varicella-zoster virus, human T-cell lymphocytotropic virus (HTLV-I, HTLV-III), and *Schistosoma mansoni* [37–39]. Myelitis has well-known etiologic associations with preceding viral illnesses such as measles, varicella, smallpox vaccinations, rabies, and typhoid [40]. A small number of patients with multiple sclerosis may present initially with a myelitis-type picture as their initial symptom.

Clinical presentation is somewhat variable. Some patients present rapidly over several hours or days with severe back pain, paresthesia, and leg weakness [40]. Another subset of patients may present with a stuttering onset over 1 to 4 weeks [41]. The most common presentation is gradual ascending paresthesia and/or leg weakness that evolves over 1 to 2 weeks [41].

MRI may be normal in the acute phase of the disease. In the subacute stage, MRI generally shows an enlarged cord which may be isointense, hypointense, or slightly hyperintense compared with the adjacent normal cord on T1-weighted images. T2-weighted images show increased signal intensity (Fig. 7.13). The abnormality is best seen on T2-weighted images, where it often appears quite extensive, extending over several segments [36]. There also may be skip areas of uninvolved cord. Contrast-enhanced scans tend to show three patterns of enhancement: diffuse intramedullary enhancement, peripheral enhancement, or speckled enhancement [36, 42]. Large nodular areas of enhancement are generally not seen, which helps to distinguish this entity from tumor [36]. Multiple sclerosis (MS) usually presents with smaller, more focal lesions that may be multiple. White matter lesions in the brain increase the probability that the patient will eventually manifest as MS [43].

CT is generally not helpful. CT performed after myelography (MCT) can show enlargement of the cord but is not specific.

Meningeal and Other Spinal Infections

Meningeal infections in the spinal canal most often represent spread of an intracranial meningitis from the basilar cisterns. For this reason, cranial symptoms generally predominate [2]. Organisms that cause a chronic basilar meningitis are, not surprisingly, those most frequently encountered. The meningitis may be caused by bacterial, fungal, or parasitic agents. The most common organisms are *M. tuberculosis*, *Coccidioides immitis*, and *Taenia solium* (cysticercosis) [2]. In some cases, the exact cause of the meningitis is not determined.

Pachymeningitis cervicales hypertrophica is a rare disease which causes chronic inflammatory hypertrophy of the dura (pachymeningitis). A number of organisms have been implicated, but the etiology is not definitely known. The disease is seen most frequently in the cervical and thoracic regions but also has been reported in the posterior fossa and lumbar spine. Symptoms generally progress from local pain to radiculopathy to cord compression. Gross changes at pathology show thickening of the dura, which also may involve the leptomeninges [44]. The appearance at MRI is that of a severe arachnoiditis [2]. On CT examination, calcification within the spinal canal may be a striking feature (Fig. 7.14).

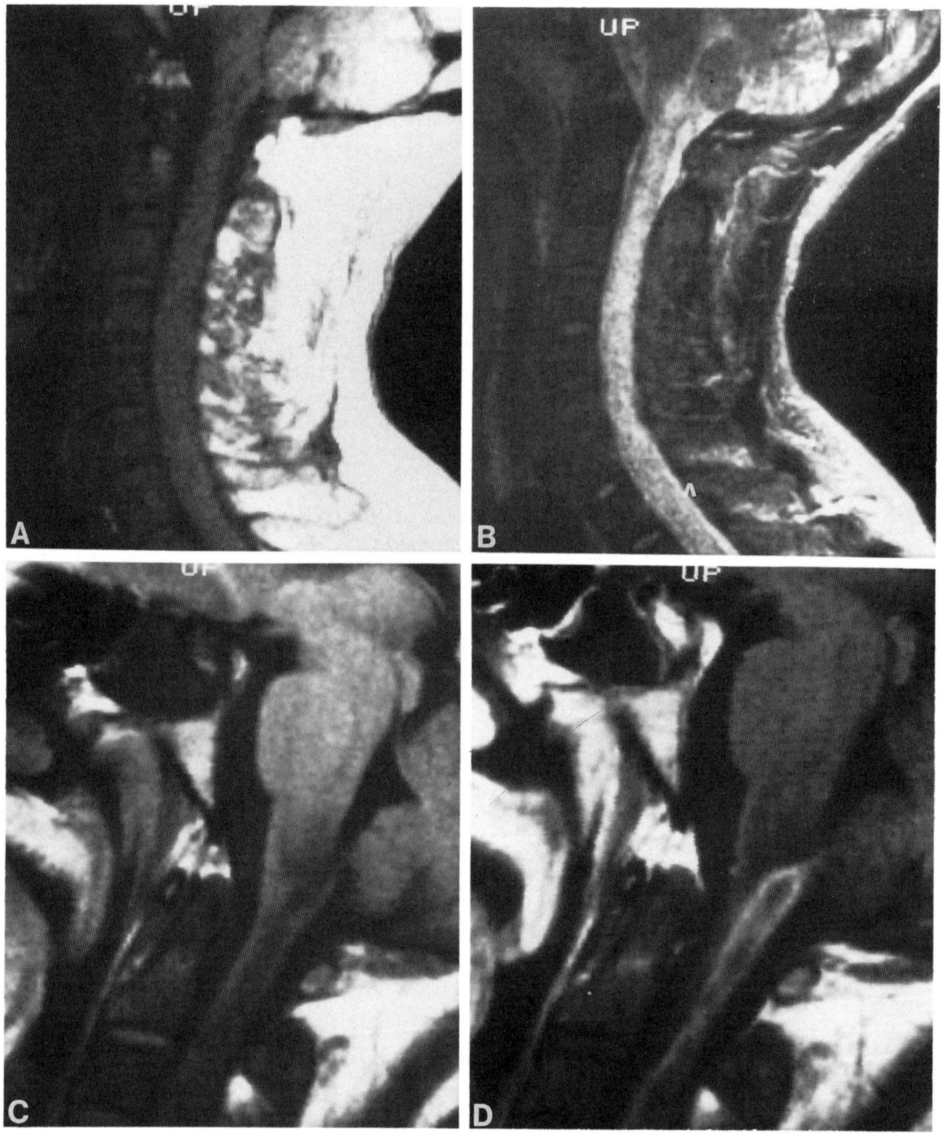

Figure 7.12. *Staphylococcus aureus* and group D enterococcus spinal cord abscess in a 34-year-old man with progressive paraplegia and numbness. **A,** Sagittal T1-weighted scan (SE 500/17) shows enlargement and decreased signal intensity of the cervical cord. **B,** Sagittal T2-weighted scan (SE 2500/90) shows increased signal intensity of the medulla and cervical and upper thoracic cord. **C,D,** Follow-up T1-weighted scans (SE 500/17) before and after Gd-DTPA performed 3 weeks later show more diffuse and marked decreased signal intensity within the medulla and cervical cord. The central cord shows an enhancing rim around an abscess cavity which at autopsy extended from the medulla to the lower thoracic cord. **E,F,** Sagittal T1-weighted scans (SE 500/17) before and after Gd-DTPA performed 3 weeks after the initial studies show two ring-enhancing lesions consistent with cerebral abscesses.

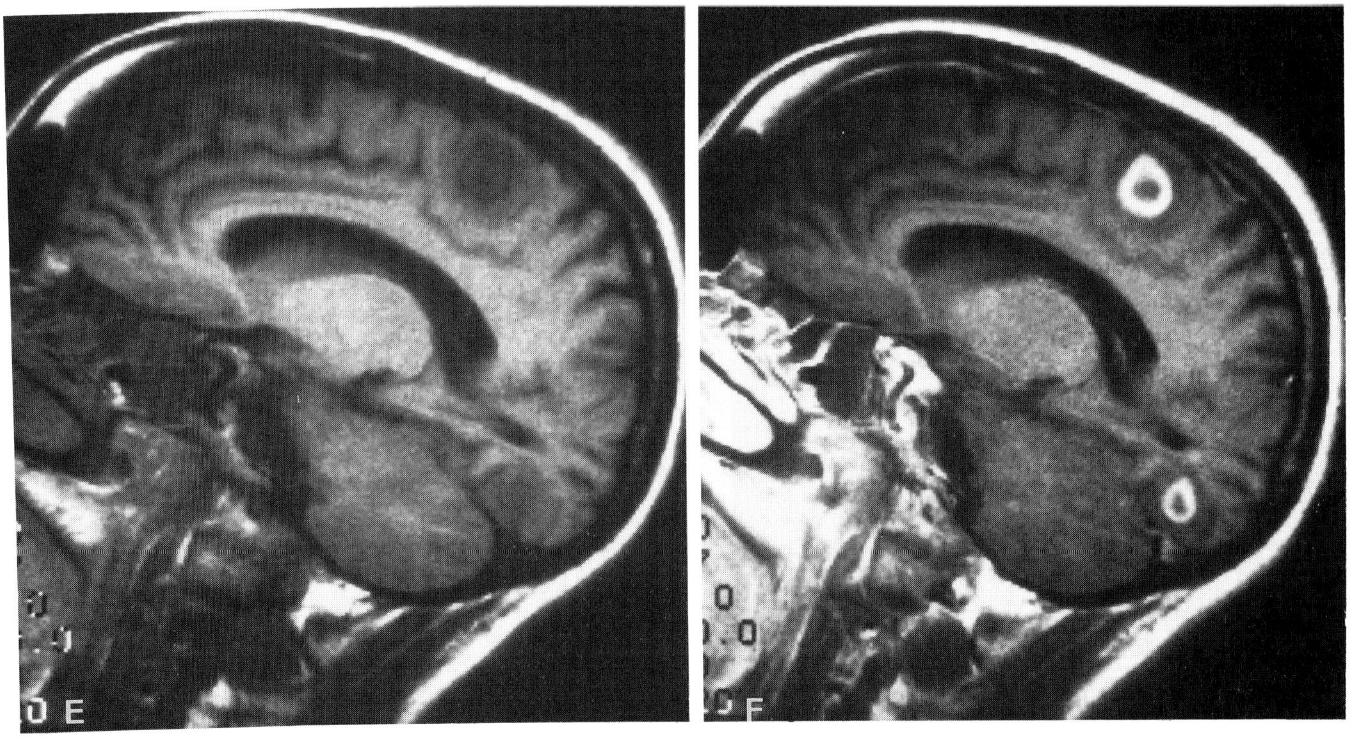

Figure 7.12–*continued*

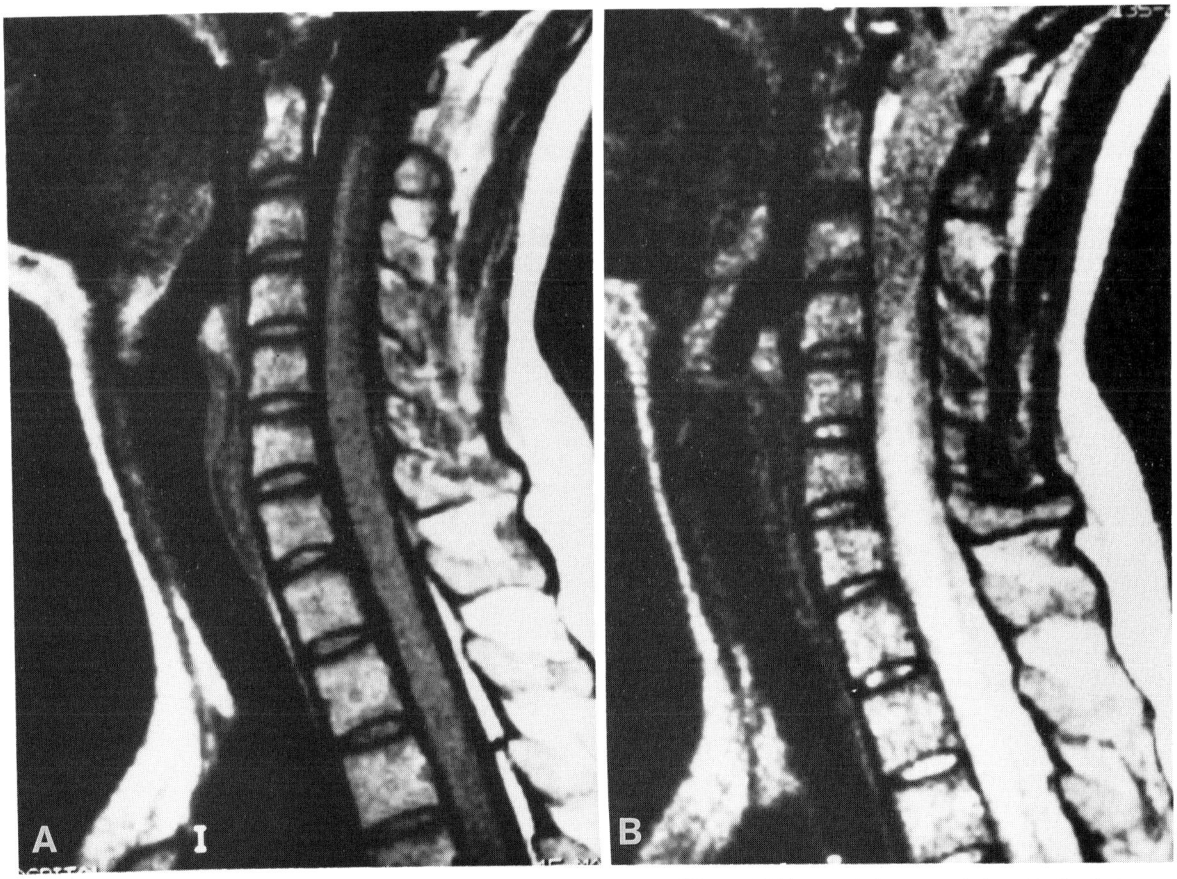

Figure 7.13. Transverse myelitis in a 28-year-old woman with a 2-week history of progressive weakness. Images obtained on a 0.5-T magnet. **A,** Sagittal T1-weighted scan (SE 500/40) shows mild enlargement of the lower cervical cord. **B,** Sagittal T2-weighted image (SE 2000/80) reveals increased signal in the lower cervical and upper thoracic cord. The history and MRI findings are consistent with transverse myelitis.

IMAGING

MRI with contrast is the best way to evaluate meningeal disease of the spine. Unenhanced scans are either normal in appearance or show nonspecific abnormalities ranging from linear stranding in the spinal canal to complete obliteration of the spinal canal (Fig. 7.15). However, after gadolinium is given, lesions can be much more readily seen and localized. Gero et al. [36] reported three patterns of enhancement: (*a*) linear enhancement on the surface of the cord and/or nerve roots, (*b*) nodular enhancement on the surface of the cord and/or nerve roots, and (*c*) diffuse, thick intradural enhancement of the leptomeninges completely surrounding the spinal cord.

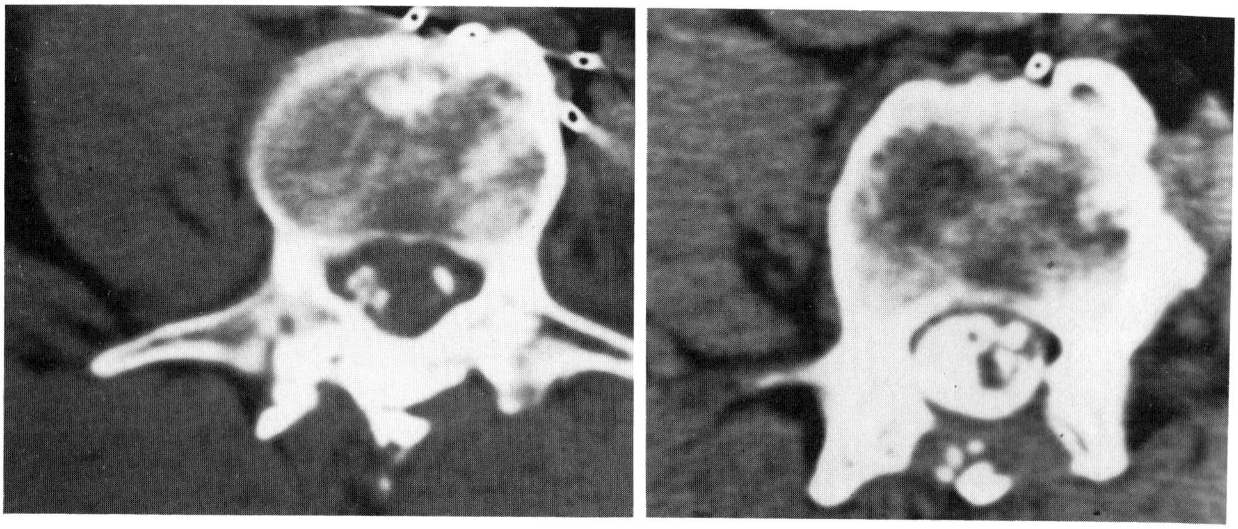

Figure 7.14. Pachymeningitis cervicalis hypertrophica. A,B, Images from a lumbar spine CT scan performed without intrathecal contrast material show extensive calcification with almost complete obliteration of the spinal canal.

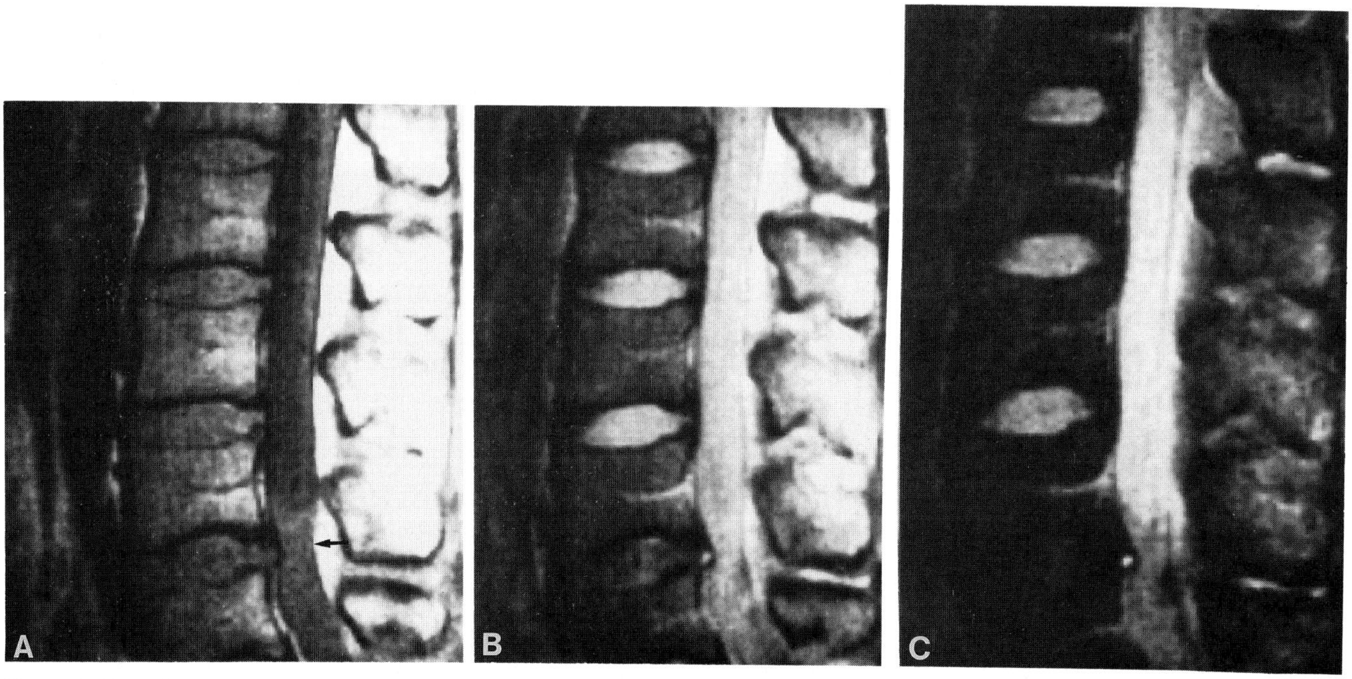

Figure 7.15. Cryptococcal meningitis in a 33-year-old man who developed abdominal pain following placement of a ventriculoperitoneal shunt. **A,** Sagittal T1-weighted scan (SE 500/17) shows an ill-defined "mass" (*arrow*) at the L4 and L5 levels which in part represents clumped nerve roots. **B,** Sagittal proton-density scan (SE 2500/35) does not clearly show the lesion. **C,** Sagittal T2-weighted scan (SE 2500/90) shows decreased intensity of the "mass" relative to CSF. The disk is decreased in signal intensity, consistent with degenerative change.

They found no correlation between the enhancement pattern and the severity of the presenting symptoms.

While MRI is an excellent test to evaluate for abnormality of the meninges, it is not specific. Leptomeningeal tumor spread can present with an identical appearance. Arachnoiditis from noninfectious causes also must be included in the differential diagnosis. Clinical history and CSF evaluation are often helpful.

A rare infection of the spine is cysticercosis. It is caused by infection from the larval stage of the pork tapeworm, *Taenia solium*. It is much more common in and around the brain than in the spine. In fact, most cases involving the spine probably originated in the subarachnoid space around the brain before migrating into the spinal subarachnoid space. These mobile cysts which migrate into the spinal subarachnoid space can cause an arachnoiditis. Less commonly, the cysts can involve the cord itself, probably from hematogenous dissemination.

Intramedullary cysts are most common in the thoracic region. They present most commonly with progressive spastic paraplegia, but they also may present with urinary and rectal sphincter dysfunction, radiating pain, sensory or motor deficits, or sexual impotence. MRI imaging shows a cystic-appearing lesion in the spinal cord with associated spinal cord expansion [45]. Myelography shows focal cord enlargement with or without a block. Differential diagnosis includes cystic cord tumor or localized syrinx. The presence of intracranial lesions and history of being from an endemic area are important clues in arriving at the correct diagnosis.

The subarachnoid form of the disease can produce symptoms of lower extremity weakness, paresthesia, or bowel and bladder dysfunction. MRI shows cysts within the spinal subarachnoid space. Because they may have exactly the same signal intensity as CSF, they may be identifiable only by noting localized deformity of the cord. Myelography better demonstrates their mobility. Imaging studies also may show findings of arachnoiditis. Once again, the presence of intracranial lesions and history of being from an endemic area are helpful in arriving at the correct diagnosis.

References

1. Post MJD, Quencer RM, Montalvo BM, et al. Spinal infection: evaluation with MR imaging and intraoperative US. Radiology 1988;169:765–771.
2. Enzmann DR. Infection and inflammation. In: Enzmann DR, DeLaPaz RL, Rubin JB, eds. Magnetic resonance of the spine. St. Louis: Mosby, 1990.
3. Burke DR, Brant-Zawadzki M. CT of pyogenic spine infection. Neuroradiology 1985;27:131–137.
4. Resnick D, Niwayama G. Osteomyelitis, septic arthritis and soft tissue infection: the axial skeleton. In: Resnick D, Niwayama G, eds. Diagnosis of bone and joint disorders. Philadelphia: Saunders, 1981.
5. Carson HW. Acute osteomyelitis of the spine. Br J Surg 1930;18:400.
6. Waldvogel FA, Vasey H. Osteomyelitis: the past decade. N Engl J Med 1980;303:360–370.
7. Sapico FL, Montgomerie JZ. Pyogenic vertebral osteomyelitis: report of nine cases and review of the literature. Rev Infect Dis 1979;1:754–776.
8. Smith AS, Blasser SI. Infectious and inflammatory processes of the spine. Radiol Clin North Am 1991;29:809–827.
9. Goldman AB, Freiberger RH. Localized infectious and neuropathic diseases. Semin Roentgenol 1979;14:19.
10. Case Records of the Massachusetts General Hospital (Case 24-1989). N Engl J Med 1989;320:1610–1618.
11. Weaver P, Lifeso RM. The radiological diagnosis of tuberculosis of the adult spine. Skeletal Radiol 1984; 12(3):178–186.
12. Smith AS, Weinstein MA, Mizushima A, et al. MR imaging characteristics of tuberculous spondylitis vs vertebral osteomyelitis. AJNR 1989;10:619–625.
13. Coventry MB, Ghormley RK, Kernohan JW. The intervertebral disk: its micro-scopic anatomy and pathology. J Bone Joint Surg 1945;27A:105.
14. Mineiro JD. Coluna vertebral humana: alguns aspectos da sua estrutura e vascularizacao. Lisboa 1: Dissertacao de Doctoramento, 1965.
15. Modic MT, Feiglin DH, Piraino DW, et al. Vertebral osteomyelitis: assessment using MR. Radiology 1985;157: 157–166.
16. Thrush A, Enzmann D. MR imaging of infectious spondylitis. AJNR 1990;11:1171–1180.
17. Post MJD, Sze G, Quencer RM, et al. Gadolinium-enhanced MR in spinal infection. J Comput Assist Tomogr 1990;14:721–729.
18. duLac P, Panuel M, Demed P, et al. MRI of disk space infection in infants and children. Pediatr Radiol 1990; 20:175–178.
19. Kopecky KK, Gilmor RI, Scott JA, et al. Pitfalls in computed tomography in diagnosis of discitis. Neuroradiology 1985;27:57–66.
20. Bielecki DK, Sartoris P, Resnick D, et al. Intraosseous and intradiscal gas in association with spinal infection: report of three cases. AJR 1986;147:83–86.
21. Resnick D, Niwayama G, Guerra J, et al. Spinal vacuum phenomena: anatomical study and review. Radiology 1981;139:341–348.
22. Baker AS, Ojemann RG, Swartz MN, et al. Spinal epidural abscess. N Engl J Med 1975;293:463–468.
23. Lasker BR, Harter DH. Cervical epidural abscess. Neurology 1987;37:1747.
24. Whelan MA, Schonfeld S, Post JD, et al. Computed tomography of nontuberculous spinal infection. J Comput Assist Tomogr 1985;9:280–287.
25. Marks WA, Bodensteiner JB. Anterior cervical epidural abscess with pneumococcus in an infant. J Child Neurol 1988;3:25–29.
26. Curling OD Jr, Gower DJ, McWhorter JM. Changing concepts in spinal epidural abscess: a report of 29 cases. Neurosurgery 1990;27:185–192.
27. Leys D, Lesoin F, Viaud C, et al. Decreased morbidity from acute bacterial spinal epidural abscesses using computed tomography and nonsurgical treatment in selected patients. Ann Neurol 1985;17:350–355.

28. Mampalam TJ, Rosegay H, Andres BT, et al. Nonoperative treatment of spinal epidural infections. J Neurosurg 1989;71:208–210.
29. Messer HD, Lenchner GS, Brust JCM, Resor S. Lumbar spinal abscess managed conservatively: case report. J Neurosurg 1977;46:825–829.
30. Stewart IC, Ford MJ, Heading RC, et al. Acute spinal epidural abscess: cause of meningism. Scott Med J 1981;26:348–349.
31. Sandhu FS, Dillon WP. Spinal epidural abscess: evaluation with contrast enhanced MR imaging. AJNR 1991; 12:1087–1093.
32. Dandy WE. Abscesses and inflammatory tumors in the spinal epidural space (so-called pachymeningitis externa). Arch Surg 1926;13:477–494.
33. Koppel BS, Daras M, Duffy KR. Intramedullary spinal cord abscess. Neurosurgery 1990;26:145–146.
34. Mehren M, Burns PJ, Mamani F, et al. Toxoplasmic myelitis mimicking intramedullary spinal cord tumor. Neurology 1988;38:1638–1650.
35. Woolsey RM, Chambers TJ, Chung HD, McGarry JD. Mycobacterial meningomyelitis associated with human immunodeficiency virus infection. Arch Neurol 1988;45: 691–693.
36. Gero B, Sze G, Sharif H. MR imaging of intradural inflammatory diseases of the spine. AJNR 1991;12:1009–1019.
37. Dupuis MJM, Atrouni S, Dooms GC, Gonsette RE. MR imaging of shistosomal myelitis. AJNR 1990;11:782–783.
38. Shabas D, Gerard G, Cunha B, et al. MR imaging of AIDS myelitis. AJNR 1989;10:551–552.
39. Tucker T, Dix RD, Katzen C, et al. Cytomegalovirus and herpes simplex virus ascending myelitis in a patient with acquired immune deficiency syndrome. Ann Neurol 1985;18:74–79.
40. Merine D, Wang H, Kumar AJ, et al. CT myelography and MR imaging of acute transverse myelitis. J Comput Assist Tomogr 1987;11:606–608.
41. Roper AH, Poskanzer DC. The prognosis of acute and subacute transverse myelopathy based on early signs and symptoms. Ann Neurol 1978;4:51–59.
42. Barakos JA, Mark AS, Dillon WP, Norman D. MR imaging of acute transverse myelitis and AIDS myelopathy. J Comput Assist Tomogr 1990;14:45–50.
43. Miller DH, Ormerod IEC, Rudge P, Kendall BE, Moseley IF, McDonald WI. The early risk of multiple sclerosis following isolated acute syndromes of the brainstem and spinal cord. Ann Neurol 1989;26:635–639.
44. Rosenfeld JV, Kaye AH, Davis S, Gonzales M. Pachymeningitis cervicalis hypertrophica. J Neurosurg 1987;66: 137–139.
45. Castillo M, Quencer R, Post MJD. MR of intramedullary spinal cysticercosis. AJNR 1988;9:393–395.

CHAPTER / 8

Trauma

P. Davis / K. C. V. G. Rao

PEDIATRIC SPINE TRAUMA

Patricia Davis

In most series, children account for approximately 5 to 10% of significant spinal injuries [1–4]. Cervical spinal injuries outnumber injuries of the lower spine, particularly in children less than 8 to 12 years of age. Children from birth to about 8 years old have a greater susceptibility to traumatic upper cervical spine injuries (occiput to C3) than older children and adults. Factors responsible for differences in spinal injuries in children include less mineralization of the child's vertebral column, greater flexibility of the nonbony musculoskeletal structures, poorly developed musculature, and the greater proportion by weight of the cranium relative to the rest of the body [1–12]. The cervical facet joints are more horizontally oriented in younger children, thereby facilitating horizontal displacement of the vertebral column without fracture. Mechanisms of injury also differ. Unlike older children and adults, young children are more often involved in pedestrian–motor vehicle accidents (MVAs) or bicycle–MVAs. Older children are often injured in sports-related injuries, nonpedestrian passenger–MVAs, or by firearms [1–12]. Other causes include falls, birth injuries, and battered child syndrome [1–14]. Males outnumber females by about 2:1 [1]. Vertebral bony injuries often heal well without surgical intervention in children, who may undergo spontaneous restoration of lost vertebral body height over time [2]. Spontaneous vertebral fusion after trauma is unusual, a finding that has been attributed to the low incidence of disk space injury in children [5]. Those with neurologic deficits and/or spinal instability may have persistent disability and/or spinal deformity [2].

Traumatic spinal cord injury in children occurs either with or without bony spinal column fracture and/or dislocation [6, 7]. Other injuries are often present, particularly facial or head trauma. Although some pediatric spinal trauma series suggest a low incidence of neurologic deficit, these studies generally exclude cord injuries without bony injury and/or cord injuries in infants and newborns [1, 2, 8, 9]. Spinal cord injury without radiographic abnormality demonstrable by plain film and/or myelography (SCIWORA) occurs predominantly in children and is frequently accompanied by significant neurologic deficits [6, 7]. In subsequent sections, patterns of spinal cord and column injury and their sequelae in children will be discussed.

Radiographic Evaluation

Evaluation of the child suspected of an acute spinal injury begins with immobilization and medical stabilization, followed by plain films of the appropriate spinal segments [9–12, 15]. The initial study is usually a cross-table lateral view in neutral position timed to coincide with inspiration, followed by an anteroposterior or frontal view. In the cervical region, an open-mouth view of the odontoid is added.

Lateral plain films should demonstrate both the craniocervical and cervicothoracic junctions. If alignment is intact, three craniocaudal lines may be drawn on lateral views without any focal irregularities or discontinuities [16]. These include a line drawn along the clivus, posterior margin of the odontoid, and posterior vertebral bodies; one from the posterior foramen magnum along the posterior spinolaminar line; and one anteriorly along the anterior vertebral body margins from C1 to T1 [16] (Fig. 8.1). The paired facet joints overlap each other on a well-positioned lateral view. An abrupt change in obliquity or overlap of the facet joints should be viewed with suspicion for a unilateral facet dislocation. The intervals between the spinous processes should be similar, although the interspinous interval between C1 and C2 posteriorly is often somewhat wider than other levels (Fig. 8.2). The frontal view should demonstrate intact vertebral alignment with a symmetrical central location of each spinous process (Fig. 8.3). The lateral masses and facet joints appear on this view as a gently undulating symmetrical complex lateral to the vertebral bodies with no focal areas of asymmetry or abrupt step-off. The open-mouth view shows the odontoid process centered between the C1–2 facet joints. The occiput–C1 and C1–2 facet joints should be symmetrical without step-offs or alignment abnormalities. The clinical assessment and findings on this initial series of plain films determine the necessity for additional imaging studies; about 70 to 90% of cervical fractures are identified on plain films [16–19].

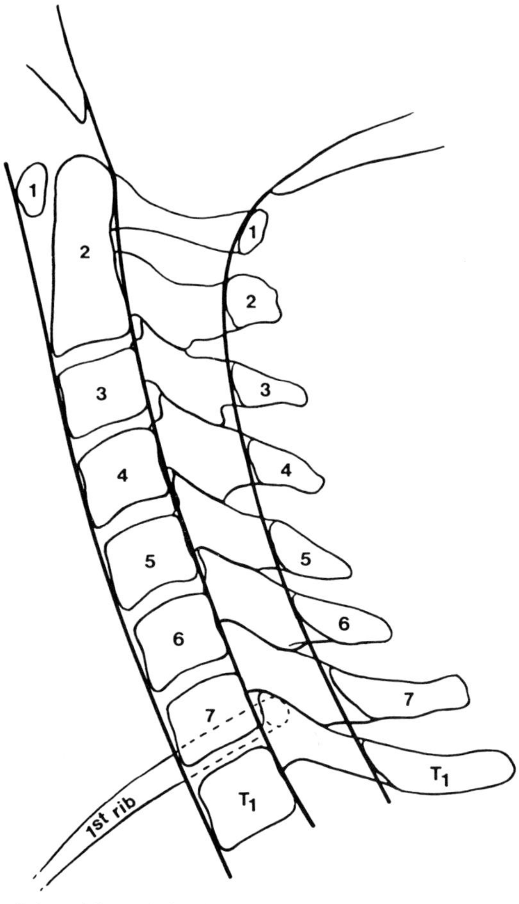

Figure 8.1. A lateral drawing of the cervical spine indicates three lines used to confirm alignment. A smoothly curving line is drawn along the anterior vertebral body margins from C1 to T1. The second line smoothly curves from the posterior clivus along the posterior dens and the posterior vertebral bodies. The third line curves smoothly from the posterior lip of the foramen magnum caudally along the spinolaminar line from C1 to C7.

The atlantodental interval (ADI), or the distance between the posterior margin of the anterior ring of C1 and the anterior margin of the odontoid, is normally larger in children than adults; nevertheless, it should not exceed 5 mm [20, 21] (Fig. 8.4). Widening of the ADI is not usually a sequela of trauma in children; the transverse ligamentous complex is so strong that the odontoid process tends to fracture before the ligament is ruptured. Instead, a wide ADI usually accompanies arthritic or inflammatory erosion of the transverse ligament or congenital anomalies of the atlantoaxial region [20, 21]. The prevertebral soft tissues may be thicker in children than in adults, with normal measurements of less than 7 mm in children under 15 years of age measured anterior to C2 [20]. The retrotracheal space is larger due to the esophageal soft tissues; in children it ranges from 5 to 14 mm in thickness [22]. Measurements above this suggest abnormal soft-tissue thickening, which may represent prevertebral hematoma, inflammatory disease, or poor distension of the pharynx on an expiratory film. Prevertebral swelling suggests significant spinal injury, although this is not present uniformly in association with acute spinal injury.

Pseudosubluxation with up to 4 mm of anterior displacement of the vertebral body of C2 on C3 has been described as a normal variant in children up to about 7 years of age [23] (see Figs. 8.2 and 8.4). In all cases, the smooth curve and alignment of the posterior spinolaminar line should be maintained. With flexion, the axis body relationship to the body of C3 must be the same as the change in position of the spinolaminar line in order to exclude instability [23]. The posterior cervical line is drawn from the posterior spinolaminar line of C1 to the posterior spinolaminar line of C3. This line should touch, cross, or be within 1 mm of the posterior arch of C2 [23].

Additional plain film views include oblique or pillar views to further define the neural arches or a swimmer's view for visualization of C7–T1. If instability is suspected or confirmed, assurance of immobilization during additional filming is essential. Flexion and extension plain film views should be considered only with careful supervision in an awake child who has no evidence of an unstable fracture. These are used in order to confirm stability and exclude a ligamentous injury. Tomography, computed tomography (CT), myelography followed by CT (MCT), and magnetic resonance imaging (MRI) are then performed to provide additional anatomic detail of suspected or confirmed injuries.

Linear tomography or polytomography provides

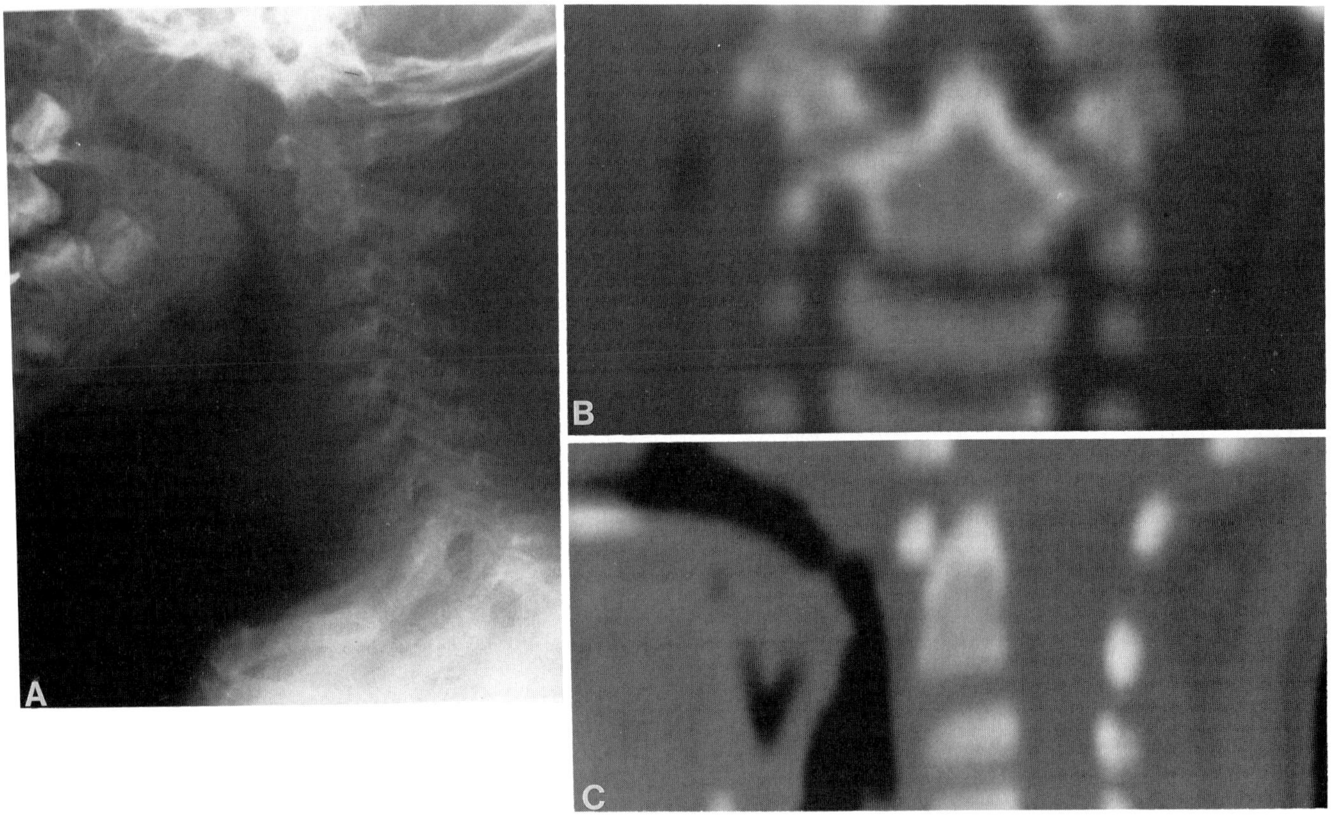

Figure 8.2. **A,** A lateral plain film of a young boy evaluated following an MVA demonstrates the normally prominent soft tissues of childhood. Note the relatively wide posterior interspinous distance between C1 and C2. C2 is slightly subluxed anteriorly relative to C3. The posterior spinolaminar line is intact, indicating a pseudosubluxation. Following a head CT for associated cranial injury, thin axial CT images of the upper cervical spine were obtained in order to exclude fracture. Re-formatted coronal (**B**) and sagittal images (**C**) demonstrate intact alignment in neutral position. Supervised flexion and extension plain films would then be considered to exclude a ligamentous injury.

additional bone detail not afforded by plain films, but lateral decubitus positioning required for sagittal tomography may be problematic for the critically ill patient or for one with an unstable spinal fracture. Bone detail is generally inferior to that provided by CT, and tomography provides no soft-tissue information. CT has generally replaced plain film tomography except for selected patients suspected of having a fracture that is primarily transversely oriented or in those with instrumentation or support devices which create artifacts on CT or MRI.

CT should be the examination of choice for further evaluation of known spinal column injuries and for confirmation or exclusion of suspected spinal fractures [24–26]. Advantages include widespread availability, quick scan time, capability to re-format images into a coronal or sagittal plane, and superb bone detail. Slice thickness must be appropriately small relative to the level(s) to be evaluated (i.e., 2- to 3-mm slices). At least one vertebral body above and below the level(s) of suspected or known injury should be scanned in order to exclude additional fractures. Images are then photographed with both bone and soft-tissues windows. A small overlap (i.e., 1 mm) between slices results in superior re-formatted images, and re-formatting of CT images in a coronal and sagittal plane is performed routinely as part of all spinal trauma studies (see Fig. 8.2). The re-formatted images are useful for evaluation of canal impingement, subluxations, and alignment abnormalities, especially with subtle or no bony injury. High-quality re-formatted images require a motionless patient; sedation may be necessary to evaluate the spinal injuries successfully. CT does not provide optimal images of associated intraspinal soft-tissue injuries (disk herniation, spinal cord contusion/transection, ligamentous tears) or axial plane fractures. Artifacts from metallic stabilization devices may degrade scan quality, and long segments of the spinal canal cannot be readily screened with CT within a reasonable scan time.

Myelography followed by CT (MCT) has been used for demonstration of spinal cord injury and exclusion of cord compression by disk herniation or extradural hematoma. However, this role is being replaced in many cases by MRI. A "minidose" myelogram should be substituted for conventional myelography if spinal column instability is suspected.

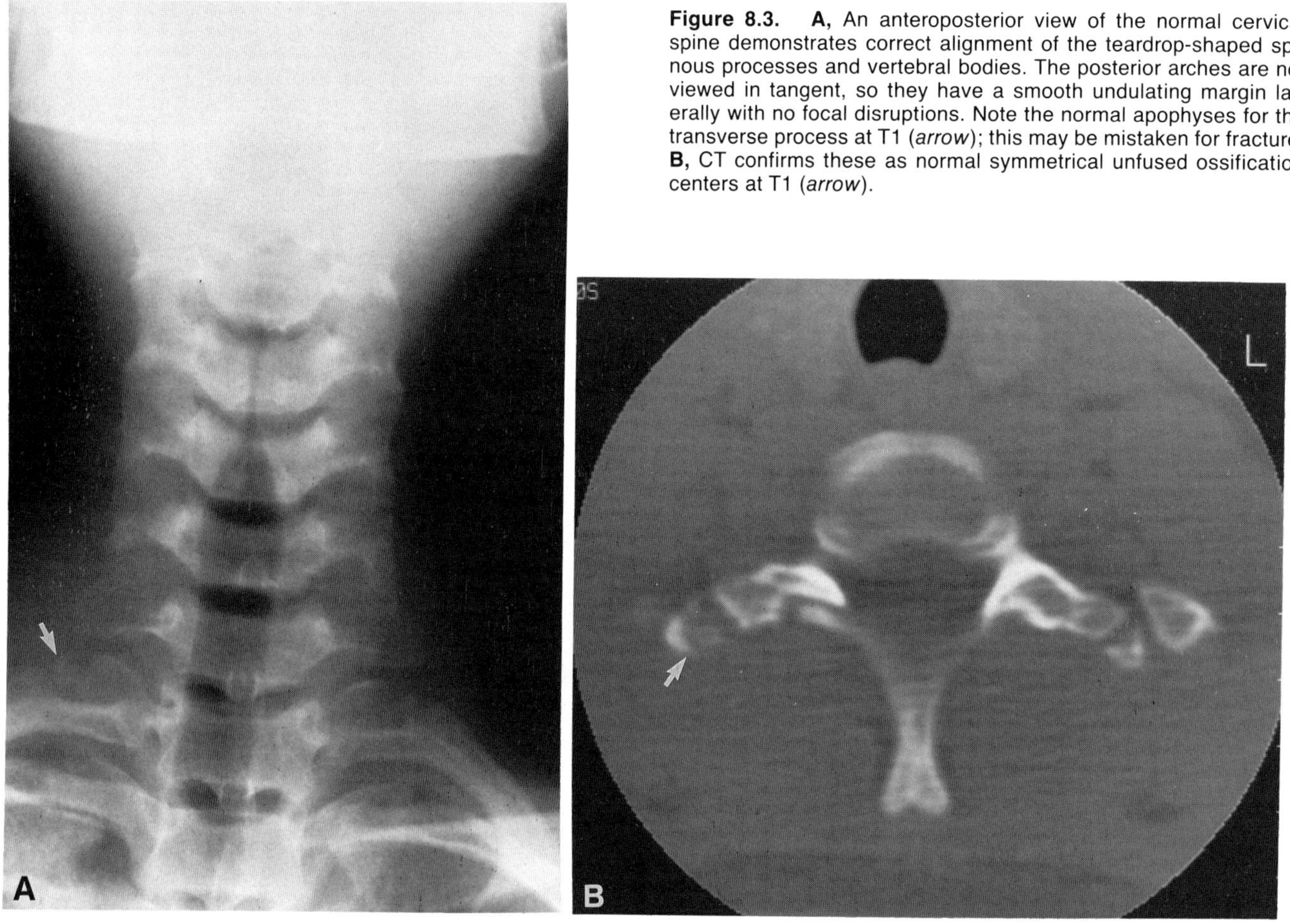

Figure 8.3. **A,** An anteroposterior view of the normal cervical spine demonstrates correct alignment of the teardrop-shaped spinous processes and vertebral bodies. The posterior arches are not viewed in tangent, so they have a smooth undulating margin laterally with no focal disruptions. Note the normal apophyses for the transverse process at T1 (*arrow*); this may be mistaken for fracture. **B,** CT confirms these as normal symmetrical unfused ossification centers at T1 (*arrow*).

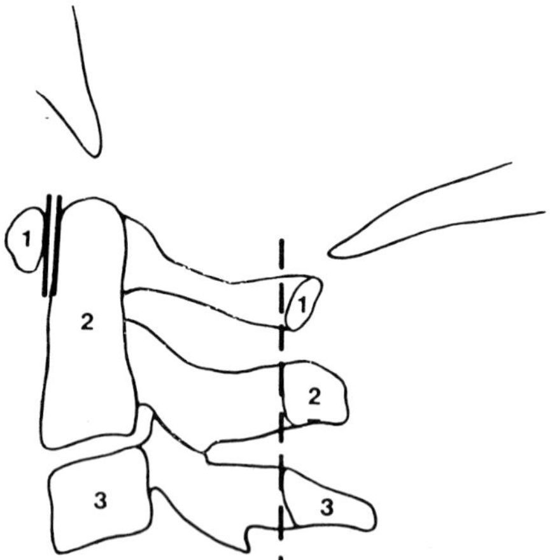

Figure 8.4. The normal atlantodental interval (ADI) is less than or equal to 5 mm in children (*solid lines*). A straight line drawn from the posterior spinolaminar line of C1 to the posterior spinolaminar line of C3 should intersect or be within 1 mm of the posterior spinolaminar line of C2. This (*dashed line*) is useful for differentiation of physiologic pseudosubluxation from traumatic subluxation (23). [Swischuk LE. Anterior displacement of C2 in children: Physiologic or pathologic? Radiology 1977: 122:759]

This procedure entails introduction of 3 to 5 cc of water-soluble contrast material in the spinal subarachnoid space followed by CT scanning without conventional postmyelographic plain films. With appropriate immobilization and support, a lumbar spinal puncture can be performed with the patient either prone or in the decubitus position. Alternatively, an older and cooperative patient may undergo a C1–2 puncture in a supine, prone, or lateral decubitus position. Myelography is required infrequently; the major indications include dural tears, nerve root avulsion, or as a substitute for children who cannot undergo successful MRI due to a foreign body or incompatible medical life-support devices.

MRI is being utilized with greater frequency for evaluating injuries in children [27–29], as has been described previously for adults [30–40]. After initial stabilization and evaluation of bony injuries using plain films and/or CT, MRI offers a noninvasive technique for visualization and localization of spinal cord injuries and exclusion of cord compression by bone or disk fragments. MRI can be performed safely and early, provided that appropriate planning and personnel are available. Indications for MRI include evaluation of spinal cord injury and/or integrity and noninvasive demonstration of disk herniation, cord compression, epidural hematoma, ligamentous injury, and possibly brachial plexus injury [11, 29].

MRI examination of the acutely injured child necessitates MRI-compatible immobilization devices that support the unstable or potentially unstable spinal column [29–38]. All life-support devices must be MRI-compatible or temporarily withheld if clinically appropriate during the MRI examination. The child should be monitored continuously (observation, pulse oximetry, cardiac tracing, respiratory rate), and personnel experienced in evaluation of acutely ill pediatric patients should be in attendance [41–42].

Surface coils are tailored to the level(s) of suspected injury, the child's size, and the curvature of the child's spine. Technology that provides the capability of imaging the entire spinal canal quickly without moving the patient facilitates a thorough and rapid MRI examination. Techniques for accomplishing this include use of decoupled surface coils so that multiple surface coils may be positioned under the child initially. This eliminates repositioning of the child or the spinal surface coils during the remainder of the examination [29]. Other techniques include placement of the child on a Plexiglas support board which encloses a movable surface coil; the coil can be repositioned as needed without repositioning the child. Adolescent or larger patients can be studied using a segmented or phased-array spine coil; the cervical support for the adult lordotic cervical spine of this coil is not optimal for the flatter and shorter spine of the younger child [43].

Rapid scanning of one segment or of the entire spinal canal with sequences tailored to the child's age, area of suspected injury, and stability are important. Choices of the field of view, slice thickness (3 to 5 mm), interslice interval (0 to 2 mm), number of signal averages, and pulse sequences are dictated by the child's size and level of suspected injury. Typical initial sequences include a T1- and a T2-weighted sagittal spin-echo sequence through the level of suspected injury. T2-weighted gradient-echo sequences may improve detection of acute hemorrhage, and fast spin-echo T2-weighted sequences are a reasonable alternative to shorten scan times [30–37]. Additional planes of imaging (axial or coronal) with spin-echo or gradient-echo sequences may be added if the child is clinically stable so as to demonstrate the extent of spinal injury.

Normal Spine Development and Variants

An understanding of normal spinal development and recognition of normal growth patterns of the spinal canal in children are essential for evaluation of the child with a spinal injury [5, 11, 12, 15, 44]. Measurements for the prevertebral and retrotracheal soft tissues of the neck in children are described above, although these may be accentuated with flexion or poor inspiration during filming. The sagittal spinal canal diameter in children is similar to that of adults, with 10 mm being a reasonable point at which canal compromise and/or spinal cord compression is to be expected [45].

Other normal findings include those associated with maturation of the spinal column. The infant spine at birth characteristically has three demonstrable ossification centers for each vertebral body [2, 44]. These consist of the centrum, which forms the vertebral body proper, and two paired centers posterolaterally, which form the neural arches. Bipartite anterior and posterior ossification centers for the centrum usually disappear by about 1 year, although a coronal cleft may remain radiographically visible up to age 4 years [15, 20]. The ossification centers between the centrum and neural arches fuse between the ages of 3 and 6 years, and the posterior arches fuse in a caudal to cranial direction from lumbar to cervical at about 3 years of age [44]. The dorsal arches of C2 and the sacrum fuse later, at 4 to 6 years and 6 to 7 years, respectively, although persistent nonfused posterior arches have been noted in 1 to 9% of adults as a normal variant at the cranial and caudal extremes of the spinal canal [44, 46, 47] (Fig. 8.5). Ring apophyses appear as a complete ring at the superior and inferior vertebral body margins by 6 to 8 years in girls and 7 to 9 years in boys, with fusion to the vertebral body completed by 15 years of age [11, 15, 20, 44]. Apophyses for the transverse processes and spinous process tips develop later and may mimic fracture if not recognized as normal developmental findings [47] (see Fig. 8.3). Vascular clefts transversely across the vertebral bodies may persist into adulthood.

The transitional levels of C1 and C2 develop uniquely; the central axis develops from three ossification centers. These consist of a centrum which ossifies earliest, a center for the dens base, and one for the dens tip. The dens tip is last to ossify but is generally visible radiographically by 2 years of age [2]. The dens base and centrum fuse by about 3 years, although a visible remnant may persist radiographically into adulthood [15, 20] (Fig. 8.6). This synchondrosis projects on a lateral film below the plane of the facet joints; this is a frequent site of odontoid fracture in children. The dens tip fuses by about age 12, although it may persist unfused as an os odontoideum on a developmental, ischemic (i.e., avascular necrosis), or in some cases posttraumatic basis [48, 49]. Paired ossification centers posterolaterally form the neural arches of C1 and C2. One or more ossification centers for the anterior arch of C1 may occasionally be visible at birth; the anterior

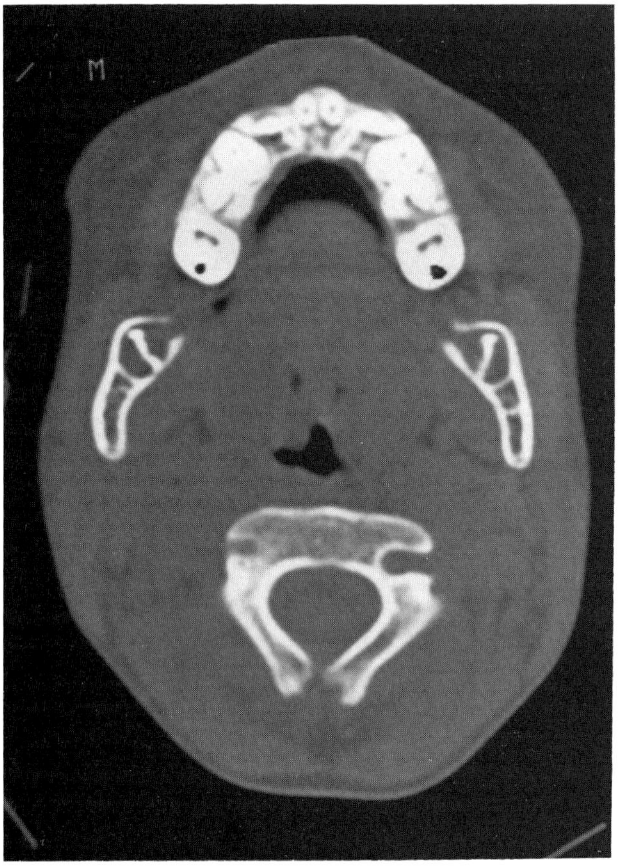

Figure 8.5. CT demonstrates posterior failure of midline fusion of the neural arches, a normal variant that may persist into adulthood.

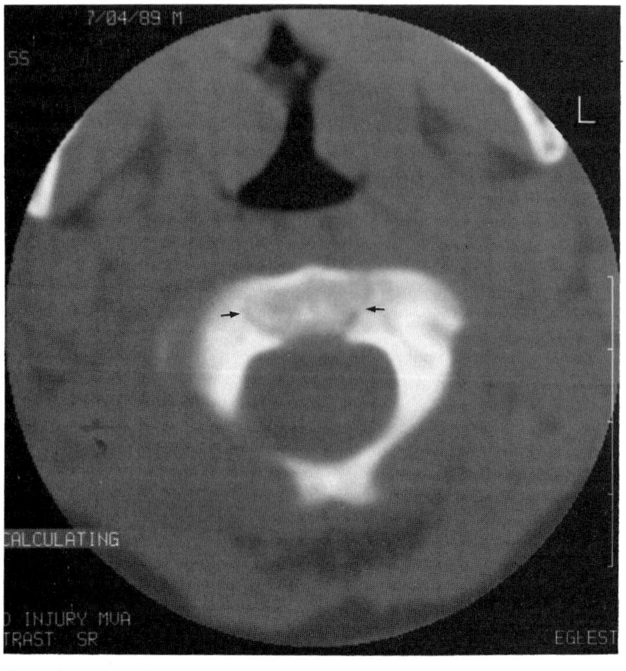

Figure 8.6. The incompletely fused synchondrosis between the ossification centers of the centrum and neural arches is located at the posterolateral margin of the vertebral body; this should not be mistaken for fracture (*arrows*).

arch fuses by 10 years of age [20] (Figs. 8.7 and 8.8). The transverse diameter of the ring of C1 may be larger than that of C2 in children less than 4 years of age; this mimics lateral displacement of the lateral masses of C1, as might be seen with C1 fractures. This is attributed to greater growth of C1 relative to C2 at this age, resulting in a "pseudo-Jefferson" variant [50] (Fig. 8.9). The anterior ring of C1 may override the dens in normal infants imaged in extension. The ossification center for the dens tip may be undeveloped or mimic an avulsion fracture in young children.

The normal MRI appearance of the pediatric spine changes markedly as the spinal column and inter-

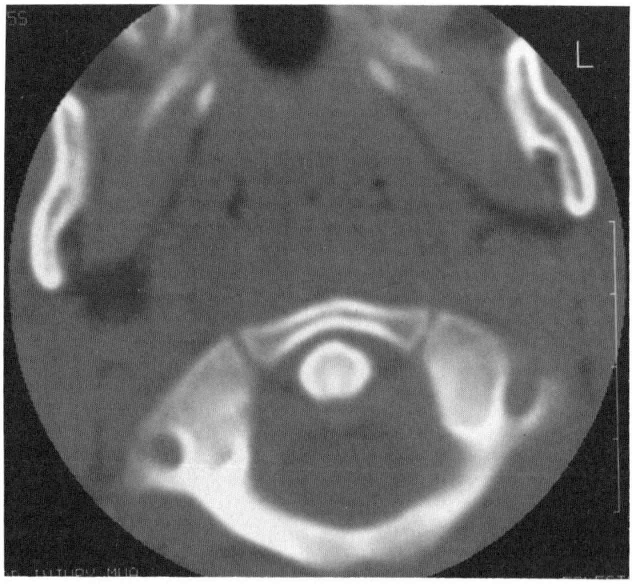

Figure 8.7. The anterior arch of C1 has one or more ossification centers which fuse with those of the neural arches by age 10. Note the intact cortex and smooth margins of the synchondrosis.

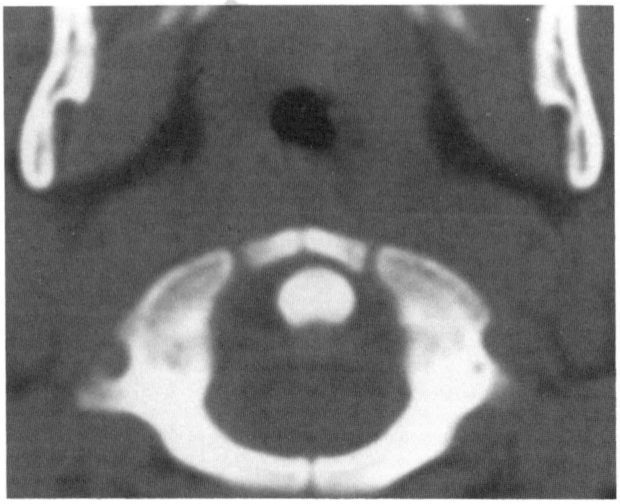

Figure 8.8. This young child has bipartite ossification centers for the anterior ring of C1, a normal variant. Also note the age-appropriate incomplete fusion posteriorly of the neural arch.

vening disk spaces mature (Figs. 8.10 and 8.11). The infant spine is characterized by very low signal intensity within the relatively small vertebral bodies on both T1- and T2-weighted sequences. The signal intensity of the disk and endplate cartilage are similar, resulting in the appearance of a large distance between vertebral bodies [51]. This appearance is a reflection of the small size of the ossified portion of the vertebral body of infants in combination with red bone marrow, which is predominant at this age. The vertebral body intensities gradually increase over the next 6 months with diminishing prominence of the cartilaginous endplates, followed by formation of a rectangular-shaped vertebral body and further diminution in endplate cartilage [51–54]. Yellow bone marrow predominates in the normal vertebral body by early adulthood, and its signal intensity should be greater than that of the disk on T1-weighted images [54]. Previous reports suggesting that the conus medullaris ascends progressively during childhood have not been supported by recent studies. Wilson and Prince found that the normal conus medullaris terminates at the L2–3 level or higher in children and adults [55] (see Fig. 8.10).

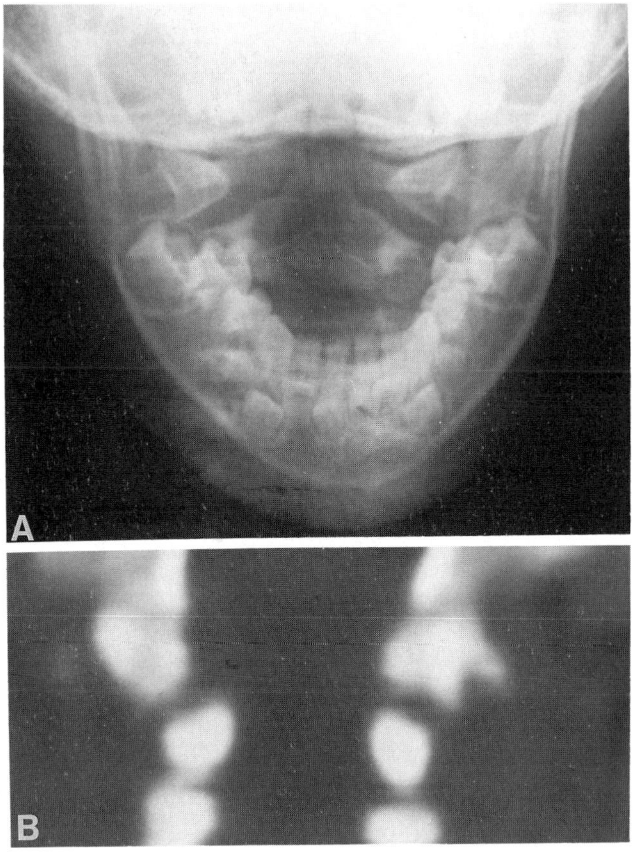

Figure 8.9. An open-mouth view of the odontoid illustrates the difficulty in obtaining this view in young children. **A,** The lateral offset on the right of C1 on C2 prompted CT scanning. **B,** Re-formatted image demonstrates intact alignment from the foramen magnum to C3. The lateral offset seen on plain films was due to rotation.

Classification of Spinal Injuries

Classifications of spinal injuries generally reflect the spinal level and likely direction of force that produced the injury [56–60]. The goal of most classification systems is to group injuries into recognizable types with similar characteristics in order to predict stability and clinical course. Although the mechanism of any given injury is often uncertain or mixed, these classifications are of some value in predicting spinal stability and likely associated fractures. In part, the type of injury also reflects the ligamentous strength of the spinal soft tissues and the angle of the facet joints of the injured segment. For example, the more steeply angled facet joints of the thoracic and lumbar spine result in fewer subluxations when compared with the cervical spine.

Classifications typically include flexion, extension, rotation, axial loading or compression, and shear-type injuries [56–60]. Milder flexion injuries include simple wedge or compression fractures of the vertebral body. More severe injuries may include fracture or ligamentous disruption of the facet joints and posterior neural arches, resulting in instability. The most severe flexion injuries are those with vertebral body fracture and fracture-dislocation of the facet joints. These injuries are often associated with cord compression and secondary neurologic deficits. Extension injuries are less frequent. Isolated fractures of the posterior neural arches are generally stable. With severe extension injury, the anterior longitudinal ligament and/or anterosuperior vertebral body may fracture, resulting in an unstable extension teardrop fracture. Compression or burst injuries are attributed to axial loading and/or marked flexion injuries. Those with fractures limited to the vertebral body are generally stable due to intact ligaments; burst vertebral body fractures accompanied by fractures of the posterior arches are unstable. Often a combination of rotation, flexion, and extension occurs coincidentally, rendering exact classification difficult. Shear injuries include injuries with a transverse vector of force and a horizontal fracture plane through the vertebral body and posterior elements. The classic injury of this type is the Chance or seatbelt injury, which may be associated with neurologic deficit and spinal instability. Occipitoatlantal distraction-dislocation and SCIWORA are injuries which occur characteristically in children; mechanisms for these injuries are discussed below.

Spinal *stability* or *instability* are ill-defined terms which are used as though they have a precise and standardized meaning [60]. *Instability* is used in some instances to indicate an injury that requires surgical fusion, although unstable injuries in children may heal with immobilization. According to White, *instability* refers to the body's inability to

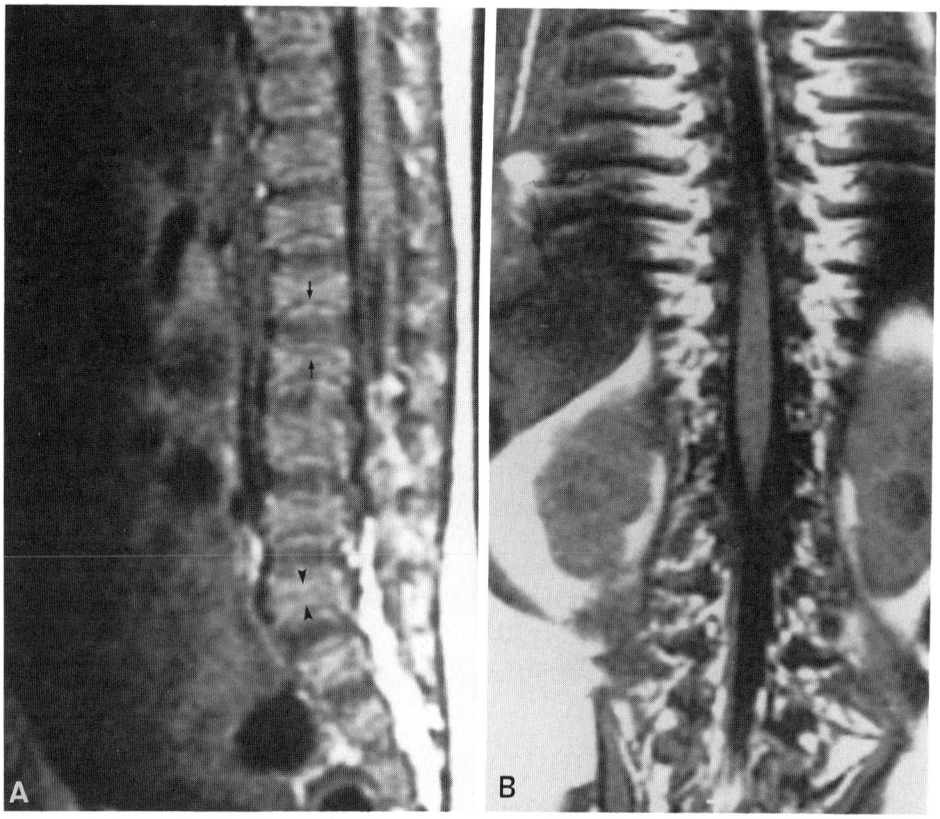

Figure 8.10. Spinal cord appearance in infants. The normal appearance of the spine changes rapidly during childhood on MRI. **A,** In infants, the small centrum (*arrows*) is surrounded by large intervening tissues of similar intensity consisting of the cartilaginous endplate and disk (*arrowheads;* TR 600, TE 25, 3.0 mm). **B,** The conus in infants, as in older children, has a smoothly tapered appearance and usually terminates at the L2–3 disk space or higher (TR 500, TE 25, 3.0 mm). A conus that terminates below this level should raise the possibility of cord tether.

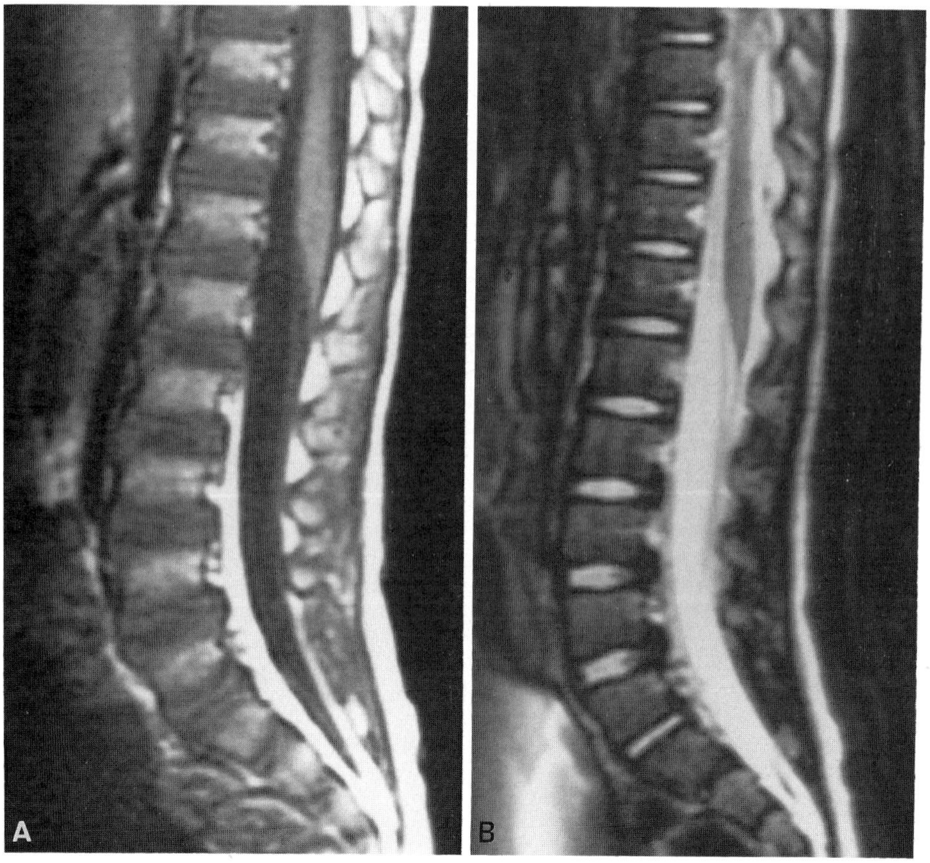

Figure 8.11. Vertebral marrow changes of aging. By about 1 year of age, the vertebral bodies are larger and more rectangular in appearance, and the disk space is inseparable from the endplates, as shown in this 4-year-old boy. The bone marrow intensity is lower on T1- and T2-weighted images due to predominance of hematopoietic marrow in childhood (**A:** TR 500, TE 11, 4.0 mm; **B:** TR 4000, TE 112, 4.0 mm).

maintain normal spinal relationships between the vertebrae without development of neurologic symptoms (radicular or spinal cord, pain, etc.) or spinal deformity [56]. Descriptions of vertebral injuries as unstable or stable are often imprecise. A two-column concept is often used for cervical injuries; disruption of both the anterior column (vertebral body, disk, anterior longitudinal ligament, annulus fibrosus, posterior longitudinal ligament) and the posterior column (neural arches and ligaments) is necessary for instability [56, 57]. Signs of potential cervical instability adapted from Swischuk [15] include

1. Vertebral body dislocation or distraction (occiput–C1)
2. Abnormal disk space width
3. Focally wide joints of Luschka
4. Facet joint dislocation
5. Spinous process distraction
6. Teardrop fractures
7. Vertebral compression fractures with retropulsed fragments and burst fractures
8. Unilateral or bilateral displacement of the ring of C1
9. Dens fractures.

The thoracolumbar spinal integrity is best assessed in terms of a three-column concept [58]. The columns consist of

Anterior column: anterior longitudinal ligament and anterior vertebral body
Middle column: posterior longitudinal ligament, posterior vertebral body, and posterior annulus fibrosus
Posterior column: posterior pillars, laminae, and spinous process [58]

Based on the preceding classification, the integrity of the middle column is the key to stability in burst or shear fractures of the thoracolumbar spine [58].

Spinal Cord Injuries

Cord injuries are described as *complete* (no motor, sensory, or autonomic innervation below the injury), *central cord* (quadriparesis with upper extremities more affected than lower), *Brown-Sequard* (ipsilateral motor deficit with contralateral hypesthesia to pain and temperature), and *anterior cord syndrome* (greater paresis of lower than upper extremities with intact proprioception, touch, and deep pressure sensation) [11, 20, 44].

Spinal cord injuries in children are infrequent but disabling injuries that occur either with or without spinal column disruption (up to 30%); about one-third have associated head injuries [1]. Cervical spinal cord injuries predominate in children, with children younger than 8 to 12 years of age being particularly prone to high cervical injury with or without fracture and to cervical subluxations [1, 4, 6, 7]. This difference reflects the relatively greater size and weight of the calvarium in children with a resultant higher center of gravity, ligamentous laxity and elasticity, immature ossification, and structural weakness of the unfused spinal column. Additionally, the cervical facet joints are more horizontal in the upper spine, contributing to the tendency to sublux with trauma. Pediatric spinal cord injuries include birth trauma, SCIWORA, infarction, craniocervical junction injuries, and sequelae of fracture-dislocation.

Most spinal cord injuries are precipitated by compression and contusion; frank transection is unusual. Over the first days to hours after injury, diminished blood flow with resultant ischemia is postulated as a major factor in the propagation of spinal injuries above and below the site of compression [11, 61]. The early foci of hemorrhagic necrosis evolve over a week or so into extensive areas of ischemia, hematomyelia, and necrosis which propagate in animal models over three to five segments above and below the injury [61]. White matter edema peaks at 2 to 3 days after injury [61]. Based on animal models of cord trauma, hypointense foci on T2-weighted MRI sequences correlate well with hematomyelia; hyperintense foci correlate with white matter edema, necrosis, and myelin degeneration [61]. Although a number of surgical and pharmacologic interventions have been studied, limited success at best has been achieved in limiting or reversing cord injury [11].

The prognosis for recovery of children with spinal injury and neurologic deficits is similar or slightly better than that of adults with similar injuries. There are reports of up to a 30% incidence of delayed partial recovery in children with complete deficits acutely [1, 2]. Onset of posttraumatic neurologic deficit in children may be immediate or delayed by 3 to 4 days [10]. Those with incomplete deficits may regain function either with or without surgical intervention [1]. Indications for surgical exploration include progression in neurologic deficit, cord compression due to bone or disk herniation, open or penetrating spinal injuries, or continued spinal column alignment abnormality with cord compression in patients with an incomplete neurologic deficit [2, 11]. Realignment of the spinal column is reasonable in children with complete spinal deficits in order to limit secondary scoliosis and remove any hindrance to recovery of neurologic function [2].

CORD INJURIES IN INFANTS

Spinal cord injury has been recognized as a complication of difficult breech or cephalic deliveries for many years and accounts for up to 10% of deaths in term infants [11, 62–73]. Most birth-related injuries affect the medulla and cervical spine, although cord lesions have been demonstrated as caudal as T10. Downward traction also may result in injuries to the cerebellum [11]. Most birth-related spinal cord injuries are presumed to result from traction or a combination of traction and torsion on the spinal cord [65, 69, 71]. Associated obstetrical factors are shoulder dystocia, torsional head rotation in cephalic deliveries, forceps rotation in cephalic deliveries, and head hyperextension in breech deliveries [69, 73]. Leventhal demonstrated at necropsy that the infant spinal cord can tolerate only about 1/4 inch of longitudinal stretch without significant injury, while the spinal column may tolerate up to 2 inches of stretch [69]. This is one explanation for the frequency in which birth injuries result in significant spinal cord injury and minimal or absent bony injury.

The spectrum of neurologic deficits and prognosis reflect the extent of cord injury and include apnea with ventilator dependence, quadriparesis, cranial nerve palsies, and hypoxic/ischemic brain injury. Although many birth-related cord injuries have no bony disruption, spinal abnormalities that have been described include atlanto-occipital distraction-dislocation, atlantoaxial instability, dens fractures, intraspinal hemorrhages (epidural, subdural, or subarachnoid), hematomyelia, cord transection, vertebral artery occlusion, and meningeal and nerve root tears [20, 71–73]. Cephalic presentation injuries tend to injure the upper cervical spine, with breech presentation injuries more often affecting the lower cervical spine [71–73].

Initial evaluation includes plain films followed by CT scans as needed to demonstrate or clarify potential sites of fracture and confirm spinal alignment. Although neurosurgical decompression has met with limited success, in the acute setting, MRI demonstrates the level and extent of spinal cord injury and excludes extramedullary cord compression (Fig. 8.12). Posttraumatic hemorrhagic cord contusions in other settings have a poorer prognosis than nonhemorrhagic lesions [32, 33]. Whether this likewise applies to birth injury awaits further MRI experience. Delayed MRI reveals the sequelae of spinal cord injury, including cord atrophy and myelomalacia, transection, syringomyelia, or hemosiderin deposition at sites of previous hemorrhage [27–29] (Fig. 8.13). Although the prognosis for survival following these injuries is poor, in a series of children suspected of having progressive infantile spinal muscular atrophy (Werdnig-Hoffmann disease), about two-thirds were found instead to have evidence of prior spinal cord injury [73].

Other etiologies for spinal cord injuries in the infant include battered child syndrome, brachial plexus stretch/avulsion, and other traumatic events such as MVAs with or without restraints, falls, pedestrian–MVAs, and SCIWORA. While the reported experience in spinal cord injuries with battered child syndrome is limited, battering should be considered in a child with clinical or radiographic evidence of repeated trauma. Spinal column lesions in battered children include fractures, fracture-dislocations, and wedge fractures with secondary kyphosis of the thoracolumbar spine or less often of the cervical spine [13, 14]. Other associated trauma such as head injuries may mask coexisting spinal trauma, or these children may present late with delayed walking, extremity spasticity, or unexplained kyphosis [13, 14].

Brachial plexus injuries may occur with spinal injury and/or arm stretch. These are classified as Duchenne-Erb's palsy of C5–7 with upper arm and shoulder deficits or Klumpke's paralysis of the hand due to injury at C8–T1, occasionally with Horner's syndrome [60]. Brachial plexus and nerve root avulsions in infants may be suspected on MRI based on posttraumatic pseudomeningoceles of the affected cervical nerve roots; MCT is generally required to confirm the diagnosis.

SPINAL CORD INJURY WITHOUT RADIOGRAPHIC ABNORMALITY (SCIWORA)

SCIWORA refers to spinal cord injury without radiographic abnormality demonstrable by plain film or myelography. SCIWORA occurs more often in children than in adults and is frequently accompanied by significant neurologic deficits [6, 7]. In general, adult spinal cord injuries occur at the level of bony injury, and posttraumatic herniated nucleus pulposus (HNP) is common [30, 33–38]. The principal exceptions are adults with cervical cord contusion in the setting of significant cervical spondylosis [30]. In contrast, children rarely suffer traumatic HNP, perhaps because the structurally weaker junction between the endplate cartilage and the vertebral body fractures before the annulus fibrosus ruptures [5]. Children less than 9 years of age frequently experience spinal cord injury remote from sites of bony injury or in the absence of spinal column disruption, with the incidence varying from 15 to 70% [4, 11] (Fig. 8.14). Possible pathogeneses include infarction of the central spinal cord (see next section), vascular disruption of large or small penetrating vessels, traction longitudinally on the spinal cord, transient disk herniation, or ligamentous compression due to the

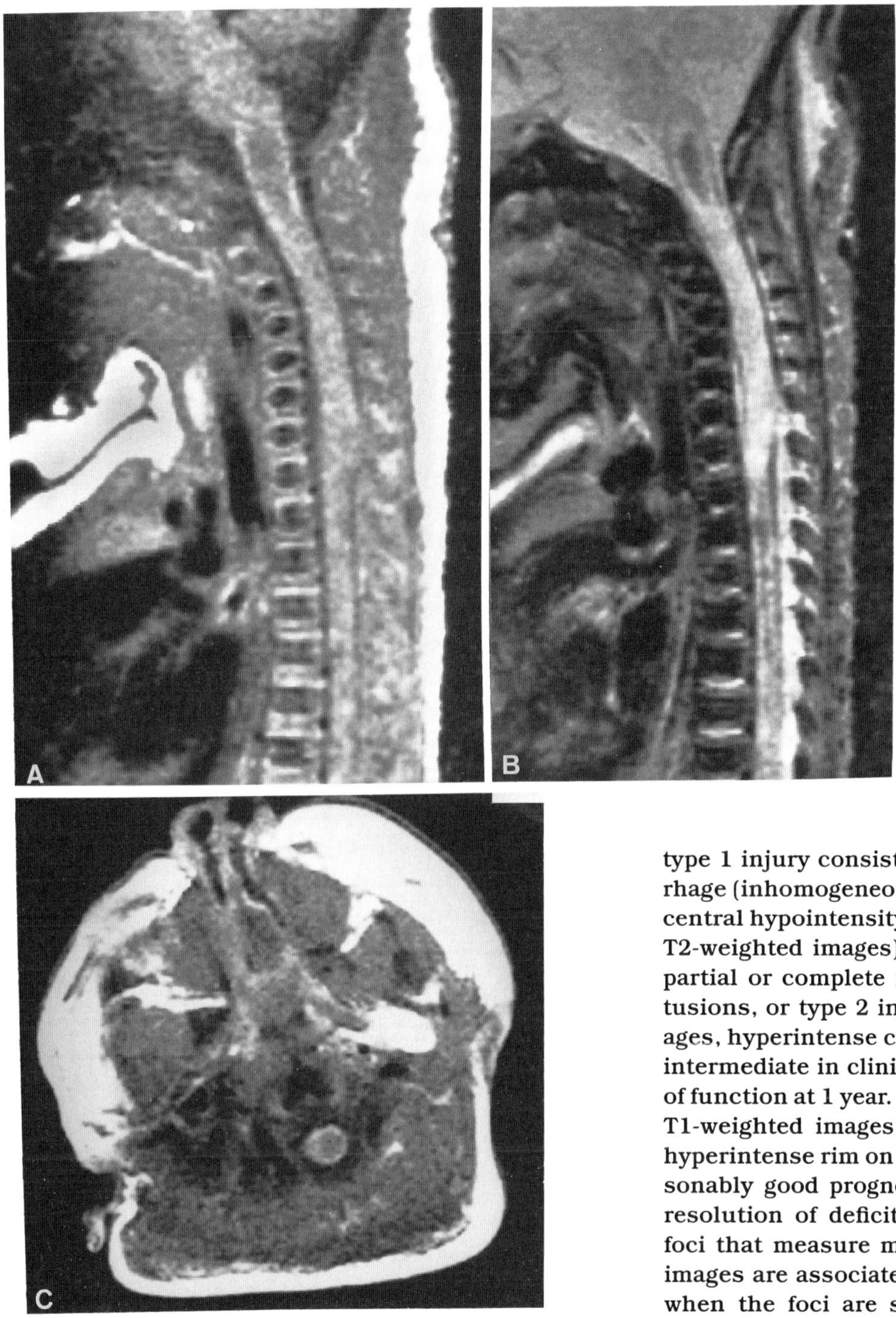

Figure 8.12. Intramedullary hematoma. Sagittal (**A**) and axial (**B**) T1-weighted MRI images of a newborn baby with birth injury demonstrate marked enlargement of the cervical cord and medulla with focal mild hypointensity centrally from the medulla to C2 (**A:** TR 602, TE 20, 3 mm; **B,** TR 650, TE 20, 4 mm). **C,** T2-weighted image confirms the intramedullary hematoma of the upper cord with extensive infarction or edema of the lower cord. Loss of signal on T2-weighted image is the earliest change on MRI in acute hemorrhage; T2-weighted gradient-echo images improve sensitivity of detection of early bleeds.

increased elasticity of the immature spinal column [74].

Spinal cord intensity patterns on MRI have prognostic value in adults with cord injury and correlate with animal models of traumatic cord injury [33–38, 61, 75]. In a series composed largely of adults, three patterns of MRI signal abnormalities that correlated well with recovery or lack of resolution of neurologic deficits at 1 year following injury have been described [32, 33]. According to this classification, a type 1 injury consisting of acute intraspinal hemorrhage (inhomogeneous on T1-weighted images, large central hypointensity with thin, hyperintense rim on T2-weighted images) has the poorest prognosis for partial or complete recovery. Cord edema and contusions, or type 2 injuries (normal T1-weighted images, hyperintense cord on T2-weighted images), are intermediate in clinical severity with some recovery of function at 1 year. Mixed or type 3 injuries (normal T1-weighted images, isointense center with thick, hyperintense rim on T2-weighted images) have a reasonably good prognosis for significant recovery or resolution of deficits [32, 33]. Large hyperintense foci that measure more than 1 cm on T2-weighted images are associated with a poorer prognosis than when the foci are smaller [30]. T2-weighted spin-echo or gradient-echo images are essential for identification of ligamentous tear, prevertebral soft-tissue hematoma, and hemorrhagic foci within the spinal cord on MRI studies performed less than 3 days after injury [30, 38]. On the basis of experimental studies, Hackney noted that cord injuries with significant recovery are often associated with petechial hemorrhages; however, current MRI technology may be inadequate for demonstration of very small hemorrhagic foci [31].

Little experience is available to date to confirm the prognostic significance of MRI findings in the pe-

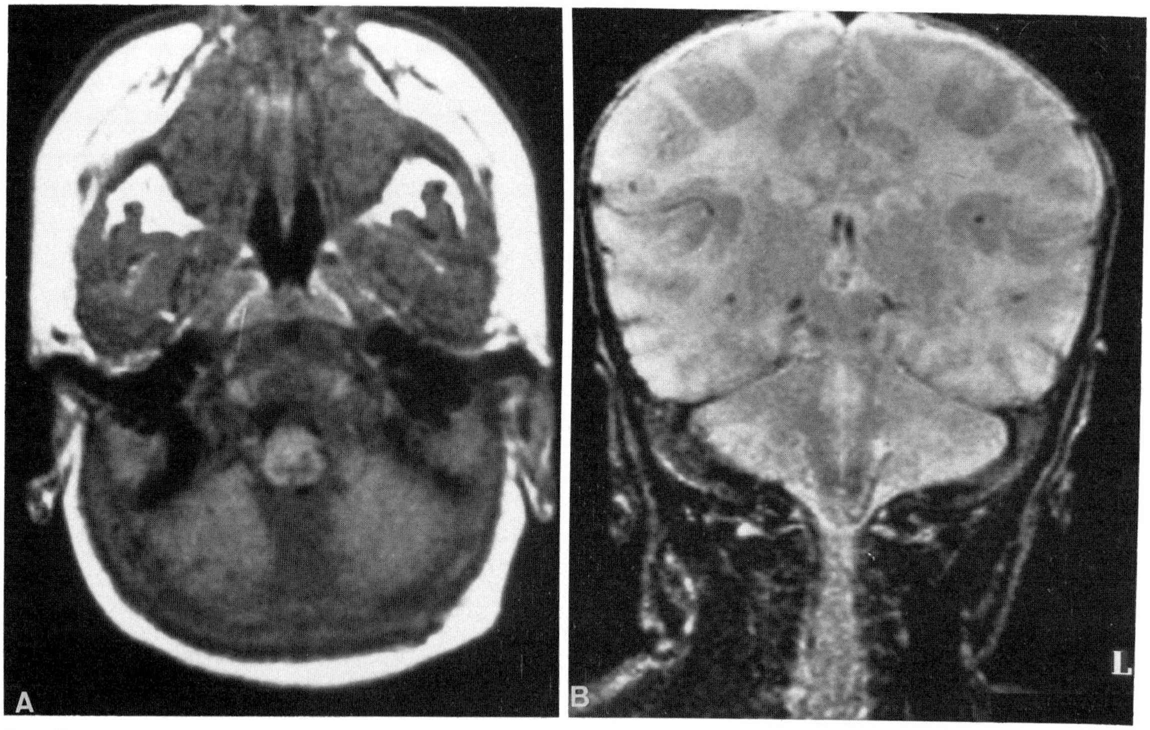

Figure 8.13. Hematomyelia due to birth injury. A 2-month-old child with palatal incoordination who was delivered in breech position and required ventilatory support for the first 5 days of life demonstrates a small focus of low signal at the cervicomedullary junction on T1- and T2-weighted sequences. This suggests hemosiderin deposition as a result of hematomyelia from birth injury; angiography was normal (**A:** TR 666, TE 30, 7 mm; **B:** TR 2000, TE 90, 3 mm).

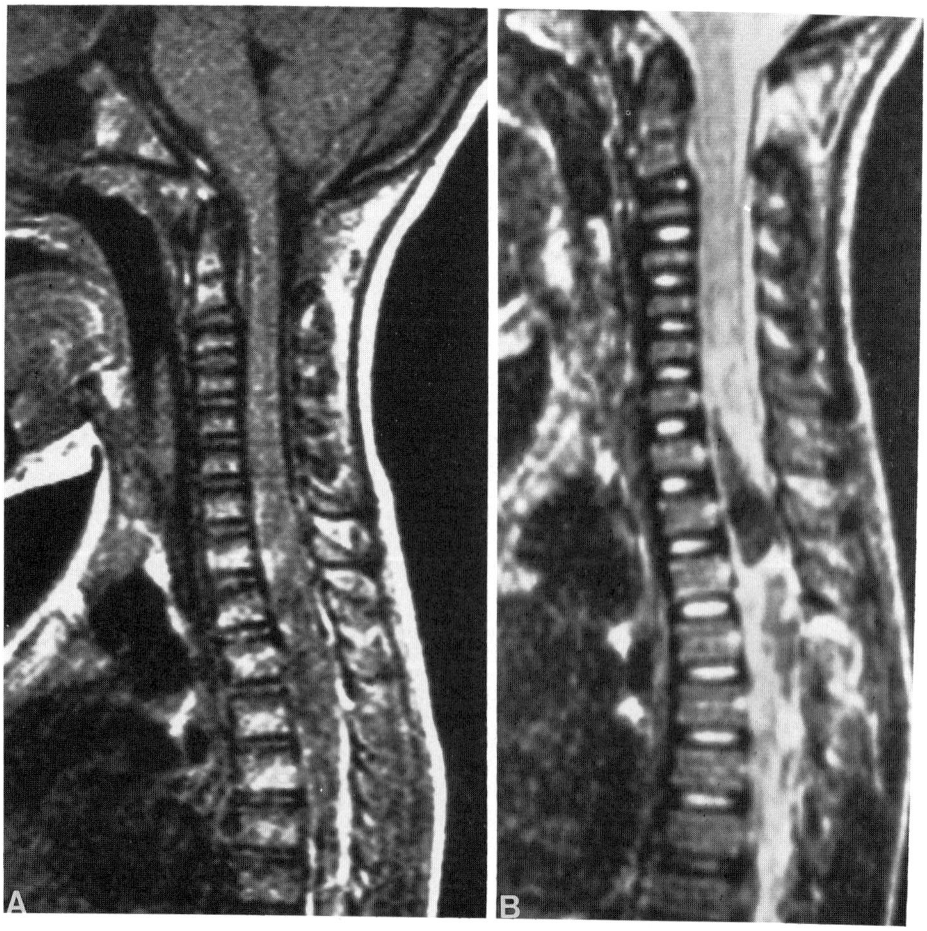

Figure 8.14. Hematomyelia secondary to trauma. A 6-year-old boy had a pulmonary contusion and immediate paraplegia (T2 level) following a pedestrian–MVA. Spinal plain films were normal. **A,** T1-weighted image (TR 363, TE 30, 5 mm) shows spinal cord enlargement from approximately C6 inferiorly into the thoracic cord. Small superficial or extramedullary focus of hyperintensity suggests methemoglobin from hemorrhage. **B,** The T2-weighted image demonstrates extensive hypointense foci over a large segment of the spinal cord corresponding with hematomyelia (TR 2000, TE 100, 5 mm).

diatric population [29, 32 33]. MRI may substitute for MCT of suspected spinal cord injury if neurosurgical intervention is contemplated. Indications include localization, confirmation, and characterization of spinal cord injury and exclusion of cord compression by bone, disk, or epidural hematoma [29]. Although the possibility of a lesion that requires emergent neurosurgical intervention has often prompted MCT, few pediatric patients require emergency decompression. In the absence of a neurosurgically correctable lesion, MRI demonstration of the presence, extent, and pattern of spinal cord injury offers diagnostic and potentially prognostic information not readily obtainable by other means. Better characterization of acute cord injuries may influence therapeutic management as newer medical and pharmacologic interventions are evaluated [76, 77].

SPINAL CORD INFARCTION

Spinal cord infarction in children occurs in association with major trauma with chest injuries; aortic, thoracic, or spinal operative complications; focal injury to the artery of Adamkiewicz; and/or hypotension [78, 79]. Infarction in some cases is the pathologic entity in children presenting with SCIWORA. Other infrequent etiologies include dissection, emboli, collagen-vascular disease, and vasculitis. The thoracic spinal cord is the region most vulnerable to infarction due to its relatively poor blood supply, although infarction may be patchy in distribution or incomplete [80]. Vascular insufficiency in association with systemic hypotension and diminished cardiac output may be a significant exacerbating factor in acute spinal trauma [81]. Onset of neurologic deficit may be delayed by hours to several days, and spinal column radiographic and myelographic abnormalities are absent. MRI reveals the location and extent of cord abnormality and excludes hematomyelia and cord compression [29]. Infarction results in abnormal hyperintensity of the spinal cord on T2-weighted images, with little or no cord enlargement (Figs. 8.15 and 8.16). Infarction is subtle or may not be apparent on T1-weighted images.

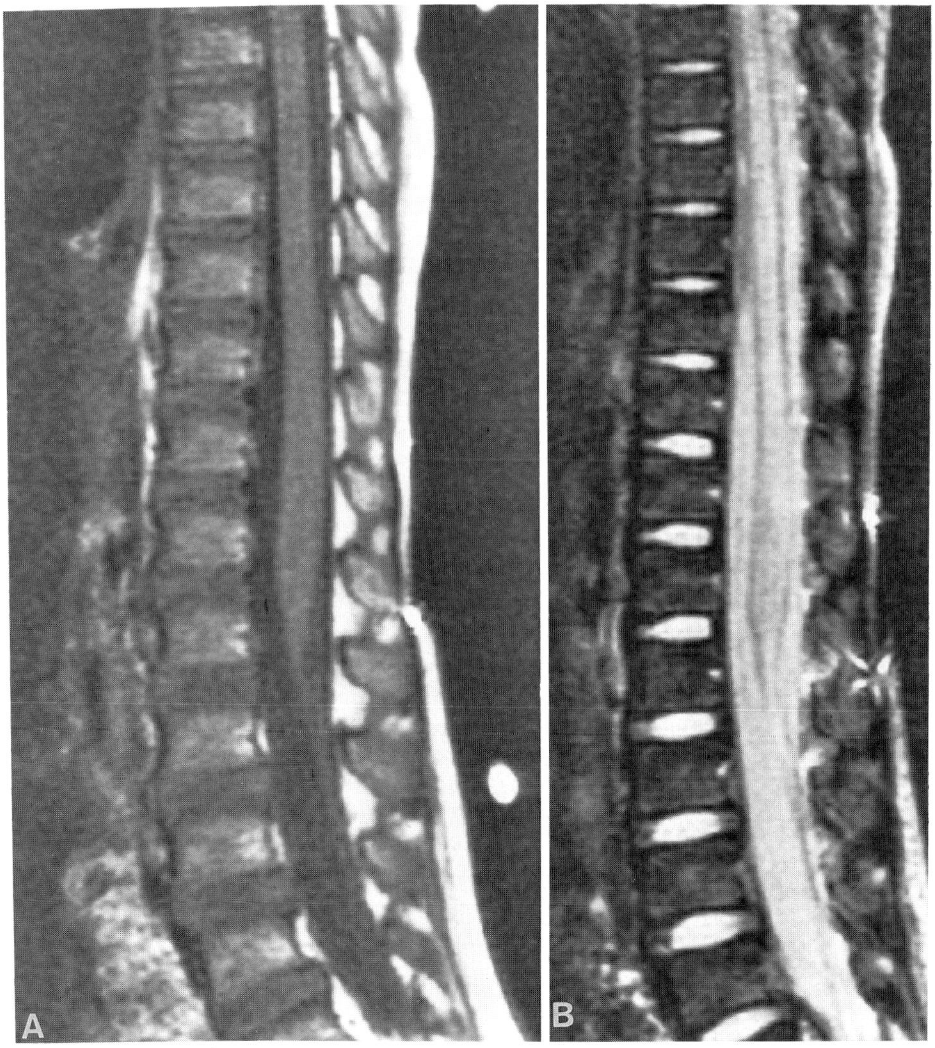

Figure 8.15. Traumatic cord ischemia. A 4-year-old injured as an unrestrained passenger in an MVA had normal spinal plain films, aortogram, and (**A**) T1-weighted MRI (TR 498, TE 20, 4 mm). **B,** T2-weighted image demonstrates extensive hyperintensity of the spinal cord without cord enlargement; this corresponded to the clinical diagnosis of infarction (TR 2000, TE 90, 4 mm).

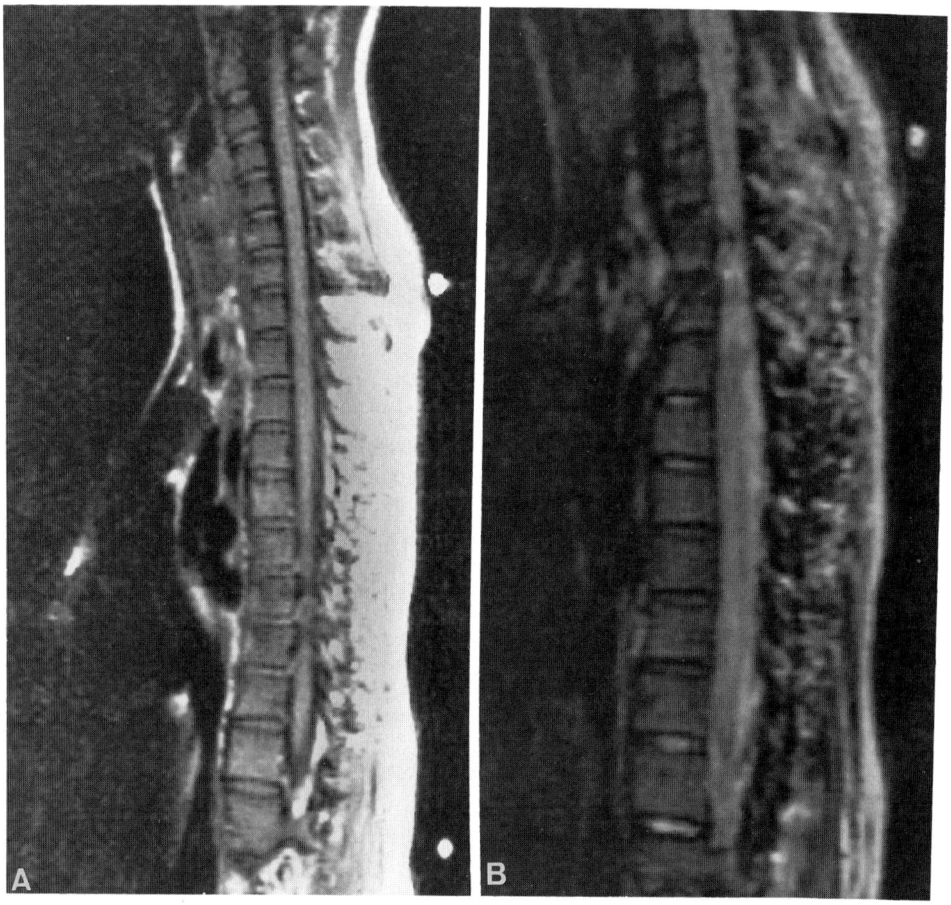

Figure 8.16. Spinal cord ischemia. An adolescent who underwent spinal instrumentation for scoliosis had leg weakness postoperatively. **A,** T1-weighted image demonstrates only scoliosis with normal cord appearance (TR 612, TE 20, 4 mm). **B,** Patchy focus of cord hyperintensity shown on T2-weighted image clinically correlates with focus of infarction (TR 1900, TE 90, 4 mm).

Cervicocranial Junction and Upper Cervical Injuries

Injuries to the upper cervical spinal column account for a significant percentage of spinal trauma in children less than 8 to 12 years of age. These injuries include occipitoatlantal dislocation-distraction, occipitoatlantal axial injuries, Jefferson burst fractures of C1, and bilateral arch fractures of C2 (hangman's fracture). Adequate immobilization and stabilization are essential to minimize cord compression with these injuries.

After the initial steps of stabilization, immobilization, and plain films, CT is the preferred imaging modality to assess craniocervical junction injuries. This study should be performed with thinly collimated overlapping slices (1.5 to 3 mm with 1-mm table incrementation) beginning just above the foramen magnum with a scan angle parallel to the C2–3 disk space. The caudal limit for imaging is determined based on clinical and plain film findings. Generally, a head CT is also required to exclude associated traumatic brain injury. Although axially oriented fractures and ligamentous injuries may be subtle or may not be apparent, CT with re-formatted images offers the best bony detail for fracture localization and canal compromise by bony or disk fragments. Re-formatting in a sagittal and coronal plane should be included routinely in order to exclude subtle subluxations or dislocations in this area. MRI may then be considered to assess cord compression by soft tissues (i.e., hematoma, disk prolapse) and to define cord injuries.

OCCIPITOATLANTAL DISTRACTION-DISLOCATION

Occipitoatlantal distraction-dislocation is described as a hyperflexion, hyperextension, or distraction injury in which the ligamentous attachments between the skull base and the ring of C1 are partially or completely disrupted. The tectorial membrane, anterior atlanto-occipital ligament, and prealar ligaments rupture between the cranium, C1, the odontoid, and the occipital condyles. Usually, the cranium is dislocated anterior to C1, and coexisting bony injuries of C2 are not uncommon [24]. Although this injury is associated with a high morbidity and mortality, prompt resuscitation and immobilization have resulted in some survivors [82–85]. Sequelae include hemiparesis or quadriparesis, absent respiratory effort or diaphragmatic breathing due to phrenic nerve injuries, lower cranial nerve palsies, and coma.

Several methods of recognition of this injury have been suggested for use with plain films and/or CT [85–88]. Lee et al. [85] suggested that an ascending line drawn from the posterior inferior edge of the body of C2 to the posterior foramen magnum should pass through the spinolaminar line of C1; a descending line from the clivus to the spinolaminar line at mid-C2 should touch or be within 5 mm of the dens [85]. The Powers et al. ratio [86] is based on two lines drawn on lateral plain films as follows: line *BC* is drawn from the basion (anterior lip to the foramen magnum) to the midposterior spinolaminar line of C1; line *OA* is drawn from the opisthion (posterior margin of the foramen magnum) to the midpoint of the posterior surface of the anterior arch of the atlas [86]. The normal ratio of *BC* to *OA* is less than 1; a ratio of greater than 1 suggests anterior cranial dislocation (Fig. 8.17). An interval of more than 5 mm between the occipital condyles and C1 also suggests this injury [87] (Figs. 8.18 and 8.19). Immobilization may result in realignment of the craniocervical junction, rendering this diagnosis more difficult. CT with sagittal and coronal re-formatting confirms the degree and direction of dislocation and demonstrates associated fractures. Injuries with ligamentous damage only are more difficult to diagnose on CT and plain films, since alignment and distraction may be minimized by immobilization. Suggestive CT findings include an otherwise unexplained prevertebral hematoma, subdural/epidural hemorrhage within the upper cervical spinal canal or foramen magnum, and unilateral or bilateral subluxation or dislocation of the occipital condyle–C1 facet joint. MRI demonstrates spinal cord and medulla contusions, shear injuries or contusions of the brain parenchyma, vascular injury of the vertebral or cerebellar arteries

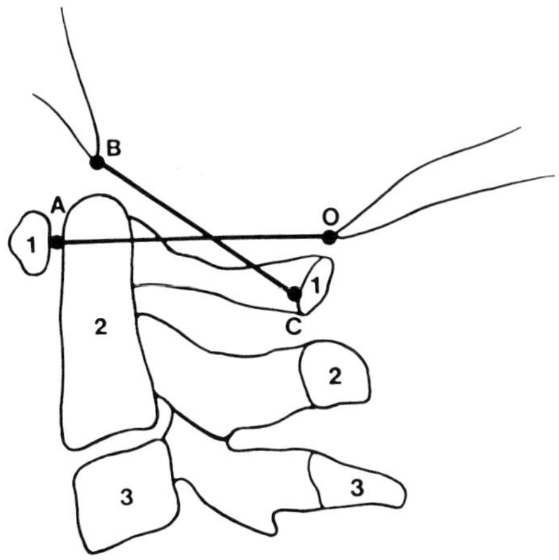

Figure 8.17. A line drawing indicating the lines that are measured for the Powers ratio. A ratio of *BC* to *OA* of greater than 1 indicates anterior displacement of the calvarium relative to C1.

with secondary inferior cerebellar infarction(s), prevertebral hematoma, and ligamentous disruption (see Fig. 8.18).

ATLANTOAXIAL FRACTURES AND DISLOCATIONS

Fractures of the atlantoaxial interval in children include Jefferson or Jefferson-variant fractures of C1, bilateral pars fractures of C2 (hangman's fracture), atlantoaxial rotary subluxation, and odontoid fractures [20]. Rupture of the transverse ligament without dens fracture is unusual in the absence of a preexisting cause for ligamentous weakening [20]. Jefferson or burst fractures of the atlas are multiple fractures of the ring of C1, attributed to axial loading or compressive forces [88]. Open-mouth views demonstrate symmetrical or asymmetrical offset of C1 laterally on C2 and one or more fracture lines through the ring of C1. Fewer than four fractures of C1 have been called Jefferson-variant fractures, and fractures confined to the anterior or posterior ring may occur with hyperflexion or hyperextension, respectively [89]. Since fusion of the ossification centers of C1 is incomplete until 7 to 10 years of age, CT is invaluable in differentiating the smoothly corticated margins of normal ossification centers and the pseudo-Jefferson variant described above from the abrupt, irregular margins associated with fracture [89] (Fig. 8.20).

Fractures of the axis in children typically occur at the synchondrosis between the dens base and centrum [20] (Fig. 8.21). This pattern is most similar to the type 2 fractures of adults, with the exception that the fracture line is caudal to the plane of the facet joints rather than at the dens base [20, 90]. Older juveniles and adolescents experience types 1 (odontoid tip) and 3 (dens and body of C2) similar to those of adults [90] (Figs. 8.22 and 8.23).

Atlantoaxial rotary subluxation is difficult to document radiographically. This occurs spontaneously following minor trauma or infection and often resolves spontaneously. CT supplements plain films for exclusion of bony injuries and demonstration of fixed (i.e., nonreducible) rotation [91]. The initial CT scan is performed to exclude fractures of the atlantoaxial junction; this is followed by a repeat study with the patient in maximal rotation in the direction opposite to the torticollis. Reduction on the repeat study indicates a transient torticollis rather than a fixed rotary subluxation. As described for fractures of this region, thin axial and re-formatted CT images are helpful for demonstration of a fixed subluxation of an occipitoatlantal facet joint prior to surgical reduction and fusion.

An awareness of developmental variants (see discussion above) and anomalies is important in the evaluation of this region. Anomalous development

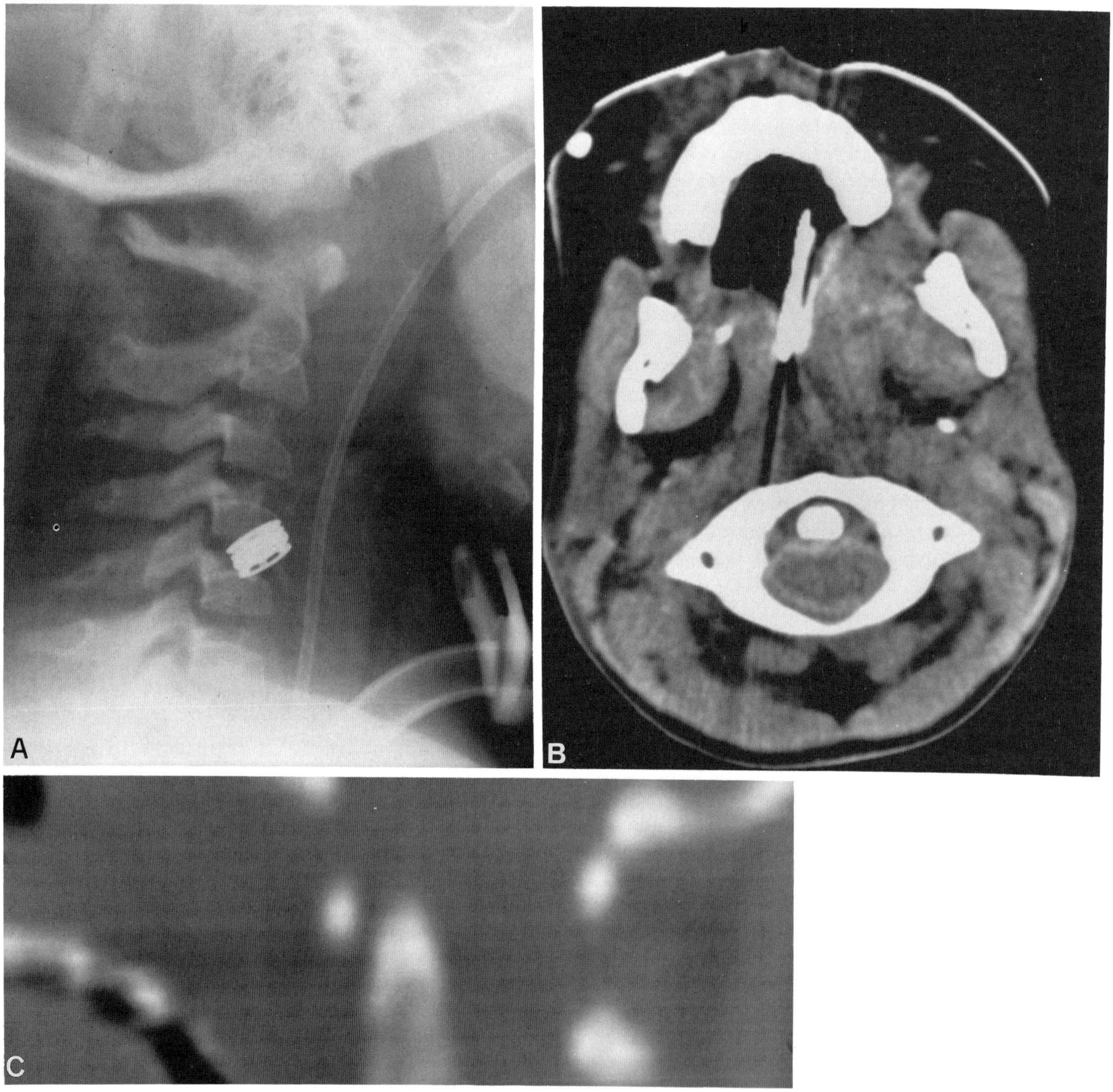

Figure 8.18. Extramedullary hematoma. A, Portable lateral view of the cervical spine demonstrates distraction of the occipitoatlantal articulation and prevertebral swelling. **B,** Axial CT scan revealed spinal canal hematoma with no fractures. **C,** Distraction, which was most severe anteriorly, was confirmed on re-formatted images. Progressive neurologic dysfunction prompted MRI. **D,** The T1-weighted image (TR 612, TE 20, 3 mm) demonstrates mild compression of the spinal cord and medulla due to a complex mass anteriorly. **E,** The prevertebral hematoma and ligamentous injury are better demonstrated on a T2-weighted image, as is the associated infarction of the inferior cerebellum. (TR 2000, TE 90, 4 mm).

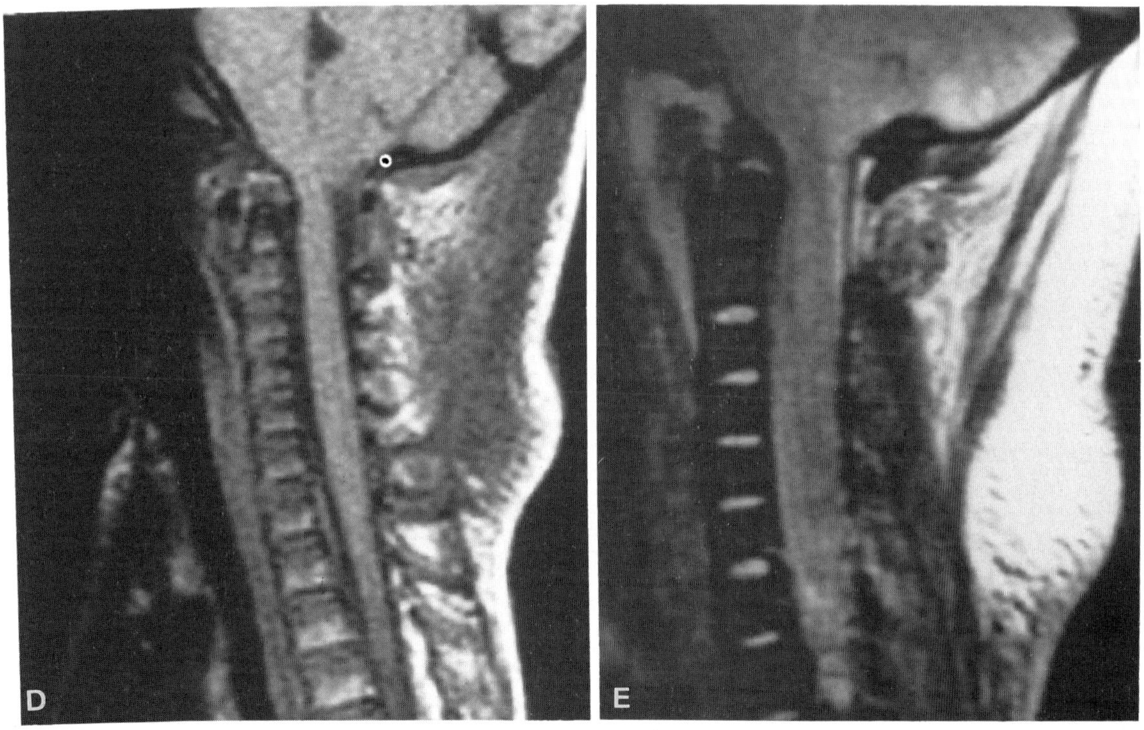

Figure 8.18–*continued*

may result in instability of the atlantoaxial junction or confusion in the setting of trauma [92, 93] (Fig. 8.24). A well-known example is the os odontoideum, or failure of fusion of the terminal ossification center of the dens. The etiology of os odontoideum may be congenital, traumatic, or vascular (avascular necrosis) and is a frequent finding with Down syndrome [48, 49, 93, 94] (Fig. 8.25).

LOWER CERVICAL FRACTURES AND DISLOCATIONS

Lower cervical injuries are typically injuries of older juveniles or adolescents [5, 20, 74]. About 60 to 65% of significant spinal injuries in older children involve the cervical spine, an incidence similar to that in adult series [20]. These occur as sequelae of sports injuries, MVAs, or falls [1–4, 20, 74]. The mechanism of injury for most cervical injuries is hyperflexion, although hyperextension and axial loading injuries are not rare. High-quality routine plain films reveal the majority of these injuries; CT and MRI are adjuncts to define bony abnormalities and soft-tissue/spinal cord lesions, respectively [9, 16–20].

Hyperflexion cervical spine injuries include simple wedge fractures involving the anterior vertebral body or more severe injuries with distraction and/or fracture of the neural arches. Disruption of the posterior longitudinal ligament accompanies all hyperflexion injuries except the clay-shoveler's fracture of the spinous process [59, 60]. More extensive injuries may be combined hyperflexion and rotary injuries and include disruption and/or dislocation of the facet joints unilaterally or bilaterally, anterior dislocation of the vertebral bodies, significant cord compression, and instability [59] (Fig. 8.26). A purely ligamentous disruption of the posterior interspinous ligaments may produce a subtle focal increase in the posterior interspinous distance or may not be apparent on routine plain films. Supervised flexion and extension images in an awake and cooperative child aid detection of this injury. This injury may be overlooked initially and result in delayed instability requiring surgical intervention [95] (Fig. 8.27).

Injuries solely due to hyperextension are uncommon, although combined hyperflexion and hyperextension injuries are not unusual. Extension injuries include fractures of the posterior neural arches, generally with sparing of the anterior longitudinal ligament and vertebral body (Fig. 8.28). Involvement of both neural arches is potentially unstable. Severe extension injuries fracture the anteroinferior vertebral body endplate, termed an *extension teardrop fracture* [59, 96]. Hyperextension-dislocation injuries may be subtle on plain films; diffuse prevertebral swelling and normal spinal column alignment may be the only plain film findings. Carefully supervised plain films positioned in extension confirm this injury [96].

Cervical burst fractures may occur with axial loading and/or hyperflexion injuries. In the absence of posterior element fractures, a burst fracture is

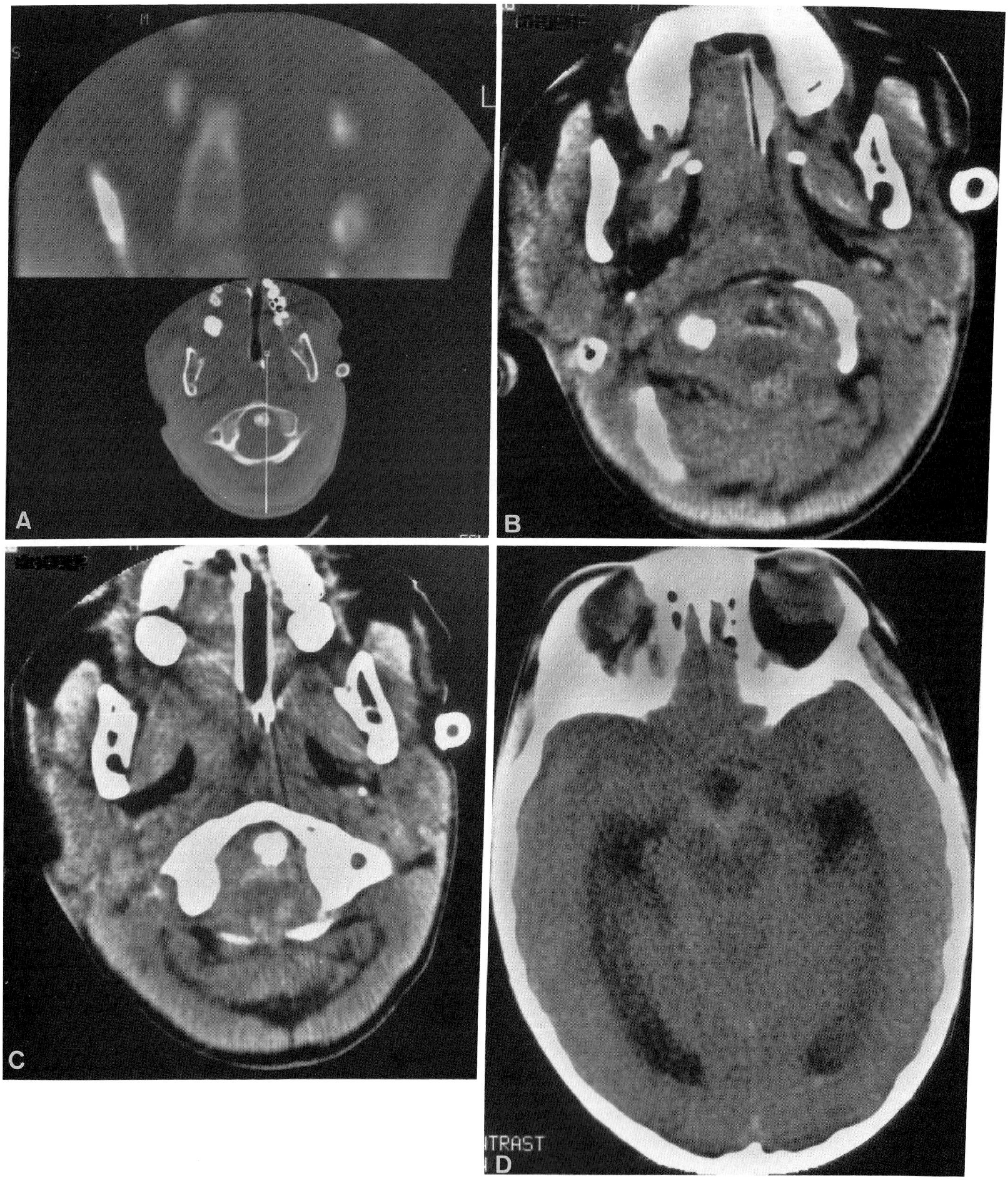

Figure 8.19. Spinal cord contusion and hematoma. This child experienced a hyperflexion injury during an MVA, with resultant anterior dislocation at the occiput-C1. **A,** After immobilization, re-formatted CT image demonstrates only distraction; the dislocation is not apparent. **B,** The absence of the normal bony structures on an axial CT scan at the craniocervical junction suggests distraction. **C,** Note the extensive intraspinal contusion and hematoma at C1. **D,** Cranial CT demonstrates subarachnoid and intraventricular hemorrhage and hydrocephalus.

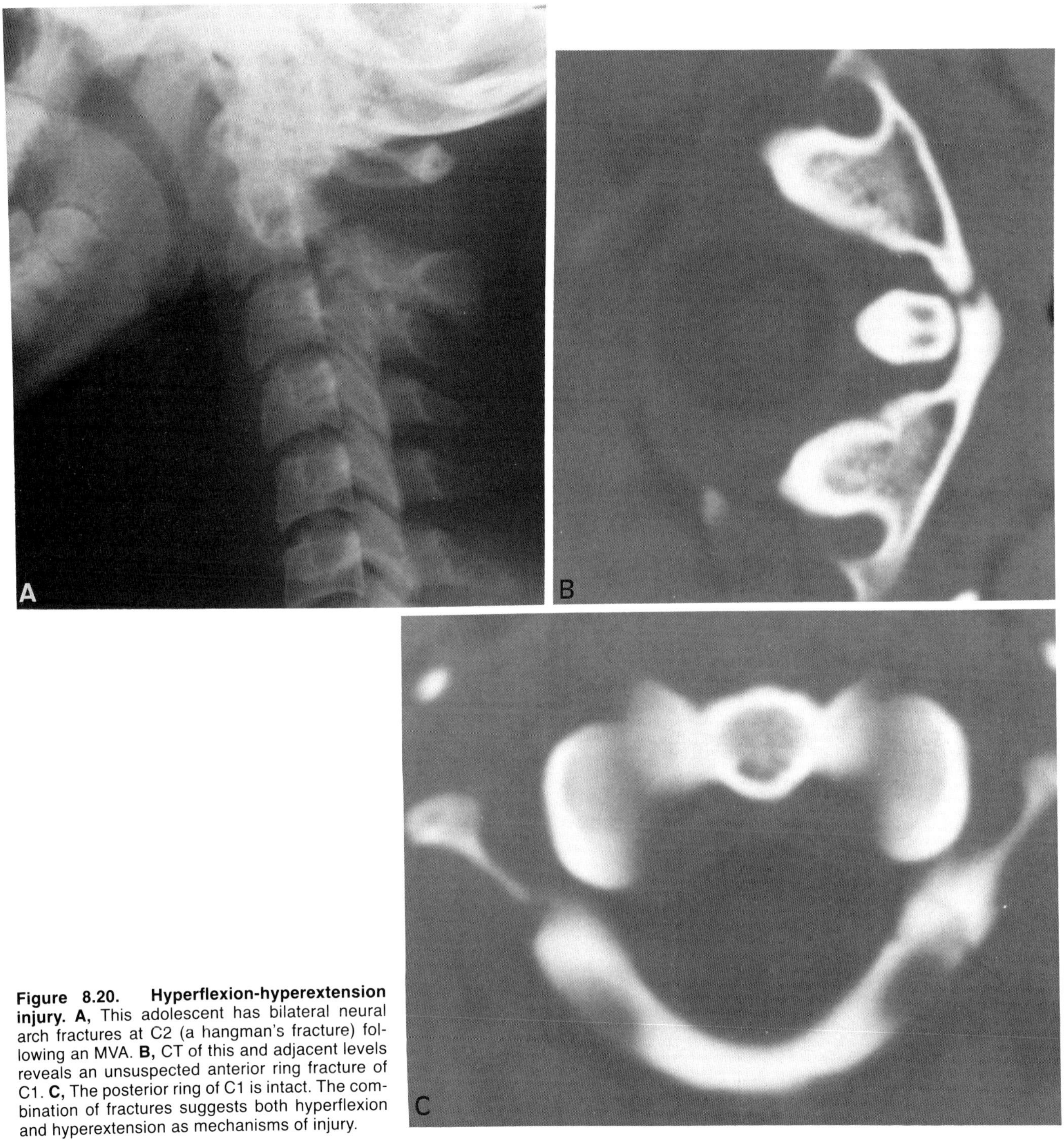

Figure 8.20. Hyperflexion-hyperextension injury. A, This adolescent has bilateral neural arch fractures at C2 (a hangman's fracture) following an MVA. **B,** CT of this and adjacent levels reveals an unsuspected anterior ring fracture of C1. **C,** The posterior ring of C1 is intact. The combination of fractures suggests both hyperflexion and hyperextension as mechanisms of injury.

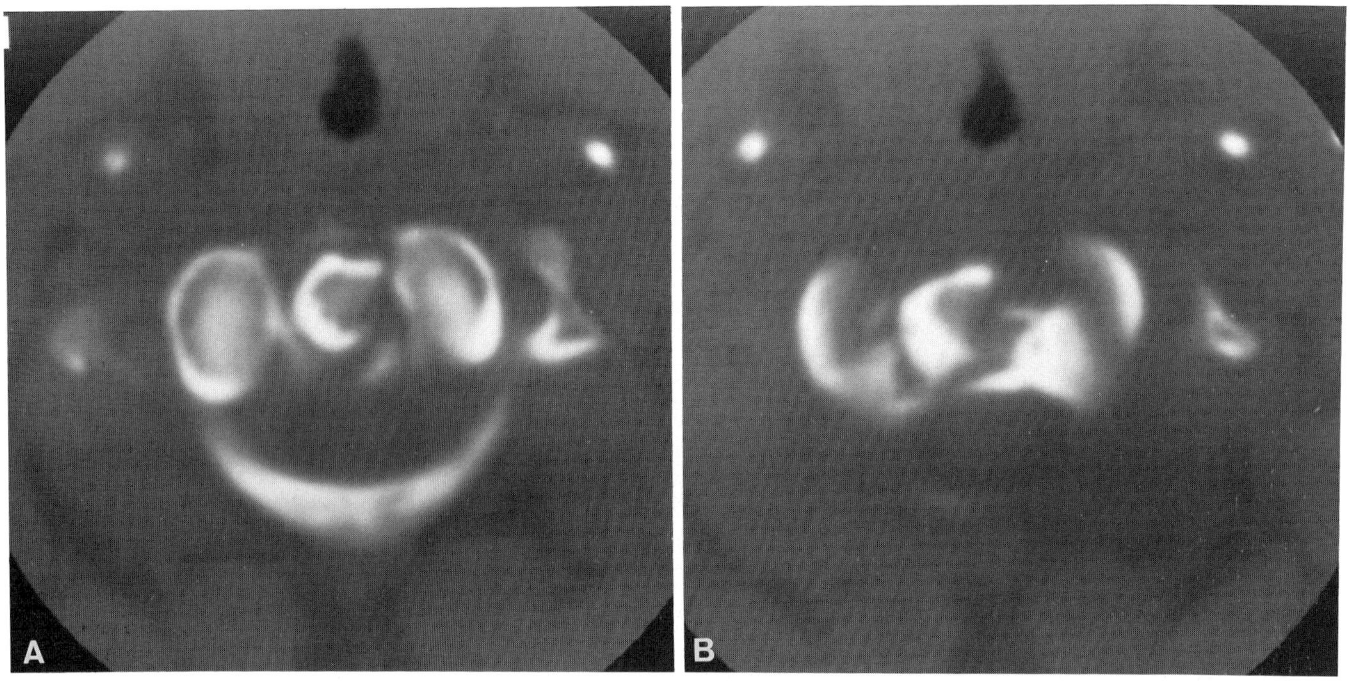

Figure 8.21. Fracture of the odontoid process. A, Odontoid fractures in younger children tend to occur at the dens-centrum synchondrosis, slightly caudal to the plane of the facet joints where type 2 odontoid fractures of older persons are located. **B,** This fracture extends into a comminuted fracture of the body of C2.

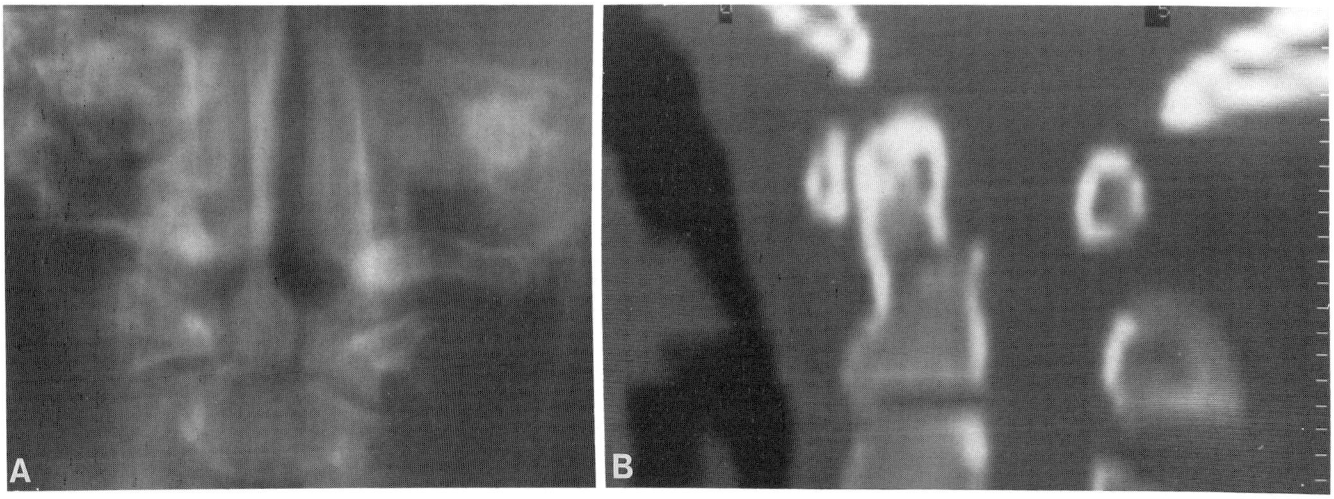

Figure 8.22. Type 3 odontoid fracture. A, Conventional tomography is sometimes preferable for fractures which are largely axial in orientation; however, bone detail is variable and artifacts can be problematic. **B,** The type 3 odontoid fracture in this case is better demonstrated using a re-formatted CT image.

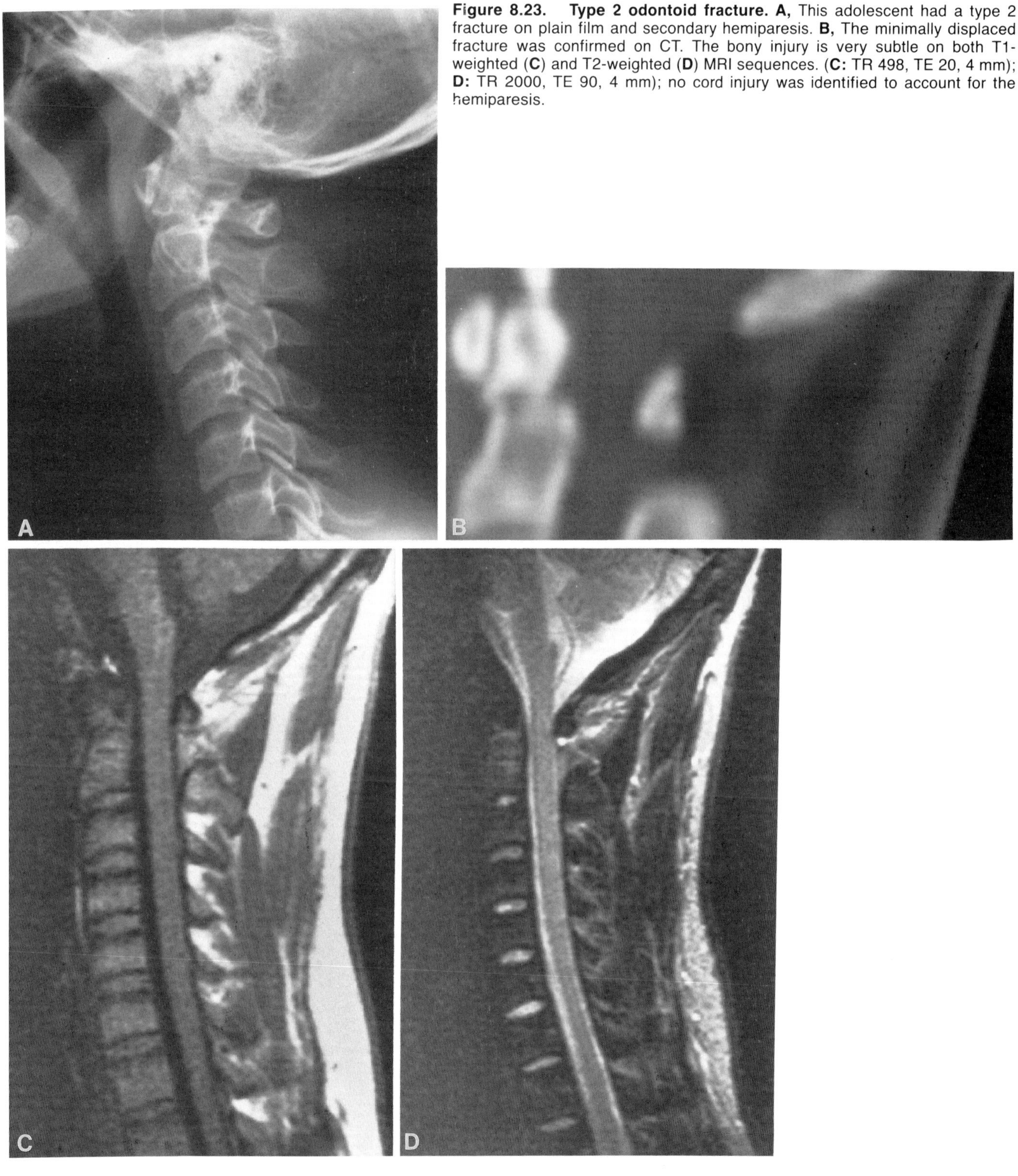

Figure 8.23. Type 2 odontoid fracture. A, This adolescent had a type 2 fracture on plain film and secondary hemiparesis. **B,** The minimally displaced fracture was confirmed on CT. The bony injury is very subtle on both T1-weighted (**C**) and T2-weighted (**D**) MRI sequences. (**C:** TR 498, TE 20, 4 mm); **D:** TR 2000, TE 90, 4 mm); no cord injury was identified to account for the hemiparesis.

Figure 8.24. Anomalous development of the ring of C1 may mimic fracture on plain films or result in instability. This child has several unfused ossification centers posteriorly at C1.

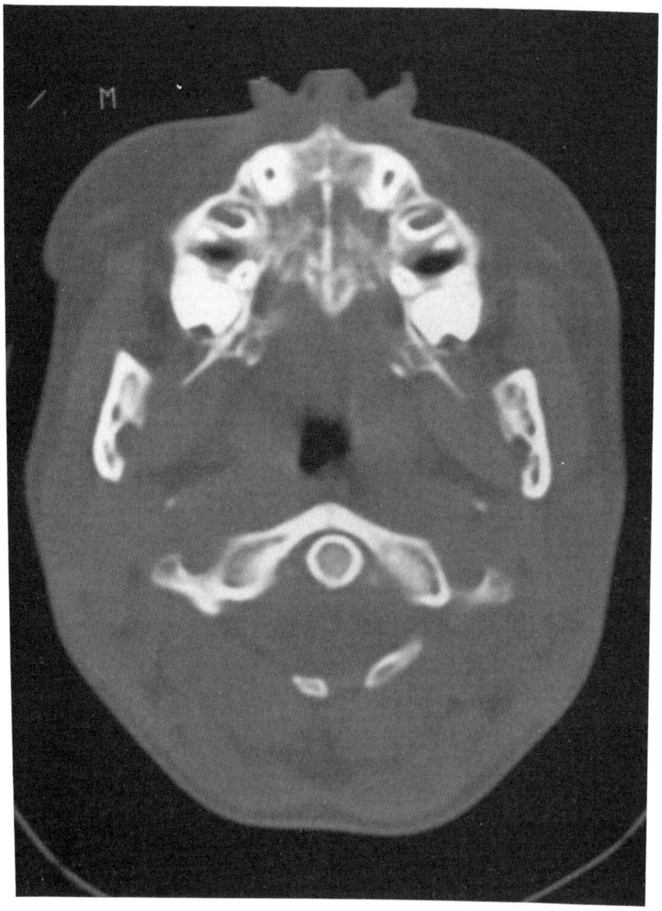

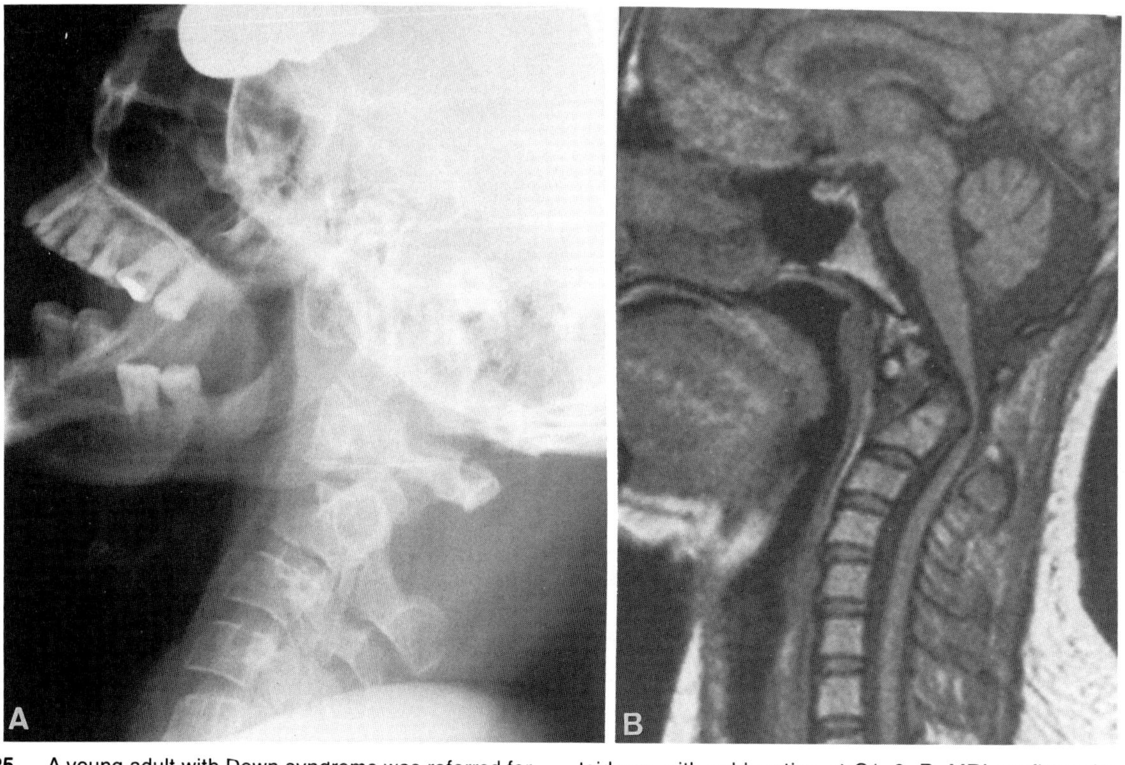

Figure 8.25. A young adult with Down syndrome was referred for screening in order to participate in sports activities. **A,** Plain film demonstrates marked alignment abnormality due to an os odontoideum with subluxation at C1–2. **B,** MRI confirms the alignment abnormality and secondary spinal stenosis with cord compression at this level (TR 700, TE 30, 5 mm).

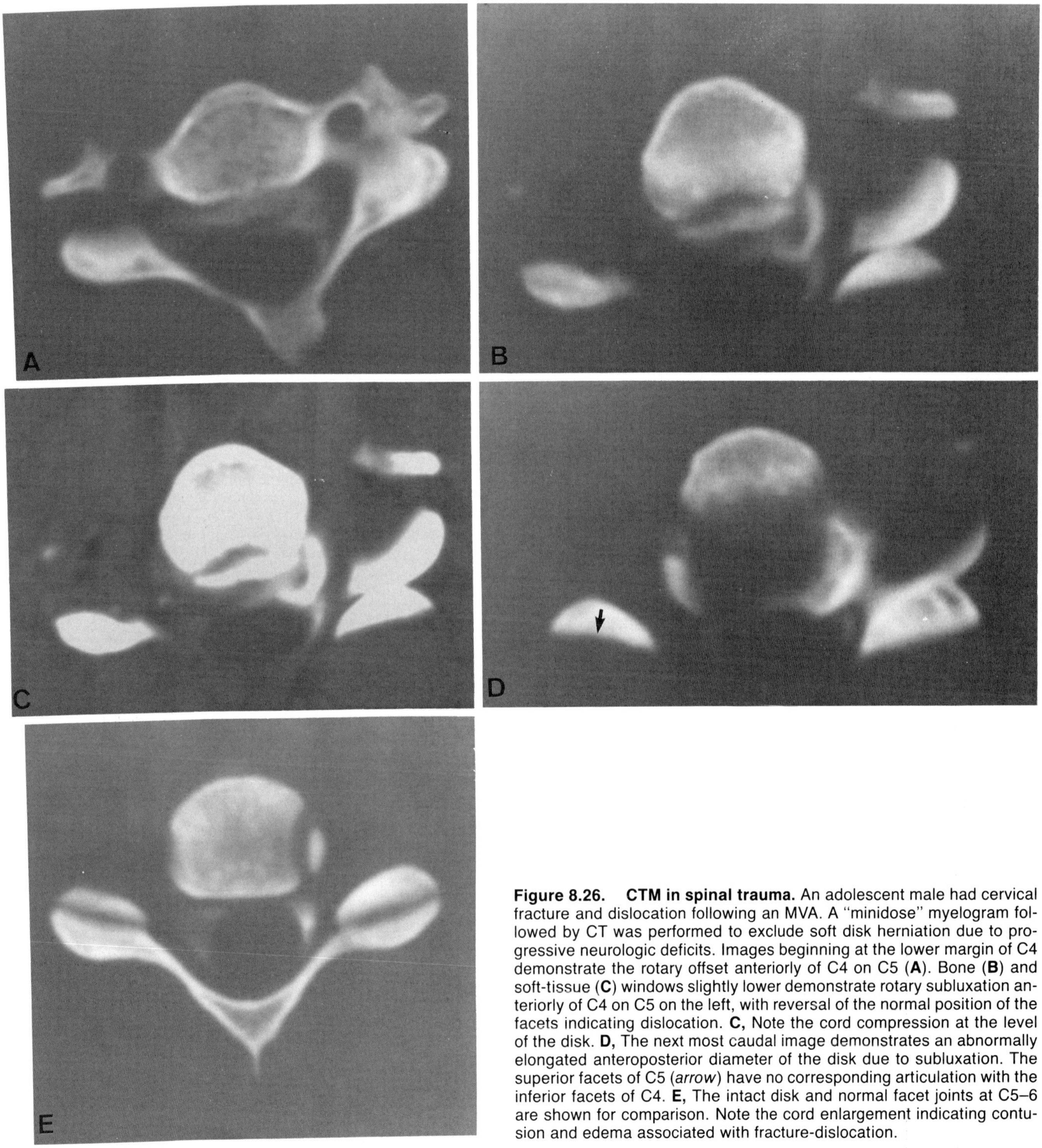

Figure 8.26. CTM in spinal trauma. An adolescent male had cervical fracture and dislocation following an MVA. A "minidose" myelogram followed by CT was performed to exclude soft disk herniation due to progressive neurologic deficits. Images beginning at the lower margin of C4 demonstrate the rotary offset anteriorly of C4 on C5 (**A**). Bone (**B**) and soft-tissue (**C**) windows slightly lower demonstrate rotary subluxation anteriorly of C4 on C5 on the left, with reversal of the normal position of the facets indicating dislocation. **C,** Note the cord compression at the level of the disk. **D,** The next most caudal image demonstrates an abnormally elongated anteroposterior diameter of the disk due to subluxation. The superior facets of C5 (*arrow*) have no corresponding articulation with the inferior facets of C4. **E,** The intact disk and normal facet joints at C5–6 are shown for comparison. Note the cord enlargement indicating contusion and edema associated with fracture-dislocation.

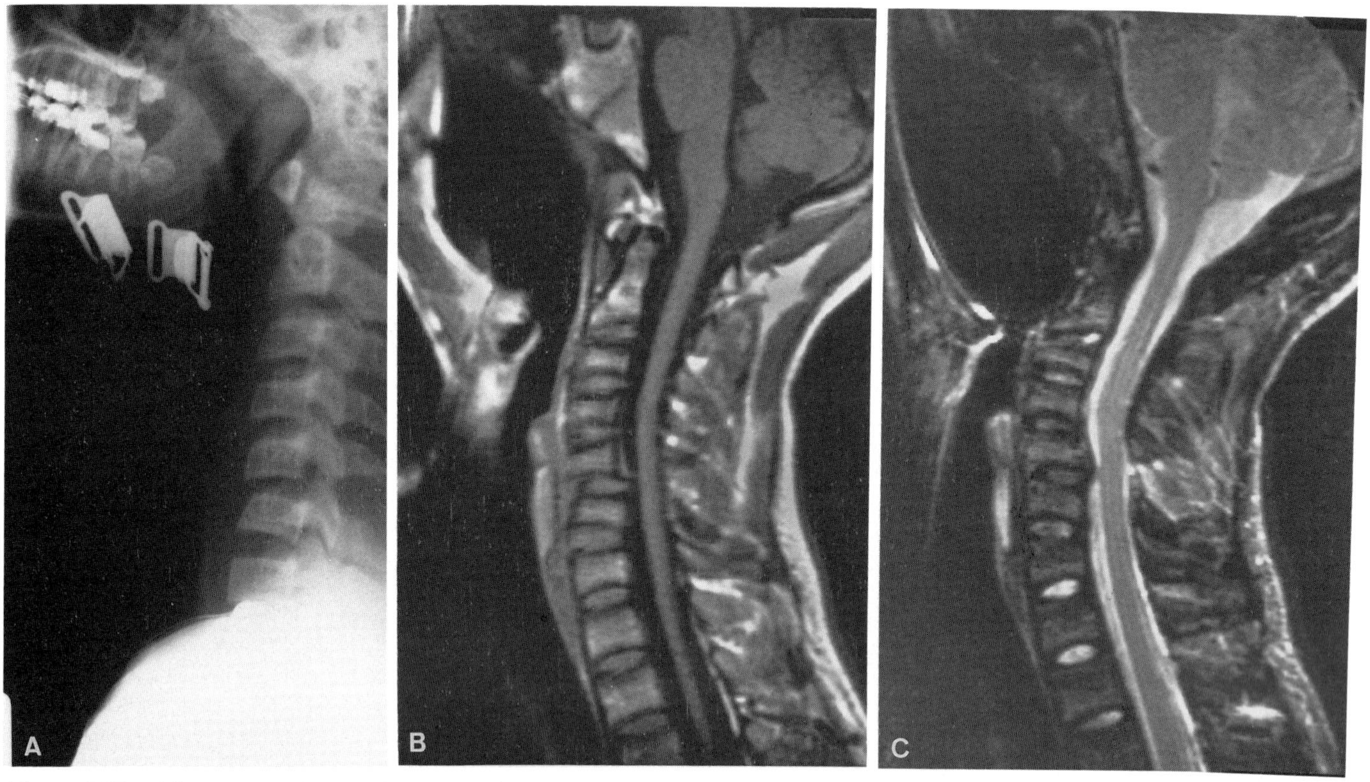

Figure 8.27. Ligamentous injury without fracture. A 15-year-old boy complained of neck pain following a wrestling injury; his initial plain films were reportedly normal. Continued symptoms prompted repeat plain films. **A,** The lateral view demonstrates focal widening posteriorly of the C5–6 posterior interspinous distance and focal widening of the posterior margin of the C5–6 disk space. **B,** A T1-weighted MRI (TR 500, TE 18, 4 mm) reveals mild canal compromise with disruption of the inferior C5 endplate and angulation of the posterior disk space at this level. There is also a suggestion of disruption of the anterior longitudinal ligament at C6–7. **C,** The T2-weighted image (TR 1714, TE 90, 4 mm) demonstrates the posterior interspinous ligament disruption and mild spinal canal compromise. Surgical fusion was performed for stabilization.

generally stable [59]. Canal compromise may result from retropulsion of bony or disk material into the spinal canal; the degree of canal compromise is a poor predictor of spinal cord deficit with this fracture [15]. Posterior element fractures and hyperflexion injuries are considered unstable fractures and require prompt immobilization.

Thoracic and Lumbar Spinal Column Injuries

Fractures of the thoracic and lumbar spine are less frequent than cervical injuries in children of all ages. The stability of the thoracic cage renders midthoracic fractures uncommon; instead, thoracic fractures occur at the cervicothoracic or thoracolumbar junction. Thoracic and lumbar fracture mechanisms in descending order of frequency are hyperflexion with or without rotation, axial loading or burst, shear or transverse, and extension injuries [11, 15, 59, 60]. An awareness of normal maturation is important for accurate interpretation of thoracic and lumbar images; the ring apophyses and ossification centers for the transverse processes are readily mistaken for fractures [11] (see Fig. 8.3).

THORACOLUMBAR FRACTURES

Thoracic and lumbar flexion injuries include the limbus vertebra, vertebral body compression or wedge fractures, and unstable injuries with both vertebral body and posterior element disruption. Simple wedge fractures of the anterior vertebral body may be a source of back pain in children with minor trauma and underlying bone abnormalities which predispose to fracture. Pathologic fractures occur in children with histiocytosis X (eosinophilic granuloma), bone marrow metastases (leukemia, neuroblastoma), disorders of calcium metabolism, osteodystrophy, and primary bone tumors [15] (Fig. 8.29). Fractures associated with repetitive trauma or battering often occur at the thoracolumbar junction, as discussed above.

Burst fractures result from axial loading or hyperflexion and are unstable if posterior element fractures accompany the comminuted vertebral body fracture [59] (Fig. 8.30). The degree of canal compromise from retropulsed bone fragments does not correlate well with the presence or degree of neurologic deficit [59]. As described for the cervical region, frontal and lateral plain films are followed by CT, MRI, and/or MCT to define fracture sites, assess canal

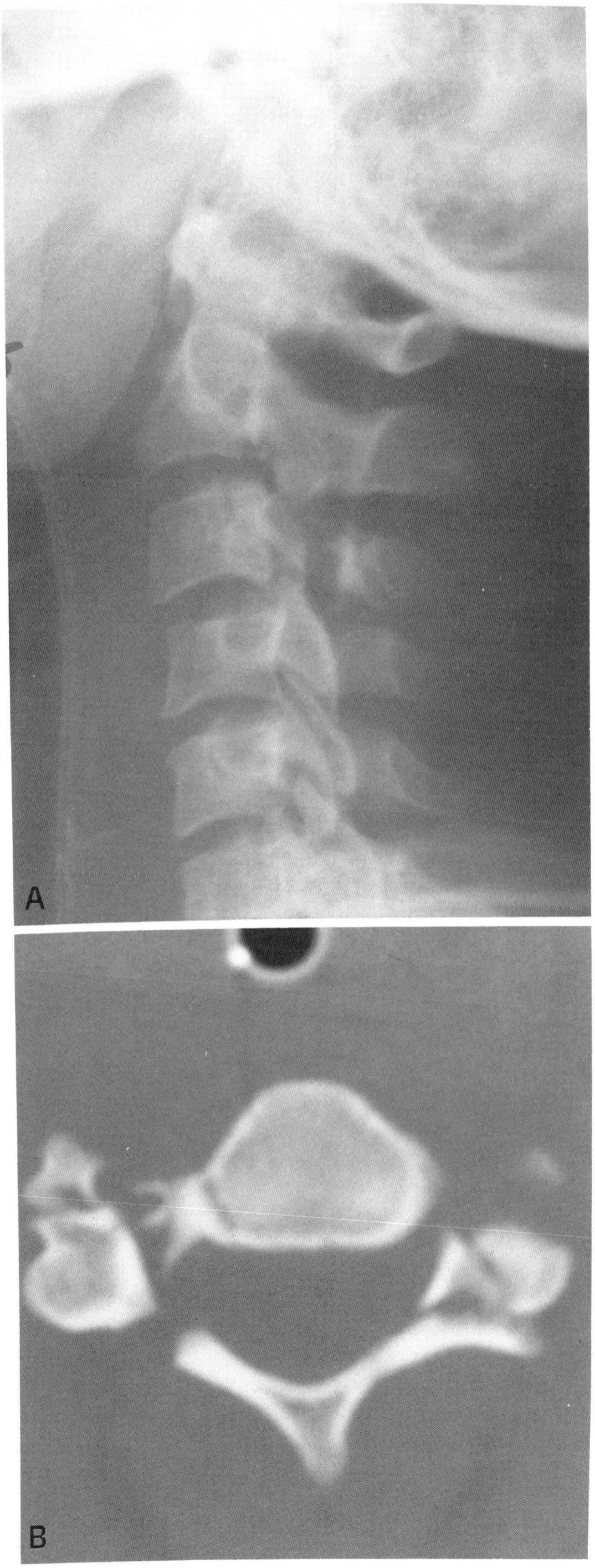

Figure 8.28. Hyperextension injury. Hyperextension injuries are less common than flexion injuries. **A,** This adolescent female has multiple fractures of the posterior elements of C3. Prevertebral swelling and a nasogastric tube are noted. **B,** CT provides superior anatomic detail of multiple comminuted fractures of the pedicle, transverse process, and foramina, pars, and lamina.

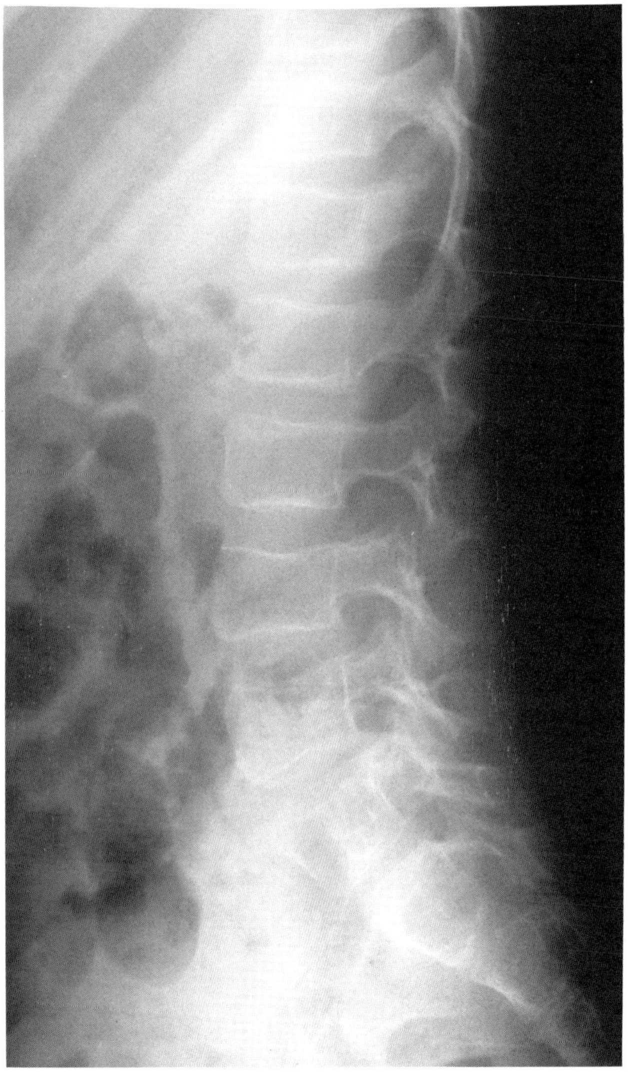

Figure 8.29. Compression fractures. Demineralization and a mild compression fracture of the vertebral body of L5 are present in this child with back pain and leukemia.

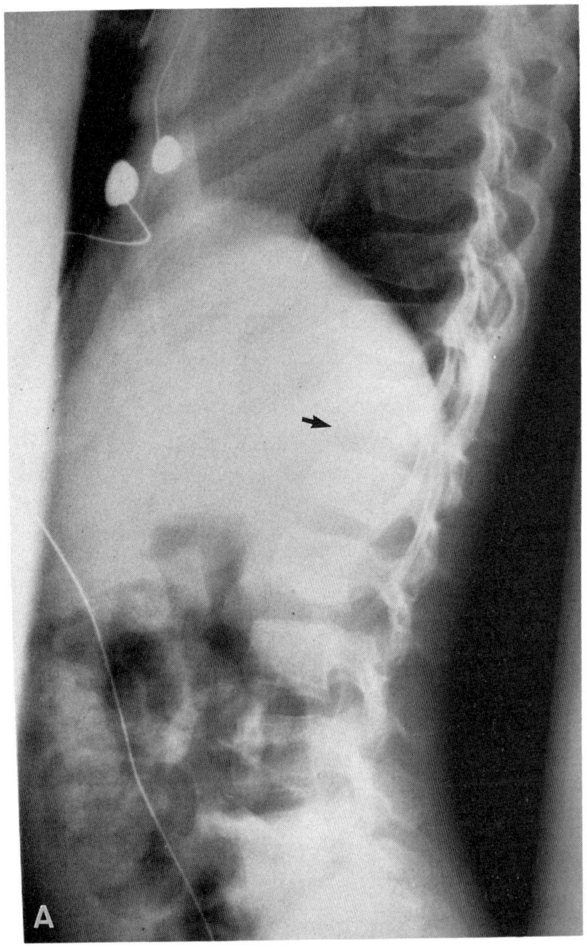

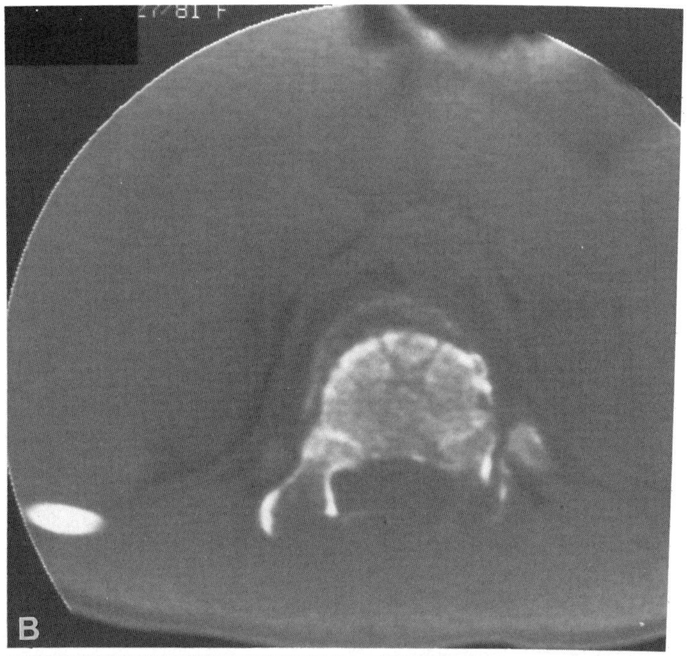

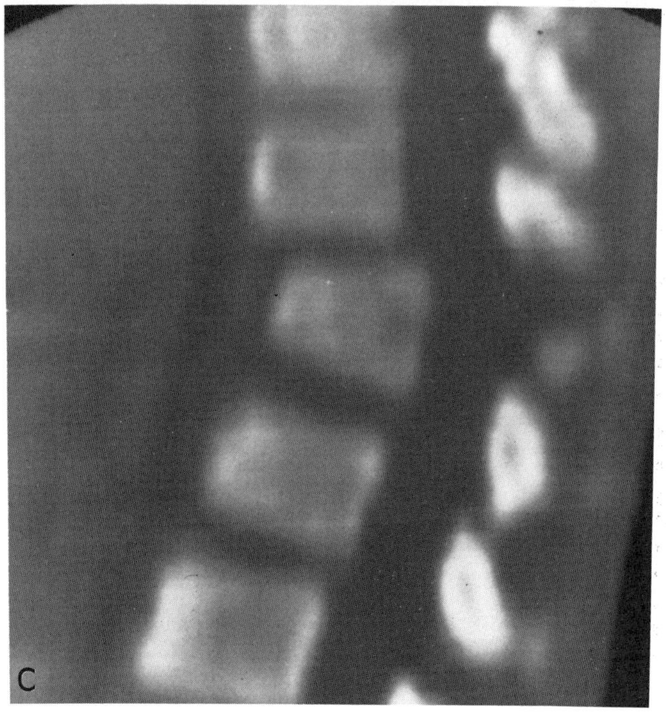

Figure 8.30. Burst fractures. A, A portable lateral film demonstrates compression fracture of T12. **B,** Axial CT scan demonstrates the comminuted T12 vertebral body fracture and disrupted bony ring posteriorly, indicating posterior element disruption and spinal instability. **C,** Re-formatted CT image shows the vertebral body compression fracture with mild retropulsion into the spinal canal and disruption of the spinous process resulting in posterior distraction from T11 to L1. The T1-weighted (**D**) and T2-weighted (**E**) MRI sequences demonstrate canal compromise with cord compression. The cord signal intensity is normal, and the child is neurologically intact (**D:** TR 612, TE 20, 4 mm; **E:** TR 2000, TE 90, 4 mm).

compromise, and exclude conus/cauda equina injuries. Extension injuries are unusual but include neural arch fractures and, rarely, extension teardrop injuries [15].

Transverse, shear, or seatbelt injuries are common in children; a subset of this group is the Chance fracture [97–103]. The *Chance fracture* refers to a horizontal fracture of the vertebral body and neural arch with separation of the posterior elements [103]. Seatbelt injuries occur at the fulcrum of flexion in passengers restrained with lap belts without shoulder restraints. These injuries occur predominantly at the thoracolumbar junction in adults. Improper lap belt positioning above the pelvis is common in children and is associated with abdominal injury [100]. In children, the combined effects of a higher center of gravity, relatively heavy cranium, improper lap belt positioning, and greater spinal flexibility result in more caudal seatbelt injuries of the mid lumbar spine [29, 100]. Abdominal injuries may overshadow manifestations of spinal injury, with a resultant delay in diagnosis of spinal trauma. Careful scrutiny of the lumbar spine on abdominal CT images of children injured in vehicular accidents with

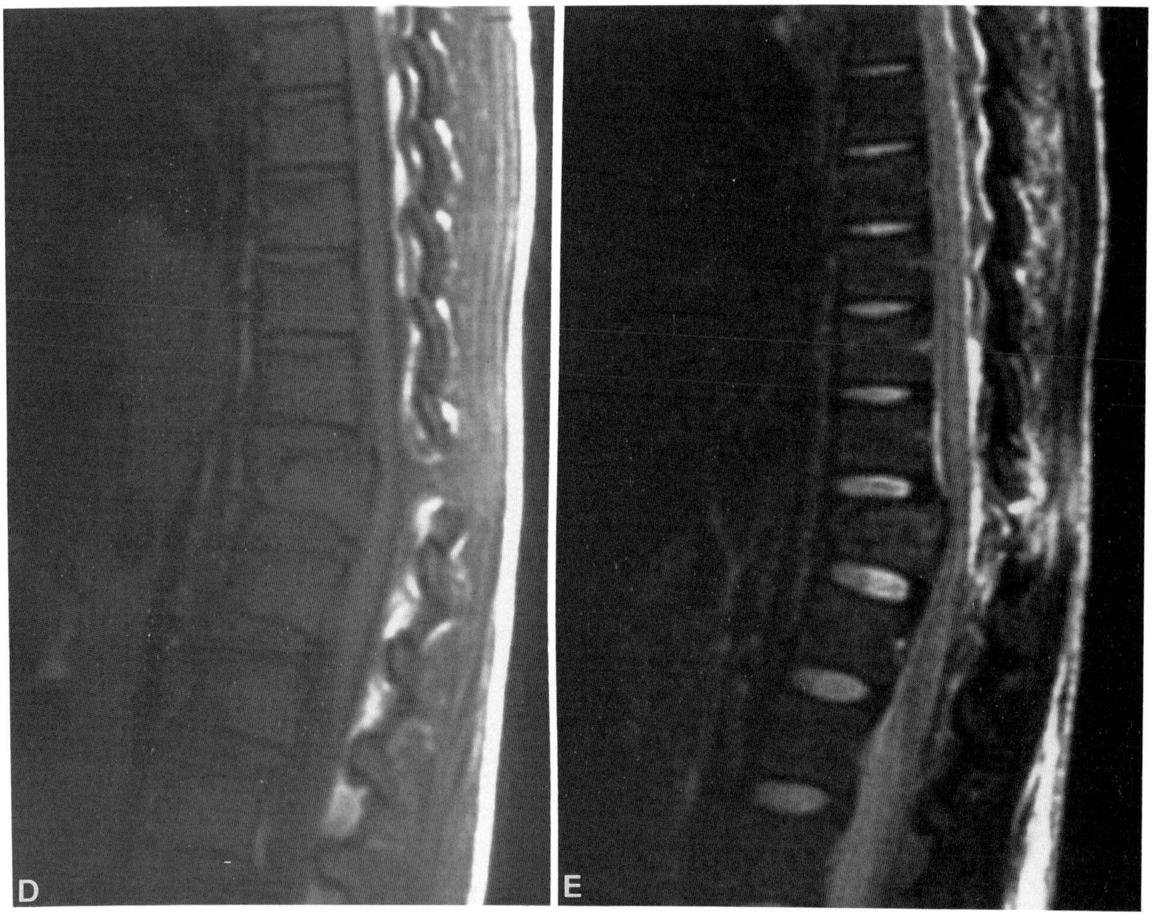

Figure 8.30–*continued*

lap belt restraints is important for early diagnosis [100].

Transverse or seatbelt-type injuries are the result of a horizontal shear force across the spinal column; included are fractures anteriorly of the vertebral body or disk space and posteriorly through the neural arches or ligaments [98, 103] (Fig. 8.31). Other specific associated abnormalities include transverse fractures of the pars, horizontal distraction or splaying of the pedicles, a wide disk space posteriorly, and vertebral fracture-dislocation. Neurologic findings range from intact to complete paraplegia due to conus/cauda equina damage [11, 59, 60]. MRI of the thoracic and lumbar spine excludes spinal canal compromise and confirms conus/lower spinal cord injury remote from the fracture site [29](Fig. 8.32).

Miscellaneous

PENETRATING INJURIES

Penetrating injuries of the spinal column and spinal cord occur in children as well as in adults. Penetrating spinal injuries are typically stable despite fractures due to incomplete bony and ligamentous disruption. Plain films demonstrate metallic fragments and fractures; CT is used to localize retained metallic fragments and associated bone and soft-tissue injuries [104] (Fig. 8.33). Metallic fragments result in significant artifacts on CT and MRI. Teitelbaum et al. [105] investigated the metallic content of various ballistic devices with regard to their safety for MRI; caution is warranted if the type of ballistic fragment is uncertain or its location critical [105]. The reported experience with MRI for evaluation of penetrating injuries is limited. Most soft-tissue abnormalities are well demonstrated on MRI, yet bony and metallic fragments are best seen on CT [29]. Since pseudomeningocele may be indistinguishable from soft-tissue hematoma on MRI, MRI serves as an adjunct to CT or MCT in penetrating injury.

DISK HERNIATION AND APOPHYSEAL PLATE FRACTURES

Herniated nucleus pulposus (HNP) is encountered uncommonly in older adolescents and teenagers in association with sports or lifting activities. Although adults with spinal injury often have an associated HNP, HNP is not usually identified with significant

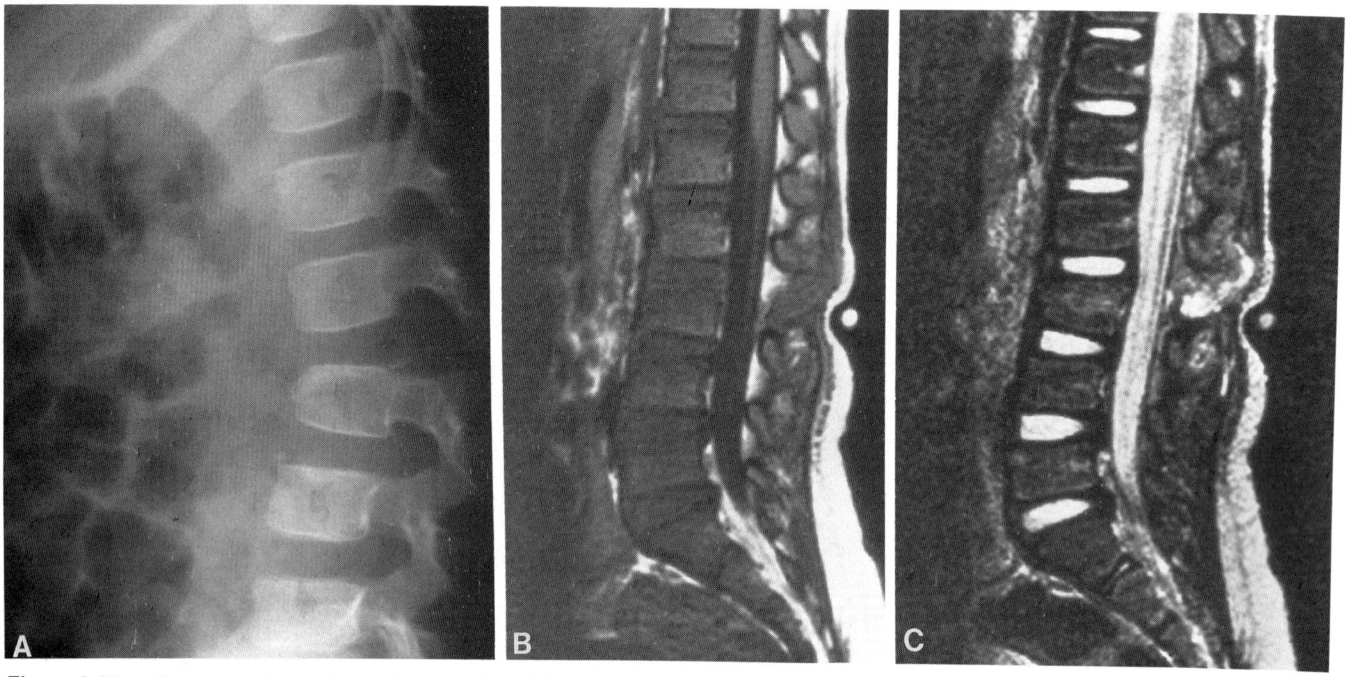

Figure 8.31. A 5-year-old sustained a lumbar injury following an MVA with lap belt restraint. **A,** Lateral plain film demonstrates disruption of the posterior disk space at L3–4 and distraction of the posterior arches. The spinal cord and canal are intact on T1-weighted image (**B**), although the soft-tissue disruption resulted in focal hyperintensity of the vertebral body and posteriorly at the site of ligamentous disruption on T2-weighted image (**C**) (**B:** TR 612, TE 20, 4 mm; **C:** TR 1900, TE 90, 4 mm).

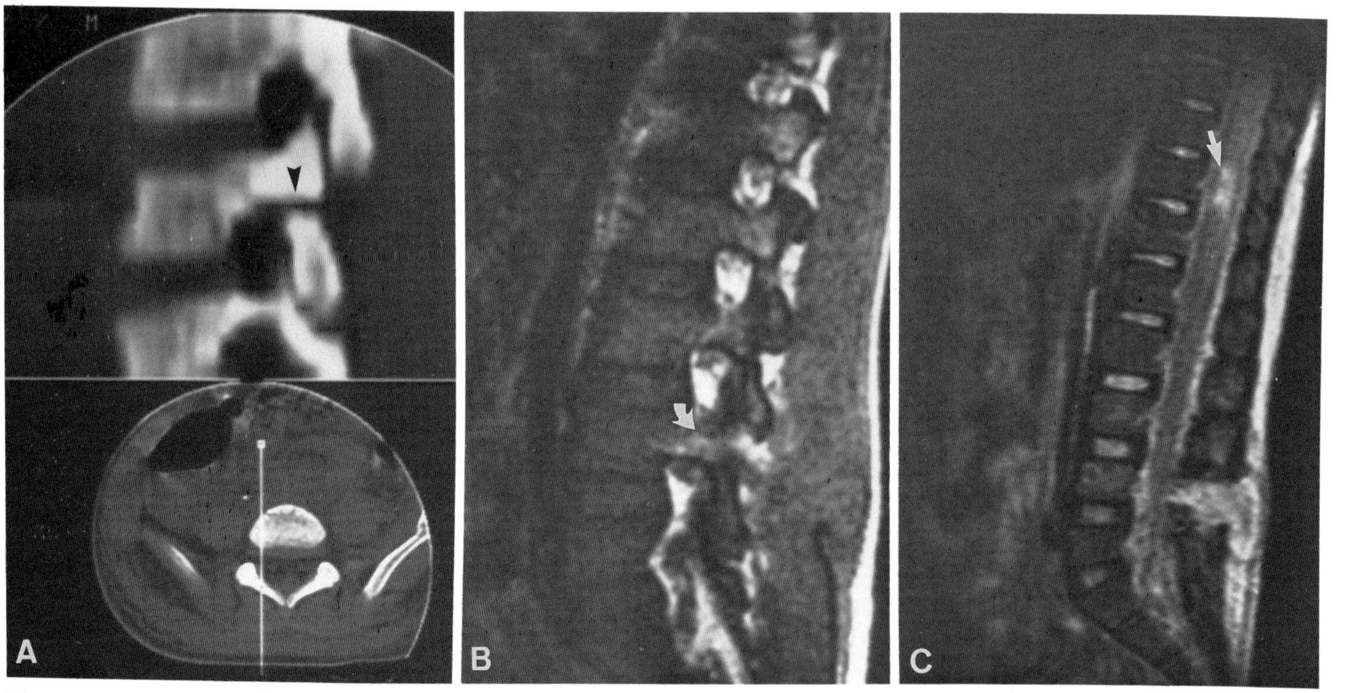

Figure 8.32. A 5-year-old boy was paraplegic following an MVA with lap belt restraints. **A,** Re-formatted CT scan reveals a transverse or shear fracture horizontally across the pars of L4; no vertebral body fracture was identified (*arrowhead*). **B,** The posterior element fracture is difficult to appreciate on T1-weighted MRI (TR 469, TE 30, 4 mm; *arrow*). **C,** T2-weighted image demonstrates edema of the L4 vertebral body, posterior interspinous hematoma, and a remote spinal cord contusion at T11–12 (TR 1800, TE 100, 4 mm; *arrow*).

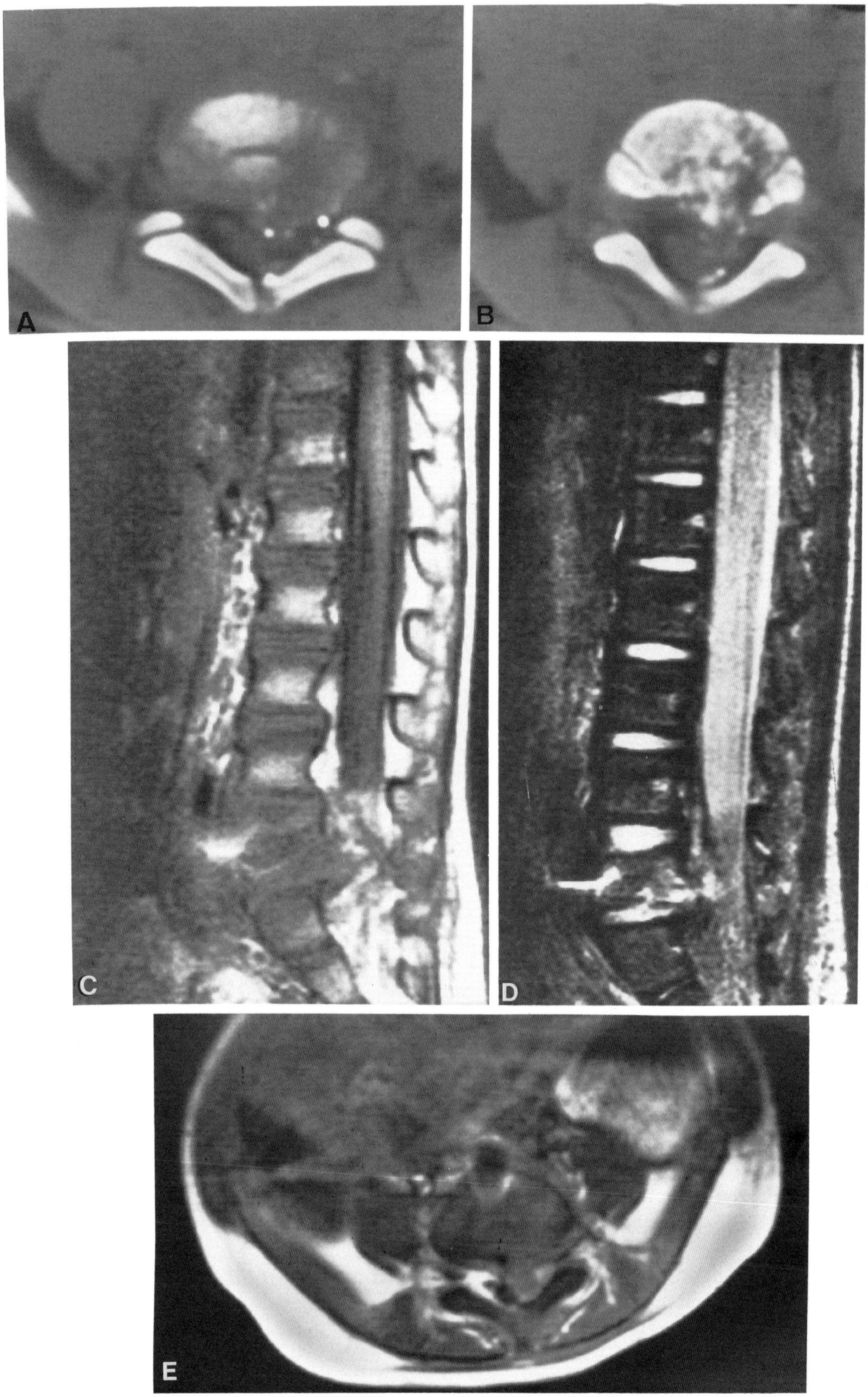

Figure 8.33. An 18-month-old child was evaluated following an abdominal gunshot wound. **A,B,** CT scans show several small metallic fragments associated with a bullet tract and lamina fracture at the L5–S1 disk space. MRI was requested in order to differentiate intradural hematoma from herniated disk fragments. **C,** The T1-weighted image shows a large spinal and paraspinal complex lesion with thecal sac compression at L5–S1 (TR 498, TE 20, 4 mm). **D,** The T2-weighted image demonstrates the mass with thecal sac compression to be composed of both hyper- and hypointense signal, indicating a partially hemorrhagic lesion. **E,** Axial T1-weighted image demonstrate both hyperintense signal of subacute hemorrhage and soft tissue compatible with disk fragments.

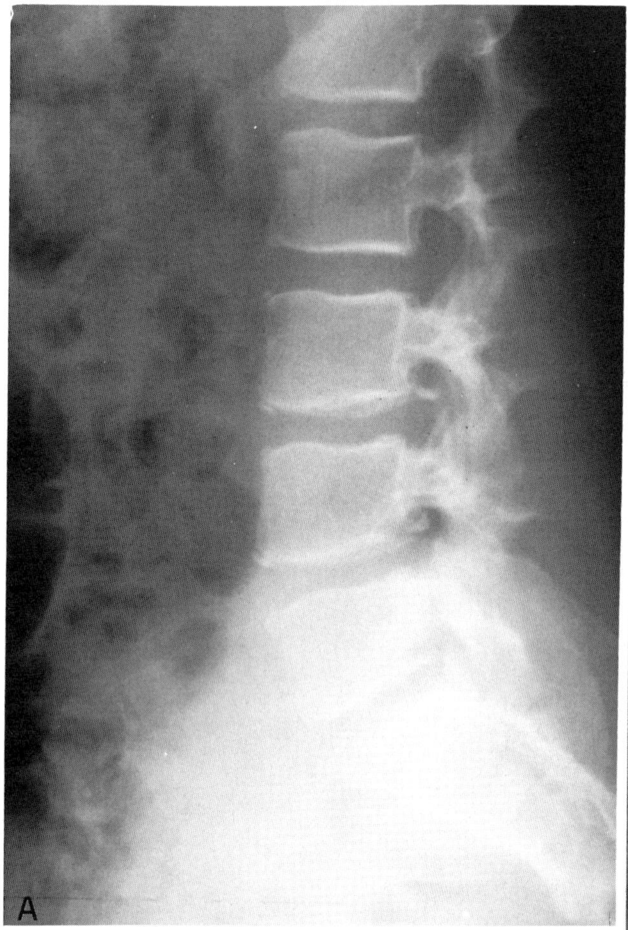

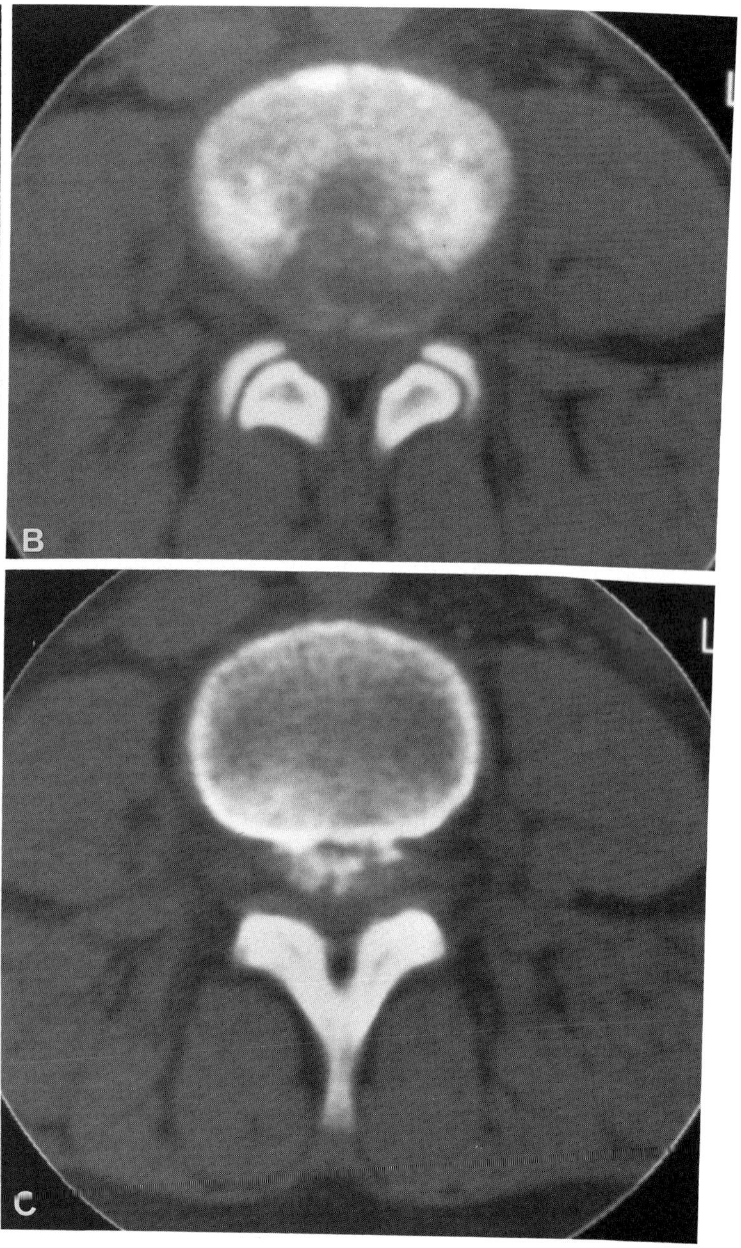

Figure 8.34. **A,** A lateral plain film in an 11-year-old girl with low back and leg pain demonstrates small calcifications which project over the spinal canal at the level of the inferior endplates of L3 and L4. **B,C,** CT scans at L3–4 show thecal sac compromise by a combined soft-tissue and calcified density inseparable from the disk and endplate. MRI images with T1-weighting (**D**) and T2-weighting (**E**) demonstrate marked spinal canal compromise at L3–4 and retropulsion of bone and disk material at the inferior L3 endplate and disk (**D:** TR 633, TE 20, 4 mm; **E:** TR 2016, TE 90, 4 mm). The thecal sac is not compromised at L4–5. Surgery confirmed an apophyseal ring fracture which was retropulsed into the spinal canal along with its attached intact disk.

spinal trauma in children [84, 106–109]. The junction between the cartilaginous endplate and the adjacent vertebral body is weaker than the annulus fibrosus, Sharpey fibers, and disk complex [109]. The result is that the endplate and ring apophysis fracture before sufficient force is generated to produce disk herniation [106–109] (Fig. 8.34). These fractures of the ring apophysis prolapse along with the adjacent cartilaginous endplate into the spinal canal and mimic disk herniation symptomatology clinically and radiographically. Although the disk may prolapse into the spinal canal along with the endplate fracture, the intact annulus indicates that a true HNP is not present. The typical endplate apophyseal fracture occurs at L4 in an adolescent male. This diagnosis is suggested by displacement posteriorly of a small bony fragment from the ring apophysis that projects into the spinal canal [108]. The ring apophysis appears at about age 6 years and fuses by 18 years of age, so this injury has a limited age of incidence [109]. Surgery is generally required; in untreated cases, this apophyseal fragment may fuse with the vertebral body in a retropulsed position. Apophyseal fractures with prolapsed disk may occur at more than one level. The apophyseal fragment may result in significant canal compromise and a myelographic block [109].

Disk space injuries in children also include prolapse of disk material into the adjacent vertebral endplate [15, 110] (Fig. 8.35). This uncommon injury mimics a destructive lesion of the vertebra, although narrowing of the disk space from which the prolapse

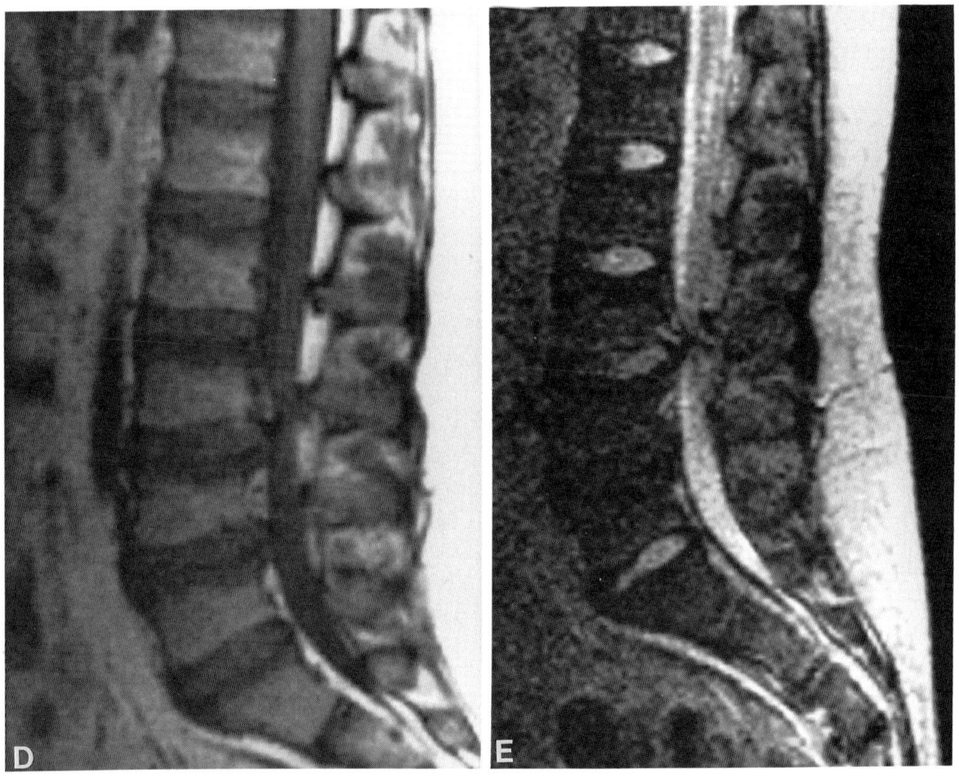

Figure 8.34–*continued*

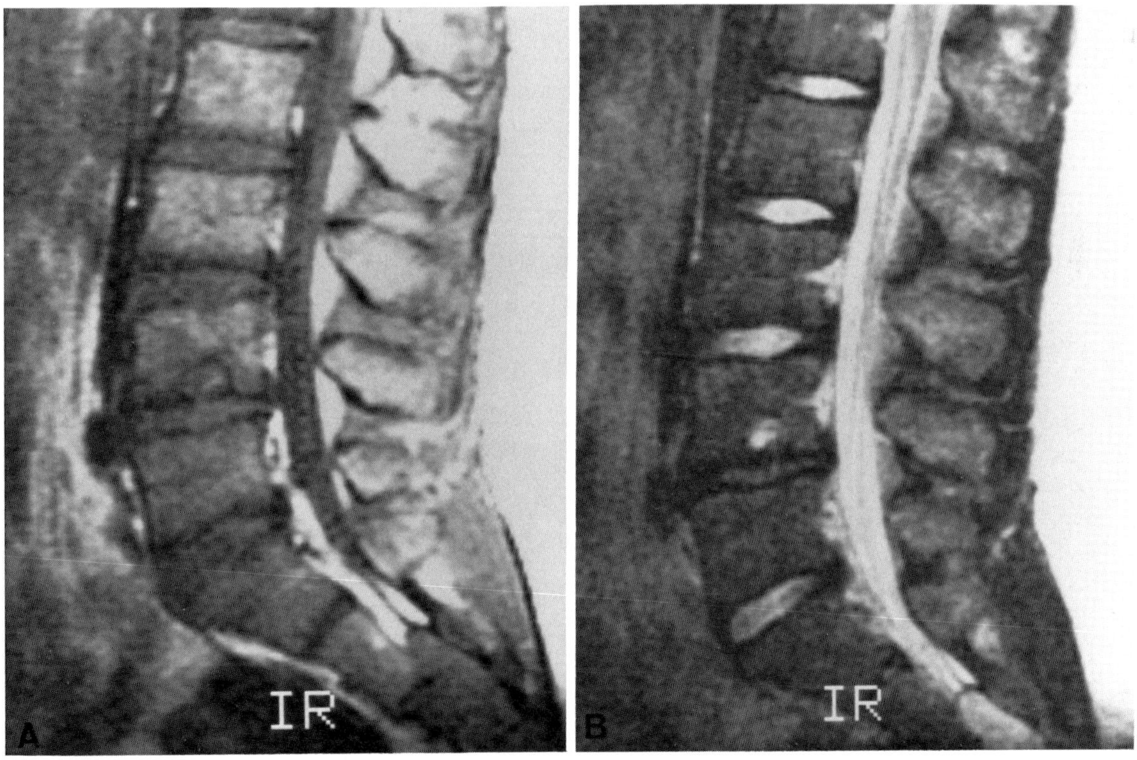

Figure 8.35. A 15-year-old boy with low back pain underwent evaluation with MRI. **A,** The T1-weighted image demonstrates a narrow L4–5 disk space with a focal endplate defect within the inferior endplate of L4 (TR 500, TE 12, 4 mm). **B,** The T2-weighted image also demonstrates the endplate defect. Although its signal intensity suggests a bone marrow lesion, its proximity to the disk space and the narrow, desiccated appearance of the disk space indicated a benign prolapse of disk material into the L4 endplate (TR 3000, TE 95, 4 mm).

originated suggests the presence of this benign process [111].

DISK CALCIFICATION

Intervertebral disk calcification has been described in children as a focal or multifocal entity of uncertain etiology; trauma or infectious/inflammatory origins are postulated [60, 111–113] (Fig. 8.36). Although adult experience might suggest that calcium disorders, premature degenerative disk disease, metabolic disorders, vitamin D hypervitaminosis, and infection are possible factors, most cases of disk space calcification in children have none of these associations. The usual presentation is that of a male child complaining of local pain and tenderness, perhaps with an elevated ESR. Symptomatic patients may have disk space calcifications and protrusions into the spinal canal; these may resolve with no residual disk space narrowing. Asymptomatic patients may have stable focal or multifocal calcifications for a prolonged time without evidence of progressive disk space narrowing or resorption of calcifications [113].

SPONDYLOLYSIS AND SPONDYLOLISTHESIS

Spondylolysis refers to a defect in the pars interarticularis which may result in subluxation of the affected vertebra. The subluxation, called *spondylolisthesis*, is graded based on the percentage of the vertebral body displacement as follows: grade 1, 25% or less subluxation; grade 2, 25 to 50% subluxation; grade 3, 50 to 75% subluxation; and grade 4, 75 to 100% subluxation [114]. Complete anterior subluxation in which the superior vertebral body is positioned anterior to the vertebral body below is referred to as *spondyloptysis*.

Although the etiology of pars defects has been debated for years, the consensus is that most arise from a focal weak region within the pars which eventually develops a fatigue fracture after repetitive microtrauma of normal or heavy activity [12, 115]. Familial occurrences suggest an underlying heritable dysplasia in some cases. The absence of an animal model with this disease and its rarity in nonambulatory children implicate upright posture and weight bearing as key factors [12, 116]. Most pars defects occur at L5, with spondylolisthesis at L5–S1; this is in contrast to degenerative spondylolisthesis of adults, which often affects L4–5. Spondylolysis and spondylolisthesis have been classified as type 1, dysplastic, anomalous lamina which allows subluxation; type 2, isthmic defects with lytic defect of the pars, elongated intact pars, or acute pars fracture; type 3, degenerative; type 4, traumatic with fracture which spares the pars interarticularis; and type 5,

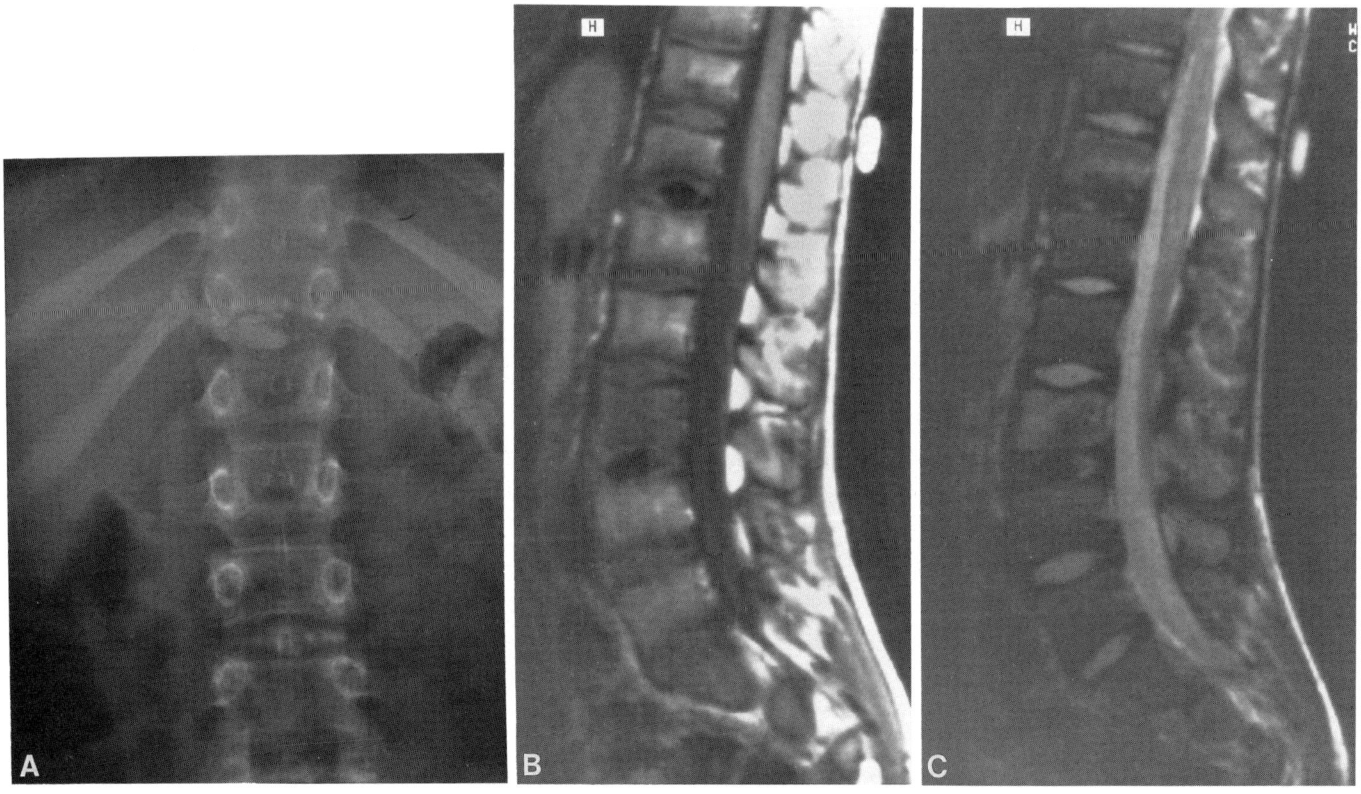

Figure 8.36. A 13-year-old boy complained of nonspecific nonradiating back pain. **A,** Plain film shows multilevel disk calcification. MRI images with T1-weighting (**B**) and T2-weighting (**C**) confirmed disk space calcifications and excluded prolapse into the spinal canal (**B:** TR 500, TE 15, 4 mm; **C:** TR 1800, TE 80, 4 mm). Clinical evaluation indicated no evidence of underlying metabolic disorder.

pathologic fracture [117]. According to this classification, the overwhelming majority of cases are categorized as type 2 defects without an acute pars fracture and are usually bilateral [117]. Spondylolysis is recognizable on plain films by 6 to 10 years of age in about 5% of normal children, although it is often asymptomatic until late childhood or adulthood [12].

Diagnosis of spondylolysis is often suggested by visualization of a defect of the pars interarticularis on lateral or oblique plain films, with spondylolisthesis best demonstrated on upright films (Fig. 8.37). The neural arch on oblique projections looks like a "Scottie dog"; a pars defect results in a lucency across the neck of the dog in this projection. CT scanning for possible spondylolysis is best performed with thin (3- to 5-mm) slices and gantry angulation parallel to the affected disk space. Contiguous scans encompass the affected level and at least one vertebral level above and below (i.e., L4 to S1). Images should include both bone and soft-tissue windows. On CT, contiguous images through a normal vertebra should demonstrate a complete bony ring at each vertebral level; images acquired parallel to the adjacent disk space demonstrate this optimally. Spondylolysis results in bilateral lytic defects across the pars and an incomplete vertebral ring. At first glance, the pars defect resembles a normal facet joint; however, its orientation is more transverse than the obliquely oriented lumbar facet joints. Bone-window scans reveal irregular sclerotic margins instead of the smooth outline of the facet joint. With spondylolisthesis, the disk at the level of the subluxation may bulge or be displaced posteriorly; sagittal re-formatted images generally demonstrate this as a pseudo-disk herniation due to the subluxation rather than a true HNP [118]. With marked subluxation, the anteroposterior canal diameter is elongated immediately above the slippage, and there is neural foraminal and spinal canal compromise at the level of subluxation.

The diagnosis of spondylolysis on MRI is difficult due to the absence of MRI signal from cortical bone. Findings include absence of a continuous high-signal intensity across the pars on T1-weighted sagittal or axial images, with or without subluxation [119, 120]. Other entities that similarly result in a low-signal intensity across the pars on T1-weighted MRI images include a developmentally thin or sclerotic neural arch, bone marrow diseases such as regenerating marrow, neoplasia, or iron overload, and the predominantly hematopoietic spinal bone marrow of children.

Sequelae of Spinal Injury

Sequelae of spinal trauma include residual spinal column deformity associated with fractures and dislocations and late effects of spinal cord injury. Young children may demonstrate significant restoration of vertebral body height following trauma with conservative measures such as immobilization. More severe injuries such as fracture-dislocation and older children with bony injury may require surgical reduction, decompression, or bony fusion [2]. Posttraumatic sequelae include scoliosis and kyphosis due to spinal column abnormalities or on a neuromuscular basis in those with neurologic deficits. Extensive surgical decompression without fusion, chest injuries, and burns may be followed by secondary kyphoscoliosis [60].

Late sequelae of spinal cord injury are best evaluated on MRI [121–123] (Fig. 8.38). Posttraumatic syringomyelia results in CSF intensity cystic spinal cord cavities detectable on MRI; these are suspected when progressive neurologic deficits occur after initial stabilization of a cord injury. Posttraumatic syringomyelia may propagate for many levels above or below the level of cord injury; surgical decompression stabilizes or improves symptoms in some cases. In contrast to syringomyelia, myelomalacia is hyperintense compared with spinal cord and CSF on proton density–weighted images [121–123]. MRI is used to exclude residual cord compression by disk herniation, bone fragments, or as a complication of operative repair. Focal atrophy, hemosiderin deposition from hematomyelia, and frank cord transection are visualized. MRI is degraded by metallic artifacts from surgical stabilization devices and clips; MCT may then be required to define the cord and spinal column at the level of injury.

Conclusion

The patterns of spinal injury in children are often unique to the pediatric population. An awareness of these patterns and of normal spinal maturation in children contributes to rapid and appropriate diagnosis and management of these injuries. Plain films in children, as for adults, should be supplemented routinely with CT for evaluation of suspected or documented bony injury. Although most spinal fractures are recognizable on plain films, the full extent and location of the fractures and of displaced fracture fragments are determined only with the addition of CT. MRI and MCT are supplemental to plain films and CT for evaluation of intraspinal soft tissues and documentation of spinal cord injury. Although MRI can be performed successfully in the acute phase of spinal injury, medical stabilization, immobilization, evaluation of the bony spinal column integrity, and attention to coexisting injuries often take precedence over assessment of spinal cord injury during the very acute phase of assess-

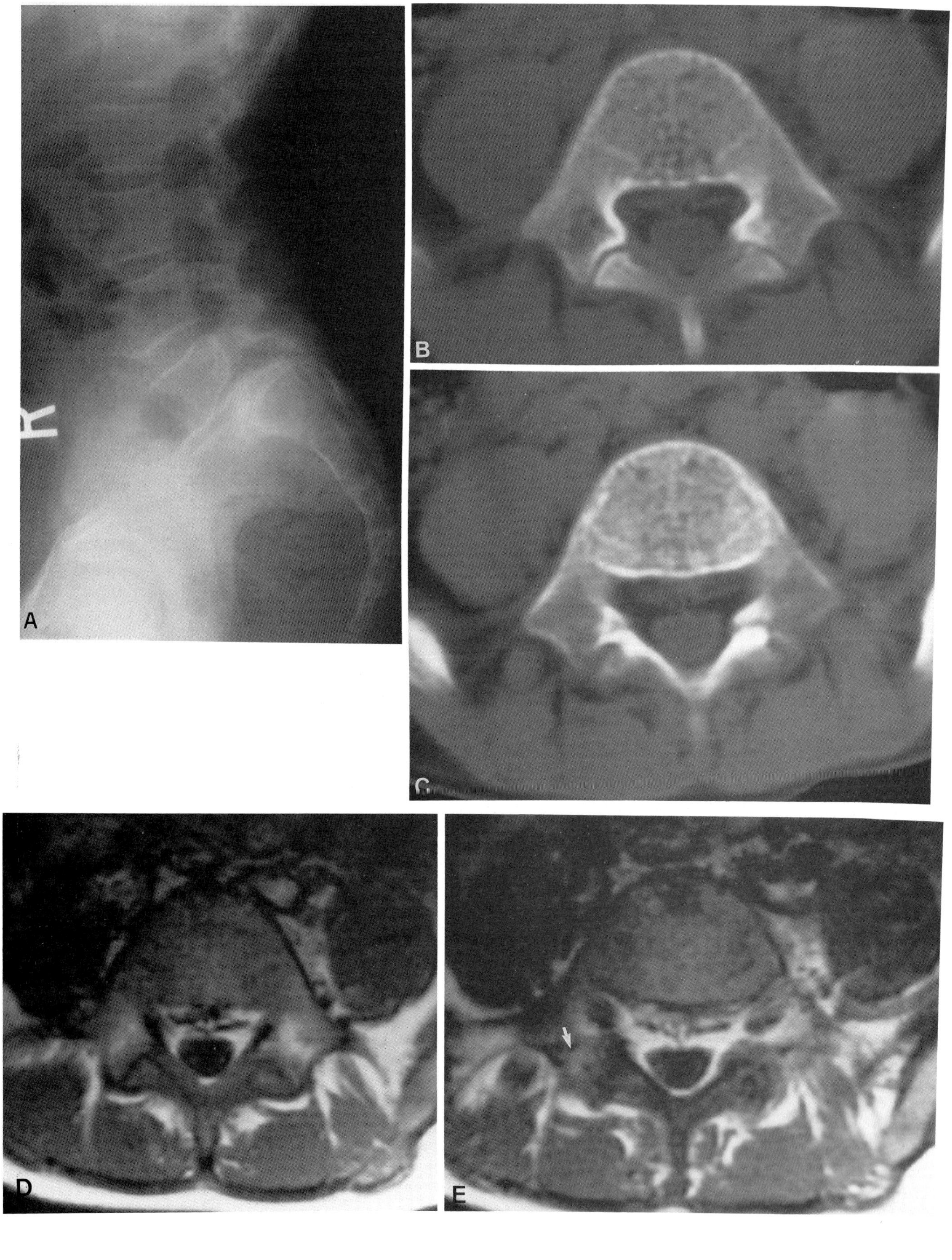
R
A
B
C
D
E

ment of the child with spinal injury. As MRI technology and availability improve with faster scanning sequences, optimal surface-coil technology, and appropriate monitoring devices, MRI may play a greater role in the early evaluation of spinal injuries in children.

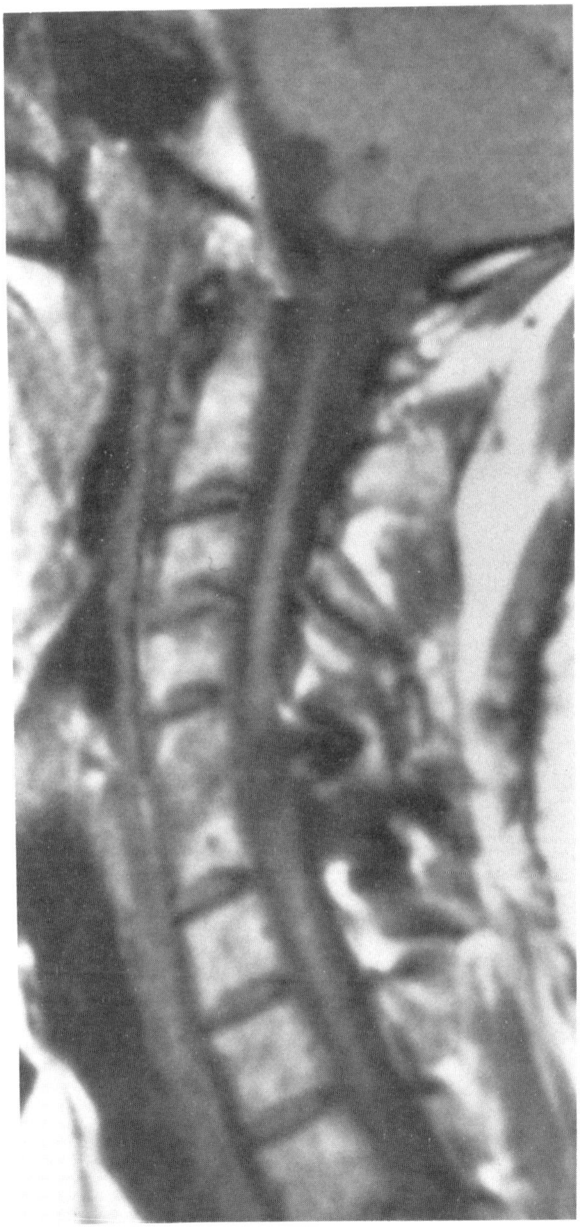

Figure 8.38. Delayed sequelae of spinal trauma. This adolescent boy was referred for MRI several months following a cervical fracture-dislocation (TR 500, TE 20, 5 mm). MRI demonstrates a focal region of marked myelomalacia versus spinal cord transection at the level of injury and surgical changes related to fusion of C5 and C6. No cystic spinal cavities were identified to indicate syringomyelia.

ADULT SPINE TRAUMA

K.C.V.G. Rao

In the adult population, spinal injuries are more often the result of blunt trauma or penetrating injuries as a result of either stab wounds or gunshot wounds. Due to an increasingly mobile society, injuries in motor vehicle accidents (MVAs), either as passengers or as pedestrians, account for a significant amount of blunt trauma to the spine. In passengers, these often involve the cervical or thoracolumbar (T11–L2) region. Another common cause of blunt spinal trauma is sports injuries, most often jumping from a height as in diving accidents, parachute jumping, or similar sports. The spinal trauma in these patients is often accompanied by fractures of the lower extremities, pelvis, or cranial vault. Penetrating injuries of the spine frequently are seen in urban communities associated with an increased incidence of crime.

Blunt Trauma

The biomechanics of spinal injury in adults are similar to those in children, although greater mechanical forces may be involved. Neurologic deficits may be encountered even without associated osseous fractures due to preexisting degenerative spondylitic disease and narrowing of the spinal canal and foramen. Degenerative changes in the ligaments and joint spaces lead to disruption of the ligaments, which may lead to subluxations not recognized in the preliminary plain radiographs taken in the supine position. Most spinal injuries are the result of indirect forces generated by movement of one segment of the body against or away from a fixed body segment. More than one vector force is usually involved, resulting in complex stresses applied to different segments of the vertebral column [123].

In most spinal trauma, the neurologic deficits due to associated spinal cord or nerve root injury usually manifest immediately at the time of the injury. However, in 5 to 10% of these injuries, the neurologic deficits may be delayed. In another 10% of patients with neurologic deficits, no radiographic evidence of vertebral fractures can be identified [123–125]. Such fractures often involve the spinal cord in the cervical region, usually following hyperextension injuries [126–128]. It is conceivable that in some cases the delayed neurologic deficits can be attributed to inadequate evaluation or nonrecognition of the extent

Figure 8.37. **A,** A late-juvenile girl with low back pain had subluxation of L5 anteriorly on S1 on a lateral plain film. **B,** The normal facet joints at L4–5 have a well-defined joint space and an oblique orientation. At a slightly more caudal level, bilateral horizontal defects are seen within the pars of L5. **C,** Adjacent image confirmed an incomplete bony ring at L5 due to bilateral pars defects. **D,** The normal facet joint is well demonstrated (TR 500, TE 20, 4 mm). **E,** However, the pars defects are very difficult to appreciate on MRI (*arrow;* TR 500, TE 20, 4 mm).

of the spinal injury in the initial clinical and radiographic assessment of the patient [127, 128].

The classification of blunt spinal injuries in adults is similar to that in the pediatric age group discussed earlier. With minor variations, the imaging strategies in both adults and children are also similar. Initial evaluation always consists of plain radiographs, obtained with proper precautions not to aggravate the bony displacement or to worsen any preexisting neurologic deficits. The indications for pleuridirectional tomography, CT, CTM, or MRI studies in most instances are similar to those described in the section on pediatric trauma. In general, CT is useful in demonstrating fractures involving the posterior elements of the vertebrae and the arch, since the same cannot be adequately evaluated with plain radiographs and lateral polytomography may be difficult in some injured patients. Retropulsed bone fragments, encroachment of the spinal canal, and in occasional patients, acute hematoma can be demonstrated on CT. Fractures of the vertebral body and some fractures involving the posterior elements per-

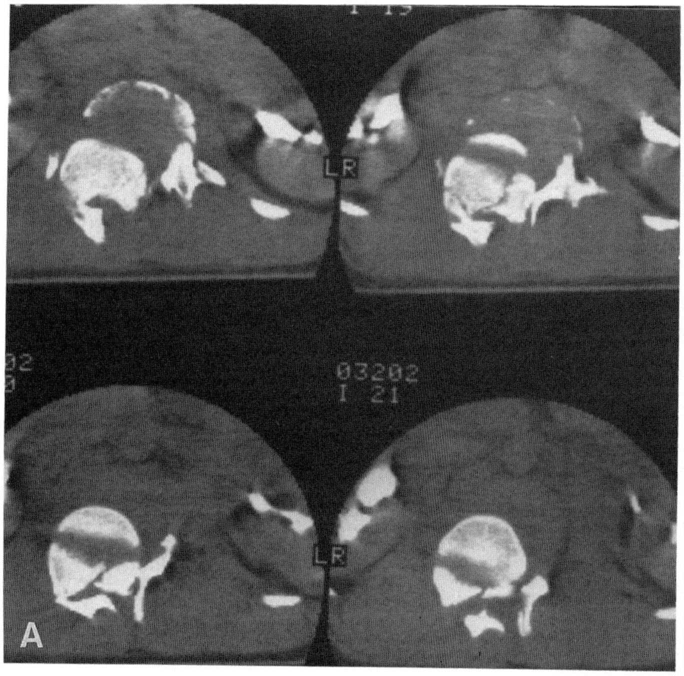

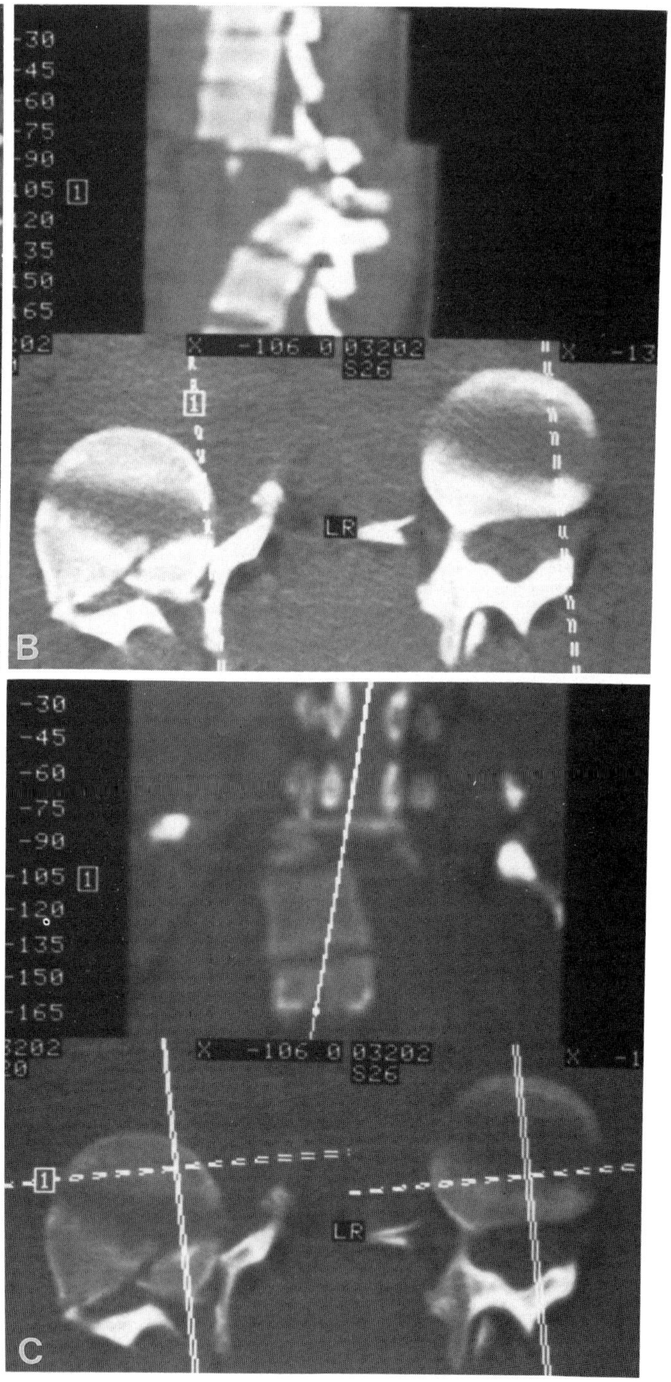

Figure 8.39. Comminuted fracture-dislocation. CT in acute trauma setting obtained following plain x-rays and prior to reduction in a patient following multiorgan trauma including spine injury. **A,** Contiguous axial CT demonstrates the spinal fracture with transection of the spinal cord. The axial section demonstrates the rotational component as well as the bony fragments within the spinal canal. Sagittal (**B**) and coronal (**C**) re-formatted images obtained from the axial sections without moving the patient demonstrate the severe malalignment of the spinal column.

pendicular to the axial CT plane may be missed, even with subsequently-reformatted CT images[25]. Similarly, minor degrees of subluxation or "perched" facets are best evaluated with lateral tomograms or CT re-formatted images. CT should always be evaluated in conjunction with the preliminary plain radiographs [26, 129–132]. In many centers, the preliminary "CT scout views" in both the frontal and lateral projections are utilized in lieu of the conventional plain radiographs.

At the present time, the role of MRI in spinal trauma is still evolving. The present indications for MRI in spinal trauma are limited to patients presenting with neurologic deficits and stable spinal fractures. CT is more useful, in acute trauma settings, when multiorgan trauma is associated with spine injury (Fig. 8.39) or when plain radiographs cannot exclude a fracture in the cervicothoracic region (C6–T2).

MRI is excellent in evaluating the spinal cord and the soft tissues both within and surrounding the spine [32–35, 38, 134, 135]. Ligamentous injury, which may be suspected from plain radiographs or CT, often can be demonstrated on MRI studies [134, 135]. Identification of the extent and type of cord injury as well as the presence of intradural or extradural hematoma often can be demonstrated on MRI [31, 33–37]. The MRI study usually consists of spin-density T1- and T2-weighted sequences. Reduction in image acquisition time and better delineation of cord edema, contusion, or hemorrhage are achieved with fast-spin-echo (FSE) T2-weighted sequences. Fat suppression when combined with an FSE sequence is useful in demonstrating osseous and ligamentous injuries. Recent availability of phased-array coils will allow evaluation of the complete spine without the need to move the patient and reposition the surface coils.

On both T1- and T2-weighted images, the cortical bone as well as the ligaments are seen as hypointense signal, whereas the soft tissues have isointense signal on T1-weighted images and isointense to varying degrees of mildly hyperintense signal on T2-weighted images. The CSF and the spinal cord can be identified easily due to the typical appearance on both the T1- and T2-weighted images. Focal cord injury resulting in swelling of the cord is seen as isointense or mildly hyperintense signal on T1-weighted images (Fig. 8.40). Hyperintensity of signal within the cord on T1-weighted images usually indicates hemorrhage. This is often seen beyond 72 hours following the trauma. On T2-weighted sequences, both edema and ischemia or hemorrhage due to trauma have a hyperintense signal in the subacute settings. Edema and ischemic changes may extend in a cephalocaudad fashion, usually involve the dorsal region, and usually decrease with time but do not disappear completely. On the other hand, hemorrhage on T2-weighted images may be isointense or hypointense (deoxyhemoglobin) initially, later changing to hyperintense (methemoglobin), and finally appearing as low signal foci or a band (hemosiderin). Subluxed or retropulsed bony fragments in fractures can be defined, and compromise of the spinal canal as well as cord compression also can be demonstrated with sagittal MRI sections on T1-weighted, T2-weighted, and GRE images. In the acute stage, marrow edema is seen as a hypointense signal on T1-weighted sequence and mildly hyperintense or heterogeneous signal on STIR pulse sequence (see Fig. 8.40).

These signal changes reflect the decreased mar-

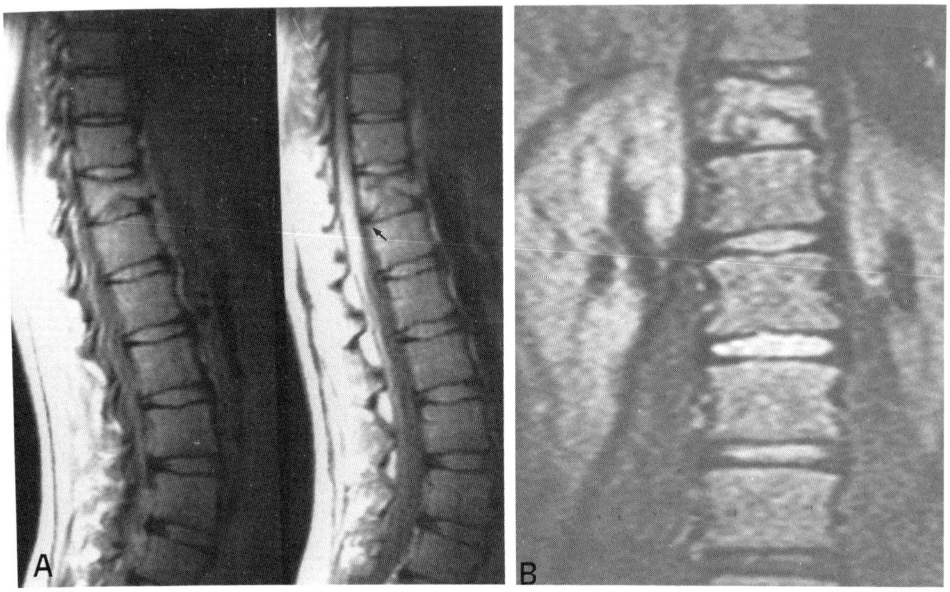

Figure 8.40. Cord contusion secondary to fracture. Acute paraplegia following a soccer injury. Plain x-rays showed loss in height of T-11 vertebral body. Minimal recovery following stabilization. Sagittal T1-weighted (**A**) and coronal STIR sequence (**B**) MRI. Compression of the T11 vertebral body with bright signal within the marrow space from edema or hemorrhage. There is a long segment of hyperintense signal involving the distal cord and conus (*arrowhead*) from cord contusion.

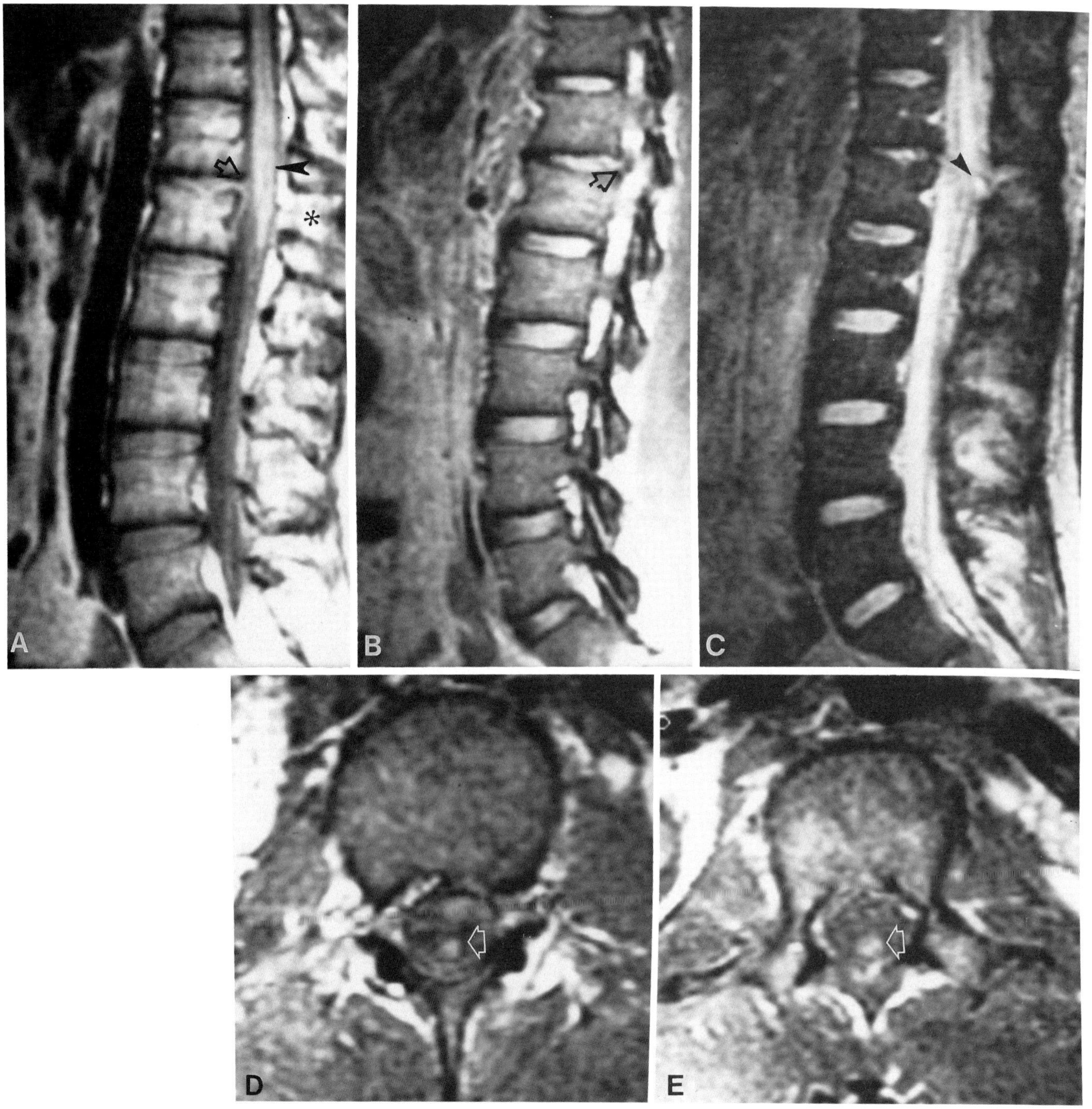

Figure 8.41. Torn ligament with cord injury. Paraplegia following a parachute jumping accident. MRI study 5 days following the injury. Adjacent sagittal T1-weighted (**A,B,**) and T2-weighted (**C**) images demonstrate the cord contusion (*arrowhead*), minimal loss in height of the posterior aspect of the vertebral body, and adjacent narrowing of the disk space. The focal ill-defined bright signal in anterior epidural space (*small arrow*) probably represents subligamentous hematoma and acute herniated disk. Increased space between the spinous process (*open arrow*) probably represents edema or hemorrhage within the disrupted interspinous ligament. Both the T1- and T2-weighted images show mild increased brightness of the vertebral body. T1-weighted axial sections through the disk level (**D**) and the adjacent T-11 vertebral body (**E**) demonstrate the bright signal within the distal conus (*open arrow*).

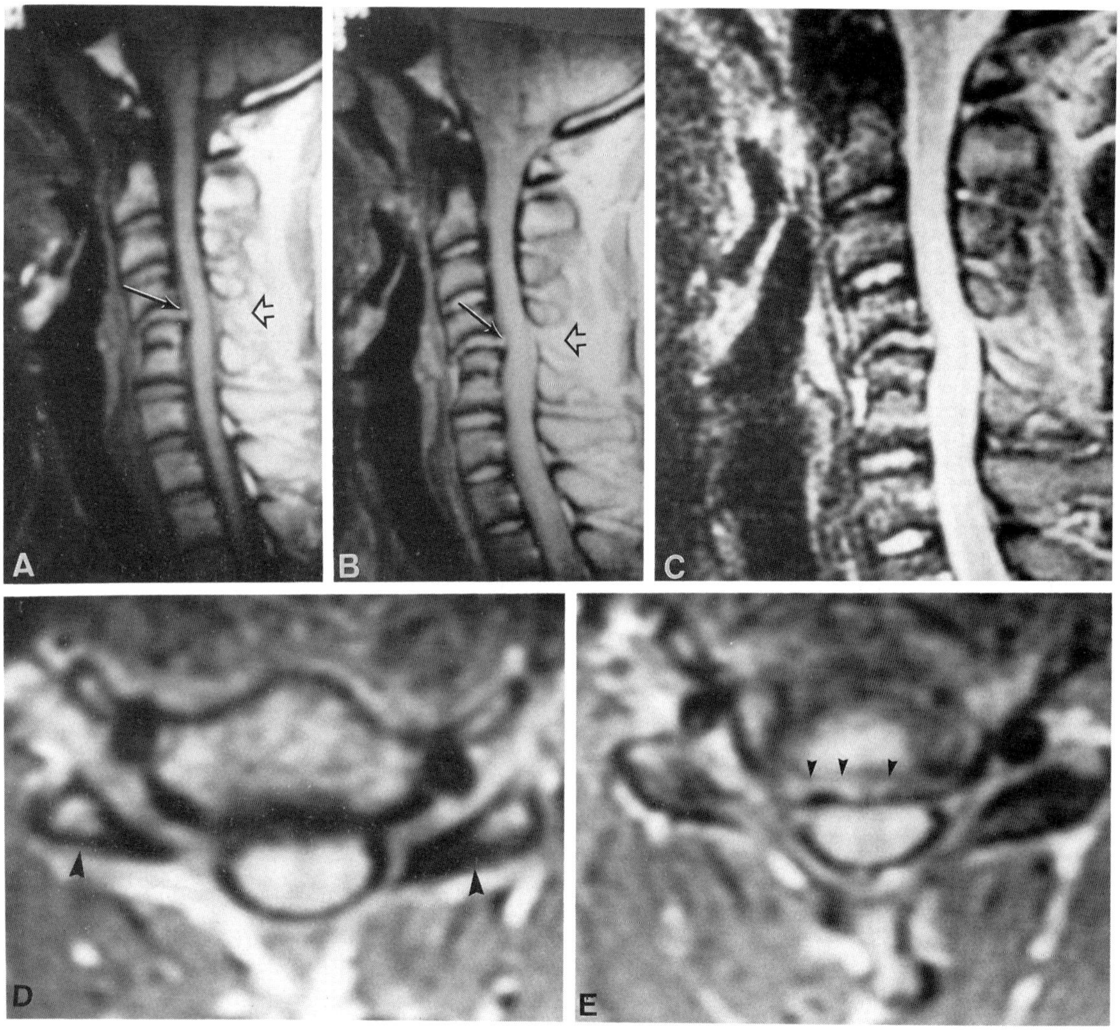

Figure 8.42. Cervical flexion injury. Adjacent sagittal T1-weighted (**A**), proton-density (**B**), and T2-weighted (**C**) MRI sequences of the cervical spine. The C4 vertebral body is subluxed anterior to the fused C5–6 vertebrae. There is bright signal within the posterior longitudinal ligament (*arrow*). The distance between the spinous process of C4 and C5 is increased (*open arrow*) due to torn ligament. Axial sections through the disk level (**D**) and adjacent vertebral body (**E**) demonstrate the naked facet (*arrowhead*) and the acute herniated disk (*small arrowheads*).

row and increased water content. Demonstration of disrupted or torn ligaments is difficult with CT or plain radiographs but should be suspected when there is more than 30% subluxation or there is an increased space between the adjacent vertebral segments or within the facet joint. On MRI, torn ligaments should be suspected when there is discontinuity within the low signal intensity of the ligaments, normally seen on all pulse sequences, with a hyperintense signal on the T2-weighted and GRE images (Fig. 8.41). The hyperintense signal probably represents a combination of hemorrhage and edema in the disrupted ligaments. Acute traumatic disk herniation has a hyperintense signal on T2-weighted images (see Fig. 8.41). The role of MRI in acute spinal trauma is primarily as an adjunct to plain radiography and CT, but it does not replace these studies. MRI is useful in excluding spinal cord contusion or hemorrhage in patients with neurologic deficits and normal plain radiographs and CT.

Cervical Region

FLEXION INJURIES

Depending on their severity, flexion injuries are always associated with varying degrees of compression of the vertebral body. These usually involve the C3 to C7 vertebral segments. In adults, because of the decreased tensile strength of the ligaments, disruption of the interspinous ligament and fractures of the laminae or spinous processes are not uncommon. Plain radiographs may show an increased dis-

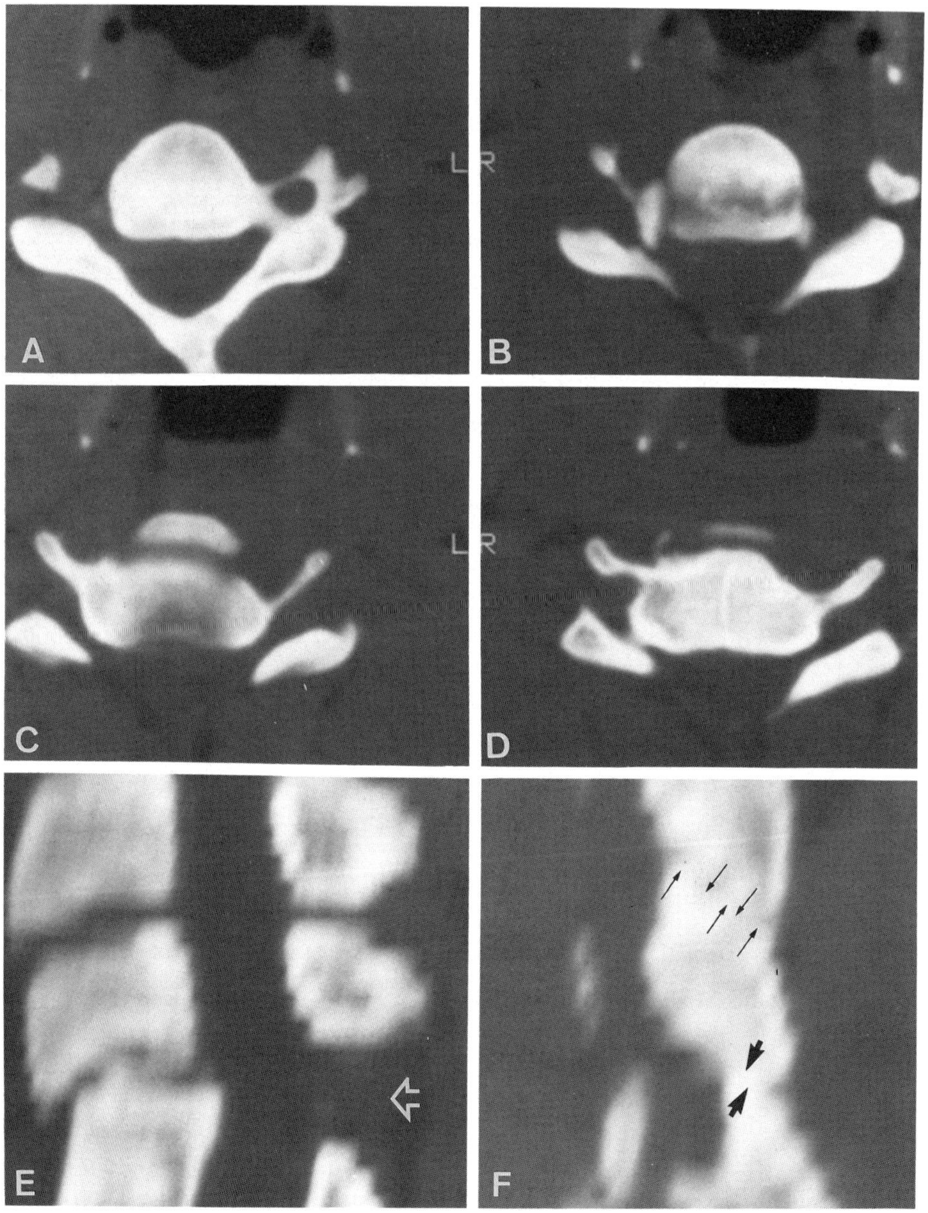

Figure 8.43. Severe flexion injury with instability of the facet joints. Following initial reduction of bilateral locked facets, follow-up x-rays demonstrated further anterior subluxation of C4 over C5 with increased interspinous distance. Contiguous CT sections (**A–D**) and re-formatted sagittal sections through the midline (**E**) and parasagittal plane (**F**). On the axial section through the disk space, the adjacent facets of the facet joint are not seen. The sagittal re-formatted images demonstrate the anterior subluxation of C4 over C5. There is increased interspinous distance (*open arrow*). The lateral sagittal section through the facet joint demonstrates the unstable facet joint (*opposing arrows*) compared with the normal alignment of the facet joint (*small arrows*). Parasagittal T1-weighted (**G,H,**) and T2-weighted (**I**) MRI sequences demonstrate the relationship of the adjacent facet surfaces (*open arrow*). The T2-weighted section also demonstrates an avulsion fracture (*open arrow*).

tance between the spinous processes. Unless there is fracture of the lamina, the torn intraspinous ligament may not be appreciated. Associated subtle or obvious fractures that may involve the lamina, pedicle, or extension into the neural foramen can be confirmed with polytomography or CT in the axial and re-formatted planes. When cord contusion, nerve root impingement, or a herniated disk is suspected, these can be better identified on MRI (Fig. 8.42). Hyperflexion sprain should be suspected if there is increased signal intensity within the interspinous space associated with increased distance between the spinous processes. This is usually the result of edema or hemorrhage. In more severe injury there is disruption of the posterior longitudinal ligament and the ligamentum flavum often associated with subluxation. Wedge compression of the anterior aspect of the vertebral bodies or avulsion of the anterior endplates of the vertebrae are signs of severity of the trauma. Compression with posterior avulsion usually results in acute cord symptoms. Depending on the extent of the ligamentous injury and fractures involving the posterior elements, these fractures can be unstable.

Severe flexion injuries often result in disruption of the ligaments and the capsular joints in addition to varying degrees of compression of the vertebral bodies or avulsion fractures. The torn ligaments may lead to bilateral facet dislocation, resulting in (*a*) *locked facets*, wherein the inferior facet of the superior subjacent vertebra is displaced from its normal relationship and wedged between the superior facet and the vertebral body of the vertebra below, or (*b*) *perched facets*, wherein the inferior facet is above and anterior to the superior facet of the adjacent vertebra below. Perched facets may result from the primary trauma or may follow "reduction" of the locked facet with traction, since in the process of reduction there is often further disruption of the capsular ligaments leading to instability even with slight flexion movement (Fig. 8.43). The instability is due to disruption of the strong ligaments in the back as well as torn capsular ligaments. The diagnosis and management with appropriate traction can be monitored by plain radiographs. Inability to reduce the locked facets with appropriate traction may be secondary to bone fragments impinging between the articular facets. Small spicules of bone which may be difficult to

Figure 8.43–*continued*

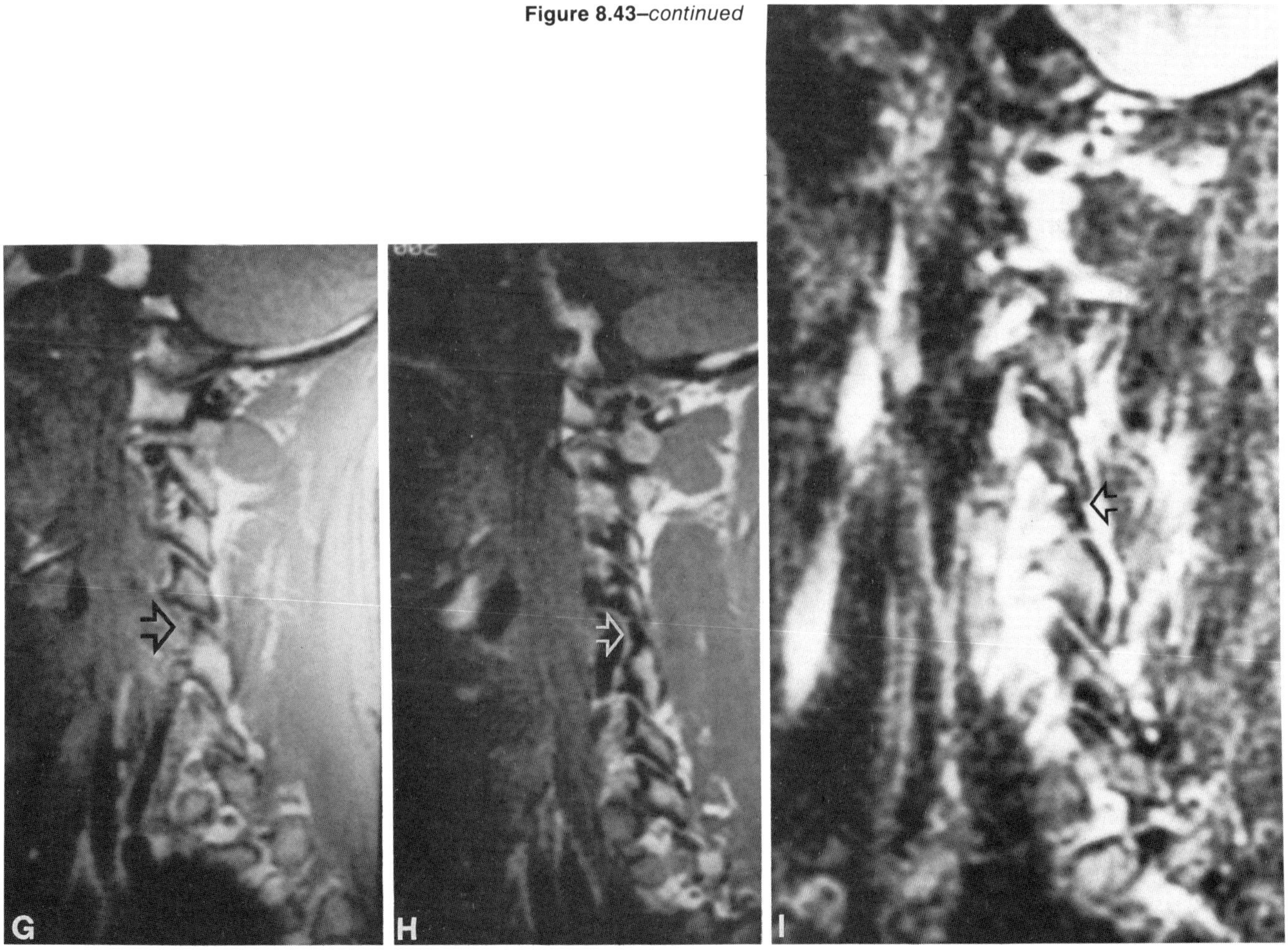

recognize on plain radiographs, linear tomography, or MRI can be demonstrated with thin axial CT sections (Figs. 8.44 and 8.45). Axial CT sections with subsequently re-formatted images are also useful in demonstrating the alignment of the facet joints and bone fragments which may compress the cord or nerve root (Fig. 8.46). MRI is usually indicated when there are neurogenic symptoms suggestive of root or cord compression following reduction. On both CT and MRI, the perched facet should be suspected in the presence of the "naked facet" sign. Even in the absence of an obvious fracture, this is a sign of instability.

Unilateral facet dislocation or *unilateral locked facet* is due to a combination of flexion and rotational stress on the adjacent vertebral segments. The inferior facet on the affected side is dislocated into the intervertebral foramen. Apart from the ligamentous injury, there is usually transverse fracture of the adjacent pillar. Several plain radiographic

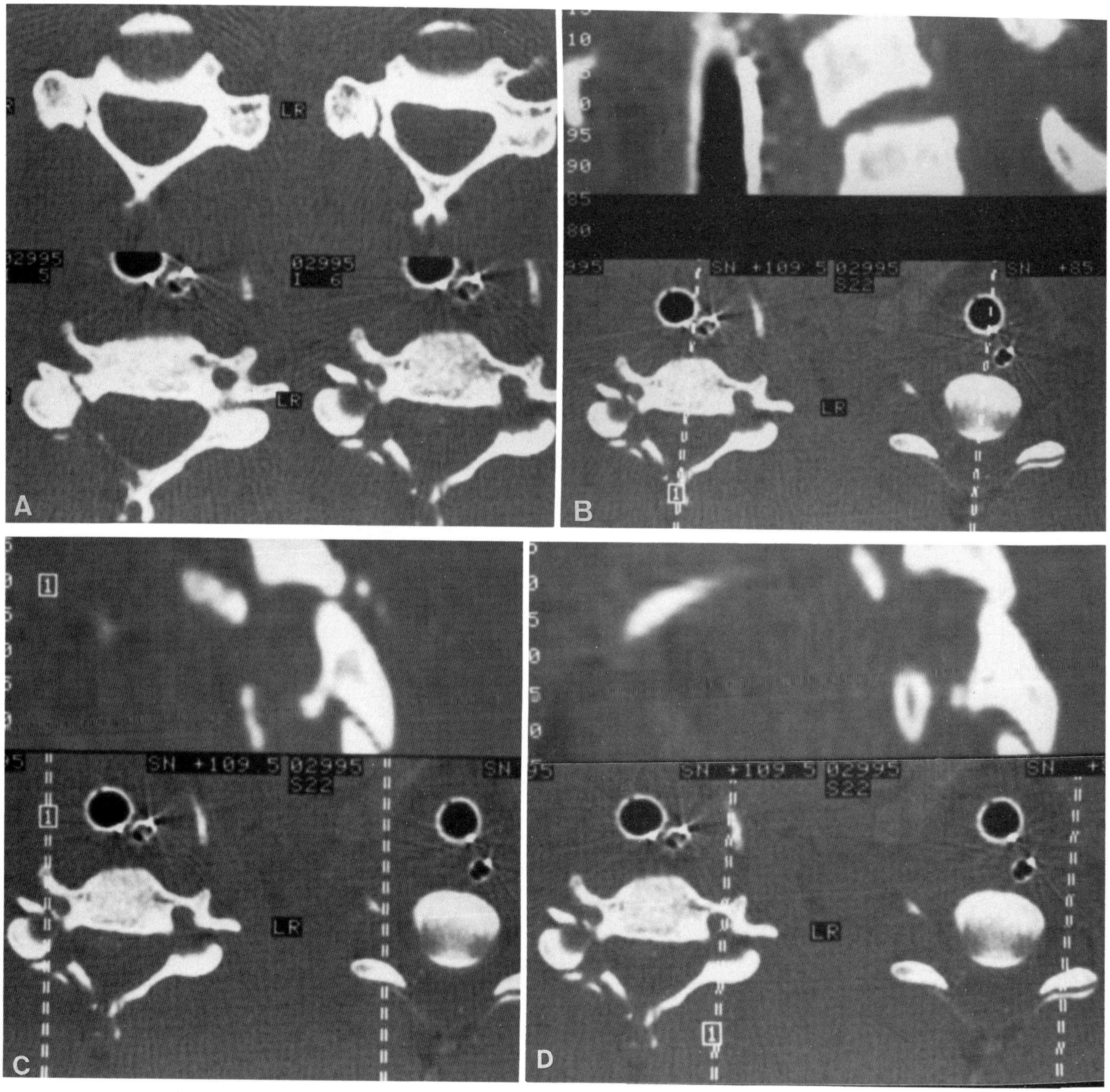

Figure 8.44. Severe flexion injury. CT scans obtained following plain radiograph to assess the extent of the fracture and bone impingement into the canal. **A,** Contiguous axial section demonstrates the fracture through the pillar on the right side, as well as fracture through the lamina. The spinal canal does not appear to be significantly encroached by the bony fragments. **B,** Sagittal midline section demonstrates the angulation and anterior subluxation of the vertebral body with increased interspinous distance. **C,D,** Note the distraction of the facet joints on both sides. The combination of these findings is indicative of an unstable fracture.

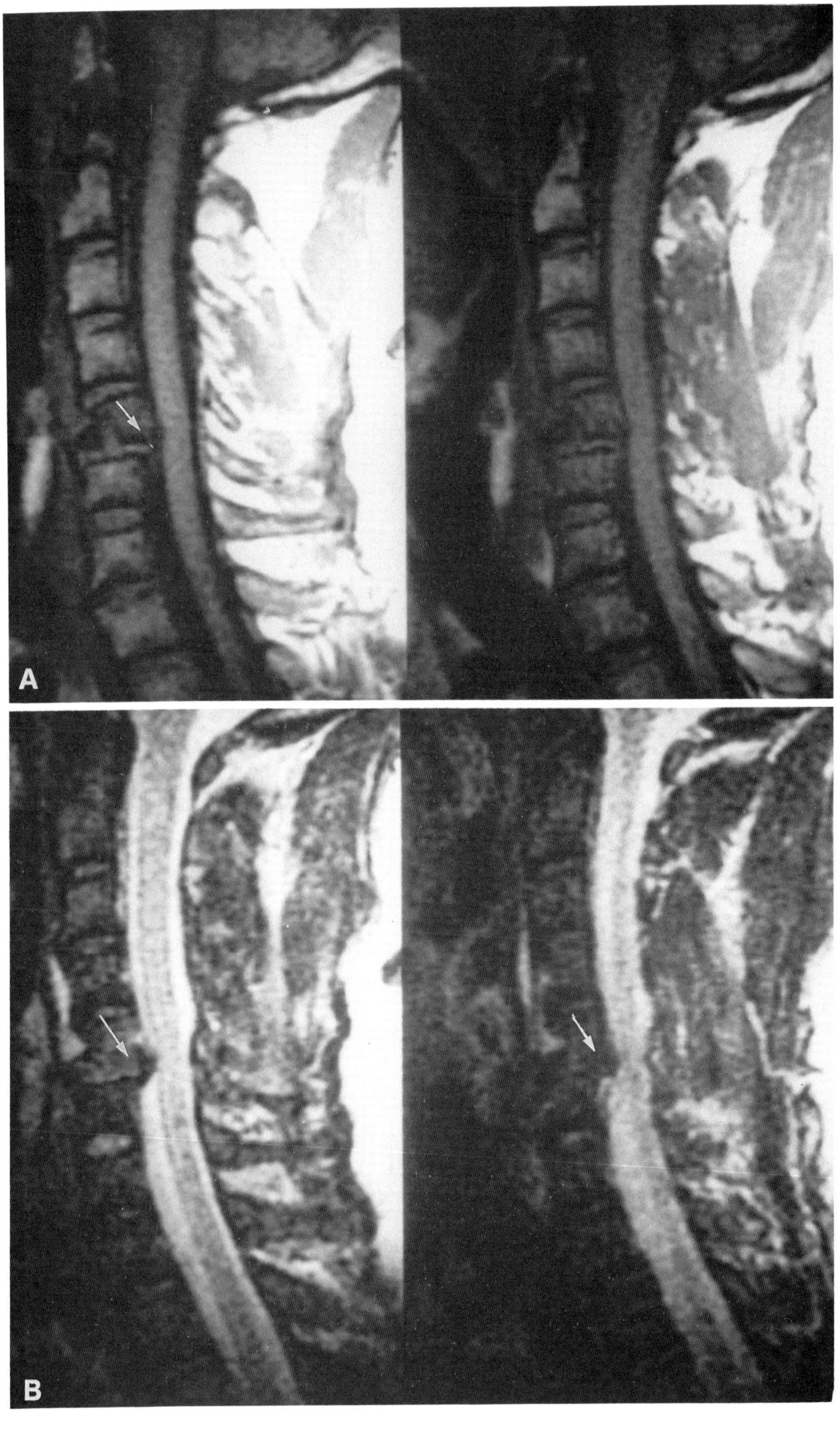

Figure 8.45. Severe flexion injury with cord compression. Sagittal T1-weighted (**A**), T2-weighted (**B**), and axial T2-weighted (**C–J**) MRI sections. The sagittal images demonstrate an avulsion fracture of the C5 vertebral body with the post segment of the vertebral body indenting the spinal cord (*arrow*). The axial sections demonstrate a linear hyperintense signal (*arrowhead*) within the vertebral body which could be mistaken for a prominent basivertebral vein. There is a fracture through the lamina (*arrow*). There is also mild focal stenosis (*open arrow*). Subsequent CT scans in the axial (**K–N**) and sagittal (**O–P**) planes demonstrate the extensive comminuted fracture of the adjacent vertebral bodies and the laminae (*arrowhead*). There is distraction of the facet joint (*small arrow*).

Figure 8.45–*continued*

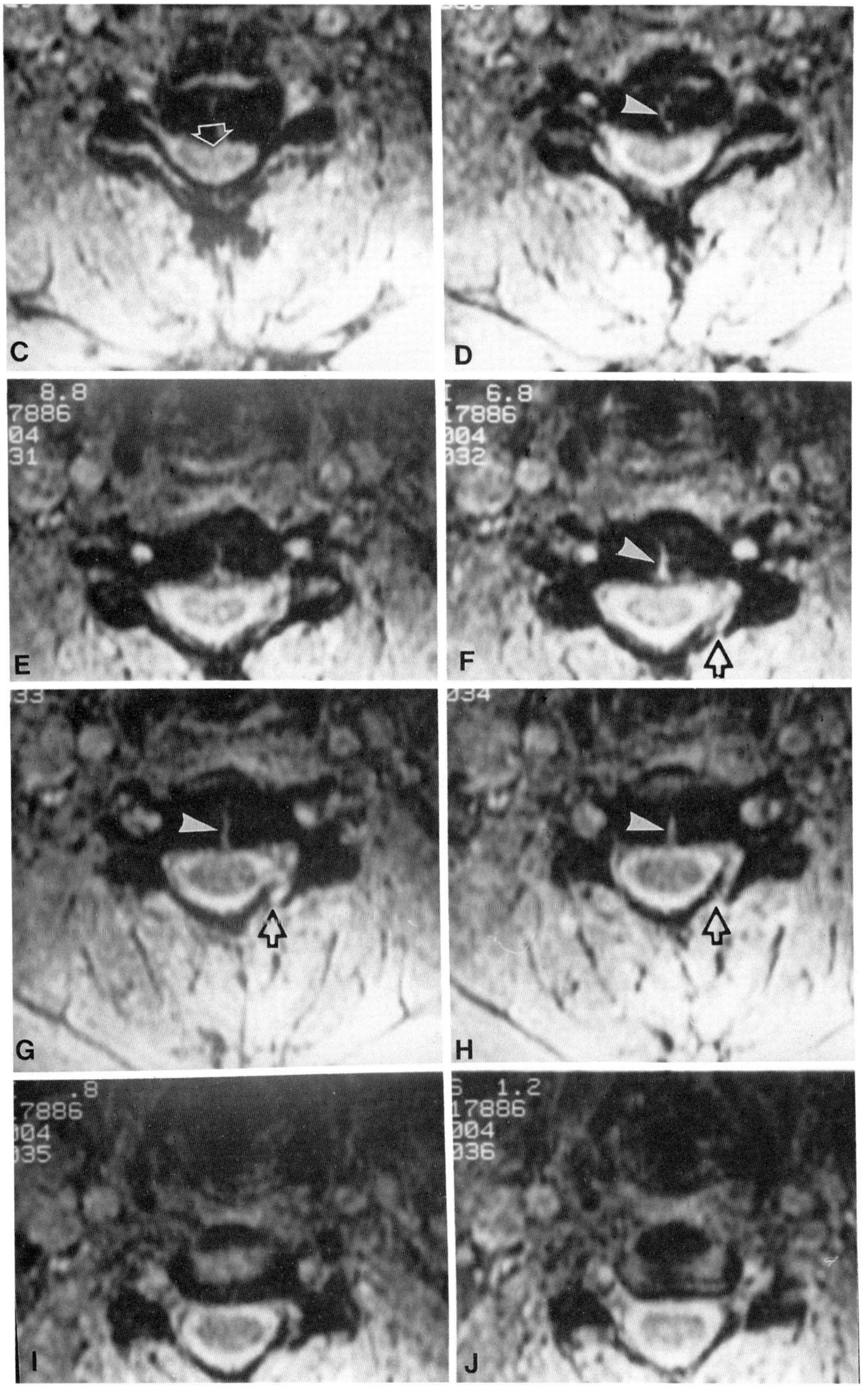

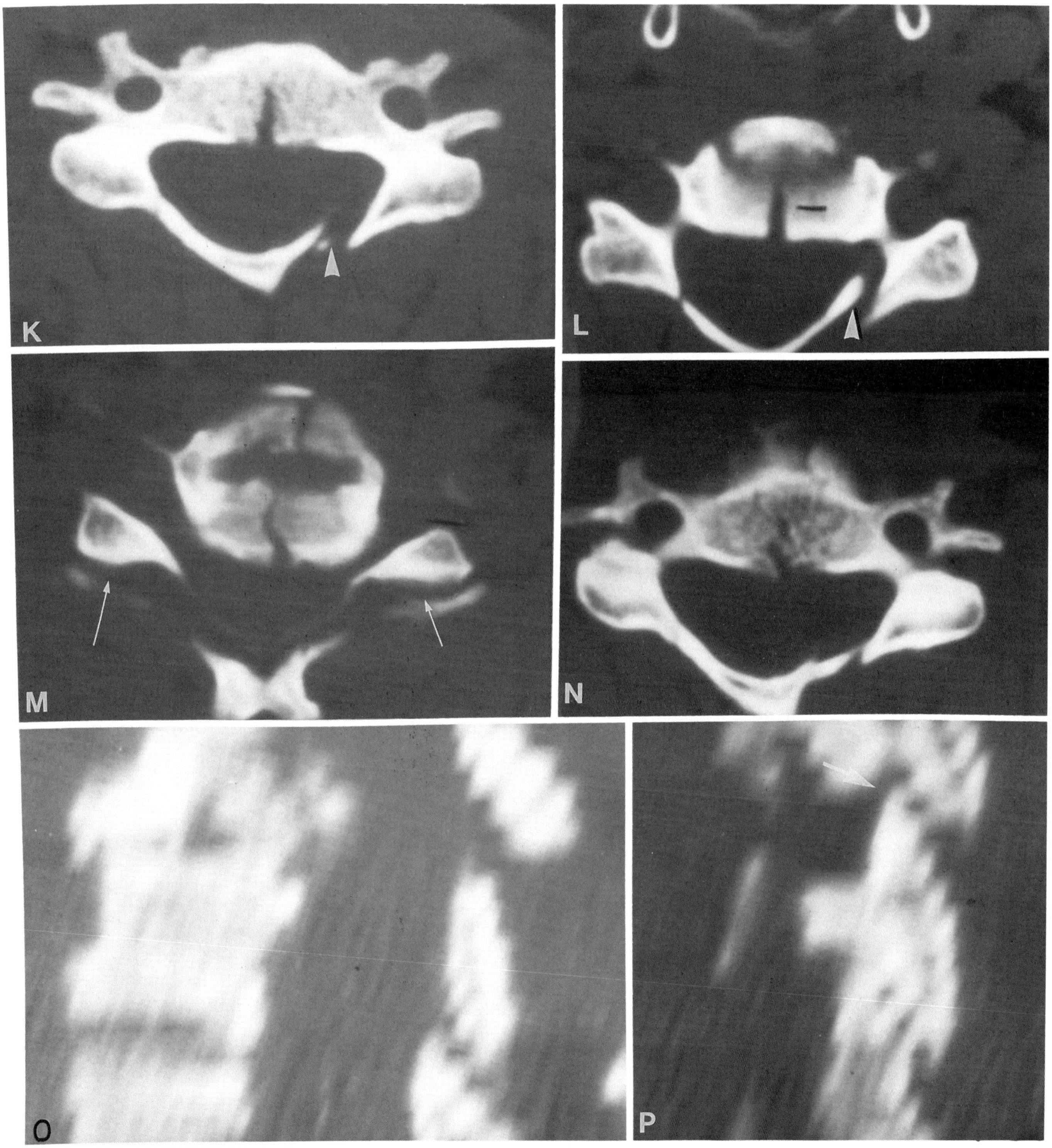

Figure 8.45–*continued*

findings have been described, the most useful in subtle cases being abrupt transition in the laminar space at the site of the dislocation on the lateral view and misalignment of the spinous process in the anteroposterior view. Polytomography or CT with sagittal and coronal re-formatted images is useful in demonstrating the foraminal encroachment by small avulsed fragments or to demonstrate the associated pillar fracture (Fig. 8.47). The pillar fracture is often part of the original trauma but rarely may follow reduction of the locked facet by traction.

EXTENSION INJURIES

In adults, extension injuries of the cervical spine are often seen in association with MVAs. They result from extension of the weight of the head away from a rigid trunk. The trauma may result in disruption of the fibers of the anterior longitudinal ligament, the adjacent annulus, or at the insertion of the Sharpey's fibers. This results in avulsion fractures of the anterosuperior and inferior aspects of the vertebral body. In more severe extension injuries or when direct force is applied to the back of the neck, associ-

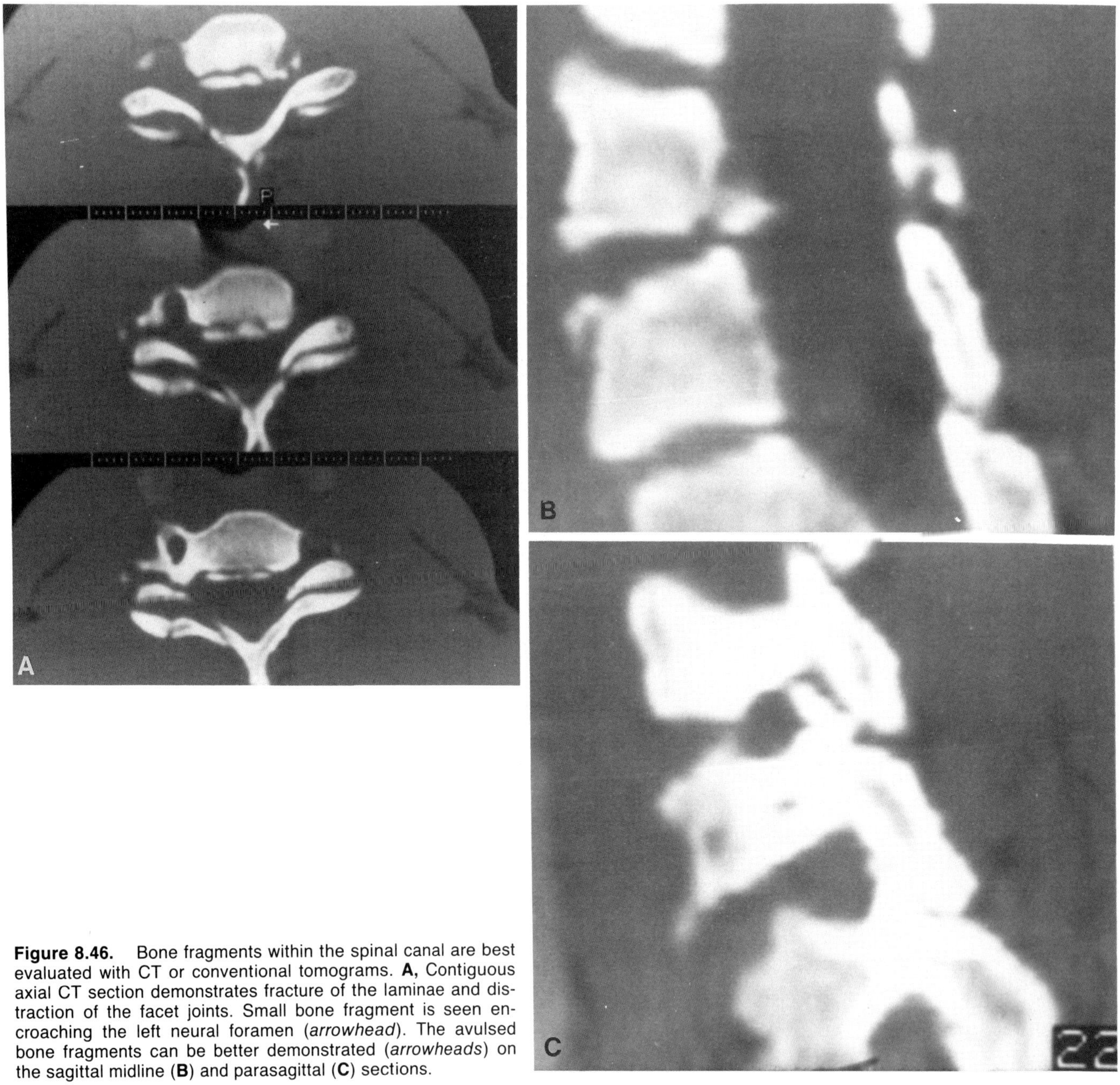

Figure 8.46. Bone fragments within the spinal canal are best evaluated with CT or conventional tomograms. **A,** Contiguous axial CT section demonstrates fracture of the laminae and distraction of the facet joints. Small bone fragment is seen encroaching the left neural foramen (*arrowhead*). The avulsed bone fragments can be better demonstrated (*arrowheads*) on the sagittal midline (**B**) and parasagittal (**C**) sections.

ated fractures of the lamina, facets, and spinous process are not uncommon.

Most hyperextension injuries of the cervical spine can be detected by plain radiographs. These include widening of the disk space due to a torn anterior longitudinal ligament, the occasional presence of a vacuum disk at the level of the injury due to avulsion of the endplate and annulus, and bony disruption as seen with teardrop fractures. CT or MRI is useful in hyperextension injuries when following stabilization there is persistence or worsening of the neurologic symptoms. CT demonstrates small avulsion fractures and their relationship to the vertebral body. In more severe injuries, often seen with hyperextension injuries with a compressive element (burst fracture), the fracture through the vertebral body and its impingement on the spinal canal can be evaluated with plain radiographs or CT. MRI is useful in assessing the presence of intrinsic cord lesions.

WHIPLASH INJURY

Whiplash injury results from a combination of both hyperflexion and hyperextension of the cervical

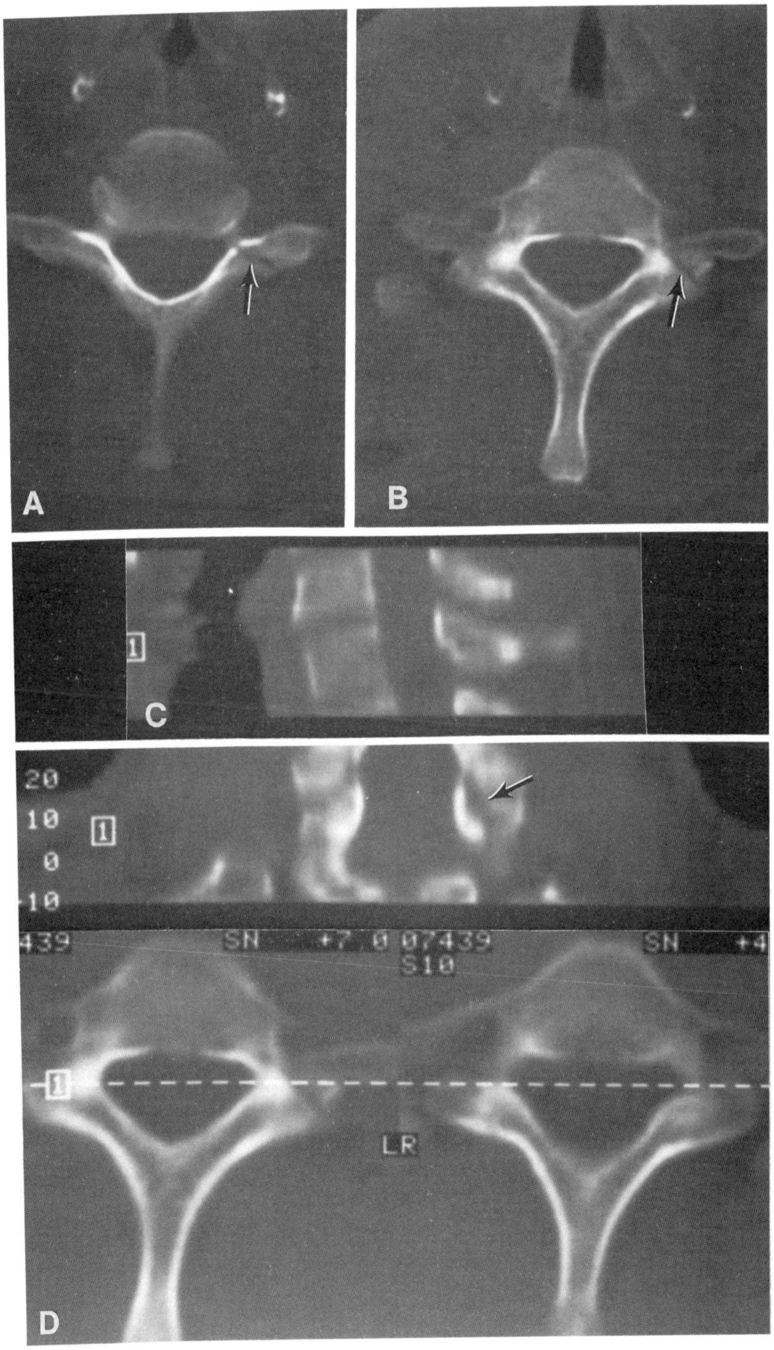

Figure 8.47. Pillar fracture. Axial (**A,B**), midline sagittal (**C**), and coronal (**D**) CT sections demonstrate a pillar fracture (*arrowhead*) on the left side.

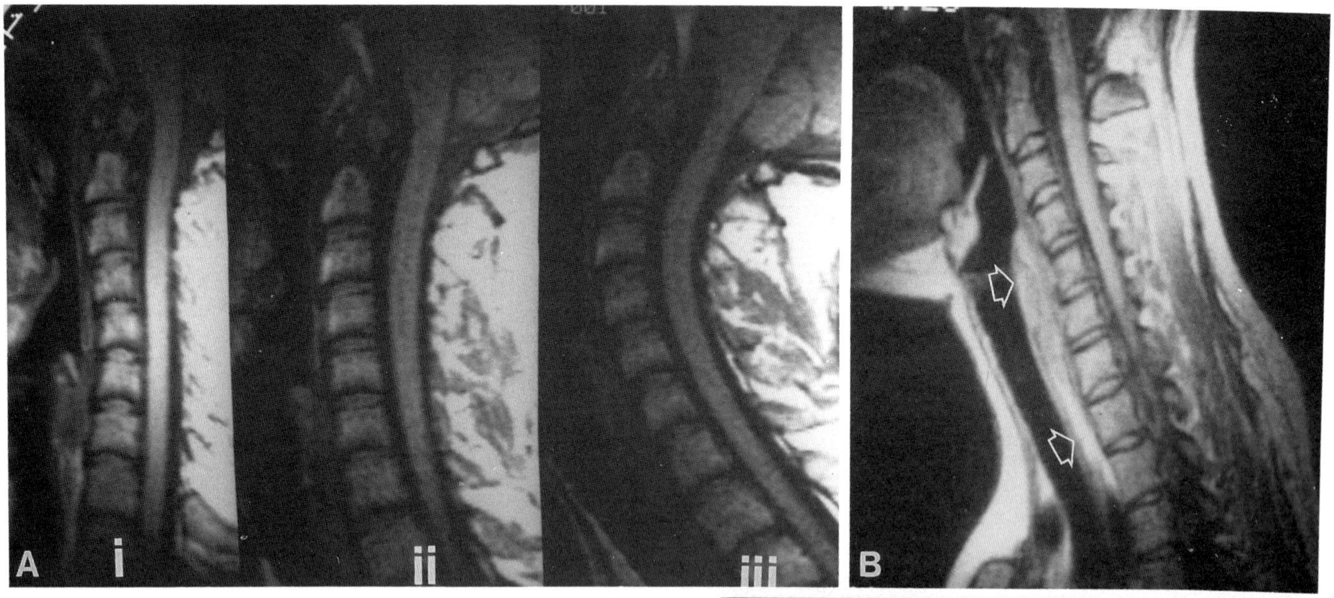

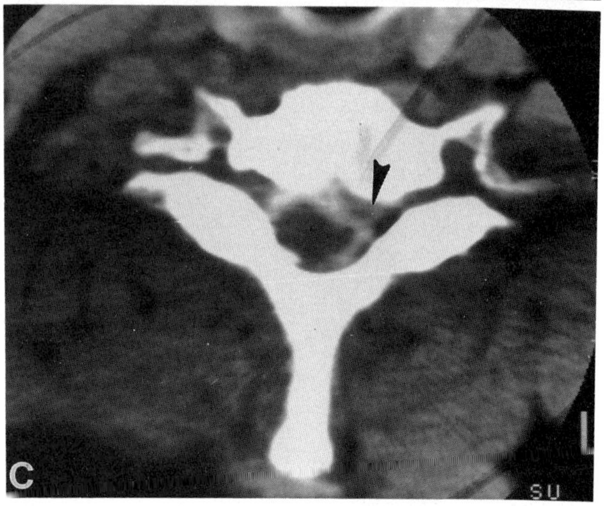

Figure 8.48. Whiplash (flexion-extension) injury. A, Similar to plain x-rays, T1-weighted midsagittal MRI section can be obtained in (**i**) flexion, (**ii**) neutral, and (**iii**) extension. Torn or disrupted ligaments, instability of the spine, disk herniation, or cord contusion can be demonstrated in a noninvasive manner. **B,** MRI performed following MVA-related whiplash injury. T1-weighted midsagittal MRI. The alignment of the vertebral bodies are normal. There is prominent prevertebral soft-tissue edema (*arrowheads*). **C,** Axial CTM a few days following MRI demonstrates a herniated disk (*arrowhead*). IVCT (**D**) and T1-weighted (**E**) and T2-weighted (**F**) midsagittal MRI sections in another patient with subligamentous epidural hematoma 6 days following whiplash injury. On the axial CT section, the enhancing margin demarcates the hematoma (*arrow*) from the thecal sac. On MRI, the focal epidural hematoma (*arrowhead*) has a bright signal on both pulse sequences.

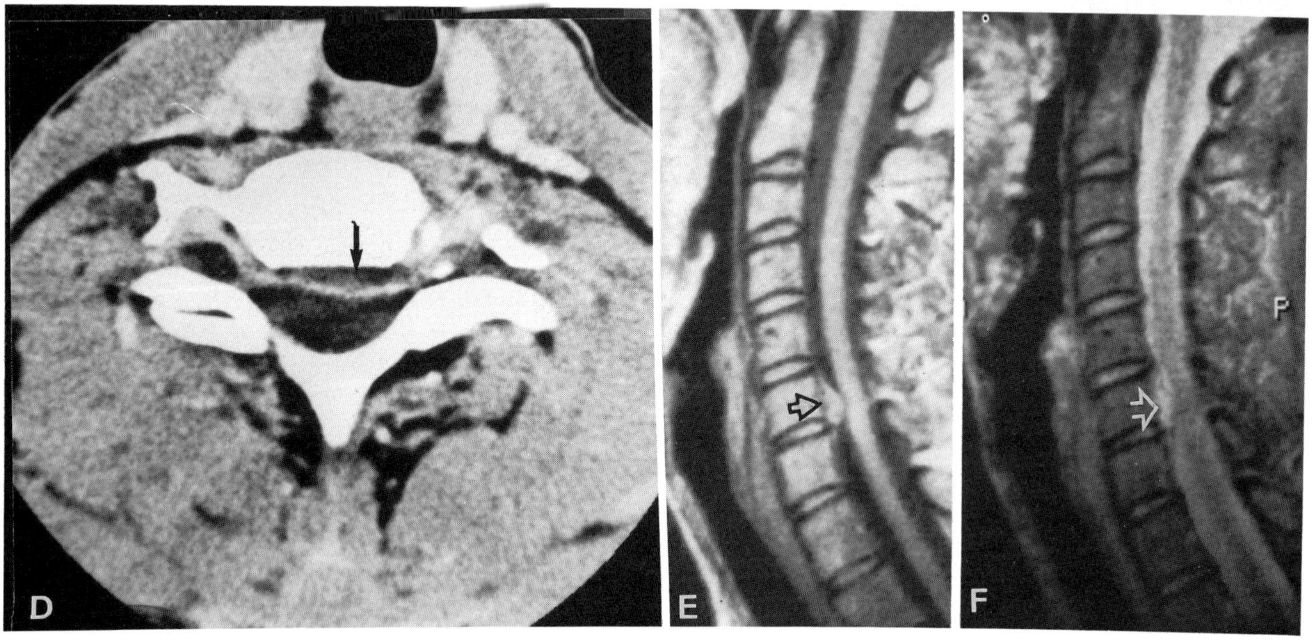

spine, usually from MVAs and most often encountered in the adult population. Depending on the severity of the vector forces involved, fractures may be identified on plain radiographs. Plain x-rays should include flexion-extension views. CT is of limited value, although prevertebral soft-tissue swelling may be identified in occasional cases. MRI is ideal in evaluating patients with whiplash injuries presenting with neurologic symptoms. Sagittal T1-weighted MRI sequences in neutral, flexion, and extension views allow evaluation of the soft-tissue trauma as well as the osseous component (Fig. 8.48). In the acute stage, the prevertebral soft-tissue swelling is seen as hyperintense signal on T2-weighted or GRE images. Similar hyperintense signal may be present within the posterior longitudinal ligament. Differentiation of focal subligamentous hematoma from focal epidural hematoma may be difficult. Edema, contusion, or hemorrhage within the spinal cord also can be identified (Fig. 8.49). Posttraumatic disk herniation

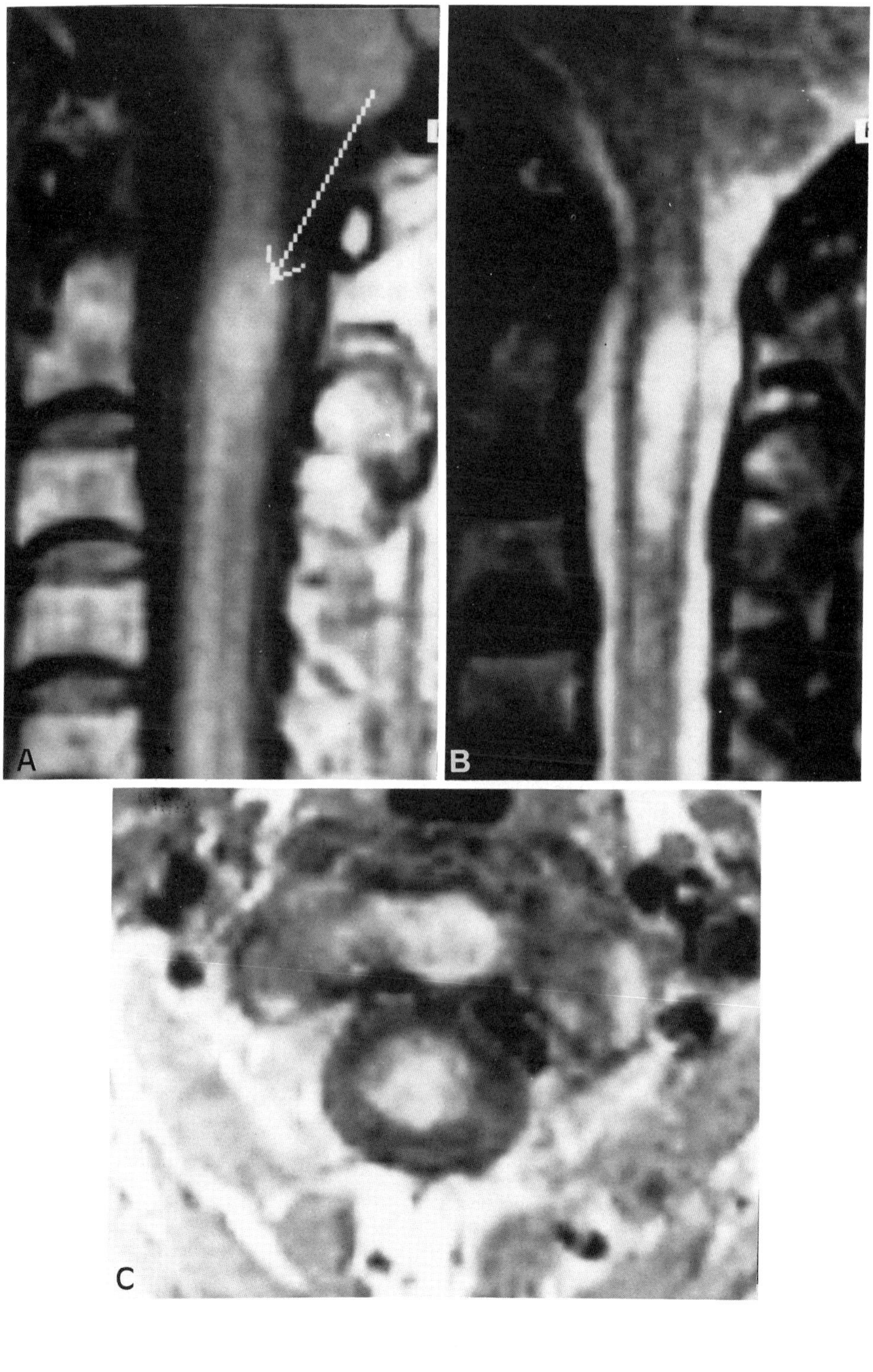

Figure 8.49. Cord contusion following severe whiplash injury. This 48-year-old woman developed a progressive neurological deficit 4 days following an MVA. Plain radiographs were normal. T1-weighted (**A**), T2-weighted (**B**), and axial T1-weighted (**C**) MRI sections. A focal hemorrhagic contusion of the cord is present. The alignment of the vertebrae appears normal. There was minimal prevertebral soft-tissue swelling.

with cord or nerve root compression is not uncommon. Posttraumatic disk herniation is often associated with subligamentous focal hematoma or epidural hematoma in the initial assessment of injury, which is best evaluated with MRI (Fig. 8.50). The disk herniation may occur immediately following the whiplash injury or a few days or weeks later.

In the aged population with preexisting cervical spondylitic disease and superimposed osteoporosis, even minor trauma may result in neurologic symptoms. Compression of the vertebral bodies and thickening of the ligaments is best evaluated with MRI. However, it is difficult to differentiate the focal bright signal associated with cord contusion from the focal bright signal seen in chronic cord compression with myelomalacia.

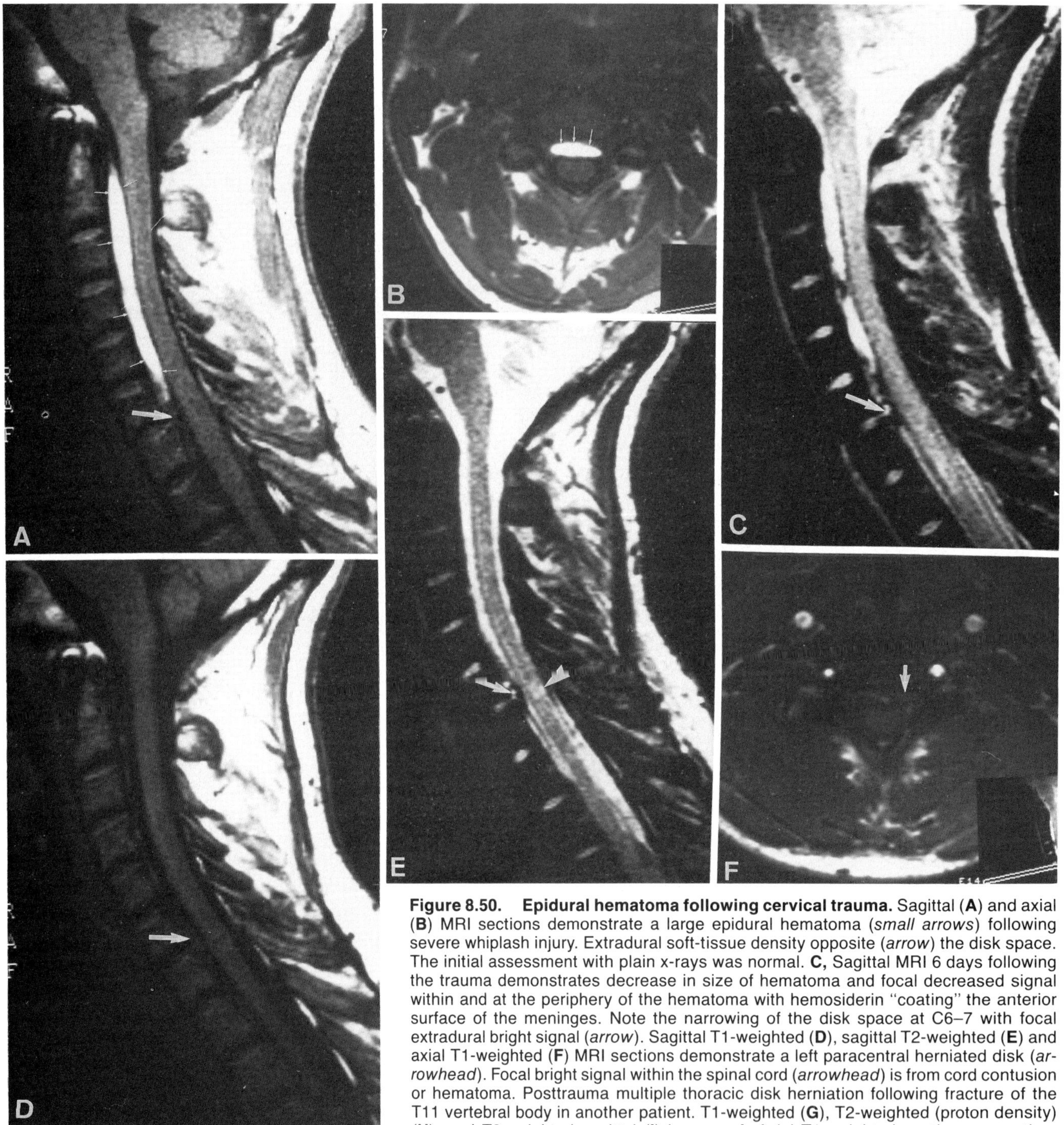

Figure 8.50. Epidural hematoma following cervical trauma. Sagittal (**A**) and axial (**B**) MRI sections demonstrate a large epidural hematoma (*small arrows*) following severe whiplash injury. Extradural soft-tissue density opposite (*arrow*) the disk space. The initial assessment with plain x-rays was normal. **C,** Sagittal MRI 6 days following the trauma demonstrates decrease in size of hematoma and focal decreased signal within and at the periphery of the hematoma with hemosiderin "coating" the anterior surface of the meninges. Note the narrowing of the disk space at C6–7 with focal extradural bright signal (*arrow*). Sagittal T1-weighted (**D**), sagittal T2-weighted (**E**) and axial T1-weighted (**F**) MRI sections demonstrate a left paracentral herniated disk (*arrowhead*). Focal bright signal within the spinal cord (*arrowhead*) is from cord contusion or hematoma. Posttrauma multiple thoracic disk herniation following fracture of the T11 vertebral body in another patient. T1-weighted (**G**), T2-weighted (proton density) (**H**), and T2-weighted sagittal (**I**) images. **J,** Axial T1-weighted contiguous section. Note the herniated disks (*arrows*) both above and below the vertebral body fracture.

ATLANTOAXIAL DISLOCATION

Atlantoaxial dislocation without fracture is usually the result of severe distraction forces associated with hyperflexion and extension. Fractures usually involve the dens (Fig. 8.51). However, if the transverse ligament is weakened from any other cause, disruption or tear of the ligament from a flexion-rotation type of injury results in *rotary subluxation*. This type of injury is uncommon in adults. Normal rotation of the neck at the atlantoaxial joint is about 45 degrees on each side. If the rotational forces exceed 45 degrees, unilateral or bilateral dislocation of the articular process occurs. If suspected on plain radiographs, it is best evaluated by tomography or CT with sagittal and coronal re-formatted images [136]. If the transverse ligament is disrupted on only one side, this results in *rotary fixation injury*, with the patient presenting with torticollis and an inability to turn the neck. Rotary fixation may be idiopathic and not related to trauma. Thin-section CT scans with the head in a neutral position and turned to either side should be obtained. In normal individuals and in patients with spasmodic torticollis, there is a decrease in the soft-tissue space between the dens and the lateral mass of C1 on the side the head is turned. With rotary fixation injury, the increased space in the soft tissues between the dens and the

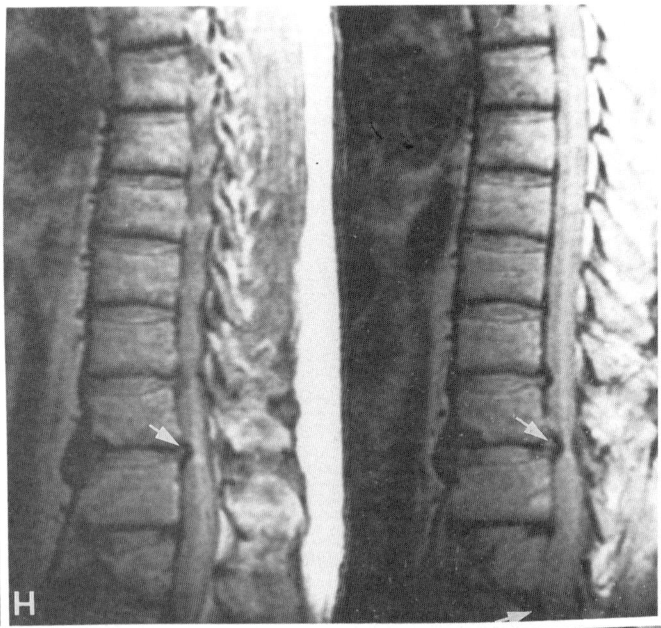

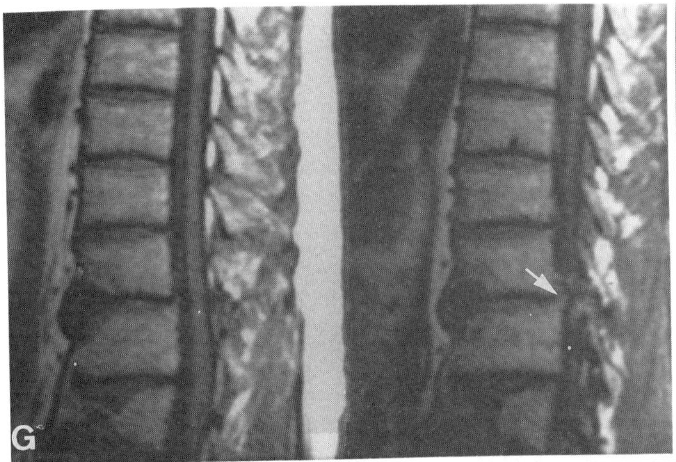

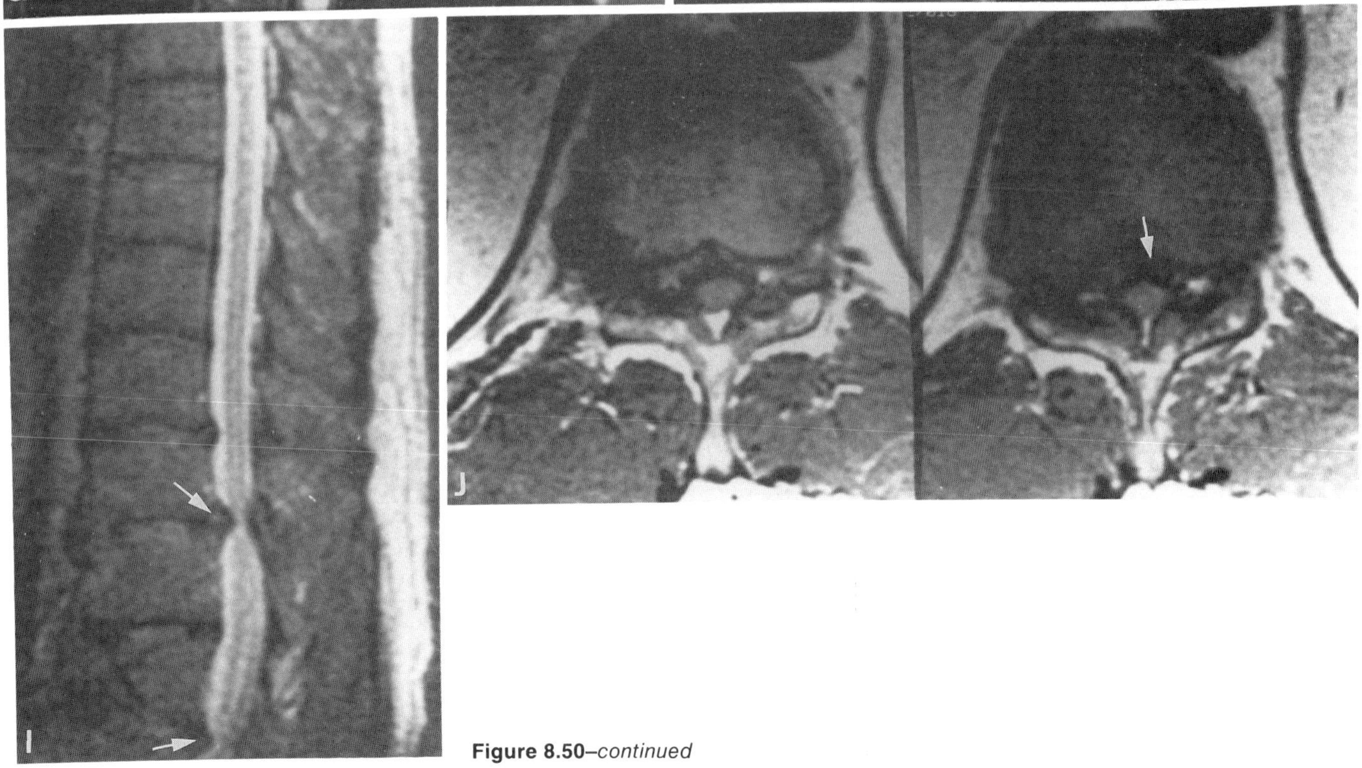

Figure 8.50–*continued*

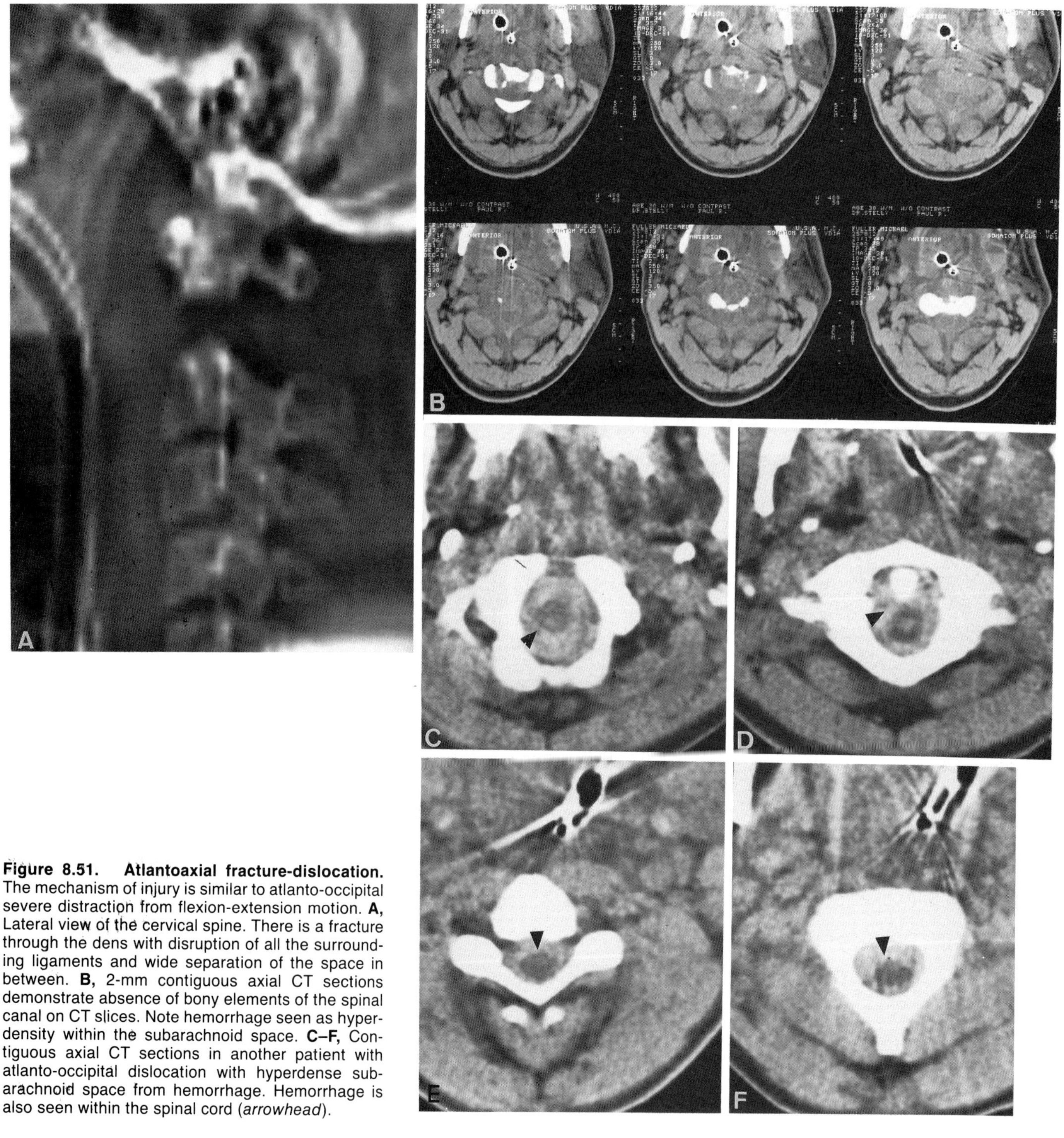

Figure 8.51. Atlantoaxial fracture-dislocation. The mechanism of injury is similar to atlanto-occipital severe distraction from flexion-extension motion. **A,** Lateral view of the cervical spine. There is a fracture through the dens with disruption of all the surrounding ligaments and wide separation of the space in between. **B,** 2-mm contiguous axial CT sections demonstrate absence of bony elements of the spinal canal on CT slices. Note hemorrhage seen as hyperdensity within the subarachnoid space. **C–F,** Contiguous axial CT sections in another patient with atlanto-occipital dislocation with hyperdense subarachnoid space from hemorrhage. Hemorrhage is also seen within the spinal cord (*arrowhead*).

arch of the atlas remains unchanged. Pleuridirectional tomography in the coronal plane or an axial CT section is ideal in demonstrating the ligamentous injury.

ATLANTO-OCCIPITAL DISLOCATION

Atlanto-occipital dislocation is a severe form of hyperextension-flexion injury resulting in disruption of the strong ligaments between the lateral masses of the atlas and the occipital condyles. It is often fatal, although a few patients may survive. Plain radiographs demonstrate the fracture-dislocation, and in most circumstances, this injury does not require evaluation with CT or MRI. However, if CT is the preliminary imaging study, atlanto-occipital dislocation usually demonstrates absence of vertebral segments in contiguous axial sections. This is usually associated with extensive hemorrhages in the meninges surrounding the spinal cord or medulla (Fig. 8.51).

Thoracic and Lumbar Spine

Trauma to the thoracic and lumbar regions results from vector forces acting either in *hyperextension* or *hyperflexion*, usually combined with *rotational*, *shearing*, or *compressive* elements. Such injuries are more common in the young athletic individual or as a result of MVAs in other age groups.

Compression fractures are usually secondary to axial loading injuries such as in diving accidents. In most cases, such fractures are fatal, often associated with head injuries. Compression fractures often involve the thoracolumbar region. *Rotational forces* often associated with other vector forces result in torn ligaments as well as fractures involving the posterior column, such as the articular facets and laminae. Rotational forces when combined with *shearing* often result in fracture-dislocation in the thoracolumbar region (Fig. 8.52).

The most common site of fracture in the thoracolumbar region is between T11 and L3 [137–138]. In nearly half these patients there is injury to other organ systems or fractures at other levels. Apart from plain radiographs, thoracic and thoracolumbar fractures are best evaluated with CT or CTM [138–139]. In severe fracture-dislocation or when more than one segment of the vertebrae are involved, such injuries are usually associated with chest or abdominal trauma.

Hyperflexion injuries are usually associated with some element of compression resulting in anterior

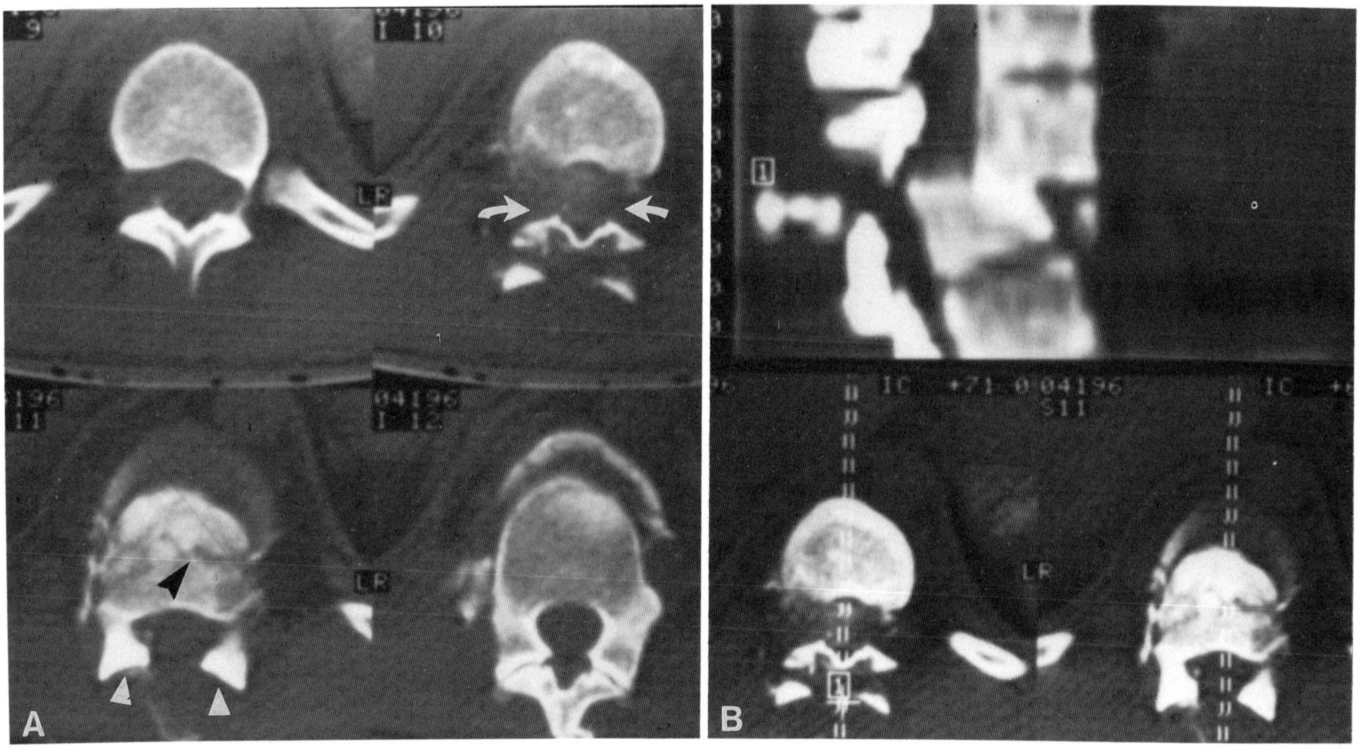

Figure 8.52. Compression with rotational element in thoracolumbar region. A, Contiguous axial CT sections demonstrate fracture through the vertebral body (*arrowhead*). There is apparent increase in the width of the neural foramen (*small arrow*). The dislocated adjacent vertebral body is displaced. Naked facet sign (*white arrowhead*). **B,** Sagittal re-formatted midline section demonstrates the unstable fracture dislocation of the T12 vertebrae and impingement within the spinal canal. Although this information can be obtained with cross-table lateral x-rays, thin-section CT with reformatted image allows better evaluation of the osseous component.

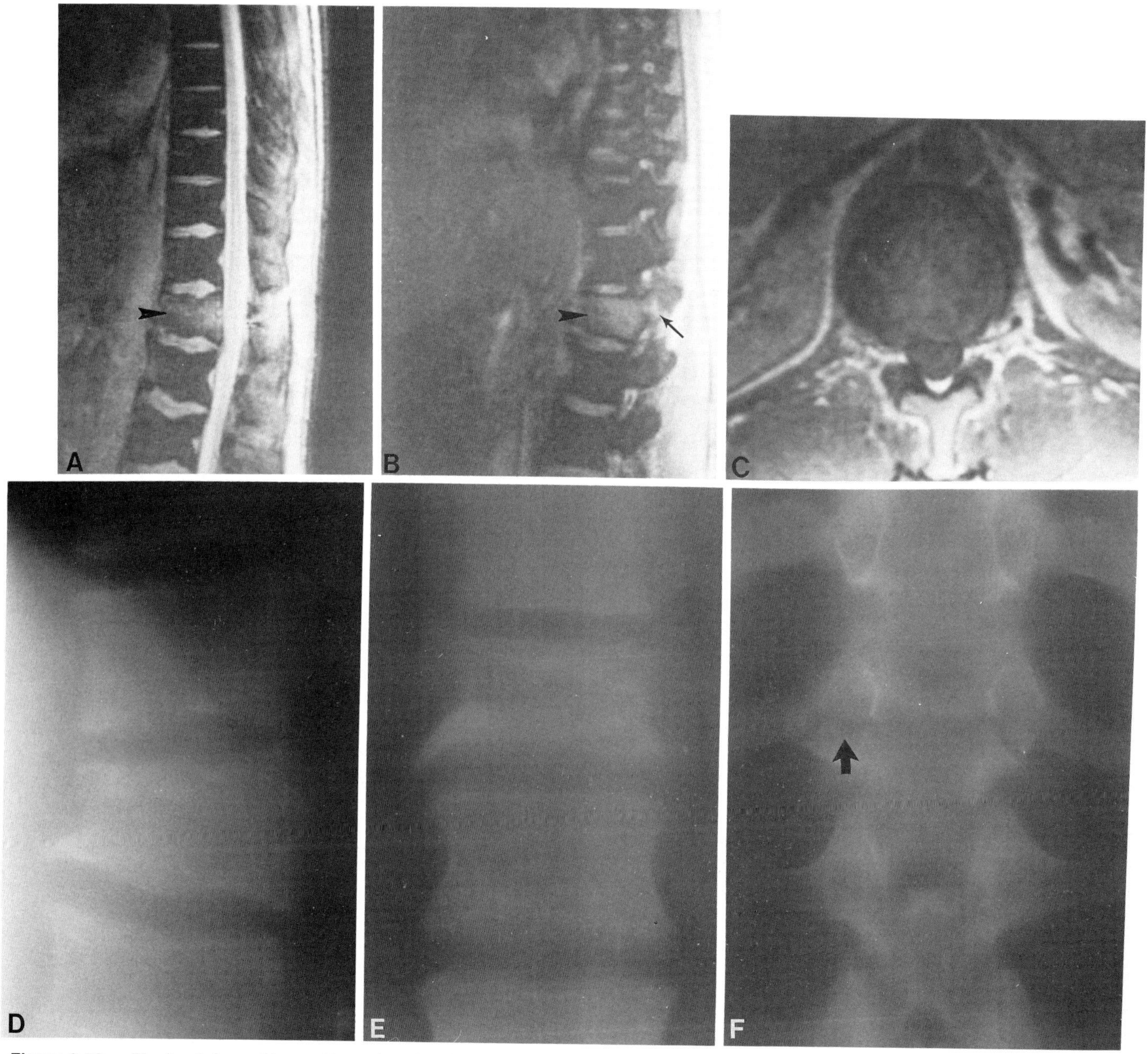

Figure 8.53. Flexion injury of lower thoracic spine. Initial plain x-rays demonstrated mild anterior wedging of the T12 vertebral body. Sagittal (**A**) and parasagittal (**B**) MRI sections demonstrate the anterior wedging of the vertebral body with mild hyperintense signal within the marrow spaces (*arrowhead*) on the T2-weighted images. Focal bright signal within the region of the pedicle (*arrow*) is suggestive of fracture through the pars involving the articular facets. **C,** T1-weighted axial MRI demonstrates mild impingement of the cord by the expanded vertebral body. **D,** Tomography in the sagittal plane demonstrates the less than 50% loss in height of the anterior aspect of the vertebral body. Coronal sections through the vertebral body (**E**) and through the posterior elements (**F**) demonstrate the fracture through the pedicle (*arrow*). Axial CTM (**G**) and sagittal re-formatted image (**H**) clearly demonstrate the stable compression fracture and the extent of cord compression.

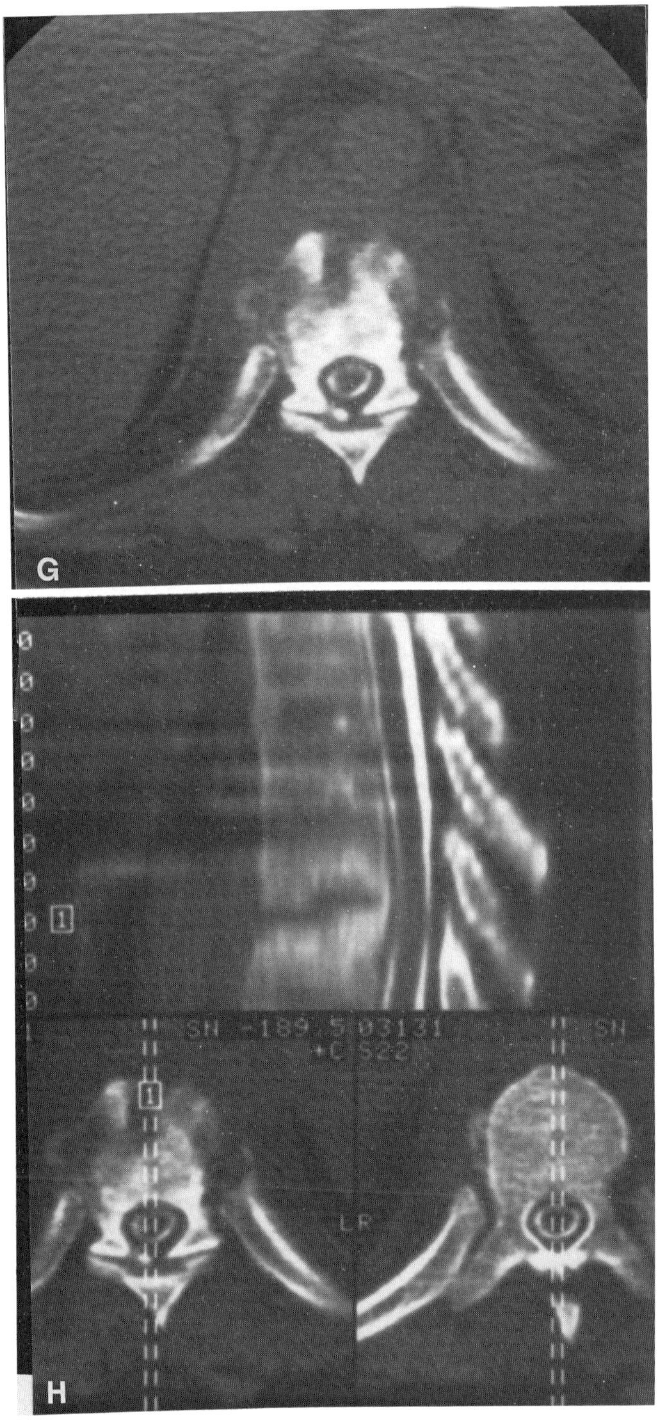

Figure 8.53–*continued*

wedging of the vertebral body. These injuries are more common in the lower lumbar and midthoracic vertebral bodies. More severe compressive forces also may lead to posterior wedging. Most of these fractures are stable if the posterior longitudinal ligament is intact and there is minimal, if any, facet distraction. Plain radiographs are adequate, although CT or MRI is useful if there is persistent pain or root symptoms or if on follow-up studies there is progressive compression of the vertebral body. In these circumstances, sagittal and coronal images allow evaluation of the retropulsed midportion of the vertebral segment or of bony fragments encroaching on the spinal canal (Fig. 8.53).

Depending on the extent of compression and retropulsion of the bone fragments within the spinal canal, burst fractures have been classified into three types: stage 1, when there is less than 50% loss in height of the anterior segment without any retropulsion; stage 2, when the anterior compression results in loss in height of more than 50%; and stage 3, when there is more than 50% loss in height with encroachment of the spinal canal [140]. The stability of the spinal column depends on an intact middle complex such as the pedicles and the posterior longitudinal ligament. The extent of retropulsion of the bony fragments encroaching within the spinal canal and the absence of involvement of the middle segment of the vertebrae are best evaluated with axial CT sections (Fig. 8.54). CTM allows identification of the soft tissues within the spinal canal. If there is no contraindication, this is complemented with MRI studies. Both these studies allow evaluation of the distal cord and the nerve roots.

Hyperflexion injuries with shearing or rotational vector forces result not only in fractures involving the vertebral bodies in the horizontal or vertical plane but also in fractures extending into the pedicle and adjacent facet joint. Plain radiographs in the frontal projection demonstrate the lateral displacement of segments of the vertebral body, with the posterior elements being splayed apart. Because of involvement of the middle column, most of these fractures are unstable. Invariably, in most cases the posterior longitudinal ligament is torn or disrupted with severe neurologic deficits. There is also associated visceral injury. The initial assessment and stabilization are based on plain radiographic studies and CT scans (see Fig. 8.54). In the majority of the cases, postoperative evaluation of residual or progressive neurologic deficits are best achieved with CTM, since MRI images may be degraded by artifacts from metallic prosthetic devices.

Penetrating and Avulsion Injuries

The common factor in both these types of injuries is contusion or transection of the spinal cord or nerve roots and disruption of the meninges. Associated dural tears are common, leading to CSF leak. Penetrating injuries are the result of stab wounds, which usually result in trauma to the soft tissues within the canal and less often are associated with fractures. Avulsion injuries are the result of distracting forces tearing the meninges with associated cord or root trauma. In both these injuries the osseous

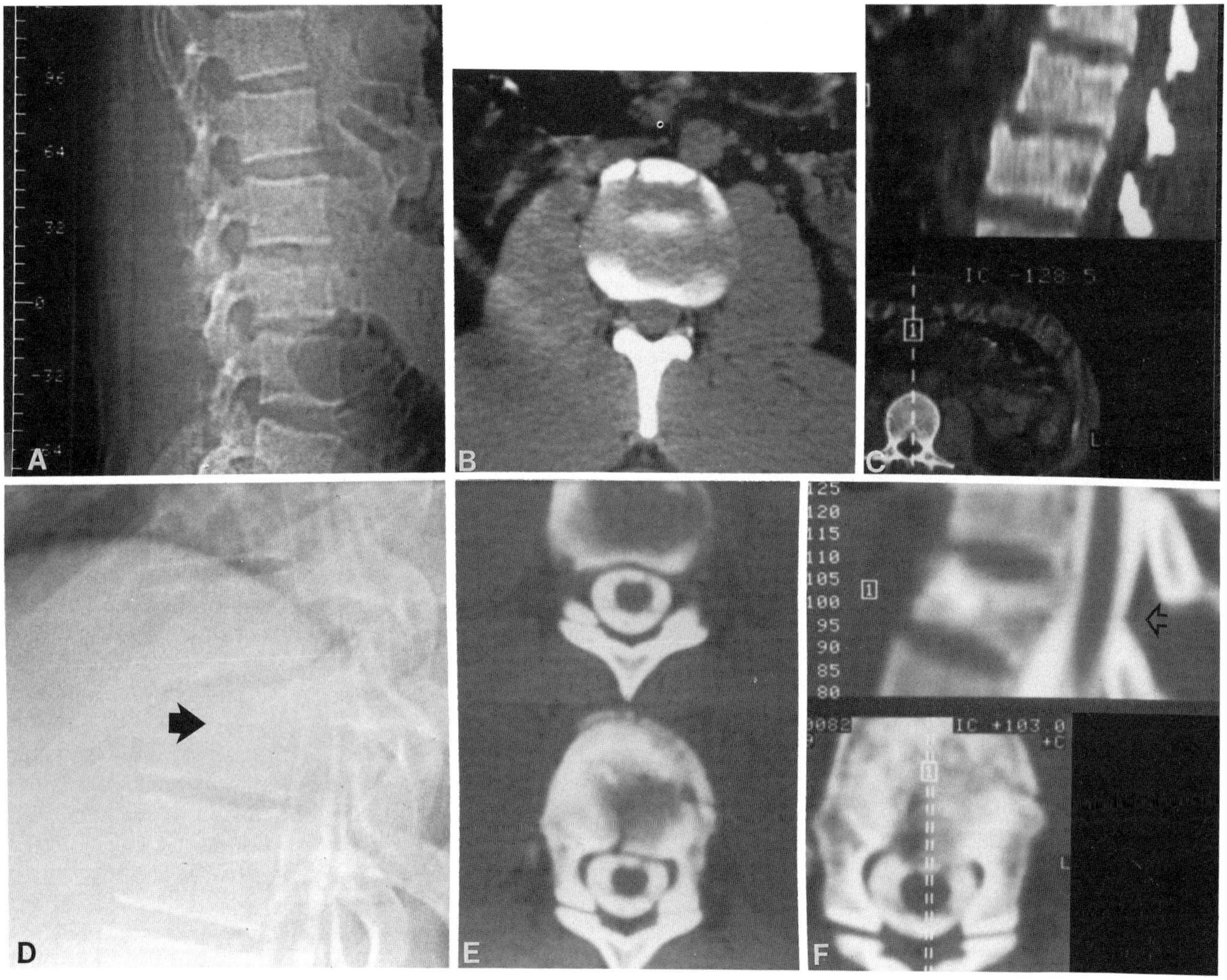

Figure 8.54. Burst fractures due to flexion and compression vector forces. Type 1 fracture. **A,** CT scout view demonstrates mild loss in height of the L3 vertebral body. **B,** Mild anterior disruption of the endplate and cortical margin of the vertebrae. **C,** Sagittal re-formatted image demonstrates the anteroposterior expansion of the vertebral body due to the compressive force. Type 2 fracture with more than 50% loss in height. Lateral view of myelogram (**D**), contiguous axial CTM (**E**), sagittal re-formatted image (**F**) demonstrate the compression fracture of the vertebral body with sparing of the middle and posterior thirds of the vertebral elements. The interspinous ligament (*arrowhead*) is torn. Type 3 burst fracture with fracture extending posteriorly and bone fragments within the spinal canal. Contiguous axial CT sections (**G**), sagittal (**H**) and coronal (**I**) re-formatted images demonstrating the comminuted nature of the fracture and the extent of compromise of the spinal canal.

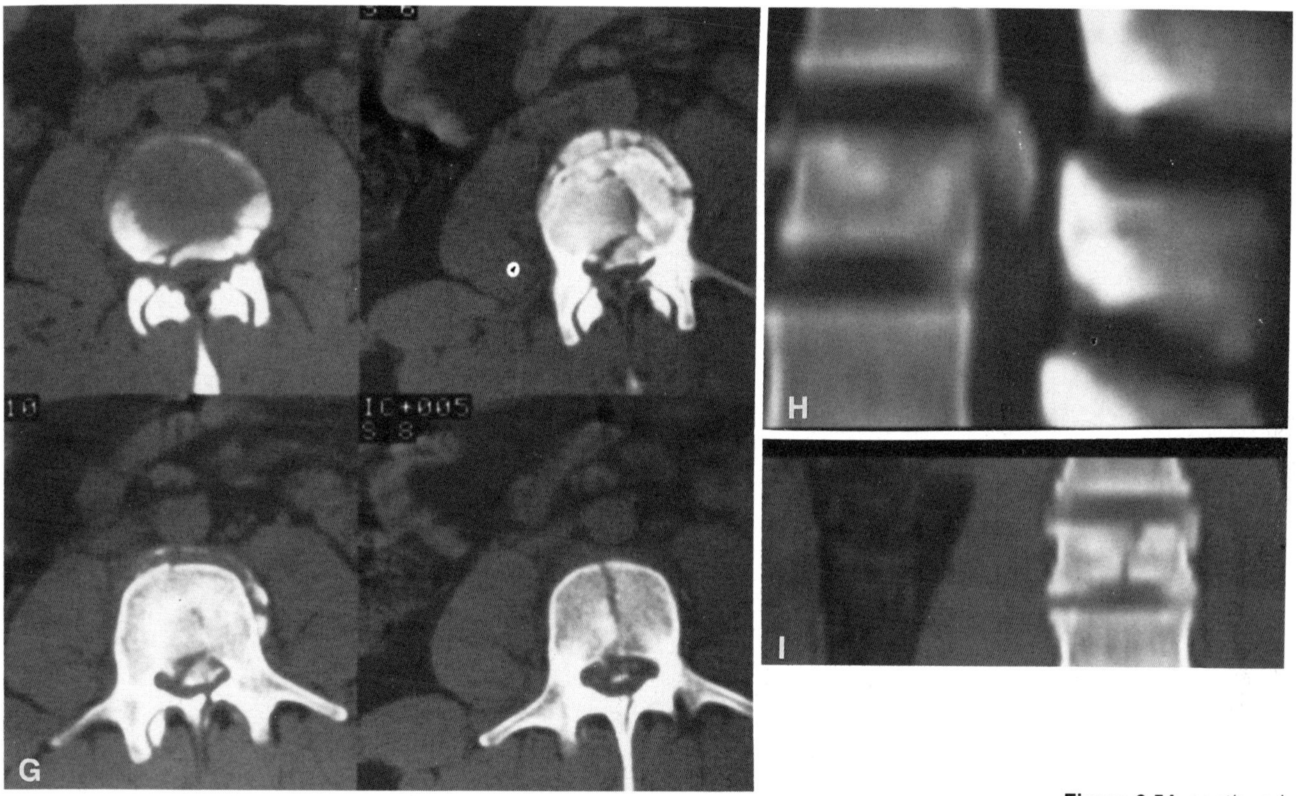

Figure 8.54–*continued*

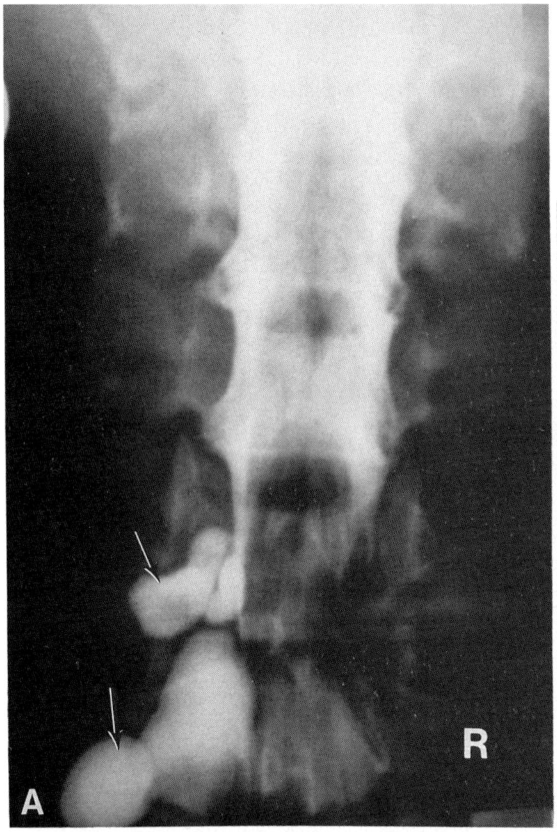

Figure 8.55. Dural tear. A, Frontal view of cervical myelogram demonstrates extravasation of contrast material from tear of the dura resulting from avulsion injury to the brachial plexus in a patient involved in brachial plexus injury from distraction forces. Axial CTM (**B**) and Gd-DTPA–enhanced T1-weighted MRI (**C**) demonstrating the dural tear (*arrow*). **D,** Myelogram in another patient with stab wound in the lumbar region demonstrates an extradural collection of contrast materials. **E,** Axial CTM shows the extravasation of contrast material.

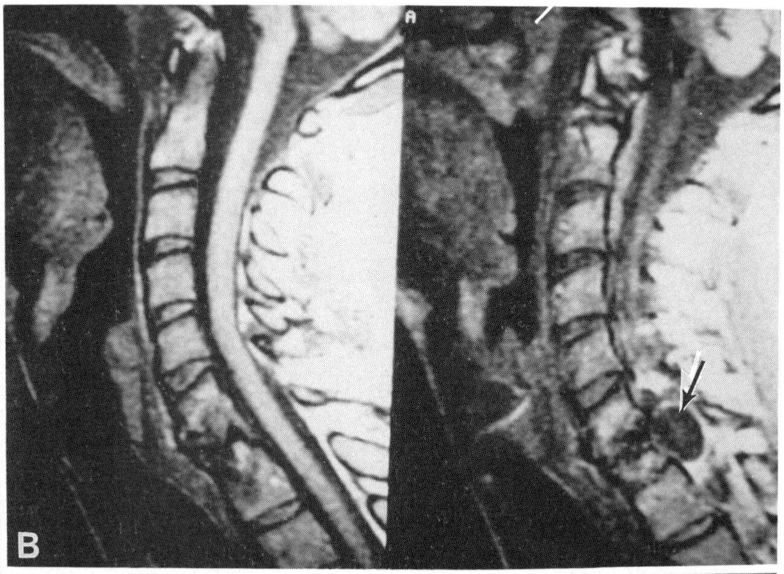

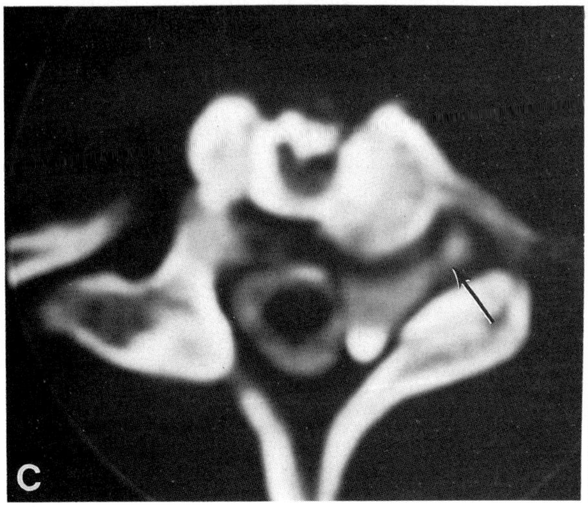

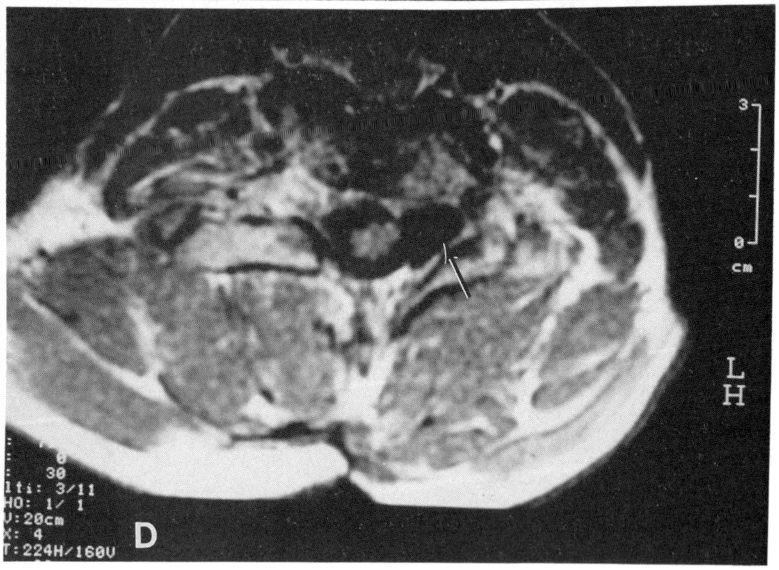

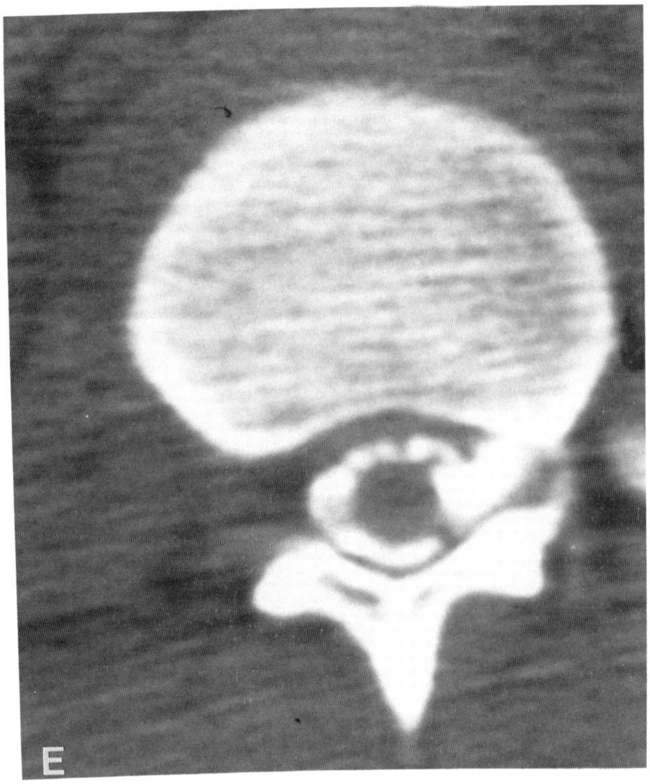

Figure 8.55–*continued*

spinal canal may be intact. Both types of injuries lead to hematoma, laceration of the spinal cord or nerve roots, or dural tears. Dural tears as a result of avulsion injuries most often involve the lower cervical region. Dural tears are often due to severe distraction of arm, as seen in motor cycle accidents. Overall, they are a rare cause of spinal trauma.

Dural tears and nerve root avulsion are best evaluated with myelography followed by CT (CTM). The myelographic study allows evaluation of the total segments of the spine. Cord compression from epidural or subdural hematoma can be localized, which on occasion may not correspond to the site of the penetrating injury. Extravasation of contrast material can be identified on myelography or CTM even with small dural tears (Fig. 8.55). CTM also allows identification of cord compression from the hematoma or presence of bone fragments.

On MRI, a dural tear should be suspected when there is deformity of the spinal cord and a subdural collection of CSF can be identified. The subdural CSF usually has an isointense or mildly hyperintense signal (Fig. 8.56). This may be due to lack of CSF pulsation or a combination of CSF mixed with blood products. In the subacute and later stages of brachial plexus injury, MRI is indicated if the CTM studies are equivocal or when cord hematoma, contusion, transection, or laceration is suspected. The dural tear may heal without any deformity of the thecal sac or may present as a pseudomeningocele, either within the spinal canal or extending beyond the canal through the neural foramen. This is usually filled with CSF but appears brighter on the T2-weighted images because of lack of pulsatile CSF motion. The avulsed nerve root may be adherent to the dura and may appear as a soft-tissue mass, mimicking a neuroma or even a herniated disk depending on its location within the neural foramen. Both CTM and MRI studies are complementary to each other.

Gunshot wounds to the spine as a civilian injury are less common, except in urban society. Depending on the bullet's trajectory and the type of firearm involved, the spinal injury may result in comminuted fractures or a nondisplaced linear fracture of a vertebral segment. In addition, there is damage to the cord and its surrounding coverings. Following plain radiographs, the extent of the intraspinal injury is evaluated in a similar fashion as with other penetrating injuries. CTM and re-formatted images are essential in demonstrating the bone fragments (Fig. 8.57). Metallic fragments may produce streak artifacts. The artifacts can be reduced to some extent by utilizing wide window settings. MRI is usually contraindicated in the initial assessment of the cord damage.

Traumatic Hemorrhage

Hemorrhage and hematoma within the spinal canal are often present with severe spinal trauma and in whiplash injuries of the cervical spine. The hemorrhage may be subligamentous, within the meningeal compartments, or within the spinal cord. Subligamentous hematoma is usually focal. Epidural hematoma often extends over several vertebral segments. Subarachnoid hemorrhage is usually associated with severe or fatal spinal trauma such as in atlanto-occipital dislocation (see Fig. 8.51). Demonstration of hematoma on CT depends on the extent of the hemorrhage and the volume of hematoma. Epidural hematomas are easier to recognize (see Fig. 8.50) compared with hemorrhage within the subarachnoid space (see Figs. 8.51 and 8.56). Due to beam-hardening artifact, smaller hematomas can be missed on CT but often identified with MRI. Identification of the hematoma is achieved with FSE and GRE sequences and should include sagittal and axial sections. Hemorrhage within the epidural space and the spinal cord follow the signal characteristics of hematoma seen with intracerebral hemorrhage, whereas similar changes are not seen with subarachnoid blood because it is mixed with CSF. Although in most instances intraspinal hematomas may not require surgical intervention, their

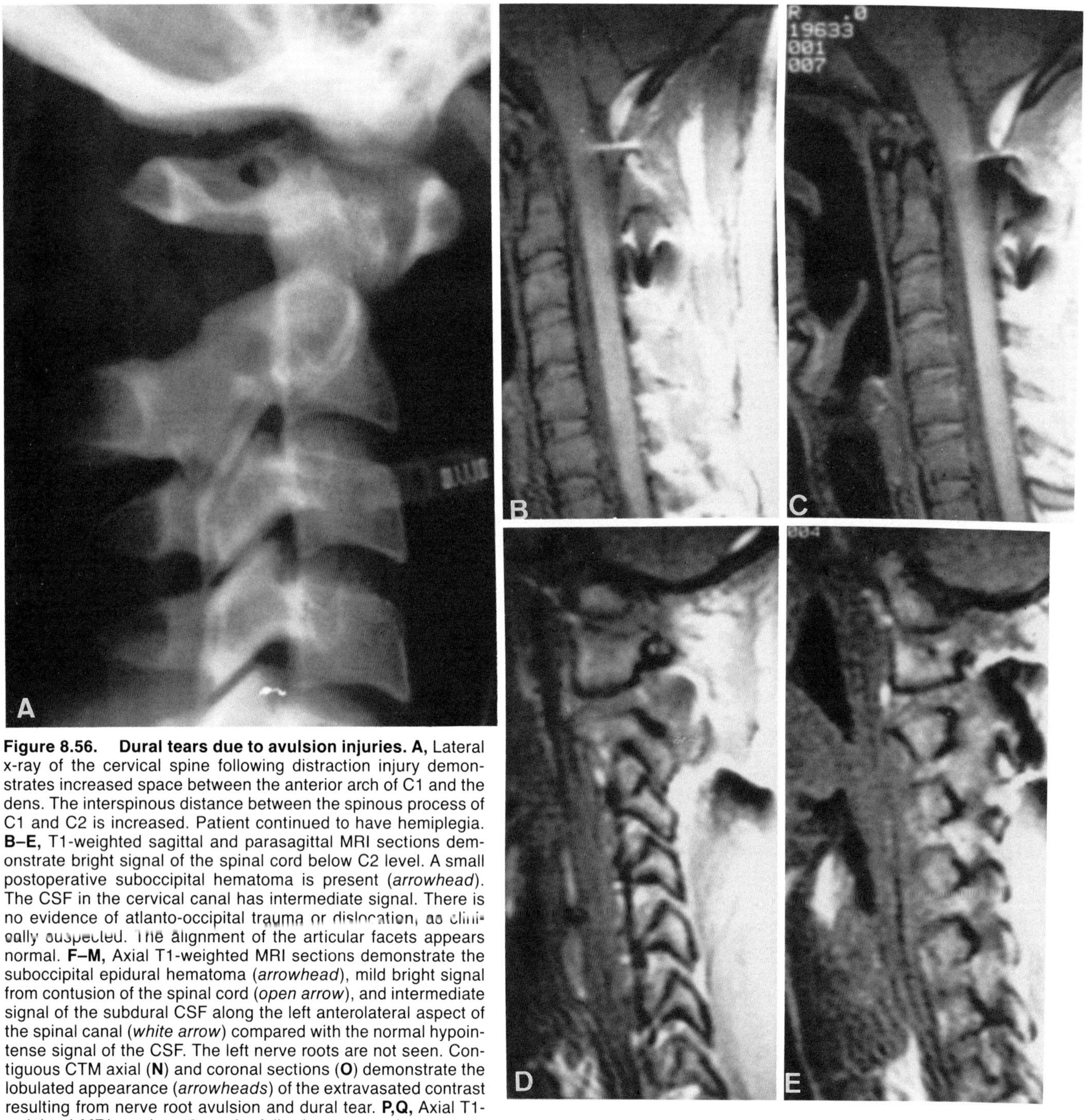

Figure 8.56. Dural tears due to avulsion injuries. A, Lateral x-ray of the cervical spine following distraction injury demonstrates increased space between the anterior arch of C1 and the dens. The interspinous distance between the spinous process of C1 and C2 is increased. Patient continued to have hemiplegia. **B–E,** T1-weighted sagittal and parasagittal MRI sections demonstrate bright signal of the spinal cord below C2 level. A small postoperative suboccipital hematoma is present (*arrowhead*). The CSF in the cervical canal has intermediate signal. There is no evidence of atlanto-occipital trauma or dislocation, as clinically suspected. The alignment of the articular facets appears normal. **F–M,** Axial T1-weighted MRI sections demonstrate the suboccipital epidural hematoma (*arrowhead*), mild bright signal from contusion of the spinal cord (*open arrow*), and intermediate signal of the subdural CSF along the left anterolateral aspect of the spinal canal (*white arrow*) compared with the normal hypointense signal of the CSF. The left nerve roots are not seen. Contiguous CTM axial (**N**) and coronal sections (**O**) demonstrate the lobulated appearance (*arrowheads*) of the extravasated contrast resulting from nerve root avulsion and dural tear. **P,Q,** Axial T1-weighted MRI sections 2 weeks following trauma demonstrate the spinal cord and avulsed nerve root being displaced by the loculated CSF collection on the left side.

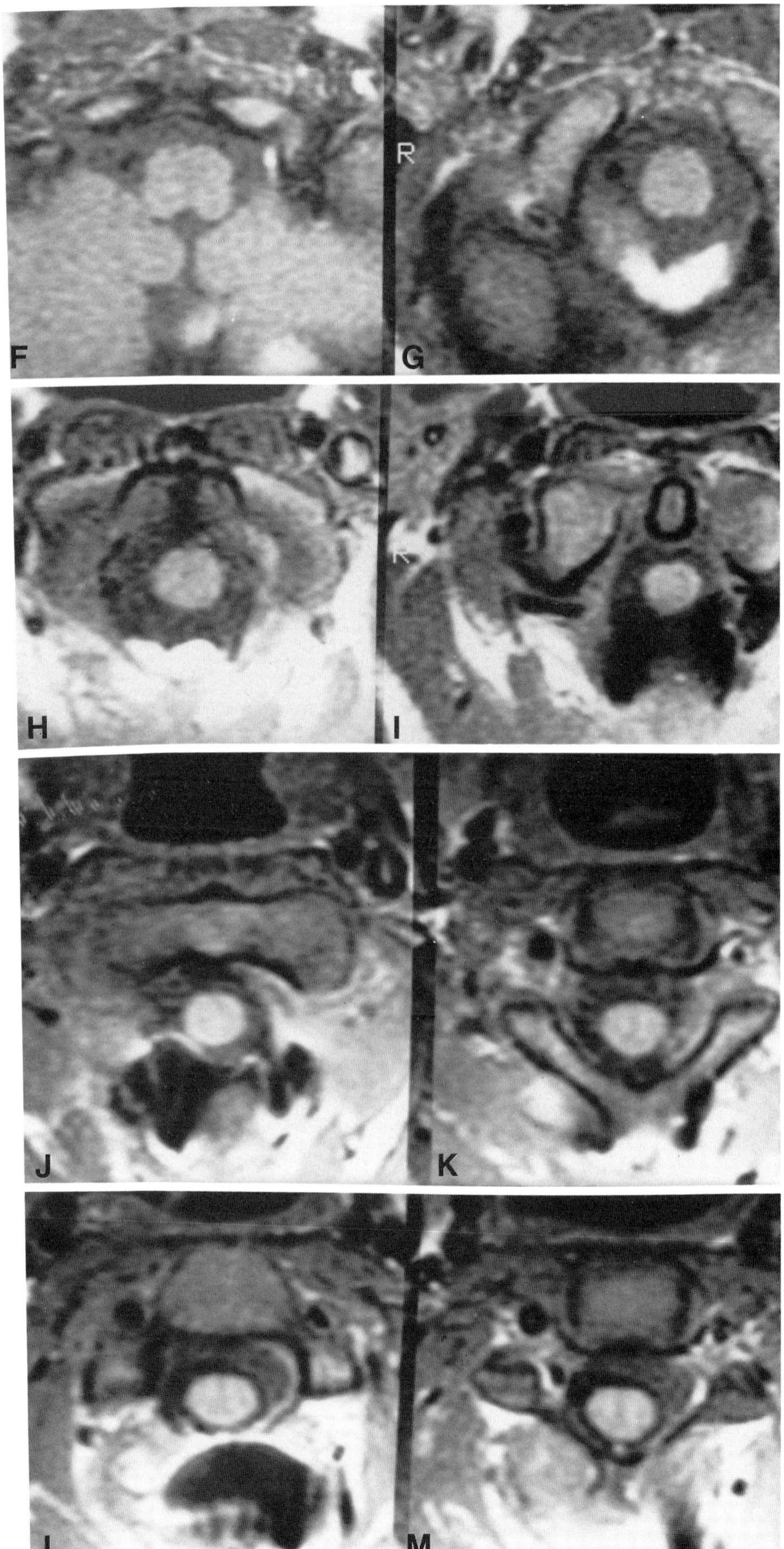

Figure 8.56–*continued*

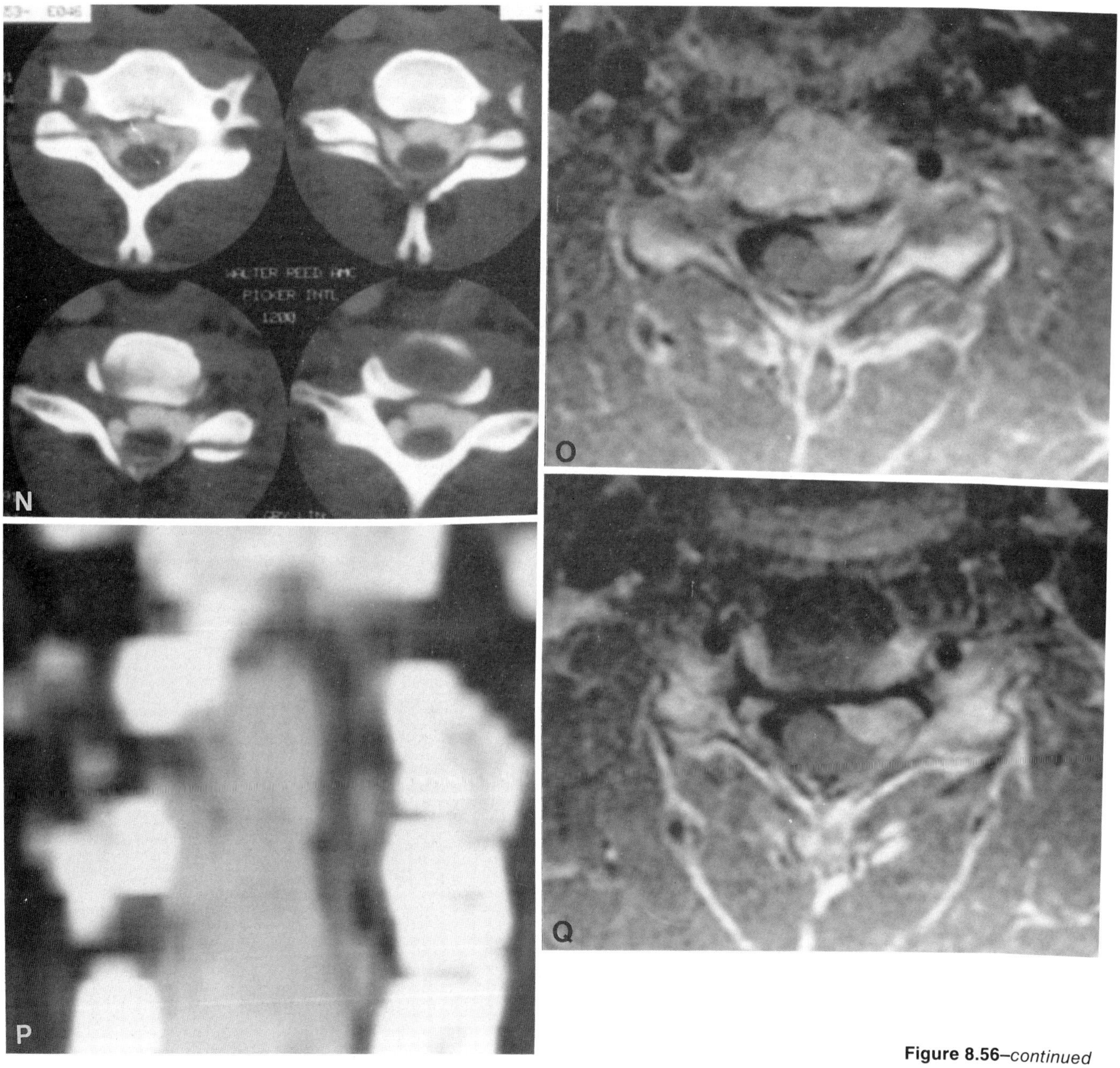

Figure 8.56–*continued*

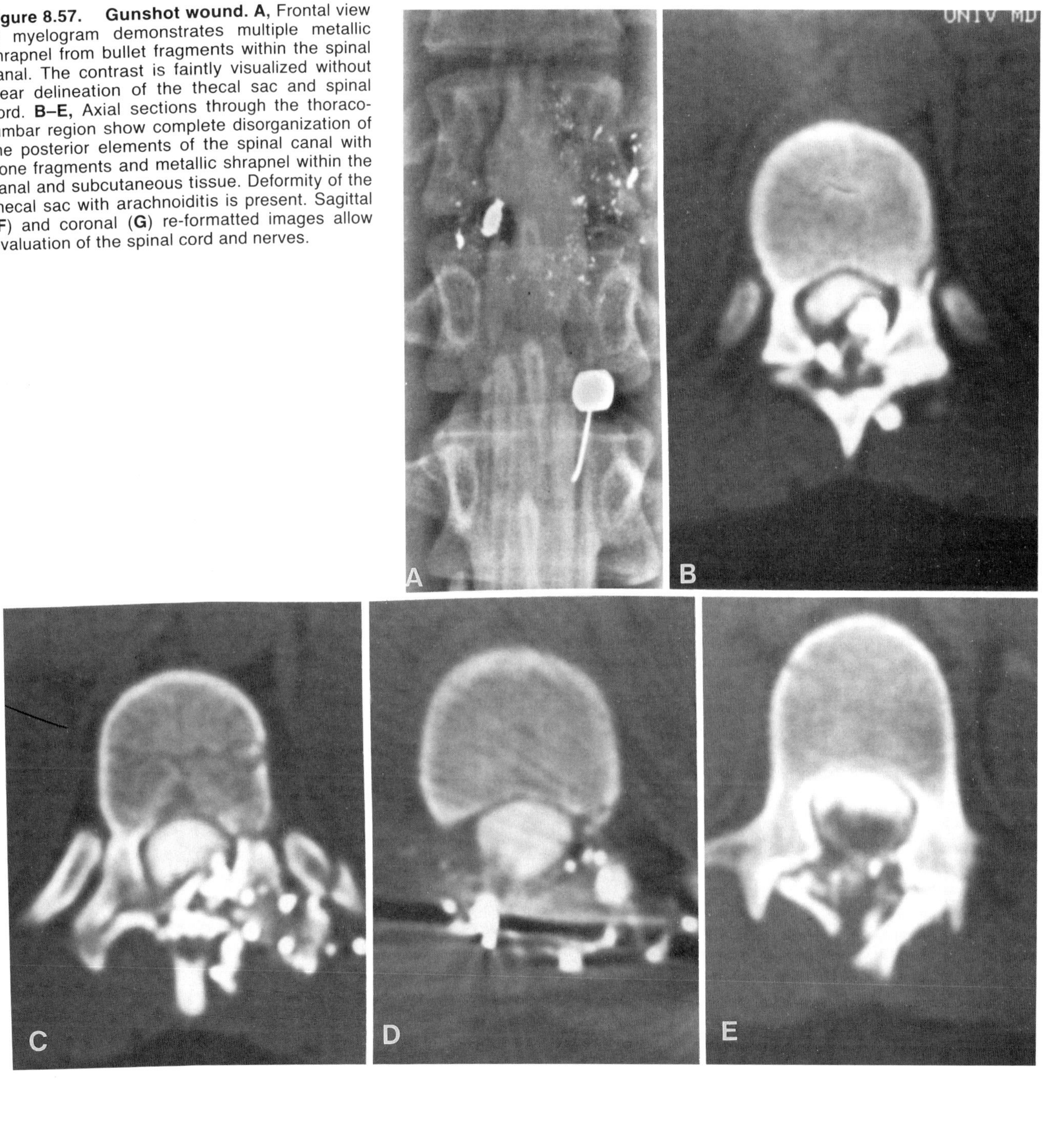

Figure 8.57. Gunshot wound. A, Frontal view of myelogram demonstrates multiple metallic shrapnel from bullet fragments within the spinal canal. The contrast is faintly visualized without clear delineation of the thecal sac and spinal cord. **B–E,** Axial sections through the thoracolumbar region show complete disorganization of the posterior elements of the spinal canal with bone fragments and metallic shrapnel within the canal and subcutaneous tissue. Deformity of the thecal sac with arachnoiditis is present. Sagittal (**F**) and coronal (**G**) re-formatted images allow evaluation of the spinal cord and nerves.

Figure 8.57–*continued*

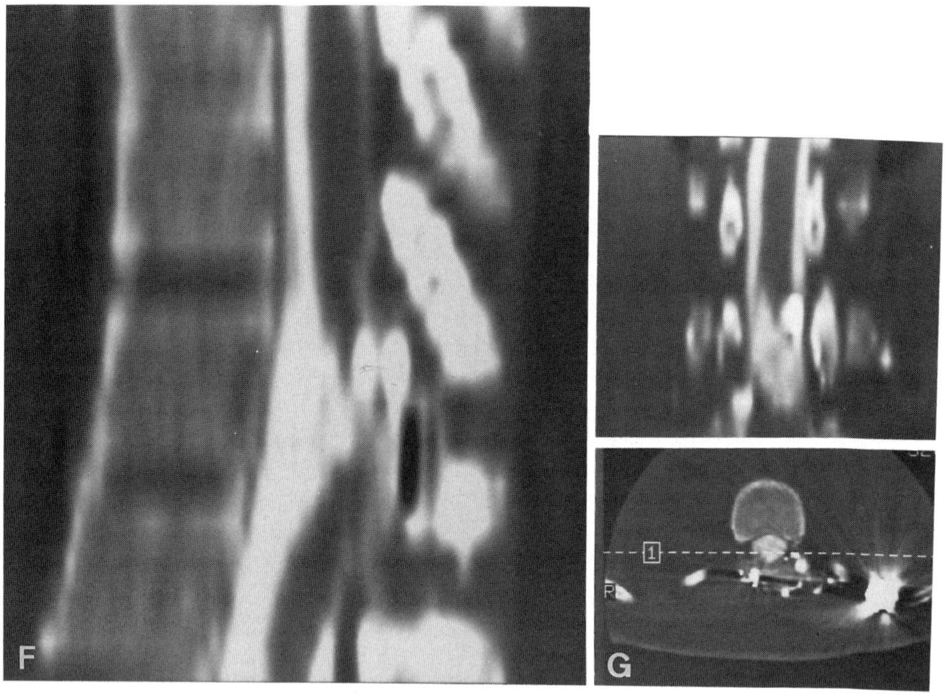

recognition is useful in defining the extent and severity of cord injury.

Spinal Cord Injury

Although osseous injuries to the spine can be stabilized over time with appropriate measures, injuries to the spinal cord often result in neurologic deficits. The neurologic deficit may be minimal and fixed, but more often progression of the deficit with time is common. Progressive deterioration of the neurologic symptoms may occur months or years after the initial injury. Imaging considerations consist of evaluating cord morphology and potential CSF spaces. In the past, this was limited because of restricted imaging capabilities. Both CTM and MRI have improved our capability in evaluating the morphologic changes in the cord resulting from trauma (Fig. 8.58). Unless contraindicated, MRI is ideal in evaluating the spinal cord and the CSF dynamics without introduction of intrathecal contrast material [32–35, 37, 38, 135] and in evaluating the chronically injured spines of trauma patients [39, 40].

Traumatic injury to the spinal cord results in cord contusion and ischemic or hemorrhagic necrosis with subsequent cavitation. The pathogenesis was discussed earlier. Cord contusions may result in complete recovery, with normal appearance of the cord. More often cord injuries are associated with peticheal hemorrhages. In the acute setting, both may have a similar appearance on MRI—as focal swelling with isointensity or mild hyperintensity on T1-weighted images and hyperintense signal on T2-weighted images. These probably represent a combination of edema and microcystic areas of necrosis. Similar discrepancy also may be present with incisional injury to the cord, where the cord signal may appear normal even though there is significant clinical neurologic deficit [141].

Sequelae of Spinal Cord Injury

The sequelae of spinal cord injury are being better understood because of correlations of the MRI studies with clinical progression [33, 34]. Within a few weeks following injury to the cord, the hyperintense foci within the cord may resolve or may become more localized and result in formation of a myelomalacic cavity or syrinx (Figs. 8.58 and 8.59). This may appear as a focal or linear band of hypointense signal on T1-weighted images and hyperintense signal on T2-weighted images. The myelomalacic cavities may remain stationary or may expand over a period of time associated with worsening of the neurologic symptoms [39, 142].

Injury to the cord and meninges also may result in cord atrophy (Fig. 8.60), arachnoidal adhesions, and formation of arachnoidal cysts (Fig. 8.61). The cysts may communicate with the cavitations within the cord. With time these cysts may expand, either outside the cord, resulting in cord compression, or within the cord, resulting in progression of paresthe-

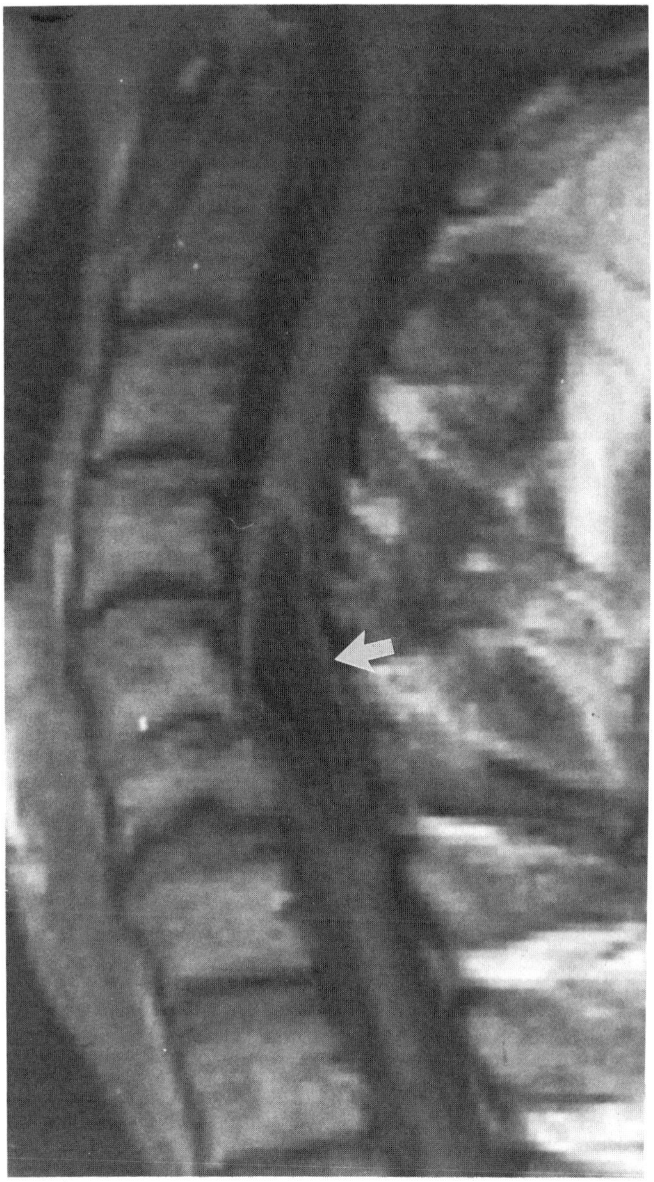

Figure 8.58. Sagittal T1-weighted MRI in a patient with quadriparesis 2 years following compression fracture of C6. Cystic cavitation within the cord (*arrow*) is present.

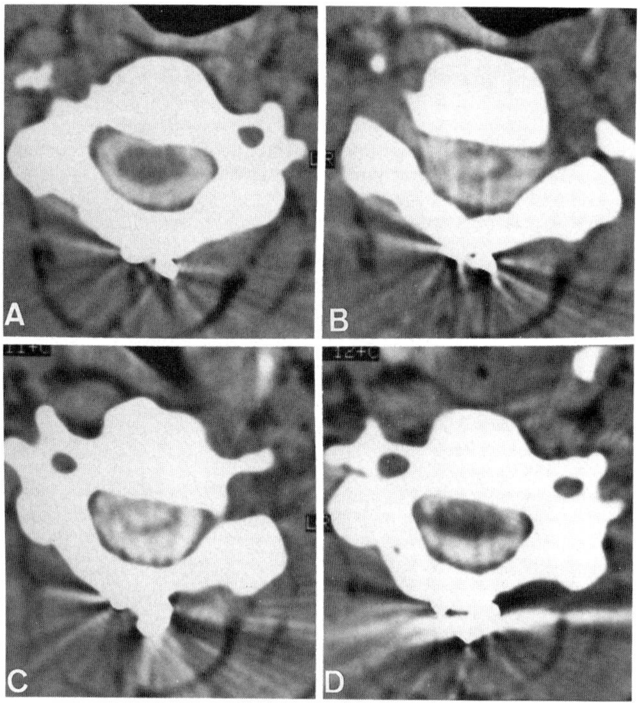

Figure 8.59. Contiguous axial CTM scans in a patient with quadriparesis following gunshot wound to the cervical spine. Contrast material is noted within the posttraumatic syrinx within the cord.

Figure 8.60. Posttraumatic cord atrophy. T1-weighted MRI. **A,** Sagittal section demonstrates cord atrophy involving the cervical and thoracic cord with a small syrinx (*arrow*) within the atrophic cord. **B,** Axial section through the level of the syrinx. **C,** Flattened atrophic thoracic cord.

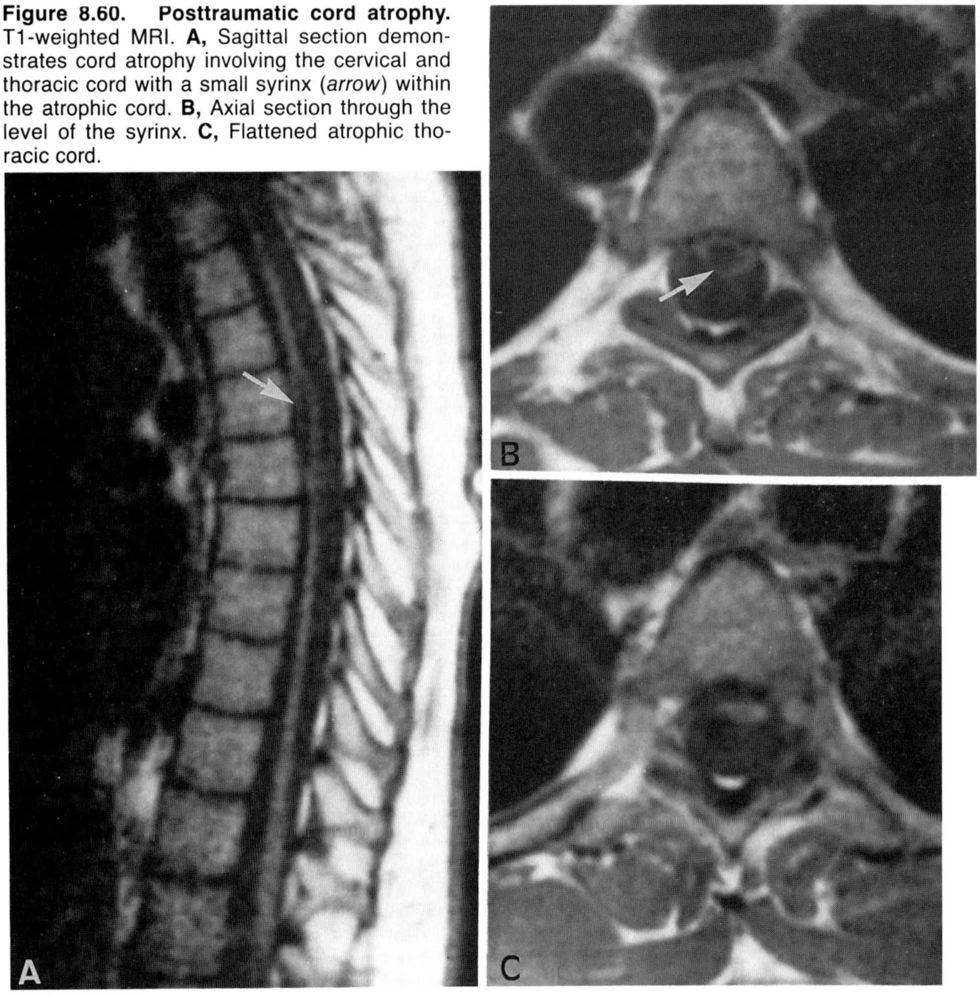

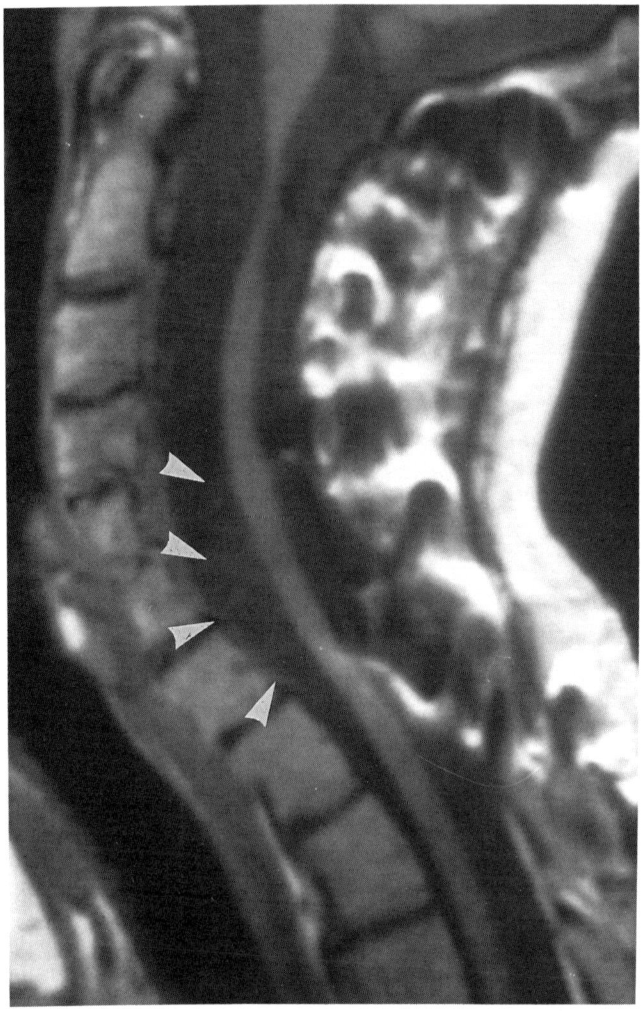

Figure 8.61. Posttraumatic subarachnoid cyst. Sagittal T1-weighted MRI demonstrates segmental atrophy of the spinal cord with a focal subarachnoid cyst. This was confirmed with CT following myelography.

sia, hyperesthesia, and occasionally motor symptoms. Arachnoidal adhesions, subarachnoid cysts, and spinal cord cysts communicating with extramedullary subarachnoid cysts have been identified as delayed sequelae of trauma [143]. These are usually associated with penetrating trauma to the spine, although they may occur following blunt trauma to the spine (see Fig. 8.61). The increase in size and volume of these cysts over a period of time is probably related to the CSF pulsatile pressure. Their proper evaluation often requires a combination of studies including MRI, myelography, and subsequent CT scans [144–146]. Occasionally, delayed CTM may be necessary to demonstrate an arachnoid cyst separate from the atrophic spinal cord. Arachnoid adhesions may result in tethering of the cord. MRI following Gd-DTPA is useful in demonstrating the site of arachnoidal adhesions and the tethered cord (see also Chap. 11).

References

1. Kewalramani LS, Tori JA. Spinal cord trauma in children: neurologic patterns, radiologic features, and pathomechanics of injury. Spine 1980;5:11.
2. Hubbard D. Injuries of the spine in children and adolescents. Clin Orthop 1974;100:56.
3. Anderson M, Schutt AH. Spinal injury in children: a review of 156 cases seen from 1950 through 1978. Mayo Clin Proc 1980;55:499.
4. Hadley MN, Zabramski JM, Browner CM, Rekate H, Sonntag VKH. Pediatric spinal trauma: review of 122 cases of spinal cord and vertebral column injuries. J Neurosurg 1988;68:18.
5. Denton JR. Trauma and the adolescent spine. In: Keim HA, ed. The adolescent spine. 2nd ed. New York: Springer Verlag, 1982:63.
6. Pang D, Wilberger JE Jr. Spinal cord injury without radiographic abnormalities in children. J Neurosurg 1982;57:114.
7. Matsumura A, Meguro K, Tsurushima H, Kikuchi Y, Wada M, Nakata Y. Magnetic resonance imaging of spinal cord injury without radiologic abnormality. Surg Neurol 1990;33:281.
8. Dietrich AM, Ginn-Pease ME, Bartkowski HM, King DR. Pediatric cervical spine fractures: predominately subtle presentation. J Pediatr Surg 1991;26:995.
9. Apple JS, Kirks DR, Merten DF, Martinez S. Cervical spine fractures and dislocations in children. Pediatr Radiol 1987;17:45.
10. Menezes AH, Godersky JC, Smoker WRK. Spinal cord injury. In: McLaurin RL, Venes JL, Schut L, Epstein F, eds. Pediatric neurosurgery. 2nd ed. Philadelphia: Saunders, 1989:298.
11. Barnes PD. Acquired abnormalities of the spine and spinal neuraxis. In: Wolpert SM, Barnes PD, eds. MRI in pediatric neuroradiology. St. Louis: Mosby–Year Book, 1992:412.
12. Ozonoff MB. The spine. In: Pediatric orthopedic radiology. 2nd ed. Philadelphia: Saunders, 1992: 1.
13. Swischuk LE. Spine and spinal cord trauma in battered child syndrome. Radiology 1969;92:733.
14. Cullen J. Spinal lesions in battered babies. J Bone Joint Surg 1975;57B:364.
15. Swischuk LE. The spine and spinal cord. In: Emergency radiology of the acutely ill or injured child. 2nd ed. Baltimore: Williams & Wilkins, 1986:556.
16. Hudgins PA, Hudgins RJ. Radiology of cervical spine trauma. In: Black PMcL, ed. Clinical neurosurgery: congress of neurological surgeons. Vol 37 Baltimore: Williams & Wilkins, 1991:571.
17. Streitwieser DR, Knopp R, Wales LR, et al. Accuracy of standard radiographic views in detecting cervical spine fractures. Ann Emerg Med 1983;12:538.
18. Vandemark RM. Radiology of the cervical spine in trauma patients: practice pitfalls and recommendations for improving efficiency and communication. AJR 1990;155:465.
19. Wales LR, Knopp RK, Morishima MS, et al. Recommendations for evaluation of the acutely injured cervical spine: a clinical radiologic algorithm. Ann Emerg Med 1980;9:422.
20. Allen BL, Ferguson RL. Cervical spine trauma in children. In: Bradford DS, Hensinger RM, eds. The pediatric spine. New York: Thieme, 1985:89.
21. Locke GR, Gardner JI, Van Epps EK. Atlas-dens interval in children. AJR 1966;97:135.
22. Wholey MH, Breuwer AJ, Baker HL. Lateral roentgeno-

gram of the neck with comments on atlanto-odontoid-basion relationship. Radiology 1958;71:350.
23. Swischuk LE. Anterior displacement of C2 in children: physiologic or pathologic? Radiology 1977;122:759.
24. Pathria MN, Petersilge CA. Spinal trauma. Radiol Clin North Am 1991;29:867.
25. Pech P, Kilgore DP, Pojunas KW, et al. Cervical spinal fractures: CT detection. Radiology 1985;157:117.
26. Faerber EN, Wolpert SM, Scott RM, Belkin SC, Carter BL. Computed tomography of spinal fractures. J Comput Assist Tomogr 1979;3:657.
27. Lanska MJ, Roessmann U, Wiznitzer M. Magnetic resonance imaging in cervical cord birth injury. Pediatrics 1990;85:760.
28. Mendelsohn DB, Zollars L, Weatherall PT, Girson M. MR of cord transection. J Comput Assist Tomogr 1990; 14:909.
29. Davis PC, Reisner A, Hudgins PA, Davis WE, O'Brien MS. Spinal injuries in children: role of MR. AJNR 1992;
30. Flanders AE, Schaefer DM, Doan HT, Mishkin MM, Gonzalez CF, Northrup BE. Acute cervical spine trauma: correlation of MR imaging findings with degree of neurologic deficit. Radiology 1990;177:25.
31. Hackney DB. Denominators of spinal cord injury. Radiology 1990;177:18.
32. Kulkarni MV, McArdle CB, Kopanicky D, et al. Acute spinal cord injury: MR imaging at 1.5 T. Radiology 1987;164:837.
33. Kulkarni MV, Bondurant FJ, Rose SL, Narayana PA. 1.5 Tesla magnetic resonance imaging of acute spinal trauma. Radiographics 1988;8:1059.
34. Bondurant FJ, Cotler HB, Kulkarni MV, McArdle CB, Harris JH. Acute spinal cord injury: a study using physical examination and magnetic resonance imaging. Spine 1990;15:161.
35. Beers GJ, Raque GH, Wagner GG, et al. MR imaging in acute cervical spine trauma. J Comput Assist Tomogr 1988;12:75.
36. Zwimpfer TJ, Bernstein M. Spinal cord concussion. J Neurosurg 1990;72:894.
37. Hackney DB, Asato R, Joseph PM, et al. Hemorrhage and edema in acute spinal cord compression: demonstration by MR imaging. Radiology 1986;161:387.
38. Mirvis SE, Geisler FH, Jelinek JJ, Joslyn JN, Gellad F. Acute cervical spine trauma: evaluation with 1.5-T MR imaging. Radiology 1988;166:807.
39. Quencer RM, Sheldon JJ, Post MJD, et al. Magnetic resonance imaging of the chronically injured cervical spinal cord. AJNR 1986;7:457.
40. Yamashita Y, Takahashi M, Matsuno Y, et al. Chronic injuries of the spinal cord: assessment with MR imaging. Radiology 1990;175:849.
41. Cohen MD. Pediatric sedation. Radiology 1990;175: 611.
42. Fisher DM. Sedation of pediatric patients: an anesthesiologist's perspective. Radiology 1990;175:613.
43. Youssem DM, Schnall MD. MR examination for spinal cord compression: impact of a multicoil system on length of study. J Comput Assist Tomogr 1991;15:598.
44. O'Rahilly R, Benson DR. The development of the vertebral column. In: Bradford DS, Hensinger RM, eds. The Pediatric Spine. New York: Thieme, 1985:3.
45. Hinck C, Hopkins CE, Savara BS. Sagittal diameter of the lumbar spinal canal in children and adults. Radiology 1962;79:97.
46. Sarwar M, Kier EL, Virpongse C. Development of the spine and spinal cord. In: Newton TH, Potts DG, eds. Modern neuroradiology, Vol. 1: Computed tomography of the spine and spinal cord. San Anselmo: Clavadel Press, 1983.
47. Schmorl G, Junghanns H. The human spine in health and disease. 2nd ed. New York: Grune and Stratton, 1971:2.
48. Fielding JW, Hensinger RN, Hawkins RJ. Os odontoideum. J Bone Joint Surg 1980;62A:376.
49. Ricciardi JE, Kaufer H, Louis DS. Acquired os odontoideum following acute ligament injury: report of a case. J Bone Joint Surg 1976;58A:410.
50. Suss RA, Zimmerman RD, Leeds NE. Pseudospread of the atlas: false sign of Jefferson fracture in young children. AJNR 1983;4:183.
51. Ho PSP, Yu S, Sether LA, Wagner M, Ho KC, Haughton VM. Progressive and regressive changes in the nucleus pulposus: I. The neonate. Radiology 1988;169:87.
52. Kricun MR. Red-yellow marrow conversion: its effect on the location of some solitary bone lesions. Skeletal Radiol 1985;14:10.
53. Sze G, Baierl P, Bravo S. Evolution of the infant spinal column: evaluation with MR imaging. Radiology 1991; 181:819.
54. Dooms GC, Fisher MR, Hricak H, Richardson M, Crooks LI, Genant HK. Bone marrow imaging: magnetic resonance studies related to age and sex. Radiology 1985; 155:429.
55. Wilson DA, Prince JR. MR imaging determination of the location of the normal conus medullaris throughout childhood. AJR 1989;152:1029.
56. White AA, Johnson RM, Punjabi MM, Southwick WO. Biomedical analysis of clinical stability in the cervical spine. Clin Orthop 1975;109:85.
57. Punjabi MD, White AA, Johnson RM. Cervical spine mechanics as a function of transection of components. J Biomech 1975;8:327.
58. Denis F. The three column spine and its significance in the classification of acute thoracolumbar spinal injuries. Spine 1983;8:817.
59. Harris JH, Edeiken-Monroe B. The radiology of acute cervical spine trauma, 2nd ed. Baltimore: Williams & Wilkins, 1987.
60. Bradford DS, Lonstein JE, Moe JH, Ogilvie JW, Winter RB, eds. Moe's textbook of scoliosis and other spinal deformities. 2nd ed. Philadelphia: Saunders, 1987:41.
61. Weirich SD, Cotler HB, Narayana PA, et al. Histopathologic correlation of magnetic resonance imaging signal patterns in a spinal cord injury model. Spine 1990; 15:630.
62. Crothers B, Putnam MC. Obstetrical injuries of the spinal cord. Medicine 1927;6:41.
63. Stern WE, Rand RW. Birth injuries to the spinal cord. Am J Obstet Gynecol 1959;78:498.
64. Towbin A. Spinal cord and brainstem injury at birth. Arch Pathol Lab Med 1964;77:620.
65. Yates PO. Birth trauma to the vertebral arteries. Arch Dis Child 1959;311:436.
66. Warwich M. Necropsy findings in newborn infants. Am J Dis Child 1921;21:488.
67. Allen JP, Myers GG, Condon VR. Laceration of the spinal cord related to breech delivery. JAMA 1969;208: 1019.
68. Byers RK. Spinal-cord injuries during birth. Dev Med Child Neurol 1975;17:103.
69. Leventhal HR. Birth injuries of the spinal cord. J Pediatr 1960;56:447.
70. Gaufin LM, Goodman SJ. Cervical spine injuries in infants. J Neurosurg 1975;42:179.
71. Stanley P, Duncan AW, Isaacson J, Isaacson AS. Ra-

diology of fracture-dislocation of the cervical spine during delivery. AJR 1985;145:621.

72. Schulman ST, Madden JD, Esterly JR, Shanklin DR. Transection of the spinal cord: a rare obstetrical complication of cephalic delivery. Arch Dis Child 1971; 46:291.
73. Allen JP. Birth injury to the spinal cord. Northwest Med 1979;69:323.
74. Hill SA, Miller CA, Kosnik EJ, Hunt WE. Pediatric neck injuries: a clinical study. J Neurosurg 1984;60:700.
75. Schouman-Claeys E, Frija G, Cuenod CA, Begon D, Paraire F, Martin V. MR imaging of acute spinal cord injury: results of an experimental study in dogs. AJNR 1990; 11:959.
76. Bracken MB, Shepard MJ, Collins WF, et al. A randomized, controlled trial of methylprednisolone or naloxone in the treatment of acute spinal-cord injury. N Engl J Med 1990;322:1405.
77. Jansen L, Hansebout RR. Pathogenesis of spinal cord injury and newer treatments: a review. Spine 1989; 14:23.
78. Ahmann PA, Smith SA, Schwartz JF, et al. Spinal cord infarction due to minor trauma in children. Neurology 1975;25:301.
79. Choi JU, Hoffman HF, Hendrick EB, Humphreys RP, Keith WS. Traumatic infarction of the spinal cord in children. J Neurosurg 1986;65:608.
80. Lazorthes G, Gouaze A, Zadeh JO, Santini JJ, Lazorthes Y, Burdin P. Arterial vascularization of the spinal cord: recent studies of the anastomotic substitution pathways. Neurosurg 1971;35:253.
81. Tator CH, Fehlings MG. Review of the secondary injury theory of acute spinal cord trauma with emphasis on vascular mechanisms. J Neurosurg 1991;75:15.
82. Kaufman RA, Dunbar JS, Botsford JA, et al. Traumatic longitudinal atlanto-occipital distraction injuries in children. AJNR 1982;3:415.
83. Aufdermaur M. Spinal injuries in juveniles: necropsy findings in 12 cases. J Bone Joint Surg 1974;3B:513.
84. Woodring JH, Selke AC, Duff DE. Traumatic atlantooccipital dislocation with survival. AJR 1981;137:21.
85. Lee C, Woodring JH, Goldstein SJ, et al. Evaluation of traumatic atlantooccipital dislocations. AJNR 1987;8: 19.
86. Powers B, Miller MD, Kramer RS, et al. Traumatic anterior atlantooccipital dislocation. Neurosurgery 1979;4:12.
87. Kaufman RA, Carroll CD, Buncher CR. Atlantooccipital junction: standards for measurement in normal children. AJNR 1987;8:995.
88. Jefferson G. Fracture of the atlas vertebra: report of four cases, and a review of those previously recorded. Br J Surg 1920;7:407.
89. Lee C, Woodring JH. Unstable Jefferson variant atlas fractures: an unrecognized cervical injury. AJR 1992; 158:113.
90. Anderson LD, D'Alonzo RT. Fractures of the odontoid process of the axis. J Bone Joint Surg 1974;56A:1663.
91. Kowalski HM, Cohen WA, Cooper P, Wisoff JH. Pitfalls in the CT diagnosis of atlantoaxial rotary subluxation. AJNR 1987;8:697.
92. Dawson EG, Smith L. Atlanto-axial subluxation in children due to vertebral anomalies. J Bone Joint Surg 1979;61A:582.
93. Pueschel SM, Scola FH, Perry CD, Pezzullo JC. Atlantoaxial instability in children with Down's syndrome. Pediatr Radiol 1981;10:129.
94. El-Khoury GY, Clark CR, Dietz FR, Harre RG, Tozzi JE, Kathol MH. Posterior atlantooccipital subluxation in Down syndrome. Radiology 1986;159:507.
95. Cheshire DJ. The stability of the cervical spine following the conservative treatment of fractures and fracture-dislocations. Paraplegia 1969;7:193.
96. Edeiken-Monroe B, Wagner LK, Harris JH. Hyperextension dislocation of the cervical spine. AJNR 1986;7: 135.
97. Hayes CW, Conway WF, Walsh JW, Coppage L, Gervin AS. Seat belt injuries: radiologic findings and clinical correlation. Radiographics 1991;11:23.
98. Hoffman MA, Spence LJ, Wesson DE, Armstrong PF, Williams JI, Filler RM. The pediatric passenger: trends in seatbelt use and injury patterns. J Trauma 1987; 27:974.
99. Woefel GD, Moore EE, Cogbill TH, Van Way CW. Severe thoracic and abdominal injury in seatbelt wearers. J Trauma 1983;23:1007.
100. Taylor GA, Eggli KD. Lap-belt injuries of the lumbar spine in children: a pitfall in CT diagnosis. AJR 1988;150:355.
101. Sivit CJ, Taylor GA, Newman KD, et al. Safety-belt injuries in children with lap-belt ecchymosis: CT findings in 61 patients. AJR 1991;157:111.
102. Chance GQ. Note on the type of flexion fracture of the spine. Br J Radiol 1948;21:452.
103. Hayes CW, Conway WF, Walsh JW, Coppage L, Gervin AS. Seat belt injuries: radiologic findings and clinical correlation. Radiographics 1991;11:23.
104. Gellald FE, Paul KS, Geisler FH. Early sequelae of gunshot wounds of the spine: radiologic diagnosis. Radiology 1988;167:523.
105. Teitelbaum G, Yee C, Von Horn D, et al. Metallic ballistic fragments: MR imaging. Radiology 1990;175:855.
106. Afshani E, Kuhn JP. Common causes of low back pain in children. Radiographics 1991;11:269.
107. Stanitski CL. Management of sports injuries in children and adolescents. Orthop Clin North Am 1988; 14:689.
108. Banerian KG, Wang A, Samberg LC, Kerr HH, Wesolowski DO. Association of vertebral end plate fracture with pediatric lumbar intervertebral disk herniation: value of CT and MR imaging. Radiology 1990;177: 763.
109. Handel SF, Twiford TW, Reigel DH, Kaufman HH. Posterior lumbar apophyseal fractures. Radiology 1979; 130:629.
110. Clarke NMP, Cleak DK. Intervertebral lumbar disc prolapse in children and adolescents. J Pediatr Orthop 1983;3:202.
111. Swick HM. Calcification of intervertebral discs in childhood. J Pediatr 1975;86:364.
112. Urso S, Colajacoma M, Migliorini A, et al. Calcifying discopathy in infancy in the cervical spine: evaluation of vertebral alteration over a period of time. Pediatr Radiol 1987;17:387.
113. Sonnabend DH, Taylor TKF, Chapman GK. Intervertebral disc calcification syndromes in children. J Bone Joint Surg 1982;64B:25.
114. Meyerding HW. Spondylolisthesis. Surg Gynecol Obstet 1932;54:371.
115. White LL, Widell EH, Jackson DW. Fatigue fracture: the basic lesion in isthmic spondylolisthesis. J Bone Joint Surg 1975;57A:17.
116. Albanese M, Pizzutillo PK. Family study of spondylolysis and spondylolisthesis. J Pediatr Orthop 1982;2: 496.
117. Wiltse LL, Newman PH, Macnab I. Classification of

spondylolysis and spondylolisthesis. Clin Orthop 1976; 117:23.
118. Teplick JG, Laffey PA, Berman A, Haskin ME. Diagnosis and evaluation of spondylolisthesis and/or spondylolysis on axial CT. AJNR 1986;7:479.
119. Johnson DW, Farnum GN, Latchaw RE, Erba SM. MR imaging of the pars interarticularis. AJR 1989;152:327.
120. Grenier N, Kressel HY, Schiebler ML, Grossman RI. Isthmic spondylolysis of the lumbar spine: MR imaging at 1.5 T. Radiology 1989;170:489.
121. Kochan JP, Quencer RM. Imaging of cystic and cavitary lesions of the spinal cord and canal: the value of MR and intraoperative sonography. Radiol Clin North Am 1991;29:867.
122. Holdsworth FW. Fractures, dislocations, and fracture dislocations of the spine. J Bone Joint Surg 1970; 52A:1534–1551.
123. Taylor AR, Blackwood W. Paraplegia in hyperextension cervical injuries with normal radiographic appearances. J Bone Joint Surg 1948;30B:245–248.
124. Schneider RC, Cherry J, Pantek H. The syndrome of acute cervical spinal cord injury. J Neurosurg 1954; 11:546–557.
125. Scher AT. Cervical spinal cord injury without evidence of fracture or dislocation: an assessment of the radiological features. S Afr Med J 1976;50:962–965.
126. Olsson O. Fractures of the upper thoracic and cervical vertebral bodies. Acta Chir Scand 1951;102:87–92.
127. Rogers WA. Fractures and dislocations of the cervical spine: an end-result study. AJR 1951;102:87–92.
128. Castellano V, Bocconi FL. Injuries of the cervical spine with spinal cord involvement (myelic fractures): statistical considerations. Bull Hosp Joint Dis 1970;31:188–194.
129. Brant-Zawadzki M, Miller EM, Federle MP. CT in the evaluation of spine trauma. AJR 1981;136:369–375.
130. Rao KCVG, Goshhazra K. CT in spinal trauma. J Comput Assist Tomogr 1980;4:271–280.
131. Keene JS, Goletz TH, Lilleas F, et al. Diagnosis of vertebral fractures: a comparison of conventional radiography, conventional tomography and computed tomography. J Bone Joint Surg 1982;64A:586–594.
132. Post MJD, Greene BA, Quencer RM, et al. The value of computed tomography in spinal trauma. Spine 1982;7: 417–431.
133. Handel SF, Lee Y. Computed tomography of spinal fractures. Radiol Clin North Am 1981;19:69–89.
134. Tarr RW, Drolshagen LF, Kerner TC, et al. MR imaging of recent spinal trauma. J Comput Assist Tomogr 1987;11:412–417.
135. McArdle CB, Crafford MJ, Mirfakhree M, et al. Surface coil MR of spinal trauma: preliminary experience. AJNR 1986;6:885–893.
136. Fielding JW, Hawkins RJ. Atlanto-axial rotary fixation: fixed rotary subluxation of the atlanto-axial joint. J Bone Joint Surg 1983;65A:461–473.
137. Atlas SW, Regenbogen V, Rogers LF, et al. The radiographic characterization of burst fractures of the spine. AJNR 1986;7:675–682.
138. Brant-Zawadzki M, Jeffery RB, Minagi H, et al. High resolution CT of thoracolumbar fractures. AJR 1982;138: 699–704.
139. McAfee PC, Yuan HA, Lasda NA. The unstable burst fracture. Spine 1982;7:365–373.
140. Ferguson RL, Allen BL. A mechanistic classification of thoracolumbar spine fractures. Clin Orthop 1984;189: 77–88.
141. Mayer JS, Kulkarni MV. MR imaging of incisional spinal cord injury. AJNR 1987;8:925–927.
142. Gebarski SS, Maynard FW, Gabrielson TO, et al. Posttraumatic progressive myelopathy. Radiology 1985; 157:379–385.
143. Osborn DRS, Vavoulis G, Nashold BS Jr, et al. Late sequelae of spinal cord trauma: myelographic and surgical correlation. J Neurosurg 1982;57:18–23.
144. Quencer RM, Green BA, Eismont FJ: Posttraumatic spinal cord cysts: clinical features and characterization with metrizamide computed tomography. Radiology 1983;146:415–423.
145. Seibert CE, Driesbach JN, Swanson WB, et al. Progressive posttraumatic cystic myelopathy: neuroradiologic evaluation. AJNR 1981;2:115–119.
146. Sklar E, Quencer RM, Green BA, et al. Acquired spinal subarachnoid cysts: evaluation with MR, CT myelography, and intraoperative sonography. AJNR 1989;10: 1097–1104.

CHAPTER / 9

Tumors

R. S. Williams / J. P. Williams

Approximately 15 to 20% of central nervous system tumors involve the spinal cord or canal [1, 2]. Primary tumors of the spinal cord and canal are relatively rare, and the true incidence of spine neoplasms is uncertain. Regional population studies suggest that the average annual incidence is probably 1 or 2 per 100,000, with a prevalence rate of approximately 13 per 100,000 [3, 4]. These studies do not include primary bone tumors of the spine nor metastatic involvement. Large series of spine tumors may be significantly skewed by population variables such as referral patterns, surgical interest, or preferential selection for autopsy series. To further complicate matters, metastatic involvement of the spine is often excluded.

Primary spine tumors occur primarily in young or middle-aged adults and are less frequently encountered in children and the elderly [2]. The sex incidence is roughly equal, although there is a definite female predilection for meningiomas. In general, spine tumors are most frequently found in the thoracic spine, but this is explained largely by the longer length of the thoracic spine relative to the remaining spine [2]. Metastatic involvement of the spine is seen primarily in patients between the ages of 50 and 70 years and is only rarely encountered in either younger or older age groups [5]. The thoracic spine is most frequently involved.

The differential diagnosis of spine neoplasms is primarily based on location of the lesion relative to the spinal cord and the age, sex, and clinical presentation [6, 7]. Tumors within the cord are referred to as *intramedullary* and account for 11 to 25% of spine tumors [2, 8]. Intramedullary neoplasms are usually associated with fusiform enlargement of the cord with narrowing of the adjacent subarachnoid space.

Tumors between the cord and the dura are referred to as *intradural extramedullary* and account for nearly 60% of primary spine tumors [2, 8]. The great majority of these are neurofibromas and meningiomas. These tumors are associated with displacement of the spinal cord away from the tumor, with widening of the ipsilateral subarachnoid space and narrowing of the contralateral space [6].

Extradural tumors account for less than 25% of primary spine neoplasms [8]. However, when metastatic disease is included, extradural tumors may account for over half of all spine neoplasms [7]. When these involve the spinal canal, there may be displacement of the cord with narrowing of the subarachnoid space both ipsilateral and contralateral to the lesion [6].

As a general rule, intramedullary tumors are more common in children and extramedullary tumors are more common in adults [8]. Mass lesions of the spinal canal usually produce symptoms by involving long tracts of the spinal cord or by involving nerve roots [9]. Pain is a common presentation and may help with localization. Sensory disturbances predominate with intramedullary tumors, motor symptoms with meningiomas, and root symptoms with nerve sheath tumors [9]. Slowly growing tumors such as meningiomas and neurofibromas may produce a spastic paresis. Rapidly growing tumors such as metastases may result in acute transection syndromes with flaccid paresis and loss of reflexes [9].

Imaging of spine neoplasms has markedly improved in the last decade due to several technological advances. The intrathecal introduction of water-soluble contrast material followed by computed tomography (CT) allows much more accurate localization of tumors [10] than either myelography or unenhanced CT. Reformatted CT images can be obtained in cooperative patients in both the sagittal and coronal planes and are useful in assessing the full extent of the lesion, especially in the cephalocaudad dimension. CT without intrathecal contrast material has limited value in the diagnosis of intramedullary and intradural extramedullary lesions but is quite useful in evaluation of extradural lesions. Intrathecal contrast material is generally required to evaluate spinal canal involvement but may be contraindicated in patients with subarachnoid block because of the significant risk of neurologic worsening [11]. The use of intravenous contrast material usually allows good distinction between an extradural soft-tissue tumor and adjacent normal soft tissues.

Magnetic resonance imaging (MRI) is generally superior to CT in the evaluation of spine neoplasms [12–14]. This is particularly evident in the intradural and intramedullary spaces but also generally applies to the extradural compartment [15]. In the evaluation of primary extradural bone tumors, CT and MRI are often complimentary. The main advantages of MRI compared with CT (even with intrathecal contrast material) are lack of ionizing radiation, superior soft-tissue contrast, lack of bone artifacts, no requirement for intrathecal contrast material, and multiplanar imaging capabilities [13]. The ability to image long segments of the spine reduces the need for accurate preimaging localization. In general, at least two imaging planes should be obtained for accurate localization of the tumor. Since MRI visualizes the cord directly, distinction between solid intramedullary tumors and other causes of cord expansion such as syringomyelia is facilitated. Disadvantages of MRI include increased susceptibility to motion artifact due to relatively longer scan times and limited sensitivity to calcification or small bone fragments that may aid in the differential diagnosis of some neoplasms.

Table 9.1.
Intramedullary Neoplasms

Ependymomas
- Most common intramedullary tumor
- Cauda and distal cord
- 4 to 8 cm length
- Often associated with syringomyelia
- CT: iso- to hypodense, enhance
- MRI: iso- to hypointense on T1, hyperintense on T2

Astrocytomas
- Most common intramedullary tumor in children
- Cervical > thoracic location
- Longer segment involved than ependymomas, may involve whole cord
- Often associated with syringomyelia
- CT and MRI: similar to ependymoma, may be more inhomogeneous

Hemangioblastoma
- 80% solitary
- One-third associated with von Hippel–Lindau disease
- Cervical involvement more common
- One-half associated with syrinx—look for tumor nodule
- Cord enlargement out of proportion to tumor size
- Meningeal varicosities, especially dorsal surface
- CT: marked homogeneous enhancement of hypodense mass
- MRI: hypointense on T1, hyperintense on T2—look for adjacent "flow voids" due to AV shunts

Intramedullary Tumors

Intramedullary tumors account for about 16 to 25% of neoplasms that involve the spinal cord or canal [2, 7, 9]. Gliomas account for perhaps 95% of all intramedullary neoplasms [7]. Of these, approximately 65% are ependymomas, 30% are astrocytomas, and 3% are oligodendrogliomas [7] (Table 9.1). Intramedullary metastases are rare [16]. In children, there is a slight predominance of astrocytomas relative to ependymomas. In adults, there is a slight predominance of astrocytomas in the cervical region. The incidence of astrocytomas and ependymomas in the thoracic region is roughly equal, and there is a great predominance of ependymomas in and below the conus medullaris [7].

Intramedullary tumors generally produce a fusiform cord enlargement over at least several segments and may involve the entire cord [17]. There is usually no associated spinal subarachnoid block [9]. Because of the long segment involvement, the signs and symptoms of intramedullary tumors are more variable than those of the extramedullary tumors [8].

Pain is a common presenting feature in 68% of patients with intramedullary tumors [2]. The pain is often nonspecific, and its significance may not be recognized. Pain may initially be due to involvement of or traction on the dorsal nerve roots [2]. As tumor size increases, pain also may be secondary to pressure on the anterolateral tract [9]. Its localization value is limited, since it may mimic other conditions such as intercostal neuralgia, cholecystitis, or sciatica. Some patients may complain of vague back pain intermittently over several years which may be either persistent or associated with long symptom-free periods. Findings suspicious for neoplasm include pain that awakens the patient at night, as well as pain that is aggravated by coughing, sneezing, or straining [2].

Motor or sensory disturbances are the first symptom of cord tumor in slightly less than 30% of patients [2]. Less than 3% have bowel or bladder dysfunction as the initial complaint. Motor findings include weakness, clumsiness, stiffness, or limping. These findings are related to disturbances of the corticospinal tract (spasticity, hyperreflexia, clonus, and absence of muscular atrophy) or damage to the motor neurons themselves (hypotony, loss of reflexes, signs of muscle atrophy in affected muscle distribution). The motor neuron signs are helpful in localization of the neoplasm.

Sensory findings include numbness, tingling, a cold feeling, or a complete anesthesia in a specific area [2]. Posterior column involvement may lead to ataxia and loss of position and vibratory sense. The upper level of involvement is most helpful in localization, but this is not always clear. Although bowel and bladder dysfunction is rarely the initial complaint, it not uncommonly develops fairly early with cauda equina lesions [8].

GLIOMAS

The great majority of gliomas are either ependymomas or astrocytomas [7]. Since imaging characteristics are similar, these two neoplasms will be considered together. The differential diagnosis between ependymomas and astrocytomas is based on size, location, and patient age.

Ependymomas are the most common intramedullary tumor and account for roughly 13% of all spinal cord tumors [9]. These are much more common [2] in the distal segment of the cord. Nearly 50% involve the cauda equina [2]. Cervical or thoracic ependymomas are generally fusiform shaped, but lesions involving the cauda equina or filum terminale may appear as bulky, extramedullary masses. These may occasionally fill the entire lumbar canal [7]. The classic ependymoma is 4 to 8 cm in length [7, 18] and rarely extends greater than five vertebral segments in length. Symptoms often begin in the midthirties, and there is a distinct male (60%) predilection [18]. Rarely, ependymomas may present acutely with subarachnoid hemorrhage and sciatica (Fincher's syndrome). Unlike intracranial ependymomas, cord ependymomas are extremely rare in children under age 4 [18].

Astrocytomas are less frequent than ependymomas, accounting for 30% of intramedullary tumors [2]. However, in the pediatric population, astrocytomas are more common than ependymomas [19]. Astrocytomas tend to involve a longer segment of cord than ependymomas and may extend over greater than five vertebral segments [7]. In the pediatric population, a cervical location is more common [19]. The tumor may be associated with holocord enlargement [17]. Although generally solitary, multiple lesions may be present in central neurofibromatosis [20]. There is a small male predominance with the more benign histologic grades (grades 1 and 2), and the sex incidence is equal for the malignant variety [2].

Bony changes are similar for both these slowly growing neoplasms and are better demonstrated with CT (Fig. 9.1). Plain film findings are evident in less than a third of patients. These include enlargement of the vertebral canal, posterior scalloping, and pedicle or lamina erosion [18]. Progressive scoliosis in a child is also a useful finding [18]. Calcification is extremely rare and should suggest other diagnoses. Myelography may show fusiform cord enlargement and help in localization but lacks in specificity [6].

Unenhanced CT is generally of little value (Fig. 9.2). The mass is usually isodense or only slightly hypodense compared with the normal thecal sac [21]. Hyperdense areas may represent hemorrhage and less often calcification. Intravenous contrast-enhanced CT (CECT) is useful because these tumors may show varying degrees of enhancement. CECT is useful in identification of the neoplasm and in distinguishing solid from cystic components [22]. Enhancement patterns are nonspecific, with reports of homogeneous, patchy, and ringlike enhancement [22]. Enhancement near the central canal is reportedly more suggestive of ependymoma [21]. CECT is also useful in evaluating recurrent tumor.

Neoplasms involving the cauda equina tend to be large and bulky and are often associated with a complete block on myelography [6]. Postmyelography CT (CTM) will often demonstrate some opacification proximal to the mass, allowing for accurate localization. Lesions involving the cord are associated with fusiform enlargement of the involved segments with associated thinning of the opacified subarachnoid space. Syrinxes may have a similar appearance. Nodularity or asymmetry of expansion favors tumor over syrinx [7]. Delayed CT in 6 to 12 hours is useful in identifying associated syrinxes, which may be opacified in 50% of cases. However, this is nonspecific because contrast material also may penetrate tumoral cavities [24].

Although myelography and CTM are both useful in demonstrating cord enlargement, neither reliably discriminates between solid and cystic components of a neoplasm [25]. MRI has largely replaced myelography in the evaluation of intramedullary lesions. MRI demonstrates the contour abnormality of the cord and allows direct visualization of associated cysts and tumor nodules [25–27] (Figs. 9.3 through 9.6). Intramedullary lesions typically have prolongation of both T1 and T2 values; that is, tumors are hypointense to cord on T1-weighted images and hyperintense on T2-weighted images. However, the T1 values are not always significantly different. The signal intensity of the tumor may be relatively isointense to cord on the T1-weighted images [25]. The signal changes in ependymoma tend to be more inhomogeneous than with astrocytomas [25]. Signal changes associated with some ependymomas may be relatively minimal, and the nodule may only be

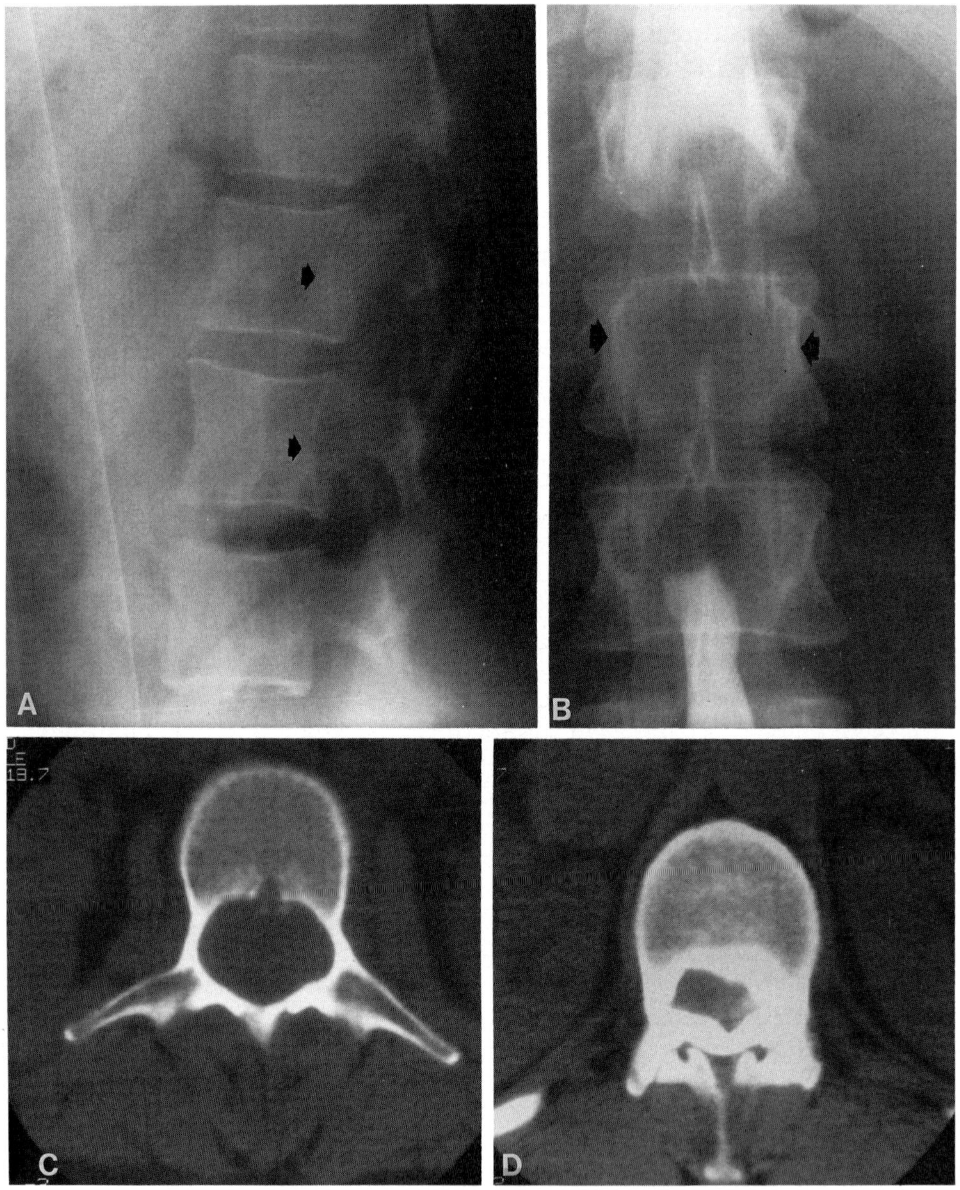

Figure 9.1. Lumbar ependymoma. A 27-year-old woman with back pain. **A,** Posterior scalloping of L1 and L2 vertebral body. **B,** Intradural mass associated with complete block. Contrast injected from above and below. Note the pedicle thinning (*arrows*). **C,** Unenhanced CT demonstrating thinning of the pedicles and lamina as well as some posterior vertebral body scalloping. Mass is isodense with thecal sac and cannot be discerned. **D,** CT with intrathecal contrast demonstrating a lobulated mass centrally in the canal. **E,F,** Proton (TR1700/TE30) and T2-weighted (TR1700/TE80) sagittal images demonstrating the lobulated mass to extend from L1 through L3 vertebral levels. **G,** Contrast-enhanced sagittal T1-weighted image (TR517/TE20) demonstrating prominent enhancement of the ependymoma. Note the thickening of the filum terminale distally (*arrows*).

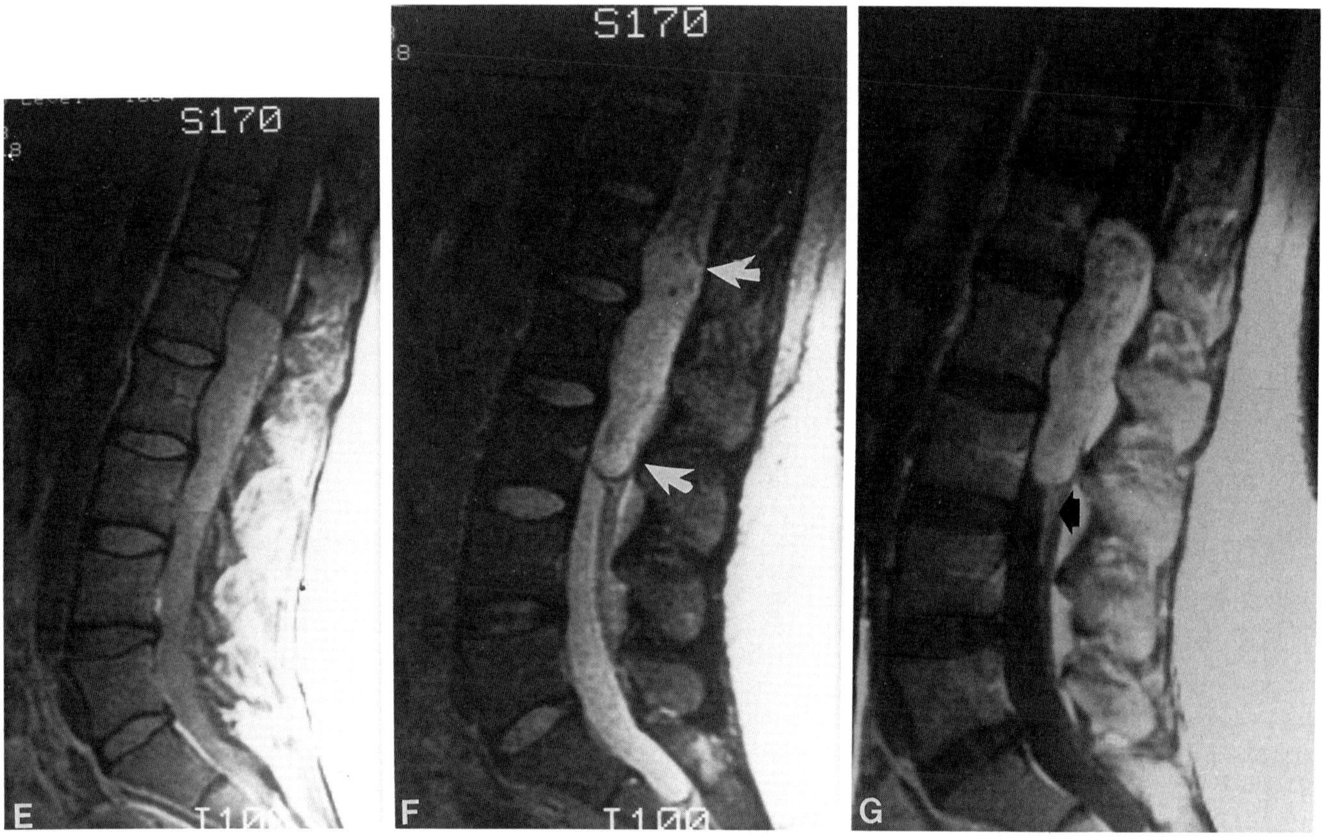

Figure 9.1–*continued*

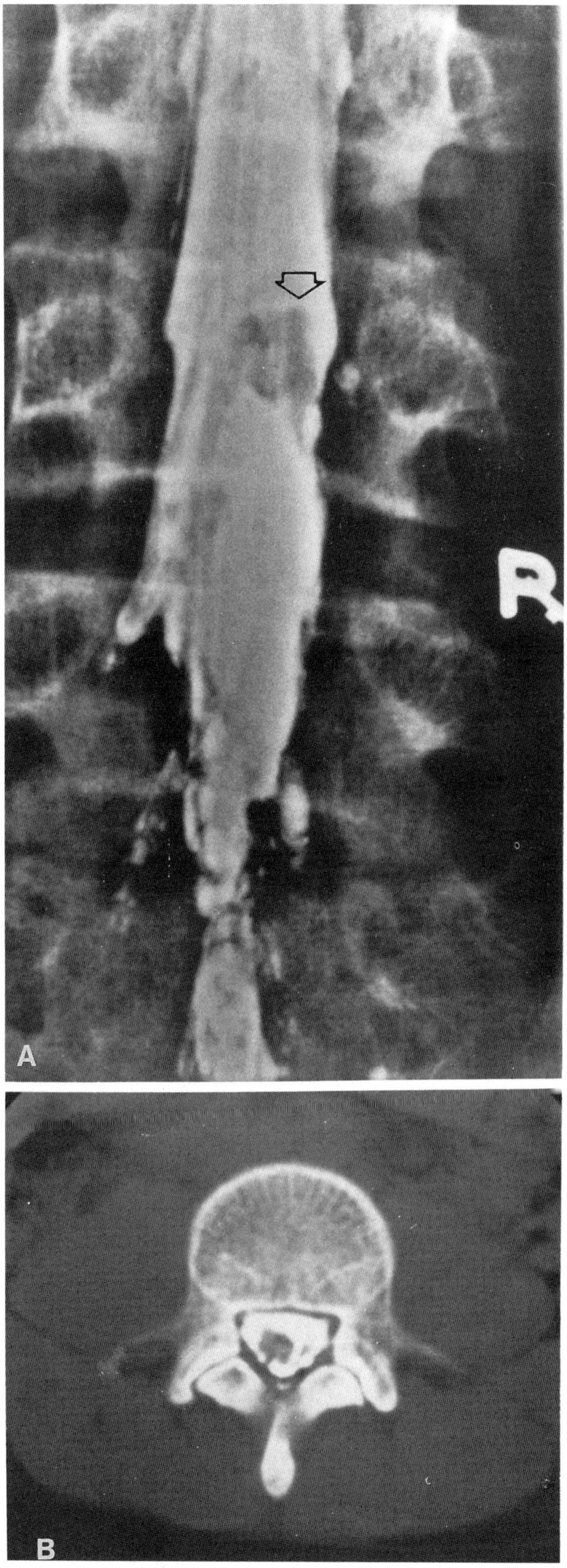

Figure 9.2. Ependymoma. A 17-year-old girl with normal unenhanced CT. **A,** Myelogram demonstrates an intradural mass at L4. **B,** Postmyelogram CT better demonstrates a lobulated mass within the thecal sac.

recognized due to the surrounding cystic fluid which has prolonged T1 and T2 values.

No specific morphologic or signal characteristics reliably distinguish ependymoma from astrocytoma. Both can be solid, cystic with one or more nodules, or of variable length [25]. However, some generalizations can be made. A solid tumor involving almost all the spinal cord with a prolonged T2 relaxation time in a young patient is more likely to be an astrocytoma. A less extensive tumor that is inhomogeneous as far as T1 and T2 relaxation times is most likely to be an ependymoma [25]. Paramagnetic contrast material improves the sensitivity and specificity in the evaluation of spinal neoplasms [15, 27, 28]. The majority of intramedullary neoplasms demonstrate enhancement [28]. Enhancement allows more accurate assessment of lesion size and configuration and precise localization of the lesion. Contrast enhancement also may help identify additional lesions and is useful in distinguishing tumor from edema.

There is a well-known association between intramedullary tumors and spinal cord cavitation [29]. Poser [29] noted that 31% of intramedullary tumors were associated with "syringomyelia." In another series, 46% of the intramedullary ependymomas and 38% of the astrocytomas were associated with "syringomyelia" [2]. *Syringomyelia* refers to abnormal longitudinally oriented cysts which are glial lined and are often associated with tumor, trauma, or arachnoiditis.

The cavitation associated with intramedullary lesions such as ependymoma may be relatively small, located within the tumor or involving only one or two segments above or below the tumor. These cavities or intratumoral cysts result from transudation from the tumor [30]. However, larger cavities (*syringomyelia*) may extend well away from the tumor and involve most of the cord. Distinguishing between the intratumoral cysts and associated nonmalignant syringomyelia is important for surgical planning, since the associated syrinx requires only decompression. In patients with an expanded cord, the combination of distinct margins, uniform signal intensity, and isointensity of cyst fluid with cerebrospinal fluid (CSF) is characteristic of hydromyelia or syringohydromyelia [31]. Since intramedullary tumors tend to be infiltrative and associated with edema, margins tend to be poorly defined. In addition, findings associated with malignancy, such as necrosis, cystic degeneration, or hemorrhage, lead to an inhomogeneous MRI appearance. Highly proteinaceous fluid within zones of cystic degeneration may produce hyperintensity or heterogeneous signal intensity on MRI [31]. The combination of poorly defined margins, signal inhomogeneity, and lack of isointensity is suggestive of neoplasm. The lack of contrast enhancement is also reassuring when evaluating syringomyelia.

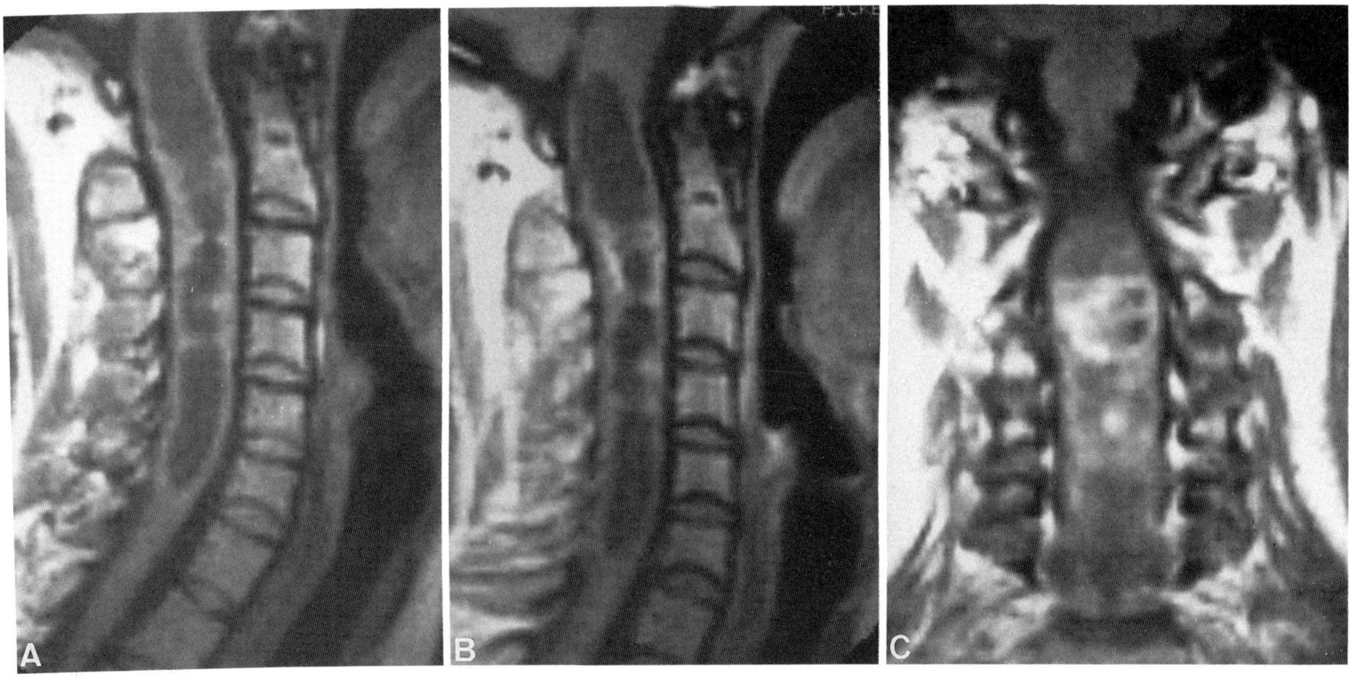

Figure 9.3. Glioma. A 34-year-old woman with neurofibromatosis and history of bilateral acoustic neuromas. **A,** Sagittal T1-weighted image (TR800/TE22). Fusiform enlargement of the cervical cord from C1 through C6 associated with syringomyelia. Irregular margin between syrinx and cord is suspicious for neoplasm, particularly evident at C2 and C3. **B,** Sagittal contrast-enhanced T1-weighted image (TR800/TE22). Abnormal enhancement is demonstrated at C2 and C3, compatible with neoplasm. **C,** Coronal postcontrast T1-weighted image (TR800/TE22) also demonstrates abnormal enhancement at the C2–3 level.

Subarachnoid seeding with intracranial metastases have been reported following surgical resection of malignant cord astrocytomas [37]. Radiation therapy seems to improve prognosis and functional recovery only in the category of patients who have had incomplete surgical resections [9, 34].

OLIGODENDROGLIOMA

Oligodendrogliomas account for 3 to 4% of intramedullary tumors and tend to present later in life [7, 21], usually in the fourth or fifth decade. They are clinically indistinguishable from other gliomas [21]. There is considerable variability in vascularity, and loops of vessels may be seen around necrotic areas [20]. Rupture of degenerated vessels may lead to hemorrhage, and areas of calcification are not infrequent.

Calcification and hemorrhage are best demonstrated on unenhanced CT [7]. On T1-weighted MRI sequences, hemorrhage in the methemoglobin stage has high signal [7], whereas older areas of hemorrhage may show marked hypointensity on T2-weighted images due to hemosiderin deposition. MRI is not useful in differentiating these neoplasms from other intraspinal gliomas [13].

Metastases from oligodendroglioma usually spread via the CSF [20]. Visceral metastases are rare. Oligodendrogliomas have a tendency to recur, and both recurrent tumor and metastases may dedifferentiate into glioblastoma.

HEMANGIOBLASTOMA

Hemangioblastomas account for 1 to 3% of intramedullary tumors, although the actual incidence may be higher [38]. Symptoms often begin in the fourth decade, but neurologic findings may be minimal and the lesion may never be diagnosed [21]. About 80% are solitary, but tumors may be present at multiple sites when associated with von Hippel–Lindau disease [7]. Although cerebellar hemangioblastomas are common in von Hippel–Lindau disease, the reported incidence of spinal cord involvement can range from 5 [39] to 33% [21].

Cervical involvement is more common than thoracic, but even the cauda equina may be involved [40]. About 60% are intramedullary, but a significant number may be both intramedullary and intradural extramedullary [21]. Most are located dorsally in the spinal canal [21, 40]. Lesions may be cystic, and approximately 50% are associated with syringomyelia [21]. When cystic, a tumor nodule can often be identified [7].

Since these are slowly growing neoplasms, plain films may demonstrate evidence of spinal canal enlargement. Over half the cases will show meningeal varicosities on myelography or angiography [21]. These serpentine vascular structures are usually seen over the dorsal surface of the cord [41]. Enhanced CT demonstrates marked homogeneous enhancement of a hypodense or isodense mass and also may demonstrate the adjacent enlarged vessels

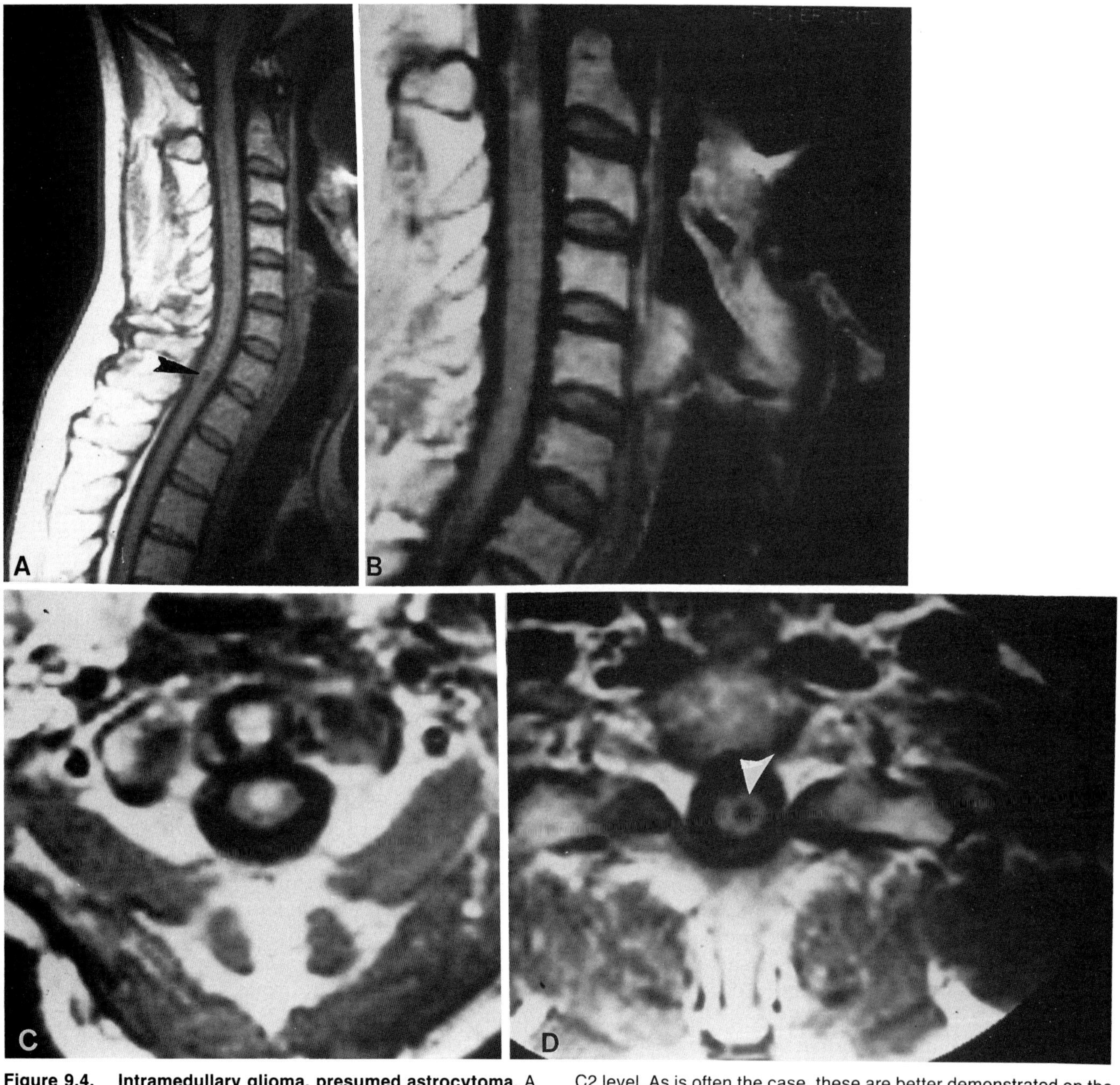

Figure 9.4. Intramedullary glioma, presumed astrocytoma. A 47-year-old neurofibromatosis patient with a history of acoustic neuromas. **A,** Sagittal T1-weighted image (TR800/TE22) demonstrates mild fusiform expansion of the cord from C2 through C7 with a small intramedullary fluid collection at C7–T1 level (*arrowhead*). **B–D,** Corresponding contrast-enhanced sagittal T1-weighted images. Three small areas of intramedullary enhancement are noted at the C2 level. As is often the case, these are better demonstrated on the corresponding axial images (**C**). Additional postcontrast axial images through the intramedullary fluid collection at the C7–T1 level (**D**) demonstrate no abnormal enhancement. This, associated with the CSF signal intensity of the intramedullary fluid, suggests that this is a syrinx rather than an intratumoral cyst.

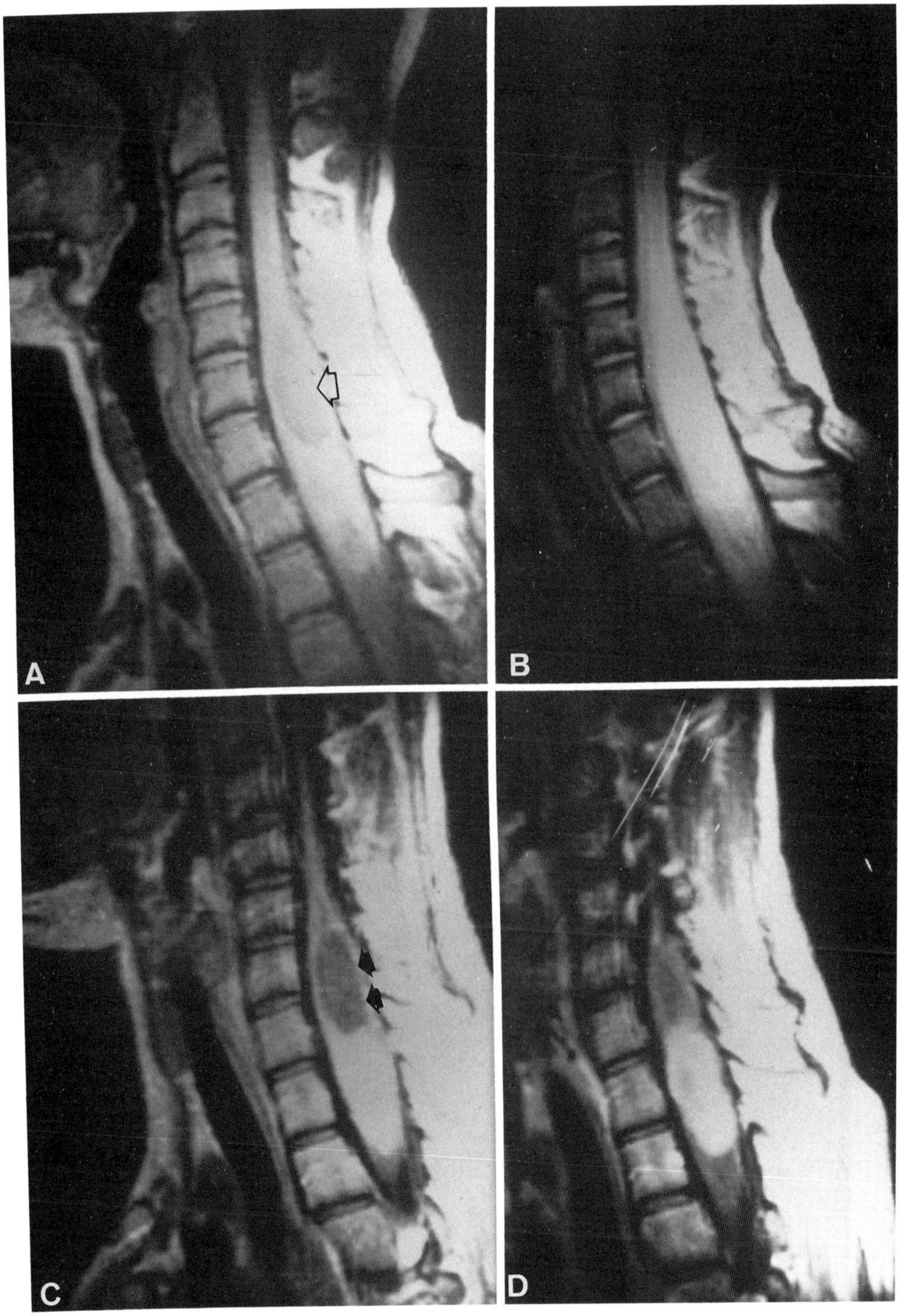

Figure 9.5. Cervical ependymoma with associated syrinx. A 21-year-old woman. **A,** Sagittal T1-weighted image (TR800/TE20). Fusiform cord enlargement extending from C5 through T2. Isointense mass associated with intramedullary fluid collection at the C6–7 level. **B,** Proton-density image (TR2000/TE30) demonstrates hyperintensity of both cord and syrinx with loss of distinction. **C,D,** Adjacent postcontrast sagittal T1-weighted images (TR800/TE20) show homogeneous enhancement of the ependymoma located at C7 through the T2 level. Note the smoothly defined margin between the syrinx and expanded cord at the C6 level as well as absence of enhancement at this level. This suggests associated syrinx rather than intratumoral cyst (*arrow*).

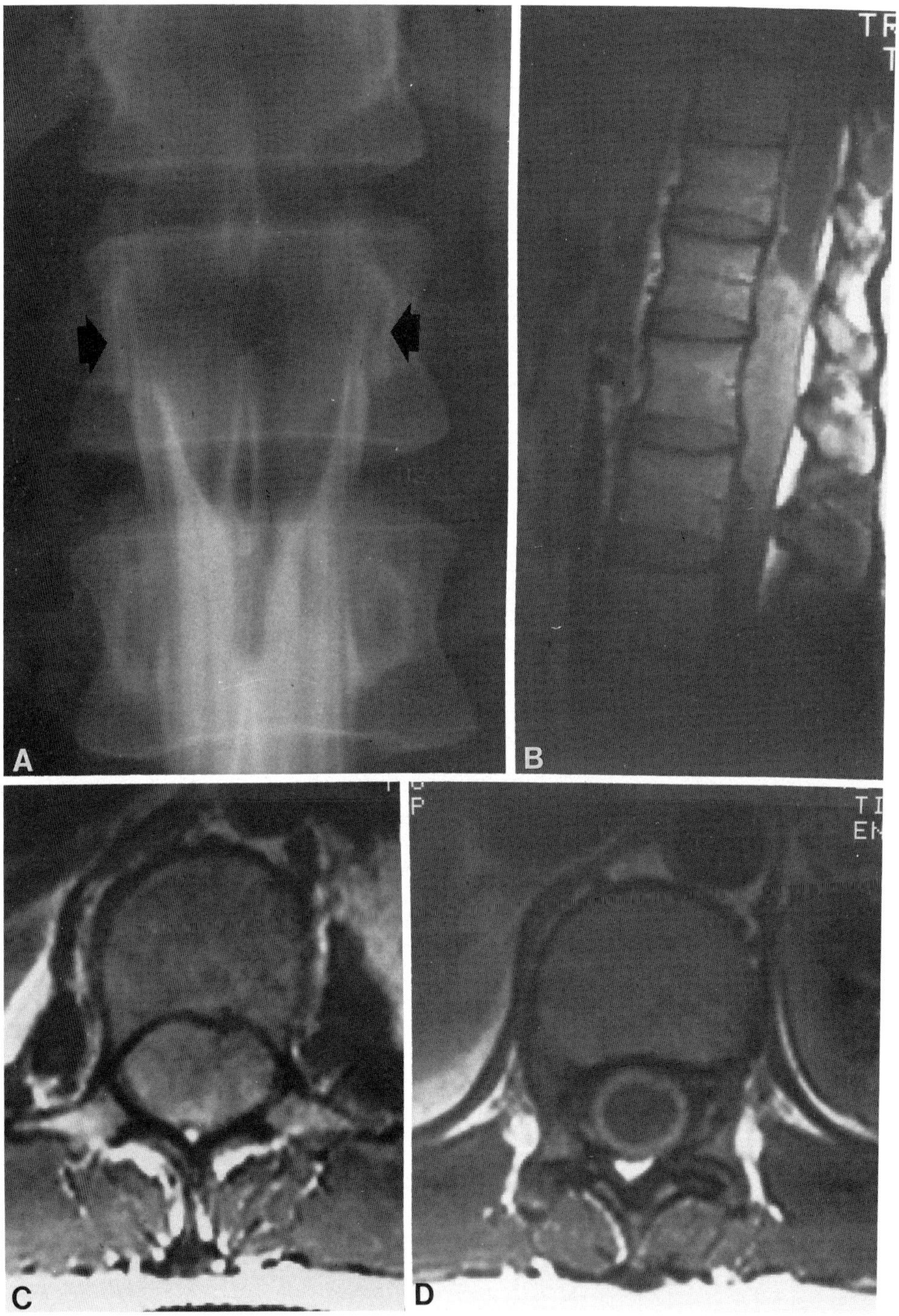

Figure 9.6. Conus ependymoma with large associated syrinx. A 25-year-old woman. **A,** Myelogram demonstrates a smoothly defined intradural mass at the L1–2 disk level which is associated with a complete block. Note the pedicle thinning (*arrows*). **B,** Sagittal T1-weighted image (TR500/TE29). Lobulated, isointense (to cord) mass of conus medullaris. Note the smoothly defined syrinx involving the thoracic cord. Mild posterior vertebral body scalloping is also present. **C,** Axial T1-weighted image (TR900/TE29) through mass again demonstrates isointense mass with scalloping more evident. **D,** Axial T1-weighted image (TR900/TE29) through the lower thoracic cord syrinx demonstrates a smooth margin between the expanded cord and the syrinx. Note that the intramedullary fluid collection is isointense with the subarachnoid CSF.

[21] (Fig. 9.7). Contrast enhancement may allow distinction between the vascular tumor nodule and adjacent syrinx [27].

MRI is well suited in the evaluation of hemangioblastomas and may eliminate the need for time-consuming angiography. Multiple spinal lesions or concurrent involvement of the posterior fossa can be detected (Fig. 9.8). These tumors are associated with prolongation of both the T1 and T2 relaxation times (hypointense on T1- and hyperintense on T2-weighted images) [41]. Cystic components have higher signal intensity than CSF on the intermediate or T2-weighted image. The high-flow arteriovenous shunts are associated with dilated afferent and efferent vessels which may be seen as adjacent tubular areas of "flow void" [39, 41].

Cord enlargement out of proportion to the actual size of the hemangioblastoma is an unexplained phenomenon that appears to be unique to this type of tumor [38]. The cord enlargement extends beyond the associated cyst. Theories include diffuse cord edema secondary to marked venous congestion associated with the prominent arteriovenous shunting, as well as possible production of some edema-producing factor by the tumor [38].

Since these tumors are highly vascular, intense enhancement with paramagnetic contrast material is common [27]. Contrast-enhanced MRI is useful in identifying both the vascular nodule of the tumor and the presence of other lesions along the entire extent of the cord or posterior fossa. In the absence of a cystic component with nodular enhancement, differentiation from a high-flow type of spinal vascular anomaly may be difficult.

INTRAMEDULLARY METASTASES

Intramedullary spinal cord metastases are rare [16, 42], with less than 5% of central nervous system (CNS) metastases involving the cord [43]. Intramedullary metastases are usually associated with epidural metastases. Solitary intramedullary involvement is quite rare [44].

Intramedullary metastases usually result from leptomeningeal spread of primary CNS lesions or hematogenous spread of other primaries. Fifty percent of intramedullary cord metastases are from bronchogenic carcinoma [16]. Breast cancer is the next most frequent primary neoplasm, accounting for 15%. Others include lymphoma, colorectal, head and neck, renal cell, and melanoma. Pain and weakness were the most frequent presenting complaints, and neurologic examination revealed weakness or paralysis in all cases. In 34% of cases, neurologic symptoms were the initial manifestation of systemic malignancy [16].

Plain films are usually normal due to the rapid growth of these lesions [46]. Since clinically significant cord lesions may be quite small with limited edema, nearly half the myelograms may be normal [16]. Nonspecific fusiform cord enlargement may be seen. CT reports are limited, but hypo- or hyperdense enhancing masses have been reported [7, 16, 45]. Postmyelography CT may show cord enlargement but is of limited value in smaller lesions with little enhancement. Diffusion of intrathecal contrast material into the lesion may result in enhancement either acutely or delayed up to 20 hours [44].

As with the other cord lesions, MRI is the imaging modality of choice. The entire cord can be evaluated and small areas of signal abnormality identified. As with other intramedullary lesions, the metastatic lesions are generally hypointense on T1- and hyperintense on T2-weighted images. Contrast enhancement aids in detection of small lesions (see Fig. 9.31).

The prognosis for patients with intramedullary metastases is grave, and the role of screening MRI in detection of clinically occult spinal metastases is still not clear. Over half such patients die within 1 month of the initial medical intervention for the neurologic dysfunction. Fewer than 20% are alive at 3 months [16].

Intramedullary Pseudotumors

The differential diagnosis of an intramedullary lesion includes primary neoplasm, metastatic neoplasm, syringomyelia or hydromyelia, nonneoplastic lesions such as multiple sclerosis and sarcoidosis, radiation myelitis, and the vascular malformations. As with neoplasms, MRI is the most useful imaging modality in sorting out these differential possibilities.

The distinction between neoplasm and syringomyelia or hydromyelia may occasionally be difficult, especially since syringomyelia and neoplasm may coexist [31–33]. In general, the combination of distinct margins, uniform signal intensity, and isointensity of the fluid within the syrinx with cerebral spinal fluid (CSF) in a patient with an expanded cord suggests the diagnosis of syringo- or hydromyelia [31]. In contrast, cysts associated with tumor are often ill-defined because of infiltration and edema. The neoplastic fluid within the cord often has inhomogeneous signal because of necrosis, cystic degeneration, or hemorrhage. Therefore, poorly defined margins, inhomogeneity of adjacent cord, and lack of uniform signal isointensity are more suggestive of neoplasm [31].

Multiple sclerosis (MS) frequently involves the spinal cord and can be demonstrated on MRI. These

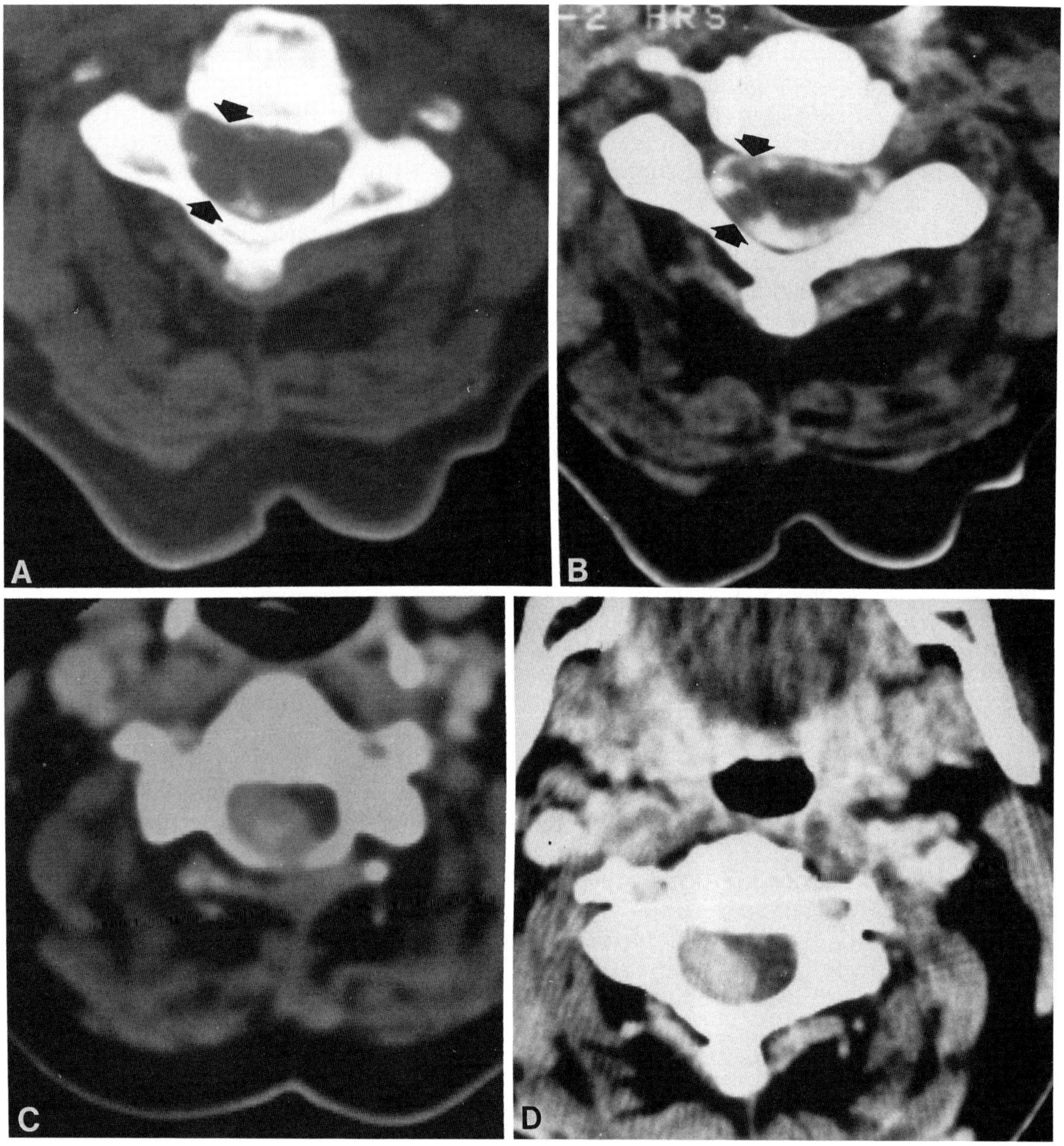

Figure 9.7. Hemangioblastoma. A 54-year-old woman with history of cerebellar hemangioblastoma resected 10 years prior. **A,B,** CT scans with intrathecal contrast. Involvement is primarily intradural extramedullary, and a well-defined mass isointense to the cord displaces the cord to the left (*arrows*). **C,D,** Intravenous contrast-enhanced CT scans. There is prominent enhancement of the surgically confirmed hemangioblastoma. **E,F,** Early and late arterial phase lateral vertebral angiograms demonstrating dense abnormal vascular blush from C1 through C4 level.

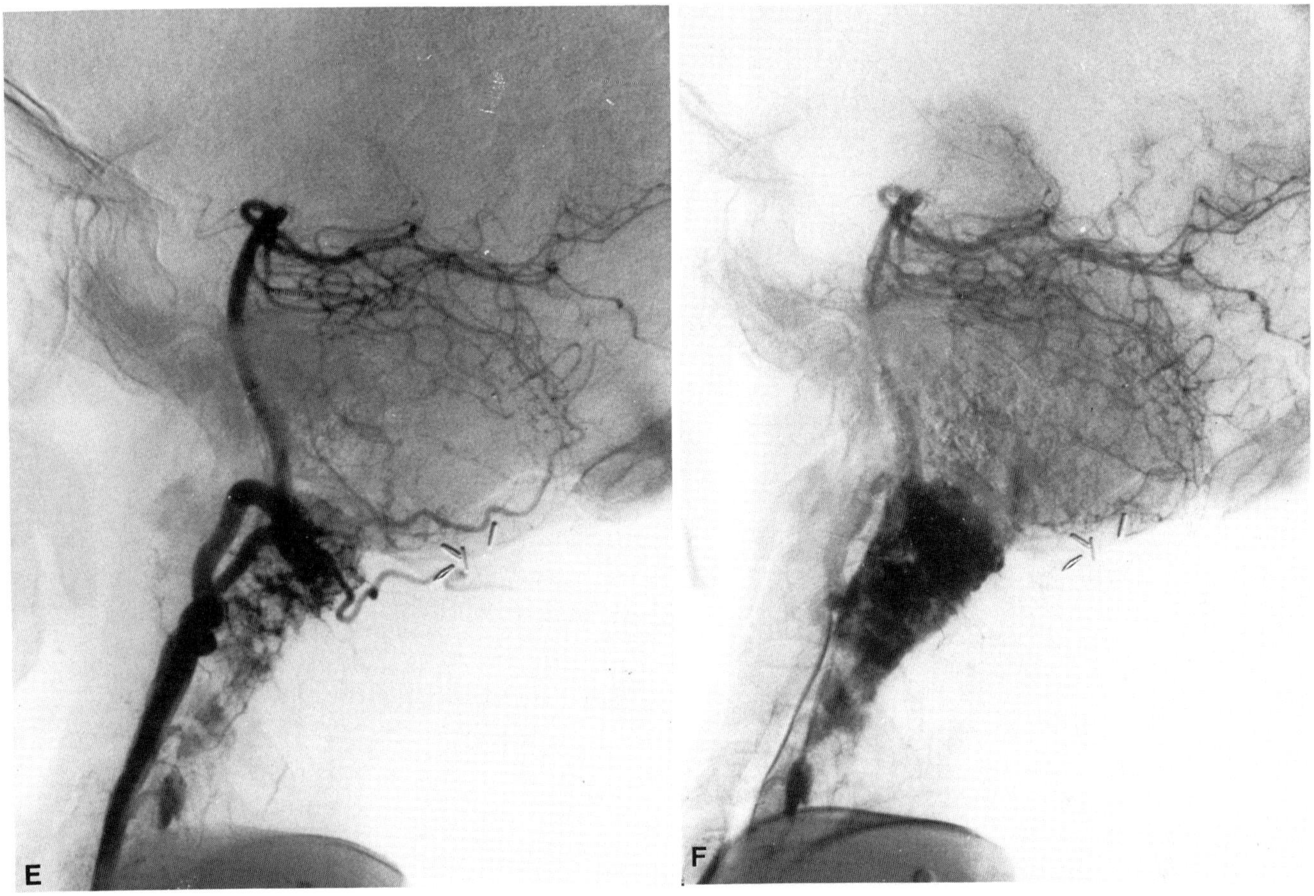
E
F

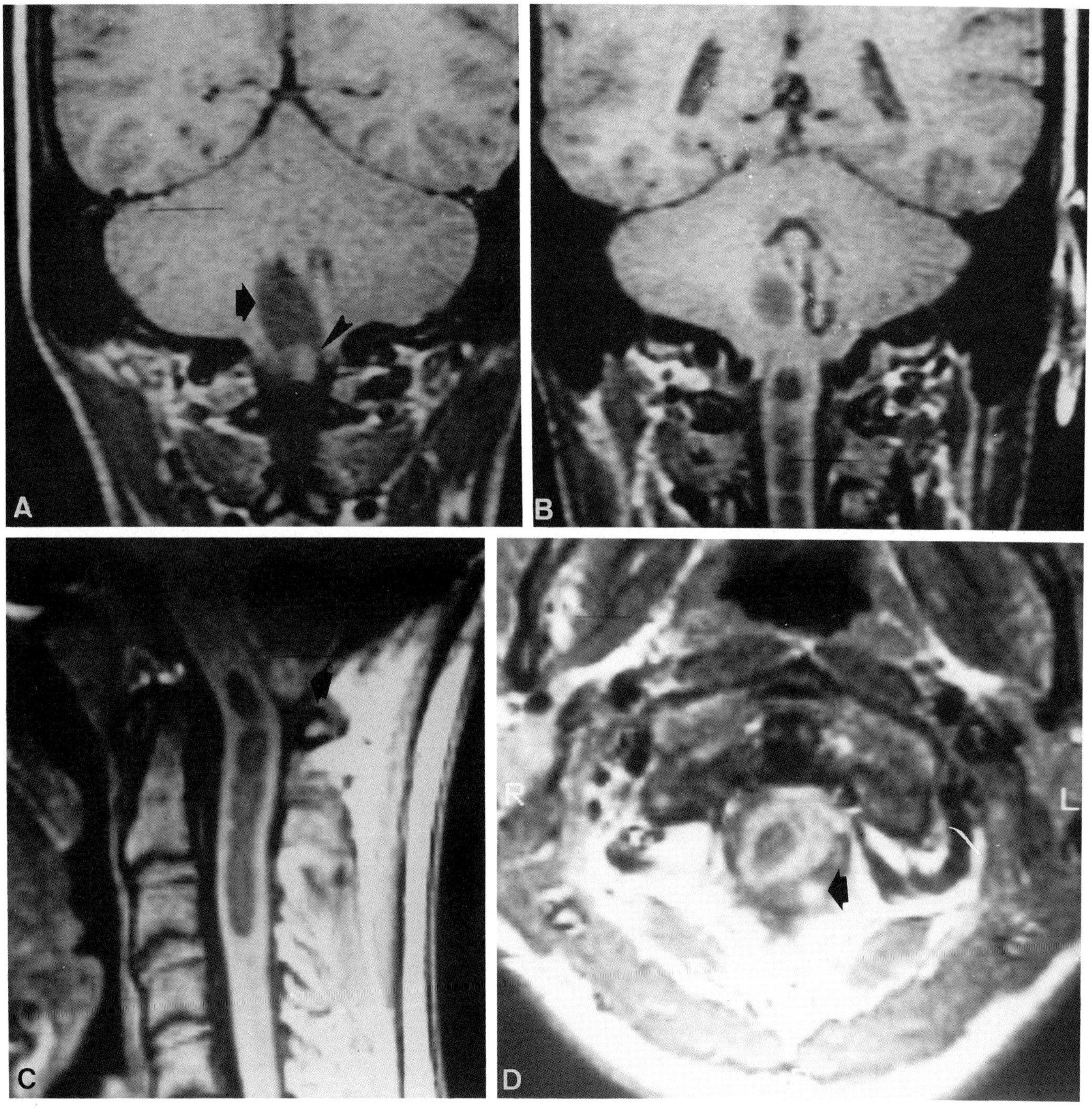

Figure 9.8. Cerebellar hemangioblastoma with cervical syrinx. A 20-year-old woman. **A,B,** 1.5-T coronal T1-weighted images (TR600/TE20). Cystic cerebellar mass (*arrow*) with tumor nodule (*arrowhead*). Note associated cervical syrinx with suggestion of multiloculation. **C,** Sagittal T1-weighted image (TR700/TE30) demonstrating mild cervical cord expansion by syrinx. **D,** Axial T1-weighted image (TR1000/TE20) demonstrating eccentric syrinx. Note the cerebellar tumor nodule evident in both **C** and **D** (*arrow*). Paramagnetic contrast would have been useful in determining whether the cervical syrinx was secondary to a cerebellar hemangioblastoma or a concurrent cervical hemangioblastoma.

patients often present with long tract symptoms but are frequently distinguished from neoplasm patients by the remitting course, signs and symptoms of more than one lesion, or characteristic CSF laboratory findings [8, 47]. However, when MS presents as a solitary focal progressive lesion, it may mimic a tumor. Approximately half of MS patients will have areas of abnormal T2 hyperintensity during acute exacerbation. There may be focal swelling of the cord, and active plaques tend to enhance with paramagnetic contrast material [47, 48] (Fig. 9.9). Patients with long-standing lesions may have distal cord atrophy. In these patients, MRI of the brain should be obtained, since most MS patients also will have the

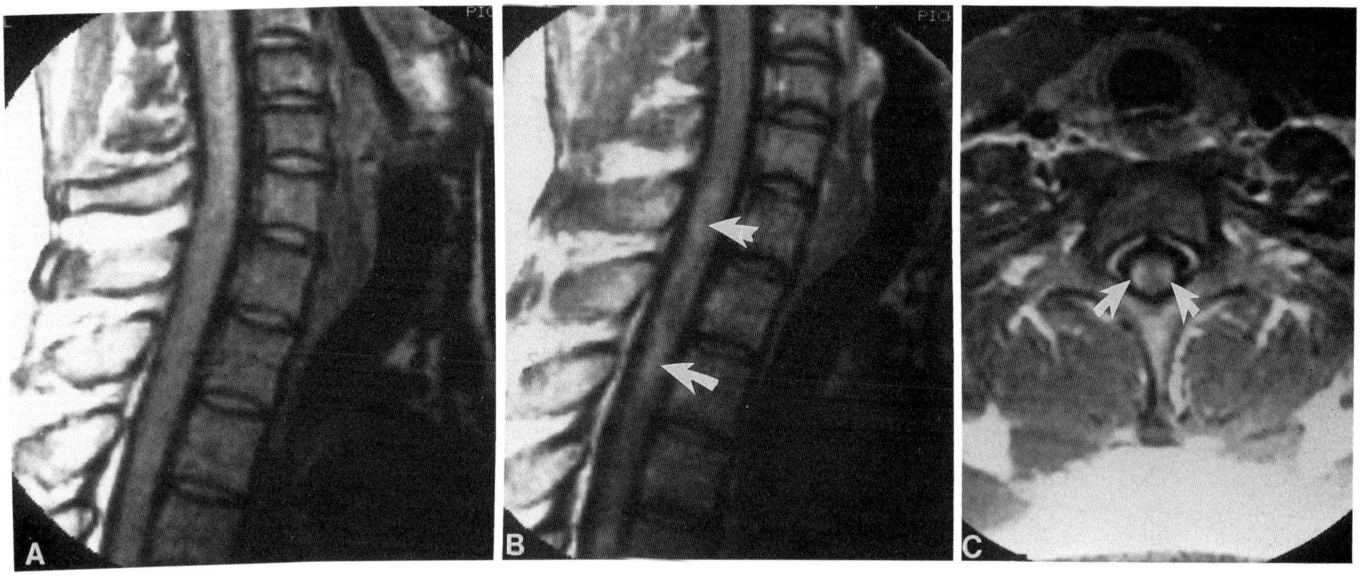

Figure 9.9. Multiple sclerosis. A 28-year-old black man presented with progressive ascending lower extremity weakness. **A,** Sagittal T1-weighted image (TR800/TE22) demonstrates a normal-appearing cord. A proton-density image (not shown) suggests some vaguely increased signal at the cervical thoracic junction. **B,** Postcontrast sagittal T1-weighted image (TR800/TE22). Abnormal enhancement at the cervicothoracic junction. **C,** Axial T1-weighted image (TR800/TE22) postcontrast image demonstrates fairly symmetrical abnormal cord enhancement at the cervicothoracic junction. Initial brain MRI was normal, but characteristic periventricular plaques have subsequently developed, and the diagnosis of multiple sclerosis has been clinically confirmed. When associated with cord expansion, multiple sclerosis may be difficult to distinguish from neoplasm on a single examination.

characteristic periventricular white matter lesions that are hyperintense on T2-weighted images. However, approximately 12% of patients with MS of the spinal cord may have normal brain MRI examinations [47]. Diagnosis may require follow-up MRI.

Primary involvement of the spinal cord by sarcoidosis is rare but may present as a solid intramedullary mass and may resemble a neoplasm [49]. There is a predilection for the cervical and upper thoracic cord. MRI often demonstrates a long, expansile, infiltrative intramedullary lesion. Approximately a third of cases are intramedullary, with the remainder being either extramedullary or a combination of intra- and extramedullary [49]. Paramagnetic contrast–enhanced images may demonstrate areas of patchy multifocal parenchymal enhancement of the intramedullary lesion. In addition, unusual areas of linear peripheral enhancement have been reported, presumably due to leptomeningeal involvement [49]. Leptomeningeal involvement would be unusual for astrocytoma. The patchy infiltrative cord involvement and cervicothoracic junction location would be atypical for ependymoma. The peripheral enhancement pattern is unusual for either tumor, since these generally involve the full diameter of the cord [49].

Other than sarcoidosis and multiple sclerosis, radiation myelopathy and paraneoplastic necrotizing myelopathy may occasionally need to be differentiated from spinal metastasis. Paraneoplastic necrotizing myelopathy, although not the result of cancer, is usually seen in patients with systemic cancer. This myelopathy accounts for less than 2% of the neurologic complications associated with cancer. Usually it precedes or follows the diagnosis of systemic cancer. The disease processes classified under this group include, among others, subacute necrotizing myelopathy and amyotrophic lateral sclerosis (see Chap. 10).

Radiation myelopathy is one of the more serious neurologic complication of radiation therapy. This may present as *acute transient myelopathy* or *acute transverse myelitis*. Acute transient myelopathy usually occurs within a few weeks following radiation therapy with a positive Lhermitte's sign. It lasts for a few months with complete recovery and is presumed to be due to reversible demyelination. Clinically, acute transient myelopathy needs to be differentiated from cord compression resulting from further progression of the metastatic disease. MRI is useful in demonstrating the absence, further progression, or new site of cord or root compression.

Acute transverse myelitis following radiation treatment is rare. The clinical presentation and MRI findings are similar to those of acute cord infarction. The cord infarction is presumed to be related to arterial occlusion. The most common complication of radiation therapy is *chronic progressive myelopathy*, which usually presents 6 to 12 months following the treatment. It is most common in patients with head and neck tumors, with only 6% of radiation myelopathy occurring in patients with lung or breast cancer. Progressive paraparesis is common. Although in occasional cases CTM may show cord swelling, the most common finding is that of atrophy. MRI demonstrates relative hypointensity on T1- and hyperintensity on T2-weighted images of the cord [50]. Even when the history of previous radiation therapy is unavailable, the diagnosis is based on the hyperintensity of the adjacent bone marrow on T1-weighted images due to postradiation fatty replacement [72] and the sharp margin of demarcation of the radiation portal.

Intradural Extramedullary Neoplasms

This category of tumors is primarily benign and consists of the most common primary tumors of the spinal canal, representing about 55% of all primary spinal neoplasms [2, 7] (Table 9.2). Most of these arise from either the nerve sheath (neurinoma/schwannoma or neurofibroma) or from the spinal cord meninges (meningioma) [7]. Exceptions include drop metastases which may seed the subarachnoid space and the occasional embryonal tumor. Imaging characteristics, patient age, sex, and tumor location all aid in the differential diagnosis.

These neoplasms are situated in the subarachnoid space, between the pia mater covering of the cord and the adjacent arachnoid mater. Expansion of these masses may displace and compress the cord. The cord displacement results in widening of the subarachnoid space on the side of the lesion and narrowing of the space on the opposite side. The mass itself is generally well-defined [6]. These characteristic features form the basis for the myelographic diagnosis, which can be applied equally to CTM and MRI.

Table 9.2.
Intradural Extramedullary Neoplasms

Meningiomas
- Marked female and thoracic spine predilection
- More common posterolateral
- Usually solitary (1–2% multiple, suspect NF)
- CT: often shows calcification, mass iso- to hyperdense, enhances
- MRI: iso- to hypointense mass with broad-based dural attachment, T2 variable, iso- to hyperintense to cord, enhances

Neurofibromas
- No age or sex predilection
- Often multiple
- One-third have extradural component (hourglass or dumbbell tumors)
- Calcifications rare
- CT: hypo- to isodense mass, enhances
- MRI: slightly hypointense on T1, hyperintense on T2

Lipomas
- Most are intradural
- CT: characteristic fat density mass
- MRI: hyperintense on T1, hypointense on T2

MENINGIOMAS

Meningiomas account for approximately 25% of all primary spine tumors. Although the overall number of nerve sheath tumors is higher at 29.5%, a significant number of these are extradural, and therefore, meningiomas are the most common intradural extramedullary tumor [2, 22]. These neoplasms have a marked female and thoracic spine predilection and tend to occur later in life than the nerve sheath tumors [1]. Meningiomas are unique among spine tumors in that there is a strong female predilection, with the sex ratio being greater than 4:1 [1, 51]. These neoplasms generally present in the fifth and sixth decades, with a mean age presentation of 50 years [1]. About 75% present between the ages of 40 and 70, and less than 10% have symptoms before age 30. These tumors are extremely rare in patients under 15 years of age [1].

Multiple meningiomas are found in less than 1 to 2% of patients and are associated with neurofibromatosis [21]. Approximately 80% of spinal meningiomas are found in the thoracic spine. However, the preference for the thoracic spine is found only in females [2, 51]. Only 47% of males demonstrated a thoracic location, whereas most of the remaining meningiomas are found in the cervical spine. Meningiomas in the lumbar region are rare.

Meningiomas are usually intradural, accounting for 85% of the total. About 7% have both intra- and extradural components, and about 8% are completely extradural [1]. Meningiomas may occur anywhere around the circumference of the spinal canal, but most have a posterolateral attachment to the insertion of the dentate ligament [1, 7]. Unlike meningiomas in the thoracic region, located posteriorly, these tumors are usually located anteriorly in the cervical region [51].

The average duration of symptoms is 23 months [51]. Pain is the most common presenting symptom, often preceding other signs and symptoms for months to years [51]. The pain is often nonspecific and is confused with other disorders such as multiple sclerosis, arthritis, arachnoiditis, and disk disease [51]. More than half the patients have local pain, and about 20% have radicular pain. Cord sensory tract damage may produce a painful dysesthesia at sites far removed from the cord lesion. These are termed *funicular pains* and are generally a manifestation of significant cord compression [1]. This deep, nonspecific pain is usually seen in the lower extremities and may prompt inappropriate therapy.

Motor disturbances are generally paresis (75%)

and less commonly complete paralysis. Motor disturbances are bilateral in about 80% [1]. The initial objective findings are usually related to long tract cord involvement. The pyramidal tracts are particularly susceptible to ischemia, and therefore, motor disturbances are seen early. The patient may complain of fleeting or slowly progressive weakness of the extremity or of increasing tiredness in them. Reflex changes are seen in 79%, usually hyperreflexia. Slowly growing tumors, such as meningiomas and neurofibromas, are generally associated with a spastic paresis.

Sensory disturbances usually start in the distal lower extremities and then slowly progress proximally to the level of the lesion. Such a progression is reportedly a pathognomonic presenting symptom of a spinal meningioma [1]. Early complaints include paresthesias and numbness, but objective sensory findings are seen relatively late.

Plain film abnormalities are evident in less than 10% and include bone erosion and widening of the interpedicular distance [7]. Adjacent bony sclerosis is occasionally present but is less frequent than that produced by intracranial meningiomas. Plain films may detect calcification in 3% [22].

CT is much more sensitive in detection of both calcification and adjacent hyperostosis [21]. The mass is generally isodense or hyperdense relative to the cord and often demonstrates calcification [7, 21 52] (Fig. 9.10). There is homogeneous enhancement with contrast material [52, 53]. Since the mass is often difficult to localize on the basis of clinical findings, myelography may be required to localize the tumor, prior to CT [7, 54]. CT typically demonstrates an intradural extramedullary mass with displacement of the cord away from the mass, ipsilateral widening (in adjacent sections) and contralateral narrowing of the subarachnoid space, and the characteristic acute interface between the mass and the subarachnoid space [6, 54]. Without contrast material, small tumors may be difficult to distinguish from dura and spinal cord [52].

MRI is useful in that neither a localizing myelogram nor a specific clinical level is necessary for successful imaging (Fig. 9.11). Unlike CT, MRI has been relatively insensitive in detection of calcification, a key differential point in distinguishing these tumors from nerve sheath neoplasms [55]. These studies should therefore be considered complementary.

The multiplanar capabilities of MRI usually demonstrate the relationship of the mass to the dura. A broad-based dural attachment is suspicious for meningioma [56]. Meningiomas are generally isointense or slightly hypointense compared with the cord on T1-weighted images [7, 57] (Fig. 9.12). On T2-weighted images, the mass usually remains isointense with the cord and is hypointense relative to the bright CSF [57, 58] (Fig. 9.13). This T2 pattern is somewhat variable, and the tumor may occasionally be hyperintense relative to the cord [7]. Although uncommon, syringomyelia related to associated cord

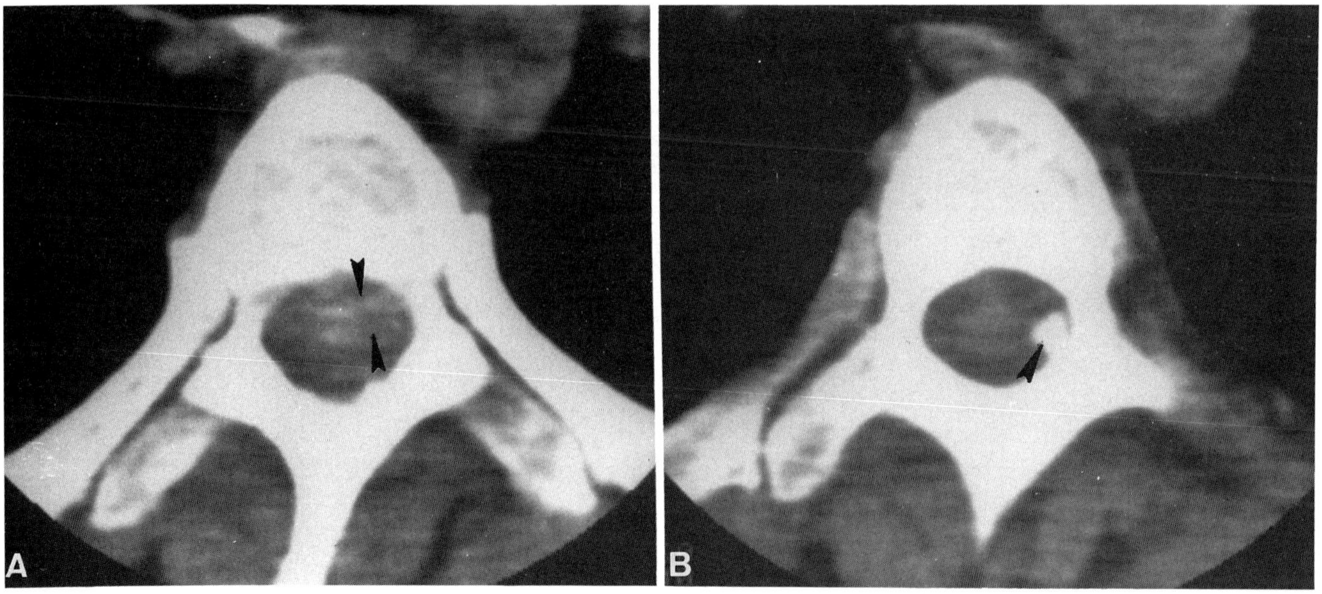

Figure 9.10. Meningioma. A 35-year-old woman with neurofibromatosis, including a past history of acoustic neuroma. **A,** Unenhanced CT image through T4 level. Isodense to slightly hyperdense mass (*arrowheads*) in the left side of the canal displaces the cord to the right but is difficult to define without either intravenous or intrathecal contrast. **B,** Calcification within the mass (*arrowheads*) strongly suggests the correct diagnosis of meningioma.

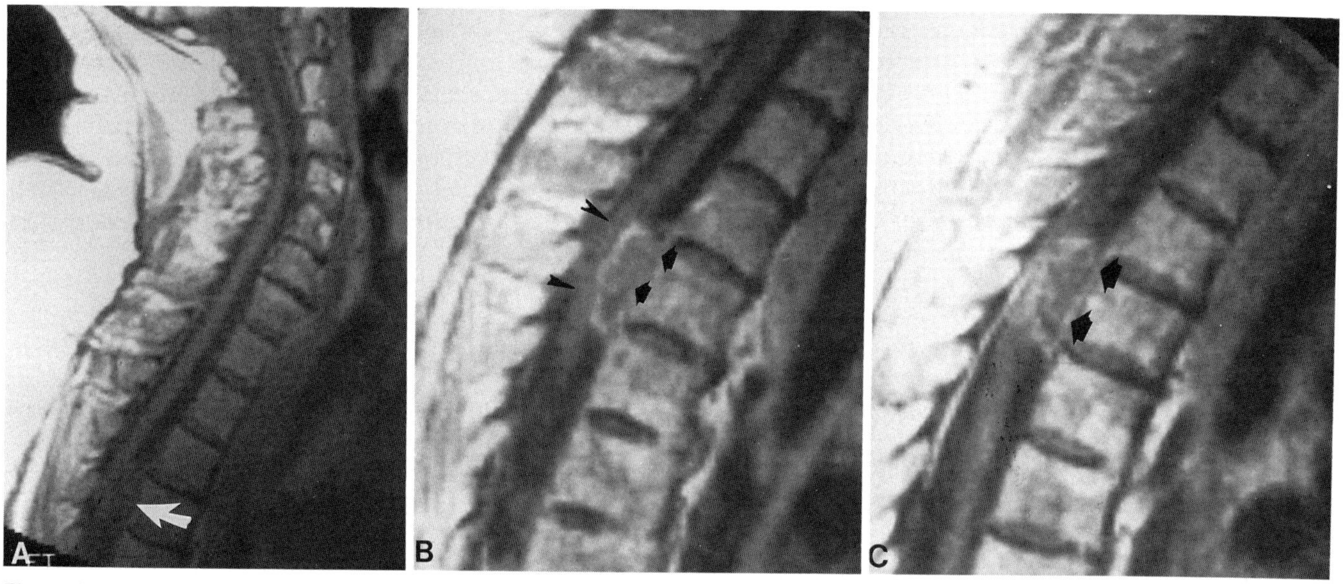

Figure 9.11. Meningioma. A 60-year-old man with lower extremity paresthesias and a history of cervical spondylosis evaluated for disk disease. **A,** Sagittal T1-weighted image (TR800/TE22) demonstrates not only cervical spondylosis but also a ventrally located intradural mass at the T4 level (*arrow*). Note the progressive widening of the anterior subarachnoid space as the displaced cord approaches the level of the lesion. **B,C,** Adjacent sagittal T1-weighted image (TR600/TE22) demonstrate a broad dural-based mass located ventrally at the T4 level (*arrow*). Rim enhancement is demonstrated, but there is only minimal inhomogeneous enhancement of the remaining portion of the meningioma. Meningiomas are usually associated with more homogeneous enhancement. Note the effacement of the adjacent thoracic cord (*arrowheads*).

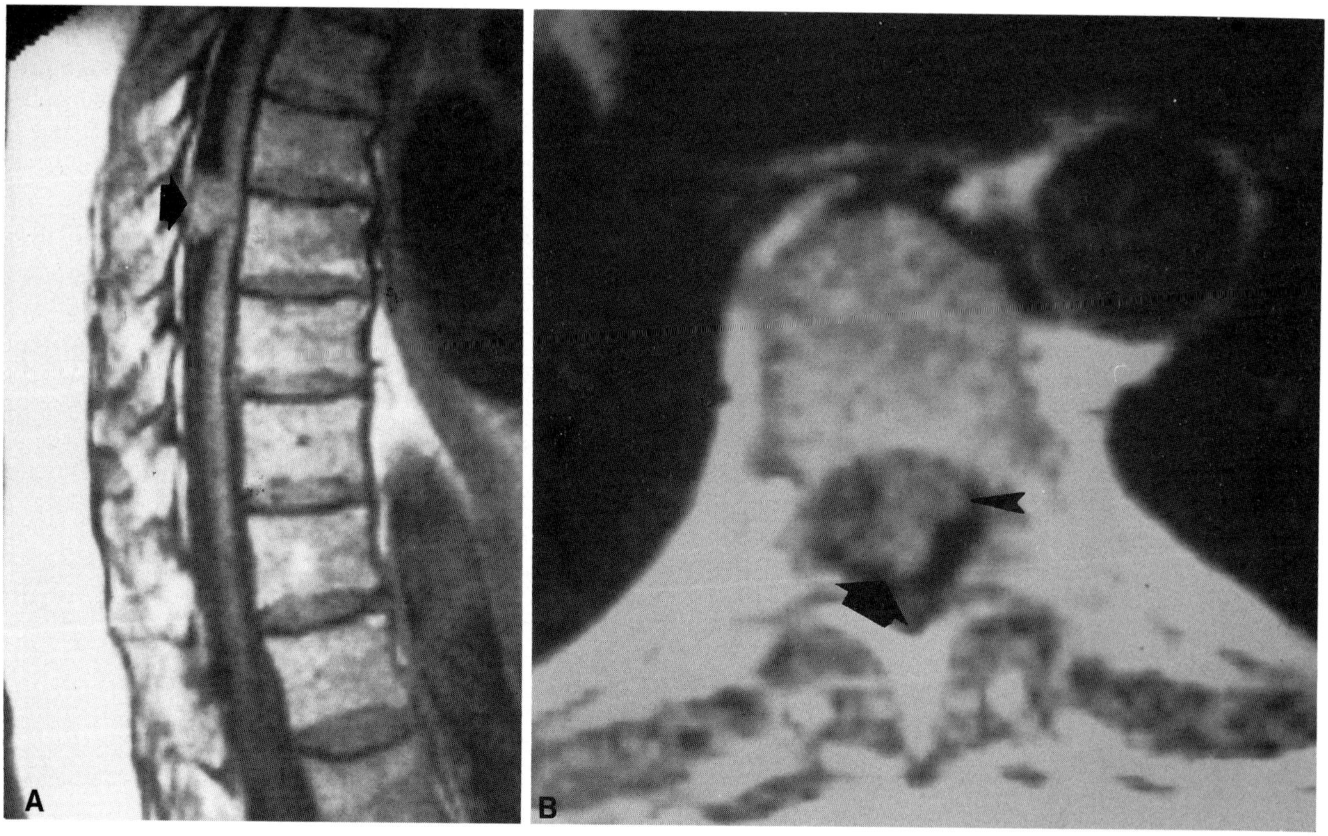

Figure 9.12. Meningioma. A 68-year-old woman who presented with a Brown-Sequard syndrome. **A,** Sagittal T1-weighted image (TR800/TE22). T8 level broad-based intradural extramedullary mass (*arrow*) is isointense with cord. Note the displacement and effacement of the thoracic cord with widening of the subarachnoid space ipsilateral to the mass. **B,** Axial T1-weighted image (TR600/TE22) through T8 level. Meningioma (*arrow*) displaces the cord (*arrowhead*) anteriorly and to the left.

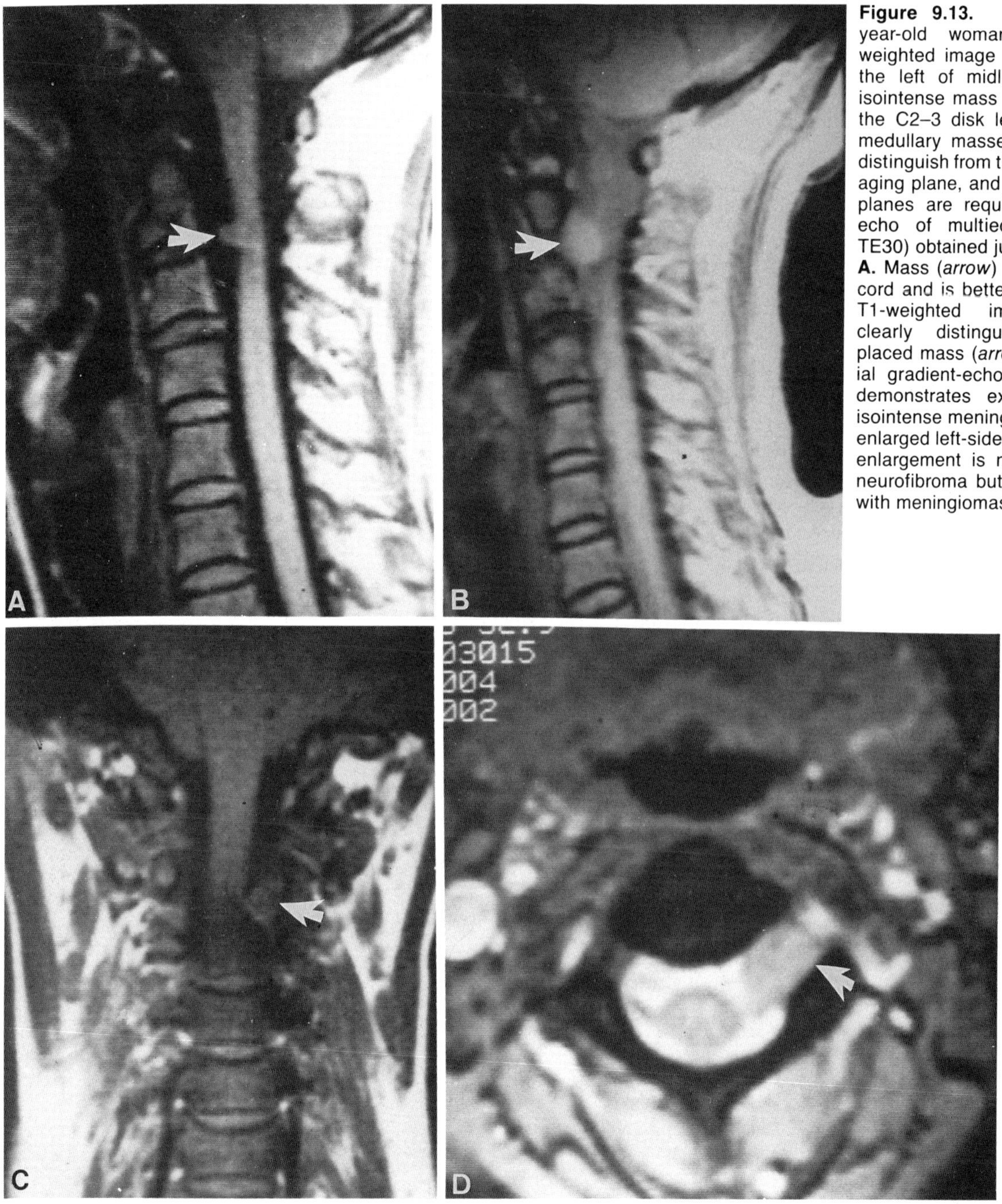

Figure 9.13. Meningioma. A 44-year-old woman. **A,** Sagittal T1-weighted image (TR500/TE20) just to the left of midline demonstrates an isointense mass ventral to the cord at the C2–3 disk level. Intradural extramedullary masses may be difficult to distinguish from the cord on a single imaging plane, and generally at least two planes are required. **B,** Sagittal first echo of multiecho exam (TR2000/TE30) obtained just to the left of image **A.** Mass (*arrow*) remains isointense to cord and is better defined. **C,** Coronal T1-weighted image (TR300/TE20) clearly distinguishes the laterally placed mass (*arrow*) from cord. **D,** Axial gradient-echo (TR25/TE13) image demonstrates extension of the still isointense meningioma (*arrow*) into the enlarged left-sided foramina. Foraminal enlargement is much more typical of neurofibroma but is seen occasionally with meningiomas.

compression by extramedullary tumors such as meningioma has been reported [62].

Meningiomas usually enhance following intravenous injection of Gd-DTPA [58, 59] (Fig. 9.14). This allows differentiation of tumor from cord and detection of smaller lesions. Although meningiomas consistently enhance compared with neurofibromas, this cannot be used in isolation as a method of differentiation. The prognosis for patients with meningioma is generally favorable. Tumors with an extradural component are more likely to be invasive and have a more rapid course [51]. The incidence of metastases is less than 1 in 1000 meningiomas [60]. These are more frequent in males, and about 60% are pulmonary [60, 61].

NERVE SHEATH TUMORS

Nerve sheath tumors are the most common primary neoplasms of the spine [2]. These are divided into two basic categories: *schwannomas* and *neurofibromas* [23]. Schwannomas are also referred to as *neurilemomas* and *neurinomas*. The principal cell type in both categories of neoplasm is the Schwann cell. The primary histologic difference is that the neurofibromas also include fibroblasts and often include nerve fibers as well [23]. Clinically, the distinction may be difficult. Schwannomas tend to be solitary, whereas neurofibromas tend to be multiple and usually are associated with neurofibromatosis. Nerve sheath tumors account for 29.5% of all primary spine tumors [2]. However, only two-thirds of the nerve sheath tumors are strictly intradural, so meningiomas are the most common intradural extramedullary tumor.

Unlike meningiomas, there is no distinct sex predilection [1]. There is an almost even spread in age distribution between the second and fifth decades [1]. Males tend to be affected earlier, in the 10- to 25-year age group. Females tend to develop the tumors between 25 and 40 years and also predominate in the early childhood and old age categories. With minor variations the mean age at presentation in both sexes is in the fifth decade.

Nerve sheath tumors may involve any segment of the spinal canal [63]. Thoracic involvement is most common, due to the relatively longer length [1]. These tumors usually involve the posterior and caudal sensory roots. Motor root involvement is uncommon. The position of the tumor in the canal is therefore usually dorsolateral, and only about 10% are seen ventrally [1]. Tumors vary in size from a few millimeters to greater than 10 cm. These neoplasms tend to be soft and smoothly defined and may have a sausage shape. Unlike meningiomas, a dural attachment is usually not evident [6]. "Dumbbell" or "hourglass" tumors result when the neoplasm extends through a neural foramen. These predominate in the cervical and upper thoracic spine. Dumbbell-shaped tumors favor the diagnosis of nerve sheath tumor rather than meningioma [1].

Pain is the presenting symptom in 75% of patients, the highest incidence of any primary spinal neoplasm [1]. The incidence of pain increases with

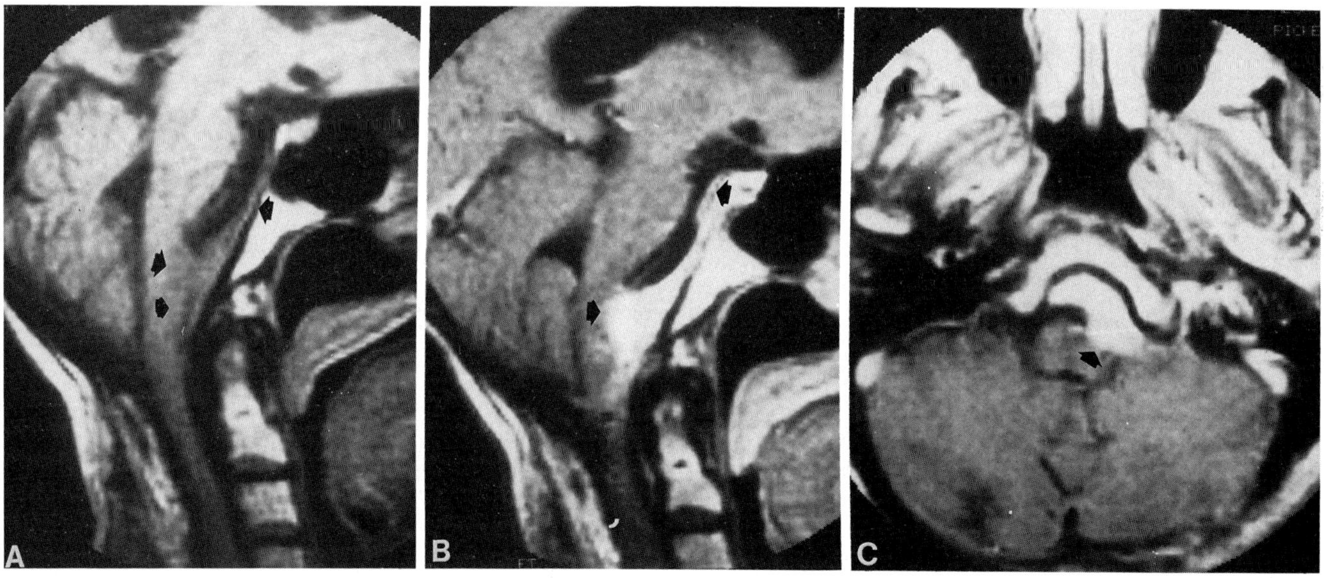

Figure 9.14. Meningioma. A 54-year-old man with foramen magnum meningioma extending into the upper cervical canal. Meningioma is the most common tumor of the foramen magnum. **A,** Unenhanced sagittal T1-weighted image (TR700/TE22). Isointense mass (*arrow*) extends along the clivus down to the level of the odontoid tip. Distinction between tumor and medulla is difficult. **B,** Sagittal contrast-enhanced T1-weighted image (TR700/TE22). There is dense homogeneous enhancement of the meningioma. Extent of the neoplasm is more obvious (*arrow*). **C,** Axial contrast-enhanced T1-weighted image (TR700/TE22). Distinction between the densely enhancing meningioma and medulla is improved.

more caudally located lesions. Neuralgic pain at the level of the lesion is the first sign of root damage and may be the only sign for a long time. The pain is often described as knifelike or lancinating. This segmental pain is usually unilateral and dependent on posture, being worse at night and aggravated by coughing or sneezing. With time, this pain may improve or resolve due to loss of nerve conduction. Long tract signs from cord compression may then predominate, making the diagnosis more difficult. Bowel and bladder dysfunction is seen in 55%. About half of patients come to surgery in the first 2 years following initial manifestations. The history extends between 2 and 7 years in about a third of patients [1].

Nerve sheath tumors are the most common primary spine neoplasm to produce plain film findings, which may be noted in 40 to 65% [1]. Kyphoscoliosis is a frequent skeletal abnormality in neurofibromatosis, occurring in 10 to 40% [6]. Larger tumors produce pedicle erosions, increased interpediculate distance, thinning of the lamina, or scalloping of the posterior vertebral body [6, 64]. Neurofibromatosis is the most frequent cause of posterior scalloping of one or more vertebral bodies [64]. This is usually secondary to dural ectasia, although it may be related to adjacent neoplasm [64]. The dumbbell tumors often enlarge the neural foramen and may erode adjacent ribs [1, 64]. A soft-tissue mass may be evident, but calcification is rare. Foraminal enlargement also may be secondary to meningoceles, especially in the thoracic spine. This protrusion of the meninges through the foramina may or may not be associated with bony changes, but there is a high association with neurofibromatosis [65]. Lateral thoracic meningoceles are the most common cause of a paraspinal mass in patients with neurofibromatosis [22]. Other causes of foraminal enlargement include meningioma, lymphoma, neural crest tumors such as ganglioneuroma, a tortuous vertebral artery, or even chordoma [66] (Fig. 9.15). Kyphoscoliosis is seen in 85% of neurofibromatosis patients with intrathoracic meningoceles [65, 67]. However, the usual small intradural tumor does not result in plain film findings.

Myelography was the gold standard in the past, yielding positive results in about 90% of neurinomas [54] (Fig. 9.16). However, over 50% of cases were associated with a spinal block, which limited visualization of the superior aspect of the mass [54]. Postmyelography CT with water-soluble contrast material allows much better assessment of the neoplasm in that CT is quite sensitive to even small amounts of contrast material that may pass proximal to the tumor. This often allows demonstration of the entire mass, as well as identification of additional masses. The need for cervical puncture is therefore diminished.

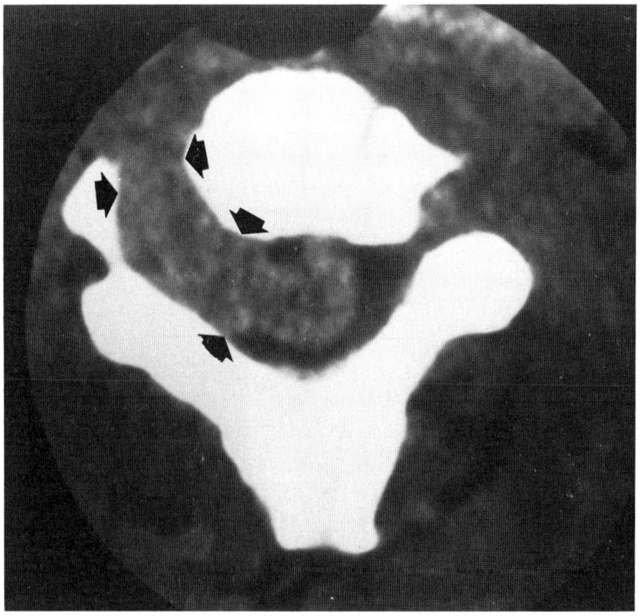

Figure 9.15. **Ganglioneuroma.** Contrast-enhanced CT through the C2–3 level. Enhancing tumor (*arrow*) associated with prominent foraminal enlargement. Benign tumors such as ganglioneuromas and neurofibromas more commonly produce pressure erosion, while malignant tumors tend to produce irregular bone destruction. Imaging characteristics do not distinguish this neoplasm from either meningioma or neurofibroma.

On non-contrast-CT, the mass tends to be hypodense to hyperdense relative to the cord [7, 54]. Density measurements are in the 40 to 60 H.U. range, and the mass is usually homogeneous, although cystic areas may be seen rarely [54]. Both nerve sheath tumors and meningiomas usually enhance homogeneously, with measurements in the 80 to 120 H.U. range (15 of 17 neurinomas enhanced in one series) [68]. A second contrast dose does not improve enhancement [69].

CT is also useful in distinguishing neurofibromas from intrathoracic meningocele [67, 70]. Meningoceles are sharply defined paravertebral CSF-density masses which communicate with the spinal canal and are not associated with intravenous contrast enhancement [70]. Intrathecal contrast material may be useful in equivocal cases, but delayed CT images may be required because meningoceles may not opacify initially [70]. Meningoceles are often associated with foraminal enlargement involving several adjacent vertebral segments, whereas neurofibromas are usually associated with isolated foraminal widening [67].

MRI is superior to CT or CTM in evaluating soft-tissue masses [71]. MRI allows for better demonstration of multiple masses and improves assessment of extradural involvement and the extent of cord compression [7, 77].

Neurofibromas have slightly brighter signal intensity than cord and paraspinal soft tissue on

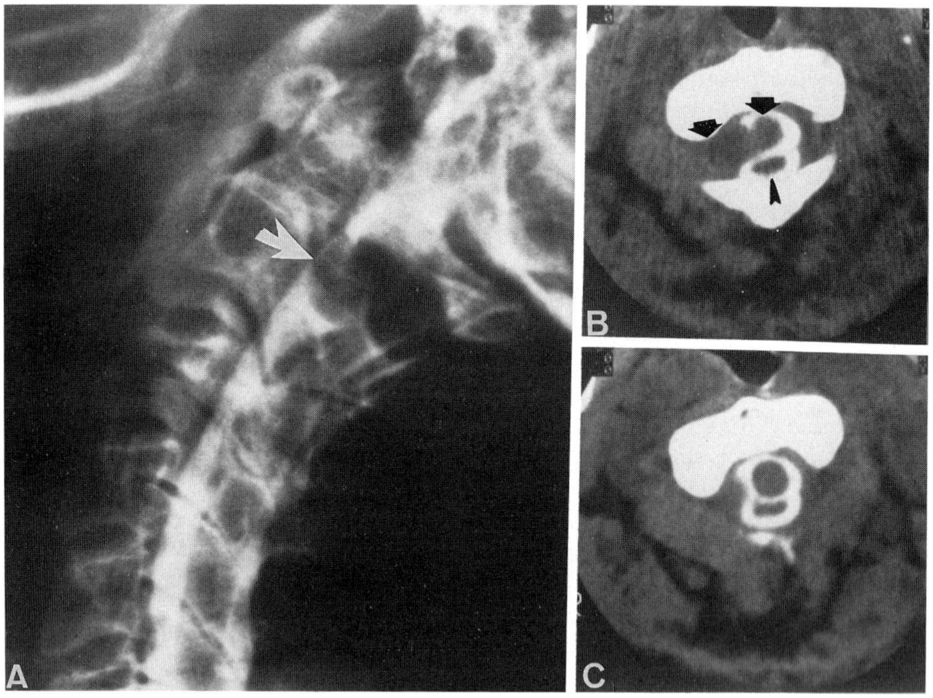

Figure 9.16. Neurofibroma. A, Lateral myelogram film demonstrates well-defined round intradural extramedullary mass (*arrow*) ventrally in the canal at C2. Note prominent posterior displacement and effacement of the cervical cord with prominent widening of anterior subarachnoid space, characteristic of an intradural extramedullary lesion. **B,** Postmyelogram CT. Extension of the ventrally located neurofibroma into the right neural foramina is demonstrated (*arrows*). Note the significant cord effacement (*arrowhead*). **C,** Sagittal postmyelogram CT reconstruction. Ventrally located neurofibroma is well defined (*arrowheads*).

T1-weighted images and are hyperintense on T2-weighted images [73]. Occasionally, these may be relatively isointense to cord on both sequences [57, 74]. The T1-weighted image gives the best anatomic image, especially of the cord, but T2-weighted images are more sensitive in detecting peripheral nerve tumors [73]. At least two imaging planes are generally required to establish the relationship of a mass to the cord and dura (Fig. 9.17). Coronal images are usually superior in demonstrating the dumbbell-shaped lesions extending through the neural foramina [73] (Fig. 9.18). Axial images are more useful in examining the paraspinal soft tissues. Meningoceles are isointense with CSF and are best distinguished on T1-weighted images [73]. Intermediate (long-TR, short-TE) scans also demonstrate these to be isointense with CSF, whereas neurofibromas tend to be hyperintense [50]. Neurofibromas frequently have well-defined central areas of decreased signal intensity on T2-weighted images, attributed to increased amounts of endoneural myxoid matrix [73]. Similar low-signal areas secondary to necrosis may be seen in neurofibrosarcomas, and this is a nonspecific finding [75, 76]. However, the presence of hypointensity on T2-weighted images distinguishes neurofibromas from the uniform T2 hyperintensity of meningoceles [50].

Paramagnetic contrast material greatly improves conspicuity and demarcation of the lesion from the cord [58].

Developmental Tumors

This group includes lipomas, teratomas, and dermoids. Lipomas are much more common and are often considered as three types: (*a*) lipomas of the filum terminale, (*b*) intramedullary lipomas, and (*c*) lipomyelomeningoceles [78]. Dysraphic states may be suggested by cutaneous stigmata such as subcutaneous lipoma, hyperpigmentation, skin dimple or sinus tract, or an unusual tuft of hair.

These fatty tumors also include abnormally increased amounts of fibrous connective tissue and often appear at least partially encapsulated [79–81]. These may invade or adhere to the cord, meninges, or cauda equina. Lipomas are somewhat more common in the intradural space but may be either extradural or both intra- and extradural. The intradural lipoma may be confined to the filum terminale, may be mixed with the nerve roots of the cauda equina, may be associated with a bone spur in diastematomyelia, or may be a distinct, well-defined mass posterior or posterolateral to the cord [81]. These intradural lesions are usually adherent to the cord or meninges, which may result in "tethering" of the cord.

Most patients are neurologically and urologically normal at birth [78]. Symptoms often appear during periods of rapid growth. Symptoms may be referable to cord tethering with resultant cord stretching and ischemia during rapid growth periods. If the mass is large enough, cord compression also may result in

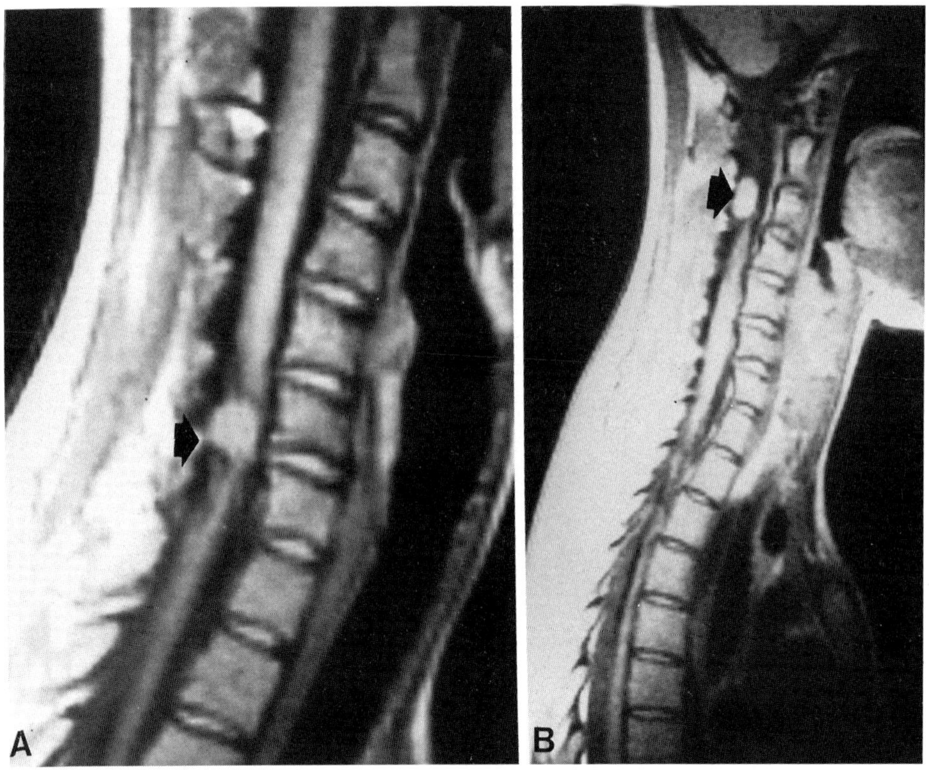

Figure 9.17. Neurofibromatosis. A 19-year-old woman without cutaneous stigmata of neurofibromatosis presented with an axillary mass which was subsequently confirmed to be a schwannoma. She has now been followed for 3 years without development of any signs or symptoms referable to her spine. All images shown are T1-weighted image (TR800/TE22). **A,B,** Sagittal unenhanced images demonstrate intradural extramedullary masses at C3 and C6 (*arrows*). **C,D,** Sagittal images of the thoracolumbar spine demonstrate a mass at the level of the conus (*arrow*) and two additional masses involving the cauda equina at the L4 level (*arrowheads*). Note that upper masses are isointense to cord. **E,F,** Sagittal postcontrast examination demonstrates homogeneous enhancement of the right-sided C6 neurofibroma which extends into the foramina (*arrows*). **G–I,** Consecutive coronal 5-mm enhanced images. Enhancement of bilateral dumbbell lesions at C6–7 (*large arrows*). Inhomogeneous enhancement of the intradural extramedullary mass at C3–4 presumed to represent neurofibroma (*small arrows*). Abnormal enhancement in the left carotid sheath is presumed to represent a vagus nerve neuroma (*arrowhead*) and is unchanged over 3 years. **J,** Sagittal enhanced image of lumbar spine. Although there has been considerable delay while the remaining spine has been imaged following contrast injection, conspicuity of lesions is still notably improved when compared with precontrast examination. Note that additional lesions are now evident (*arrowheads*).

symptoms. Steroid administration may precipitate symptoms, apparently related to swelling of the lipomatous tissue [78, 82]. The patient may present initially for evaluation of scoliosis.

Plain film evaluation may demonstrate nonspecific spinal canal widening with thinning of the pedicle [83]. However, both CT and MRI findings are characteristic and diagnostic (Fig. 9.19). CT demonstrates a lucent mass with density measurements in the fat range (generally 0 to -100 H.U) [84]. CT also may demonstrate widening of the canal with flattening of the adjacent cord [83]. CT is superior to MRI in evaluating for associated bony changes [85]. Reformatted sagittal CT with intrathecal contrast material is helpful in establishing anatomic relationships. However, to provide adequate spatial resolution, very thin (1.5 mm) axial images are required. This is a distinct disadvantage in evaluating extensive lesions because of both radiation and imaging time concerns.

The multiplanar imaging capabilities and superior soft-tissue contrast of MRI make this the imaging method of choice for evaluating the lipoma and its complex relationship with adjacent soft-tissue structures [85, 86]. Multiple planes of the entire cord and spinal canal can be obtained. Sagittal views are superior to reformatted sagittal CT views in demonstrating tethering of the cord or filum terminale [85]. MRI is ideal in demonstrating the relationship of the lipoma to the dura and may demonstrate extension of the mass through the dura [81]. Coronal MRI images are especially useful in evaluating the conus medullaris and cauda equina.

T1-weighted images are preferred in evaluation of lipomas because of the superior anatomic definition and the characteristic hyperintense appearance of fat (Fig. 9.20). When present, fibrous components may be demonstrated as linear bands of decreased signal intensity either within or encapsulating the lipoma [81]. T2-weighted images are less useful

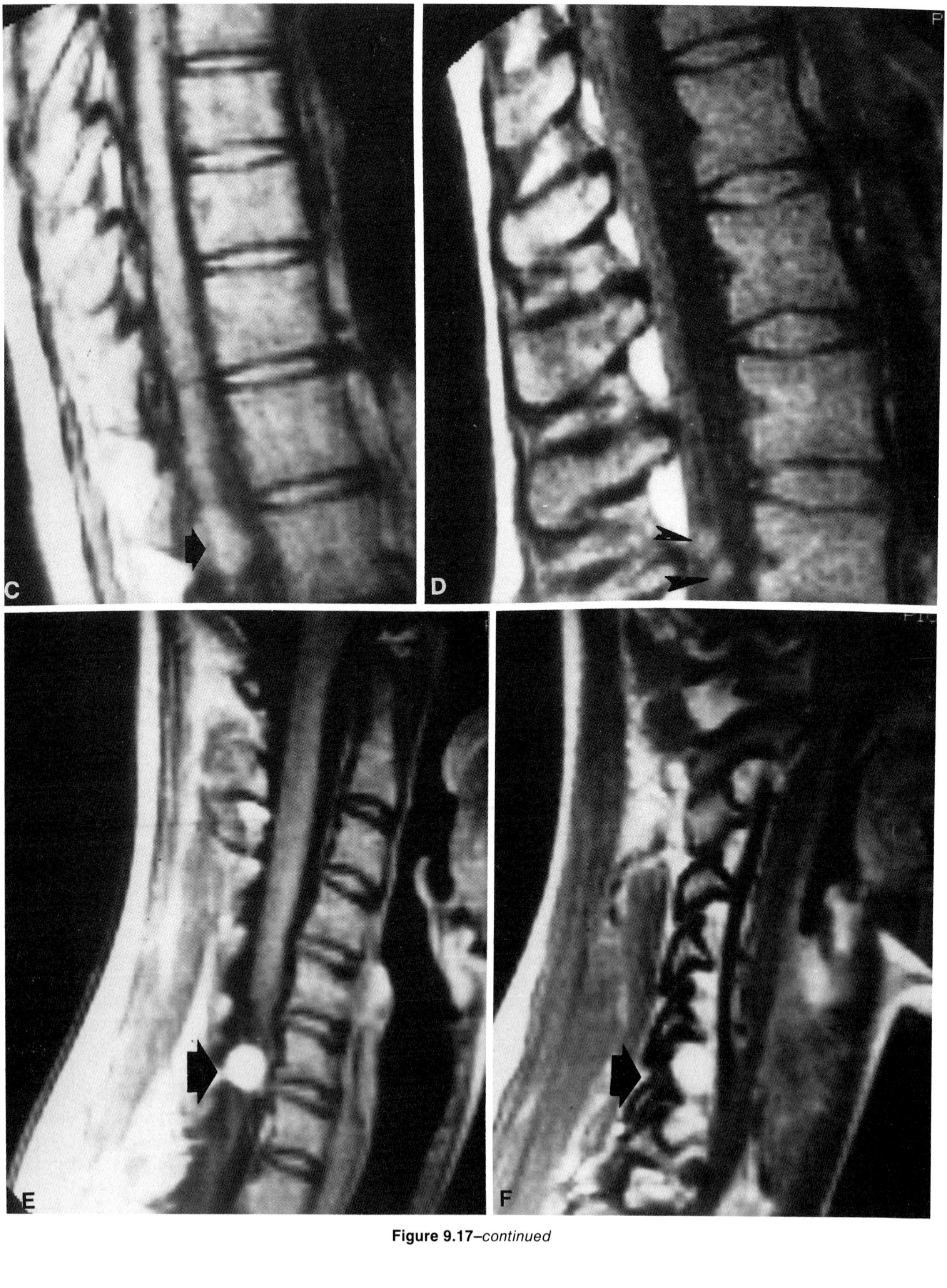

Figure 9.17–*continued*

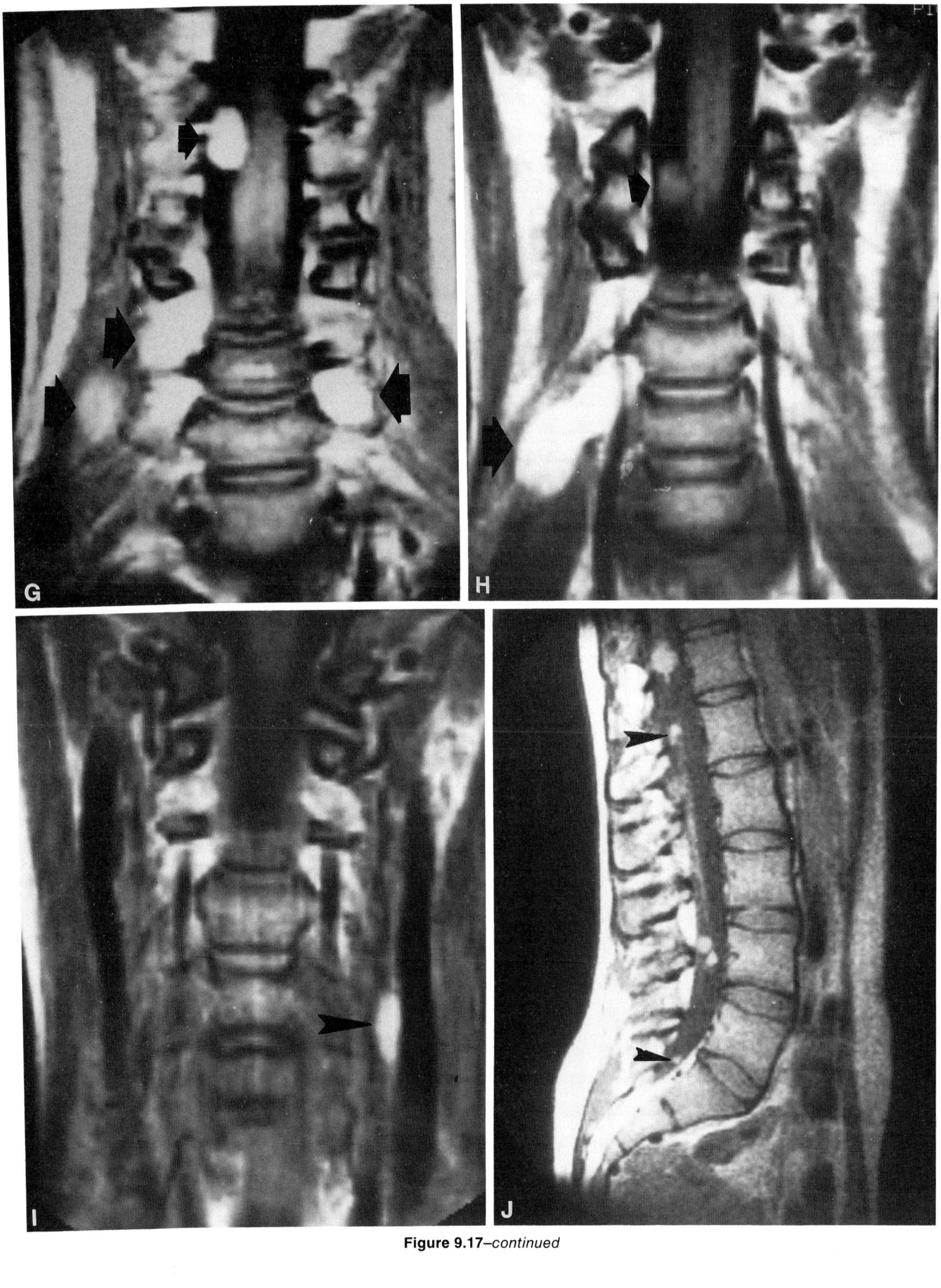

Figure 9.17–*continued*

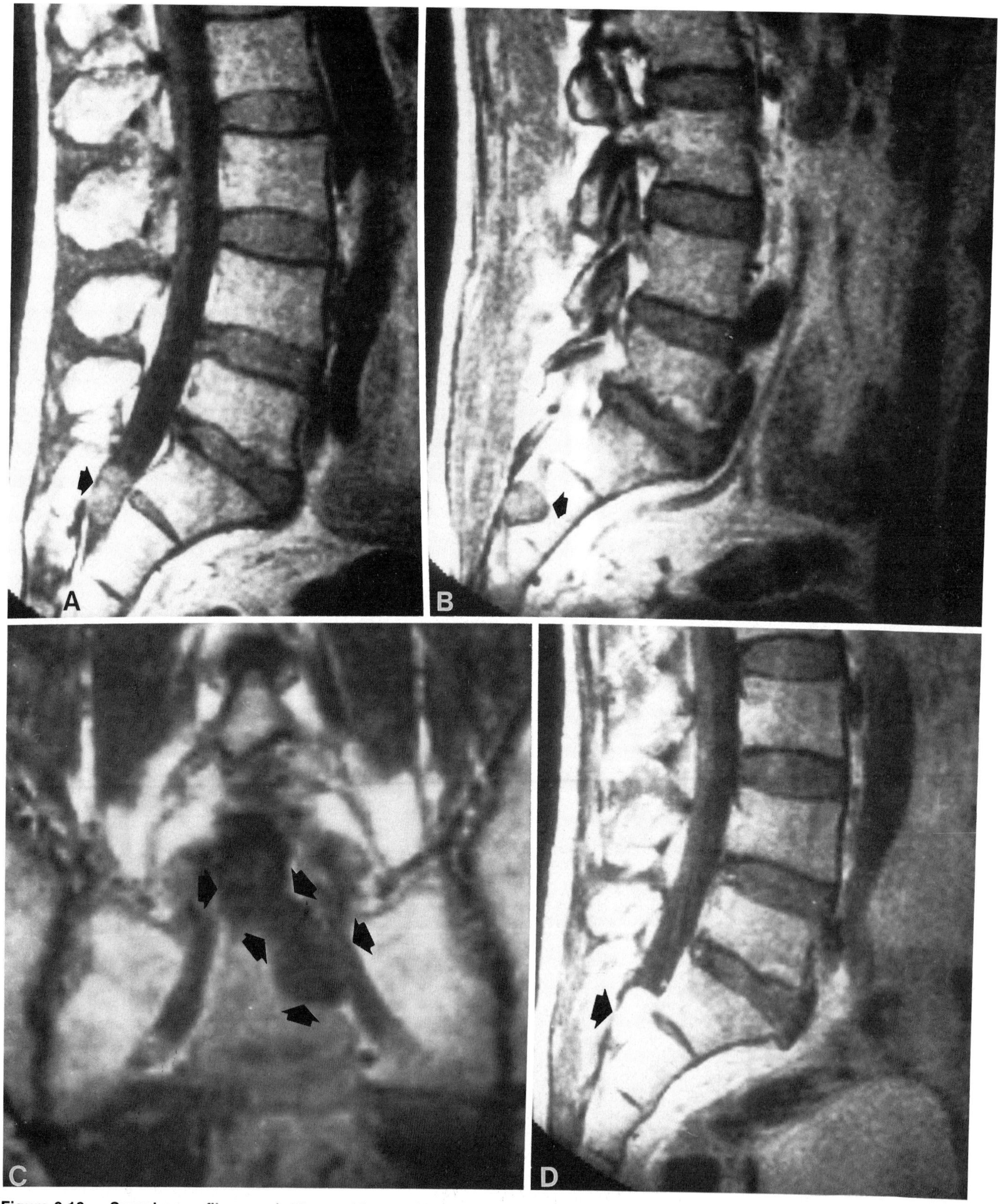

Figure 9.18. Sacral neurofibroma. A 45-year-old man with long history of neurofibromatosis with multiple skin lesions, presumed cervical astrocytoma, and bilateral acoustic neuromas. Sagittal (**A**) and left parasagittal (**B**) T1-weighted image (TR800/TE22) demonstrate a mass in the caudal thecal sac (*arrows*) which extends into the left sacral foramina. Note that mass is isointense to the distal cord. **C,** Coronal T1-weighted image (TR550/TE22) demonstrates dumbbell-shaped lesion (*arrows*) extending through the left S1 foramina. The characteristic dumbbell shape strongly suggests that this is a neurofibroma rather than a drop metastasis from the patient's cervical astrocytoma. **D,** Sagittal contrast-enhanced T1-weighted image (TR800/TE22). There is dense homogeneous enhancement.

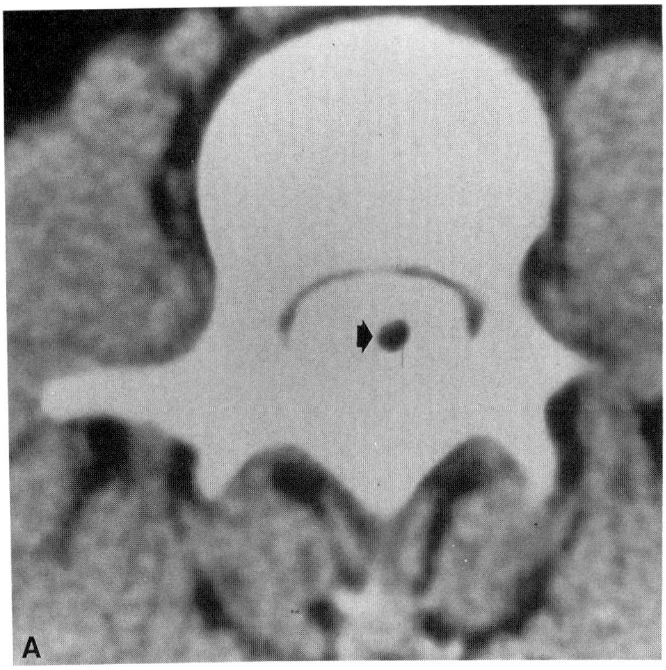

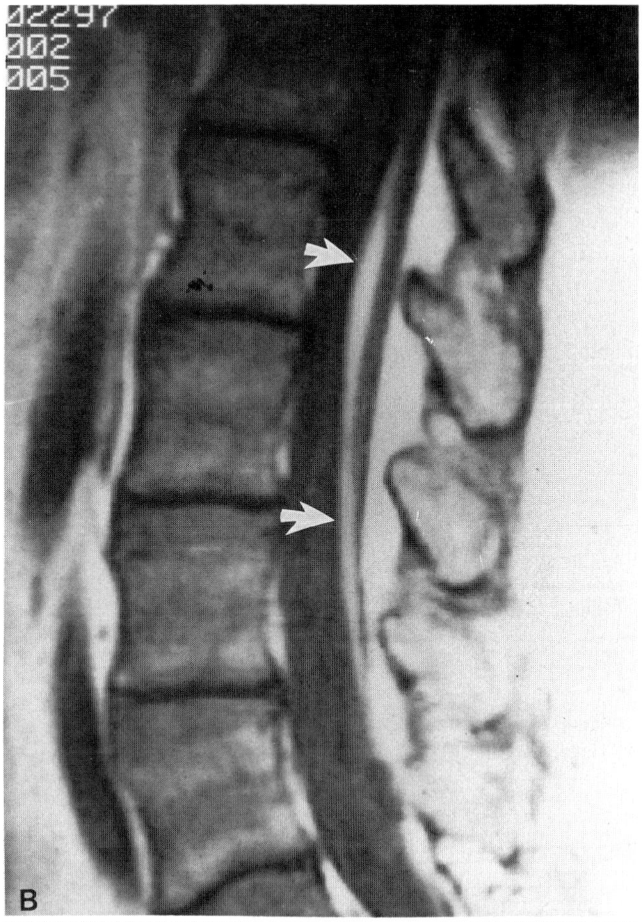

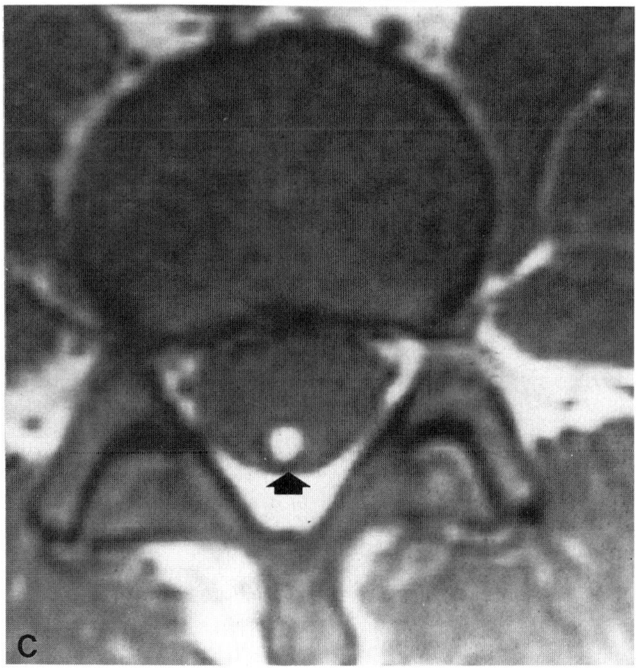

Figure 9.19. Lipoma of the filum terminale. A, Axial postmyelogram CT image demonstrates the characteristic fat-density lipoma (*arrow*) to be circumscribed by the opacified subarachnoid space. Lipoma measures −82 H.U. **B,** Sagittal T1-weighted image (TR600/TE22). Thickening and abnormal hyperintensity of the filum terminale are diagnostic of a filum lipoma (*arrows*). **C,** Axial T1-weighted image (TR700/TE22). Characteristic hyperintensity of the filum lipoma (*arrows*).

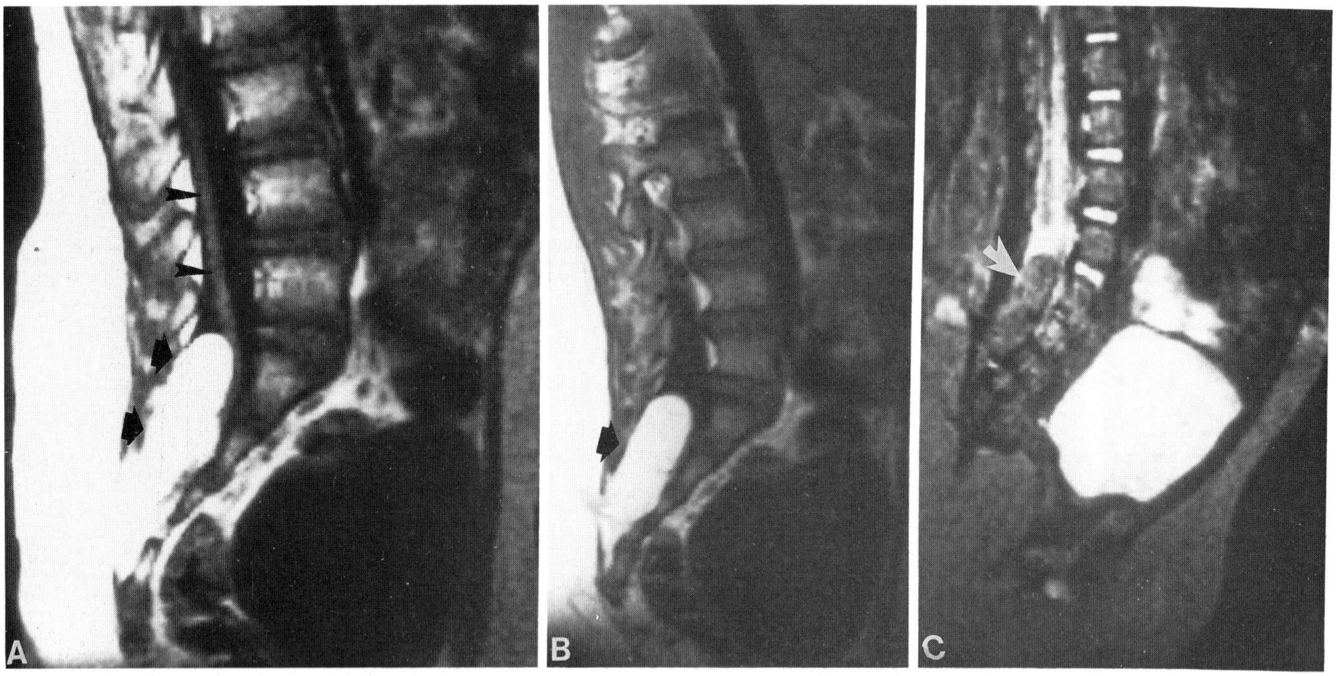

Figure 9.20. Lipoma. A 4-year-old boy with spinal dysraphism, large lipoma, cutaneous stigmata, and tethered cord. **A,B,** Sagittal T1-weighted image (TR450/TE22). Large hyperintense lipoma (*arrows*) in the caudal sac associated with a tethered cord (*arrowheads*). Note the enlarged spinal canal as well as the markedly distended urinary bladder. **C,** T2-weighted (TR2800/TE120). Decreased signal in the lipoma (*arrows*), which remains isointense with the subcutaneous fat.

because fat signal intensity diminishes and CSF signal intensity increases, effectively diminishing contrast. T2-weighted images provide a lower signal-to-noise ratio, and anatomic boundaries are somewhat obscured [85]. Chemical-shift or fat-suppression images occasionally may be useful, especially when the differential diagnosis includes hemorrhage (see Fig. 9.20). T2-weighted images are useful in distinguishing Pantopaque from fat because signal intensity of Pantopaque diminishes more rapidly than that of fat with progressive T2 weighing [87]. However, plain film correlation is both faster and easier.

Incidental fibrolipomas of the filum terminale are frequently encountered and may be seen in up to 6% of "normal" individuals at autopsy [79]. They are significant only when large enough to compress the cauda or if associated with tethering [79]. Treatment of spinal lipomas is controversial, with some advocating early surgical intervention, even in asymptomatic infants [78]. The time of onset or progression of a deficit cannot be predicted and may be precipitous [78]. Progressive neurologic deficits are frequently encountered, with an incidence of 56% associated with intraspinal lipoma [81]. Surgical intervention stabilizes a fixed deficit; however, only 40% show improvement in motor deficits [78]. Return of urinary continence is even less likely.

Other developmental tumors such as teratomas and dermoids are much less frequent than lipomas [88]. These are more frequent in the dorsal aspect of the lower cervical canal and at the thoracolumbar junction. There may be associated bony abnormalities such as spina bifida occulta and diastematomyelia [88]. The imaging characteristics are quite variable and are dependent on the relative amounts of fat or cystic components. Both CT and MRI are well suited to evaluate the fatty component of the mass. T1-weighted images or proton-density images are useful in distinction between the different tissue types [89].

Intradural Extramedullary Metastases

Metastatic involvement of the meninges is relatively uncommon and may be due to either seeding from intracranial primary neoplasms or direct metastases from systemic malignancies (Fig. 9.21). Seeding from primary intracranial neoplasms is the most common cause of subarachnoid space metastases and is most frequently seen in children [7]. This is seen in aggressive CNS neoplasms such as medulloblastoma, the "primitive neuroectodermal tumors," ependymomas, and glioblastoma multiforme [7]. More than one-third of medulloblastoma patients will have metastatic spinal canal involvement either at the time of initial intracranial presentation or at the first recurrence [7, 90, 91, 100]. An interesting observation is that patients who do not undergo surgery for their medulloblastoma actually appear to have a lower incidence of spinal metas-

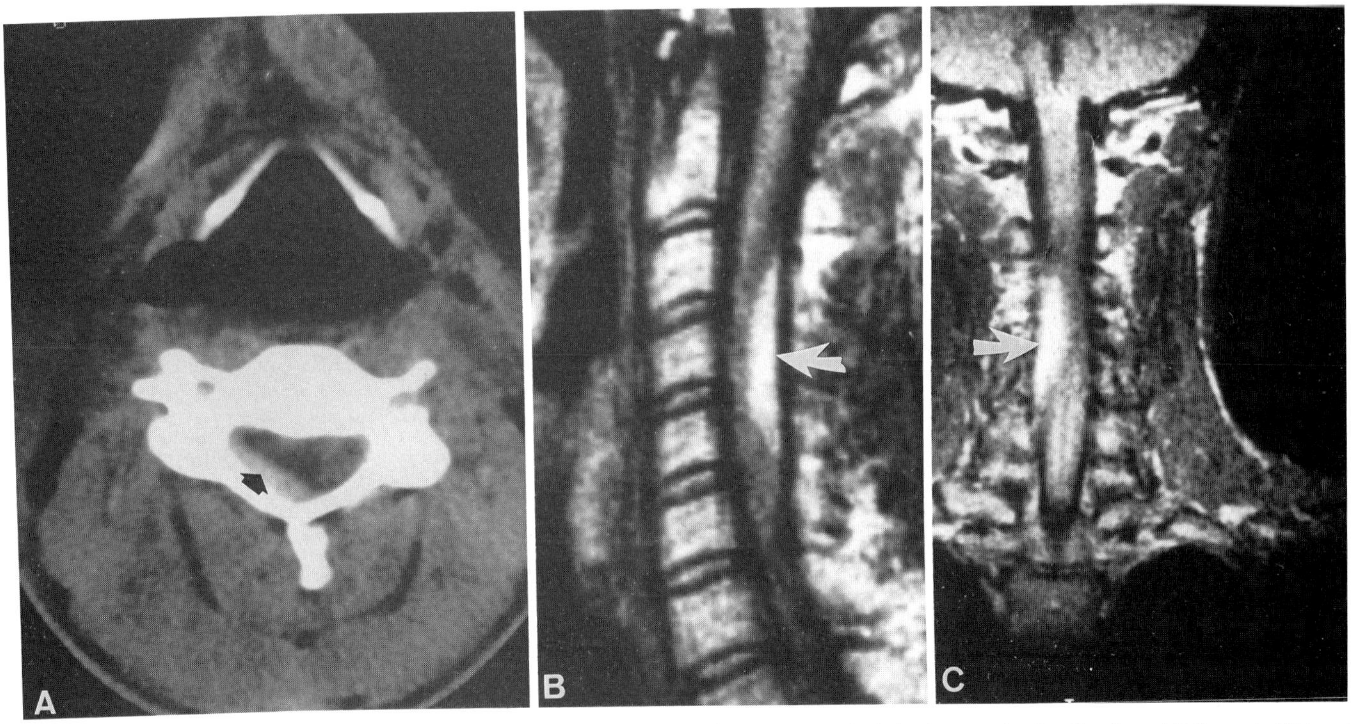

Figure 9.21. Epidural hemorrhage. A, Unenhanced CT. Characteristic hyperdense epidural hemorrhage (*arrow*). **B,** Sagittal T1-weighted image (TR500/TE40). Hyperintense epidural hemorrhage is seen posteriorly in the canal extending from C3 through C5 level (*arrow*). **C,** Coronal T1-weighted image (TR1000/TE40) shows hyperintense hemorrhage (*arrow*).

tases, suggesting that surgery may be a possible mechanism for subarachnoid dissemination of tumor [90, 91]. Most of these metastases are found in the lumbosacral portion of the canal.

The relatively high incidence of subarachnoid involvement and its consequent alteration of the treatment plan necessitate accurate screening. This is all the more important because early clinical signs and symptoms may be insidious and nonspecific [92]. Subarachnoid space involvement may eventually produce spinal root signs and symptoms, including back and lower extremity pain, motor deficits, and both urinary and bowel disturbances [93]. CSF cytology results are inaccurate in predicting spinal involvement [91].

The value of myelography in evaluation of subarachnoid metastases is well recognized [91, 93]. Nodular filling defects within the subarachnoid space as well as a crowding together of nerve roots of the cauda equina or irregular enlargement of individual nerve roots are typical findings [93]. Postmyelography CT improves conspicuity of small lesions and allows a better definition of focally thickened nerve roots of the cauda equina [94] (Fig. 9.22). This is particularly useful in demonstrating a small lesion adherent to the cord. Postmyelography CT also may salvage myelographic examinations that are compromised by poor opacification or other technical limitations. Unenhanced CT is of little value. Occasional reports have demonstrated enhancement of subarachnoid metastases following intravenous contrast administration, but again, this is relatively insensitive [7, 92, 93].

Unenhanced MRI in screening patients with meningeal carcinomatosis has been disappointing [7, 28, 93, 94]. Occasionally, multiple nodular lesions which are slightly hyperintense on T1-weighted images may be seen. Contrast-enhanced MRI has considerably improved conspicuity and detection of neoplasms in the subarachnoid space [95–98]. Even small drop metastases of 2 to 3 mm may be seen [97]. T1-weighted images are generally most useful, although occasionally intermediate or more T2-weighted images provide additional information. T2-weighted images are limited due to the increased signal intensity of CSF, long acquisition times, and a decrease in the signal-to-noise ratio. In the evaluation of subarachnoid metastases, pre- and postcontrast T1-weighted images are satisfactory [99]. Axial images are often complementary to the usual sagittal sequences when evaluating for subarachnoid metastases. Careful comparison with precontrast images will prevent false-positive results related to the normally present hyperintense extradural fat in the lumbosacral canal.

Extradural Tumors

Extradural neoplasms account for 30 to 50% of spinal tumors, depending on the series and patient

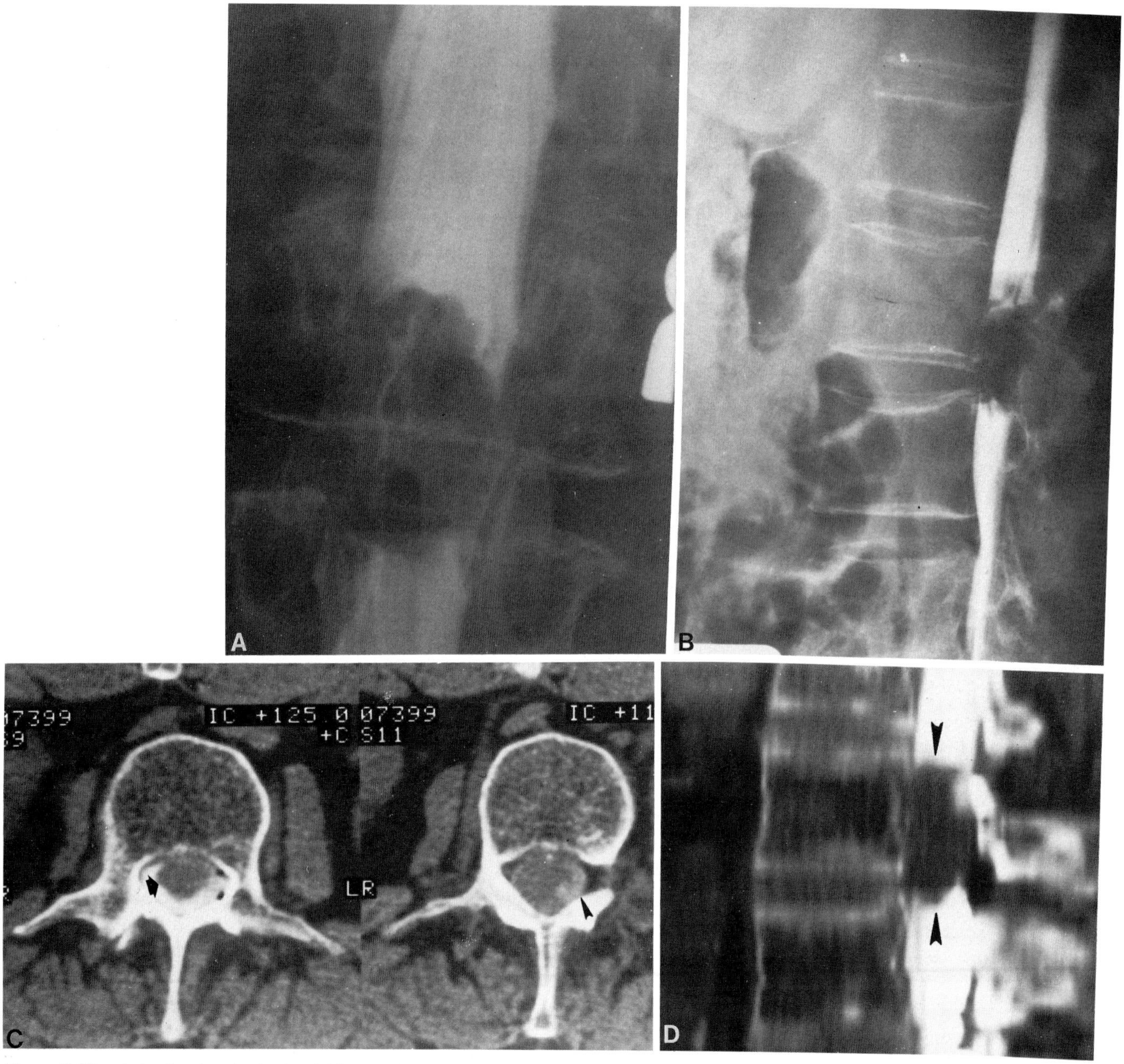

Figure 9.22. Intradural metastasis. A 67-year-old man with metastatic carcinoid. **A,** frontal, **B,** lateral views of myelogram demonstrate a lobulated intradural mass at the L2–3 level which nearly completely fills the canal. **C,** Two consecutive images from a post-myelogram CT demonstrating an intradural mass (*arrow*). The second image demonstrates only a small amount of contrast to the left of the mass (*arrowhead*). CT without either intrathecal or intravenous contrast is of little value in evaluating lesions in either the intramedullary or intradural extramedullary space. **D,** Sagittal reconstruction following CTM demonstrates characteristic intradural mass (*arrowhead*).

population studied [7, 8, 50, 101]. The great majority of these are due to metastatic disease and are therefore malignant. Extradural metastases are estimated to be three to four times more common than primary vertebral tumors [101]. These extradural tumors primarily involve the vertebral column but also may involve the dura or paravertebral soft tissues (Table 9.3). Occasionally, primarily intradural tumors such as neurofibromas or meningiomas may be seen in the extradural space. The spinal cord or exiting nerve roots may be secondarily involved by metastases, myeloma, primary bone tumors, or direct extension of paravertebral neoplasms or malignant adenopathy.

METASTASES

Metastases are the most common neoplasm of the spine and are seen primarily in patients over age 40 [22, 111]. Approximately one-third of all metastatic bone disease is found in the spine [102, 103]. How-

Table 9.3.
Extradural Neoplasms

Metastases
- Most common neoplasm of spine
- Lung, breast, prostate most common
- Pain is nearly always initial symptom
- Pedicles involved earlier than in myeloma
- CT: lytic or sclerotic, margins may be indistinct, no matrix calcifications
- MRI: hypointense on T1, variable hyperintensity on T2, contrast may decrease conspicuity

Myeloma
- Most common primary malignant bone neoplasm
- Age > 40, 69% male, bone pain especially in T and L spine
- Pedicles involved late
- CT: multiple small, irregular lytic foci throughout marrow, no sclerotic rim
- MRI: small hypointense foci on T1, variable hyperintensity on T2

Chordoma
- Most frequent primary sacral neoplasm
- Mean age 56 years, 2:1 male
- Lytic bone destruction, often with sclerotic rim
- May involve disk and multiple vertebral bodies
- CT: nearly all have calcification in soft-tissue mass, many also have cystic areas
- MRI: hypointense on T1, hyperintense on T2—look for cystic areas, calcifications, and hypointense pseudocapsule

Osteoid osteoma
- Size usually 1 cm or less
- Well-defined, lucent nidus, often with surrounding sclerosis
- Teenage males most common, lumbar predilection
- Painful scoliosis in two-thirds
- Usually occur near base of transverse process
- CT usually required to see nidus, may have central calcific density

Osteoblastoma
- 40% involve spine, more common on right
- Usually involve base of transverse process region
- Teenage males most common
- Painful scoliosis in one-third
- May resemble osteoid osteoma, but central nidus much larger, usually > 1.5 cm
- Destructive, expansile, lytic lesions but usually well-circumscribed, may have sclerotic rim
- About one-half have epidural extension

Leukemia
- Plain films: osteopenia or focal lytic lesions
- CT: nonspecific bone destruction, with thecal contrast, may see nerve root thickening, masses are iso- to hyperdense
- MRI: nonspecific marrow replacement (hypointense on T1), T2 variable, may be iso- or hyperintense to marrow

Lymphoma
- Plain films usually normal, may be lytic or blastic
- CT: paravertebral nodes with anterior scalloping of vertebrae, foramina often involved, epidural mass often extends for more than one level
- MRI: marrow replacement hypointense on T1, variable hyper- to isointense on T2, foraminal and epidural involvement seen better

Sarcomas
- Primary spine involvement rare
- Plain films still best in guessing probable histology
- Chondrosarcomas usually have calcification, in older age groups
- Ewing's usually in first three decades, lytic, may have reactive sclerosis, rarely have calcium, soft-tissue mass may be huge
- Osteosarcomas of spine usually in older age groups, may see new bone formation

Eosinophilic Granuloma
- Three-quarters of patients < age 20
- Vertebra plana more common in younger children
- Vertebral body most frequently involved, most commonly lumbar
- CT shows bone destruction and/or vertebra plana
- MRI shows either vertebra plana or nonspecific marrow replacement, best for demonstrating multiple sites and epidural extension

Giant cell tumors
- Distinct female predilection, ages 20 to 40
- Sacrum much more frequent than spine
- Less than one-third involve pedicle or arch
- Large, well-defined expansile lytic area, often eccentric
- May involve S1 joint or disk space
- Sacral lesions often have large soft-tissue mass

Aneurysmal bone cyst
- Great majority in second decade
- Spine involvement much more common than sacrum
- Spinous process and neural arch most frequent location
- Expansile, lytic, well-defined, often involves more than one level
- CT, MRI show sharp margins, occasionally fluid levels of varying density

Hemangiomas
- Slowly growing vascular tumor, usually in vertebral body
- Symptoms rare but more likely in young females with thoracic lesion
- Plain films: characteristic coarse vertical striations, "corduroy cloth"
- CT: "polka dot" appearance due to coarse trabeculae surrounded by fatty matrix, enhances
- MRI: mottled hyperintensity on T1, marked hyperintensity on T2, enhances

ever, prognosis is poor if neurologic compromise is already present [104].

Extradural metastases may arise via hematogenous dissemination, direct extension from metastatic nodes or paravertebral masses (Figs. 9.23 and 9.24), or direct transcortical extension of tumor from the vertebral marrow into the epidural space [7]. The location of hematogenous metastases is dependent on the distribution of red marrow in the vertebrae, as well as the vascular supply [105]. The abundant vascular red marrow of the vertebral body is therefore involved more often than the pedicles or posterior elements. However, early pedicle involvement is also frequent in metastatic disease because of the rich vascular supply to this area [105]. This is in contrast to multiple myeloma, which involves the pedicle relatively late.

Although any tumor may metastasize to bone, lung, breast, and prostate are the most common sources [106]. Nearly 50% of patients with lung cancer develop bone metastasis during the course of their disease [103]. As many as 40% of breast cancer and 29% of prostate cancer patients have evidence of bone metastasis at the time of diagnosis [106]. Many

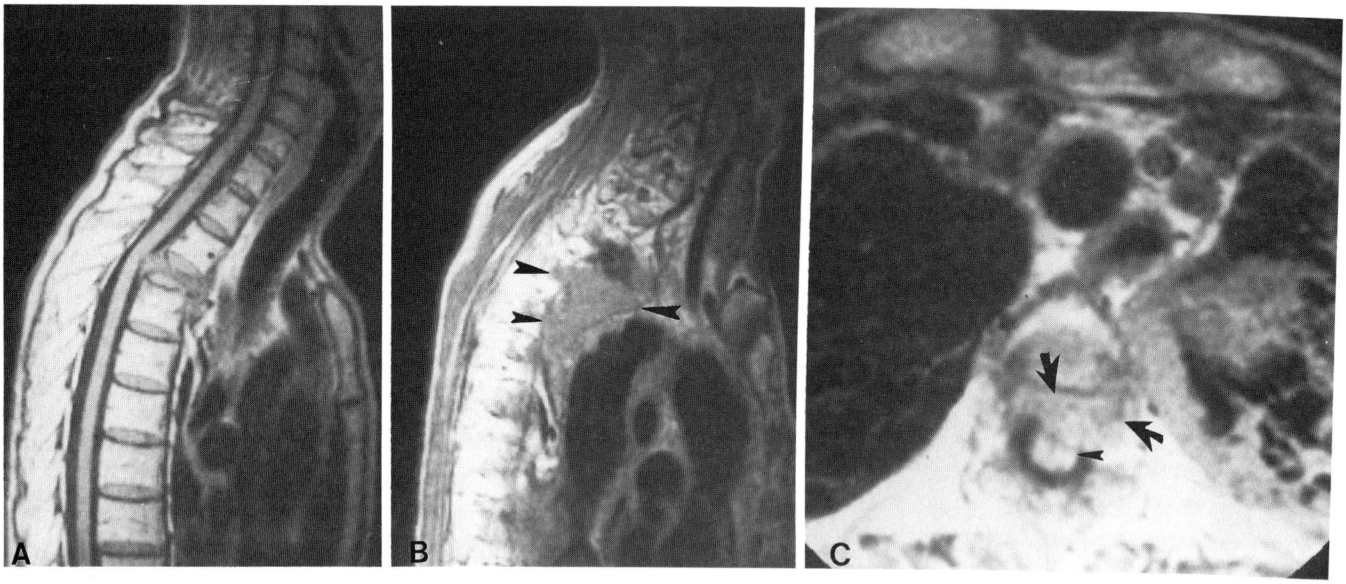

Figure 9.23. Pancoast tumor with direct extension into spinal canal. A 72-year-old with known left apical squamous cell carcinoma presents with signs of cord compression. **A,** Sagittal T1-weighted image (TR800/TE22). Tumor replacement of the collapsed T4 vertebral body with extension into the epidural space and compression of the cord. Note also the anterior paravertebral extension as well as lack of adjacent disk involvement. **B,** Left parasagittal T1-weighted image (TR800/TE22). Extensive tumor is seen anterior and lateral to the vertebral column (*arrowheads*). **C,** Axial T1-weighted image (TR800/TE22). Direct extension of the left apical neoplasm into the spinal canal (*arrows*) with displacement and effacement of the thoracic cord (*arrowheads*).

develop bone metastases within 12 to 18 months of initial diagnosis [103]. Other common primary tumors include myeloma, lymphoma, and renal cell carcinoma. Thyroid carcinoma has a distinct spinal predilection but is infrequently seen because of its relatively lower overall incidence [7]. Neuroblastoma is the most frequent metastatic lesion under the age of 1 year but is rare in adults [105, 107] (Figs. 9.25, 9.26, and 9.27).

Spinal cord compression due to metastatic tumor is a common neurologic complication [108]. The thoracic spine is involved most frequently (68%), with the remainder fairly evenly divided between the cervical and lumbosacral spine. Lung and breast carcinomas metastasize more frequently to the thoracic spine; colon and cervix carcinomas more frequently involve the lumbar spine [108, 109]. As with other spine neoplasms, metastatic tumors usually present with pain, which may be local or radicular in nature and exacerbated by movement, coughing, and straining.

Radionuclide bone imaging provides an ideal screening examination for metastases [103, 110]. However, this study has only a limited ability to detect myeloma and is also limited by lack of specificity [102, 110, 112]. Degenerative disorders, osteophytes, Paget's disease, and healing fractures are examples of possible false-positive results [110]. Specificity can be considerably improved by a follow-up plain film examination of the involved areas. If a bone scan is positive and the plain films are normal, further evaluation is warranted [102, 112].

CT is much more sensitive than plain film examination and is more specific than bone scintigraphy in the detection of metastasis [102, 112, 113]. CT is useful in demonstrating osteolytic lesions of the vertebral body, as well as destruction of cortical bone and extension of the tumor into the epidural space or paravertebral soft tissues [7] (Fig. 9.28). Conversely, secondary involvement of the vertebral body or spinal canal by an adjacent paraspinal mass or malignant adenopathy is also well demonstrated [109].

Osteolytic lesions such as lung carcinoma metastases generally appear as one or more osteolytic areas within the vertebral body or pedicle. Metastatic lesions usually do not have a sclerotic rim or matrix calcification [22]. Reactive bone sclerosis is often seen with metastases from prostate, lymphoma, and occasionally breast [22]. Contrast enhancement of an epidural or paravertebral soft-tissue mass may be demonstrated [7]. If cord compression is suspected, MRI evaluation is preferred [114]. If MRI is unavailable, postmyelography CT often allows adequate demonstration of spinal canal involvement [115]. This is particularly useful when the myelogram demonstrates what appears to be a complete block. The superior sensitivity of CT often allows a small amount of contrast material proximal to the tumor to be demonstrated, allowing a more accurate assessment of the extent of the neoplasm for radiation planning [7].

MRI has replaced myelography and MCT as the procedure of choice in the diagnosis of spinal metastases and possible cord compression [114, 116].

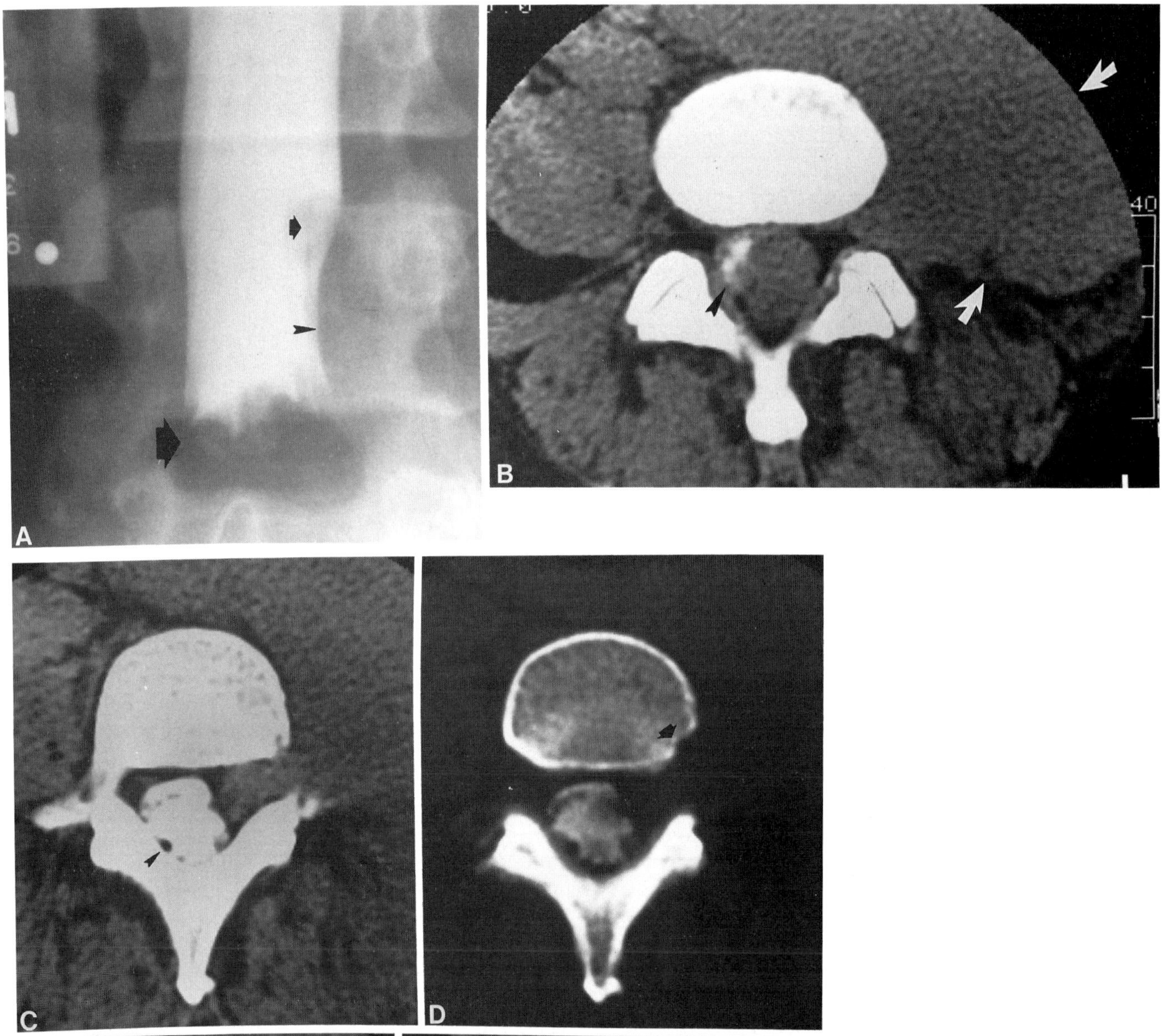

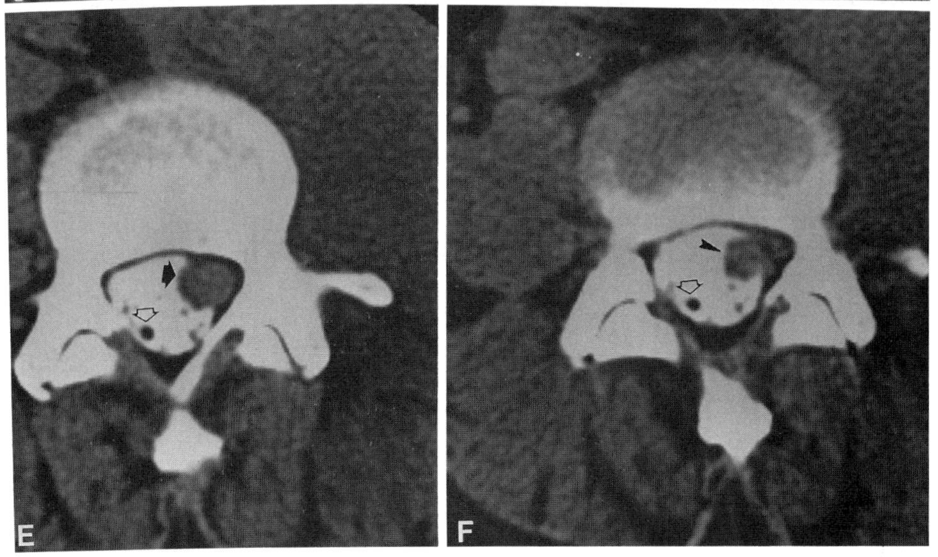

Figure 9.24. Retroperitoneal sarcoma with dural extension. A 35-year-old black man with rapidly progressive lower extremity weakness and systemic malaise. Same patient as Figure 9.19, films obtained 6 months later. **A,** AP myelogram. Epidural block at L4–5 disk level (*large arrow*). Scalloping due to extradural extension of tumor (*arrowhead*) with some intradural extension of tumor (*small arrow*). **B,** Huge left psoas mass (*arrows*) and large epidural mass with marked compression of opacified thecal sac (*arrowhead*). **C,** Foraminal extension with sac effacement. Filum lipoma (*arrowhead*) has been displaced to right. **D,** Vertebral body erosion best appreciated on bone window image (*arrow*). **E,F,** Extradural mass (*arrow*) extends into intradural space (*arrowhead*). Incidental filum terminale lipoma (*open arrow*).

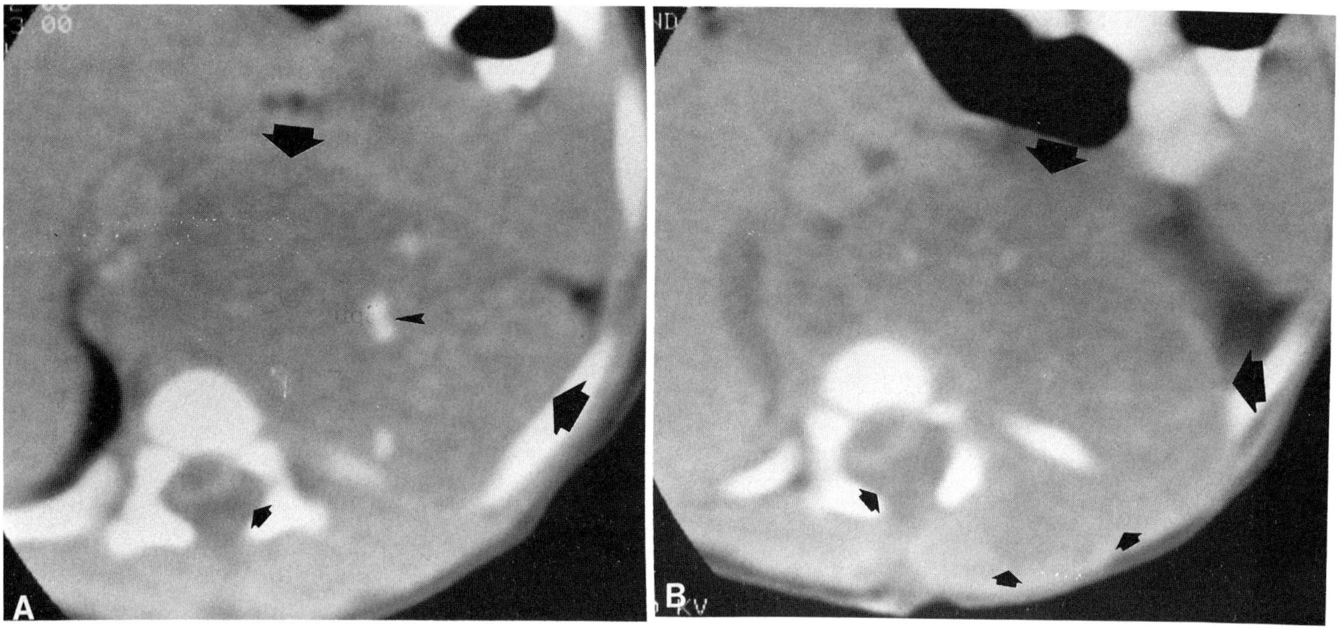

Figure 9.25. Neuroblastoma. A 1-month-old girl. **A,B,** Axial post-myelogram CT scans through T12 level. Large paravertebral mass (*large arrows*) with characteristic calcifications (*arrowheads*). Note the extension into the posterior paravertebral musculature and into the posterior epidural space (*small arrows*). **C,D,** Sagittal T1-weighted image (TR500/TE30). Huge mass anterior to the spine with extensive involvement of the spinal canal (*arrows*) and paravertebral musculature. **E,** Axial T2-weighted image (TR1800/TE50). Hyperintensity of tumor significantly improves tissue contrast (*arrowheads*) and better defines full extent of tumor. Note the extension through the left sided neural foramen.

This is due to both MRI's superior imaging characteristics and the relative dangers of myelography in a patient with a complete subarachnoid block. There is a significant incidence of rapid neurologic deterioration following lumbar puncture in patients with complete block [11].

Additional advantages of MRI include sensitivity in evaluating bone marrow replacement and superior soft-tissue contrast, as well as the ability to image relatively long segments of the spine [7, 113, 114, 116]. Approximately 20% of patients with spinal metastatic disease have more than two sites of involvement at some point. MRI allows detection of widely separated lesions that may be missed on CT-myelography due to lack of subarachnoid space compression or location between two areas of myelographic block [114]. When a mass is associated with a subarachnoid block, T1-weighted images may demonstrate some hyperintensity of the CSF distal to the block (Fig. 9.29). This finding is suspicious for a subarachnoid block and is presumably related to obstruction of CSF pulsations and an increase in protein concentration in the trapped distal CSF [7].

Normal marrow is usually hyperintense on T1-weighted images because of the high fat content [117]. When this fat is replaced by even small amounts of tumor, there is some loss of the signal intensity on the T1-weighted images [7, 113, 114, 117] (Fig. 9.30). Any process that alters or replaces marrow or fat will decrease the T1 signal [117] (Fig. 9.31). T2 signal is more variable and depends on the type of tissue that has replaced the normal marrow. Areas of sclerosis or fibrosis result in low signal intensity on both the T1- and T2-weighted images, whereas areas of necrotic tissue, hemorrhage, or inflammatory debris tend to have higher T2 signal than neoplastic tissue [7, 117]. In general, vertebral metastases are relatively hyperintense to unaffected marrow on T2-weighted images [118]. Extensive tumor infiltration may result in an abnormal heterogeneous appearance [118]. Sclerotic metastases may be hypointense on T2-weighted images [118].

MRI may be helpful in distinguishing benign from malignant compression fractures and may add information unavailable with other imaging modalities [227, 231]. Complete replacement of normal marrow in malignant vertebral body compression fractures is often observed [231]. This suggests that most vertebral metastases do not result in compression fracture until the entire vertebral body is replaced by tumor. Conversely, in the majority of benign fractures (without a history of trauma), there is always preservation of normal bone marrow. Irregular marrow replacement is suspicious for malignancy, but differentiation between benign and malignant fractures is not always possible. Another feature that may be helpful is preservation of normal T1 signal intensity in the dorsal one-third of the compressed vertebral body. This "posterior band sign" is reportedly seen in benign, but not metastatic, vertebral body frac-

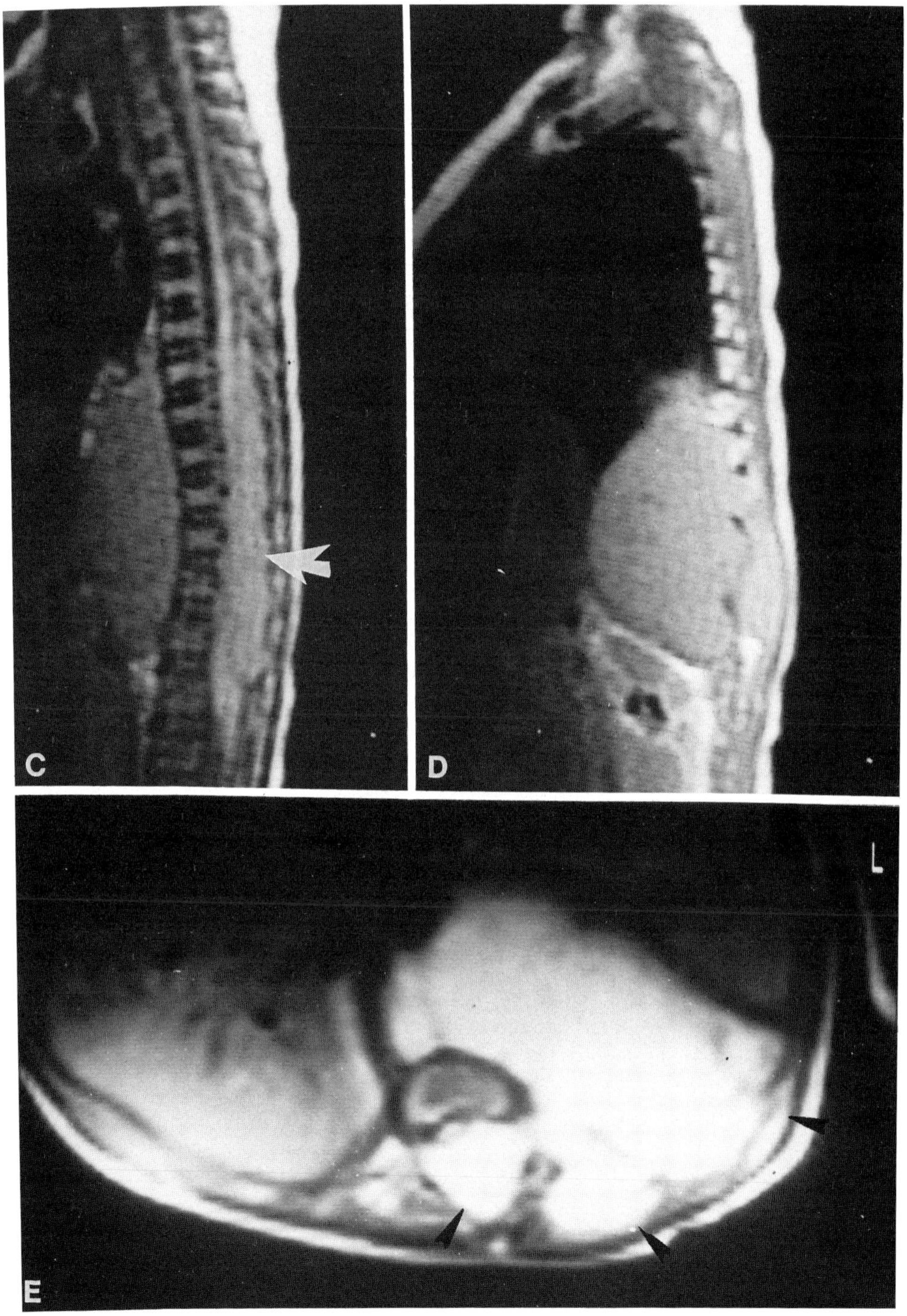

Figure 9.25–*continued*

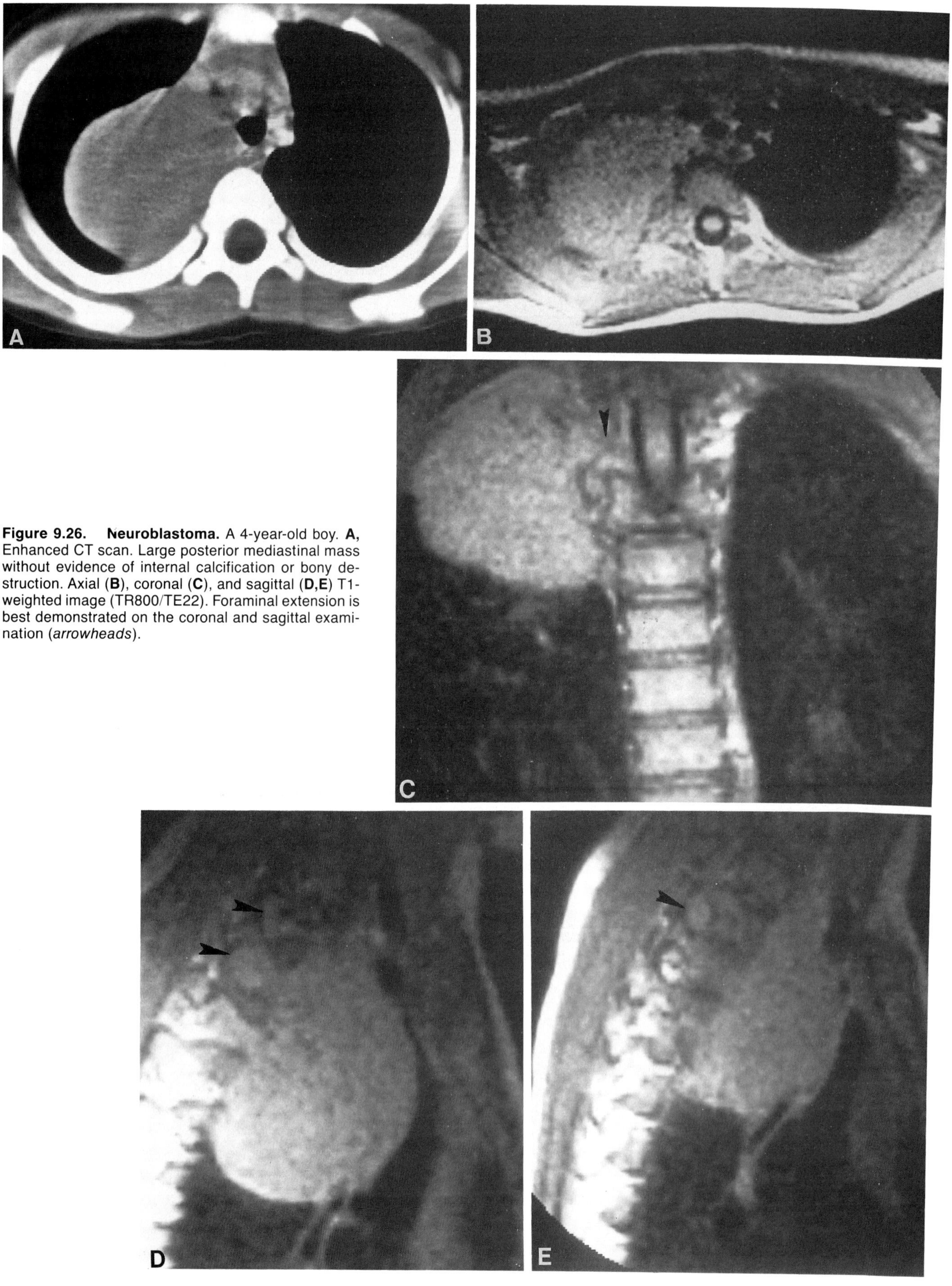

Figure 9.26. Neuroblastoma. A 4-year-old boy. **A,** Enhanced CT scan. Large posterior mediastinal mass without evidence of internal calcification or bony destruction. Axial (**B**), coronal (**C**), and sagittal (**D,E**) T1-weighted image (TR800/TE22). Foraminal extension is best demonstrated on the coronal and sagittal examination (*arrowheads*).

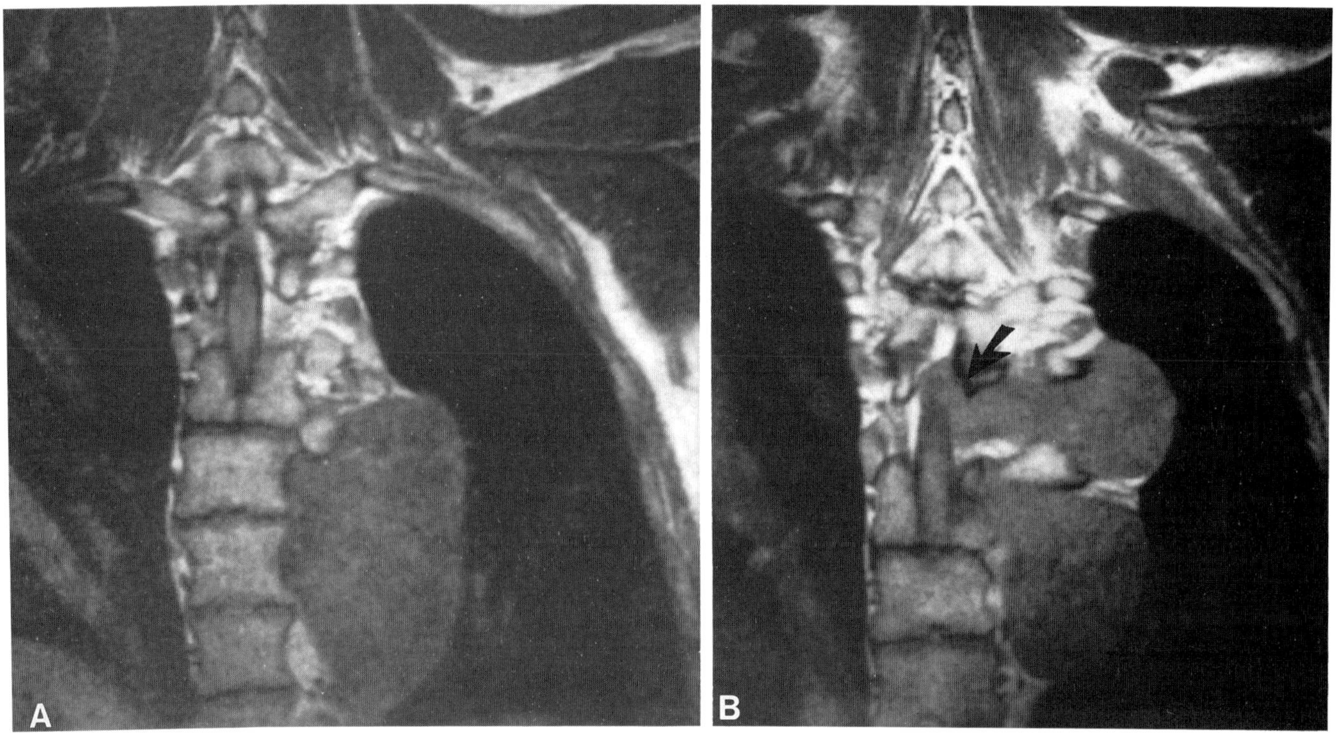

Figure 9.27. Ganglioneuroma. An 18-year-old woman with back pain. Coronal T1-weighted image. Left paraspinal mass extends into the spinal canal via the T6 and T7 neural foramina. Note the epidural extension of tumor cephalad to the T6 foramen (*arrow*) with some cord displacement. Calcification is less frequent in benign ganglioneuromas.

tures [227]. However, this is not always reliable (see Fig. 9.32).

Paramagnetic contrast material is not as useful in the extradural space as it is elsewhere in the spine [118]. Before contrast administration, marrow replacement is evident as areas of relatively decreased signal intensity relative to normal marrow fat. Paramagnetic contrast enhancement of these neoplastic areas may actually decrease relative contrast, and the tumor may become relatively isointense with the normal marrow [118]. This can be resolved by utilizing fat-suppression techniques, resulting in decreased signal from fatty tissue such as vertebral marrow. Contrast between the suppressed-fat signal and the intermediate signal of the neoplasm may be increased. This difference in contrast between normal and pathologic marrow may be further improved by injection of paramagnetic contrast material [116].

Contrast enhancement may help to better define the configuration of the lesion. This is useful when evaluating extension of tumor into the spinal canal [118]. Epidural extension of the tumor becomes hyperintense relative to the low signal CSF. Postcontrast MRI is also useful when localizing tumor for biopsy or in the unusual situation where epidural tumor needs to be distinguished from disk herniation [118].

MYELOMA

Multiple myeloma is a plasma cell malignancy characterized by diffuse multicentric involvement of the bone marrow. Myeloma is the most common primary malignant neoplasm of bone, accounting for 45% of malignant bone tumors in one large series [111]. Myeloma is rarely encountered in patients less than age 40, and 69% of cases are seen in males. Bone pain is the most frequent symptom, presenting in approximately 68% of patients [119]. The pain is generally in the thoracic or lumbar area. Most patients also have weakness, weight loss, and a moderate to severe anemia. Hypercalcemia is seen in 20 to 50% [111]. Serum electrophoresis is generally abnormal, although minimal or no abnormalities are seen in approximately 15% [111]. Diagnosis is established by bone marrow aspiration.

Myeloma involves the hematopoietic red marrow, and therefore, vertebral body location is quite common. The pedicles and posterior elements contain predominantly fatty marrow and are generally not involved early. However, the anemia associated with replacement disorders such as myeloma may prompt conversion of the fatty marrow to hematopoietic red marrow [105]. Because of this change, the pedicles and posterior elements may be involved subsequently. This relatively late involvement of the

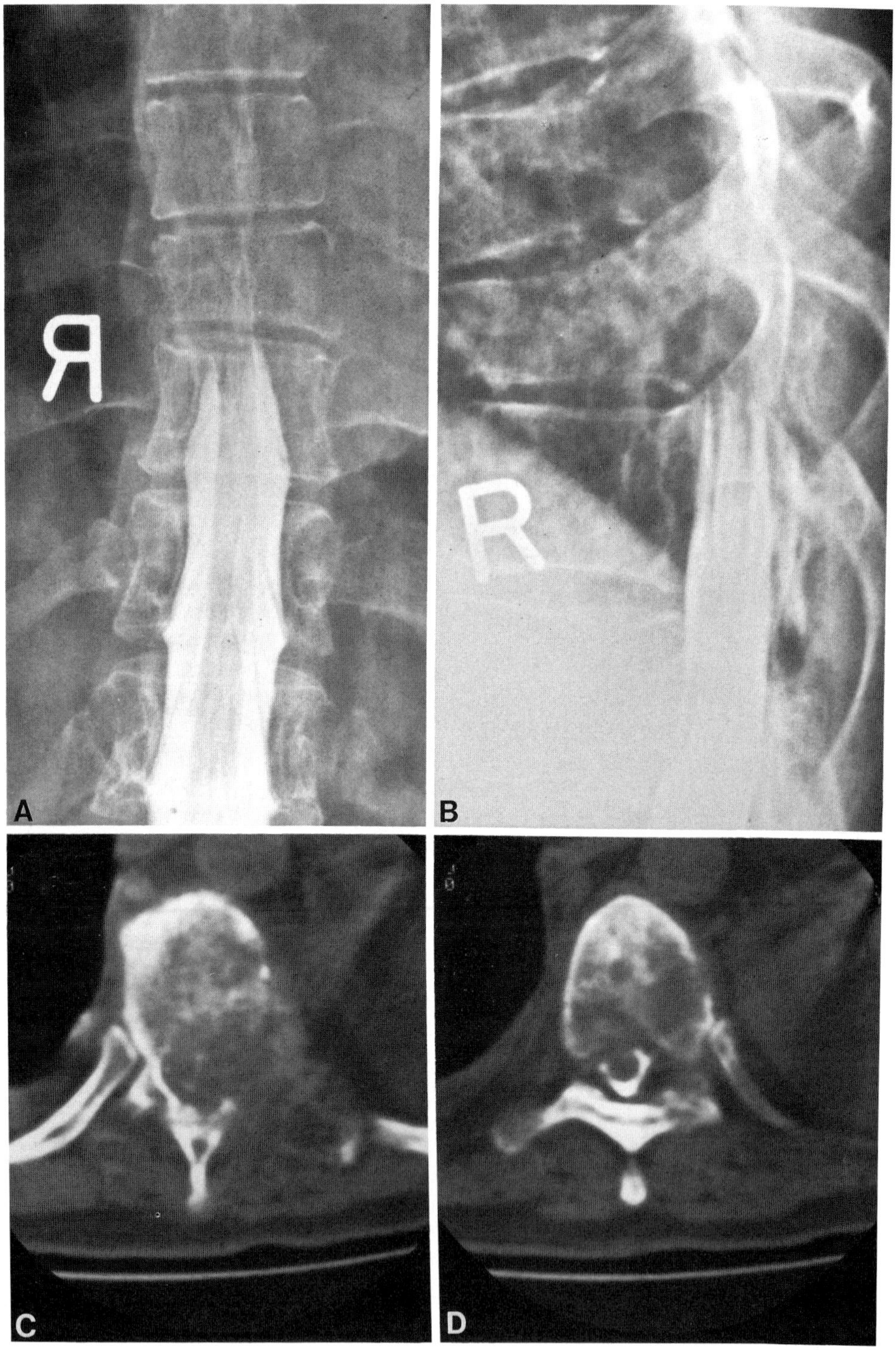

Figure 9.28. Metastatic renal cell carcinoma. A 62-year-old man with back pain. **A,** CT scan with intrathecal contrast. Destructive, expansile mass replaces much of the vertebral body and extends into the right transverse process and into the spinal canal. **B,** Sagittal T1-weighted image (TR800/TE22). Tumor replacement of the normal vertebral marrow in the T7 and T8 vertebrae results in relative hypointensity (*arrows*) relative to the normal marrow in the adjacent vertebrae. There is some destruction of the T7–8 disk, unusual for metastases and actually more suggestive of infection. Note the extension of tumor into the anterior epidural space. **C,** Contrast-enhanced sagittal T1-weighted image (TR800/TE22). There is some enhancement of the tumor with the T7 and T8 vertebral bodies, resulting in relatively decreased contrast when compared with the adjacent normal vertebrae. **D,** Axial CT demonstrates epidural and osseous metastases involving the thoracic vertebrae with cord compression. There are metastatic deposits within the rib. **E,** Coronal re-formatted image demonstrating the circumferential narrowing of the thecal sac from the epidural metastasis. **F,** Sagittal. T1-weighted image demonstrates the vertebral body metastasis compressing the thoracic spinal cord. Note the decreased fluid (*arrows*) involving the vertebral bodies. **G,H,** Contiguous axial CT scans demonstrating posterior epidural metastasis with involvement of the pedicle, lamina, and adjacent ribs in a patient with renal cell carcinoma.

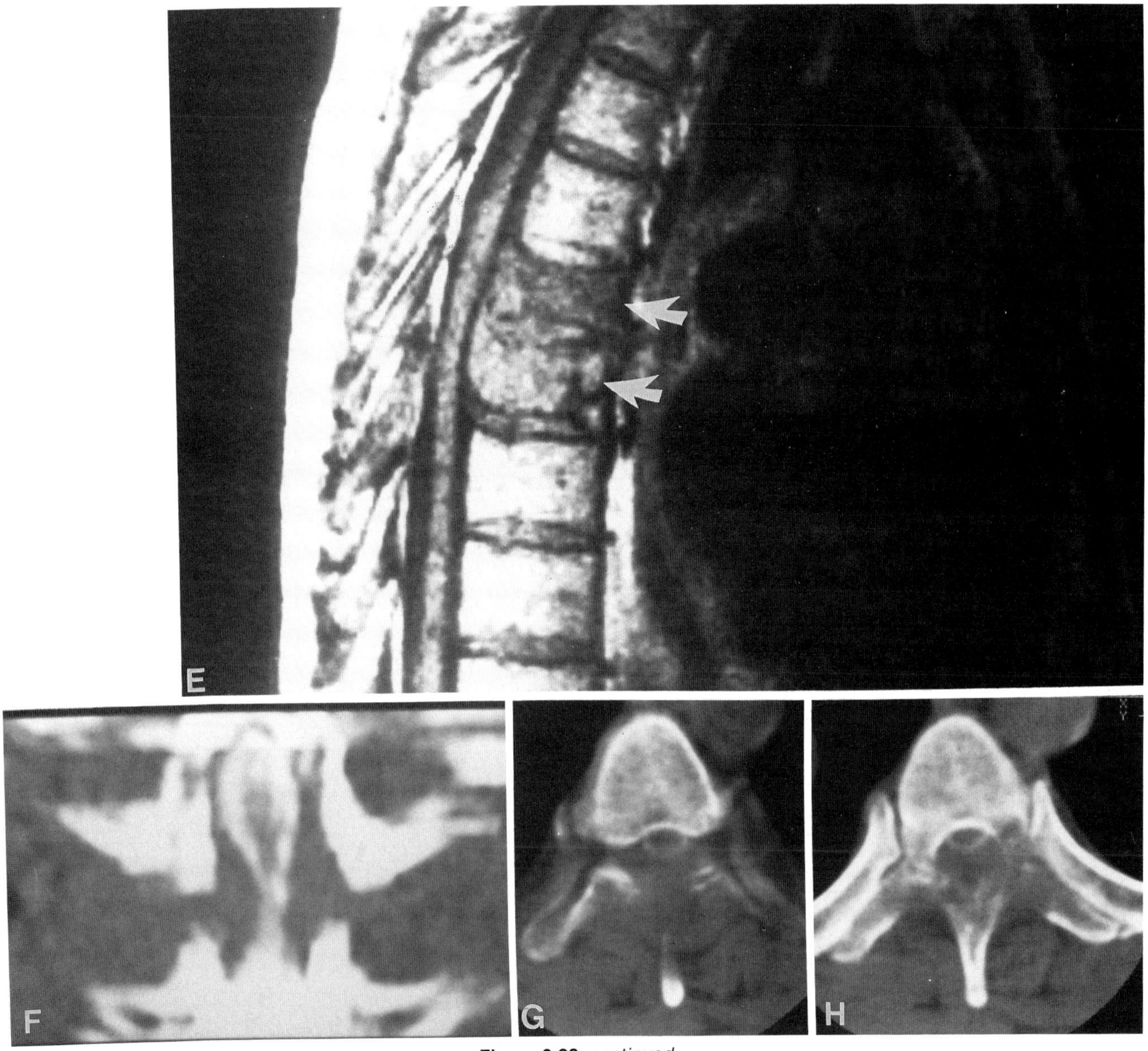

Figure 9.28–*continued*

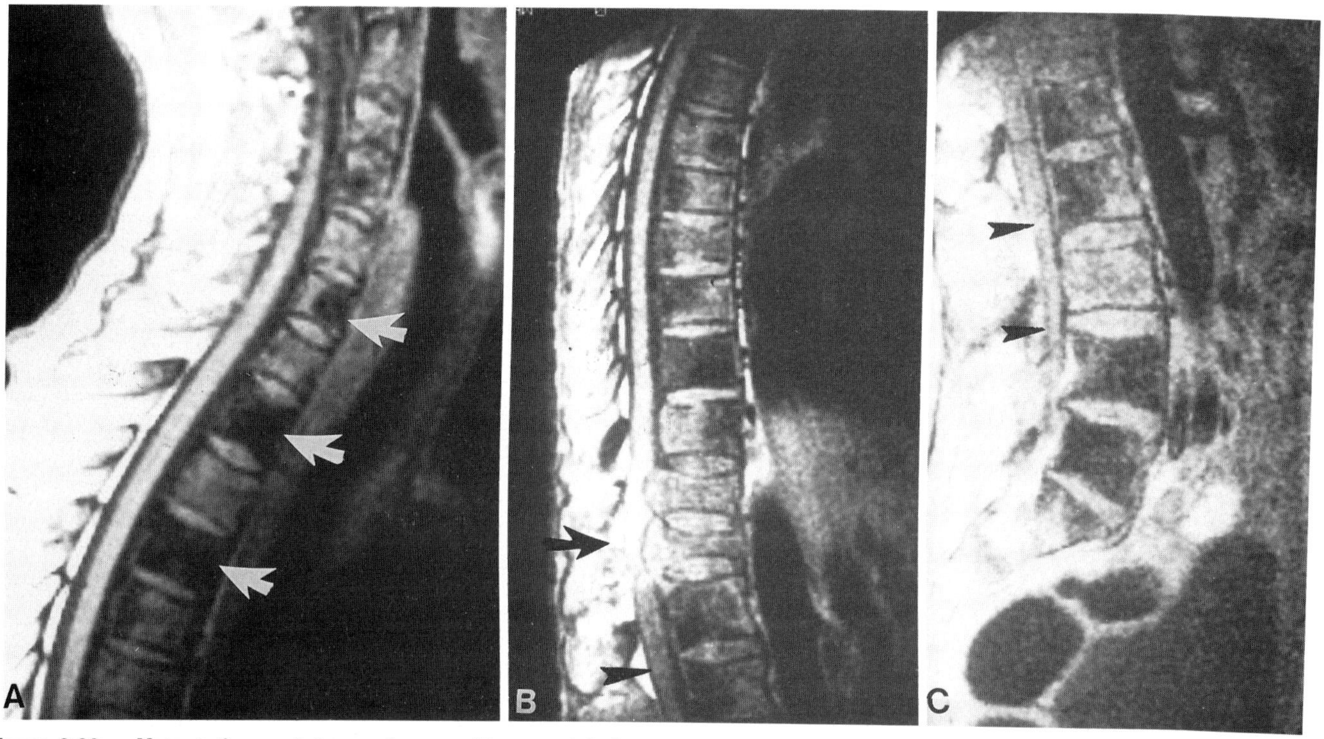

Figure 9.29. Metastatic prostate carcinoma with associated subarachnoid block. A 60-year-old man with diffuse prostate metastases. **A,** Sagittal T1-weighted image (TR800/TE22). Multiple areas of abnormal marrow replacement (*arrows*) corresponding to sclerotic metastases noted on plain film exam (not shown). Note the normal-appearing cervical thoracic cord and the normal-appearing hypointense CSF. **B,C,** Sagittal T1-weighted midline images through the thoracic and lumbar spine. Multiple areas of abnormal marrow replacement of varying signal intensity. Note the posterior extension of tumor into the spinal canal at the T11 and T12 levels with cord compression (*arrows*). The thecal sac distal to this complete block is markedly hyperintense (*arrowheads*) when compared with the normal CSF signal seen proximal to the level of the block. This abnormal T1 hyperintensity of the CSF is presumably due to a combination of increased protein and absent CSF pulsations and is suspicious for a subarachnoid block.

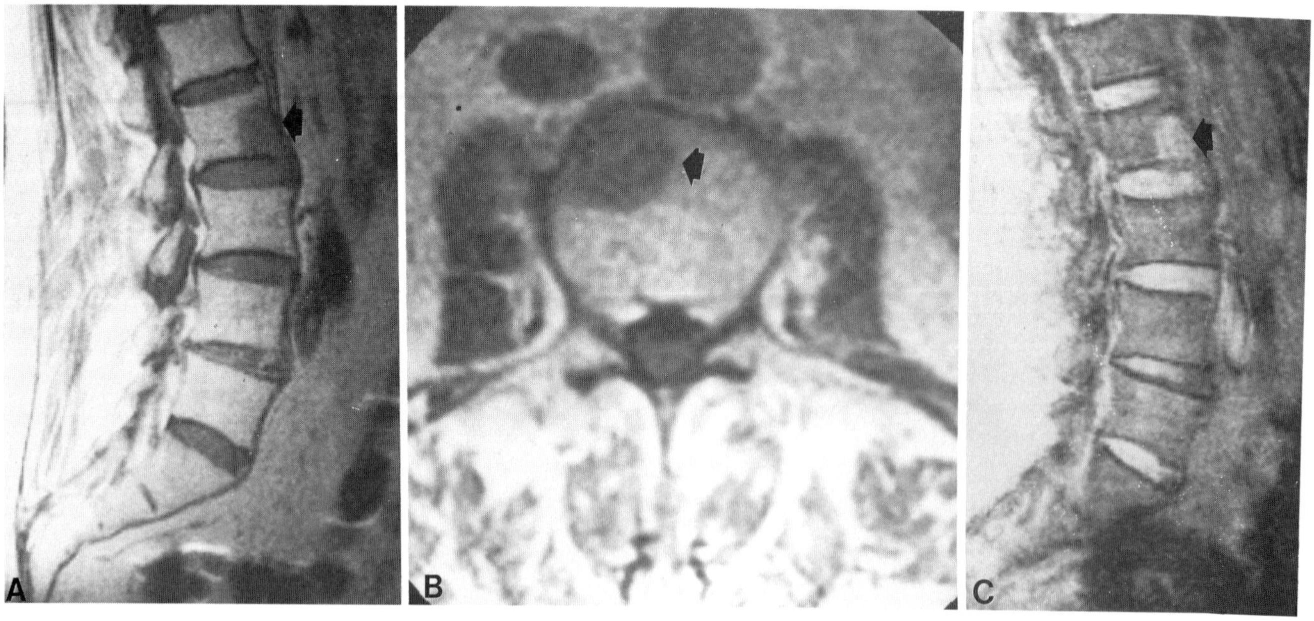

Figure 9.30. Prostate metastasis to L2 vertebral body. A 64-year-old man with back pain and normal plain film examination. **A,** Sagittal T1-weighted image (TR800/TE22). Rounded 2-cm hypointense area of marrow replacement in the anteroinferior L2 vertebral body (*arrow*). **B,** Axial T1-weighted image (TR900/TE22). Well-defined area of marrow replacement (*arrows*). **C,** Sagittal T2 sequence (free induction decay). Metastasis is relatively hyperintense on T2-weighted image (*arrow*).

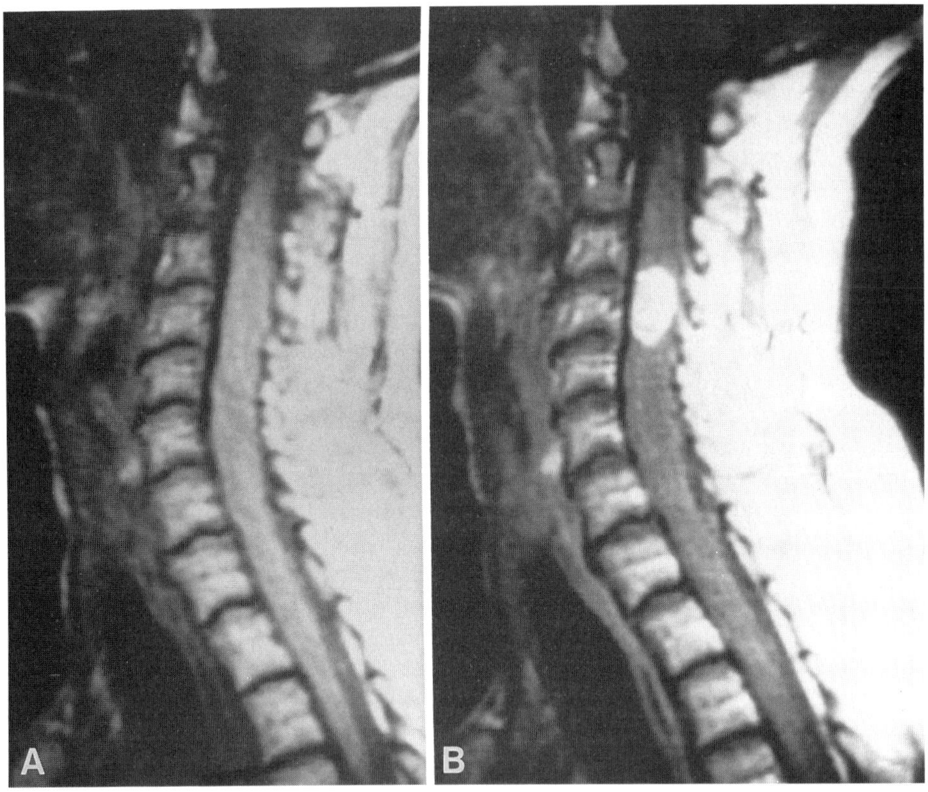

Figure 9.31. Intramedullary metastasis. A 50-year-old man with a widely metastatic bronchogenic carcinoma. **A,** Sagittal T1-weighted image shows a fusiform enlargement of the cervical cord which is relatively isointense. **B,** Contrast-enhanced sagittal T1-weighted image shows intense homogeneous enhancement of the intramedullary metastasis.

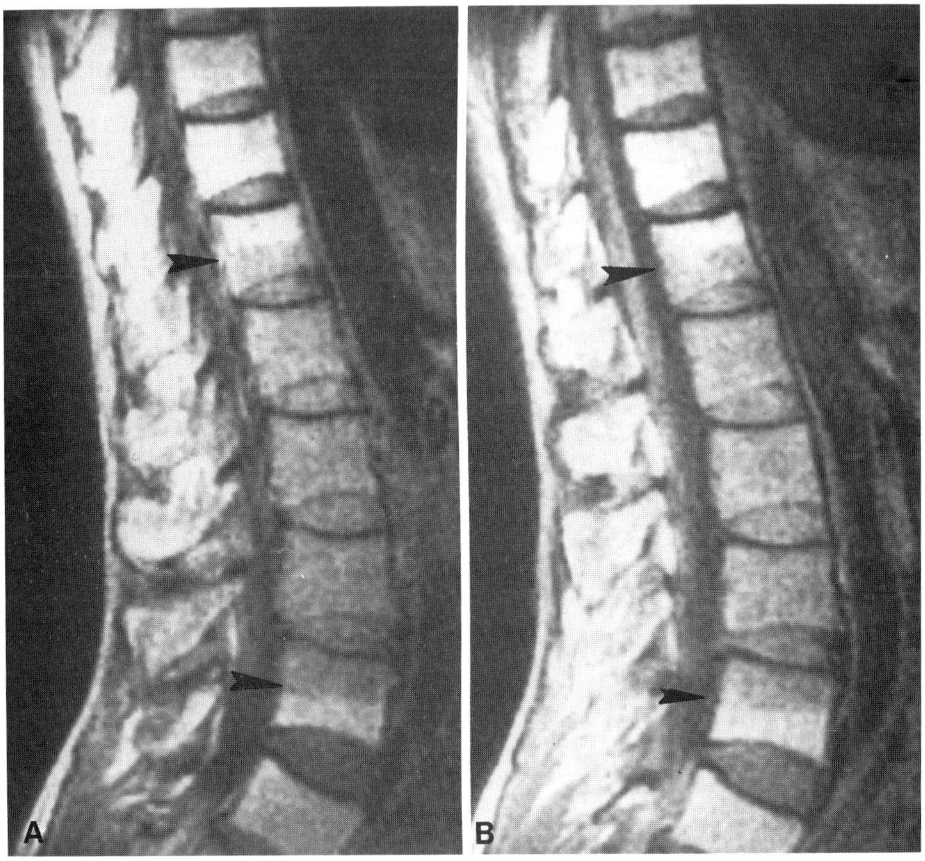

Figure 9.32. Radiation change. A 61-year-old man with two previous radiation ports for metastatic disease. **A,B,** Adjacent sagittal T1-weighted image (TR800/TE22). Prior radiation is associated with fatty replacement of marrow resulting in marked hyperintensity on T1. The previous radiation ports extended caudally from mid-L4 vertebral body and superiorly from the mid-T12 vertebral body. Note the sharp delineation between the associated fatty replaced marrow (*arrowheads*) and the normal marrow between T12 and L4. Relative hypointensity of the normal marrow could be confused with an infiltrative process if the characteristic radiation changes are not recognized.

pedicles and posterior elements is a distinguishing feature from metastatic disease, which tends to involve the pedicles earlier [7, 105].

Vertebral involvement may lead to early pathologic fracture with abrupt onset of symptoms [111]. Neurologic signs and symptoms are frequently due to either a pathologic fracture or extraosseous extension of neoplastic tissue. The rare sclerotic form of myeloma has a higher association with peripheral polyneuropathy [111].

Bone scans are of limited use in evaluation of myeloma because of the significant number of false-negative results, which exceed 50% [117]. Bone scintigraphy also underestimates the extent of skeletal involvement [120]. Plain films are of limited use because approximately 50% of cancellous bone is replaced before the abnormality is demonstrated [121]. However, plain films are abnormal in approximately three-fourths of patients, and the diagnosis is usually straightforward [120]. The plain film findings result from myelomatous masses replacing bone and classically include "punched out" lytic areas of bone destruction that vary from several millimeters to 5 cm in diameter [111]. There is no surrounding zone of sclerosis, although the lytic lesions are typically well defined [111]. Generalized osteoporosis and pathologic fractures also may be encountered [111, 122]. Disk narrowing may be encountered due to vertebral endplate collapse with subsequent intravertebral disk herniation (Schmorls node) [22]. Expansion of the vertebral body may produce a "ballooned out" appearance similar to that seen in the ribs [111].

Solitary plasmacytomas are isolated lesions which involve the spine in nearly 50% of cases [111]. The lesions are destructive, expansile, and may be multicystic. Hematologic and bone marrow studies are often normal. Although a majority of these patients may develop multiple myeloma subsequently, there is a latent period of 5 to 10 years [111].

The lesions of myeloma occasionally become sclerotic following therapy. Rarely, sclerotic forms of myeloma are encountered in untreated patients [122]. The sclerosis apparently results from production of reactive new bone, but what triggers the myelomatous deposits to produce this new bone is unclear. Interestingly, about three-fourths of patients with the blastic form of myeloma present with the chief complaint of radicular pain radiating down one extremity [122]. The sclerotic form of myeloma is often associated with the POEMS syndrome [123]. The acronym for this multisystem disorder stands for *p*olyneuropathy, *o*rganomegaly, *e*ndocrine abnormalities, *M* protein, and *s*kin changes.

Plain films are generally sufficient for the diagnosis of multiple myeloma, and additional imaging is not required [120]. CT is more sensitive in evaluation for myeloma in patients with bone pain and normal plain x-rays [120]. CT also demonstrates associated soft-tissue components within the spinal canal and neural foramina in patients with neurologic complications related to their myeloma [120]. However, intrathecal contrast material is generally required.

Both bone and soft-tissue window images should be obtained. CT demonstrates replacement of the normal fatty bone marrow with dense plasma cell infiltrate as multiple, small, irregular lytic foci replacing the bony trabeculae. Cortical involvement in the way of pathologic fractures also may be demonstrated. The lytic areas are generally 1 to 5 mm in diameter, but confluent lytic areas greater than 1 cm are not infrequent [119] (Fig. 9.33). Margins are generally well defined but not sclerotic [119–121]. CT demonstrates myelomatous involvement of the vertebrae earlier than plain films and also generally demonstrates the extent to be greater than sus-

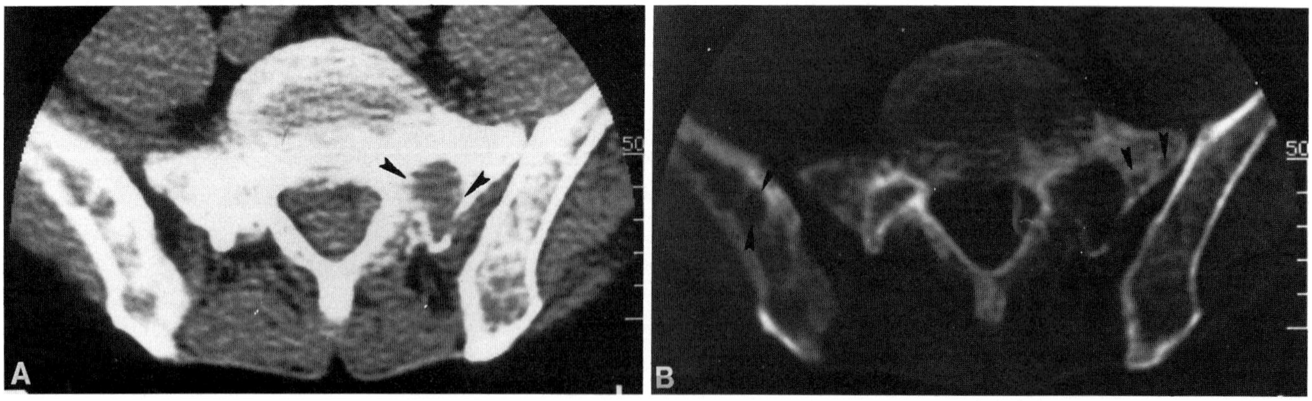

Figure 9.33. Multiple myeloma. A 40-year-old man with back pain. **A,** Postmyelogram CT image through the sacrum. Lytic area of bony destruction (*arrowheads*) associated with cortical destruction of the sacral component of the left SI joint. The distal thecal sac is also involved and no intrathecal contrast extends caudal to the L4–5 disk level. **B,** CT bone window image demonstrates the lytic lesion on the left and also demonstrates additional small lytic foci in the left vertebral body (*arrowheads*) and in the right iliac bone (*arrowheads*).

pected by plain film evaluation [117, 119, 120]. The disadvantage of CT is related to the difficulty in screening large segments.

Although CT is more sensitive than plain films in evaluation of myeloma, early involvement may be difficult to distinguish from osteoporosis [117, 121]. MRI is well suited for evaluation of myeloma. Benefits include a large field of view, excellent delineation of the adjacent spinal canal and neural foramina, superior soft-tissue contrast, and the high signal intensity of normal bone marrow, which facilitates identification of infiltrative processes [117, 124]. MRI is also helpful in evaluating for extension of the neoplastic tissue into the spinal canal with possible cord or nerve root compression [7, 117, 124].

MRI may be useful in differentiating between lesions that are stabilized by radiotherapy and progressive myelomatous bone disease [124]. Diffuse inhomogeneity of the bone marrow is probably caused by small foci of myeloma [124]. Diffuse bone marrow infiltration is best demonstrated on the T1-weighted images.

Normal bone marrow contains a significant amount of fat and therefore has high signal intensity on T1-weighted images. Normal signal intensity usually diminishes on the more T2-weighted images. The sensitivity of MRI is based on replacement of the normal marrow by the myelomatous tissue, resulting in a decrease in the signal intensity on the T1-weighted images [117, 124] (Fig. 9.34). The longer T1 relaxation time of neoplastic tissue results in a hypointense area on the T1-weighted image. In patients with untreated myeloma, these foci typically become relatively hyperintense on the more T2-weighted images. However, as with other spine tumors, this appearance is somewhat variable and depends on the amount of hemorrhage and necrotic tissue. Some patients may not demonstrate focal lesions and will show only nonspecific inhomogeneity of the bone marrow on the T1-weighted image [124]. Patients with treated tumors may show areas of decreased signal intensity on both the T1- and T2-weighted images. This is apparently related to sclerotic regeneration after irradiation [124].

Despite the excellent sensitivity of MRI, this modality lacks specificity, since many systemic diseases such as lymphoma, leukemia, and metastases can have a similar appearance [7, 124]. Correlation with plain films is imperative, since old fractures or degenerative changes that produce hypertrophic bone may have a low signal intensity on the T1-weighted image. Another pitfall to be avoided is an apparent diffuse decrease in T1-weighted signal intensity of the marrow related to distance from the surface coil. This is a problem often encountered with obese patients and may be overcome by additional imaging with a body coil [124]. Focal fat deposition should not be confused with myeloma. Although both may be hyperintense on a T2-weighted image, the focal fatty replacement of marrow is markedly hyperintense on the T1-weighted image (unlike myeloma). Differential considerations also include vertebral hemangiomas, but these are typically more hyperintense on both T1- and T2-weighted images [124].

MRI is not required in most routine cases of myeloma. It is useful in establishing the diagnosis in patients with back pain with both negative bone scans and negative plain radiographs. Although data are preliminary, MRI will probably be useful in assessment of therapy and may help differentiate between lesions stabilized by radiotherapy and active progressive myelomatous involvement [124].

LEUKEMIA

Leukemia is a myeloproliferative disorder that may initially cause complaints referable to the musculoskeletal system [125]. Diffuse marrow hyperplasia is generally found in all forms of leukemia, but the frequency and character of the surrounding bony changes vary greatly and are dependent on both the age of the patient and the histologic type of tumor [126].

Leukemia is the most frequent malignancy of childhood and may even be encountered during infancy [127]. In children, most cases are from acute lymphoblastic leukemia. Bone pain and demonstrable skeletal lesions are common. Pain is secondary to proliferation of hematopoetic tissue within the medullary cavities, especially in the long bones and vertebral bodies [125]. Patients occasionally present with back pain that may be secondary to either collapse of a vertebral body or secondary to leukemic infiltration [125]. Approximately one-half of patients present with radiographic abnormalities at the time of diagnosis, and diffuse osteopenia is frequently noted in the spine [125, 126]. Occasionally, lytic or sclerotic lesions can be identified [125]. Plain films may demonstrate zones of radiolucency adjacent to the cortex of the inferior and superior borders of the vertebral bodies which may progress to demineralization during the course of the disease. These nonspecific zones of lucency correspond to the "leukemic" lines seen in metastases of the long bones, but these are generally not well defined after a few months of age [127].

In adults, leukemic bony lesions are much less frequent, occurring in less than 10% of cases [128]. The most common findings are diffuse osteoporosis or focal areas of osteolysis. The generalized osteoporosis particularly affects the spine, ribs, skull, and pelvis [128]. Lytic areas may be either due to local infiltration or secondary to infarction related to vas-

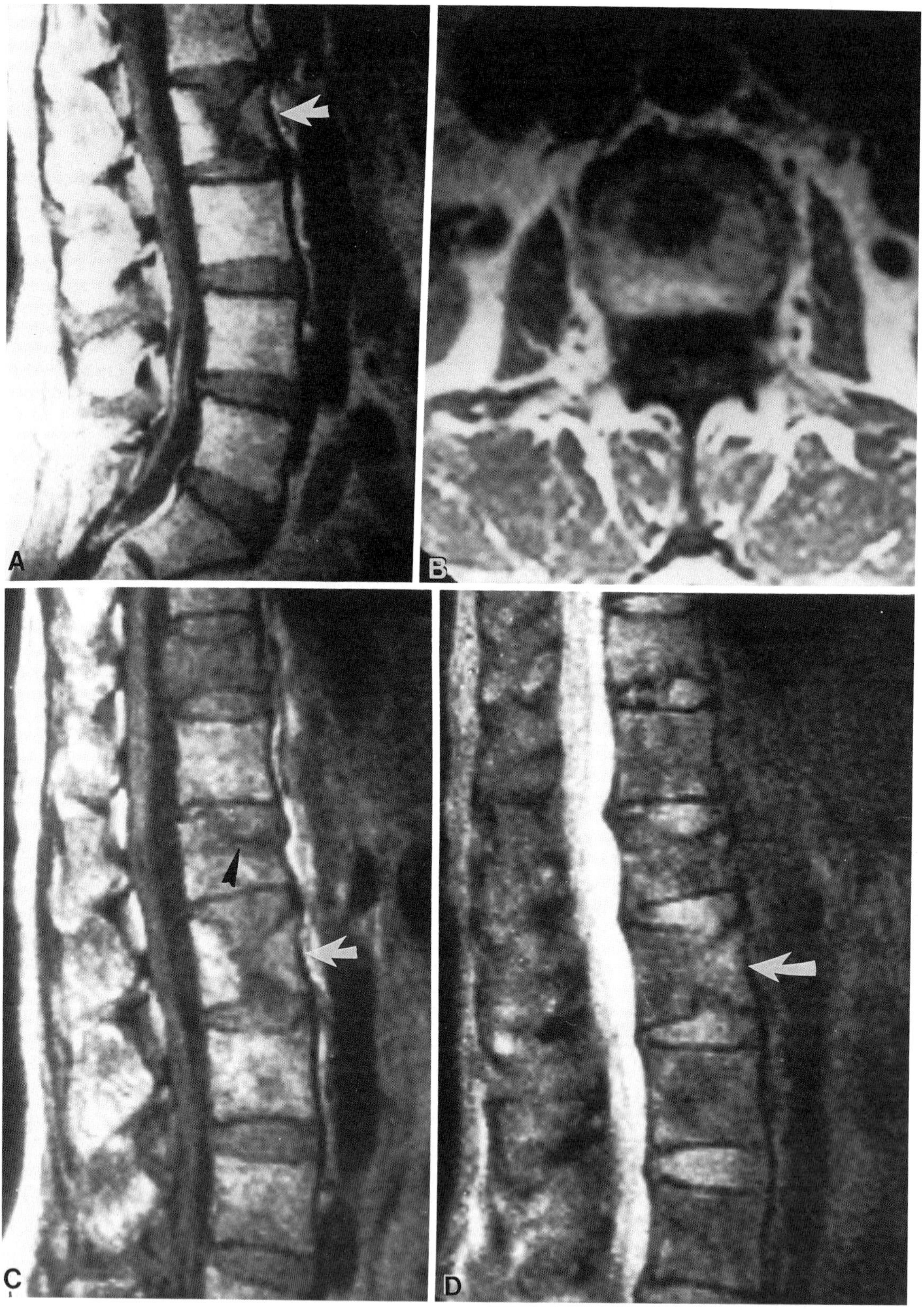

Figure 9.34. Multiple myeloma. A 60-year-old woman with back pain. **A,** Sagittal T1-weighted image (TR800/TE22). Abnormal replacement involving the mid-L2 vertebral body which is relatively hypointense compared with adjacent normal marrow. There is intravertebral disk herniation of the L1–2 disk through the superior vertebral endplate of L2. The area of marrow replacement was isointense on the T2 (FID) sequence (not shown). **B,** Axial T1-weighted image (TR800/TE22) through L2 vertebra. Well-defined area of marrow replacement seen centrally in the vertebral body. **C,** Sagittal T1-weighted image (TR800/TE22) obtained 18 months later. Intravertebral disk herniation through both the superior and inferior vertebral endplates of L2 (*arrow*). Marrow replacement in the L1 vertebral body and T11 vertebral body is now evident. There is also now herniation of the T12–L1 disk through the superior vertebral endplate of L1 (*arrowhead*). **D,** Sagittal T2 sequence (FID) at the same follow-up visit. Abnormal T2 hyperintensity is seen anteriorly in T2 body (*arrow*).

cular occlusion [126]. Rarely, an osteosclerotic lesion may be encountered and should suggest the possibility of myeloid metaplasia [128].

Acute leukemia is less frequent in adults, and plain film findings are also less frequent than seen in children. Chronic lymphocytic leukemia is more common in the elderly and is generally associated with diffuse osteoporosis [126]. Chronic myelogenous leukemia is also encountered in adults. Skeletal symptoms are relatively infrequent, being generally associated with acute blastic transformation [126]. Both these chronic forms of leukemia may be associated with osteoporosis, although it is less frequently encountered in chronic myelogenous leukemia. Focal osteolytic areas may be more suggestive of myeloma.

Chloromas are solid extramedullary tumors composed of early myeloid precursors which are occasionally encountered in the myelogenous leukemias [129]. These locally invasive masses are interchangeably referred to as either *chloromas* or *granulocytic sarcomas;* the distinction is based on demonstration of green pigmentation on pathologic examination in chloromas [130]. Chloromas are somewhat more frequent in the acute myelogenous leukemias, although these are encountered in less than 10% of patients [130].

Chloromas may be diagnosed concurrently with myelogenous leukemia or may be discovered during the course of the disease, even when the patient is in clinical remission. Younger patients are affected more commonly than older patients [130]. Although chloromas may occur anywhere, spinal involvement is common and frequently symptomatic. These are often associated with nerve root or cauda equina compression [131]. Pain and symptoms related to mass effect predominate, and acute paraplegia may be encountered [130].

Chloromas are usually extradural but may occasionally be either intradural or both intra- and extradural (dumbbell-type) [130]. There may be prominent destruction of adjacent bone and pathologic fracture [126]. Posterior epidural involvement is more common [130, 131]. Subarachnoid involvement may be associated with nerve root thickening and, if extensive, can produce an intradural block [131]. Multiple spinal tumors may be encountered, and either neurofibromatosis or lymphoma may be suspected initially [130].

As with other lytic tumors, CT is superior to plain films in demonstrating both the extent of bony destruction and the associated soft-tissue mass (Fig. 9.35). CT with intrathecal contrast material is useful in demonstration of abnormally thickened nerve roots due to leukemic infiltration [130, 131]. Both intra- and extradural masses can be demonstrated. Occasional "dumbbell" lesions can be identified, strongly resembling neurofibromatosis [131]. Nodules may be iso- to hyperdense [132, 131] and have a variable appearance that may mimic lymphoma, neurofibroma, or metastatic neuroblastoma. CT is valuable in evaluating associated paraspinal soft-tissue inflammatory processes that may secondarily involve the spinal canal with neurologic impairment [133]. Homogeneous enhancement of leukemic masses in the CNS has been reported [134].

The MRI features in leukemia are nonspecific and need to be correlated with other imaging studies as well as the clinical and pathologic studies. Marked decrease in marrow signal on T1-weighted images is noted in the blastic phase of chronic myelogenous leukemia. It is probably related to the complete replacement of marrow fat by the leukemic infiltrate as well as marrow edema. On T2-weighted images, the leukemic infiltrate within the marrow is isointense or mildly hyperintense.

Other processes such as old fractures or degenerative changes can be differentiated on the basis of decreased signal intensity in both the T1- and T2-weighted images, apart from loss in height of the vertebral bodies and marked endplate sclerosis. An inflammatory or necrotic process involving the marrow appears as hypointense signal on T1-weighted images and hyperintense signal on T2-weighted images.

Leukemic infiltration of the nerve roots and adjacent intra- and paraspinal soft tissues can be better demonstrated on MRI. As with CT, paramagnetic contrast material aids in demonstrating the leukemic infiltration of the soft tissues [118]. In addition, the multiplanar capabilities of MRI allow visualization of larger segments of the spine in coronal and sagittal planes.

MRI is also useful in the diagnosis of recurrent disease, confirmation of remission, and as a screening test in detecting focal lesions, providing a guide for direct needle biopsy of the marrow site [135]. Similar to CT, MRI is also useful in the detection of chloroma. Recognition of these lesions is important because they are extremely sensitive to radiation, which may be preferred over or combined with chemotherapy [129, 132]. Surgery is usually reserved for symptomatic spinal cord compression [129].

LYMPHOMA

Lymphoma is a frequent cause of spinal malignancy and may occur in any age group, although it is rare in the very young [111]. There is a 3:2 male-to-female ratio. Patients generally present with pain, which occasionally is long-standing. Neurologic symptoms are common when spinal involvement is present. In general, prognosis is better than with other malignancies.

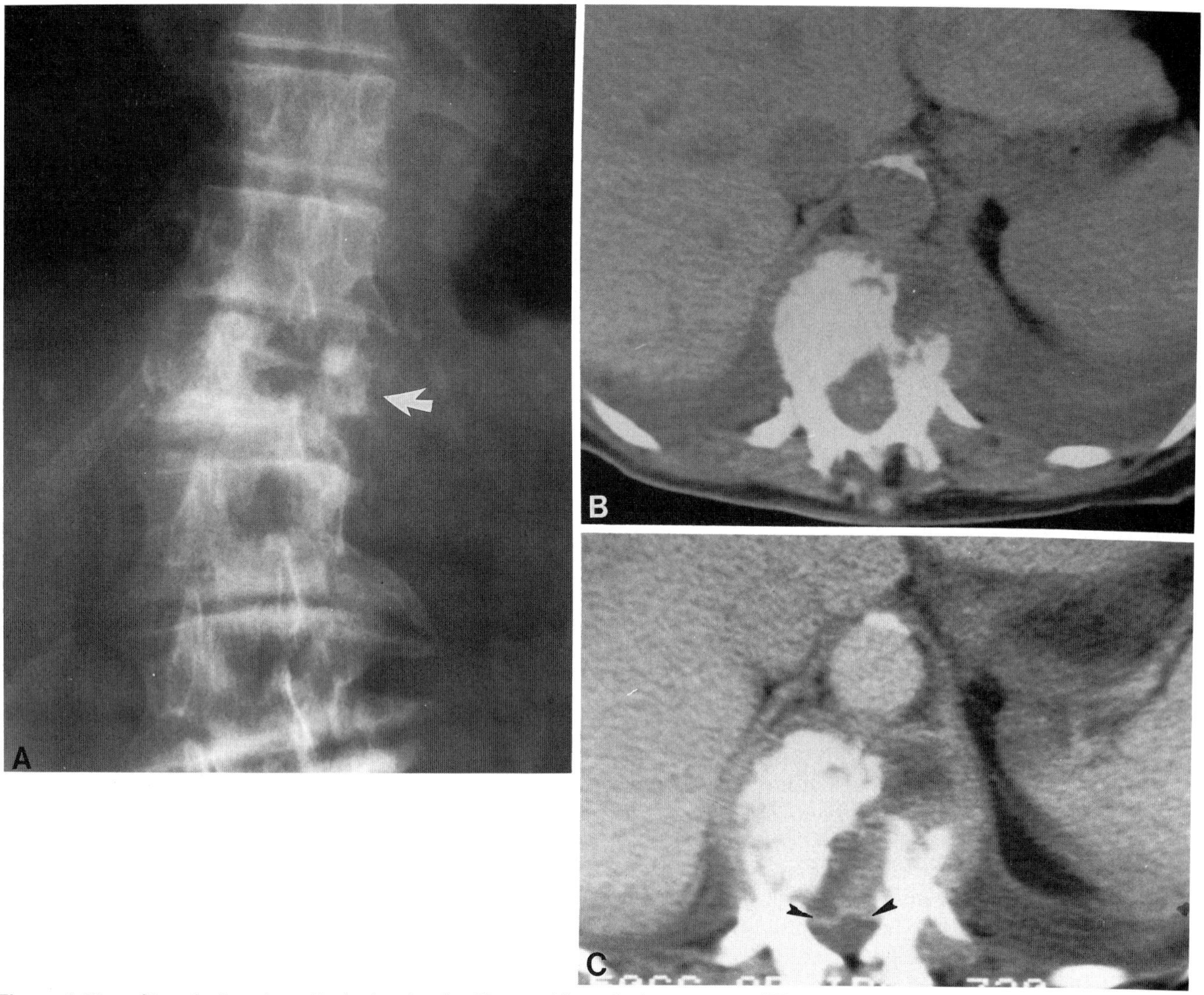

Figure 9.35. Chronic lymphocytic leukemia. An 85-year-old woman with systemic disease and back pain. **A,** AP radiograph. Mixed lytic and blastic destructive process at T12 (*arrow*). **B,** Unenhanced CT scan. Leukemic infiltration associated with prominent bony destruction and extension into the paravertebral region and epidural space. Note bilateral pleural effusions. **C,** Enhanced CT scan. Enhancement of the leukemic infiltrate better defines epidural extension (*arrowheads*). **D,** First and second echo of a sagittal multiecho sequence (TR1000/TE30/80). Epidural extension of the T12 leukemic infiltrate is best demonstrated on the first echo (*arrowhead*), as is the associated cord displacement. A true T1-weighted image should demonstrate this leukemic infiltrate to be hypointense relative to normal marrow. **E,** Sagittal contrast-enhanced T1-weighted image (TR600/TE30). Prominent homogeneous enhancement of the leukemic infiltrate is seen both superiorly and inferiorly in the canal (*arrows*).

Lymphoma is classified as either Hodgkin's disease or non-Hodgkin's lymphoma, and bony involvement may be either primary or secondary. These two clinical subtypes are radiographically indistinguishable [137]. Between 15 and 25% of lymphoma patients have demonstrable bone involvement on plain films [138–141]. The thoracolumbar spine and pelvis are the most frequent sites, and the lesions are usually lytic or mixed lytic and blastic [139]. Bone destruction in the aggressive histologic type is more likely to be permeative [140, 141], and the transition zone is generally indistinct. Occasionally, the vertebral body may be sclerotic, producing the "ivory" vertebrae [138, 140, 141]. Anterior scalloping of the vertebral body is commonly encountered due to enlargement of adjacent lymph nodes. These also may account for the periosteal reaction [140, 141]. The periosteal reaction is seen most frequently in the lateral and inferior margins of the vertebral bodies [137].

Lymphoma involves bone either by direct extension from adjacent lymphadenopathy or via hematogenous spread to the marrow (Fig. 9.36). Paravertebral soft-tissue masses are frequently demonstrated. These often extend through the neural foramina to involve the spinal canal [7, 137, 142, 143]. This may result in extradural compression of the spinal cord or cauda equina. The dura is generally an

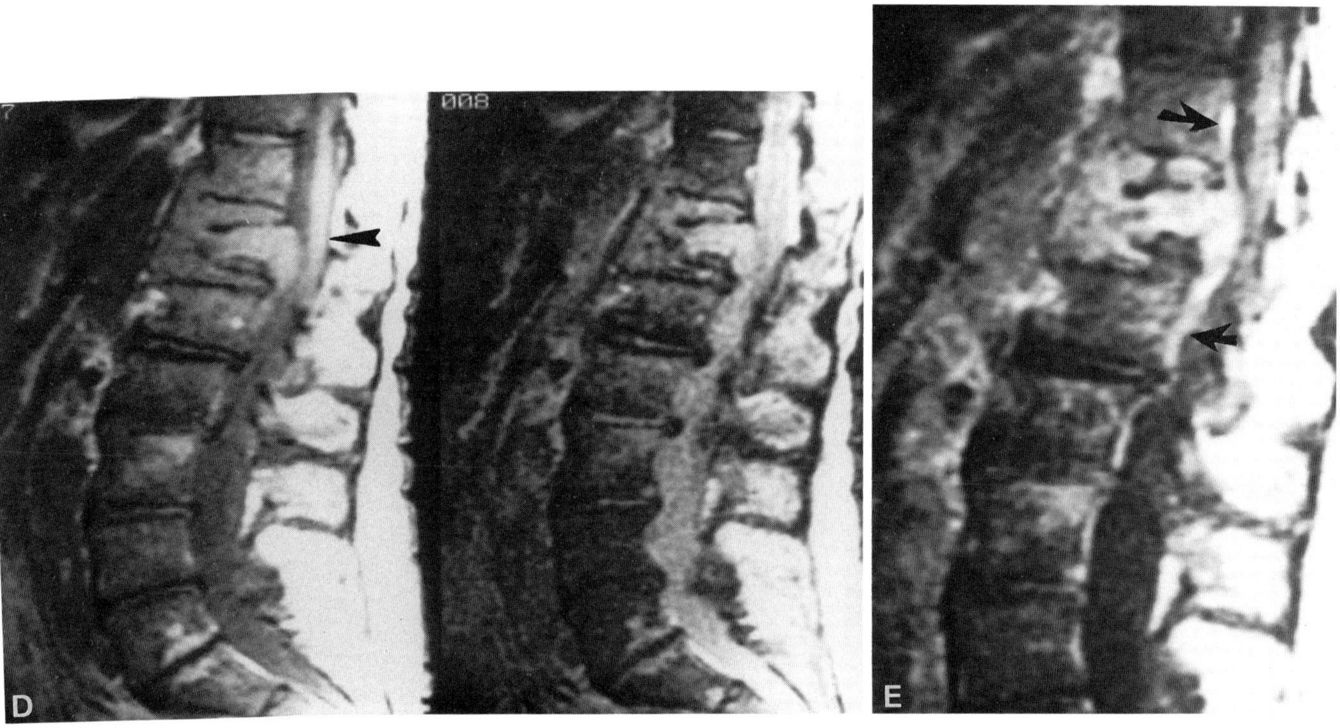

Figure 9.35–*continued*

effective barrier against intrathecal extension [7]. Vertebral body destruction with resultant compression of the thecal sac is less common than foraminal extension of tumor. Vertebral collapse is more often seen with the more aggressive forms of non-Hodgkin's lymphoma [7].

CT is useful in demonstrating extent of involvement as well as the adjacent paravertebral mass. When performed with intrathecal contrast material, foraminal and epidural involvement is well demonstrated. The soft-tissue mass has a homogeneous density ranging from 40 to 100 H.U. Calcifications or areas of central necrosis are uncommon [142]. Associated bony changes such as destruction or sclerosis, as well as erosion or compression fracture, also may be demonstrated. The epidural mass often extends beyond one vertebral segment and is suggestive of lymphoma [142].

MRI is more sensitive than either plain films or scintigraphy in evaluation of lymphomatous bony involvement [143, 144]. This is due to the permeative growth pattern of lymphoma, which is occasionally associated with little bone destruction or bony response. Foraminal extension of a paravertebral mass into the epidural space does not require intrathecal contrast material (Fig. 9.37). MRI is superior to CT in that long segments of the spine can be evaluated. This facilitates identification of multiple areas of involvement, critical in planning for radiation therapy. As with most other marrow infiltrative lesions, MRI generally shows areas of involvement to be hypointense on T1-weighted images and relatively hyperintense to isointense on the T2-weighted images [145]. Epidural involvement, which may extend over several levels, can be defined clearly with MRI [145]. Although nonspecific, Hodgkin's lymphoma occasionally will be detected as "diffuse marrow infiltration with superimposed focal areas of even lower signal intensity, reflecting the nodular nature of Hodgkin's" [135]. Leptomeningeal enhancement with both iodinated contrast material and Gd-DTPA has been reported [96].

The distinction of lymphoma from metastatic disease or other extradural malignancy is difficult (Fig. 9.38). When there is no bony destruction or significant lymphadenopathy, distinction between epidural lymphoma and disk herniation also may be difficult, although lymphoma may have more indistinct margins [146]. As in lymphoma, neurofibromas also may cause foramina erosion but are generally localized. Lymphoma is a likely consideration when bony destruction with a paravertebral soft-tissue mass and epidural extension over several levels is present [142].

OSTEOSARCOMA

Osteosarcoma of the spine is rare, accounting for only 3% of all primary osteosarcomas [147]. Osteosarcoma is most often seen in the first three decades of life, with a slight male predominance. However, in the spine, no sex predilection is evident, and these

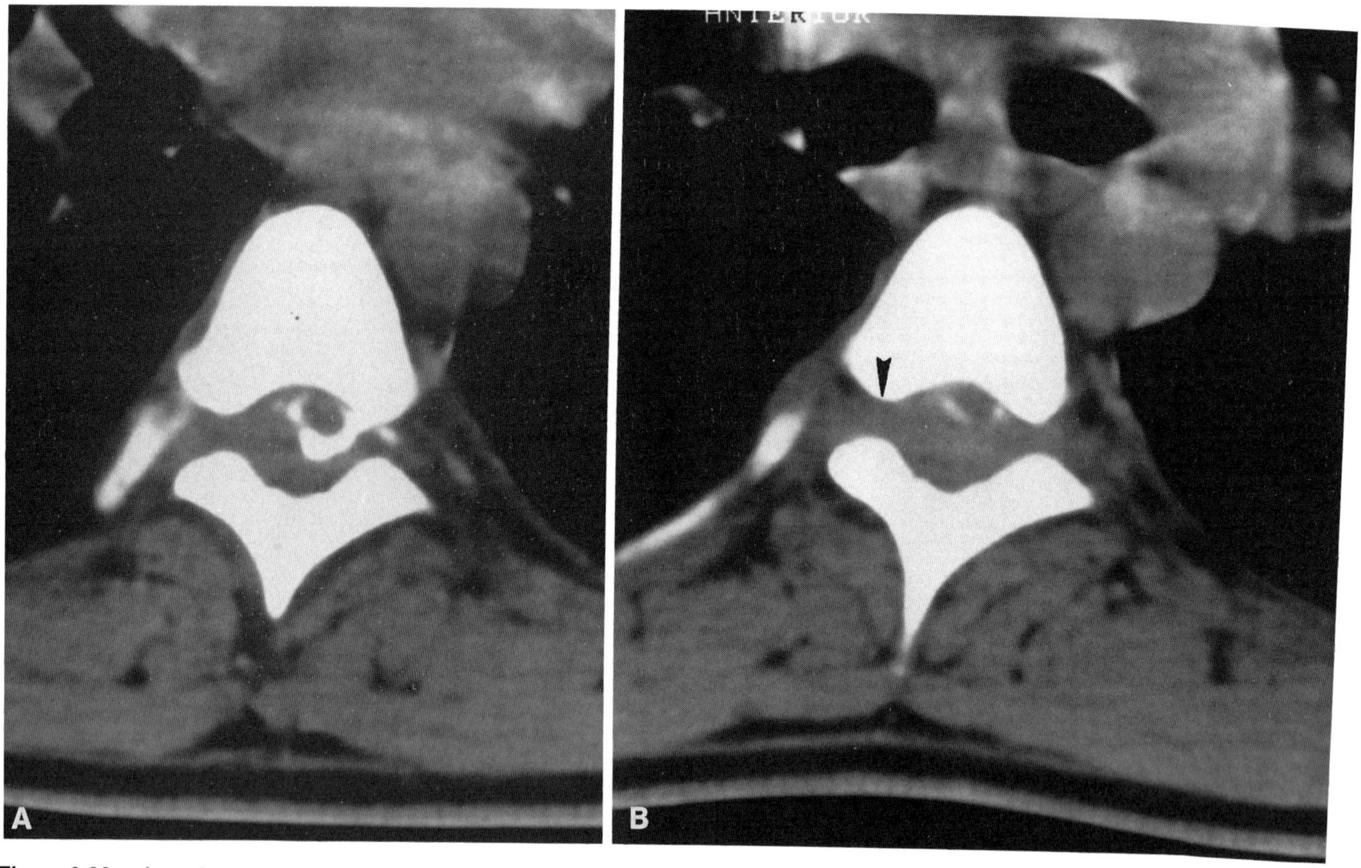

Figure 9.36. Lymphoma. A 27-year-old HIV^{+} man with systemic lymphoma. Myelogram (not shown) demonstrated complete epidural block at T6 level. **A–C,** Postmyelogram CT scans. Nonspecific epidural soft-tissue mass located primarily posteriorly and to the right. Note the foraminal involvement (*arrowheads*) as well as the compression of the opacified thecal sac. Lymphoma may extend a considerable distance within the spinal canal, but the dura is generally an effective barrier. **D,** Noncontrast axial CT scan. **E,** Coronal re-formatted CT scan. There is a bulky soft-tissue area in the prevertebral space extending into the spinal canal. This involves the lymph nodes and displaces the abdominal aorta. The contrast image demonstrates the typical sclerosis of the vertebrae usually associated with Hodgkin's disease involving the vertebrae.

tumors tend to occur in the older age group [147]. Occasionally, a history of either radiation or Paget's disease can be obtained. The median latent period following previous spinal radiation is 10 years, although the range is from 6 to 30 years [148]. There is a well-known relationship between Paget's disease and sarcomatous transformation, but this tends to involve the long bones with relative sparing of the vertebral column [147].

Local pain is the usual presenting symptom, although 70% of patients also will have neurologic signs and symptoms [147]. These tumors are fairly evenly distributed throughout the spine and sacrum. Most involve the vertebral body either alone or in association with posterior element involvement. When only the posterior elements are involved, osteoblastoma may be impossible to exclude radiographically, especially in the cervical spine. Lesions may be either predominantly lytic or blastic, depending on the amount of ossification or calcification within the neoplasm. Most demonstrate a combination of these features [111]. Cortical involvement is usually evident, with a gradual transition between the neoplasm and adjacent uninvolved bone, producing indistinct borders [111]. An associated soft-tissue mass is frequently evident on plain film examination [147]. The neoplasm may eventually involve adjacent vertebrae [149].

Radiographic findings are nonspecific and are dependent on the amount of ossification or calcification within the neoplasm and may therefore be osteosclerotic, osteolytic, or mixed in appearance [149]. CT is superior to MRI in demonstration of soft-tissue calcification and tumor matrix mineralization [150]. Large foci of calcification appear as discrete areas of signal void on MRI, and smaller calcifications occasionally result in a somewhat inhomogeneous appearance. However, some noncalcified areas of tumor also have a hypointense appearance on MRI [150].

Both MRI and CT are accurate in demonstrating involvement of cortical bone. Cortical involvement is identified on MRI by replacement of the normal hypointense cortex with the relatively hyperintense

Figure 9.36–continued

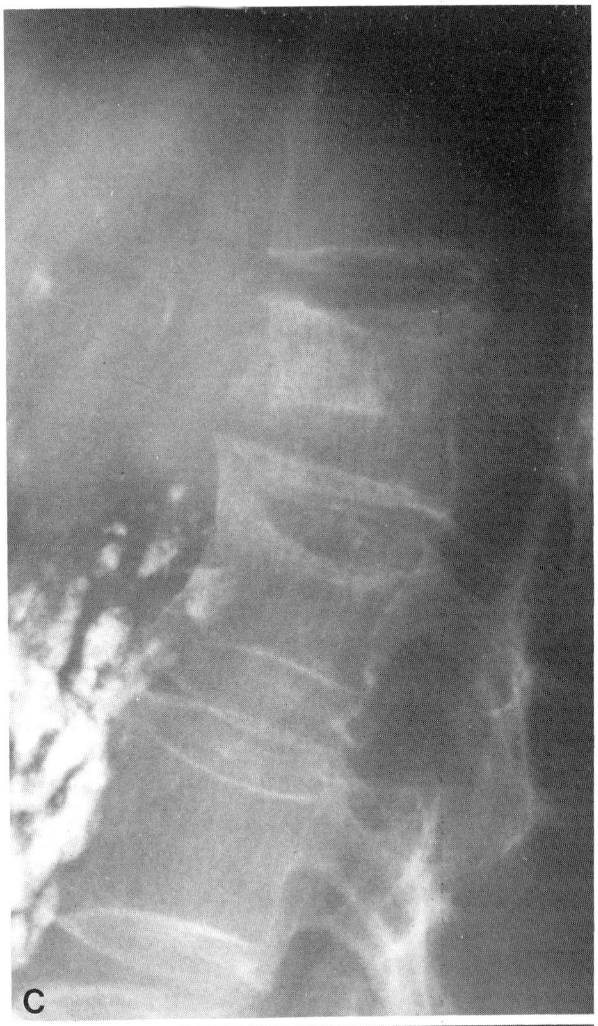

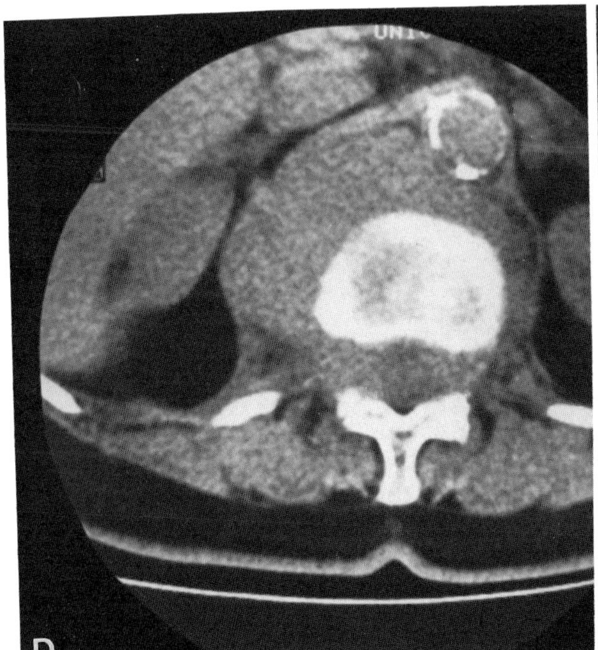

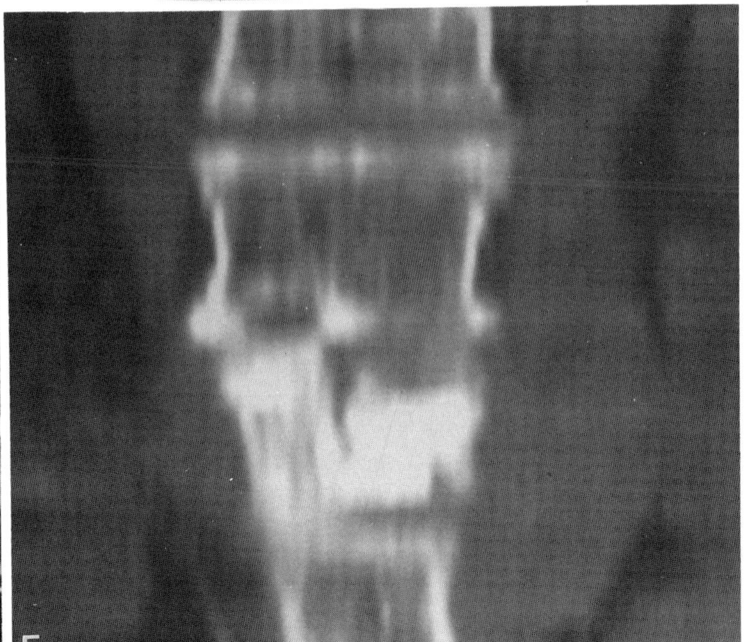

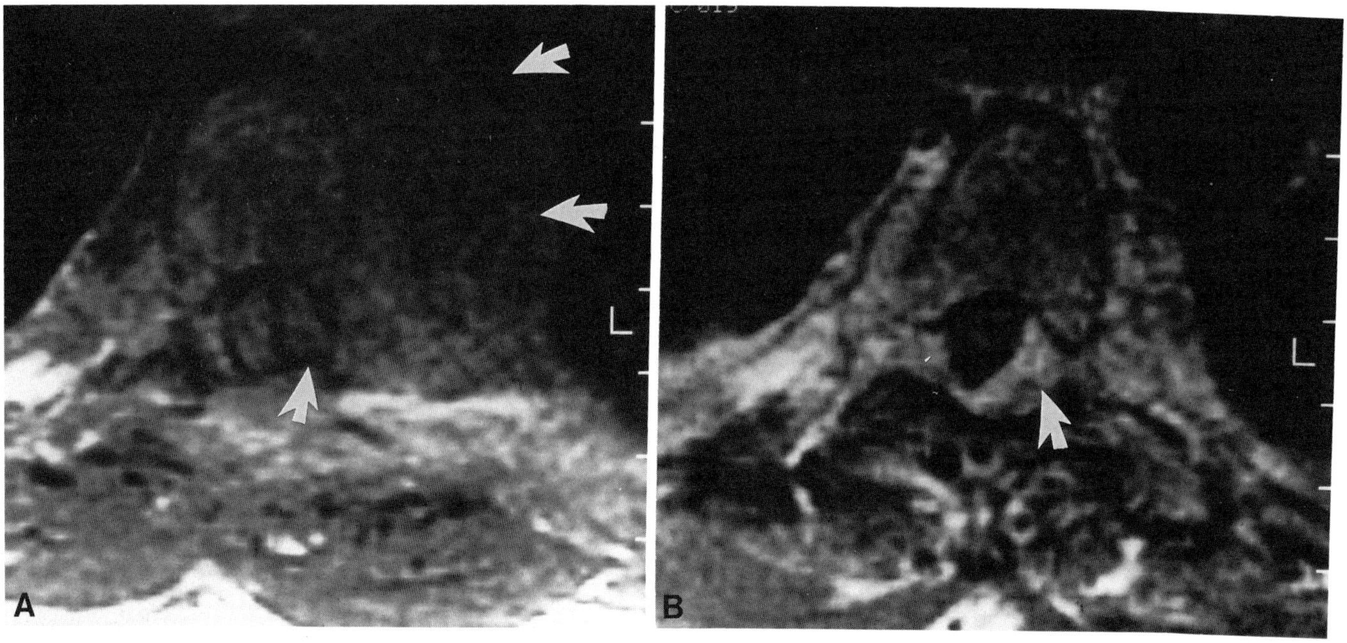

Figure 9.37. Lymphoma. Large paravertebral mass with foraminal extension and epidural compression of the cord. **A,** Axial T1-weighted image demonstrates a large left paravertebral mass (*arrows*) with extension into the left side of the spinal canal. **B,** Contrast-enhanced axial T1-weighted image demonstrating enhancement of the paravertebral mass bilaterally as well as the epidural extension of the mass (*arrows*).

signal of the invading neoplasm [150]. CT generally demonstrates a hyperdense bony tumor that may extend into the spinal canal or adjacent paravertebral musculature. Heterogeneous contrast enhancement may be demonstrated [149].

Involvement of the spinal canal is equally well demonstrated by either MRI or CT with intrathecal contrast material. CT findings of spinal canal involvement include obliteration of the epidural fat, deformity of the opacified subarachnoid space, cord compression, and compromise of the neural foramina [151]. These findings are often not apparent without intrathecal contrast material.

MRI is superior in demonstrating extent of marrow abnormality as well as the relationship of the neoplasm to surrounding structures [152–154]. Both T1- and T2-weighted sequences should be obtained. On T1-weighted images, the normally hyperintense marrow is replaced by the more hypointense neoplasm.

Marrow involvement is better shown by MRI [150, 154]. Clear demarcation between the normally high-signal marrow and the hypointense areas of neoplastic replacement is frequently demonstrated. On T2-weighted images, findings are more variable, with areas of hyperintensity that may be very similar to that of marrow or areas of marked hypointensity which correspond to osteoblastic activity within the marrow [150, 155]. Involvement of adjacent musculature is better demonstrated on T2-weighted images, where the neoplasm becomes relatively hyperintense compared with the normal musculature [150]. On T1-weighted images, the invading neoplasm may occasionally be isointense with adjacent musculature [150]. In general, axial images are best for demonstrating the extent of paraspinal involvement, and coronal or sagittal images are useful in depicting the superior or inferior extent of the neoplasm [151, 154].

Dynamic Gd-DTPA studies are useful in assessing chemotherapeutic response [228]. Reduction in signal intensity on T2-weighted images obtained during therapy is a highly specific but not very sensitive criterion indicating a favorable response. Hyperintensity on contrast-enhanced T1-weighted images following chemotherapy is sensitive but not specific for detection of viable tumor regions [228].

The prognosis is poor, with or without surgical intervention [147]. Hematogenous metastases result in pulmonary involvement. Early widespread bony metastasis also may be encountered [111].

EWING'S SARCOMA

Ewing's sarcoma is a highly malignant neoplasm that only rarely affects the spinal column. Most lesions are in the sacrum or less commonly in the lumbar spine [111]. In general, Ewing's sarcoma is associated with a distinct male predilection, but this is not evident in vertebral column involvement [156]. Most cases are encountered in the first three decades of life. Although primary involvement of the spinal column is rare, metastatic involvement is frequent [156].

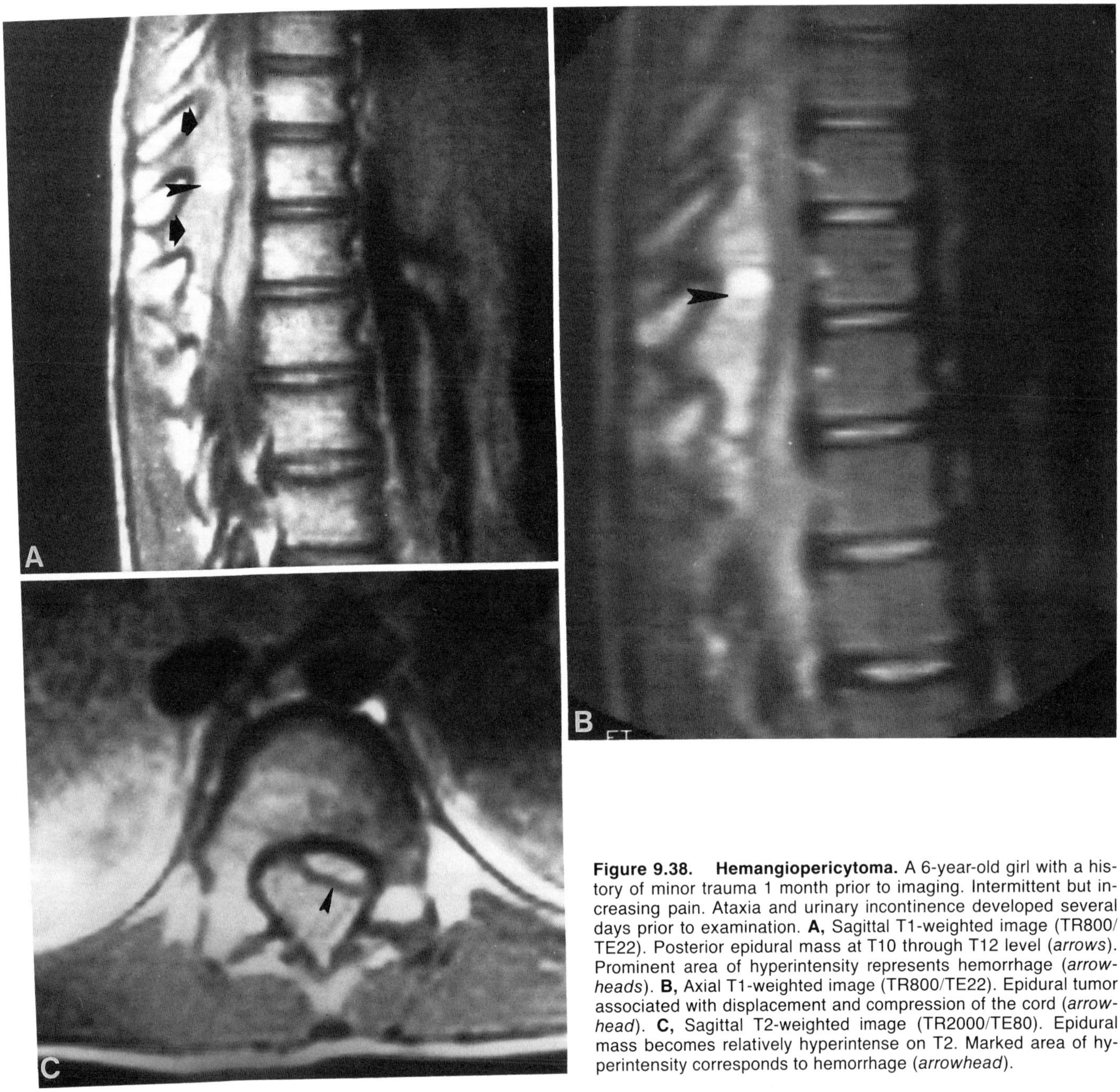

Figure 9.38. Hemangiopericytoma. A 6-year-old girl with a history of minor trauma 1 month prior to imaging. Intermittent but increasing pain. Ataxia and urinary incontinence developed several days prior to examination. **A,** Sagittal T1-weighted image (TR800/TE22). Posterior epidural mass at T10 through T12 level (*arrows*). Prominent area of hyperintensity represents hemorrhage (*arrowheads*). **B,** Axial T1-weighted image (TR800/TE22). Epidural tumor associated with displacement and compression of the cord (*arrowhead*). **C,** Sagittal T2-weighted image (TR2000/TE80). Epidural mass becomes relatively hyperintense on T2. Marked area of hyperintensity corresponds to hemorrhage (*arrowhead*).

Ewing's sarcoma accounts for approximately 5% of skeletal neoplasms and usually arises in medullary bone. Rarely, extraosseous Ewing's sarcoma involving the epidural space may be encountered either primarily or secondarily [157]. This is more common in the thoracolumbar region. Approximately 75% of Ewing's sarcomas present before age 20 and 90% before age 30 [158].

Pain radiating to the lower extremities is common and may be confused with disk herniation [158]. The extraosseous component of the neoplasm may be quite extensive, especially in the distal spine and sacrum, where dimensions of 10 to 20 cm have been encountered [156].

Plain films remain the most useful modality in assessing probable histologic diagnosis [159]. In the spine, the lesion may be lytic, sclerotic, or mixed [160]. The vertebral body is typically involved, although arch involvement also may be demonstrated. Calcifications within the tumor are rarely encountered, and the tumor rarely extends to an adjacent vertebrae [111]. Occasionally, a loss of intervertebral disk height is noted.

The lytic or sclerotic aspect of the neoplasm is well demonstrated by CT, which also allows differentiation between the sclerosis resulting from buildup of periosteal bone and direct intratumor sclerosis [160]. The sclerosis is apparently a result

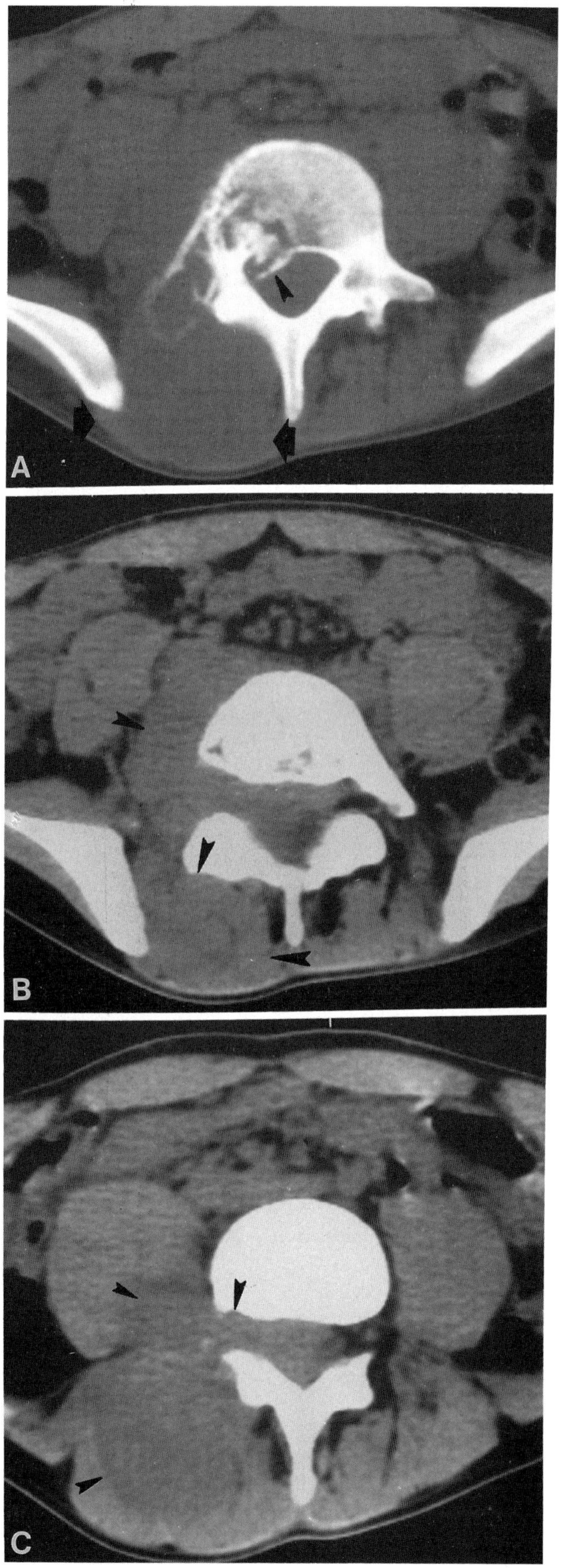

of osteogenesis, occurring from contact of the neoplasm with adjacent bone and not from formation of bone by the neoplasm itself. CT also demonstrates the associated soft-tissue mass, which may be huge, in the presacral region (Fig. 9.39). Occasionally, CT also will demonstrate some widening of the spinal canal which may not be visible with plain films [160].

MRI may provide more information than CT regarding the extension of neoplasm within bone marrow as well as soft-tissue involvement of the paravertebral region and spinal canal [159] (Fig. 9.40). In general, bone marrow is best assessed on T1-weighted images, and soft-tissue involvement is best demonstrated on the T2-weighted images [150]. MRI may show cortical thickening and tumor infiltration on either T1- or T2-weighted images which is usually manifest by replacement of the normally hypointense cortex signal with comparatively increased signal intensity of this aggressive neoplasm [150]. Periosteal reaction may be identified with either CT or MRI. Cortical involvement is best demonstrated on axial images.

MRI does not distinguish Ewing's sarcoma from osteosarcoma [150]. Both demonstrate hypointense marrow replacement on T1-weighted images and may have a variable appearance on T2-weighted images. Ewing's sarcoma is usually hyperintense on T2-weighted images, but this hyperintensity is not specific for neoplasm and may be seen with associated inflammatory changes or necrosis. Occasionally, Ewing's sarcoma will be hypointense on T2-weighted images [159]. Gadolinium enhancement occasionally is useful in distinguishing between tumor and necrosis but is nonspecific because both tumor and reactive changes may enhance [159].

Although MRI is inferior to plain films in initial diagnosis and no better than CT in evaluating the effectiveness of chemotherapy, it is the best single study for preoperative evaluation of the extent of neoplastic involvement and follow-up of patients with Ewing's sarcoma [159]. Local recurrence is frequent following radiotherapy; therefore, accurate imaging to determine tumor extent before both surgery

Figure 9.39. Ewing's sarcoma. A 17-year-old boy presented with back and radicular pain extending into the right hip and leg. There was no fever or systemic complaint. **A,** CT through L5 level viewed at bone window. There is a destructive lytic lesion involving the right side of the body extending into the right transverse process. Note the pathologic fracture involving the posterior vertebral body (*arrowhead*). **B,C,** Soft-tissue window images demonstrate a huge right paraspinal and paravertebral mass (*arrowheads*) which displaces the psoas muscle laterally. Extension through the right foramina into the canal is also well demonstrated (Courtesy of William R. Scott, M.D., West Florida Regional Medical Center, Pensacola, Florida.)

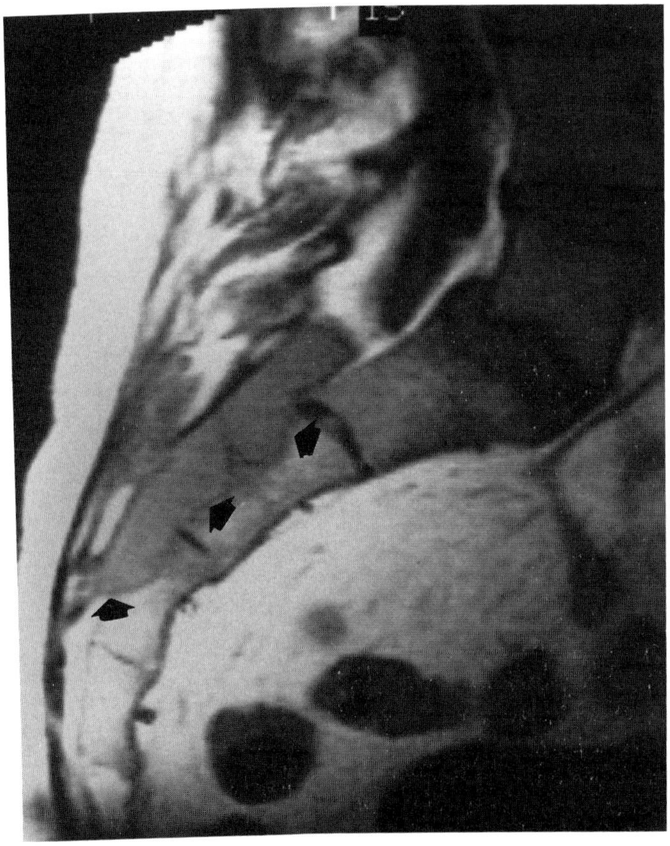

Figure 9.40. Ewing's sarcoma. A 38-year-old man with low back pain radiating to right leg associated with some right leg numbness. Sagittal T1-weighted image (TR450/TE26). Soft-tissue signal mass located in the caudal spinal canal. There is prominent scalloping of the posterior aspect of the sacrum (*arrows*). Lesions in this location are generally not included on routine CT examinations. Myelograms frequently miss these lesions because the thecal sac will appear to terminate at the lower S1 level and therefore the lack of opacification distally cannot be appreciated. MRI also may miss these lesions, especially if multiecho sequences are used (the mass may appear isointense with the thecal sac).

and radiotherapy is important [159]. Long-term follow-up after chemotherapy suggests that a hypointense appearance on T2-weighted image is a sign of tumoral sterilization [159]. However, this is rarely encountered, since most patients also have radiotherapy. Hyperintensity on contrast-enhanced T1-weighted images following chemotherapy is highly sensitive but not specific in the detection of viable tumor regions [228]. A hyperintense appearance on T2-weighted images is nonspecific, since the distinction between recurrent and reactive change is usually not possible [159].

Ewing's sarcoma metastasizes to the lungs, bones, and lymph nodes. In the past, Ewing's sarcoma was highly lethal, with a 5-year survival of less than 15%. This has improved to as high as 75% with the use of radiation and chemotherapy and may be as high as 85% when local surgery can be performed [158].

CHONDROSARCOMA

Chondrosarcomas are cartilaginous tumors that infrequently involve the vertebral column. These arise primarily or subsequent to malignant degeneration of a solitary cartilaginous tumor such as an enchondroma or osteochondroma [161]. Unlike the osteosarcomas and Ewing's sarcomas, chondrosarcomas generally involve older adults, with more than half of patients greater than 40 years of age. However, the secondary chondrosarcomas are seen in a younger age group, typically in the late twenties or early thirties [162]. Chondrosarcomas are rare in children and adolescents.

Approximately 6% of chondrosarcomas involve the vertebral axis [161]. The neoplasm usually involves the neural arch or centra and may extend from one level to another [161]. Chondrosarcomas may occur anywhere in the spine but are most common in the thoracic spine and sacrum [111]. These tumors are relatively slow growing. Malignant change should be suspected when pain develops at an osteochondroma site. The pain is often described initially as dull, but with time it becomes almost unbearable. Osteochondromas may grow to a large size prior to clinical presentation, and the patient may present with symptoms of spinal cord compression or with masses of the neck or trunk [22, 161]. Approximately 60% of patients are males.

Plain films are highly suggestive of the diagnosis and may be pathognomonic. Prominent bony destruction associated with mottled densities from calcification is usually noted. Bony expansion and cortical destruction may be evident [111]. Calcifications are frequently extensive and may be punctate or ringlike, although they may appear somewhat amorphous in more aggressive tumors [22]. Occasionally, the entire tumor is osteolytic. The osteolytic area often has a sclerotic margin which with time becomes less defined [161]. Soft-tissue extension of the tumor usually contains calcific debris [161, 163].

CT easily assesses the extent of the lytic process (Fig. 9.41) and is superior in demonstration of the matrix calcifications as well as the associated soft-tissue mass. This superior sensitivity compared with plain films is particularly important when evaluating pelvic and other central lesions in which the soft-tissue component is often difficult to detect [163]. CT is also useful in demonstrating the extent of vertebral body destruction and canal involvement [161]. The tumor is usually hyperdense and enhances with intravenous contrast material [164]. Indistinct tumor margins and a large concentric soft-tissue mass with amorphous calcification on CT may be suggestive of a higher-grade malignancy. The presence and amount of calcification are reportedly inversely related to the degree of malignancy [162].

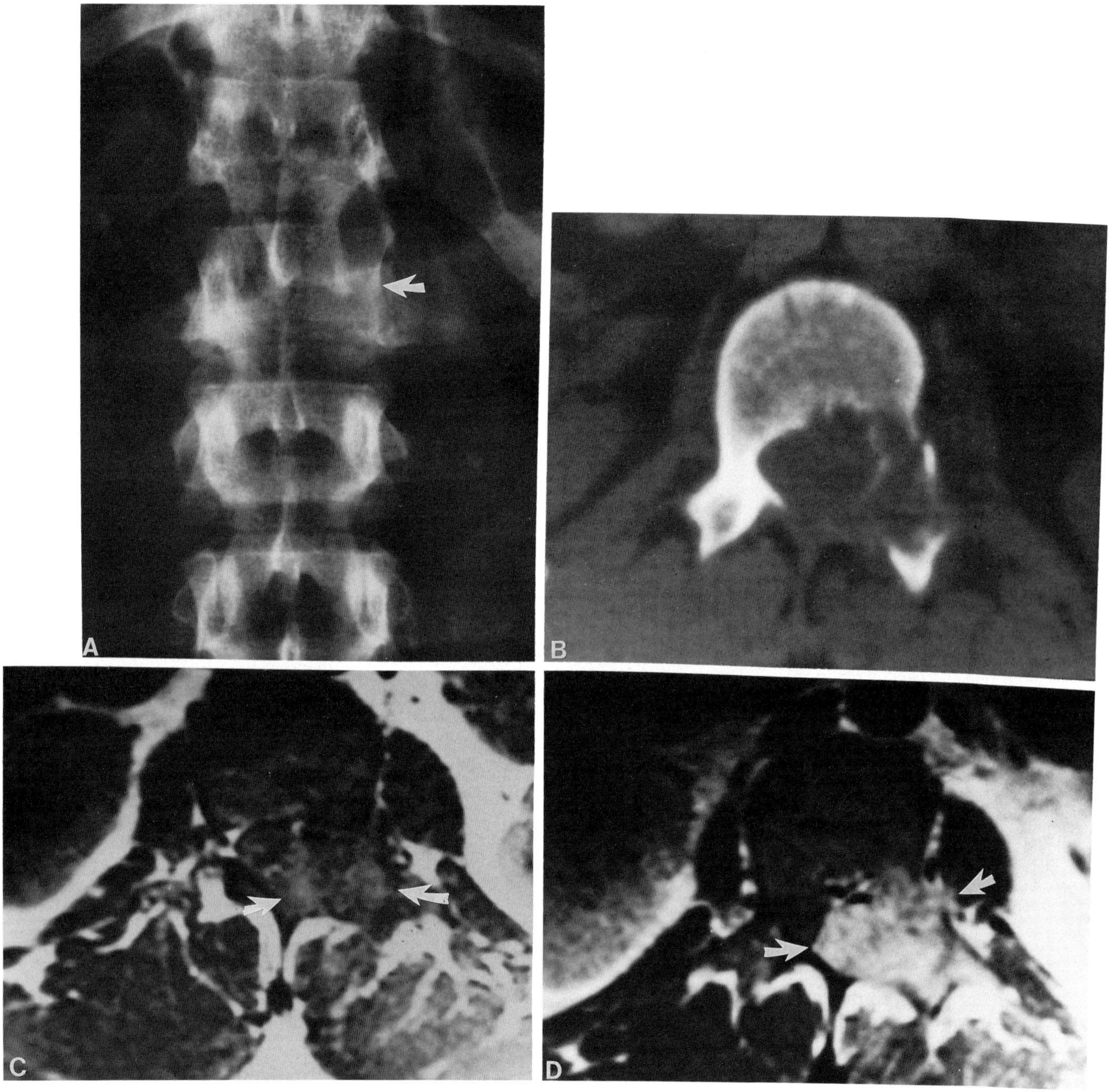

Figure 9.41. Mesenchymal chondrosarcoma. A 20-year-old man with back pain. **A,** Plain film shows absent pedicle on the left at L1 (*arrow*). **B,** Axial CT at L1. Destructive expansile lesion of the left pedicle. Lack of calcification is atypical for chondrosarcoma, although this may occur in the dedifferentiated forms. **C,** Axial T1-weighted image (TR700/TE20). Soft-tissue signal mass in the left pedicle (*arrows*) impinges on thecal sac. **D,** Axial contrast enhanced T1-weighted image (TR700/TE20). Enhancement of the left pedicle mass (*arrows*). Extent of canal involvement is more evident. **E,** Contiguous axial CT section demonstrates the calcified osteochondroma arising from the pedicle and transverse process of the C6 vertebral body. There is minimal encroachment into the spinal canal. The fluffy calcification with areas of hypodensity is characteristic of osteochondromas. (Case courtesy of Dr. Margaret A. Stoll, Georgetown University, and Dr. Mark Kransdorf, Armed Forces Institute of Pathology, Washington, D.C.)

Approximately 10 to 20% of chondrosarcomas may undergo sarcomatous transformation and are referred to as *dedifferentiated chondrosarcomas* [165]. This is very rare in the vertebral column but can be recognized as a lytic intraosseous area devoid of the usual calcifications associated with chondrosarcoma. The associated soft-tissue mass is usually large and is also devoid of calcification. Recognition of this complication is important because these are much more aggressive neoplasms. CT is particularly useful in determining the appropriate biopsy sites [165].

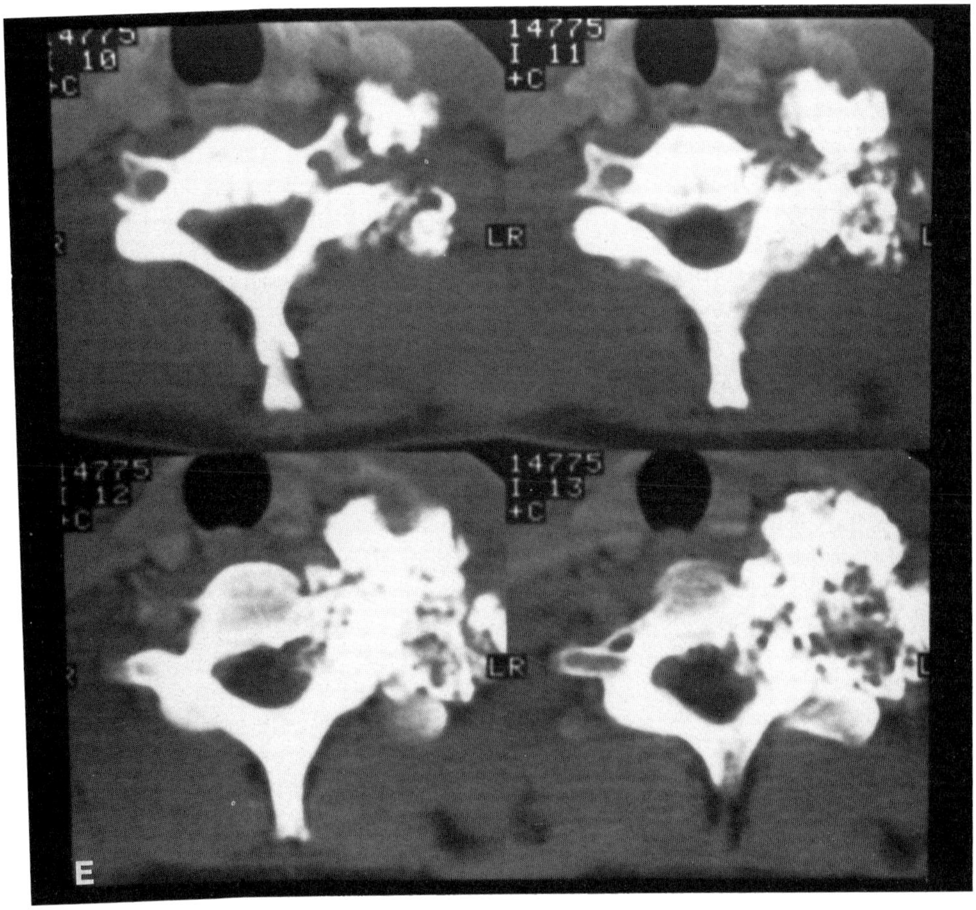

Figure 9.41–*continued*

MRI is useful in delineating the associated soft-tissue mass and canal extension of the tumor but is inferior to CT in detection of the calcifications. On MRI, chondrosarcomas tend to be hypointense on T1- and hyperintense on T2-weighted images. The mass may have either homogeneous or heterogeneous signal on T2-weighted images [165]. Signal intensity characteristics are nonspecific and do not allow differentiation between chondrosarcoma and chordoma. In general, chordomas tend to be midline, whereas chondrosarcomas are more likely off midline [164].

The surgical inaccessibility of most chondrosarcomas of the spine results in a high incidence of local recurrence, and the overall prognosis is poor [161]. Hematogenous metastases to the lung are rarely encountered early in the course of disease. However, recurrent or partially resected tumors do present a threat of early metastasis due to hematogenous spread [164].

CHORDOMA

Chordomas are slowly growing, locally aggressive midline neoplasms that develop from remnants of the primitive notochord [111, 166]. These uncommon tumors account for between 1 to 4% of all malignant tumors of bone but account for approximately 40% of primary sacral neoplasms [166]. The sacrum is the most common site for chordoma and accounts for just over 50% of all chordomas [167, 168]. The skull base accounts for approximately 35%, and the remaining 15% are found in the spine. The cervical spine is involved in approximately one-half of spinal chordomas; the thoracic spine is the least commonly involved spinal location [111, 167]. Chordomas affect males approximately twice as frequently as females [111, 168]. The mean age of patients with sacrococcygeal tumors is 56 years, and the mean age for vertebral chordoma presentation is 46 years [168]. Chordomas are distinctly uncommon in children and are usually limited to the skull base and cervical spine [166].

These tumors are slowly growing but locally invasive. Sacrococcygeal chordomas may reach a huge size before the diagnosis is established [166, 169]. Pain is the most frequent presenting complaint but is nonspecific and is often present for months to years before the diagnosis is established. Approximately one-fourth of patients have neurologic complaints such as numbness, radicular pain, or bowel/bladder incontinence. Sacrococcygeal tumors often have huge anterior soft-tissue components and frequently are associated with rectal dysfunction or constipation [168]. The presacral extension of these tumors can almost always be detected on rectal examination.

Chordomas are predominantly destructive neoplasms that arise from ectopic notochordal remnants within the vertebral bodies. The vertebral body is primarily involved, with progressively less frequent involvement of the pedicle, lamina, and spinous process [167]. Contiguous vertebrae also may be destroyed, and there may be involvement of the intervertebral disk space, a unique characteristic of this malignancy [167, 170]. Chordomas are characterized by destructive changes which may be lytic, blastic, or a combination of both [171]. Secondary sclerotic changes are present in more than one-half of cases and are a helpful differential point [168, 170]. There is often an associated anterior or anterolateral soft-tissue mass which is out of proportion to the degree of bony destruction. Soft-tissue calcification is frequent, but plain films demonstrate calcific debris in only 15 to 44% of patients [166, 168, 172]. A common radiologic finding in sacral chordomas is destruction of several vertebral segments associated with an anterior soft-tissue mass [168]. Expansion of the anteroposterior diameter of the sacrum is present in 50% of cases [167].

CT demonstrates bony destruction in 90% of patients. The margins of the neoplasms are usually well defined and sclerotic in nearly one-half of cases [166, 169, 170, 172]. Approximately 20% of untreated patients will demonstrate pseudoloculation resulting in a honeycombed appearance [166].

On CT, the soft-tissue component is usually out of proportion to the amount of vertebral destruction and may be greater than 10 cm in diameter. In nearly 50% of cases the solitary or multiple septated areas of hypodensity [173] in the soft tissue correspond with the areas of cystic degeneration noted on pathologic specimens. The cystic areas are generally not very extensive but may produce an inhomogeneous appearance in larger masses. CT usually shows a sharply defined tissue margin which corresponds to the fibrous pseudocapsule noted pathologically [173].

Soft-tissue calcification may be present in 86% of sacrococcygeal neoplasms. Calcification is less frequent in spinal chordomas and is noted in approximately 40% of studies [173]. Calcifications are generally coarse and irregular and may represent sequestrated necrotic bone [166]. These are usually located peripherally but also may be seen centrally [171].

Spinal chordomas characteristically involve the epidural space [169, 170]. CT satisfactorily demonstrates this epidural involvement and is further complemented when combined with myelography (Fig. 9.42). The encroaching soft-tissue mass may distort or displace the epidural fat, thecal sac, or nerve roots [169] (Fig. 9.43).

In the absence of MRI, contrast-enhanced CT allows detection of epidural extent and is useful in assessing the associated soft-tissue component, which may be relatively isodense with adjacent muscle prior to contrast administration [169]. Contrast enhancement also accentuates cystic areas of necrosis [169].

MRI is superior to CT due to its ability to view in orthagonal planes and evaluate the soft-tissue component. Tumor tissue is generally associated with markedly prolonged T1 and T2 values. Chordomas tend to be isointense to muscle on T1-weighted images and hyperintense on T2-weighted images (175–178). T1-weighted images may demonstrate the cystic areas separated by strands of low signal intensity, with the mass often circumscribed by a fibrous pseudocapsule of low signal intensity [176]. T2-weighted images (long TR and TE) provide the best contrast between tumor and surrounding soft tissues. This high contrast between tumor and the adjacent soft tissue makes muscle invasion much easier to perceive [175]. Lesions with higher mucoid content or necrosis have higher signal intensity on T2-weighted images [177]. Bone marrow replacement by tumor and CSF signal changes due to epidural extension are well demonstrated by MRI [177].

The primary differential considerations are myeloma, metastases, chondrosarcoma, giant cell tumor, lymphoma, and granulomatous osteomyelitis [166, 169, 170, 172]. With the exception of the sacral chondrosarcoma, the other neoplasms are generally not associated with soft-tissue masses containing large, amorphous calcifications or fragments of bone [170]. Chordomas are nearly always midline lesions, while giant cell tumors are often eccentric [170, 179]. Although of limited value in very large sacral masses, the sacrococcygeal chordomas almost always arise from the lower sacral segments, whereas tumors such as giant cell tumor, chondrosarcoma,

Figure 9.42. Sacral chordoma. A 60-year-old man with low back pain which had been unsuccessfully evaluated with plain films, bone scan, and myelography. **A,** Plain films initially interpreted as normal. In retrospect, lytic area of lower sacrum can be identified (*arrowheads*). **B,** Posterior image from bone scan remarkable only for large amount of activity within the urinary bladder. Sacral lesions may be associated with urinary retention. **C,** Myelogram also interpreted as normal. Normal opacification of the thecal sac. The sacral lesion is located caudal to the thecal sac and was therefore not demonstrated. **D,E,** Axial and sagittal CT reconstructions of the sacrum. Soft-tissue mass (*arrows*) within the inferior sacrum associated with cortical breakthrough anteriorly (*arrowhead*). Sacral chordomas are often associated with very large anterior presacral masses by the time they are diagnosed.

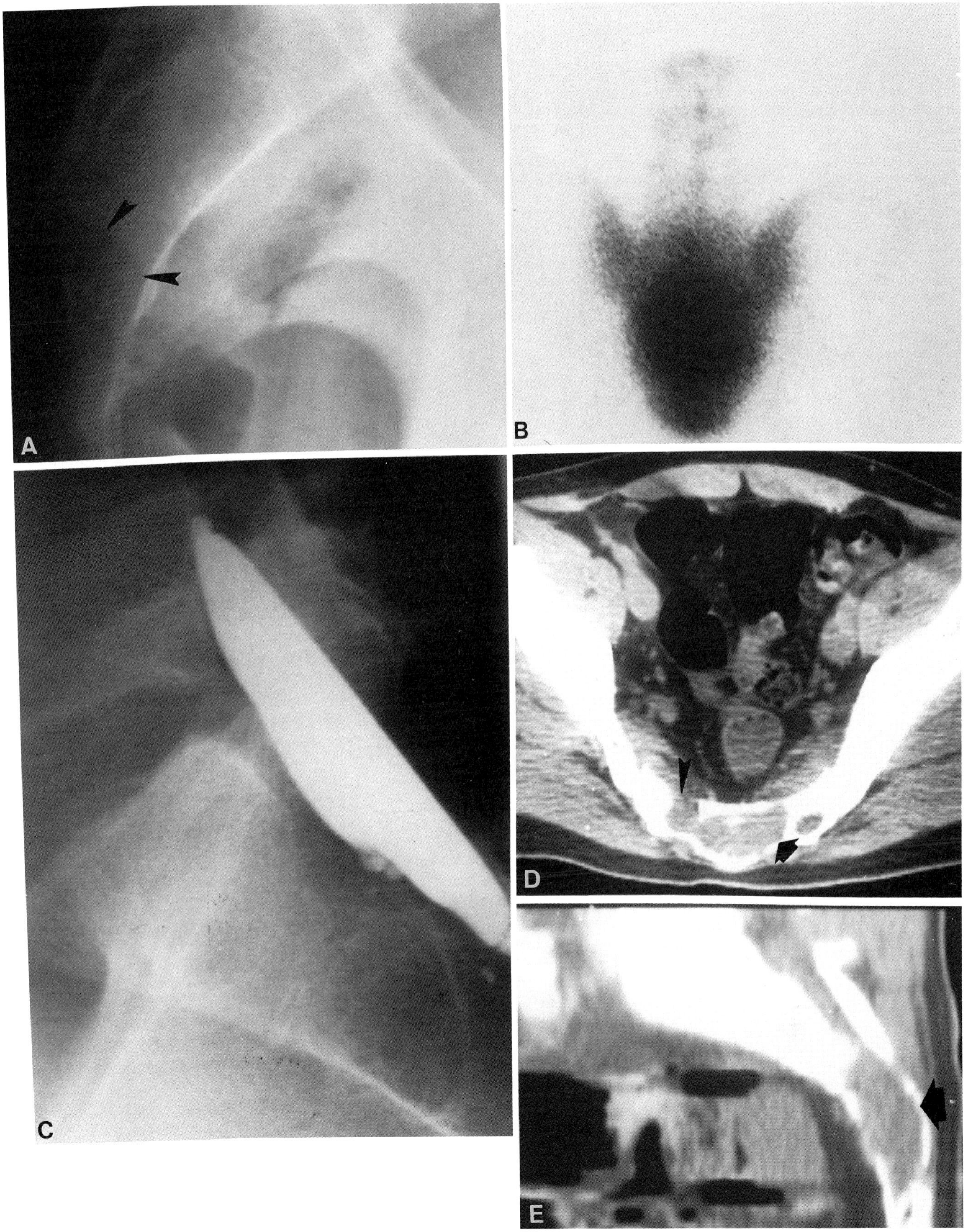
A
B
C
D
E

Figure 9.43. Thoracic chordoma. A 59-year-old man with back pain. **A,** AP myelogram demonstrates sclerosis of the T8 vertebral body (*arrowhead*) associated with a right-sided epidural mass primarily at T8 but also at the T7 level (*arrows*). Paravertebral soft-tissue mass is also evident bilaterally. **B,** Postmyelogram CT through T8 level. Paravertebral mass is seen anteriorly and primarily to the right (*arrow*). Epidural mass is effacing the thecal sac (*arrowheads*). **C,** Sagittal postmyelogram CT reconstruction at T8 level. Prominent sclerosis of the T8 vertebral body. Epidural soft-tissue mass extends into the anterior canal (*arrowhead*) and is associated with posterior displacement of the thoracic cord. Thoracic location is the least common for chordomas, and with the absence of soft-tissue calcifications, chordoma is an unlikely prebiopsy diagnosis.

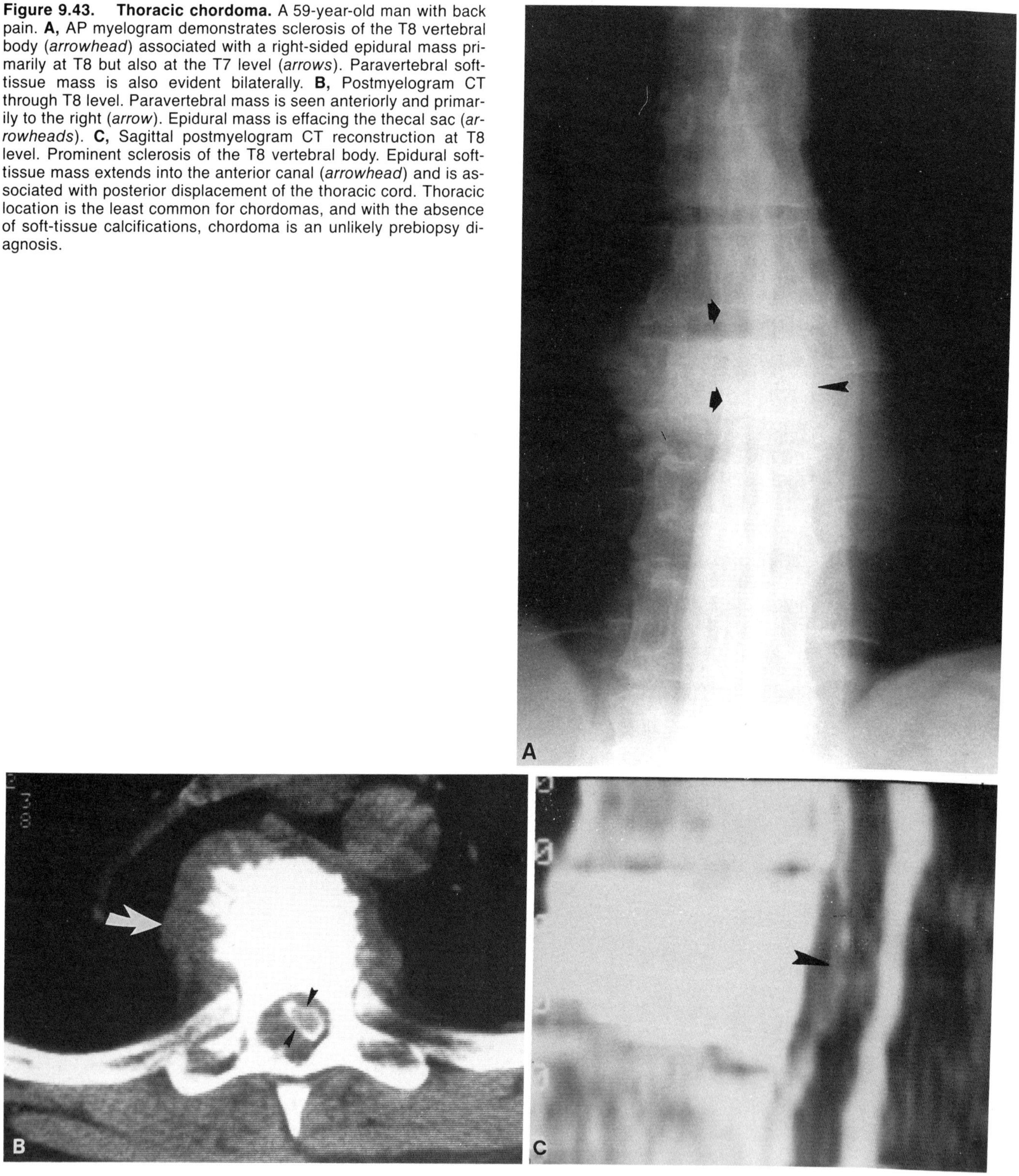

and osteosarcoma more often involve the upper segments [166]. When present, osteosclerosis is a helpful radiologic sign, although this also may be seen in myeloma and lymphoma. Although only occasionally present, a unique characteristic of chordoma is the presence of bony destruction involving multiple adjacent vertebral bodies with involvement of the disk spaces [169, 170]. However, since multilevel involvement with disk destruction is more characteristic of osteomyelitis, biopsy may be necessary to confirm the diagnosis.

Local recurrence is common and often involves the posterior elements. The 5-year survival for sacrococcygeal tumors is 66%, and 5-year survival for patients with vertebral chordomas is approximately 50% [168]. Approximately one-fourth of patients will at some time have distant metastases, but these are commonly asymptomatic [166, 168]. Metastases are by the hematogenous route, and the most frequent sites are the lungs, spine, skin, and lymph nodes [168, 180]. Intradural metastases also have been reported [181]. The presence of metastases generally does not affect survival, and death is generally related to secondary effects of the paraplegia produced by the primary tumor (sepsis, urinary tract infection, pneumonia) [168].

OSTEOID OSTEOMA AND OSTEOBLASTOMA

These histologically similar neoplasms are occasionally indistinguishable radiographically but can be separated generally on the basis of size and biologic behavior [111, 182, 183]. Osteoid osteomas are confined and self-limited in growth and generally do not exceed 1 cm in size. Osteoblastomas are more extensive and increase in size with time.

Osteoid osteomas are much more common than osteoblastomas, but both show a predilection for the vertebral column, with approximately 40% of osteoblastomas and 10% of osteoid osteomas localized to the spine [185]. More than half these neoplasms occur in the first two decades of life, predominantly in teenagers [185]. Between 60 and 70% of patients are male [183]. More than half of osteoid osteomas occur in the lumbar spine, whereas osteoblastomas are more equally distributed throughout the spine [183]. The majority of spinal osteoid osteomas and osteoblastomas occur in the region of the base of the transverse process [186]. Both tumors typically involve either the lamina or the pedicle, but osteoblastomas involve more than one element of the neural arch three times more frequently than do osteoid osteomas. Osteoblastomas are twice as frequent on the right side of the spine, but there is no side preference for osteoid osteomas [183].

Histology, size, radiographic appearance, and clinical features help distinguish osteoid osteoma from osteoblastoma in most cases. The distinction is useful because of the differences in clinical course. However, there is considerable overlap in the preceding finding, and occasionally, an arbitrary diameter of 1.5 cm is used to separate these two neoplasms [182].

Nearly all patients present with pain which is usually worse at night and aggravated by motion [186]. A unique characteristic is the relief provided by aspirin, noted in 25 to 40% of patients [183]. Painful scoliosis is encountered frequently, and the apex of the curve corresponds well to the site of the lesion [186]. This scoliosis is present in 66% of patients with osteoid osteoma and in 36% of patients with osteoblastoma. The convexity of the curve is away from the involved side of the vertebrae and is induced by painful spasms [183]. Neurologic findings are more often associated with osteoblastoma, since these frequently involve the neural canal.

Osteoid osteomas are benign lesions characterized by a small oval or round mass referred to as a *nidus*. When located in cortical bone, this nidus is often associated with a prominent surrounding zone of sclerosis [111]. The zone of sclerosis is much less prominent or absent when the nidus is located within cancellous bone. The nidus is generally 1 cm or less in diameter. Plain films may be normal or demonstrate an area of sclerosis but infrequently demonstrate the nidus [185, 187]. Radioisotope bone scans demonstrate an area of increased activity which is useful in directing the CT examination to the appropriate level [185].

CT frequently demonstrates the radiolucent nidus associated with osteoid osteoma (Fig. 9.44). Accurate preoperative localization is important because it permits a less radical resection without the risk of missing the nidus [187]. The inner surface of this nidus is smooth and well defined [188]. Occasionally, the nidus will contain a calcified density, which is usually round, smooth, and located in the center of the lucent nidus [188]. The CT characteristics are valuable in the differential diagnosis from osteomyelitis because the sequestrum of osteomyelitis tends to be irregular and eccentric and the inner surface of the radiolucent focus also is irregular. In addition, osteomyelitis is often associated with a soft-tissue mass adjacent to the area of destruction [188].

CT is currently preferred in the evaluation of osteoid osteoma. Experience with MRI of spinal osteoid osteoma is extremely limited, and findings may be confusing [229]. MRI should demonstrate the lesion, but signal characteristics will be dependent on the amount of calcification within the nidus. If the calcification is prominent, then the nidus may be markedly hypointense on both T1- and T2-weighted im-

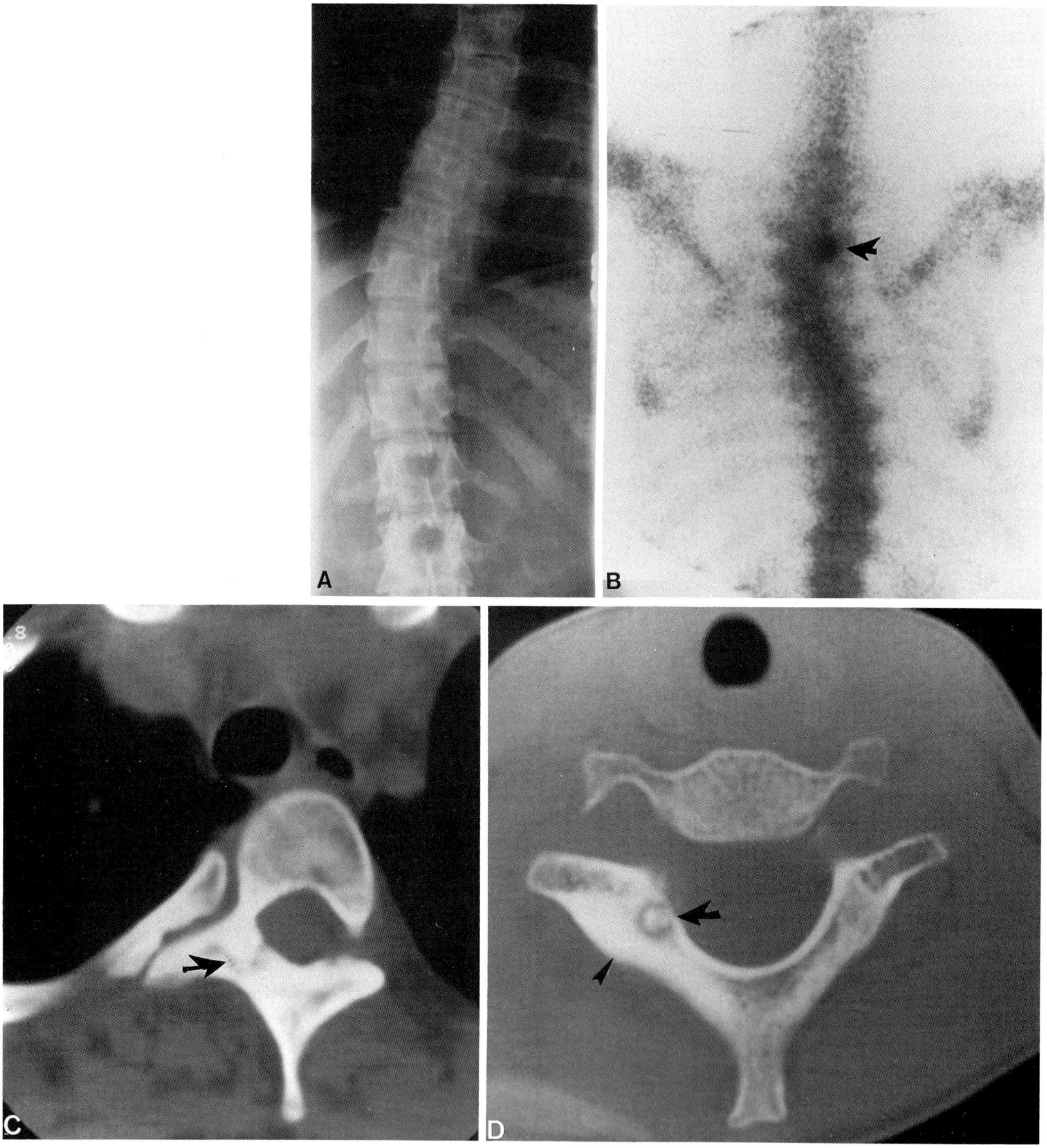

Figure 9.44. Osteoid osteoma. An 18-year-old with painful scoliosis. **A,** Prominent scoliosis, but lesion not seen. **B,** Posterior bone scan. Focal increased activity along the concave margin of T3 (*arrow*). **C,** Unenhanced CT scans. Sclerosis involving the right lamina, particularly anteriorly (*arrow*). The radiolucent nidus is sometimes difficult to demonstrate. **D,** Another patient with osteoid osteoma. Note the typical appearance of the radiolucent nidus which contains a central calcific or ossific component (*arrow*). Also note the prominent adjacent reactive sclerosis (*arrowhead*). (Case courtesy of J. Powell Williams, III, New Orleans, La.)

ages (Fig. 9.45). Adjacent bone may be hypointense on T1- and hyperintense on T2-weighted images, probably secondary to adjacent inflammation and edema [229].

The radiographic features of osteoblastoma are much less characteristic than those of osteoid osteoma. The central nidus in osteoblastoma is almost always many times larger than that seen in osteoid osteoma [111]. In some cases, only fairly well-circumscribed bone destruction is observed, and this may not appear to be a benign process. The majority of spinal osteoblastomas appear as destructive, expansile lytic lesions that are well circumscribed even when the cortex has been expanded [182, 183, 189] (Fig. 9.46). Although most are confined to the posterior elements, 24% involve both the body and the posterior elements and 14% involve the vertebral body only [182]. Definite cortical destruction may be seen occasionally [183]. The appearance may be osteolytic, sclerotic and mottled, or mixed sclerotic and lytic [185]. The inhomogeneous appearance is due to the lytic and calcified portion of this bone-forming tumor [189]. The adjacent bone is often normal [56%] but may have either a surrounding sclerotic rim or occasionally considerable surrounding sclerosis [182].

Either CT with intrathecal contrast material or MRI should be performed in all cases of osteoblastoma [185] and in selected cases of osteoid osteoma. About one-half of patients with osteoblastoma have epidural extension, while osteoid osteomas do not extend beyond the site of origin [183]. When performed with intrathecal contrast material, CT can define the extent of bony destruction, the well-defined nature of the tumor, and both paraspinal and intraspinal components, as well as the relationship of the mass to the cord [189].

MRI findings in osteoblastoma are similar to those in other osseous neoplasms. Osteoblastomas are hypointense on T1- and intermediate to hyperintense on T2-weighted images [230].

The differential diagnosis of osteoid osteoma and osteoblastoma generally include osteomyelitis, aneurysmal bone cyst, Ewing's sarcoma, and osteosarcoma [185]. Although the majority of osteoblastomas have a destructive and expansile appearance, these findings are often not obvious on plain radiographs and can be easily overlooked. A benign-appearing, expansile lesion in the posterior elements of the spine with a mineralized matrix most likely represents osteoblastoma [182]. However, when the matrix is lucent, distinction between osteoblastoma and aneurysmal bone cyst may be difficult [190]. Aneurysmal bone cyst has a similar location but tends to be larger, more osteolytic, and more expansile than osteoblastoma. The sharp delineation of the tumor would be atypical for osteosarcoma or chondrosarcoma. Osteoblastoma generally involve the posterior elements and may extend secondarily into the vertebral body, whereas both osteomyelitis and Ewing's sarcoma predominate in the vertebral body. However, when Ewing's sarcoma does occasionally arise in the posterior elements, the differential diagnosis with osteoblastoma is quite difficult [185]. Osteomyelitis generally involves the adjacent disk space. Vertebral osteosarcomas in children are generally secondary to prior irradiation [185].

The distinction between osteoid osteoma and osteoblastoma is based on size, since osteoid osteomas rarely exceed 1 to 1.5 cm in diameter [182], the degree of osteosclerosis, and the presence of osteolysis or of a nidus. The expansile nature of osteoblastoma without evidence of vertebral collapse or significant soft-tissue component is atypical for both metastasis and solitary plasmacytoma [191]. Giant cell tumors are quite rare in the spine and are also frequently associated with vertebral collapse and the presence of a paravertebral soft-tissue mass [191].

The prognosis for both osteoid osteoma and osteoblastoma is generally excellent. Surgical removal often results in immediate relief of pain and early return of spinal mobility [183, 186]. However, a delay in treatment in the growing child may result in some residual scoliosis. Persistent pain suggests residual tumor [184]. If the nidus of the tumor is not removed, or if there is a multifocal nidus, recurrence of symptoms is likely, and reoperation may be necessary [183].

HISTIOCYTOSIS X

Histiocytosis X is a reticuloendothelial system disorder characterized by the abnormal proliferation of histiocytes [111]. There are three overlapping clinical presentations. Letterer-Siwe disease is the acute, fulminant form with a poor prognosis. This accounts for about 10% of cases, is seen primarily in children under age 2, and involves liver, spleen, nodes, skin, and bone. Hand-Schuller-Christian disease is a relatively severe chronic form, which also involves bone and visceral sites. This chronic form accounts for nearly a third of cases, and the peak incidence is in children between age 5 and 10 years [192, 193].

Eosinophilic granuloma accounts for over one-half of cases of histiocytosis X and is more often monostotic than polyostotic [194]. Three-quarters of cases occur in patients less than 20 years of age, and there is a slight male preponderance [194]. Although the cause is unknown, there is speculation that it may be an infectious or an autoimmune disorder [194].

Spinal involvement in histiocytosis X is often in-

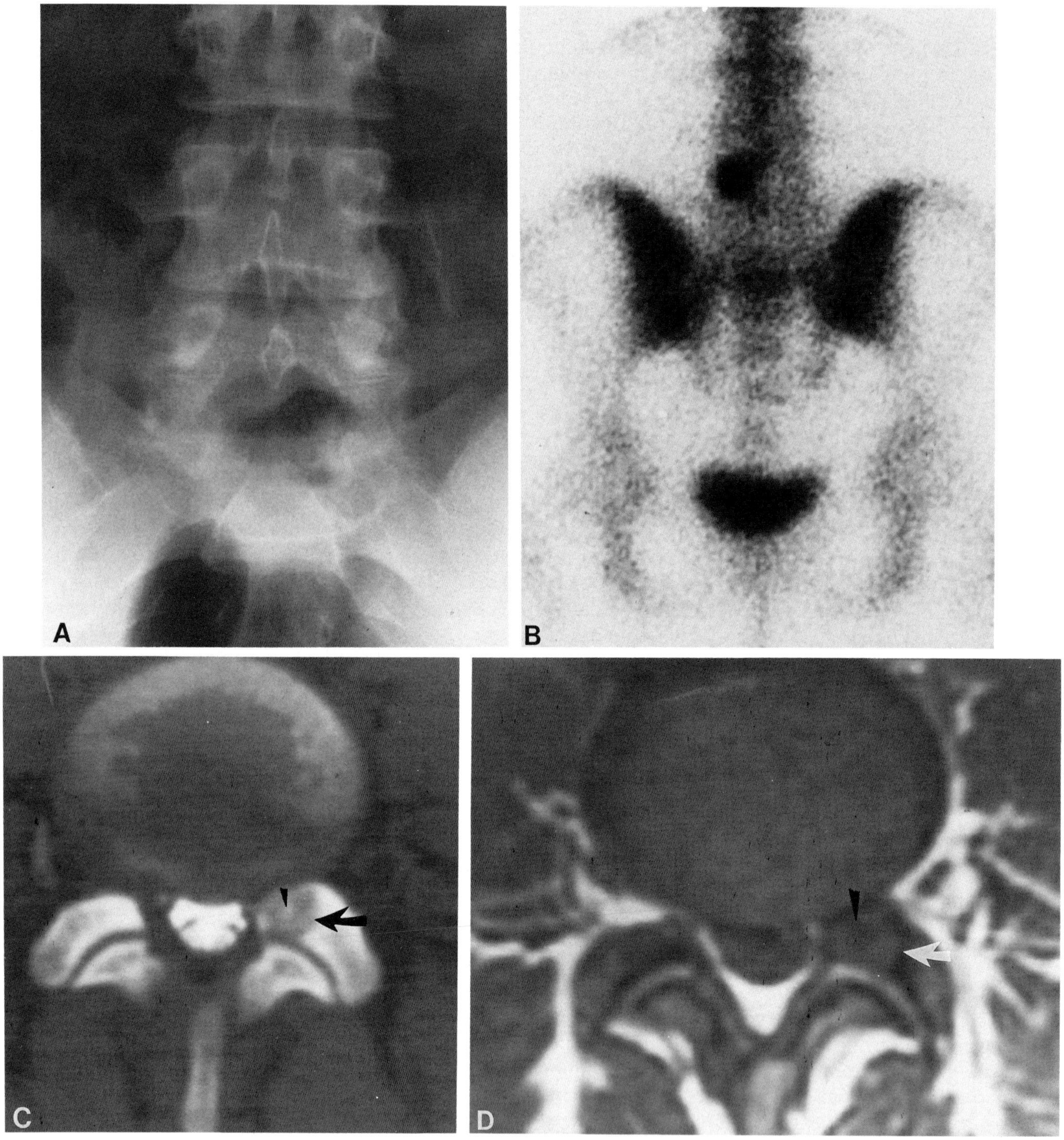

Figure 9.45. Osteoid osteoma. A 17-year-old girl with back and left leg pain of 8 months' duration. **A,** Plain film shows sclerosis of the left pedicle of L5. **B,** Posterior bone scan. Marked focal uptake in the left pedicle and facet region of L5. **C,** CT myelogram. Lucency in the medial aspect of the superior articular facet of L5 (*arrow*) which contains small central foci of calcification (*arrowhead*). **D,** Axial 1.5-T T1-weighted image shows the nidus (*arrow*) in the central calcification (*arrowhead*). **E,** Sagittal 1.5-T intermediate and T2-weighted image (TR2000/TE30–75). Note expansion of the superior articular facet of L5 (*arrow*). (Case courtesy of Dr. Margaret A. Stoll, Georgetown University, and Dr. Mark Kransdorf, Armed Forces Institute of Pathology, Washington, D.C.)

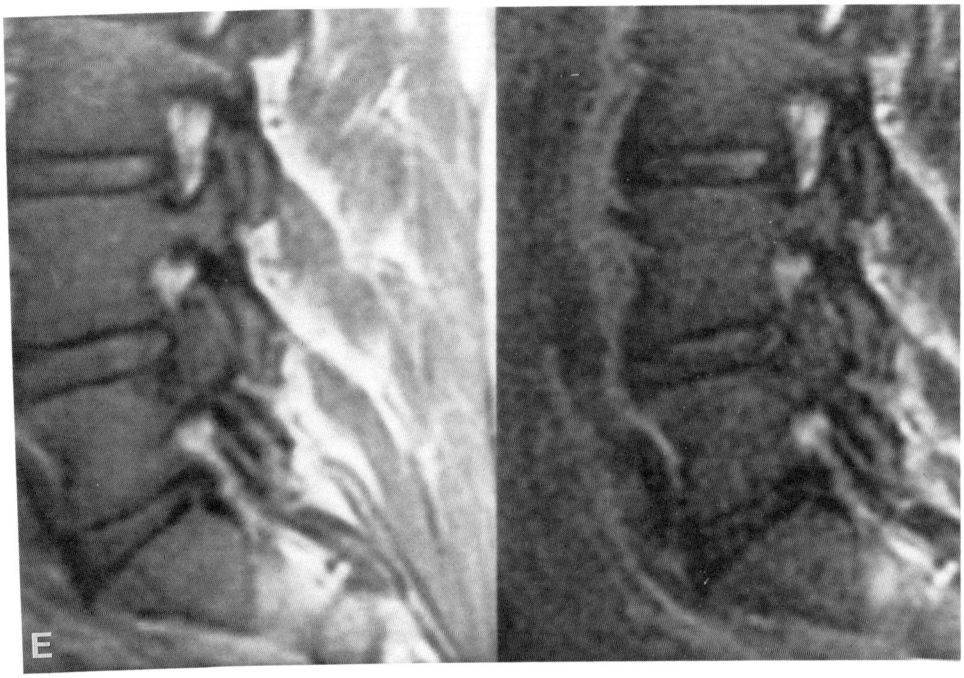

Figure 9.45–*continued*

cidentally discovered during a skeletal survey in patients with a known diagnosis [195]. Occasionally, spinal lesions may be the first manifestation of histiocytosis X when they produce pain, muscle stiffness, or neurologic symptoms [195]. Symptomatic patients often present with pain that may be either sudden or gradual in onset but which is often localized to the area of vertebral involvement [195]. Approximately one-half of patients who present with symptomatic eosinophilic granuloma (EG) of the spine will have neurologic symptoms such as paresis [195]. The mean age of patients with symptomatic EG of the spine is 12 years, with over two-thirds of patients seen in the first three decades of life. The pain and restriction of motion are similar to what would be expected with other compression fractures, and the degree of compressive collapse is dependent on the extent of involvement of the vertebral body [196]. In younger children, most of the body is often affected, and the result is the classic wafer-like collapse known as *vertebra plana.* Older children often have relatively smaller areas of involvement resulting in a more segmental type of collapse [196].

Most lesions involve the vertebral body, and the pedicles are infrequently involved [195] (Fig. 9.47). The disk spaces are usually preserved, and the height of the disk space may actually appear to be increased. Histiocytosis X more often involves the lumbar spine and least often affects the cervical spine [195]. However, patients who present with symptomatic EG most frequently have cervical spine involvement and less often thoracic spinal involvement [195]. The higher incidence of cervical involvement in symptomatic patients may be due to greater mobility in this area, predisposing to pain and muscle stiffness [195]. Scoliosis or torticollis is often present. In patients with thoracic involvement, there is often kyphosis with the apex over the affected vertebrae [197]. Spinal cord involvement may be caused by either vertebral collapse with bony impingement or extradural extension of the lesion [198].

Although EG is the most frequent cause of vertebra plana, it is not pathognomonic because it also can be encountered in malignant disease. EG also may involve the spine without causing vertebra plana [198]. Earlier diagnosis often allows identification of osteolytic lesions within the vertebral body without evidence of collapse. This was encountered in over half the patients in one series of symptomatic vertebral lesions, and the authors concluded that histiocytosis X should be included in the differential diagnosis of all patients with osteolytic vertebral lesions, regardless of age [195].

Most bone lesions can be detected on plain radiographs. Both CT and MRI are helpful in providing better delineation of a lesion, and both are superior to plain films in demonstration of extraosseous extension. MRI is preferred for evaluating canal encroachment, since intrathecal contrast material is not required [194]. The multiplanar imaging capabilities of MRI, the high soft-tissue contrast, the ability to demonstrate additional unsuspected areas of involvement, and the absence of ionizing radiation also favor MRI in both the initial evaluation and follow-up examinations in children with spinal involvement [199]. MRI demonstrates nonspecific prolongation of

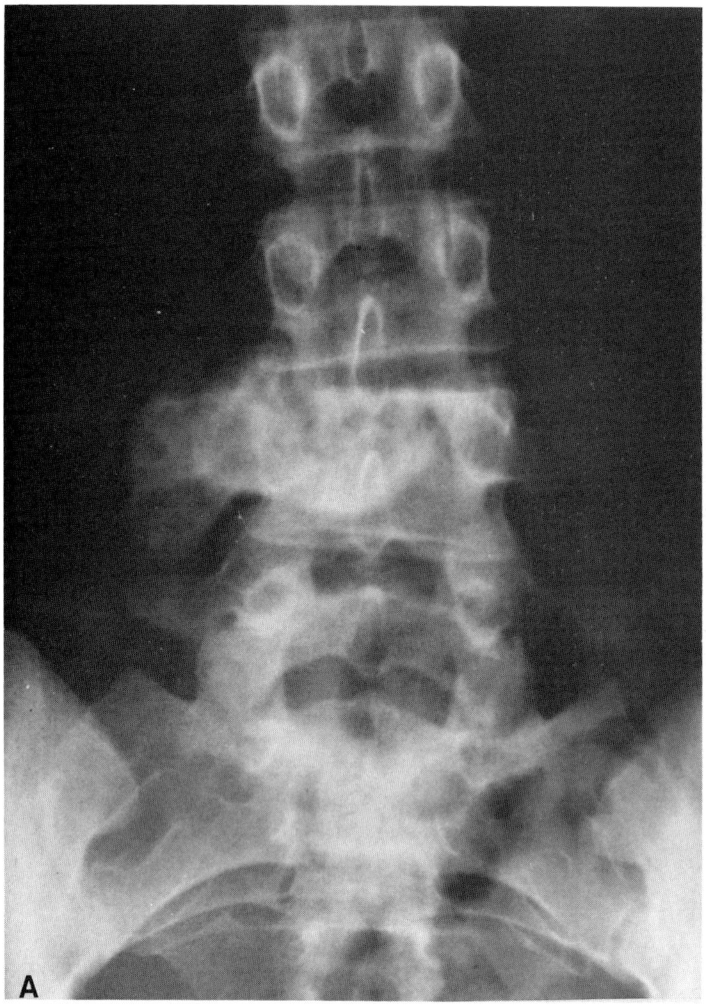

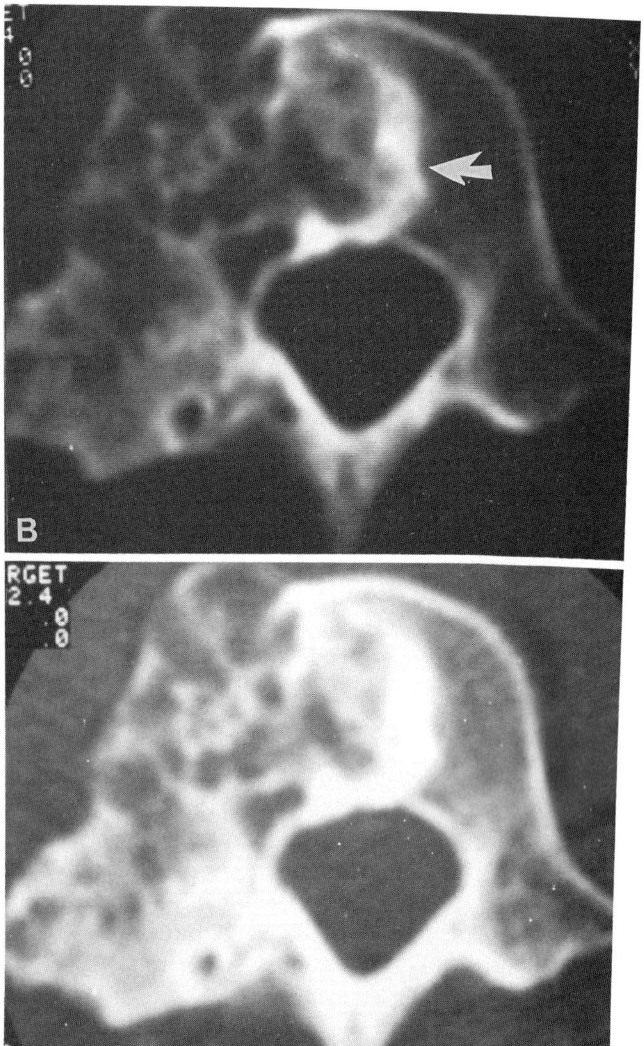

Figure 9.46. Osteoblastoma. A 21-year-old man with low back pain. **A,** Plain film. Loss of definition of the right pedicle of L4 with expansion of the right transverse process. Note that lesion is along the concave aspect of the scoliotic curve. **B,C,** CT of L4 level. Expansion of the transverse process and vertebral body and to a much lessor degree the right lamina. Note that the lesion is well circumscribed. The sclerotic rim along the medial aspect of the vertebral body is suggestive of osteoblastoma, but the overall appearance and the patient age are otherwise indistinguishable from aneurysmal bone cyst (*arrow*). Sharp tumor delineation is atypical for sarcoma. **D,** Axial T1-weighted image (TR700/TE26). **E,** T2-weighted image (TR2000/TE80). Corresponding MRI images demonstrate the tumor to be relatively hypointense on T1-weighted image (*arrow*) and somewhat mottled on a T2-weighted image. Note again the prominent transverse process expansion. Osteoblastomas often involve the base of the transverse process and are more common on the right. **F,** Axial CT of another patient with osteoblastoma involving the vertebral body and pedicles. It has the characteristic sclerotic rim with an expansile mass and punctate calcification. (Cases courtesy of Dr. Margaret A. Stoll, Georgetown University, and Dr. Mark Kransdorf, Armed Forces Institute of Pathology, Washington, D.C.)

both T1 and T2 values (hypointense on T1, hyperintense on T2), similar to other infiltrative disorders [193].

Osseous histiocytosis X is considered a benign self-limited disorder when systemic disease is absent [192]. The bone lesions of histiocytosis X tend to resolve spontaneously at a rate unaffected by the mode of therapy [192]. The treatment of localized spinal EG is controversial and is generally reserved for patients with neurologic involvement [198, 199]. Treatment modalities have included radiation therapy, chemotherapy, local steroid injection, or surgical intervention [198]. Unnecessary radiation may reduce repair capacity, particularly in children, in whom impairment of growth potential is possible. Bone heals very slowly following radiation or use of chemotherapeutic agents [192].

Varying degrees of reconstitution of vertebral height have been reported, irrespective of the treatment. The younger the patient, the greater is this potential [196]. Long-term follow-up shows reconstitution of vertebral height equivalent to 48 to 95% of

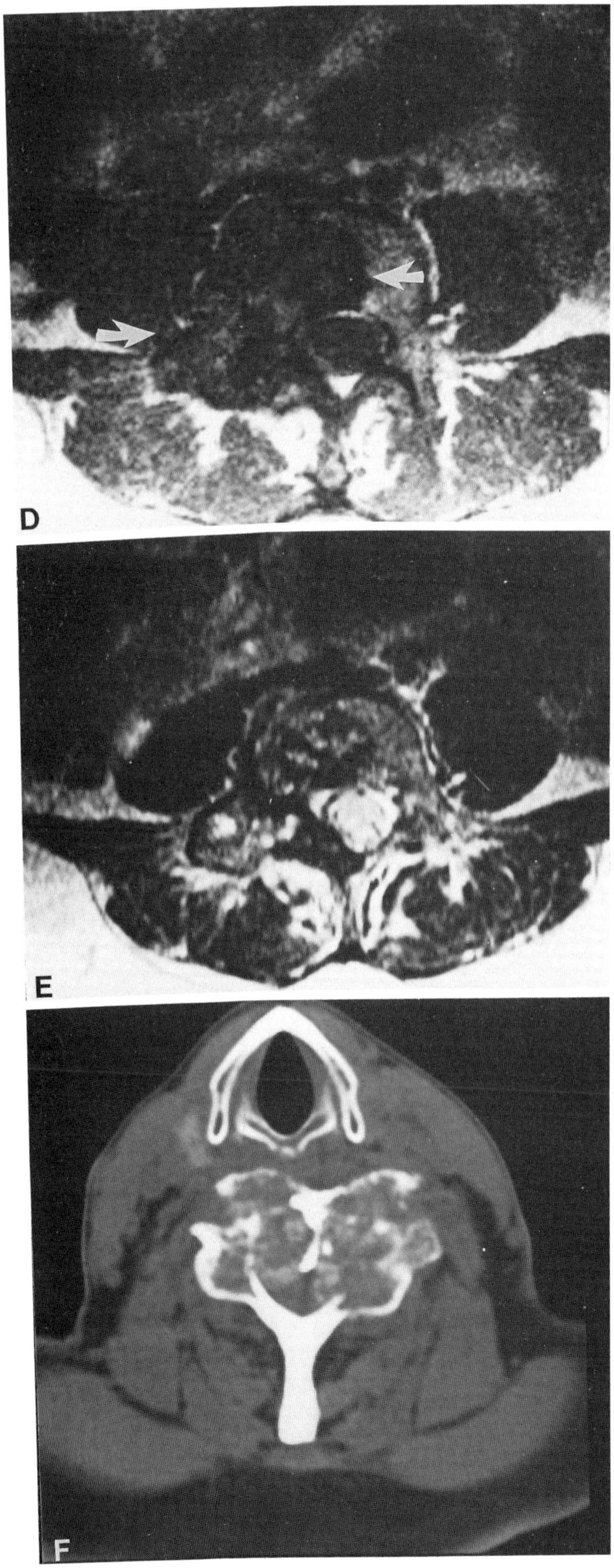

Figure 9.46–*continued*

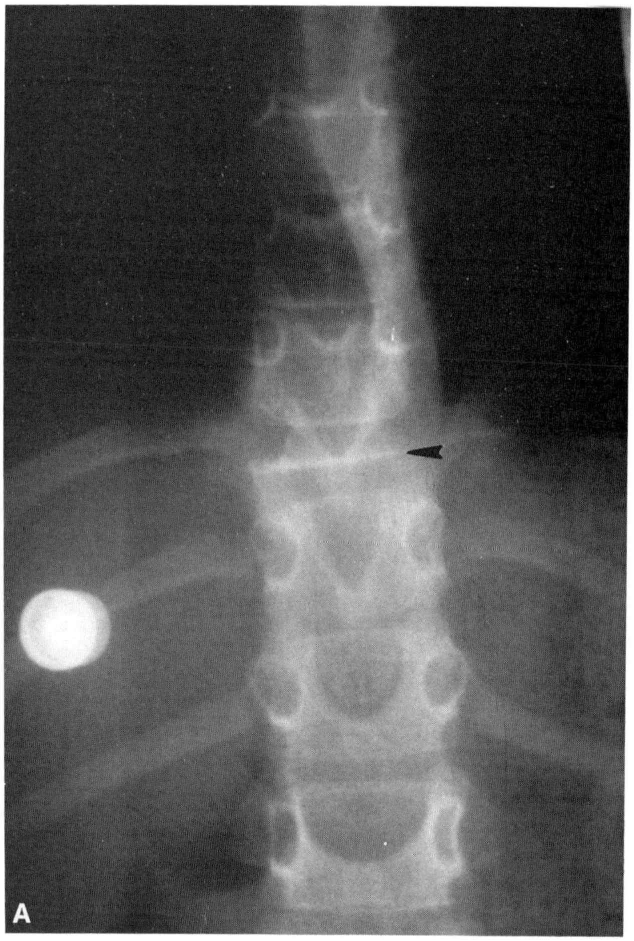

Figure 9.47. Histiocytosis X. A 10-year-old boy with eosinophilic granuloma. **A,** AP radiograph. T10 vertebra plana (*arrowhead*). **B,** Sagittal T1-weighted image (TR800/TE22). Vertebra plana at T10 (*arrowhead*). **C,** Coronal T2-weighted image (TR2000/TE80) demonstrates increased height and hyperintensity of the disks adjacent to the T10 vertebra plana (*arrowhead*). Sagittal T1-weighted and (**D**) and T2-weighted (**E**) images obtained 11 months later. There has been no change in vertebral plana (*arrowhead*). T1-weighted image demonstrates little, if any, loss of signal intensity at T12, but prominent hyperintensity at this level was noted on the T2-weighted sequence (*arrow*). **F,** Follow-up sagittal T1 examination several months later. There has been interval collapse of the T12 vertebral body (*arrow*). **G,** Sagittal T1-weighted image of cervical spine demonstrates vertebra plana at T1 with some epidural proliferation (*arrowheads*).

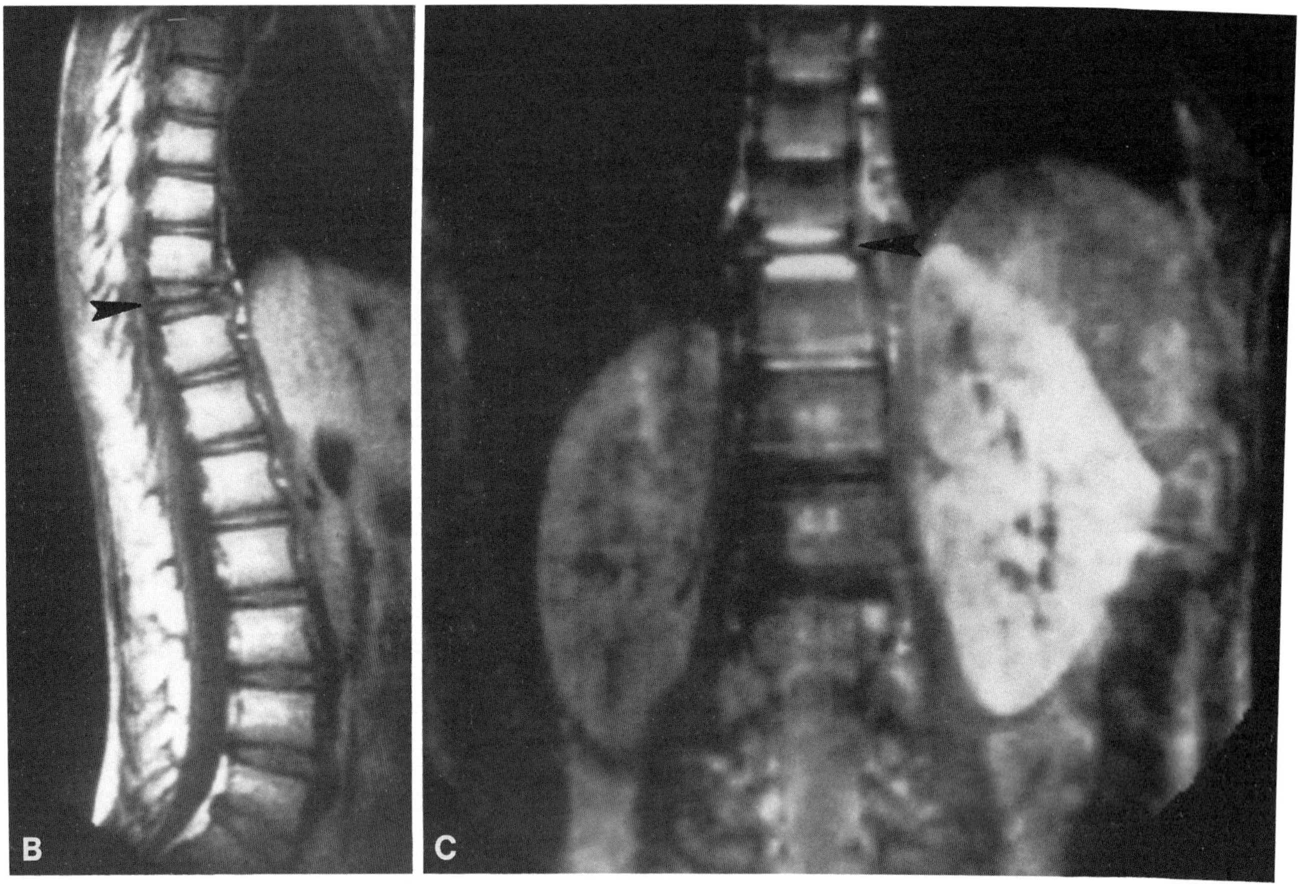

Figure 9.47–*continued*

normal, with no sign of osteoarthritis [197]. Long-term follow-up also may demonstrate return of the thickened adjacent vertebral disk to a normal height [197]. Residual scoliosis may be present.

GIANT CELL TUMORS

Giant cell tumors, or osteoclastomas, are locally aggressive vascular tumors that are generally benign [200, 201]. As many as 50% of these lesions recur after initial therapy, but less than 10% are actually malignant on initial presentation [200, 201]. The great majority of these histologically controversial tumors are located in the epiphysis of the long bones, and less than 10% are found in the sacrum and spine [201, 202]. The sacrum is involved much more frequently than the spine, and giant cell tumor is second only to chordoma as the most common primary tumor of the sacrum [200]. Giant cell tumor is the most common "benign" tumor of the sacrum [200].

Giant cell tumors of the spine and sacrum usually present in the third and fourth decades of life, and there is a distinct female preponderance [200, 203, 204]. Patients generally present with localized back pain with occasional radiation to an extremity. The majority are symptomatic for 3 to 4 months before diagnosis [200]. Constipation and urinary frequency are also encountered in sacral tumors because of the bulky soft-tissue component [200]. Local tenderness is also frequent.

In the sacrum, the tumors are large at presentation. There is no predominant site of origin within the sacrum, and occasionally, the entire sacrum may be involved. Rarely, extension across the sacroiliac joint or the L5–S1 disk may be demonstrated [200]. The characteristic appearance of giant cell tumor of the sacrum is a large, prominent lytic area which is usually fairly well delineated, although occasionally the margins are indistinct. No internal calcification or ossification is present in untreated tumors, unless complicated by response to pathologic fracture [200, 201]. Reactive sclerosis is usually absent in untreated cases [201]. Trabeculation or pseudotrabeculation due to some remnants of nearly destroyed bone is only occasionally encountered [200]. This pseudotrabeculation may result in a multiloculated and somewhat mottled appearance [205]. Gradual sclerosis of the neoplasm may be demonstrated following radiation therapy [200].

In the spine, most giant cell tumors involve the vertebral body. Less than one-third involve either the

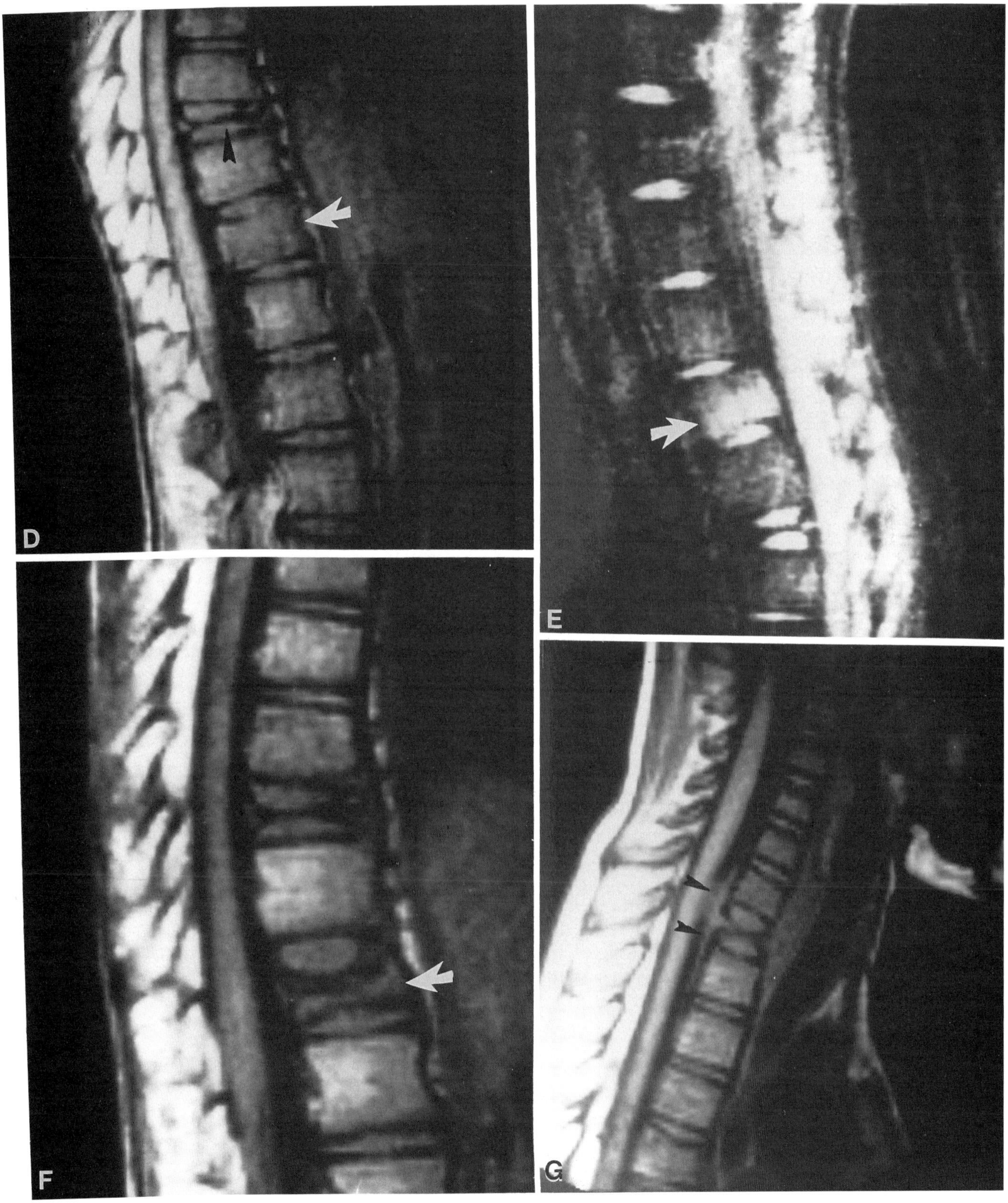

Figure 9.47–*continued*

pedicle or the vertebral arch. This is in contrast to both aneurysmal bone cysts and osteoblastomas, which almost always arise in the posterior elements and only infrequently arise in the vertebral body [203]. Plain film examinations are often limited by overlying bowel contents, especially in patients who have constipation as a result of their sacral lesion. Plain film findings in the spine are nonspecific and usually demonstrate a lytic area which may be expansile. Occasionally, a multiloculated or "soap bubble" appearance may be encountered, but this is usually not distinctive [205]. The initial cystlike lesions may eventually produce cortical expansion and may result in vertebral body collapse. There is often a large paravertebral soft-tissue mass, but this may be difficult to demonstrate on plain films [206].

CT demonstrates an expansile, lytic, destructive mass often associated with a large paravertebral component [21, 201, 202]. This soft-tissue mass may be huge in sacral lesions. Expanded cortex may simulate a sclerotic margin. Lesions are frequently eccentric [179]. CT is useful in young adults with persistent sacral or vertebral pain in whom plain films are normal and subsequent bone scan is positive [200]. CT is also useful in demonstrating the extent of spinal canal involvement.

MRI is superior in demonstrating both the paravertebral component and the spinal canal involvement (Fig. 9.48). In addition, extension of the lesion across the disk is better demonstrated with sagittal and coronal MRI than with axial imaging. As with other tumors of the spine and sacrum, T1-weighted images demonstrate bone marrow replacement by relatively hypointense tumor. Cortical bone thinning and destruction also can be demonstrated [207]. Sagittal images are particularly useful in assessing vertebral collapse and associated canal involvement. Both CT and MRI are useful in assessing soft-tissue extension and are particularly useful in demonstrating its effect on associated pelvic structures such as the bladder and rectum [200].

Ten percent of giant cell tumors are malignant (Dahlin). Most of the malignant giant cell tumors are associated with a previous history of irradiation [201]. Despite the usual "benign" histology, the prognosis for giant cell tumor of the sacrum is much worse than that for giant cell tumors in other sites, including other areas of the spine [205]. Recurrence and death rates are much higher [201]. This is related to a combination of delayed diagnosis, resulting in larger tumor size, and inability to completely resect the tumor [200, 208]. Incomplete surgical removal is considered the most significant factor in recurrence [204]. Recurrent tumors behave more aggressively and are associated with an increased risk of malignancy [201].

Approximately 2% of giant cell tumors will metastasize to the lung despite their benign histology [111]. Resection of these pulmonary metastases is associated with a good prognosis [200]. Radiation of giant cell tumors is generally contraindicated because of relative radioresistance and the significant risk of sarcomatous degeneration [200, 201, 205]. Radiation therapy is occasionally utilized when complete surgical resection cannot be obtained [208].

Differential diagnosis for an expansile, radiolucent lesion of the vertebral body or sacrum should include aneurysmal bone cyst, osteoblastoma, Brown tumor, and fibrous dysplasia. In older patients, myeloma and metastases are also considered. Aneurysmal bone cysts and osteoblastomas usually arise in the posterior elements and infrequently arise in the vertebral body, unlike giant cell tumors [203]. In addition, aneurysmal bone cysts are generally seen in younger patients. Brown tumors may have an identical appearance but often can be identified because of the clinical or biochemical evidence of hyperparathyroidism. Myeloma and metastases generally occur in an older age group than that affected by giant cell tumors.

ANEURYSMAL BONE CYSTS

An aneurysmal bone cyst is a nonneoplastic, lytic, expansile lesion which may resemble a neoplasm radiographically. It is neither an aneurysm nor a bone cyst [209]. The lesion consists of unclotted venous blood under pressure which fills communicating bone spaces causing expansion and ballooning of the involved portion of bone [210]. The etiology is unknown but may be related to other reactive nonneoplastic processes [111]. An alteration in local

Figure 9.48. Giant cell tumor. A 31-year-old man presented with back pain. **A,** AP radiograph. Partial collapse of T11 vertebrae, more prominent on right. Note right pedicle involvement (*arrow*). **B,** Lateral film. Lytic lesion involves much of T11 vertebral body (*arrow*). Note partial collapse of inferior vertebral endplate of T11 (*arrowheads*) secondary to bony replacement. Sagittal (**C**) and parasagittal (**D**) T1-weighted image (TR600/TE20). Hypointense tumor signal replaces the normal marrow of T11. Note prominent extension into the spinal canal (*arrow*) with cord involvement. Also note prominent involvement of the posterior elements (*arrowhead*), unusual for giant cell tumor. There is also involvement of the T11 disk anteriorly (*open arrow*). **E,** Sagittal T2-weighted image (TR2000/TE75). Tumor is relatively hyperintense compared with adjacent normal marrow. **F,** Axial T1-weighted image (TR1000/TE20) through level of T11–12 disk. Associated soft-tissue mass with canal encroachment and cord displacement (*arrowhead*). Contiguous axial CTM (**G**) and sagittal (**H**) and parasagittal (**I**) re-formatted CTM sequences in another patient with giant cell tumor. Expansile vertebral body mass with disruption of the cortex. There is extension into the pedicle and the epidural space with cord compression. (Courtesy of William R. Scott, M.D., West Florida Regional Medical Center, Pensacola, Florida.)

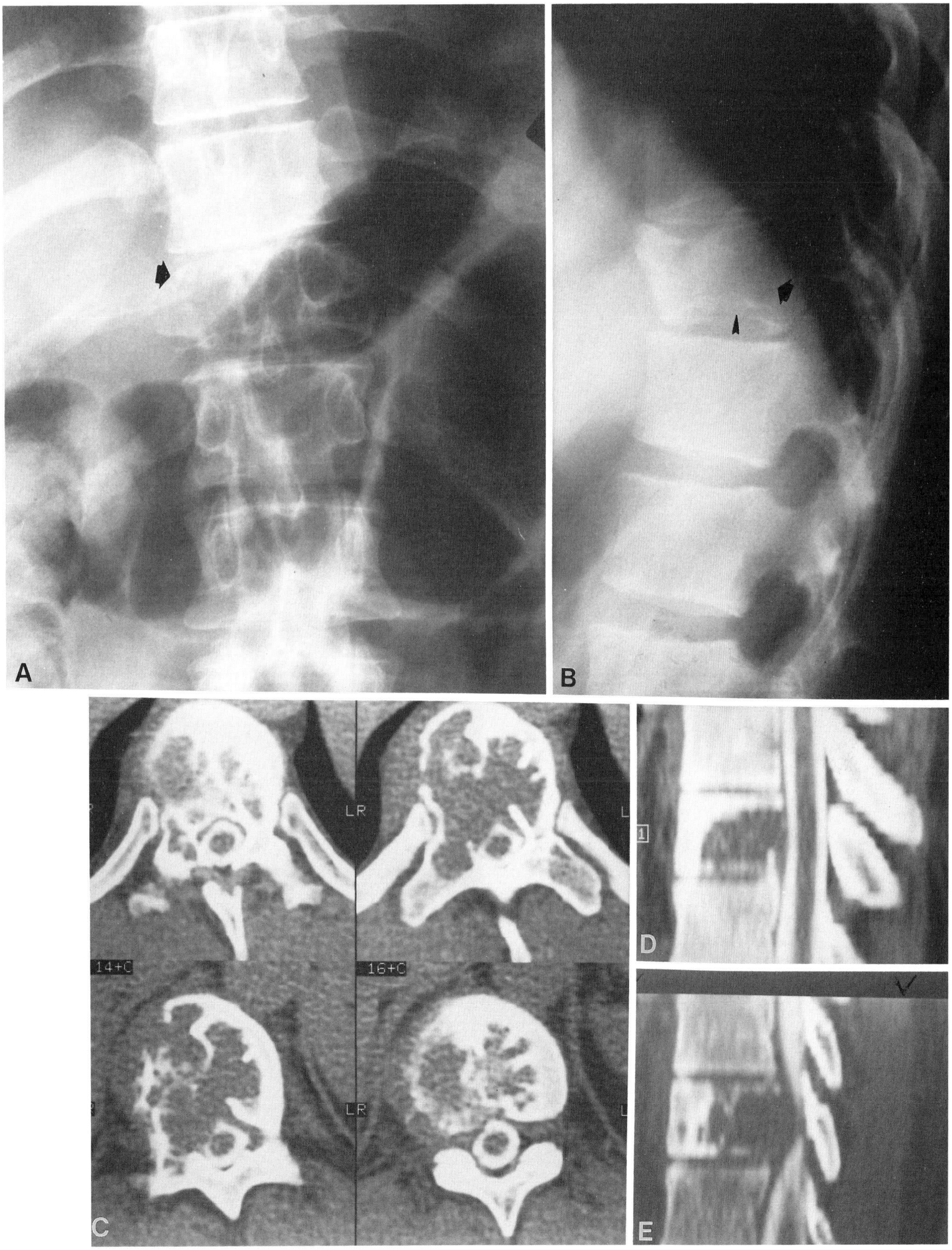
A
B
LR
14+C
16+C
LR
C
1
D
E

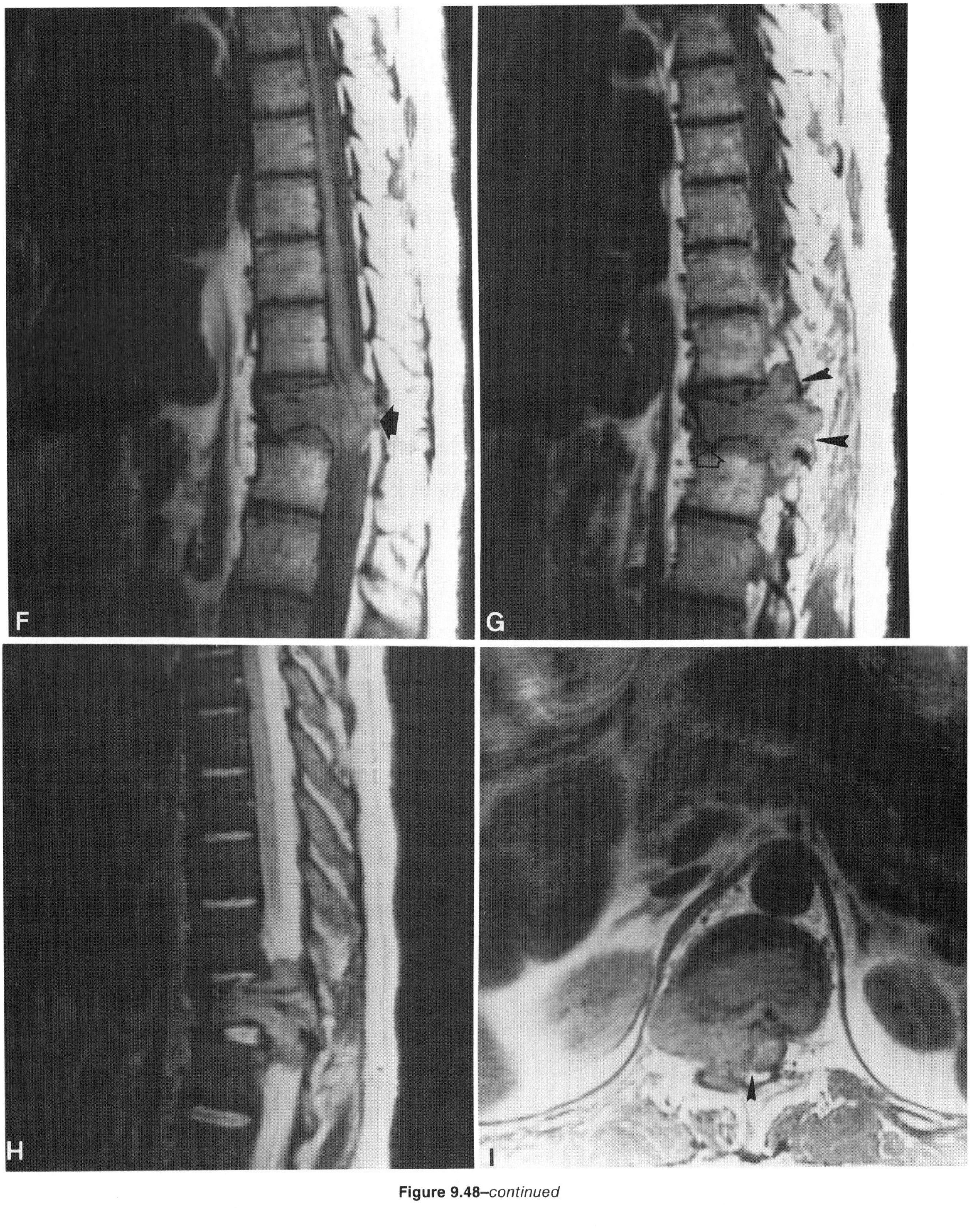

Figure 9.48–*continued*

hemodynamics related to venous obstruction or arteriovenous fistula also has been postulated [211]. This may be triggered by trauma or the development of other pathology in the area which interferes with venous drainage [210].

Aneurysmal bone cyst is primarily a condition of the second decade, with an average patient age of 16.6 years [209]. Between 80 and 90% of cases occur in patients below age 20 [210]. About 20% of cases involve the spine. Local pain is the presenting symptom in approximately three-fourths of patients. Approximately one-quarter will report a palpable mass, and another one-quarter of patients will present with significant long tract signs resulting in some degree of paraplegia. Occasionally, patients will have symptoms referable to nerve root involvement [209]. The lesion usually grows rapidly, with the duration of symptoms ranging from 4 to 8 months [209, 210]. Pain is usually slight and intermittent at first but may become severe and constant when the lesion expands rapidly [210].

There is a predilection for the lumbar spine, although aneurysmal bone cysts are seen throughout the cervical and thoracic spine [209]. Sacral involvement is much less common. The lesions often involve the neural arch, particularly the spinous process. The vertebral body also may be affected, usually due to extension from a lesion in the posterior elements, although primary vertebral body involvement is not unusual [209, 210]. Aneurysmal bone cysts often involve more than one contiguous vertebrae [209].

The classic plain film finding is an expansile, osteolytic cavity, often containing fine strands of bone surrounded by an eggshell-like, blown-out cortex [209]. The bony shell is characteristically continuous with the cortex of the adjacent bone. The thin periosteal shell of bone is often interrupted or only partially visible on plain film examination [212]. More advanced lesions may show a partial collapse or erosion of contiguous vertebral bodies or adjoining ribs. The cyst usually begins eccentrically in the posterior arch or body and may grow into contiguous parts of the neural arch, often extending far beyond the original contours of the affected bone with involvement of the adjacent soft tissues [213]. The bone cyst is confined by fibrous periosteum or a thin shell of periosteal bone and tends not to invade surrounding tissues.

CT with sagittal reconstruction will demonstrate most of the characteristic features of aneurysmal bone cysts of the spine. CT may show a thin rim of cortical bone not evident on plain films and is helpful in excluding a calcified tumor matrix [214].

Both CT and MRI may demonstrate the characteristic, although not pathognomonic, fluid levels associated with aneurysmal bone cysts. Fluid levels within the lesion may be seen in approximately one-third of patients and are best demonstrated when viewed at very narrow window widths (100 to 150 H.U.) with the window level set at +40 to +80 H.U. [212]. The dependent layers are visibly denser than the nondependent layers. These fluid levels correspond with hemorrhagic fluid collections contained within the variably-sized cavities of the lesion and are not associated with contrast enhancement. However, there is a variable amount of fibrous vascular tissue within aneurysmal bone cysts. This varies from thin membranes to considerable areas of solid tissue. This tissue surrounds the cystic cavities and is associated with contrast enhancement [216]. Since these fluid levels are the result of layering out of the various components of blood, sufficient immobilization time must be allowed prior to scanning. This is especially critical with the new quick-scan capabilities. A time interval of at least 10 minutes has been suggested for improving the sensitivity of this important differential point [216].

The combination of CT with MRI provides the best preoperative information [215]. MRI is useful in evaluating the associated soft-tissue component. The multiplanar capability of MRI allows direct sagittal imaging which facilitates demonstration of multilevel involvement (Figs. 9.49 and 9.50). Spinal cord encroachment is best demonstrated with MRI.

MRI is often more specific than CT [216, 217]. MRI findings include fluid levels, well-defined cystic spaces with a potential wide range of signal intensities, diverticular-like soft-tissue projections, and a hypointense circumscribing rim at both the bone-tumor interface and the soft-tissue interface [216, 217]. MRI demonstrates the characteristic expansile appearance of the lesion. The soft-tissue component of the lesion is generally hypointense on T1- and hyperintense on T2-weighted images. A thin, well-defined, low-signal rim of fibrous tissue circumscribing the lesion can be demonstrated. This has a low signal on both T1- and T2-weighted images [211, 214, 216].

Small, diverticula-like projections in the walls of larger cysts have been described [216]. These lobulations are of soft-tissue signal intensity. MRI generally shows septa within the aneurysmal bone cyst, and adjacent loculations may have markedly different signal characteristics because of the varying age of the hemorrhagic components [216]. Although acute hemorrhage has little signal, within a few days methemoglobin effects predominate, and the lesions are hyperintense on T1-weighted images [216]. Because of the significant T1 effect, even the T2-weighted images may be hyperintense at this time. Like CT, MRI also may demonstrate characteristic fluid levels, which are probably due to settling of degraded blood products within the cyst [211]. When

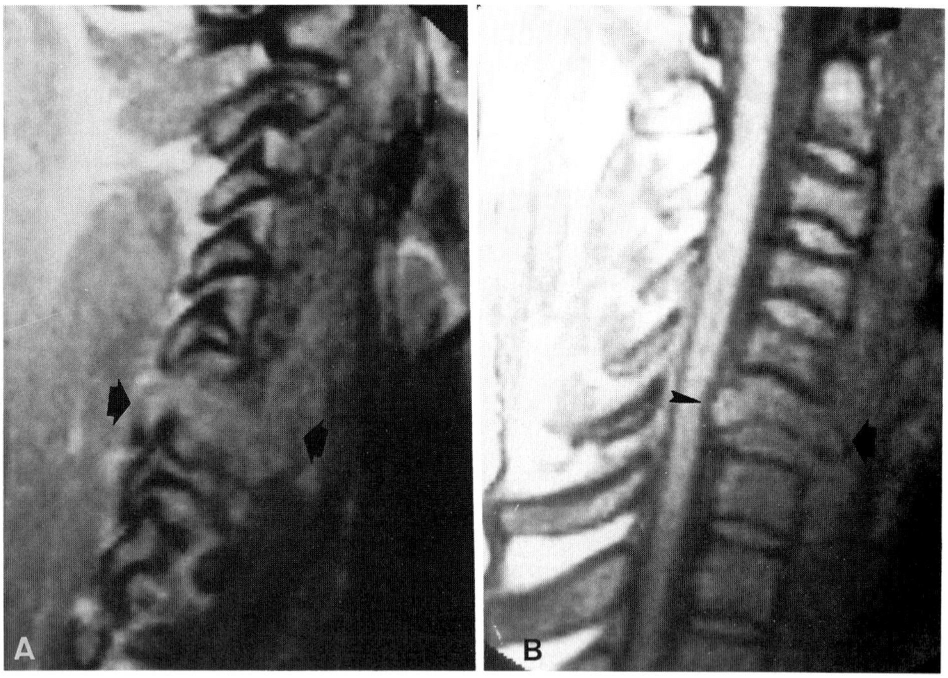

Figure 9.49. Aneurysmal bone cyst. An 18-year-old man with neck pain. Initial MRI. **A,** Parasagittal T1-weighted image (TR800/TE22). There is destruction of the right facet of C6 which is replaced by soft-tissue signal mass (*arrows*). **B,** Sagittal midline T1-weighted image (TR800/TE22). There is also involvement of the vertebral body with extension of the lesion both anteriorly (*arrow*) and posteriorly (*arrowhead*) into the canal with cord compression.

fluid levels are present, the dependent layer is generally of brighter signal intensity on T1-weighted images, presumably due to T1 shortening by methemoglobin [214]. The distinctness of the fluid levels seen within the cysts may vary with the amount of T2 weighting and may not be evident if only a single sequence is obtained. In general, the T1-weighted images are best for demonstration of fluid levels [214].

Although the fluid levels are highly suggestive of aneurysmal bone cysts, they are not pathognomonic because they also may be seen in telangiectatic osteosarcoma, simple cyst with fracture, giant cell tumor, and even chondroblastoma [211, 214]. However, the thin, well-defined margins of aneurysmal bone cyst should help to distinguish it from these other lesions. CT is useful in narrowing the differential considerations, since the lack of osteoid formation would be atypical for osteoblastoma and the lack of internal calcifications would be atypical for chondroblastoma [218]. The lack of a large soft-tissue mass makes osteomyelitis less likely.

The giant cell tumor is seen rarely in patients younger than age 20 and, unlike aneurysmal bone cysts, occurs most frequently in the sacrum. In addition, large cavernous vascular spaces are atypical [210]. Vertebral hemangioma can be distinguished by its characteristic plain film appearance. The cavernous form typically affects the vertebral body and does not demonstrate ballooning or involvement of the neighboring vertebrae. Hydatid disease can be distinguished by CT because the cysts contain lower-density fluid [210]. Osteosarcoma and metastatic lesions generally produce more pain, involve older age groups, and are not associated with the "aneurysmal" appearance or the characteristic thin rim of bone or fibrous tissue at the periphery of the lesion [210].

Prognosis for surgical resection is generally excellent [209]. Even with incomplete resection, some lesions will regress spontaneously. However, the recurrence rate of aneurysmal bone cysts is 30% [213]. Incomplete removal is associated with a 70% recurrence rate. There is a risk of significant hemorrhage during surgery, which occurs in 22% of cases, mostly in the flat bones and spine [213]. Recur-

Figure 9.50. Aneurysmal bone cyst. Recurrence several months following surgery, same patient as in preceding figure. **A,B,** Plain films demonstrate expansile, lytic lesion now involving both C5 and C6 vertebrae (*arrows*) with obliteration of C5–6 disk. **C,** Contrast-enhanced CT scan. Enhancing, expansile lesion is circumscribed by bony rim (*arrows*). Note cystic areas (*arrowheads*). Sagittal (**D**) and parasagittal (**E**) T1-weighted image (TR800/TE22). Lesion now involves both the C5 and C6 vertebrae with obliteration of the disk space. Note posterior element involvement (*arrowhead*). **F,** Sagittal T2-weighted image (TR2000/TE80). Lesion is markedly hyperintense on T2-weighted image. Note associated prevertebral hemorrhage (*arrow*). Aneurysmal bone cyst in another patient. **G,** Lateral plain x-ray of the cervical spine demonstrates lucency of the C2 vertebral body with disruption of the corticle margin, suggestive of a fracture. **H,** Axial CT section demonstrating an expansile lytic lesion involving the vertebral body with disruption of the corticle margin both along its anterior and posterior aspect. Sagittal (**I**) and coronal (**J**) re-formatted images provide evaluation of the tumor in all three phases.

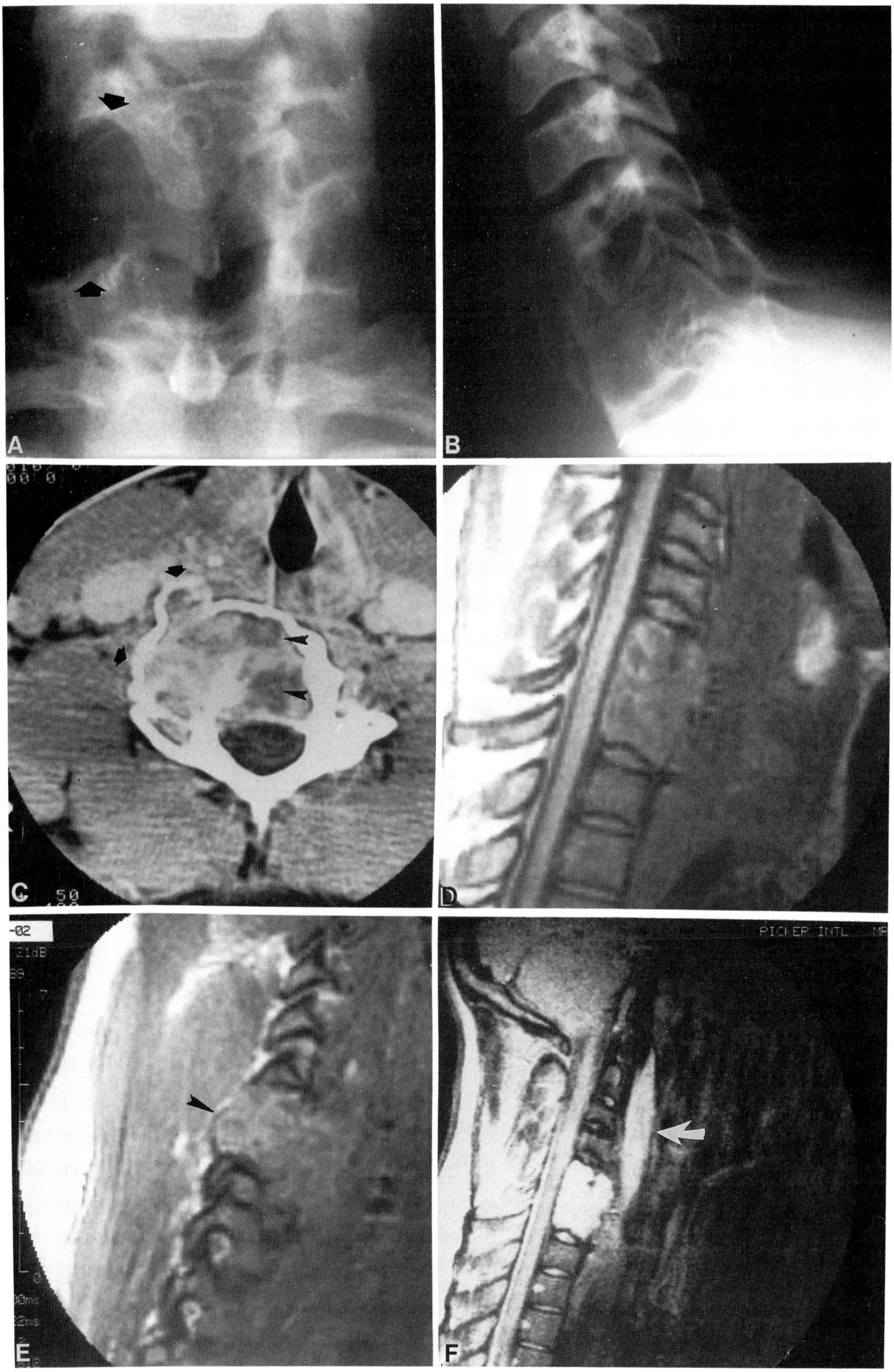
A
B
C
D
E
F
PICKER INTL

Figure 9.50–*continued*

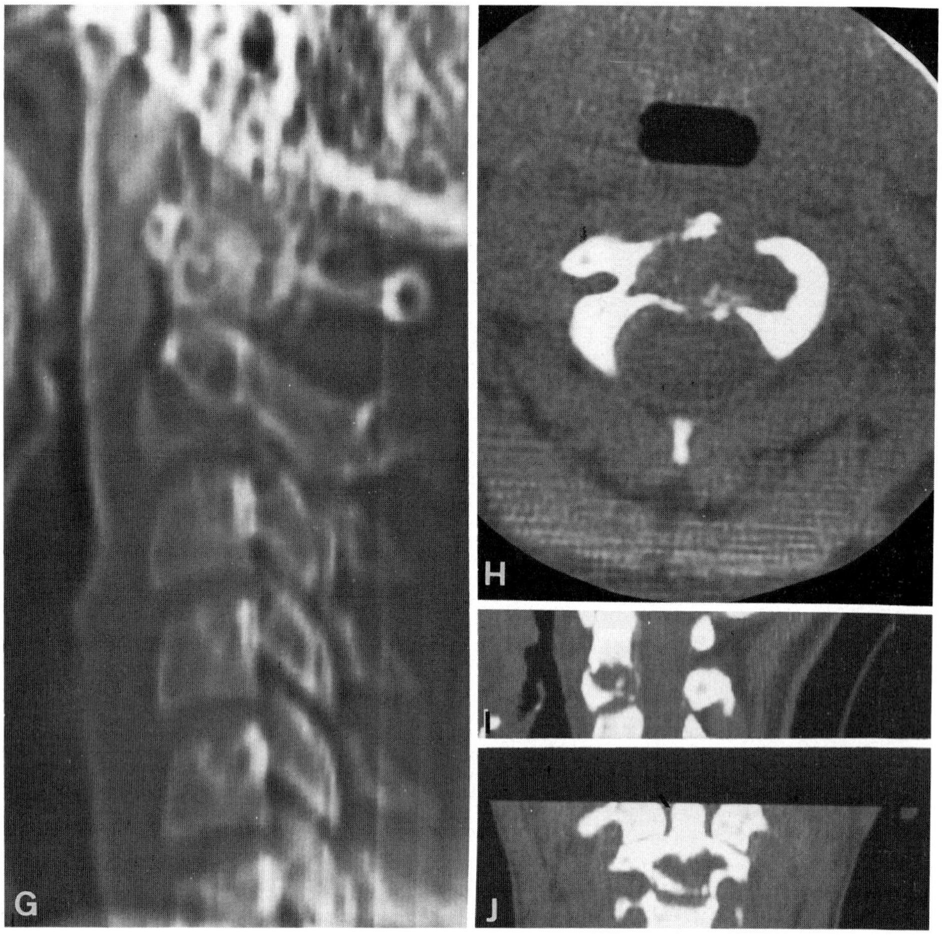

rences usually occur rapidly, with an average interval of 4 months following surgery [209]. Radiotherapy is very effective in arresting the lesion and can induce rapid ossification and provide rapid relief of symptoms [215]. However, this cannot be advocated because of the significant risk of sarcomatous change [209, 215]. Embolotherapy may be useful in preoperative reduction of blood supply to the lesion. Embolotherapy also may be used as definitive therapy for inaccessible lesions [214].

VERTEBRAL HEMANGIOMAS

Hemangiomas are slowly growing vascular tumors which are usually solitary and asymptomatic. Although hemangiomas are encountered in about 10% of autopsy series, only a small percentage of these are detected on plain radiographs [219]. These are often demonstrated incidentally on lateral chest or spine examinations. The most common skeletal locations are the calvarium and spine [111]. Thoracic involvement is more common than lumbar involvement, and cervical involvement is quite rare [220]. The median patient age is 40 years, and there is a 3:2 female predominance [219].

Symptoms may result from pathologic fracture or epidural extension with resulting cord or root impingement [20]. When present, symptoms vary from localized or radicular pain to weakness or paralysis of an extremity [221]. Symptomatic hemangiomas are more frequent in younger females [220] and have been reported as a complication of pregnancy [101]. Midthoracic lesions are more prone to produce compression symptoms [220]. This is probably related to the larger size of the cord in relation to the spinal canal at this level as well as the dorsal kyphosis.

Spinal hemangiomas generally involve only the vertebral body, although occasionally the posterior elements may be involved [219]. Patients with posterior element involvement are more prone to develop cord compression [219, 222]. Cord compression may result from hypertrophy and ballooning of the vertebrae leading to a narrower canal, subperiosteal growth resulting in epidural extension, compression fracture, or epidural hemorrhage [222]. The epidural extension from subperiosteal growth generally involves the posterior aspect of the vertebral body [222] (Fig. 9.51).

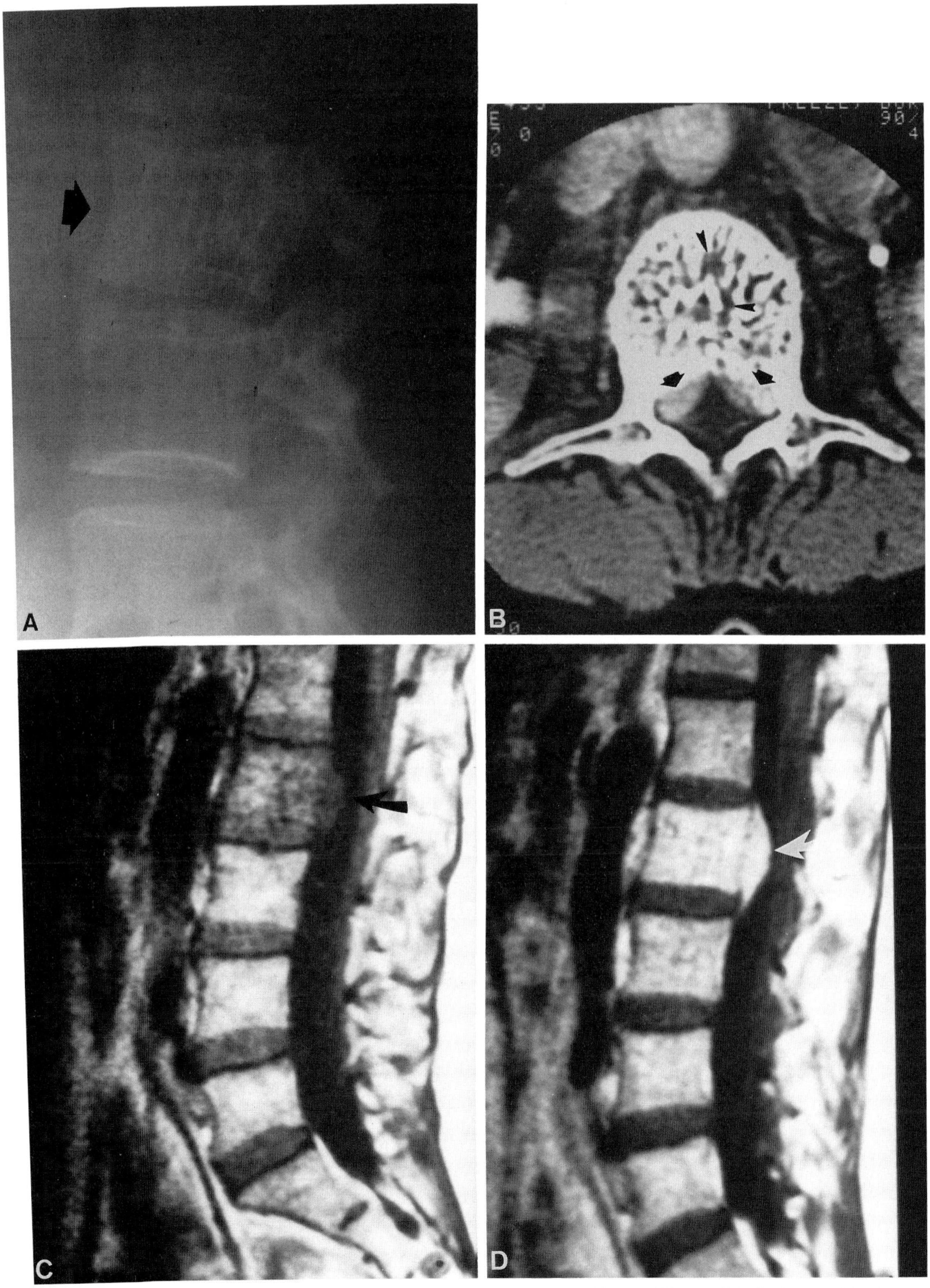

Figure 9.51. Hemangioma. A 58-year-old woman. **A,** Coarse vertical striations in L2 vertebral body (*arrow*). **B,** Contrast-enhanced CT scan. There is epidural extension of the tumor into the anterior spinal canal bilaterally (*arrows*). Note the fatty matrix (*arrowheads*) interspersed between the coarsened vertical trabeculae. **C,** Sagittal T1-weighted image (TR616/TE15). Vertebral body marrow of L2 is somewhat mottled. Note tumor extension into the spinal canal (*arrow*). Sagittal (**D**) and axial (**E**) contrast-enhanced T1-weighted image (TR616/TE15). Prominent homogeneous enhancement of the vertebral body and epidural extension of tumor (*arrow*).

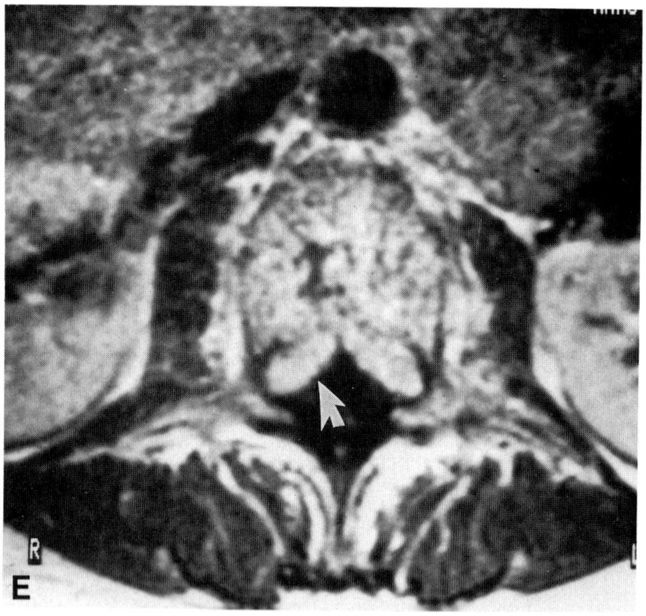

Figure 9.51–*continued*

Characteristic plain film findings include coarse vertical striations in the vertebral body separated by more lucent zones [220]. These lucent zones result from a loss of trabeculae which are resorbed by the dilated venous channels. The associated mechanical stress response results in thickening of adjacent trabeculae, which results in a "honeycombed" or "corduroy cloth" appearance [22].

The high sensitivity of CT has resulted in increased detection of incidental vertebral hemangiomas [219, 222]. The course vertical striations are well demonstrated and often have a "polka dot" appearance [22, 223]. The fatty matrix between these thickened trabeculae often has negative CT attenuation values compatible with fat. In some patients, this matrix tissue may actually be soft-tissue density [224]. Rarely, local bony expansion or "ballooning" may be demonstrated [221]. This cortical expansion is much more frequent in patients with symptomatic vertebral hemangiomas [219]. Cortical expansion in symptomatic patients generally involves either the pedicles or the posterior or lateral cortex of the body. Swelling of the adjacent soft tissues is also often demonstrated in symptomatic patients [219]. These associated soft-tissue masses may be due to local extension of the lesion [225] or epidural hemorrhage into the canal [222].

Intravenous contrast enhancement often confirms the highly vascular nature of the lesion and also may demonstrate some enhancement of the intratrabecular channels of the vertebral body [225]. Intrathecal contrast material is quite useful in demonstrating possible canal encroachment as well as in planning surgical intervention [222].

MRI is both highly sensitive and specific for vertebral hemangiomas. T1-weighted images demonstrate areas of abnormally increased signal intensity within the involved vertebrae relative to adjacent normal marrow [224]. These hyperintense foci tend to be small and scattered, producing a somewhat mottled appearance. These areas of hyperintensity correspond with the hypodense foci on CT and represent the fat component of the tumor matrix. This also has been confirmed with chemical-shift imaging and histopathologic findings [224]. The abnormally thickened, coarse trabeculae are best seen on axial images and are hypointense to isointense on T1-weighted images [224].

On T2-weighted images, abnormal increased signal within the involved vertebral bodies, with the coarse trabeculae remaining relatively hypointense, results in a somewhat mottled appearance [224]. Associated paravertebral or epidural extension is demonstrated as intermediate soft-tissue signal on T1-weighted images and is hyperintense on T2-weighted images [226].

The presence of hyperintense foci on T1- and marked hyperintensity on T2-weighted images, coupled with the mottled appearance of the vertebral bodies, is considered relatively specific for vertebral hemangioma. Neoplastic, inflammatory, or ischemic processes may be hyperintense on T2- but would not be expected to be hyperintense on T1-weighted images, unless there are areas of hemorrhage or thrombosis. Plain film or CT evaluation may help in these instances. Marrow conversion in the elderly may result in focal hyperintensity on T1-weighted images within the vertebral bodies, but this is not associated with marked hyperintensity on the T2-weighted images [226].

Symptomatic vertebral hemangiomas may require surgical decompression. There is a significant risk of intraoperative hemorrhage, which may be minimized with preoperative embolization [221]. Radiation may be useful in cases where surgery is not feasible [221].

References

1. Nittner K. Spinal meningiomas, neurinomas and neurofibromas and hourglass tumours. In: Vinken PJ, Bruin GW, eds. Handbook of clinical neurology. Vol. 20, part 2. Amsterdam: North-Holland, 1976:177–322.
2. Sloof JL, Kernohan JW, MacCarty CS. Primary intramedullary tumors of the spinal cord and filum terminale. Philadelphia: Saunders, 1964.
3. Kurland LT. The frequency of intracranial and intraspinal neoplasms in the resident population of Rochester, Minnesota. J Neurosurg 1958;15:627–641.
4. Guomundsson KR. A survey of tumours of the central nervous system in Iceland during the ten-year period 1954–1963. Acta Neurol Scand 1970;46:538–552.

5. Chade HO. Metastatic tumours of the spine and spinal cord. In: Vinken PJ, Bruin GW, eds. Handbook of clinical neurology. Vol. 20, part 2. Amsterdam: North-Holland, 1976:415–433.
6. Shapiro R. Myelography. 3rd ed. Chicago: Year Book Medical Publishers, 1975.
7. Zimmerman RA, Bilaniuk LT. Imaging of tumors of the spinal canal and cord. Radiol Clin North Am 1988; 26:965–1007.
8. Stein BM. Spinal tumors. In: Rowland LP, ed. Merritt's textbook of neurology. 7th ed. Philadelphia: Lea & Febiger, 1984:264–274.
9. Connolly ES. Spinal cord tumors in adults. In: Youmans JR, ed. Neurological surgery. Vol. 5, 2nd ed. Philadelphia: Saunders, 1982:3196–3214.
10. Dorwart RH, LaMasters DL, Watanabe TJ. Tumors. In: Newton TH, Potts DG, eds. Modern neuroradiology. Vol 1: Computed tomography of the spine and spinal cord. San Anselmo, CA: Clavadel Press, 1983:115–147.
11. Hollis PH, Malis LI, Zappulla RA. Neurological deterioration after lumbar puncture below complete spinal subarachnoid block. J Neurosurg 1986;64:253–256.
12. Kucharczyk W, Brant-Zawadzki M, Sobel D, et al. Central nervous system tumors in children: detection by magnetic resonance imaging. Radiology 1985;155:131–136.
13. Di Chiro G, Doppman JL, Dwyer AJ, et al. Tumors and arteriovenous malformations of the spinal cord: assessment using MR. Radiology 1985;156:689–697.
14. Bradley WG Jr, Waluch V, et al. Comparison of CT and MR in 400 patients with suspected disease of the brain and cervical spinal cord. Radiology 1984;152: 695–702.
15. Bydder GM, Brown J, Niendorf HP, Young IR. Enhancement of cervical intraspinal tumors in MR imaging with intravenous gadolinium-DTPA. J Comput Assist Tomogr 1985;9(5):847–851.
16. Grem JL, Burgess J, Trump DL. Clinical features and natural history of intramedullary spinal cord metastasis. Cancer 1985;56:2305–2314.
17. Epstein F, Epstein N. Surgical treatment of spinal cord astrocytomas of childhood. J Neurosurg 1982;57:685–689.
18. Fischer G, Tommasi M. Spinal ependymomas. In: Vinken PJ, Bruin GW, eds. Handbook of clinical neurology. Vol. 20, part 2. Amsterdam: North-Holland, 1976:353–387.
19. Epstein F, Epstein N. Intramedullary tumors of the spinal cord. In: Pediatric neurosurgery: surgery of the developing nervous system. New York: Grune & Stratton, 1982:529–540.
20. Arendt A. Spinal leiomas. In: Vinken PJ, Bruin GW, eds. Handbook of clinical neurology. Vol. 20, part 2. Amsterdam: North-Holland, 1976:323–352.
21. Green BA, Diaz RD, Donovan Post J. The diagnosis of spinal column and spinal cord tumors with emphasis on the value of computed tomography. In: Donovan Post MJ, ed. Computed tomography of the spine. Baltimore: Williams & Wilkins, 1984:659–695.
22. Kricun ME. Computed tomography. In: Imaging modalities in spinal disorders. Philadelphia: Saunders, 1988: 376–467.
23. Rubinstein LJ. Tumors of the central nervous system. Washington: Armed Forces Institute of Pathology, 1972.
24. Kan S, Fox A, Vinuela F. Delayed CT metrizamide enhancement of syringomyelia secondary to tumor. AJNR 1983;4:73–78.
25. Scotti G, Scialfa G, Columbo N, Landoni L. Magnetic resonance diagnosis of intramedullary tumors of the spinal cord. Neuroradiology 1987;29:130–135.
26. Goy AM, Pinto RS, Rashavendra BN, et al. Intramedullary spinal cord tumors: MR imaging, with emphasis on associated cysts. Radiology 1986;161:381–386.
27. Slasky BS, Bydder GM, Neindorf HP, Young IR. MR imaging with gadolinium-DTPA in the differentiation of tumor, syrinx, and cyst of the spinal cord. J Comput Assist Tomogr 1987;11(5):845–850.
28. Sze G, Stimac GK, Bartlett C, et al. Multicenter study of gadopentetate dimeglumine as an MR contrast agent: evaluation in patients with spinal tumors. AJNR 1990; 11:967–974.
29. Poser CM. The relationship between syringomyelia and neoplasm. In: American lectures in neurology (American Lecture Series No. 262). Springfield, Ill.: Charles C. Thomas, 1956.
30. Blackwood W, Corsellis JAN. Diseases of the spine and spinal cord. In: Greenfield's neuropathology. London: Edward Arnold Publishing, 1976:668–674.
31. Williams AL, Haughton VM, Pojunas KW, et al. Differentiation of intramedullary neoplasms and cysts by MR. AJR 1987;149:159–164.
32. Lee BCP, Zimmerman RD, Manning JJ, Deck MDF. MR imaging of syringomyelia and hydromyelia. AJNR 1985; 6:221–228.
33. Sherman JL, Barkovich AJ, Citrin CM. The MR appearance of syringomyelia: new observations. AJR 1987;148: 381–391.
34. Guidetti B, Mercuri S, Vagnozzi R. Longterm results of the surgical treatment of 129 intramedullary spinal gliomas. J Neurosurg 1981;54:323–330.
35. Cooper PR, Epstein F. Radical resection of intramedullary spinal cord tumors in adults: recent experience in 29 patients. J Neurosurg 1985;63:492–499.
36. DeSousa AL, Kalsbeck JE. Intraspinal tumors in children: a review of 81 cases. J Neurosurg 1979;51:437–445.
37. Johnson DL, Schwarz S. Intracranial metastases from malignant spinal cord astrocytoma: case report. J Neurosurg 1987;66:621–625.
38. Solomon RA, Stein BM. Unusual spinal cord enlargement related to intramedullary hemangioblastoma. J Neurosurg 1988;68:550–553.
39. Sato Y, Waziri M, Smith W, et al. Hippel-Lindau disease: MR imaging. Radiology 1988;166:241–246.
40. Umbach W, Kunft HD. Vascular tumours of the spinal cord. In: Vinken PJ, Bruin GW, eds. Handbook of clinical neurology. Vol. 20, part 2. Amsterdam: North-Holland, 1976:435–480.
41. Kaffenberger DA, Shah CP, Murtagh FR, et al. MR imaging of spinal cord hemangioblastoma associated with syringomyelia. J Comput Assist Tomogr 1988; 12(3):495–498.
42. Costigan DA, Winkelman MD. Intramedullary spinal cord metastasis: a clinicopathological study of 13 cases. J Neurosurg 1985;62:227–233.
43. Chason JL, Walker FB, Landers JW. Metastatic cancer in CNS and dorsal root ganglia. Cancer 1963;16:781–787.
44. Reddy SC, Vijayamohan G, Rao GR. Delayed CT myelography in spinal intramedullary metastasis. J Comput Assist Tomogr 1984;8(6):1182–1185.
45. Post MJD, Quencer RM, Green BA, et al. Intramedullary spinal cord metastases, mainly of nonneurogenic origin. AJR 1987;148:1015–1022.
46. Puljic S, Batnitzky S, Hilton P, et al. Metastases to the

medulla of the spinal cord: myelographic features. Radiology 1975;117:89–91.
47. Honig LS, Sheremata WA. Magnetic resonance imaging of spinal cord lesions in multiple sclerosis. J Neurol Neurosurg Psychiatry 1989;52:459–466.
48. Kermode AG, Tofts PS, et al. Heterogeneity of blood-brain barrier changes in multiple sclerosis: an MR study with gadolinium-DTPA enhancement. Neurology 1990;40:229–235.
49. Nesbit GM, Miller GM, Baker HL, et al. Spinal cord sarcoidosis: a new finding at MR imaging with Gd-DTPA enhancement. Radiology 1989;173:839–843.
50. Masaryk TJ. Spine tumors. In: Modic MT, Masaryk TJ, Ross JS, eds. Magnetic resonance imaging of the spine. Chicago: Year Book Medical Publishers, 1989.
51. Levy WJ, Bay J, Dohn D. Spinal cord meningioma. J Neurosurg 1982;57:804–812.
52. Wang AM, Gall CM, Shillito J, et al. CT demonstration of extradural thoracic meningioma. J Comput Assist Tomogr 1988;12(3):536–538.
53. Wood JB, Wolpert SM. Lumbosacral meningioma. AJNR 1985;6:450–451.
54. Boisserie-Lacroix M, Kien P, Caille JM. Imaging of intradural extramedullary tumors: neurinomas and meningiomas. J Neuroradiol 1987;14:66–81.
55. Zimmerman RD, Fleming CA, Saint-Louis LA, et al. Magnetic resonance imaging of meningiomas. AJNR 1985;6:149–157.
56. Epstein BS. The spine radiological text and atlas. 2nd ed. Philadelphia: Lea & Febiger, 1962:540.
57. Scotti G, Scialfa G, Columbo N, Landoni L. MR imaging of intradural extramedullary tumors of the cervical spine. J Comput Assist Tomogr 1985;9(6):1037–1041.
58. Schroth G, Thron A. Magnetic resonance imaging of spinal meningiomas and neurinomas: improvement of imaging by paramagnetic contrast enhancement. J Neurosurg 1987;66:695–700.
59. Bydder GM, Kingsley DPE, Brown J, et al. MR imaging of meningiomas including studies with and without gadolinium-DTPA. J Comput Assist Tomogr 1985;9(4):690–697.
60. Som PM, Sacher M, Strenger SW, et al. "Benign" metastasizing meningiomas. AJNR 1987;8:127–130.
61. Dastur KJ, Raji MR, Smith WI. Pulmonary metastasis from intraspinal meningioma. AJNR 1984;5:483–484.
62. Castillo M, Quencer RM, Green BA, Montalvo BM. Syringomyelia as a consequence of compressive extramedullary lesions: postoperative clinical and radiological manifestations. AJNR 187;8:973–978.
63. Lombardi G, Passerini A. Spinal cord tumors. Radiology 1961;76:381–391.
64. Yaghmai I. Spine changes in neurofibromatosis. Radiographics 1986;6(2):261–285.
65. Kornberg M, Rechtine GR, Dupy TE. Thoracic vertebral erosion secondary to an intrathoracic meningocele in a patient with neurofibromatosis. Spine 1984;9(8):821–824.
66. Wang AM, Jochim CL, Shillito J, et al. Cervical chordoma presenting with intervertebral foramen enlargement mimicking neurofibroma: CT findings. J Comput Assist Tomogr 1984;8(3):529–532.
67. So CB, Li KB. Anterolateral cervical meningocele in association with neurofibromatosis: MR and CT studies. J Comput Assist Tomogr 1989;13(4):692–695.
68. Nakagawa H, Huang YP. Computed tomography of intraspinal and paraspinal neoplasms. J Comput Assist Tomogr 1977;1:377–389.
69. Lapointe JS, Graeb BA, Nugent RA, Robertson WD. Value of intravenous contrast enhancement in the CT evaluation of intraspinal tumors. AJNR 1985;6:939–943.
70. Maiuri F, Corriero G. Lateral thoracic meningocele. Surg Neurol 1986;26:409–412.
71. Weeks RG, Berquist TH, McLeod RA, Zimmer WD. Magnetic resonance imaging of soft tissue tumors: comparison with computed tomography. Magnet Reson Imag 1985;3:345–352.
72. Ramsey RG, Zacharias CE. MR imaging of the spine after radiation therapy: easily recognizable effects. AJR 1985;144:1131–1135.
73. Burk DL, Brunberg JA, Kanal E, et al. Spinal and paraspinal neurofibromatosis: surface coil MR imaging at 1.5 T. Radiology 1987;162:797–801.
74. Lewis TT, Kingsley DPE. Magnetic resonance imaging of multiple spinal neurofibromatoma–neurofibromatosis. Neuroradiology 1987;29:562–564.
75. Levine E, Huntrakoon M, Wetzel LH. Malignant nerve sheath neoplasms in neurofibromatosis: distinction from benign tumors by using imaging techniques. AJR 1987;149:1059–1064.
76. Coleman BG, Arger PH, Dalinka MK, et al. CT of sarcomatous degeneration in neurofibromatosis. AJR 1983; 140:383–387.
77. Mayer JS, Kulkarni MV, Yeakley JW. Craniocervical manifestations of neurofibromatosis: MR versus CT studies. J Comput Assist Tomogr 1987;11(5):839–844.
78. McLone DG, Naidich TP. Laser resection of 50 spinal lipomas. Neurosurgery 1986;18(5):611–615.
79. Naidich TP, McLone DG. Congenital pathology of the spine and spinal cord. In: Taveras JM, Ferrucci JT, eds. Radiology Diagnosis-Imaging-Intervention, Vol. 3: Neuroradiology and Radiology of the Head and Neck. Philadelphia: Lippincott, 1990.
80. Naidich TP, McLone DG, Mutluer S. A new understanding of dorsal dysraphism with lipoma (lipomyeloschisis): radiologic evaluation and surgical correction. AJR 1983;140:1965–1078.
81. Brunberg JA, Latchaw RE, Kanal E, et al. Magnetic resonance imaging of spinal dysraphism. Radiol Clin North Am 1988;26(2):181–198.
82. Neblett CR. Comment on laser resection of 50 spinal lipomas. Neurosurgery 1986;18:615.
83. Wood BP, Harwood-Nash DC, Berger P, Goske M. Intradural spinal lipoma of the cervical cord. AJR 1985;145: 174–176.
84. Schroeder S, Lackner K, Weiand G. Lumbosacral intradural lipoma. J Comput Assist Tomogr 1981;5(2): 274.
85. Wippold FJ, Citrin C, Barkovich AJ, Sherman JS. Evaluation of MR in spinal dysraphism with lipoma: comparison with metrizamide computed tomography. Pediatr Radiol
86. Barnes PD, Lester PD, Yamanashi WS, Prince JR. Magnetic resonance imaging in infants and children with spinal dysraphism. AJNR 1986;7:465–472.
87. Suojanen J, Wang AM, Winston KR. Pantopaque mimicking spinal lipoma: MR pitfall. J Comput Assist Tomogr 1988;12(2):346–348.
88. Smoker WRK, Moore SA, Beck DW, Hart MN. Intradural spinal teratoma: case report and review of the literature. AJNR 1986;7:905–910.
89. Monajati A, Spitzer RM, Wiley JL, Heggeness L. MR imaging of a spinal teratoma. J Comput Assist Tomogr 1986;10(2):307–310.

90. Dorwart RH, Wara WM, Norman D, Levin VA. Complete myelographic evaluation of spinal metastases from medulloblastoma. Radiology 1981;139:403–408.
91. Deutsch M, Reigel DH. The value of myelography in the management of childhood medulloblastoma. Cancer 1980;45:2194–2197.
92. Hoffman SM, Baker TW, El Gammal T, Brooks BS. Hematogenous spinal leptomeningeal metastasis: a unique CT enhancement pattern. Pediatr Radiol 1987; 17:147–148.
93. Krol G, Sze G, Malkin M, Walker R. MR of cranial and spinal meningeal carcinomatosis: comparison with CT and myelography. AJNR 1988;9:709–714.
94. Davis PC, Friedman NC, Fry SM, et al. Leptomeningeal metastasis: MR imaging. Radiology 1987;163:449–454.
95. Dillon WP, Norman D, Hans Newton T, Bolla K, Mark A. Intradural spinal cord lesions: Gd-DTPA–enhanced MR imaging. Radiology 1989;170:229–237.
96. Berns DH, Blaser S, Ross JS, Masaryk TJ, Modic MT. MR imaging with Gd-DTPA in leptomeningeal spread of lymphoma. J Comput Assist Tomogr 1988;12(3):499–500.
97. Sze G, Abramson A, Krol G, et al. Gadolinium-DTPA in the evaluation of intradural extramedullary spinal disease. AJNR 1988;9:153–163.
98. Lim V, Sobel DF, Zyroff J. Spinal cord pial metastases: MR imaging with gadopentetate dimeglumine. AJNR 1990;11:975–982.
99. Barloon TJ, Yuh WTC, Yang CKC, Schultz DH. Spinal subarachnoid tumor seeding from intracranial metastasis: MR findings. J Comput Assist Tomogr 1987;11 (2):242–244.
100. Erlich SS, Davis RL. Spinal subarachnoid metastasis from primary intracranial glioblastoma multiforme. Cancer 1978;42:2854–2864.
101. Paillas JE, Alliez B, Pellet W. Primary and secondary tumors of the spine. In: Vinken PJ, Bruin GW, eds. Handbook of clinical neurology. Vol. 20, part 2. Amsterdam: North-Holland, 1976:19–54.
102. Redmond J, Spring DV, Munderloh SH, et al. Spinal computed tomography scanning in the evaluation of metastatic disease. Cancer 1984;54:253–258.
103. McNeil BJ. Rationale for the use of bone scans in selected metastatic and primary bone tumors. Semin Nucl Med 1978;8(4):336–345.
104. Miller F, Whitehill R. Carcinoma of the breast metastatic to the skeleton. Clin Orthop 1984;184:121–127.
105. Kricun ME. Red-yellow marrow conversion: its effect on the location of some solitary bone lesions. Skeletal Radiol 1985;14:10–19.
106. Schocker JD, Brady LW. Radiation therapy for bone metastasis. Clin Orthop 1982;169:38–43.
107. Roberts JL, Magid D. Case report 493. Skeletal Radiol 1988;17:375–377.
108. Gilbert RW, Kim JH, Posner JB. Epidural spinal cord compression from metastatic tumor: diagnosis and treatment. Ann Neurol 1978;3:40–51.
109. Fisher MS. Lumbar spine metastasis in cervical carcinoma: a characteristic pattern. Radiology 1980;134: 631–634.
110. Mettler FA, Guiberteau MJ. Essentials of nuclear medicine imaging. 2nd ed. New York: Grune & Stratton, 1986.
111. Dahlin DC, Unni KK. Bone tumors: general aspects and data on 8542 cases. 4th ed. Springfield, Ill.: Charles C. Thomas, 1986.
112. Muindi J, Coombes RC, Golding S, et al. The role of computed tomography in the detection of bone metastases in breast cancer patients. Br J Radiol 1983; 56:233–236.
113. Algra PR, Bloem JL. Detection of vertebral metastases: comparison between MR imaging and bone scintigraphy. Radiographics 1991;11:219–232.
114. Smoker WRK, Godersky JC, Knutzon RK, et al. The role of MR imaging in evaluating metastatic spinal disease. AJNR 1987;8:901–908.
115. Resjo IM, Harwood-Nash DC, Fitz CR, Chuang S. CT metrizamide myelography for intraspinal and paraspinal neoplasms in infants and children. AJR 1979;132:367–372.
116. Kamholtz R, Sze G. MRI of spinal metastases. MRI Decisions November 1990:2–15.
117. Daffner RH, Lupetin AR, Dash N, et al. MRI in the detection of malignant infiltration of bone marrow. AJR 1986;146:353–358.
118. Sze G, Krol G, Zimmerman RD, Deck MDF. Malignant extradural spinal tumors: MR imaging with Gd-DTPA. Radiology 1988;167:217–223.
119. Solomon A, Rahamani R, Seligsohn U, Ben-Artzi F. Multiple myeloma: early vertebral involvement assessed by computerized tomography. Skeletal Radiol 1984;11: 258–261.
120. Schreiman JS, McLeod RA, Kyle RA, Beabout JW. Multiple myeloma: evaluation by CT. Radiology 1985; 154:483–486.
121. Helms CA, Genant HK. Computed tomography in the early detection of skeletal involvement with multiple myeloma. JAMA 1982;248:2886–2887.
122. Edelman RR, Kaufman H, Kolodny GM. Case report 350. Skeletal Radiol 1986;15:160–163.
123. Resnick D, Greenway GD, Bardwick PA, et al. Plasma-cell dyscrasia with polyneuropathy, organomegally, endocrinopathy, M-protein, and skin changes: the POEMS syndrome. Radiology 1981;140:17–22.
124. Fruehwald FXJ, Tscholakoff D. Magnetic resonance imaging of the lower vertebral column in patients with multiple myeloma. Invest Radiol 1988;23:193–199.
125. Rogalsky RJ, Black GB, Reed MH. Orthopaedic manifestations of leukemia in children. J Bone Joint Surg 1986;68A:494–501.
126. Schabel SI, Tyminski L, Holland RD, Rittenburg GM. The skeletal manifestations of chronic myelogenous leukemia. Skeletal Radiol 1980;5:145–149.
127. Nixon GW, Gwinn JL. The roentgen manifestations of leukemia in infancy. Radiology 1973;107:603–609.
128. De Boeck M, Peeters O, Van Camp B, et al. Monocytic leukaemia associated with myeloid metaplasia resembling metastatic bone disease. Skeletal Radiol 1984; 11:9–12.
129. Stork JT, Cigtay OS, Schellinger D, Jacobson RJ. Recurrent chloromas in acute myelogenous leukemia. AJR 1984;142:778–778.
130. Eelkema E, Johnson DW, Latchaw RE. Multiple spinal granulocytic sarcomas simulating neurofibromatosis. AJNR 1989;10:S42–S44.
131. McAllister MD, O'Leary DH. CT myelography of subarachnoid leukemic infiltration of the lumbar thecal sac and lumbar nerve roots. AJNR 1987;8:568–569.
132. Pomeranz SJ, Hawkins HH, Towbin R, et al. Granulocytic sarcoma (chloroma): CT manifestations. Radiology 1985;155:167–170.
133. Khardori N, McCredie K, Bodey GP. Brown-Sequard syndrome secondary to soft tissue infection in a pa-

tient with acute lymphocyte leukemia. Neuroradiology 1988;30:348–350.
134. Pagani JJ, Libshitz HI, Wallace S, Hayman LA. Central nervous system leukemia and lymphoma: computed tomographic manifestations. AJR 1981;137:1195–1201.
135. Olson DO, Shields AF, Scheurich CJ, et al. Magnetic resonance imaging of the bone marrow in patients with leukemia, aplastic anemia and lymphoma. Invest Radiol 1986;21:540–546.
136. McKinstry CS, Steiner RE, Young AT, et al. Bone marrow in leukemia and aplastic anemia: MR imaging before, during, and after treatment. Radiology 1987;162:701–707.
137. Edeiken-Monroe B, Edeiken J, Kim EE. Radiologic concepts of lymphoma of bone. Radiol Clin North Am 1990;28:841–864.
138. Grossman H, Winchester PH, Bragg DG, et al. Roentgenographic changes in childhood Hodgkin's disease. AJR 1970;108(2):354–364.
139. Beachley MC, Lau BP, King ER. Bone involvement in Hodgkin's disease. AJR 1972;114(3):559–563.
140. Braunstein DM, White SJ. Non-Hodgkin lymphoma of bone. Radiology 1980;135:59–63.
141. Braunstein DM. Hodgkin disease of bone: radiographic correlation with the histological classification. Radiology 1980;137:643–646.
142. Beres J, Pech P, Berns TF, et al. Spinal epidural lymphomas: CT features in seven patients. AJNR 1986;7:327–328.
143. Holtas SL, Kido DK, Simon JH. MR imaging of spinal lymphoma. J Comp Assist Tomogr 1986;10(1):111–115.
144. Weaver GR, Sandler MP. Increased sensitivity of magnetic resonance imaging compared to radionuclide bone scintigraphy in the detection of lymphoma of the spine. Clin Nucl Med 1987;12:333–334.
145. Greco A, Jelliffe AM, Maher EJ, Leung AWL. MR imaging of lymphomas: impact on therapy. J Comp Assist Tomogr 1988;12(5):785–791.
146. Williams AL, Haughton VM, Daniels DL, Grogan JP. Differential CT diagnosis of extruded nucleus pulposus. Radiology 1983;148:141–148.
147. Shives TC, Dahlin DC, Sim FH, et al. Osteosarcoma of the spine. J Bone Joint Surg 1986;68A:660–668.
148. Sundaresan N, Huvos AG, Krol G, et al. Post radiation sarcoma involving the spine. Neurosurgery 1986;18:721–724.
149. Miled KB, Siala M, Hamza KR, et al. Primary osteosarcoma of the cervical spine: one case. J Neuroradiol 1988;15:294–300.
150. Boyko OB, Cory DA, Kohen MD, et al. MR imaging of osteogenic and Ewing's sarcoma. AJR 1987;148:317–322.
151. Beltran J, Noto AM, Chakeres DW, Christoforidis AJ. Tumors of the osseous spine: staging with MR imaging versus CT. Radiology 1987;162:565–569.
152. Aisen AM, Martel W, Braunstein EM, et al. MRI and CT evaluation of primary bone and soft tissue tumors. AJR 1986;146:749–756.
153. Bloem JL, Taminiau AHM, Eulderink F, et al. Radiologic staging of primary bone sarcoma: MR imaging, scintigraphy, angiography, and CT correlated with pathologic examination. Radiology 1988;169:805–810.
154. Zimmer WD, Berquist TH, McLeod RA, et al. Bone tumors: magnetic resonance imaging versus computed tomography. Radiology 1985;155:709–718.
155. Sundaram M, McGuire MH, Herbold DR. Magnetic resonance imaging of osteosarcoma. Skeletal Radiol 1987;16:23–29.
156. Pilepich MV, Vietti TJ, Nestbit ME, et al. Ewing's sarcoma of the vertebral column. Int J Radiat Oncol Biol Phys 1981;7:27–31.
157. Scheithauer BW, Egbert BM. Ewing's sarcoma of the spinal epidural space: report of two cases. J Neurol Neurosurg Psychiatry 1978;41:1031–1035.
158. Kornberg M. Primary Ewing's sarcoma of the spine: a review and case report. Spine 1986;11(1):54–57.
159. Frouge C, Vanel D, Coffre C, et al. The role of magnetic resonance imaging in the evaluation of Ewing's sarcoma. Skeletal Radiol 1988;17:387–392.
160. Vanel D, Contesso G, Couanet D, et al. Computed tomography in the evaluation of 41 cases of Ewing's sarcoma. Skeletal Radiol 1982;9:8–13.
161. Camins MB, Duncan AW, Smith J, Marcove RC. Chondrosarcoma of the spine. Spine 1978;3(3):202–209.
162. Hermann G, Wagner LD, Klein MJ, et al. Case report 541. Skeletal Radiol 1989;18:241–244.
163. Mayes CB, Wallace S, Bernardino ME. Computed tomography of chondrosarcoma. CT 1981;5:345–348.
164. Oot RF, Melville GE, New PFJ, et al. The role of MR and CT in evaluating clival chordomas and chondrosarcomas. AJNR 1988;9:715–723.
165. de Lange EE, Pope TL, Fechner RE. Dedifferentiated chondrosarcoma: radiographic features. Radiology 1986;160:489–492.
166. Smith J, Ludwig RL, Marcove RC. Sacrococcygeal chordoma: a clinicoradiological study of 60 patients. Skeletal Radiol 1987;16:37–44.
167. Utne JR, Pugh DG. The roentgenologic aspects of chordoma. AJR 1955;74:593–608.
168. Sundaresan N, Galicich JH, Chu FCH, Huvos AG. Spinal chordomas. J Neurosurg 1979;50:312–319.
169. Firooznia H, Golimbu C, Rafii M, et al. Computed tomography of spinal chordomas. CT 1986;10:45–50.
170. Firooznia H, Pinto RS, Lin JP, Baruch HH, Zausner J. Chordoma: radiologic evaluation of 20 cases. AJR 1976;127:797–805.
171. Krol G, Sundaresan N, Deck M. Computed tomography of axial chordomas. J Comput Assist Tomogr 1983;7(2):286–289.
172. de Bruine FT, Kroon HM. Spinal chordoma: radiologic features in 14 cases. AJR 1988;150:861–863.
173. Meyer JE, Lepke RA, Lindfors KK, et al. Chordomas: their CT appearance in the cervical, thoracic and lumbar spine. Radiology 1984;153:693–696.
174. Yuh WTC, Lozano RL, Flickinger FW, et al. Lumbar epidural chordoma: MR findings. J Comput Assist Tomogr 1989;13(3):508–510.
175. Rosenthal DI, Scott JA, Mankin HJ, et al. Sacrococcygeal chordoma: magnetic resonance imaging and computed tomography. AJR 1985;145:143–147.
176. Sze G, Uichanco LS, Brant-Zawadzki MN, et al. Chordomas: MR imaging. Radiology 1988;166:187–191.
177. Yuh WTC, Flickinger FW, Barloon TJ, Montgomery WJ. MR imaging of unusual chordomas. J Comput Assist Tomogr 1988;12(1):30–35.
178. Oot RF, Melville GE, New PFJ, et al. The role of MR and CT in evaluating clival chordomas and chondrosarcomas. AJNR 1988;9:715–723.
179. Whelan MA. Computed tomography of the sacrum. In: Donovan-Post MR, ed. Computed tomography of the spine. Baltimore: Williams & Wilkins, 1980:94–118.
180. Resnick CS, Young JWR, Levine AM, Aisner SC. Case report 544. Skeletal Radiol 1989;18:303–305.

181. Krol G, Sze G, Arbit E, Marcove R, Sundaresan N. Intradural metastases of chordoma. AJNR 1989;10:193–195.
182. McLeod RA, Dahlin DC, Beabout JW. The spectrum of osteoblastoma. AJR 1976;126(2):321–335.
183. Janin Y, Epstein JA, Carras R, Kahn A. Osteoid osteomas and osteoblastomas of the spine. Neurosurgery 1981;8:31–38.
184. Epstein N, Benjamin V, Pinto R, Budzilovich G. Benign osteoblastoma of a thoracic vertebra. J Neurosurg 1980;53:710–713.
185. Azouz EM, Kozlowski K, Marton D, et al. Osteoid osteoma and osteoblastoma of the spine in children. Pediatr Radiol 1986;16:25–31.
186. Kirwan EO, Hutton TAN, Pozo JL, Ransford AO. Osteoid osteoma and benign osteoblastoma of the spine: clinical presentation and treatment. J Bone Joint Surg 1984;66B:21–26.
187. Wedge JH, Tchang S, Macfadyen DJ. Computed tomography in localization of spinal osteoid osteoma. Spine 1981;6:423–427.
188. Mahboubi S. CT appearance of nidus in osteoid osteoma versus sequestration in osteomyelitis. J Comput Assist Tomogr 1986;10(3):457–459.
189. Omojola MF, Fox AJ, Vinuela FV. Computed tomographic metrizamide myelography in the evaluation of thoracic spinal osteoblastoma. AJNR 1982;3:670–673.
190. Vade A, Wilbur A, Pudlowski R, Ghosh L. Case report 566. Skeletal Radiol 1989;18:475–480.
191. Abdelwahab IF, Frankel VH, Klein MJ. Case report 351. Skeletal Radiol 1986;15:164–169.
192. Sartoris DJ, Parker BR. Histiocytosis X: rate and pattern of resolution of osseous lesions. Radiology 1984; 152:679–684.
193. Graif M, Pennock JM. MR imaging of histiocytosis X in the central nervous system. AJNR 1986;7:21–23.
194. David R, Oria RA, Kumar R, et al. Radiologic features of eosinophilic granuloma of bone. AJR 1989;153:1021–1026.
195. Sanchez RL, Llovet J, Moreno A, Galito E. Symptomatic eosinophilic granuloma of the spine: report of two cases and review of the literature. Orthopedics 1984;7 (11):1721–1726.
196. Seimon LP. Eosinophil granuloma of the spine. J Pediatr Orthop 1981;1:371–376.
197. Ippolito E, Farsetti P, Tudisco C. Vertebra plana. J Bone Joint Surg 1984;66A:1364–1368.
198. Green NE, Robertson WW, Kilroy AW. Eosinophilic granuloma of the spine with associated neural deficit. J Bone Joint Surg 1980;62A:1198–1202.
199. Haggstrom JA, Brown JC, Marsh PW. Eosinophilic granuloma of the spine: MR demonstration. J Comput Assist Tomogr 1988;12(2):344–345.
200. Smith J, Wixon D, Watson RC. Giant cell tumor of the sacrum: clinical and radiologic features in 13 patients. J Can Assoc Radiol 1979;30:34–39.
201. Dahlin DC. Giant cell tumor of bone: highlights of 407 cases. AJR 1985;144:955–960.
202. Wuisman P, Roessner A, Harle A, et al. Case report 503. Skeletal Radiol 1989;17:592–597.
203. Schwimmer SR, Bassett LW, Mankooso AA, et al. Giant cell tumor of the cervicothoracic spine. AJR 1981; 136:63–67.
204. McDonald DJ, Sim FH, McLeod RA, Dahlin DC. Giant-cell tumor of bone. J Bone Joint Surg 1986;68A:235–242.
205. Di Lorenzo N, Spallone A, Nolletti A, Nardi P. Giant cell tumors of the spine: a clinical study of six cases with emphasis on the radiological features, treatment and follow-up. Neurosurgery 1980;6:29–34.
206. Edwards CC. The diagnosis and surgical management of spinal tumors using computed tomography. In: Donovan-Post MJ, ed. Computed tomography of the spine. Baltimore: Williams & Wilkins, 1984:704–737.
207. Brady TJ, Gebhardt MC, Pykett IL, et al. NMR imaging of forearms in healthy volunteers and patients with giant-cell tumor of bone. Radiology 1982;144:549–552.
208. Seider MJ, Rich TA, Ayala AG, Murray JA. Giant cell tumors of bone: treatment with radiation therapy. Radiology 1986;161:537–540.
209. Hay MC, Paterson D, Taylor TKF. Aneurysmal bone cysts of the spine. J Bone Joint Surg 1978;60B:406–411.
210. Ameli NO, Abbassioun K, Saleh H, Eslamdoost A. Aneurysmal bone cysts of the spine: report of 17 cases. J Neurosurg 1985;63:685–690.
211. Munk PL, Helms CA, Holt RG, et al. MR imaging of aneurysmal bone cyst. AJR 1989;153:99–101.
212. Hudson TM. Fluid levels in aneurysmal bone cyst: a CT feature. AJR 1984;141:1001–1004.
213. Haney P, Gellad F, Swartz J. Aneurysmal bone cyst of the spine: computed tomographic appearance. J Comput Assist Tomogr 1983;7:319–322.
214. Cory DA, Fritsch SA, Cohen M, et al. Aneurysmal bone cysts: imaging findings and embolotherapy. AJR 1989; 153:369–373.
215. Capanna R, Van Horn JR, Biagini R, Puggieri P. Aneurysmal bone cyst of the sacrum. Skeletal Radiol 1989; 18:109–113.
216. Beltran J, Simon DC, Levy M, et al. Aneurysmal bone cysts: MR imaging at 1.5 T. Radiology 1986;158:689–690.
217. Zimmer WD, Berquist TH, Sim FH, et al. Magnetic resonance imaging of aneurysmal bone cyst. Mayo Clin Proc 1984;59:633–636.
218. Daffner RH, Linetsky L, Zabkar JH. Case report 433. Skeletal Radiol 1987;16:428–432.
219. Laredo JD, Reizine D, Bard M, Merland JJ. Vertebral hemangiomas: radiologic evaluation. Radiology 1986; 161:183–189.
220. Mohan V, Gupta SK, Tuli SM, Sanyal B. Symptomatic vertebral hemangiomas. Clin Radiol 1980;31:575–579.
221. Baker ND, Greenspan A, Neuwirth M. Symptomatic vertebral hemangiomas: a report of four cases. Skeletal Radiol 1986;15:458–463.
222. Schnyder P, Fankhauser H, Mansouri B. Computed tomography in spinal hemangioma with cord compression. Skeletal Radiol 1986;15:372–375.
223. Leehey P, Naseem M, Every P, et al. Vertebral hemangioma with compression myelopathy: metrizamide CT demonstration. J Comput Assist Tomogr 1985;9(5): 985–986.
224. Ross JS, Masaryk TJ, Modic MT, et al. Vertebral hemangiomas: MR imaging. Radiology 1987;165:165–169.
225. Yu R, Brunner R, Rao KCVG. Role of computed tomography in symptomatic vertebral hemangiomas. CT 1984;8:311–315.
226. Hajek PC, Baker LL, Goobar JE, et al. Focal fat deposition in axial bone marrow: MR characteristics. Radiology 1987;162:245–249.
227. Horowitz SW, Fine M, Azar-Kia B. Differentiation of benign versus metastatic vertebral body compression fractures. Neuroradiology 1991;33(Suppl.):114–116.
228. Erlemann R, Sciuk J, Bosse A, et al. Response of osteosarcoma and Ewing's sarcoma to preoperative chemo-

therapy: assessment with dynamic and static MR imaging and skeletal scintigraphy. Radiology 1990;175:791–796.
229. Houang B, Grenier N, Greselle JF, et al. Osteoid osteoma of the cervical spine: misleading features about a case involving the uncinate process. Neuroradiology 1990;31:549–551.
230. Kroon HM, Schurmans J. Osteoblastoma: clinical and radiologic findings in 98 new cases. Radiology 1990;175:783–790.
231. Yuh WTC, Zachar CK, Barloon TJ, et al. Vertebral compression fractures: distinction between benign and malignant causes with MR imaging. Radiology 1989;172:215–218.

CHAPTER / 10

Vascular Disorders

B. C. P. Lee / K. C. V. G. Rao

Vascular lesions of the spine may manifest as intradural or extradural hemorrhage or as infarction of the spinal cord. Such lesions are often associated with vascular malformations of the spinal cord, although a number of other pathologic entities may present with similar clinical manifestations. These lesions may involve the spinal column, the meninges, or the spinal cord.

Of these, the most important are those related to vascular anomalies of the spinal cord and meninges. Vascular anomalies are primarily considered to be congenital in origin. Unlike the craniocerebral vascular anomalies, where aneurysms are common, isolated aneurysms within the spinal vasculature are extremely rare [1]. The more common congenital vascular lesions of the spinal cord include arteriovenous malformations, cavernous hemangiomas, and arteriovenous fistulas. Included among vascular lesions of the spine are certain tumors such as hemangioblastoma and other spinal cord tumors. These may resemble malformations either by virtue of their intrinsic vascularity or because of venous compression.

Venous congestion and dilatation of the epidural and intradural veins may mimic an intraspinal vascular anomaly and result from extrinsic compression by epidural masses, vascular lesions within the vertebral bodies or enlargement of the epidural venous plexus (Batson's plexus) following obstruction in the systemic venous pathways. Clinical manifestations of spinal vascular lesions result from hemorrhage, ischemia, myelomalacia, and cord compression.

Spinal Vascular Malformations

Vascular malformations of the spine are categorized histologically as (*a*) *arteriovenous malformations* (AVM), in which the vessels are arteries and veins which often communicate directly with one another, (*b*) *capillary hemangiomas*, in which the caliber of the abnormal vessels is smaller and only identified on pathologic studies, and (*c*) *cavernous hemangioma*, which consists of dilated venous space.

Arteriovenous malformations (AVMs) and AV fistulas may be differentiated on selective spinal angiography [2–8]. Discrete arteries feed an AVM, whereas a single direct communication between an artery and a vein without an intervening network of malformed vessels is present in an AV fistula. Abnormal vessels are not visible angiographically in capillary and cavernous malformations.

AVMs are usually intramedullary and rarely extramedullary; the latter may be intradural or extradural [2]. The precise location has an important bearing on the surgical management: Extramedullary lesions may be excised, whereas those within the spinal cord are seldom amenable to surgery. It is important to differentiate AV fistulas from AVMs, since nearly all fistulas are extramedullary and are readily curable by embolization or surgical ligation.

Clinical Manifestations

Spastic weakness of the lower and sometimes the upper extremities is the most common presenting symptom of AVMs and fistulas. The weakness is usually progressive, but it may be recurrent and occasionally acute and devastating. Rarely, malformations present as subarachnoid hemorrhages.

It has been postulated that the spastic paraparesis or quadraparesis is secondary to venous congestion from transmission of the arterial pressure directly to the veins. The often dramatic improvement of the symptoms following therapeutic occlusion of a high-flow AV fistula is thought to be related to relief of this congestion [7]. The lesser improvement following extirpation of an AVM may have similarly explanation [6, 8]. Sudden deterioration or weakness may be due to hemorrhage into the spinal cord or subarachnoid space. The increased venous pressure also may account for subarachnoid hemorrhages and less commonly for bleeding into the spinal cord. Whether ischemia plays a role in producing clinical symptoms is unclear.

Although the vessels visible on MRI may be arteries or veins, the latter are more readily detected because they have larger calibers. Veins enlarge because of transmitted arterial pressure in a true arteriovenous malformation or arteriovenous fistula. True venous malformations are probably uncommon, and these, together with capillary or telangiectatic malformations, are diagnosed by the presence of hemorrhage within the spinal cord rather than by the presence of vessels. Venous engorgement and enlargement are common with severe spinal cord compression from intrinsic or extrinsic causes and should not be confused with the enlargement of vascular malformations [8].

Cavernous malformations are occult vascular malformations that have different clinical presentations from AVMs and AV fistulas. These malformations are not under higher arterial pressure and do not cause venous engorgement. They may rarely result in spinal subarachnoid hemorrhage. They often bleed into the spinal cord, but the hemorrhage tends to be insidious rather than catastrophic. Neurologic symptoms are probably the result of such intrinsic hemorrhages, and they may be recurrent.

The management of AVMs and AV fistulas is different from that of occult malformations. Definable arterial feeders to an AVM may be embolized, followed by excision of the malformations, and a fistula may be definitively treated by embolization of the supplying artery. Occult malformations may not require treatment, since they cause less severe neurologic defects and may have a relatively benign course. Treatment, when required, is by surgical excision of the hematoma and the malformation.

Imaging Techniques

The indications for imaging studies are to demonstrate the presence of hemorrhage, to identify if it is associated with a vascular lesion, and to delineate the vascular anatomy. The latter is best achieved with spinal angiography. The purpose of such relatively noninvasive imaging studies as CT and MRI is to demonstrate the presence of hematoma in an acute or subacute setting and the presence of enlarged vascular channels.

CT has very limited application in the demonstration of intradural vascular lesions (Fig. 10.1), since the majority of these are small. Acute hematomas, especially if they are of moderate size, can be detected on CT. IVCT has been utilized in the past to demonstrate the prominent vascular channels within the spinal canal (Fig. 10.2). Intramedullary enhancement, when associated with prominent epidural and intradural vascular structures, is suggestive of spinal cord vascular anomalies; however, a similar appearance may be seen with intradural vascular tumors (hemangioblastoma) or extradural masses compressing the spinal cord (hemangiopericytoma, meningioma, or epidural extension of vertebral hemangioma). MRI should be the primary screening examination in suspected spinal vascular lesions [9–12].

MRI detects hemorrhages and vessels within the spinal cord but is insensitive in visualizing abnormal vessels confined to the subarachnoid and epidural spaces. Artifacts within the spinal canal are common on MRI scans and are often mistaken for vascular malformations within the subarachnoid space (Fig. 10.3). Despite these disadvantages, MRI should still be the initial examination because it is entirely noninvasive.

Myelography is sensitive in detecting vascular malformations on the surface of the spinal cord and subarachnoid space and should always be performed when the MRI is negative. Myelography should be followed by CT scanning. CTM provides essentially the same information but without the artifacts of MRI. Selective spinal angiography is mandatory whenever surgical or embolization therapy is contemplated, since none of the preceding examinations is capable of revealing the detailed anatomy of vascular lesions.

The technique of MRI is standard spin-echo sagittal T1-, T2-weighted sequences, supplemented by axial T1- and T2-weighted images of the appropriate region. Artifact-suppression techniques such as presaturation of cardiac, arterial, and respiratory movements, gradient-motion correction, and phase and frequency wrap corrections should be implemented

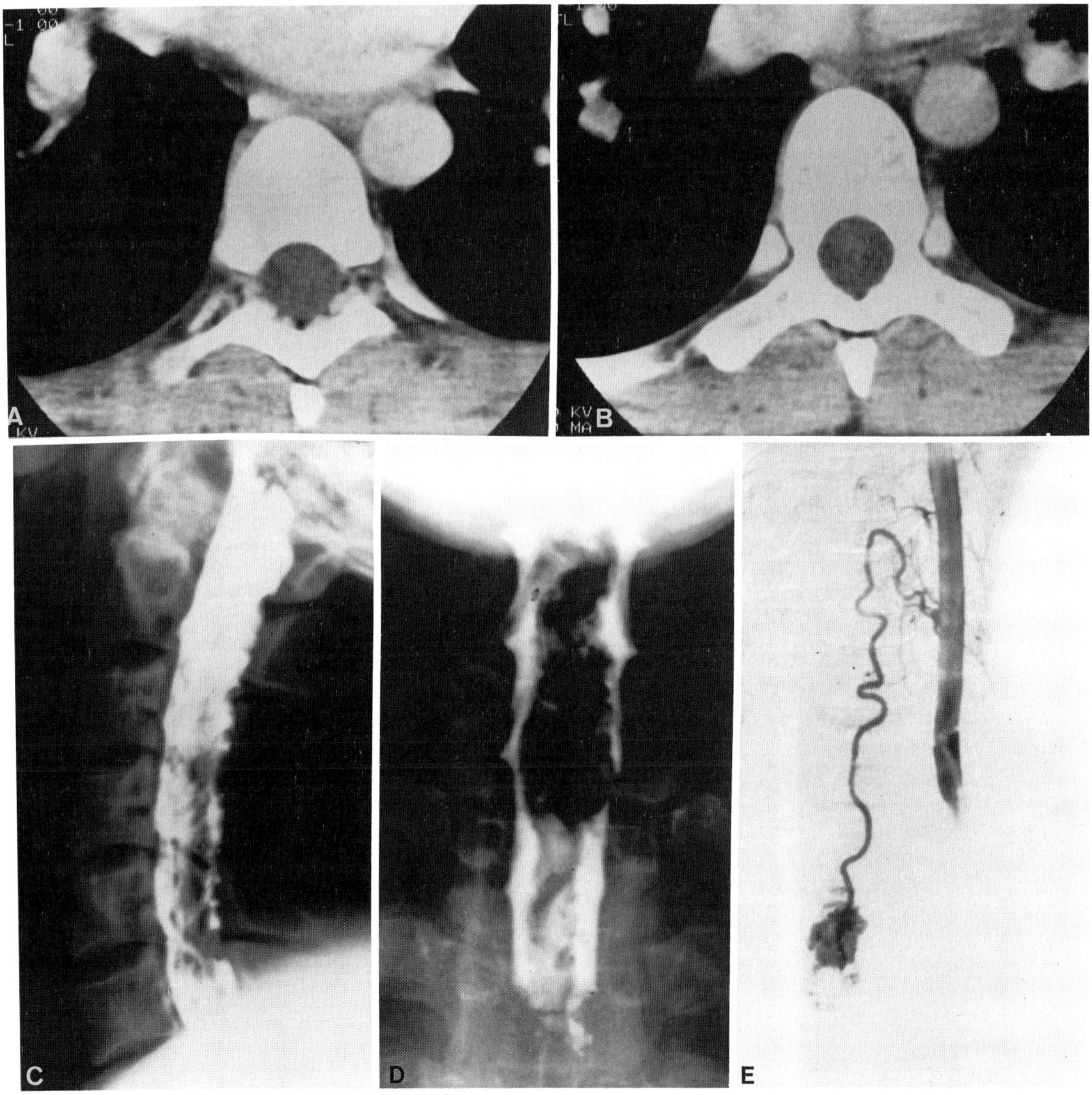

Figure 10.1. Intramedullary vascular anomaly. Non-contrast CT (NCCT) (**A**) and intravenous enhanced CT (IVCT) (**B**). A faint enhancement is noted within the center of the cord (*arrow*). Lateral (**C**) and anteroposterior (**D**) views of the myelogram demonstrate a large irregular filling defect with a serpentine tubular filling defect (*arrowhead*). **E,** Vertebral arteriogram demonstrates a vascular nidus (*arrowhead*) being fed by a tortuous dilated spinal artery.

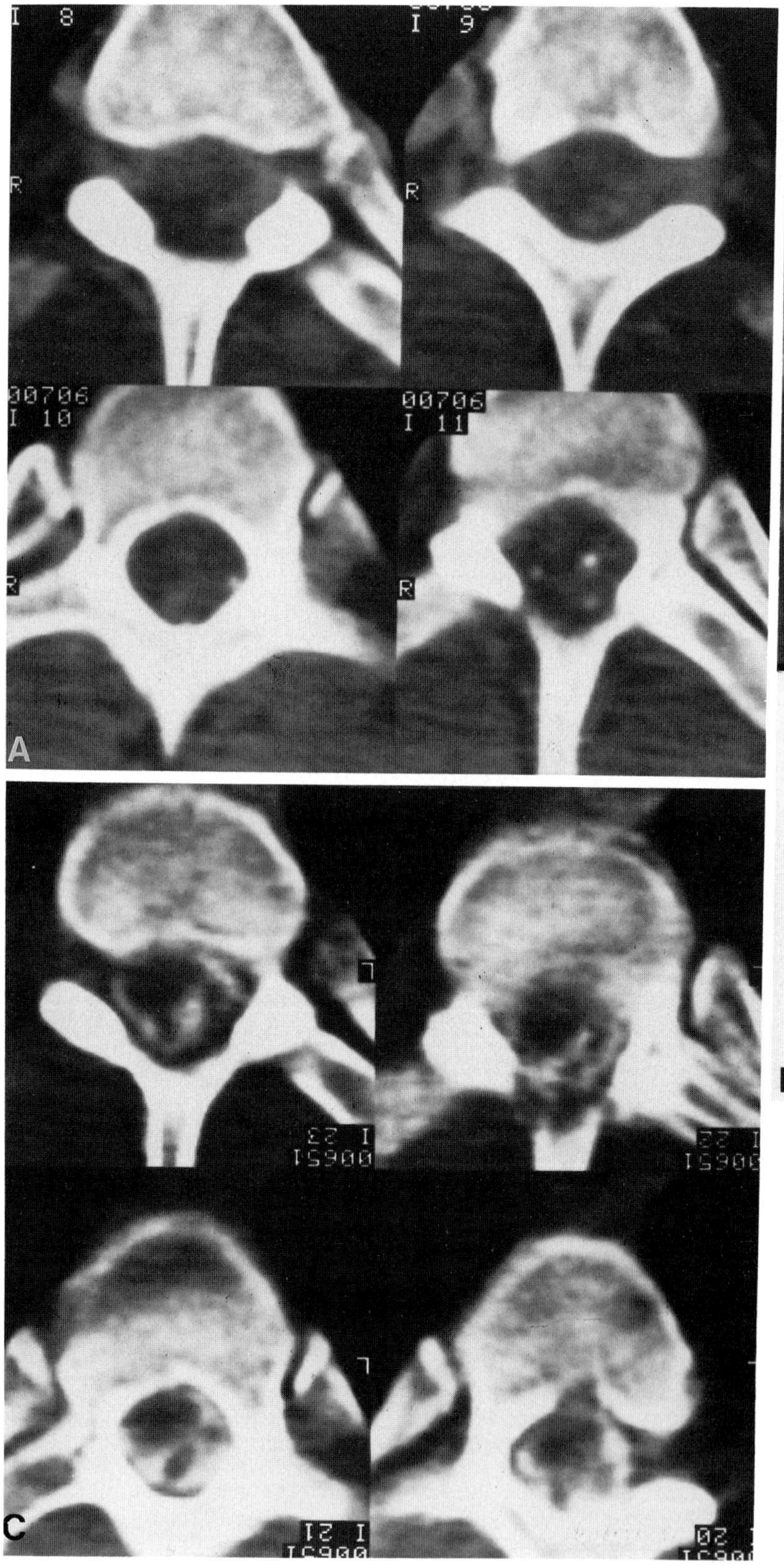

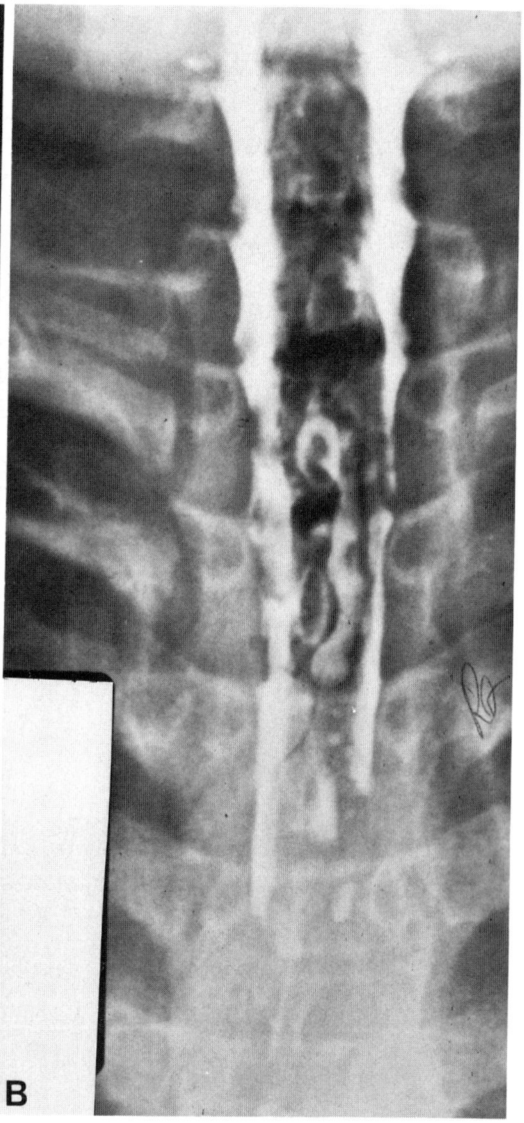

Figure 10.2. Foix-Alajouanine syndrome. A 47-year-old man presented with acute exacerbation of pain and further progression of paraplegia, with a clinical level localized to the midthoracic region (T8–9 root level). **A,** Adjacent CT sections following intravenous contrast show enhancement of the dorsolateral extramedullary compartment with extension into the right neural foramen (*arrowheads*). Few punctate hyperdense foci represent contrast material in vessel (*arrow*). **B,** Frontal view of the myelogram demonstrates serpigenous filling defects with an apparent block. **C,** Contiguous CT following the myelogram demonstrates intradural extramedullary filling defects with epidural component highly suggestive of dilated vessels and indicative of a vascular malformation. **D,** Screening digital subtraction spinal angiography failed to demonstrate the suspected arteriovenous malformation. Patient subsequently underwent surgery. Large dilated thrombosed veins were identified without evidence of arterial feeders.

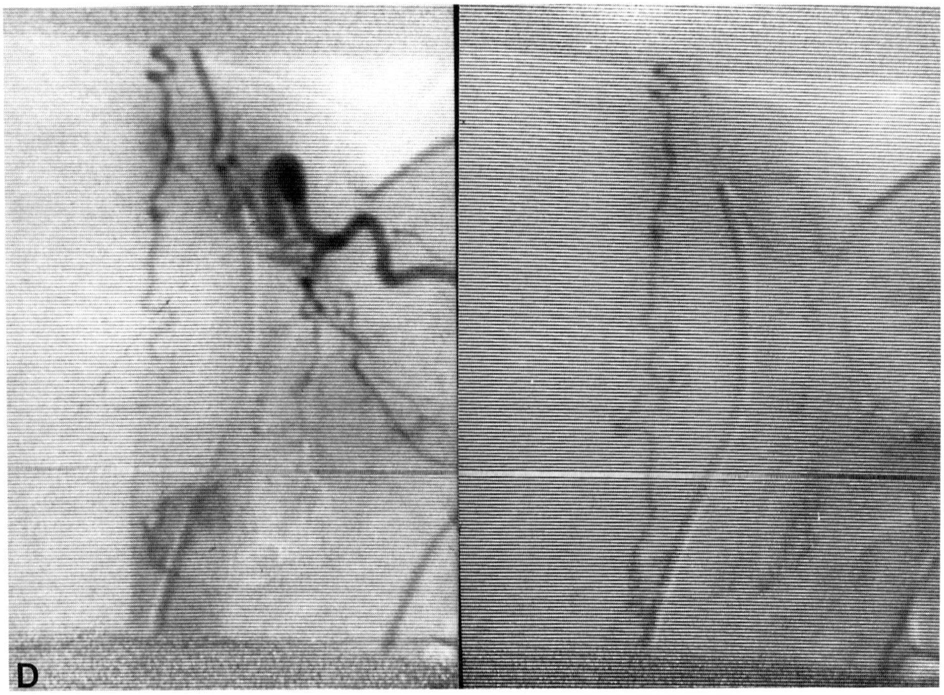

Figure 10.2–*continued*

routinely. Cardiac and respiratory gating may be needed whenever the artifacts are not suppressed by the preceding correction techniques. Gradient-echo sequences are usually helpful in differentiating blood flow from thrombosed vessels; signal void of flowing blood on spin-echo sequences usually shows increased signal of "flow-related enhancement" on the gradient-echo sequences. Intravenous gadolinium should be administered routinely [13, 14].

Diagnosis

MRI is useful as a screening study to detect vascular lesions and determine their locations relative to the spinal cord and subarachnoid and epidural spaces. Hemorrhage and other complications arising from these lesions often can be identified. Cavernous hemangiomas often have a distinct appearance, but it is more difficult to differentiate AVMs from fistulas. Other nonvascular pathologic lesions may give rise to similar appearances and may be confused with intrinsic vascular malformations.

ARTERIOVENOUS MALFORMATIONS

Dilated vessels, thrombosis, hemorrhage, and myelomalacia are often associated with vascular anomalies, on MRI [9–13]. The abnormal vessels are seen as "flow void" on both T1- and T2-weighted images. Most of these vessels are dilated veins, which may appear as varices and may occasionally be aneurysmal in size. It is sometimes possible to detect the arterial feeders and the abnormal vessels constituting the malformation by their smaller calibers (Fig. 10.4). Sagittal and axial views are helpful in determining the locations of the flow void; vessels within the spinal cord usually constitute the malformation, and those in the subarachnoid and epidural spaces are likely to be dilated draining veins [12] (Fig. 10.5). Intradural extramedullary vessels may compress and produce a "scalloping" appearance on the surface of the spinal cord (see Fig. 10.4). The vascular malformation and draining veins do not enhance because of the rapid flow. Thrombosis within the vessels and hemorrhage into the spinal cord that give rise to increased signal on T1-weighted images are the result of methemoglobin (see Fig. 10.5). Signal void due to susceptibility effect is present in the chronic stage. Increased signal within the vessels on contrast-enhance T1-weighted images may be related to flow-related enhancement. Flow void on gradient-echo images differentiates this from intraluminal thrombosis. The appearance of subarachnoid hemorrhage depends on the quantity and age of the blood. Small amounts of blood mixed with cerebrospinal fluid (CSF) may show evidence of increased protein as a mild increase in signal on T1-weighted and proton-density images; larger amounts usually show the typical increased signal of methemoglobin on T1-weighted images (see Fig. 10.5). The signal on T2-weighted images depends on susceptibility; acute blood is dark because of deoxyhemoglobin, whereas subacute blood has a similar signal to cerebrospinal fluid. Chronic blood is always

Figure 10.3. MRI artifact. A 16-year-old girl presented with sudden onset of paraplegia. Sagittal T1-weighted (**A,B**), sagittal T2-weighted (30/1800) and (**C,D**), contiguous T2-weighted (30/2000) axial (**E–H**) sections. Large flow void signal (*arrowheads*) is identified on the axial and sagittal images which was suspected to represent a large intradural AVM. Subsequent spinal angiography proved to be normal. Patient had a gradual but slow recovery.

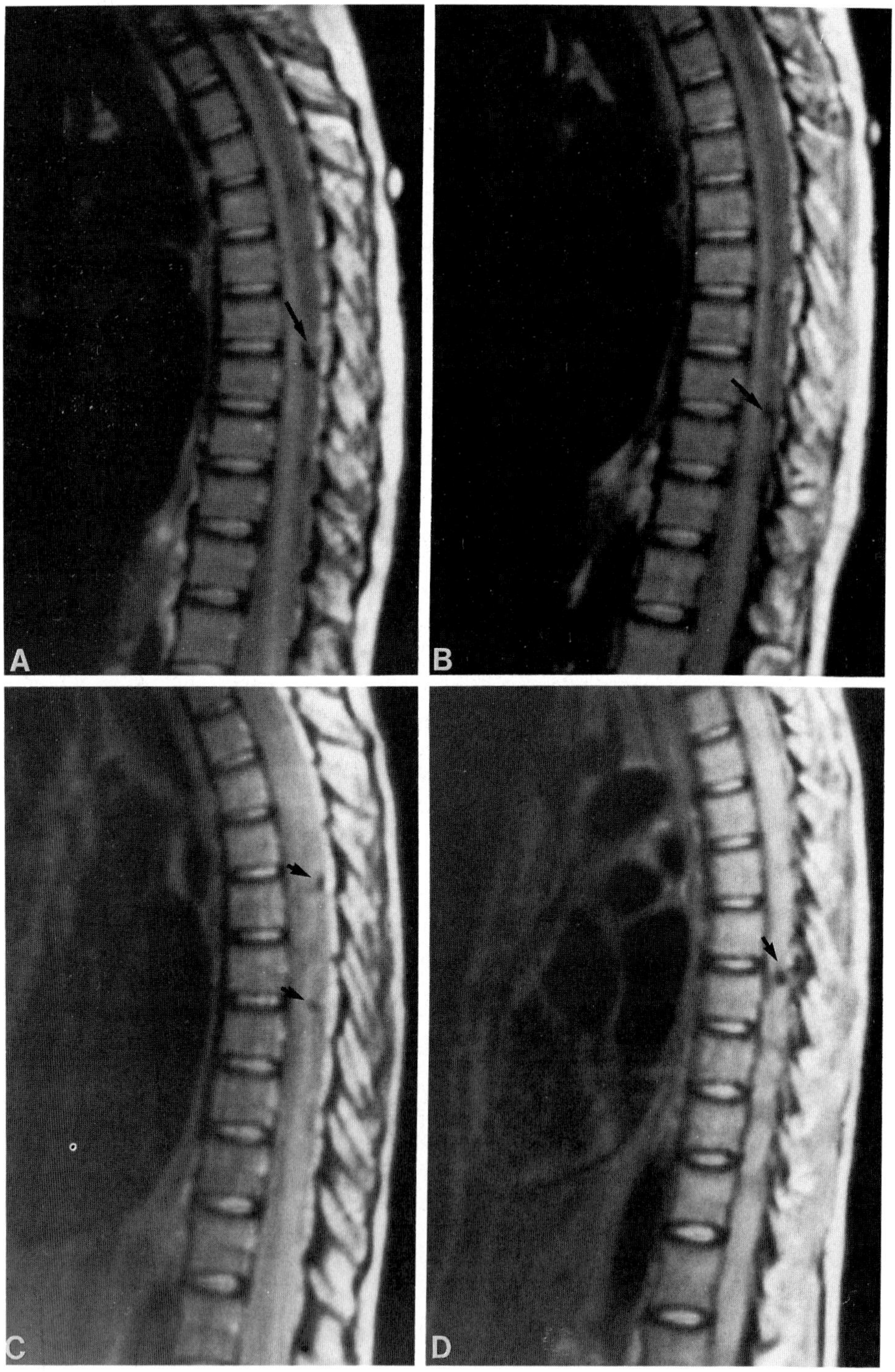

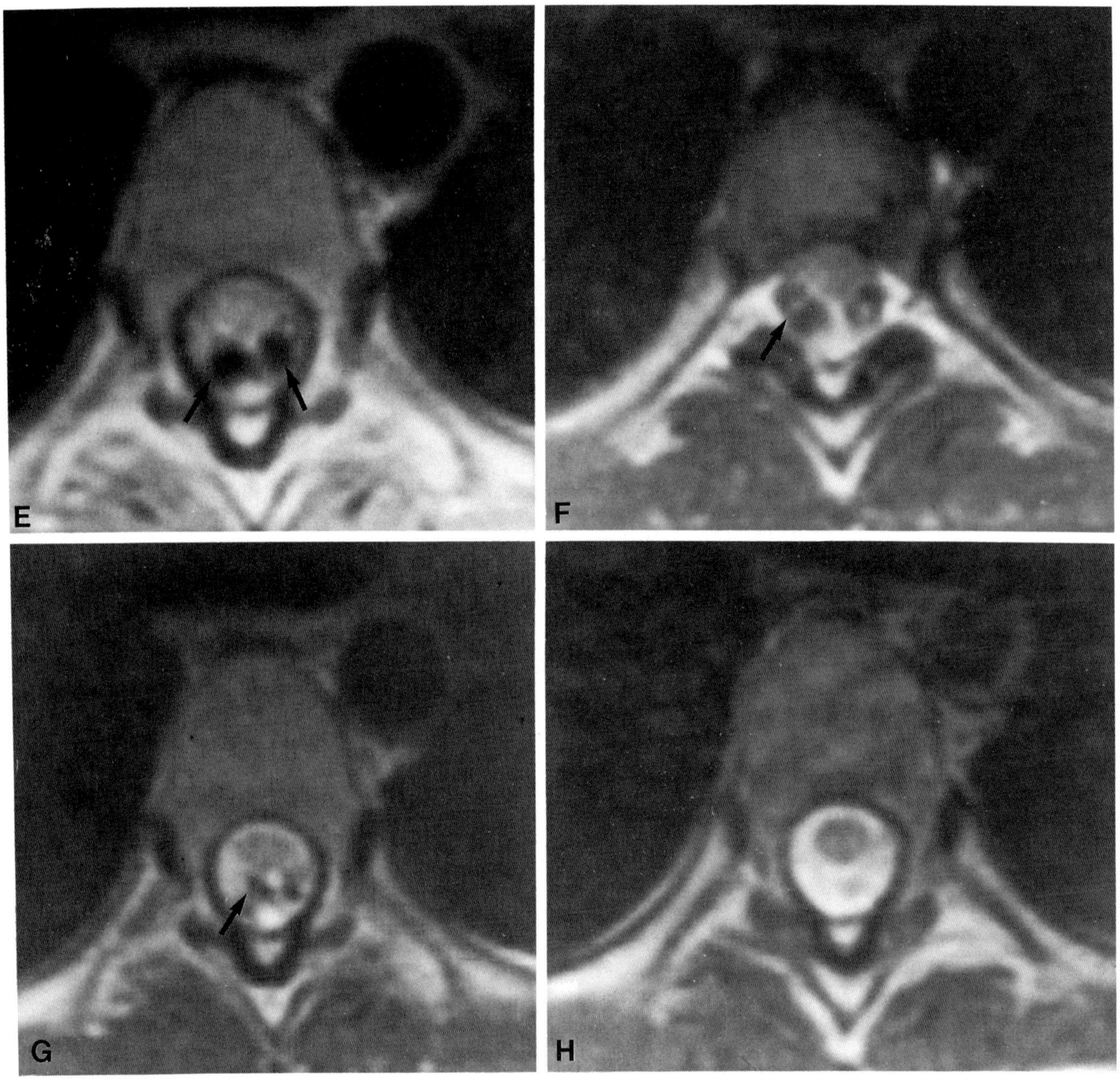

Figure 10.3–*continued*

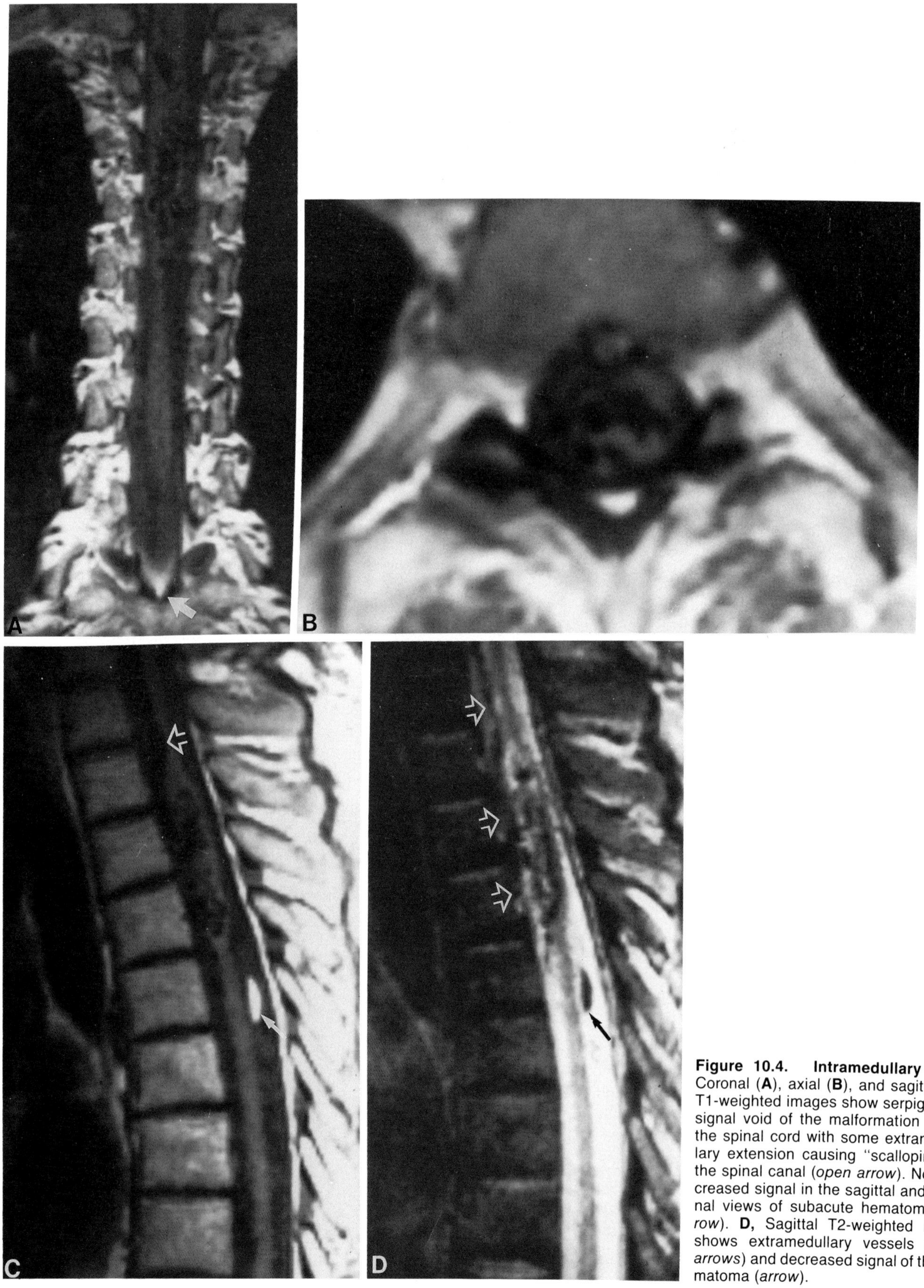

Figure 10.4. Intramedullary AVM. Coronal (**A**), axial (**B**), and sagittal (**C**) T1-weighted images show serpigenous signal void of the malformation within the spinal cord with some extramedullary extension causing "scalloping" of the spinal canal (*open arrow*). Note increased signal in the sagittal and coronal views of subacute hematoma (*arrow*). **D,** Sagittal T2-weighted image shows extramedullary vessels (*open arrows*) and decreased signal of the hematoma (*arrow*).

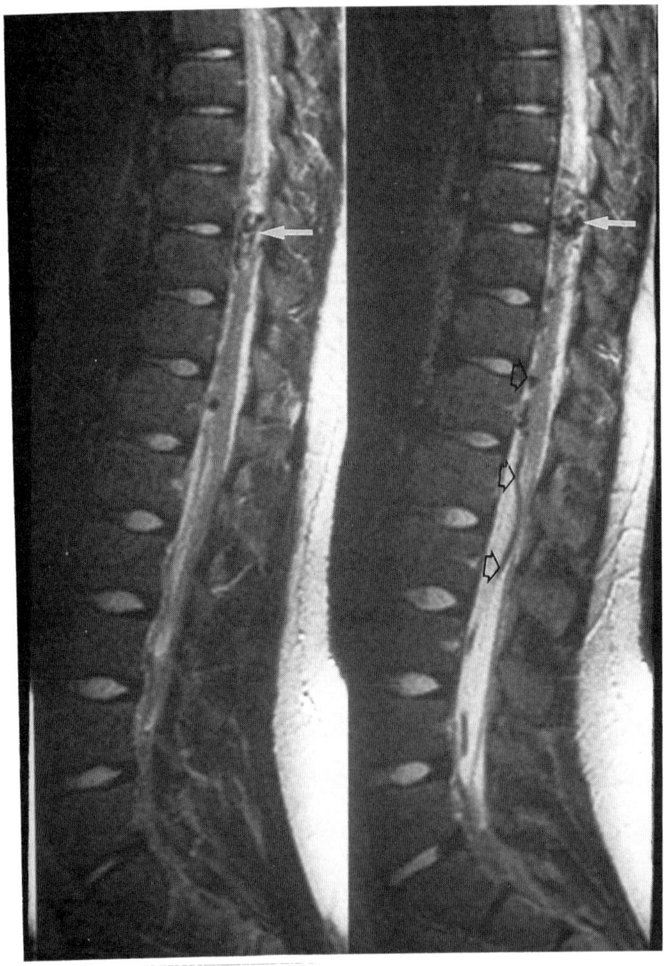

Figure 10.5. Intra/extramedullary AVM. Sagittal T2-weighted image shows malformation within and outside the spinal cord (*arrow*), with dilated veins extending a considerable distance caudally (*open arrows*). (Courtesy of D. A. Nichols and the Mayo Clinic.)

absorbed. Increased signal within the spinal cord adjacent to the abnormal vessels is sometimes seen on T2-weighted images and probably represents myelomalacia. Whether these changes are reversible is unknown (Fig. 10.6).

Two unusual types of AVMs deserve mention: The *Foix-Alajouanine abnormality* represents an advanced form of radiculomeningeal AVM which results in spontaneous thrombosis and may be a cause of myelomalacia [15, 16] (see Fig. 10.1). The MRI appearance reflects these changes. The *Cobb syndrome* is a metameric form in which the AVM involves the spinal cord, canal, and corresponding vertebral segment [12, 16].

AV FISTULA

Depending on the size of the arteriovenous communication, a single fistula may give rise to numerous dilated veins. The fistula is usually dural and originates from the radicular arteries, rarely from the anterior or posterior spinal arteries directly. Although the fistula itself is almost never identified, the dilated draining veins are almost always seen. The dilated veins may be aneurysmal in size, are numerous, and often extend a considerable distance into the intradural and epidural spaces. The surface of the spinal cord may be "scalloped" by subarachnoid vessels (Figs. 10.7 and 10.8). The veins cannot be differentiated from those draining an AVM. They do not enhance with contrast material unless flow is slow (Fig. 10.9).

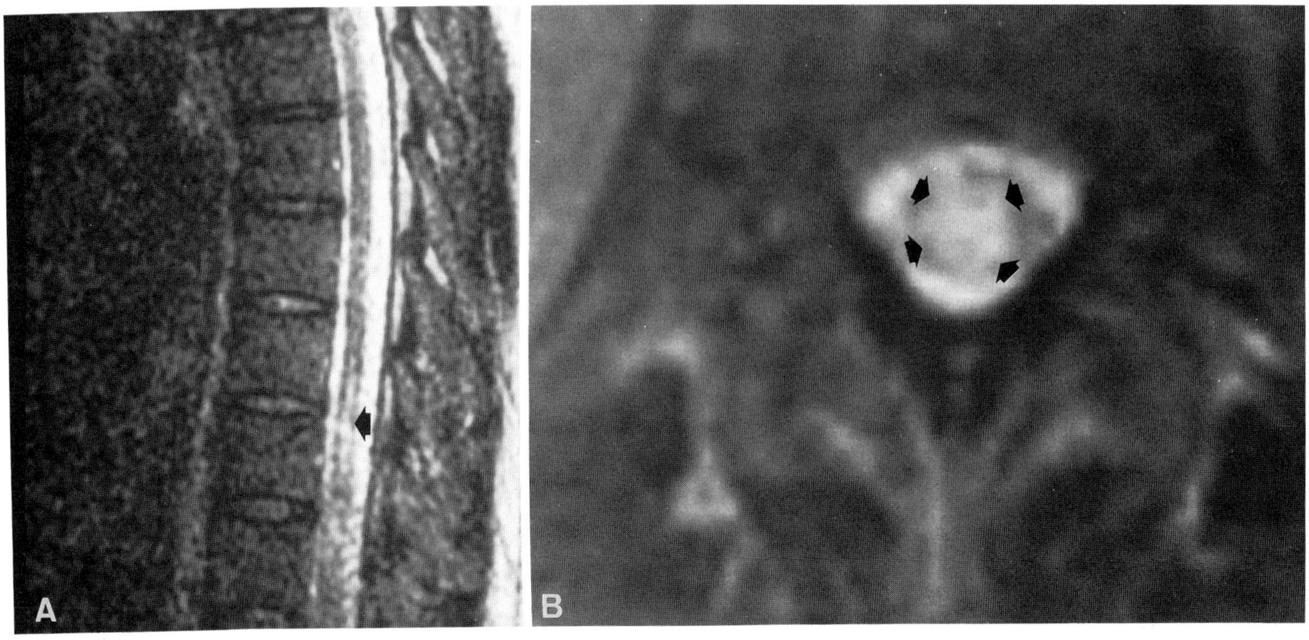

Figure 10.6. Myelomalacia. Sagittal (**A**) and axial (**B**) T2-weighted images show an area of increased signal related to an AVM (*arrow*). (Courtesy of D. A. Nichols and the Mayo Clinic.)

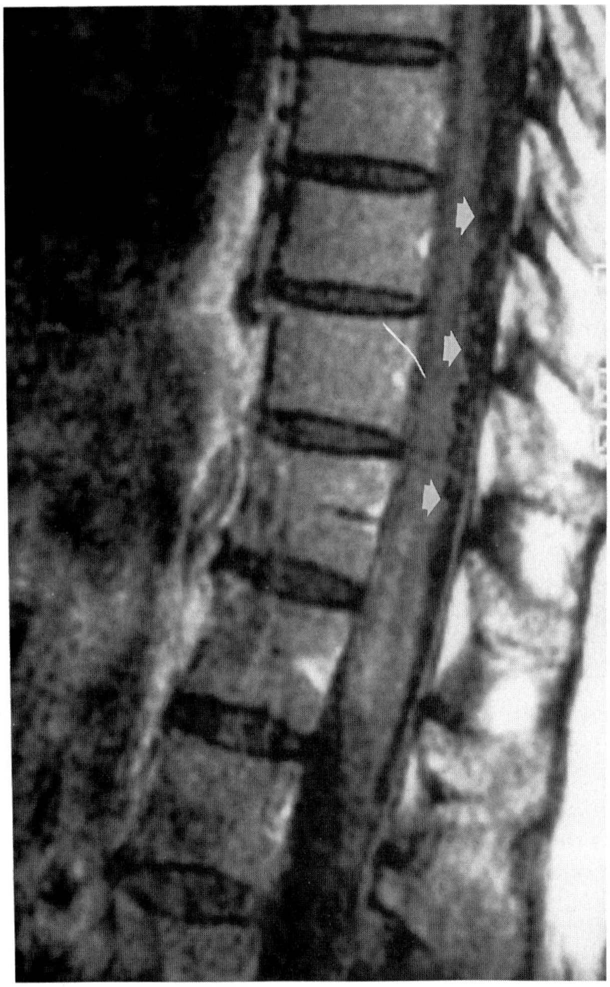

Figure 10.7. AV fistula. Draining veins extending a considerable distance in the subarachnoid space behind the spinal cord and causing "scalloping" of the surface of the spinal cord (*open arrows*). (Courtesy of D. A. Nichols and the Mayo Clinic.)

CAVERNOUS HEMANGIOMA

Cavernous angiomas are vascular anomalies consisting of venous sinusoids and slow flow. They constitute 5 to 12% of all spinal vascular anomalies [1]. They are difficult to demonstrate angiographically because of the small feeding arteries. Calcification within the angioma is not unusual and probably represents a dystrophic calcium deposition secondary to old hemorrhage. They are probably a larger variant of capillary angiomas. Multiplicity of the angiomas is common within both the brain and the spinal cord. As such, in the presence of spinal cord angioma, the brain parenchyma also should be imaged. Cavernous angiomas have a tendency to bleed within the parenchyma, although they may be symptomatic due to their size and location within the spinal cord.

Cavernous angiomas are difficult to detect on CT but are easily identified on MRI. This may explain the higher detection rates at the present time. On CT, cavernous angioma may be mistaken for an ependymoma due to the small size, punctate calcification, and enhancement (Fig. 10.10). The MRI appearance of this lesion is similar to that of angiomas in the brain and is characterized by signal changes due to recurrent hemorrhages [18, 19]. Subacute hemorrhage is shown as increased signal on T1-weighted images, and there are usually areas of decreased signal on T2-weighted images which represent the susceptibility effect of hemosiderin from chronic bleeds (Figs. 10.11 and 10.12). Some of the hypointense signal within the cavernous angiomas are due to calcification. Two observations are helpful in differentiating hemosiderin from flow void within the angiomas. First, the susceptibility effect is much more pronounced on T2-weighted and gradient-echo images compared with flow void, which should change little between imaging sequences (Fig. 10.13). Second, contrast enhancement is often seen in regions of flow void because of the slow flow. The spinal cord is variably enlarged, depending on the size of the hemorrhage. Increased signal within the spinal cord adjacent to the hemangioma is less commonly seen than with AVMs.

Telangiectasia. Spinal cord telangectasias or capillary angiomas have a similar appearance to cavernous hemangiomas. They are identified as telangectasias in the presence of linear hypointense signal resulting from hemosiderin deposition or telangiectic vessels. Absence of abnormal and dilated vessels helps in differentiating this lesion from AVMs and fistulas.

False-Positive/Negative Diagnosis

Signal void within the spinal canal due to vascular pulsation in an otherwise sharp image is the most common cause for false diagnosis of vascular malformations. These artifacts are more prominent with high- compared with low-field scanners and are seen despite the routine application of motion-suppression techniques. Conversely, vascular malformations may be present despite a completely normal MRI. The incidence of such false-negative findings is unknown. Myelography or occasionally spinal angiography is required to clarify the diagnostic dilemma in both instances.

Differential Diagnosis

Increased vascularity due to venous obstruction is seen in the spinal canal with vascular spinal cord tumors as well as lesions causing obstruction of the spinal canal (Fig. 10.14). The MRI sequences should be scrutinized for intramedullary, intradural extramedullary, and extradural masses which cause such obstructions. Intramedullary tumors may some-

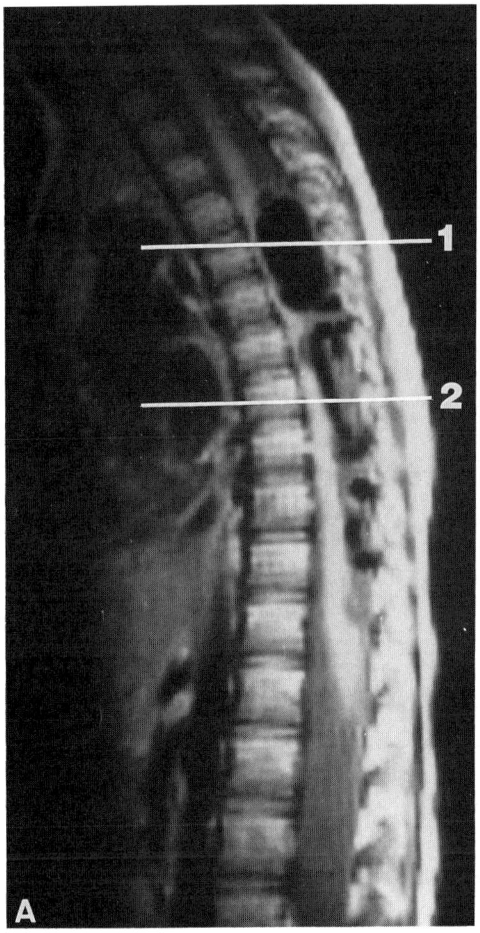

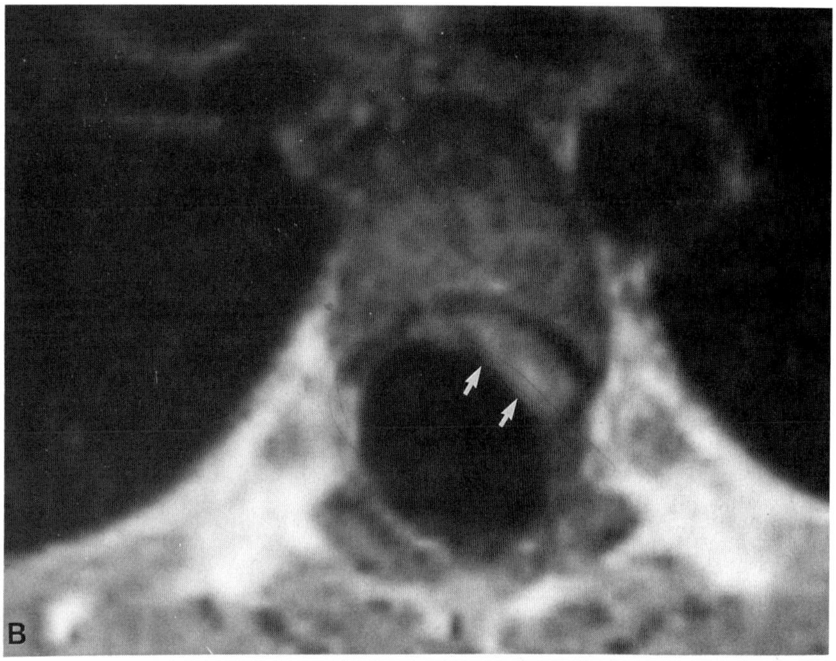

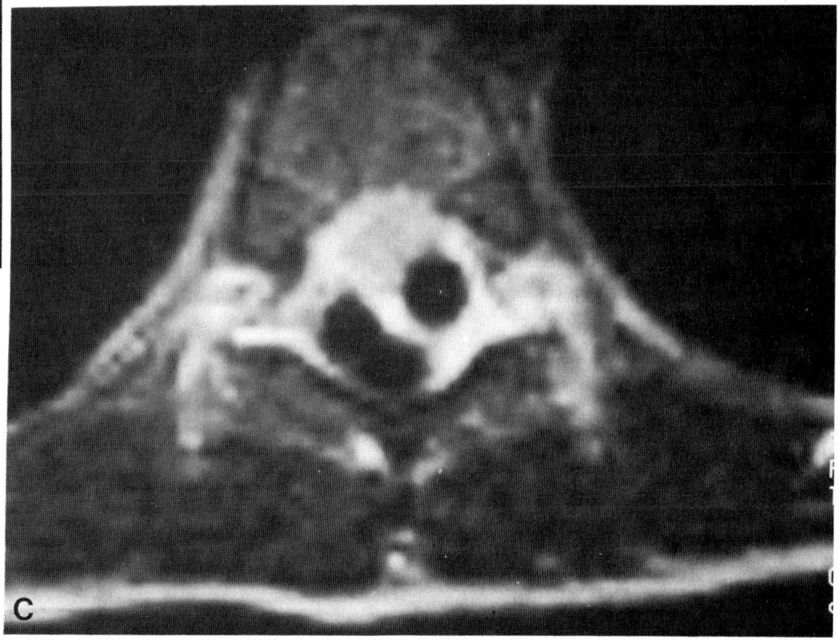

Figure 10.8. AV fistula. A, Sagittal proton-density image shows flow void of dilated veins. **B,** Axial T1-weighted image at level 1 shows aneurysmal dilatation with compression of the spinal cord (*arrows*). **C,** Axial T2-weighted image at level 2 shows two dilated veins. **D,** Coronal gradient-echo image shows flow-related enhancement of the dilated veins. **E,** Selective angiogram demonstrates the fistula and the dilated draining veins.

Figure 10.8–*continued*

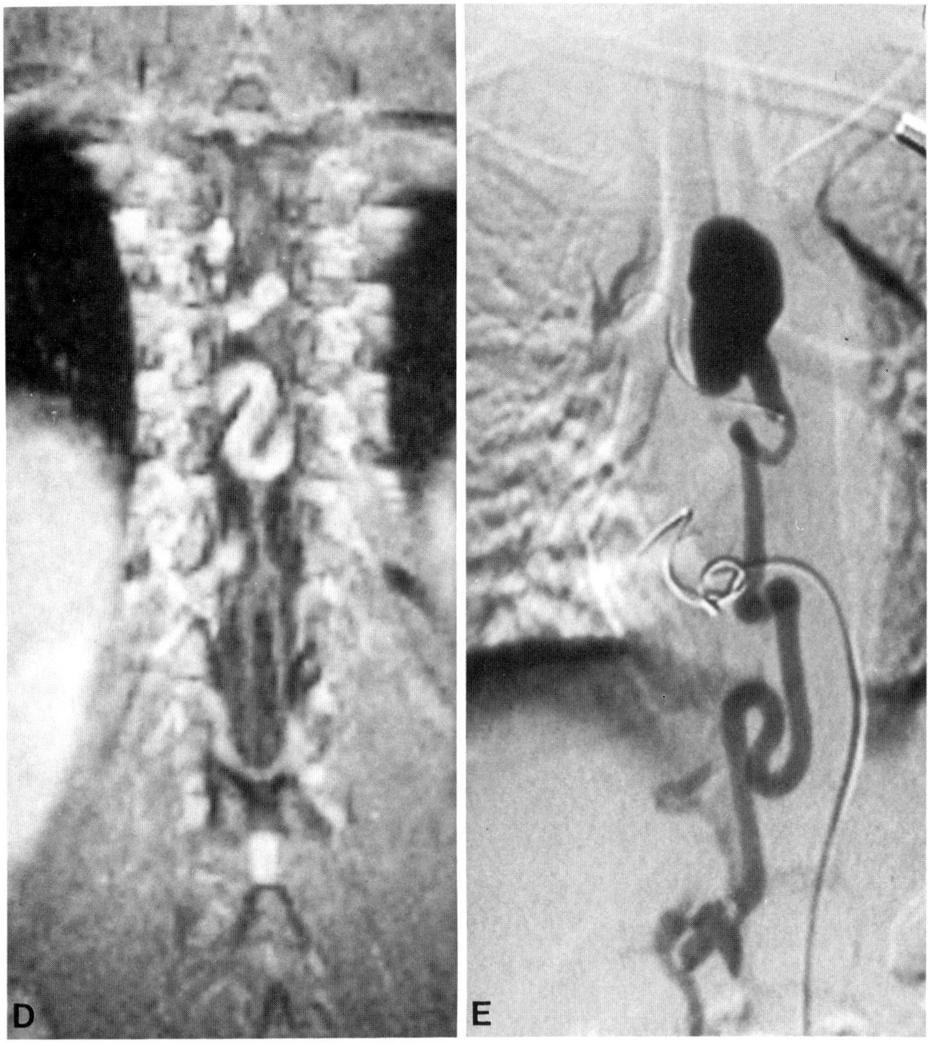

times hemorrhage, and bleeding may be associated with trauma, cord infarction, and bleeding diathesis.

Hemangioblastoma is a vascular tumor with vessels that may appear similar to those of AVMs. Since it is a true tumor, the spinal cord is always enlarged in the affected region, and the tumor has increased signal on T2-weighted images. The tumor mass enhances with intravenous contrast material. Syrinx cavities are often associated with this abnormality [20–22] (Fig. 10.15). Hemangioblastoma in the cord or overlying the cord and even in the nerve roots may occur only within the spinal canal (Fig. 10.16) or be associated with similar lesions in the cerebellum and also may be part of the Osler-Rendau-Weber syndrome.

Other primary tumors, such as vascular astrocytomas and ependymomas, may have enlarged vessels on MRI. The veins on the surface of the cord may be more prominent in the presence of obstruction and compression. Hemorrhage into the cord is uncommon, but contrast enhancement is almost invariable.

Metastases from hypervascular primary tumors such as renal and thyroid tumors, as well as from choriocarcinoma and melanoma, may be hemorrhagic. It is possible to distinguish these tumors from AVMs and AV fistulas by the absence of discrete vessels; the hemorrhages may be identical to those of cavernous hemangiomas. Contrast enhancement is usually more marked than in vascular malformations. Because melanin is a paramagnetic substance, the increased signal on T1-weighted images is indistinguishable from methemoglobin of hemorrhage.

SPINAL CORD INFARCTION

Spinal cord infarction is rare, and its diagnosis both on a clinical basis and by imaging studies is mostly by excluding other causes. Infarction is usually suspected in patients with severe atherosclerosis of the vessels and an associated history of diabetes. Spinal cord infarction may follow surgical procedures involving the heart and aorta and aortic dissection [23–25]. The diagnosis of spinal cord in-

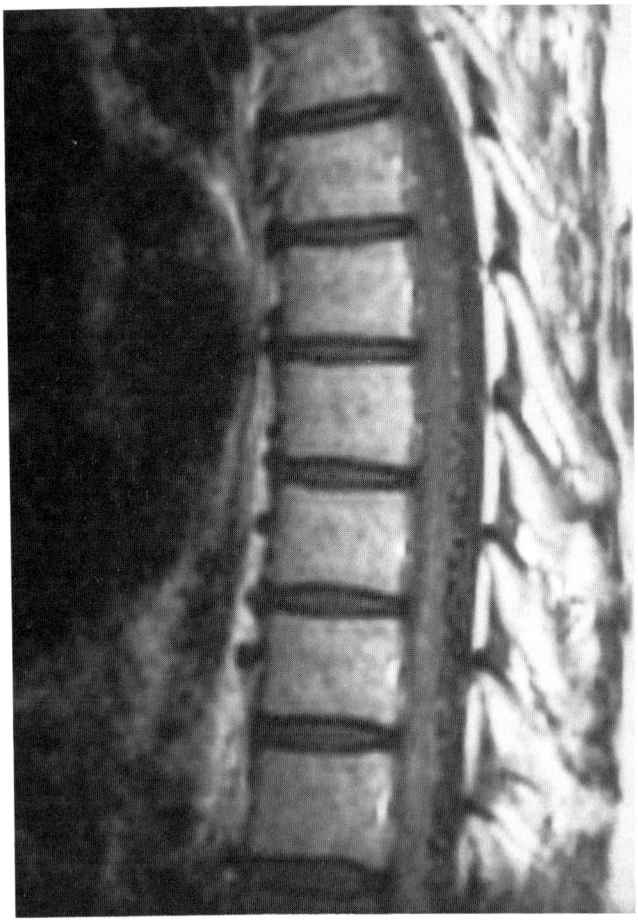

Figure 10.9. AV fistula. Postcontrast T1-weighted image shows enhancement of some of the vessels on the surface of the spinal cord because of slow flow. (Courtesy of D. A. Nichols and the Mayo Clinic.)

farction is made on the basis of clinical setting, since the imaging findings in the acute stage of the infarction are often normal. In the subacute stage, the focal or diffuse swelling of the spinal cord may be identified on CTM (Fig. 10.17). MRI is ideal in evaluating suspected cord infarction [23–30]. On MRI, the cord swelling may be seen on the T1-weighted image; however, the T2-weighted sequence usually shows bright signal with a longitudinal flame-shaped appearance (Fig. 10.18). This often persists for a long time. In the subacute stage, this represents vasogenic edema and later probably gliosis. Enhancement with contrast material may be observed if the area of infarction is significant. In general, in the acute stage, the imaging characteristics are similar to those of acute myelitis and in the chronic stage to cord atrophy. Spinal cord infarcts usually are focal and most often involve the watershed zones of the spinal cord. This is usually around the upper thoracic region or in the region of the conus. Infarction of the spinal cord may be associated with infarction of the adjacent vertebral body [26, 29]. Although rare, spinal cord infarction following embolization of the medullary artery which supplies the vertebrae and the spinal cord from herniated disk or fibrocartilaginous material may result in both spinal cord and vertebral body infarction. In short, the presence of vertebral body marrow signal changes in association with cord swelling is indicative of spinal cord infarction rather than other causes of the cord abnormality.

FOIX-ALAJOUANINE SYNDROME

This syndrome consists of progressive paraplegia with initial spasticity and later flaccidity, usually in an older age group and more often in males. It is presumed to be the result of progressive thrombosis of the epidural and leptomeningeal veins with resulting infarction of the cord. Dilated tortuous, thrombosed veins overlie the necrotic spinal cord [15] (see Fig. 10.1). In a sense, this may represent thrombophlebitis with venous infarction involving the spinal cord. Similar findings also may be present following thrombosis of a dural AV fistula [30]. Imaging studies may show a swollen or atrophic cord with prominent vascular channels on the epidural surface which often enhances with contrast.

SPINAL HEMATOMA

Hemorrhage within the spinal canal may be within the epidural space or in the subdural space. Subarachnoid hemorrhage may spill over from the intracranial subarachnoid bleed. Primary spinal subdural hemorrhage is often traumatic or results from a bleeding diathesis [31]. A few instances of spontaneous subdural hematomas have been reported [32]. Clinical signs of meningismus and acute onset of low back pain are the usual clinical presentations. On CT, small amounts of subdural hematoma may not be detected. Larger amounts of blood products are seen as hyperdensity compressing the thecal sac or surrounding the thecal sac. MRI is useful in the subacute stage to demonstrate the localized hematoma. Spinal epidural hematoma other than from trauma is rare. It is presumed to be venous in origin, although arterial hemorrhage has been reported. "Spontaneous" epidural hematoma is often the result of minor trauma or may result from increased flow in Batson's plexus secondary to increased abdominal pressure or venous obstruction in the inferior vena cava. Other causes which may result in spontaneous hematoma include bleeding disorders, hypertension, and the vasculitides. Most hematomas are located in the thoracic and thoracolumbar regions, usually along the dorsal aspect. Clinical presentation is of an acute onset of burning pain that is radicular in distribution, with subsequent paraparesis or quadriparesis.

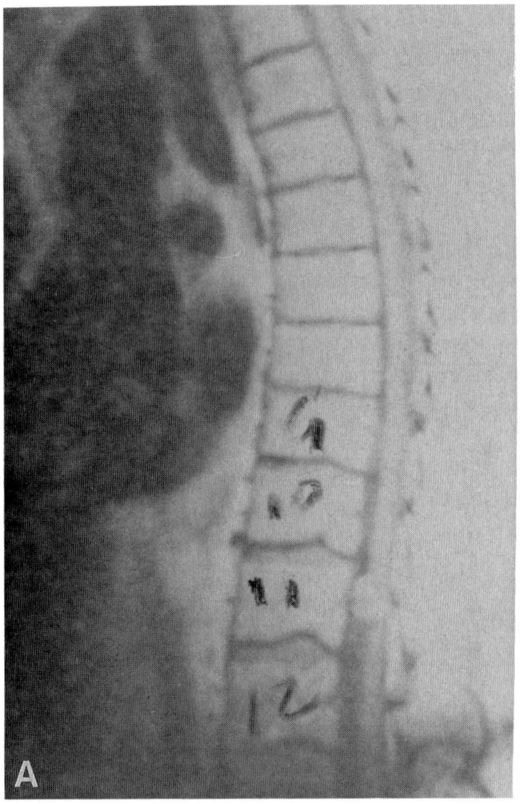

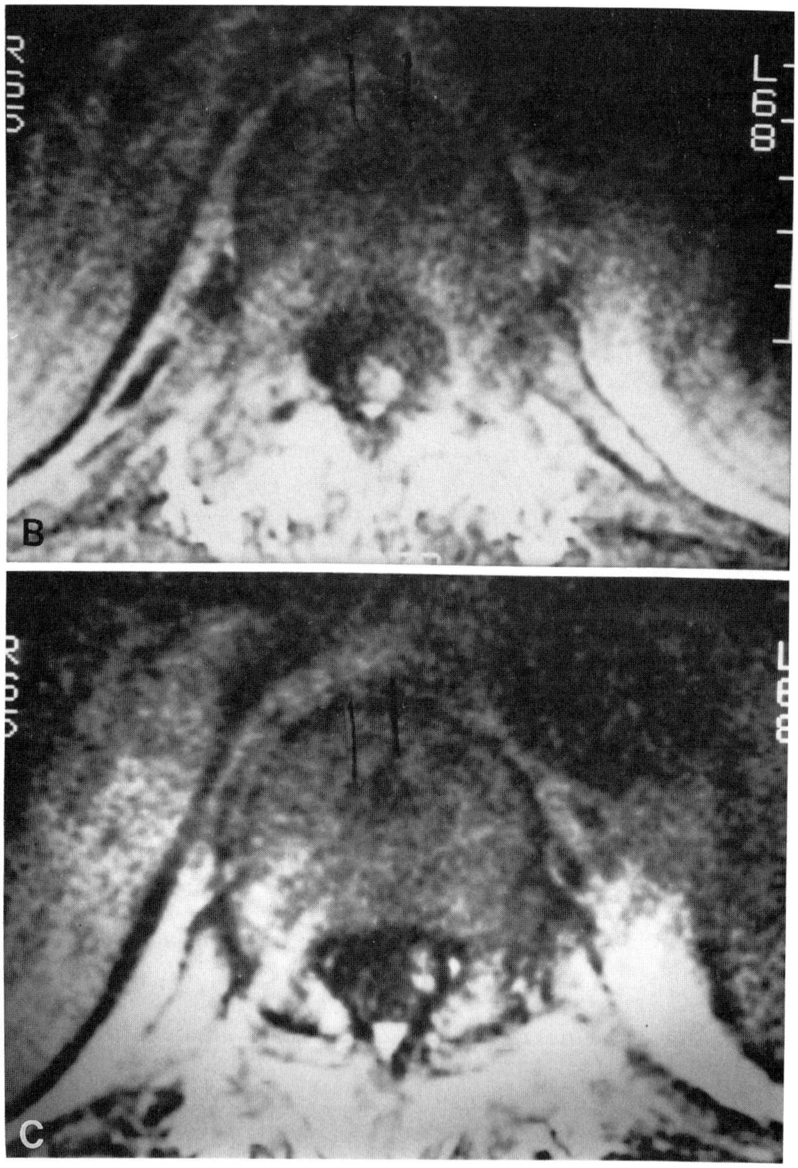

Figure 10.10. Cavernous hemangioma. A 26-year-old man with low back pain. Sagittal (**A**) and axial Gd-DTPA–enhanced (**B,C**) T1-weighted MRI sequences demonstrate an exophytic enhancing intradural lesion (*arrowhead*) opposite the T11 vertebral body. Frontal (**D**) and lateral (**E**) views of the myelogram demonstrate focal swelling of the spinal cord (*arrow*). Subsequent axial CT sections (**F,G**) demonstrate the exophytic cavernous angioma.

Figure 10.10–*continued*

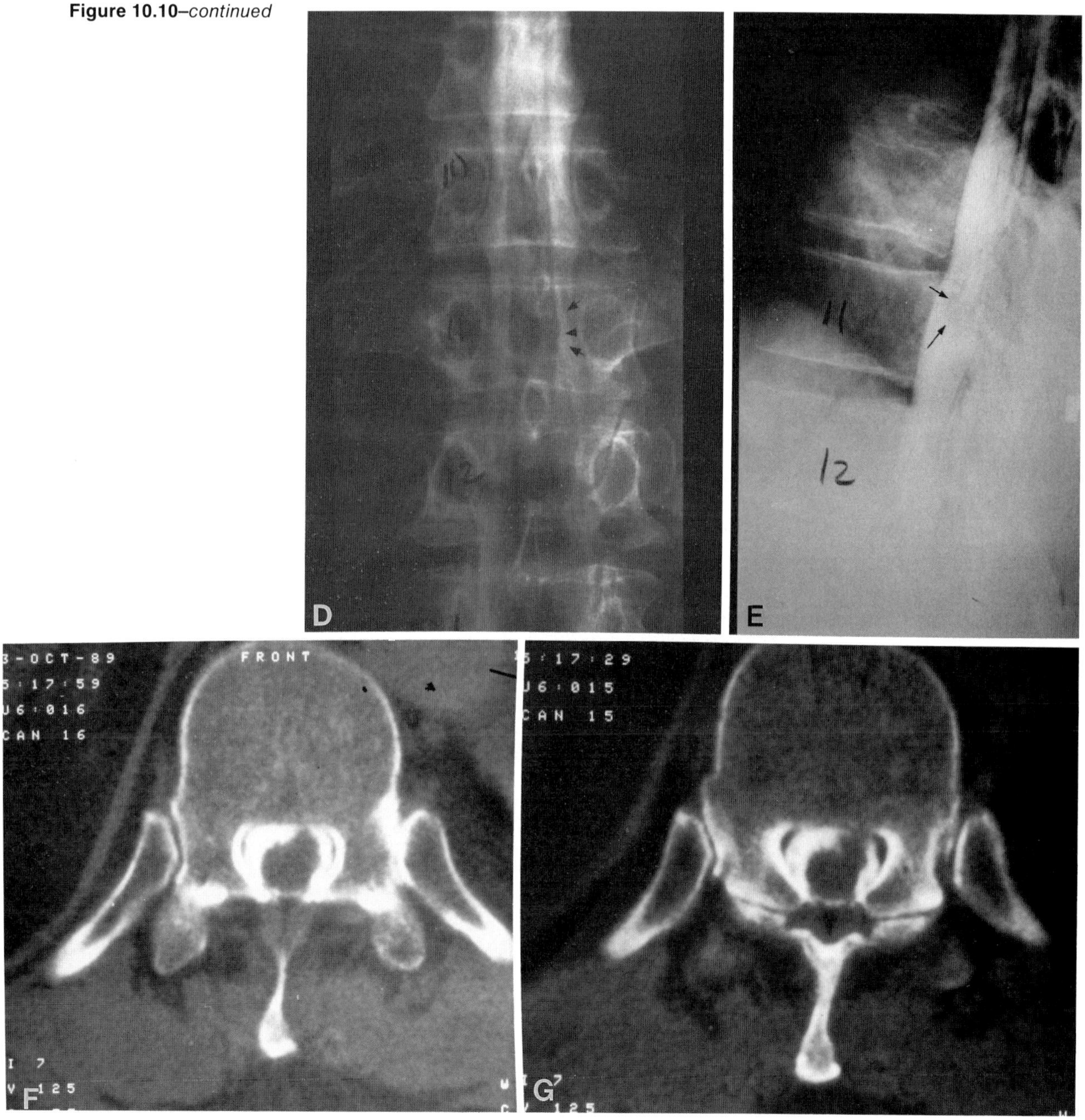

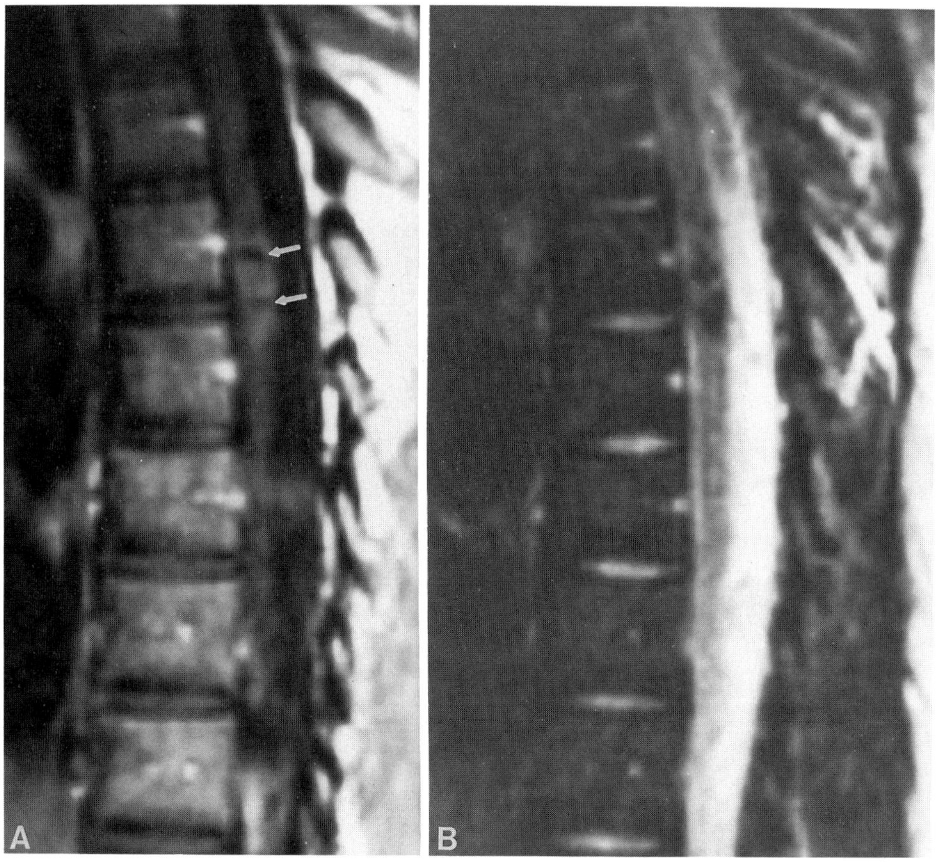

Figure 10.11. Cavernous hemangioma. A, Sagittal T1-weighted image shows an area of slightly increased signal with two regions of decreased signal (*arrows*) within the spinal cord. **B,** Sagittal T2-weighted image shows decreased signal within these regions (*open arrows*). The signals are characteristic of a hematoma with methemoglobin and susceptibility effect.

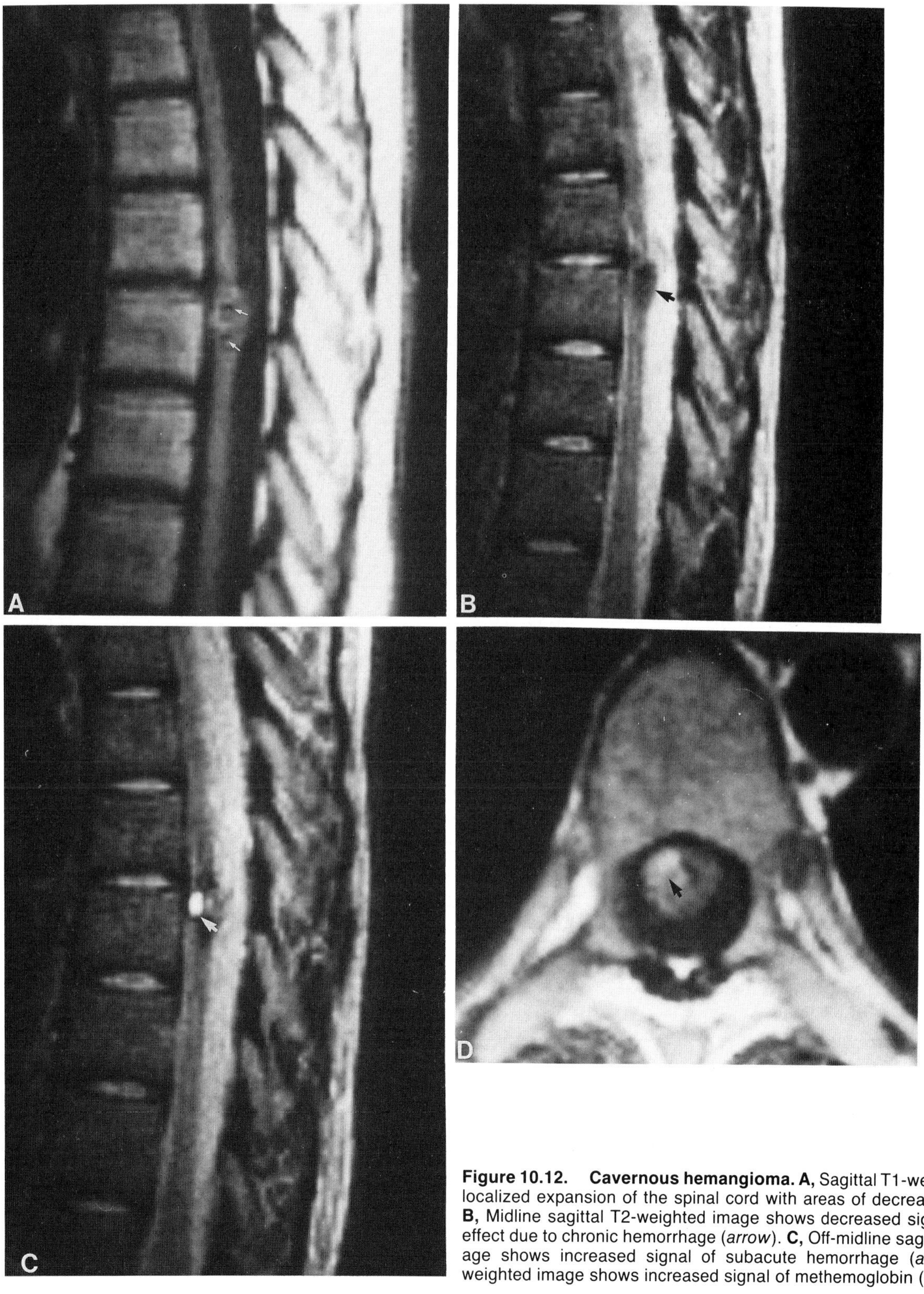

Figure 10.12. Cavernous hemangioma. A, Sagittal T1-weighted image shows localized expansion of the spinal cord with areas of decreased signal (*arrows*). **B,** Midline sagittal T2-weighted image shows decreased signal of susceptibility effect due to chronic hemorrhage (*arrow*). **C,** Off-midline sagittal T2-weighted image shows increased signal of subacute hemorrhage (*arrow*). **D,** Axial T1-weighted image shows increased signal of methemoglobin (*arrow*).

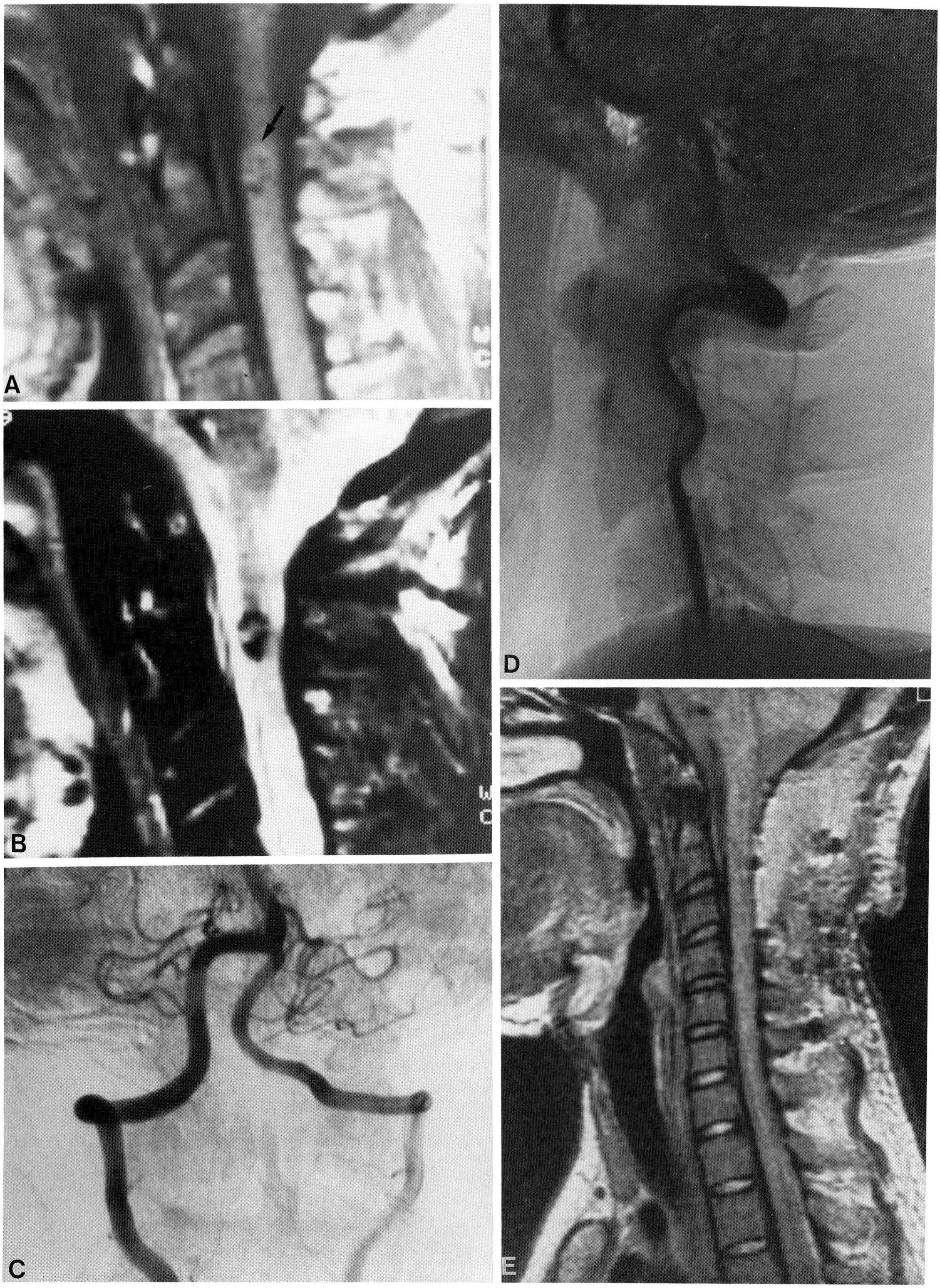
A
B
C
D
E

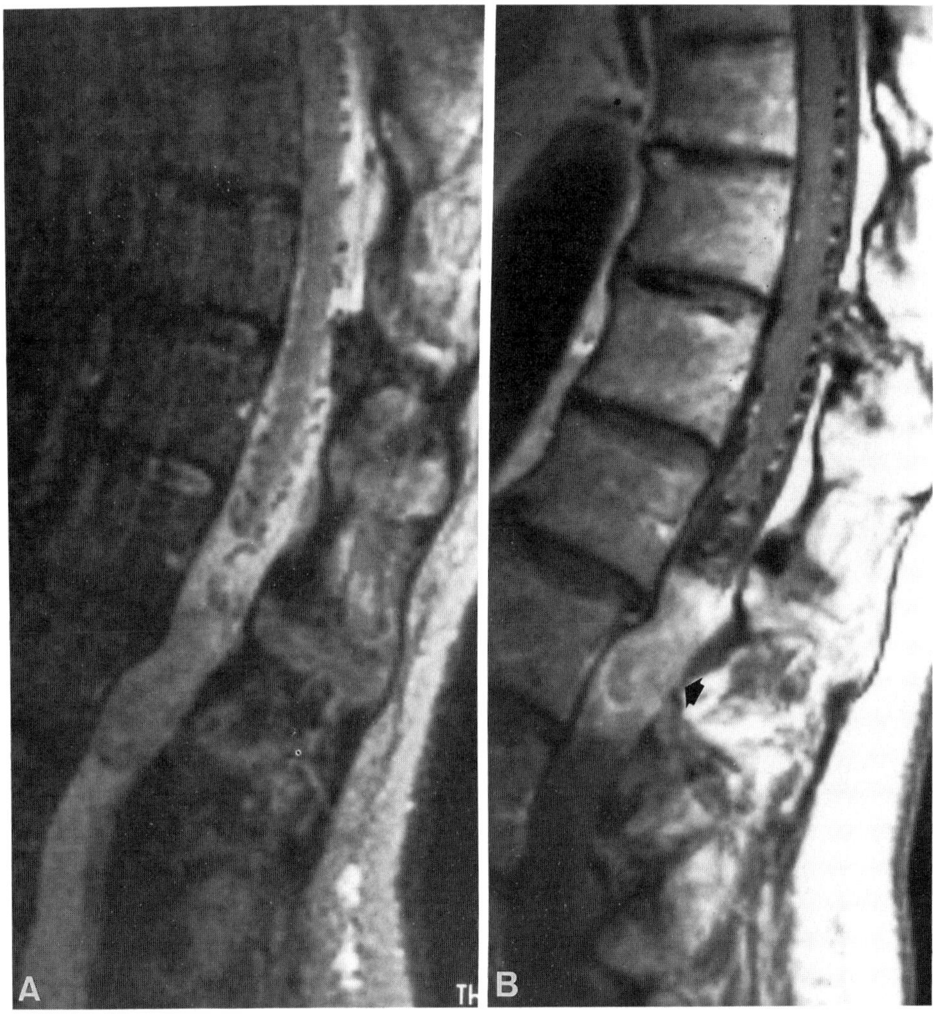

Figure 10.14. Ependymoma. A, Sagittal T2-weighted image shows numerous vessels causing signal void in the subarachnoid space. **B,** Sagittal T1-weighted image after contrast shows increased signal in many of these vessels, indicating slow venous flow. The enhancing ependymoma of the cauda is demonstrated (*arrow*).

Figure 10.13. Cavernous hemangioma. 27-year-old man came with clinical symptoms of subarachnoid hemorrhage with xanthochromic CSF and normal CT of the head. **A,** Sagittal T1-weighted image shows minimal cord expansion with areas of bright and decreased signal within the cervical cord (*arrow*). **B,** Sagittal gradient-echo image demonstrates the typical MRI characteristics of hemosiderin deposits within this lesion. Frontal (**C**) and lateral (**D**) views of vertebral arteriogram does not demonstrate a vascular blush or nidus. **E,** Postoperative T1-weighted image demonstrates complete excision of the lesion.

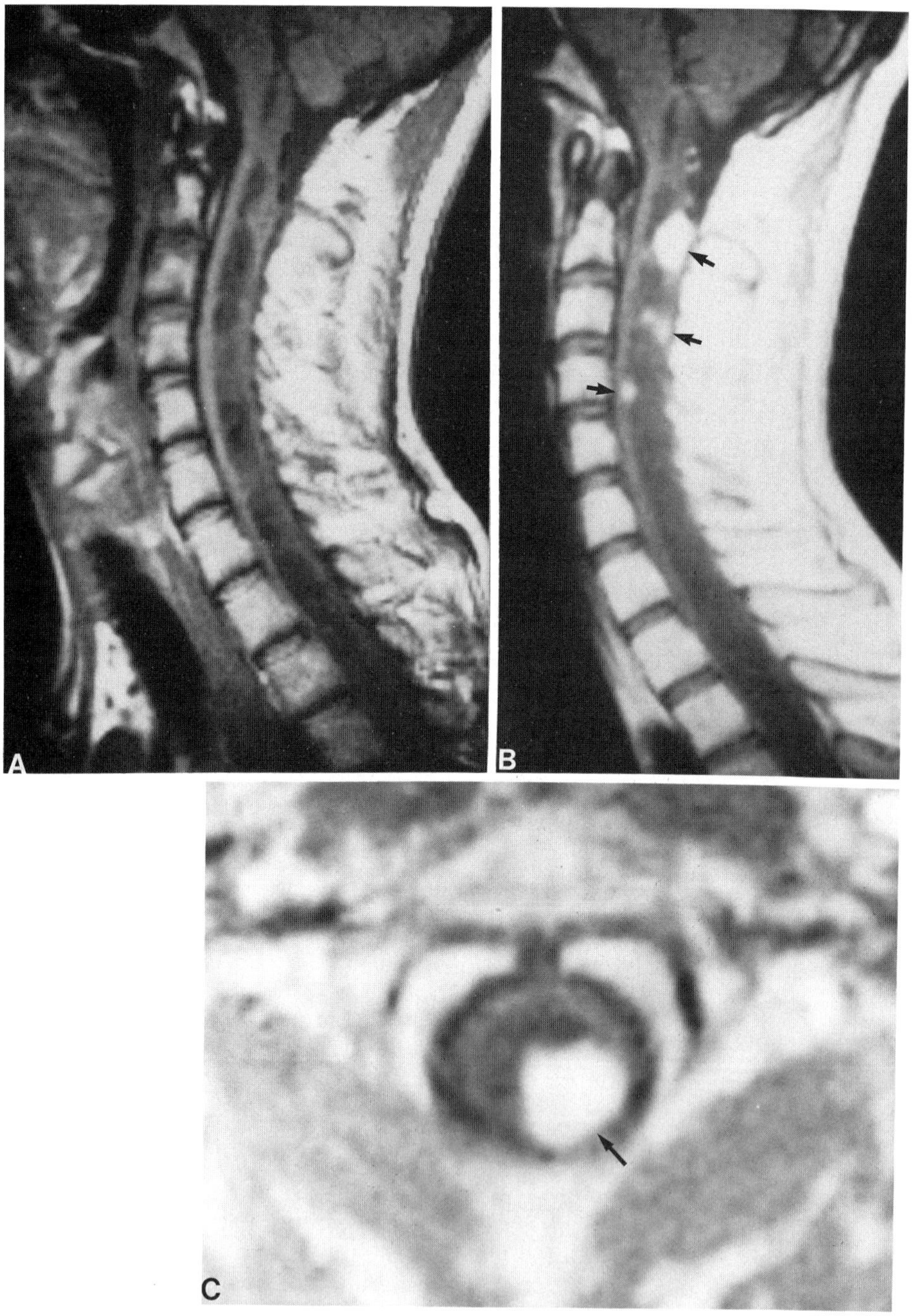

Figure 10.15. Hemangioblastoma. Sagittal T1-weighted images without (**A**) and with (**B**) contrast material shows intramedullary cysts and enhancing nodules (*arrows*). **C,** Axial T1-weighted image with contrast material shows relationship of one of the enhancement masses to the cyst.

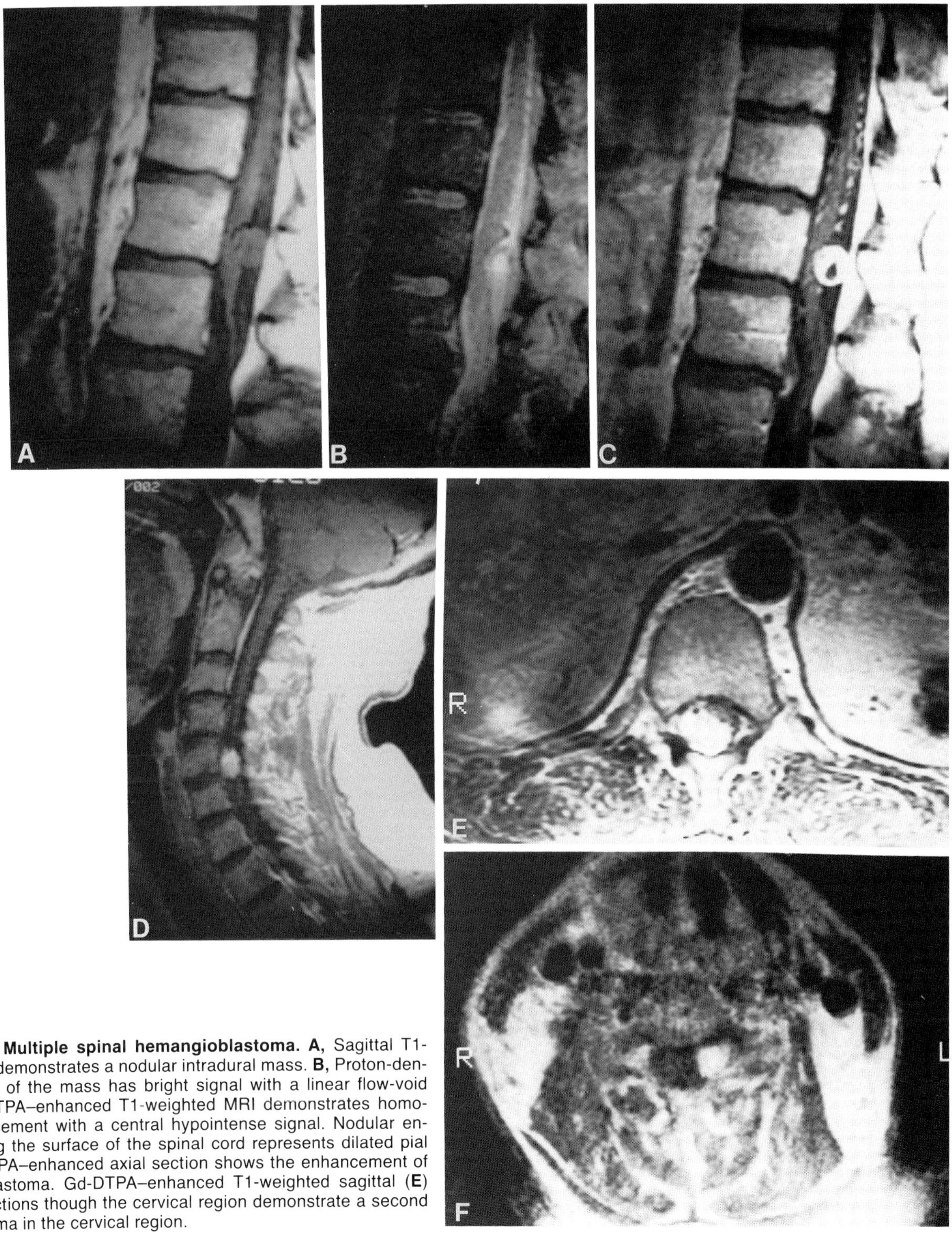

Figure 10.16. Multiple spinal hemangioblastoma. A, Sagittal T1-weighted image demonstrates a nodular intradural mass. **B,** Proton-density sagittal MRI of the mass has bright signal with a linear flow-void signal. **C,** Gd-DTPA–enhanced T1-weighted MRI demonstrates homogeneous enhancement with a central hypointense signal. Nodular enhancement along the surface of the spinal cord represents dilated pial veins. **D,** Gd-DTPA–enhanced axial section shows the enhancement of the hemangioblastoma. Gd-DTPA–enhanced T1-weighted sagittal (**E**) and axial (**F**) sections though the cervical region demonstrate a second hemangioblastoma in the cervical region.

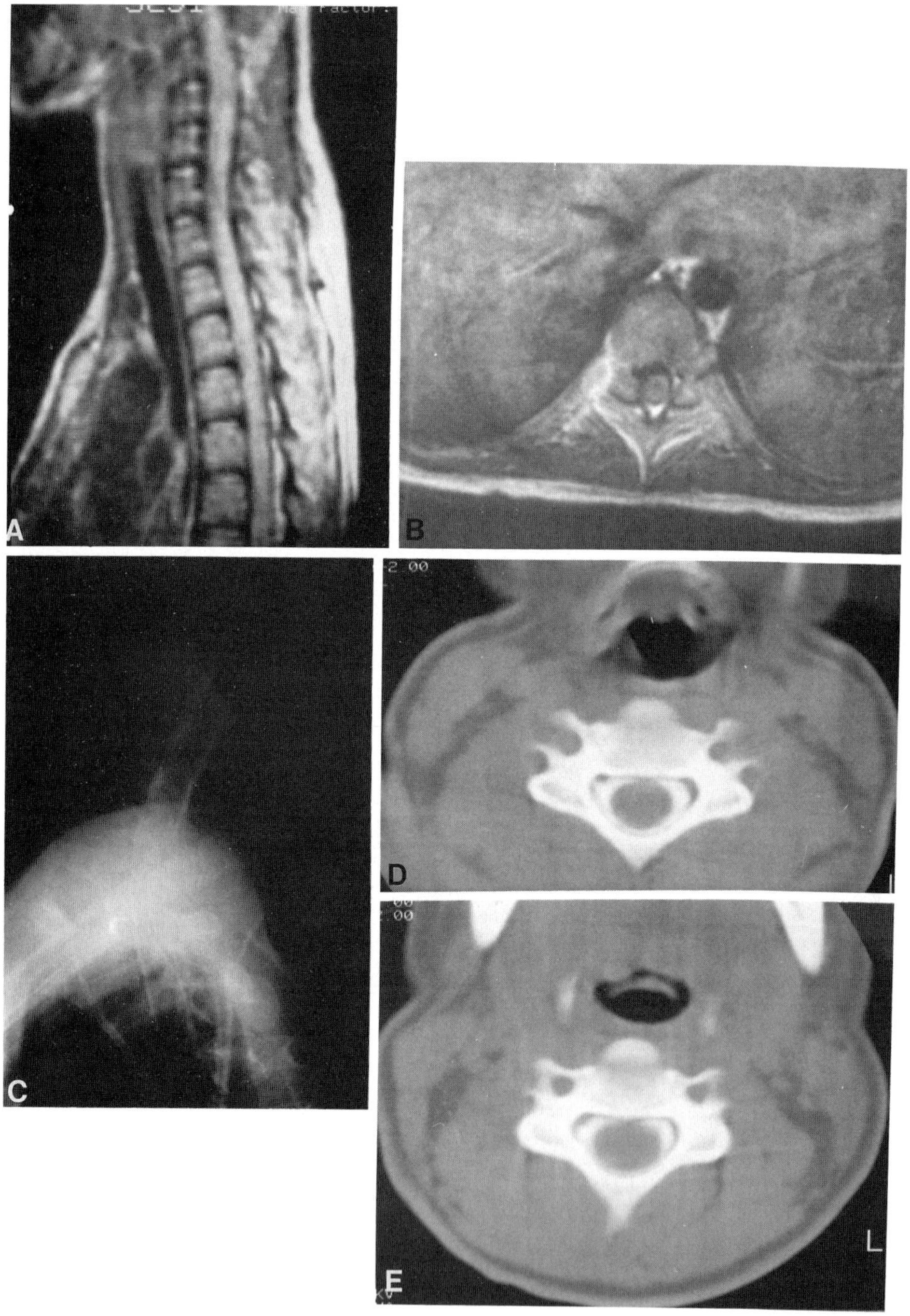

Figure 10.17. Cord infarction. A 9-year-old girl with acute onset of quadriplegia without antecedent history of viral disease. Sagittal (**A**) and axial (**B**) T1-weighted MRI sequences demonstrate diffuse marked swelling of the cord involving the cervical and thoracic segments. **C,** Lateral view of the cervical myelogram demonstrates diffuse cord swelling. **D,E,** Axial CT scans demonstrate similar findings indicative of cord ischemia.

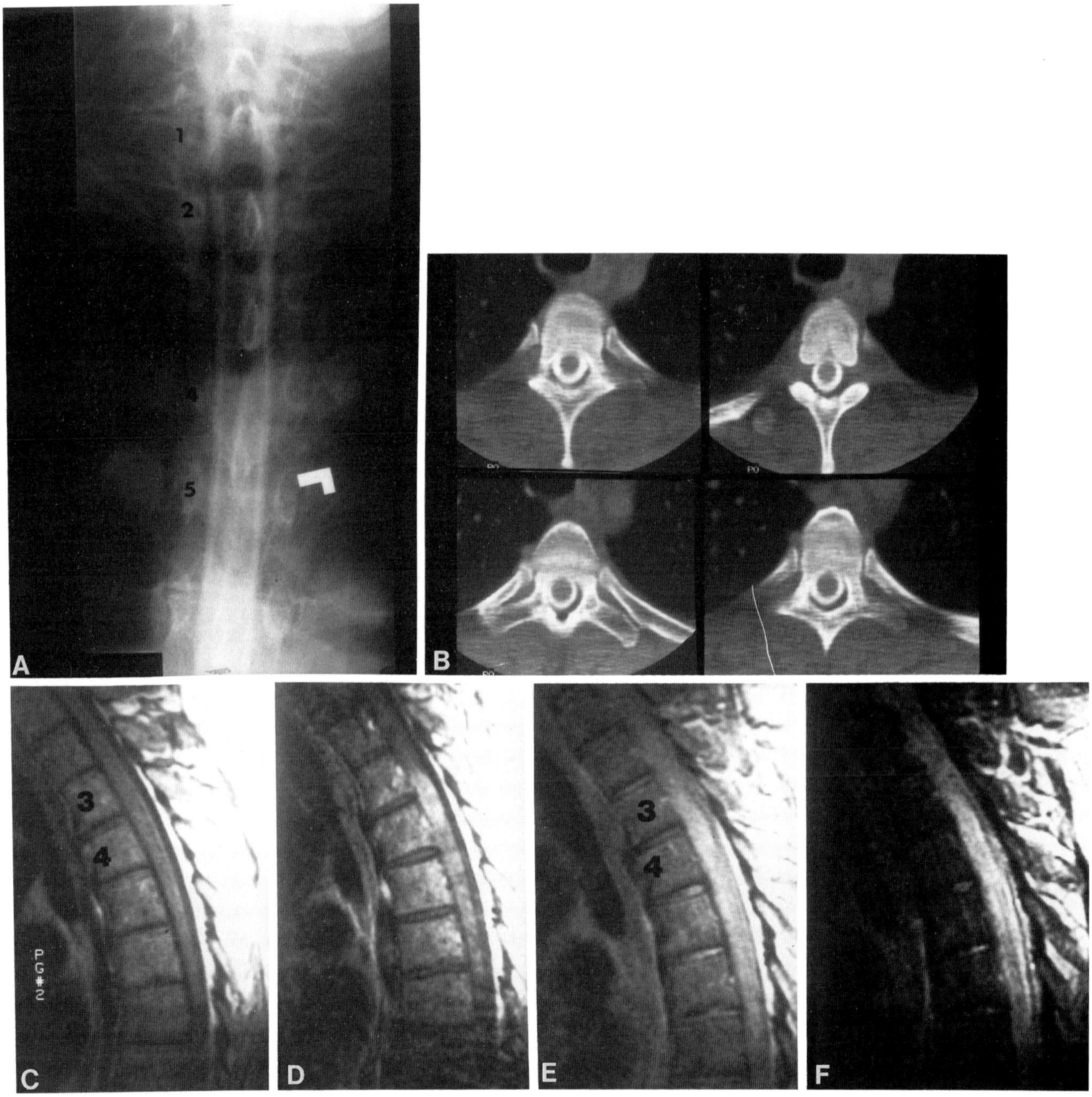

Figure 10.18. Spinal cord and vertebral infarction. A 56-year-old man with acute onset of chest pain later confined to the back. **A,** Frontal view of the myelogram demonstrates diffuse cord swelling. **B,** Contiguous axial CT demonstrates similar findings. T1-weighted precontrast (**C**), T1-weighted Gd-DTPA–enhanced (**D**), and T2-weighted (30/2000) (**E**) and (80/2000) (**F**) images demonstrate focal mild cord swelling with enhancement and focal flame-shaped bright signal on the T2-weighted images localized to T3–5 level. **G,H,** Post Gd-DTPA–enhanced axial sections demonstrate enhancement primarily in the ventral aspect of the cord (*arrow*). Some enhancement of the vertebral marrow is also present in **D.** There was minimal clinical improvement. Repeat MRI performed 4 months later. Sagittal T1-weighted precontrast (**I**), T1-weighted post Gd-DTPA (**J**), and axial sections (**K**), demonstrate decrease in cord swelling; however, there is continued focal enhancement within the cord as well as the adjacent vertebral marrow.

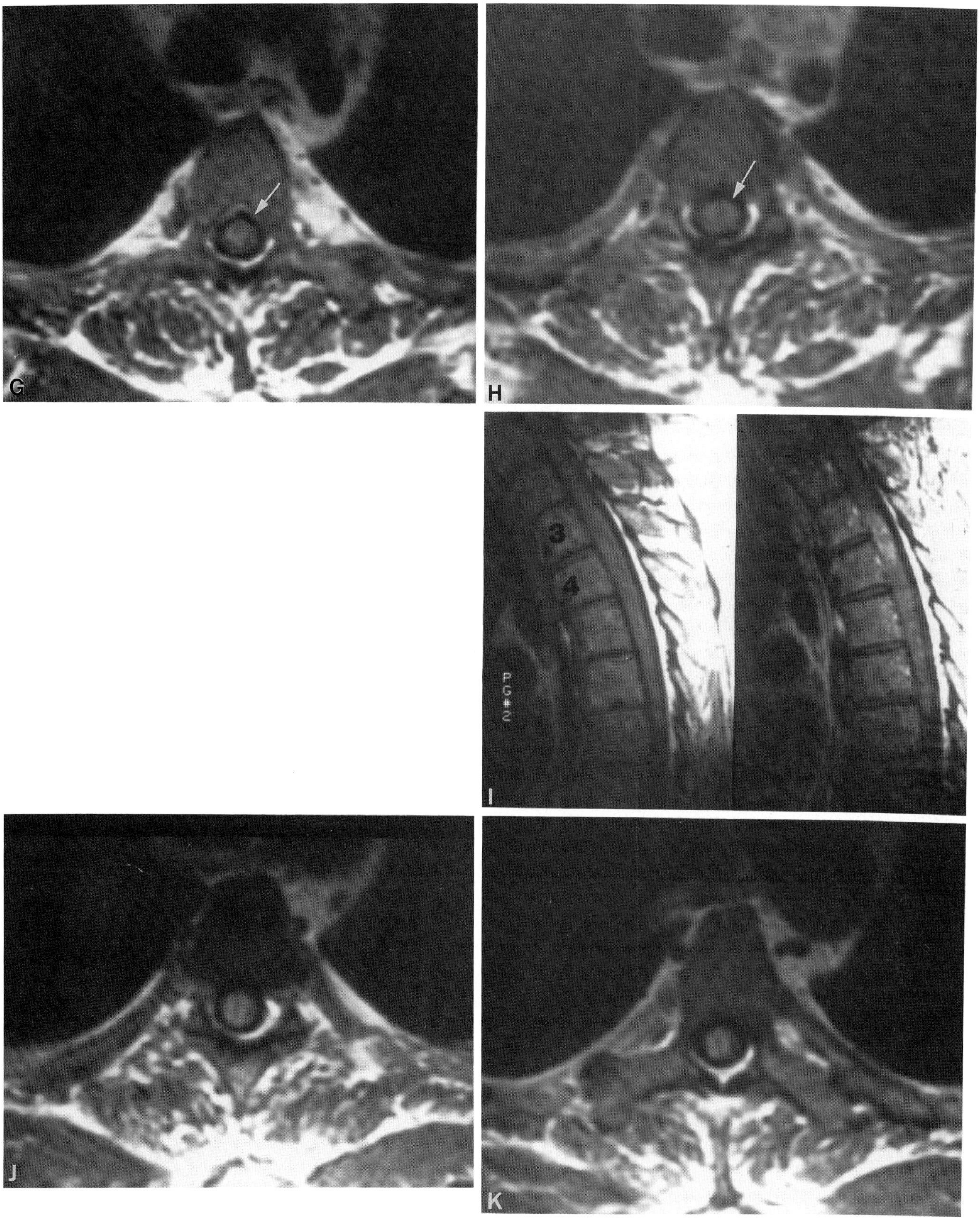

Figure 10.18–*continued*

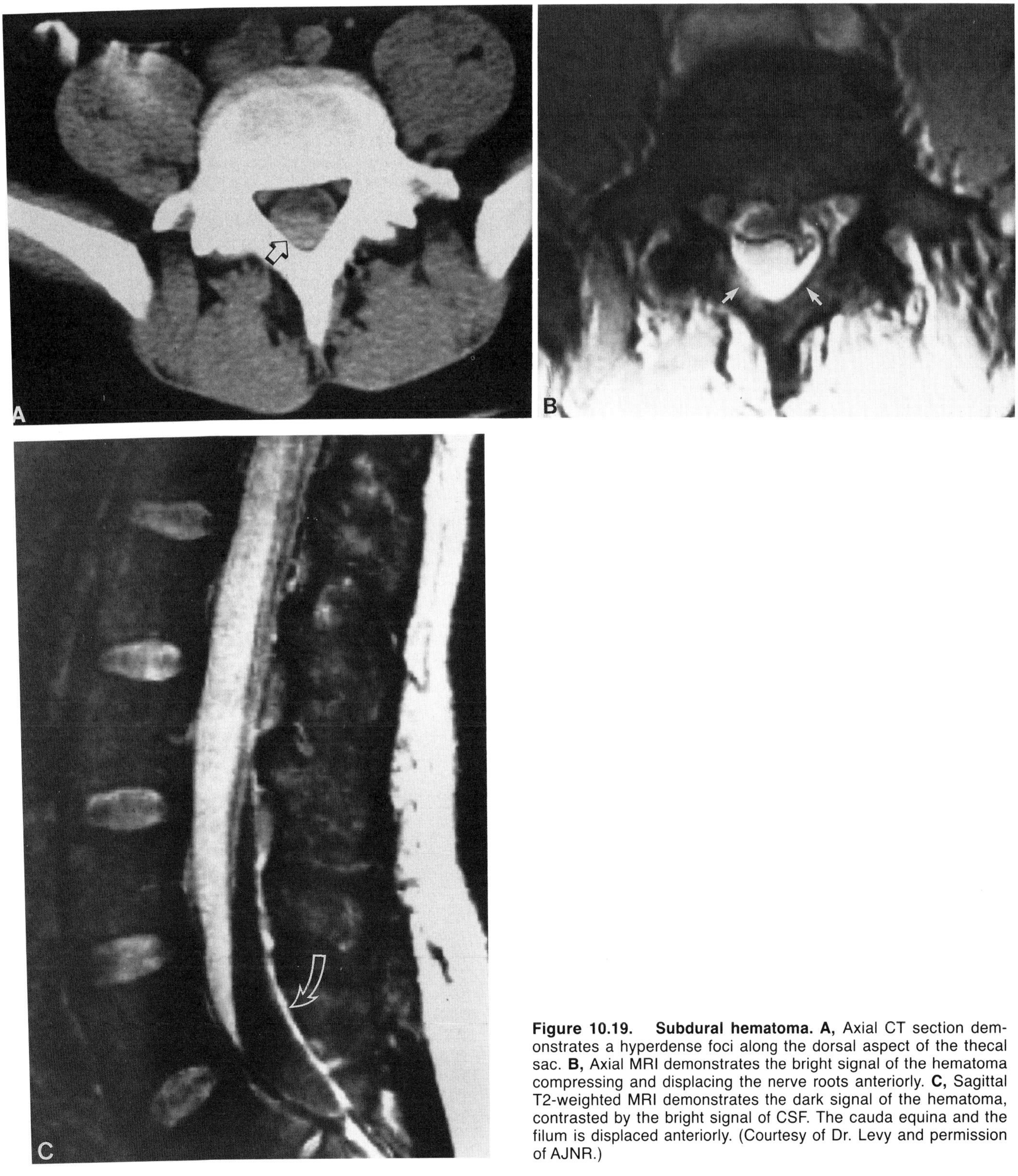

Figure 10.19. Subdural hematoma. A, Axial CT section demonstrates a hyperdense foci along the dorsal aspect of the thecal sac. **B,** Axial MRI demonstrates the bright signal of the hematoma compressing and displacing the nerve roots anteriorly. **C,** Sagittal T2-weighted MRI demonstrates the dark signal of the hematoma, contrasted by the bright signal of CSF. The cauda equina and the filum is displaced anteriorly. (Courtesy of Dr. Levy and permission of AJNR.)

CT is useful in the acute stage in demonstrating epidural hematomas, since they are localized and appear hyperdense. On MRI, epidural or subdural hematomas (Fig. 10.19) may be identified even in the subacute stage, and even the small hematomas can be identified. On T1-weighted images they may be mildly hyperintense or more often isointense. On T2-weighted images they follow the signal changes associated with aging of hematomas.

EXTRADURAL SPINAL VASCULAR LESIONS

The most common extradural spinal vascular lesions are those associated with cavernous hemangiomas (cavernomas) of the vertebrae. They may extend into the epidural space and occasionally may extend through the dura. Dural extension often follows surgery.

Epidural extension usually is associated with radicular symptoms. Both IVCT and MRI are useful in demonstrating the intraspinal extension of these vascular tumors (see Chap. 9).

References

1. Jellinger K. Pathology of spinal vascular malformations and vascular tumors. In: Pia HW, Djindjian R, eds. Spinal angiomas: Advances in diagnosis and therapy. New York: Springer, 1978:9–201.
2. Djindjian R. L'angiographie de la moelle epiniere. Paris: Masson, 1970.
3. Di Chiro G, Doppman J, Ommaya AK. Selective arteriography of arteriovenous aneurysm of spinal cord. Radiology 1967;88:1065–1077.
4. Di Chiro G, Doppman J, Ommaya AK. Angiographic and surgical aspects of spinal cord vascular anomalies. Radiology 1969;88:1078–1085.
5. Kendall BE, Logue V. Spinal epidural angiomatous malformations draining into intrathecal veins. Neuroradiology 1977;13:181–189.
6. Merland JJ, Riche MC, Chiras J. Intraspinal extramedullary arteriovenous fistulae draining into the medullary veins. J Neuroradiol 1980;7:271–320.
7. Aminoff MJ, Barnard RO, Valentine L. The pathophysiology of spinal vascular malformations. J Neuroradiol Sci 1974;23:255–263.
8. Di Chiro G, Doppman JL, Dwyer AJ, et al. Tumors and arteriovenous malformations of the spinal cord. Radiology 1985;156:689–697.
9. Kulkarni MV, Burks DD, Price AC, Cobb C, Allen JH. Diagnosis of spinal arteriovenous malformation in a pregnant patient by MR imaging. J Comput Assist Tomogr 1985;9:171–173.
10. Masaryk TJ, Ross JS, Modic MT, et al. Radiculomeningeal vascular malformations of the spine: MR imaging. Radiology 1987;164:845–849.
11. Dormont D, Gelbert F, Assouline E, et al. MR imaging of spinal cord arteriovenous malformations at 0.5T: study of 34 cases. AJNR 1988;9:833–838.
12. Minami S, Sagoh T, Nishimura K, et al. Spinal arteriovenous malformation: MR imaging. Radiology 1988;169: 109–115.
13. Terwey B, Becker H, Thron A, et al. Gadolinium-DTPA enhanced MR imaging of spinal dural arteriovenous fistulae. J Comput Assist Tomogr 1989;13:30–37.
14. Dillon WP, Norman D, Newton TH, et al. Intradural spinal cord lesions: Gd-DTPA enhanced MR imaging. Radiology 1989;170:229–237.
15. Wirth FP, Post KD, Di Chiro D, et al. Foix-Alajouanine disease: spontaneous thrombosis of a spinal cord arteriovenous malformation—a case report. Neurology 1970; 20:1115–1118.
16. Symon L, Kuyama H, Kendall B. Dural arteriovenous malformations of the spine: clinical features and surgical results in 55 cases. J Neurosurg 1984;60:238–247.
17. Riche MC, Reizine D, Melki JP, Merland JJ. Classification of spinal cord malformations. Radiat Med 1985;3: 17–24.
18. Heimberger K, Schnaberth G, Koos W, et al. Spinal cavernous hemangioma (intradural-extramedullary) underlying repeated subarachnoid hemorrhage. J Neurol 1982; 226:289–293.
19. Fontaine S, Melanson D, Cosgrove R, Bertrand G: Cavernous hemangiomas of the spinal cord: MR imaging. Radiology 1988;16:839–842.
20. Rebner M, Gebarski SS. Magnetic resonance imaging of spinal cord hemangioblastoma. AJNR 1985;6:287–289.
21. Sato Y, Waziri M, Smith WL, et al. Hippel-Lindau disease: MR imaging. Radiology 1988;166:241–246.
22. Silbergeld J, Cohen WA, Maravilla KR, et al. Supratentorial and spinal cord hemangioblastomas: gadolinium-enhanced MR appearance with pathologic correlation. J Comput Assist Tomogr 1989;13:1048.
23. Laguna J, Cravioto H. Spinal cord infarction secondary to occlusion of the anterior spinal artery. Arch Neurol 1973;28:134–136.
24. Mawad ME, Rivera V, Crawford ES, et al. Spinal cord ischemia after resection of thoracoabdominal aortic aneurysms: MR findings in 24 patients, AJNR 1990;11:987–992.
25. Sandson TA, Friedman JH. Spinal cord infarction: report of 8 cases and review of the literature. Medicine 1989; 68:282–292.
26. Yuh WTC, Marsh EE, Wang AK, et al. MR of spinal cord and vertebral body infarction. AJNR 1992;13:145–154.
27. Brown E, Virapongse C, Gregorios JB. MR imaging of cervical spinal cord infarction. J Comput Assist Tomogr 1989;13:920–922.
28. Casselman JW, Jolie E, Dehaene I, et al. Gadolinium enhanced MR imaging of infarction of the anterior spinal cord. AJNR 1991;12:561.
29. Mikulis DJ, Ogilvy CS, McKee A, et al. Spinal cord infarction and fibrocartilagenous emboli. AJNR 1992;13:155–160.
30. Larsson E-M, Desai P, Hardin CW, et al. Venous infarction of the spinal cord resulting from dural arteriovenous fistula: MR imaging findings. AJNR 1991;12:739.
31. Tantana S, Pilla TS, Luisiri A. Computed tomography of acute spinal subdural hematoma. J Comput Assist Tomogr 1986;10:891–893.
32. Levy JM. Spontaneous lumbar subdural hematoma. AJNR 1990;11:780–781.

CHAPTER / 11

Syrinx and Cysts

J. Sherman / B. C. P. Lee

Cystic lesions within the spinal canal may involve the spinal cord or may be located outside the spinal cord but within the dural tube. Those within the spinal cord are termed as *syringomyelia* or *hydromyelia*, whereas intradural extramedullar cysts may have varied etiologies. In both these conditions, the symptom complex is associated with progressive neurologic deficits and periods of remission. Extramedullary cystic lesions often present with symptoms of cord compression, whereas intramedullary cystic lesions frequently present with a combination of sensory and motor deficits. Imaging studies are important in differentiation of the location of these lesions. At the present time, MRI is the primary imaging modality, occasionally combined with CTM.

Syringomyelia

Syringomyelia is a chronic disorder involving the spinal cord. Pathologically, it is characterized by the presence of longitudinally oriented cavities and gliosis. If the cavitation and gliosis are limited to the brainstem, the term *syringobulbia* is applicable. The term *hydromyelia* has been used to describe the appearance of dilatation of the central canal of the spinal cord, while the term *syringomyelia* has been reserved for cavities independent of the central canal [1]. From a practical viewpoint, it is impossible to differentiate most cases of true hydromyelia from true syringomyelia. Consequently, recent literature tends to unite the two terms (*syringohydromyelia*) [2] or to use the terms *syringomyelia* or *syrinx* [3–5] in a general manner to represent any spinal cord cyst, and we shall use these terms in the same general sense in this chapter.

Pathogenesis of Syringomyelia

Syrinx cavities contain fluid which may have many sources. The means by which the fluid accumulates and the cavities extend are various. Syringomyelia is thus not one entity but has many causes, just as cystic spaces in the brain may be of diverse origins. The shape and structure of the spinal cord and the spinal canal have a strong influence on the appearance of the spinal cord cystic spaces, which makes the cavities and the clinical signs similar, regardless of the etiology.

Syringomyelia may be extensions from a true hydromyelia but may not communicate with the central canal. When the intramedullary cavity extends to the medulla, it is known as *syringobulbia*. Cerebellar tonsillar ectopia or the Chiari I malformation is commonly found on MRI scans in patients presenting with spinal symptoms of syringomyelia [6–10]. Major neurologic defects in addition to those of syringomyelia are commonly present with the more severe Chiari II malformation [11–14]. Other congenital abnormalities of the craniocervical junction, such as platybasia, basilar invagination, and occipitalization of the atlas, may be present concurrently.

Although most of the intramedullary cavities are idiopathic and presumably congenital (Table 11.1), less often they may be secondary to trauma, infection, or inflammation and intramedullary tumors. The precise mechanism of idiopathic syringomyelia is unknown, but the frequent association of cerebellar ectopia has prompted suggestions of a causal relationship between these abnormalities [6, 7]. It is thought that the pathogenesis of hydromyelia depends on whether the fourth ventricle communicates with the central canal [15] (Fig. 11.1). Gardner

Table 11.1.
Etiology of Spinal Cord Cysts

1.	Idiopathic
2.	Congenital
	Chiari Anomaly
	Dandy-Walker cyst
	Ventriculus terminalis
3.	Chronic cord compression from degenerative disk disease
4.	Trauma
	Acute: Hematomyelia/Partial cord transection
	Chronic: Delayed effects of cord injury
5.	Tumor
	Hemangioblastoma
	Ependymoma
	Astrocytoma
6.	Infection
7.	Vascular
	Cord infarction
	Vascular anomalies
8.	Post surgical

Differentiation of the etiologic cause for the cyst is based on prior history, the location within the spinal cord, the extent of the cyst, enhancement characteristics on MRI, and its appearance on early and delayed CTM. CTM is indicated in spinal cord cysts associated with trauma, and when syringomyelia is suspected with appearance of an atrophic spinal cord on MRI.

[5, 16, 17] postulates that the primary defect is a lack of permeability of the fourth ventricular exit which hinders exit of cerebrospinal fluid (CSF) into the subarachnoid space. The arterial pulsations of the choroid plexuses of the lateral and third ventricles exert a "water hammer" effect which pushes the cerebellar tonsils below the foramen magnum and propels ventricular CSF into the central canal of the spinal cord [5, 16, 17] (Fig. 11.1*A*). Williams [18] postulates that the hydromyelia occurs because of differences in the hydrostatic pressure of the intracranial and spinal subarachnoid CSF (Fig. 11.1*B*). Normally, CSF exits through the fourth ventricular openings into cranial and caudal subarachnoid spaces. The hydrostatic pressure of the subarachnoid spaces are the same in these compartments. In the presence of cerebellar tonsillar ectopia, the intracranial egress of the CSF space is impeded, thus lowering the intracranial subarachnoid pressure. Egress into the spinal subarachnoid space, however, is unimpeded, but eventually the direction of the flow is reversed and CSF reenters the fourth ventricle. At this stage, the ventricular CSF, instead of exiting into the spinal subarachnoid space, flows caudally into the central canal. A difference in the hydrostatic pressure within the central canal and the spinal subarachnoid is said to account for the differences in size of the hydromyelia at different locations [19–22]. As the pressure within the central canal increases, CSF may escape into the parenchyma of the spinal cord, giving rise to true fissures or "syrinxes" which communicate with the dilated canal. The preceding two hypotheses may explain formation of hydromyelia in the presence of the the Chiari malformation, but not when this anomaly is absent.

Abouker [23] believes that obstruction at the foramen magnum causes increased spinal CSF pressure by inhibiting upward movement of CSF to sites of absorption (Fig. 11.1*C*). This is in contradistinction to Williams, who believes that foramen magnum obstruction prevents the downward passage of CSF. Abouker theorizes that increased pressure drives CSF through the spinal parenchyma or via a pathway along the posterior roots into the spinal cord, where chronic edema gradually undergoes cavitation. The same basic mechanism is invoked by Ball and Dayan [24], except that they propose that the CSF passes into the central canal by tracking under pressure into the spinal cord along perivascular (Virchow-Robin) spaces (Fig. 11.1*C*). This mechanism is often invoked to explain the development of syringomyelia in patients who have suffered trauma, arachnoiditis, or cord compression (Fig. 11.2).

In noncommunicating syringomyelia there is thought to be no direct communication with the central canal, fourth ventricle, or subarachnoid space. Noncommunicating syringomyelia usually occurs following trauma or is associated with tumor or arachnoiditis.

Syringomyelia communicating with the central canal is more frequently seen than true hydromyelia. One explanation is that these lesions originate as fissures from the ependymal lining of a true hydromyelia which enlarge and extend into the adjacent spinal cord parenchyma [5, 15–17, 19–22]. The explanation of the progressive enlargement of the syringomyelia is unclear. True hydromyelia and isolated syringomyelia are seen commonly with congenital spinal cord anomalies, such as tethered cord and diastematomyelia (see "Congenital Abnormalities").

Mechanisms of Syrinx Enlargement

The mechanisms of syrinx enlargement are not completely understood. Some patients have syrinx cavities that remain stable throughout the life of the patient. In other patients, the syrinx undergoes craniocaudad expansion, leading to progressive disability. Syrinx movement is probably dependent on transmitted pressure from the subarachnoid space. Pulsations in the subarachnoid space are transmitted to the fluid in the syrinx by generating waves in the walls. Marked fluid shifts are seen in coughing and other Valsalva-like maneuvers [17, 24]. These pulsations are thought to be caused by engorgement of the epidural venous plexus. The effect of these fluid shifts is to cause the cavity to extend into the part of the cord offering least resistance (see Fig. 11.1). Craniocaudad pulsation is most prominent, but the movement of CSF within the cavity is uneven and turbulent. There is usually less movement of fluid at the rostral and caudal ends of a syrinx than in the middle of the syrinx. Larger syrinx cavities usually demonstrate more prominent fluid motion. Gliotic ridges and bands may form around the margins of the cavity and offer temporary or permanent resistance to syrinx progression. These glial ridges produce the "beaded" or "stack of coins" appearance of syringomyelia cavities. Passive filling of preexisting cavities or patent segments of the central spinal canal may occur. Adhesions around the spinal cord caused by arachnoiditis may pull the cord open and also prevent normal craniocaudad spinal cord movement. Hemorrhage into the syrinx (Gowers syringal hemorrhage) is a rare cause for

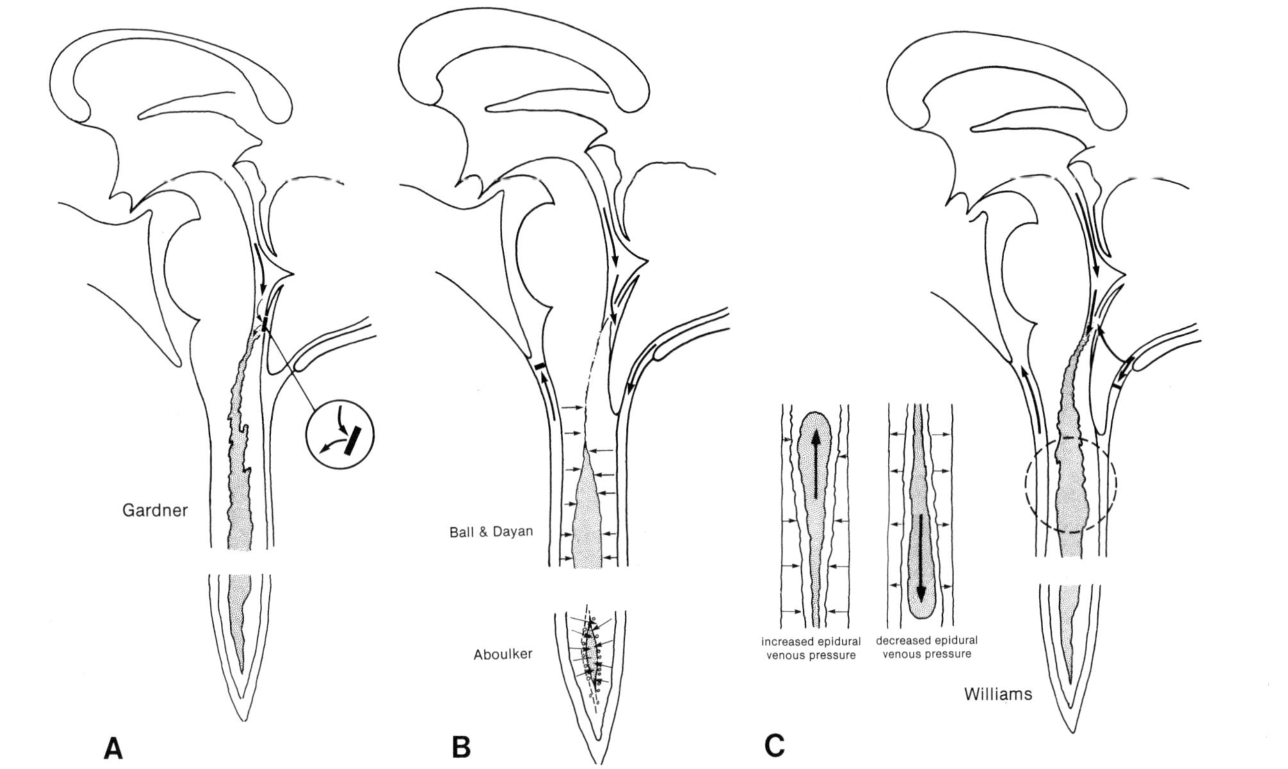

Figure 11.1. Theories of pathogenesis of syringomyelia (see text). **A,** Diagram of syringomyelia illustrating Gardner's theory of syrinx pathogenesis. Gardner theorizes that fourth ventricle outlet foramina are developmentally hypoplastic or obstructed in prenatal development allowing continual transmission of pressure into central canal of the spinal cord. **B,** Diagram of syringomyelia illustrating Williams' theory of syrinx pathogenesis. Williams emphasizes intermittent ball-valve theory of obstruction at foramen magnum as initiating factor. He theorizes that "sloshing" fluid in syrinx leads to propagation of the cavity in the spinal cord. **C,** Diagram of syringomyelia illustrating the theory of Aboulker and Ball. They postulate that fluid enters spinal cord along the perivascular spaces of Virchow-Robin or along perineural spaces of posterior roots.

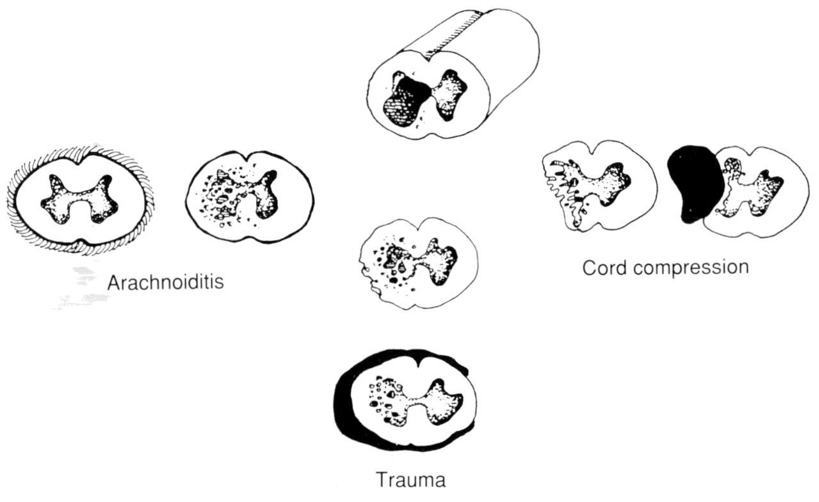

Figure 11.2. Diagram of pathogenetic mechanisms of syringomyelia due to arachnoiditis, trauma, or cord compression. In arachnoiditis, leptomeningeal granulation tissue and thickening damage the microcirculation of the spinal cord and disrupt normal flow of CSF. Microcystic myelomalacia develops and may progress to true syrinx formation. In trauma, the cord may be damaged by direct compression and indirectly by hemorrhagic arachnoiditis. Cord edema, myelin breakdown, and hemorrhage together with damage to the cord microcirculation lead to cyst development. A chronic compressive lesion causes permanent cord deformity and damages the cord microcirculation and the surrounding leptomeninges. When the compression is relieved, the perivascular spaces enlarge, and CSF may be forced into the cord, forming small cysts which subsequently coalesce into larger cysts.

acute syrinx extension [25]. Spontaneous reduction and eventual disappearance of a syrinx can occur [26], possibly due to autodecompression through a slit or fissure in the spinal cord.

Clinical Manifestations

The neurologic deficits caused by syringomyelia and hydromyelia depend on their location within the spinal cord. A slowly evolving hydromyelia may be virtually asymptomatic until it is very large, or neurologic deficits may be produced by the associated abnormalities such as the tonsil ectopia and spinal cord tethering. A syrinx within a portion of the spinal cord affecting motor or sensory long tracts will be clinically apparent at an early stage. In the classic case, the manifestations of syringomyelia begin in late adolescence or early adulthood and progress irregularly with long periods of stability. The clinical picture is variable, depending on the cross-sectional and vertical extent of cord destruction. Syringomyelia may be suspected when the patient reports that superficial injuries or burns of the arm and hand have caused no pain. Neuropathic arthropathy (Charcot's joint) may develop secondary to the acquired loss of sensibility to deep pain. Advanced shoulder joint destruction may be the presenting sign of long-standing syringomyelia (Fig. 11.3). Examination may disclose segmental anesthesia of the dissociated type (loss of pain and temperature sense with preservation of the sense of touch) in a capelike distribution over the neck, shoulders, and arms. There may be segmental weakness and atrophy of the hands and arms with loss of tendon reflexes. The principal site of the intramedullary lesion is the anterior commissure, but involvement of the ventral and dorsal areas of the cord is common. When the anterior gray matter is involved, segmental amyotrophy occurs. With dorsal lesions, segmental pain may develop in addition to other sensory disorders. Pain is often severe in one-third of patients and may be most intense in an analgesic limb (anesthesia dolorosa). Some of the cavities extend upward into the lateral tegmentum of the medulla (syringobulbia) or higher levels of the brainstem, causing analgesia and thermoanesthesia of one or both sides of the face and other disorders of lower cranial nerves. It is thus important not only to diagnose the abnormality but also to accurately determine the precise location and configuration of the syrinx relative to the long tracts within the spinal cord. Scoliosis, which is a common association, is probably the result of asymmetrical involvement of the motor tracts by the syringomyelia or hydromyelia.

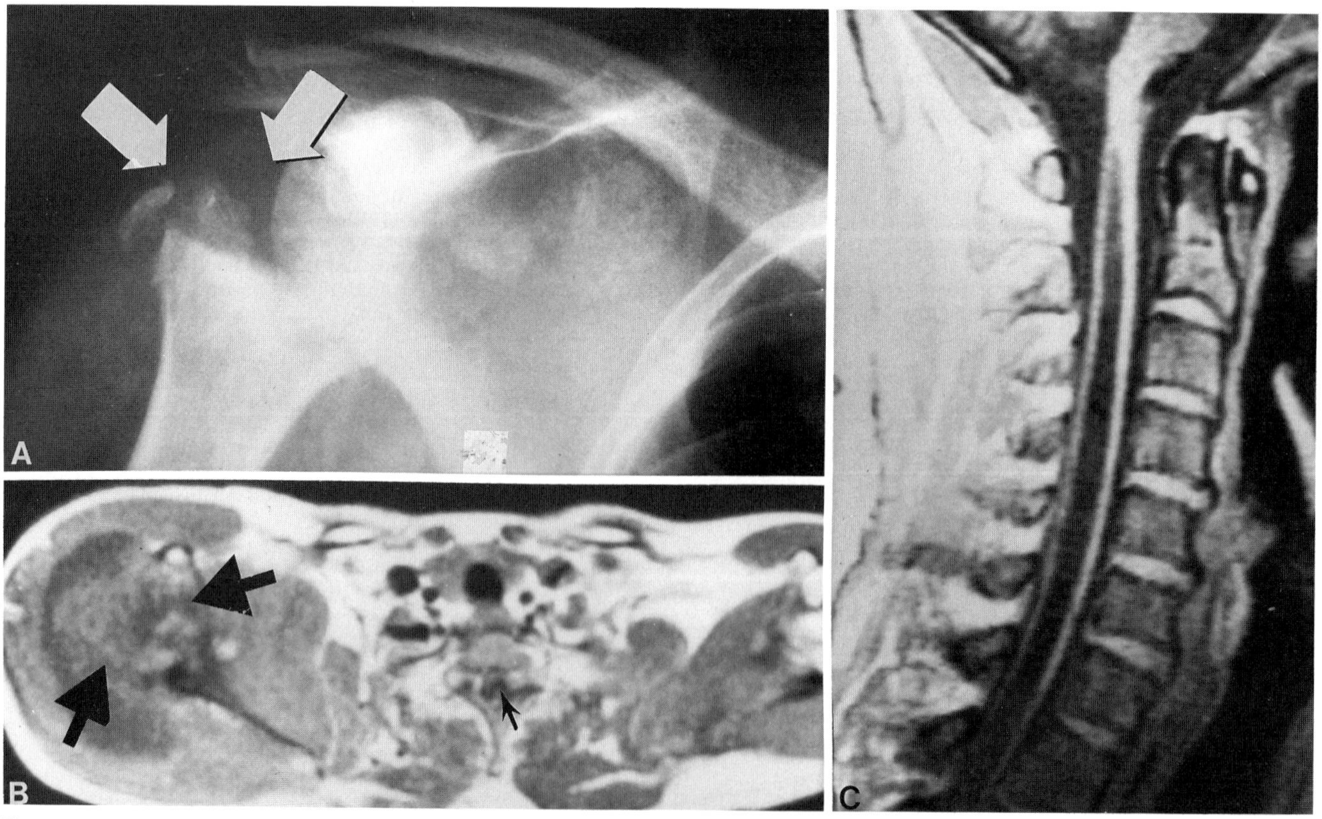

Figure 11.3. Neuropathic joint caused by syringomyelia. A, Radiograph of shoulder showing markedly destructive neuropathic joint (Charcot's joint). **B,** MRI of upper thorax including right shoulder showing destructive hypertrophic changes in neuropathic joint (*large arrows*). Syrinx cavity present in cord (*small arrow*). **C,** Sagittal MRI showing idiopathic syringomyelia.

Imaging of Syringomyelia

COMPUTED TOMOGRAPHY

Non-contrast-enhanced computed tomography (NCCT) may show a large spinal cord cavity as an area of low density (Fig. 11.4); however, medium- and small-sized cavities can be easily missed on NCCT. CT following intravenous contrast administration (IVCT) may demonstrate an enhancing spinal cord tumor associated with syringomyelia [27]. CT scanning with intrathecal injection of water-soluble contrast medium is necessary to demonstrate the cord size and may show a syrinx cavity on delayed scans performed 6 to 24 hours after injection [28, 29]. Opacification of syrinx cavities may occur regardless of the etiology of the cavity (Fig. 11.5). The mechanism of delayed opacification is probably direct diffusion of contrast material into the cord via perineural or perivascular pathways [30].

In general, the reliability of CT studies in evaluating suspected syrinx depends on an early CT scan following intrathecal contrast injection and, if the syrinx cavity is not visualized, even if the cord appears atrophic, a delayed CT 6 to 24 hours later, when the syrinx may be opacified with contrast material. This method has been superseded by high-resolution MRI because of the more sensitive contrast resolution of the latter technique [31–37].

MAGNETIC RESONANCE IMAGING

T1-weighted images in the sagittal and axial planes are the ideal sequence for evaluating intramedullary cavities. The scans should be performed with a surface or Helmholtz coil, and the sections should be no more than 4 mm thick. Sagittal T2-weighted double-echo sequences are useful for evaluating the signal of the fluid within the cavities compared with the surrounding subarachnoid space. T1-weighted images after intravenous gadolinium contrast administration either in the sagittal or axial planes are necessary for distinguishing primary syringomyelia from syrinxes secondary to tumors, vascular malformations, and inflammatory causes.

The total length and diameter of the syrinx should be evaluated by using a combination of sagittal and axial T1-weighted images, as well as T2-weighted sagittal sections. The syrinx cavity should appear as a CSF-intensity fluid collection within the cord. Small cavities may be overlooked if transaxial images are not obtained. The cord is enlarged in about 50% of patients and atrophic in about 16% of patients [37]. On long-TR images, the syrinx fluid should be isointense compared with CSF, but signal differences may be present due to fluid pulsation. Some cavities appear "tense" and contain nonpulsatile fluid. The fluid in some syrinx cavities is very pulsatile and leads to prominent low signal intensity (CSF flow void) on long-TR images obtained without motion-suppression techniques (Fig. 11.6). The low signal intensity is due to a combination of phase-shift effects and time-of-flight effects. Truncation and movement artifacts may resemble cavities and may lead to a false-positive diagnosis of syringomyelia. The truncation artifact is seen as dark and bright bands parallel to the spinal cord–subarachnoid space interface; the dark band within the spinal cord is mistaken for a syrinx involving the entire seg-

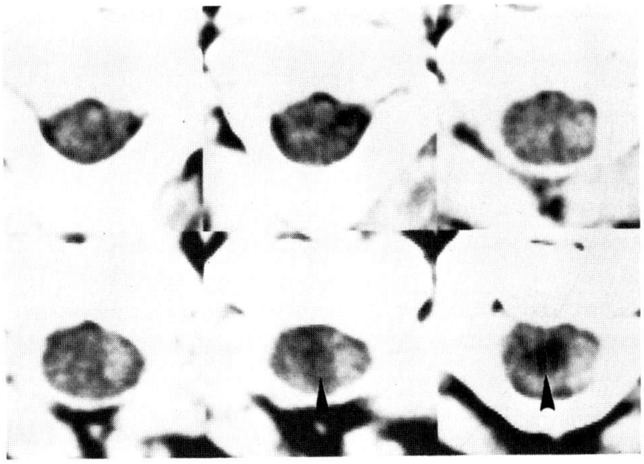

Figure 11.4. Syringomyelia on NCCT. Contiguous 5-mm CT section demonstrates a large syrinx (*arrowhead*) within the cervical cord. Small syrinx cavities can be easily missed.

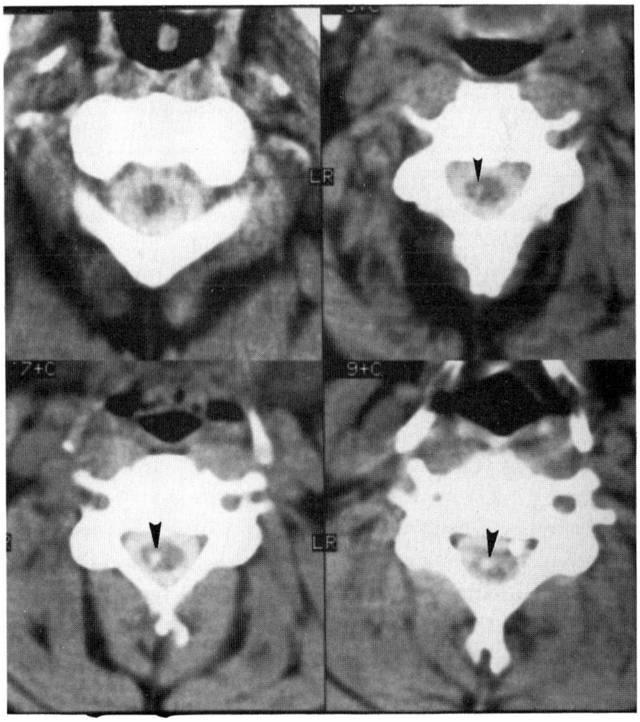

Figure 11.5. Syringomyelia on CTM. Axial CT section following myelography demonstrates an irregular collection of hyperdense contrast material within the substance of the cord in a patient with posttraumatic syrinx. This was not detected on myelography.

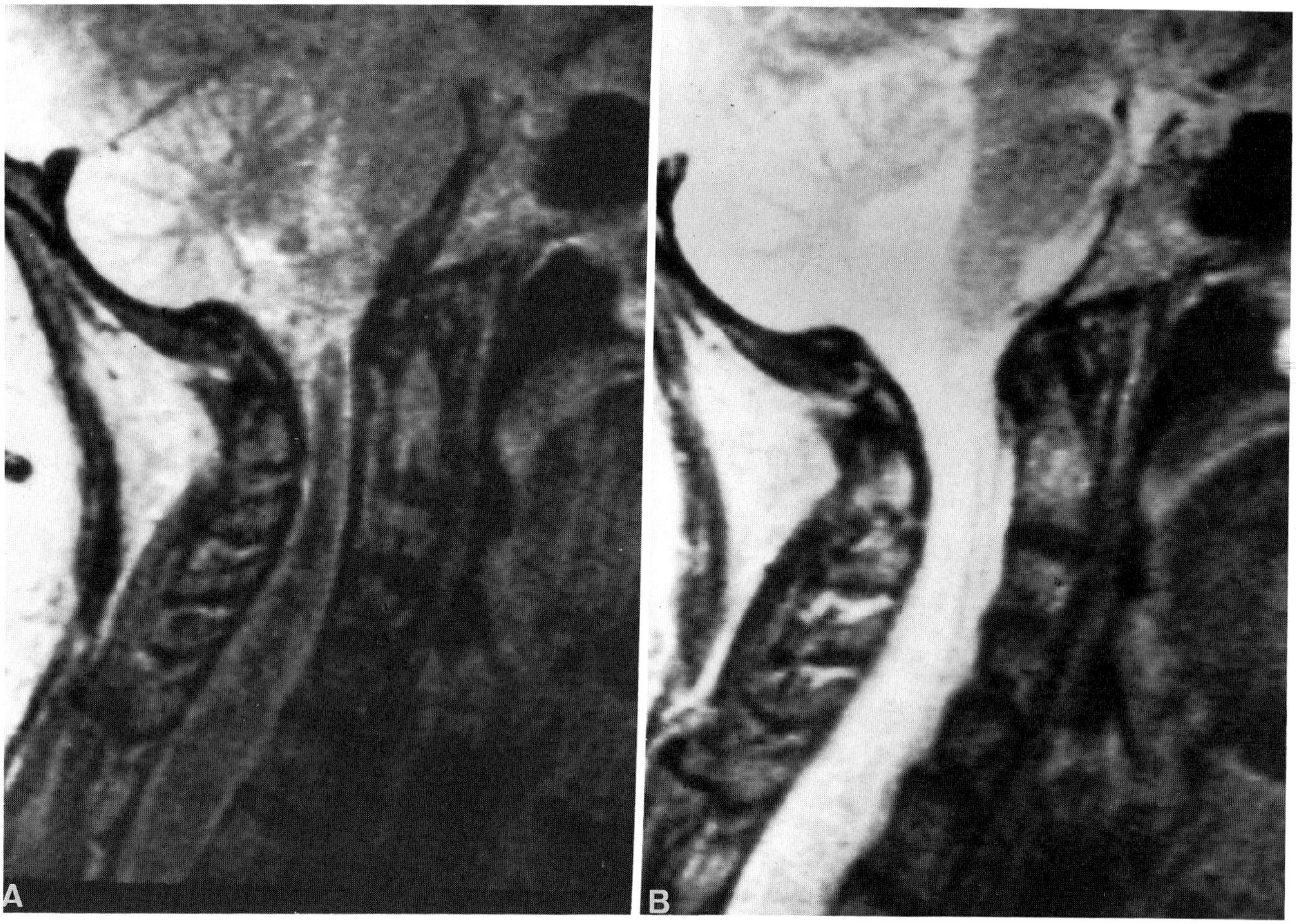

Figure 11.6. Syringomyelia with demonstration of the effect of pulsatile CSF and use of motion artifact suppression technique (MAST). A, Sagittal SE 2000/80 image (without MAST) of brain and cervical cord with syringomyelia. Pulsatile fluid in syrinx causes marked hypointensity in cavity compared with CSF in brain. Syrinx cavity enlarges the spinal cord. **B,** Sagittal SE 2000/80 MAST image of same location. Note uniform high signal appearance of CSF in brain, spine, and syrinx. Overall image "quality" has improved, but syrinx cavity is no longer apparent.

ment of the cord imaged. This artifact is due to aliasing or undersampling of data in the phase-encoding direction [39, 40]. The artifact may be corrected by decreasing the field of view using a localized surface coil and by increasing the data sampling. Increasing such sampling in the frequency readout corrects the artifact in this direction and is implemented routinely because it requires no additional scanning time. Increasing the data sampling in the phase-encoding direction, on the other hand, prolongs the scanning time and is impractical in most instances. Artifacts secondary to vascular pulsation and patient movement are also in the phase-encoding direction and have a similar appearance to truncation artifacts. Unlike truncation artifacts, which are restricted to the spinal canal, movement artifacts are also visible outside the anatomic boundaries of the body. The latter artifacts may be corrected by gradient compensation and cardiac gating implementable in most MRI scanners [41–45].

The appearance of a syrinx cavity on long-TR images changes significantly depending on the type of gating or motion suppression applied (Fig. 11.7). In some cases, an actual change in cavity diameter can be documented by comparing cardiac gated T1-weighted images in end diastole with those in end systole (Fig. 11.8). Quantitative methods for CSF flow velocity measurements are now available. These techniques typically use a phase-contrast cine MRI pulse sequences to generate velocity-encoded images during the cardiac cycle. This technique shows direction of fluid pulsations as well as velocity and also can be used to evaluate brain and spinal cord motion [38]. Myelomalacia and gliosis are frequently present around a syrinx cavity and appear hyperintense compared with the syrinx fluid on density-weighted images (short TE, long TR).

Idiopathic Syringomyelia

Idiopathic syringomyelia and hydromyelia have to be differentiated from syrinx from other causes. The

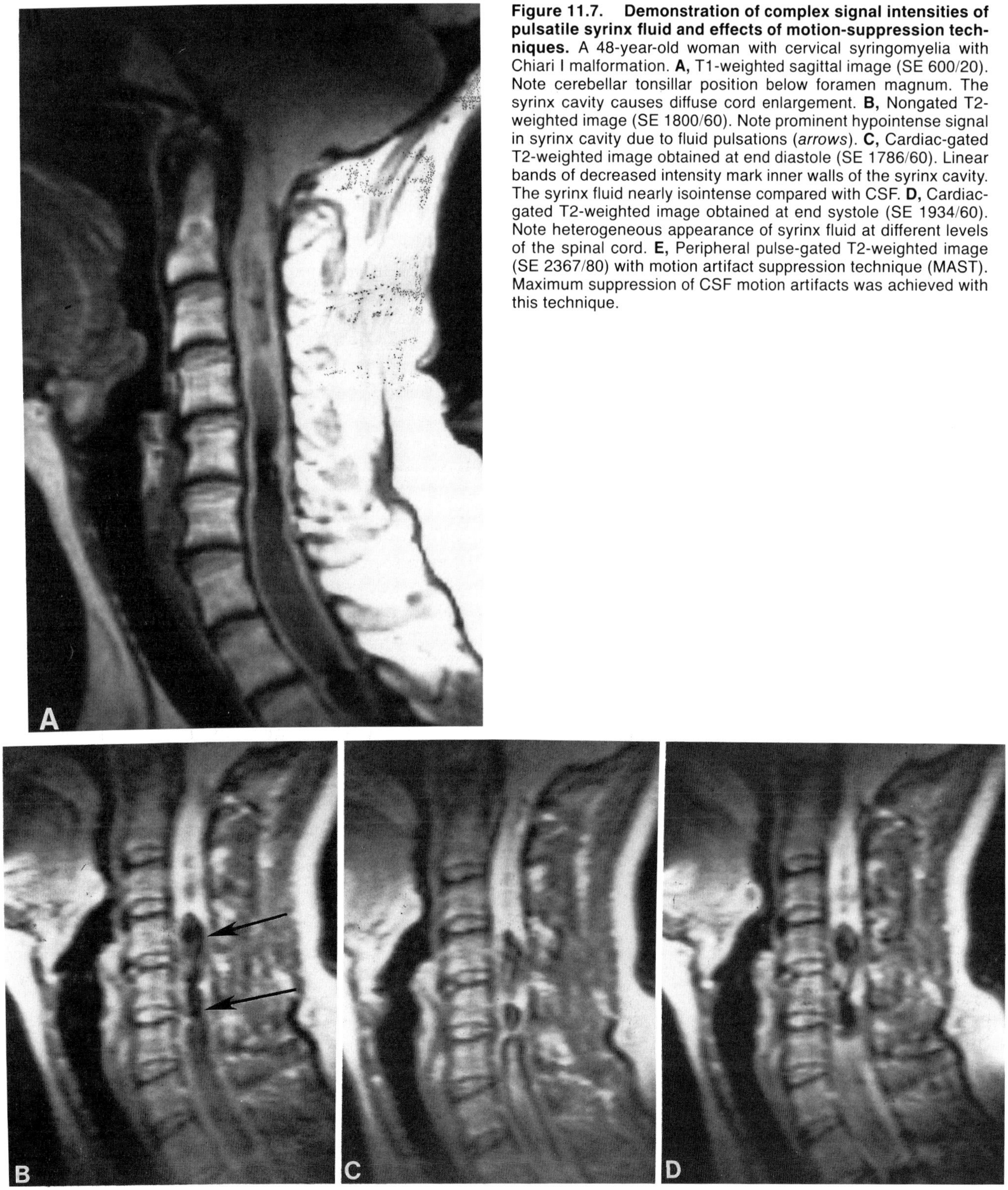

Figure 11.7. Demonstration of complex signal intensities of pulsatile syrinx fluid and effects of motion-suppression techniques. A 48-year-old woman with cervical syringomyelia with Chiari I malformation. **A,** T1-weighted sagittal image (SE 600/20). Note cerebellar tonsillar position below foramen magnum. The syrinx cavity causes diffuse cord enlargement. **B,** Nongated T2-weighted image (SE 1800/60). Note prominent hypointense signal in syrinx cavity due to fluid pulsations (*arrows*). **C,** Cardiac-gated T2-weighted image obtained at end diastole (SE 1786/60). Linear bands of decreased intensity mark inner walls of the syrinx cavity. The syrinx fluid nearly isointense compared with CSF. **D,** Cardiac-gated T2-weighted image obtained at end systole (SE 1934/60). Note heterogeneous appearance of syrinx fluid at different levels of the spinal cord. **E,** Peripheral pulse-gated T2-weighted image (SE 2367/80) with motion artifact suppression technique (MAST). Maximum suppression of CSF motion artifacts was achieved with this technique.

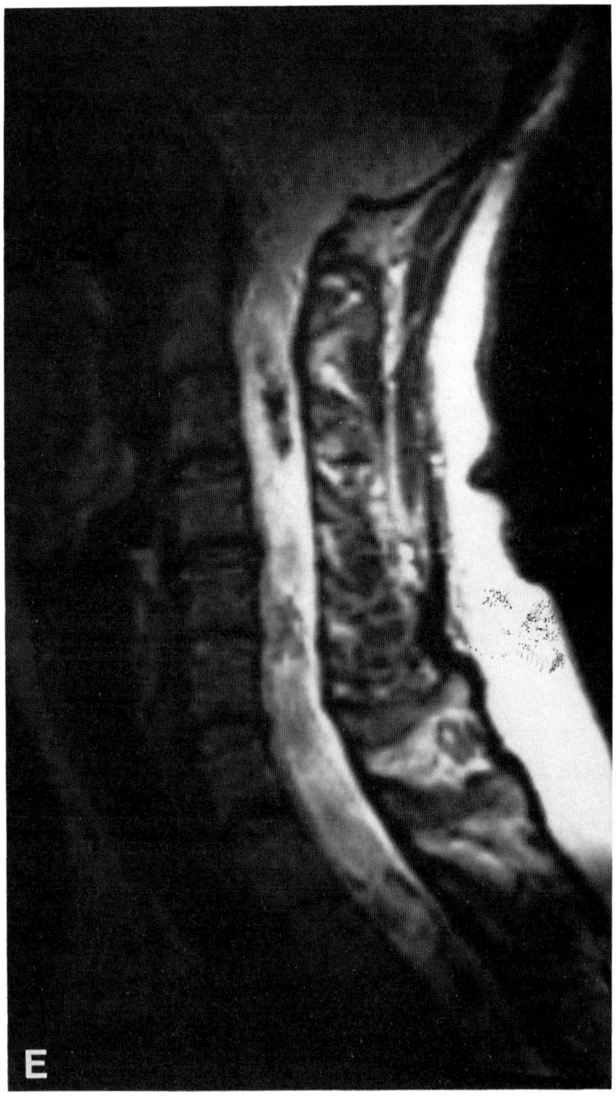

Figure 11.7–*continued*

exact location of the syrinx and the extent of involvement of the spinal cord and brainstem should be demonstrated. Abnormalities of the cerebellum and the spinal cord such as cerebellar tonsillar ectopia and spinal cord tethering also should be clearly shown. Bony congenital anomalies of the craniocervical junction and caudal region also should be documented (see Fig. 11.21). The contiguity of multiple syrinxes and the etiology of the cavities are important considerations in the management of syringomyelia.

In general, intramedullary cavities should be termed *syringomyelia* unless there is unequivocal evidence that they represent a dilated central canal in axial views. The caudal and cranial extents of the cavity are best evaluated in the lateral T1-weighted view. The shape depends on their size; small and moderate-sized syrinxes are widest in the midportion and taper at the extremities (Fig. 11.9). Large lesions are more uniform in size, may extend the full length of the spinal cord, and are often blunted at the extremities (Fig. 11.10). The degree of enlargement of the spinal cord is proportional to the size of the syrinx. Syringomyelias involving the brainstem (syringobulbias) usually have a tapered appearance (Fig. 11.11). These lesions are potentially life threatening and may require urgent surgical therapy. Multiple septa may be visible and may appear to completely divide the cavity into compartments (Fig. 11.12). Axial views not only are used in assessing the configuration of large cavities but also are sensitive in detecting small cavities not visible on lateral views because of the partial volume effect due to the thickness of the sagittal sections. A cavity should be considered a hydromyelia only when it is central at the site of the central canal location and is completely symmetrical (Figs. 11.13 and 11.14). Some syrinxes may be fairly symmetrical and still appear contiguous with the central canal (Fig. 11.15). Septa of various thicknesses may separate the syrinx into compartments in the sagittal and axial planes (Figs. 11.12 and 11.16). These septa are usually incomplete, and the entire syringomyelia nearly always collapses with surgical decompression. The presence of a signal void on T2-weighted images has been used by some authors to differentiate idiopathic communicating syringomyelias from those secondary to other causes [33, 37, 46]. This feature is not always reliable because of the current practice of using motion-compensatory techniques for routine spinal cord imaging (Fig. 11.17). However, an increase in the signal on T1-weighted and proton-density images above CSF suggests proteinaceous content, and a primary cause for the syringomyelia should be sought.

Differences in the size and configuration of syringomyelias have been explained by collapse and distension of portions of a continuous syringo- or hydromyelia (Figs. 11.11 and 11.18). Air myelography in the past has demonstrated the ability of syrinx to collapse. In this now-obsolete examination, the entire subarachnoid CSF accumulation is replaced by air. When the patient is inclined, the elevated portion of the spinal cord containing the syringo- or hydromyelia collapses and is small in caliber compared with the dependent lower portion. The reverse occurs when the dependent part is then elevated. This phenomenon, which is due to the higher hydrostatic pressure of the CSF within the syrinx/hydromyelia compared with the air introduced into the subarachnoid space, cannot explain differences in syrinx size seen on MRI [47]. It has been postulated that differences in the venous pressure in the subarachnoid space and the spinal cord may exert the same effect, i.e., increase in caudal subarachnoid venous pressure compresses the cor-

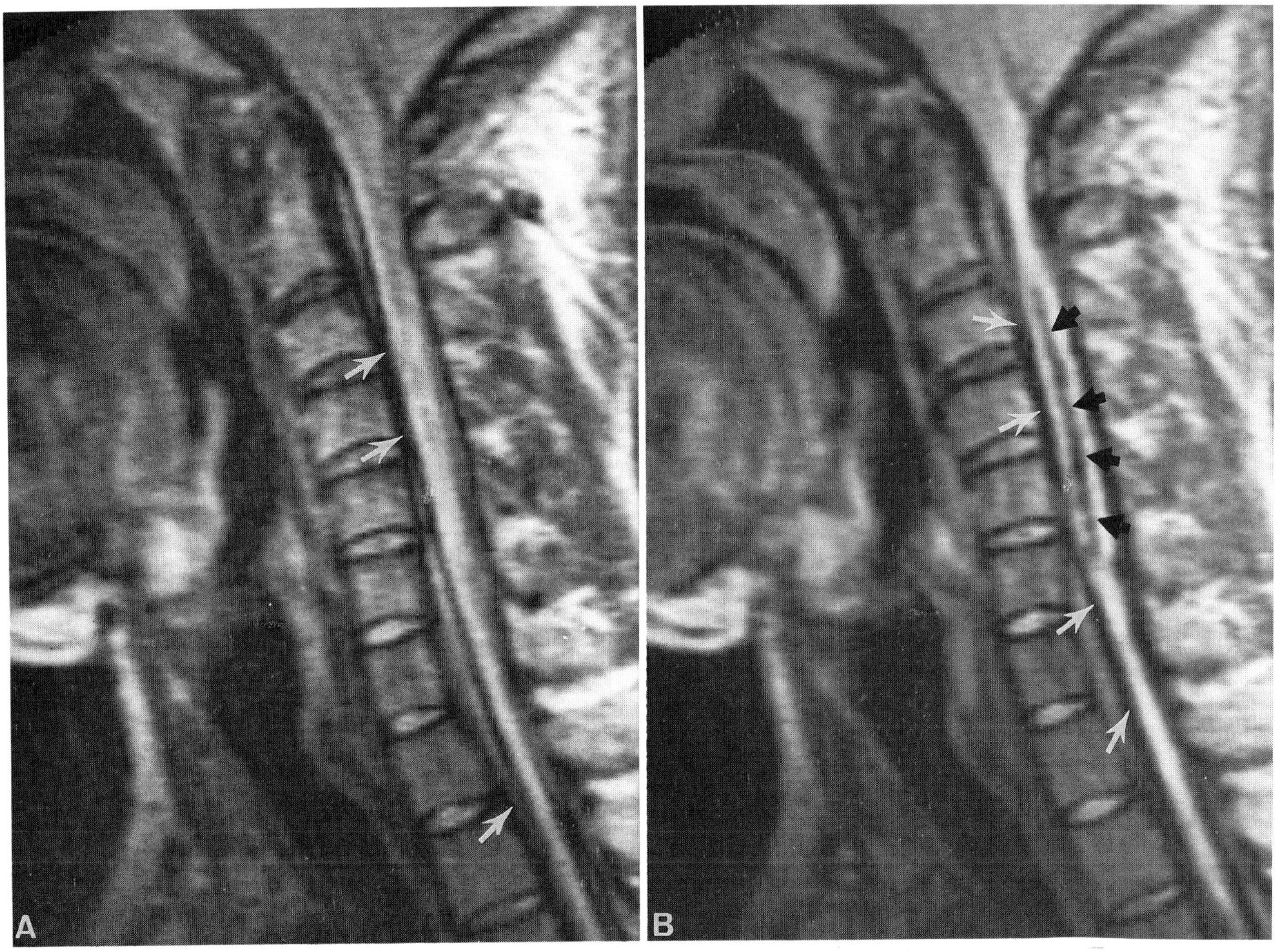

Figure 11.8. Demonstration of syrinx cavity pulsation. A 28-year-old man with syringomyelia and Chiari I malformation. **A,** Cardiac-gated image obtained at end diastole (SE 1554/60). Hypointense CSF is seen along the margin of the spinal cord (*white arrows*) but not in the syrinx cavity. **B,** Cardiac-gated image obtained at end systole (SE 1766/60) at identical slice location. Note expansion of cavity (*black arrows*) and increased visibility of the low-signal-intensity syrinx fluid. Fluid in subarachnoid space is also more visible because of signal loss (*white arrows*).

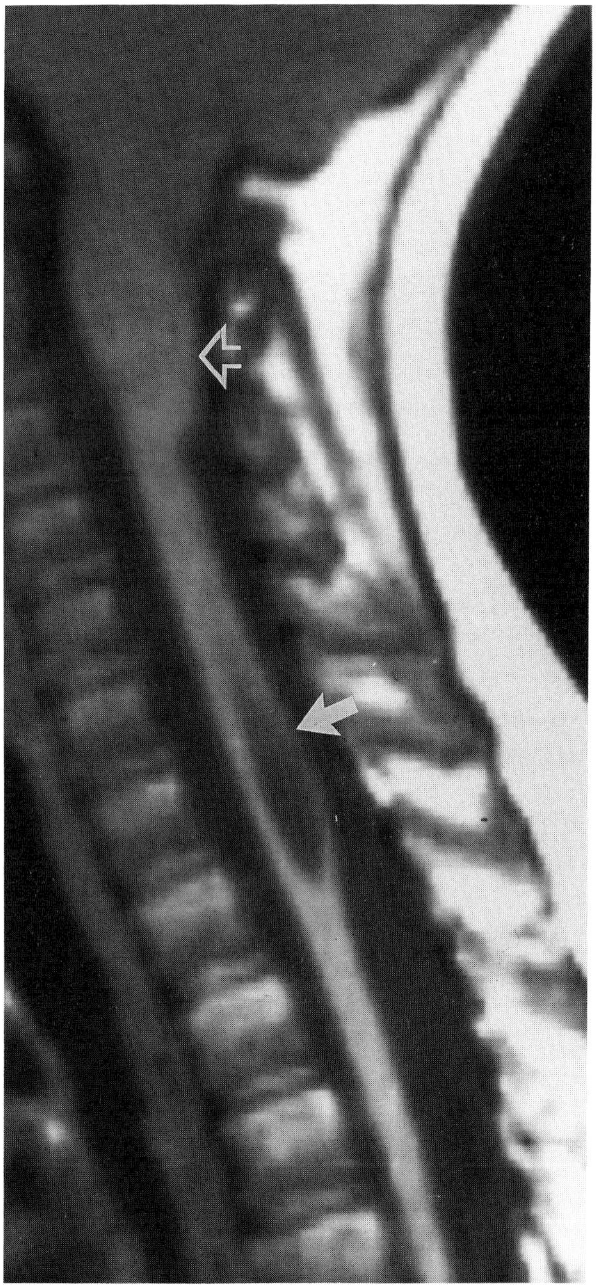

Figure 11.9. A syrinx is present in the midcervical region with tapering ends causing proportionate enlargement of the spinal cord (*arrow*). Note cerebellar ectopia (*open arrow*).

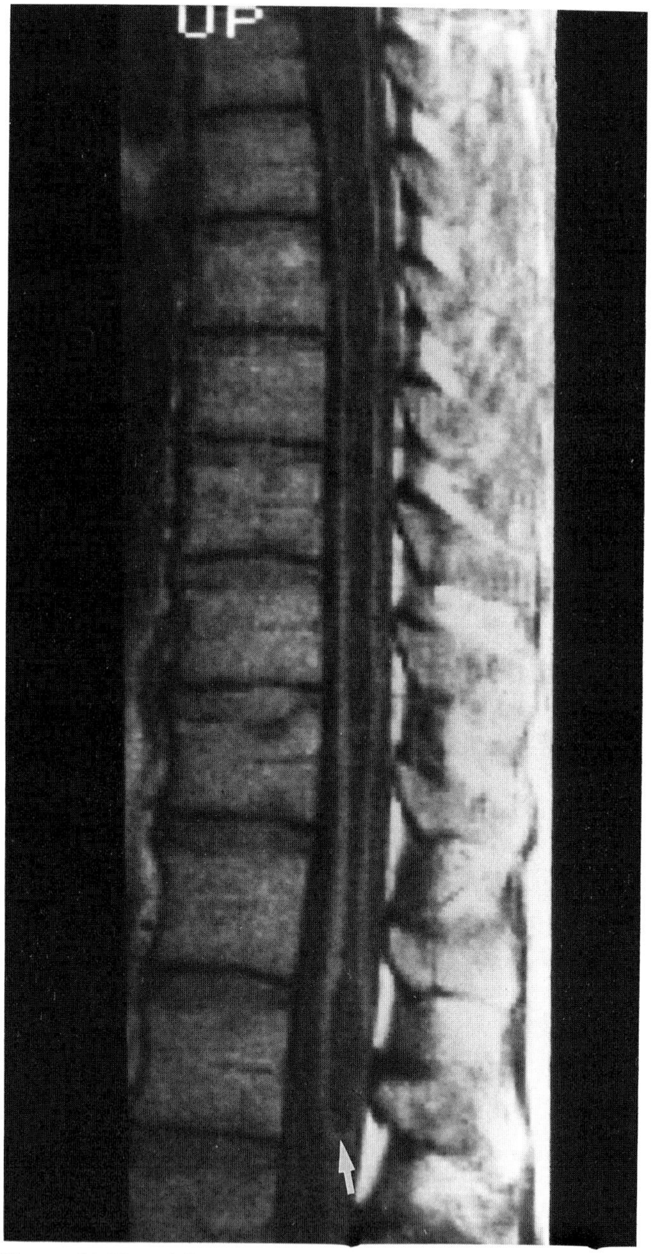

Figure 11.10. A large syrinx involves the entire thoracic spinal cord and conus medullaris. Note the blunted appearance of the caudal end of the cavity (*arrow*).

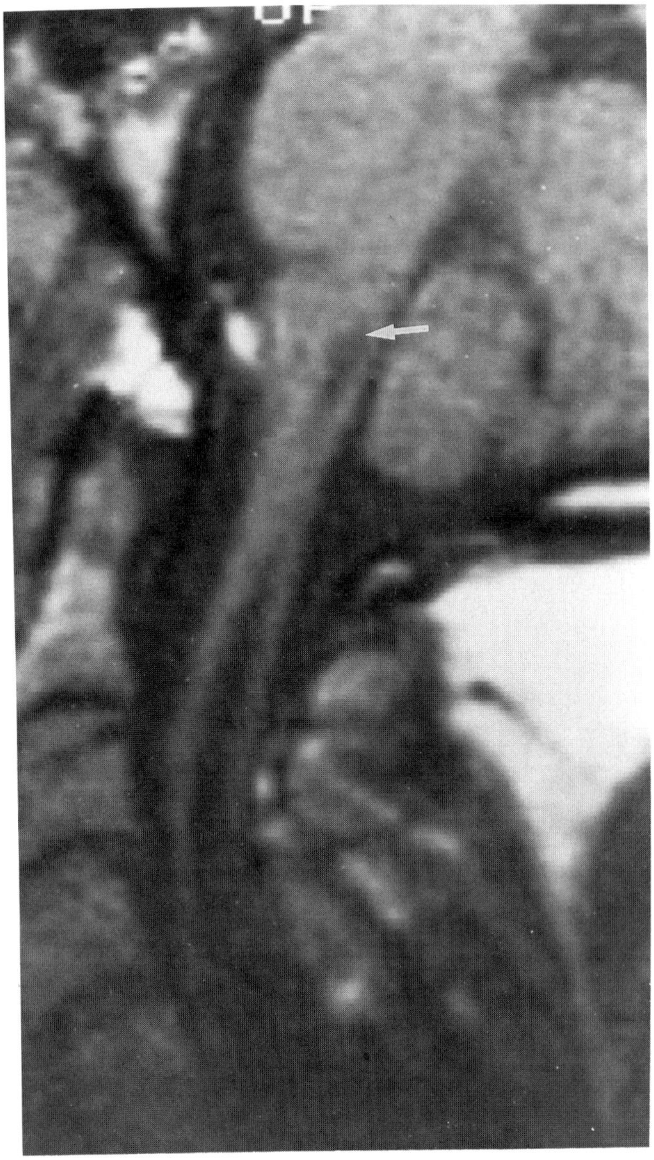

Figure 11.11. Cervical syrinx extends into the medulla. Note the tapered appearance of the cranial extension (*arrow*) and apparent lack of continuity between the two portions of the syrinx.

responding portion of the syringomyelia and causes the intramedullary fluid to be displaced into the cranial portion part of the spinal cord [18, 20, 21]. Multiple intramedullary cavities on MRI sequences should not be considered as separate syrinxes and usually represent portions of a single continuous cavity with intervening collapsed segments. It is seldom necessary to introduce contrast material directly into the intramedullary cavities to demonstrate continuity of the syringomyelia.

Congenital Syringomyelia

Congenital syringomyelia is by definition also referred to as *communicating syringomyelia.* This form is associated with Chiari I or Chiari II malformation and/or basilar invagination. The cerebellar tonsils average 8 to 9 mm below the foramen magnum (Figs. 11.7, 11.8, 11.12 and 11.18 to 11.21) in congenital syringomyelia, while in normal patients the cerebellar tonsils should lie no more than 3 mm below the level of the foramen magnum as measured on sagittal images [48]. The position of the cerebellar tonsils is evaluated best on sagittal T1-weighted images. Cerebellar tonsils with a normal configuration extending slightly below the foramen magnum are of little significance and should not be confused with true cerebellar tonsillar ectopia. The configuration of the tonsils and the available CSF space at the craniocervical junction are important considerations in this diagnosis. Ectopic cerebellar tonsils may appear as tongue-shaped extensions with a clear plane of cleavage from the upper cervical spinal cord, or they may be inseparable from the posterior aspect of the spinal cord (see Fig. 11.9). Solid masses can be simulated in the Chiari I malformation by arachnoiditis, which can cause the tonsils to become closely adherent to the dorsal surface of the upper cervical cord. In some cases, a prominent dorsal bulge occurs at the cervicomedullary junction. A "vermian peg" has been described in Chiari II malformation that also may simulate a localized cervicomedullary mass [49]. The spinal cord is focally or diffusely enlarged by the syrinx cavity in 50% of patients, focally atrophic in about 25% of patients, and normal in diameter in 25% of patients. Lobulation of the syrinx cavities often produces a "beaded" appearance (see Figs. 11.8 and 11.19). Eccentric cavities are present in one or more areas in about 75% of patients (see Figs. 11.14 and 11.16). The cisterna magna is very small or nonexistent in about 70% of patients. There is ventriculomegaly in about 50% of patients, but significant hydrocephalus is uncommon.

A firm diagnosis may be made on noncontrast MRI scans. Intravenous gadolinium should be administered in atypical cases, e.g., where the syrinxes are small, the spinal cord enlargement is inappropriately greater than the size of the syrinx, and the signal of the cystic fluid suggests increased protein content. Gadolinium-enhanced MRI is also indicated when the T2-weighted images reveal an abnormally increased signal intensity in the cord. Although in most cases this bright signal represents gliosis, occasional cases may demonstrate a coexisting tumor. Lack of contrast enhancement is used to distinguish primary syringomyelia from those secondary to tumors and vascular and inflammatory lesions.

Traumatic Syrinx

Intramedullary cavities in the spinal cord secondary to trauma may be divided into those occurring at

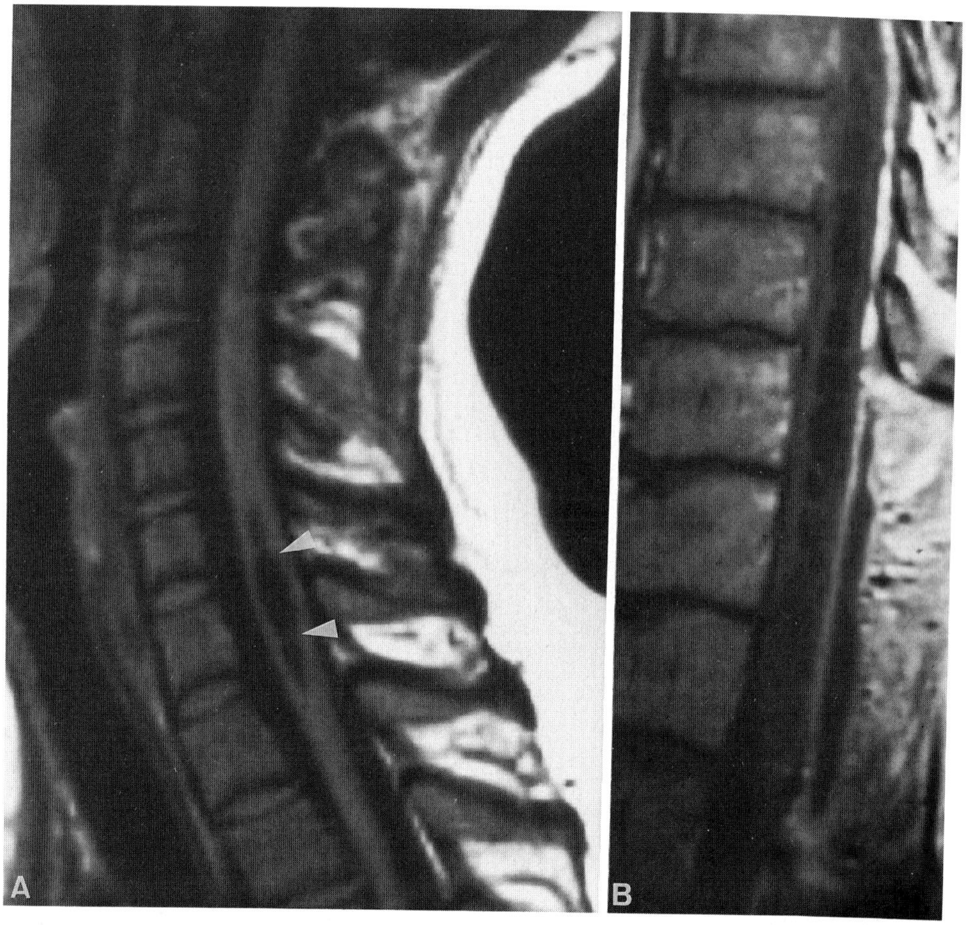

Figure 11.12. Cervical (**A**) and thoracic (**B**) spinal cord shows multiple septa within the syringomyelia (*arrowheads*).

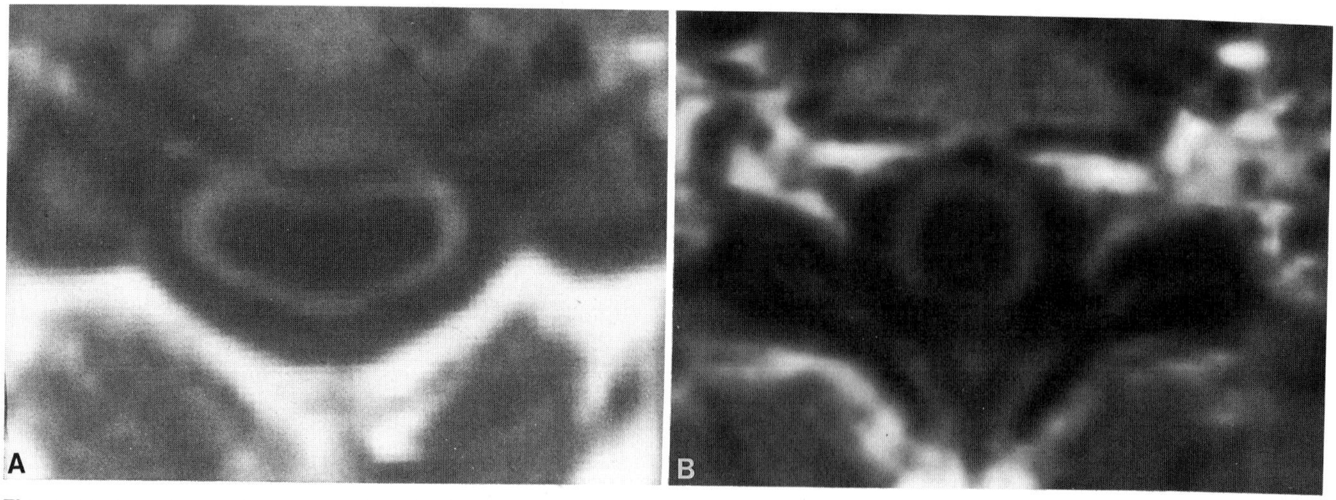

Figure 11.13. The cervical (**A**) and thoracic (**B**) spinal cord is enlarged by a hydromyelia which is central and completely symmetrical. **C–G,** Arachnoid cyst in another patient following blunt trauma. T1-weighted sagittal (**C**) and T2-weighted sagittal (**D**) images demonstrate focal anterior displacement of spinal cord (*arrows*). The cord appears to be mildly atrophic. **E,** Frontal view of the myelogram demonstrates the focal intradural defect caused by the arachnoid cyst (*arrow*). **F,** Contiguous axial CT scan demonstrates the dorsally placed arachnoid cyst (*arrow*) with ventral displacement of the mildly atrophic cord. **G,** Delayed CT scan demonstrates the intravasation of the contrast material within the cyst (*arrow*).

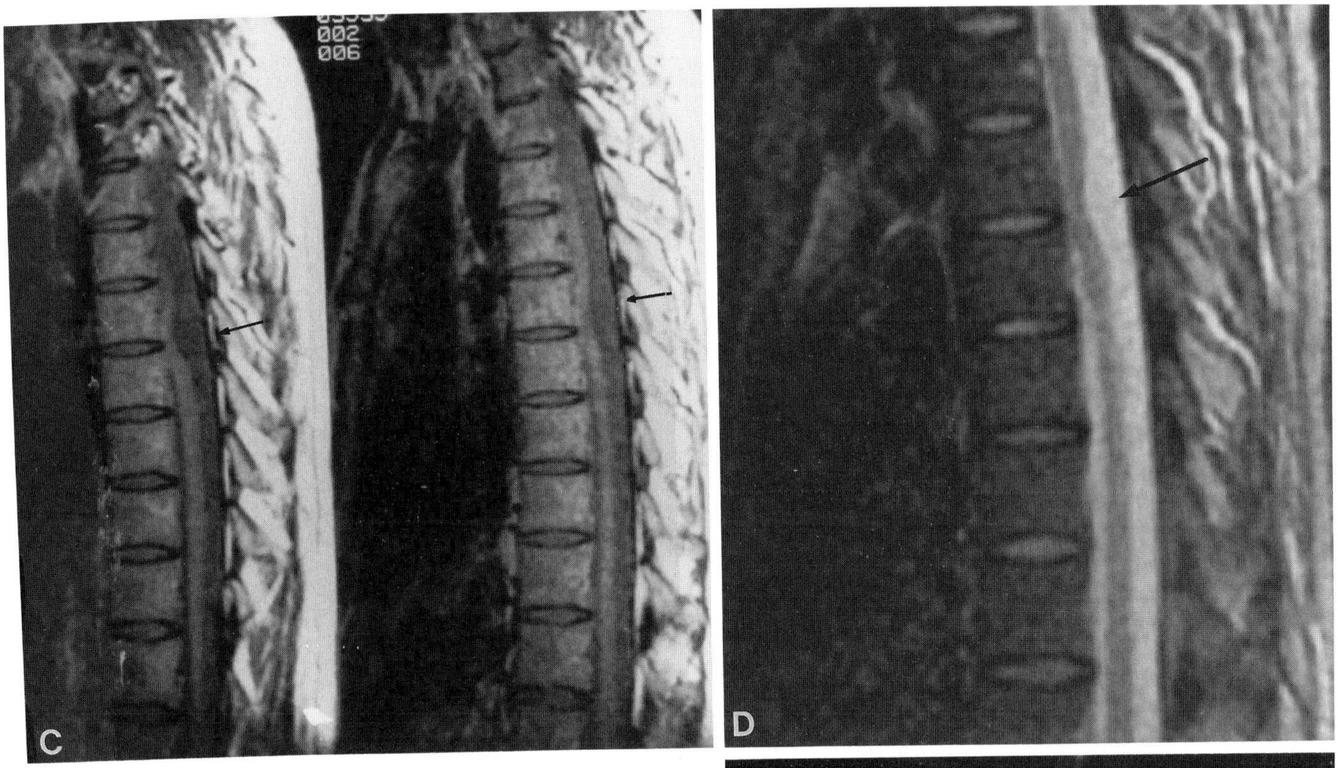

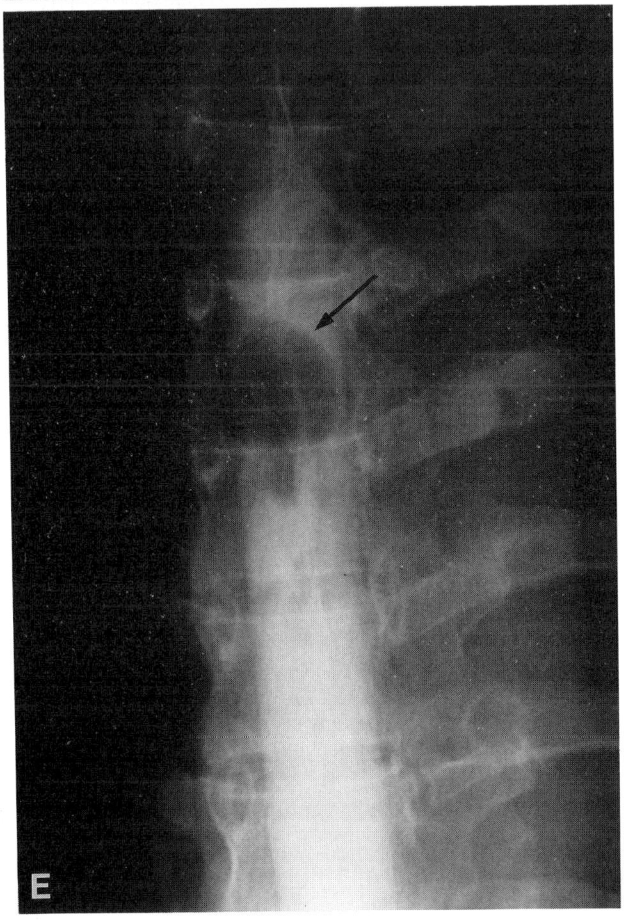

Figure 11.13–*continued*

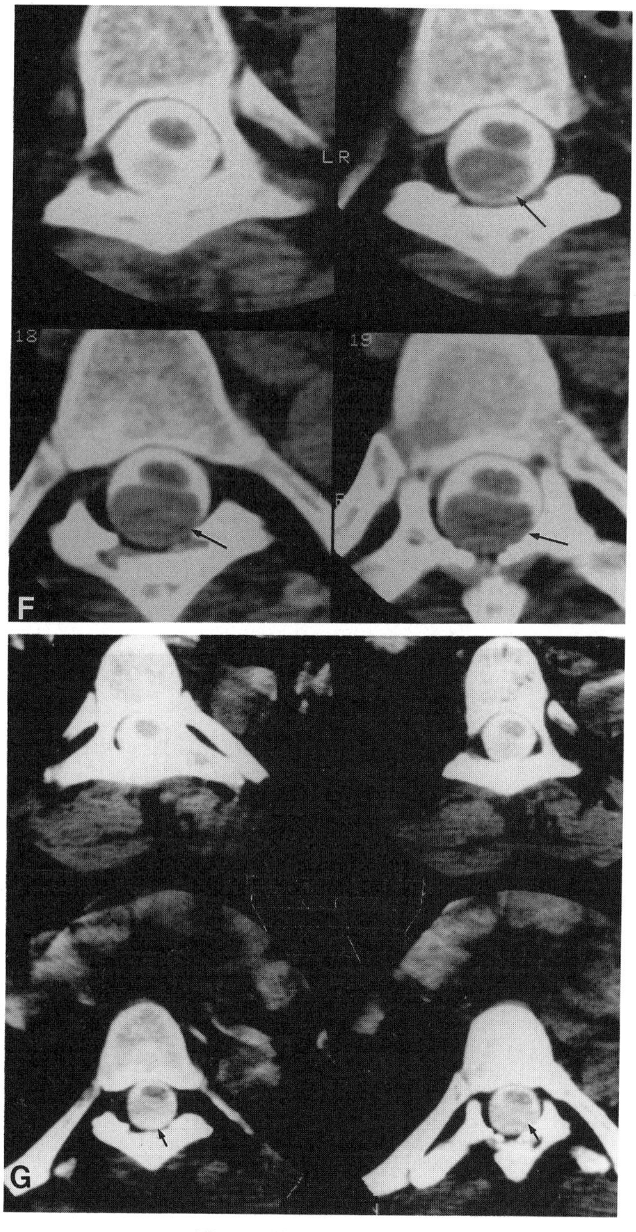

Figure 11.13–*continued*

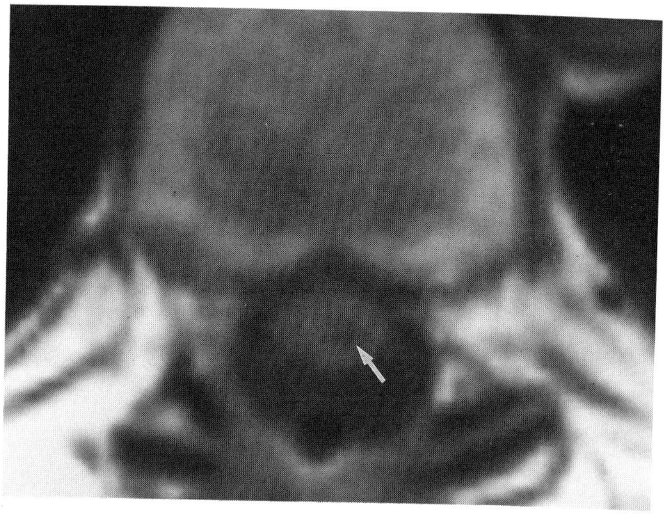

Figure 11.14. A small syrinx cavity is located to the left of the midline (*arrow*).

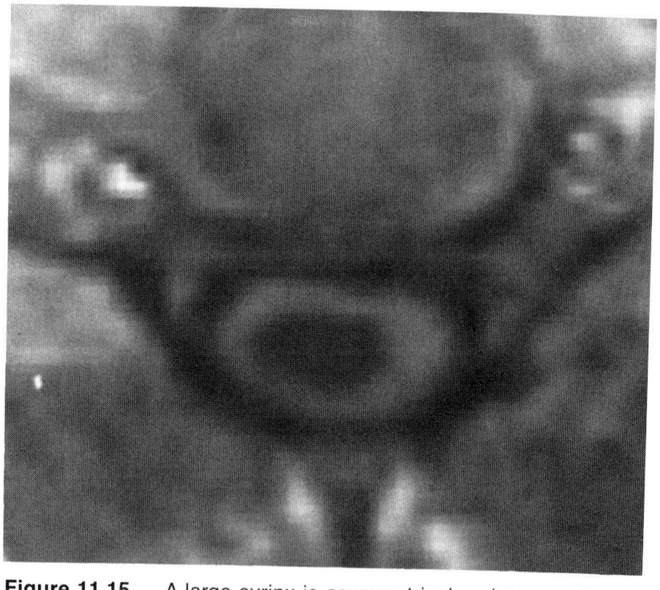

Figure 11.15. A large syrinx is asymmetrical and communicates with the central canal.

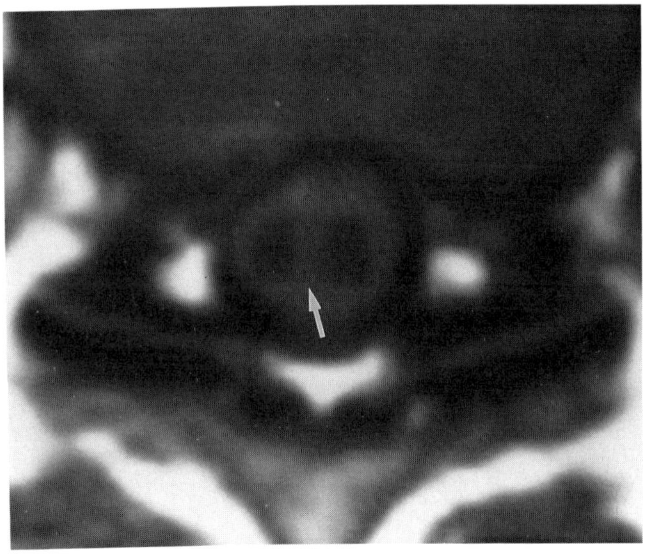

Figure 11.16. The syrinx is separated by a midline septum (*arrow*).

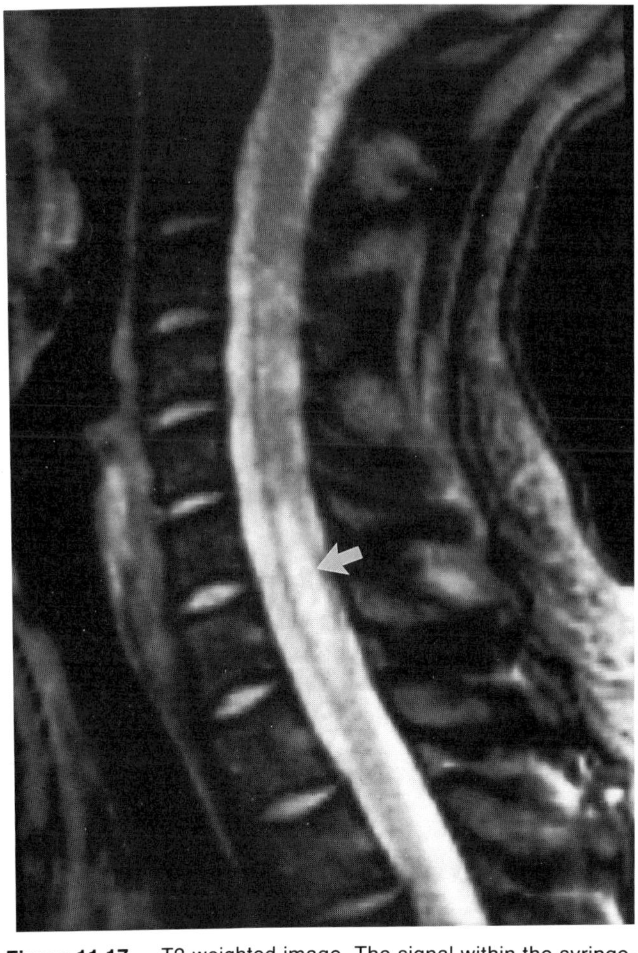

Figure 11.17. T2-weighted image. The signal within the syringomyelia is the same as CSF (*arrow*).

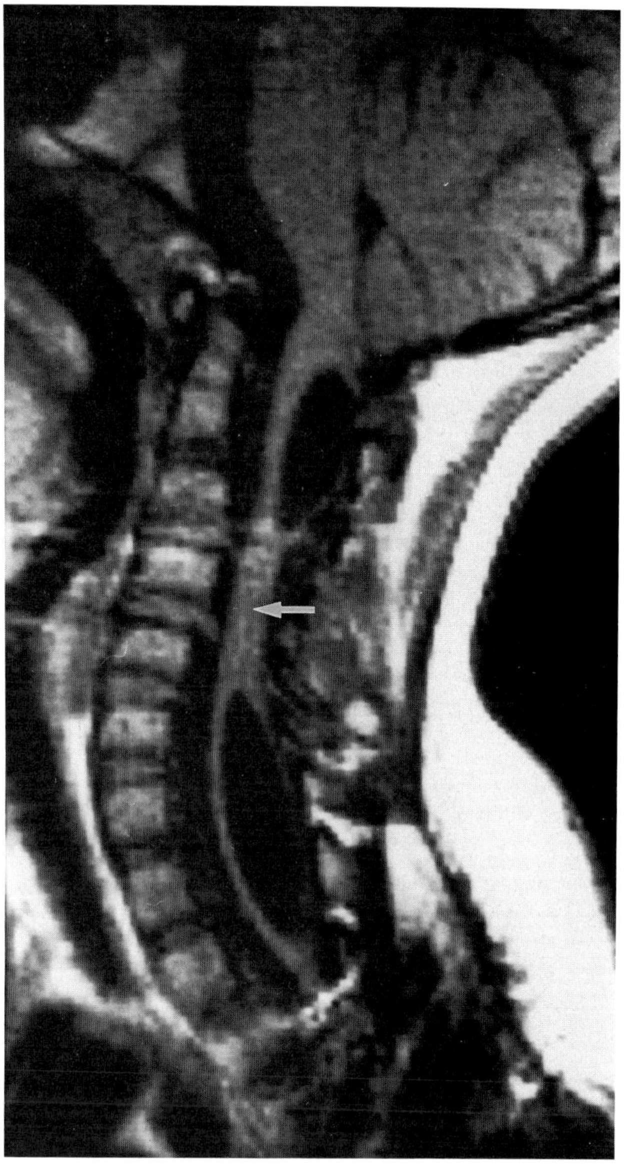

Figure 11.18. A syringomyelia with an apparently "collapsed" portion (*arrow*).

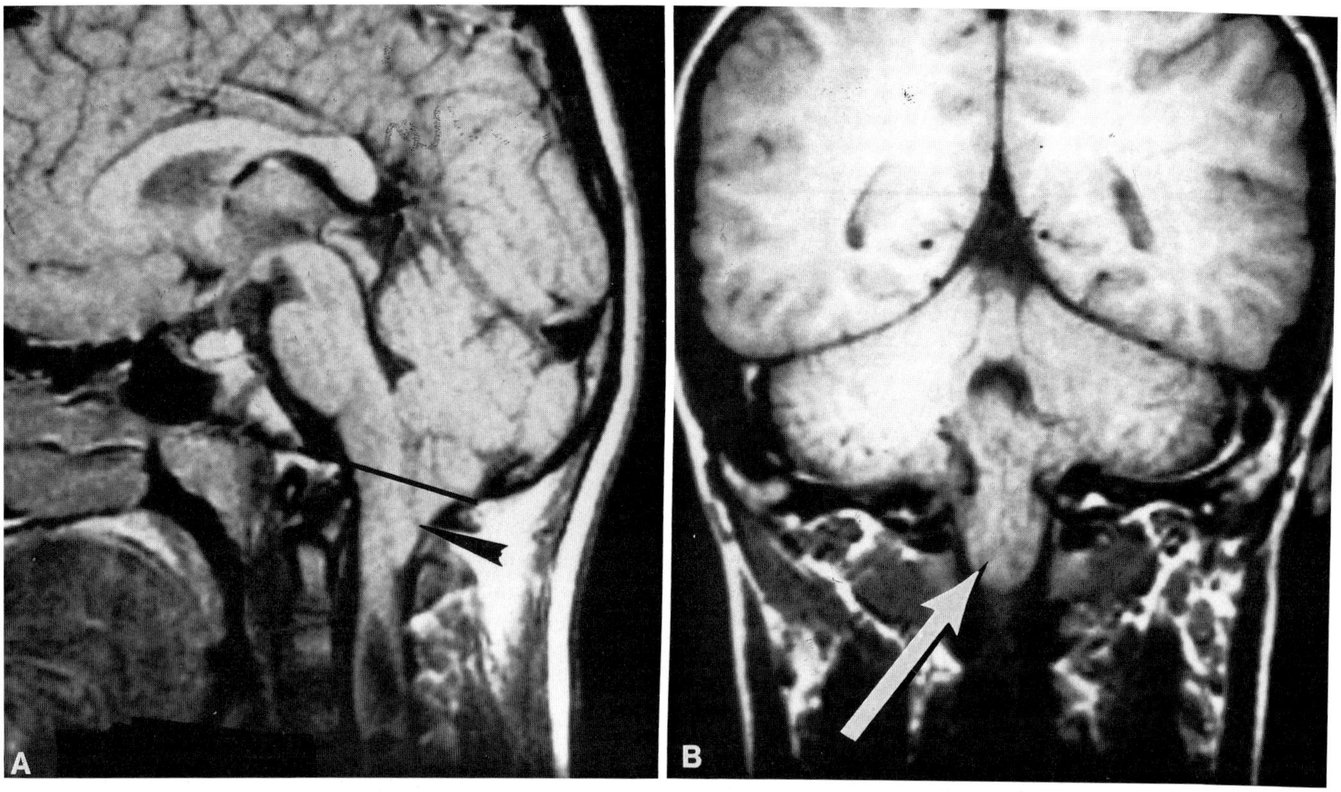

Figure 11.19. Chiari I malformation and syringomyelia. A 15-year-old boy with left body paresthesias. **A,** Sagittal image demonstrating position of cerebellar tonsils (*arrowhead*) below McRae's line (*black line*). Note absence of cisterna magna. Syrinx cavity is seen at C2 level. **B,** Coronal image showing nodular appearance of both cerebellar tonsils (*arrow*) compressed together and extending below foramen magnum.

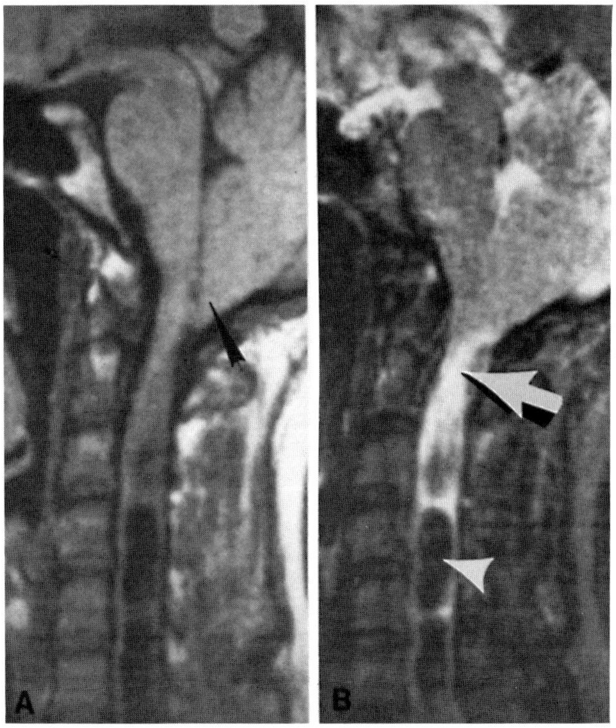

Figure 11.20. Syringomyelia and Chiari I malformation. A 34-year-old woman. **A,** Sagittal T1-weighted image. Note tonsillar ectopia (*arrowhead*) and "beaded" syrinx cavity. **B,** Sagittal T2-weighted image without motion-suppression technique. Note hyperintense area of gliosis at C1–2 level (*arrow*) and hypointense appearance of pulsatile fluid (*arrowhead*).

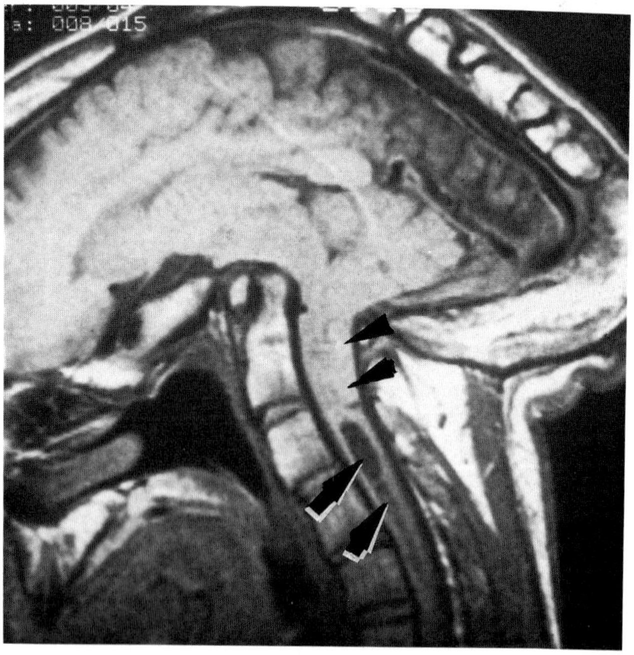

Figure 11.21. Syringomyelia (*arrows*) with severe congenital basilar invagination and platybasia. Note compressed cerebellar tonsils extending down to the C2–3 level (*arrowheads*).

the time of the injury and those developing later. Because the history of severe trauma is always clear-cut, there is seldom difficulty in attributing the abnormality to this cause. MRI is used to confirm the diagnosis and to determine the size and location of the syrinx within the normal functioning spinal cord cranial to the site of injury [50].

Spinal cord injury produces a spectrum of cord pathology ranging from microcystic myelomalacia to expansile, pulsatile syringomyelia cavities extending cephalad or caudad from the level of injury. Trauma causes cord edema, myelin breakdown, axonal loss, and possible hemorrhage [51, 52]. Trauma also damages the microcirculation, inducing secondary ischemia, and also causes arachnoiditis. Microcysts are formed initially and later coalesce, forming a syrinx [53, 54].

Acute Injury. In severe injuries, the spinal cord may be partially or totally transacted, but less severe trauma results in contusion (Fig. 11.22). Spinal cord contusion consists of necrosis and hemorrhage with edema at the site of injury. The edema resolves, but necrosis and hemorrhage may result in syrinx within an atrophic spinal cord. Clinical symptoms are the result of the spinal cord contusions and transection; the syrinxes do not produce symptoms unless they enlarge and extend later to involve the normal portions of the spinal cord. The configuration and degree of myelomalacia and spinal cord atrophy are evaluated on both T1- and T2-weighted images. After the acute stage, syrinxes may be seen on sagittal and axial T1-weighted images but are difficult to detect on T2-weighted images, since there is usually increased signal in the spinal cord as a result of gliosis.

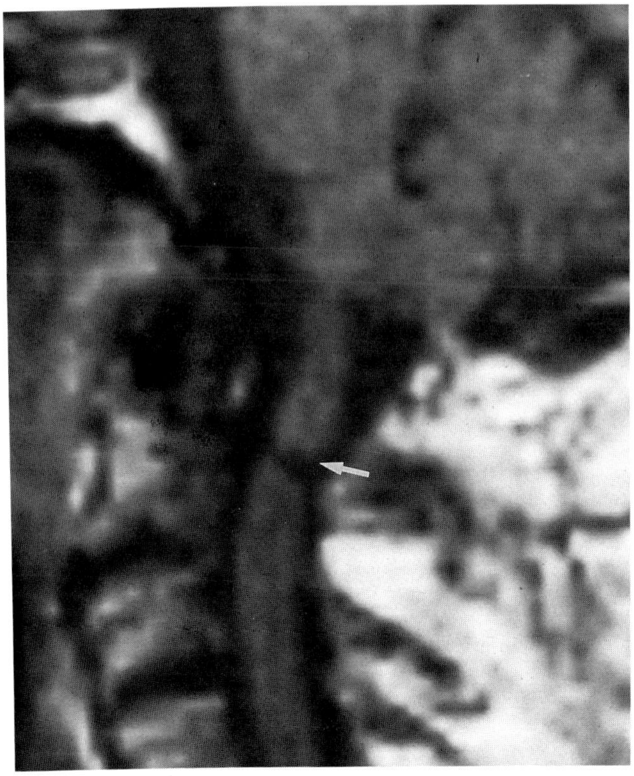

Figure 11.22. Spinal cord avulsion involving the cervical cord. T1-weighted image demonstrates an area of focal hypointense signal (*arrow*). This may be confused with a small syrinx.

Delayed Syrinx. Severe spinal cord injury, usually resulting in transection, may give rise to large syrinxes which become symptomatic months or even years later. Traumatic syrinxes are true cavities and not merely dilatations of the central canal. It is thought that the syrinxes represent enlargement and extension of the intramedullary cavities which develop at the time of the acute injury [55–57]. The syrinxes may extend both caudal and cranial to the site of injury. However, since the spinal cord is functionless below the transection, caudal syrinxes are irrelevant; cranial extension, on the other hand, may produce important clinical symptoms. Paraplegic patients depend on power in their arms for mobilization, so weakness of these limbs as a result of cranial ascent of a syrinx may be severely disabling. MRI has been found useful in evaluating progressive posttraumatic myelopathy to detect cavities which may be amenable to shunting [50, 67]. The MRI features of traumatic syrinxes are similar to those of syringomyelia on sagittal views; however, on axial sections they may be seen to be placed eccentrically (Fig. 11.23). The fluid within the syrinxes may have slightly increased signal on T1-weighted and proton-density images due to the higher protein content compared with CSF. The cord is usually slightly enlarged by the syrinx, and no enhancement is seen after intravenous gadolinium administration. It is important to try to determine the precise cranial extent of the syrinx, since bulbar involvement is a serious complication that requires prompt decompression of the cyst.

Remote Syrinx. Rarely, syrinx cavities are found at locations remote from the site of injury. The cause of these lesions is unclear; they may arise from intramedullary hemorrhages, which become cavities after the blood has absorbed, or they may result from ischemia of the spinal cord at border zones between major vascular supplies [55] (Fig. 11.24).

Neoplastic Syrinx

Cavities secondary to intramedullary tumors may be situated adjacent to the tumor or within the tumor mass and rarely represent dilatation of the central canal (hydromyelia). It is estimated that about 20 to 40% of patients with intramedullary tumors have associated cysts [56, 67]. Neoplastic syringomyelia cavities are termed *caudal* or *rostral* when they delineate the inferior or superior extent of the lesion

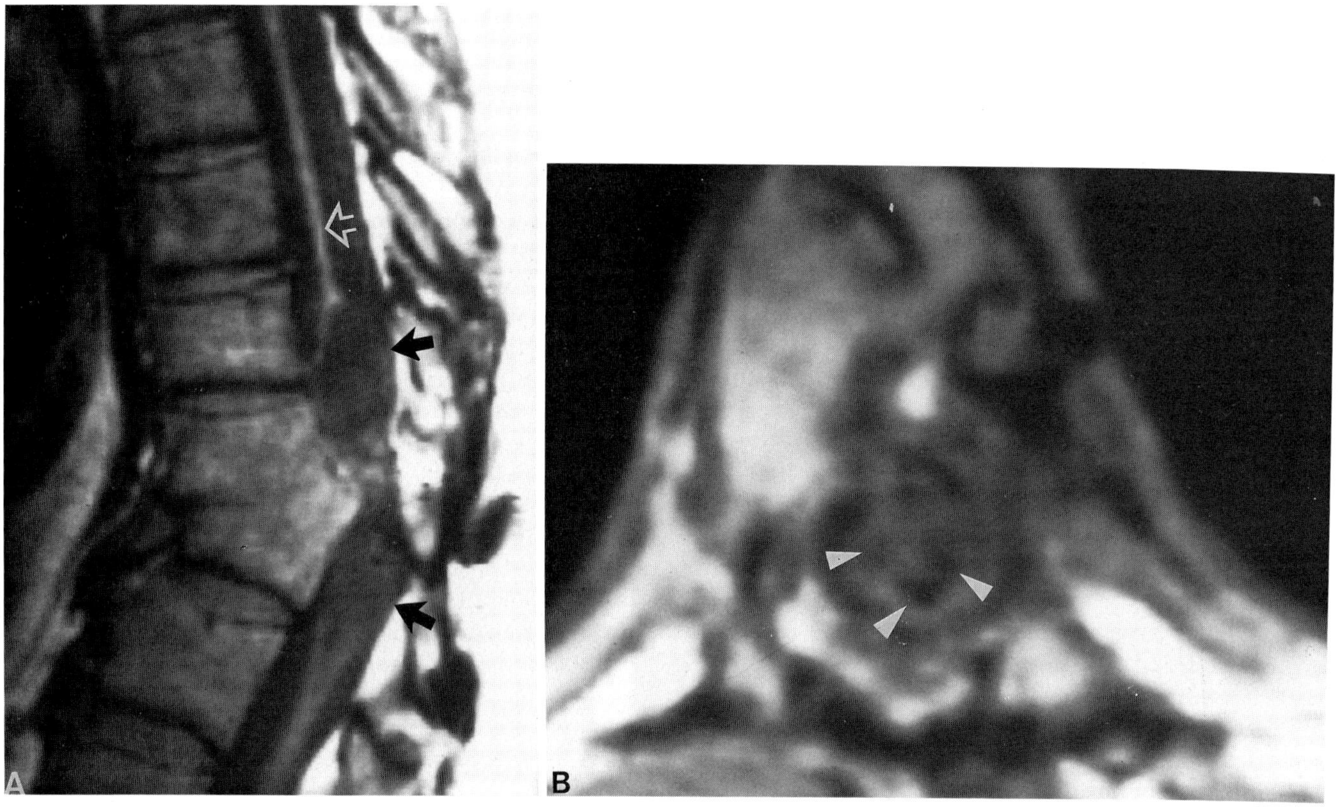

Figure 11.23. **A,** Syrinxes at the site of myelomalacia behind the gibbus of a compression fracture extends caudally and cranially (*arrow*). Note atrophy of the cranial thoracic spinal cord (*open arrow*). **B,** Axial views shows multiple syrinx cavities (*arrowheads*) as a result of myelomalacia from trauma.

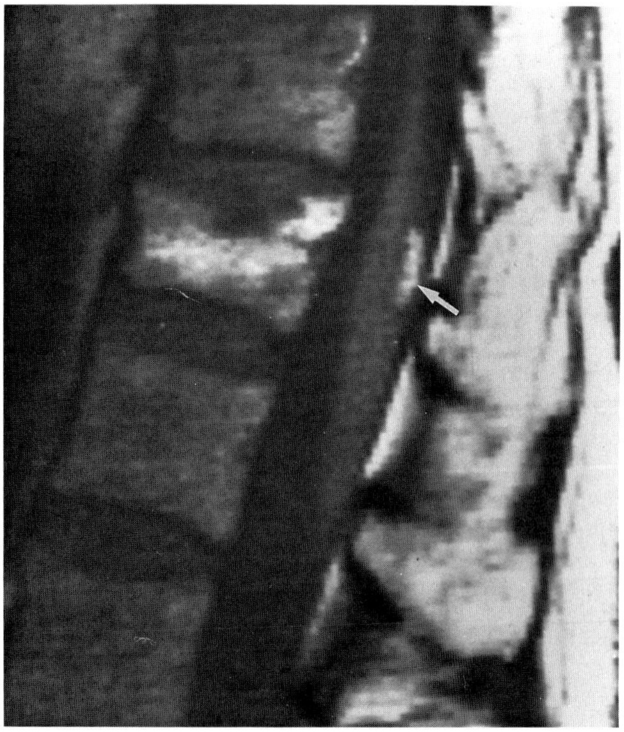

Figure 11.24. Hemorrhage in the posterior conus medullaris may eventually give rise to a syrinx cavity (*arrow*). Note hemorrhagic compression fracture.

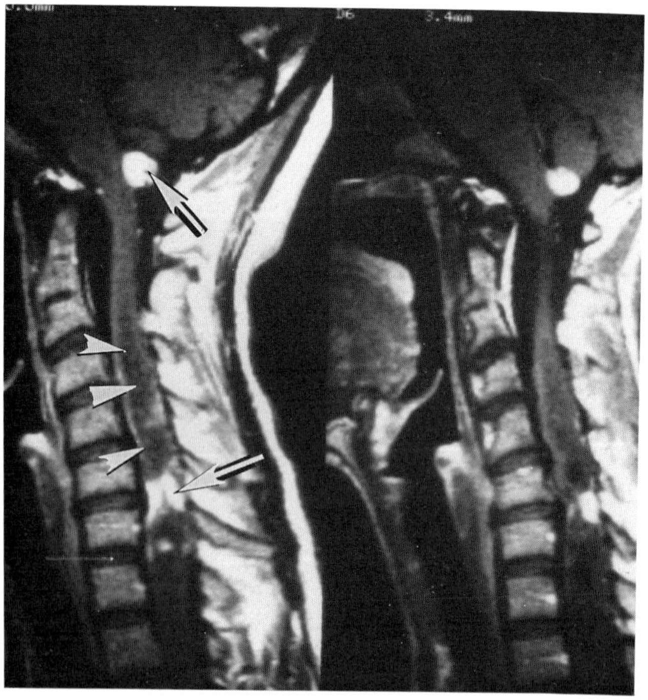

Figure 11.25. **Hemangioblastoma with syringomyelia.** T1-weighted sagittal image with gadolinium-DTPA. Note enhancing tumor nodules (*arrows*) and secondary syringomyelia (*arrowheads*).

and are not of tumorous origin. They are designated as *intratumoral cysts* when they are surrounded by solid neoplasm [58].

Syringomyelia associated with extramedullary tumors is rare [59, 60], unless surgery has been performed [61]. Extramedullary tumor at the craniocervical junction may produce syringomyelia by causing craniospinal pressure dissociation, similar to patients with craniovertebral anomalies. Any type of tumor may be associated with syringomyelia [62]. The tumors most commonly associated with intramedullary cavities are ependymomas, astrocytomas, and hemangioblastomas [62–66]. Hemangioblastoma (Fig. 11.25) has the highest incidence of associated cysts (40 to 60% of patients), but its incidence is low. Tumor nodules of hemangioblastoma may be very small, but they enhance consistently on the postcontrast MRI, and this may be the only differentiating feature from idiopathic syringomyelia (Fig. 11.26). The association of spinal ependymomas with syringomyelia is well known, especially in the context of neurofibromatosis type II. Ependymomas are more frequently associated with syringomyelia (50% of patients) than astrocytomas and often contain areas of hemorrhage. Astrocytoma is less frequently associated with syringomyelia (25% of patients) but is more likely to contain an intratumoral cyst (Figs. 11.27 and 11.28). Although these tumors cannot be definitely differentiated by MRI, the finding of hemosiderin deposition at the superior and inferior borders of a centrally located tumor with an associated syrinx is most suggestive of an ependymoma. Intramedullary cavities are formed as a result of hemorrhage and necrosis within or adjacent to the tumors. True hydromyelia may be secondary to obstruction of flow of CSF within the central canal by a critically situated tumor and occurs sometimes with tumors at the craniocervical junction.

Sagittal and axial T1-weighted views with and without intravenous contrast material and T2-weighted images are required for evaluation of neoplastic syrinxes. Unlike syringomyelia, neoplastic syrinxes are usually irregular, eccentric, and seldom extend far from the tumor masses (Figs. 11.29 and 11.30). The enlargement of the spinal cord is usually related more to the tumor mass and less to the cysts themselves. The margins of the syrinx may be smooth when they are true cysts or may be irregular

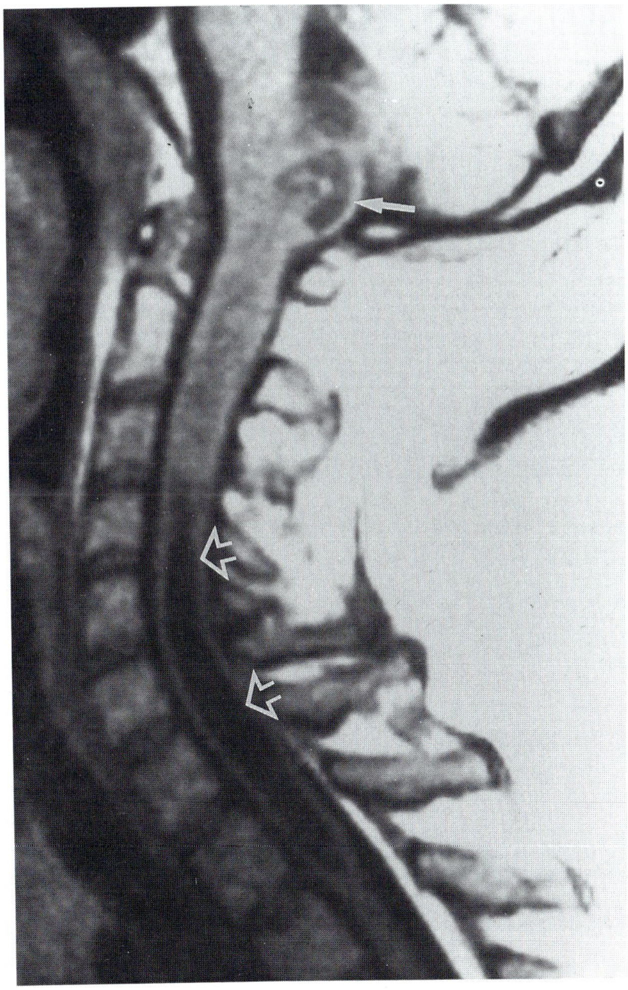

Figure 11.26. Syrinx (*open arrow*) extends below an intramedullary astrocytoma of the spinal cord. Another cystic cavity is present within the tumor mass (*arrow*).

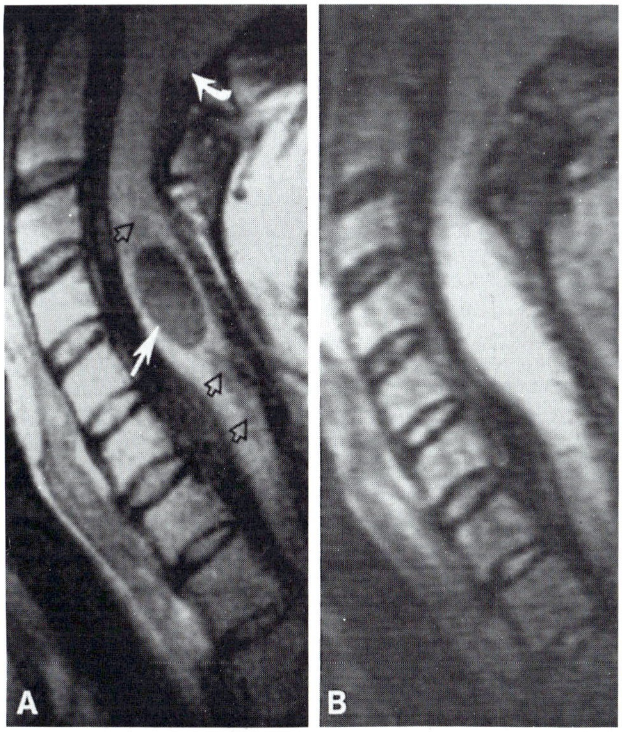

Figure 11.27. Astrocytoma with a large tumor cyst. A, T1-weighted sagittal image. Tumor cysts enlarge the cord (*white arrow*), and small areas of syringomyelia (*open arrows*) are noted above and below the cyst. Chiari I malformation is also present. Note cerebellar tonsil position (*curved arrow*). **B,** T2-weighted sagittal image.

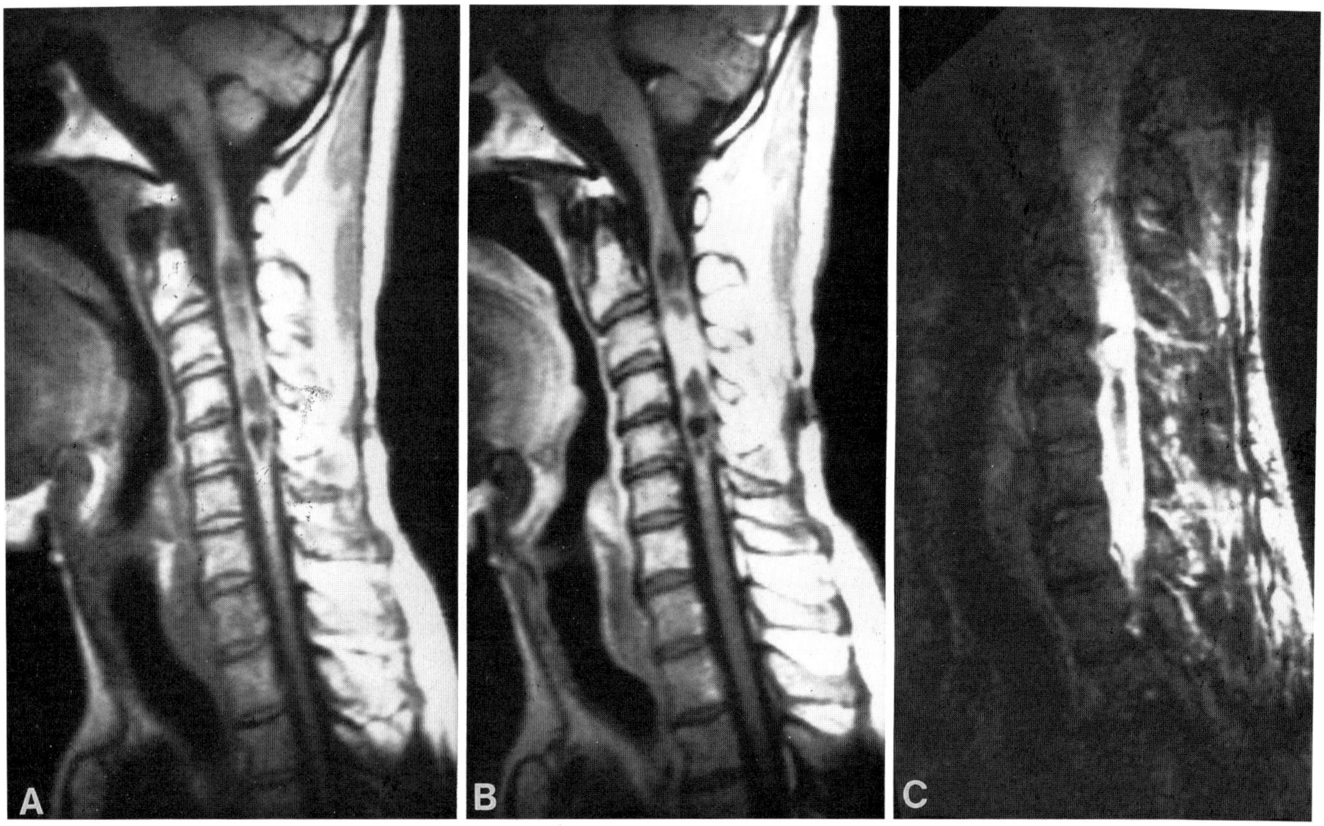

Figure 11.28. Astrocytoma with syringomyelia. A, Precontrast T1-weighted sagittal image. Cord tumor cannot be differentiated from area of gliosis or myelomalacia. **B,** Postcontrast T1-weighted sagittal image. Enhancement of solid tissue indicates location of tumor (*arrows*). **C,** T2-weighted sagittal image shows diffuse increased intensity in tumor and syrinx with small area of flow-related signal hypointensity in part of cavity.

Figure 11.29. **A,** Non-contrast-enhanced scan shows multiple syrinxes within an intramedullary tumor (*arrowheads*). These syrinxes have signal intensity higher than CSF. **B,** Contrast scan of another case shows syrinxes with the enhancing tumor (*arrow*).

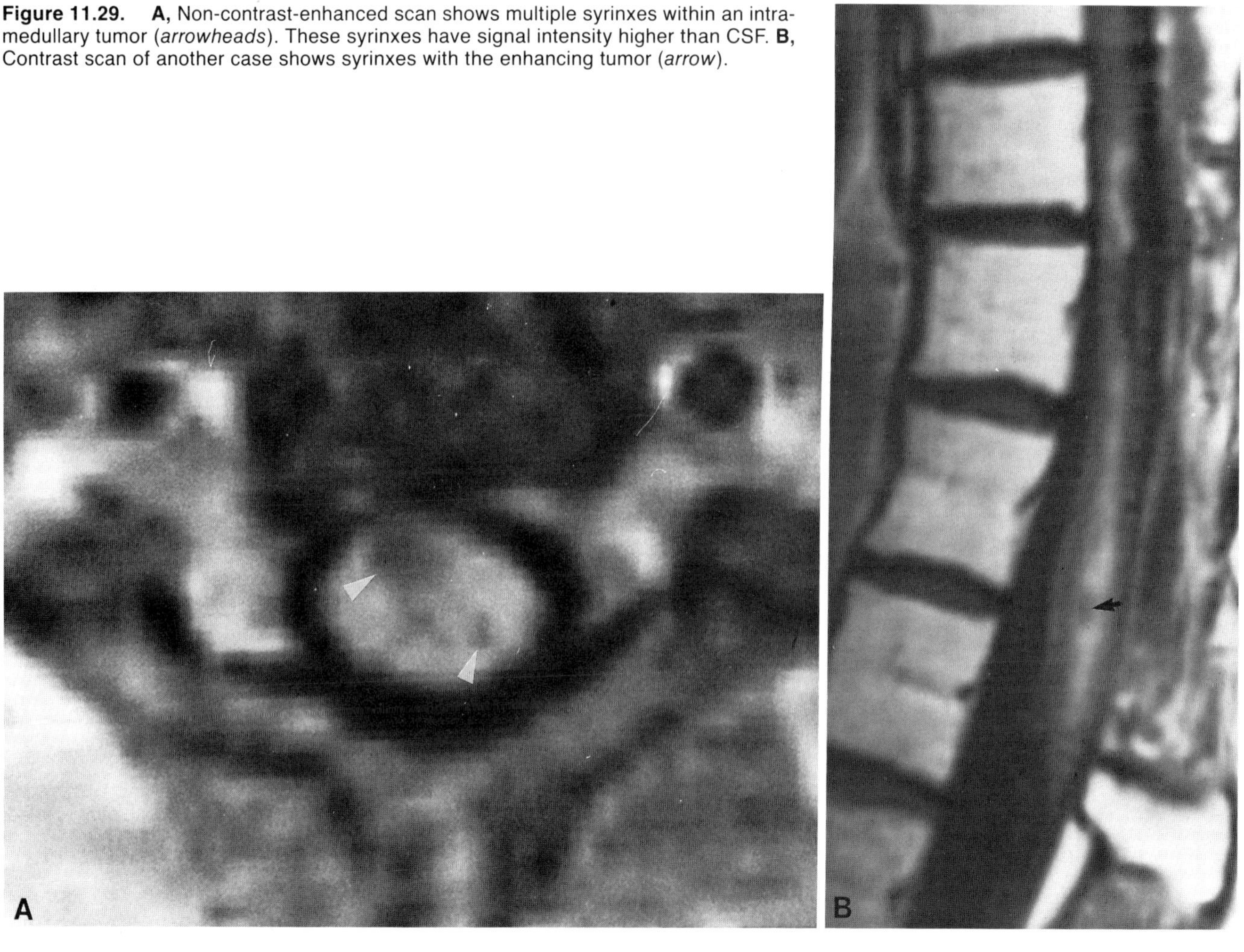

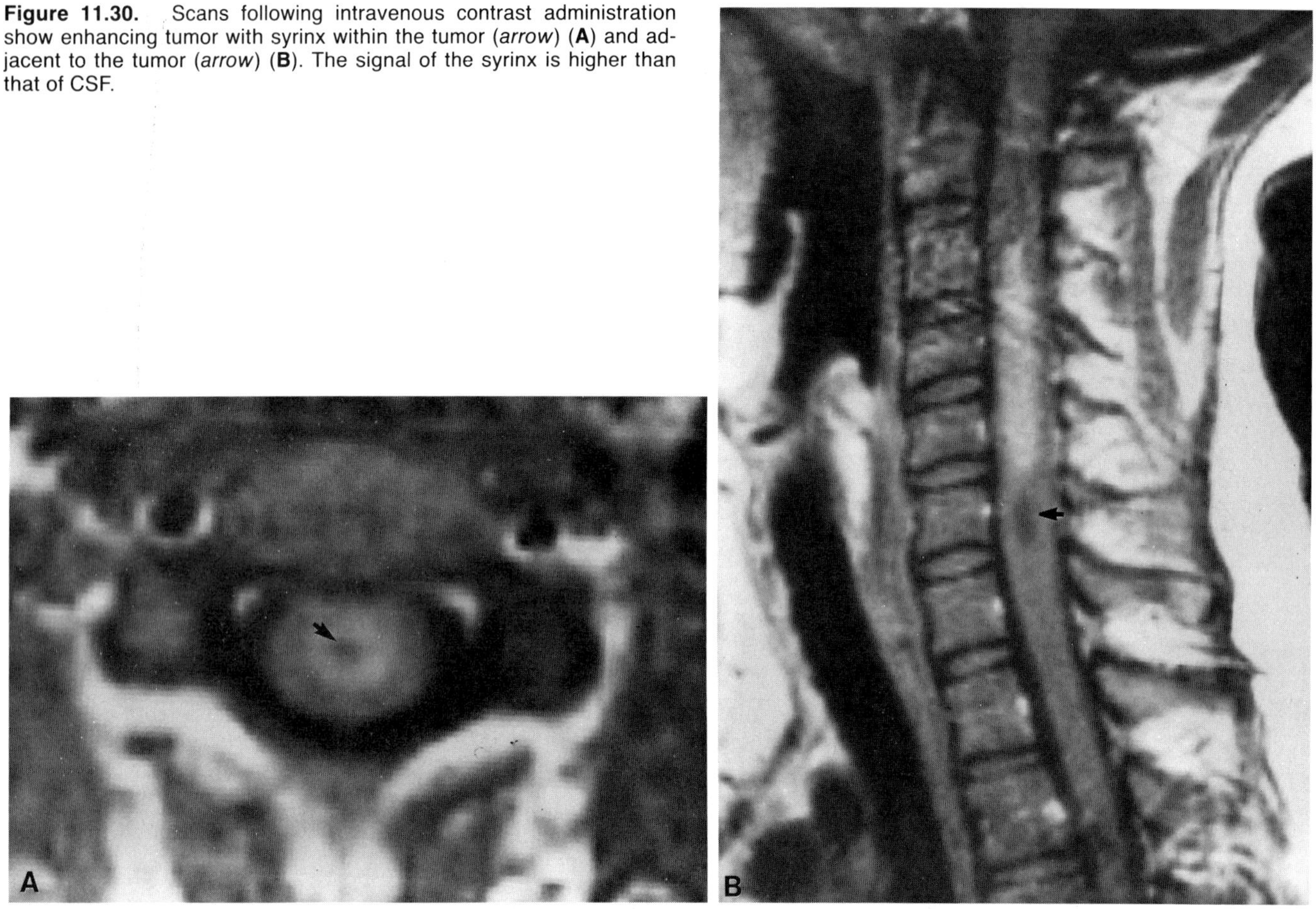

Figure 11.30. Scans following intravenous contrast administration show enhancing tumor with syrinx within the tumor (*arrow*) (**A**) and adjacent to the tumor (*arrow*) (**B**). The signal of the syrinx is higher than that of CSF.

when they are the result of necrosis. The signal within the syrinxes reflects their high protein content; it tends to be slightly higher in intensity in T1-weighted and proton-density images but variable in T2-weighted images compared with CSF. These syrinxes may be difficult to diagnose when they are small [32, 63] (see Figs. 11.29 and 11.30). Signal void, seen sometimes in pulsatile primary syringomyelia, is not seen in tumor syrinxes. Hemorrhages, when present, may show increased signal of methemoglobin on T1-weighted images and either increased or decreased signal of hemosiderin and intracellular blood products on T2-weighted images. Rarely, cavities with characteristics of hydromyelia may be seen with tumors situated at the craniocervical junction.

The most important criterion in deciding that syrinxes are secondary to an intramedullary tumor is visualization of the tumor itself. Such a tumor usually enlarges and causes localized expansion and deformity of the spinal cord in excess of that due to the syrinx alone. The tumor mass may have different signals from the spinal cord [32, 58, 63, 64]. Almost all intramedullary tumors enhance with intravenous contrast [58, 64–66] (see Figs. 11.28, 11.29, and 11.30). This may be seen within the tumor mass itself, within the entire lining, or as distinct nodules along the wall of the cystic cavities. The configurations of intramedullary tumors, on both non-enhanced and enhanced scans, are not usually specific for histologic types.

Arachnoiditis and Syringomyelia

Chronic adhesive arachnoiditis may be caused by infection, hemorrhage, granulomatous disease, surgery, and a variety of agents injected into the subarachnoid space. The resulting arachnoidal and dural scarring may lead to syringomyelia, nerve root and spinal cord tethering, or extramedullary cyst formation [68–70]. The clinical presentation of patients is often complex and may be the result of the arachnoiditis, cord compression, or syrinx formation. Most patients have a history of trauma or spinal surgery, but the symptoms of syringomyelia usually develop months to years after the event that incited the development of the arachnoiditis. Three different forms of arachnoiditis have been associated with syrinx formation: basal cistern arachnoiditis, diffuse adhesive spinal arachnoiditis, and focal spinal meningeal cicatrix formation [71].

Most cases of syringomyelia due to arachnoiditis develop in the thoracic spinal cord. The combination

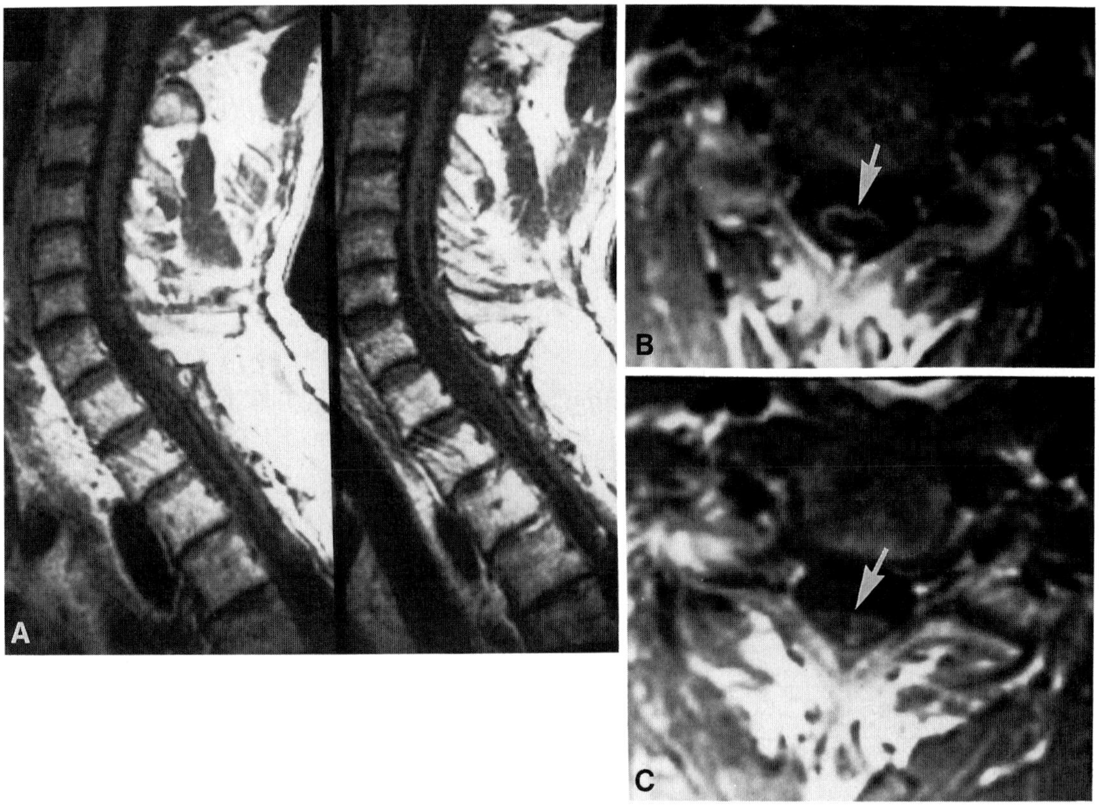

Figure 11.31. Arachnoiditis with syringomyelia and cord compression due to a subarachnoid cyst. A, Contiguous sagittal and parasagittal T1-weighted images. **B,C,** Axial T1-weighted images showing syrinx cavity (*arrow*). **C,** Section at lower level shows posterior displacement of cord (*arrowhead*) and cord compression caused by a loculated ventral subarachnoid cyst (*arrow*).

of the pathologic features of arachnoiditis and syringomyelia causes this type of syringomyelia to have significantly different MRI characteristics compared with those of other etiologies. The subarachnoid space may be obliterated or distorted due to adhesions and cyst formation. Extramedullary (subarachnoid cysts) commonly develop in patients with arachnoiditis and frequently coexist with syringomyelia. It is important to evaluate the entire spinal subarachnoid space and cord in patients with arachnoiditis and symptoms of myelopathy, since an extramedullary loculated cyst may compress the spinal cord regardless of whether or not there is associated syringomyelia (Fig. 11.31).

Surgical drainage of an extramedullary cyst may be required in addition to syrinx shunting. A ventral subarachnoid cyst will cause posterior displacement and compression of the cord (see Fig. 11.31). In the thoracic area, the cord is normally situated along the ventral margin of the canal, and posterior cord positioning should therefore raise suspicion of a ventral extramedullary cyst. The contents of the cyst are usually identical to CSF, and therefore, the differentiation of cyst from CSF space enlargement due to atrophy or cord tethering may be difficult. With severe arachnoiditis, the margins of the spinal cord are usually indistinct (Fig. 11.32). There may be tethering of the cord, nerve root thickening, clumping, and soft-tissue "mass" formation. The spinal cord cavity more often appears septated and irregular, and the "beaded" or "haustral" pattern is usually not present. Focal adhesive arachnoiditis with associated syringomyelia may simulate an intramedullary tumor [65]. Contrast enhancement in arachnoiditis is variable and does not correlate with the severity of the disease [72]. Gadolinium-DTPA may improve visualization of the thickened meninges in many cases, especially in granulomatous disease. The MRI studies need to be complemented with CTM to correlate the morphologic patterns and signal changes seen on the T1- and T2-weighted images. Although CSF dynamics by cine MRI are being evaluated, clinical application is still under investigation. Despite all the findings from the multimodality imaging studies, operative sonography is usually required to identify the true cystic spaces within the cord.

Chronic Compressive Syringomyelia

Cervical spondylosis is the most common cause of benign chronic extramedullary cord compression (Fig. 11.33). Small, irregular slitlike cavities develop in a small number of patients with spinal stenosis. Rarely, a syrinx may be caused by an extramedullary

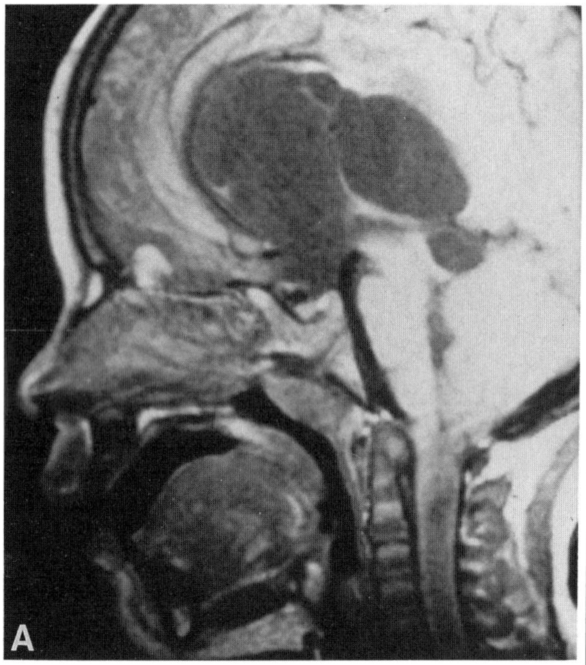

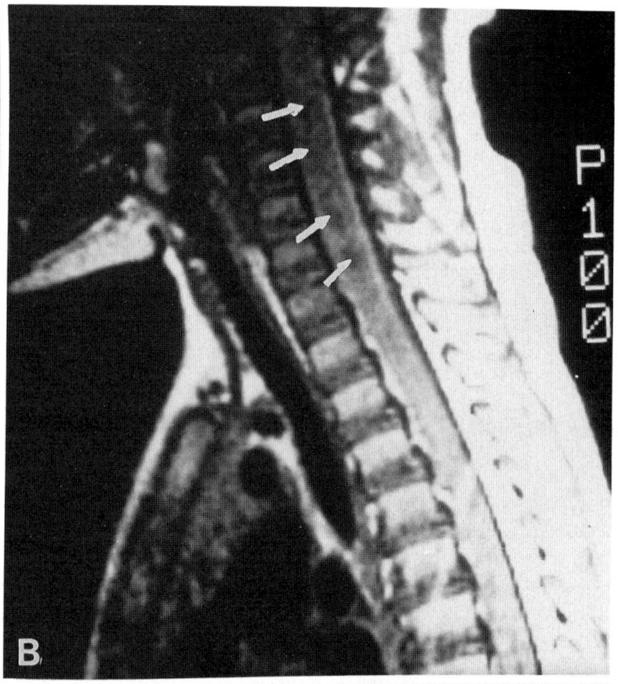

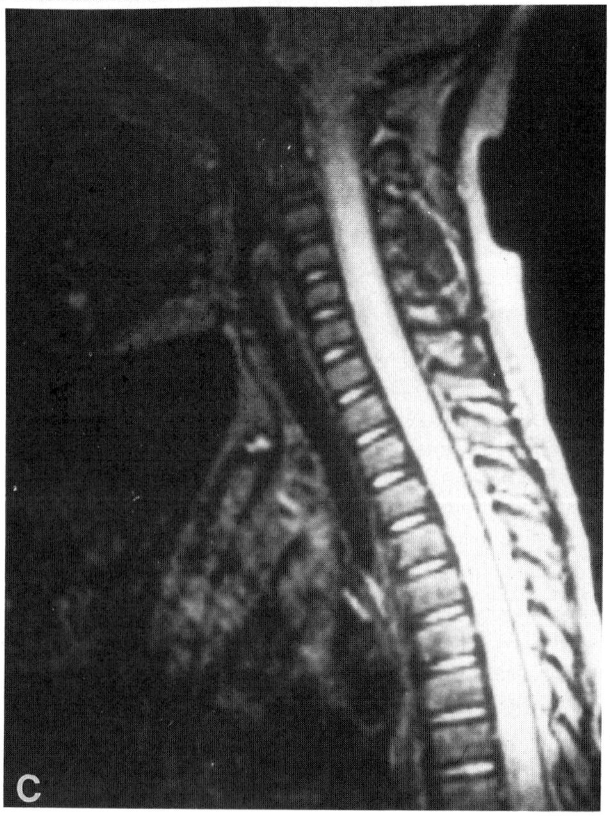

Figure 11.32. Arachnoiditis due to Listeria monocytogenes causing syringomyelia and hydrocephalus. A, Sagittal T1-weighted image of brain and upper spine. Note hydrocephalus due to arachnoiditis. **B,** Sagittal T1-weighted image of cervicothoracic cord. Margin of cord is indistinct due to arachnoidal thickening and adhesions. There is poorly defined hypointensity in cord due to cavitation and myelomalacia (*arrows*). **C,** Sagittal T2-weighted image. Note diffuse increased intensity in cord. **D,** Axial images at two different levels reveal eccentric cavitation (*arrows*) and central edema/myelomalacia.

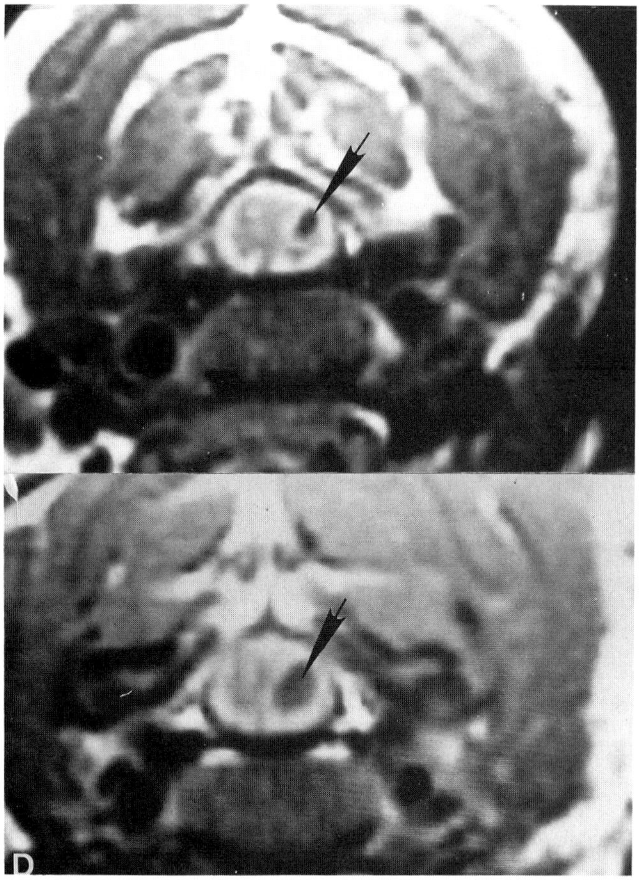

Figure 11.32–*continued*

intradural mass and extend above or below the level of obstruction. It has been suggested to be secondary to ischemic damage of vessels supplying the spinal cord [69] and that the cavitation develops from a combination of chronic repetitive trauma and vascular insults in such cases [73, 74]. CSF may penetrate the damaged spinal cord via enlarged Virchow-Robin perivascular spaces. The CSF may be forced into the cord by abnormal CSF pulsations associated with the obstruction [56]. Syringomyelia may develop after surgical removal of a chronic compressive extramedullary lesion such as a tumor [76].

The presence of syringomyelia may contribute to the development or progression of spondylosis and spinal degenerative disease in some patients. However, syringomyelia does not have a consistent, major role in the development of cervical spondylosis [75].

The MRI appearances are nonspecific, and the signal characteristics depend on the contents of the cystic fluid. There is no available experience with contrast enhancement.

Cysts of Uncertain Origin

Intramedullary cysts have been described in association with posterior fossa cysts, arachnoid cysts, and following excision of extramedullary lesions [75, 77, 78]. Occasionally, the MRI appearances of intramedullary cysts are not explained as a primary syringomyelia, nor by trauma, neoplasm, or infection (Fig. 11.34). In these cases, the MRI scans should be scrutinized for evidence of prior hemorrhage. Hematomyelia may be caused by an intramedullary vascular malformation, generalized blood diathesis, and anticoagulation therapies. Focal infarcts from emboli or severe hypotension also may give rise to syrinxes.

POSTOPERATIVE APPEARANCE OF SYRINGOMYELIA

A number of surgical procedures have been advocated for treatment of syringomyelia. These treatments have been directed at drainage of the syrinx cavity or correction of the pathologic mechanism that directs the filling of the cavity (Fig. 11.35). The treatments have included ventricular shunting, suboccipital craniectomy and cervical laminectomy, syrinx puncture and aspiration, syringosubarachnoid shunting, syringoperitoneal shunting, syrinx marsupiliazation, and conus "snipping" (terminal ventriculostomy). Successful treatment of the syrinx decreases pain sensation, halts progression of neurologic deterioration, and results in marked reduction in the size of the cavity. When the distension of the cord is relieved, the spinal cord will be shown to be atrophic in many patients (Fig. 11.36).

Occipital-cervical decompression (foramen magnum decompression) seeks to relieve compression of the cerebellar tonsils and cervicomedullary junction, reestablish normal CSF flow pathways, and normalize craniospinal pressures [80]. MRI reveals a widened foramen magnum with truncation of the occipital bone posterior to the normal position of the opisthion. Ideally, CSF will be seen around the cerebellar tonsils and brainstem (Figs. 11.37 to 11.39). In some cases, the cerebellum descends caudally into the defect, increasing the hindbrain herniation and compression of the posterior fossa structures. This has been referred to as cerebellar "slumping" and may be associated with worsening of symptoms [46].

Shunt tubes are frequently placed in syrinx cavities, usually shunting fluid directly into the subarachnoid space but occasionally shunting fluid into the peritoneal cavity. The shunt tubes are usually very small but can be detected in about 75% of patients. Obstruction of the syringostomy tube causes reexpansion of the syrinx cavity. Arachnoid adhesions may cause localized extramedullary cyst formation, which can compress the spinal cord. Tethering of the cord can occur after surgery and can simulate the symptoms of a syrinx.

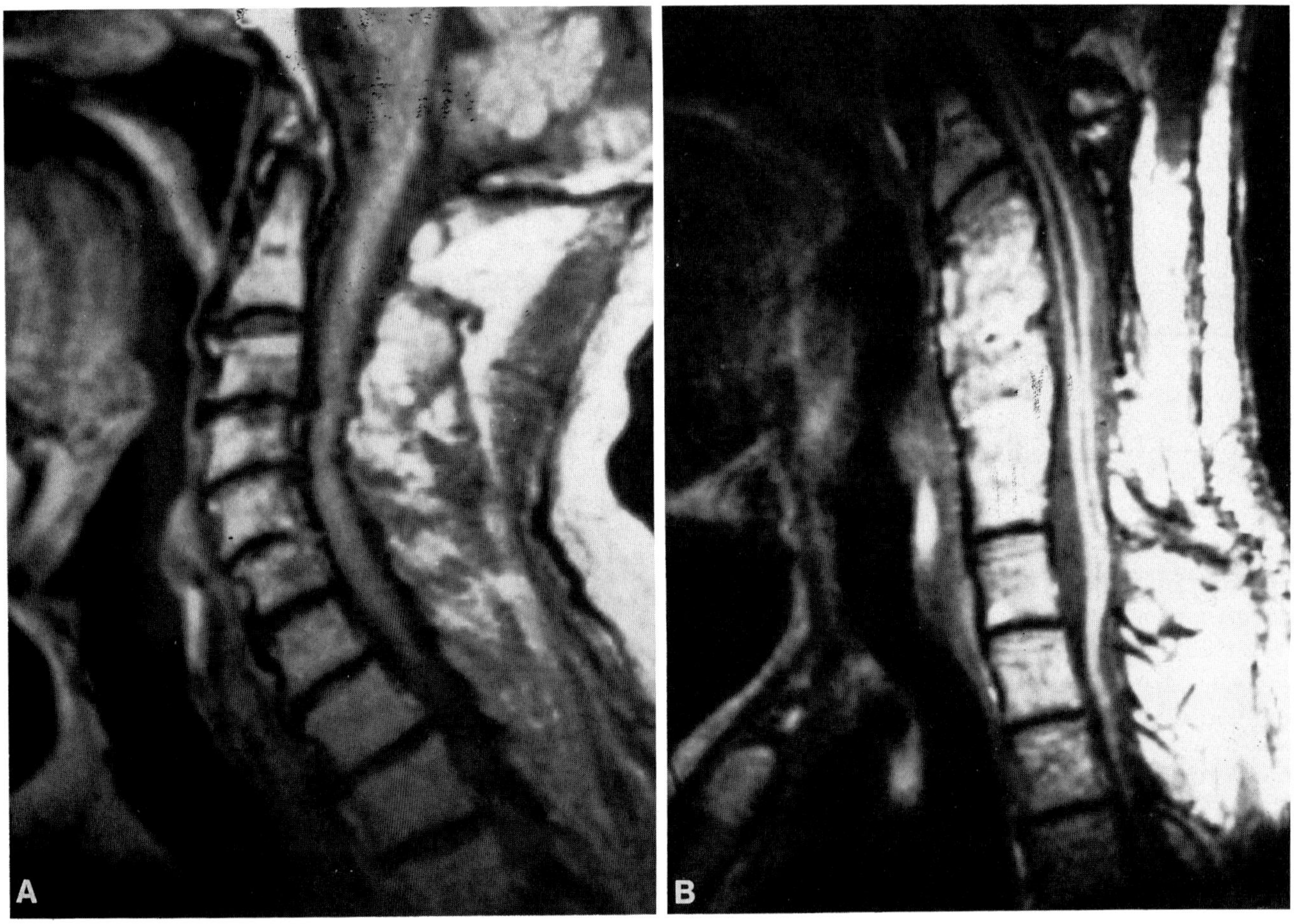

Figure 11.33. Postcompressive syringomyelia. A 57-year-old man with chronic cord compression due to stenosis. **A,** Preoperative sagittal T1-weighted image showing spondylosis and cord compression. **B,** Postoperative sagittal image showing syringomyelia, laminectomy, and anterior vertebral fusion.

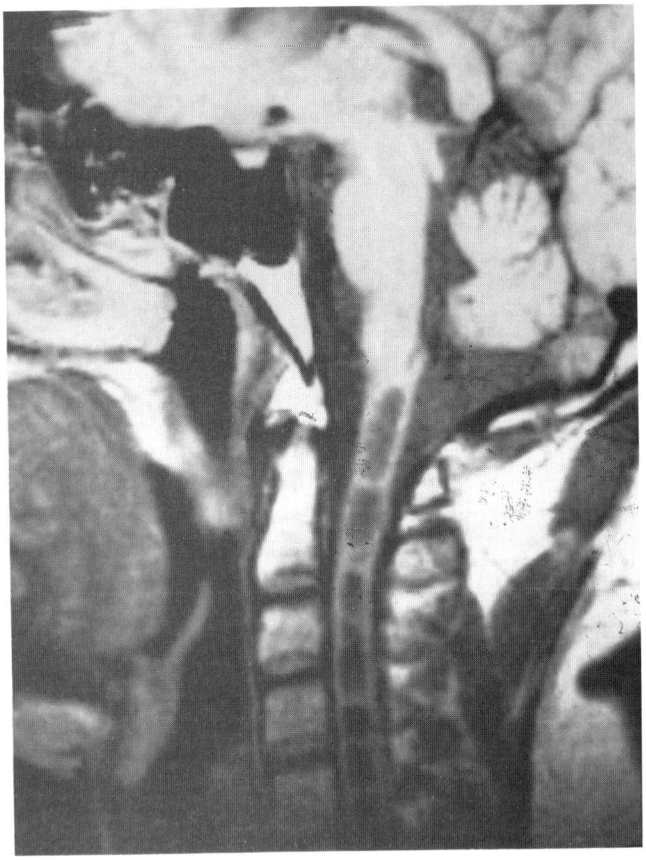

Figure 11.34. **Idiopathic syringomyelia/syringobulbia.** A 45-year-old man with chronic paresthesias and increasing muscle weakness. Sagittal image. Note syrinx cavity extending into cervicomedullary junction. There was no history of trauma or infection and no evidence of Chiari malformation or neoplasm.

Differential Diagnosis of Syringomyelia

Diagnosis of syringomyelia and differentiation from other causes of intramedullary cavities on MRI scans are generally not difficult. Rarely, acute myelitis and intramedullary abscess may have similar MRI appearances.

Acute myelitis is an acute inflammation and not a true cavity of the spinal cord. It is seen on MRI scans as regions of decreased signal on T1-weighted images and increased signal on T2-weighted images with variable degrees of enlargement of the spinal cord. The signal is due to edema which has poorly defined margins and a T1 signal that is higher than the fluid of syringomyelia. Acute lesions may enhance with intravenous contrast material, unlike true cysts and syrinx, which do not.

Intramedullary abscesses are rare and are associated with systemic infection and emboli. They have a similar appearance to abscesses in other parts of the central nervous system; they have a center of decreased signal on T1-weighted images, and the cavity and surrounding spinal cord may have increased signal on T2-weighted images. The signal within the abscess is greater on T1- and less on T2-weighted images compared with CSF, reflecting a higher protein content [79]. Enhancement is seen in the wall of the abscess. When meningitis is present, enhancement of the meningitis also may be seen. Abscesses are rarely confused with syringomyelia but may have similar appearances to cystic degeneration of intramedullary tumors.

MISCELLANEOUS SPINAL CANAL CYSTS

Isolated noncommunicating cysts can develop in the spinal cord or within the canal and need to be distinguished from syringomyelia or hydromyelia. The differential diagnosis of spinal intramedullary and intradural cystic lesions includes neuroepithelial cyst, epidermoid cyst, dermoid cyst, and enterogenous cyst [81]. These cysts are distinguishable by pathologic analysis, and their nature may be suspected on MRI because of the cyst contents and associated anomalies.

Isolated noncommunicating neuroepithelial cysts contain fluid which is similar or identical to CSF [82]. A focal cyst in the conus medullaris may represent congenital cystic dilatation of the ventriculus terminalis (Fig. 11.40). The ventriculus terminalis is a small ependymal-lined cavity of the conus medullaris [83]. Isolated delation of the ventriculus terminalis may be due to acquired or developmental blockage between the central canal of the upper spinal cord and the central canal of the lower spinal cord and conus [84]. These ependymal-lined cysts are isointense compared with adjacent CSF and do not enhance after intravenous contrast administration. They do not undergo progressive enlargement, as do syrinx cavities in this region.

Dermoid cysts contain heterogeneous material consisting of the secretions of sebaceous glands and desquamated epithelium. A dermal sinus may be present. Epidermoid cysts contain homogeneous fluid which is very variable in content, sometimes simulating CSF but more often of higher intensity than CSF due to the presence of cholesterol crystals. Enterogenous cysts are lined by mucus-secreting cells. About half the cases are associated with congenital abnormalities such as spina bifida or fused vertebrae, hemivertebrae, or anterior spina bifida. Enterogenous cysts may be intramedullary or extramedullary, and if extramedullary, they will typically cause spinal cord compression [85].

ARACHNOID CYST

Arachnoidal or subarachnoid cyst is the most common intradural extramedullary cyst within the spinal canal. Based on a recent classification of spinal meningeal cysts (86), these are also classified

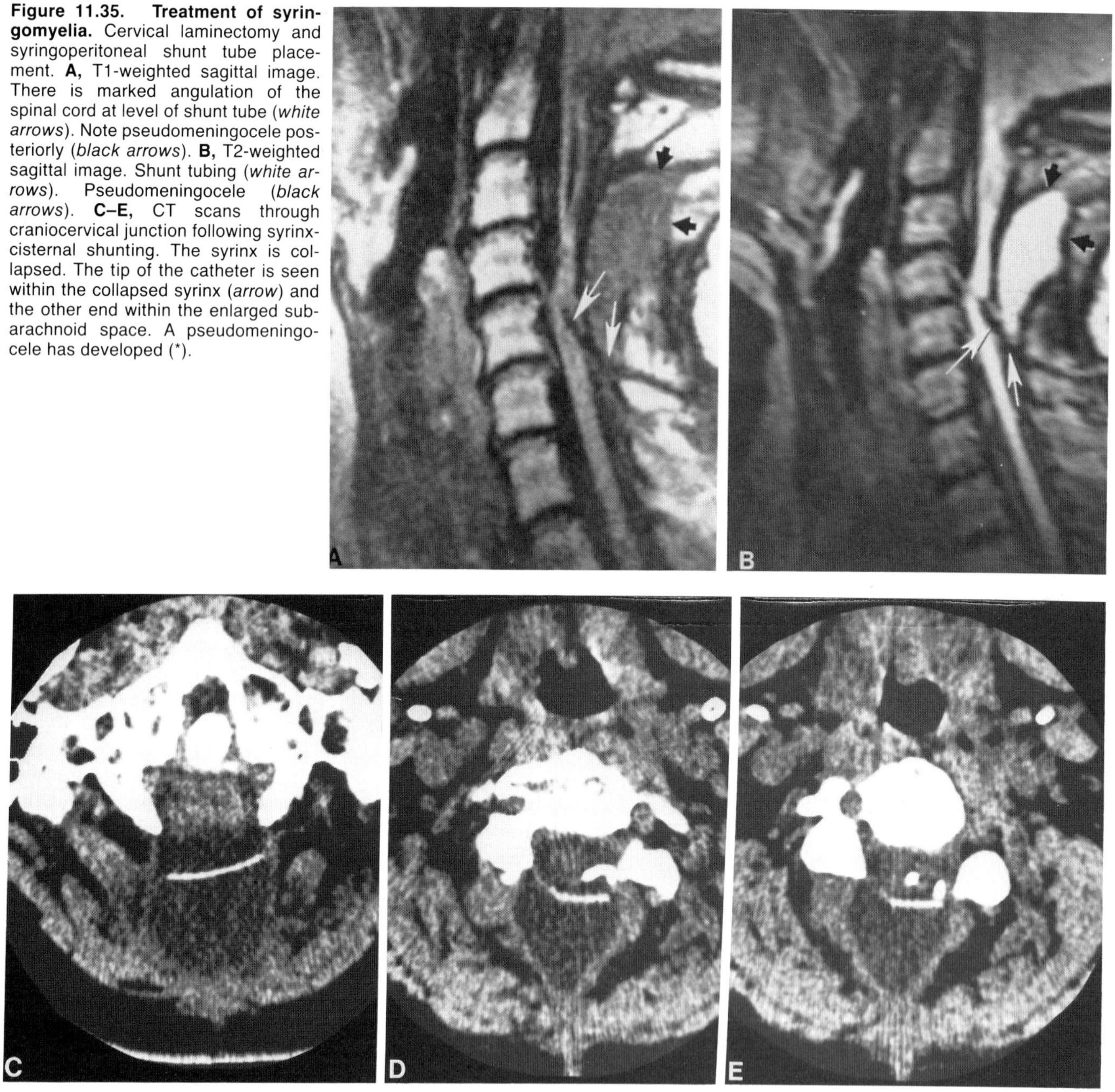

Figure 11.35. Treatment of syringomyelia. Cervical laminectomy and syringoperitoneal shunt tube placement. **A,** T1-weighted sagittal image. There is marked angulation of the spinal cord at level of shunt tube (*white arrows*). Note pseudomeningocele posteriorly (*black arrows*). **B,** T2-weighted sagittal image. Shunt tubing (*white arrows*). Pseudomeningocele (*black arrows*). **C–E,** CT scans through craniocervical junction following syrinx-cisternal shunting. The syrinx is collapsed. The tip of the catheter is seen within the collapsed syrinx (*arrow*) and the other end within the enlarged subarachnoid space. A pseudomeningocele has developed (*).

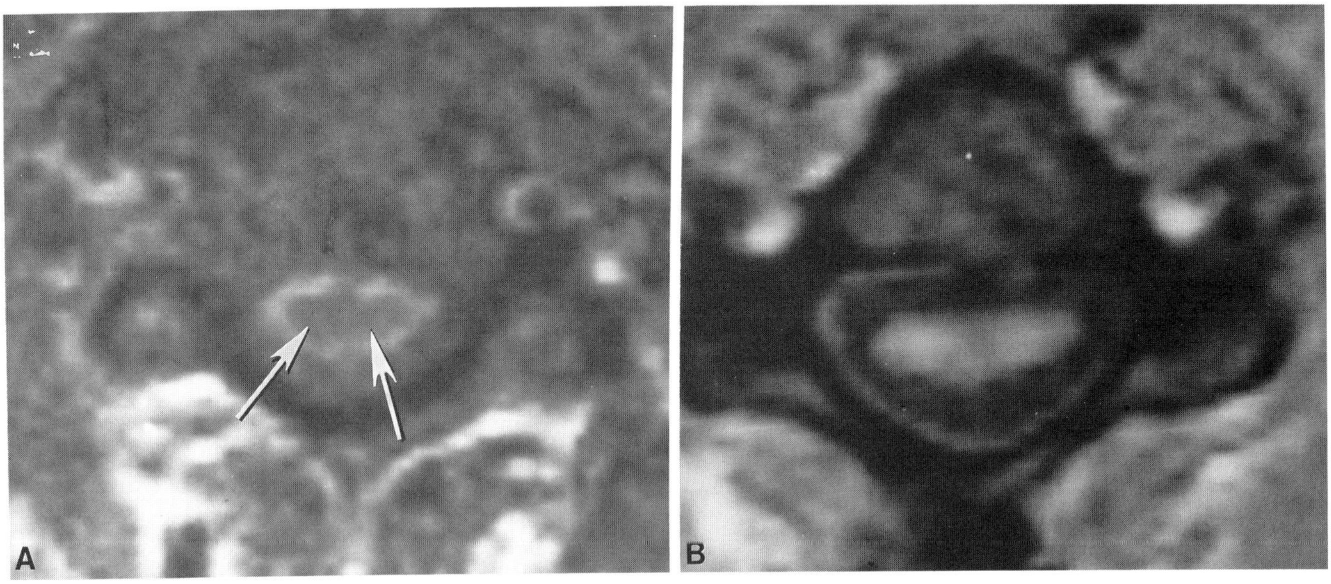

Figure 11.36. Spinal cord atrophy after syrinx treatment. A 61-year-old man with Chiari I malformation and syringomyelia. **A,** Axial T1-weighted image before treatment. Note syrinx cavity (*arrows*). **B,** Axial T1-weighted image after treatment. The atrophic spinal cord is flattened in the anteroposterior diameter and the syrinx cavity has been completely drained.

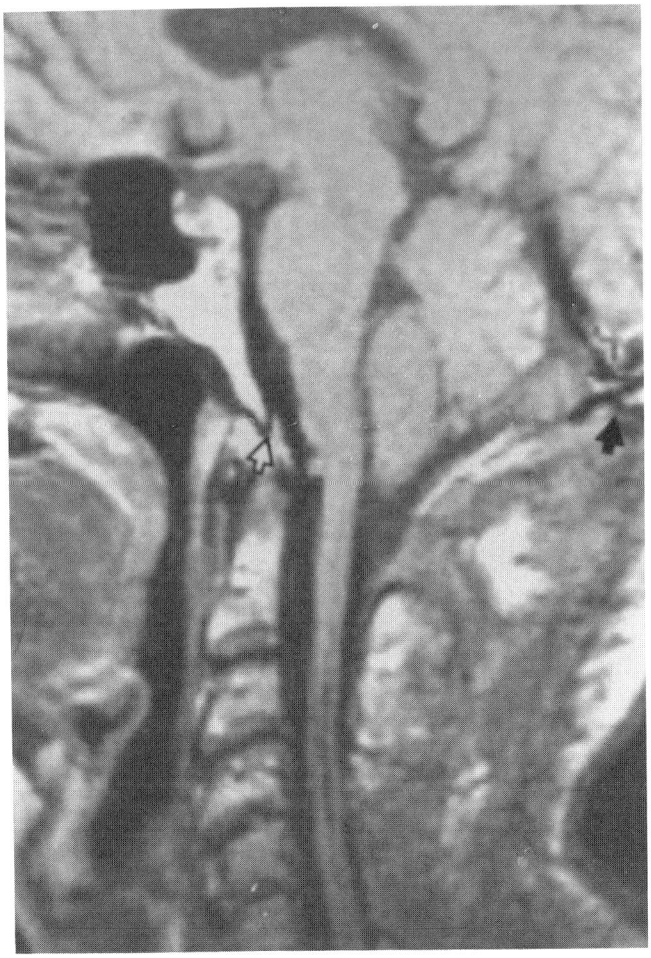

Figure 11.37. Foramen magnum decompression. Decompression was obtained by suboccipital craneictomy. The posterior lip of the foramen magnum (*solid arrow*) is now located posterior to its normal position. There is CSF between the basion (*open arrow*) at the anterior margin of the foramen magnum.

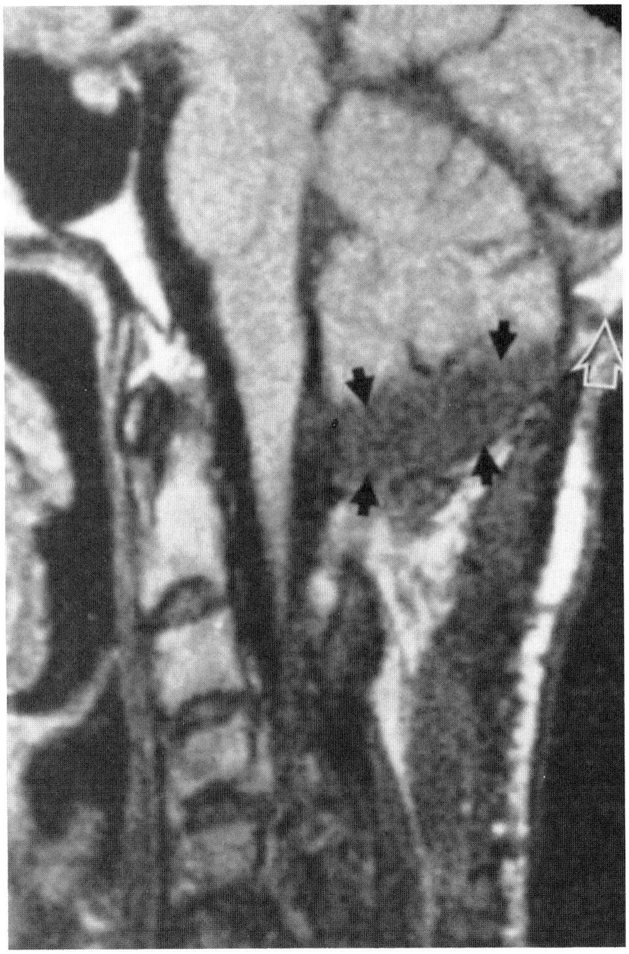

Figure 11.38. Foramen magnum decompression with duroplasty. Note increased CSF (*arrows*) below cerebellum (*arrows*). There is slight flaring of the occipital bone at the margin of the occipital craniectiomy (*open arrow*) which was of no clinical significance.

Figure 11.39. Foreman magnum decompression for treatment of syringomyelia which resulted in exacerbation of symptoms referable to the craniovertebral junction. Note "slumping" of cerebellum through surgically enlarged foramen magnum. The original plane of the foramen magnum is indicated by solid line. The cervicomedullary junction appears kinked (*small arrow*). The edge of occipital bone indents the cerebellum (*long arrow*). There is a small residual syrinx cavity in the lower cervical spinal cord.

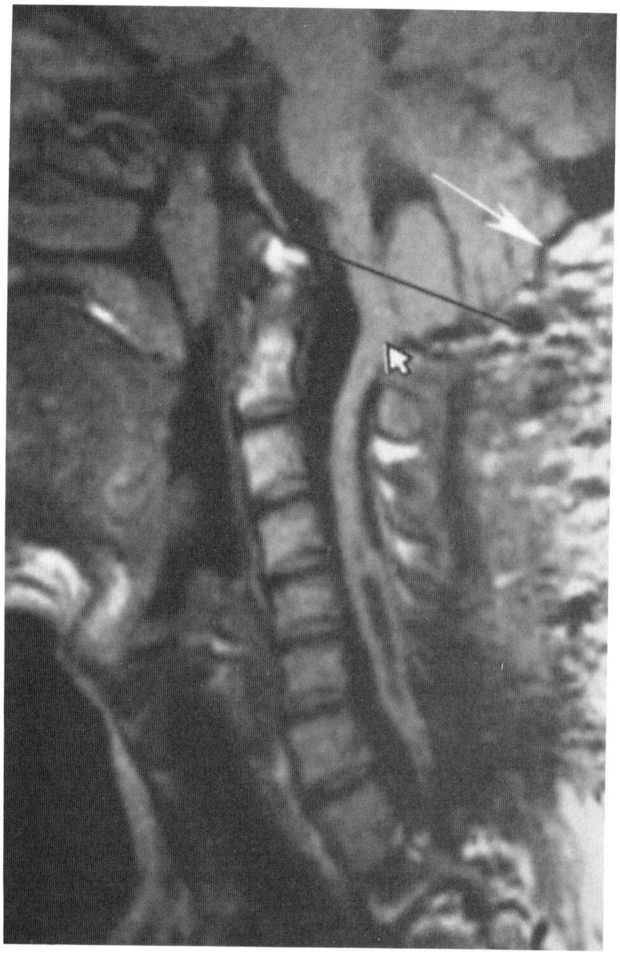

Figure 11.40. Benign focal cyst in the conus medullaris probably due to cystic dilatation of the ventriculus terminalis in a 71-year-old woman. A, T1-weighted sagittal image with gadolinium. Note absence of enhancement and well-defined cyst (*arrows*). **B,** Proton-weighted sagittal image. Cyst contents parallel CSF intensity (*arrows*). **C,** T2-weighted sagittal image. Cyst contents parallel CSF intensity (*arrows*).

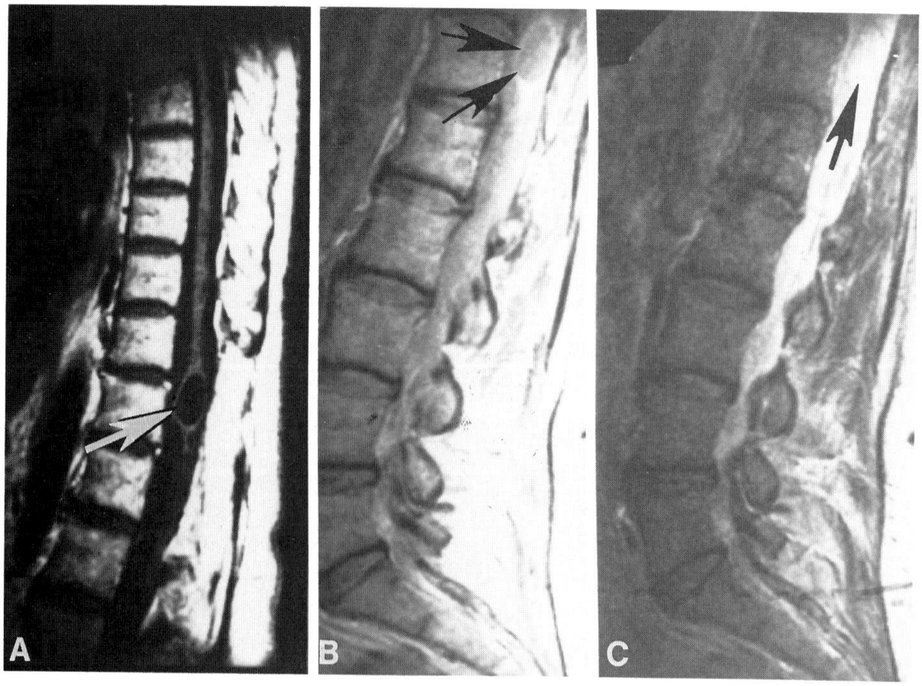

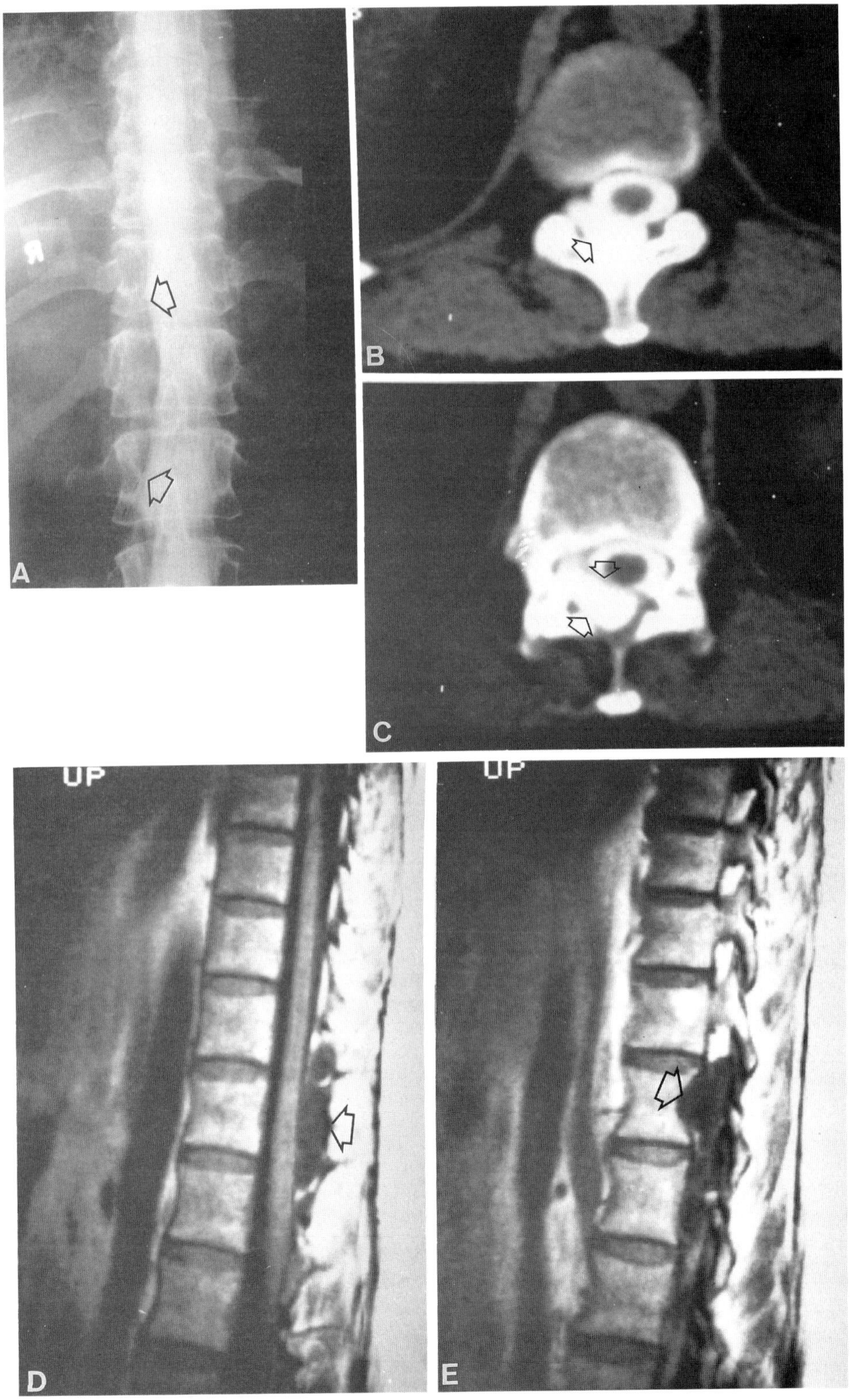

Figure 11.41. Intraspinal arachnoid cyst. A, Frontal view of the myelogram shows a large extradural mass deforming the thecal sac between T11 and L2 (*arrows*). There is erosion of the adjacent pedicle on the right side. **B,C,** Axial CTM sections show the contrast collection along the dorsal aspect of the cord extending toward the right neural foramen (*arrows*). **D,E,** Sagittal and parasagittal T1-weighted MRI sequences show the arachnoid cyst has same signal as CSF and appears lobulated along the dorsal surface of the spinal cord. Note the scalloping of the vertebral body and the wide neural foramen (*arrows*).

as intradural meningeal cysts. They need to be differentiated from cystic outpouchings along the nerve root sleeves (Tarlov's cyst, common in the lumbosacral region) or the meningeal cyst common in the anterior sacral region, presenting as a pelvic mass (see Chap. 3).

Arachnoid or subarachnoid cysts may occur at any level within the spinal canal but are most often seen in the thoracolumbar region. Most are idiopathic in origin, although a few may develop following infection or trauma. Those of infectious or traumatic origin (acquired) may develop at any level and may be located anteriorly or posteriorly. The idiopathic variety are usually seen in the subarachnoid space along the posterior aspect of the spinal cord. The subarachnoid cyst may herniate through a defect in the dura, resulting in an extradural arachnoid cyst. Whether intradural or extradural, the cyst often communicates with the rest of the subarachnoid space through either an ostium or a synichea in the acquired variety. Both the extradural and intradural subarachnoid cysts are within the confines of the spinal canal and neural foramen. They enlarge with time due to the ball-valve mechanism resulting from the pulsatile flow of CSF in the rest of subarachnoid space.

Intradural Arachnoid Cyst

Intradural arachnoid cyst usually presents in patients in their fourth or fifth decade; most are acquired, although a few may be congenital. They become symptomatic at an early stage compared with the extradural arachnoid cyst. As a result, it is unusual to see remodeling of the bone. They are presumed to arise by loculation of the arachnoid trabeculae adjacent to the septum posticum [87]. As a result, most of these cysts are located along the dorsal surface of the spinal cord. They are usually elongated and large, extending over several vertebral segments.

On T1-weighted MRI images, the cyst has a hypointense signal and is associated with a focal enlargement of the subarachnoid space (Fig. 11.41). On density-weighted and T2-weighted images, it has a signal intensity slightly brighter than CSF, which should raise the suspicion of a subarachnoid cyst. The cyst may extend laterally and rarely projects into the neural foramen. Extradural extension of subarachnoid cysts should be suspected when apart from a focal increase in the extradural space, there is enlargement of the neural foramen at one or more levels. A CSF flow-velocity-aided sequence or gradient-echo sequence is useful in separating the cyst from the rest of the subarachnoid space. The flow velocity of CSF is absent or decreased in the cyst relative to the pulsatile flow in the rest of the subarachnoid space. Gradient-echo images not only may show the brighter signal within the cyst but also allow identification of the cyst wall and occasionally hemosiderin deposition in the wall of the cyst.

The MRI findings can be confirmed with CTM. This often demonstrates the dural defect and the communication between the two compartments. Occasionally, when the myelogram demonstrates an extradural filling defect and a similar finding is noted on CTM, a delayed CT will demonstrate the subarachnoid cyst and the dural defect, even after the contrast material in the subarachnoid space is diluted or absorbed.

Extradural Arachnoid Cyst

Extradural arachnoid cysts are congenital lesions usually associated with enlargement of the neural foramen, and marked expansion of the spinal canal is common. Congenital fusion anomalies are common. They are seen most often in the sacral region and truly represent simple meningocele (see Chap. 4). Involvement of the thoracic or cervical region is uncommon but not rare.

References

1. Greenfield JG. Syringomyelia and syringobulbia. In: Blackwood W, McMenemy WH, Mayer A, Norman RM, Russell DS, eds. Greenfield's neuropathology. Baltimore: Williams & Wilkins, 1963:331–333.
2. Harwood-Nash DC, Fitz CR. Myelography and syringohydromyelia in infancy and childhood. Radiology 1974; 113:661–669.
3. Yeats A, Brant-Zwadzki M, Norman D, et al. Nuclear magnetic resonance imaging of syringohydromyelia. AJNR 1983;4:234–237.
4. Peerless SJ, Durward QJ. Management of syringomyelia: a pathophysiologic approach. Clin Neurosurg 1983;30: 531–576.
5. Gardner WJ. Hydrodynamic mechanism of syringomyelia: its relationship to myelocele. J Neurol Neurosurg Psychiatry 1965;28:247–259.
6. Logue V. Fourteenth Crookshank lecture. Syringomyelia: a radiodiagnostic and radiotherapeutic saga. Clin Radiol 1972;22:2–16.
7. Logue V, Edwards MR. Syringomyelia and its surgical treatment: an analysis of 75 patients. J Neurol Neurosurg Psychiatry 1981;44:273–284.
8. Lee BCP, Deck MDF, Kneeland JB, et al. MR imaging of craniocervical junction. AJNR 1985;6:209–213.
9. Spinos E, Laster DW, Moody DM, et al. MR evaluation of Chiari I malformations at 0.15 T. AJR 1985;146:1143–1148.
10. Samuelsson L, Bergstrom K, Thuomas K-A, et al. MR imaging of syringohydromyelia and Chiari malformation in myelomeningocele patients with scoliosis. AJNR 1987; 8:539–546.
11. Cameron AH. The Arnold-Chiari malformation and other neuroanatomical malformations associated with spina bifida. J Pathol 1957;73:195–211.
12. Peach B. Arnold-Chiari malformation: anatomic features of 20 cases. Arch Neurol 1965;12:613–621.

13. Naidich TP, McLone DG, Fulling KH. The Chiari II malformation: IV. The hindbrain deformity. Neuroradiology 1983;25:179–197.
14. Curnes JT, Oakes WJ, Boyko OB. MR imaging of hindbrain deformity in Chiari II patients with and without symptoms of brainstem compression. AJNR 1989;10: 293–302.
15. Foster JB, Hudgson P. The pathogenesis of communicating syringomyelia. In: Barnett MJM, Foster JB, Hudgson P, eds. Syringomyelia. London: Saunders, 1973; 104–123.
16. Gardner WJ. Hydrodynamic mechanism of syringomyelia: its relation to myelocele. J Neurol Neurosurg Psychiatry 1965;28:247–259.
17. Gardner WJ, Angel J. The mechanism of syringomyelia and its surgical correction. Clin Neurosurg 1975;6:131–141.
18. Williams B. The distending force in the production of "communicating syringomyeli." Lancet 1969;2:189–193.
19. du Boulay G, Shah SH, Currie JC, et al. The mechanism of hydromyelia in Chiari type I malformation. Br J Radiol 1974;47:579–587.
20. Williams B. Current concepts of syringomyelia. Br J Hosp Med 1970;79:331–341.
21. Williams B. Pathogenesis of syringomyelia. Lancet 1972;1:142–143.
22. Williams B. On the pathogenesis of syringomyelia: a review. J R Soc Med 1980;73:798–806.
23. Abouker J. La syringomyelie et les liquides intrarachidiens. Neurochirugie 1979;25(Suppl 1):396–403.
24. Williams B. Cerebrospinal fluid pressure changes in response to coughing. Brain 1976;99:331–346.
25. Perot P, Feindel W, Lloyd-Smith D. Hematomyelia as a complication of syringomyelia: Gower's hemorrhage. J Neurosurg 1966;25:447–451.
26. Sudo K, Dol S, Maruo Y, et al. Syringomyelia with spontaneous resolution. J Neurol Neurosurg Psychiatry 1990;5:437–438.
27. Nakagawa H, Huang YP, Malis Li, et al. Computer tomography of intraspinal and paraspinal neoplasms. J Comput Assist Tomogr 1978;2:226–228.
28. Deeb ZL, Daffner RH, Rothfus WE, et al. Syringomyelia: myelography, computed tomography and magnetic resonance imaging. CT 1988;12:1–6.
29. Aubin ML, Vigneaud J, Jardin C, Bar D. Computed tomography in 75 clinical cases of syringomyelia. AJNR 1981;2:199–204.
30. Kan S, Fox A, Vinuela F, et al. Delayed metrizamide enhancement of syringomyelia secondary to tumor. AJNR 1980;4:77–80.
31. Lee BCP, Zimmerman RD, Manning JJ, Deck MDF. MR imaging of syringomyelia and hydromyelia. AJNR 1985; 6:221–228.
32. Williams AL, Haughton VM, Pojumas KW, et al. Differentiation of intramedullary neoplasms and cysts by MR. AJNR 1987;8:527–532.
33. Aubin ML, Baleriaux D, Cosnard G, et al. MRI in syringomyelia of congenital, infectious, traumatic or idiopathic origin: a study of 143 cases. J Neuroradiol 1987;14:313–336.
34. Abe K, Sawada H, Fujiwara M, et al. MR imaging of syringobulbia with giant syrinx. Neuroradiology 1988; 30:442–443.
35. Houng MTW, Stern M, Brew B, et al. Magnetic resonance imaging (MRI) appearances of syringohydromyelia. Australas Radiol 1988;32:172–177.
36. Samuelsson L, Saaf J, Wahlund L-O, et al. Evaluation of syrinx and Chiari malformations using ultra-low magnetic resonance imaging. Magn Reson Imag 1990;8: 123.
37. Sherman JL, Barkovich AJ, Citrin CM. The MR appearance of syringomyelia: new observations. AJNR 1986; 7:985–995.
38. Bradley WG, Nitz W, O'Sullivan RM, et al. CSF velocity imaging in the spine. Presented at the 29th annual meeting of American Society of Neuroradiology, Washington, D.C., 1991.
39. Bronskill MJ, McVeigh ER, Kucharczyk W, Henkelman RM. Syrinx-like artifacts on MR images of the spinal cord. Radiology 1988;166:451–454.
40. Levy LM, Di Chiro G, Brooks RA, et al. Spinal cord artifacts from truncation errors during MR imaging. Radiology 1988;166:479–481.
41. Rubin JB, Enzmann DR, Wright A. CSF-gated MR imaging of the spine: theory and clinical implementation. Radiology 1987;163:784–795.
42. Rubin JB, Enzmann DR. Optimizing conventional MR imaging of the spine. Radiology 1987;163:777–783.
43. Holtas SL, Plewes DB, Simon JH, et al. Technical aspects on magnetic resonance imaging of the spine at 1.5 tesla. Acta Radiol 1987;28:375–383.
44. Hyman RA, Gorey MT. Imaging strategies for MR of the spine. Radiol Clin North Am 1988;26:505–520.
45. Edelman RR, Atkinson DJ, Silver MS, et al. FRODO pulse sequences: a new means of eliminating motion, flow and wraparound artifacts. Radiology 1988;166:231–238.
46. Barkovich AJ, Sherman JL, Citrin CM, Wippold FJ II. MR of postoperative syringomyelia. AJNR 1987;8:319–327.
47. Westberg G. Gas myelography and percutaneous puncture in the diagnosis of spinal cord cysts. Acta Radiol Suppl (Copenh) 1966;252:1–67.
48. Barkovich AJ, Wippold FJ, Sherman JL, et al. Significance of cerebellar tonsillar position on MR. AJNR 1986;7:795–799.
49. Venes JL, Black KL, Latack JT. Preoperative evaluation and surgical management of the Arnold-Chiari malformation. J Neurosurg 1986;64:363–370.
50. Rossier AB, Foo D, Shilleto J, et al. Post traumatic cervical syringomyelia. Brain 1985;108:439–461.
51. Kulkarni MV, McArdale CB, Kopanicky J, et al. Acute spinal cord injury: MR imaging at 1.5 T. Radiology 1987; 164:837–843.
52. Quencer RM, Egnor M, Bunge R, et al. Acute posttraumatic cord syndrome: new observations and challenges to traditional concepts. Presented at the 76th annual meeting of the Radiological Society of North America, Chicago, 1990.
53. Cohen WA, Young W, DeCrescito V, et al. Posttraumatic syrinx formation: experimental study. AJNR 1985;6: 823–827.
54. Macdonald RL, Findlay JM, Tator CH. Microcystic spinal cord degeneration causing post traumatic myelopathy. J Neurosurg 1988;68:466–471.
55. Barnett HJM, Jousse AT, Ball MJ. Pathology and pathogenesis of progressive cyst myelopathy as a late sequel to spinal cord injury. In: Barnett MJM, Foster JB, Hudgson P, eds. Syringomyelia. London: Saunders, 1973: 179–219.
56. Quencer RM, Sheldon J, Post MJD, et al. Magnetic resonance imaging of chronically injured cervical spinal cord: correlation with delayed metrizamide CT, intraoperative spinal sonography and surgery. AJNR 1986; 7:464–475.

57. Yamashita Y, Takahashi M, Matsuno Y, et al. Chronic injuries of the spinal cord: assessment with MR imaging. Radiology 1990;175:849–854.
58. Goy AMC, Pinto RS, Raghavendra BN, et al. Intramedullary spinal cord tumors: MR imaging, with emphasis on associated cysts. Radiology 1986;161:381–386.
59. Hirata Y, Matsukado, Kaku M. Syringomyelia associated with foramen magnum meningioma. Surg Neurol 1985; 23:291–294.
60. Quencer RM, Gammal TE, Cohen G. Syringomyelia associated with intradural extramedullary masses of the spinal canal. AJNR 1986;7:143–148.
61. Castillo M, Quencer RM, Green BA, et al. Syringomyelia as a consequence of compressive extramedullary lesions: postoperative, clinical and radiological manifestations. AJR 1988;150:391–396.
62. Barnett HJM, Rawcastle NB. Syringomyelia and tumors of the nervous system. In: Barnett MJM, Foster JB, Hudgson P, eds. Syringomyelia. London: Saunders, 1973;261–301.
63. Rubin JM, Aisen AM, DiPietro MA. Ambiguities in MR imaging of tumoral cysts in the spinal cord. J Comput Assist Tomogr 1986;10:395–398.
64. Slasky BS, Bydder GM, Niendorf HP, et al. MR imaging with gadolinium-DTPA in the differentiation of tumor, syrinx and cyst of the spinal cord. J Comput Assist Tomogr 1987;11:840–845.
65. Parizel PM, Baleriuax D, Rodesch G, et al. Gd-DTPA enhanced MR imaging of spinal tumors. AJNR 1989;10: 249–258.
66. Bydder GM, Brown J, Niendorf HP, et al. Enhancement of cervical intraspinal tumors in MR imaging with IV Gd-DTPA. J Comput Assist Tomogr 1985;9:847851.
67. Tator CH, Meguro K, Rowed DW. Favourable results with syringosubarachnoid shunts for treatment of syringomyelia. J Neurosurg 1982;20:339–345.
68. Mackay R. Chronic adhesive spinal arachnoiditis. JAMA 1939;112:802–808.
69. Barnett JM. The pathogenesis of syringomyelic cavitation associated with arachnoiditis localized to the spinal cord. In: Barnett MJM, Foster JB, Hudgson P, eds. Syringomyelia. London: Saunders, 1973:245–260.
70. Sklar E, Quencer RM, Green BA, et al. Acquired spinal subarachnoid cysts: evaluation with MR, CT, myelography and intraoperative sonography. AJNR 1989;10: 1097–1104.
71. Caplan LR, Noronhna AN, Amico LL, Syringomyelia and arachnoiditis. J Neurol Neurosurg Psychiatry 1990; 53:106–113.
72. Johnson CE, Sze G. Benign lumbar arachnoiditis: MR imaging with gadopentate dimeglumine. AJNR 1990; 11:763–770.
73. Wilkinson M. Cervical spondylosis—its early diagnosis and treatment. London: Heinmann, 1971.
74. Jinkins JR. Focal syringomyelia in compressive cervical spinal stenosis with associated distant cranio-caudad cavitary changes of the spinal cord. Presented at the 24th annual meeting of ASNR, San Diego, January 1986.
75. Castillo M, Quencer RM, Green BA, et al. Syringomyelia as a consequence of compressive extramedullary lesions. AJNR 1987;8:973–978.
76. Yu YL, Mosley IF. Syringomyelia and cervical spondylosis: a clinicoradiological investigation. Neuroradiology 1987;29:143–151.
77. Banna M. Syringomyelia in association with posterior fossa cysts. AJNR 1988;9:867–873.
78. Andrews BT, Weinstein PR, Rosenbaum ML, et al. Intradural arachnoid cysts of the spinal canal associated with intramedullary cysts. J Neurosurg 1988;68:544–549.
79. Castillo M, Quencer RM, Post MJD, et al. MR of intramedullary spinal cysticercosis. AJR 1989;9:393–395.
80. Williams B. Progress in syringomyelia. Neurol Res 1986;8:130–145.
81. Russell DS, Rubenstein L, eds. Pathology of tumors of the nervous system. 5th ed. Baltimore: Williams & Wilkins, 1989:702.
82. Rousseau M, Lesion F, Comnbellas G, et al. Intramedullary ependymal cyst in a 71 year old woman. Neurosurgery 1983;13:52–54.
83. Kernohan JW. The ventriculus terminalis: its growth and development. J Comp Neurol 1924;38:10–125.
84. Sigal R, Denys A, Halimi P, et al. Ventriculus terminalis of the conus medullaris: MR imaging in four patients with congenital dilatation. AJNR 1991;12:733–737.
85. Miyagi K, Mukawa J, Meekaru S, et al. Enterogenous cyst in the cervical spinal canal. J Neurosurg 1988;68:292–296.
86. Nabors MW, Piat Tg, Naz K, et al. Updated assessment and current classification of spinal meningeal cysts. J Neurosurg 1987;68:366–372.
87. Perret G, Green D, Keller J. Diagnosis and treatment of intradural arachnoid cysts of the thoracic spine. Radiology 1962;79:425–429.

CHAPTER / 12

Nonneoplastic Cord Lesions

J. Sherman

Nonneoplastic disorders of the spinal cord constitute a variety of disease processes, including, among others, demyelinating and degenerative diseases, vasculitides (including collagen-vascular disease), occult vascular malformations, and the results of acute and chronic ischemia from a variety of causes. The evaluation of these entities has been difficult prior to the development of MRI. MRI is now the preferred method for evaluation of the spinal cord. Myelography is restricted to the evaluation of diseases that displace, compress, or expand the spinal cord (i.e., contour abnormalities). CT myelography (CTM) is excellent for the evaluation of degenerative disk disease and, prior to MRI, had a limited role primarily in evaluating cord swelling or atrophy and in demonstrating the cavitating lesions, including syringomyelia. Intravenous contrast–enhanced computed tomography (IVCT) has had very limited success in the evaluation of cord parenchyma. Spinal cord angiography is reserved for the few patients with spinal vascular malformations and hypervascular tumors.

Clinical Overview

An understanding of the various clinical presentations is useful when evaluating imaging studies and in excluding intrinsic cord pathology from that resulting from extramedullary causes. The importance of the clinical examination in determining the differential diagnosis in this group of disorders is based on five basic neurologic abnormalities and eight spinal cord syndromes that encompass the majority of spinal cord diseases [1] (Table 12.1). The five neurologic abnormalities include pain, motor dysfunction, sensory abnormalities, abnormalities of reflexes and muscle tone, and bladder symptoms. These neurologic abnormalities each have many different expressions and variations but can be grouped broadly into eight neurologic syndromes.

1. *Acute complete cord syndrome* (paraplegia or quadriplegia with local pain, abnormal reflexes, nonselective sensory loss, and bladder dysfunction). This syndrome is caused by cord infarction, acute trauma, spinal hemorrhage, or acute compression from a large disk.
2. *Subacute complete cord syndrome* (subacute disorders encompass those developing over hours, days, or weeks). The differential diagnosis includes transverse myelitis and multiple sclerosis. Epidural tumors and abscesses are included but are more slowly progressive than myelitis.
3. *Chronic spastic paraplegia/quadraplegia and ataxia without sensory signs.* Includes amyotrophic lateral sclerosis, primary lateral sclerosis, familial spastic paraplegia, and the myelopathy of cervical spondylosis.
4. *Chronic dorsal cord syndrome.* These patients have chronic spastic paraplegia or quadriplegia, ataxia, and disorders of position and vibratory sensation.
5. *Chronic anterior cord syndrome and Brown-Sequard syndrome.* Patients with these syndromes have chronic spastic paraplegia/quadriplegia and ataxia with loss of pain sensation bilaterally (anterior cord syndrome) or unilaterally (Brown-Sequard syndrome).
6. *Chronic atrophic paralysis of the hands with sensory disturbances.* Sensory disturbances may

Table 12.1.
Common Causes of Spinal Cord Syndromes

1. Acute complete cord syndrome
Cord infarction
Acute trauma
Spinal hemorrhage
Acute cord compression from disk herniation
2. Subacute complete cord syndrome
Transverse myelitis
Multiple sclerosis
Epidural tumor
Epidural abscess
3. Chronic spastic paraplegia/quadriplegia and ataxia without sensory signs
Amyotrophic lateral sclerosis
Familial spastic paraplegia
Myelopathy of cervical spondylosis
4. Dorsal cord syndrome
Arachnoiditis
Spinal multiple sclerosis
Myelopathy of cervical spondylosis
Arteriovenous malformation
Subacute combined degeneration of the spinal cord
Meningioma and schwannoma
Spinal cord glioma
Vacuoloar myelopathy of AIDS
Spinocerebellar degeneration
Spinal cord injury
5. Chronic anterior cord syndrome and Brown-Sequard syndrome
Meningioma and schwannoma
Myelopathy of cervical spondylosis
Radiation myelopathy
Spinal cord injury
6. Chronic atrophic paralysis of hands *with* sensory disturbances
Intrinsic spinal cord tumors
Syringomyelia
Spinal cord injury
7. Chronic atrophic paralysis of hands *without* sensory disturbances
Amyotrophic lateral sclerosis
Cervical spondylosis
Extrinsic tumors of the upper cervical spinal cord
8. Cauda equina syndrome
Cauda equina tumor
Lumbar spinal stenosis
Herniated lumbar disk
Arachnoiditis
Trauma with cauda equina injury

involve only pain (central cord syndrome) or all sensory modalities.

7. *Chronic atrophic paralysis of hands without sensory disturbances.*
8. *Cauda equina syndrome.* This syndrome comprises chronic paraplegia, low back pain, areflexia, "saddle anesthesia," and loss of bladder function.

Imaging Considerations

At the present time, unless there is a contraindication, MRI should be the imaging modality of choice. If CT is employed, it is usually following myelography with nonionic contrast material. The CT sections are usually obtained through the region found abnormal on the myelogram. The sections should not exceed 5 mm in thickness. Obvious cord swelling or cord atrophy can be easily demonstrated. This is best achieved utilizing wide window settings. If abnormal cord morphology is identified, delayed CT after an interval of 6 to 12 hours will allow identification of syrinx and in other circumstances intravasation of contrast material. Axial CT sections are also useful in precisely demonstrating the various patterns of cord atrophy.

The goal of spinal cord MRI is to produce consistently clear and detailed images of the spinal cord which allow evaluation of contour abnormalities, tissue changes, or changes in the normal differentiation of spinal gray matter and white matter. This requires optimization of all components of the examination. The techniques may vary, depending on the capabilities of the different MRI imagers. Standard technique usually requires T1-weighted, T2-weighted, and proton-weighted (balanced) images. Differentiation of white matter and gray matter requires proton-weighted or T2-weighted images in the axial plane. T2-weighted spin-echo and/or gradient-echo sequences may be employed, depending on the imager. Properly optimized gradient recalled echo imaging techniques adequately reveal pathology and can often replace T2-weighted SE images in the spine, especially for axial images [2].

Spinal cord detail can be obscured by artifact, including edge artifacts due to chemical shift, motion, and Fourier transformation truncation artifact (Gibbs' phenomenon). Motion artifact occurs predominantly in the phase-encoded direction. Truncation artifact is represented by bands of low and high signal intensity paralleling cord-CSF interfaces. Truncation artifact may cause factitious increased width of the CSF and therefore cause the cord to appear atrophic. The artifact can be reduced by decreasing pixel size [3]. Artifacts from CSF pulsations may cause ill-defined streaks along the spinal cord contour or a heterogeneous loss of signal intensity in the CSF. It is advisable to use all available methods to attempt to reduce or nullify motion. Such methods include motion-correction gradient waveforms [4], e.g., low compensation (FC), motion artifact suppression technique (MAST), optimal placement of the phase-encoding gradient, peripheral pulse gating, cardiac gating, and presaturation slabs. Patient motion can be reduced if the patient is comfortable [5, 6].

Normal Spinal Cord

A brief overview of the morphologic appearance of the spinal cord and its vasculature is useful when evaluating the pathologic changes seen on cross-sectional imaging. The spinal cord is approximately cylindrical in shape, tapering caudally to become the conus medullaris. The surface of the cord is longitudinally indented by the median fissure, which penetrates the cord for a depth of up to 3 mm and contains branches of the anterior spinal artery and vein (Fig. 12.1). The posterior median sulcus is a shallow groove that marks the position of the dorsal median septum, which extends into the cord and reaches the deep-lying gray matter. The dorsolateral sulcus marks the position of the entrance of the dorsal nerve roots, while the emergence of the ventral roots is marked by the less distinct ventrolateral sulcus.

The blood supply of the spinal cord is dependent on three longitudinal arterial trunks: a single anterior spinal artery and paired posterior spinal arteries [7]. The anterior radiculomedullary arteries supplying the anterior spinal artery vary from 2 to 17 in number but average 9. The artery of Adamkiewicz is the largest anterior medullary feeder and supplies the thoracolumbar region. The posterior medullary feeders (radiculomedullary arteries) supplying the posterior spinal artery are more numerous, usually ranging from 10 to 23 in number. In most cases, the

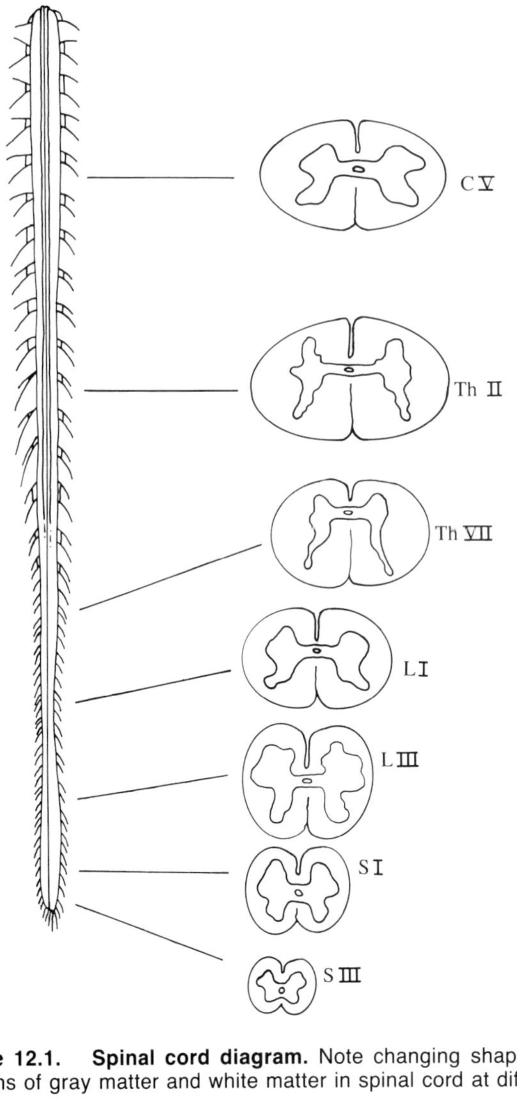

Figure 12.1. Spinal cord diagram. Note changing shape and patterns of gray matter and white matter in spinal cord at different levels.

artery of Adamkiewicz supplies the entire lumbosacral cord, including the posterior arteries. The radiculomedullary arteries form two distinct systems of cord vascularization. The first system is a posterolateral and peripheral plexus supplying the outer 50 to 75% of the spinal cord. This system is formed primarily by the two posterior spinal arteries and their anastomoses running longitudinally along the posterolateral spinal cord sulcus [8]. The second system supplies the central gray matter and adjacent white matter, including the corticospinal tracts. This system is formed by numerous central arteries that arise from the anterior spinal artery within the central sulcus of the spinal cord. Blood from the intraspinal capillary net is drained through radial veins into the external venous perispinal venous net and then into anterior and posterior radicular veins. The venous network is more dense posteriorly than anteriorly, while the arterial network is denser anteriorly [9].

The central spinal cord gray matter comprises nerve cell bodies, unmyelinated fibers, and dendrites. Large motor neurons are situated in the ventral horns of gray matter. The receptor part of the spinal cord is in the dorsal gray horns. The gray matter consists of various cell masses forming columns within the cord (Fig. 12.2). These can be broadly separated into three subdivisions: dorsal columns, intermediate columns, and ventral columns. The anterior and lateral marginal areas of the spinal cord are formed by long ascending and descending white matter fiber tracts. The white matter is divided into three main columns known as *funiculi*. The dorsal funiculi are the prominent columns of white matter that carry the largest ascending tracts to the cerebral cortex. The lateral funiculi contain several ascending and descending fasciculi. The corticospinal tract is found in the posterior portion of the lateral funiculus. The smaller ventral corticospinal tract (uncrossed fibers) is found in the anterior funiculus.

The spinal cord in cross section has the shape of an ellipse that varies with the level of section (Fig. 12.3). The spinal cord has two enlargements, the cervical and lumbar. These enlargements occur because of the increased neural tissue required for the brachial and lumbar plexi. At the highest cord levels, the cord is a near-circular ellipse. It gradually enlarges in both transverse and anteroposterior planes to its maximum area at C4 or C5 and then decreases [14] (Table 12.2 and Figs. 12.4 to 12.6). In vivo measurements of the spinal cord in the cervical and thoracic regions have been performed using pneumomyelography, oily contrast myelography, postmortem myelography, and CT-assisted myelography [10–13]. These measurements were each affected by technique-dependent errors and thus are not reliable. Myelographic measurements [10] are approximately

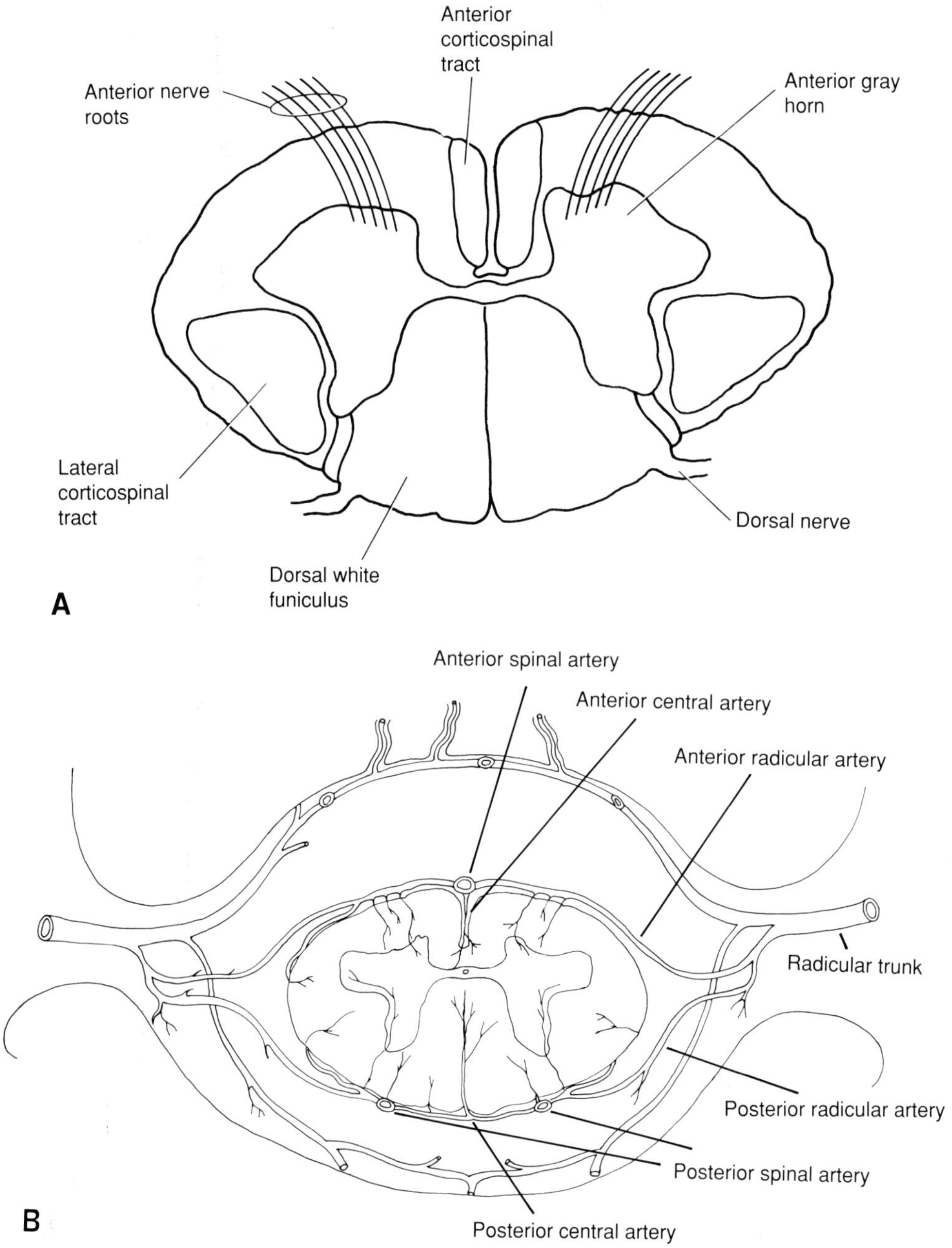

Figure 12.2. Spinal cord diagrams. A, Diagram of cross section of cervical cord. **B,** Diagram of cross section of spinal cord showing arterial blood supply.

2 mm greater and CT myelography measurements [11, 12] are 1 to 2 mm less than MRI data [14]. There is no correlation between cord size and patient weight, age, and sex. Each spinal cord level is slightly different, and the measurements are only valid in the absence of stenosis or cord compression.

Differentiation of the gray matter and white matter can be made on the axial MRI images (see Fig. 12.3). The butterfly-shaped central gray matter of the cord is the most conspicuous landmark on high-resolution axial MRI sections. Gray matter has higher spin density and thus appears "brighter" on proton-weighted images of the spinal cord and brain, whereas on T1-weighted images there is poor differentiation of gray and white matter [15, 16]. The central gray matter is situated posteriorly near the obex in the medulla. Sagittal images reveal the hyperintense gray matter "stripe." An area of "discontinuity" appears at the level of the dens due to the decussation.

Proton-weighted MRI images reveal increased intensity in the gray matter. At the cervicomedullary junction, the spinal trigeminal nuclei, located pos-

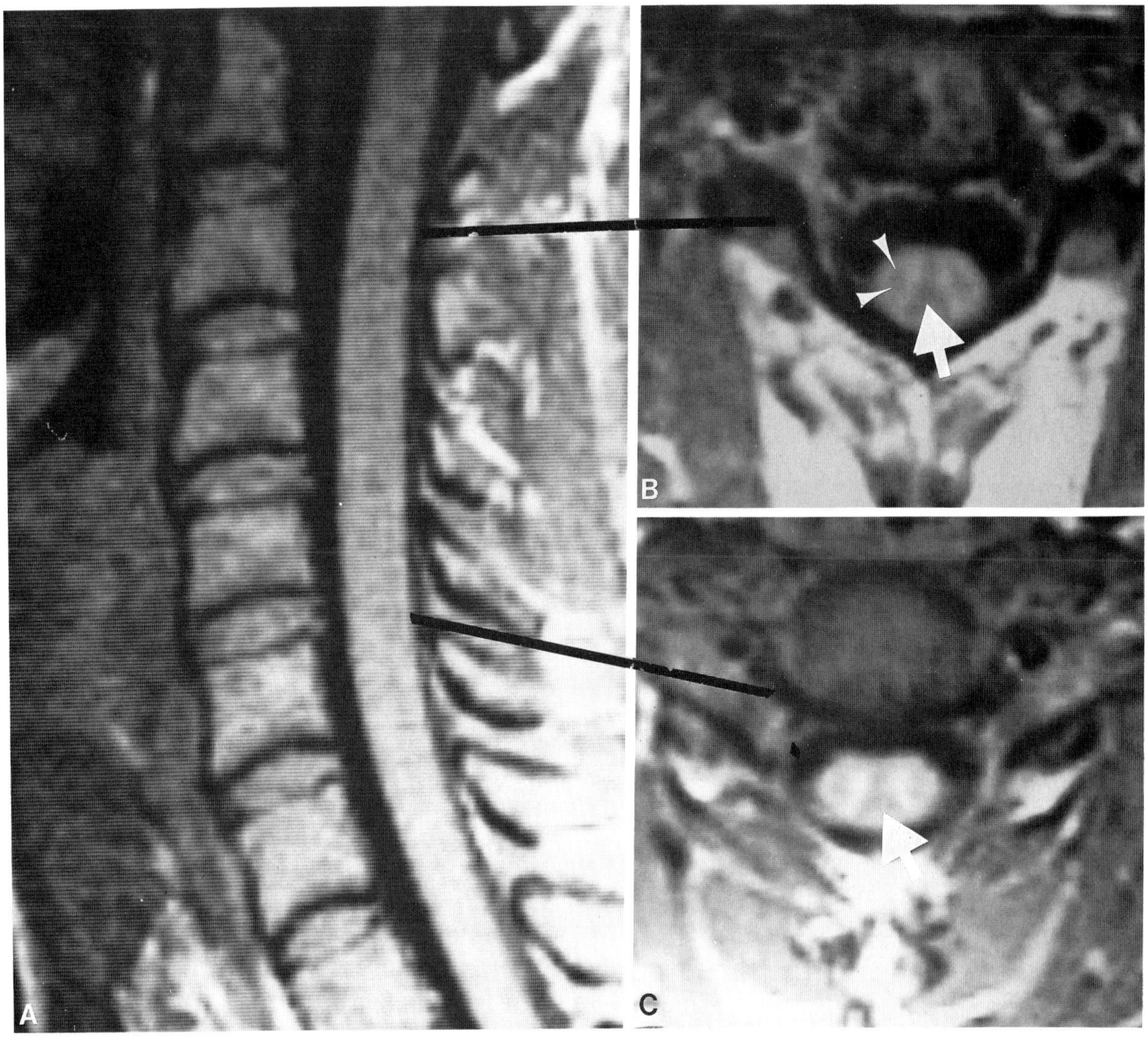

Figure 12.3. Normal spinal cord. A, Sagittal T1-weighted image. **B,** Axial proton-weighted image (SE 1000/20) at C3 level. Note rounded contour, dorsal white funiculi (*large arrow*), lateral corticospinal tract (*arrowheads*). **C,** Axial image at C5 level. Note difference in shape compared with C3 level and dorsal white funiculi (*large arrow*).

terolaterally and peripherally in the cord, and the supraspinal nuclei, located anterolaterally, appear as high-signal-intensity areas similar to the central gray matter. The pyramidal tracts and the fasciculus gracilis and cuneatus are hypointense compared with the gray matter.

Diagnosis of the intrinsic cord abnormalities discussed in this chapter depends on two features: (*a*) the morphologic change in the size of the cord, which can be evaluated with either CTM or T1-weighted MRI, and (*b*) the intrinsic signal change seen on T2-weighted MRI or in the gadolinium-enhanced T1-weighted MRI. The imaging findings, when combined with the clinical presentation and laboratory findings, are useful in most instances and dictate the need for additional imaging studies or follow-up studies.

Table 12.2.
Calculated Average Spinal Cord Measurements and Standard Deviations

	Average AP (SD)	Average Transverse (SD)	Average ACA (SD)
C1	9.3 mm (0.9)	11.3 mm (0.9)	104.1 mm^2 (13.8)
C2	8.8 mm (0.9)	12.4 mm (0.9)	110.0 mm^2 (15.0)
C3	8.6 mm (0.8)	13.2 mm (1.0)	113.6 mm^2 (16.0)
C4	8.7 mm (0.9)	14.0 mm (1.1)	121.9 mm^2 (17.5)
C5	8.3 mm (0.9)	13.9 mm (1.0)	115.4 mm^2 (15.7)
C6	7.9 mm (0.8)	13.2 mm (1.0)	104.6 mm^2 (15.0)
C7	7.4 mm (0.9)	11.4 mm (1.2)	84.6 mm^2 (15.7)
T1	7.4 mm (0.7)	10.3 mm (1.1)	76.0 mm^2 (12.7)
T2	7.4 mm (1.1)	9.8 mm (0.9)	73.7 mm^2 (15.4)
T3	7.3 mm (0.8)	9.8 mm (0.9)	72.2 mm^2 (10.8)

These average measurements and the corresponding standard deviations were calculated from original data that was rounded to the nearest millimeter.

Figure 12.4. Measurements of the anteroposterior diameter of the spinal cord from C1 to T3 in millimeters.

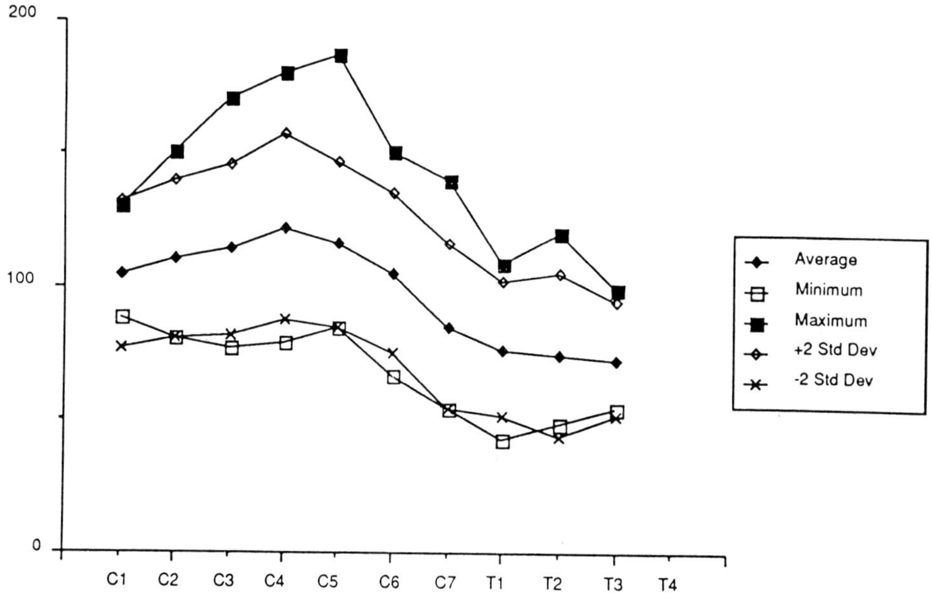

Figure 12.5. Measurements of the transverse diameter of the spinal cord from C1 to T3 in millimeters. Note the cord enlargement that peaks at the C4 level.

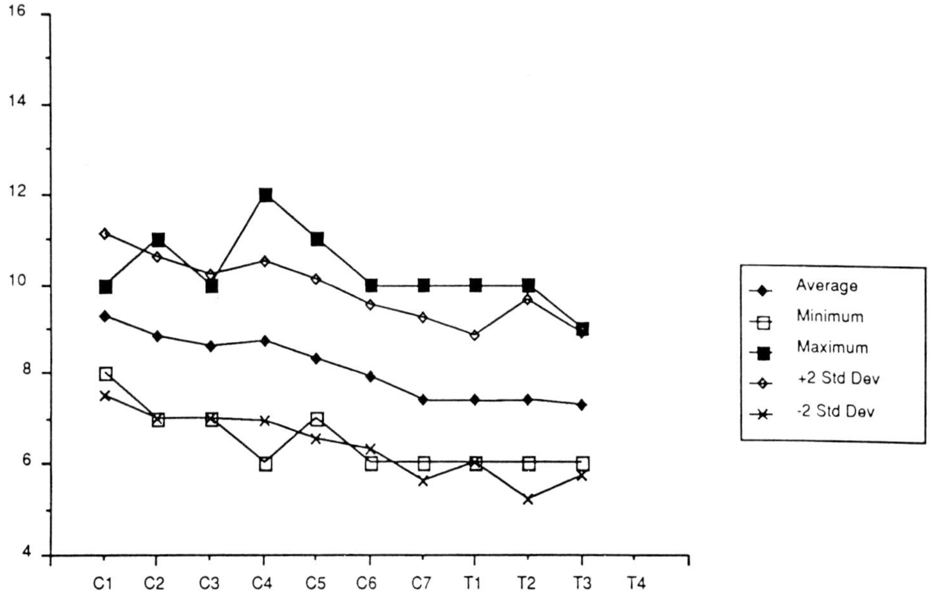

Figure 12.6. Graph of the approximate cord area (ACA) values from C1 to T3. These numbers are obtained by multiplying the anteroposterior diameter by the transverse diameter. These values do not represent true area measurements of the spinal cord because the shape of the cord is a variable ellipse.

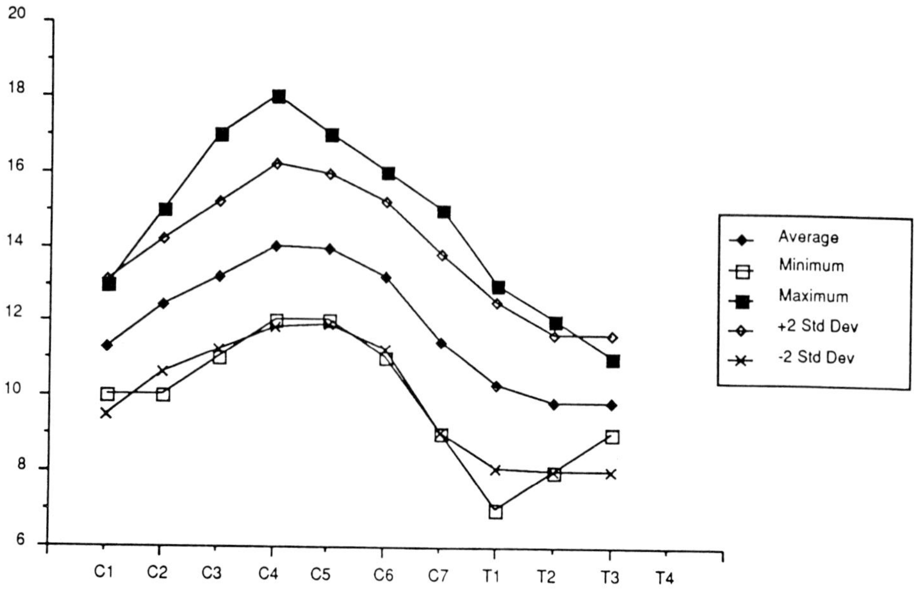

Demyelinating Disease

Multiple sclerosis (MS) is a demyelinating disease of the central nervous system (CNS). MS is characterized by fluctuating multifocal symptoms of the sensory and motor systems [17]. As a general clinical rule, the diagnosis of MS is not secure unless there is a history of remission and relapse and evidence on examination of more than one discrete lesion of the CNS [18]. The disease is more frequent in the Western Hemisphere, found in northern Europe, Great Britain, Canada, and the United States. The age of onset is between 10 and 50 years. Pathologically, MS plaques are defined as perivenous inflammatory lesions that cause destruction of myelin while preserving axons [19]. In the subacute stages of the disease process (6 to 12 weeks after clinical onset of symptoms), there is reactive astrocytosis and infiltration by lymphocytes and plasma cells. Active demyelination may persist at the edges of the plaque. Chronic inactive plaques have intact axons surrounded by reactive astrocytosis.

Patients with MS have autopsy findings of demyelinating lesions in brain and spinal cord with almost equal frequency [19]. However, there have been few accounts of the myelographic or CT appearance of MS. MRI is clearly superior to other imaging methods for demonstration of MS plaques [20, 21]. Spinal cord MRI is positive in 75 to 86% of patients in whom MS of the cord is clinically suspected [22, 23]. The ability of MRI to demonstrate multiple lesions in the brain, as well as in the cord, helps to establish the diagnosis by showing dissemination in space and by revealing "silent" lesions that may reflect prior involvement. The occurrence of a normal brain MRI and a positive spinal cord MRI occurs in 16 to 50% of patients [23, 24]. The MS plaques are usually not visible on T1-weighted images. However, long-TR, long-TE spin-echo T2-weighted images show the MS plaques as areas of increased signal intensity compared with the adjacent normal white matter. The hyperintensity may be quite subtle and frequently is only recognized as a loss of differentiation between the gray matter and the normally low signal intensity of the dorsal white columns. T2* low-flip-angle gradient-echo images can effectively demonstrate spinal cord MS, particularly when axial images are acquired [25]. Gradient-echo images often have superior resolution, are more robust, and are easier to obtain than their spin-echo counterparts. However, optimized spin-echo T2-weighted images are generally considered superior to optimized gradient-echo T2* images.

Patients with MS have plaques in the major white matter tracts, especially the lateral columns and dorsal white funiculi [26], because they comprise the largest white matter areas (Fig. 12.7). Lesions in the cervical cord are about twice as common as lesions at lower levels. Spinal cord demyelination in multiple sclerosis may occur as solitary lesion or as multiple lesions (Fig. 12.8). As in the brain, acute lesions are sometimes associated with surrounding edema leading to focal cord enlargement. Cavitation in MS plaques in the spinal cord can occur [27] and is presumed to result from severe inflammatory changes occurring in some acute plaques (Fig. 12.9). Large acute lesions may be associated with severe edema and cord swelling leading to vascular compression, ischemia, and acute necrotic myelopathy [19]. Devic's disease (neuromyelitis optica) is often considered a variant of MS in which an acute severe attack of demyelination affects the optic nerve and spinal cord within an interval of days or weeks [28] (Fig. 12.10). However, other diseases also have been implicated (see section on Devic's disease). Chronic lesions cannot be differentiated from acute or subacute lesions on non-contrast-enhanced images unless there is a change in cord size. Gd-DTPA enhancement occurs in acute lesions, but the duration of enhancement is variable, correlating with the variable course of disease in patients. MS plaque enhancement is not likely in patients with clinically stable disease [29]. T2-weighted images should always be obtained because inactive lesions are not seen on T1-weighted enhanced or nonenhanced images. Areas of cord enlargement and enhancement

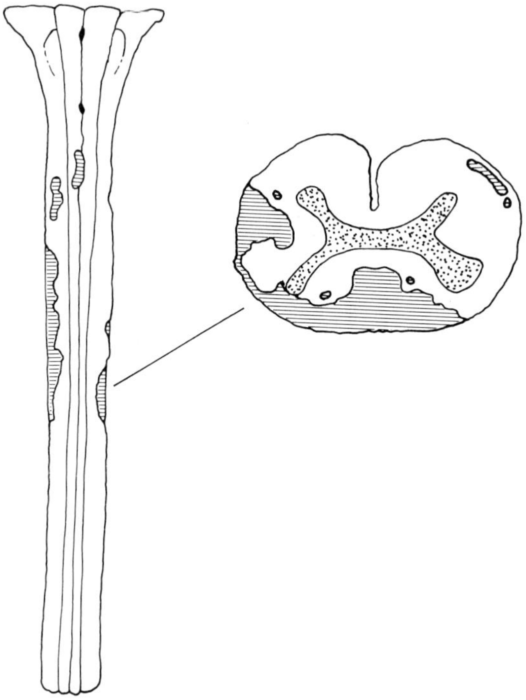

Figure 12.7. Diagram of spinal cord multiple sclerosis in a patient with bilateral lower extremity paresthesias and mild paraparesis. Note elongated peripheral pattern of demyelination in craniocaudad orientation and wedge-shaped pattern on cross-sectional view. There is sparing of central gray matter.

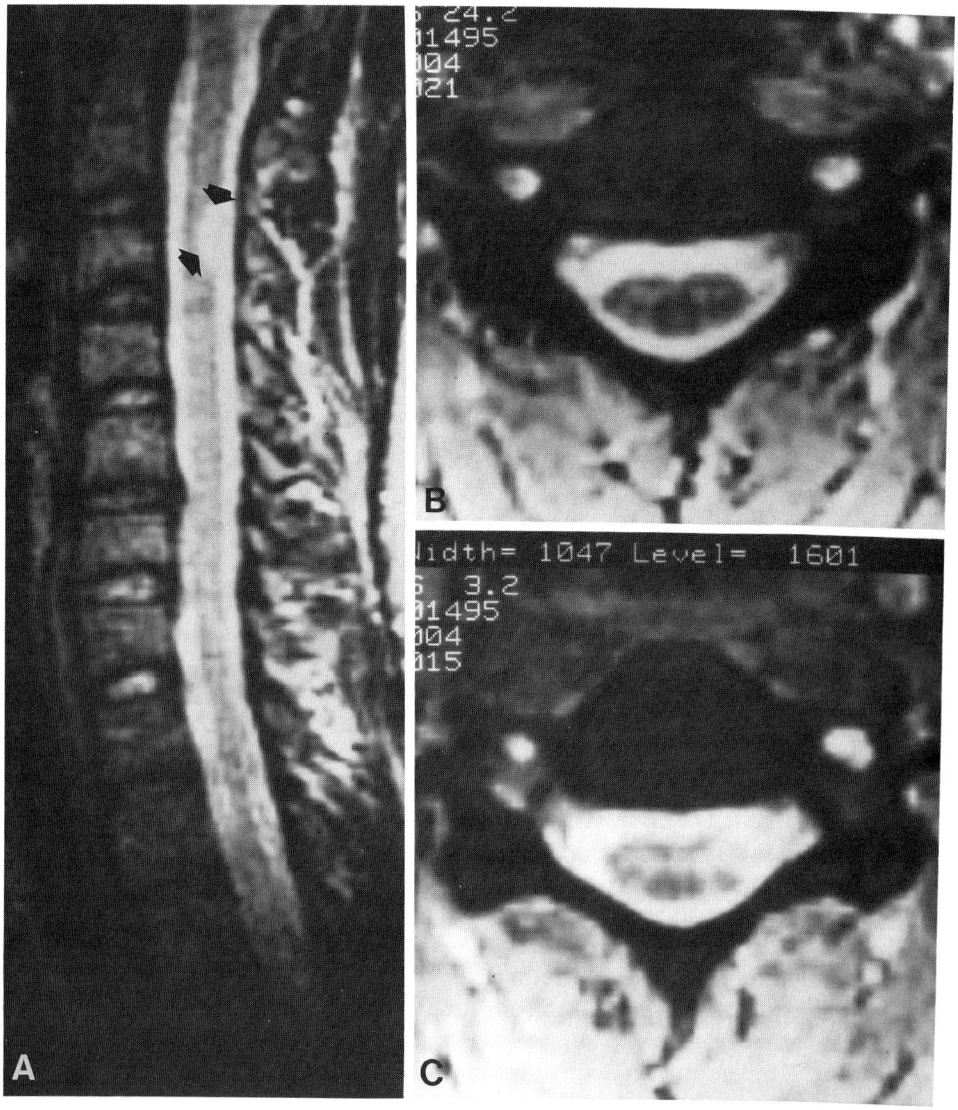

Figure 12.8. Spinal cord MS with gradient-echo imaging. A, Sagittal 2DFT T2*-weighted gradient-echo section (FE 600/30/12°) showing hyperintense focus of demyelination in spinal cord (*arrow*). **B,** Normal 3DFT T2*-weighted gradient-echo axial section (FE60/15/5°) showing cord at unaffected level, for comparison with next section. **C,** Demyelinating disease affects the spinal cord at this section. Note heterogeneous areas of increased intensity.

due to MS should diminish after about 3 months. If progressive enlargement has occurred or there has been no change, then neoplasm may be present, and biopsy should be considered.

Focal or diffuse spinal cord atrophy may occur in chronic MS. Short segment focal atrophic areas result from involution of a localized plaque (Fig. 12.11). These areas often have a persistent high signal abnormality on T2-weighted MRI sequences. Diffuse cord atrophy (Fig. 12.12) is a well-known end result of chronic MS and frequently will not show any signal abnormality [30].

Myelitis

ACUTE TRANSVERSE MYELITIS

Acute transverse myelitis (ATM) is a clinical entity affecting children [31] and adults. The clinical diagnosis is determined according to the following criteria [32]: (*a*) acutely developing paraparesis affecting motor and sensory systems, (*b*) spinal segmental level of sensory disturbance, (*c*) stable, nonprogressive clinical course, (*d*) no clinical or laboratory evidence of spinal cord compression, and (*e*) absence of other known neurologic disease. ATM may be caused by a long list of etiologic factors. There are well-known etiologic associations with viral diseases, including measles, varicella, and smallpox vaccinations. Antecedent upper respiratory viral infections in many patients and pathologic similarity to acute disseminated encephalomyelitis suggest an autoimmune mechanism in some cases [33]. Transverse myelitis occurs in approximately 4% of patients with lupus erythematosus and rarely in patients with other collagen-vascular diseases [34, 35].

MS may present initially with a myelopathy suggestive of transverse myelitis, but detection of typical "silent" brain lesions on MRI frequently indicates the diagnosis of MS [36]. The pathologic changes involve short or long segments of the cord,

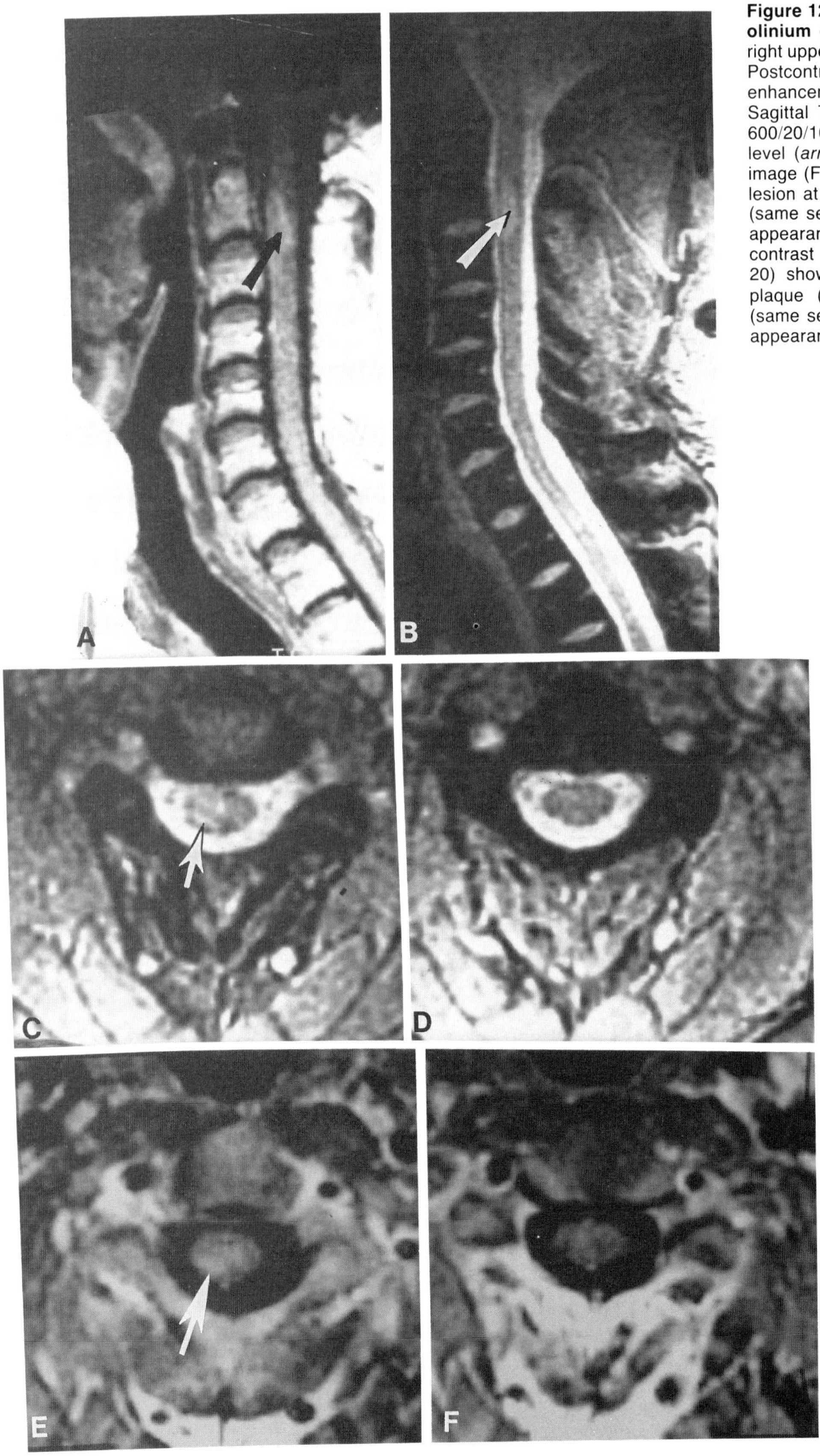

Figure 12.9. Acute spinal cord MS with gadolinium enhancement. A 45-year-old man with right upper extremity and truncal paresthesias. **A,** Postcontrast sagittal T1-weighted MRI showing enhancement in lesion at C2 level (*arrow*). **B,** Sagittal T2*-weighted gradient-echo image (FE 600/20/10°) showing hyperintense focus at C2 level (*arrow*). **C,** Transaxial T2*-weighted 3DFT image (FE 50/15/5°/4.0 mm) showing right sided lesion at C2 level (*arrow*). **D,** Transaxial section (same sequence) at lower level showing normal appearance of spinal cord for comparison. **E,** Postcontrast axial T1-weighted MRI image (SE600/20) showing slight enhancement of acute MS plaque (*arrow*). **F,** Postcontrast axial section (same sequence) at lower level showing normal appearance of spinal cord for comparison.

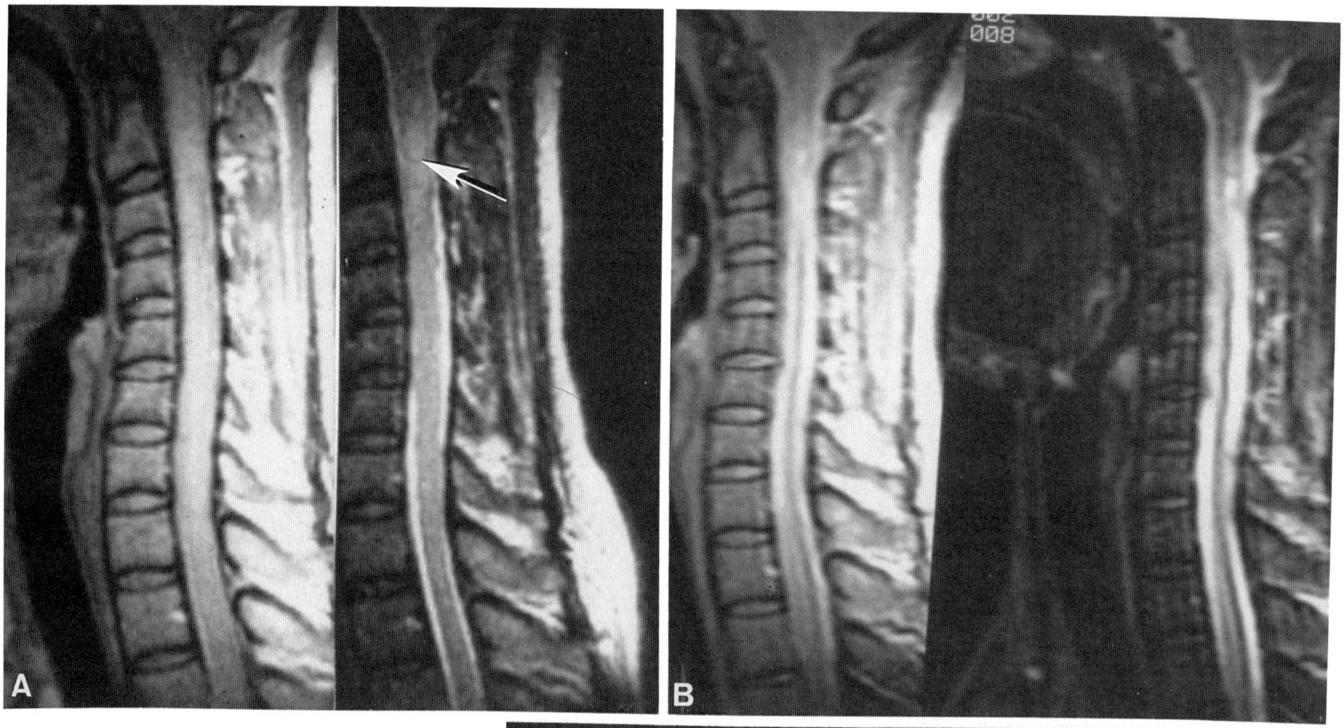

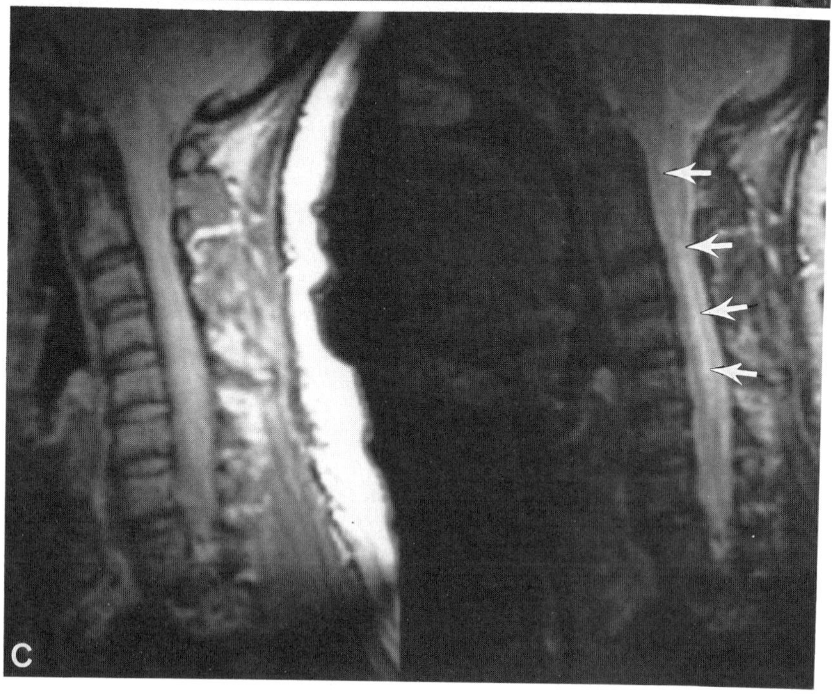

Figure 12.10. Devic's disease. Progressive spinal cord MS in a 32-year-old woman with optic neuritis and bilateral upper extremity paresthesias progressing to paraparesis. **A,** MRI at time of presentation. Proton-weighted and T2-weighted sagittal images (SE 2100/30,80). Note subtle lesion in spinal cord at C2 level (*arrow*). **B,** Sagittal follow-up MRI 9 months later (SE 2500/30,80). Note progression of disease. **C,** Sagittal follow-up MRI 15 months after initial study (SE 2100/30,80). Note involvement at C1–2 and C3 through C6 (*arrows*).

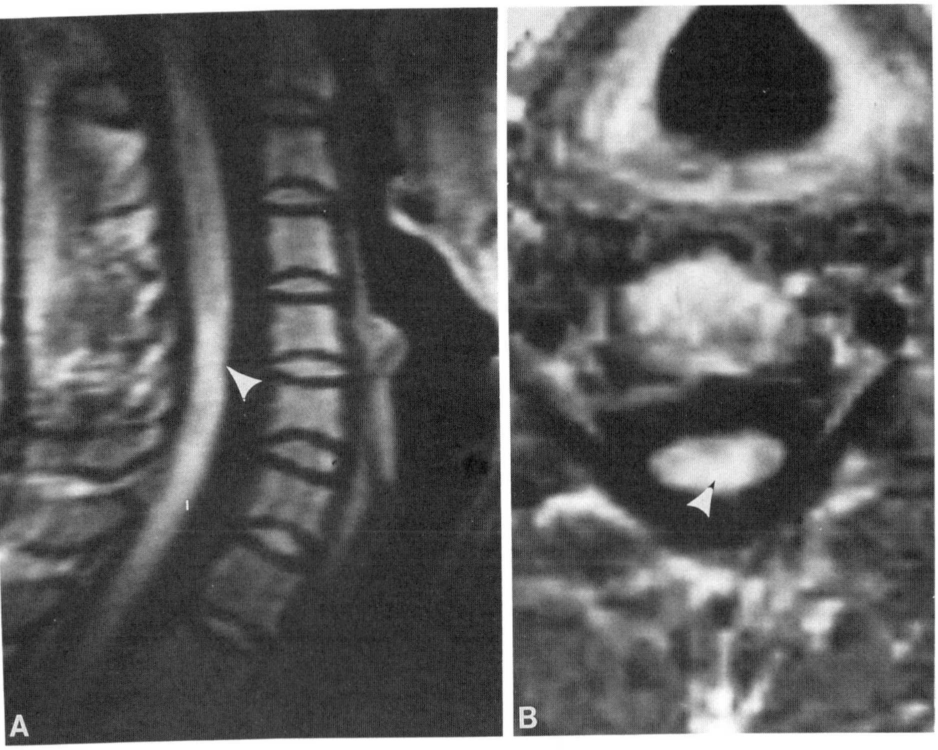

Figure 12.11. Chronic spinal cord MS with focal atrophy. A, Sagittal proton-weighted image (SE 1500/60) shows focal cord narrowing at location of hyperintense plaque (*arrowhead*). **B,** Axial proton-weighed image (SE 1500/40) shows demyelination primarily involving dorsal white funiculi (*arrowhead*).

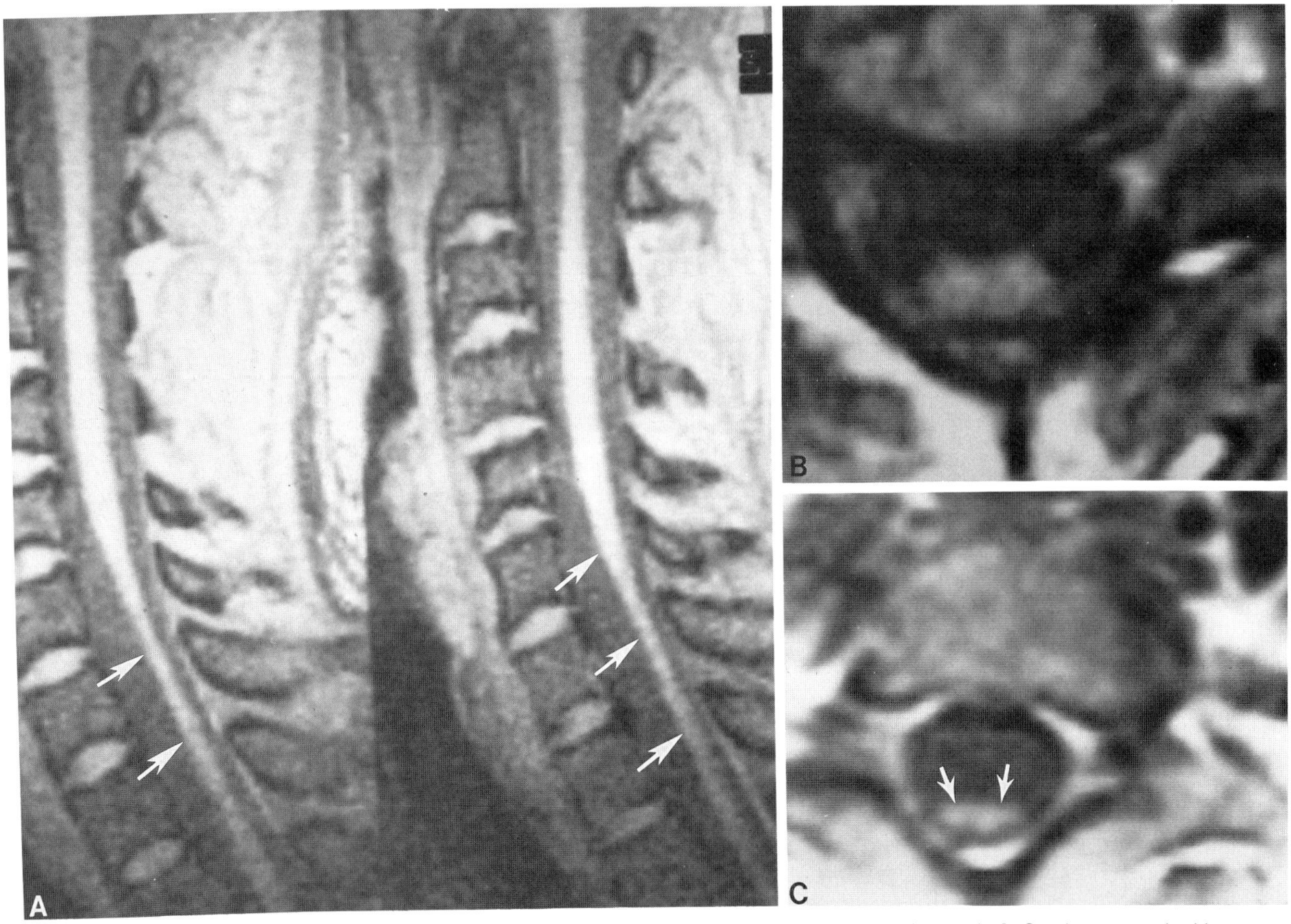

Figure 12.12. Chronic spinal cord MS with diffuse atrophy. A 38-year-old woman with severe combined motor and sensory dysfunction in trunk and lower extremities. Note diffuse lower cervical and thoracic cord atrophy (*arrows*). **A,** Contiguous sagittal images. **B,** Axial image at C3 level. **C,** Axial image at T1 level.

usually in the thoracic area. Lymphocytic infiltration and demyelination probably occur in all cases, but in some patients the disease progresses, with the development of polymorphonuclear cell infiltration, edema, secondary ischemia, and cord necrosis [37].

Collagen-vascular disease and paraneoplastic syndromes are other postulated causes of ATM.

The purpose of imaging studies is primarily to demonstrate any morphologic changes in the spinal cord or extrinsic disease process. This can be achieved with myelography, CTM, or MRI (Fig. 12.13). Myelography may miss subtle cord swelling. CTM is slightly better, since focal enlargement of the cord can be identified. The cord may have a relative hypodense appearance due to the edema. However, in the clinical findings suggestive of acute transverse myelitis, MRI should be the imaging study of choice.

The MRI may be normal in some patients, fulfilling the clinical criteria of the disease. The MRI may appear falsely "negative" if performed very early in the disease [38]. When positive, the MRI appearance of ATM is variable and nonspecific. The affected area of the cord may appear swollen or normal in size. There

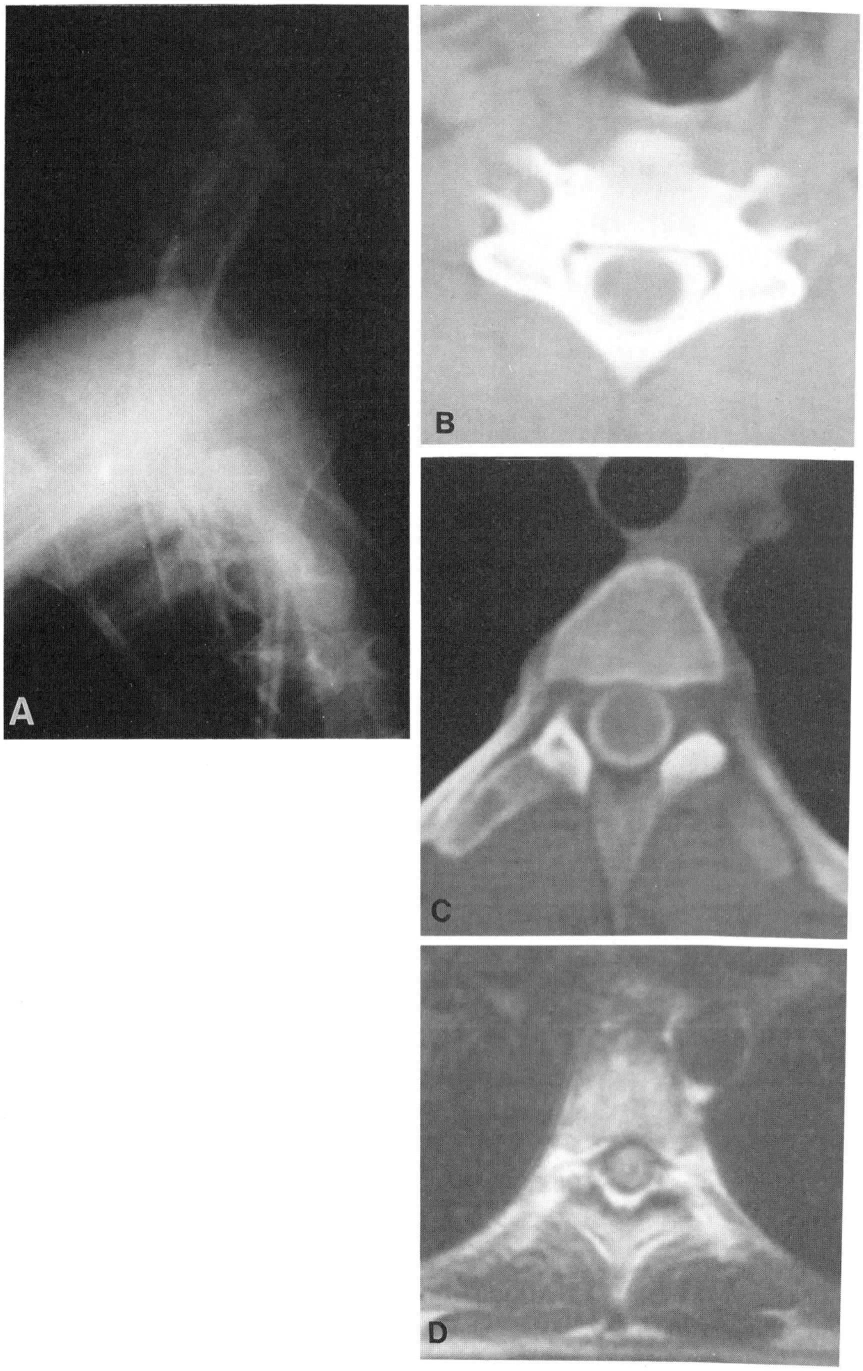

Figure 12.13. Acute transverse myelitis in a 12-year-old girl about 2 weeks following upper respiratory viral infection. A, Lateral view of the myelogram demonstrates diffuse cord swelling with almost complete obliteration of the subarachnoid space. CTM scans through upper cervical region and (**B**) thoracic region (**C**) demonstrating diffuse symmetrical swelling of the spinal cord. Note the obliteration of the midline sulci in the cervical segment of the cord. **D,** T1-weighted gadolinium-enhanced axial MRI section. Cord swelling is noted without any focal enhancement.

is no cavitation, but hypointensity may be seen T1-weighted images. T2-weighted images show abnormal increased signal intensity. Short or long segments of the cord may be affected (Figs. 12.14 through 12.16). Mild to moderate enhancement may be seen with gadolinium-DTPA [39, 40], in contrast to the more intense enhancement noted with cord tumors. Complete resolution of the T2-weighted changes and cord swelling is expected as the patient recovers. In some patients, recovery is poor or incomplete, and the spinal cord becomes atrophic [41].

SUBACUTE ENCEPHALOMYELITIS

This disease may develop as variant of acute transverse myelitis (ATM) (Fig. 12.17) or as a secondary degenerative/inflammatory process in association with a spinal dural arteriovenous fistula (see section on Foix-Alajouanine syndrome). It is distinguished from ATM by the progressive development of spinal cord and brainstem dysfunction. MRI before treatment with high-dose steroids reveals spinal cord swelling, increased intensity on T2-weighted images, and enhancement following gadolinium-DTPA injection. The etiology of the disease is unknown, but histologic analysis reveals inflammatory changes without infection, similar to findings in patients with ATM [42].

AIDS MYELITIS

AIDS-related spinal cord disorders include a range of infectious, neoplastic, and idiopathic etiologies. AIDS patients presenting with chronic myelopathy may have normal MRI of the cord or images showing cord atrophy. Pathologically, these patients often have vacuolar myelopathy primarily involving the lateral and posterior columns, resembling that of subacute combined degeneration of vitamin B_{12} deficiency. The myelopathy may be a remote or indirect effect of HIV or other infectious agent [43]. It is most common in patients with opportunistic infections. Myelitis due to *Mycobacterium* [44], *Toxoplasma* [45, 46], and viral infections with varicella-zoster, herpes simplex, cytomegalovirus, and HTLV-III viruses has been reported [47–49]. Increased intensity with or without cord swelling has been seen on T2-weighted MRI sequences [46, 50, 51] (Fig. 12.18). The MRI appearance of AIDS myelitis on pre- and post-contrast images is indistinguishable from that of

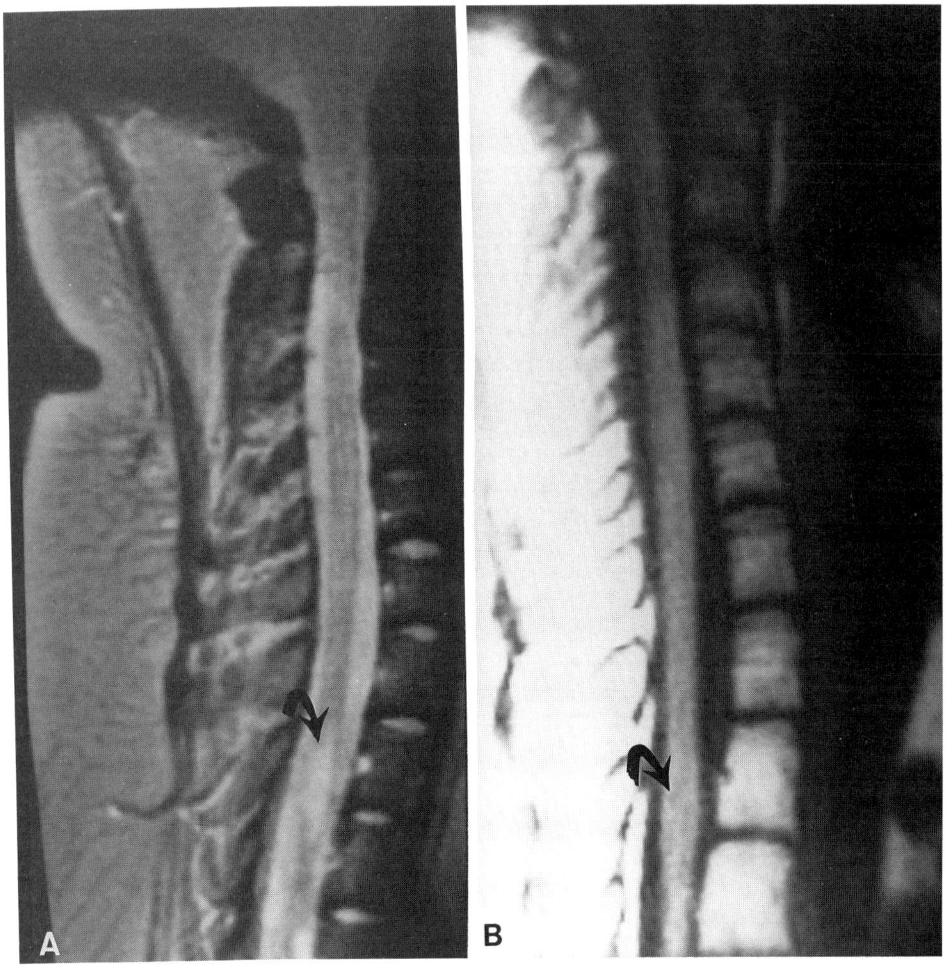

Figure 12.14. Acute transverse myelitis in a 23-year-old woman. A, T2-weighted sagittal image. Note increased intensity involving upper thoracic cord (*arrow*). **B,** T1-weighted sagittal image (without gadolinium) shows cord enlargement (*arrow*).

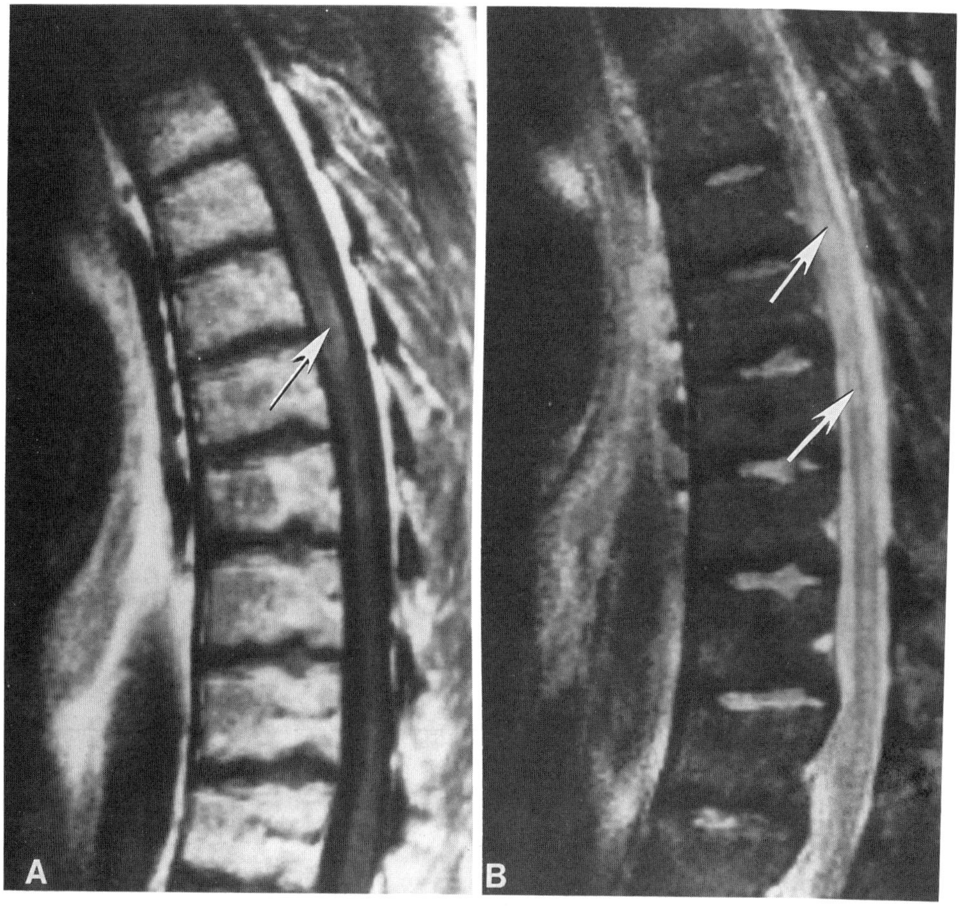

Figure 12.15. Acute transverse myelitis in thoracic cord. A 45-year-old man with acute lower extremity paresthesias. **A,** Contrast-enhanced sagittal T1-weighted image. Note areas of enhancement in cord without significant cord enlargement (*arrow*). **B,** T2-weighted image. Increased signal is present in several areas (*arrows*).

acute transverse myelitis [40]. Myelopathy due to viruses in immunocompromised patients may be caused by primary or secondary mechanisms. Secondary mechanisms include immune-mediated demyelination of the spinal cord and parainfectious vasculitis with subsequent cord ischemia [47].

TUBERCULOSIS

TB is a rare cause of spinal arachnoiditis. It frequently involves the spinal cord. Gd-DTPA–enhanced MRI sequences reveal enhancing nodules, enhancement of the leptomeninges and dura, and segmental enhancement of the spinal cord due to TB myelitis or vasculitis with infarction [52]. Intramedullary tuberculomas are rare and may cause focal cord enlargement and decreased or increased signal intensity on T2-weighted images [53]. Gadolinium-DTPA–enhanced MRI is recommended for evaluation. Spinal sarcoidosis and tuberculosis are indistinguishable on MRI.

NEUROSYPHILIS

Spinal syphilis is classified into (*a*) syphilitic meningomyelitis, (*b*) spinal vascular syphilis, (*c*) syphilitic spinal pachymeningitis, (*d*) spinal cord compression due to gumma, and (*e*) tabes dorsalis. The most common presentation now is meningomyelitis, sometimes associated with superimposed endarteritic cord ischemia or infarction [54]. Spinal cord enlargement and dural thickening are seen on myelography and MRI. Tabetic neurosyphilis (tabes dorsalis) occurs in 10% of improperly treated males with early syphilis, but it is presently very uncommon. It is associated with spinal cord atrophy and demyelination in the posterior columns and dorsal roots.

POLIOMYELITIS

Poliomyelitis has largely been eradicated by vaccination programs against poliovirus. The disease caused degeneration in the anterior horn cells of the spinal cord and anterior spinal cord atrophy. The postpolio syndrome is of more current interest. Patients who had poliomyelitis early in life and whose neurologic deficit had been stable may note increasing weakness decades after the initial attack. Spinal cord atrophy may be seen. This syndrome is probably caused by age-related loss of motor units rather than reactivation of the disease or amyotrophic lateral sclerosis.

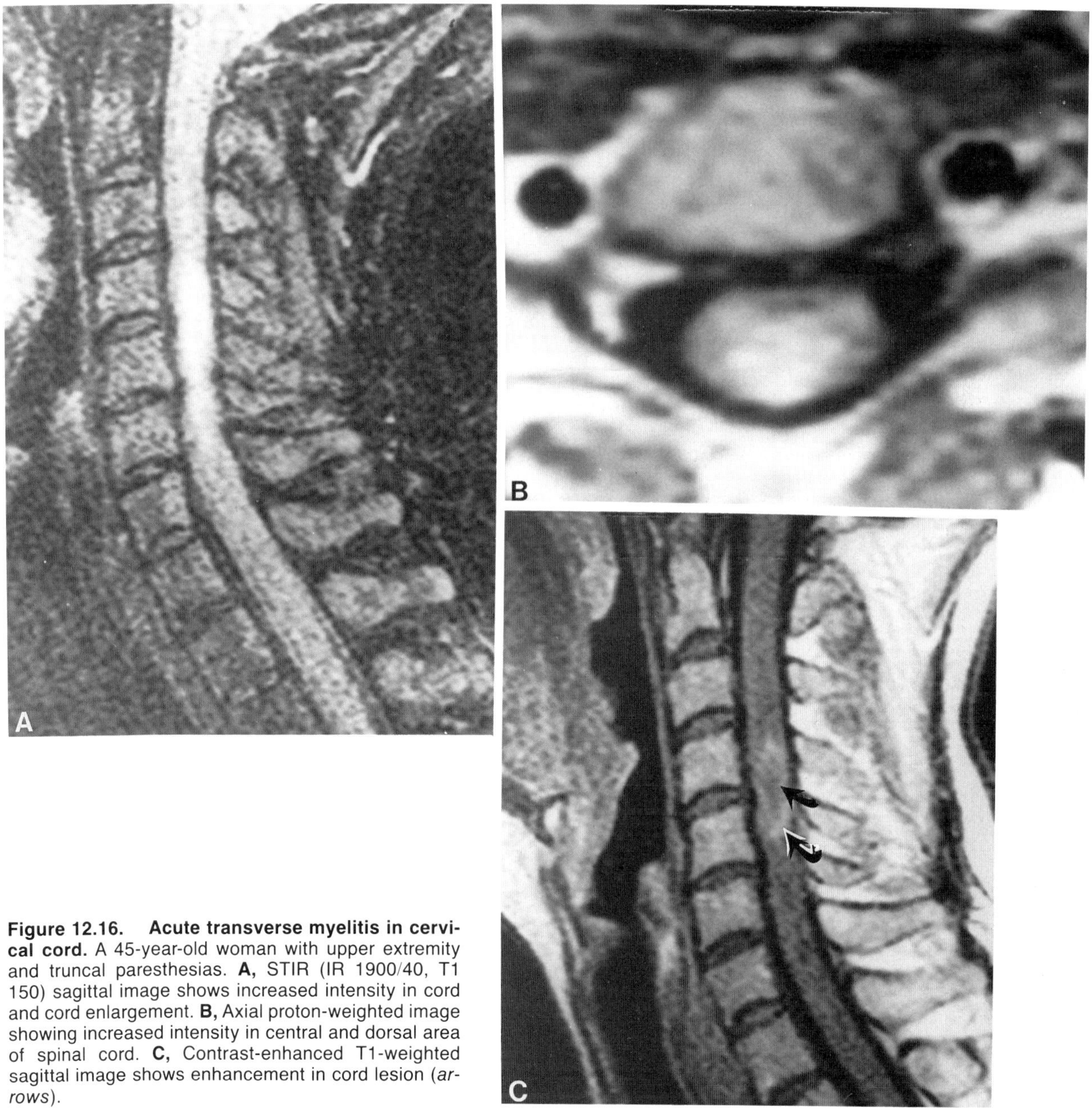

Figure 12.16. Acute transverse myelitis in cervical cord. A 45-year-old woman with upper extremity and truncal paresthesias. **A,** STIR (IR 1900/40, T1 150) sagittal image shows increased intensity in cord and cord enlargement. **B,** Axial proton-weighted image showing increased intensity in central and dorsal area of spinal cord. **C,** Contrast-enhanced T1-weighted sagittal image shows enhancement in cord lesion (*arrows*).

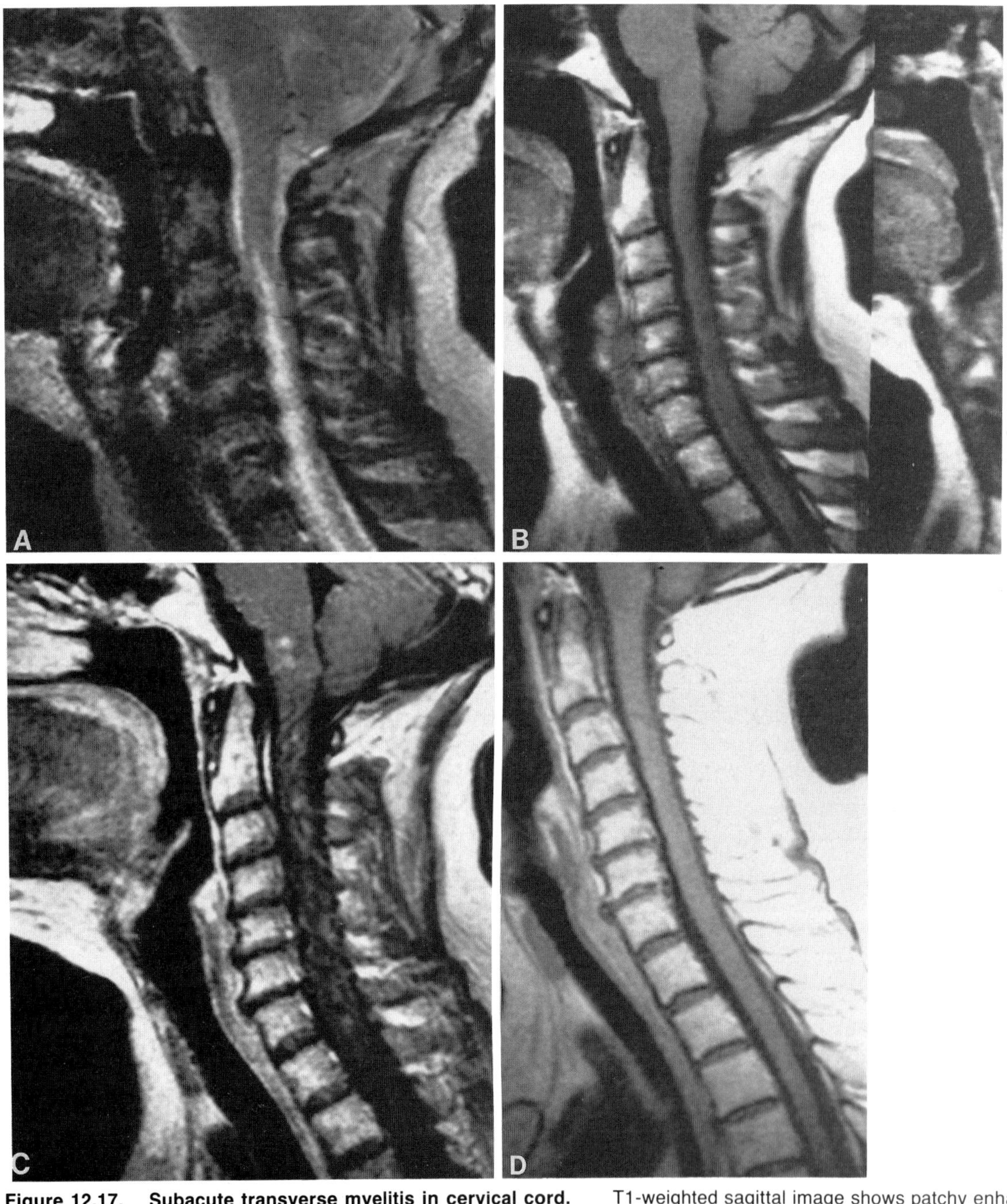

Figure 12.17. Subacute transverse myelitis in cervical cord. A 56-year-old woman with agitation and progressive sensory and motor symptoms. **A,** T2-weighted sagittal image reveals diffuse increased intensity in cord from C2 through T1. **B,** T1-weighted sagittal image shows mild diffuse cord swelling. **C,** Contrast-enhanced T1-weighted sagittal image shows patchy enhancement in cord. **D,** Follow-up study 3 weeks after therapy was initiated, patient asymptomatic at this time. Contrast-enhanced T1-weighted sagittal images show resolution of cord swelling and enhancement.

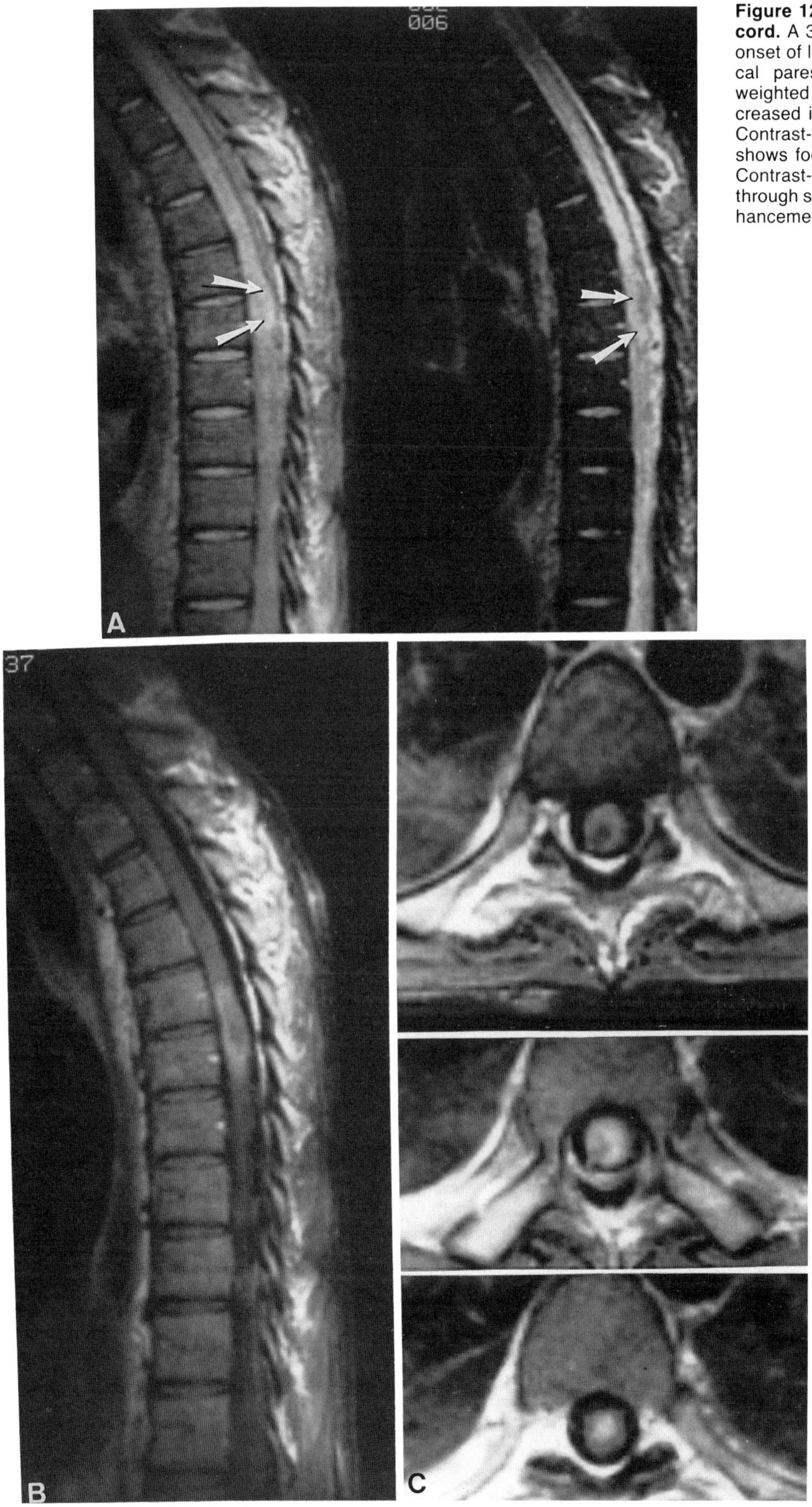

Figure 12.18. AIDS acute myelitis in thoracic cord. A 36-year-old male AIDS patient with acute onset of left leg paresthesias, weakness, and truncal paresthesias. **A,** Proton-weighted and T2-weighted sagittal image pair showing subtle increased intensity in midthoracic cord (*arrows*). **B,** Contrast-enhanced T1-weighted sagittal image shows focal enhancing lesion in thoracic cord. **C,** Contrast-enhanced T1-weighted axial images through spinal cord lesion. Note heterogeneous enhancement and cord enlargement.

TROPICAL SPASTIC PARAPARESIS (TSP)

Tropical spastic paraparesis (TSP) is a slowly advancing myelopathy that occurs in high prevalence in the equatorial countries of the world. TSP is associated with chronic HTLV-I infection. The primary lesion involves inflammatory demyelination and axonal damage leading to severe spinal cord atrophy [55]. MRI in the active inflammatory stage reveals a high-signal-intensity lesion. The thoracic spinal cord is most affected, particularly the lateral and anterior funiculi [55]. Diagnosis requires a positive HTLV-I antibody test and evidence of ongoing infection.

SARCOIDOSIS

The pathologic manifestations of CNS sarcoidosis consist of granulomatous dural invasion with plaquelike or focal masses [56], leptomeningitis, parenchymatous infiltration especially along the perivascular spaces of Virchow-Robin [57], and coalescence of granulomas in the spinal cord, rarely producing an intramedullary "tumor" [58]. Edema

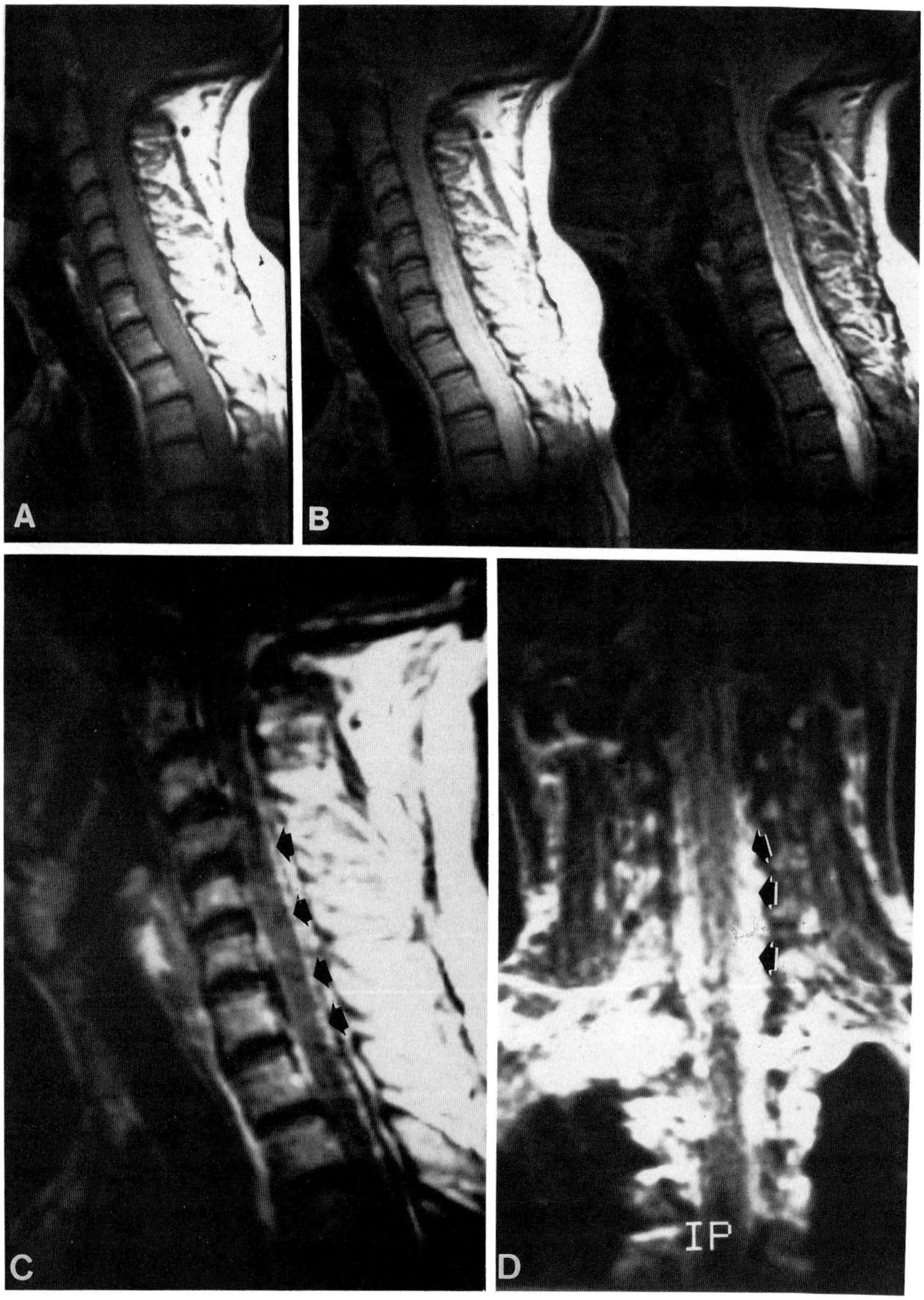

Figure 12.19. Sarcoidosis of the spinal cord in a 42-year-old man. A, Sagittal T1-weighted image. Note loss of differentiation between spinal cord and CSF. **B,** Proton-weighted and T2-weighted sagittal image pair. There is increased intensity in cord at cervicothoracic junction. **C,** Contrast-enhanced T1-weighted sagittal image reveals diffuse leptomeningeal thickening and enhancement (*arrows*). **D,** Contrast-enhanced T1-weighted coronal image through cord showing diffuse leptomeningeal enhancement (*arrows*).

accompanies infiltration of the cord and extends along white matter fascicles (vasogenic edema) [59–61]. Heterogeneous areas of abnormal increased intensity may be seen on T2-weighted images (Fig. 12.19). Cord enlargement is variable. Enhanced MRI is the preferred imaging technique for the evaluation of CNS sarcoidosis [62] and provides additional information in all but the most chronic phase of the disease process. Postcontrast T1-weighted images reveal patchy, multifocal parenchymal enhancement and areas of leptomeningeal enhancement [63, 64] (Figs. 12.19 and 12.20). The diagnosis can be confirmed if there is concurrent evidence of sarcoidosis on chest x-rays (Fig. 12.21). Spinal cord atrophy may be seen in the chronic phase, probably as a result of ischemia and the direct disruption of neural pathways [65].

VASCULITIS

Collagen-vascular disease may occasionally involve the spinal cord. Systemic lupus erythematosus (SLE) (Fig. 12.22) is the most likely disease of this category to be recognized clinically. Myelopathy is the presenting feature in 15 to 20% of patients with SLE [66]. The disease is usually one of rapid progression, but it may be characterized by remissions and relapses. Vasculitis and the associated complications of this disease and its treatment may affect the spinal cord, including infarction and transverse myelitis and one or more foci of subtotal coagulative or liquefactive necrosis [67]. SLE patients may develop vacuolar myelopathy similar to that seen in subacute combined degeneration [68].

Spinal cord enlargement and hyperintensity on T2-weighted images with rapid regression after steroid and cytoxan therapy has been reported [69]. There is variable enhancement with gadolinium. In most instances, the spinal cord is affected at or below thoracic levels, but 25% of patients have involvement in the cervical region.

Behçet's disease is an uncommon multisystem immune-related vasculitis of unknown origin which may involve the CNS, especially affecting the brainstem. The disease is more frequent in the Orient and the Middle East. This condition may simulate multiple sclerosis [70, 71].

DEVIC'S DISEASE

Devic's disease (neuromyelitis optica) classically consists of a combination of blindness and paraplegia (Fig. 12.23). Although multiple sclerosis [28, 72] is the most likely explanation for this symptom complex, there have been many other diseases that have been associated with Devic's disease, including acute disseminated encephalomyelitis, varicella-zoster infection, mumps, rubeola, infectious mononucleosis, systemic lupus erythematosus, pulmonary tuberculosis, and clioquinol intoxication [73].

MISCELLANEOUS MYELITIS/MYELOPATHY

The differential diagnosis of myelopathies poses a clinical problem. Multiple sclerosis is often suspected but accounts for only about 10 to 32% of idiopathic myelopathy [37, 74]. Myelitis can be caused by a wide variety of infectious and toxic agents. Although uncommon, spinal cord infections may be

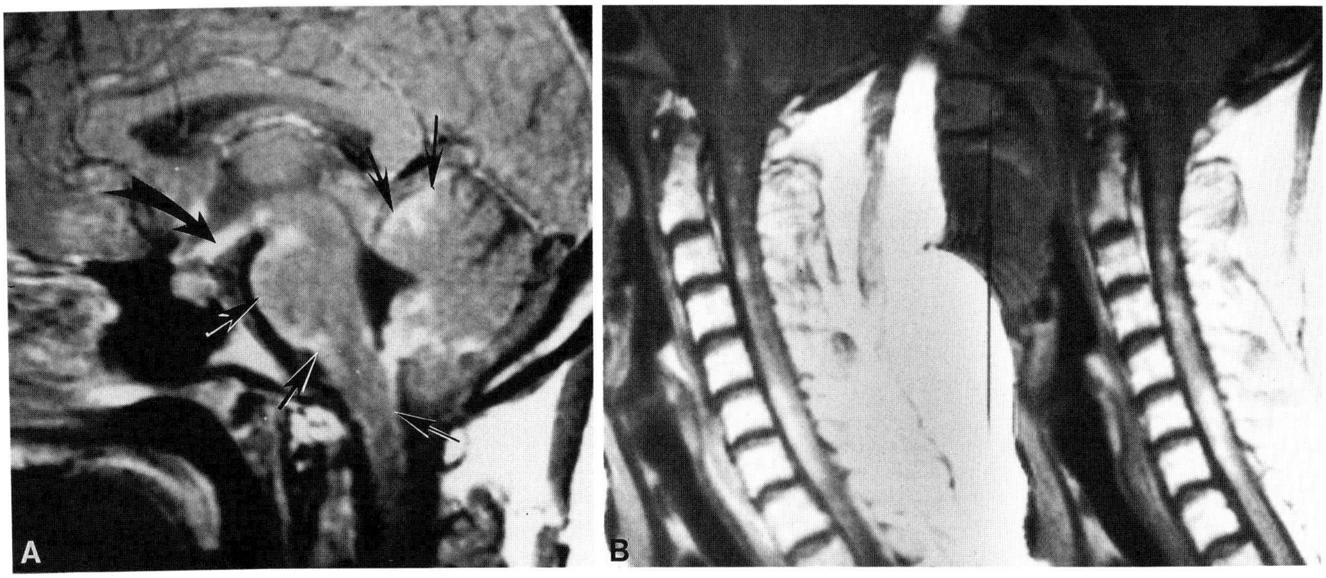

Figure 12.20. Sarcoidosis of the brain and spinal cord. A 37-year-old woman with decreasing vision due to optic chiasm involvement. Patient has mild complaints of paresthesias, apparently due to spinal cord involvement. **A,** T1-weighted sagittal MRI of brain. Note abnormal optic chiasm (*curved arrow*) and leptomeningeal enhancement (*arrows*). **B,** T1-weighted sagittal MRI of spinal cord. Note diffuse enhancement of cord.

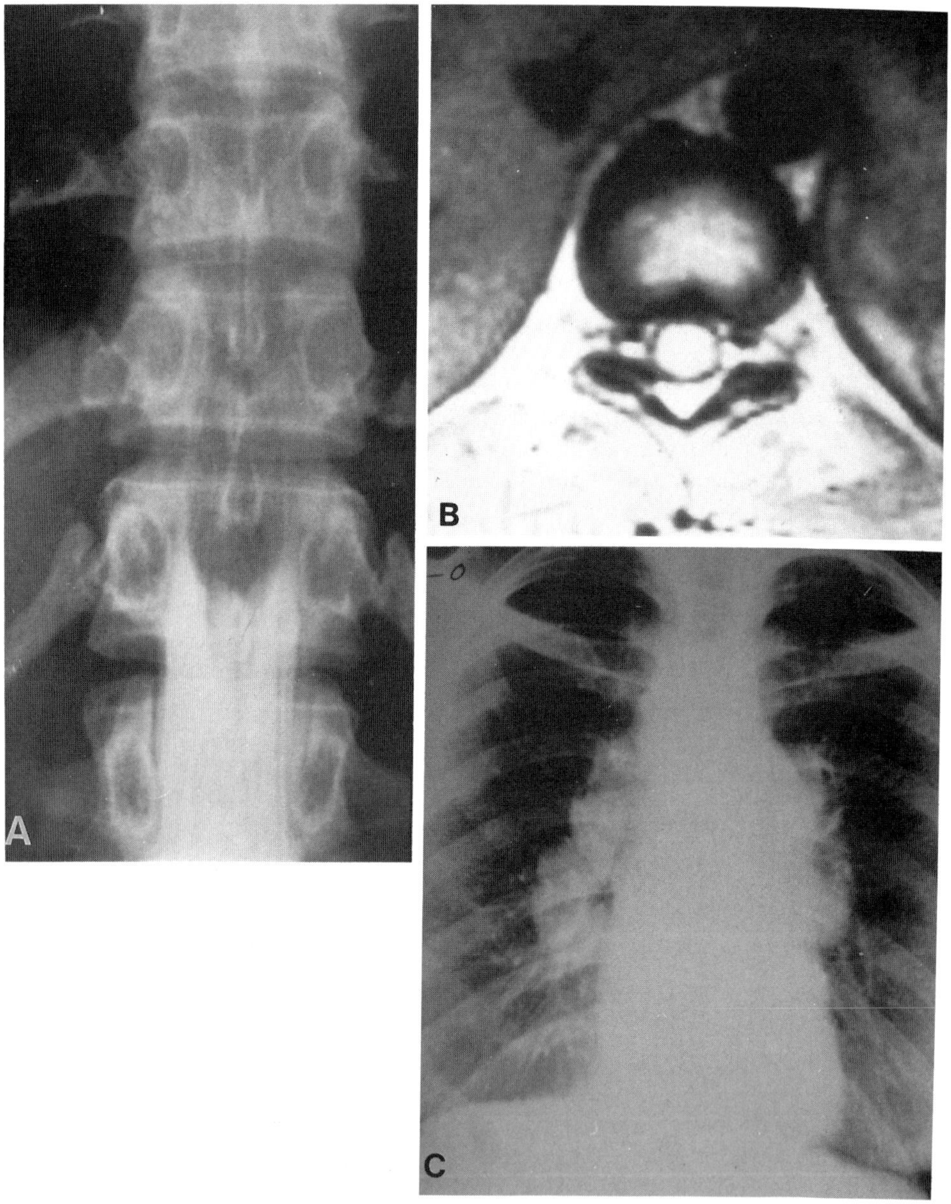

Figure 12.21. Subacute transverse myelitis secondary to sarcoidosis in a 42-year-old woman presenting with paraplegia. A, Anteroposterior view of the lumbar myelogram demonstrates almost complete block of the contrast column at the level of the conus. **B,** Axial MRI section at a similar level demonstrates nonspecific, diffuse symmetrical, swelling of the distal cord. This was focal involving the distal cord and conus. **C,** Frontal view of the chest x-rays shows characteristic hilar adenopathy due to sarcoidosis.

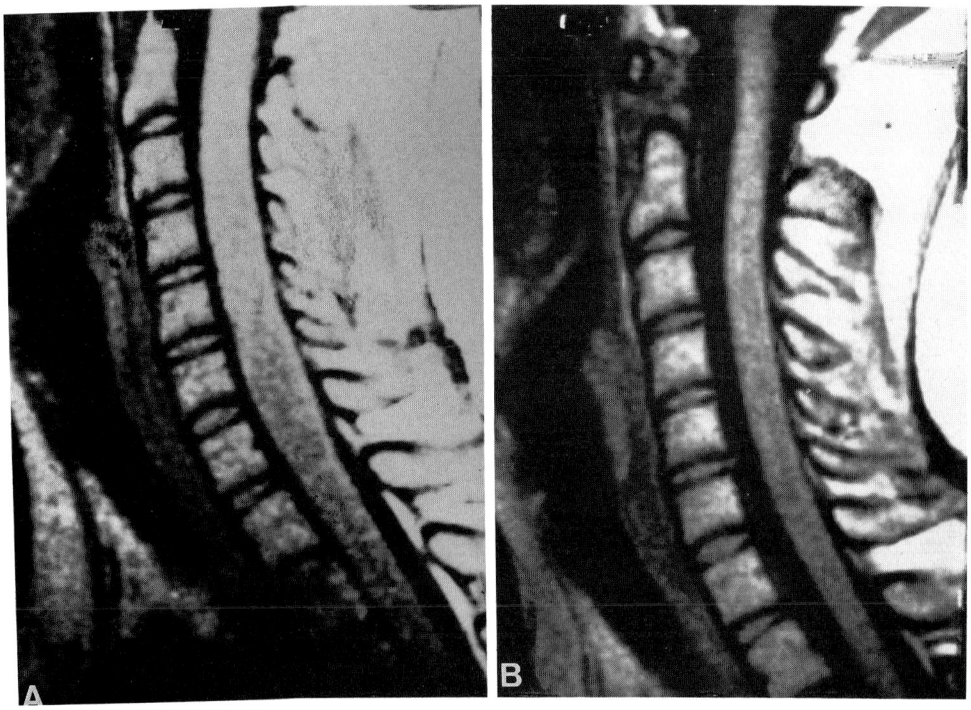

Figure 12.22. Systemic lupus erythematosus myelitis. A, T1-weighted sagittal image shows diffuse cord swelling. **B,** Follow-up examination shows complete resolution of cord swelling.

caused by bacteria, viruses, fungi (Fig. 12.24), or parasites, particularly in patients who are intravenous drug abusers or immunocompromised [75–78] (Fig. 12.25). There have been a few reports of spinal cord toxoplasmosis in AIDS patients, in whom the disease caused well-localized cord expansion [45, 50]. MRI with gadolinium-DTPA may be helpful by indicating that the primary disease site is either intramedullary or meningeal.

Spinal cysticercosis may present with single or multiple cord cysts [79]. MRI of one patient with intramedullary schistosomiasis of the conus medullaris showed marked conus enlargement and irregular enhancement after Gd-DTPA administration [80]. Spinal echinococcosis is rare but may occur as a result of direct extension from pulmonary lesions into the spinal canal, causing extradural cord compression [81]. Lyme disease (spirochete *Borrelia burgdorferi*) diversely affects the CNS and may therefore simulate multiple sclerosis clinically and on MRI studies [82–84].

RADIATION MYELOPATHY

Radiation myelopathy is a recognized complication of therapeutic irradiation. However, its pathogenesis is unknown but believed to be due to small-vessel and capillary damage [85]. The diagnosis of radiation myelopathy is based on the following criteria: spinal cord within the irradiation field, late onset of the clinical disorder (36 months), concordance between the level of clinical signs and that of the spinal cord radiation port, and absence of clinical and imaging evidence of metastasis [86]. The dose of ionizing radiation has ranged from 2500 to 6500 rads, but frequently the amount of ionizing radiation received by the spinal cord has not been measured accurately. Chronic radiation myelopathy has a variable appearance on MRI [87, 88]. The MRI image may be normal or reveal diffuse atrophy, or abnormal areas of increased intensity in the cord may be seen on T2-weighted images. The areas of hyperintensity correspond to radiation ports and may be associated with either a normal-sized cord or cord enlargement. Gadolinium enhancement may be seen in these patients.

Degenerative Diseases

AMYOTROPHIC LATERAL SCLEROSIS (ALS)

Amyotrophic lateral sclerosis is a relentless degenerative disease of the motor system leading to death in half the cases within 3 years from onset of the disease [89]. There is degeneration of the lateral and anterior corticospinal tracts in the spinal cord. Massive neuronal cell loss with gliosis is present in the anterior gray horns, especially in the cervical and lumbar enlargements [90]. The lack of involvement of the posterior white columns stands out prominently compared with the lateral and anterior corticospinal tracts in sporadic ALS [91, 92]. Symmetrical patches of posterior column demyelination occur in familial ALS [90].

Imaging studies early in the course of the disease may reveal no abnormality, but cord atrophy is frequently seen in patients evaluated later in the course of the disease. This probably reflects both the

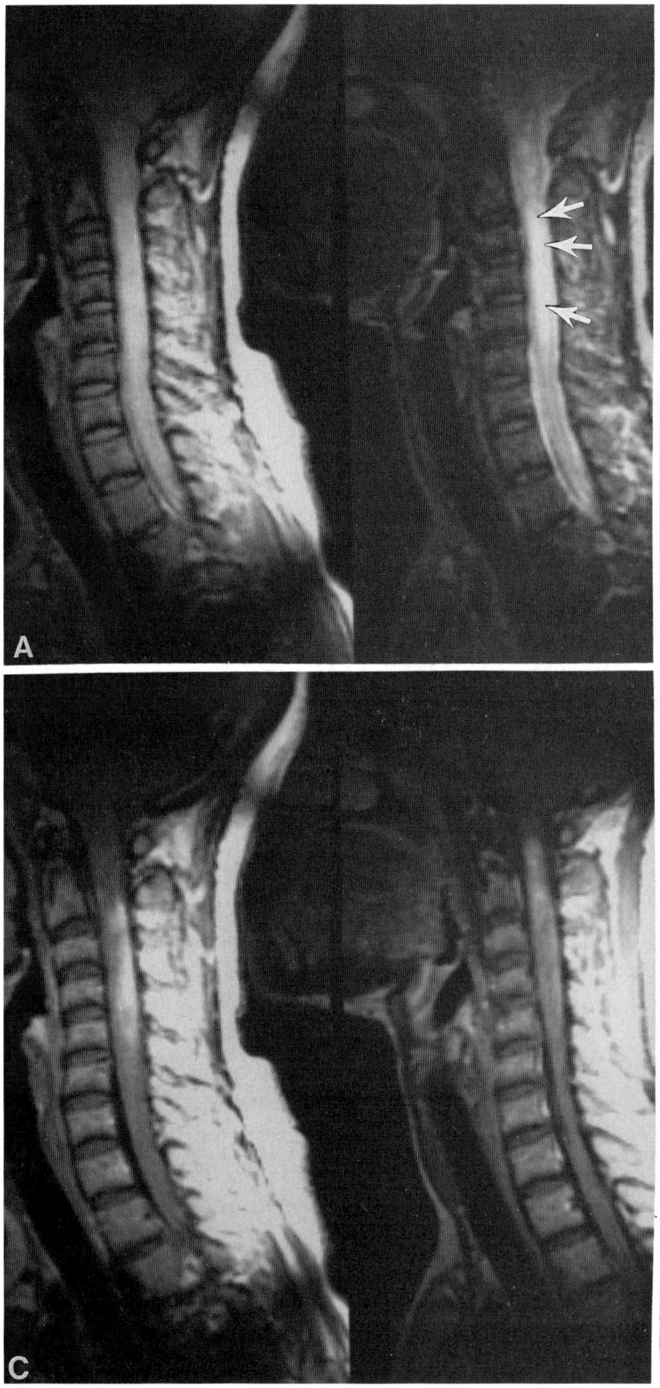

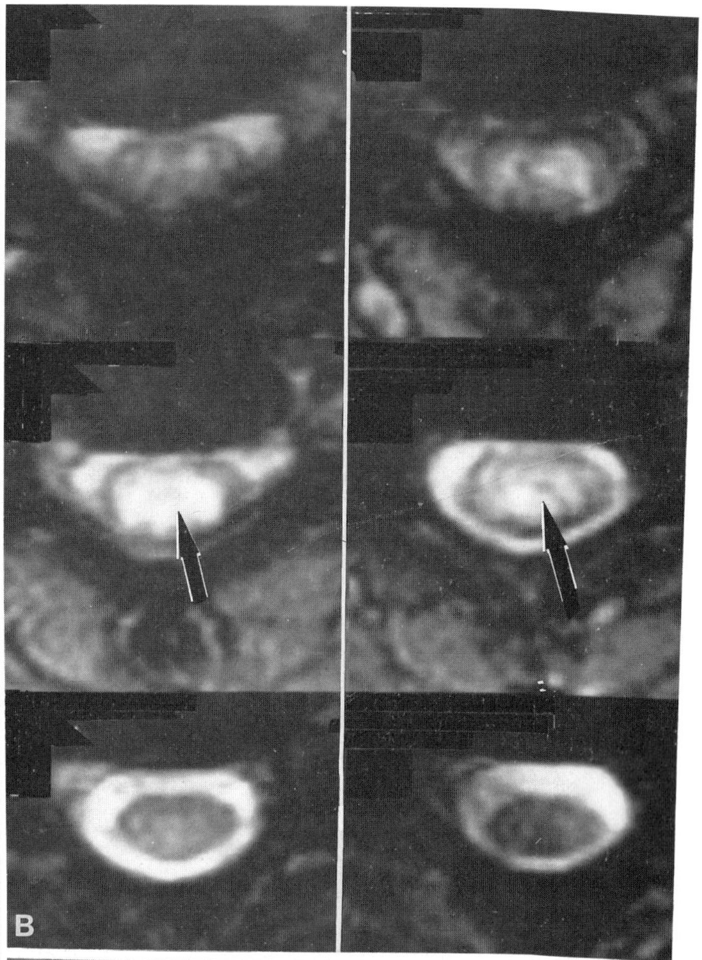

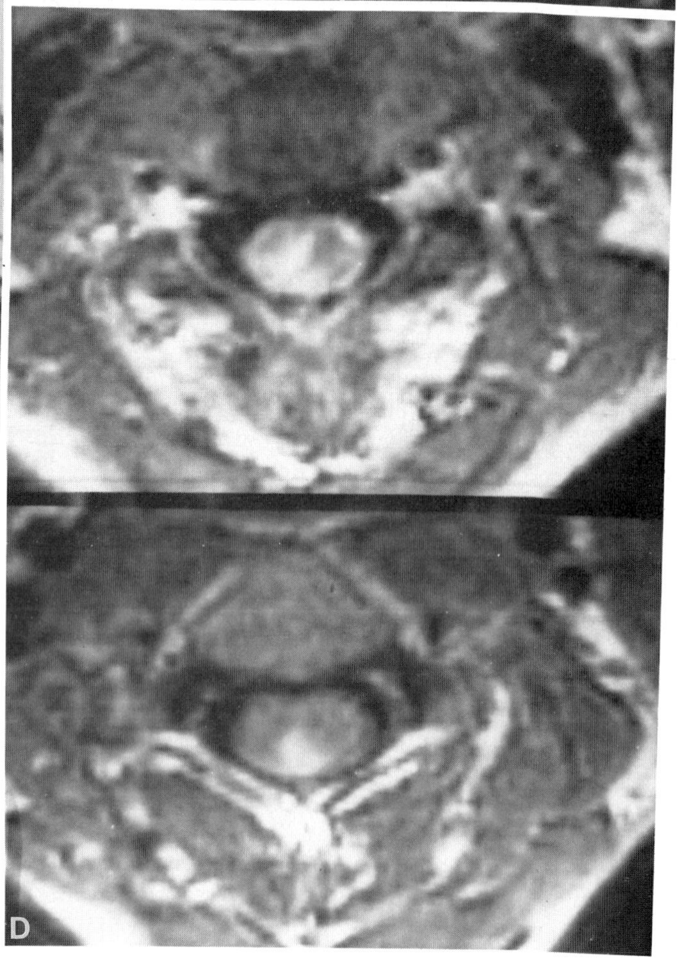

Figure 12.23. Devic's syndrome. A 28-year-old woman with sudden onset of vision loss and cervical myelopathy. There was no history of demyelinating disease. **A,** Long-TR spin-echo sagittal images (SE 2400/30,80) show hyperintense area in cord from C2 through C6 (*arrows*). **B,** T2*-weighted gradient-echo axial images show hyperintensity at affected level (*arrows*). **C,** Gadolinium-DTPA–enhanced sagittal T1-weighted images show enhancement of acute demyelination between C2 and C6. **D,** Gadolinium-DTPA–enhanced axial images showing abnormal enhancement in cord. **E,** Gadolinium-DTPA enhanced coronal images. Note enhancement in both central and peripheral areas of the spinal cord. **F,** STIR 2500/30/150 coronal view of orbits. Note hyperintense focus in left optic nerve (*arrow*).

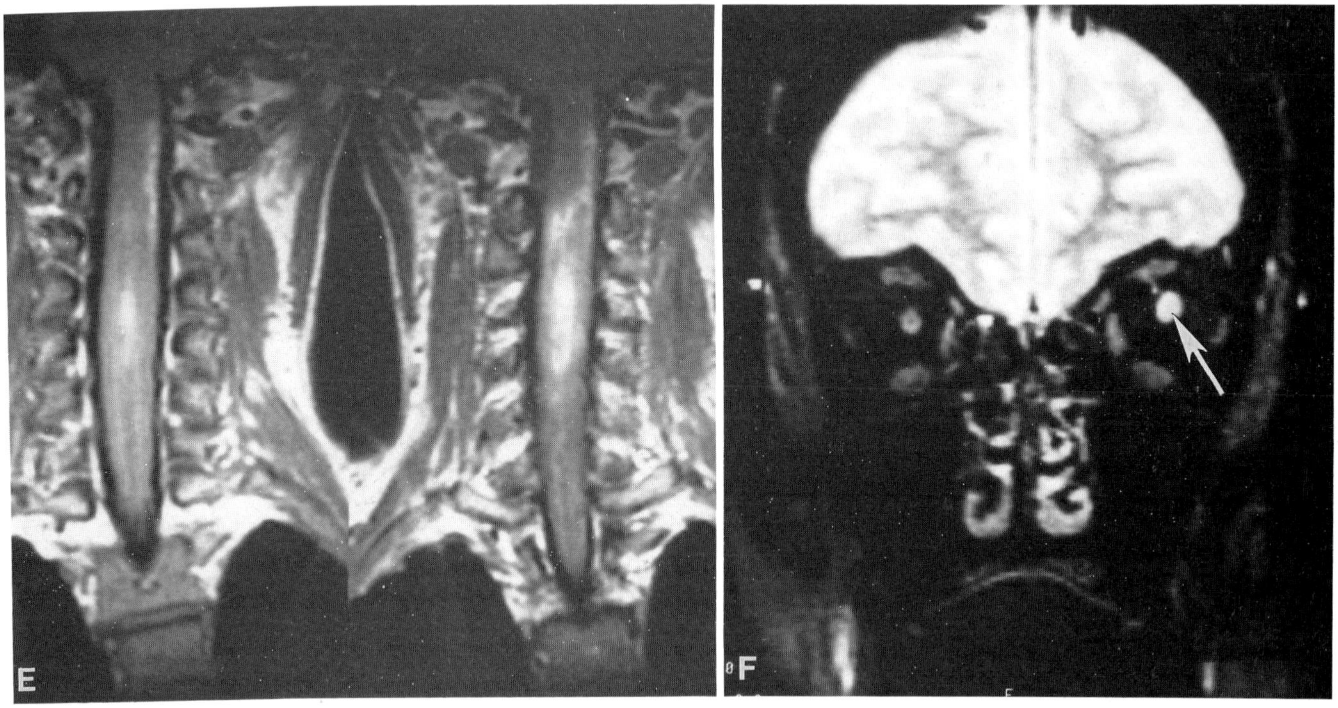

Figure 12.23–*continued*

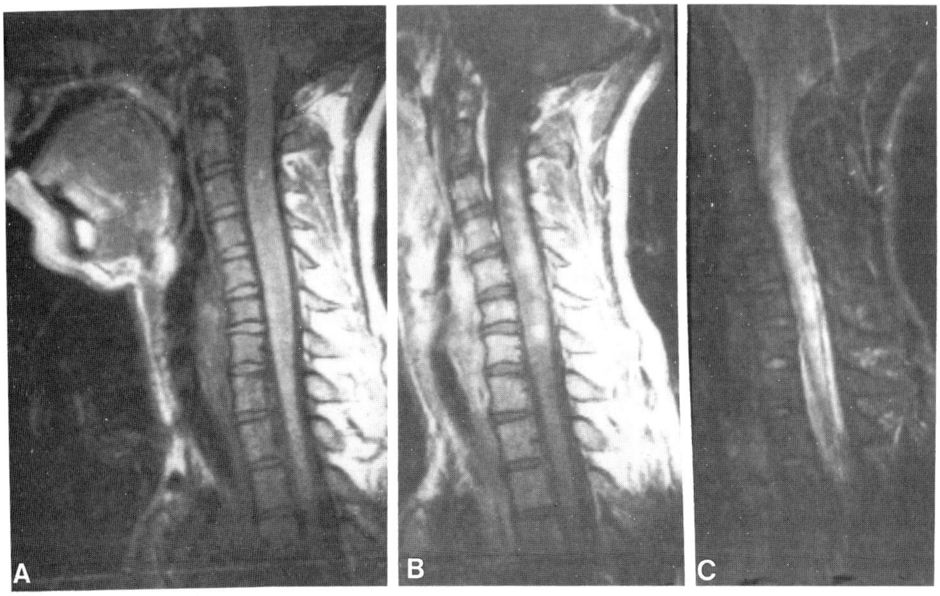

Figure 12.24. Histoplasmosis of the spinal cord. A, Sagittal T1-weighted image of lower thoracic area reveals minimal irregularity of cord contour. **B,** Sagittal T2-weighted image reveals patchy increased intensity in cord (*arrows*) with a small "bulls-eye" hypointense area (*arrowhead*). **C,** Postcontrast (Gd-DTPA) image reveals enhancement of histoplasma granuloma in cord. (Case courtesy of Dr. J.R. Jinkins.)

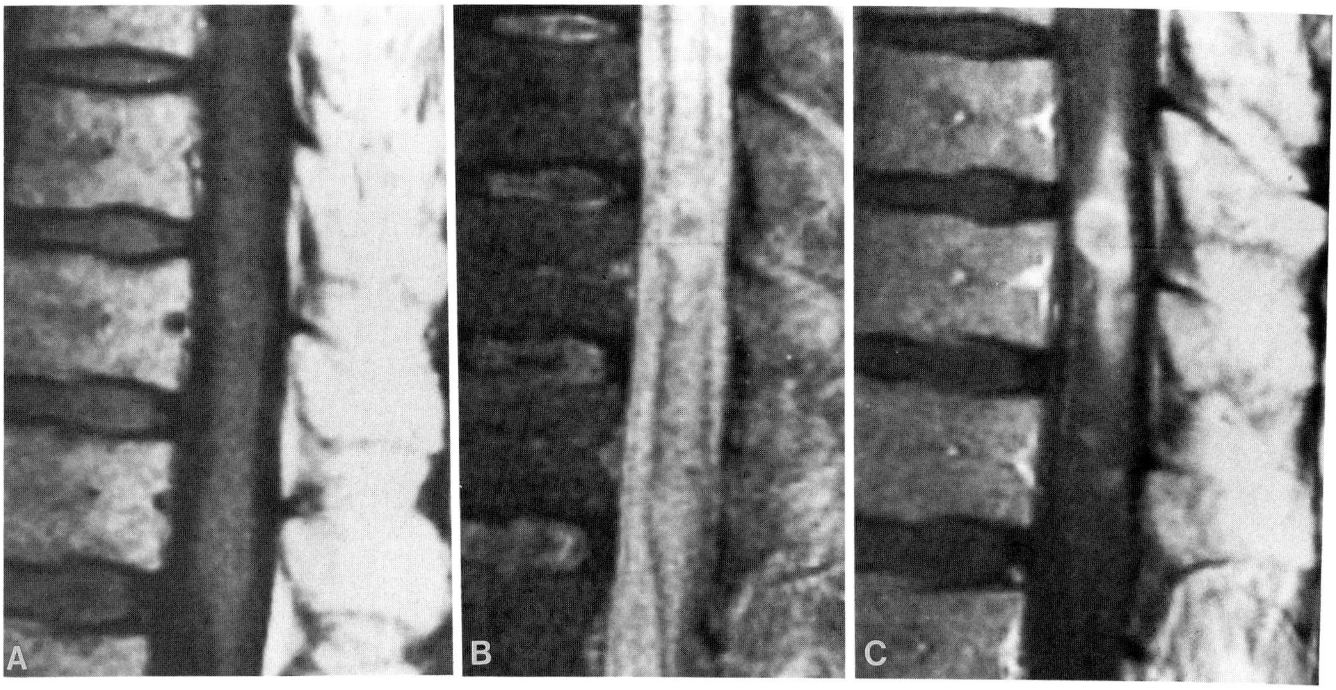

Figure 12.25. A 22-year-old man with chronic granulomatous disease with acute bacterial myelitis. A, T1-weighted sagittal image shows cord enlargement. **B,** Contrast-enhanced (Gd-DTPA) T1-weighted sagittal image reveals patchy intramedullary enhancement of inflammatory disease in spinal cord. **C,** T2-weighted sagittal image shows hyperintensity due to inflammatory disease and reactive vasogenic edema. (Case courtesy of Nick Patronas, M.D.)

loss of motor neurons and corticospinal tract degeneration. Cord atrophy (Figs. 12.26 and 12.27) may be diffuse or selectively affect either the cervical or thoracic levels. In about 50% of ALS patients with cord atrophy, the anterior and lateral columns of the spinal cord are affected, with preservation of the posterior white funiculi [30]. Degenerative changes in the axons of the lateral corticospinal tracts may cause increased intensity in these areas on proton-weighted MRI sequences [93].

BENIGN FOCAL AMYOTROPHY

Benign focal amyotrophy [94] (monomelic amyotrophy) is a degenerative disease of unknown etiology characterized by unilateral weakness and muscular atrophy of one extremity, usually an upper extremity (Fig. 12.28). Pathologically, there is unilateral degeneration of anterior horn cells in a short segment of the spinal cord. Unlike ALS, benign focal amyotrophy has a benign course. The disease has an insidious onset, progresses slowly over 2 to 3 years, and then stabilizes. A variation of this disease affecting young males is characterized by muscular atrophy of the hand and forearm and is termed *juvenile amyotrophy.* Focal unilateral atrophy of the cervical cord may be demonstrated by MRI [95] or CT myelography [96, 97].

SPINOCEREBELLAR DEGENERATION

Spinocerebellar degeneration (SCD) comprises a group of progressive degenerative diseases of the cerebellum and its connections characterized by ataxia (Fig. 12.29). The hereditary ataxias are a heterogeneous group, and the classification of these disorders is controversial. Morphologic analysis is an important consideration in the determination of the specific type of SCD [88–100]. Friedreich's ataxia is the most frequent and the best characterized of the group of SCD disorders. The peak age of this autosomal recessive disease is about 10 years, with death occurring most commonly in the third and fourth decades. Pathologically, there is degeneration and atrophy of the posterior columns, pyramidal tracts, and spinocerebellar tracts [90]. There are other types of recessive and dominant hereditary ataxias associated with variable degrees of severity of spinal cord atrophy in which the onset of presenting symptoms and course of the disease vary from the first to the eighth decades [101]. There may be profound spinal cord atrophy in some of these patients.

HEREDITARY SPASTIC PARAPLEGIA

Hereditary spastic paraplegia is a heterogeneous disease primarily causing a loss of motor cortex

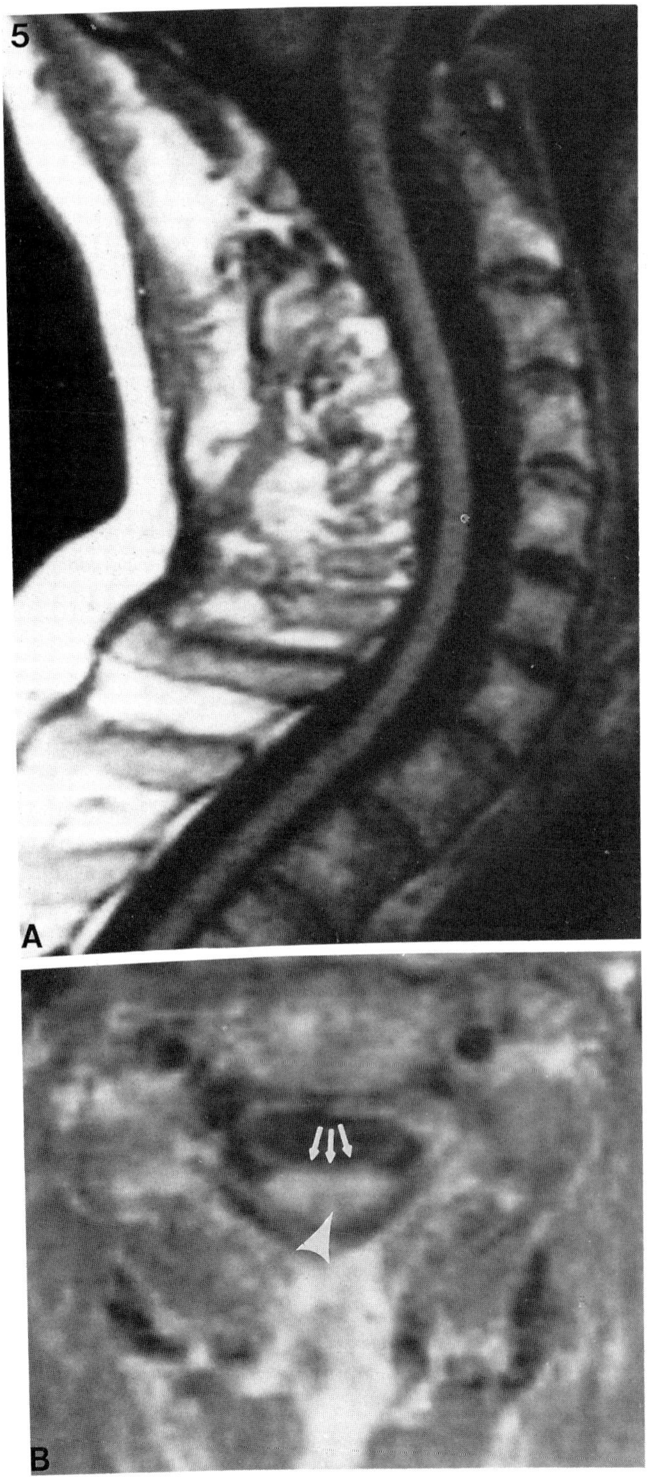

Figure 12.26. Amyotrophic lateral sclerosis in a 54-year-old man. A, Sagittal image shows cervical and thoracic cord atrophy. **B,** Axial image shows anterior cord atrophy (*arrows*) and loss of visualization of lateral corticospinal tracts but preservation of normal dorsal white funiculi (*arrowhead*).

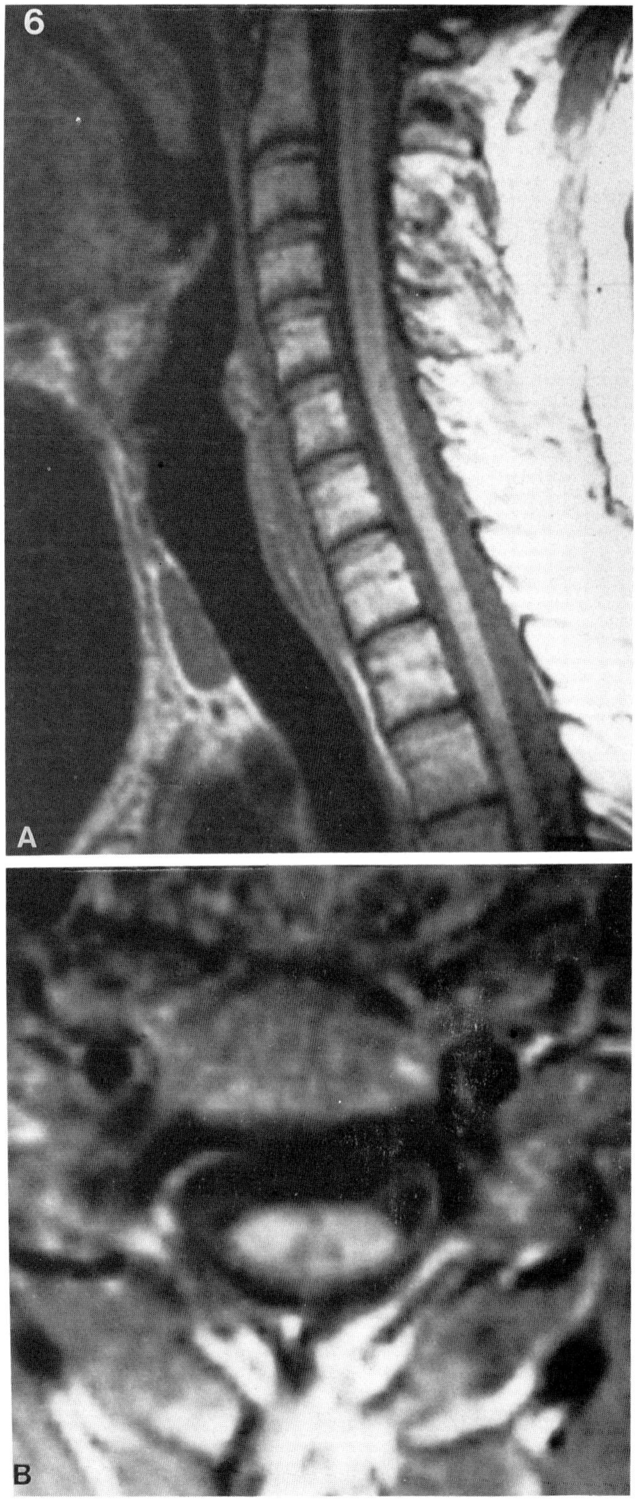

Figure 12.27. Amyotrophic lateral sclerosis in a 23-year-old man. A, Sagittal image suggests mild cord atrophy. **B,** Axial image shows anterior cord atrophy with preservation of dorsal white funiculi.

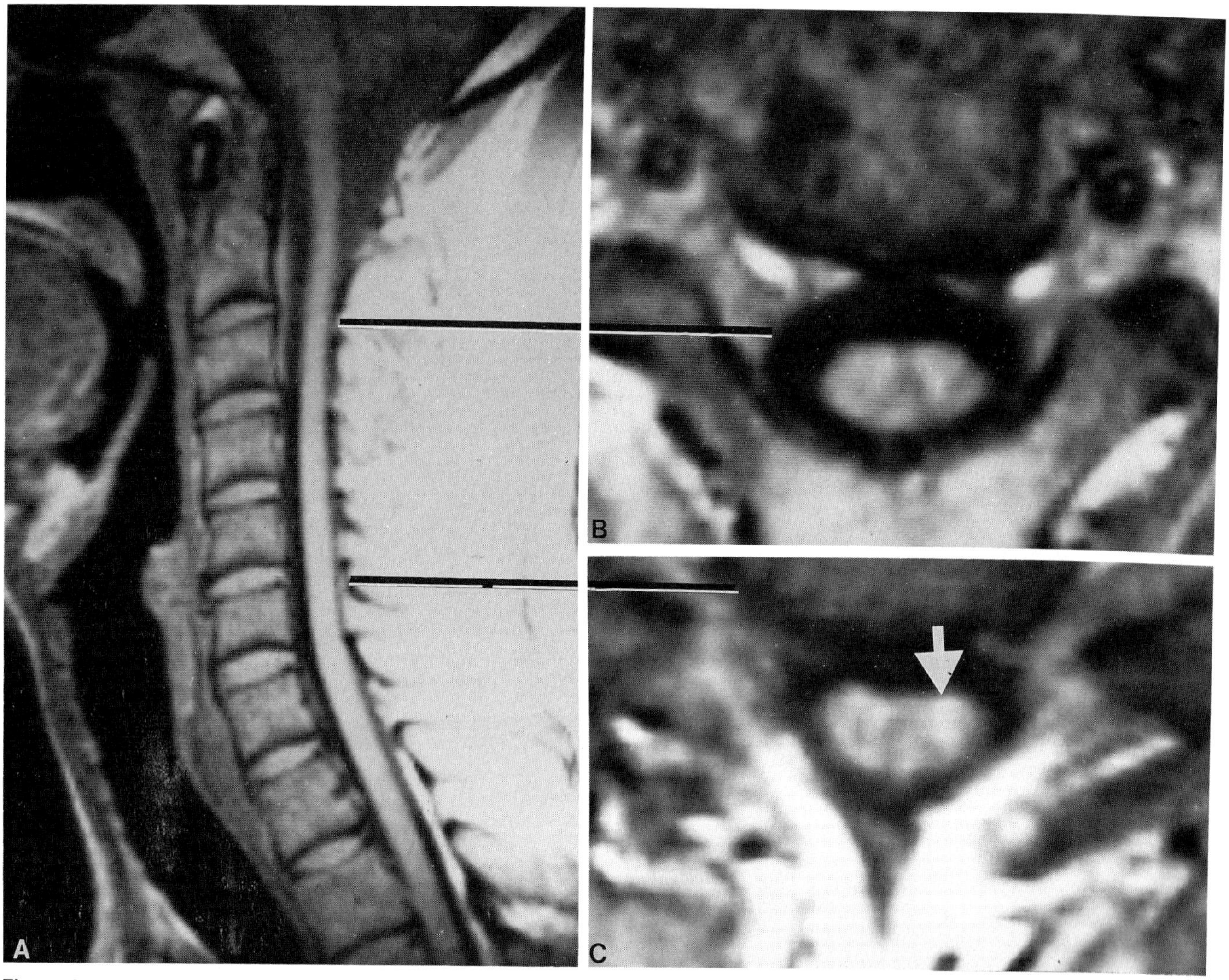

Figure 12.28. Focal amyotrophy. A 38-year-old man with a 12-year history of left arm weakness. **A,** Sagittal image. **B,** Axial image at C2–3 level shows normal spinal cord contour. **C,** Axial image at C5–6 level shows left sided ventral cord atrophy (*arrow*).

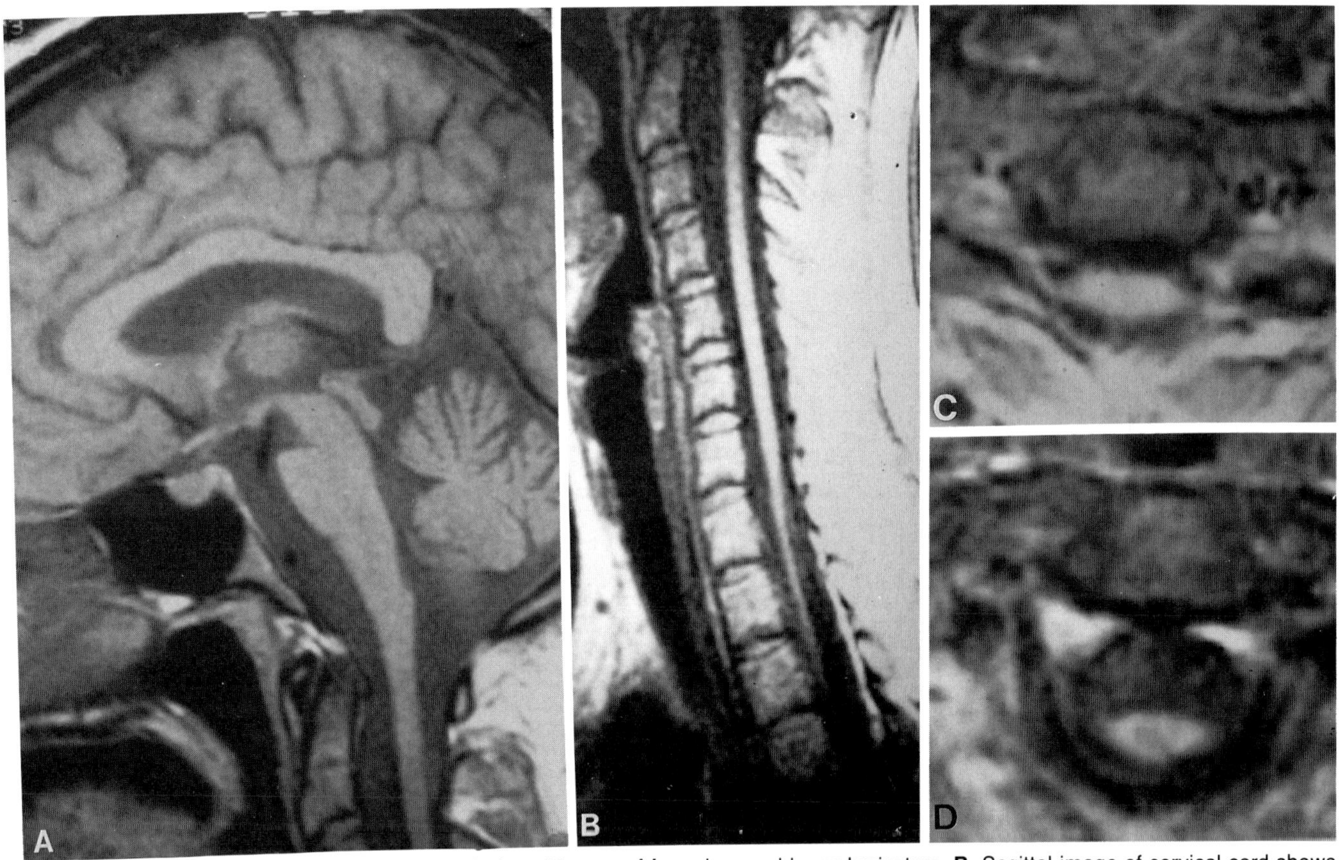

Figure 12.29. Familial spinocerebellar ataxia in a 25-year-old woman. A, Sagittal image of brain. Note diffuse atrophy of cerebellum and lower brainstem. **B,** Sagittal image of cervical cord shows diffuse atrophy. **C–D,** Axial images of cord show diffuse atrophy.

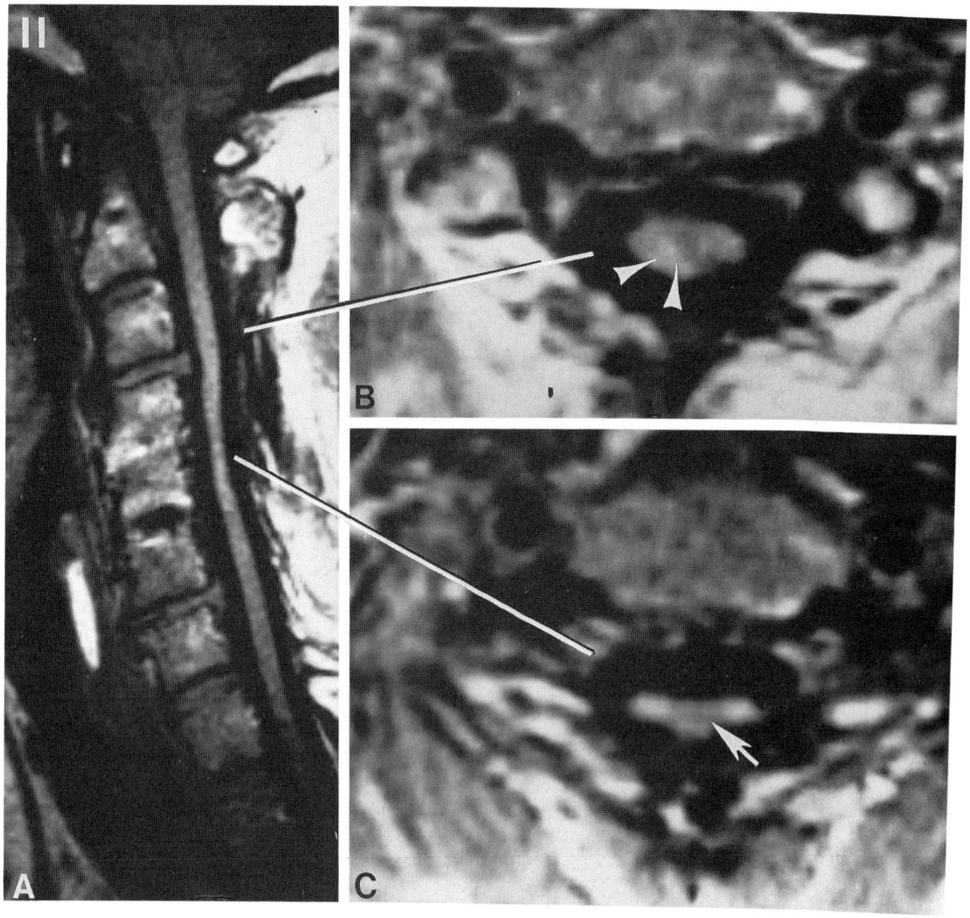

Figure 12.30. Cord atrophy due to infarction. Patient had surgery for cervical stenosis. **A,** Sagittal T1-weighted image shows focal atrophy and cord deformity. **B,** Axial images show myelomalacia in lower right quadrant of cord (*arrowheads*) at C3 level. **C,** "Ribbon-like" atrophy of cord at C5 level due to anterior spinal artery ischemia. Note relative sparing of dorsal funiculus (*arrow*).

neurons and corticospinal tract degeneration in the spinal cord [102].

ADRENOMYELONEUROPATHY

Adrenoleukodystrophy (ALD) is a sex-linked recessive genetic disorder of childhood which causes CNS demyelination and adrenal insufficiency [103]. There is considerable phenotypic variability. Adrenomyeloneuropathy (AMN) is a variant of ALD manifesting in adulthood as a slowly progressive spastic paraparesis, distal symmetrical polyneuropathy, Addison's disease, and hypogonadism. AMN is characterized by myelin destruction, loss of axons and myelin throughout the entire length of the lateral corticospinal tracts, dorsal spinocerebellar tracts, and gracile tracts, inflammation and gliosis, and atrophy in the spinal cord [104]. MRI findings of diffuse thoracic spinal cord atrophy have been reported in some patients [105–106] and in some asymptomatic individuals with the ALS-AMN complex.

PARANEOPLASTIC SYNDROMES

Neurologic paraneoplastic syndromes affecting the spinal cord have been described in some patients with systemic neoplasms [107, 108]. The typical clinical course is one of fairly rapidly progressive neurologic dysfunction that reflects involvement of the spinal cord at any level. The condition affects some patients with lymphoma or carcinoma of the breast or lung. Pathologically, there is patchy multifocal myelomalacia or massive coagulative necrosis affecting both the gray and white matter over several spinal segments. The pathogenesis of the disorder is unknown, but herpes simplex virus type 2 has been demonstrated immunohistochemically and by electron microscopy in a few cases [109].

Subacute paraneoplastic anterior horn cell disease is seen primarily in patients with lymphoma, especially Hodgkin's disease. Striking degeneration of the anterior horns occurs which may simulate ALS.

NUTRITIONAL DEFICIENCIES AND METABOLIC DISORDERS

Vitamin B_{12} deficiency (pernicious anemia) causes characteristic clinical and pathologic findings that have been named *subacute combined degeneration of the spinal cord* [110]. The disease involves mainly the midthoracic segments of the cord but may extend into the cervical and medullary levels. Degenerative changes typically affect the center of the posterior columns but advance to involve the lateral

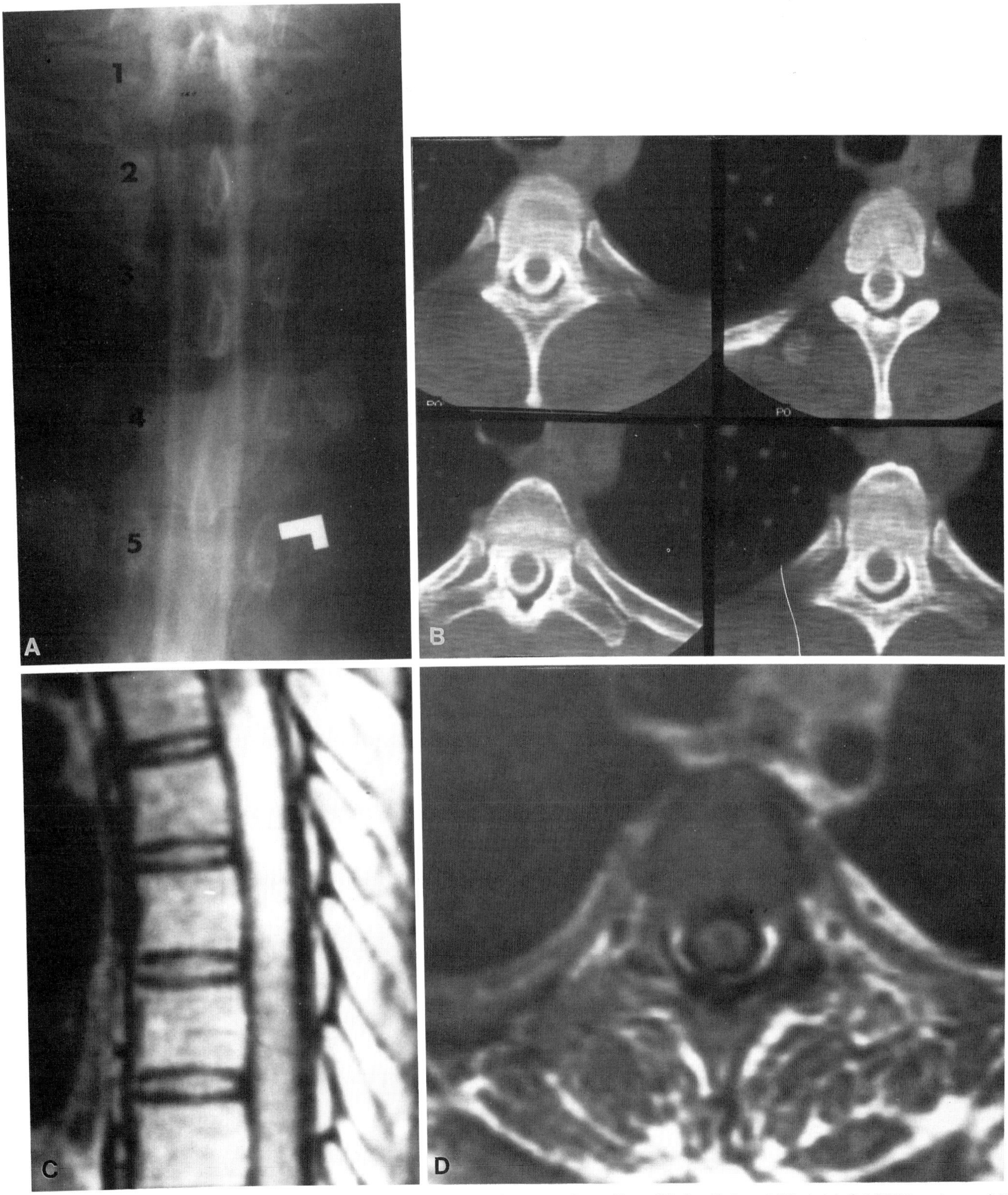

Figure 12.31. Thoracic cord infarction. A 37-year-old man with acute spinal cord infarction following cardiac arrest and sudden onset of paraplegia. (**A**) Frontal view of the myelogram demonstrates apparent cord swelling. (**B**) Subsequent CT sections demonstrates cord swelling. (**C**) Sagittal and (**D**) Axial Gd-DTPA enhanced MR demonstrates the swollen cord with enhancement. The MR study was obtained about a week following the myelogram.

columns and eventually, all the spinal cord tracts [102]. Demyelination occurs in early lesions, and remyelination may occur with suitable treatment. Variable degrees of neuroglial scarring are typically observed on pathologic examination. When untreated, the damage to the cord is irreversible [111]. Posterior column degeneration is also seen in deficiencies of nicotinic acid (pellagra), riboflavin, and pantothenic acid.

INFARCTION

Spinal cord infarction is a rare although devastating cause of myelopathy (Figs. 12.30 to 12.32). There is no obvious explanation for why spinal strokes occur less frequently than do cerebral strokes. One hypothesis is the presence of extraspinal anastomoses which may provide functionally adequate redistribution of flow in instances of occlusion of spinal or segmental vessels. These are most evident in the cervical region and less well defined caudally, a possible explanation for the relatively greater incidence of caudal ischemia [112]. When ischemia occurs, the most vulnerable region of the spinal cord is the grey matter [113] because of its higher metabolic rate. This explains the cases in which there is severe paraparesis with little or no sensory involvement.

Any pathologic process that interferes with the crucial blood supply of the spinal cord may result in infarction. Most spinal cord infarctions occur at the upper thoracic region or thoracolumbar junction. The upper thoracic cord is in a watershed area between major feeding arteries [114]. Upper thoracic medullary feeder arteries are sparse, the anterior spinal artery is smaller, and sulcus arteries are fewer than in the cervical area. Thoracolumbar infarctions often are the result of occlusion of the artery of Adamkiewicz, which is frequently the only blood supply to the thoracolumbar junction [115]. Spinal cord infarction may be caused by aortic dissection, aortic occlusion, aortic surgery, atherosclerosis, acute disk herniation, subarachnoid hemorrhage, infection, vasculitis, sickle cell anemia, decompression sickness, radiation, and rarely, emboli, epidural spinal anesthesia, and arteriography [116–121]. Cord infarction is a well-recognized complication of spontaneous aortic dissection and may be the most common cause of vascular cord injury. The dissections appear to cause occlusion of the mouth of segmental arteries. Ischemic myelopathy is most likely to occur in aortic dissections involving the descending thoracic and upper abdominal aorta [120].

Atherosclerosis commonly affects the aorta and the orifices of the segmental arteries but may account for only 1% of all instances of ischemic myelopathy [122]. Fibrocartilaginous embolization of disk tissue has been recognized as an unusual cause of cord infarction, but pathologic diagnosis is difficult, and this disease may be more common than suspected [118, 123].

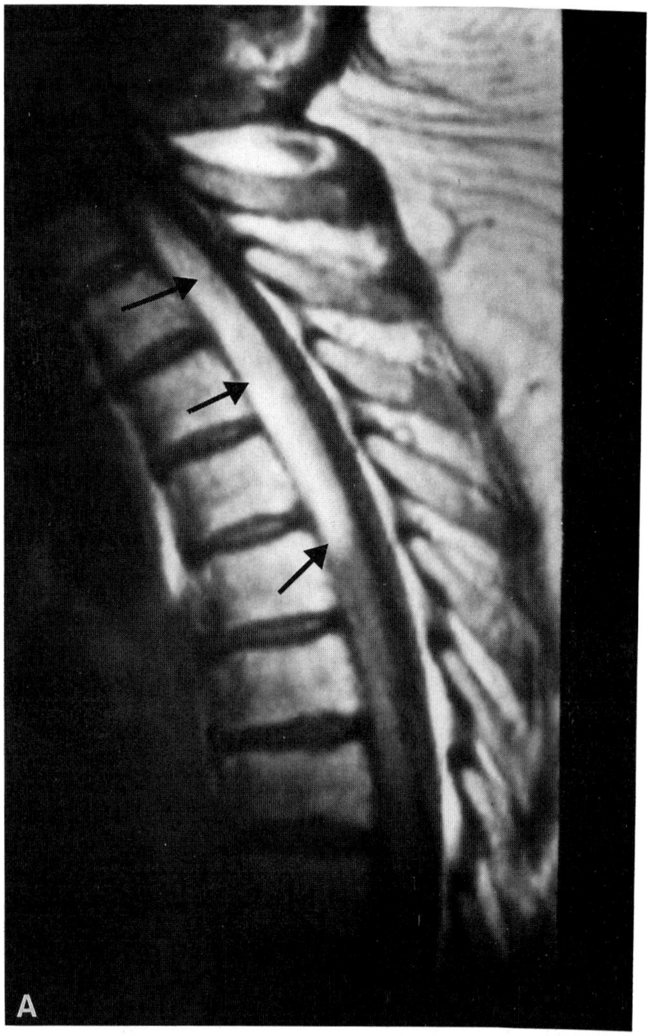

Figure 12.32. Thoracic cord infarction (two patients). A, A 55-year-old woman with acute spinal cord infarction due to dissecting aneurysm of the descending aorta. Infarct enhances intensely with gadolinium-DTPA (*arrows*). **B,** A 25-year-old paraplegic female with congenital heart disease. Cardiac arrest and spinal cord infarction occurred during open-heart surgery. Sagittal images show cord atrophy between T6 and T10. **C,** Axial images show rounded, markedly atrophic cord at T8 level.

The MRI appearance of cord infarction parallels that of cerebral infarction. Normal or decreased signal on T1-weighted images and increased signal on T2-weighted images are the dominant signal changes in spinal cord infarction [117]. The MRI signal intensity may be normal within the first 24 hours of an infarct [124]. Slight spinal cord enlargement may be seen. Gadolinium-DTPA may be helpful spinal cord infarction, but its efficacy has not been determined (Fig. 12.32). The timing of enhancement in cord infarction is probably similar to that in cerebral infarction, where lack of enhancement in the acute

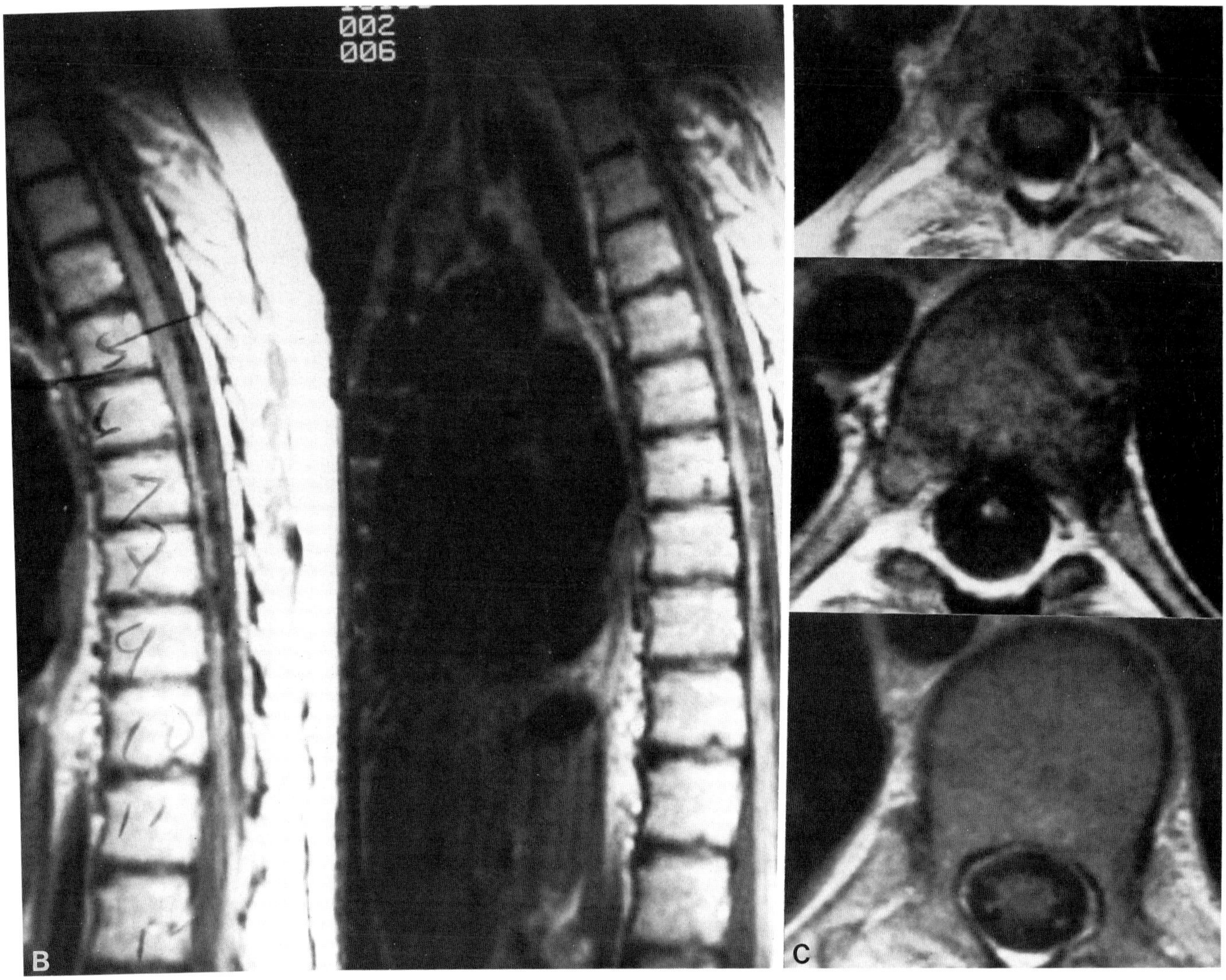

Figure 12.32–*continued*

phase is followed by enhancement in the subacute phase. An important ancillary finding is the presence of abnormal increased signal on T2-weighted images in one or more vertebral bodies due to vertebral bone marrow infarction. Yuh et al. [117] describe abnormal bone marrow signal predominantly in the anterior one-half of the vertebra or in multiple areas near the endplate and/or deep medullary portion of the vertebral body involving several vertebrae. Abnormal signal or, more important, evolution of signal abnormality in a vertebral body associated with a nonexpansile or slightly expansile cord lesion should suggest the possibility of spinal cord infarction [118] and decreases the likelihood of other diagnoses such as neoplasm, infection, and vascular malformation.

FOIX-ALAJOUANINE SYNDROME

Foix-Alajouanine syndrome (subacute necrotic myelitis) is a slowly progressive spastic paraparesis with variable sensory loss and sphincter dysfunction (Fig. 12.33). Flaccid paralysis and occasional fasciculations are also seen. Men are affected more than women, usually in the sixth and seventh decades of life, which parallels the prevalence of spinal dural AVFs [125]. There is spinal cord necrosis associated with dilated, tortuous, and thick-walled vessels found in the subarachnoid space. The disease may be a manifestation of a dural arteriovenous fistula. The fistula causes stagnation in the spinal medullary veins and venous congestion in and around the spinal cord. The resultant thrombosis may lead to venous infarction and advancing necrosis [126, 127]. Prolonged cord enhancement may be seen on Gd-DTPA–enhanced T1-weighted images. It is believed that persistent enhancement of the spinal cord on serial MRIs in patients with dural arteriovenous fistula probably results from chronic progressive venous ischemia [128]. The MRI appearance is difficult to distinguish from acute transverse myelitis or neoplasm [129].

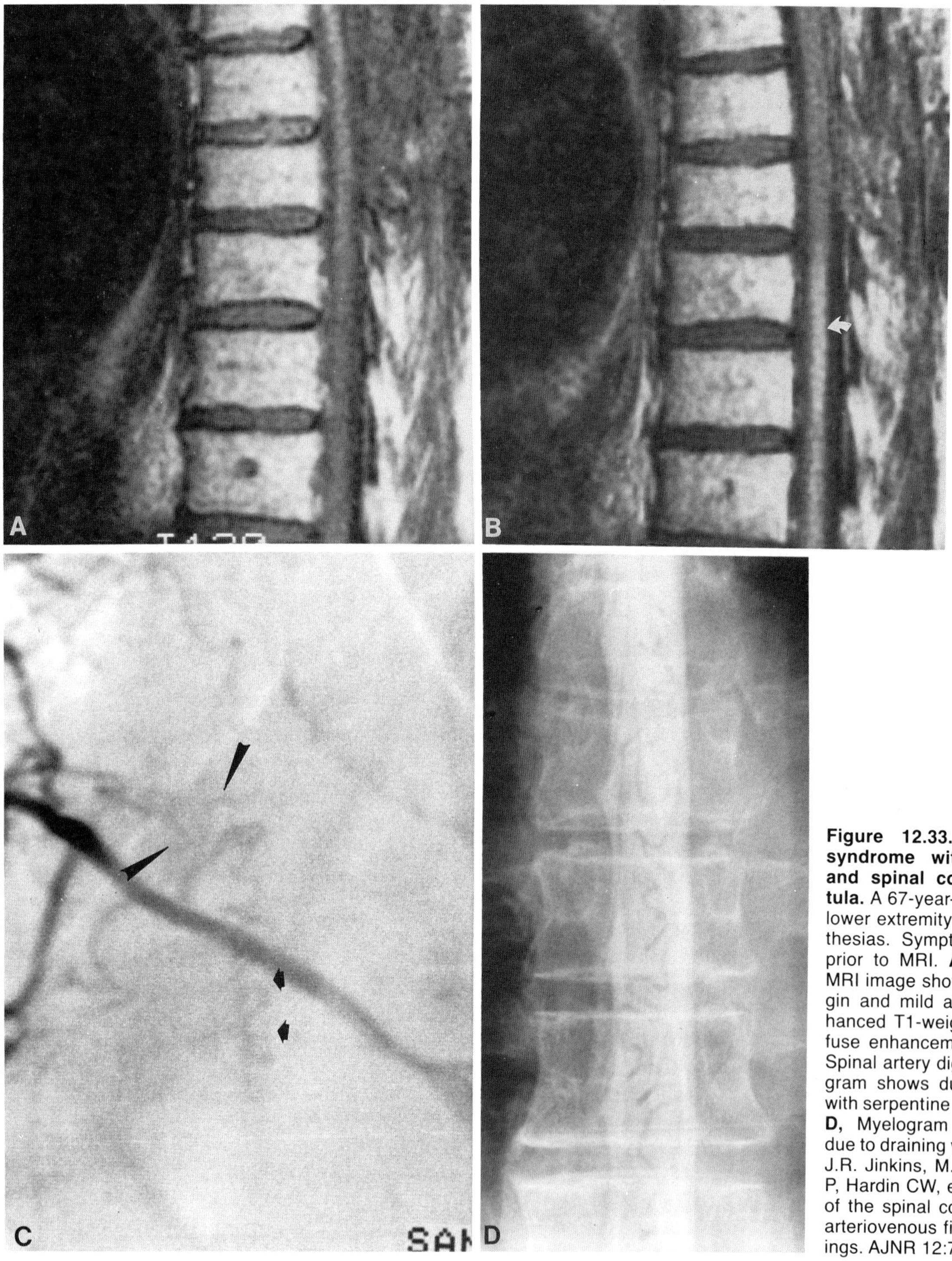

Figure 12.33. Foix-Alajounanine syndrome with venous infarction and spinal cord arteriovenous fistula. A 67-year-old man with persistent lower extremity paraparesis and paresthesias. Symptoms started 5 months prior to MRI. **A,** T1-weighted sagittal MRI image shows blurring of cord margin and mild atrophy. **B,** Contrast-enhanced T1-weighted image shows diffuse enhancement of cord (*arrow*). **C,** Spinal artery digital subtraction arteriogram shows dural AVM (*arrowheads*) with serpentine draining veins (*arrows*). **D,** Myelogram demonstrates defects due to draining veins. (Case courtesy of J.R. Jinkins, M.D.) Larsson EM, Desai P, Hardin CW, et al: Venous infarctions of the spinal cord resulting from dural arteriovenous fistula: MR imaging findings. AJNR 12:739–743, 1991.

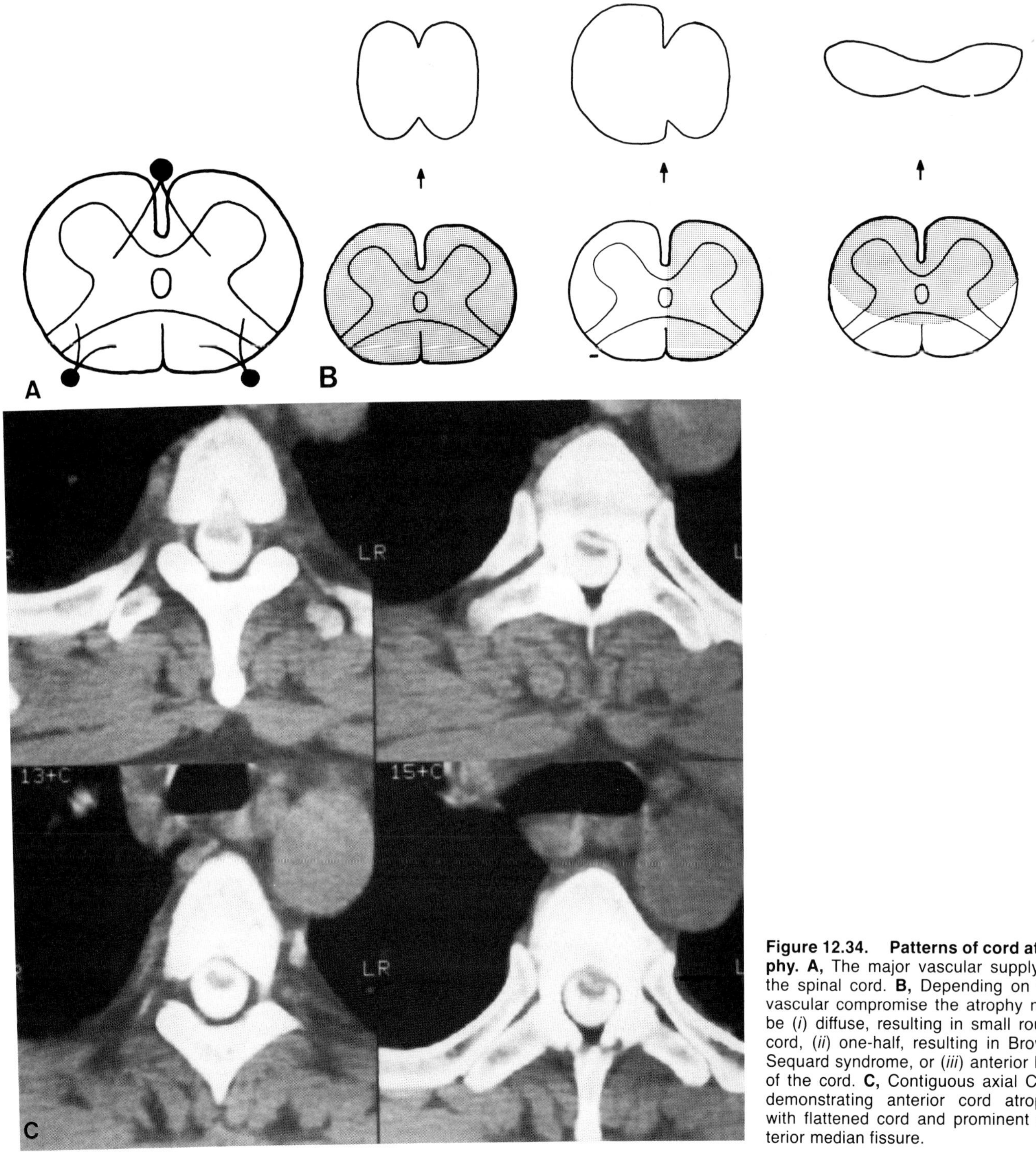

Figure 12.34. Patterns of cord atrophy. A, The major vascular supply of the spinal cord. **B,** Depending on the vascular compromise the atrophy may be (*i*) diffuse, resulting in small round cord, (*ii*) one-half, resulting in Brown-Sequard syndrome, or (*iii*) anterior half of the cord. **C,** Contiguous axial CTM demonstrating anterior cord atrophy with flattened cord and prominent anterior median fissure.

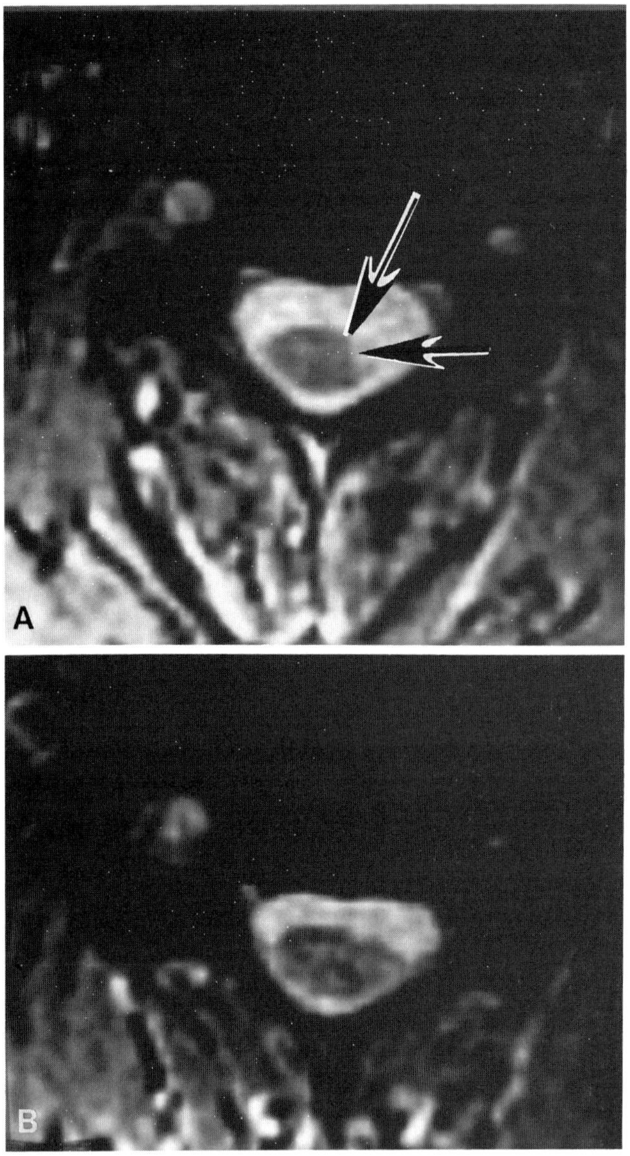

Figure 12.35. Brown-Sequard syndrome (idiopathic). T2* gradient-echo images. **A,** Midcervical level. Note left hemiatrophy of cervical spinal cord. **B,** Lower cervical level.

CORD ATROPHY

Spinal cord atrophy may be thought of as the final common end point of all myelopathies (Fig. 12.29). Degenerative processes of the spinal cord affect a number of specific neuroanatomic fiber tract systems. For example, the anterior horn cells and the corticospinal tracts are specifically affected in ALS, whereas the posterior columns (dorsal white funiculi) are notably unaffected in ALS. Although frequently nonspecific, spinal cord atrophy can occur in distinctive patterns suggestive of certain disease processes or the regions of the cord involved (see Fig. 12.34). The most common patterns are diffuse atrophy, in which a small round spinal cord (see Fig. 12.29) is visualized but the anterior median fissure is not distinct, and hemiatrophy (Brown-Sequard syndrome), recognizable because one-half the cord appears normal while the other half is atrophic (Fig. 12.35). The anterior median fissure delineates the midline. Anterior cord atrophy should be suspected when the cord appears flattened and the anterior median fissure is widened. Posterior atrophy occurs when the posterior white funiculi are selectively involved [130].

References

1. Woolsey RM, Young RR. The clinical diagnosis of disorders of the spinal cord. Neurol Clin 1991;9:573–583.
2. Katz BH, Quencer RM, Hinks RS. Comparison of gradient-recalled-echo and T2-weighted spin-echo pulse sequences in intramedullary spinal lesions. AJNR 1989;10:815–822.
3. Yousem DM, Janick PA, Atlas SW, Grossman RI. Pseudo-atrophy of the cervical spinal cord: a manifestation of truncation artifact. Presented at the 27th annual meeting of the American Society of Neuroradiology, Orlando, 1989.
4. Szeverenyi NK, Kieffer SA, Cacayorin ED. Correction of CSF motion artifact on MR images of the brain and spine by pulse sequence modification: clinical evaluation. AJNR 1988;9:1069–1074.
5. Bronskill MJ, McVeigh ER, Kucharczyk W, Henkelman RM. Syrinx-like artifacts on MR images of the spinal cord. Radiology 1988;166:485–488.
6. Curtin AJ, Chakeres DW, Bulas R, Boesel CP, Finneran M, Flint E. MR imaging artifacts of the axial internal anatomy of the cervical spinal cord. AJNR 1989;10:19–26.
7. El-Toraei I, Juler G. Ischemic myelopathy. Angiology 1979;30:81–94.
8. Gillilan L. The arterial supply of the human spinal cord. J Comp Neurol 1958;110:75–103.
9. Schoenen J. Clinical anatomy of the spinal cord. Neurol Clin 1991;9(3):503–532.
10. DiChiro G, Fisher RI. Contrast radiography of the spinal cord. Arch Neurol 1964;11:125–131.
11. Thijssen HOM, Keyser A, Horstink MWM, Meijer E. Morphology of the cervical spinal cord on computed myelography. Neuroradiology 1979;18:57–62.
12. Yu YL, du Boulay GH, Stevens JM, Kendall BE. Morphology and measurements of the cervical spine cord in computer-assisted myelography. Neuroradiology 1985;27:399–402.
13. Gellad FE, Rao KCVG, Joseph PM. Morphology of the thoracic cord on metrizamide computer assisted tomography. AJNR 1983;4:614–617.
14. Sherman JL, Nassaux PY, Citrin CM. Measurements of the normal cervical spinal cord on MR imaging. AJNR 1990;11:369–372.
15. Carvilin MJ, Asato R, Hackney DB, Kassab E, Joseph PM. High-resolution MR of the spinal cord in humans and rats. AJNR 1989;10:13–17.
16. Ho PSP, Yu S, Czervionke LF, et al. MR appearance of gray and white matter at the cervicomedullary region. AJNR 1989;10:1051–1055.
17. Dejong RN. Multiple sclerosis: history, definition and general considerations. In: Vinken PJ, Bruyn GW, eds. Handbook of clinical neurology. Amsterdam: North-Holland, 1978.

18. Adams RD, Victor M, eds. Principles of neurology. 4th ed. New York: McGraw-Hill, 1981:693–718.
19. Allen IV. Demyelinating disease. In: Adams JH, Corsellis AN, Duchen LW, eds. Greenfield's neuropathology. 4th ed. New York: Wiley, 1984.
20. Maravilla KR, Weinreb JC, Suss R, et al. Magnetic resonance demonstration of multiple sclerosis plaques in the cervical cord. AJNR 1984;5:685.
21. Floris R, Bianco F, Rossi F, et al. The role of spinal cord MRI in the diagnosis of multiple sclerosis: Riv Neurol 1987;57:170.
22. Enzmann DR, Rubin JB. Cervical spine: MR imaging with a partial flip angle, gradient refocused pulse sequence: II. Spinal cord disease. Radiology 1988;166: 473–479.
23. Nilsson O, Larsson E-M, Holtas S. Myelopathy patients studied with magnetic resonance for multiple sclerosis plaques. Acta Neurol Scand 1987;76:272.
24. Honig LS, Sheremata A, Shapiro L, et al. Spinal cord abnormalities without cerebral lesions: magnetic resonance imaging study of demyelinating disease (abstract). Ann Neurol 1986;20:152.
25. Enzmann DR, Rubin JB. Cervical spine: MR imaging with a partial flip angle, gradient-refocused pulse sequence: II. Spinal cord disease. Radiology 1988;166: 473–478.
26. Miller DH, McDonald WI, Blumhardt LD, et al. Magnetic resonance imaging in isolated noncompressive spinal cord syndromes. Ann Neurol 1987;22:714–723.
27. Ransohoff RM, Whitman G, Weinstein M. Detection of cavitary spinal cord lesions in multiple sclerosis patients by magnetic resonance imaging. Neurology 1989; 39(Suppl 1):333.
28. Lumsden CE. The neuropathology of multiple sclerosis. In: Vinken PH, Bruyn GW, eds. Handbook of clinical neurology. Vol. 9. Amsterdam: North-Holland, 1978.
29. Larsson E-M, Holtas S, Nilsson O. Gd-DTPA–enhanced MR of suspected spinal multiple sclerosis. AJNR 1989; 10:1071–1076.
30. Sherman JL. MR patterns in spinal cord degenerative disease. Presented at the XIV Symposium Neuroradiologicum, London, 1990.
31. Adams C, Armstrong D. Acute transverse myelopathy in children. Can J Neurol Sci 1990;17:40–45.
32. Berman M, Feldman S, Alter M, Zilber N, Kahana E. Acute transverse myelitis: incidence and etiologic considerations. Neurology 1981;31:966–971.
33. Abramsky O, Teitelbaum D. The autoimmune features of acute transverse myelopathy. Ann Neurol 1977;2:36–40.
34. Kenik JGG, Krohn K, Kelly RB, et al. Transverse myelitis and optic neuritis in systemic lupus erythematosis: a case report with magnetic resonance imaging findings. Arthritis Rheum 1987;30:947.
35. Propper DJ, Bucknall RC. Acute transverse myelopathy complicating systemic lupus erythematosus. Ann Rheum Dis 1989;48:512–515.
36. Miller DH, Ormerod EC, Rudge P, Kendall BE, Moseley IF, McDonald WI. The early risk of multiple sclerosis following isolated acute syndromes of the brainstem and spinal cord. Ann Neurol 1989;26:635–639.
37. Ropper AH, Poskanzer DC. The prognosis of acute and subacute transverse myelopathy based on early signs and symptoms. Ann Neurol 1978;4:51.
38. Merine D, Wang H, Kumar AJ, Zinreich SJ, Rosenbaum AE. CT myelography and MR imaging of acute transverse myelitis. J Comput Assist Tomogr 1987;11:606–608.
39. Appel B, Moens E, Lowenthal A. MRI of the spine and spinal cord: infectious and inflammatory pathology. J Neuroradiol 1988;15:325–334.
40. Barakos JA, Mark AS, Dillon WP, Norman D. MR imaging of acute transverse myelitis and AIDS myelopathy. J Comput Assist Tomogr 1990;14:45–50.
41. Awerbuch G, Feinberg WM, Ferry P, Komar NN, Clements J. Demonstration of acute post-viral myelitis with magnetic resonance imaging. Pediatr Neurol 1987; 3:367–369.
42. McCombe PA, Chalk JB, Searle JW, Tannenberg AEG, Smith JJ. Progressive encephalomyelitis with rigidity: a case report with magnetic resonance imaging findings. J Neurol Neurosurg Psychiatry 1989;52:1429–1431.
43. Grafe MR, Wiley CA. Spinal cord and peripheral nerve pathology in AIDS: the roles of cytomegalovirus and human immunodeficiency virus. Ann Neurol 1989; 25:561–566.
44. Woolsey RM, Cambers TJ, Chung HD, McGarry JD. Mycobacterial meningomyelitis associated with human immunodeficiency virus infection. Arch Neurol 1988; 45:691–693.
45. Herskovitz S, Siegel SE, Schneider AT, Nelson SJ, Goodrich JT, Lantos G. Spinal cord toxoplasmosis in AIDS. Neurology 1989;39:1552–1553.
46. Harris TM, Smith RR, Bognanno JR, Edwards MK. Toxoplasmic myelitis in AIDS: gadolinium-enhanced MR. J Comput Assist Tomogr 1990;14:809–811.
47. Britton CB, Miller JR. Neurologic complications in acquired immunodeficiency syndrome. Neurology 1986; 36:912–916.
48. Tucker T, Dix RD, Katzen C, et al. Cytomegalovirus and herpes simplex virus ascending myelitis in a patient with acquired immune deficiency syndrome. Ann Neurol 1985;18:74–79.
49. Ho DD, Rota TR, Schooley RE, et al. Isolation of HTLV-III from cerebrospinal fluid and neural tissues of patients with neurologic syndromes related to the acquired immunodeficiency syndrome. N Engl J Med 1985;24:1493–1497.
50. Mehren M, Burns PJ, Mamami F, Levy CS, Laureno R. Toxoplasmic myelitis mimicking intramedullary spinal cord tumor. Neurology 1988;9:797–799.
51. Sze G, Brant-Zawadzki MN, Norman D, Newton TH. The neuroradiology of AIDS. Semin Roentgenol 1987;22: 42–53.
52. Chang KH, Han MH, Choi YW, Kim IO, Han MC, Kim CW. Tuberculous arachnoiditis of the spine: findings on myelography, CT, and MR imaging. AJNR 1989;10: 1255–1262.
53. MacDonell AH, Baird RW, Bronze MS. Intramedullary tuberculomas of the spinal cord: case report and review. Rev Infect Dis 1990;12:432–439.
54. Silber MH. Syphilitic myelopathy. Genitourinary Med 1989;65:338–341.
55. Moore GRW, Taugott U, Scheinberg LC, Raine CS. Tropical spastic paraparesis: an inflammatory myelopathy with remyelination (a) J Neuropathol Exp Neurol 1988;47:69.
56. Osenbach RK, Blumenkopf B, Ramirez H, Gutierrez J. Meningeal neurosarcoidosis mimicking convexity en-plaque meningioma. Surg Neurol 1986;26:387–390.
57. Mirfakhraee M, Crofford MJ, Guinto FC, Nauta HJ, Weedn VW. Virchow-Robin space: a path of spread in neurosarcoidosis. Radiology 1986;158:715–720.
58. Kumpe DA, Krishna Rao CVG, Garcia JH, et al. Intracranial neurosarcoidosis. J Comput Assist Tomogr 1979; 3:324–330.

59. Klatzo I. Neuropathological aspects of brain edema. J Neuropathol Exp Neurol 1967;26:1–11.
60. Klatzo I, Suzuki R, Orzi R, Schuier F, Nitsch C. Pathomechanisms of ischemic brain edema. In: Proceedings of the fifth international symposium on brain edema. Gronningen, The Netherlands, 1982. New York: Plenum Press, 1984:1–10.
61. Sage MR. Blood-brain barrier: phenomenon of increasing importance to the imaging clinician. AJR 1982; 136:887–898.
62. Sherman JL, Stern BJ. Sarcoidosis of the CNS: comparison of unenhanced and enhanced MR images. AJNR 1990;11:15–23.
63. Nesbit GM, Miller GM, Baker HL, Ebersold MJ, Scheithauer B. MR demonstration of spinal cord sarcoidosis: new finding with Gd-DTPA enhancement. Presented at the 75th Radiological Society of North America, Chicago, 1989.
64. Kelly AB, Johnson CE, Zimmerman RD, Deck MDF. Gadolinium-enhanced MR of spinal cord sarcoidosis (a). AJNR 1989;10:887.
65. Junger SS, Stern BJ, Levine SR, Sipos E, Marti-Masso JF. Intramedullary spinal sarcoidosis. Presented at the 41st Meeting of the American Academy of Neurology, Chicago, 1989.
66. Al-Husaini A, Jamal GA. Myelopathy as the main presenting feature of systemic lupus erythematosus. Eur Neurol 1985;24:94–106.
67. Johnson RT, Richardson EP. The neurological manifestation of systemic lupus erythematosus: a clinical-pathological study of 24 cases and review of the literature. Medicine 1968;47:337–369.
68. Devinsky O, Petito CK, Alonso DR. Clinical and neuropathological findings in systemic lupus erythematosus: the role of vasculitis, heart emboli, and thrombotic thrombocytopenic purpura. Ann Neurol 1988; 23:380–384.
69. Tobben PJ, Hodges FJ, Erickson KK. Lupus myelitis: MR findings before and after treatment. Presented at the 29th A.S.N.R., Washington, D.C., 1991.
70. Banna M, El-Ramahi K. Neurologic involvement in Behçet disease: imaging findings in 16 patients. AJNR 1991;12:791–796.
71. Lindenberg R, Haymaker W. Tissue reactions in the gray matter of the central nervous system. In: Haymaker W, Adams RD, eds. Histology and histopathology of the nervous system. Springfield, Ill.: Charles C. Thomas, 1982:1037–1228.
72. Stansbury FC. Neuromyelitis optica (Devic's disease): presentation of five cases, with pathologic study, and review of literature. Arch Ophthalmol 1949;43:292–335.
73. Kim RC. Necrotizing myelopathy (commentary). AJNR 1991;12:1084–1086.
74. Marti-Febregas J, Martinez JM, Illa Ii, Escartin A. Myelopathy of unknown etiology. Acta Neurol Scand 1989;80:455–460.
75. Johnson RT: Viral infections of the nervous system. New York: Raven Press, 1982.
76. Dupuis MJM, Atrouni S, Dooms GC, Gonsette RE. MR imaging of schistosomal myelitis. AJNR 1990;11:782–783.
77. Gero B, Sze G, Sharif MDH. MR of intradural inflammatory disease of the spine. Presented at the XIV Symposium Neuroradiologicum, London, 1990.
78. Koppel BS, Daras M, Duffy KR. Intramedullary spinal cord abscess. Neurosurgery 1990;26:145–146.
79. Castillo M, Quencer RM, Post MJD. MR of intramedullary spinal cysticercosis. AJNR 1988;9:393–395.
80. Silbergleit JA, Brunberg EM Spickler. MR of intramedullary schistosomiasis. Presented at the 29th A.S.N.R., Washington, D.C., 1991.
81. Jena A, Tripathi RP. Primary spinal echinococcosis causing paraplegia: case report with MR and pathologic correlation. AJNR 1991;12:560.
82. Peterman SB, Hoffman JC. Lyme disease simulating multiple sclerosis. Presented at the annual meeting of the American Society of Neuroradiology, Orlando, Fla., March 1989.
83. Kohler J. Lyme borreliosis: a case of transverse myelitis with syrinx cavity. Neurology 1989;39:1553–1554.
84. Fernandez RE, Rothberg M, Ferencz G, Wujack D. Lyme disease of the CNS: MR imaging findings in 14 cases. AJNR 1990;11:479–481.
85. Delattre JY, Rosenblum MK, Thaler HT, Mandell L, Shapiro WR, Posner JB. A model of radiation myelopathy in the rat: pathology, regional capillary permeability changes and treatment with dexamethasone. Brain 1988;989:1319–1336.
86. Palmer JJ. Radiation myelopathy. In: Vinken PJ, Bruyn GW, eds. Handbook of Clinical Neurology, Amsterdam: North-Holland, 1976:81–95.
87. Sze G, Russel E, Lee D. MR imaging of chronic radiation myelopathy. Presented at the 75th Radiological Society of North America, Chicago, 1989.
88. deToffol B, Cotty P, Calais G, et al. Chronic cervical radiation myelopathy diagnosed by MRI. J Neuroradiol 1989;16:251–253.
89. Tandan R, Bradley WG. Amyotrophic lateral sclerosis: 1. Clinical features, pathology, and ethical issues in management. Ann Neurol 1985;18:271–280.
90. Oppenheimer DR. Diseases of the basal ganglia, cerebellum and motor neurons. In: Adams JH, Corsellis JA, Duchen LW, eds. Greenfield's neuropathology. 4th ed. Baltimore: Williams & Wilkins, 1984:699–747.
91. Lawyer T, Netsky MG. Amyotrophic lateral sclerosis: clinicoanatomic study of 53 cases. Arch Neurol Psychiatry 1953;69:171–192.
92. Brownell B, Oppenheimer DR, Hughes JT. The central nervous system in motor neurone disease. J Neurol Neurosurg Psychiatry 1970;33:338–357.
93. Sherman JL, Clawson L, Citrin CM, et al. MR evaluation of amyotrophic lateral sclerosis (ALS) (a) AJNR 1987; 8:941.
94. Adornato BT, Engel WK, Kucera J, Bertorini TE. Benign focal amyotrophy. Neurology 1978;28:399.
95. Biondi A, Dormont D, Weitzner I, Bouche P, Chaine P, Bories J. MR imaging of the cervical cord in juvenile amyotrophy of distal upper extremity. AJNR 1989;10: 263–268.
96. Metcalf JC, Wood JB, Bertorini TE. Benign focal amyotrophy: metrizamide CT evidence of cord atrophy: case report. Muscle Nerve 1987;10:338–345.
97. Oryema J, Ashby P, Spiegel S. Monomelic atrophy. Can J Neurol Sci 1990;17:124–130.
98. Babatame H, Fukuyama H, Akiguchi I, et al. Spinocerebellar degeneration: qualitative and quantitative MR analysis of atrophy. J Comput Assist Tomogr 1988; 12:298–303.
99. Ramos A, Quintana F, Diez C, Leno C, Berciano J. CT findings in spinocerebellar degeneration. AJNR 1987; 8:635–640.
100. Duvoisin RC. The olivopontocerebellar atrophies. In: Marsden CD, Fahn S, eds. Movement disorders. 2nd ed. London, Butterworth, 1987:249–271.
101. Konigsmark BW, Weiner LP. The olivopontocerebellar atrophies: a review. Medicine 1970;19:227–241.
102. Adams RD, Salam Adams M: Chronic nontraumatic

diseases of the spinal cord. Neurol Clin 1991;9:605–624.

103. Schaumburg HH, Powers JM, Raine CS, et al. Adrenoleukodystrophy: a clinical and pathological study of 17 cases. Arch Neurol 1975;32:577–591.

104. Schaumburg HH, Powers JM, Raine CS, et al. Adrenomyeloneuropathy: a probable variant of adrenoleukodystrophy: II. General pathologic neuropathologic and biochemical aspects. Neurology 1977;27:1114–1119.

105. Kumar AJ, Naidu S, Wang H, Zinreich SJ, Bryan RN, Moser H. MRI of spinal cord in adrenomyeloneuropathy. Presented at the XIV Symposium Neuroradiologicum, London, 1990.

106. Snyder RD, King JN, Keck GM, Orrison WW. MR imaging of the spinal cord in 23 subjects with ALD-AMN complex. AJNR 1992;12:1095–1098.

107. Henson RA, Urich H. Cancer and the nervous system: the neurological manifestations of systemic malignant disease. Oxford, England: Blackwell, 1982.

108. Handforth A, Nag S, Sharp D, Robertson DM. Paraneoplastic subacute necrotic myelopathy. Can J Neurol Sci 1983;10:204–207.

109. Iwamasa T, Utsumi Y, Sakuda H, et al. Two cases of necrotizing myelopathy associated with malignancy caused by herpes simplex virus type 2. Acta Neuropathol 1989;78:252–257.

110. Russell JSR, Batten FE, Collier J. Subacute combined degeneration of the spinal cord. Brain 1900;23:39–110.

111. Duchen WL, Jacobs JM. Nutritional deficiencies and metabolic disorders. In: Adams JH, Corsellis JAN, Duchen LW, eds. Greenfield's neuropathology. 4th ed. New York: Wiley, 1984:573–626.

112. Thron AK. Vascular anatomy of the spinal cord. New York: Springer-Verlag, 1988;78–103.

113. Marcus ML, Heistadd DD, Ehrhardt JC, et al. Regulation of the total regional spinal cord blood flow. Circ Res 1977;xx:128–136.

114. Sandson TA, Friedman JH. Spinal cord infarction: report of 8 cases and review of the literature. Medicine 1989;68:282–292.

115. Doppman JL, DiChiro G, Ommaya AK. Selective arteriography of the spinal cord. St. Louis: Warren H. Green, 1969.

116. Elksnis SM, Hogg JP, Cunningham ME. MR imaging of spontaneous spinal cord infarction. J Comput Assist Tomogr 1991;15:228–232.

117. Yuh WTC, Marsh E, Wang AK, et al. MR imaging of spinal cord and vertebral body infarction. AJNR 1992;13:145–154.

118. Mikulis DJ, Ogilivy CS, McKee A, Davis KR, Ojeman RG. Spinal cord infarction and fibrocartilagenous emboli: spinal cord infarction and fibrocartilagenous emboli. AJNR 1992;13:155–160.

119. Ackerman WE, Juneja MM, Knapp RK. Maternal paraparesis after epidural anesthesia and cesarean section. South Med J 1990;83:695–700.

120. Dawson DM, Potts F. Acute nontraumatic myelopathies. Neurol Clin 1991;3:585–603.

121. Szilagyi DE, Hageman JH, Smith RF, et al. Spinal cord damage in surgery of the abdominal aorta. Surgery 1978;83:38–50.

122. Jellinger K. Spinal cord arteriosclerosis and progressive vascular myelopathy. J Neurol Neurosurg Psychiatry 1967;30:195–208.

123. Naiman JL, Donhue WL, Prichard JS. Fatal nucleus pulposus embolism of spinal cord after trauma. Neurology 1961;11:83–87.

124. Casselman JW, Jolie E, Dehaeni I, et al. Gadolinium-DTPA enhanced MR imaging of infarction of the anterior spinal cord. AJNR 1991;12:561.

125. Crisuolo GR, Oldfield EH, Doppman JL. Reversible acute and subacute myelopathy in patients with dural arteriovenous fistulas: Foix-Alajouanine syndrome reconsidered. J Neurosurg 1989;70:354–359.

126. Minami S, Sagoh T, Nishimura K, et al. Spinal arteriovenous malformation: MR imaging. Radiology 1988;169:109–115.

127. Wrobel CJ, Oldfield EH, Dichiro G, Tarlov EC, Baker RA, Dopman JL. Myelopathy due to intracranial dural arteriovenous fistulas draining intrathecally into spinal medullary veins: report of three cases. J Neurosurg 1989;69:934–939.

128. Larson E-M, Desai P, Hardin CW, Story J, Jinkins JR. Venous infarction of the spinal cord resulting from dural arteriovenous fistula: MR imaging findings. AJNR 1991;12:739–743.

129. Mirich DR, Kucharczyk W, Keller MA, Deck J. Subacute necrotizing myelopathy: MR imaging in four pathologically proved cases. AJNR 1991;12:1077–1083.

130. Mawad ME, Hilal SK, Fetell MR, Silver AJ, Ganti SR, Sane P. Patterns of spinal cord atrophy by metrizamide CT. AJNR 1983;4:611–613.

SUBJECT INDEX

E